6th Edition

CARPENTRY

Floyd **Vogt**

Australia • Brazil • Mexico • Singapore • United Kingdom • United States

Carpentry, Sixth Edition
Floyd Vogt

Vice President, Career and Professional
 Editorial: Dave Garza

Director of Learning Solutions: Sandy Clark

Senior Acquisitions Editor: Jim DeVoe

Managing Editor: Larry Main

Senior Product Manager: Jennifer A. Starr

Product Manager: Jennifer Jacobson,
 Ohlinger Publishing Services

Editorial Assistant: Aviva Ariel

Vice President, Marketing: Jennifer Baker

Marketing Director: Deborah Yarnell

Senior Market Development Manager:
 Erin Brennan

Marketing Coordinator: Danielle Yannotti

Brand Manager: Kristin McNary

Senior Production Director: Wendy A. Troeger

Production Manager: Mark Bernard

Content Project Manager: David Barnes

Senior Art Director: Casey Kirchmayer

Technology Project Manager: Joe Pliss

Cover Image: ©iStockphoto.com/
 Skip ODonnell ©iStockphoto.com/Elif Gunyeli

For product information and technology assistance, contact us at
Cengage Customer & Sales Support, 1-800-354-9706
For permission to use material from this text or product,
submit all requests online at **www.cengage.com/permissions**
Further permissions questions can be e-mailed to
permissionrequest@cengage.com

Library of Congress Control Number: 2012947574

ISBN-13: 978-1-133-60736-6

ISBN-10: 1-133-60736-5

Cengage
20 Channel Center Street
Boston, MA 02210
USA

Cengage is a leading provider of customized learning solutions with office locations around the globe, including Singapore, the United Kingdom, Australia, Mexico, Brazil, and Japan. Locate your local office at **www.cengage.com/global**

Cengage products are represented in Canada by Nelson Education, Ltd.

To learn more about Cengage platforms and services, register or access your online learning solution, or purchase materials for your course, visit **www.cengage.com**.

Printed in Mexico
Print Number: 09   Print Year: 2019

Contents

Section 1

Section 2

Section 3

Section 4

Preface

Welcome to *Carpentry, 6th Edition,* a modern approach to residential construction. Designed for students enrolled in carpentry courses at secondary and two-year, four-year postsecondary schools, *Carpentry, 6th Edition* walks the student step-by-step through the various principles and practices associated with constructing a residential building.

APPROACH

A unique blend of traditional and up-to-date construction practices, the author focuses topics on need-to-know information in four comprehensive sections: **Section 1—Tools and Materials, Section 2—Rough Carpentry, Section 3—Exterior Finish, Section 4—Interior Finish.** Beginning with the layout of the building and finishing with trim carpentry, each section features step-by-step procedures of key carpentry jobs, critical safety information, tips of the trade, and insight into the construction industry.

Section 1

This section describes wood as a building material in its many forms, building fasteners, hand and power tools, jobsite safety, and prints and codes. Three units explore wood in many forms from boards to engineered sheets and beams. A unit on fasteners used to construct residential buildings and three units on the tools necessary to shape and secure building products are included. A unit concentrating on understanding jobsite safety in terms of personal safety equipment, tools used on the job, and the hazards of being on the job is included. A unit devoted to prints and codes helps students to understand the written language of residential and commercial construction.

Section 2

Beginning with the basics, this section explains how to locate a building on a site using surveying equipment. It then explores various types of residential foundations, provides concrete applications for foundations, slabs, and stairs, and walks students through the framing of floors, walls, roofs, and stairs. Two units devoted to roof framing are designed to give the student an introduction to the subject and an exposure to advanced framing methods. In addition, a section on energy conservation considerations for insulation and ventilation helps students keep pace with industry standards. Engineering the framing member sizes for joists and rafters is also included.

Section 3

Exterior carpentry jobs, including applying various roof and siding finishes, installing windows and doors, and constructing decks with rails and fences, are covered in this section. There is also a focus on protecting the building from water intrusion during vertical rain and wind-blown horizontal rain.

Section 4

Trim work involved in finishing a residential building is described in this section. The section begins with the application of drywall and various finishes on floors, walls, and ceilings. Next, fitting trim to windows, doors, and walls is explained. Also, constructing and finishing a staircase with balustrade is covered. Lastly, to finish the job, the text explains how to install manufactured cabinets and construct countertops, drawers, and doors.

A **Success Story** opens each of the sections in the book, providing a look at the day-to-day job of a carpenter and the successes accomplished through dedication and education.

Green Tip

Orienting a building to address the sun will affect a lifetime of heating and cooling dollars. Warm climates should include natural shading and northerly exposures. Cooler areas should orient longer building sides and larger glass toward southerly exposures.

Green Tips highlight environmental resources and green building techniques, focusing on sustainability and eco-friendly architecture.

Crawl Space and Basement Foundations

The crawl space is the area enclosed by the foundation between the ground and the floor above. A minimum distance of 18 inches from the ground to the floor and 12 inches from the ground to the bottom of any beam is required. The ground is covered with a plastic sheet, called a vapor barrier, to prevent moisture rising from the ground from penetrating into the floor frame

*IRC
R317.1 above.*

A foundation enclosing a basement is simi-

International Residential Code references highlight content related to the requirements of the 2012 IRC.

Estimating Formula can be found at the end of applicable chapters, and combines an example with shorthand formulas to help complete calculations and apply chapter content to the actual estimation process.

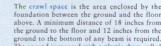

SAFETY REMINDER

Concrete is a universal building material used for support of various structures. It must be handled and properly supported during placement. This is particularly important because concrete can be worked for only a short time before it sets up. Avoid prolonged contact with fresh concrete or wet cement because of possible skin irritation. Wear protective clothing when working with newly mixed concrete. Wash skin areas that have been exposed to wet concrete as soon as possible.

CAUTION

- Be trained and competent in the use of stationary power tools before attempting to operate them without supervision.
- Make sure power is disconnected when changing blades or cutters. Ensure that safety guards are in place and that all guides are in proper alignment and secured.
- Make sure saw blades are sharp and suitable for the operation.
- Wear eye and ear protection. Wear appropriate, properly fitted clothing.
- Keep the work area clear of scraps that might present a tripping hazard.
- Keep stock clear of saw blades before starting a machine.
- Do not allow your attention to be distracted by others or be distracting to others while operating power tools.
- Turn off the power and make sure the machine has stopped before leaving the area.

Safety precautions that apply to specific operations are given when those operations are described in this unit.

Safety Reminders and Cautions prominently feature the latest safety considerations so readers can avoid the dangers of certain procedures, tools, and equipment in order to stay safe on the job.

Step-by-Step Procedures walk readers through the key tasks associated with specific residential building tasks, while **On the Job** tips of the trade provide practical advice.

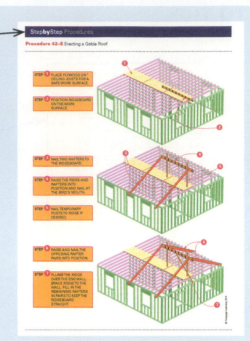

On the Job

Safety is a team sport. It requires every worker to play by the rules.

Deconstruct This features illustrate situations where a problem may or may not exist in the construction process, giving students the opportunity to consider what is right and wrong and and develop their critical thinking skills.

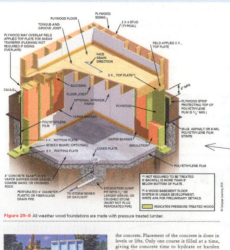

Visually intensive content featuring photo-realistic drawings and full-color jobsite photos, drawn from live construction sites in geographically diverse locations in the United States, bridge the gap between the classroom and the jobsite.

NEW TO THIS EDITION

The sixth edition offers a realistic representation of residential construction from which learners can make practical connections between descriptions and application of concepts.

Technical Content

1   Estimating math and introductory concepts have been added.

2   Throughout the text, the author has streamlined content and cut material based on the latest innovations in residential construction. Organizational changes include moving the scaffold chapter into a new safety chapter, which appears before the units of actual construction. Scaffold content from former Chapters 39 and 40 is now in Chapters 20 and 21. The unit on roof framing has been expanded from one to two units. First, Unit 15 Roof Framing is introductory, covering roof types (gable, shed, and flat roofs) and trusses. Unit 16 Advanced Roof Framing covers hip, intersecting, and complex roof framing details. Complex roof framing includes dormers, gambrel rafters, knee walls, meeting rafters with differing, and birds mouth heights. Also included is the use of roof theory to cut gable end sheathing and roof sheathing to hip and valley rafters without using the trial-and-error cut-and-fit method.

3   A new step-by-step procedure for operating a circular saw has been added to Chapter 14.

4   A new safety unit (Unit 8 Jobsite Safety and Scaffolds) has been added. Topics include jobsite attitude, personal protective equipment, activity around a jobsite and its hazards, jobsite housekeeping and its importance, electrical hazards and proper use, hazardous materials, and fire protection. Also included are instruction methods for erecting scaffolds and using various jobsite aids.

5   Understanding the differences in drawings for residential and commercial buildings has been woven throughout Unit 9 Architectural Plans and Building Codes. This will better prepare the student to work on light commercial sites.

6   Information on the total station has been added to surveying equipment in Chapter 28 on Leveling and Layout Tools.

7   Unit 11 explores various types of foundations used nationwide. Masonry materials and their styles for each foundation are included as well as the methods and materials of damp proofing a foundation.

8   Engineering the size and type of framing members has been included. Structural terms, strengths of materials, and span charts are included. These appear in Chapter 34 for joist and girders and Chapter 41 for rafters.

9   Structural steel framing has been expanded (in Unit 14) to include light commercial applications.

10   Unit 18 Insulation and Ventilation has been expanded to include discussions on heat loss, thermal conductivity, and the insulation requirements of the 2012 International Energy Conservation Code (IECC).

11   The roofing unit (Unit 19) now includes safety considerations for working a roof. Content on flashing a chimney has been reworked to include a drawing and a step-by-step description of flashing pieces.

12   Determining the swing of exterior and interior doors has been added to Chapter 56 on exterior doors. Also included is a step-by-step procedure on installing a prehung door and hardware often installed on doors such as stops, closers, and security hinges.

13   Deck installation in Unit 23 now includes IRC requirements for fastening the ledger to a building as it pertains to bolt size and locations.

14   Unit 24 Drywall Construction has been expanded to include automatic tools, more on drywall finishing, causes and repair of defects, and patching techniques.

15   Chapter 85 Countertop and Cabinet Construction now includes details on cabinet materials, frames, parts, and assembly.

Learning Enhancements

- At the end of each unit, a **Deconstruct This** feature includes a photograph or figure with open-ended questions for students to consider. This feature provides opportunities for students to think critically about the situation presented. These photos show situations where a problem may or may not exist in a construction process. This will give the students an opportunity to ponder what is right and what is wrong.

- At the end of some chapters, a description of an estimating process is followed by an **Estimating Formula**—a figure that summarizes in a table the shorthand estimating formulas and includes an example. The summarizing graphic makes the estimating content more user-friendly and provides an application opportunity.

- **Green** Tips, scattered across each unit, highlight environmental resources and green building techniques, focusing on sustainability and eco-friendly architecture.

- Icons in the margins draw students' attention to places in the 2012 International Residential Code (IRC) that relate to the text. This will help the student navigate the pages of code manuals.

SUPPLEMENTAL PACKAGE

Included with the sixth edition of *Carpentry* is an extensive supplemental package to aid student learning, and help instructors prepare lessons and evaluations:

An *Instructor's Guide* provides many helpful tools to facilitate effective classroom presentations of material—comprehensive *Lesson Plans,* integrated PowerPoint references and Teaching Tips, a review of important math, blueprint reading to help evaluate student knowledge at the start of the course, and *Answers to Review Questions* and *Workbook Questions.* The *Instructor's Guide* is also available in electronic format on the accompanying Instructor Resources CD.

A *Workbook* for students provides a wide range of practice problems to reinforce concepts learned in each chapter, as well as to prepare students for exams. Question types include multiple choice, completion, and identification as well as critical math problems and soft skill activities. Instructors may assign these questions as homework, to ensure full comprehension of the material.

Instructor Resources CD—a complete CD-ROM for the instructor, this electronic resource contains several components for classroom prep:

- An important *Math Review* as well as *Lab Projects,* including brief lessons and practice problems for students, allows instructors to review important concepts to ensure that their students are up to speed prior to introduction of the material, as well as providing an opportunity

to practice their skills with additional hands-on activities.

- The *Instructor's Guide* in Word format allows instructors to customize the lesson plans, or add their own notes to further enhance classroom presentation. Answers to the Review Questions in the book and workbook are also provided.

- *PowerPoint presentations* include an outline of each of the chapters in the text, and features drawings, photos, and procedures from the book.

- A *Testbank* available in ExamView format provides 1,500 questions, covering the thirty units in the text, for evaluating student comprehension of important concepts.

- An *Image Library* contains a multitude of photos, drawings, and procedures included in the book, allowing instructors to supplement presentations in class. A set of *AutoCAD* drawings of the house plan presented in the book is also included.

- *Step-by-Step Procedures* found in the book are also available in electronic format to enable instructors to highlight specific skills and enhance classroom presentations.

A *Companion Website* is also available to students and instructors and offers an online alternative to the *Instructor Resources CD*.

- A *CourseMate,* featuring an ebook, PowerPoint, interactive games and activities, quizzing, video, and more, provides a digital and engaging learning experience for the student. Featuring an Engagement Tracker, instructors can monitor student progress throughout the course, and pinpoint areas requiring further review. To learn more about this resource, please visit www.cengagebrain.com. At the cengagebrain.com homepage, search for the ISBN your title using the search box at the top of the page. This will take you to the product page, where your resources can be found.

Acknowledgments

The publisher and author wish to sincerely thank those who contributed to the *Carpentry* book and helped to enhance the text for the sixth edition.

Our gratitude is extended to those reviewers who contributed to the revision and previous editions of the *Carpentry* book—your insights and recommendations were invaluable:

Greg Blaney, Dawson Technical Institute of Kennedy King College

Gary Brackett
State University of New York at Delhi

Dwight Belles
Northwest Kansas Technical College

Stephen Hollman
Owensboro Community and Technical College

Louis Bermudes
Instructor
San Jose City College
San Jose, California

Richard Cappelmann
Technical College of the Low Country
Beaufort, South Carolina

Carl Gamarino
Industrial Technical Education Center
Brockport, Pennsylvania

Kirk Garrison
Department Co-Chair
Portland Community College
Portland, Oregon

Rick Glanville
Camosun College
Victoria, British Columbia, Canada

Robert Gresko
Instructor, Construction Technology
Pennsylvania College of Technology
Williamsport, Pennsylvania

Ihab Habib
Austin Peay State University

Douglas A. Holman
East Tennessee Carpenters Joint Apprenticeship and Training Program
Knoxville, Tennessee

Larry Kness
Southeast Community College
Milford, Nebraska

Jim Loosle
Kearns High School
Taylorsville, Utah

John E. Mackay
Training Coordinator
New Jersey Carpenters Technical Training Centers
Kenilworth, New Jersey

David Neu
University of Montana College of Technology

Oscar Ortiz
Orange Coast College

Dave Rainforth
Southeast Community College
Milford, Nebraska

Thomas Roever
Guilford Technical Community College

Wm. David Sanders
Fayetteville Technical Community College

Terry Schaefer
Western Wisconsin Technical College
LaCrosse, Wisconsin

Kathy Swan
Carpenter's Training Trust of Western Washington
Renton, Washington

David Vancise
Master Carpentry Instructor
Indian River Community College
Fort Pierce, Florida

Robert Wilcke
Carpentry Instructor
Western Iowa Tech Community College
Sioux City, Iowa

Kim Zupan
University of Montana College of Technology

Special thanks also to Gary Brackett for researching the IRC code references included in many chapters; to Jennifer Jacobson for her continuous, collaborative support; and to Terrel Broiles for his tireless effort to bring the drawings contained within these pages to life.

And to the Delmar Cengage Learning team, whose dedication to the project produced quality learning materials for aspiring carpenters everywhere—Jim DeVoe, Senior Acquisitions Editor; Larry Main, Managing Editor; Jennifer A. Starr, Senior Product Manager; Aviva Ariel, Editorial Assistant; David Barnes, Content Project Manager; and Casey Kirchmayer, Senior Art Director.

And thanks finally to all the numerous tradespeople who were willing to share their trade techniques and pause in their tasks for pictures. Last and most, thanks to my wife, Pamela, for her ubiquitous and untiring assistance and encouragement.

Floyd Vogt is a sixth-generation carpenter/builder. He was raised in a family with a business devoted to all phases of home construction, and began working in the family business at age fifteen.

After completing a B.A. in chemistry from the State University of New York at Oneonta, Mr. Vogt returned to the field as a self-employed remodeler. In 1985, he began teaching at State University of New York at Delhi in Delhi, New York, **www.delhi.edu**, in the Carpentry program., He has taught many courses, including Light Framing, Advanced Framing, Math, Energy Efficient Construction, Finish Carpentry, Finish Masonry, and Estimating. Currently, Mr. Vogt is a professor in construction design-build management, bachelor's degree program at Delhi. Course responsibilities include Residential Construction, AutoCAD, Construction Seminar, and Physical Science Applications. He has served as a carpentry regional coordinator for Skills-USA and postsecondary Skills-USA student advisor. He is currently a member of a local town planning board. E-mail *vogtfh@delhi.edu*

INTRODUCTION

The history of construction is long. It encompasses building materials, engineering, building techniques, and machinery. The earliest structures date back 500,000 years. The only remnants of these structures are rock and stone because any wood material has long since decayed. Structures dating back 12,000 years ago, such as GÖbekli Tepe, erected at the top of a mountain ridge in southeastern Turkey, were able to be built without metal tools (*Figure I–1*). Stone Age Europeans 8,000 B.C.E built rectangular timber houses more than 100 feet long—proving the existence of carpentry even at this early date.

The Egyptians used copper woodworking tools as early as 4000 B.C.E. By 2000 B.C.E. they had developed bronze tools and were proficient in the drilling, dovetailing, mitering, and mortising of wood.

The Greek mathematicians introduced the principals of pulleys that lead to the ability to lift heavy stone. They also had excellent surveying skills. These newer construction techniques made the construction of the Parthenon possible in 438 B.C.E. (*Figure I–2*)

In the Roman Empire, two-wheeled chariots, called *carpentum* in Latin, were made of wood. A person who built such chariots was called a *carpentarius,* from which the English word *carpenter* is derived. Roman carpenters handled iron adzes, saws, rasps, awls, gouges, and planes (*Figure I–3*).

The Romans improved lime mortar by adding volcanic ash. Its greater strength allowed them to build on a massive scale. During this time, they also developed methods for heating and ventilation that were mainly used in the thermal baths.

During the Middle Ages, most carpenters were found in larger towns where work was plentiful. They would also travel with their tools to outlying villages or wherever there was a major construction

© Vincent J. Musi/National Geographic Society/Corbis

Fig I–1 Ruins of Gobekli Tepe erected at the top of a mountain ridge in Turkey.

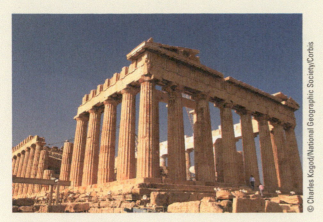

Fig I–2 The Parthenon was built in Greece in 438 B.C.E..

project in progress. By this time, they had many efficient, steel-edged hand tools. Water-wheel mills could make wood planks easier to produce. During this period, skillful carpentry was required for the building of timber churches and castles (*Figure I–4*).

In the 1100s, carpenters banded together to form guilds. The members of the guild were divided into masters, journeymen, and apprentices. The master was a carpenter with much experience and knowledge who trained apprentices. The apprentice lived with the master and was given food, clothing, and shelter and worked without pay. After a period of five to nine years, the apprentice became a journeyman who worked and traveled for wages. Eventually, a journeyman could become a master. Guilds were the forerunners of the modern labor unions and associations.

Starting in the 1400s, carpenters used great skill in constructing the splendid buildings of the Renaissance period and afterward. The 1600's was considered the birth of modern science, the first plate glass manufactured, and an increased use of iron as a building material. The triangulated roof truss was introduced during this time.

In the 1800s the steam engine, explosives, and optical equipment made larger projects possible. The introduction of the balloon frame began to replace the slower mortise-and-tenon framing style. Glass became less expensive to manufacture and was no longer considered a luxury item. In 1873, electric power was used for the first time to drive machine tools. The first electric hand drill was developed in 1917, and in 1925, electric portable saws were being used.

At present, many power tools are available to the carpenter to speed up the work. Although the volume of the carpenter's work has been reduced by the use of manufactured parts, some of the same skills carpenters used in years past are still needed for the intricate interior finish work in buildings.

Figure I–3 Tools used by workers in wood during ancient times of the Roman Empire.

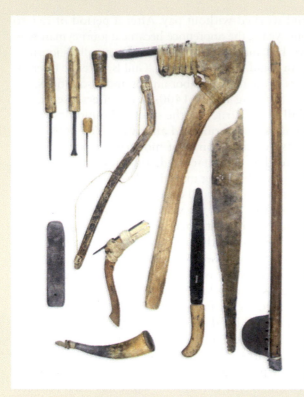

Courtesy of Art Resource

Figure I–4 Carpentry tools from the Middle Ages.

CARPENTERS

Today, a carpenter is a tradesperson that works with wood as well as metal, plastic, masonry and wood composite materials. They construct and repair structures and their parts using wood, plywood, and other building materials. They lay out, cut, fit, and fasten the materials to erect the framework and apply the finish. Carpenters build houses, factories, banks, schools, hospitals, churches, bridges, dams, and other structures. In addition to new construction, a large part of our industry is engaged in remodeling and repair of existing buildings.

The majority of workers in the construction industry are carpenters (*Figure I–5*). They are usually the first trade workers on the job, laying out for the excavation and building lines. They take part in every phase of the construction, working below the ground, at ground level, or at great heights. They can be the last to leave the job when they put the key in the door lock.

JOB OPPORTUNITES IN CONSTRUCTION

The residential construction industry is one of the biggest sectors of the American economy. When the economy is doing well so is the housing industry, and job opportunities are plentiful. The types of job in construction are many, involving people who are good at using a shovel, a tape measure, or a computer. There are those who do the physical work, those who create the ideas for the building, and those who organize the process. These occupational areas include management and design, unskilled and semiskilled labor, and skilled trades.

© Cengage Learning 2014

Figure I–5 Carpenters make up the majority of workers in the construction industry.

Management and Design

Architecture, engineering, contracting, and construction management are management and design professions. These professions require more than four years of college and often a license to practice. Many contractors have less than four years of college, but they often operate at a very high level of business, influencing millions of dollars, and so they are included with the professions here. These construction professionals spend more time in offices than on the jobsite.

Architects.   Architects usually have a strong background in art, so they are well prepared to design attractive and functional buildings. A typical architect's education includes a four-year degree and a master's degree in architecture. Most of their construction education comes during the final years of work on the architecture degree.

Engineers.   Engineers generally have more background in math and science, so they are prepared to analyze conditions and calculate structural characteristics. There are many specialties within engineering, but civil engineers are most commonly found in construction. Some civil engineers work mostly in road layout and building. Other civil engineers work mostly with structures in buildings. They are sometimes referred to as structural engineers.

Contractors.   Contractors are the owners of the businesses that do the building. Some contractors are referred to as general contractors and others as subcontractors. The general contractor is the principal construction company hired by the owner to construct the building. In larger construction firms, the contractor may be more concerned with running the business than with supervising actual construction. A general contractor might have only a skeleton crew, relying on subcontractors for most of the actual construction. The general contractor may have a superintendent who coordinates the work of all the subcontractors. Subcontractors are smaller companies concerned with only a portion of the building, such as framing or electrical work.

Most states require contractors to have a license to do contracting in their state. Requirements vary from state to state, but a contractor's license usually requires several years of experience in the trade and a test on both trade information and the contracting business.

Construction managers.   Construction managers are those who arrange, guide, coordinate, and financially plan the construction process. Large construction jobs are too complicated for one person to manage. Construction managers oversee the entire job from start to finish. They are hired by the owner as their agent to work with the architect and general contractor. They are often responsible for tracking down all licenses and permits and for watching that building and safety codes are met. The job titles include project managers, construction superintendents, project engineers, and construction supervisors (*Figure I–6*).

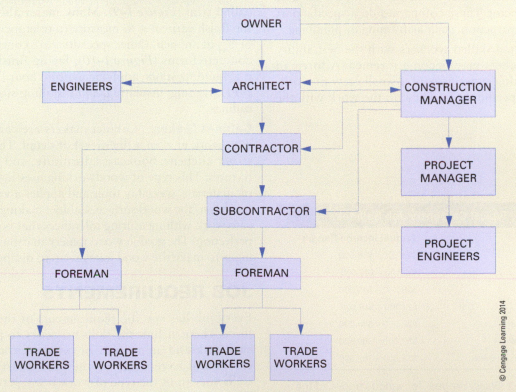

Figure I–6 An organizational chart showing the path of responsibilities for the construction industry.

© Cengage Learning 2014

Unskilled or Semiskilled Labor

Construction by its very nature is labor intensive. The industry needs many kinds of workers. Construction workers with limited skills are called laborers. Laborers are assigned tasks of moving materials and working under the close supervision of a skilled worker. Their work is strenuous, and so construction laborers must be in excellent physical condition.

Construction laborers are often construction workers who have not yet reached a high level of skill in a particular trade. Many laborers go on to acquire trade skills and become skilled workers. Laborers often specialize in working with a particular trade, such as mason's tenders or carpenter's helpers. Laborers who specialize in a particular trade are often paid slightly more than completely unskilled laborers. Some laborers are skilled construction workers opting to work in a role with reduced responsibility and increased simplicity.

Skilled Trades

A skilled tradesperson works with tools and materials to build structures. They are collectively referred to as members of the building trades. These trades are among the highest paying of all skilled occupations (*Figure I–7*). Generally plenty of work is available to provide a comfortable living for a good worker. It is quite common for a successful tradesperson to start his or her own business as a subcontractor. These companies do a specific part of the construction, such as framing or plumbing.

Foremen.   Skilled workers with the best ability to lead others may become foremen. A foreman is a working supervisor of a small crew of workers in a specific trade. They do the work with the assistance of other workers and see that the work is done properly and efficiently. Foremen earn their responsibilities from trade experience and personal leadership skills.

There are many jobs and duties in construction. Go to **www.bls.gov/oco/cg/cgs003.htm** for more information on construction careers at the U.S. Department of Labor, Bureau of Labor Statistics.

Trade Specializations

In less densely populated communities, where the volume of construction is lighter, carpenters often perform tasks in all areas of the trade. They come to the site when the foundation is complete and stay until the house is completed. They are known as general carpenters. In more densely populated areas where work is abundant, carpenters tend to specialize in one particular phase of carpentry. The general carpenter needs a more complete knowledge of the trade than the specialist does.

Rough Carpentry.   Rough carpentry is work involving erecting structural frames, scaffolding, and concrete formwork (*Figure I–8*). For the most part, this type of work is either taken apart later or covered with layers of finish. Rough carpentry does not mean that the workmanship is crude. They typically measure to within 1/16 inch. Just as much care is taken in the rough work as in any other work.

Finish Carpentry.   Finish carpenters specialize in applying exterior and interior finish, sometimes called trim (*Figure I–9*). Many materials require the finish carpenter to measure to tolerances of $\frac{1}{32}$ or $\frac{1}{64}$ of an inch. Other specialties are constructing concrete forms (*Figure I–10*), laying finish flooring, building stairs, applying gypsum board (*Figure I–11*), roofing, insulating, and installing suspended ceilings.

Cabinet Maker.   Cabinet makers are carpenters who work at even a finer level of detail. They construct kitchens, bookcases, furniture, and built-in fixtures. This type of woodworking uses high quality materials, speciality tools and applies a variety of finishes to the wood surface. Cabinet making is intricate work with measuring tolerances that border on perfection. The goal in wood joinery in cabinet making is to make the wood joints seem to disappear.

JOB REQUIREMENTS

Every job has specific requirements of the workforce. These include the skills necessary to perform each task and an attitude or mind-set the workers should have while on the job. Construction skills vary to fit the type of work being performed. They tend to be clearly defined, and it is easy to

Typical Carpentry Job Salaries	
Title	**Annual Income Ranges**
Laborer	$23,500 – $38,500
Framer	$27,500 – $43,000
Roofer	$28,500 – $46,000
Floorer	$27,000 – $48,500
Cabinetmaker	$29,500 – $44,000
Drywall Finisher	$31,000 – $50,500
Construction Foremen	$39,500 – $62,000

Cengage Learning 2014

Figure I–7 Salary ranges for various trades of the construction industry.

Figure I–8 Some carpenters specialize in framing.

Figure I–9 Carpenters may be specialists in applying interior trim.

© Cengage Learning 2014

Figure I–10 Erecting concrete formwork may be a specialty.

© Cengage Learning 2014

Figure I–11 Some carpenters choose to do only gypsum board application.

determine if a worker possesses these skills. The construction attitude is less tangible, but no less important. While work may continue if the workers have only the required skills, minor or severe slowdowns are inevitable if workers do not have the proper attitude.

Skills

The skills of a successful carpenter are many. They include technical, mental, communication, and physical skills.

Technical Skills.   Carpenters need to know how to use and maintain hand and power tools. They need to know the kinds, grades, and characteristics of the materials with which they work—how each can be cut, shaped, and most satisfactorily joined. Carpenters must be familiar with the many different fasteners available and choose the most appropriate for each task.

Carpenters should know how to lay out and frame floors, walls, stairs, and roofs. They must know how to install windows and doors, and how to apply numerous kinds of exterior and interior finish. They must use good judgment to decide on proper procedures to do the job at hand in the most efficient and safest manner.

Physical Skills.   Carpenters need to be in good physical condition because much of the work is done by hand and sometimes requires great exertion. They must be able to lift large sheets of plywood, heavy wood timbers, and bundles of roof shingles; they also have to climb ladders and scaffolds.

Mental Skills   Carpenters need reading and math skills. Much of construction begins as an idea put on paper. A carpenter must be able to interpret these ideas from the written form to create the desired structure. This is done by reading prints and using a ruler. The quantity of material needed must be estimated using math and geometry. Accurate measurements and calculations speed the construction process and reduce wasted materials.

Communication Skills.   Communication skills are also very important. Carpenters must communicate with many people during the construction process. Work to be done as determined by the owner or architect must be accurately understood; otherwise costly delays and expenses may result. Efficiency of work relies heavily on workers' understanding of what others are doing and what work must be done next. Communication is vital for a jobsite to be safe.

Attitude

The proper construction attitude is not as clearly defined as job skills. Regional variations and requirements will affect the expected jobsite attitude. For example, in regions where heat is a concern, workers develop a steady rhythm to their work that survives the heat. In regions where winters are harsh, workers develop a faster style of work when the weather is warm knowing they may have downtime in the winter. Other attitudes, such as having a good work ethic, are universal to all jobsites.

Work Ethic.   A good work ethic is not easily defined in one sentence. It involves the person as a whole, the way he or she approaches life and their work. A person with a good work ethic has respect and lets it show. Respect for the jobsite, fellow workers, tools and materials, and oneself reveals care and concern for the construction process. Workers demonstrating this form of respect are safe and more pleasant to work alongside.

Professional.   Workers with a good work ethic show up fifteen minutes early, not a few minutes late. That is being a professional. When finished with a task, they look for something else to do that promotes the job, even if it is using a broom. They perform their tasks as well as expected, up to the standard required for that application. They finish the task, not leaving something undone for someone else to fix. They are interested in learning, looking around at the work of others for ways to improve their own skills. They cooperate with other workers and tradespeople to make the jobsite a pleasant workplace. They are also honest with the material and their time, never cheating on the accepted method. They feel that nothing less than a first-class job is acceptable.

Jobsite Humor.   Jobsite humor makes work easier to do and more fun. Some tasks, by their very nature, are boring and unpleasant. Humor can make difficult jobs seem to get done faster. Unfortunately, humor has a bad side, because humor for one person can be pain for another. Jobsite humor can be tasteless and just plain mean. It can single out a person, making them feel alienated. This type of humor can severely affect jobsite safety. If someone feels like a victim, they defend themselves, and do not concentrate on what they are doing. During these times, unintended things can happen, and someone may get hurt. Keep jobsite humor suitable for everyone present.

Teamwork.   There is no replacement for teamwork on the jobsite (*Figure I–12*). Someone once said, "Two people working together can outperform three people working alone." Teamwork is

Figure I–12 Teamwork makes any job more safe and efficient.

©Richard Levine/Alamy

why construction is so much fun. It can be seen when one person holds material to be fastened or cut for another. It exists when a heavy object is lifted into place by four people. Like the old saying goes, "Many hands make light work."

But the difference between a team and a group of individuals is more dramatic when each member of the team anticipates the next move of the other. For example, while holding a board that is being fastened or cut, a good team member looks to see what will be the next step. Is more material needed? Does the horse or ladder have to be moved into a better position? Will there be another tool needed? How can the tool best be handed to the fellow worker? It can be easy to miss this type of teamwork when it is done without words and without being asked.

Another way a team works better is in the mental energy used on the job. Many pairs of eyes working on one task can ward off errors and mistakes. better than one. Two minds can find the better, faster, safer method. Teamwork is a major reason why people stay in construction for a lifetime.

TRAINING

Vocational training in carpentry is offered by many high schools for those who become seriously interested at an early age. For those who have completed high school, carpentry training programs are offered at many postsecondary vocational schools and colleges. Because there is so much to learn, it can take years to learn carpentry.

Some schools participate in programs in which students go to school part-time and work as on-the-job apprentices part-time. Upon graduation, they may continue with the same employer as full-time apprentices.

Apprenticeship training programs (usually four years in length) are offered by the United Brotherhood of Carpenters and Joiners of America, a carpenters' union, and by contractors' associations, such as the National Association of Home Builders of the United States (NAHB) and the Associated General Contractors of America (AGC) in cooperation with the Bureau of Apprenticeship and Training of the U.S. Department of Labor (*Figure I–13*). In Canada, apprenticeships are administered by the

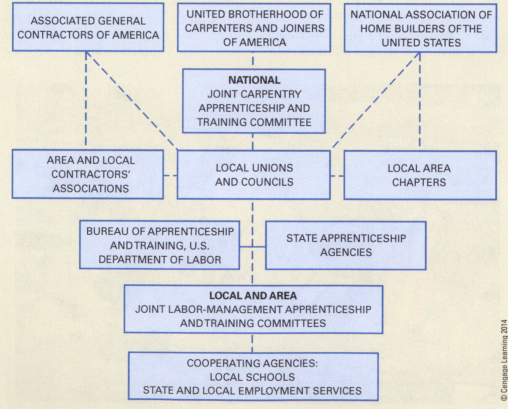

APPRENTICESHIP AND TRAINING SYSTEM OF THE CARPENTRY TRADE

© Cengage Learning 2014

Figure I–13 Industry, labor, and government work together in carpentry apprenticeship programs.

provincial governments with a final exam called the interprovincial examination written for national accreditation. Usually, these organizations give apprenticeship credit for completion of previous vocational school training in carpentry, resulting in a shortened apprenticeship and a higher starting wage.

The apprentice (*Figure I–14*) must be at least seventeen years of age, in some areas eighteen, and must learn the trade while working on the job with experienced journeymen. Under the apprenticeship agreement, basic standards provide, among other things, that there can be only a certain number of apprentices hired by a particular contractor in relation to the number of journeymen. This is to ensure that apprentices receive proper supervision and training on the job. Basic standards also provide that there is a progressively increasing schedule of wages. Starting pay for the apprentice carpenter is usually about 50 percent of the journeyman carpenter's rate. Due to periodic increases about every six months, the apprentice should receive 95 percent of the journeyman's wage during the last six months of the apprenticeship.

Under the terms of agreement, the apprentice is required to attend classes for a certain number of hours each year. A minimum of 144 hours per year is normally considered necessary. These classes are usually held at local schools, twice a week for about thirty-six weeks during the school year, for the length of the apprenticeship. The apprentice becomes accepted as a journeyman carpenter when the training is completed, and is awarded a Certificate of Completion of Apprenticeship. The newly

graduated apprentice is now expected, within reason, to do jobs required of the journeyman.

Although it is in the best interests of the apprentice to be indentured, that is, to have a written contract with an organization, with conditions of the apprenticeship agreed on, it is possible to learn the trade as a helper until enough skills have been acquired to demand the recognition and rewards of a journeyman carpenter. However, self-discipline is required to gain knowledge of both the practical and theoretical aspects of the trade. There may be time on the job to explain to a helper or apprentice how to do a certain task, but there usually is not time to explain concepts.

Many opportunities exist for the journeyman carpenter. Advancement depends on dependability, skill, productivity, and ingenuity, among other characteristics. Carpentry foremen, construction superintendents, and general contractors usually rise from the ranks of the journeyman carpenters. Many who start as apprentice carpenters eventually operate their own construction firms. A survey revealed that 90 percent of the top officials (presidents, vice presidents, owners, and partners) of construction companies who replied began their careers as apprentices. Many of the project managers, superintendents, and craft supervisors employed by these companies also began as apprentices.

National student organizations like SkillsUSA, formally VICA (Vocational Industrial Clubs of America, Inc.), offer students training in leadership, teamwork, citizenship, and character development. SkillsUSA helps build and reinforce work attitudes, self-confidence, and communications skills for the future workforce. It emphasizes total quality at work, high ethical standards, superior work skills, and pride in the dignity of work. The organization also promotes involvement in community service activities. Yearlong student activities culminate with skill competitions where local winners move up to regional, state, and then national and international levels.

SAFETY AT THE WORK SITE

Much of the work performed in construction carries risk with it. Although the risk will vary with the type of work, all construction workers, at some time, may be at risk of being maimed or killed on the job.

Tools are, by their very nature, dangerous to use. Each must be operated in a fashion suited to the design of the tool. Some tools can cause cuts, while others can kill. Safety cannot be taken for granted and must be built into the methods and process being used for any particular task.

© Cengage Learning 2014

Figure I–14 Many opportunities lie ahead for the apprentice carpenter.

Unusual situations leading to accidents can happen at any time. For example, a roofing carpenter can be thrown to the ground by a gust of wind, cleanup workers can be injured by falling objects from a scaffold, and trim carpenters can fall through a stairway opening that is under construction. Safety is like air; it must be continuous and everywhere for people to survive.

Jobsite safety is like team sportsmanship. Everyone on the team must work together and play by the same rules or someone is at risk of being injured (*Figure I–15*). Safety is like silence; just as everyone in a room must agree to be silent for silence to exist, so too everyone on the job must agree to be safe for safety to exist. One person can create a situation where dozens of workers will be at risk. Safety programs are successful only because all workers join the team.

Safe Work Practices

Safe methods and practices learned correctly today will develop into habits that will become second nature for a lifetime. Always approach new tasks thoughtfully and carefully. If a new tool is being used, become familiar with its requirements. Adapt to the tool, because it cannot adapt to you. Always read and follow the directions associated with

Figure I–15 Safety is a team sport.

the tool or task. Read the manufacturer's recommendations for use and installation. Manufacturers clearly define the risks and recommended uses. Failure to follow their instructions is foolish and short-sighted. Always use the safety devices designed into a tool. They are there to protect the operator, not to make it harder to use the tool. Always remember where you are and the possible dangers of the work site. Horseplay and practical jokes are distracting to everyone around. When horseplay happens, the work environment is less safe. Workers have suffered permanent back injuries while lifting an object too fast because someone touched them inappropriately. Respect is the key to safety. Always listen and communicate fully. Think of possible ways for misunderstanding to develop and ensure they are avoided. The only stupid question is the one not asked.

Listening   Construction efficiency and jobsite safety require good listening skills. Listening to what is asked and performing that task efficiently is one form of good listening. It is just as important to listen to the sounds of the job. Most sounds are normal; some are not. A saw cutting wood, a mixer mixing mortar, an air compressor providing power, and the bangs, taps, and thuds of construction are examples of normal noises. As workers become acquainted and accustomed to these sounds, it is important to be able to tell when these sounds change. A change in sound can be an early warning signal that something is wrong. It could be that something is about to break or someone is in trouble and needs assistance.

Safe Work Conditions

Safe work conditions are easy to create, yet they can be just as easily neglected. Always maintain clear areas for working and walking. Store materials in an orderly fashion, and dispose of waste materials as they are produced. If you stumble and struggle for a place to stand, then safety risks have increased. Always work to keep the area as dry as possible. Sweep water puddles away or make small trenches to divert the water. Keep electrical cords and tools dry. Always maintain tools and equipment according to manufacturer's recommendations. Equipment needs may include lubrication, cleaning, drying, or inspection. Always be aware of the effect of air temperature on safety. Cold or frozen areas can be slippery to work on. Hot areas can have soft material that is easily marred.

Always understand the risks associated with building materials. All workers are to have access to information on the hazards of materials they are

1. Product and manufacturer information	9. Physical and chemical properties
2. List of any hazardous ingredients in the product	10. Reactivity and stability of product
3. Physical characteristics of the product	11. Health hazards
4. First Aid Measures	12. Ecological and environmental information
5. Fire and explosive nature of product	13. Disposal concerns
6. Spill cleanup procedures	14. Transport information
7. Safe handling and storing information	15. U.S. Regulatory information
8. Personal exposure concerns	16. Special Precautions

© Cengage Learning 2014

Figure I–16 MSDS (Material Safety Data Sheets) contain much information.

working with through the **Material Safety Data Sheet (MSDS)**. Topics of information that may be included in MSD sheets are shown in *Figure 1–16*. Read them. Always look for ways to keep fires from starting. Any possible heat source should be carefully studied and isolated to prevent fires from starting.

Always wear personal protective devices. Eyes, ears, and airways should be protected when injury is possible. It is easier to understand why eyes are important to protect. Most of our perception of the world comes through our eyes. But ear, lung, and sinus damage often occurs slowly. Most times these effects do not show up for years. Personal protection is a personal responsibility.

OSHA

OSHA (Occupational Safety and Health Administration) was created in 1970 to improve workplace safety. OSHA's goal is to save lives, prevent injuries, and protect the health of America's workers. They have been given the authority to work with state agencies to enforce acceptable work site safety standards. OSHA provides research information, education, and training in the field of occupational safety and health. Their efforts, and safety in general, will succeed only if every worker agrees to be a team player and does their best to maintain safe work habits and site conditions.

SUMMARY

Carpentry is a trade in which there is a great deal of self-satisfaction, pride, and dignity associated with the work. It is an ancient trade and the largest of all trades in the building industry.

Skilled carpenters who have labored to the best of their ability can take pride in their workmanship, whether the job was a rough concrete form

or the finest finish in an elaborate staircase (**Figure I–17**). At the end of each workday, carpenters can stand back and actually see the results of their labor. As the years roll by, the buildings that carpenters' hands had a part in creating still can be viewed with pride in the community.

Courtesy of L. J. Smith, Inc.

Figure I–17 After completing a complicated piece of work, such as this intricate staircase, carpenters can take pride in and view their accomplishment.

Tools and Materials

© iStockphoto.com/mattjeacock

Courtesy Michael E.C. Surguy

Michael E. C. Surguy
Title: Owner
Company: Michael E. C. Surguy Carpenter Contractors Inc., New Providence, NJ

EDUCATION

Michael graduated from high school in 1987 and admits that he did not apply himself while there. Next, he attended vocational school studying carpentry, then started framing, and worked in the field for several years.

When he was twenty-six years old, with several years of labor experience to build on, Michael decided to take two years off and study. He researched schools, finding that few had carpentry programs. He entered the State University of New York at Delhi as the oldest student in the program. "I went there because I wanted to learn," he explained. "Going to class was one thing, but I brought blueprints from old employers so I could study them, and my instructors helped me out with the math." Michael obtained his Associate of Occupation Studies degree from SUNY—Delhi, having developed his specialized carpentry and woodworking skills.

HISTORY

Today, Michael is the sole owner of Michael E. C. Surguy Carpenter Contractors Inc. Michael's company does jobs ranging from basic framing to complex remodeling and additions, working on $3-million to $4-million homes. Recognized by the Community Builders Association of New Jersey, in 2004, as a recipient of their Subcontractor of the Year Award, Surguy is one of New Jersey's specialized carpentry contractors.

ON THE JOB

In a typical day, Michael is on the job. "In my business and doing the types of jobs I do, I need to be on the job. I don't just set up my guys and leave. I have a tool belt on every day. People pay me to be there and work. If I'm not there, the work doesn't get done." Michael expresses his disapproval of those competitors who subcontract out their work, believing this is not the right thing to do. "If you get the job, you need to be there and run the job."

Michael takes his responsibility seriously. "I work in expensive homes, and I need to be there every day. I've learned that I need to be there so I am there."

BEST ASPECTS

Michael has a deep passion for carpentry. He reported learning at a National Home Builders Association meeting that only one in seventy-five people make it, and his theory for why this is the case is what the job requires emotionally. "You really need to enjoy it and be passionate about it. I don't think it's with every type of job. In carpentry, you are using your head and your body. Every 2 × 10–24 feet piece of wood you need to carry with passion. If you show up for work in the morning, and you're thinking that it's too heavy or it's too cold outside, you won't do it."

CHALLENGES

Michael knows that his job isn't for everyone. While he shares that the business has been good to him, he explains the stress it has involved and describes the growing pains of building a business. "When I started out, I lived at my parents' house. I was lucky. I made nothing for the first three years, but my guys always got paid. I got my tools and paid my insurance. If I hadn't lived at home, there's no way I could have done it. Nobody tells you this," he explains. Michael described building his reputation by doing jobs that he thought he could do and being honest with his bids, breaking each job down, figuring out the cost of materials and employees. At first, Michael took on small jobs that turned into bigger jobs. He started out as sole proprietor and then became a corporation with employees. "Now it's a different ball game. You have to have liability insurance. It's a business." Currently in his twelfth year, he is confident about running the business and where to find good workers. "I know what I'm doing now."

After many years of custom framing dominating his business, Michael reports, "You can't just frame now, not in this economy. There's not enough work." Four years ago, Surguy Construction was doing seven to eight custom houses each year.

"Now we're only doing three to four a year." He added, "I know a lot of people who are not working. Unemployment is about 7.5 percent." In response to the economic downturn, Michael has diversified and is now doing exterior detail and cedar siding. "We reframe them, do all the exterior stuff, all the soffit work and bevel siding."

IMPORTANCE OF EDUCATION

Michael persevered in school, committed to learning what he knew he needed to learn. "When I finished Delhi, I did a lot of studying on my own with trigonometry. I studied a lot at night on my own and took night classes. I bought carpentry books and studied out of those." When he was starting his business, he prepared himself to be a one-man operation. "I knew that when I started out, I wouldn't have a crew. I can probably pull a five-guy job with just two guys. I kept a smaller crew but produced just as much work."

FUTURE OPPORTUNITIES

Michael expressed a great deal of concern over what this economy has done and will continue to do to the home-building business, and named several businesses that are closing stores, laying off workers, and reducing employee hours. Michael is hoping to make it through these tough times. "In the last four years, I started making money. Last year was a bad year, but this year has started off well. No one knows what is going to happen."

WORDS OF ADVICE

Michael recalls the advice of his favorite instructor at Delhi: "You can either learn the right way. Or you can learn the wrong way. But if you learn the wrong way, you will always work the wrong way." And to this he adds advice from his own experience: "You have to do work to get work. If you don't do work, people won't know you."

Interview courtesy of Michael E.C. Surguy

UNIT 1
Wood and Lumber

CHAPTER 1 Wood
CHAPTER 2 Lumber

The construction material most often associated with a carpenter is wood. Wood has properties that make it the first choice in many applications in home construction. Wood is easy to tool and work with, is pleasing to look at and smell, and has strength that will last a long time.

Lumber is manufactured from the renewable resources of the forest. Trees are harvested and sawn into lumber in many shapes and sizes having a variety of characteristics. It is necessary to understand the nature of wood to get the best results from the use of it. With this knowledge, the carpenter is able to protect lumber from decay, select it for appropriate use, work it with proper tools, and join and fasten it to the best advantage.

Wood comes from many tree species having many different characteristics. A good carpenter knows the different applications for specific wood species.

OBJECTIVES

After completing this unit, the student should be able to:

- name the parts of a tree trunk and state their functions.

- describe methods of cutting the log into lumber.

- define hardwood and softwood, give examples of some common kinds, and list their characteristics.

- explain moisture content at various stages of seasoning, tell how wood shrinks, and describe some common lumber defects.

- state the grades and sizes of lumber and compute board measure.

CHAPTER 1

Wood

The carpenter works with wood more than any other material and must understand its characteristics in order to use it intelligently. Wood is a remarkable substance. It can be more easily cut, shaped, or bent into almost any form than just about any other structural material. It is an efficient insulating material. It takes almost 6 inches of brick, 14 inches of concrete, or more than 1,700 inches of aluminum to equal the insulating value of only 1 inch of wood.

There are many kinds of wood, and they vary in strength, workability, elasticity, color, grain, texture, and smell. It is important to keep these qualities in mind when selecting wood. For instance, baseball bats, diving boards, and tool handles are made from hickory and ash because of their greater ability to bend without breaking (elasticity). Oak and maple are used for floors because of their beauty, hardness, and durability. Redwood, cedar, cypress, and teak are used in exterior situations because of their resistance to decay (*Figure 1–1*). Cherry, mahogany, and walnut are typically chosen for their beauty.

With proper care, wood will last indefinitely. It is a material with beauty and warmth that has thousands of uses. Wood is one of our greatest natural resources. With wise conservation practices, wood will always be in abundant supply. It is fortunate that we have perpetually producing forests that supply this major building material from which we can construct homes and other structures that last for hundreds of years. When those structures have served their purpose and are torn down, the wood used in their construction can be salvaged and used again (recycled) in new building, remodeling, or repair. Wood is biodegradable, and when it is considered not feasible for reuse, it is readily absorbed back into the earth with no environmental harm.

STRUCTURE AND GROWTH

Wood is a material cut from complex living organisms called trees. Trees are made up of many different kinds of cells and growth areas that are visible in the cross-sectional view (*Figure 1–2*). Wood is made up of many hollow cells held together by a natural substance called lignin. The size, shape, and arrangement of these cells determine the strength, weight, and other properties of wood. Tree growth takes place in the cambium layer, which is just inside the protective shield of the tree called the *bark*. The tree's roots absorb water, which passes upward through the sapwood to the leaves, where it is combined with

Figure 1–1 Redwood is often used for exterior trim and siding.

Courtesy of California Redwood Association

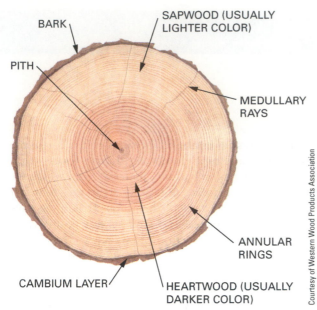

BARK

PITH

SAPWOOD (USUALLY LIGHTER COLOR)

MEDULLARY RAYS

ANNULAR RINGS

CAMBIUM LAYER

HEARTWOOD (USUALLY DARKER COLOR)

Courtesy of Western Wood Products Association

Figure 1–2 A cross section of a tree showing its structure.

carbon dioxide from the air. Sunlight causes these materials to change into food, which is then carried down to the root. Sap is distributed toward the center of the trunk through the medullary rays.

As the tree grows outward from the pith (center), the inner cells become inactive and turn into heartwood. This older section of the tree is the central part of the tree and usually is darker in color and more durable than sapwood. The heartwood of cedar, cypress, and redwood, for instance, is extremely resistant to decay and is used extensively for outdoor furniture, patios, and exterior siding. Used for the same purposes, sapwood decays more quickly.

Each spring and summer, a tree adds new layers to its trunk. Wood grows rapidly in the spring; it is rather porous and light in color. In summer, tree growth is slower; the wood is denser and darker, forming distinct rings. Because these rings are formed each year, they are called annular rings. By counting the dark rings, the age of a tree can be determined. Periods of abundant rainfall and sunshine or periods of slow growth can be discerned by studying the width of the rings. Some trees, like the Douglas fir, grow rapidly to great heights and have very wide and pronounced annular rings. Mahogany, which grows in a tropical climate where the weather is more constant, has annular rings that do not contrast as much and sometimes are hardly visible.

HARDWOODS AND SOFTWOODS

Woods are classified as either hardwood or softwood. There are different methods of classifying these woods. The most common method of classifying wood is by its source. Hardwood comes from deciduous trees, which shed their leaves each year. Softwood is cut from coniferous, or cone-bearing, trees, commonly known as evergreens (*Figure 1–3*). In this method of classifying wood, some of the softwoods may actually be harder than the hardwoods. For instance, fir, a softwood, is harder and stronger than basswood, a hardwood. There are other methods of classifying hardwoods and softwoods, but this method is the one most widely used.

© Cengage Learning 2014

Figure 1–3 Softwood is from cone-bearing trees, hardwood from broad-leaf trees.

Some common hardwoods are ash, birch, cherry, hickory, maple, mahogany, oak, and walnut. Some common softwoods are pine, fir, hemlock, spruce, cedar, cypress, and redwood.

Wood can also be divided into two groups according to cell structure. Open-grained wood has large cells that show tiny openings or pores in the surface. To obtain a smooth finish, these pores must be filled with a specially prepared paste wood filler. Examples of open-grained wood are oak, mahogany, and walnut. Some close-grained hardwoods are birch, cherry, maple, and poplar. All softwoods are close-grained. (See *Figure 1–4* for common types of softwoods and their characteristics and *Figure 1–5* for common hardwoods.)

Identification of Wood

Identifying different kinds of wood can be very difficult because some closely resemble each other. For instance, ash and white oak are hard to distinguish from each other, as are some pine, hemlock, and spruce. Not only are they the same color, but the grain pattern and weight are about the same. Only the most experienced workers are able to tell the difference.

It is possible to get some clues to identifying wood by studying the literature, but the best way to learn the different kinds of wood is by working with them. Each time you handle a piece of wood, examine it. Look at the color and the grain; feel if it is heavy or light, if it is soft or hard; and smell it for a characteristic odor. Aromatic cedar, for instance, can always be identified by its pleasing, moth-repelling odor, if by no other means. After studying the characteristics of the wood, find out the kind of wood you are holding, and remember it. In this manner, after a period of time, those kinds of wood that are used regularly on the job can be identified easily. The identification of kinds of wood that are seldom worked with can be accomplished in the same manner, but, of course, the process will take a little longer.

Softwoods									
Kind	**Color**	**Grain**	**Hardness**	**Strength**	**Work Ability**	**Elasticity**	**Decay Resistance**	**Uses**	**Other**
Red Cedar	Dark Reddish Brown	Close Medium	Soft	Low	Easy	Poor	Very High	Exterior	Cedar Odor
Cypress	Orange Tan	Close Medium	Soft to Medium	Medium	Medium	Medium	Very High	Exterior	
Fir	Yellow to Orange Brown	Close Coarse	Medium to Hard	High	Hard	Medium	Medium	Framing Millwork Plywood	
Ponderosa Pine	White with Brown Grain	Close Coarse	Medium	Medium	Medium	Poor	Low	Millwork Trim	Pine Odor
Sugar Pine	Creamy White	Close Fine	Soft	Low	Easy	Poor	Low	Pattern-making Millwork	Large Clear Pieces
Western White Pin	Brownish White	Close Medium	Soft to Medium	Low	Medium	Poor	Low	Millwork Trim	
Southern Yellow Pine	Yellow Brown	Close Coarse	Soft to Hard	High	Hard	Medium	Medium	Framing Plywood	Much Pitch
Redwood	Reddish Brown	Close Medium	Soft	Low	Easy	Poor	Very High	Exterior	Light Sapwood
Spruce	Cream to Tan	Close Medium	Medium	Medium	Medium	Poor	Low Odor	Siding Subflooring	Spruce Odor

Figure 1–4 Common species of softwood and their characteristics.

Hardwoods									
Kind	Color	Grain	Hardness	Strength	Work Ability	Elasticity	Decay Resistance	Uses	Other
Ash	Light Tan	Open Coarse	Hard	High	Hard	Very High	Low	Tool Handles, Oars, Baseball Bats	
Basswood	Creamy White	Close Fine	Soft	Low	Easy	Low	Low	Drawing Boards, Veneer Core	Imparts No Taste or Odor
Beech	Light Brown	Close Medium	Hard	High	Medium	Medium	Low	Food Containers, Furniture	
Birch	Light Brown	Close Fine	Hard	High	Medium	Medium	Low	Furniture Veneers	
Cherry	Light Reddish Brown	Close Fine	Medium	High	Medium	High	Medium	Furniture	
Hickory	Light Tan	Open Medium	Hard	High	Hard	Very High	Low	Tool Handles	
Lauan	Light Reddish Brown	Open Medium	Soft	Low	Easy	Low	Low	Veneers Paneling	
Mahogany	Russet Brown	Open Fine	Medium	Medium	Excellent	Medium	High	Quality Furniture	
Maple	Light Tan	Close Medium	Hard	High	Hard	Medium	Low	Furniture Flooring	
Oak	Light Brown	Open Coarse	Hard	High	Hard	Very High	Medium	Flooring Boats	
Poplar	Greenish Yellow	Close Fine	Medium Soft	Medium Low	Easy	Low	Low	Furniture Veneer Core	
Teak	Honey	Open Medium	Medium	High	Excellent	High	Very High	Furniture Boat Trim	Heavy Oily
Walnut	Dark Brown	Open Fine	Medium	High	Excellent	High	High	High Quality Furniture	

Figure 1–5 Common species of hardwood and their characteristics.

© Cengage Learning 2014

CHAPTER 2

Lumber

MANUFACTURE OF LUMBER

When logs arrive at the sawmill, the bark is removed first. Then a huge band saw slices the log into large planks, which are passed through a series of saws. The saws slice, edge, and trim the planks into various dimensions, and the pieces become **lumber.**

Once trimmed of all uneven edges, the lumber is stacked according to size and grade and taken outdoors where **stickering** takes place. Stickering is the process of restacking the lumber on small cross-sticks that allow air to circulate between the pieces. This air-seasoning process may take six months to two years due to the large amount of water found in lumber.

Following air-drying, the lumber is dried in huge ovens. Once dry, the rough lumber is surfaced to standard sizes and shipped (*Figure 2–1*).

The long, narrow surface of a piece of lumber is called its edge; the long, wide surface is termed

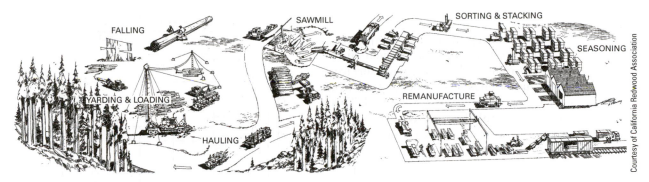

Figure 2–1 Manufacturing path of lumber from forest to finished product.

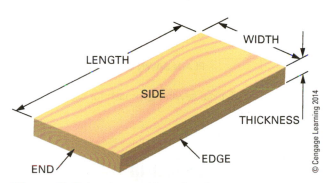

Figure 2–2 Lumber and board surfaces are distinguished by specific names.

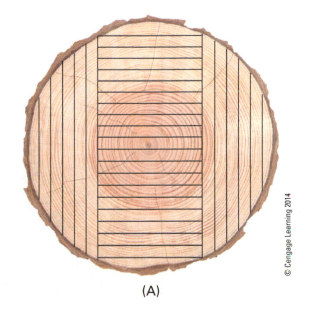

(A)

its side; and its extremities are called ends. The distance across the edge is termed its thickness, across its side is called its width, and from end to end is termed its length (*Figure 2–2*). The best-appearing side is called the face side, and the best-appearing edge is called the face edge.

In certain cases, the surfaces of lumber may acquire different names. For instance, the distance from top to bottom of a beam is called its depth, and the distance across its top or bottom may be called its width or its thickness. The length of posts or columns, when installed, may be called their height.

Plain-Sawed Lumber

A common way of cutting lumber is called the plain-sawed method, in which the log is cut tangent to the annular rings. This method produces a distinctive grain pattern on the wide surface (*Figure 2–3*). This method of sawing is the least expensive and produces greater widths. However, plain-sawed lumber shrinks more during drying and warps easily. Plain-sawed lumber is sometimes called slash-sawed lumber.

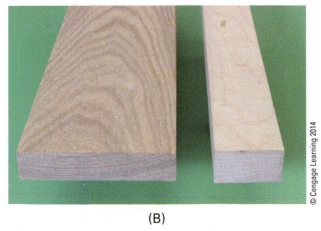

(B)

Figure 2–3 (A) Typical sawing approach for plain-sawed lumber. (B) Surfaces of plain-sawed lumber.

Quarter-Sawed Lumber

Another method of cutting the log, called quarter-sawing, produces pieces in which the annular rings are at or almost at right angles to the wide surface. Quarter-sawed lumber has less tendency to warp, and shrinks less and more evenly when dried. This type of lumber is durable because the wear is on the edge of the annular rings. Quarter-sawed lumber is frequently used for flooring.

A distinctive and desirable grain pattern is produced in some wood, such as oak, because the lumber is sawed along the length of the medullary rays. Quarter-sawed lumber is sometimes called vertical-grain or edge-grain (*Figure 2–4*).

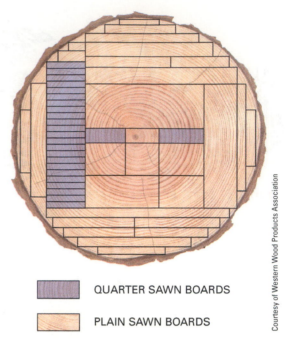

QUARTER SAWN BOARDS

PLAIN SAWN BOARDS

Figure 2–5 Typical sawing approach for combination-sawed lumber.

Courtesy of Western Wood Products Association

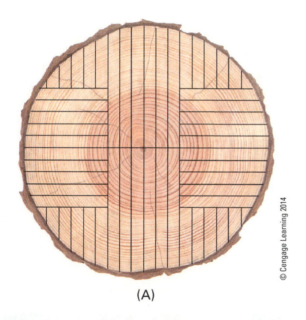

(A)

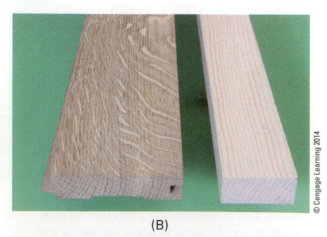

(B)

Figure 2–4 (A) Typical sawing approach for quarter-sawed lumber. (B) Surfaces of quarter-sawed lumber.

© Cengage Learning 2014

Combination Sawing

Most logs are cut into a combination of plain-sawed and quarter-sawed lumber. With computers and laser-guided equipment, the sawyer determines how to cut the log with as little waste as possible in the shortest amount of time to get the desired amount and kinds of lumber (*Figure 2–5*).

MOISTURE CONTENT AND SHRINKAGE

When a tree is first cut down, it contains a great amount of water. Lumber, when first cut from the log, is called green lumber and is very heavy because most of its weight is water. A piece 2 inches thick, 6 inches wide, and 10 feet long may contain as much as $4\frac{1}{4}$ gallons of water weighing about 35 pounds (*Figure 2–6*).

Green Tip

Manufacturing wood products from logs uses less energy than does manufacturing metal or masonry products.

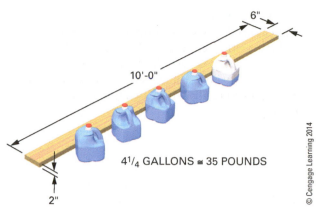

Figure 2–6 Green lumber contains a large amount of water.

Figure 2–7 Wood cells are long, hollow tubes that fill with water during tree growth.

Green lumber should not be used in construction because it shrinks as it dries to the same moisture content as the surrounding air. This shrinking is considerable and unequal because of the large amount of water that leaves it. When it shrinks, it usually warps, depending on the way it was cut from the log. The use of green lumber in construction results in cracked ceilings and walls, squeaking floors, sticking doors, and many other problems caused by shrinking and warping of the lumber as it dries. Therefore, lumber must be dried to a suitable degree before it can be surfaced and used.

Green lumber is also subject to decay. Decay is caused by *fungi,* low forms of plant life that feed on wood. This decay is commonly known as dry rot because it usually is not discovered until the lumber has dried. Decay will not occur unless wood moisture content is in excess of 19 percent. Wood construction maintained at moisture content of less than 20 percent will not decay. It is important that lumber with an excess amount of moisture be exposed to an environment that will allow the moisture to evaporate. Seasoned lumber must be protected to prevent the entrance of moisture, which allows the growth of fungi and decay of the wood. (The subject of fungi and wood decay is discussed in detail in Chapter 33.)

Moisture Content

The amount of liquid water in wood is significant because of the nature of wood cells. They grow in columns of tubes (*Figure 2–7*). Water fills these tubes during the growing process. The moisture content (MC) of lumber is expressed as a percentage and indicates how much of the weight of a wood sample is actually water. It is derived by determining the difference in the weight of the sample before and after it has been oven-dried and dividing that number by the dry weight.

$$MC = \frac{\text{wood wet wt} - \text{wood dry wt}}{\text{wood dry wt}} \times 100\%$$

For example, if a wood sample weighs 16 ounces prior to drying and 13 ounces after drying, we assume that there were $16 - 13 = 3$ ounces of water in the wood. To determine the moisture content of the sample before drying,

$$MC\% = \frac{16 - 13}{13} \times 100\% = 23.0769\%$$

Thus, the moisture content of the sample before drying was roughly 23 percent. Lumber used for framing and exterior finish preferably should have an MC of 15 percent, not to exceed 19 percent. For interior finish, an MC of 10 to 12 percent is recommended.

Green lumber has water in the hollow part of the wood cells as well as in the cell walls. When wood starts to dry, the water in the cell cavities, called *free water,* is first removed. When all of the free water is gone, the wood has reached the fiber-saturation point, approximately 30 percent MC. No noticeable shrinkage of wood takes place up to this point.

As wood continues to dry, the water in the walls of the cells is removed and the wood starts to shrink. It shrinks considerably from its size at the fiber-saturation point to its size at the desired MC of less than 19 percent. The actual shrinkage

Lumber Size	Actual Width	Width at 19% MC	Width at 11% MC	Width at 8% MC
2 × 4	3½"	3½"	3⁷⁄₁₆"	3⅜"
2 × 6	5½"	5½"	5⅜"	5⁵⁄₁₆"
2 × 8	7½"	7¼"	7⅛"	7¹⁄₁₆"
2 × 10	9¼"	9¼"	9¹⁄₁₆"	9"

Source: *U.S. Span Book for Major Wood Species* [available from Canadian Wood Council, (800)463-5091; http://www.cwc.ca/publications/US_Span_book].

Figure 2–8 Lumber dimensions change with the moisture content of the wood.

of the lumber will vary with regional climate conditions and the MC of the lumber when delivered (*Figure 2–8*). Lumber at this stage is called *dry* or seasoned and now must be protected from getting wet.

It is important to understand not only that wood shrinks as it dries, but also how it shrinks. Little shrinkage occurs along the length of lumber; therefore, shrinkage in that direction is not considered. Most shrinkage occurs along the length of each annular ring, with the longer rings shrinking more than the shorter ones. When viewing plainsawed lumber in cross section, it can be seen that the piece warps as it shrinks because of the unequal length of the annular rings. A cross section of quarter-sawed lumber shows annular rings of equal length. Therefore, although the piece shrinks, it shrinks evenly with little warp (*Figure 2–9*). Wood warps as it dries according to the way it was cut from the tree. Cross-sectional views of the annular rings are different along the length of a piece of lumber; therefore, various kinds of warps result when lumber dries and shrinks.

When the moisture content of lumber reaches that of the surrounding air (about 10 to 12 percent MC),

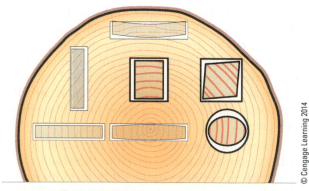

Figure 2–9 As lumber dries, the annular rings become shorter, sometimes causing wood to deform.

© Cengage Learning 2014

it is at equilibrium moisture content. At this point, lumber shrinks or swells only slightly with changes in the moisture content of the air. Realizing that lumber undergoes certain changes when moisture is absorbed or lost, the experienced carpenter uses techniques to deal with this characteristic of wood (*Figure 2–10*).

Drying Lumber

Lumber is either air-dried or kiln-dried, or a combination of both. In air-drying, the lumber is stacked in piles with spacers, which are called stickers, placed between each layer to permit air to circulate through the pile (*Figure 2–11*). Kiln-dried lumber is stacked in the same manner but is dried in buildings, called kilns, which are like huge ovens (*Figure 2–12*).

Kilns provide carefully controlled temperatures, humidity, and air circulation to remove moisture. First, the humidity level is raised and the temperature is increased; the humidity is then gradually decreased. Kiln-drying has the advantage of drying lumber in a shorter period of time, but is more expensive than air-drying.

The recommended moisture content for lumber to be used for exterior finish at the time of installation is 12 percent, except in very dry climates, where 9 percent is recommended. Lumber with low moisture content (8 to 10 percent) is necessary for interior trim and cabinet work.

The moisture content of lumber is determined by the use of a moisture meter (*Figure 2–13*). Points on the ends of the wires of the meter are driven into the wood, and the moisture content is read off the meter.

Experienced workers know when lumber is green (because it is much heavier than dry lumber), and they can estimate fairly accurately the moisture content of lumber simply by lifting it.

Lumber is brought to the planer mill, where it is straightened, smoothed, and uniformly sized. This process can be done when the lumber is dry or green. Most construction lumber is surfaced on four sides (S4S) to standard thicknesses and widths. Some may be surfaced on only two sides (S2S) to required thicknesses.

Lumber Storage

Lumber should be delivered to the job site so materials are accessible in the proper sequence; that is, those that are to be used first are on the top and those to be used last are on the bottom.

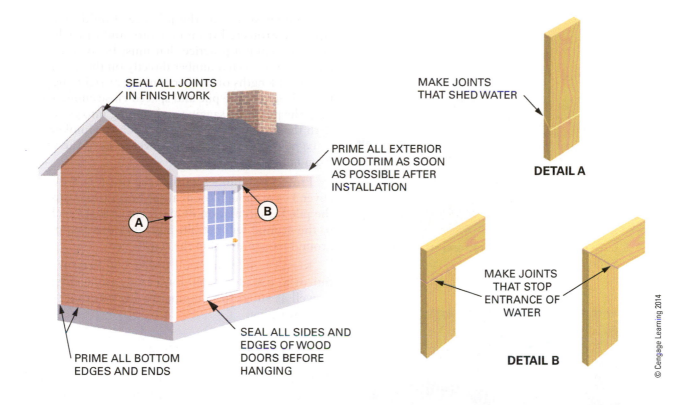

SEAL ALL JOINTS IN FINISH WORK

PRIME ALL EXTERIOR WOOD TRIM AS SOON AS POSSIBLE AFTER INSTALLATION

A

B

PRIME ALL BOTTOM EDGES AND ENDS

SEAL ALL SIDES AND EDGES OF WOOD DOORS BEFORE HANGING

MAKE JOINTS THAT SHED WATER

DETAIL A

MAKE JOINTS THAT STOP ENTRANCE OF WATER

DETAIL B

© Cengage Learning 2014

© Cengage Learning 2014

On the Job

Techniques to prevent moisture from entrance into wood.

Figure 2–10 Techniques to prevent water from getting in behind the wood surface.

Figure 2–11 Lumber is stickered and stacked to allow air drying.

Figure 2–12 Air-dried lumber is placed in a kiln to further reduce moisture content.

Figure 2–13 A moisture meter's prongs are inserted into wood to determine its moisture content.

Lumber stored at the job site should be adequately protected from moisture and other hazards. A common practice that must be avoided is placing unprotected lumber directly on the ground. Use short lengths of lumber running at right angles to the length of the pile and spaced close enough to keep the pile from sagging and coming into contact with the ground. The base on which the lumber is to be placed should be fairly level to keep the pile from falling over.

Protect the lumber with a tarp or other type of cover. Leave enough room at the bottom and top of the pile for circulation of air. A cover that reaches to the ground will act like a greenhouse, trapping ground moisture within the stack.

Keep the piles in good order. Lumber spread out in a disorderly fashion can cause accidents as well as subject the lumber to stresses that may cause warping.

LUMBER DEFECTS

A defect in lumber is any fault that detracts from its appearance, function, or strength. One type of defect is called a **warp**. Warps are caused by, among other things, drying lumber too fast, careless handling and storage, or surfacing the lumber before it is thoroughly dry. Warps are classified as **crooks**, **bows**, **cups**, and **twists** (*Figure 2–14*).

Splits in the end of lumber running lengthwise and across the annular rings are called **checks** (*Figure 2–15*). Checks are caused by faster drying of the end than of the rest of the stock. Checks can be prevented to a degree by sealing the ends of lumber with paint, wax, or other material during the drying period. Cracks that run parallel to and between the annular rings are called **shakes** and may be caused by weather or other damage to the tree.

The *pith* is the spongy center of the tree. It contains the youngest portion of the lumber, called **juvenile wood**. Juvenile wood is the portion of wood that contains the first seven to fifteen growth rings. The wood cells in this region are not well aligned and are therefore unstable when they dry. They shrink in different directions, causing internal stresses. If a board has a high percentage of juvenile wood, it will warp and twist in remarkable ways. **Knots** are cross sections of branches in the trunk of the tree. Knots are not necessarily defects unless they are loose or weaken the piece. **Pitch pockets** are small cavities that hold pitch, which sometimes oozes out. A **wane** is bark on the edge of lumber or the surface from which the bark has fallen. *Pecky* wood has small grooves or channels running with the grain. This is common in cypress. Pecky cypress

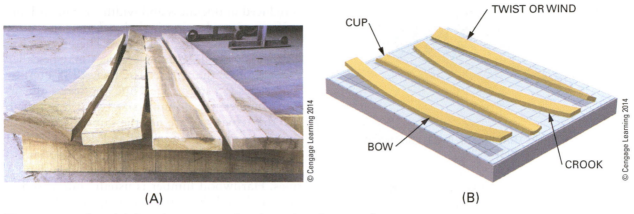

(A)   (B)

Figure 2–14 Board defects have names that depend on the type of warp.

Figure 2–15 Examples of common wood defects.

is often used as an interior wall paneling when that effect is desired. Some other defects are *stains, decay,* and *wormholes.*

LUMBER GRADES AND SIZES

Lumber grades and sizes are established by wood products associations, of which many wood mills are members. Member wood mills are closely supervised by the associations to ensure that standards are maintained. The grade stamp of the association is assurance that lumber grade standards have been met.

Member mills use the association grade stamp to indicate strict quality control. A typical grade stamp is shown in *Figure 2–16* and includes the association trademark, the mill number, the lumber grade, the species of wood, and whether the wood was green or dry when it was planed.✱

✱IRC
R502.1

Softwood Grades

The largest softwood association of lumber manufacturers is the Western Wood Products

A) ASSOCIATION'S TRADEMARK (WESTERN WOOD PRODUCTS ASSOCIATION)

B) MILL NUMBER

C) GRADE OF LUMBER (IN THIS CASE IT IS STANDARD)

D) KIND OF WOOD (DOUGLAS FIR)

E) MOISTURE CONTENT (S-DRY STANDS FOR 19 PERCENT M.C.)

Figure 2–16 Typical softwood lumber grade stamp.

Courtesy of Western Wood Products Association

Association (WWPA), which grades lumber in three categories: boards (under 2 inches thick), dimension (2 to 4 inches thick), and timbers (5 inches and thicker). The board group is divided into boards, sheathing, and form lumber. The dimension group is divided into light framing, studs, structural light framing, and structural joists and planks. Timbers are divided into beams and stringers. The three main categories are further classified according to strength and appearance as shown in *Figure 2–17*.

Hardwood Grades

Hardwood grades are established by the National Hardwood Lumber Association. Firsts and seconds (FAS) is the best grade of hardwood and must yield about 85 percent clear-cutting. Each piece must be at least 6 inches wide and 8 feet long. The next best grade is called *select*. For this, the minimum width is 4 inches and the minimum length is 6 feet. No. 1 common allows even narrower widths and shorter lengths, with about 65 percent clear-cutting.

Note that the price and quality of lumber are often linked. Some large home centers sell lumber at lower prices because it is of a lower quality. Understanding the information on lumber grade stamps will help the builder wisely purchase material.

Lumber Sizes

Rough lumber that comes directly from the sawmill is close in size to what it is called, that is, nominal size. There are slight variations to nominal size because of the heavy machinery used to cut the log into lumber. When rough lumber is planed, it

is reduced in thickness and width to standard and uniform sizes. Its nominal size does not change even though the actual size does. Therefore, when *dressed* (surfaced), although a piece may be called a 2 × 4, its actual size is $1^1/_2$ inches (38 mm) by $3^1/_2$ inches (89 mm). The same applies to all surfaced lumber; the nominal size (what it is called) and the actual size are not the same. *Figure 2–18* shows the standard nominal and dressed sizes of softwood lumber. *Figure 2–19* gives descriptions of grades and standard lumber sizes based on WWPA rules. Hardwood lumber is usually purchased in the rough and then straightened, smoothed, and sized as needed by the purchaser.

> **CAUTION**
>
> It is important for workers in construction to know actual sizes of dimension lumber.

BOARD MEASURE

Softwood lumber is usually purchased by specifying the number of pieces—thickness (″) × width (″) × length (′) (e.g., 35 pieces, 2″ × 6″ × 16′)—in addition to the grade. This is referred to as material list form. Sometimes, when no particular lengths are required, the thickness, width, and total number of linear feet (length in feet) are ordered. The length of the pieces then may vary, and these are called *random* lengths. Another method of purchasing lumber is by specifying the thickness, width, and total number of *board feet* (*Figure 2–20*). Lumber purchased in this manner may also contain random lengths.

Hardwood lumber is purchased by specifying the grade, thickness, and total number of board feet. Large quantities of both softwood and hardwood lumber are priced and sold by the board foot.

A board foot is a measure of lumber volume. It is defined as the volume of wood equivalent to a piece of wood that measures 1 inch thick by 12 inches wide by 1 foot long. It allows for a comparison of different-size pieces. For example, the volume of wood in a 2 × 6 that is 1 foot long is also one board foot (*Figure 2–21*).

To calculate board feet, use the following formula:

$$\text{B.F} = \frac{\#PC \times \text{Thick}'' \times \text{Width}'' \times \text{Length}}{12}$$

This formula uses the dimensions used to identify the board, for example, 2 × 6 −12 feet. Always use

Grade Selector Charts

Boards

HIGHEST QUALITY APPEARANCE GRADES	**SELECTS**	B & BETTER (IWP—SUPREME)* C SELECT (IWP—CHOICE) D SELECT (IWP—QUALITY)	**Specification Checklist** ☐ Grades listed in order of quantity. ☐ Include all species suited to project. ☐ Specify lowest grade that will satisfy job requirement.
	FINISH	SUPERIOR PRIME E	☐ Specify surface texture desired. ☐ Specify moisture content suited to project.
	PANELING	CLEAR (ANY SELECT OR FINISH GRADE) NO. 2 COMMON SELECTED FOR KNOTTY PANELING NO. 3 COMMON SELECTED FOR KNOTTY PANELING	☐ Specify Ⓦ grade stamp. For finish and exposed pieces, specify stamp on back or ends.
	SIDING (Bevel, Bungalow)	SUPERIOR PRIME	

BOARDS SHEATHING & FORM LUMBER	NO. 1 COMMON (IWP—COLONIAL) NO. 2 COMMON (IWP—STERLING) NO. 3 COMMON (IWP—STANDARD) NO. 4 COMMON (IWP—UTILITY) NO. 5 COMMON (IWP—INDUSTRIAL)	**Western Red Cedar**
		FINISH CLEAR HEART PANELING A AND CEILING B
	ALTERNATE BOARD GRADES SELECT MERCHANTABLE CONSTRUCTION STANDARD UTILITY ECONOMY	CLEAR—V.G. HEART BEVEL A—BEVEL SIDING SIDING B—BEVEL SIDING C—BEVEL SIDING

*Idaho white pine (IWP) carries its own comparable grade designations.

Dimension/All Species   2″ to 4″ thick (also applies to finger-jointed stock)

STRUCTURAL LIGHT FRAMING 2″ to 4″Thick 2″ to 4 ″Wide	SELECT STRUCTURAL NO. 1 NO. 2 NO. 3	These grades are designed to fit those engineering applications where higher bending strength ratios are needed in light framing sizes. Typical uses would be for trusses, concrete pier wall forms, etc.
LIGHT FRAMING 2″ to 4″Thick 2″ to 4″ Wide	CONSTRUCTION STANDARD UTILITY	This category is used where high strength values are NOT required; such as studs, plates, sills, cripples, blocking.
STUDS 2″ to 4″ Thick 2″ and Wider	STUD	An optional all-purpose grade. Characteristics affecting strength and stiffness values are limited so that the "Stud" grade is suitable for vertical framing members, including load-bearing walls.
STRUCTURAL JOISTS & PLANKS 2″ to 4″ Thick 5″ and Wider	SELECT STRUCTURAL NO. 1 NO. 2 NO. 3	These grades are designed especially to fit in engineering applications for lumber 5 inches and wider, such as joists, rafters, and general framing uses.

Timbers   5″ and thicker

BEAMS & STRINGERS 5″ and thicker Width more than 2″ greater than thickness	DENSE SELECT STRUCTURAL* DENSE NO. 1* DENSE NO. 2* SELECT STRUCTURAL NO. 1 NO. 2** NO. 3**	**POSTS & TIMBERS** 5″ × 5″ and larger Width not more than 2″ greater than thickness	DENSE SELECT STRUCTURAL * DENSE NO. 1* DENSE NO. 2* SELECT STRUCTURAL NO. 1 NO. 2** NO. 3**

*Douglas Fir or Douglas Fir–Larch only.

Figure 2–17 Softwood lumber grades.

Nominal versus Actual Sizes of Lumber

Nominal Size	Actual Size	Nominal Size	Actual Size
1 × 4	¾″ × 3½″	2 × 4	1½″ × 3½″
1 × 6	¾″ × 5½″	2 × 6	1½″ × 5½″
1 × 8	¾″ × 7¼″	2 × 8	1½″ × 7¼″
1 × 10	¾″ × 9¼″	2 × 10	1½″ × 9¼″
1 × 12	¾″ × 11¼″	2 × 12	1½″ × 11¼″

© Cengage Learning 2014

Figure 2–18 Actual sizes of lumber are smaller than nominal sizes.

the nominal dimensions, not the actual dimensions. For example, how many board feet are in 16 pieces of 2 × 4 −8 feet?

$$\frac{16 \times 2 \times 4 \times 8}{12} = 85.333 = 85\frac{1}{3} \text{ 1/3 board feet}$$

Note that this formula seems to ignore the basic rules of algebra by multiplying feet and inches together. This is because it calculates what is called board feet, not cubic feet or cubic inches.

Standard Lumber Sizes/Nominal, Dressed, Based on WWPA Rules

Product Description	Thickness In.	Nominal Size — Width In.	Nominal Size — Surfaced Dry	Dressed Dimensions — Thicknesses and Widths In. Surfaced Unseasoned	Dressed Dimensions — Lengths Ft.	
DIMENSION	S4S . . . . Other surface combinations are available. See "Abbreviations" below.	2 3 4	2 3 4 5 6 8 10 12 Over 12	1½ 2½ 3½ 4½ 5½ 7½ 9½ 11½ ¾ off normal	1⁹⁄₁₆ 2⁹⁄₁₆ 3⁹⁄₁₆ 4⅝ 5⅝ 7½ 9½ 11½ Off ½	6′ and longer, generally shipped in multiples of 2′
SCAFFOLD PLANK	Rough Full Sawn or S4S . . . . (Usually shipped unseasoned)	1¼ & Thicker	8 and Wider	If Dressed refer to "DIMENSION" sizes		6′ and longer, generally shipped in multiples of 2′
TIMBERS	Rough or S4S . . . . (Shipped unseasoned)	5 and Larger		½ off nominal (S4S) See 3.20 of WWPA Grading Rules for Rough		6′ and longer, generally shipped in multiples of 2′

Product Description		Nominal Size — Thickness In.	Nominal Size — Width In.	Dressed Dimensions — Thickness In.	Dressed Dimensions — Width In.	Lengths Ft.
DECKING	2″ Single T&G	1	5 6 8 10 12	1½	4 5 6¾ 8¾ 10¾	6′ and longer, generally shipped in multiples of 2′
	3″ and 4″ Double T&G	3 4	6	2½ 3½	5¼	
FLOORING	(D & M), (S2S & CM)	⅜ ½ ⅝ 1 1¼ 1½	2 3 4 5 6	⁵⁄₁₆ ⁷⁄₁₆ ⁹⁄₁₆ ¾ 1 1¼	1⅛ 2⅛ 3⅛ 4⅛ 5⅛	4′ and longer, generally shipped in multiples of 2′
CEILING AND PARTITION	(S2S & CM)	⅜ ½ ⅝ ¾	3 4 5 6	⁵⁄₁₆ ⁷⁄₁₆ ⁹⁄₁₆ ¹¹⁄₁₆	2⅛ 3⅛ 4⅛ 5⅛	4′ and longer, generally shipped in multiples of 2′
FACTORY AND SHOP LUMBER	S2S	1 (4/4) 1¼ (5/4) 1½ (5/4) 1¾ (7/4) 2 (8/4) 2½ (10/4) 3 (12/4) 4 (16/4)	5″ and wider (except 4″ and wider in 4/4 No. 1 Shop and 4/4 No. 2 Shop, and 2″ and wider in 5/4 & Thicker No. 3 Shop)	¾ (4/4) 1⁵⁄₃₂ (4/4) 1¹³⁄₃₂ (6/4) 1¹⁹⁄₃₂ (7/4) 1¹³⁄₁₆ (8/4) 2⅜ (10/4) 2¾ (12/4) 3¾ (16/4)	Usually sold random width	6′ and longer, generally shipped in multiples of 2′

Courtesy of Western Wood Products Association

Figure 2–19 Softwood lumber sizes.

Abbreviations

Abbreviated descriptions appearing in the size table are explained below.

S1S—Surfaced one side.
S2S—Surfaced two sides.

S4S—Surfaced four sides.
S1S1E—Surfaced one side, one edge.
S1S2E—Surfaced one side, two edges.
CM—Center matched.

D & M—Dressed and matched.
T & G—Tongue and grooved.
Rough Full Sawn—Unsurfaced green lumber cut to full specified size.

Product Classification

Nailing Diagram

	thickness in.	width in.		thickness in.	width in.
board lumber	1″	2″ or more	beams & stringers	5″ and thicker	more than 2″ greater than thickness
light framing	2″ to 4″	2″ to 4″	posts & timbers	5″ × 5″ and larger	not more than 2″ greater than thickness
studs	2″ to 4″	2″ to 6″ 10′ and shorter	decking	2″ to 4″	4″ to 12″ wide
structural light framing	2″ to 4″	2″ to 4″	siding	thickness expressed by dimension of butt edge	
structural joists & planks	2″ to 4″	5″ and wider	mouldings	size at thickest and widest points	

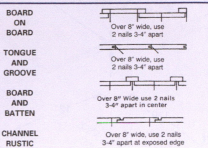

BOARD ON BOARD — Over 8″ wide, use 2 nails 3-4″ apart

TONGUE AND GROOVE — Over 8″ wide, use 2 nails 3-4″ apart

BOARD AND BATTEN — Over 8″ Wide use 2 nails 3-4″ apart in center

CHANNEL RUSTIC — Over 8″ wide, use 2 nails 3-4″ apart at exposed edge

Lengths of lumber generally are 6 feet and longer in multiples of 2′

Figure 2–19 *(continued)*

Figure 2–20 Board foot measure allows lumber to be referred to in a volume of varying width and length boards.

BOARD FEET = NUMBER OF PIECES X THICKNESS" X WIDTH" X LENGTH' ÷ 12

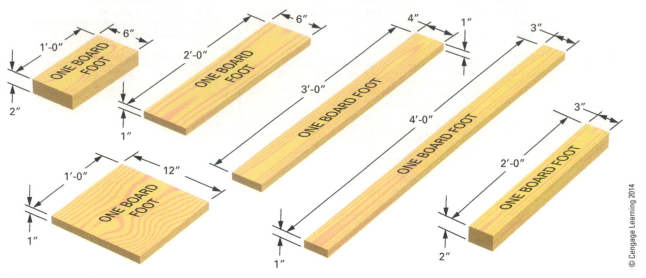

Figure 2–21 One board foot of lumber is a volume of wood; it can have many shapes.

DECONSTRUCT THIS

Carefully study *Figure 2-22* and think about what is wrong and/or what is right. Consider all possibilities. What construction practice or method is different in your area of the country?

Figure 2–22 Here we see a wood exterior corner board.

KEY TERMS

air-dried lumber

annular rings

board foot

boards

bows

cambium layer

checks

close-grained

coniferous

crooks

cups

deciduous

dimension lumber

dry rot

edge-grain

equilibrium moisture content

fiber-saturation point

firsts and seconds

grade

grain

green lumber

hardwood

heartwood

juvenile wood

kiln-dried

knots

lignin

lumber

lumber grades

medullary rays

moisture content

moisture meter

No. 1 common

nominal size

open-grained

pitch pocket

pith

plain-sawed

quarter-sawed

sapwood

sawyer

seasoned

shakes

softwood

stickering

timbers

twists

wane

warp

REVIEW QUESTIONS

Select the most appropriate answer.

1. The center of a tree in a cross section is called the

   a. heartwood.
   b. lignin.
   c. pith.
   d. sapwood.

2. New wood cells of a tree are formed in the

   a. heartwood.
   b. bark.
   c. medullary rays.
   d. cambium layer.

3. Tree growth is faster in the

   a. spring.
   b. summer.
   c. fall.
   d. winter.

4. Quarter-sawed lumber is sometimes called

   a. tangent-grained.
   b. slash-sawed.
   c. vertical-grained.
   d. plain-sawed.

5. Lumber is called "green" when

   a. it is stained by fungi.
   b. the tree is still standing.
   c. it is first cut from the log.
   d. it has reached equilibrium moisture content.

6. Wood will not decay unless its moisture content is in excess of

   a. 15 percent.
   b. 19 percent.
   c. 25 percent.
   d. 30 percent.

7. When all of the free water in the cell cavities of wood is removed and before water is removed from cell walls, lumber is at the

   a. framing lumber moisture content.
   b. fiber-saturation point.
   c. equilibrium moisture content.
   d. shrinkage commencement point.

8. Air-dried lumber cannot be dried to less than

   a. 8 to 10 percent moisture content.
   b. equilibrium moisture content.
   c. 25 to 30 percent moisture content.
   d. its fiber-saturation point.

9. A commonly used and abundant softwood is

   a. ash.
   b. fir.
   c. basswood.
   d. birch.

10. The wood species most naturally resistant to decay is

   a. pine.
   b. spruce.
   c. cypress.
   d. hemlock.

UNIT 2
Engineered Panels

Growing concern over the efficient use of forest resources led to the development of reconstituted wood. The resulting wood products are referred to as *engineered lumber*. One form of engineered lumber is manufactured from peeled logs and reconstituted wood into large sheets referred to as *engineered panels*, commonly called plywood. The tradesperson must know the different kinds, sizes, and recommended applications to use these materials to the best advantage.

Plywood is a general term used to cover a variety of materials. A carpenter must understand the uses for the various materials to make optimum use of the engineered panel.

OBJECTIVES

After completing this unit, the student should be able to describe the composition, kinds, sizes, grades, and several uses of:

- plywood
- oriented strand board
- composite panels
- particleboard
- hardboard
- medium-density fiberboard
- softboard

CHAPTER 3
Rated Plywood and Panels

The term **engineered panels** refers to human-made products in the form of large reconstituted wood sheets, sometimes called **panels**. In some cases, the tree has been taken apart and its contents have been redistributed into sheet or panel form. The panels are widely used in the construction industry. They are also used in the aircraft, automobile, and boat-building industries, as well as in the making of road signs, furniture, and cabinets.

With the use of engineered panels, construction progresses at a faster rate because a greater area is covered in a shorter period of time (*Figure 3–1*). These panels, in certain cases, present a more attractive appearance and give more protection to a surface than does solid lumber. It is important to know the kinds and uses of various engineered panels in order to use them to the best advantage.

APA-RATED PANELS

Many mills belong to associations that inspect, test, and allow mills to stamp the product to certify that it conforms to government and industrial standards. The grade stamp assures the consumer that the product has met the rigid quality and performance requirements of the association.

The trademarks of the largest association of this type, the Engineered Wood Association (formerly called the American Plywood Association and still known by the acronym APA), appear only on products manufactured by APA member mills (*Figure 3–2*). This association is concerned with quality supervision and testing not only of **plywood** (cross-laminated wood veneer) but also of **composites** (veneer faces bonded to reconstituted wood cores) and nonveneered panels, commonly called **oriented strand board (OSB)** (*Figure 3–3*).

Plywood

One of the most extensively used engineered panels is plywood. Plywood is a sandwich of wood. Most plywood panels are made up of sheets of veneer (thin pieces) called **plies**. These plies, arranged in layers, are bonded under pressure with glue to form a very strong panel. The plies are glued together so that the grain of each layer is at right angles to the next one. This cross-graining results in a sheet that

Courtesy of APA–The Engineered Wood Association

Figure 3–1 Sheet material covers a greater area in a shorter period of time than does solid lumber.

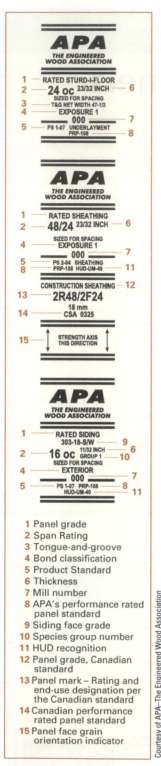

1 Panel grade
2 Span Rating
3 Tongue-and-groove
4 Bond classification
5 Product Standard
6 Thickness
7 Mill number
8 APA's performance rated panel standard
9 Siding face grade
10 Species group number
11 HUD recognition
12 Panel grade, Canadian standard
13 Panel mark – Rating and end-use designation per the Canadian standard
14 Canadian performance rated panel standard
15 Panel face grain orientation indicator

Figure 3–2 The grade stamp is assurance of a high-quality, performance-rated panel.

is as strong as or stronger than the wood it is made from. Plywood usually contains an odd number of layers so that the face grain on both sides of the sheet runs in the direction of the long dimension

Figure 3–3 APA performance-rated panels.

ARROWS SHOW DIRECTION
OF GRAIN IN EACH LAYER

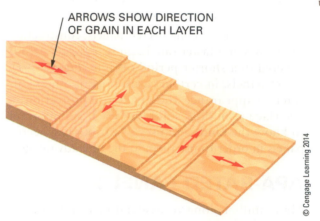

Figure 3–4 In the construction of plywood, the grains of veneer plies are placed at right angles to each other.

of the panel (*Figure 3–4*). Softwood plywood is commonly made with three, five, or seven layers. Because of its construction, plywood is more stable under changes of humidity and is more resistant to shrinking and swelling than are wood boards.

Manufacture of Veneer Core Plywood.   Specially selected "peeler logs" are mounted on a huge lathe in which the log is rotated against a sharp knife. As the log turns, a thin layer is peeled off like paper unwinding from a roll (*Figure 3–5*). The entire log is used. The small remaining spindles are utilized for making other wood products.

The long ribbon of veneer is then cut into desired widths, sorted, and dried to a moisture content of 5 percent. After drying, the veneers are fed through glue spreaders that coat them with a uniform thickness. The veneers are then assembled to make panels (*Figure 3–6*). Large presses bond the assembly under controlled heat and pressure. From the presses, the panels are left unsanded, touch-sanded, or smooth-sanded, and then are cut to size, inspected, and stamped.

Figure 3–5 The veneer is peeled from the log like paper unwinding from a roll.

Figure 3–6 Gluing and assembling plywood veneers into panels.

VENEER GRADES

A Smooth, paintable. Not more than 18 neatly made repairs, boat, sled, or router type, and parallel to grain, permitted. Wood or synthetic repairs permitted. May be used for natural finish in less demanding applications.

B Solid surface. Shims, sled or router repairs, and tight knots to 1 inch across grain permitted. Wood or synthetic repairs permitted. Some minor splits permitted.

C Plugged Improved C veneer with splits limited to 1/8-inch width and knotholes or other open defects limited to 1/4 x 1/2 inch. Wood or synthetic repairs permitted. Admits some broken grain.

C Tight knots to 1-1/2 inch. Knotholes to 1 inch across grain and some to 1-1/2 inch if total width of knots and knotholes is within specified limits. Synthetic or wood repairs. Discoloration and sanding defects that do not impair strength permitted. Limited splits allowed. Stitching permitted.

D Knots and knotholes to 2-1/2-inch width across grain and 1/2 inch larger within specified limits. Limited splits are permitted. Stitching permitted. Limited to Exposure 1.

Figure 3–7 Veneer letter grades define veneer appearance.

veneer faces are always sanded smooth. Some panels, such as APA-Rated Sheathing, are unsanded because their intended use does not require sanding. Other panels used for such purposes as subflooring and underlayment require only a touch-sanding to make the panel thickness more uniform.

Strength Grades. Softwood veneers are made of many different kinds of wood. These woods are classified in groups according to their strength (*Figure 3–8*). Group 1 is the strongest. Douglas fir and southern pine are in Group 1 and are used to make most of the softwood plywood. The group number is also shown in the grade stamp.

Oriented Strand Board

Oriented strand board (OSB) is a nonveneered, performance-rated structural panel composed of small oriented (lined-up) strandlike wood pieces arranged in three to five layers with each layer at right angles to the other (*Figure 3–9*). The cross-lamination of the layers achieves the same advantages of strength and stability as in plywood.

Manufacture of OSB. Logs from specially selected species are debarked and sliced into strands that are between $25/1000$ and $30/1000$ of an inch thick,

Green Tip

The plywood manufacturing process uses the best-quality veneer logs for outer surfaces, while lesser-quality logs are used for the inner cores. This efficiency results in the best appearance while maintaining superior strength.

Veneer Grades. In declining order, the letters *A*, *B*, *C plugged*, *C*, and *D* are used to indicate the appearance quality of panel veneers. Two letters are found in the grade stamp of veneered panels. One letter indicates the quality of one face, while the other letter indicates the quality of the opposite face. The exact description of these letter grades is shown in *Figure 3–7*. Panels with B-grade or better

CLASSIFICATION OF SPECIES

Group 1	Group 2		Group 3	Group 4	Group 5
Apitong [a][b]	Cedar, Port Orford	Maple, Black	Alder, Red	Aspen	Basswood
Beech, American	Cypress	Mengkulang [a]	Birch, Paper	Bigtooth	Poplar, Balsam
Birch	Douglas-fir 2 [c]	Meranti, Red [a][d]	Cedar, Alaska	Quaking	
Sweet	Fir	Mersawa [a]	Fir, Subalpine	Cativo	
Yellow	Balsam	Pine	Hemlock, Eastern	Cedar	
Douglas-fir 1 [c]	California Red	Pond	Maple, Bigleaf	Incense	
Kapur [a]	Grand	Red	Pine	Western Red	
Keruing [a][b]	Noble	Virginia	Jack	Cottonwood	
Larch, Western	Pacific Silver	Western White	Lodgepole	Eastern	
Maple, Sugar	White	Spruce	Ponderosa	Black (Western Poplar)	
Pine	Hemlock, Western	Black	Spruce	Pine	
Caribbean	Lauan	Red	Redwood	Eastern White	
Ocote	Almon	Sitka	Spruce	Sugar	
Pine, Southern	Bagtikan	Sweetgum	Engelmann		
Loblolly	Mayapis	Tamarack	White		
Longleaf	Red Lauan	Yellow Poplar			
Shortleaf	Tangile				
Slash	White Lauan				
Tanoak					

[a] Each of these names represents a trade group of woods consisting of a number of closely related species.

[b] Species from the genus Dipterocarpus marketed collectively: Apitong if originating in the Philippines, Keruing if originating in Malaysia or Indonesia.

[c] Douglas-fir from trees grown in the states of Washington, Oregon, California, Idaho, Montana, Wyoming, and the Canadian Provinces of Alberta and British Columbia shall be classed as Douglas-fir No. 1. Douglas-fir from trees grown in the states of Nevada, Utah, Colorado, Arizona and New Mexico shall be classed as Douglas-fir No. 2.

[d] Red Meranti shall be limited to species having a specific gravity of 0.41 or more based on green volume and oven dry weight.

Courtesy of APA—The Engineered Wood Association

Figure 3–8 Plywood is classified into five groups according to strength and stiffness, with Group 1 being the strongest.

Courtesy of Louisiana Pacific Corporation

Figure 3–9 Oriented strand board is often used for sheathing.

between $^3/_4$ and 1 inch wide, and between $2^1/_2$ and $4^1/_2$ inches long. The strands are dried, loaded into a blender, coated with liquid resins, formed into a mat consisting of three or more layers of systematically oriented wood fibers, and fed into a press where, under high temperatures and pressure, they form a dense panel. As a safety measure, the panels have one side textured to help prevent slippage during installation.

Nonveneered panels have previously been sold with such names as waferboard, structural particleboard, and others. At present, almost all panels manufactured with oriented strands or wafers are called oriented strand board. Various manufacturers have their own particular brand names for OSB, such as Oxboard, Aspenite, and many others.

Composite Panels

Composite panels are manufactured by bonding veneers of wood to both sides of reconstituted wood panels. More efficient use of wood is thus allowed with this product while retaining the wood grain appearance on both faces of the panel.

Composite panels rated by the American Plywood Association are called *COM-PLY* and are manufactured in three or five layers. A three-layer panel has a reconstituted wood core with wood veneers on both sides. A five-layer panel has a wood veneer in the center as well as on both sides.

PERFORMANCE RATINGS

A performance-rated panel meets the requirements of the panel's end use. The three end uses for which panels are rated are single-layer flooring, exterior siding, and sheathing for roofs, floors, and walls.

Names given to designate end uses are *APA-Rated Sheathing, Structural I, APA-Rated Sturd-I-Floor,* and *APA-Rated Siding* (**Figure 3–10**). Panels are tested to meet standards in areas of resistance to moisture, strength, and stability.

GUIDE TO APA PERFORMANCE RATED PANELS[a][b]
FOR APPLICATION RECOMMENDATIONS, SEE FOLLOWING PAGES.

APA RATED SHEATHING
Typical Trademark

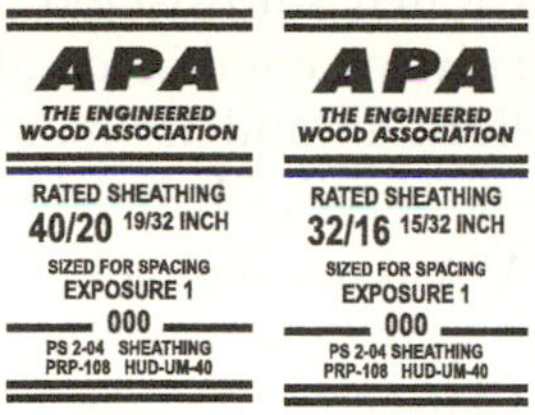

APA THE ENGINEERED WOOD ASSOCIATION
RATED SHEATHING
40/20 19/32 INCH
SIZED FOR SPACING
EXPOSURE 1
000
PS 2-04 SHEATHING
PRP-108 HUD-UM-40

APA THE ENGINEERED WOOD ASSOCIATION
RATED SHEATHING
32/16 15/32 INCH
SIZED FOR SPACING
EXPOSURE 1
000
PS 2-04 SHEATHING
PRP-108 HUD-UM-40

Specially designed for subflooring and wall and roof sheathing. Also good for a broad range of other construction and industrial applications. Can be manufactured as OSB, plywood, or other wood-based panel. BOND CLASSIFICATIONS: Exterior, Exposure 1. COMMON THICKNESSES (in.): 3/8, 7/16, 15/32, 1/2, 19/32, 5/8, 23/32, 3/4.

APA STRUCTURAL I RATED SHEATHING[c]
Typical Trademark

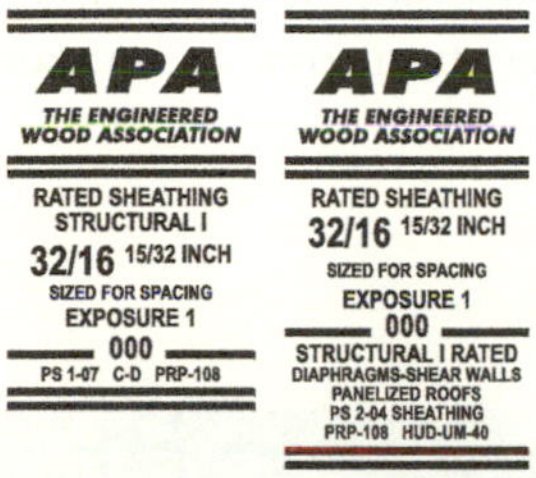

APA THE ENGINEERED WOOD ASSOCIATION
RATED SHEATHING
STRUCTURAL I
32/16 15/32 INCH
SIZED FOR SPACING
EXPOSURE 1
000
PS 1-07 C-D PRP-108

APA THE ENGINEERED WOOD ASSOCIATION
RATED SHEATHING
32/16 15/32 INCH
SIZED FOR SPACING
EXPOSURE 1
000
STRUCTURAL I RATED
DIAPHRAGMS-SHEAR WALLS
PANELIZED ROOFS
PS 2-04 SHEATHING
PRP-108 HUD-UM-40

Unsanded grade for use where shear and cross-panel strength properties are of maximum importance, such as panelized roofs and diaphragms. Can be manufactured as OSB, plywood, or other wood-based panel. BOND CLASSIFICATIONS: Exterior, Exposure 1. COMMON THICKNESSES (in.): 3/8, 7/16, 15/32, 1/2, 19/32, 5/8, 23/32, 3/4.

APA RATED STURD-I-FLOOR
Typical Trademark

APA THE ENGINEERED WOOD ASSOCIATION
RATED STURD-I-FLOOR
24 oc 23/32 INCH
SIZED FOR SPACING
T&G NET WIDTH 47-1/2
EXPOSURE 1
000
PS 2-04 SINGLE FLOOR
PRP-108 HUD-UM-40

APA THE ENGINEERED WOOD ASSOCIATION
RATED STURD-I-FLOOR
20 oc 19/32 INCH
SIZED FOR SPACING
T&G NET WIDTH 47-1/2
EXPOSURE 1
000
PS 1-07 UNDERLAYMENT
PRP-108

Specially designed as combination subfloor-underlayment. Provides smooth surface for application of carpet and pad and possesses high concentrated and impact load resistance. Can be manufactured as OSB, plywood, or other wood-based panel. Available square edge or tongue-and-groove. BOND CLASSIFICATIONS: Exterior, Exposure 1. COMMON THICKNESSES (in.): 19/32, 5/8, 23/32, 3/4, 1, 1-1/8.

APA RATED SIDING
Typical Trademark

APA THE ENGINEERED WOOD ASSOCIATION
RATED SIDING
24 oc 19/32 INCH
SIZED FOR SPACING
EXTERIOR
000
PS 1-07 PRP-108
HUD-UM-40

APA THE ENGINEERED WOOD ASSOCIATION
RATED SIDING
303-18-S/W
16 oc 11/32 INCH
GROUP 1
SIZED FOR SPACING
EXTERIOR
000
PS 1-07 PRP-108
HUD-UM-40

For exterior siding, fencing, etc. Can be manufactured as plywood, as other wood-based panel or as an overlaid OSB. Both panel and lap siding available. Special surface treatment such as V-groove, channel groove, deep groove (such as APA Texture 1-11), brushed, rough sawn and overlaid (MDO) with smooth- or texture-embossed face. Span Rating (stud spacing for siding qualified for APA Sturd-I-Wall applications) and face grade classification (for veneer-faced siding) indicated in trademark. BOND CLASSIFICATION: Exterior. COMMON THICKNESSES (in.): 11/32, 3/8, 7/16, 15/32, 1/2, 19/32, 5/8.

(a) Specific grades, thicknesses and bond classifications may be in limited supply in some areas. Check with your supplier before specifying.

(b) Specify Performance Rated Panels by thickness and Span Rating. Span Ratings are based on panel strength and stiffness. Since these properties are a function of panel composition and configuration as well as thickness, the same Span Rating may appear on panels of different thickness. Conversely, panels of the same thickness may be marked with different Span Ratings.

(c) For some Structural I plywood panel constructions, the plies are special improved grades. Panels marked PS 1 are limited to Group 1 species. Other panels marked Structural I Rated qualify through special performance testing.

Courtesy of APA—The Engineered Wood Association

Figure 3–10 Guide to APA performance-rated panels.

Exposure Durability

APA performance-rated panels are manufactured in three exposure durability classifications: *Exterior, Exposure 1,* and *Exposure 2.* Panels marked Exterior are designed for permanent exposure to the weather or moisture. Exposure 1 panels are intended for use where long delays in construction may cause the panels to be exposed to the weather before being protected. Panels marked Exposure 2 are designed for use when only moderate delays in providing protection from the weather are expected. The exposure durability of a panel may be found in the grade stamp.

Span Ratings

The span rating in the grade stamp on APA-Rated Sheathing appears as two numbers separated by a slash, such as $^{32}/_{16}$ or $^{48}/_{24}$. The left number denotes the maximum recommended spacing of supports when the panel is used for roof or wall sheathing. The right number indicates the maximum recommended spacing of supports when the panel is used for subflooring. In both cases, the long dimension of the panel must be placed across three or more supports. A panel marked $^{32}/_{16}$, for example, may be used for roof sheathing over rafters not more than 32 inches on center, or for subflooring over joists not more than 16 inches on center.

The span ratings on APA-Rated Sturd-I-Floor and APA-Rated Siding appear as a single number. APA-Rated Sturd-I-Floor panels are designed specifically for combined subflooring-underlayment applications and are manufactured with span ratings of 16, 20, 24, and 48 inches.

APA-Rated Siding is produced with span ratings of 16 and 24 inches. The rating applies to vertical installation of the panel. All siding panels may be applied horizontally direct to studs 16 to 24 inches on center provided horizontal joints are blocked.

CHAPTER 4
Nonstructural Panels

PLYWOOD

All the rated products discussed in the previous chapter may be used for nonstructural applications. In addition, other plywood products, grade-stamped by the APA–The Engineered Wood Association, are available for nonstructural use. They include sanded and touch-sanded plywood panels (*Figure 4–1*) and specialty plywood panels (*Figure 4–2*).

Nonstructural panels include hardwood plywood, particleboard, and fiberboards (*Figure 4–3*).

Hardwood Plywood

Plywood is available with hardwood face veneers, of which the most popular are birch, oak, and lauan. Beautifully grained hardwoods are sometimes matched in a number of ways to produce interesting face designs. Hardwood plywood is used in the interior of buildings for such things as wall paneling, built-in cabinets, and fixtures.

Particleboard

Particleboard is a reconstituted wood panel made of wood flakes, chips, sawdust, and planer shavings (*Figure 4–4*). These wood particles are mixed with an adhesive, formed into a mat, and pressed into sheet form. The kind, size, and arrangement of the wood particles determine the quality of the board.

The highest-quality particleboard is made of large wood flakes in the center. The flakes become gradually smaller toward the surfaces where finer particles are found. This type of construction results in an extremely hard board with a very smooth surface. Softer and lower-quality boards contain the same size particles throughout. These boards usually have a rougher surface texture. In addition to the size, kind, and arrangement of the particles, the quality of the board is determined by the method of manufacture.

GUIDE TO APA SANDED AND TOUCH-SANDED PLYWOOD PANELS[(a)(b)(c)]
FOR APPLICATION RECOMMENDATIONS, SEE FOLLOWING PAGES.

APA A-A Typical Trademark (mark on panel edge) `A-A • 3/4 IN. • G-1 • EXTERIOR-APA • 000 • PS 1-07`	Use where appearance of both sides is important for interior applications such as built-ins, cabinets, furniture, partitions; and exterior applications such as fences, signs, boats, shipping containers, tanks, ducts, etc. Smooth surfaces suitable for painting. BOND CLASSIFICATIONS: Exposure 1, Exterior. COMMON THICKNESSES (in.): 1/4, 11/32, 3/8, 15/32, 1/2, 19/32, 5/8, 23/32, 3/4.
APA A-B Typical Trademark (mark on panel edge) `A-B • 1/4 IN. • G-1 • EXPOSURE 1-APA • 000 • PS 1-07`	For use where appearance of one side is less important but where two solid surfaces are necessary. BOND CLASSIFICATIONS: Exposure 1, Exterior. COMMON THICKNESSES (in.): 1/4, 11/32, 3/8, 15/32, 1/2, 19/32, 5/8, 23/32, 3/4.
APA A-C Typical Trademark APA THE ENGINEERED WOOD ASSOCIATION A-C 23/32 INCH GROUP 1 EXTERIOR 000 PS 1-07	For use where appearance of only one side is important in exterior or interior applications, such as soffits, fences, farm buildings, etc.[(d)] BOND CLASSIFICATION: Exterior. COMMON THICKNESSES (in.): 1/4, 11/32, 3/8, 15/32, 1/2, 19/32, 5/8, 23/32, 3/4.
APA A-D Typical Trademark APA THE ENGINEERED WOOD ASSOCIATION A-D 11/32 INCH GROUP 1 EXPOSURE 1 000 PS 1-07	For use where appearance of only one side is important in interior applications, such as paneling, built-ins, shelving, partitions, flow racks, etc.[(d)] BOND CLASSIFICATION: Exposure 1. COMMON THICKNESSES (in.): 1/4, 11/32, 3/8, 15/32, 1/2, 19/32, 5/8, 23/32, 3/4.
APA B-B Typical Trademark (mark on panel edge) `B-B • 19/32 IN. • G-2 • EXTERIOR-APA • 000 • PS 1-07`	Utility panels with two solid sides. BOND CLASSIFICATIONS: Exposure 1, Exterior. COMMON THICKNESSES (in.): 1/4, 11/32, 3/8, 15/32, 1/2, 19/32, 5/8, 23/32, 3/4.
APA B-C Typical Trademark APA THE ENGINEERED WOOD ASSOCIATION B-C 23/32 INCH GROUP 1 EXTERIOR 000 PS 1-07	Utility panel for farm service and work buildings, boxcar and truck linings, containers, tanks, agricultural equipment, as a base for exterior coatings and other exterior uses or applications subject to high or continuous moisture.[(d)] BOND CLASSIFICATION: Exterior. COMMON THICKNESSES (in.): 1/4, 11/32, 3/8, 15/32, 1/2, 19/32, 5/8, 23/32, 3/4.
APA B-D Typical Trademark APA THE ENGINEERED WOOD ASSOCIATION B-D 15/32 INCH GROUP 2 EXPOSURE 1 000 PS 1-07	Utility panel for backing, sides of built-ins, industry shelving, slip sheets, separator boards, bins and other interior or protected applications.[(d)] BOND CLASSIFICATION: Exposure 1. COMMON THICKNESSES (in.): 1/4, 11/32, 3/8, 15/32, 1/2, 19/32, 5/8, 23/32, 3/4.

Continued on next page

Courtesy of APA—The Engineered Wood Association

Figure 4–1 Guide to APA sanded and touch-sanded plywood panels.

continued

GUIDE TO APA SANDED AND TOUCH-SANDED PLYWOOD PANELS[a][b][c]
FOR APPLICATION RECOMMENDATIONS, SEE FOLLOWING PAGES.

APA UNDERLAYMENT Typical Trademark		For application over structural subfloor. Provides smooth surface for application of carpet and pad and possesses high concentrated and impact load resistance. For areas to be covered with resilient flooring, specify panels with "sanded face."[e] BOND CLASSIFICATION: Exposure 1. COMMON THICKNESSES[f] (in.): 1/4, 11/32, 3/8, 15/32, 1/2, 19/32, 5/8, 23/32, 3/4.
APA C-C PLUGGED[g] Typical Trademark		For use as an underlayment over structural subfloor, refrigerated or controlled atmosphere storage rooms, pallet fruit bins, tanks, boxcar and truck floors and linings, open soffits, and other similar applications where continuous or severe moisture may be present. Provides smooth surface for application of carpet and pad and possesses high concentrated and impact load resistance. For areas to be covered with resilient flooring, specify panels with "sanded face."[e] BOND CLASSIFICATION: Exterior. COMMON THICKNESSES[f] (in.): 11/32, 3/8, 15/32, 1/2, 19/32, 5/8, 23/32, 3/4.
APA C-D PLUGGED Typical Trademark		For open soffits, built-ins, cable reels, separator boards and other interior or protected applications. Not a substitute for Underlayment or APA Rated Sturd-I-Floor as it lacks their puncture resistance. BOND CLASSIFICATION: Exposure 1. COMMON THICKNESSES (in.): 3/8, 15/32, 1/2, 19/32, 5/8, 23/32, 3/4.

(a) Specific plywood grades, thicknesses and bond classifications may be in limited supply in some areas. Check with your supplier before specifying.

(b) Sanded Exterior plywood panels, C-C Plugged, C-D Plugged and Underlayment grades can also be manufactured in Structural I (all plies limited to Group 1 species).

(c) Some manufacturers also produce plywood panels with premium N-grade veneer on one or both faces. Available only by special order. Check with the manufacturer. For a description of N-grade veneer, refer to the APA publication *Sanded Plywood,* Form K435.

(d) For nonstructural floor underlayment, or other applications requiring improved inner ply construction, specify panels marked either

"plugged inner plies" (may also be designated plugged crossbands under face or plugged crossbands or core); or "meets underlayment requirements."

(e) Also available in Underlayment A-C or Underlayment B-C grades, marked either "touch sanded" or "sanded face."

(f) Some panels 1/2 inch and thicker are Span Rated and do not contain species group number in trademark.

(g) Also may be designated APA Underlayment C-C Plugged.

Courtesy of APA–The Engineered Wood Association

Figure 4–1 *(continued)*

Many nonstructural panels are not designed to be wet. Keep these panels dry.

Particleboard Grades. The quality of particleboard is indicated by its density (hardness), which ranges from 28 to 55 pounds per cubic foot. Nonstructural particleboard is used in the construction industry for the construction of kitchen cabinets and countertops, and for the core of veneer doors and similar panels.

Fiberboards

Fiberboards are manufactured as *high-density, medium-density,* and *low-density* boards.

Hardboards

High-density fiberboards are called hardboards and are commonly known by the trademark *Masonite* regardless of the manufacturer. The hardboard industry makes almost complete use of the great natural resource of wood by utilizing wood chips and board trimmings, which were once considered waste.

Wood chips are reduced to fibers, and water is added to make a soupy pulp. The pulp flows onto a traveling mesh screen where water is drawn off to form a mat. The mat is then pressed under heat to weld the wood fibers back together by utilizing lignin, the natural adhesive in wood.

Some panels are tempered (coated with oil and baked to increase hardness, strength, and water resistance). Carbide-tipped saws trim the panels to standard sizes.

GUIDE TO APA SPECIALTY PLYWOOD PANELS(a)
FOR APPLICATION RECOMMENDATIONS, SEE FOLLOWING PAGES.

APA Decorative
Typical Trademark

```
APA
THE ENGINEERED
WOOD ASSOCIATION

DECORATIVE
3/8 INCH
GROUP 2
EXTERIOR
000
PS 1-07
```

Rough-sawn, brushed, grooved, or striated faces. For paneling, interior accent walls, built-ins, counter facing, exhibit displays. Can also be made by some manufacturers in Exterior for exterior siding, gable ends, fences and other exterior applications. Use recommendations for Exterior panels vary with the particular product. Check with the manufacturer. BOND CLASSIFICATIONS: Exposure 1, Exterior. COMMON THICKNESSES (in.): 5/16, 3/8, 1/2, 5/8.

APA High Density Overlay (HDO)(b)
Typical Trademark (mark on panel edge)

`HDO • INDUSTRIAL • A-A • 3/4 IN. • G-1 • EXTERIOR-APA • 000 • PS 1-07`

Has a hard semi-opaque resin-fiber overlay on both faces. Abrasion resistant. For concrete forms, cabinets, countertops, signs, tanks. Also available with skid-resistant screen-grid surface. BOND CLASSIFICATION: Exterior. COMMON THICKNESSES (in.): 3/8, 1/2, 5/8, 3/4.

APA Medium Density Overlay (MDO)(b)
Typical Trademark

```
APA
THE ENGINEERED
WOOD ASSOCIATION

MDO GENERAL
B-C  23/32 INCH
     GROUP 1
EXTERIOR
000
PS 1-07
```

Smooth, opaque, resin-fiber overlay on one or both faces. Ideal base for paint, both indoors and outdoors. For exterior siding, paneling, shelving, exhibit displays, cabinets, concrete forms, signs. BOND CLASSIFICATION: Exterior. COMMON THICKNESSES (in.): 11/32, 3/8, 15/32, 1/2, 19/32, 5/8, 23/32, 3/4.

APA Marine
Typical Trademark (mark on panel edge)

`MARINE • A-A • 5/8 IN. • EXTERIOR-APA • 000 • PS 1-07`

Ideal for boat hulls. Made only with Douglas-fir or western larch. Subject to special limitations on core gaps and face repairs. Also available with HDO or MDO faces. BOND CLASSIFICATION: Exterior. COMMON THICKNESSES (in.): 1/4, 3/8, 1/2, 5/8, 3/4.

APA Plyform Class I(b)
Typical Trademark

```
APA
THE ENGINEERED
WOOD ASSOCIATION

PLYFORM
B-B  23/32 INCH
     CLASS 1
EXTERIOR
000
PS 1-07
```

Concrete form grades with high reuse factor. Sanded both faces and mill-oiled unless otherwise specified. Special restrictions on species. Also available in HDO or MDO for very smooth concrete finish, and with special overlays. BOND CLASSIFICATION: Exterior. COMMON THICKNESSES (in.): 19/32, 5/8, 23/32, 3/4.

APA Plyron
Typical Trademark (mark on panel edge)

`PLYRON • 3/4 IN. • EXPOSURE 1-APA • 000`

Hardboard face on both sides. Faces tempered, untempered, smooth or screened. For countertops, shelving, cabinet doors, flooring. BOND CLASSIFICATIONS: Exposure 1, Exterior. COMMON THICKNESSES (in.): 1/2, 5/8, 3/4.

(a) Specific plywood grades, thicknesses and bond classifications may be in limited supply in some areas. Check with your supplier before specifying.

(b) Can also be manufactured in Structural I (all plies limited to Group 1 species).

Courtesy of APA–The Engineered Wood Association

Figure 4–2 Guide to APA specialty plywood panels.

a - 3/4" MAPLE PLYWOOD

b - 1/2" BIRCH APPLE PLY

c - 1/4" LUAN PLYWOOD

d - 3/4" OAK VENEERED MDF

e - 3/4" MDF

f - 3/4" PARTICLE BOARD

© Cengage Learning 2014

Figure 4–3 Various types of nonstructural panels.

Courtesy of Duraflake Division, Williamette Industries, Inc.

Courtesy of Duraflake Division, Williamette Industries, Inc.

Figure 4–4 Particleboard is made from wood flakes, shavings, resins, and waxes.

Sizes of Hardboard. The most popular thicknesses of hardboard range from $\frac{1}{8}$ to $\frac{3}{8}$ inch. The most popular sheet size is 4 feet by 8 feet, although sheets can be ordered in practically any size.

Classes and Kinds of Hardboard. Hardboard is available in three different classes: tempered, standard, and service tempered (*Figure 4–5*). It may be obtained smooth-one-side (S1S) or smooth-two-sides (S2S). Hardboard is available in many forms, such as perforated, grooved, and striated.

Uses of Hardboard. Hardboard may be used inside or outside. It is used for exterior siding and interior wall paneling. It is also used extensively for cabinet backs and drawer bottoms. It can be used wherever a dense, hard panel is required. Because of the composition of hardboard, it is important to seal all sides and edges in exterior applications.

Because it is a wood-based product, hardboard can be sawed, routed, shaped, and drilled with standard woodworking tools. It can be securely fastened with glue, screws, staples, or nails.

Medium-Density Fiberboard

Medium-density fiberboard (MDF) is manufactured in a manner similar to that used to make hardboard except that the fibers are not pressed as tightly together. The refined fiber produces a fine-textured, homogeneous board with an exceptionally smooth surface. Densities range from 28 to 65 pounds per cubic foot. It is available in thicknesses ranging from $\frac{3}{16}$ to $1\frac{1}{2}$ inches and comes in widths of 4 and 5 feet. Lengths run from 6 to 18 feet.

MDF can be used for case goods, drawer parts, kitchen cabinets, cabinet doors, signs, and some interior wall finish.

Softboard

Low-density fiberboard is called softboard. Softboard is very light and contains many tiny air spaces because the particles are not compressed tightly.

The most common thicknesses range from $\frac{1}{2}$ to 1 inch. The most common sheet size is 4 feet by 8 feet, although many sizes are available.

Uses of Softboard. Because of their lightness, softboard panels are used primarily for insulating or sound control purposes. They are used extensively as decorative panels in suspended ceilings and as ceiling tiles (*Figure 4–6*). Softboard can be used for exterior wall sheathing. This type may be coated or impregnated with asphalt to protect it from moisture during construction.

Softboard panels can easily be cut with a knife, handsaw, or power saw. They cannot be handplaned with any satisfactory results. Wide-headed nails, staples, or adhesive are used to fasten softboards in place, depending on the type and use.

Class	Surface	Nominal Thickness
1 Tempered	S1S and S2S	$\frac{1}{8}$ $\frac{1}{4}$
2 Standard	S1S and S2S	$\frac{1}{8}$ $\frac{1}{4}$ $\frac{3}{8}$
3 Service tempered	S1S and S2S	$\frac{1}{8}$ $\frac{1}{4}$ $\frac{3}{8}$

© Cengage Learning 2014

Figure 4–5 Kinds and thicknesses of hardboard.

Figure 4–6 Softboards are used extensively for decorative ceiling panels.

Courtesy of Armstrong World Industries, Inc.

OTHER

The preceding chapters have been limited to engineered wood panels and boards. Many other building materials are used in the construction industry besides those already mentioned. It is recommended that the student study *Sweet's Architectural File* to become better acquainted with the thousands of building material products on the market. This reference is well known by architects, contractors, and builders, and is revised and published annually. *Sweet's* may be found online at http://www.sweets.com. Two publications, *Products for General Building and Renovations* and *Products for Home Building and Remodeling,* may be available at the school or city library.

DECONSTRUCT THIS

Carefully study **Figure 4–7** and think about what is wrong and/or what is right. Consider all possibilities. What construction practice or method is different in your area of the country?

© Cengage Learning 2014

Figure 4–7 Framing in the springtime.

KEY TERMS

composites	hardboard	particleboard	softboard
engineered panels	oriented strand board (OSB)	plies	tempered
fiberboard	panels	plywood	veneer

REVIEW QUESTIONS

Select the most appropriate answer.

1. Plywood usually contains

   a. three layers.
   b. four layers.
   c. an even number of layers.
   d. an odd number of layers.

2. Strength of plywood is classified by

   a. veneer grades
   b. group numbers 1 to 5
   c. density
   d. exposure

3. The best-appearing face veneer of a softwood plywood panel is indicated by the letter

   a. *A.*
   b. *B.*
   c. *E.*
   d. *Z.*

4. Which of those listed below is the best selection of plywood for exterior wall framing?

   a. APA Underlayment
   b. APA A-C, Exterior
   c. APA-Rated Sheathing, Exposure 1
   d. APA C-D Plugged

5. The fraction $^{32}/_{16}$ on an APA grade stamp refers to

   a. panel thickness.
   b. maximum rafter spacing.
   c. maximum wall stud spacing.
   d. maximum wall stud and floor joist spacing.

6. Particleboard not rated as structural may be used for

   a. countertops.
   b. subflooring.
   c. wall sheathing.
   d. roof sheathing.

7. Hardboard may be used in

   a. interior applications only.
   b. exterior and interior applications.
   c. applications protected from moisture.
   d. cabinet and furniture work only.

8. Much of the softboard is used in the construction industry for

   a. underlayment for wall-to-wall rugs.
   b. roof covering.
   c. decorative ceiling panels.
   d. interior wall finish.

9. For use as an underlayment over structural subflooring where continuous or severe moisture may be present, select

   a. A-C, Group 1, Exterior plywood.
   b. B-D, Group 2, Exposure 1 plywood.
   c. Underlayment, Group 1, Exposure 1 plywood.
   d. C-C Plugged, Group 2, Exterior plywood.

10. The recommended panel where the appearance of only one side is important for interior applications such as built-ins and cabinets is

   a. APA A-A plywood.
   b. APA MDO.
   c. APA A-D plywood.
   d. APA Plyron.

UNIT 3
Engineered Lumber Products

An increased demand for lumber and a diminishing supply of large old-growth trees have forced the wood industry to come up with solutions to ensure the survival of our natural resources. Structural lumber previously sawn from large logs is now produced from reconstituted wood in many shapes and sizes. These products are collectively referred to as engineered lumber.

OBJECTIVES

After completing this unit, the student should be able to describe the manufacture, composition, uses, and sizes of:

- laminated veneer lumber
- parallel strand lumber and laminated strand lumber
- wood I-beams
- glue-laminated beams

SAFETY REMINDER

Engineered wood is stronger than wood sawn from logs. Yet it must be handled, cut, and fastened properly to achieve the safe and desirable end result.

©iStockphoto.com/Curt Pickens

CHAPTER 5
Laminated Veneer Lumber

Old-growth trees are large, tall, tight-grain trees that take more than two hundred years to mature. The lumber produced from these trees is of the highest quality. Large-size lumber can be efficiently cut from these trees. Due to centuries of logging, the number of old-growth trees is decreasing. Second- and third-growth trees as well as trees planted during reforestation efforts are more abundant. These are smaller and produce fewer large-size pieces. Lumber from these trees sometimes has undesirable wood characteristics, such as a tendency to warp.

An inevitable result of the decreasing supply of large old-growth trees and the abundance of smaller trees is the development and use of engineered lumber products.

Engineered lumber products (ELPs) are reconstituted wood products and assemblies designed to replace traditional structural lumber. Engineered lumber products consume less wood and can be made from smaller trees than can traditional lumber. Traditional lumber processes typically convert 40 percent of a log to structural solid lumber. Engineered lumber processes convert up to 75 percent of a log into structural lumber. In addition, the manufacturing processes of engineered lumber consume less energy than those of solid lumber. Also, some ELPs make use of abundant, fast-growing species not currently harvested for solid lumber.

The final engineered lumber product has greater strength and consequently can span greater distances. It is predicted that engineered lumber will be used more than solid lumber in the near future. It is important that present and future builders be thoroughly informed about engineered lumber products.

This unit describes how engineered lumber products are made, where they are used, and what sizes are available. Construction details for ELPs are shown in later units on floor, wall, and roof framing.

LAMINATED VENEER LUMBER

Laminated veneer lumber (LVL) is one of several types of engineered lumber products (*Figure 5–1*). It was first used to make airplane propellers

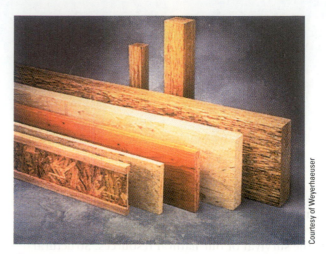

Figure 5–1 There are several types of engineered lumber.

Courtesy of Weyerhaeuser

during World War II. The world's first commercially produced LVL for building construction was patented as MICRO-LAM laminated veneer lumber in 1970. LVL is now widely used in wood frame construction.

Manufacture of LVL

Like plywood, LVL is a wood veneer product. The grain in each layer of veneer in LVL runs in the same direction, parallel to its length (*Figure 5–2*). This is unlike plywood, in which each layer of veneer is laid with the grain at right angles to each other.

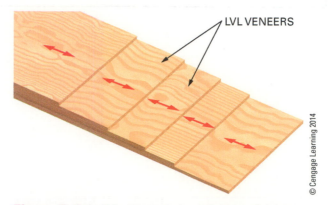

LVL VENEERS

Figure 5–2 In LVL, the grain in each layer of veneer runs in the same direction.

© Cengage Learning 2014

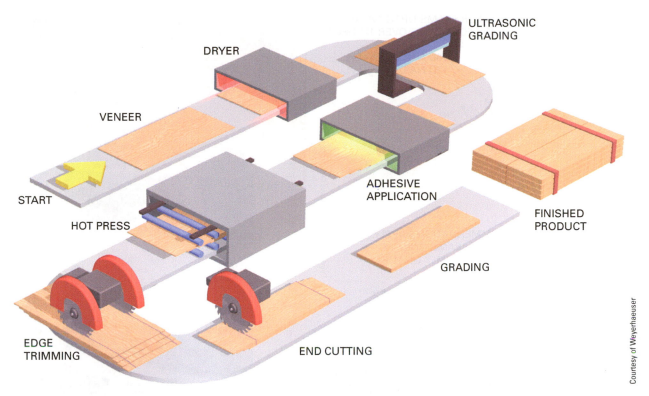

ULTRASONIC
GRADING

DRYER

VENEER

START

HOT PRESS

ADHESIVE
APPLICATION

FINISHED
PRODUCT

GRADING

EDGE
TRIMMING

END CUTTING

Courtesy of Weyerhaeuser

Figure 5–3 LVL manufacturing process, simplified.

Laminated veneer lumber is made from sheets of veneer peeled from logs, similar to the first step in the manufacture of plywood (see Figure 3–5). Douglas fir or southern pine is used because of its strength and stiffness. The veneer is peeled in widths of 27 or 54 inches and from $\frac{1}{10}$ to $\frac{3}{16}$ inch in thickness. It is then dried, cut into sheets, ultrasonically graded for strength, and sorted (*Figure 5–3*).

The veneer sheets are laid in a staggered pattern so that the ends overlap. They are then permanently bonded together with an exterior-type adhesive in a continuous press under precisely controlled heat and pressure. Unlike plywood, the LVL veneers are densified; that is, the thickness is compressed and made more compact. Fifteen to twenty layers of veneer make up a typical $1\frac{3}{4}$-inch-thick beam (*Figure 5–4*). The edges of the bonded veneers are then edge-trimmed to specified widths and end-cut to specified lengths.

LVL Sizes and Uses

Laminated veneer lumber is manufactured up to 3½ inches thick, 24 inches wide, and 80 feet long. The usual thickness is 1¾ inches. Doubling

© Cengage Learning 2014

Figure 5–4 Close-up view of LVL.

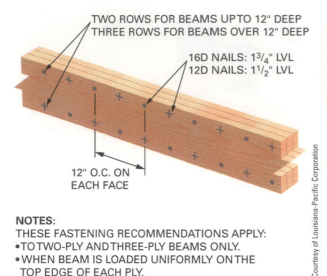

TWO ROWS FOR BEAMS UP TO 12" DEEP
THREE ROWS FOR BEAMS OVER 12" DEEP

16D NAILS: 1³/₄" LVL
12D NAILS: 1¹/₂" LVL

12" O.C. ON
EACH FACE

NOTES:
THESE FASTENING RECOMMENDATIONS APPLY:
•TO TWO-PLY AND THREE-PLY BEAMS ONLY.
•WHEN BEAM IS LOADED UNIFORMLY ON THE
 TOP EDGE OF EACH PLY.

Figure 5–5 Recommended nailing pattern for fastening LVL beams together.

Courtesy of Louisiana-Pacific Corporation

the 1³/₄-inch thickness equals the width of nominal 2 × 4 framing. Typical LVL widths are 9¹/₂, 11¹/₂, 11⁷/₈, 14, 16, and 18 inches. LVL beams may be fastened together to make a thicker and stronger beam (*Figure 5–5*).

Laminated veneer lumber is intended for use as high-strength, load-carrying beams to support the weight of construction over window and door openings, and in floor and roof systems (*Figure 5–6*). Although LVL is heavier than dimension lumber, it is still easy to handle and suitable

© Cengage Learning 2014

Figure 5–6 LVL is designed to be used for load-carrying beams.

for use in concrete forming in manufactured housing, and in many other specialties where a light-weight, strong beam is required. It can be cut with regular tools and requires no special fasteners.

CHAPTER 6
Parallel Strand Lumber and Laminated Strand Lumber

PARALLEL STRAND LUMBER

Parallel strand lumber (PSL), commonly known by its brand name, Parallam (*Figure 6–1*), was developed, like all engineered lumber products, to meet the need of the building industry. PSL provides large-dimension lumber (beams, planks, and posts). PSL also utilizes small-diameter, second-growth trees, thus protecting the diminishing supply of old-growth trees.

Manufacture of PSL

Parallam PSL is manufactured by peeling veneer of Douglas fir and southern pine from logs in much the same manner as for plywood and LVL. The veneer

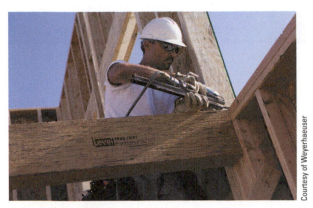

Figure 6–1 Parallel strand lumber is commonly called Parallam and is used as beams and posts to carry heavy loads.

is then dried and clipped into *strands* (narrow strips) up to 8 feet in length and $\frac{1}{8}$ or $\frac{1}{10}$ inch in thickness (*Figure 6–2*). Small defects are removed, and the strands are then coated with a waterproof adhesive. The oriented strands are fed into a rotary belt press and bonded using a patented microwave pressing process. The result is a continuous timber, up to 11 inches thick by 17 inches wide, which can then be factory-ripped into widths and thicknesses to fit builders' needs (*Figure 6–3*). The four surfaces are sanded smooth before the product is shipped.

PSL Sizes and Uses

PSL comes in many thicknesses and widths and is manufactured up to 66 feet long. PSL is available in square and rectangular shapes for use as posts and beams. Beams are sold in a convenient $1\frac{3}{4}$-inch thickness for installation of single and multiple laminations. Solid $3\frac{1}{2}$-inch thicknesses are compatible with 2 × 4 wall framing. A list of beam and post sizes is shown in *Figure 6–4*. Also available is Parallam 269, which measures $2\frac{11}{16}$ inches thick.

Green Tip

Engineered lumber maximizes the efficiency of wood resources by reducing manufacturing waste. Sawn lumber has about 50 percent waste, while OSB production, for example, has about 8 percent waste.

Parallel strand lumber can be used wherever there is a need for a large beam or post. The differences between PSL and solid lumber are many. Solid lumber beams may have defects, such as

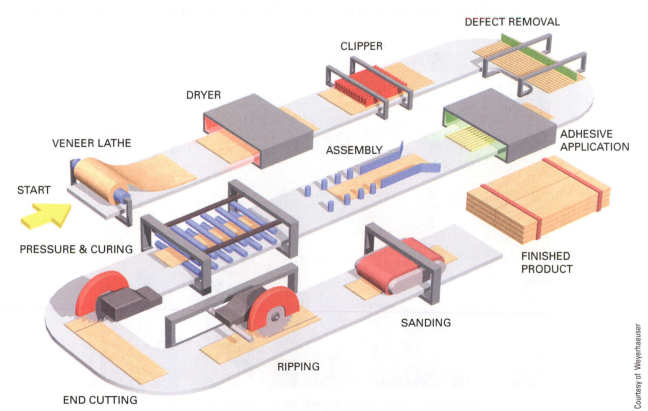

Figure 6–2 The manufacture of parallel strand lumber, simplified.

Figure 6–3 Close-up view of PSL.

© Cengage Learning 2014

PSL Column & Post Sizes		PSL Beam Sizes	
Thickness	Width	Thickness	Depths
3½"	3½"	1¾"	7¼"
3½"	5¼"	3½"	9¼"
3½"	7"	5¼"	11¼"
5¼"	5¼"	7"	11½"
5¼"	7"		11⁷⁄₈"
7"	7"		12"
			12½"
			14"
			16"
			18"

© Cengage Learning 2014

Figure 6–4 Available sizes of PSL used for posts and *beams*.

knots, checks, and shakes, which weaken them, whereas PSL is consistent in strength throughout its length. PSL is readily available in longer lengths, and its surfaces are sanded smooth, eliminating the need to cover them by boxing the beams.

LAMINATED STRAND LUMBER

A registered brand name of laminated strand lumber (LSL) is *TimberStrand* (*Figure 6–5*). While LVL and PSL are made from Douglas fir and southern pine,

LSL can be made from very small logs of practically any species of wood; its strands are much shorter than those of parallel strand lumber. At present, LSL is being manufactured from surplus, overmature aspen trees that usually are not large, strong, or straight enough to produce ordinary wood products.

PSL

LSL

© Cengage Learning 2014

Figure 6–5 Laminated strand lumber is used for non-load-bearing situations such as rim joists.

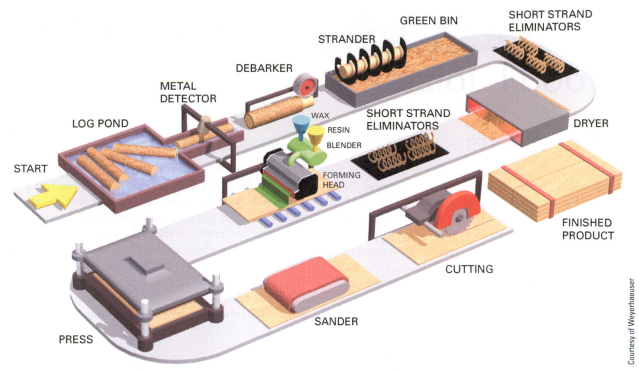

Figure 6–6 Laminated strand lumber can be manufactured from practically any species of wood.

Manufacture of LSL

The TimberStrand LSL manufacturing process begins by cleaning and debarking 8-foot aspen logs (*Figure 6–6*). The wood is then cut into strands up to 12 inches long, dried, and treated with a resin. The treated strands are aligned parallel to each other to take advantage of the wood's natural strength. The strands are pressed into solid *billets* (large blocks) up to $5\frac{1}{2}$ inches thick, 8 feet wide, and 35 feet long (*Figure 6–7*). Scraps from the process fuel the furnace that provides heat and steam for the plant.

LSL Sizes and Uses

The long billet is resawn and sanded to sizes as required by customers. It is used for a wide range of millwork, such as doors, windows, and virtually any product that requires high-grade lumber. It is also used for truck decks, manufactured housing, and some structural lumber, such as window and door headers.

LSL is made from wood that is less strong and stiff than the wood used to make PSL. For this reason LSL is not designed to be a structural framing member that will carry heavy loads over long spans. It is often used as a rim joist attached to engineered floor joists.

Figure 6–7 Close-up view of LSL.

CHAPTER 7
Wood I-Joists

WOOD I-JOISTS

The wood I-beam joist was invented in 1969, as a substitute for solid lumber joists (*Figure 7–1*). Wood I-joists are engineered wood assemblies that utilize an efficient "I" shape, common in steel beams, which gives them tremendous strength in relation to their size and weight. Consequently, they are able to carry heavy loads over long distances while using considerably less wood than solid lumber of a size necessary to carry the same load over the same span.

The flanges of the joist may be made of laminated veneer lumber or specially selected finger-jointed solid wood lumber (*Figure 7–2*). The web of the beam may be made of plywood, laminated veneer lumber, or oriented strand board.

The Manufacture of Wood I-Joists

Regardless of who makes them, the manufacturing process of wood I-joists consists of gluing top and bottom flanges to a connecting center web. First the web material is ripped to a specified width, and then the edges and ends are shaped for joining to flanges and adjacent web sections. The ends of the flange material are finger-jointed for gluing end to end. One side of the flange material is grooved to receive the beam's web. Flanges and webs are then assembled with waterproof glue by pressure-fitting the web into

© Cengage Learning 2014

Courtesy of Louisiana-Pacific Corporation

Courtesy of Louisiana-Pacific Corporation

Figure 7–1 Wood I-joists are available in many sizes.

© Cengage Learning 2014

Figure 7–2 Finger-joints are used to join the ends of short pieces of lumber to make a longer piece.

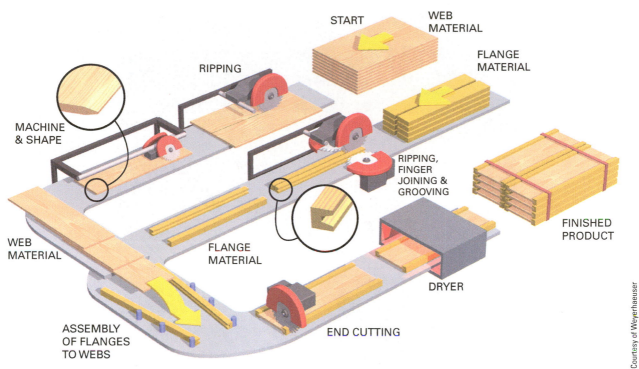

START
WEB MATERIAL
FLANGE MATERIAL
RIPPING
MACHINE & SHAPE
RIPPING, FINGER JOINING & GROOVING
FINISHED PRODUCT
WEB MATERIAL
FLANGE MATERIAL
DRYER
ASSEMBLY OF FLANGES TO WEBS
END CUTTING

Courtesy of Weyerhaeuser

Figure 7–3 Manufacture of wood I-joists, simplified.

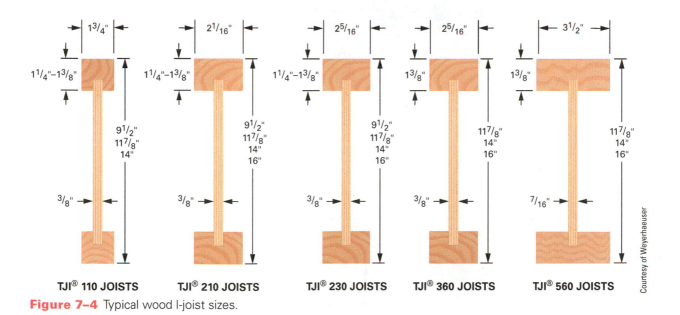

TJI® 110 JOISTS
TJI® 210 JOISTS
TJI® 230 JOISTS
TJI® 360 JOISTS
TJI® 560 JOISTS

Courtesy of Weyerhaeuser

Figure 7–4 Typical wood I-joist sizes.

the flanges. The wood I-joist is end-trimmed and the adhesive cured in an oven or at room temperature (*Figure 7–3*). As with most engineered wood products, wood I-joists are produced to an approximate equilibrium moisture content.

Wood I-Joist Sizes and Uses

Wood I-joists may have webs of various thicknesses, and flanges may vary in thickness and width, depending on intended end use and the manufacturer. Joist depths are available from $9\frac{1}{4}$ to 24 inches (*Figure 7–4*). Joists with larger webs and flanges are designed to carry heavier loads. Wood I-joists are available up to 80 feet long.

Wood I-joists are intended for use in residential and commercial construction as floor joists and roof rafters (*Figure 7–5*).

Courtesy of Louisiana-Pacific Corporation

(A)

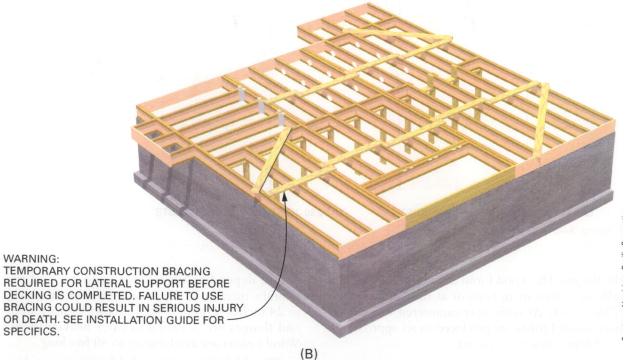

WARNING:
TEMPORARY CONSTRUCTION BRACING
REQUIRED FOR LATERAL SUPPORT BEFORE
DECKING IS COMPLETED. FAILURE TO USE
BRACING COULD RESULT IN SERIOUS INJURY
OR DEATH. SEE INSTALLATION GUIDE FOR
SPECIFICS.

Courtesy of Louisiana-Pacific Corporation

(B)

Figure 7–5 Wood I-joists are used as (A) roof rafters, (B) for floor joists and window and door headers.

CHAPTER 8

Glue-Laminated Lumber

GLUE-LAMINATED LUMBER

Glue-laminated lumber, commonly called *glulam*, is constructed of solid lumber glued together, side against side, to make beams and joists of large dimensions that are stronger than natural wood of the same size (*Figure 8–1*). Even if it were possible, it would not be practical to make solid wood beams as large as most glulams. Glulams are used for structural purposes, but architectural-appearance-grade glulams are decorative as well and, in most cases, their surfaces are left exposed to show the natural wood grain.

Manufacture of Glulams

Glulam beams are made by gluing stacks of lam stock into large beams, headers, and columns (*Figure 8–2*). Lams are the individual pieces in the glued-up stack. The lams, which are at approximate equilibrium moisture content, are glued together with exterior adhesives. Thus glulams are rated in terms of weather resistance.

Lam Layup. The lams are arranged in a certain way for maximum strength and minimum shrinkage. Different grade lams are placed where they will do the most good under load conditions. High-grade tension lams are used in tension faces, and high-grade compression lams are used in compression faces.

Figure 8–1 Glue-laminated lumber is commonly called glulam.

Courtesy of APA–The Engineered Wood Association

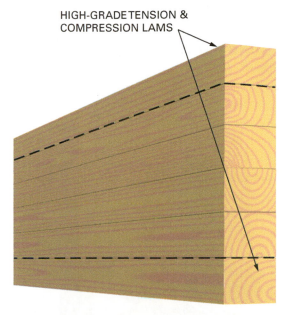

HIGH-GRADE TENSION & COMPRESSION LAMS

THE SEQUENCE OF LAM GRADES, FROM BOTTOM TO TOP OF A GLULAM, IS REFERRED TO AS A LAM LAYUP AND IS A VITALLY IMPORTANT FACTOR IN GLULAM PERFORMANCE.

© Cengage Learning 2014

Figure 8–2 Glulams are made by gluing stacks of solid lumber.

Tension is a force applied to a member that tends to increase its length, while compression is a force tending to decrease its length (*Figure 8–3*). When a load is imposed on a glulam beam that is supported on both ends, the topmost lams are in compression while those at the bottom are in tension.

More-economical lams are used in the lower-stressed middle sections of the glulam. The sequence of lam grades, from bottom to top of a glulam, is referred to as *lam layup*, and it is a vitally important factor in glulam performance. Because of this lam layup, glulam beams come with one edge stamped "TOP." Always remember to install glulam beams with the "TOP" stamp facing toward the sky (*Figure 8–4*).

Glulam Grades

The American Plywood Association–Engineered Wood Systems trademark, APA-EWS (*Figure 8–5*), appears on all beams manufactured by American

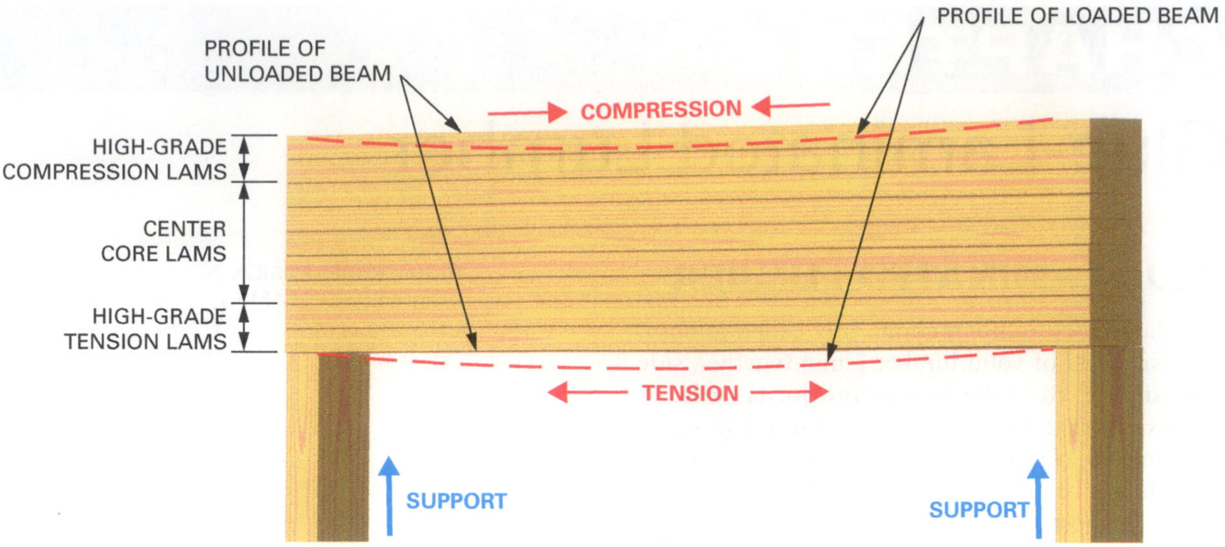

PROFILE OF LOADED BEAM

PROFILE OF
UNLOADED BEAM

COMPRESSION

HIGH-GRADE
COMPRESSION LAMS

CENTER
CORE LAMS

HIGH-GRADE
TENSION LAMS

TENSION

SUPPORT

SUPPORT

Courtesy of Bohemia, Inc.

Figure 8–3 The load on a beam places lams in tension and compression.

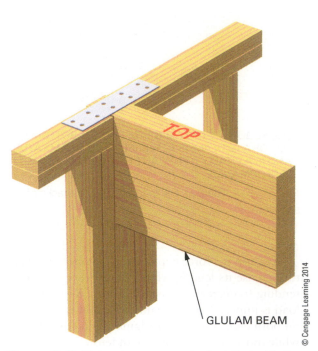

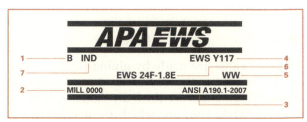

© Cengage Learning 2014

GLULAM BEAM

Figure 8–4 Glulam beams must be installed with the edge stamped "TOP" pointed toward the sky.

1) INDICATES STRUCTURAL USE:
   B - SIMPLE SPAN BENDING MEMBER.
   C - COMPRESSION MEMBER.
   T - TENSION MEMBER.
   CB- CONTINUOUS OR CANTILEVERED SPAN BENDING MEMBER.
2) MILL NUMBER.
3) IDENTIFICATION OF ANSI STANDARD A190.1, STRUCTURAL GLUED LAMINATED TIMBER.
4) CODE RECOGNITION OF AS A QUALITY ASSURANCE AGENCY FOR GLUED STRUCTURAL MEMBERS.
5) APPLICABLE LAMINATING SPECIFICATION.
6) APPLICABLE COMBINATION NUMBER.
7) SPECIES OF LUMBER USED.
8) DESIGNATES APPEARANCE GRADE (INDUSTRIAL ARCHITECRAL, PREMIUM).

Courtesy of APA–The Engineered Wood Association

Figure 8–5 The grade stamp assures that the beam has met all the necessary requirements.

Wood Systems (AWS) member mills. The AWS is a related corporation of the APA. The trademark guarantees that the glulams meet all the requirements of the American National Standards Institute (ANSI).

Glulams are manufactured in three grades for appearance. The *industrial-appearance grade* is used in warehouses, garages, and other structures in which appearance is not of primary importance or when the beams are not exposed. *Architectural-appearance grade* is for projects where appearance is important, and the *premium-appearance grade*

is used when appearance is critical (*Figure 8–6*). There is no difference in strength among the different appearance grades. All glulams with the same design values are rated for the same loadings, regardless of appearance grade.

Glulam Sizes and Uses

The dimensions of glulam beams are indicated by width, depth, and length (*Figure 8–7*). Widths range from 2½ to 8¾ inches, depths from 6 to 28½ inches,

Figure 8–6 Glulam beams are often manufactured for strength and appearance.

and lengths are generally available from 10 to 40 feet in 2-foot increments (*Figure 8–8*).

Various wood species are used to produce straight, curved, arched, and special shapes for all structures—from elegant homes and churches to large malls, warehouses, and civic centers. In all, the American Institute of Timber Construction (AITC) recognizes more than one hundred glulam beam combinations, most of which are for specialized applications.

Green Tip

Engineered lumber is more dimensionally stable than sawn lumber from the effects of shrinking and swelling due to moisture. This reduces callbacks.

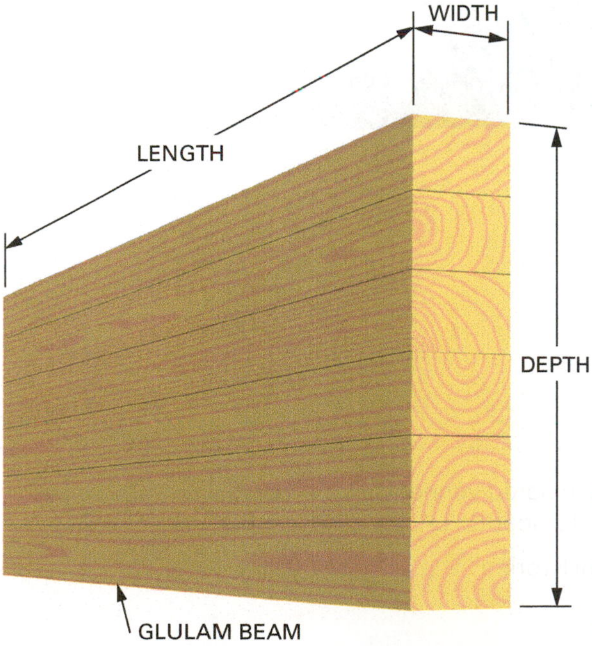

Figure 8–7 The dimensions of glulam beams are indicated by width, depth, and length.

Depth in Inches	Width						
	2½"	3⅛"	3½"	5⅛"	5½"	6¾"	8¾"
6"		*	*				
7½"	*	*	*				
9"	*	*	*	*	*		
10½"	*	*	*	*	*		
12"	*	*	*	*	*	*	*
13½"	*	*	*	*	*	*	*
15"	*	*	*	*	*	*	*
16½"	*	*	*	*	*	*	*
18"	*	*	*	*	*	*	*
19½"	*	*	*	*	*	*	*
21"	*	*	*	*	*	*	*
22½"	*	*	*	*	*	*	*
24"				*	*	*	*
25½"				*	*	*	*
27"						*	*
28½"						*	*

Sizes generally available from 10 to 40 feet long in 2'-0" increments

Figure 8–8 Available glulam beam sizes.

DECONSTRUCT THIS

Carefully study *Figure 8–9* and think about what is wrong and/or what is right. Consider all possibilities. What construction practice or method is different in your area of the country?

© Cengage Learning 2014

Figure 8–9 This photo shows a wood I-joist.

KEY TERMS

engineered lumber
   products (ELPs)

finger-jointed

glue-laminated lumber

headers

joist

laminated strand
   lumber (LSL)

laminated veneer
   lumber (LVL)

millwork

parallel strand lumber
   (PSL)

rim joist

REVIEW QUESTIONS

Select the most appropriate answer.

1. Engineered lumber is produced today because of

   a. the decreasing supply of old-growth trees.
   b. a need for stronger, more stable building material.
   c. a need to make efficient use of natural resources.
   d. all of the above.

2. Unlike plywood, the veneers of LVL are

   a. deciduous.
   b. densified.
   c. diversified.
   d. double-faced.

3. Laminated veneer lumber is often used for

   a. high-strength load-carrying beams.
   b. posts and columns.
   c. low-strength rim joists.
   d. floor joists.

4. Parallel strand lumber is made from

   a. Alaskan cedar and California redwood.
   b. Douglas fir and southern pine.
   c. Idaho pine and eastern hemlock.
   d. Englemann spruce and western pine.

5. Parallel strand lumber is often used for

   a. high-strength load-carrying beams.
   b. window and door headers.
   c. low-strength rim joists.
   d. floor joists.

6. Laminated strand lumber is often used for

   a. high-strength load-carrying beams.
   b. posts and columns.
   c. low-strength rim joists.
   d. floor joists.

7. The web of wood I-beams may be made of

   a. hardboard.
   b. particleboard.
   c. solid lumber.
   d. oriented strand board.

8. The flanges of wood I-beams are often made from

   a. glue-laminated lumber.
   b. laminated veneer lumber.
   c. parallel strand lumber.
   d. laminated strand lumber.

9. The lams used in glue-laminated lumber are at approximately

   a. the fiber-saturation point.
   b. 19 percent moisture content.
   c. equilibrium moisture content.
   d. 30 percent moisture content.

10. In glulam beams, specially selected tension lams are placed

    a. at the top of the beam.
    b. at the bottom of the beam.
    c. in the center of the beam.
    d. throughout the beam.

UNIT 4
Fasteners

©iStockphoto.com/Curt Pickens

The simplicity of fasteners can be misleading to students of construction. It is easy to believe only that nails are driven, screws are turned, and sticky stuff is used to glue. Although this tends to be true, joining material together so it will last a long time is more challenging. Many times, a fastener is used for just one type of material. Some fasteners should never be used with certain other materials. The fastener selected often separates a quality job from a shoddy one.

Fasteners have been evolving for centuries. Today they come in many styles, shapes, and sizes requiring different fastening techniques. It is important for the carpenter to know what fasteners are available, which securing technique should be employed, and how to wisely select the most appropriate fastener for various materials under different conditions.

There are many kinds and styles of fasteners made of various materials. Selecting the proper fastener is important for strength and durability.

OBJECTIVES

After completing this unit, the student should be able to name and describe the following commonly used fasteners and select them for appropriate use:

- nails
- screws and lag screws
- bolts
- solid- and hollow-wall anchors
- adhesives

CHAPTER 9

Nails, Screws, and Bolts

Nails, screws, and bolts are the most widely used of all fasteners. They come in many styles and sizes. The carpenter must know what is available and wisely select those most appropriate for fastening various materials under different conditions.

NAILS

There are hundreds of kinds of nails manufactured for just about any kind of fastening job. They differ according to purpose, shape, material, coating, and in other ways. Nails are made of aluminum, brass, copper, steel, and other metals. Some nails are hardened so that they can be driven into masonry without bending. Only the most commonly used nails are described in this chapter (*Figure 9–1*).

Uncoated steel nails are called bright nails. Various coatings may be applied to reduce corrosion, increase holding power, and enhance appearance. To prevent rusting, steel nails are coated with *zinc*. These nails are called galvanized nails. They may be coated by being dipped in molten zinc (*hot-dipped galvanized* nails), or they may be *electroplated* with corrosion-resistant metal (*plated* nails). Hot-dipped nails have a heavier coating than plated nails and are more resistant to rusting. Some building material manufacturers specify that their products be fastened with hot-dipped nails because of the heavier coating.

Figure 9-2 Improper choice of fasteners can cause dark stains to form on wood surfaces.

Green Tip

Use the correct fastener style and size for a particular application. This will ensure that the building performs as expected and lasts a long time.

When fastening metal that is going to be exposed to the weather, use nails of the same material. For example, when fastening aluminum, copper, or galvanized iron, use nails made of the same metal. Otherwise, a reaction with moisture and the two different metals, called electrolysis, will cause one of the metals to disintegrate over time.

When fastening some woods, such as cedar, redwood, and oak, that will be exposed to the weather, use stainless steel fasteners. Otherwise, a reaction between the acid in the wood and fastener will cause dark stains to appear around the fasteners (*Figure 9–2*).

Nail Sizes

The sizes of some nails are designated by the penny (d) system. The origin of this system of nail measurement is not clear. Although many people think

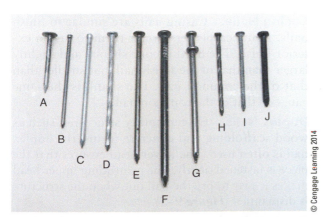

Figure 9-1 Kinds of commonly used nails: (A) roofing, (B) finish, (C) galvanized, (D) galvanized spiral, (E) box, (F) common, (G) duplex, (H) spiral, (I) coated box, and (J) masonry.

PENNY (D) SIZES

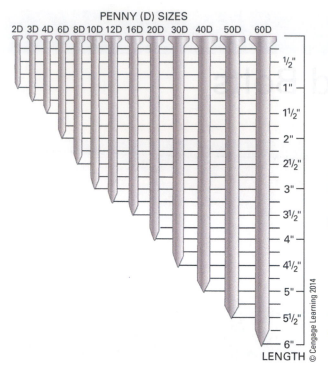

Figure 9-3 Most nails are sized according to the penny system.

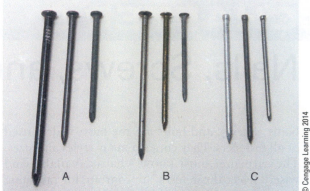

Figure 9-4 The most widely used nails are the (A) common, (B) box, and (C) finish nails.

of metal. The most widely used wire nails are the common, box, and finish nails (*Figure 9–4*).

Common Nails. Common nails are made of wire, are of heavy gauge, and have a medium-size head. They have a pointed end and a smooth shank. A barbed section just under the head increases the holding power of common nails.

Box Nails. Box nails are similar to common nails, except they are thinner. Because of their small gauge, they can be used close to edges and ends with less danger of splitting the wood. Many box nails are coated with resin cement to increase their holding power.

Finish Nails. Finish nails are of light gauge with a very small head. They are used mostly to fasten interior trim. The small head is sunk into the wood with a nail set and covered with a filler. The small head of the finish nail does not detract from the appearance of a job as much as would a nail with a larger head.

Casing Nails. Casing nails are similar to finish nails. Many carpenters prefer them to fasten exterior finish. The head is cone shaped and slightly larger than that of the finish nail, but smaller than that of the common nail. The shank is the same gauge as that of the common nail.

Duplex Nails. On temporary structures, such as wood scaffolding and concrete forms, the duplex nail is often used. The lower head ensures that the piece is fastened tightly. The protruding upper head makes it easy to pry the nail out when the structure is dismantled (*Figure 9–5*).

Brads. Brads are small finishing nails (*Figure 9–6*). They are sized according to length in inches and gauge. Usual lengths are from $\frac{1}{2}$ inch to $1\frac{1}{2}$ inches, and gauges run from #14 to #20. The higher the gauge number, the thinner the brad.

it should be discarded, it is still used in the United States. Some believe it originated years ago when nails cost a certain number of pennies for a hundred of a specific length. Of course, the larger nails cost more per hundred than smaller ones, so nails that cost eight pennies per hundred were larger than those that cost four pennies per hundred. The symbol for penny is *d*; perhaps it is the abbreviation for *denarius,* an ancient Roman coin.

In the penny system, the shortest nail is 2d and is 1 inch long. The longest nail is 60d and is 6 inches long (*Figure 9–3*). A six-penny nail is written as 6d and is 2 inches long. Experienced carpenters can determine the penny size of nails just by looking at them.

The thickness or diameter of the nail is called its gauge. In the penny system, gauge depends on the kind and length of the nail. The gauge increases with the length of the nail. Long nails, 20d and over, are sometimes called *spikes*. The length of nails not included in the penny system is designated in inches and fractions of an inch, and the gauge may be specified by a number.

Kinds of Nails

Most nails, cut from long rolls of metal wire, are called wire nails. *Cut nails,* used only occasionally, are wedge-shaped pieces stamped from sheets

Figure 9-5 Duplex nails are used for temporary fastening. (Note, second nail was left out for clarity.)

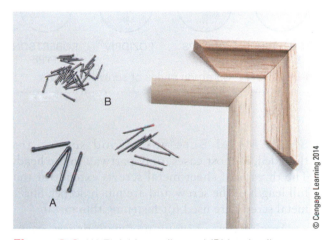

Figure 9-6 (A) Finishing nails, and (B) brad nails.

Brads are used for fastening thin material, such as small molding.

Roofing Nails. Roofing nails are short nails of fairly heavy gauge with wide, round heads. They are used for such purposes as fastening roofing material and softboard wall sheathing. The large head holds thin or soft material more securely.

Some roofing nails are coated to prevent rusting. Others are made from noncorrosive metals such as aluminum and copper. The shank is usually barbed to increase holding power. Usual sizes run from $\frac{3}{4}$ inch to 2 inches. The gauge is not specified when ordering (Figure 9–1).

Masonry Nails. Masonry nails may be cut nails or wire nails (*Figure 9–7*). These nails are made from hardened steel to prevent them from bending when being driven into concrete or other masonry. The cut nail has a blunt point that tends to prevent

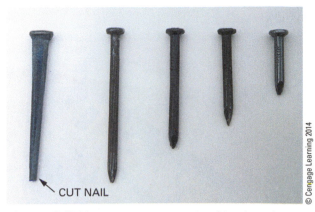

CUT NAIL

Figure 9-7 Masonry nails are made of hardened steel and may chip apart when driven.

splitting when it is driven into hardwood. Some masonry and flooring nails have round shanks of various designs for better holding power.

> **CAUTION**
>
> Masonry nails are made of hardened steel that is brittle. Care should be taken when driving them to avoid shattering the head. WEAR SAFETY GLASSES.

Staples. Staples are U-shaped fasteners used to secure a variety of materials. They come in a number of sizes and are designated by their length. Sold in boxes of one thousand or more, they are driven by several types of tools, including the squeeze stapler, hammer tacker, and electric stapler (*Figure 9–8*). They are used to fasten thin material.

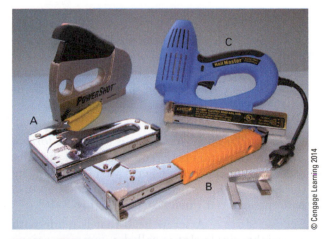

Figure 9-8 Staples can be driven by a variety of guns. (A) squeeze stapler, (B) hammer tacker stapler, and (C) electric stapler.

Figure 9-9 Using a power stapler to fasten wood.

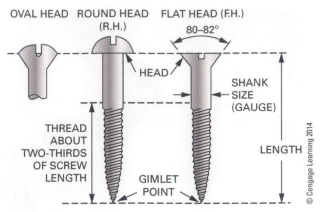

Figure 9-10 Common styles of slotted screws and screw terms.

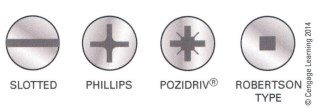

SLOTTED   PHILLIPS   POZIDRIV®   ROBERTSON TYPE

Figure 9-11 Common styles of screwhead slots.

Heavy-duty staples are similar in design only larger in size and gauge. These types of staples are often used to install some roofing materials or during cabinet construction (*Figure 9–9*).

SCREWS

Wood screws are used when greater holding power is needed and when the work being fastened must at times be removed. For example, door hinges must be applied with screws because nails would pull loose after a while, and the hinges may, at times, need to be removed. Screws cost more than nails and require more time to drive. When ordering screws, specify the length, gauge, type of head, coating, kind of metal, and screwdriver slot.

Kinds of Screws

A wood screw is identified by the shape of the screw head and screwdriver slot. For instance, a screw may be called a *flat-head Phillips* or a *round-head* screw. Three of the most common shapes of screw heads are the *flat head, round head,* and *oval head.*

The pointed end of a screw is called the *gimlet point*. The threaded section is called the *thread*. The smooth section between the head and thread is called the *shank*. Screw lengths are measured from the point to that part of the head that sets flush with the material when fastened (*Figure 9–10*).

Screw Slots.   A screw head that is made with a straight, single slot is called a common screw. A Phillips head screw has a crossed slot. There are many other types of screwdriver slots, each with a different name (*Figure 9–11*).

Sheet Metal Screws.   Wood screws are threaded, in most cases, only partway to the head. The threads of sheet metal screws extend for the full length of the screw and are much deeper. Sheet metal screws are used for fastening thin metal.

Another type of screw, used with power screwdrivers, is the self-tapping screw, which is used extensively to fasten metal framing. This screw has a cutting edge on its point to eliminate the need to predrill a hole (*Figure 9–12*). It is important that the drilling process be completed before the threading process begins. Drill points are available in

Figure 9-12 The self-tapping screws have pierce-point or drill-point tips.

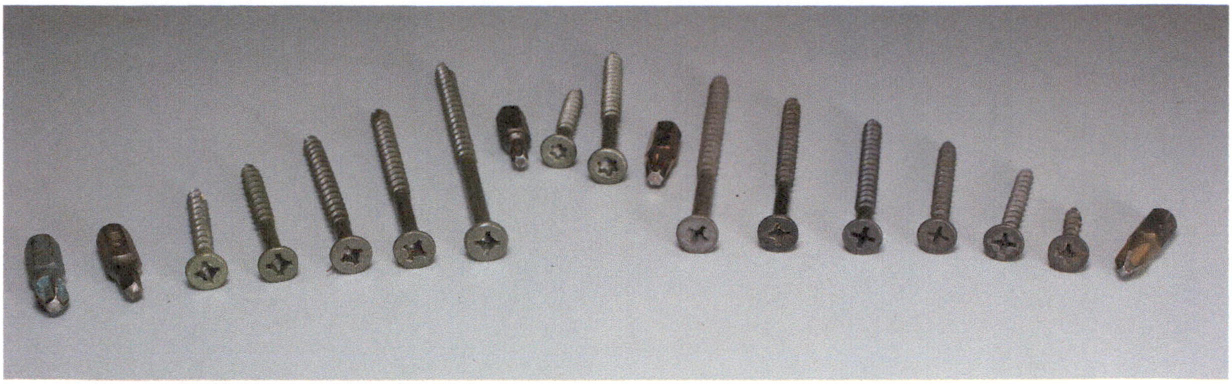
© Cengage Learning 2014

Figure 9-13 Typical drywall and deck screws.

various lengths and must be equal to the thickness of the metal being fastened.

Drywall screws are designed to fasten drywall to wood or metal framing members. The head is shaped like a bugle with threads extending nearly to the head. Drywall screws come in coarse or fine threads and sharp or drill points. Deck screws are similarly shaped but have an epoxy coating to resist corrosion (***Figure 9–13***). Screw heads vary depending on screw manufacturer.

Many other screws are available that are designed for special purposes. Like nails, screws come in a variety of metals and coatings (***Figure 9–14***). Steel screws with no coating are called bright screws.

Screw Sizes

Screws are made in many different sizes. Usual lengths range from $\frac{1}{4}$ inch to 4 inches. Gauges run from 0 to 24 (***Figure 9–15***). Unlike some nails, the higher the gauge number, the greater the diameter of the screw. Screw lengths are not available in every gauge. The lower gauge numbers are for shorter, thinner screws. Higher gauge numbers are for longer screws.

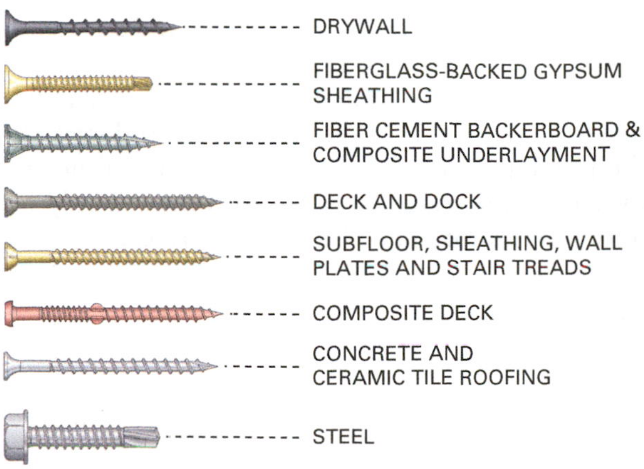
DRYWALL
FIBERGLASS-BACKED GYPSUM SHEATHING
FIBER CEMENT BACKERBOARD & COMPOSITE UNDERLAYMENT
DECK AND DOCK
SUBFLOOR, SHEATHING, WALL PLATES AND STAIR TREADS
COMPOSITE DECK
CONCRETE AND CERAMIC TILE ROOFING
STEEL
Courtesy of Simpson Strong-Tie Company

Figure 9-14 Screws come in a variety of styles for different applications.

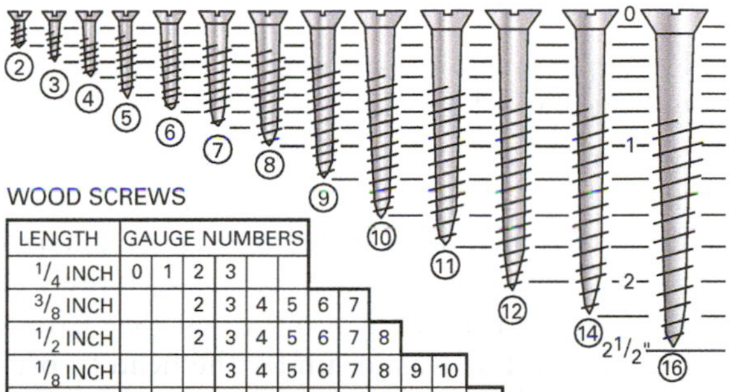
WOOD SCREWS
2
3
4
5
6
7
8
9
10
11
12
14
16
0
1
2
2 1/2"

LENGTH	GAUGE NUMBERS																	
1/4 INCH	0	1	2	3														
3/8 INCH			2	3	4	5	6	7										
1/2 INCH			2	3	4	5	6	7	8									
1/8 INCH				3	4	5	6	7	8	9	10							
3/4 INCH					4	5	6	7	8	9	10	11						
7/8 INCH							6	7	8	9	10	11	12					
1 INCH							6	7	8	9	10	11	12	14				
1 1/4 INCH								7	8	9	10	11	12	14	16			
1 1/2 INCH							6	7	8	9	10	11	12	14	16	18		
1 3/4 INCH									8	9	10	11	12	14	16	18	20	
2 INCH									8	9	10	11	12	14	16	18	20	
2 1/4 INCH										9	10	11	12	14	16	18	20	
2 1/2 INCH													12	14	16	18	20	
2 3/4 INCH														14	16	18	20	
3 INCH															16	18	20	
3 1/2 INCH																18	20	24
4 INCH																18	20	24

WHEN YOU BUY SCREWS, SPECIFY (1) LENGTH, (2) GAUGE NUMBER, (3) TYPE OF HEAD – FLAT, ROUND, OR OVAL, (4) MATERIAL – STEEL, BRASS, BRONZE, ETC., (5) FINISH – BRIGHT, STEEL, CADMIUM, NICKEL, OR CHROMIUM PLATED.

© Cengage Learning 2014

Figure 9-15 Wood screw sizes.

Lag Screws

Lag screws (***Figure 9–16***) are similar to wood screws except that they are larger and have a square or hex head designed to be turned with a wrench instead of a screwdriver. This fastener is

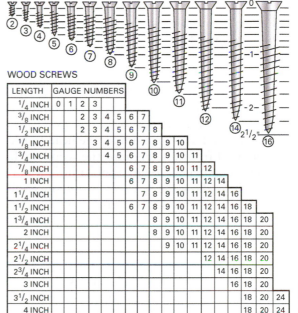

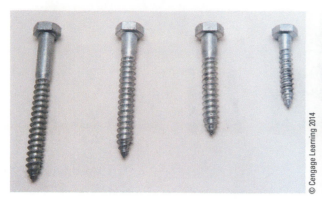

Figure 9-16 Lag screws are large screws with a square or hex head.

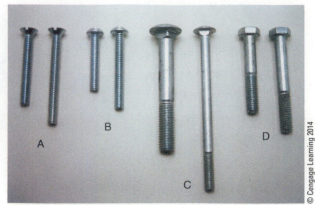

Figure 9-17 Commonly used bolts include (A) flat-head stove, (B) round-head stove, (C) carriage, and (D) machine.

used when great holding power is needed to join heavy parts and where a bolt cannot be used.

Lag screws are sized by diameter and length. Diameters range from $\frac{1}{4}$ inch to 1 inch, with lengths from 1 inch to 12 inches and up. Shank and pilot holes are predrilled to receive lag screws in the same manner as for wood screws. (See Chapter 13.) Place a flat washer under the head to prevent the head from digging into the wood as the lag screw is tightened down. Apply a little wax to the threads to allow the screw to turn more easily and to prevent the head from twisting off.

BOLTS

Most bolts are made of steel. To retard rusting, galvanized or stainless steel bolts are used. As with nails and screws, they are available in different kinds of metals and coatings. Many kinds are used for special purposes, but only a few are generally used. Commonly used bolts are the carriage, machine, and stove bolts (*Figure 9–17*).

Carriage Bolts

The carriage bolt has a square section under its oval head. The square section is embedded in wood or surface material and prevents the bolt from turning as the nut is tightened.

Machine Bolts

The machine bolt has a square or hex head. This is held with a wrench to keep the bolt from turning as the nut is tightened.

Stove Bolts

Stove bolts have either round or flat heads with a screwdriver slot. They are usually threaded all the way up to the head. *Machine screws* are very similar to stove bolts.

Bolt Sizes

Bolt sizes are specified by diameter and length. Carriage and machine bolts range from $\frac{3}{4}$ inch to 20 inches in length and from $\frac{3}{16}$ to $\frac{3}{4}$ inch in diameter. Stove bolts are small in comparison to other bolts. They commonly come in lengths from $\frac{3}{8}$ inch to 6 inches and from $\frac{1}{8}$ to $\frac{3}{8}$ inch in diameter.

Drill holes for bolts the same diameter as the bolt. Use flat washers under the head (except for carriage bolts) and under the nut to prevent the nut from cutting into the wood and to distribute the pressure over a wider area. Be careful not to overtighten carriage bolts. The head need only be drawn snug, not pulled below the surface.

CHAPTER 10

Anchors and Adhesives

ANCHORS

Special kinds of fasteners used to attach parts to solid masonry and hollow walls and ceilings are called anchors. There are hundreds of types available. Those most commonly used are described in this chapter.

Solid-Wall Anchors

Solid-wall anchors may be classified as heavy, medium, or light duty. Heavy-duty anchors are used to install such things as machinery, hand rails, dock bumpers, and storage racks. Medium-duty anchors may be used for hanging pipe and ductwork, securing window and door frames, and installing cabinets. Light-duty anchors are used for fastening such things as junction boxes, bathroom fixtures, closet organizers, small appliances, smoke detectors, and other lightweight objects.

Heavy-Duty Anchors. The wedge anchor (*Figure 10–1*) is used when high resistance to pullout is required. The anchor and hole diameter are the same, simplifying installation. The hole depth is not critical as long as the minimum is drilled. Proper installation requires cleaning out the hole.

The sleeve anchor (*Figure 10–2*) and its hole size are the same, and the hole depth need not be exact. After inserting the anchor in the hole, it is expanded by tightening the nut. This anchor can be used in material such as brick that may have voids or pockets.

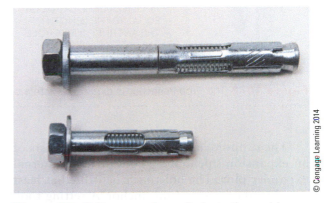

© Cengage Learning 2014

Figure 10–2 Sleeve anchors eliminate the problem of exact hole depth requirements.

INSERT – CLEAN HOLE, THEN DRIVE THE ANCHOR FAR ENOUGH INTO THE HOLE SO THAT AT LEAST SIX THREADS ARE BELOW THE TOP OF THE SURFACE OF THE FIXTURE.

DRILL – SIMPLY DRILL A HOLE THE SAME DIAMETER AS THE ANCHOR. DO NOT WORRY ABOUT DRILLING TOO DEEP BECAUSE THE ANCHOR WORKS IN A "BOTTOMLESS HOLE." YOU CAN DRILL INTO THE CONCRETE WITH THE LOAD POSITIONED IN PLACE; SIMPLY DRILL THROUGH THE PREDRILLED MOUNTING HOLES.

ANCHOR – MERELY TIGHTEN THE NUT. RESISTANCE WILL INCREASE RAPIDLY AFTER THE THIRD OR FOURTH COMPLETE TURN.

Courtesy of Porteous Fastener Co.

Figure 10–1 The wedge anchor has high resistance to pullout.

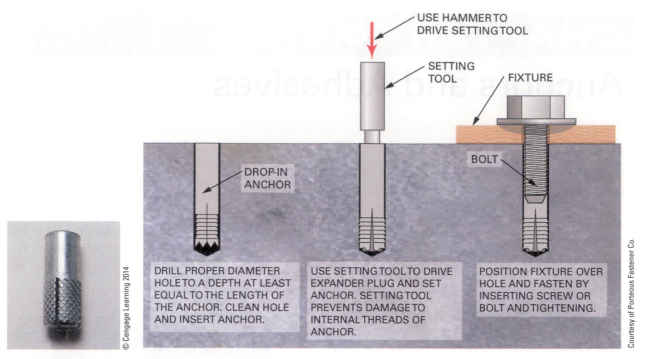

USE HAMMER TO
DRIVE SETTING TOOL

SETTING
TOOL

FIXTURE

DROP-IN
ANCHOR

BOLT

DRILL PROPER DIAMETER
HOLE TO A DEPTH AT LEAST
EQUAL TO THE LENGTH OF
THE ANCHOR. CLEAN HOLE
AND INSERT ANCHOR.

USE SETTING TOOL TO DRIVE
EXPANDER PLUG AND SET
ANCHOR. SETTING TOOL
PREVENTS DAMAGE TO
INTERNAL THREADS OF
ANCHOR.

POSITION FIXTURE OVER
HOLE AND FASTEN BY
INSERTING SCREW OR
BOLT AND TIGHTENING.

© Cengage Learning 2014

Courtesy of Porteous Fastener Co.

Figure 10–3 The drop-in anchor is expanded with a setting tool.

The **drop-in anchor** (*Figure 10–3*) consists of an expansion shield and a cone-shaped, internal expander plug. The hole must be drilled at least equal to the length of the anchor. A setting tool, supplied with the anchors, must be used to drive and expand the anchor. This anchor takes a machine screw or bolt.

Medium-Duty Anchors. **Split fast anchors** (*Figure 10–4*) are one-piece steel with two sheared expanded halves at the base. When driven, these halves are compressed and exert immense outward force on the inner walls of the hole as they try to regain their original shape. They come in both flat- and round-head styles.

Single and *double expansion anchors* (*Figure 10–5*) are used with machine screws or bolts. Drill a hole of recommended diameter to a depth equal to the length of the anchor. Place the anchor into the hole, flush with or slightly below the surface. Position the object to be fastened and bolt into place. Once fastened, the object may be unbolted, removed, and refastened, if desired.

The **lag shield** (*Figure 10–6*) is used with a lag screw. The shield is a split sleeve of soft metal, usually a zinc alloy. It is inserted into a hole of recommended diameter and a depth equal to the length of the shield plus $\frac{1}{2}$ inch or more. The lag screw length is determined by adding the length of the shield, the thickness of the material to be fastened, plus $\frac{1}{4}$ inch. The tip of the lag screw must protrude

DRILL HOLE AND DRIVE ANCHOR WITH HAMMER
THROUGH FIXTURE AND INTO HOLE UNTIL FLUSH.

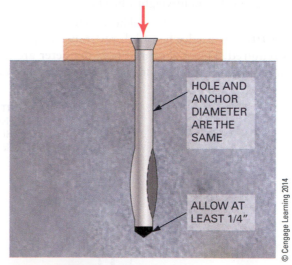

HOLE AND
ANCHOR
DIAMETER
ARE THE
SAME

ALLOW AT
LEAST 1/4"

© Cengage Learning 2014

Figure 10–4 The split fast is a one-piece, all-steel anchor for hard masonry.

from the bottom of the anchor to ensure proper expansion. As the fastener is threaded in, the shield expands tightly and securely in the drilled hole.

The *concrete screw* (*Figure 10–7*) utilizes specially fashioned high and low threads that cut into a properly sized hole in concrete. Screws come in $\frac{3}{16}$- and $\frac{1}{4}$-inch diameters and up to 6 inches in length. The hole diameter is important to the performance of the screw. It is recommended that a

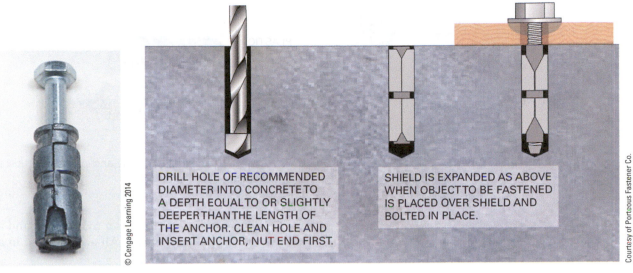

DRILL HOLE OF RECOMMENDED DIAMETER INTO CONCRETE TO A DEPTH EQUAL TO OR SLIGHTLY DEEPER THAN THE LENGTH OF THE ANCHOR. CLEAN HOLE AND INSERT ANCHOR, NUT END FIRST.

SHIELD IS EXPANDED AS ABOVE WHEN OBJECT TO BE FASTENED IS PLACED OVER SHIELD AND BOLTED IN PLACE.

© Cengage Learning 2014

Courtesy of Porteous Fastener Co.

Figure 10–5 Two opposing wedges of the double expansion anchor pull toward each other, expanding the full length of the anchor body.

© Cengage Learning 2014

Figure 10–6 Lag shields are designed for light- to medium-duty fastening in masonry.

minimum of 1-inch and a maximum of $1\frac{3}{4}$-inch embedment be used to determine the fastener length. The concrete screw system eliminates the need for plastic or lead anchors.

The *machine screw anchor* (*Figure 10–8*) consists of two parts. A lead sleeve slides over a threaded, cone-shaped piece. Using a special setting punch that comes with the anchors, the lead sleeve is driven over the cone-shaped piece to expand the sleeve and hold it securely in the hole. A hole of recommended diameter is drilled to a depth equal to the length of the anchor.

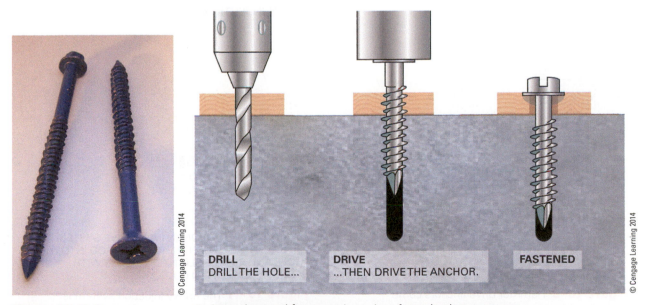

DRILL
DRILL THE HOLE...

DRIVE
...THEN DRIVE THE ANCHOR.

FASTENED

© Cengage Learning 2014

© Cengage Learning 2014

Figure 10–7 Concrete screws eliminate the need for an anchor when fastening into concrete.

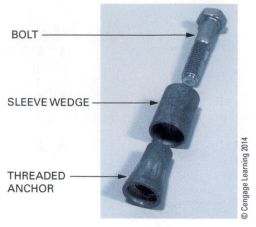

BOLT

SLEEVE WEDGE

THREADED ANCHOR

© Cengage Learning 2014

Figure 10–8 Machine screw anchors have a threaded bottom held by a sleeve wedged over the top.

Light-Duty Anchors.

Three kinds of drive anchors are commonly used for quick and easy fastening in solid masonry. They differ only in the material from which they are made. The *hammer drive anchor* (**Figure 10–9**) has a body of zinc alloy containing a steel expander pin. In the *aluminum drive anchor,* both the body and the pin are aluminum to avoid the corroding action of electrolysis.

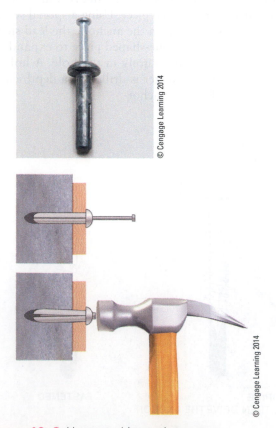

© Cengage Learning 2014

Figure 10–9 Hammer drive anchors come assembled for quick and easy fastening.

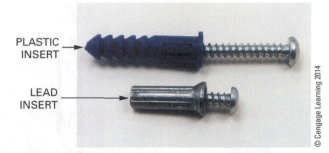

PLASTIC INSERT

LEAD INSERT

© Cengage Learning 2014

Figure 10–10 Lead and plastic anchors or inserts are used for light-duty fastening.

The *nylon nail anchor* utilizes a nylon body and a threaded steel expander pin. All are installed in a similar manner.

Lead and *plastic anchors,* also called inserts (**Figure 10–10**), are commonly used for fastening lightweight fixtures to masonry walls. These anchors have an unthreaded hole into which a screw is driven. The anchor is placed into a hole of recommended diameter and $\frac{1}{4}$ inch or more deeper than the length of the anchor. As the screw is turned, the threads of the screw cut into the soft material of the insert. This causes the insert to expand and tighten in the drilled hole. Ribs on the sides of the anchors prevent them from turning as the screw is driven.

Chemical Anchoring Systems.

Threaded studs, bolts, and rebar (concrete reinforcing rod) may be anchored in solid masonry with a chemical bond using an *epoxy resin compound.* Two types of systems commonly used are the *epoxy injection* (**Figure 10–11**) and *chemical capsule* (**Figure 10–12**).

EPOXY INJECTION

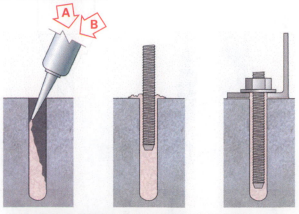

MIX THE TWO-COMPONENT ADHESIVE SYSTEM AND PLACE IN HOLE. PUSH THE ANCHOR ROD INTO THE HOLE AND ROTATE SLIGHTLY TO COAT WITH ADHESIVE. ALLOW TO CURE.

Courtesy of Adhesives Technology

Figure 10–11 The epoxy injection system is designed for high-strength anchoring.

CHEMICAL CAPSULE ANCHOR

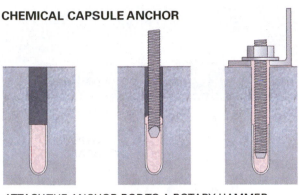

ATTACH THE ANCHOR ROD TO A ROTARY HAMMER ADAPTER. THE COMBINATION OF THE ROTATION AND HAMMERING ACTION MIXES THE CAPSULE CONTENTS TOGETHER. ALLOW TO CURE.

Figure 10–12 The chemical capsule anchoring system provides for easy, premeasured application.

Figure 10–13 Toggle bolts are used for fastening in hollow walls.

In the injection system, a dual cartridge is inserted into a tool similar to a caulking gun. The chemical is automatically mixed as it is dispensed. Small cartridges are available to accurately dispense epoxy from an ordinary caulking gun.

Chemical capsules contain the exact amount of all chemicals needed for one installation. Each capsule is marked with the appropriate hole size to be used. Drill holes according to diameter and depth indicated on each capsule. It is important to thoroughly clean and clear the hole of all concrete dust before inserting the capsule.

With all solid-masonry anchors, follow the specifications in regard to hole diameter and depth, minimum embedment, maximum fixture thickness, and allowable load on anchor.

Hollow-Wall Fasteners

Toggle Bolts. Toggle bolts (*Figure 10–13*) may have a wing or a tumble toggle. The wing toggle is fitted with springs, which cause it to open. The tumble toggle falls into its open position when passed through a drilled hole in the wall. The hole must be drilled large enough for the toggle of the bolt to slip through. A disadvantage of using toggle bolts is that, if removed, the toggle falls off inside the wall.

Plastic Toggles. The plastic toggle (*Figure 10–14*) consists of four legs attached to a body that has a hole through the center and fins on its side to prevent turning during installation. The legs collapse to allow insertion into the hole. As sheet metal screws are turned through the body, they draw in and expand the legs against the inner surface of the wall.

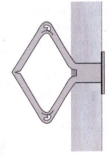

SQUEEZE TOGGLE WINGS FLAT AND PUSH INTO HOLE DRILLED IN WALL.

TAP ANCHOR IN AND FLUSH WITH WALL.

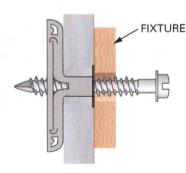

FIXTURE

PLACE FIXTURE OVER HOLE, INSERT SHEET METAL SCREW, AND TIGHTEN.

Figure 10–14 The plastic toggle hollow wall anchor.

Figure 10-15 Hollow wall expansion anchors.

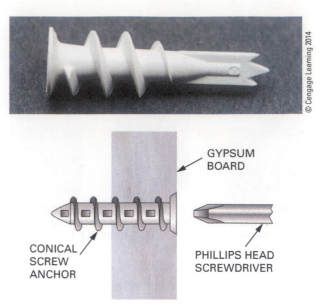

DRIVE ANCHOR IN WALL BY TURNING WITH SCREWDRIVER UNTIL HEAD IS FLUSH WITH SURFACE.

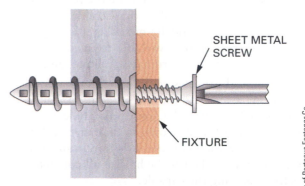

PLACE FIXTURE OVER HOLE IN ANCHOR AND FASTEN WITH PROPER SIZE SHEET METAL SCREW.

Figure 10–16 The conical screw anchor is a self-drilling, hollow wall anchor for lightweight fastenings.

Expansion Anchors. Hollow-wall expansion anchors (*Figure 10–15*) consist of an expandable sleeve, a machine screw, and a fiber washer. The collar on the outer end of the sleeve has two sharp prongs that grip into the surface of the wall material. This prevents the sleeve from turning when the screw is tightened to expand the anchor. After expanding the sleeve, the screw is removed, inserted through the part to be attached, and then screwed back into the anchor. Some types require that a hole be drilled, while other types have pointed ends that may be driven through the wall material.

Installed fixtures may be removed and refastened or replaced by removing the anchor screw without disturbing the anchor expansion. Anchors are manufactured for various wallboard thicknesses. Make sure to use the right size anchor for the wall thickness in which the anchor is being installed.

Conical Screws. The deep threads of the conical screw anchor (*Figure 10–16*) resist stripping out when screwed into gypsum board, strand board, and similar material. After the plug is seated flush with the wall, the fixture is placed over the hole and fastened by driving a screw through the center of the plug.

Nylon Plugs. The nylon plug (*Figure 10–17*) is used for a number of hollow-wall and some solid-wall applications. A hole of proper diameter is drilled. The plug is inserted, and the screw is driven to draw or expand the plug.

Connectors

Widely used in the construction industry are devices called connectors. Connectors are metal pieces formed into various shapes to join wood to wood, or wood to concrete or other masonry. Connectors are available in hundreds of shapes and styles and have specific names depending on their function. Only a few are discussed here.

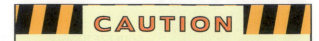

CAUTION

Framing ties and anchors must have the proper size and quantity of fasteners. Check manufacturer's recommendations.

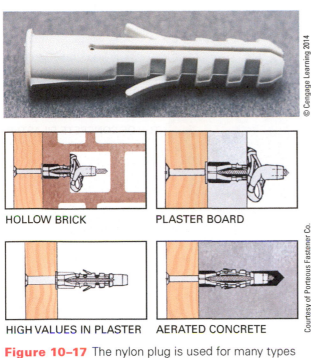

HOLLOW BRICK   PLASTER BOARD

HIGH VALUES IN PLASTER   AERATED CONCRETE

© Cengage Learning 2014

Courtesy of Porteous Fastener Co.

Figure 10–17 The nylon plug is used for many types of hollow wall fastening.

Wood-to-Wood. *Framing anchors* and *seismic* and *hurricane ties* (**Figure 10–18**) are used to join parts of a wood frame. *Post* and *column caps* and *bases* are used at the top and bottom of those members (**Figure 10–19**). *Joist hangers* and *beam hangers* are available in many sizes and styles (**Figure 10–20**). It is important to use the proper style, size, and quantity of nails in each hanger.

Wood-to-Concrete. Some wood-to-concrete connectors are *sill anchors, anchor bolts,* and *holddowns* (**Figure 10–21**). A *girder hanger* and a *beam seat* (**Figure 10–22**) make beam-to-foundation wall connections. *Post bases* come in various styles. They are used to anchor posts to concrete floors or footings.

Many other specialized connectors are used in frame construction. Some are described in the framing sections of this book. See Simpson Strong-Tie website at www.strongtie.com for more details on anchoring systems.

ADHESIVES

The carpenter seldom uses any glue in the frame or exterior finish. Glue is used on some joints and other parts of the interior finish work. A number of **mastics** (heavy, pastelike adhesives) are used throughout the construction process.

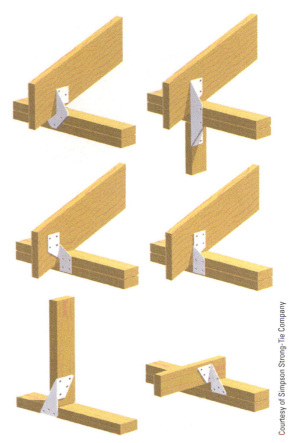

Courtesy of Simpson Strong-Tie Company

Figure 10–18 Framing ties and anchors are manufactured in many unique shapes.

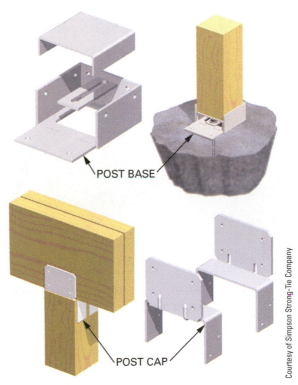

POST BASE

POST CAP

Courtesy of Simpson Strong-Tie Company

Figure 10–19 Caps and bases help fasten tops and bottoms of posts and columns.

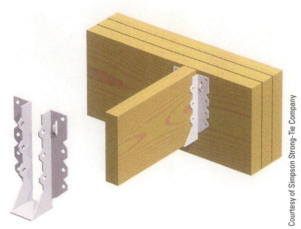

Figure 10–20 Hangers are used to support joists and beams.

Courtesy of Simpson Strong-Tie Company

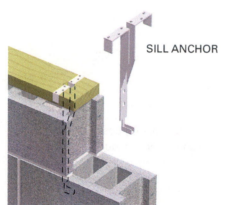

SILL ANCHOR

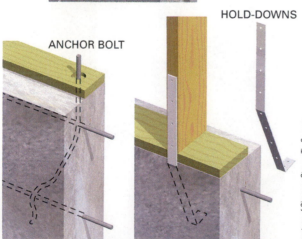

HOLD-DOWNS

ANCHOR BOLT

Courtesy of Simpson Strong-Tie Company

Figure 10–21 Sill anchors, anchor bolts, and hold-downs connect frame members to concrete.

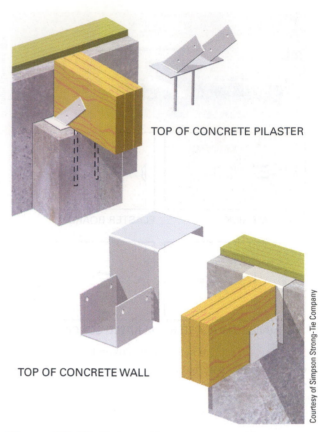

TOP OF CONCRETE PILASTER

TOP OF CONCRETE WALL

Courtesy of Simpson Strong-Tie Company

Figure 10–22 Girder and beam seats provide support from concrete walls.

Glue

White and Yellow Glue. Most of the glue used by the carpenter is the so-called white glue or yellow glue. The white glue is *polyvinyl acetate;* the yellow glue is *aliphatic resin.* Neither type is resistant to moisture. Both are fast setting, so joints should be made quickly after applying the glue. They are available under a number of trade names and are excellent for joining wood parts not subjected to moisture.

Urethane Glue. Urethane glue is a fine all-purpose glue available for bonding wood, stone, metal, ceramics, and plastics. Its strong, waterproof bond cures with exposure to moisture in material and air. It can be used for interior or exterior work and does not become brittle over time. It tends to expand while curing, filling gaps and spaces in material.

© Cengage Learning 2014

Figure 10–23 Applying panel adhesive to joists with a caulking gun for plywood subfloor.

Because urethane glue sticks to just about anything, care should be taken when working with it. It cannot be dissolved by common solvents, and cleanup can be difficult. It often requires days to scrape and rub it from skin.

Contact Cement. Contact cement is so named because pieces coated with it bond on contact and need not be clamped under pressure. It is extremely important that pieces be positioned accurately before contact is made. Contact cement is widely used to apply plastic laminates for kitchen countertops. It is also used to bond other thin or flexible material that otherwise might require elaborate clamping devices.

Mastics

Several types of mastics are used throughout the construction trades. They come in cans or cartridges used in hand or air guns and in large quantities that are troweled into place. With these adhesives, the bond is made stronger, and fewer fasteners are needed.

Construction Adhesive. One type of mastic is called *construction adhesive*. It is used in a glued floor system, described in a later unit on floor framing (*Figure 10–23*). It can be used in cold weather, even on wet or frozen wood. It is also used on stairs to increase stiffness and eliminate squeaks.

Panel Adhesive. *Panel adhesive* is used to apply such things as wall paneling, foam insulation, gypsum board, and hardboard to wood, metal, and masonry. It is usually dispensed with a caulking gun. It is important to use the adhesives matched for the material being installed.

Troweled Mastics. Other types of mastics may be applied by hand for such purposes as installing vinyl base, vinyl floor tile, or ceramic wall tile. A notched trowel is usually used to spread the adhesive. The depth and spacing of the notches along the edges of the trowel determine the amount of adhesive left on the surface.

It is important to use a trowel with the correct notch depth and spacing. Failure to follow recommendations will result in serious consequences. Too much adhesive causes the excess to squeeze out onto the finished surface. This leaves no alternative but to remove the applied pieces, clean up, and start over. Too little adhesive may result in loose pieces.

DECONSTRUCT THIS

Carefully study *Figure 10–24* and think about what is wrong and/or what is right. Consider all possibilities. What construction practice or method is different in your area of the country?

Figure 10–24 Here plywood is being attached to a framing member.

KEY TERMS

anchors	contact cement	lag shield	sleeve anchor
box nails	drive anchor	machine bolt	solid wall anchors
brads	drop-in anchor	masonry nails	split fast anchor
bright	duplex nail	mastics	staple
carriage bolt	electrolysis	nylon plug	stove bolt
casing nails	expansion anchors	penny (d)	toggle bolts
common nails	finish nails	Phillips head	urethane glue
common screw	galvanized	plastic toggle	wedge anchor
conical screw	inserts	roofing nails	wire nails
connectors	lag screws	self-tapping screw	

REVIEW QUESTIONS

Select the most appropriate answer.

1. The length of an eight-penny nail is

   a. 1$\frac{1}{2}$ inches.
   b. 2 inches.
   c. 2$\frac{1}{2}$ inches.
   d. 3 inches.

2. Fasteners coated with zinc to retard rusting are called

   a. coated.
   b. dipped.
   c. electroplated.
   d. galvanized.

3. Brads are

   a. types of screws.
   b. small box nails.
   c. small finishing nails.
   d. kinds of stove bolts.

4. When a moisture-resistant exterior glue is required, use

   a. white glue.
   b. urethane glue.
   c. yellow glue.
   d. rubber cement.

5. Many carpenters prefer to use casing nails to fasten

   a. interior finish.
   b. exterior finish.
   c. door casings.
   d. roof shingles.

6. The blunt point on the end of a cut nail helps

   a. drive the nail straight.
   b. prevent splitting of the wood.
   c. hold the fastened material more securely.
   d. start the nail in the material.

7. On temporary structures, such as concrete forms, use

   a. common nails.
   b. duplex nails.
   c. galvanized nails.
   d. spikes.

8. The common name for a fastener with tapered threads is

   a. bolt.
   b. screw.
   c. lag.
   d. all of the above.

9. An example of a solid-wall anchor is a

   a. wedge anchor.
   b. conical screw.
   c. toggle bolt.
   d. all of the above.

10. The term *connector* refers to a

   a. metal device used to fasten wood members.
   b. wire used to make nails and screws.
   c. worker using a nail or screw gun.
   d. all of the above.

UNIT 5
Hand Tools

One of the many benefits to working in the field of construction is the variety and diversity of tools available. Tools are the means by which construction happens.

Knowing how to choose the proper tool and how to keep it in good working condition is essential. A tradesperson should never underestimate the importance of tools and never neglect their proper use and care. Tools should be kept clean and in good condition. If they get wet on the job, dry them as soon as possible, and coat them with light oil to prevent them from rusting.

Carpenters are expected to have their own hand tools and to keep them in good working condition. Tools vary in quality, which is related to cost. Generally, expensive tools have better quality than inexpensive tools. For example, inferior tools cannot be brought to a sharp, keen edge and will dull rapidly. They will bend or break under normal use. Quality tools are worth the expense. The condition of a tool reveals the attitude of the owner toward his or her profession.

OBJECTIVES

After completing this unit, the student should be able to:

- identify and describe the hand tools that are commonly used by the carpenter.
- use each of the hand tools in a safe and appropriate manner.
- sharpen and maintain hand tools in suitable working condition.

SAFETY REMINDER

Carpentry as a trade was created using hand tools, some having a long history. Each tool has a specific purpose and associated risk of use.

The use of tools requires the operator to be knowledgeable about how to safely manipulate the tools.

This applies to hand tools as well as power tools. Safety is an attitude—an attitude of acceptance of a tool and all of its operational requirements. Safety is a blend of ability, skill, and knowledge—a blend that should always be present when working with tools.

CHAPTER 11

Layout Tools

LAYOUT TOOLS

Much of the work a carpenter does must first be laid out, measured, and marked. Layout tools are used to measure distances, mark lines and angles, test for depths, and align various material into the proper positions.

Measuring Tools

The ability to take measurements quickly and accurately must be mastered early in the carpenter's training. Practice reading the rule or tape to gain skill in fast and precise measuring.

Most industrialized countries use the metric system of measure. Linear metric measure centers on the meter, which is slightly larger than a yard. Smaller parts of a meter are denoted by the prefix *deci-* ($\frac{1}{10}$), which is used instead of *feet*. *Centi-* ($\frac{1}{100}$) and *milli-* ($\frac{1}{1000}$) are used instead of inches and fractions. The prefix *kilo-* represents one thousand times larger and is used instead of *miles*. For example, in metric measure a 2 × 4 is 39 mm (millimeter) × 89 mm. The metric system is easier to use than the English system because all measurements are in decimal form and there are no fractions.

Pocket Tapes. Most measuring done by tradespeople is done with pocket tapes (*Figure 11–1*). These are painted steel ribbons wound around a spool with a spring inside. The spring returns the tape after it is extended. They are made as small as 3 feet, but typical professional models are 16, 25, and 35 feet long.

Pocket tapes are divided into feet, inches, and sixteenths of an inch. They have clearly marked increments of 12 and 16 inches, the spacing for standard framing members, to speed up the layout. Markings are usually black for each 12 inches and red for every 16 inches. Some tapes also have small black dots at increments of 19.2 inches (*Figure 11–2*). This spacing is typically used only for layout of some engineered floor members.

Each inch on a tape is divided into fractions of an inch. Each fraction line has a name that must

Figure 11–1 Pocket tape.

Courtesy of The Stanley Works

be memorized (see Figure 11–2). Most measuring done by a carpenter is to the nearest sixteenth, while a cabinetmaker will work to a sixty-fourth. A carpenter should be able to read a ruler quickly and accurately.

Steel tapes in 50 and 100 foot lengths are commonly used to lay out longer measurements. They are not spring loaded, so they must be rewound by hand. The end of the tape has a steel ring with a folding hook attached. The hook may be unfolded to go over the edge of an object. It may also be left in the folded position and the ring placed over a nail when extending the tape. Remember to place the nail so that the *outside* of the ring, which is the actual end of the tape, reaches to the desired mark (*Figure 11–3*). Rewind the tape when not using it. If the tape is kinked, it will snap. Keep it out of water. If it gets wet, dry it thoroughly while rewinding.

Some tradespeople prefer the 6 or 8 foot *folding rule*. The folding rule sometimes has a metal extension on one end for making inside measurements (*Figure 11–4*).

On the Job

Experienced tradespeople can read ruler fractions quickly and accurately. Study and memorize the one-sixteenth increments of an inch by name.

BLACK BACKGROUND EVERY FOOT

1 FOOT –1INCH = 13 INCHES

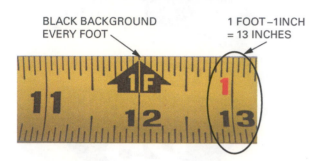

Figure 11–3 Steel tapes have ends with folding hooks. User must be aware of how to hold the tapes to maintain accuracy.

RED BACKGROUND EVERY 16 INCHES

BLACK DIAMOND EVERY 19.2 INCHES

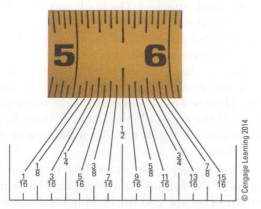

Figure 11–2 Tapes have color coded markings at 12 , 16 and 19.2 inch intervals. Each inch is typically broken into sixteenth of an inch increments.

Figure 11–4 Folding wood rulers have a sliding metal extension.

Squares

The carpenter has the use of a number of different kinds of squares to measure and lay out for square and other angle cuts.

Figure 11–5 The body and blade of a combination square are adjustable.

Combination Squares. The combination square (*Figure 11–5*) consists of a movable blade, 1 inch wide and 12 inches long, that slides along the body of the square. It is used to lay out or test 90 and 45 degree angles. Hold the body of the square against the edge of the stock and mark along the blade (*Figure 11–6*). It can function as a depth gauge to lay out or test the depth of rabbets, grooves, and dadoes. It can also be used with a pencil as a marking gauge to draw lines parallel to the edge of a board. Drawing lines in this manner is called *gauging* lines. Lines may also be gauged by holding the pencil and riding the finger along the edge of the board. Finger-gauging takes practice, but once mastered saves a lot of time. Be sure to check the edge of the wood for slivers first.

Speed Squares. Some carpenters prefer to use a triangular-shaped square known by the brand name Speed Square (*Figure 11–7*). Speed Squares are made of one-piece plastic or aluminum alloy and are available in two sizes. They can be used to lay out 90 and 45 degree angles and as guides for portable

On the Job

Watch the grain of the wood as you adjust scribing speed to avoid splinters.

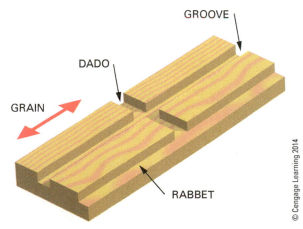

GROOVE

DADO

GRAIN

RABBET

Figure 11–6 The combination square is useful for squaring and as a marking gauge. A pencil held in one hand is a quick way to draw a parallel line.

Figure 11–7 Speed Squares are used for layout of rafters and other angles.

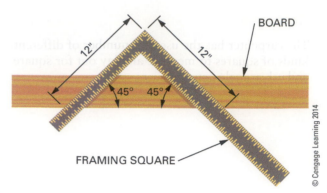

Figure 11–9 Laying out 45 degree angles with a framing square.

power saws. A degree scale allows angles to be laid out; other scales may be used to lay out rafters.

Framing Squares. The framing square, often called the *steel, or rafter square* (*Figure 11–8*), is an L-shaped tool made of thin steel or aluminum. The longer of the two legs is called the *blade* or *body* and is 2 inches (50 mm) wide and 24 inches (600 mm) long. The shorter leg is called the *tongue* and is 1½ inches (38 mm) wide and 16 inches (400 mm) long. The outside corner is called the *heel.*

The framing square is a centuries-old tool. Entire books have been written about it. Based on the use of the right triangle, many layout techniques have been devised and used. These techniques and necessary scales, tables, and graduations stamped on the square were designed to assist the carpenter in the many calculations needed. Today a pocket calculator has virtually replaced these aids. The only exception is the rafter table, which is still useful today and will be discussed in more detail in Unit 15.

The framing square is useful in laying out roof rafters, bridging, and stairs. It is also used to lay out 90 and 45 degree angles (*Figure 11–9*).

The side that has the manufacturer's name stamped on it is referred to as the *face side.* The rafter table is printed on the body. On the same side of the square, on the tongue, can be found the *octagon scale,* which is used to lay out eight-sided timbers from square ones (*Figure 11–10*).

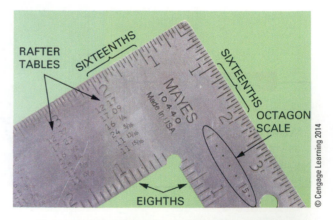

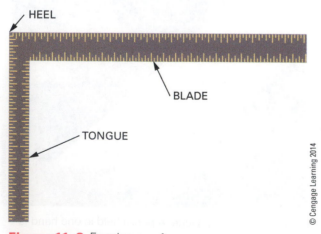

Figure 11–8 Framing or rafter square.

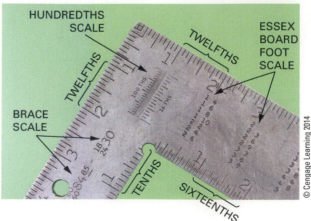

Figure 11–10 Both sides of a rafter square have tables and scales to assist the carpenter.

On the back side of the square, the *Essex board foot table* is used to calculate the number of board feet in lumber. The *brace table* is used to figure the length of diagonal braces. The *hundredths scale,* consisting of an inch divided into one hundred parts, is used to find $\frac{1}{100}$ths of an inch. This scale may be used to convert fractions to decimals and vice versa.

On the face side, the edges are divided into inches that are graduated into $\frac{1}{8}$ths on the inside and $\frac{1}{16}$ths on the outside. The edges on the back are divided into inches and $\frac{1}{12}$ths on the outside, while one inside edge is graduated into $\frac{1}{16}$ths and the other into $\frac{1}{10}$ths.

Sliding T-Bevels. The *sliding T-bevel,* sometimes called a *bevel square* or just a **bevel** (*Figure 11–11*),

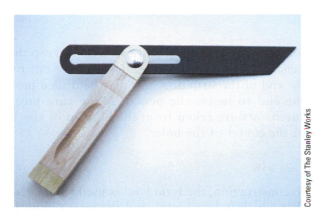

Figure 11–11 Sliding T-bevel.

Courtesy of The Stanley Works

consists of a body and a sliding blade that can be turned to any angle and locked in position. It is used to lay out or test angles other than those laid out with squares. The body of the tool is held against the edge of the stock, and the angle is laid out by marking along the blade.

Straightedges

A **straightedge** can be made of metal or wood. It can have any thickness, width, or length, as long as the size is convenient for its intended use and it has at least one edge that is absolutely straight from one end to the other. To determine if it is straight, sight along the edge. Another way is to lay the piece on its side and mark along its edge from one end to the other. Turn the piece over. Hold each end on the line just marked, and mark another line. If both lines coincide, the edge is straight (*Procedure 11–A*).

Straightedges are useful for many purposes. The framing square, the blade of the combination square, or the back of a saw could be used as a straightedge for drawing short, straight lines.

To determine a straight distance over large distances, a line (or string) and gauge blocks can be used. This method uses a taut string held away by offset blocks from the surface of the material being straightened. Another block of the same thickness as the offset blocks is then used periodically to test the material's distance from the line (*Figure 11–12*). This method is easy to do and can be very accurate.

StepbyStep Procedures

Procedure 11–A Checking a Straightedge

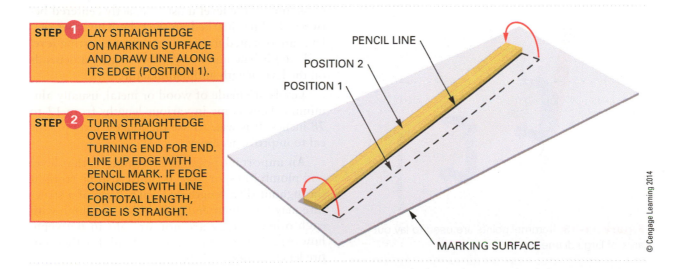

STEP **1** LAY STRAIGHTEDGE ON MARKING SURFACE AND DRAW LINE ALONG ITS EDGE (POSITION 1).

STEP **2** TURN STRAIGHTEDGE OVER WITHOUT TURNING END FOR END. LINE UP EDGE WITH PENCIL MARK. IF EDGE COINCIDES WITH LINE FOR TOTAL LENGTH, EDGE IS STRAIGHT.

PENCIL LINE

POSITION 2

POSITION 1

MARKING SURFACE

© Cengage Learning 2014

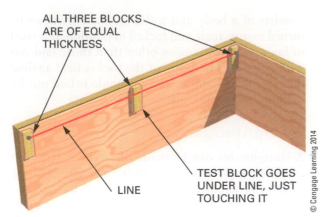

Figure 11–12 Use of a line and gauge blocks is an effective method to determine a straight line.

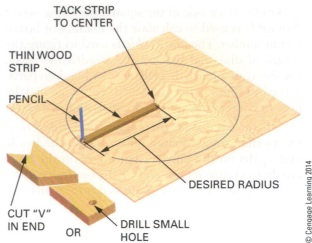

Figure 11–14 A thin strip of wood can be used to lay out circles or arcs.

Trammel Points. A pair of tools called trammel points may be used to draw circles or parts of circles, called arcs (*Figure 11–13*), which may be too large for a compass. They can be clamped to a strip of wood any distance apart according to the desired radius of the circle to be laid out. One trammel point can be set on the center while the other, which may have a pencil attached, is swung to lay out the circle or arc.

In place of trammel points, the same kinds of layouts can be made by using a thin strip of wood with a brad or small finish nail through it for a center point. Measure from the end of the strip a distance equal to the desired radius. Drive the brad through the strip until the point comes through. Set the point of the brad on the center, and hold a pencil against the end while swinging the strip to

form the circle or arc (*Figure 11–14*). To keep the pencil from slipping, a small V may be cut on the end of the strip or a hole may be drilled near the end to insert the pencil. Make sure measurements are taken from the bottom of the V or the center of the hole.

Levels

In construction, the term level is used to indicate that which is *horizontal*, and the term plumb is used to mean the same as *vertical*. The term *level* also refers to a tool that is used to achieve both level and plumb.

Carpenter's Levels. The carpenter's level (*Figure 11–15*) is used to test both level and plumb surfaces. Accurate use of the level depends on accurate reading. The air bubble in the slightly crowned glass tube of the level must be exactly centered between the lines marked on the tube. The tubes of a level are oriented in two directions for testing level and plumb. The number of tubes in a level depends on the level length and manufacturer.

Levels are made of wood or metal, usually aluminum. They come in various lengths from 12 to 78 inches. It is wise to use the longest level practical to improve accuracy.

An important point to remember is that level and plumb lines, or objects, must also be straight throughout their length or height. Parts of a structure may have their end points level or plumb with each other. If they are not straight in between, however, they are not level or plumb for their entire length (*Figure 11–16*).

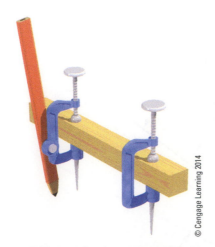

Figure 11–13 Trammel points are used to lay out arcs of large diameter.

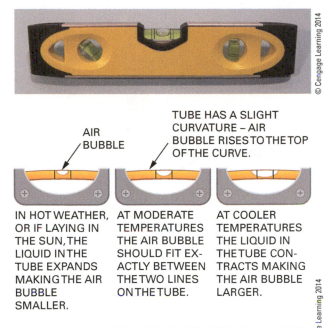

AIR BUBBLE

TUBE HAS A SLIGHT CURVATURE – AIR BUBBLE RISES TO THE TOP OF THE CURVE.

| IN HOT WEATHER, OR IF LAYING IN THE SUN, THE LIQUID IN THE TUBE EXPANDS MAKING THE AIR BUBBLE SMALLER. | AT MODERATE TEMPERATURES THE AIR BUBBLE SHOULD FIT EXACTLY BETWEEN THE TWO LINES ON THE TUBE. | AT COOLER TEMPERATURES THE LIQUID IN THE TUBE CONTRACTS MAKING THE AIR BUBBLE LARGER. |

REGARDLESS OF CONDITIONS, THE AIR BUBBLE MUST BE CENTERED BETWEEN THE TWO LINES ON THE TUBE FOR THE INSTRUMENT TO BE LEVEL OR PLUMB.

Figure 11–15 The bubble size of a carpenter's level can be affected by temperature.

To check a level for accuracy, place it on a nearly level or plumb object that is firm. Note the exact position of the level on the object. Read the level carefully and remember where the bubble is located within the lines on the bubble tube. Rotate the level along its vertical axis and reposition it in the same place on the object (*Figure 11–17*). If the bubble reads the same as the previous measurement, then the level is accurate.

Line Levels. The line level (*Figure 11–18*) consists of one glass tube encased in a metal sleeve with hooks on each end. The hooks are attached to a stretched line, which is then moved up or down until the bubble is centered. However, this is not an accurate method and gives only approximate levelness. Care must be taken that the level be attached close to the center of the suspended line because the weight of the level causes the line to sag. If the line level is off center to any great degree, the results are faulty.

Plumb Bobs. The plumb bob (*Figure 11–19*) is accurate and is used frequently for testing and establishing plumb lines. Suspended from a line, the plumb bob hangs absolutely vertical. However, it

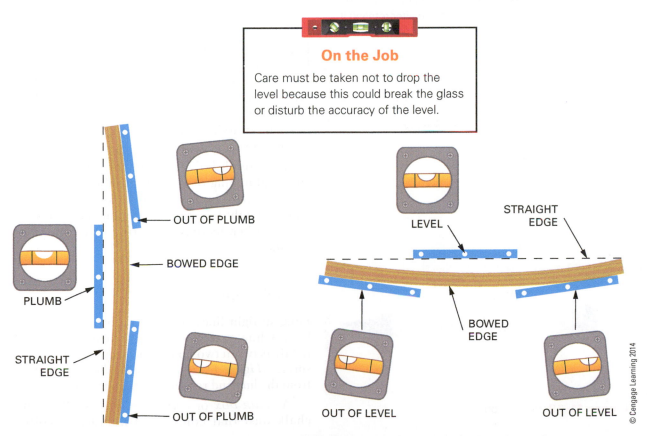

On the Job

Care must be taken not to drop the level because this could break the glass or disturb the accuracy of the level.

OUT OF PLUMB

BOWED EDGE

PLUMB

STRAIGHT EDGE

OUT OF PLUMB

LEVEL

STRAIGHT EDGE

BOWED EDGE

OUT OF LEVEL

OUT OF LEVEL

Figure 11–16 To be level or plumb for their entire length, pieces must be straight from end to end.

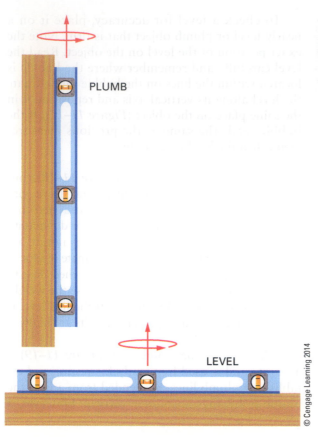

Figure 11–17 To check a level for accuracy the bubble should read exactly the same before and after rotating it.

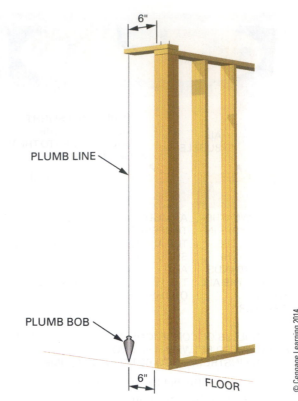

Figure 11–20 The post is plumb when the distance between it and the plumb line is the same.

Figure 11–18 Line level.

Figure 11–19 Plumb bob.

is difficult to use outside when the wind is blowing because it will move with the wind. Plumb bobs come in several different weights. Heavy plumb bobs stop swinging more quickly than lighter ones. Some have hollow centers that are filled with heavy metal to increase the weight without enlarging the size.

The plumb bob is useful for quick and accurate plumbing of vertical members of a structure (*Figure 11–20*). It can be suspended from a great height to establish a point that is plumb with another. Its only limitation is the length of the line.

Chalk Lines

Long straight lines are laid out by using a chalk line. A line coated with chalk dust is stretched tightly between two points and snapped against the surface (*Figure 11–21*). The chalk dust is dislodged from the line and remains on the surface.

A *chalk box* or *chalk line reel* is filled with chalk dust that comes in a number of colors

Figure 11–21 Snapping a chalk line on a roof deck.

sure lines are stretched tight before snapping in order to snap a straight and true line. Sight long lines by eye for straightness to make sure there is no sag in the line. If there is a sag, take it out by supporting the line near the center. Press the center of the line to the material and snap the line on both sides of the center. Keep the line from getting wet. If it does get wet, leave it outside the box until it dries.

Wing Dividers

Wing dividers can be used as a compass to lay out circles and arcs and as dividers to space off equal distances. However, this tool is used mainly for scribing and is often called a *scriber*. Scribing is the technique of laying out stock to fit against an irregular surface (*Figure 11–23*). For easier and more accurate scribing, heat and bend the end of

Figure 11–22 Chalk line reel.

(*Figure 11–22*). The most popular colors are blue, yellow, red, and white. The dust saturates the line, which is on a reel inside the box. The line is ready to be snapped when it is pulled out of the box. After several snaps, the line will need more chalk and must be reeled in to be recoated with chalk. Shaking or tapping the box helps recoat the line.

Chalk Line Techniques.   When unwinding and chalking the line, keep it off the surface until snapped. Otherwise many lines will be made on the surface, and this could be confusing. Make

Figure 11–23 Scribing is laying out a piece to fit against an irregular surface.

On the Job

Bend the leg of the dividers as shown for easier and more accurate scribing.

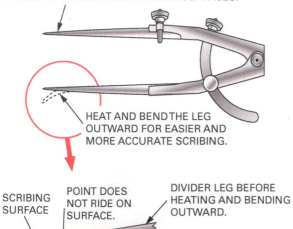

METAL LEG IS INTERCHANGEABLE WITH PENCIL AND IS USED TO SCRIBE LINES ON DARK SURFACES.

HEAT AND BEND THE LEG OUTWARD FOR EASIER AND MORE ACCURATE SCRIBING.

SCRIBING SURFACE

POINT DOES NOT RIDE ON SURFACE.

DIVIDER LEG BEFORE HEATING AND BENDING OUTWARD.

SCRIBING MAY NOT BE ACCURATE BECAUSE POINT IS NOT RIDING ON SURFACE.

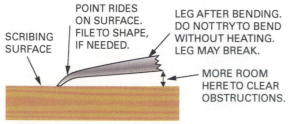

POINT RIDES ON SURFACE. FILE TO SHAPE, IF NEEDED.

LEG AFTER BENDING. DO NOT TRY TO BEND WITHOUT HEATING. LEG MAY BREAK.

SCRIBING SURFACE

MORE ROOM HERE TO CLEAR OBSTRUCTIONS.

SCRIBING IS MORE ACCURATE WHEN POINT RIDES ON SURFACE.

© Cengage Learning 2014

Figure 11–24 Adjusting one of the metal legs of a scriber makes it a more accurate tool.

the solid metal leg outward (*Figure 11–24*). Pencils are usually used in place of the interchangeable steel marking leg. Use pencils with hard lead that keep their points longer.

Butt Markers

Butt markers (*Figure 11–25*) are available in three sizes. They are often used to mark hinge gains. The marker is laid on the door edge at the hinge location and tapped with a hammer to outline the cutout for the hinge.

© Cengage Learning 2014

Figure 11–25 Butt hinge markers make mortising the hinge into a door easier.

CHAPTER 12
Boring and Cutting Tools

BORING TOOLS

The carpenter is often required to cut holes in wood and metal. Boring denotes cutting large holes in wood. Drilling is often thought of as making holes in metal or small holes in wood. Boring tools include those that actually do the cutting and those used to turn the cutting tool. The hole size and the bit size are measured according to their diameters. (See also Chapter 14.)

Bit Braces

The bit brace (*Figure 12–1*) is used to hold and turn auger bits to bore holes in wood. Its size is determined by its *sweep* (the diameter of the circle made by its handle). Sizes range from 8 to 12 inches. Most bit braces come with a ratchet that can be used when there is not enough room to make a complete turn of the handle.

Bits

Auger Bits.   Auger bits (*Figure 12–2*) are available with coarse or fine *feed screws*. Bits with coarse feed screws are used for fast boring in rough work. Fine feed bits are used for slower boring in finish work. As the bit is turned, the feed screw pulls the bit through the wood so little or no pressure on the bit is necessary. The *spurs* score the outer circle of the hole in advance of the *cutting lips*. The cutting lips lift the chip up and through the twist of the bit.

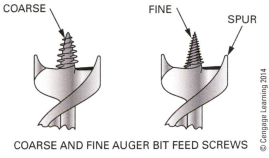

COARSE AND FINE AUGER BIT FEED SCREWS

Figure 12–2 Auger bits have feed screws and spurs.

A full set of auger bits ranges in sizes from $\frac{1}{4}$ to 1 inch, graduated in $\frac{1}{16}$ inch increments. The bit size is designated by the number of $\frac{1}{16}$ inch increments in its diameter. For instance, a #12 bit has 12 sixteenths. Therefore, it will bore a $\frac{3}{4}$ inch diameter hole.

CUTTING TOOLS

The carpenter uses many kinds of cutting tools and must know which to select for each job, as well as how to use, sharpen, and maintain them.

Edge-Cutting Tools

Wood Chisels.   The wood chisel (*Figure 12–3*) is used to cut recesses in wood for such things as door hinges and locksets and to make joints.

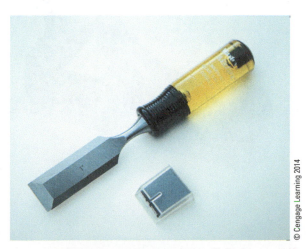

Figure 12–3 Wood chisel.

Figure 12–1 Bit brace.

Chisels are sized according to the width of the blade and are available in widths of $\frac{1}{8}$ inch to 2 inches. Most carpenters can do their work with a set consisting of chisels that are $\frac{1}{4}$, $\frac{1}{2}$, $\frac{3}{4}$, 1, and $1\frac{1}{2}$ inches in size.

Firmer chisels have long, thick blades and are used on heavy framing. *Butt chisels* are short, with a thinner blade. They are preferred for finish work.

CAUTION

Improper use of chisels has caused many accidents. When using chisels, keep both hands behind the cutting edge at all times (*Figure 12–4*). When not in use, the cutting edge should be shielded. Never put or carry chisels or other sharp or pointed tools in pockets.

Figure 12–4 Keep both hands in back of the chisel's cutting edge.

Figure 12–5 The jack plane is a general-purpose bench plane.

Bench Planes. Bench planes (*Figure 12–5*) come in several sizes. They are used for smoothing rough surfaces and bringing work down to the desired size. Large planes are used, for instance, on door edges to produce a straight surface over a long distance. Long planes will bridge hollows in a surface and cut high spots better than short plane (*Figure 12–6*). Small planes are more easily used for shorter work.

Bench planes are given names according to their length. The longest is called the *jointer*. In declining order are the *fore*, *jack*, and *smooth* planes. It is not necessary to have all the planes. The jack plane is 14 inches long and of all the bench planes is considered the best for all-around work.

Block Planes. Block planes are small planes designed to be held in one hand. They are often used to smooth the edges of short pieces and for trimming end grain to make fine joints (*Figure 12–7*).

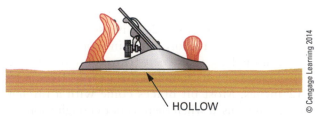

HOLLOW

Figure 12–6 Longer planes bridge hollows to allow for planing of long, straight edges.

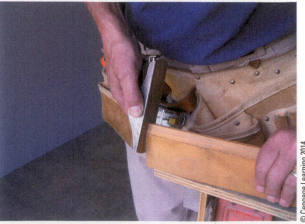

Figure 12–7 A block plane is small and often has a low blade angle.

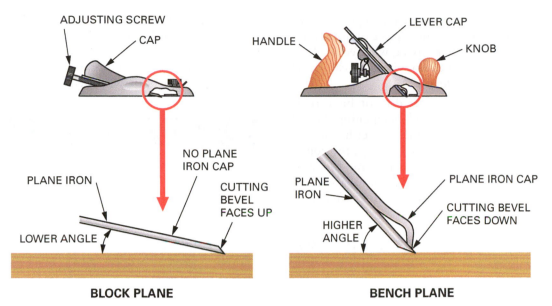

Figure 12–8 Difference between a block plane and a bench plane.

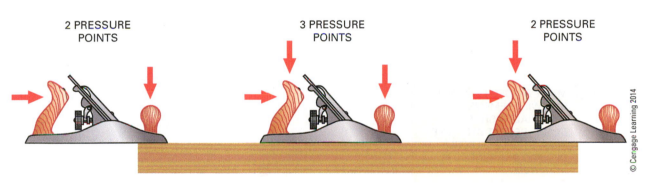

Figure 12–9 Correct method for using a plane.

Block planes are designed differently than bench planes. On bench planes, the cutting edge bevel is on the bottom side. On block planes, it is on the top. In addition, the bench plane iron has a plane iron cap attached to it, while the block plane iron has none (*Figure 12–8*).

Unlike bench planes, block planes are available with their blades set at a high angle or at a low angle. Most carpenters prefer the low-angle block plane because this type of plane cuts end grain more effectively. Block planes also have a smoother cutting action and fit into the hand more comfortably.

CAUTION

Keep plane irons sharp. When planes are not in use, place them on their sides or retract the iron.

Using Planes. When planing, have the stock securely held against a stop. When starting, push forward while applying pressure downward on the **toe** (front). When the **heel** (back) clears the end, apply pressure downward on both ends while pushing forward. When the opposite end is approached, relax pressure on the toe and continue pressure on the heel until the cut is complete (*Figure 12–9*). This method prevents tilting the plane over the ends of the stock and helps ensure a straight, smooth edge.

Sharpening Chisels and Plane Irons. To produce a keen edge, the tool must be **whetted** (sharpened) using an oilstone or waterstone. Hold the tool on a well-oiled stone so that the bevel rests flat on it. Move the tool back and forth across the stone for a few strokes. Then, make a few strokes with the flat side of the chisel or plane iron held absolutely flat on the stone. Continue whetting in this manner until as keen an edge as possible is obtained. To obtain a keener edge, repeat the procedure on a finer stone or on a piece of leather. The edge is sharp when, after having whetted the bevel and before tuning it over, no wire edge can be felt on the flat side (*Procedure 12–A*).

Chisels and plane irons do not have to be ground each time they need sharpening. Grinding is necessary only when the bevel has lost its concave shape by repeated whettings, the edge is badly nicked, or the bevel has become too short and blunt. The edge of a blade may be whetted many times before it needs grinding. After many

whettings, the bevel of the wood chisel or plane iron may need to be shaped by a grinding wheel. The bevel should have a concave surface, which is called a *hollow grind*. To obtain a hollow grind, a grinding attachment may be used.

CAUTION

Use safety goggles or otherwise protect your eyes when grinding.

Step**by**Step Procedures

Procedure 12–A Sharpening a Plane Iron

....

STEP 1   ONE OR TWO PASSES WILL REMOVE ANY BURRS FROM THE BACK SIDE. BE SURE TO KEEP THE IRON FLAT ON THE STONE.

STEP 2   MANY PASSES ON BEVELED SIDE WILL REMOVE MATERIAL TO SHARPEN BLADE. REPEAT STEP ONE THEN STEP TWO UNTIL DESIRED SHARPNESS IS ACHIEVED.

STEP 3   PULL BLADE OVER LEATHER STROP TO HONE THE EDGE. BOTH SURFACES SHOULD BE DONE IN A BACK AND FORTH MOTION.

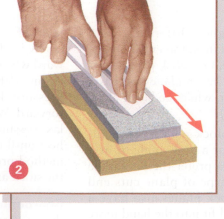

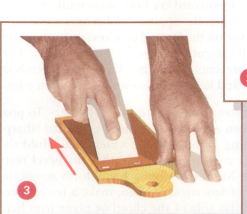

© Cengage Learning 2014

....

If a grinding attachment is not available, hold the chisel or plane iron by hand on the tool rest at the proper angle. A general rule is that the width of the bevel is approximately twice the thickness of the blade (*Figure 12–10*). Move the blade up on the grinding wheel for a longer bevel and down for a shorter bevel.

Let the index finger of the hand holding the chisel ride against the outside edge of the tool rest as the chisel is moved back and forth across the revolving wheel. Dip the blade in water frequently to prevent overheating. Do not move the position of your index finger, making sure that the tool can be replaced on the wheel at exactly the same angle to obtain a smooth hollow to the bevel. Grind the chisel or plane iron until an edge is formed (*Figure 12–11*). A burr or *wire edge* will be formed on the edge on the flat side. This can be felt by lightly rubbing your thumb along the flat side toward the edge.

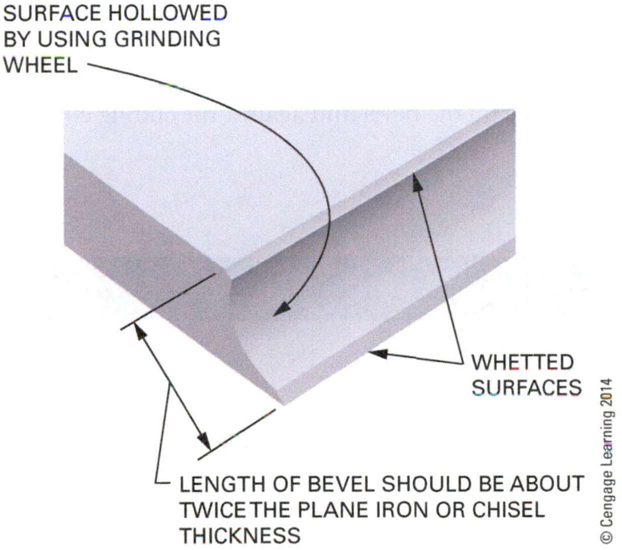

Snips. **Straight tin snips** are generally used to cut straight lines on thin metal, such as roof flashing and metal roof edging. Three styles of **aviation snips** are available for straight metal cutting and for left and right curved cuts (*Figure 12–12*). The color of the handle denotes the differences in the design of the snips. Yellow handles are for straight cuts, green are for cutting curves to the right, and red are for cutting to the left.

Hatchets. For wood shingling of side walls and roofs, among other purposes, the **shingling hatchet** is used. In addition to the shingling hatchet, many carpenters carry a slightly heavier hatchet for such uses as pointing stakes or otherwise tapering rough stock. A special drywall hatchet is also used for the installation of gypsum board (*Figure 12–13*).

SURFACE HOLLOWED BY USING GRINDING WHEEL

WHETTED SURFACES

LENGTH OF BEVEL SHOULD BE ABOUT TWICE THE PLANE IRON OR CHISEL THICKNESS

© Cengage Learning 2014

Figure 12–10 Details of the cutting edge of chisels and planes.

© Cengage Learning 2014

Figure 12–11 Grinding a wood chisel.

© Cengage Learning 2014

Figure 12–12 (A) Metal shears and (B) left-, straight-, and right-cutting aviation snips.

© Cengage Learning 2014

Figure 12–13 Shingling hatchet.

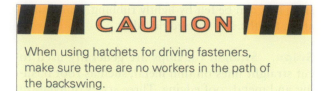

When using hatchets for driving fasteners, make sure there are no workers in the path of the backswing.

Knives. A carpenter usually has a jackknife of good quality. The jackknife is used mostly for sharpening pencils and for laying out recessed cuts for some types of finish hardware, such as door hinges. The jackknife is used for laying out this type of work because a finer line can be obtained with it than with a pencil. In addition to marking, it also scores the layout line, which is helpful when chiseling the recess (*Procedure 12–B*).

The utility knife (*Figure 12–14*) is frequently used instead of a jackknife for such things as cutting gypsum board and softboards. Replacement blades are carried inside the handle.

Figure 12–14 Utility knives.

Scrapers. The hand scraper (*Figure 12–15*) is very useful for removing old paint, dried glue, pencil, crayon, and other marks from wood surfaces. The scraper blades are reversible, removable, and replaceable. They dull quickly, but can be easily sharpened by filing on the bevel and against the cutting edge.

Step**by**Step Procedures

Procedure 12-B Chiseling Square Edges

STEP 1 SCORE LAYOUT LINE WITH KNIFE.

STEP 2 LIFT CHIP WITH CHISEL.

STEP 3 CHIP BREAKS OFF AT SCORED LINE.

STEP 4 A SHOULDER IS PROVIDED TO REST THE CHISEL AGAINST WHEN DEEPENING THE CUT.

On the Job

Score layout lines with a knife for more accurate chiseling of recesses.

Figure 12–15 Hand scraper.

> **CAUTION**
>
> Be careful when filing so that the filing hand does not come in contact with the cutting edge. Also, care should be taken to file evenly the entire cutting edge and not to hollow out the center of the blade or the outside corners.

Tooth-Cutting Tools

The carpenter uses several kinds of saws to cut wood, metal, and other material. Each one is designed for a particular purpose.

Handsaws

Handsaws (*Figure 12–16*) used to cut across the grain of lumber are called **crosscut saws**. To cut with the grain, **ripsaws** are sometimes used. The

Figure 12–16 Handsaws are still useful on the jobsite. Some handsaws are made with deeper teeth, which are designed to cut in both directions.

difference in the cutting action is in the shape of the teeth. The crosscut saw has teeth shaped like knives. These teeth cut through the wood fibers to give a smoother action and surface when cutting across the grain. The ripsaw has teeth shaped like rows of tiny chisels that cut the wood ahead of them (*Figure 12–17*). Another design, called a **shark tooth saw**, has longer teeth and is able to cut in both directions of blade travel.

To keep the saw from binding, the teeth are *set,* that is, alternately bent, to make the saw cut or *kerf* wide enough to give clearance for the blade.

A 7 point or 8 point (number of tooth points to the inch) saw is designed to cut across the grain of framing and other rough lumber. A 10 point or 11 point saw is designed for fine crosscuts on finish work. The number of points is usually stamped on the saw blade at the heel.

Using Handsaws.   Stock is handsawed with the face side up because the back side is splintered, along the cut, by the action of the saw going through the stock. This is not important on rough work. However, on finish work, it is essential to identify the face side of a piece and to make all layout lines and saw cuts with the face side up.

The saw cut is made on the waste side of the layout line by cutting away part of the line and leaving the rest. This takes some practice, especially when it is important to make thin layout lines rather than broad, heavy ones. Press the blade of the saw against the thumb when starting a cut. Make sure the thumb is above the teeth; steady it with the index finger, with the rest of the hand on the work. Move the thumb until the saw is aligned as desired and start the cut on the upstroke (*Figure 12–18*). Move the hand away when the cut is deep enough. When handsawing, hold crosscut saws at about a 45 degree angle.

> **CAUTION**
>
> Do not use a ripsaw for cutting across the grain. It can jump at the start of the cut, possibly causing injury.

Most carpenters prefer to have their saws set and filed by sharpening shops. Sharpening handsaws requires special tools, much skill, and experience to do a professional job.

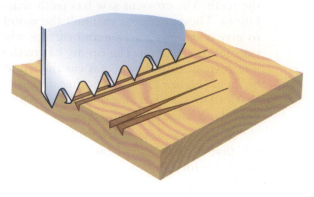

HOW A CROSSCUT SAW CUTS

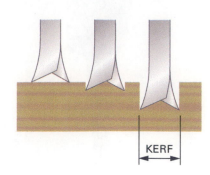

CROSS SECTION OF CROSSCUT TEETH

KERF

CROSSCUT SAW CUTS

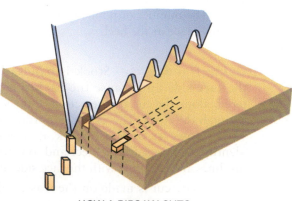

HOW A RIPSAW CUTS

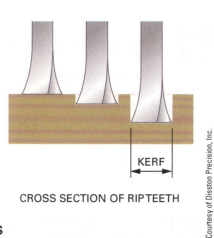

CROSS SECTION OF RIP TEETH

KERF

RIPSAW CUTS

Courtesy of Disston Precision, Inc.

Figure 12–17 Cutting action of ripsaws and crosscut saws.

© Cengage Learning 2014

Figure 12–18 Starting a cut with a handsaw.

© Cengage Learning 2014

Figure 12–19 Compass saw.

is narrower for making curved cuts of smaller diameter (*Figure 12–19*). To start the saw cut, a hole needs to be bored (except in soft material, when the point of the saw blade can be pushed through).

Coping Saws. The coping saw (*Figure 12–20*) is used primarily to cut molding to make coped joints. A *coped joint* is made by cutting and fitting the end of a molding against the face of a similar

Special-Purpose Saws

Compass and Keyhole Saws. The compass saw is used to make circular cuts in wood. The keyhole saw is similar to the compass saw except its blade

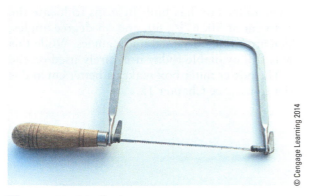

Figure 12–20 Coping saw.

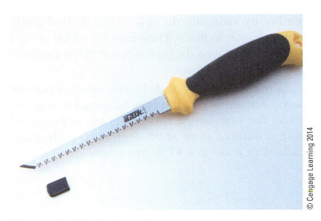

Figure 12–22 Wallboard saw.

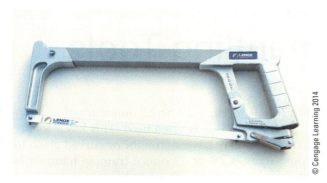

Figure 12–21 Hacksaw.

Pullsaws. The Japanese-style pullsaws are gaining popularity. They have a unique design in that they cut on the up- (pull) stroke instead of the downstroke (*Figure 12–23*). Some models cut in both directions. The pulling action of the pullsaw is actually easier to use. It takes less effort and gives more control. Pullsaws cut fast and smooth with thin kerfs. Many styles are available for cutting rough and fine work.

Miter Boxes. The miter box (*Figure 12–24*) is used to cut angles of various degrees on finish

piece. (Coping is explained in detail in Chapter 79.) The coping saw is also used to make any small, irregular curved cuts in wood or other soft material.

The coping saw blade has fine teeth that may be installed with the teeth pointing either toward or away from the handle. Which is best depends on the operator and the situation. The blade cuts only in the direction the teeth point.

Hacksaws. Hacksaws (*Figure 12–21*) are used to saw metal. Hacksaw blades are available with 18, 24, and 32 points to the inch. Coarse-toothed blades are used for fast cutting in thick metal. Fine-toothed blades are used for smooth cutting of thin metal. At least three teeth of the blade should be in contact with thin metal or cutting will be difficult. Make sure that the blades are installed with the teeth pointing away from the handle.

Wallboard Saws. The wallboard saw (*Figure 12–22*) is similar to the compass saw but is designed especially for gypsum board. The point is sharpened to make self-starting cuts for electric outlets, pipes, and other projections. Another type with a handsaw handle is also used frequently.

Figure 12–23 Pullsaws cut when they are pulled, which is the reverse of handsaws.

Figure 12–24 A handsaw miter box.

lumber by swinging the saw to the desired angle and locking it in place. These cuts are called miters. The joint between the pieces cut at these angles is called a mitered joint.

A mitered joint is made by cutting each piece at half the angle at which it is to be joined to another piece. For instance, if two pieces are to be joined with a mitered joint at 90 degrees to each other, each piece is cut at a 45 degree angle.

The miter box has built-in stops to locate the saw to cut at 90, 67½, 60, and 45 degree angles, which are commonly used miter angles. While this saw is still available today it is rarely used on the job. The power miter box makes a better cut and is easier to use. See Chapter 18.

CHAPTER 13

Fastening and Dismantling Tools

Discussed in this chapter are those tools used to drive nails and turn screws and other fasteners. Tools used to clamp, hold, pry, and dismantle workpieces are also included.

FASTENING TOOLS

The carpenter must decide which fastening tool to select and be able to use it competently and safely for the job at hand.

Hammers

The carpenter's claw hammer is available in a number of styles and weights. The claws may be straight or curved. Head weights range from 7 to 32 ounces. Most popular for general work is the 16 ounce, curved-claw hammer (*Figure 13–1*).

For rough work, a 20 or 22 ounce framing hammer (*Figure 13–2*) is often used. This has a longer handle and may have a straight or curved claw. In some areas, a 28 or 32 ounce framing hammer is preferred for extra driving power.

Hammer handles vary in styles being made of wood, steel, or fiberglass. They also come in different lengths. The longer handles allow the carpenter to drive the nail harder and faster.

The hammerheads are smooth or serrated into a waffled surface. The waffle surface keeps the head from slipping off the nail, making nailing more effective. The direction of the driven nail can even be changed slightly by twisting the wrist. These hammers should be used exclusively for framing because the wood surface is damaged by the waffle imprint that is left when the nail is seated.

Figure 13–1 Curved claw hammer.

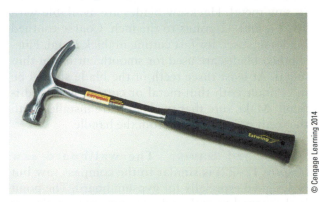

Figure 13–2 Straight claw framing hammer.

© Cengage Learning 2014

Nail Sets. Nail sets (*Figure 13–3*) are used to set nail heads below the surface. The most common sizes are $\frac{1}{32}$, $\frac{2}{32}$, and $\frac{3}{32}$ inch. The $\frac{1}{4}$ inch nail set is used to drive the large-headed nails typically used for exterior finish work. The size refers to the diameter of the tip. The surface of the tip is concave to prevent it from slipping off the nail head. If the tip becomes flattened, the nail set is more difficult to keep on the nail being driven.

Figure 13–4 Wear eye protection when driving nails.

Nailing Techniques. Hold the hammer firmly, close to the end of the handle, and hit the nail squarely. If the hammer frequently glances off the nail head, try cleaning the hammer face (*Figure 13–5*). As a general rule, use nails that are three times longer than the thickness of the material being fastened. To swing a hammer, the entire arm and shoulder are used. It is important to use the wrist too. During the latter part of the swing, as the hammer nears the nail, the wrist is rotated quickly, giving more speed to the hammerhead. This increased speed generates more nail-driving force, all with less arm effort.

Toenailing is the technique of driving nails at an angle to fasten the end of one piece to another

On the Job

To help prevent glancing off the nail head when driving nails, clean the hammer face by rubbing it back and forth on a rough surface.

SANDPAPER BLOCK, CONCRETE, OR ANY HARD ROUGH SURFACE

HAMMER FACE

Figure 13–5 Roughing up the hammerhead face helps keep the hammerhead from glancing off the nail.

Figure 13–3 Nail set.

Figure 13–6 Toenailing is the technique of driving nails at an angle.

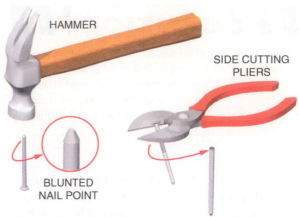

SOMETIMES BLUNTING THE POINT OF A NAIL WILL PREVENT SPLITTING THE STOCK. TAP THE POINT LIGHTLY SO AS NOT TO BEND THE NAIL.

IF A TWIST DRILL IS NOT HANDY, CUT THE HEAD OFF A NAIL OF THE SAME GAUGE AS THE NAILS BEING USED, AND USE IT TO DRILL HOLES.

Figure 13–7 Methods to avoid splitting wood.

(*Figure 13–6*). It is used when nails cannot be driven into the end, called face nailing. Toenailing generally uses smaller nails than face nailing and offers greater withdrawal resistance of the pieces joined. Start the nail about ¾ to 1 inch from the end and at an angle of about 30 degrees from the surface.

Drive finish nails almost all the way. Then set the nail below the surface with a nail set to avoid making hammer marks on the surface. Set finish nails at least ⅛ inch deep so the filler will not fall out.

In hardwood, or close to edges or ends, drill a hole slightly smaller than the nail shank to prevent the wood from splitting or the nail from bending. If a twist drill of the desired size is not available, cut the head off a finish nail of the same gauge and use it for making the hole.

Blunting or cutting off the point of the nail also helps prevent splitting the wood (*Figure 13–7*). The point spreads the wood fibers as the nail is driven, while the blunt end pushes the fibers ahead of it and reduces the possibility of splitting.

Holding the nail tightly with the thumb and as many fingers as possible while driving the nail in hardwood helps prevent bending the nail. Of course, hold the nail in this manner only as long as possible. Be careful not to glance the hammer off the nail and hit the fingers.

When nailing along the length of a piece, stagger the nails from edge to edge, rather than in a straight line. This avoids splitting and provides greater strength (*Figure 13–8*). Drive nails at an angle into end grain for greater holding power. This is called dovetail nailing. When fastening pieces side to side, nails are driven at an angle for greater strength (*Figure 13–9*). In addition, this may keep the nail points from protruding if using 12d or 3¼ inch nails to fasten 2 inch nominal stock.

Figure 13–8 Stagger and angle nails for greater strength and to avoid splitting the stock.

When fastening two pieces of stock together, the alignment is usually very important. To make this process easier, drive the first nail only to where it protrudes from the first layer slightly (*Procedure 13–A*). Then align the pieces as necessary. Push or tap the first layer with the small protruding point into the second layer. This will act as a tack, holding the proper alignment until the nail is driven.

When it is necessary to start a nail higher than you can hold it, use the nail starter located in the

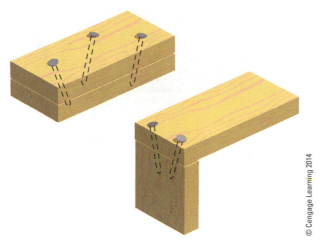

Figure 13–9 Driving nails at an angle increases holding power.

Figure 13–10 Method of starting a nail with one hand.

head of many framing hammers. If your hammer does not have one, press the nail tightly between the claws of the hammer, with the head of the nail against the handle (*Figure 13–10*). Turn the claws of the hammer toward the surface. Reach up and swing the hammer to start and hold the nail in the stock. Pull the hammer claws away from the nail, turn the hammer around, and drive the nail all the way in.

Screwdrivers

Screwdrivers are manufactured to fit all types of screw slots. The carpenter generally uses only the *slotted* screwdriver (*Figure 13–11*), which has a straight tip to drive common screws, and the *Phillips* screwdriver. Other screwdrivers include the Robertson screwdriver, which has a squared tip (see Figure 9–11).

Screwdriver Sizes. Slotted screwdrivers are sized by the length of the blade and by the type. Lengths generally run from 3 to 12 inches. Phillips screwdrivers are sized by their length and point size. Commonly used sizes are lengths that run the same as common screwdrivers and points that come in numbers 0, 1, 2, 3, and 4. The higher number indicates a point with a larger diameter.

Screwdrivers should fit snugly, without play, into the slot of the screw being driven. The screwdriver tip should not be wider than the screwhead, nor should it be too narrow (*Figure 13–12*). The correct size screwdriver helps ensure that the screw

---

Step**by**Step Procedures

Procedure 13-A Aligning Boards with a Protruding Nail

| STEP **1** | START NAIL UNTIL IT PROTRUDES SLIGHTLY THROUGH THE FIRST LAYER. |

| STEP **2** | ALIGN PIECES AND TAP THEM TOGETHER. THE NAIL SERVES AS A TACK. FINISH DRIVING THE NAIL. |

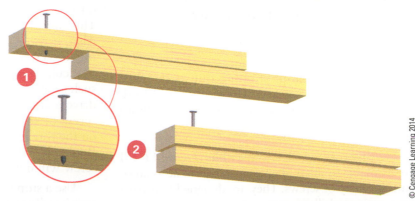
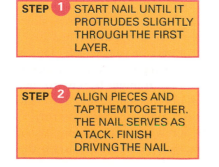

Figure 13–11 Slotted and Phillips screwdrivers.

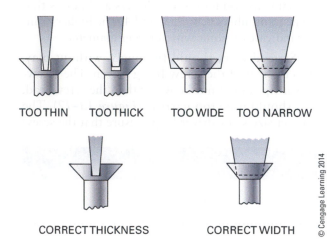

TOO THIN   TOO THICK   TOO WIDE   TOO NARROW

CORRECT THICKNESS   CORRECT WIDTH

Figure 13–12 The correct size screwdriver for the screw being driven is best.

Figure 13–13 Screwgun drive bits for various screw head styles.

Screwdriving Techniques. If possible, select screws so that two-thirds of their length penetrates the piece in which they are gripping. In preparation for driving a screw into hardwood, for example, a pilot hole and a shank hole must be drilled. The pilot hole allows the screw to be driven into place without splitting the material. The shank hole, very importantly, allows the two pieces to come together tight when the screw is driven.

In addition, if the screw has a flat head, the shank hole may be countersunk so the head will be flush or set below the surface when driven (*Procedure 13–B*).

To select the pilot hole drill bit size, hold the bit against the threaded portion of the screw. Determine by eye the bit that just covers the solid center portion of the screw while leaving the threads visible. For the shank hole, hold the bit over the shank portion of the screw, selecting the bit that is closest in size.

Select drills with great care. Smaller drills may be used for a pilot hole in softwoods. Some may advocate that in softwood no pilot hole is necessary, but it is wise to drill them anyhow. It does not take that much more time, and the screw can be driven straight and more easily. Without a pilot hole, the screw may follow the grain and go in at an undesirable angle.

If the pilot hole is too small or not deep enough, difficulty may be encountered in driving the screw. This causes slipping and damage to the screw slots. Also, if too much pressure is applied when driving the screw, the head may be twisted off. This is particularly true when driving screws of soft metal, such as aluminum or brass. It might be wise to first drive a steel screw, remove it, and then drive the screw of softer metal. Rub some wax (paraffin) on the threads of the screw to make driving easier. Remember that if the pilot hole is too large, the screw will not grip.

Use a stop when drilling the shank hole to make sure it will not be drilled too deep. This will prevent

will be driven without slipping out of the slot. When seated, the screwdriver slot should look the same as before the screw was driven, with no burred edges.

Screwdriver Bits.   Screwdriver bits (*Figure 13–13*) are available in many shapes and sizes to accommodate a variety of screws. They are designed to drive a screw using a drill or screw gun.

StepbyStep Procedures

Procedure 13-B Making a Pilot Hole

STEP 1 DETERMINE SIZE OF AND DRILL PILOT HOLE. USE STOP BLOCK, IF NEEDED, TO PREVENT GOING THROUGH. DEPTH OF PILOT HOLE MUST BE SAME OR DEEPER THAN SCREW LENGTH.

STEP 2 DRILL SHANK HOLE SAME DIAMETER AS SCREW SHANK. USE STOP BLOCK OF APPROPRIATE LENGTH TO PREVENT DRILLING SHANK HOLE TOO DEEP.

STEP 3 COUNTERSINK DEEP ENOUGH SO SCREWHEAD WILL BE SLIGHTLY BELOW SURFACE WHEN DRIVEN.

STEP 4 DRIVE SCREW WITH SCREWDRIVER OF PROPER SIZE UNTIL SCREW IS WELL SEATED.

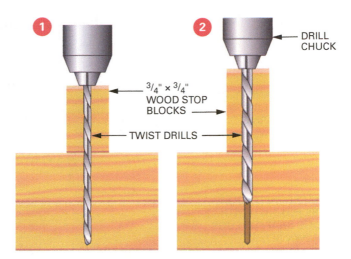

DRILL CHUCK

3/4" × 3/4" WOOD STOP BLOCKS

TWIST DRILLS

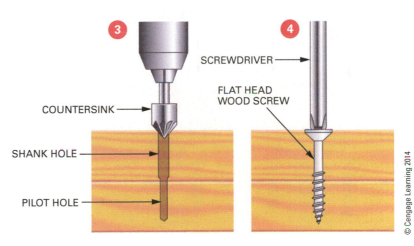

SCREWDRIVER

FLAT HEAD WOOD SCREW

COUNTERSINK

SHANK HOLE

PILOT HOLE

© Cengage Learning 2014

it from going through the material when drilling the pilot hole. A simple stop can be made by drilling a hole lengthwise through a piece of nominal 1 × 1 stock, cutting it to the desired length, and inserting it on the twist drill against the chuck.

If the material to be fastened is thick, the screw may be set below the surface by counterboring to gain additional penetration without resorting to a longer screw. To set the screwhead below the surface, bore the counterbored hole first. Drilling pilot and shank holes first leaves no stock to guide the center point of the bit used to make the counterbored hole. Use a stop block to ensure desired depth. The diameter of the hole should be equal to or slightly larger than the diameter of the screwhead. Next, drill the pilot hole; then, the shank hole (*Procedure 13–C*).

DISMANTLING TOOLS

Dismantling tools are used to take down staging and scaffolding, concrete forms, and other temporary structures. In addition, they are used for tearing out sections of a building when remodeling. Carpenters must be skilled in the use of dismantling tools, and in the work, so that the dismantled members are not damaged any more than necessary.

Hammers

In addition to fastening, hammers are often used for pulling nails to dismantle parts. To increase leverage and make nail pulling easier, place a small block of wood under the hammerhead (*Figure 13–14*).

Step**by**Step Procedures

Procedure 13-C Counterboring a Screw Hole

STEP 1 DRILL COUNTERBORED HOLE OF DESIRED DIAMETER.

STEP 2 DRILL SHANK AND PILOT HOLES AS DESCRIBED PREVIOUSLY. DRIVE SCREW UNTIL WELL SEATED.

STEP 3 COUNTERBORED HOLES MAY BE PLUGGED WITH VARIOUS SHAPED PLUGS AS SHOWN BELOW.

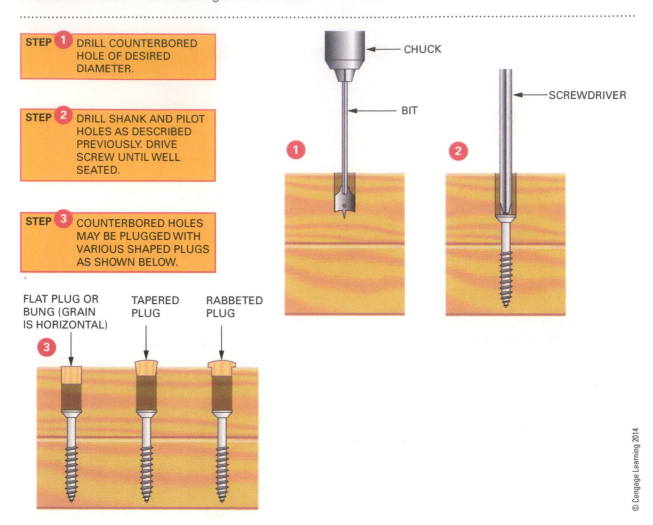

FLAT PLUG OR BUNG (GRAIN IS HORIZONTAL)   TAPERED PLUG   RABBETED PLUG

CHUCK

BIT

SCREWDRIVER

© Cengage Learning 2014

© Cengage Learning 2014

Figure 13–14 Pull a nail more easily by placing a block of wood under the hammer.

Bars and Pullers

The **wrecking bar** (*Figure 13–15*) is used to withdraw spikes and to pry when dismantling parts of a structure (*Figure 13–16*). They are available in lengths from 12 to 36 inches, with the 30 inch size often preferred for construction work.

Carpenters need a small **flat bar,** similar to those shown in *Figure 13–17,* to pry small work and pull small nails. To extract nails that have been driven home, a **nail claw,** commonly called a *cat's paw,* is used (*Figure 13–18*).

Holding Tools

To turn nuts, lag screws, bolts, and other objects, an **adjustable wrench** is often used (*Figure 13–19*). The

Figure 13-15 Wrecking bars.

Figure 13-16 Using a wrecking bar to pry stock loose.

Figure 13-17 Flat bars.

Figure 13-17 Flat bars.

Figure 13-18 Nail claw.

Figure 13-19 Adjustable wrench.

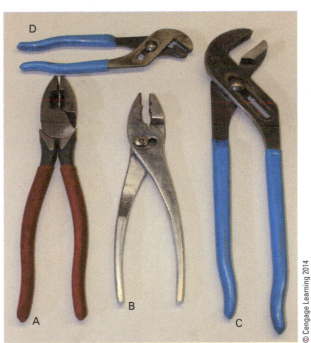

Figure 13-20 (A) Lineman pliers, (B) combination pliers, and (C & D) adjustable pliers.

wrench is sized by its overall length. The 10 inch adjustable wrench is the one most widely used.

For extracting, turning, and holding objects, a pair of pliers is often used. Many kinds are manufactured, but the combination pliers (*Figure 13-20*) is designed for general use. Lineman pliers are designed to work with wire, and adjustable pliers are designed for pipe.

Clamps come in a variety of styles and sizes (*Figure 13-21*). They are useful for holding objects together while they are being fastened, glued, and

© Cengage Learning 2014

Figure 13–21 (A) Spring clamp, (B) C-clamp, (C) quick clamp, and (D) wood screw.

used as temporary guides. *Spring clamps* are quick and easy to set. They are spring loaded for ease of closing their jaws. Simply squeezing the handles opens the jaws; releasing sets them. *C-clamps* are named for their shape. The size is designated by the throat opening. *Quick clamps* are named for the speed at which they can be adjusted and set. One side of the jaws is stationary and the other slides on the bar. After the material is placed in the jaws, the handles are squeezed to tighten and set the clamp. The small trigger is pulled to release the clamp. The *wood screw*, also called *parallel clamps*, is made of wood blocks and large screws. These clamps are used primarily for holding wood pieces while they are glued. It takes some practice to set this clamp quickly. The center screw is turned one way to tighten, and the other screw is turned the opposite way.

DECONSTRUCT THIS

Carefully study *Figure 13–22* and think about what is wrong and/or what is right. Consider all possibilities. What construction practice or method is different in your area of the country?

© Cengage Learning 2014

Figure 13–22 This photo shows a worker using a wood chisel to clean out a dado.

KEY TERMS

adjustable wrench	counterboring	miter box	shingling hatchet
auger bits	countersink	mitered joint	Speed Square
aviation snips	crosscut saws	miters	steel tapes
bench plane iron	dadoes	nail claw	straightedge
bench planes	drilling	nail set	straight tin snips
bevel	expansive bit	pilot hole	tack
bit brace	face nailing	plumb	toe
block planes	flat bar	plumb bob	toenailing
boring	framing hammer	pocket tapes	trammel points
butt markers	framing square	pullsaws	twist drills
carpenter's level	grooves	rabbets	utility knife
chalk line	hacksaws	ripsaws	wallboard saw
clamps	hand scraper	screwdriver bits	whetted
claw hammer	heel	screwdrivers	wing dividers
combination pliers	keyhole saw	scribing	wood chisel
combination square	level	shank hole	wrecking bar
compass saw	line level	shark tooth saw	
coping saw			

REVIEW QUESTIONS

Select the most appropriate answer.

1. A safe worker attitude that promotes a safe job site comes from

   a. ability.
   b. skill.
   c. knowledge.
   d. all of the above.

2. When stretching a steel tape to lay out a measurement, place the ring so the

   a. 1 inch mark is on the starting line.
   b. end of the steel tape is on the starting line.
   c. inside of the ring is on the starting line.
   d. outside of the ring is on the starting line.

3. In construction, the term *plumb* means perfectly

   a. horizontal.
   b. level.
   c. straight.
   d. vertical.

4. A large, L-shaped squaring tool that has tables stamped on it is called a

   a. framing square.
   b. Speed Square.
   c. bevel square.
   d. combination square.

5. The layout tool that may be adjusted to serve as a marking gauge is the

   a. framing square.
   b. Speed Square.
   c. bevel square.
   d. combination square.

6. To adjust a carpenter's level into a level position when the bubble is found to be touching the right line, the

   a. right side of the level should be raised.
   b. left side of the level should be raised.
   c. left side of the level should be lowered.
   d. entire level should be raised.

7. When snapping a long chalk line, care should be taken to
   a. dampen the string.
   b. keep the string from sagging.
   c. hold the string loosely.
   d. let the string touch the surface as it unwinds.

8. The tool used to mark material to conform to an irregular surface is called a
   a. pen.
   b. chisel.
   c. scriber.
   d. chalk line.

9. The name of the one-handed plane with a low blade angle is the
   a. block plane.
   b. bench plane.
   c. chisel.
   d. plane iron.

10. The color of the handle on aviation snips indicates
    a. which hand to use.
    b. the direction in which curves can easily be cut.
    c. the manufacturer.
    d. what material may be easily cut.

UNIT 6
Portable Power Tools

©iStockphoto.com/Curt Pickens

The sound of construction has changed over the years. The rhythmic whoosh of a handsaw has virtually been replaced by the whir and ring of a circular saw. Power tools have been created to increase the productivity of most job site tasks.

The number and style of power tools available today for the carpenter is vast, and the list continues to grow. Power tools enable the carpenter to do more work in a shorter time period with less effort.

OBJECTIVES

After completing this unit, the student should be able to:

- state general safety rules for operating portable power tools.

- identify, describe, and safely use the following portable power tools: circular saws, jigsaws, reciprocating (also called saber) saws, drills, hammer-drills, screwdrivers, planes, routers, sanders, staplers, nailers, and powder-actuated drivers.

SAFETY REMINDER

With increased speed and production comes an increase in personal risk. This danger can come from a spectrum of human shortcomings that range from a lack of knowledge and skill to overconfidence and carelessness. Safe operation of power tools requires knowledge and discipline.

Learn the safe operating techniques from the manufacturer's recommended instructions before operating any tool. Once you understand these procedures, follow them every time the tool is used. Don't take chances; life is too short as it is.

Being aware of the dangers of operating power tools is the first step in avoiding accidents. This begins with eye and ear protection.

Portable power tools are everywhere on the construction site today. All tools must be carefully used and properly maintained to keep both the workers using the tools and those nearby safe.

CAUTION

Safety is as important as breathing. Have a complete understanding of a tool before attempting to operate it. Read the manufacturer's operating instructions.

Maintain a proper attitude of safety at all times by following these guidelines:

- Wear eye and ear protection. Eyes and ears do not grow back.
- Do not become distracted by others or distracting to others when tools are being operated.
- Do not wear loose-fitting clothes or jewelry that might become caught in the tool.
- Make sure the material being tooled is securely held and supported.
- Remember, using sharp tools and cutters is actually safer than using dull ones.
- Stay alert and develop an attitude of care and respect for yourself and others.

Use the proper power source:

- Electricity used to power a tool can be fatal to humans. Use ground fault interrupter circuits (GFICs) at all times. These will trip before any electricity can leak out of a tool and cause a shock.
- Do not use frayed or badly worn power cords.
- Use properly sized extension cords that are rated for the power requirements of the tool.
- Avoid using a cord longer in length than necessary. Voltage to the tool drops as the cord gets longer.
- Keep extension cords out of the path of all construction traffic.
- Unplug the tool whenever touching the cutting surface of the tool.

On the Job

Safety is a team sport. It requires every worker to play by the rules.

CHAPTER 14
Saws, Drills, and Drivers

SAWS

The carpenter uses several kinds of portable saws for crosscutting and ripping, for making circular cuts, and for cutting openings in floors, walls, and ceilings.

Circular Saws

Commonly called the Skilsaw, the portable electric circular saw (*Figure 14–1*) is used often by the carpenter. The circular saw blade is driven by an electric motor. The saw has a base that rests on the work to be cut. A handle with a trigger switch is provided for the operator to control the tool. The saw is adjustable for depth of cut. A retractable safety guard is provided over the blade, extending under the base.

© Cengage Learning 2014

Figure 14–1 Using a portable electric circular saw to cut framing material.

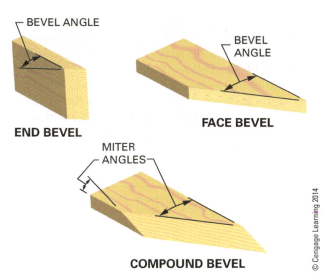

Figure 14–2 Edge, flat, and compound bevel cuts.

© Cengage Learning 2014

Courtesy of Porter Cable

Courtesy of Robert Bosch Tool Company

Figure 14–3 Direct drive and worm gear drive portable electric circular saws.

The base may be tilted for making *bevel* cuts. Bevel or miter cuts are those where the edge, end, or face of a board is cut at an angle. Edge bevels run along the length of the board or with the grain. End bevels run along the width of the board or across the grain cutting. Face bevels are angle cuts made on the face (*Figure 14–2*). Compound bevels are cuts with two angles, usually a combination of face and end bevels.

Saws are manufactured in many styles and sizes. The size is determined by the diameter of the blade, which ranges from 4½ to 16 inches. The most common size of circular saw uses a 7¼-inch-diameter blade. The handle and switch may be located on the top or in back. The blade may be driven directly by the motor or through a worm gear (*Figure 14–3*).

The forward end of the base is notched in two places to serve as guides for following layout lines. One notch is used to follow layout lines when the base is tilted to 45 degrees and the other when the base is not tilted.

> ### CAUTION
>
> Make sure the saw blade is installed with the teeth pointing in the correct direction. The teeth of the saw blade projecting below the base should point toward the base as the blade rotates.

To loosen the bolt that holds the blade in place, first unplug the saw and lock the arbor. This is done by pushing the arbor locking slide of the saw, usually found between the blade shield and the handle. While pushing the slide, rotate the blade by hand until the slide locks the arbor. With the proper wrench, turn the bolt in the same direction as the rotation of the blade. To tighten, turn the bolt in a direction opposite to the rotation of the blade.

On most models an adjustable attachment that fits into the base is used for ripping narrow pieces parallel with the edge. The saw may also be guided by tacking or clamping a straightedge to the material and running the edge of the saw base against it. Allowance must be made for the distance from the saw blade to the edge of the saw base when positioning the straightedge.

Circular Saw Blades. Circular saw blades are available in a number of styles. The shape and number of teeth around the circumference of the blade determine their cutting action. Carbide-tipped blades are used more often than high-speed steel blades. They stay sharper longer. More complete information on saw blades can be found in Unit 7.

Using the Portable Circular Saw. Always wear safety glasses when cutting. Make sure to cut on a stable work surface and that the waste piece will fall clear and not bind the saw blade (*Figure 14–4*). Adjust the depth of cut so that the blade just cuts through the work. Do not expose the blade any more than is necessary. This also reduces the amount of sawdust spray flying into the operator's face when cutting thinner material (*Figure 14–5*).

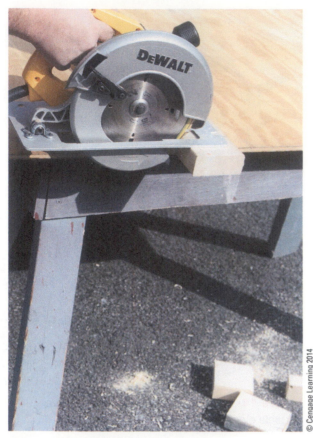

Figure 14–4 Saw cuts are made over the end of supports so the waste will fall clear and not bind the blade.

CAUTION

Make sure the blade guard operates properly. Be aware that the guard may possibly stick in the open position. Never wedge the guard back in an open position.

Figure 14–5 The blade of the saw is adjusted for depth only enough to cut through the material.

The saw blade removes material as it cuts—this is known as the *kerf*. Therefore the blade must be guided along on the waste side of the layout line (*Figure 14–6*). Follow the line closely by watching the layout line as the saw cuts for the whole length. Straight cuts are best; deviations from the line may cause the saw to bind and kick back. Sometimes a speed square is helpful to keep the saw traveling straight (*Figure 14–7*).

CAUTION

Do not force the saw forward. Listen to the saw: The sound of the saw should be the same throughout the cut. If the saw does slow and bind, stop the motor and bring the saw back to where it will run free. Restart and guide the saw by observing the line where the saw blade is cutting.

Figure 14–6 The blade must follow alongside the desired cut line.

Figure 14–7 A speed square may serve as a guide for straight cuts.

Near the end of the cut, the forward end of the base will go off the work. Keep watching the cut, releasing the switch when the cut is completed (*Procedure 14-A*).

CAUTION

Keep the saw clear of your body until the saw blade has completely stopped.

When starting cuts across stock at an angle, it may be necessary to retract the guard by hand. A handle is provided for this purpose (*Figure 14–8*). Release the handle after the cut has been started and continue as above. Compound miter cuts may be made by cutting across the stock at an angle with the base tilted.

Portable circular saws cut on the upstroke towards the saw base rotating upward through the material. As the teeth of the saw blade come through the top surface, the stock often is

Step**by**Step Procedures

Procedure 14–A Using a Circular Saw

STEP 1 Always wear safety glasses when cutting material. Mark the stock with a line as needed.

STEP 2 Rest the forward end of the base on the work and align the saw base parallel to the cut line. First align the notch of base with the cut line. Next, make the base perpendicular with the cutline. Imagine a square on the material as shown. Then bring the saw close to the material, but not touching it.

STEP 3 With the blade clear of the material, start the saw. When it has reached full speed, advance the saw into the work a short distance. Make sure to observe the blade and the cut line.

STEP 4 Cut as close to the line as possible. Pause the saw advancement into the material and check the alignment of the saw base edge and the cut line. If realignment is needed, release the switch and begin the cut over.

STEP 5 Continue following the line closely. Let the waste drop clear and release the switch. Ensure that the guard has returned to the protected position.

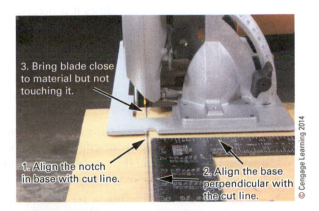

3. Bring blade close to material but not touching it.

1. Align the notch in base with cut line.

2. Align the base perpendicular with the cut line.

© Cengage Learning 2014

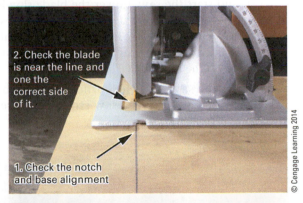

2. Check the blade is near the line and one the correct side of it.

1. Check the notch and base alignment

© Cengage Learning 2014

© Cengage Learning 2014

splintered at the layout line. The severity of the splintering depends on the style of blade used, the kind and thickness of the material being cut, and blade sharpness. More splintering occurs when cutting across grain than with the grain.

On finish work, that is, work that will ultimately be exposed to view, any splintering along the cut is unacceptable. One way to prevent this is to mark and cut from the back side. Sometimes it is not possible to cut from the back side, or both sides may be exposed to view, like cutting off the bottom of a door. To do this, mark the layout lines on the face that will be least visible. Then score with a sharp knife along the layout lines cutting deeply into the face. Make the saw cut along the waste side of the scored lines (*Figure 14–9*). This will allow the face grain to cut away without chipping.

Making Plunge Cuts. Many times it is necessary to make internal cuts in the material such as for sinks in countertops or openings in floors and walls. To make these cuts with a portable electric circular saw, the saw must be plunged into the material:

- Accurately lay out the cut to be made. Adjust the saw for depth of cut.
- Wearing eye protection, hold the guard open and tilt the saw up with the front edge of the base resting on the work. Hold the saw blade over, and in line with, the cut to be made (*Figure 14–10*).
- Make sure the teeth of the blade are clear of the work and start the saw. Lower the blade slowly into the work, applying a slight forward pressure to keep the saw from kicking back. Follow the line carefully, until the entire base rests squarely on the material.
- Advance the saw into the corner. Release the switch and wait until the saw stops before removing it from the cut.

Figure 14–9 Scoring the material before cutting reduces chipping of the face.

CAUTION

Make sure the tool comes to a complete stop before withdrawing it from the material being cut.

- Reverse the direction and cut into the corner. Again, wait until the saw stops before removing it from the cut.

CAUTION

Never move the saw backward while cutting. The direction of the turning blade will make the saw want to rise up out of the cut, jumping backward across everything in its path.

- Proceed in like manner to cut the other sides. Finish the cut into the corners with a handsaw or jigsaw.

Figure 14–8 Sometimes it is necessary to retract the guard of the portable circular saw by hand. Be sure to release it when the saw is well into the cut.

Figure 14–10 Making a plunge cut with a portable circular saw. First retract the guard, place the front edge of the saw base on the material and then pivot, running the saw slowly into the material.

Jigsaws

The **jigsaw** is widely used to make curved cuts (*Figure 14–11*). The teeth of the blade point upward when installed, so the saw cuts on the upstroke. To produce a splinter-free cut on the face side, it is best to cut on the side opposite the face of the work, if possible, or to score with a knife along the layout line.

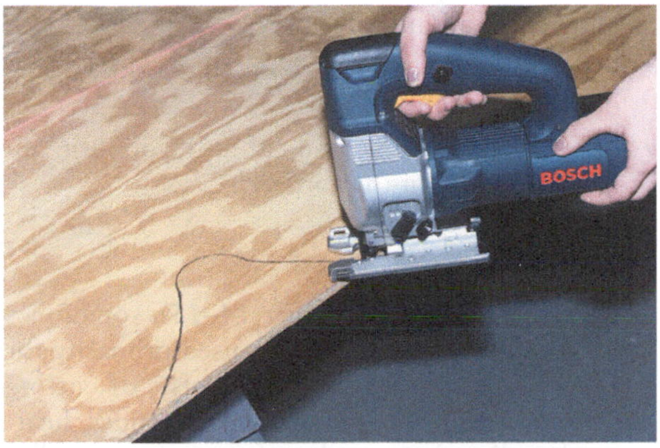
BOSCH

Figure 14–11 The jigsaw can cut along curved lines.

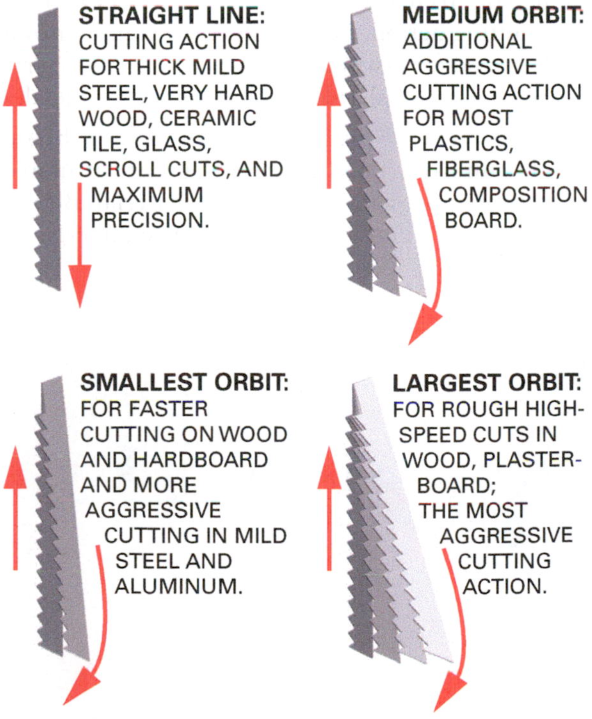
STRAIGHT LINE:
CUTTING ACTION FOR THICK MILD STEEL, VERY HARD WOOD, CERAMIC TILE, GLASS, SCROLL CUTS, AND MAXIMUM PRECISION.
MEDIUM ORBIT:
ADDITIONAL AGGRESSIVE CUTTING ACTION FOR MOST PLASTICS, FIBERGLASS, COMPOSITION BOARD.
SMALLEST ORBIT:
FOR FASTER CUTTING ON WOOD AND HARDBOARD AND MORE AGGRESSIVE CUTTING IN MILD STEEL AND ALUMINUM.
LARGEST ORBIT:
FOR ROUGH HIGH-SPEED CUTS IN WOOD, PLASTER-BOARD; THE MOST AGGRESSIVE CUTTING ACTION.

Figure 14–12 Some jigsaws have an additional orbital cutting action during each stroke of the blade.

There are many styles and varieties of jigsaws. The length of the stroke along with the amperage of the motor determine its size and quality. Strokes range from ½ to 1 inch. The longest stroke is the best for faster and easier cutting. Some saws can be switched from straight up-and-down strokes to several orbital (circular) motions to provide more effective cutting action for various materials (*Figure 14–12*). The base of the saw may be tilted to make bevel cuts.

Blade Selection. Many blades are available for fine or coarse cutting in wood or metal (*Figure 14–13*). Wood-cutting blades have teeth that are from six to twelve points to the inch. Blades with coarse teeth (fewer points to the inch) cut faster, but rougher. Blades with more teeth to the inch may cut slower, but produce a smoother cut surface. They do not splinter the work as much. Proper blade selection and use will produce the best cuts.

APPLICATION	PART NUMBER	TEETH PER INCH	OVERALL LENGTH (INCHES)	WIDTH (INCHES)
Metal cutting (High speed steel)				
Non-ferrous metal cutting 1/4" to 3/4" thick	**94612**	8	3 5/8	5/16
Metal cutting over 1/8" thick	**94613**	12	2 3/4	5/16
Metal cutting up to 1/8" thick	**94614**	21	2 3/4	5/16
Metal cutting up to 1/16" thick	**94615**	36	2 3/4	5/16
Wood cutting (Taper ground, high speed steel)				
Fast, smooth curve cutting in wood up to 2" thick	**94618**	6	3 5/8	1/4
Fast, smooth cutting in wood up to 2" thick	**94619**	6	3 5/8	5/16
Very smooth curve cutting in wood up to 1" thick	**94622**	10	3 5/8	1/4
Very smooth cutting in wood up to 1" thick	**94623**	10	3 5/8	5/16
Wood cutting (Alternate set—ground, high carbon steel)				
Fast, medium curve cutting in wood up to 2" thick	**94627**	6	3 5/8	1/4
Fast, medium cutting in wood up to 2" thick	**94628**	6	3 5/8	5/16
Wood cutting (Alternate set—high carbon steel)				
Fast, rough cutting in wood up to 2" thick	**94631**	6	3 5/8	5/16
Medium cutting in wood up to 2" thick	**94634**	8	3 5/8	5/16
General purpose (Alternate set—high speed steel)				
General purpose cutting in wood, plastic, metal, etc.	**94637**	12	3 5/8	5/16
Wood scroll cutting (Taper ground, high carbon steel)				
Smooth, intricate scroll cuts in wood up to 1 3/4" thick	**94640**	20	2 3/4	3/16

Figure 14–13 Jigsaw blade selection guide.

Using the Jigsaw. Follow a safe and established procedure:

- Outline the cut to be made. Secure the work either by hand, tacking, or clamping, or by some other method.

- Using eye protection, hold the base of the saw firmly on the work. With the blade clear, squeeze the trigger.

- Push the saw into the work, following the line closely. Make the saw cut into the waste side, and cut as close to the line as possible without completely removing it.

- Keep the saw moving forward, holding the base down firmly on the work. This will allow the saw to cut faster and more efficiently by keeping saw vibration to a minimum. Turn the saw as necessary in order to follow the line to be cut. Feeding the saw into the work as fast as it will cut, but not forcing it, finish the cut. Keep the saw clear of your body until it has stopped.

Making Plunge Cuts. Plunge cuts may be made with the jigsaw in a manner similar to that used with the circular saw:

- Tilt the saw up on the forward end of its base with the blade in line and clear of the work (*Figure 14–14*).

- Start the motor, holding the base steady. Very gradually and slowly, lower the saw until the blade penetrates the work and the base rests firmly on it.

CAUTION

Hold the saw firmly to prevent it from jumping when the blade makes contact with the material and to make a successful plunge cut. Thicker material may require a pilot hole to be drilled prior to blade insertion.

- Cut along the line into the corner. Back up for about an inch, turn the corner, and cut along the other side and into the corner.

- Continue in this manner until all the sides of the opening are cut.

- Turn the saw around and cut in the opposite direction to cut out the corners.

Reciprocating Saws

The reciprocating saw (*Figure 14–15*), sometimes called a *sawzall* or saber saw, is used primarily for *roughing-in* work. This work consists of cutting holes and openings for such things as pipes, heating and cooling ducts, and roof vents. It can be likened to a powered compass saw.

Most models have a variable speed of from 0 to 2,400 strokes per minute. Like jigsaws, some models may be switched to several *orbital* cutting strokes, taking you from a straight back-and-forth to an orbital cutting action. Ordinarily, the orbital cutting mode is used for fast cutting in wood. The normal reciprocating stroke should be used for cutting metal.

Reciprocating Saw Blades. Common blade lengths run from 4 to 12 inches. They are available for cutting practically any type of material. Blades are available to cut wood, metal, plaster, fiberglass,

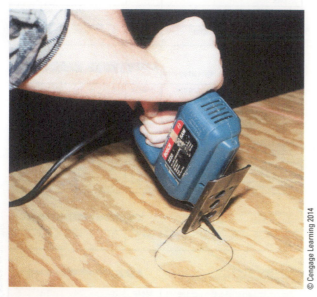

Figure 14–14 Making an internal cut by plunging the jigsaw.

© Cengage Learning 2014

Figure 14–15 Using the reciprocating saw to cut an opening in the subfloor.

© Cengage Learning 2014

ceramics, and other material. They are made of a hardened steel that allows the blades to occasionally cut nails with ease.

Using the Reciprocating Saw. The reciprocating saw is used in a manner similar to the jigsaw. The difference is that the reciprocating saw is heavier and more powerful. It can be used more efficiently to cut through rough, thick material, such as walls when remodeling. With a long, flexible blade, it can be used to cut flush with a floor or along the side of a stud.

To use the saw the base or shoe of the saw is held firmly against the work. Like the jigsaw, this reduces saw vibration and allows the saw to cut faster and more efficiently. To make cutouts, first drill a hole in the material. Then insert the blade, start the motor, and follow the layout lines. The blades can be reversed for cutting in confined areas.

DRILLS AND DRIVERS

Portable power drills, manufactured in a great number of styles and sizes, are widely used to drill holes and drive fasteners in all kinds of construction materials.

Drills

The drills used in the construction industry are classified as light duty or heavy duty. Light-duty drills usually have a *pistol-grip* handle. Heavy-duty drills have a *spade-shaped* or *D-shaped* handle (*Figure 14–16*).

Drill Sizes. The size of a drill is determined by the capacity of the *chuck*, its maximum opening. The chuck is that part of the drill that holds the cutting tool. The most popular sizes for light-duty models are ¼ and ⅜ inch. Heavy-duty drills have a ½-inch chuck or larger.

Drill Speed and Rotation. Most drills have *variable speed* and *reversible* controls. Speed of rotation can be controlled from 0 to maximum rpm (revolutions per minute) by varying the pressure on the trigger switch. Slow speeds are desirable for drilling larger holes or holes in metal. Faster speeds are used for drilling smaller holes in softer material. A reversing switch changes direction of the rotation for removing screws or withdrawing bits and drills from clogged holes.

Using Portable Electric Drills. Select the proper size and type of bit and insert it. Tighten the chuck with the chuck key or by holding the chuck while slowly spinning the chuck. Holes in hard materials may be center-punched to keep the spinning from wandering off center. Watch the direction of the hole being drilled to make sure the resulting hole is straight or plumb.

Courtesy of Porter Cable

© Cengage Learning 2014

Figure 14–16 Portable power drills are available in a number of styles.

> ⚠️ **CAUTION**
>
> Hold small pieces securely by clamping or other means. When drilling through metal especially, the drill has a tendency to hang up when it penetrates the underside. If the piece is not held securely, the hang up will cause the piece to rotate with the drill. It could then hit anything in its path and possibly cause injury before the power can be shut off.

Place the bit on the center of the hole to bedrilled. Hold the drill at the desired angle and start the motor. Apply pressure as required, but do not force the bit. Drill into the stock, being careful not to wobble the drill. Failure to hold the drill steady may result in breakage of small twist drills.

> ## CAUTION
>
> While drilling a hole, withdraw the turning bit from the material periodically to clear the shavings. This will keep the bit cooler and help it last longer. More importantly, it will keep the bit from jamming in the hole. Jamming causes the bit to stop suddenly. Personal injury may occur if the drill is powerful and the jammed bit is large. Therefore, be ready to release the trigger at any time.

Drill Bits

Electric drills are able to turn a variety of bits. Each bit is designed for different types of holes with varying results.

Twist Bits. Twist bits or drills make smaller holes in wood, metal, or plastic (*Figure 14–17*). They range in size from $\frac{1}{16}$ to 1 inch in increments of $\frac{1}{64}$ inch. High-speed twist bits are made of hardened steel that allows the bit to drill holes in various materials including steel. Regular steel bits are only designed to cut wood or plastic.

The general rule for drilling in wood is to do so at a high rotating speed with low drill pressure, but to drill in steel at slow rotating speeds with high drill pressure. Cutting oil can be used to cool the bit, keeping it sharper longer.

Speed Bores. For boring holes in rough work, the speed bore or *spade bit* is commonly used. They cut fast often leaving a rough hole. Drill will high speed and low drill pressure, particularly near the bottom of the hole. The bit may punch through the bottom and bind. To avoid this ease off the pressure near the bottom of the hole. If possible turn the stock over and drill from the other side when the bit point protrudes (*Figure 14–18*).

Forstner Bits. For holes with a cleaner edge in finish work, the Forstner bit, may be used (*Figure 14–19*). These bits make fine clean holes in wood with diameters from $\frac{1}{4}$ inch to $4\frac{1}{2}$ inches. Drills require more drill pressure to move through the stock.

Expansive Bits. To bore holes over 1 inch in diameter, an **expansive bit** may be used (*Figure 14–20*).

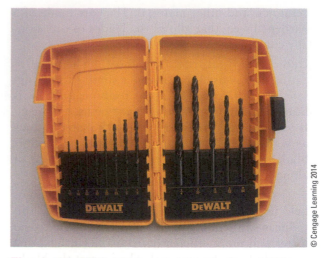

Figure 14–17 Twist bits come in sizes from 1/16" to 1" and drill a variety of materials.

Figure 14–18 Speed bores wood bits.

Figure 14–19 Forstner style wood bits.

With two interchangeable and adjustable cutters, large-diameter holes may be bored. This tool cuts slowly but the finished hole is usually smooth.

Countersinks. The **countersink** (*Figure 14–21*) spin to form a recess for a flat-head screw to set

Figure 14–20 Expansive wood bit.

Figure 14–21 Countersink boring bit.

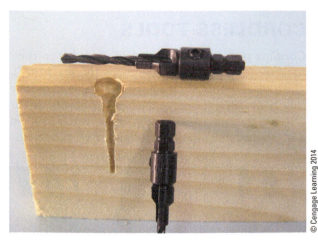

Figure 14–22 Combination drill and countersink.

flush with the surface of the material in which it is driven.

Combination Drills. *Combination drills* and *countersinks* are used to drill shank and pilot holes for screws and countersink in one operation (*Figure 14–22*).

Hole Saws. Occasionally, carpenters may use hole saws (*Figure 14–23*). These saws cut holes through material from ⅝ inch to 6 inches in diameter. One disadvantage of the hole saw is that a hole cannot be made partially through the material. Only the circumference is cut, and the waste is not expelled.

Masonry drill bits. Masonry drill bits have carbide tips for drilling holes in concrete, brick, tile, and other masonry. They are frequently used in portable power drills. They are more efficiently used in hammer-drills (*Figure 14–24*).

Boring Techniques.

To avoid splintering the back side of a piece when boring all the way through, stop when the point comes through. Finish by boring from the opposite side. This is especially important when both sides are exposed to view as in the case of a door. Care must be taken not to strike any nails or other objects that might cause blunting and shortening of the spurs. If the spurs become too short, the auger bit is ruined.

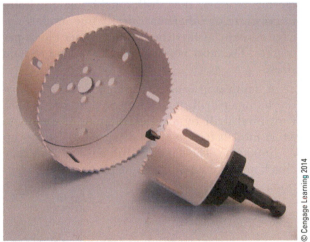

Figure 14–23 Hole saws actually saw holes in material leaving a circular center plug.

Figure 14–24 Masonry bits have a carbide tip.

Figure 14–25 The hammer-drill is used to make holes in masonry.

Hammer-drills

Hammer-drills (*Figure 14–25*) are similar to other drills. However, they can be changed to a hammering action as they drill, quickly making holes in concrete or other masonry. Some models deliver as many as fifty thousand hammer blows per minute. The most popular are the ⅜- and ½-inch sizes.

A depth stop is usually attached to the side of the hammer-drill. It can be converted to a conventional drill by a quick-change mechanism. Most models have a variable speed of from 0 up to 2,600 rpm. The hammer-drill has the same type chuck and is used in the same manner as conventional portable power drills.

Screwguns

Screwguns or *drywall drivers* (*Figure 14–26*) are used extensively for fastening gypsum board to walls and ceilings with screws. They are similar in appearance to the light-duty drills, except for the chuck. They have a pistol-type grip for one-hand operation and controls for varying the speed and reversing the rotation.

The chuck is made to receive special screwdriver bits of various shapes and sizes. A screwgun has an adjustable nosepiece, which surrounds the bit. When the forward end of the nosepiece touches the surface, the clutch is separated and the bit stops turning with a vibrating noise. Adjusting the nosepiece makes variations in the screw depth.

Accessories are available for driving screws with different heads. Hex head nutsetters are available in magnetic or nonmagnetic styles, in sizes ranging from ³⁄₁₆ to ⅜ inch.

Figure 14–26 The drywall driver is used to fasten wallboard with screws.

CORDLESS TOOLS

Cordless power tools are widely used due to their convenience, strength, and durability (*Figure 14–27*). The tools' power source is a removable, rechargeable battery usually attached to the handle of the tool. The batteries range in voltage from 9.6 to 36 volts and, in general, the higher the voltage, the stronger the tool.

CAUTION

It is easy to think that because these tools are battery powered, they are safer to use than higher voltage tools with cords. While they are safe to use, the operator should never forget the proper techniques and requirements for using the tools. Always wear personal protection equipment.

Interchangeable batteries make it cost effective to have an assortment of tools. They come with different types of battery which include nickel-cadmium and lithium-ion. Lithium-ion batteries tend to last longer

Green Tip

Batteries, particularly from portable power equipment, should be recycled. This keeps the toxic material inside batteries away of the environment and helps reduce other raw-material manufacturing costs.

between charges and are lighter than nickel-cadmium batteries. They also take longer to charge and are a little more expensive (*Figure 14–28*).

Cordless tools tend to make the job site safer by eliminating extension cords. Yet the extension cord sometimes serves as a safety line for tools that fall. Take care to set the tool down in a safe place after each use, particularly when working on scaffolds.

Cordless drills come with variable-speed reversing motors and a positive clutch, which can be adjusted to drive screws to a desired torque and depth. Some models allow for a hammer-drilling mode when drilling masonry. Cordless circular saws are powerful and able to easily cut dimension lumber as well as trim material (*Figure 14–29*). Sizes include 4½- and 6½-inch blades. Cordless jigsaws and reciprocating saws offer the same features as corded models. They also cut and handle in much the same manner with the same performance.

Courtesy of Stanley Black & Decker, Inc.

Figure 14–27 Cordless models are available for many tools.

© Cengage Learning 2014

Figure 14–28 Cordless tool batteries are typically either nickel-cadmium or lithium-ion.

© Cengage Learning 2014

Figure 14–29 Cordless circular saws easily cut dimension lumber.

CHAPTER 15

Planes, Routers, and Sanders

PORTABLE POWER PLANES

Portable power planes make planing jobs much easier for the carpenter. Planing the ends and edges of hardwood doors and stair treads, for example, takes considerable effort with hand planes, even with razor-sharp cutting edges.

Jointer Planes

The jointer plane is used primarily to smooth and straighten long edges, such as fitting doors in openings (*Figure 15–1*). It is manufactured in lengths up to 18 inches. The electric motor powers a cutter head that may measure up to 3¾ inches wide. The planing depth, or the amount that can be taken off with one pass, can be set for 0 up to ⅛ inch.

An adjustable fence allows planing of squares, beveled edges to 45 degrees, or chamfers (*Figure 15–2*). A rabbeting guide is used to cut rabbets up to ⅞ inch deep.

Operating Power Planes

The operation and feel of the power plane is similar to that of the bench plane. The major differences are the vibration and the ease of cutting with the power plane.

Figure 15–1 A portable electric jointer plane.

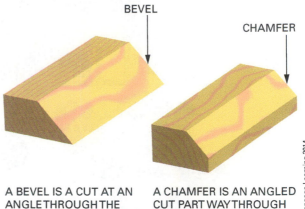

BEVEL

CHAMFER

A BEVEL IS A CUT AT AN ANGLE THROUGH THE TOTAL THICKNESS

A CHAMFER IS AN ANGLED CUT PART WAY THROUGH THE THICKNESS

Figure 15–2 A bevel and chamfer.

> ### CAUTION
>
> Extreme care must be taken when operating power planes. There is no retractable guard, and the high-speed cutterhead is exposed on the bottom of the plane. Keep the tool clear of your body until it has completely stopped. Keep extension cords clear of the tool.

- Set the side guide to the desired angle, and adjust the depth of cut.
- Hold the toe (front) firmly on the work, with the plane cutterhead clear of the work.
- Start the motor. With steady, even pressure, make the cut through the workpiece for the entire length. Guide the angle of the cut by holding the guide against the side of the stock. Apply pressure to the toe of the plane at the beginning

of the cut. Apply pressure to the heel (back) at the end of the cut to prevent tipping the plane over the ends of the workpiece.

- To plane a taper, that is, to take more stock off one end than the other, make a number of passes. Each pass should be shorter than the preceding one. Lift the plane clear of the stock at the end of the pass. Make the last pass completely from one end to the other (*Figure 15–3*).

PORTABLE ELECTRIC ROUTERS

One of the most versatile portable tools used in the construction industry is the router (*Figure 15–4*). It is available in many models, ranging from ¼ hp

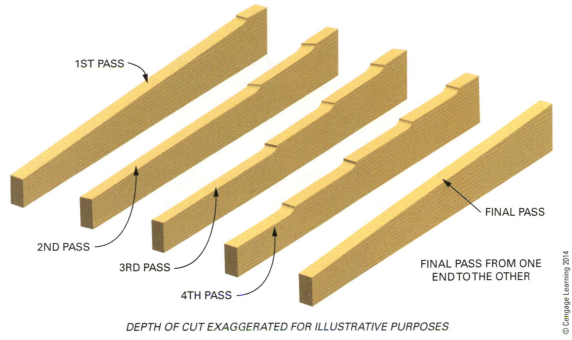

1ST PASS

2ND PASS

3RD PASS

4TH PASS

FINAL PASS

FINAL PASS FROM ONE END TO THE OTHER

© Cengage Learning 2014

DEPTH OF CUT EXAGGERATED FOR ILLUSTRATIVE PURPOSES

Figure 15–3 Technique for planing a taper.

© Cengage Learning 2014

Figure 15–4 Using a portable electric router.

(horsepower) to more than 3 hp with speeds of from 18,000 to 30,000 rpm. These tools have high-speed motors that enable the operator to make clean, smooth-cut edges.

The motor powers a ¼- or ½-inch chuck in which cutting bits of various sizes and shapes are held (*Figure 15–5*). An adjustable base is provided to control the depth of cut. A trigger or toggle switch controls the motor.

The router is used to make many different cuts, including grooves, dadoes, rabbets, and dovetails

(*Figure 15–6*). It is also used to shape edges and make cutouts, such as for sinks in countertops. It is extensively used with accessories, called templates, to cut recesses for hinges in door edges and door frames. When operating the router, it is important to be mindful of the bit at all times. Watch what you are cutting, and keep the router moving. Stalling the movement of the router will cause the bit to burn or melt the material.

Laminate Trimmers

A light-duty specialized type of router is called a laminate trimmer. It is used almost exclusively for trimming the edges of *plastic laminates* (*Figure 15–7*). Plastic laminate is a thin, hard material used primarily as a decorative covering for kitchen and bathroom cabinets and countertops. (The installation of this material is described in Chapter 85.)

Guiding the Router

Controlling the sideways motion of the router is accomplished by the following methods:

- By using a router bit with a pilot (guide) (*Figure 15–8*). The pilot may be solid and rotate with the bit or may have ball-bearing pilots. These guide the router along the uncut portion of the material being routed.

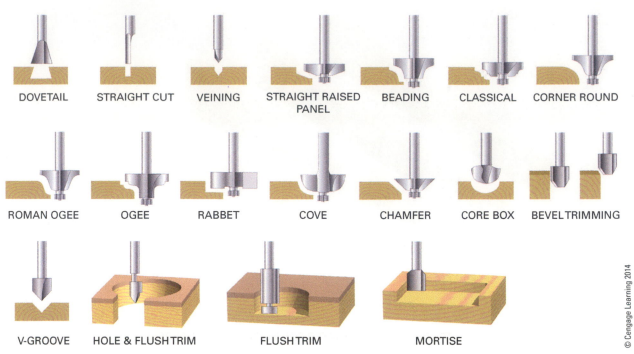

Figure 15–5 Router bit selection guide.

Figure 15–6 A dovetail cut is easily made with a router.

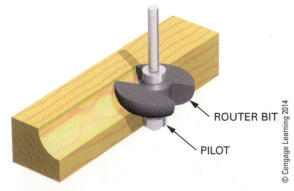

ROUTER BIT

PILOT

Figure 15–8 Router bits may have a pilot bearing to guide the cut.

Figure 15–7 The laminate trimmer is used to trim the edges of plastic laminates.

- By guiding the edge of the router base against a straightedge (*Figure 15–9*). Be sure to keep the router tight to the straightedge, and do not rotate the router during the cut, because its base may not be centered.

- By using an adjustable guide attached to the base of the router (*Figure 15–10*). The guide rides along the edge of the stock. Make sure the edge is in good condition.

- By using a template (pattern) with template guides attached to the base of the router (*Figure 15–11*). This is the method widely used for cutting recesses for door hinges. (This process is explained in detail in Chapter 57.)

STRAIGHTEDGE IS ON OPERATOR'S RIGHT. AS ROUTER IS PULLED, ROTATION OF ROUTER TENDS TO KEEP IT AGAINST STRAIGHTEDGE. IF STRAIGHTEDGE WERE ON LEFT SIDE, ROUTER WOULD HAVE TENDENCY TO PULL AWAY FROM THE STRAIGHTEDGE.

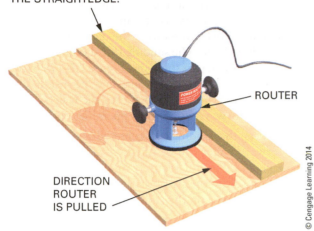

ROUTER

DIRECTION ROUTER IS PULLED

Figure 15–9 Using a straightedge to guide the router.

Figure 15–10 A guide attached to the base of the router rides along the edge of the stock and controls the sideways motion of the router.

- By freehand routing, in which the sideways motion of the router is controlled by the operator only. Care should be taken during this operation.
- To make *circular* cuts, remove the subbase. Replace it, using the same screwholes, with a custom-made one in which one side extends to any desired length. Along a centerline, make a series of holes to fasten the newly made subbase to the center of the desired arc (*Figure 15–12*).

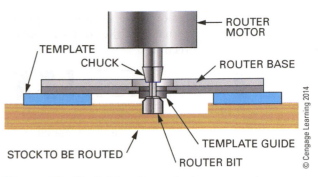

ROUTER MOTOR

TEMPLATE

CHUCK

ROUTER BASE

STOCK TO BE ROUTED

TEMPLATE GUIDE

ROUTER BIT

Figure 15–11 Guiding the router by means of a template and template guide.

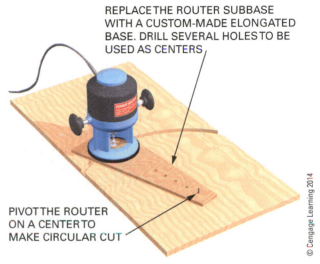

REPLACE THE ROUTER SUBBASE WITH A CUSTOM-MADE ELONGATED BASE. DRILL SEVERAL HOLES TO BE USED AS CENTERS

PIVOT THE ROUTER ON A CENTER TO MAKE CIRCULAR CUT

Figure 15–12 Technique for making arcs using a router.

Using the Router

Before using the router, make sure power is disconnected. Follow the method outlined:

- Select the correct bit for the type of cut to be made.
- Insert the bit into the chuck. Make sure the chuck grabs at least 1/2 inch of the bit. Adjust the depth of cut.
- Control the sideways motion of the router by one of the methods previously described.
- Clamp the work securely in position. Plug in the cord.
- Lay the base of the router on the work with the router bit clear of the work. Start the motor.
- Advance the bit into the cut, pulling the router in a direction that is against the rotation of the bit. On outside edges and ends, the router

is moved counterclockwise around the piece. When making internal cuts, the router is moved in a clockwise direction.

> ### CAUTION
>
> Finish the cut, keeping the router clear of your body until it has stopped. Be aware that the router bit is unguarded.

PORTABLE ELECTRIC SANDERS

Interior trim, cabinets, and other finish should be sanded before any finishing coats of paint, stain, polyurethane, or other material are applied. It is shoddy workmanship to coat finish work without sanding. In too many cases, expediency seems to take precedence over quality. Trim needs to be sanded because the grain probably has been *raised*. This happens because the stock has been exposed to moisture in the air between the time it was milled and the time of installation. Also, rotary planing of lumber leaves small ripples in the surface. Although hardly visible before a finish is applied, they become very noticeable later.

Some finishing coats require more sanding than others. If a penetrating stain is to be applied, extreme care must be taken to provide a surface that is evenly sanded with no cross-grained scratches. If paints or transparent coatings are to be applied, surfaces need not be sanded as thoroughly. Portable sanders make sanding jobs less tedious.

Portable Electric Belt Sanders

The belt sander is used frequently for sanding cabinetwork and interior finish (*Figure 15–13*). The size of the belt determines the size of the sander. Belt widths range from 2½ to 4 inches. Belt lengths vary from 18 to 24 inches. The 3-inch by 21-inch

belt sander is a popular, lightweight model. Some sanders have a bag to collect the sanding dust. Remember to wear eye protection.

Installing Sanding Belts.   Sanding belts are usually installed by retracting the forward roller of the belt sander. An arrow is stamped on the inside of some sanding belts to indicate the direction in which the belt should run. The sanding belt should run *with*, not against, the lap of the joint (*Figure 15–14*). Sanding belts joined with butt joints may be installed in either direction. Install the belt over the rollers. Then release the forward roller to its operating position.

To keep the sanding belt centered as it is rotating, the forward roller can be tilted. Stand the sander on its back end. Hold it securely and start it. Turn the adjusting screw one way or the other to track the belt and center it on the roller (*Figure 15–15*).

Using the Belt Sander.   More work has probably been ruined by improper use of the portable belt sander than any other tool. It is wise to practice on scrap stock until enough experience in its use is gained to ensure an acceptable sanded surface. Care must be taken to sand squarely on the sander's pad. Allowing the sander to tilt sideways or to ride on either roller results in a gouged surface.

> ### CAUTION
>
> Make sure the switch of the belt sander is off before plugging the cord into a power outlet. Some trigger switches can be locked in the "ON" position. If the tool is plugged in when the switch is locked in this position, the sander will travel at high speed across the surface. This could cause damage to the work and/or injury to anyone in its path.

Figure 15–13 Using a portable electric belt sander.

Figure 15–14 Sanding belts should be installed in the proper direction, as indicated by the arrow on the inside of the belt.

Figure 15–15 The belt should be centered on its rollers by using the tracking screw.

- Secure the work to be sanded. Make sure the belt is centered on the rollers and is tracking properly.
- Hold the tool with both hands so that the edge of the sanding belt can be clearly seen.
- Start the machine. Place the pad of the sander flat on the work. Pull the sander back and lift it just clear of the work at the end of the stroke.

> ## CAUTION
>
> Be careful to keep the electrical cord clear of the tool. Because of the constant movement of the sander, the cord may easily get tangled in the sander if the operator is not alert.

- Place the sander back on the material and bring the sander forward. Continue sanding using a skimming motion that lifts the sander just clear of the work at the end of every stroke. Sanding in this manner prevents overheating of the sander and helps to remove material evenly. It also allows debris to be cleared from the work, and the operator can see what has been done.
- Do not sand in one spot too long. Be careful not to tilt the sander in any direction. Always

sand with the pad flat on the work. Do not exert excessive pressure. The weight of the sander is enough. Always sand with the grain to produce a smooth finish.

- Make sure the sander has stopped before setting it down. It is a good idea to lay it on its side to prevent accidental traveling.

Finishing Sanders

The finishing sander or palm sander (*Figure 15–16*) is used for the final sanding of interior work. These tools are manufactured in many styles and sizes. They are available in cordless models.

Finishing sanders have either an orbital motion, an oscillating (straight back-and-forth) motion, or a combination of motions controlled by a switch. The orbital motion has faster action, but leaves scratches across the grain. The *random orbital* sander reduces this problem with a design that randomly moves the center of the rotating paper at high speed. This allows the paper to sand in all directions at once. The straight-line motion is slower, but produces no cross-grain scratches on the surface.

Most sanders take one-quarter or one-half sheet of sandpaper. It is usually attached to the pad by some type of friction or spring device. Some sanding sheets come precut with an adhesive backing for easy attachment to the sander pad.
To use the finishing sander, proceed as follows:

- Select the desired grit sandpaper. Attach it to the pad, making sure it is tight. A loose sheet will tear easily.
- Start the motor and sand the surface evenly, *slowly* pushing and pulling the sander with the grain. Let the action of the sander do the work.

Figure 15–16 A portable electric finishing sander.

Do not use excessive pressure because this may overload the machine and burn out the motor. Always hold the sander flat on its pad.

Abrasives. The quality of the abrasives on the sandpaper is determined by the length of time it is able to retain its sharp cutting edges. *Flint* and *garnet* are natural minerals used as abrasives. Although sandpaper made with flint is less expensive, it does not last as long as garnet. Synthetic (human-made) abrasives include *aluminum oxide* and *silicon carbide*. Sandpaper coated with aluminum oxide is probably the most widely used for wood.

Grits. Sandpaper grit refers to the size of the abrasive particles. Sandpaper with large abrasive particles is considered coarse. Small abrasive particles are used to make fine sandpaper.

Sandpaper grits are designated by a grit number. The grit numbers range from No. 12 (coarse)

Description	Grit No.
Very Fine	400, 360, 320, 280
Fine	240, 220, 180, 150
Medium	120, 100
Coarse	80, 60, 50, 40
Very coarse	36, 30, 24, 20, 16, 12

© Cengage Learning 2014

Figure 15–17 Grits of coated abrasives.

to No. 400 (fine) (*Figure 15–17*). Commonly used grits are 60 or 80 for rough sanding, and 120 or 180 for finish sanding.

Sand with a coarser grit until a surface is uniformly sanded. Do not switch to a finer grit too soon. Do not use worn or clogged abrasives. Their use causes the surface to become glazed or burned.

CHAPTER 16

Fastening Tools

Portable fastening tools included in this chapter are those called **pneumatic** (powered by compressed air) and **powder-actuated** (drive fasteners with explosive powder cartridges). They are widely used throughout the construction industry for practically every fastening job from foundation to roof.

PNEUMATIC STAPLERS AND NAILERS

Pneumatic **staplers** and **nailers** are commonly called *guns* (*Figure 16–1*). They are used widely for quick fastening of framing, subfloors, wall and roof sheathing, roof shingles, exterior finish, and interior trim. A number of manufacturers make a variety of models in several sizes for special fastening jobs. For instance, a *framing gun,* although used for many fastening jobs, is not used to apply interior trim.

Older models required frequent oiling, either by hand or by an oiler installed in the air line.

Courtesy of Paslode

Figure 16–1 Pneumatic nailers and staplers are widely used to fasten building materials.

With improvements in design, some newer models require no oiling at all.

Remember to wear eye protection. It is also recommended to wear ear protection. Prolonged exposure to loud noises will damage the ear.

Nailing Guns

The heavy-duty framing gun (*Figure 16–2*) drives smooth-shank, headed nails up to 4 inches, ring-shank nails up to 2⅜ inches, and screw shank nails up to 3 inches to fasten all types of framing. A light-duty version (*Figure 16–3*) drives smooth-shank nails up to 2⅜ inches and ring-shank nails up to 1¾ inches to fasten light framing, subfloor, sheathing, and similar components of a building. Nails come glued in strips for easy insertion into the magazine of the gun (*Figure 16–4*). Check local building codes and specifications for the correct nail styles and sizes. For example, clip-headed nails cannot be used to fasten plywood that is used as a shear (high-stress) panel.

The *finish nailer* (*Figure 16–5*) drives finish nails from 1 to 2 inches long. It can be used for the application of practically all kinds of exterior and interior finish work. It sets or flush drives nails as desired. A nail set is not required, and the possibility of marring the wood is avoided.

The *brad nailer* (*Figure 16–6*) drives both slight-headed and medium-headed brads ranging

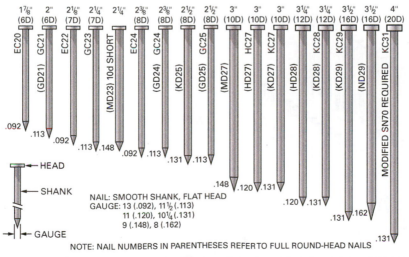

Figure 16–2 Heavy-duty framing guns are used for floor, wall, and roof framing.

Courtesy of Senco Products, Inc.

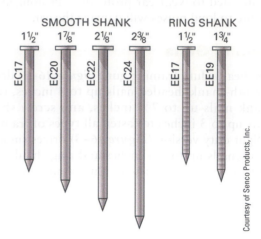

SMOOTH SHANK   RING SHANK

EC17 1½"   EC20 1⅞"   EC22 2⅛"   EC24 2⅜"   EE17 1½"   EE19 1¾"

Courtesy of Senco Products, Inc.

Figure 16–3 A light-duty nailer is used to fasten light framing, subfloors, and sheathing.

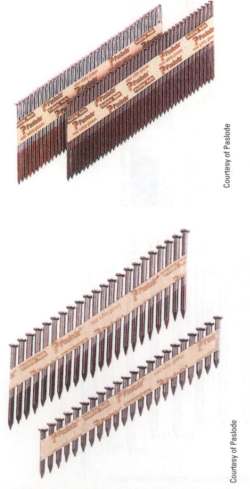

Courtesy of Paslode

Courtesy of Senco Products, Inc.

FINISH NAILS

1" (2D)   1¼" (3D)   1½" (4D)   1¾" (5D)   2" (6D)

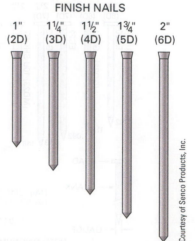

Courtesy of Senco Products, Inc.

Figure 16–4 Both headed and finish nails used in nailing guns come glued together in strips.

Figure 16–5 The finish nailer is used to fasten all kinds of interior trim.

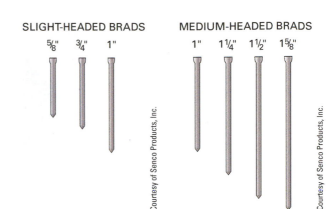

SLIGHT-HEADED BRADS   MEDIUM-HEADED BRADS

⅝" ¾" 1"   1" 1¼" 1½" 1⅝"

Courtesy of Senco Products, Inc.

Courtesy of Senco Products, Inc.

Figure 16–6 A light-duty brad nailer is used to fasten thin molding and trim.

COIL ROOFING NAILS

⅞" 1" 1¼" 1½" 1¾"

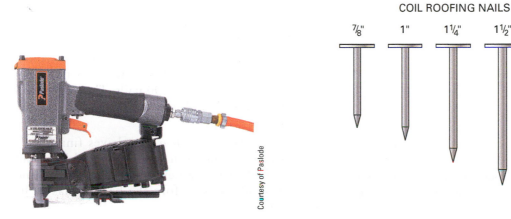

Courtesy of Paslode

Courtesy of Senco Products, Inc.

Figure 16–7 A coil roofing nailer is used to fasten asphalt roof shingles.

in length from ⅝ inch to 1⅝ inches. It is used to fasten small moldings and trim, cabinet door frames and panels, and other miscellaneous finish carpentry.

The *coil roofing nailer* (*Figure 16–7*) is designed for fastening asphalt and fiberglass roof shingles. It drives five different sizes of wide, round-headed roofing nails from ⅞ to 1¾ inches. The nails come in coils of 120 (*Figure 16–8*), which are easily loaded in a nail canister.

Staplers

Like nailing guns, *staplers* are manufactured in a number of models and sizes and are classified as light-, medium-, and heavy-duty staplers.

A popular tool is the *roofing stapler* (*Figure 16–9*), which is used to fasten roofing shingles. It comes in several models and drives 1-inch wide-crown staples in lengths from ¾ inch to 1½ inches. It can also be used for fastening other materials, such as lath wire, insulation, furniture, and cabinets. The staples, like nails, come glued

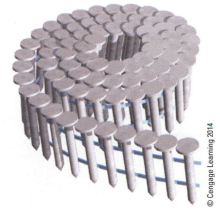

© Cengage Learning 2014

Figure 16–8 Roofing nails come in coils for use in the roofing nailer.

together in strips (*Figure 16–10*) for quick and easy reloading. Most stapling guns can hold up to 150 staples.

No single model of stapler drives all widths and lengths of staples. Models are made to drive narrow-crown, intermediate-crown, and wide-crown

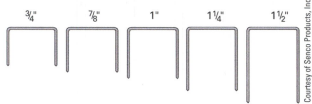

¾"   ⅞"   1"   1¼"   1½"

Figure 16–9 The wide-crown stapler, being used in this photograph to fasten roof shingles, can be used to fasten a variety of materials.

staples. A popular model drives ⅜-inch intermediate-crown staples from ¾ inch to 2 inches in length.

Cordless Guns

Conventional pneumatic staplers and nailers are powered by compressed air. The air is supplied by an air compressor (*Figure 16–11*) through long lengths of air hose stretched over the construction site. The development of cordless nailing and stapling guns (*Figure 16–12*) eliminates the need for air compressors and hoses. The cordless gun utilizes a disposable fuel cell. A battery and spark plug power an internal combustion engine that forces a piston down to drive the fastener. Another advantage is the time saved in setting up the compressor, draining it at the end of the day, coiling the hoses, and storing the equipment.

The *cordless framing nailer* drives nails from 2 to 3¼ inches in length. Each fuel cell will deliver energy to drive about 1,200 nails. The battery will last long enough to drive about 4,000 nails before recharging is required.

The *cordless finish nailer* drives finish nails from ¾ to 2½ inches in length. It will drive about 2,500 nails before a new fuel cell is needed and about 8,000 nails before the battery has to be recharged.

Figure 16–10 Staples, like nails, come glued together in strips for use in stapling guns.

Figure 16–11 Compressors supply air pressure to operate nailers and other pneumatic tools.

Courtesy of Paslode

Figure 16–12 Each cordless gun comes in its own case with battery, battery charger, safety glasses, instructions, and storage for fuel cells.

The *cordless stapler* drives intermediate-crown staples from ¾ inch to 2 inches in length. It will drive about 2,500 staples with each fuel cell and about 8,000 staples with each charge of the battery.

Using Nailers and Staplers

Because of the many designs and sizes of staplers and nailers, you should study the manufacturer's directions and follow them carefully. Use the right nailer or stapler for the job at hand. Make sure all safety devices are working properly, and always wear eye protection. A work contact switch allows the tool to operate only when this device is firmly depressed on a work surface and the trigger is pulled, promoting safe tool operation.

- Load the magazine with the desired size staples or nails.
- Connect the air supply to the tool. For those guns that require it, make sure there is an oiler in the air supply line, adequate oil to keep the gun lubricated during operation, and an air filter to keep moisture from damaging the gun. Use the recommended air pressure. Larger nails require more air pressure than smaller ones.

CAUTION

Exceeding the recommended air pressure may cause damage to the gun or burst air hoses, possibly causing injury to workers.

- Press the trigger and tap the nose of the gun to the work. When the trigger is depressed, a fastener is driven each time the nose of the gun is tapped to the work. The fastener may also be safely driven by first pressing the nose of the gun to the surface and then pulling the trigger.
- Upon completion of fastening, disconnect the air supply.

CAUTION

Never leave an unattended gun with the air supply connected. Always keep the gun pointed toward the work. Never point it at other workers or fire a fastener except into the work. Serious injury can result from improper use of the tool.

POWDER-ACTUATED DRIVERS

Powder-actuated drivers (*Figure 16–13*) are used to drive specially designed pins into masonry or steel. They are used in a manner similar to firing a gun. Powder charges of various strengths drive the pin when detonated.

Courtesy of ITW Ramset

Figure 16–13 Powder-actuated drivers are used for fastening into masonry or steel.

Drivepins

Drivepins are available in a variety of sizes. Three styles are commonly used. The *headed* type is used for fastening material. The *threaded* type is used to bolt an object after the pin is driven. The *eyelet* type is used when attachments are to be made with wire.

Powder Charges

Powder charges are color coded according to strength. Learn the color codes for immediate recognition of the strength of the charge. A stronger charge is needed for deeper penetration or when driving into hard material. The strength of the charge must be selected with great care.

Because of the danger in operating these guns, many states require the operator to be certified. Certificates may be obtained from the manufacturer's representative after a brief training course.

Using Powder-Actuated Drivers

- Study the manufacturer's directions for safe and proper use of the gun. Use eye and ear protection.

CAUTION

Treat this tool with great respect. Never use it if the end that ejects the fastener is facing toward other workers.

- Make sure the drivepin will not penetrate completely through the material into which it is driven. This has been the cause of fatal accidents.

- To prevent ricochet hazard, make sure the recommended shield is in place on the nose of the gun. A number of different shields are available for special fastening jobs.

- Select the proper fastener for the job. Consult the manufacturer's drivepin selection chart to determine the correct fastener size and style.

- Select a powder charge of necessary strength. Always use the weakest charge that will do the job and gradually increase strength as needed. Load the driver with the pin first and the cartridge second.

- Keep the tool pointed at the work. Wear safety goggles. Press hard against the work surface, and pull the trigger. The resulting explosion drives the pin. Eject the spent cartridge.

CAUTION

If the gun does not fire, hold it against the work surface for at least thirty seconds. Then remove the cartridge according to the manufacturer's directions. Do not attempt to pry out the cartridge with a knife or screwdriver; most cartridges are rim fired and could explode.

DECONSTRUCT THIS

Carefully study *Figure 16–14* and think about what is wrong and/or what is right. Consider all possibilities. What construction practice or method is different in your area of the country?

Figure 16–14 This photo shows a carpenter cutting a 2 x 6.

KEY TERMS

belt sander	hole saws	nailers	screwguns
carbide-tipped	jigsaw	pneumatic	Skilsaw
chamfers	jointer plane	powder-actuated	staplers
cordless nailing gun	laminate	powder-actuated drivers	stapling guns
dovetails	laminate trimmer	reciprocating saw	taper
grit	magazine	ripping	templates
hammer-drills	masonry drill bits	router	

REVIEW QUESTIONS

Select the most appropriate answer.

1. To use a power tool properly, the operator should always

   a. wear eye protection.
   b. wear ear protection.
   c. understand the manufacturer's recommended instructions.
   d. all of the above.

2. The guard of the portable electric saw should never be

   a. lubricated.
   b. adjusted.
   c. retracted by hand.
   d. wedged open.

3. When selecting an extension cord for a power tool,

   a. use a longer cord to keep the cord from heating up.
   b. keep the cord evenly spread out around the work area.
   c. use one with a GFCI or plugged into a GFCI outlet.
   d. all of the above.

4. When using a power tool for cutting, the operator should wear

   a. safety contact lenses and steel-toed work boots.
   b. ear and eye protection.
   c. stereo headphones.
   d. all of the above.

5. Sharp tools

   a. put less stress on the operator than dull tools.
   b. cut slower than dull tools.
   c. are more dangerous to use than dull tools.
   d. all of the above.

6. The jigsaw is used primarily for making

   a. curved cuts.
   b. compound miters.
   c. cuts in drywall.
   d. long straight cuts.

7. The saw primarily used for rough-in work is the

   a. reciprocating saw.
   b. sawzall.
   c. saber saw.
   d. all of the above.

8. The tool that is best suited for drilling metal as well as wood is the

   a. auger bit.
   b. high-speed twist drill.
   c. expansive bit.
   d. speed bit.

9. To produce a neat and clean hole in wood

   a. use a fast-spinning, sharp bit.
   b. use a slower travel speed.
   c. finish the hole by drilling from the back side.
   d. all of the above.

10. When using the router to shape four outside edges and ends of a piece of stock, the router is guided in a

    a. direction with the grain.
    b. clockwise direction.
    c. counterclockwise direction.
    d. all of the above.

11. The tool for which some codes require the operator to be certified because of the potential danger is the

    a. powder-actuated driver.
    b. hammer-drill.
    c. cordless pneumatic nailer.
    d. screwgun.

12. The saw arbor nuts that hold circular saw blades in position are loosened by rotating the nut

    a. clockwise.
    b. with the rotation of the blade.
    c. counterclockwise.
    d. against the rotation of the blade.

13. The best way to achieve a clean cut when using a circular saw is to cut

    a. on the face side of the material.
    b. on the back side of the material.
    c. with a quick travel of the saw.
    d. while periodically pausing the travel.

14. An electric power plan is most similar to a

    a. router.
    b. sander.
    c. bench plane.
    d. all of the above.

15. Tools that are driven by compressed air are referred to as

    a. air-powered tools.
    b. compressed tools.
    c. powder-actuated tools.
    d. pneumatic tools.

UNIT 7
Stationary Power Tools

©iStockphoto.com/Curt Pickens

Many kinds of stationary power woodworking tools are used in wood mills and cabinet shops for specialized work. On the building site, usually only table and miter saws are available for use by carpenters. These are not heavy-duty machines. They must be light enough to be transported from one job to another. These saws are ordinarily furnished by the contractor.

OBJECTIVES

After completing this unit, the student should be able to:

- describe different types of circular saw blades and select the proper blade for the job at hand.

- describe, adjust, and operate the radial arm saw and table saw safely to crosscut lumber to length, rip to width, and make miters, compound miters, dadoes, grooves, and rabbets.

- operate the table saw in a safe manner to taper rip and to make cove cuts.

- describe, adjust, and operate the power miter saw to crosscut to length, making square and miter cuts safely and accurately.

SAFETY REMINDER

Stationary power tools are strong and durable, designed for heavy-duty tasks. While using power tools the worker must keep a constant awareness of the risks of using the tool.

> # CAUTION
>
> - Be trained and competent in the use of stationary power tools before attempting to operate them without supervision.
> - Make sure power is disconnected when changing blades or cutters.
> - Make sure saw blades are sharp and suitable for the operation. Ensure that safety guards are in place and that all guides are in proper alignment and secured.
> - Wear eye and ear protection. Wear appropriate, properly fitted clothing.
> - Keep the work area clear of scraps that might present a tripping hazard.
> - Keep stock clear of saw blades before starting a machine.
> - Do not allow your attention to be distracted by others or be distracting to others while operating power tools.
> - Turn off the power and make sure the machine has stopped before leaving the area.
>
> Safety precautions that apply to specific operations are given when those operations are described in this unit.

CHAPTER 17

Circular Saw Blades

All of the stationary power tools described in this unit use circular saw blades. To ensure safe and efficient saw operation, it is important to know which type to select for a particular purpose (*Figure 17–1*).

CIRCULAR SAW BLADES

The more teeth a saw blade has, the smoother the cut. However, a fine-toothed blade does not cut as fast. This means that the stock must be fed more slowly. A coarse-toothed saw blade leaves a rough surface, but cuts more rapidly. Thus, the stock can be fed faster by the operator.

If the feed is too slow, the blade may overheat. This will cause it to lose its shape and wobble at high speed. This is a dangerous condition that must be avoided. An overheated saw may bind in the cut, possibly causing kickback and serious operator injury.

The same results occur when trying to cut material with a dull blade. Always use a sharp blade. Use fine-toothed blades for cutting thin, dry material, and coarse-toothed blades for cutting heavy, rough lumber.

Types of Circular Saw Blades

Many different types of circular saw blades are used for various cutting applications. Blades are made to cut wood and wood products, plastic

WOOD CROSSCUT BLADE

ABRASIVE METAL CUTTING BLADE

MASONRY DIAMOND BLADE

WOOD COMBINATION BLADE

© Cengage Learning 2014

Figure 17–1 Circular saw blades are made to cut different materials.

laminates, masonry, and steel. Each material has a different requirement for cutting and the blades are designed accordingly (see *Figure 17–1*).

Wood-cutting blades are diverse in design styles. They are loosely classified into two groups, high-speed steel blades and tungsten carbide-tipped blades. High-speed steel blades are the original generation of circular blades. Teeth may be sharpened with a file and tend not to stay sharp as long as carbide-tipped blades. The carbide-tipped blades are superior, and well suited for cutting material that contains adhesives and other foreign material that rapidly dull high-speed steel blades. Tungsten carbide is a hard metal that is sharpened with diamond-impregnated grinding wheels. Most saw blades being used today are carbide tipped. Resurfacing of the tips should be done by a professional blade sharpener.

In both groups of saw blades, the number and shape of the teeth vary to give different cutting actions according to the kind, size, and condition of the material to be cut. Masonry and steel cutting blades are made in general types and also abrasive and steel types. Abrasive blades are made of composite materials that are consumed as the blade cuts. The blade diameter gets smaller with each cut, and it is easy to tell when the blade should

be replaced. Steel blades designed to cut steel and masonry are carbide tipped or diamond coated. They tend to last longer and are more expensive than abrasive blades.

High-Speed Circular Saw Blades

High-speed steel blades are classified as *rip, crosscut,* or *combination* blades. They may be given other names, such as plywood, panel, or flooring blades, but they are still in one of the three classifications.

Ripsaws. The ripsaw blade (*Figure 17–2*) usually has fewer teeth than crosscut or combination blades. As in the hand ripsaw, every tooth of the circular ripsaw blade is filed or ground at right angles to the face of the blade. This produces teeth with a cutting edge all the way across the tip of the tooth. These teeth act like a series of small chisels that cut and clear the stock ahead of them.

Use a ripsaw for cutting solid lumber with the grain when a smooth edge is not necessary. Also, use a ripsaw when cutting unseasoned or green lumber and lumber of heavy dimension with the grain.

Crosscut Saws. The teeth of a crosscut circular saw blade are shaped like those of the crosscut handsaw (*Figure 17–3*). The sides of the teeth are

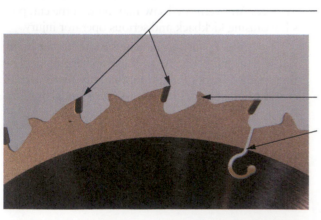

WIDELY SPACED TEETH

CUTTING-DEPTH CONTROL TOOTH

HEAT EXPANSION SLOT

© Cengage Learning 2014

Figure 17–2 The teeth of a ripsaw have square-edged cutting tips with wide spaces between them.

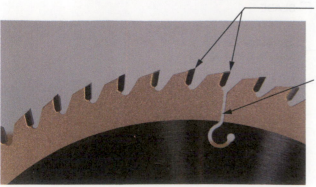

CLOSELY SPACED TEETH

EXPANSION SLOT

© Cengage Learning 2014

Figure 17–3 The teeth of a crosscut blade have beveled sides and pointed tips.

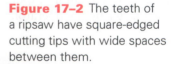

alternately filed or ground on a bevel. This produces teeth that come up to points instead of edges.

Crosscut teeth slice through the wood fibers smoothly. A crosscut blade is an ideal blade for cutting across the grain of solid lumber. It also cuts plywood satisfactorily with little splintering of the cut edge.

Combination Saws.   The combination blade is used when a variety of ripping and crosscutting is to be done. It eliminates the need to change blades for different operations.

There are several types of combination blades. One type has groups of teeth around its circumference (*Figure 17–4*). The leading tooth in each group is a rip tooth. The ones following are crosscut teeth.

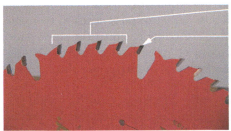

CROSS CUT
TEETH
RIP TOOTH

© Cengage Learning 2014

Figure 17–4 The teeth of a combination blade include both crosscut and rip teeth.

Carbide-Tipped Blades

Carbide Blade Component Parts.   The component parts of a carbide blade have functions that make the blade perform well (*Figure 17–5*). The teeth consist of bits of carbide and are silver soldered to the metal blade. They should be the only part to actually touch the material. Carbide is made in different hardnesses for various applications. Drilling stone and masonry often requires hammering, thus softer carbide is used. The hardest carbide is used for cutting metal. The harder the carbide material, the more brittle the blade, which should be handled gently.

> **CAUTION**
>
> The teeth of a carbide blade can break like glass if the blade is dropped.

The *gullet* is a gap created to allow for blade expansion when it heats up from normal use. The anti-vibration slots are cut in the blade with a laser to help keep the blade true while it is being run under heavy load. The antikickback teeth are designed to keep the tooth that follows it from biting too deeply into the material. Otherwise the blade might kick back at the operator. They usually are located

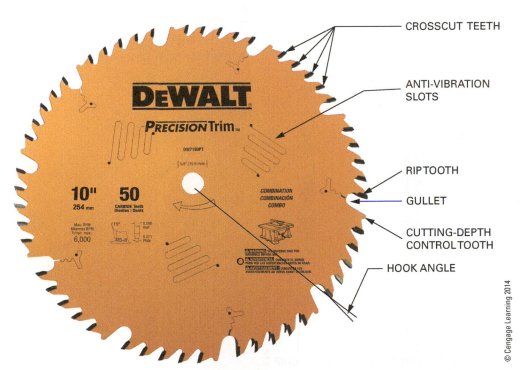

CROSSCUT TEETH

ANTI-VIBRATION
SLOTS

RIP TOOTH

GULLET

CUTTING-DEPTH
CONTROL TOOTH

HOOK ANGLE

© Cengage Learning 2014

Figure 17–5 Parts of a typical carbide combination blade.

before any large space in the blade. The hook angle is the angle at which the tooth is pitched forward. Larger hook angles cut more aggressively, making a faster but less smooth cut.

Carbide Tooth Style. The teeth of a carbide blade are ground to different shapes, giving the blade different cutting abilities (*Figure 17–6*).

Square Grind. The *square grind* is similar to the rip teeth in a steel blade. It is used primarily to cut solid wood with the grain. It can also be used on composition boards when the quality of the cut surface is not important.

Alternate Top Bevel. The *alternate top bevel* grind is used with excellent results for crosscutting solid lumber, plywood, hardboard, particleboard, fiberboard, and other wood composite products. It can also be used for ripping operations. However, the feed is slower than that of square or combination grinds.

Triple Chip. The *triple chip grind* is designed for cutting brittle material without splintering or chipping the surface. It is particularly useful for cutting an extremely smooth edge on plastic laminated material, such as countertops. It can also be used like a planer blade to produce a smooth cut surface on straight, dry lumber of small dimension.

Combination. Carbide-tipped saws also come with a combination of teeth. The leading tooth in each set is square ground. The following teeth are

ground at alternate bevels. These saws are ideally suited for ripping and crosscutting solid lumber and all kinds of engineered wood products. Because of their versatility, these saw blades are probably the most widely used by carpenters.

Carbide-tipped teeth are not set (bent slightly outward). The carbide tips are slightly thicker than the saw blade itself. Therefore, they provide clearance for the blade in the saw cut. In addition, the sides of the carbide tips are slightly beveled back to provide clearance for the tip.

Dado Blades

Dadoes, grooves, and *rabbets* can be cut using a single saw blade (*Figure 17–7*), but this would require making several passes of the material with the single blade. A *dado set,* which consists of more than one blade, is commonly used to make these cuts faster because only one pass needs to be made through the stock. One type, called the stack dado head, consists of two outside circular saw blades with several *chippers* of different thicknesses placed between (*Figure 17–8*).

Most dado sets make cuts from $\frac{1}{8}$ to $\frac{13}{16}$ inch wide. This is done by using one or more of the blades and chippers together. Wide cuts are obtained by adding shims between chippers or by making more passes through the material. When installing this type of dado set, make sure the tips of the chippers are opposite the gullets and not against the side of the blade. Chipper tips are *swaged* (made wider than the body of the chipper) to ensure a clean cut across the width of the dado or groove.

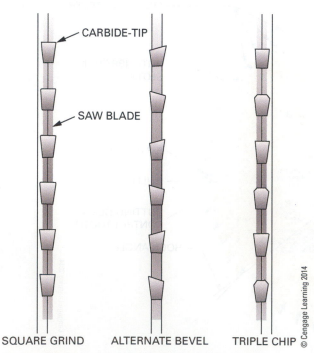

CARBIDE-TIP

SAW BLADE

SQUARE GRIND   ALTERNATE BEVEL   TRIPLE CHIP

© Cengage Learning 2014

Figure 17–6 Three main tooth styles of carbide-tipped saw blades.

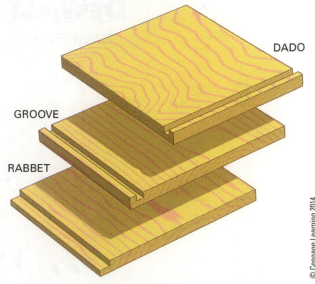

DADO

GROOVE

RABBET

© Cengage Learning 2014

Figure 17–7 Notches in wood have different names depending on where they are located.

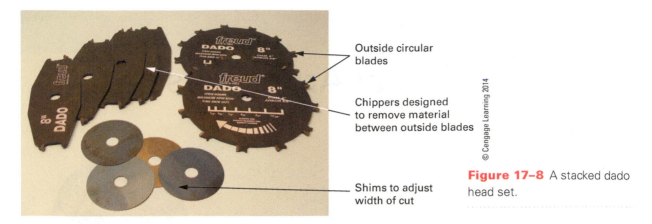

Outside circular blades

Chippers designed to remove material between outside blades

Shims to adjust width of cut

Figure 17–8 A stacked dado head set.

Figure 17–9 A wobbler head dado blade.

Another type of dado blade is the one-unit *adjustable dado head* (*Figure 17–9*), commonly called a *wobbler head.* This type can be adjusted for width of dado by rotating the sections of the head. The head can be adjusted by loosening the arbor nut and need not be removed from the saw arbor. This type of dado head does not make perfectly square-bottomed slots.

CAUTION

Always unplug tools before touching blades or cutters.

Removing and Replacing Circular Saw Blades

Carbide circular saw blades will at some time need to be sharpened or replaced. Several clues will help the operator realize that the blade is dull:

- The feeding pressure of the material into the saw will seem to be increasing. The ability to notice this takes some experience with the saw.

- Burn marks are left on the material after the cut. Note that some material, like cherry, burns easily even with a sharp blade. Keep the feed rate constant to reduce burning.

- Pitch and sawdust will build up on the blade. This buildup will cause blade drag, resulting in more heating up than normal.

- Broken or chipped teeth are the most obvious sign of a dull blade. Remember that the teeth are brittle. Protect them from being struck by or dropped on hard objects.

Saw arbor shafts may have a right- or left-hand thread for the nut, depending on which side of the blade it is located. No matter what direction the arbor shaft is threaded, the arbor nut is loosened in the same direction in which the saw blade rotates (*Figure 17–10*). The arbor nut is always tightened against the rotation of the saw blade. This design prevents the arbor nut from loosening during operation.

Figure 17–10 The saw arbor nut is loosened by turning it in the same direction in which the blade rotates.

CHAPTER 18

Miter and Radial Arm Saws

The miter saw and the radial arm saw are similar in purpose. The miter saw is specifically designed for crosscuts at various angles. Materials are placed against a fence, and the saw makes a downward cut. The radial arm saw is designed to do crosscuts and rip cuts. It is highly recommended here to use a table saw for all ripping operations and use only the crosscut operations of a radial arm saw.

POWER MITER SAW

The term power miter saw, also called a *power miter box,* refers to a tool that has a circular saw mounted above the base. It comes in variety of styles and sizes. The saw is operated by pushing it down into the material using a chopping action. These tools are used in many phases of construction. A retractable blade guard provides operator protection and should always be in place when the saw is being used.

Courtesy of DeWalt

CAUTION

- Wear eye and ear protection. Also remember that loose clothing, like gloves, is a hazard to the operator.
- Watch the line of travel of the saw. Only the material being cut should be in the cutting line.
- Do not cross your hands and arms. If the stock is on the left side of the blade, hold it with the left hand, using the right hand to pull the saw. If the stock is on the right side of the blade, hold it with the right hand, using the left hand to pull the saw.
- Bowed stock should be cut with the crown of the board down on the table or against the fence. Otherwise the bow will tend to bind the blade as it is cut.

The simplest of these designs is often referred to as a *chop box* (*Figure 18–1*). It comes in sizes with 10-, 12-, 14-, and 16-inch-diameter blades. The saw cannot be adjusted to cut angles; it cuts only in the up-and-down direction. To cut angles, the material itself must be clamped in place at the desired angle. Chop boxes are designed to cut metal, typically metal studs and pipe. An abrasive blade or a special metal-cutting carbide-toothed blade is used.

Courtesy of DeWalt

Figure 18–1 A chop saw is designed to cut with a chopping action. These saws are designed to cut steel.

Abrasive blades will cut all types and thicknesses of metal and do so with a shower of sparks. Over their life span they shrink in diameter. Carbide blades cut cooler and retain their size over their life span.

The power miter saw designed for wood has more possible adjustments (*Figure 18–2*). The saw blade may be adjusted 45 degrees to the right or left. Some models tilt to the side: left, right, or both directions. It is possible with these saws to cut compound miters. Positive stops are located in the saw angle and the tilt angle adjustments to help the operator cut common angles. Other angles are cut by reading the degree scale and locking in the desired position. Typical sizes include 8½-, 10-, and 12-inch-diameter blades. Cutting width depends on the blade diameter.

In more sophisticated models, the saw slides out on rails to increase the width of the cut. These models are referred to as *sliding compound miter saws* (*Figure 18–3*). The 12-inch models can cut a 4½-inch by 12-inch block at 90 degrees, a 4½-inch by 8½-inch block at 45 degrees, and a 3½-inch by 8½-inch block to a compound 45-degree miter. They can also be adjusted to cut compound angles that exceed 45 degrees. Some will cut up to 60 degrees in either the left or right direction. Most tilt for a bevel cut in both directions.

The sliding compound miter saw allows for cutting large crown moldings with the stock laying flat on the base of the saw. The blade is adjusted to the compound angles using the preset stops on the saw. Check the manufacturer's instructions for the saw to determine crown molding cutting procedures. Some models have a laser attachment that lays a red line on the material before the cut is made, showing where the blade will enter the material (*Figure 18–4*).

A laser light washer is also available to retrofit older saws. It is designed to replace the saw's outer

Figure 18–3 Sliding compound miter saw.

Figure 18–4 Some miter saws have lasers to assist the operator.

Figure 18–2 Power miter box.

blade washer and is activated when the blade spins (*Figure 18–5*).

The operation of all of these types of miter saws is primarily the same. Adjustments are made to the blade travel and then made secure with locking clamps. Material is held firmly against the base and

Figure 18–5 A laser washer may be purchased to retrofit an older saw.

Courtesy of Delta

Figure 18–6 Hold-down clamps are used to keep material from shifting during the cut.

fence of the saw. Some models have material hold-down clamps to ensure the piece does not move during the cut (*Figure 18–6*). The saw is started and the blade eased into the material. With sliding models the saw may be pulled toward the operator then pushed down and then back to cut the material. Rate of feed depends on the material being cut. Listen to the saw to determine the best feed rate. The saw should nearly keep its full-rpm running sound when cutting; that is, it should not slow down much.

Cutting Identical Lengths

When many pieces of identical length need to be cut, clamp or otherwise fasten a *stop block* to the table in the desired location. Cut one end of the stock. Slide the cut end against the stop block. Then cut the other end. Continue in this manner until the desired number of pieces is cut.

Be careful not to bring the stock against the stop block with too much force. Doing so will cause the stop block to move. Any movement results in pieces of unequal length. It is helpful, when cutting many identical lengths, to mark the location of the stop block. Occasionally, look to see if it has moved.

Be careful that no sawdust or wood chips are trapped between the stock and the stop block. This results in pieces shorter than desired. A stop block with a rabbeted end helps prevent sawdust or chips from being trapped between the stop block and the piece being cut (*Figure 18–7*).

RADIAL ARM SAWS

The major function of a radial arm saw (*Figure 18–8*) is crosscutting. The cutting operation is the same as for a power miter box. The stock remains stationary while the saw moves across it. The radial arm saw can also be used for ripping operations.

The size of the radial arm saw is determined by the diameter of the largest blade that can be used. The arm of an industrial saw moves horizontally in a complete circle (*Figure 18–9*). The motor unit tilts to any desired angle and also rotates in a complete circle. The depth of cut is controlled by raising or lowering the arm. This flexibility allows practically any kind of cut to be made. Extreme care should be employed to follow all setup procedures exactly.

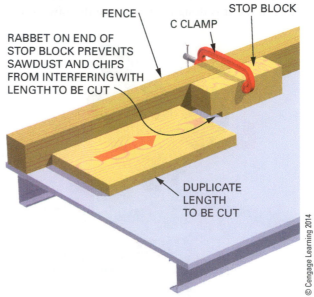

FENCE
C CLAMP
STOP BLOCK
RABBET ON END OF STOP BLOCK PREVENTS SAWDUST AND CHIPS FROM INTERFERING WITH LENGTH TO BE CUT
DUPLICATE LENGTH TO BE CUT

© Cengage Learning 2014

Figure 18–7 Technique for making a fence stop where sawdust will not interfere.

Figure 18–8 The radial arm saw.

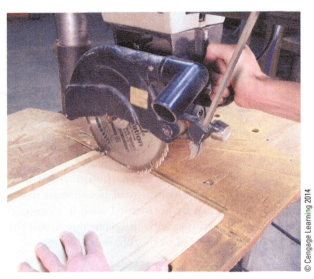

Figure 18–10 Straight crosscutting using the radial arm saw. Note a blade guard is removed for clarity.

Figure 18–9 The arm of the radial saw moves horizontally in a complete circle.

Crosscutting

For straight crosscutting, make sure the arm is at right angles to the fence. Adjust the depth of cut so that the teeth of the saw blade are about $\frac{1}{16}$ inch below the surface of the table. With the saw all the way back and all guards in place, hold the stock against the fence. Make the cut by bringing the saw forward and cutting to the layout line (*Figure 18–10*).

When crosscutting stock thicker than the capacity of the saw, cut through half the thickness. Then turn the stock over and make another cut.

> ### CAUTION
> - Pull the saw gently into the material. It will tend to feed itself into the stock, causing it to jam into the stock, so be prepared to resist this force as you pull the saw into the stock.
> - Remember, dull blades tend to self-feed, that is, run into the cut fast, more than sharp blades.

When the cut is complete, return the saw to the starting position, behind the fence. Turn off the power. As a safety precaution, all saws should automatically return to the retracted rest position if the operator should let go of the saw.

Ripping

Although it is possible to rip lumber using the radial arm saw, it should be avoided if a table saw can be used. See Chapter 19. The setup for a radial arm saw takes some time, and the ripping setup must be followed very carefully to protect both the operator and the material being cut.

> ### CAUTION
> Feed the stock against the rotation of the blade when ripping on a radial arm saw. Feeding stock in the wrong direction may cause it to be pulled from the operator's hands and through the saw with great force. This could cause serious or fatal injury to anyone in its path.

CHAPTER 19

Table Saws and Other Shop Tools

The table saw is one of the most frequently used woodworking power tools. In many cases, it is brought to the job site when the interior finish work begins. It is a useful tool because so many kinds of work can be performed with it. Common table saw operations, with different jigs to aid the process, are discussed in this chapter.

TABLE SAWS

The size of the table saw (*Figure 19–1*) is determined by the diameter of the saw blade. It may measure up to 16 inches. A commonly used table saw on the construction site is the 10-inch model.

The blade is adjusted for depth of cut and tilted up to 45 degrees by means of handwheels. A rip fence guides the work during ripping operations. A miter gauge is used to guide the work when cutting square ends and miters. The miter gauge slides in grooves in the table surface. It may be turned and locked in any position up to 45 degrees.

A guard should always be placed on the blade to protect the operator. Exceptions to this include some table saw operations like dadoes, rabbets, and cuts where the blade does not penetrate the entire stock thickness. A general rule is if the guard can be used for any operation on the table saw, it should be used.

Ripping Operations

For ripping operations, the table saw is easier to use and safer than the radial arm saw. To rip stock to width, adjust the rip fence to the desired width. To check the accuracy of the scale under the *rip fence,* measure from the rip fence to the point of a saw tooth set closest to the fence. Lock the fence in place. Adjust the height of the blade to about $\frac{1}{4}$ inch above the stock to be cut. With the stock clear of the blade, turn on the power.

> **CAUTION**
>
> - Stand to either side of the blade's cutting line. Avoid standing directly behind the saw blade. Make sure no one else is in line with the saw blade in case of kickback.
> - Be prepared to turn the saw blade off at any moment should the blade bind.
> - A splitter and antikickback devices should be used when the cutting operation allows it.

Hold the stock against the fence with the left hand. Push the stock forward with the right hand, holding the end of the stock (*Figure 19–2*). As the end approaches the saw blade, let it slip through

Figure 19–1 A table saw.

© Cengage Learning 2014

Figure 19–2 Using the table saw to rip lumber (guard has been removed for clarity).

© Cengage Learning 2014

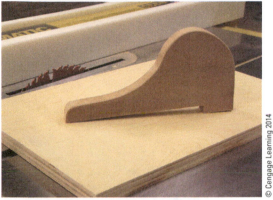

Figure 19–3 Use a push stick to rip narrow pieces (guard has been removed for clarity).

the left hand. Remove the left hand from the work. Push the end all the way through the saw blade with the right hand, if the stock is of sufficient width (at least 5 inches wide). If the stock is not wide enough, use a *push stick* (*Figure 19–3*).

CAUTION

- Make sure the stock is pushed all the way through the saw blade. Leaving the cut stock between the fence and a running saw blade may cause a kickback, injuring anyone in its path.
- Use a push stick, especially when ripping narrow pieces.
- Keep small cutoff pieces away from the running blade. Do not use your fingers to remove cutoff pieces; use the push stick or wait until the saw has stopped to do so.
- Always use the rip fence for ripping operations. Never make freehand cuts.
- Never reach over a running saw blade.

Bevel Ripping

Bevel ripping is done in the same manner as straight ripping except that the blade is tilted (*Figure 19–4*). The blade is adjustable from 0 to

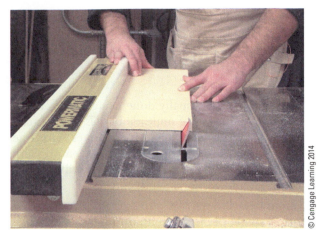

Figure 19–4 When bevel ripping, the blade is tilted (guard has been removed for clarity).

45 degrees. Take care not to let the blade touch the rip fence. Also keep the stock firmly in contact with the table. If the stock is allowed to lift off from the table during the cut, the width of the stock will vary.

Taper Ripping

Tapered pieces (one end narrower than the other) can be made with the table saw by using a taper ripping jig. The jig consists of a wide board with the length and amount of taper cut out of one edge. The other edge is held against the rip fence. The stock to be tapered is held in the cutout of the jig. The taper is cut by holding the stock in the jig as both are passed through the blade (*Figure 19–5*).

By using taper ripping jigs, tapered pieces can be cut according to the design of the jig. A handle on the jig makes it safer to use. Also, if the jig is the same thickness as the stock to be cut, then the

Figure 19–5 Using a taper ripping jig to cut identical wedges (guard has been removed for clarity).

cutout section of the jig can be covered with a thin strip of wood to prevent the stock from flying out of the jig and back toward the operator.

Rabbeting and Grooving

Making *rabbets* is usually done using a dado head, which is comprised of multiple blades and cutters. This tool makes fast, accurate grooves with one pass (*Figure 19–6*). Make sure firm even pressure is applied to the table and fence. This will provide consistent and accurate depth cuts.

Rabbets and narrow *grooves* can also be made with a single blade when a dado head is unavailable or when only a few are needed. The operation takes two settings of the rip fence, and two passes through the saw blade are required (*Figure 19–7*). For narrow grooves, one or more passes are required. Move the rip fence slightly with each pass until the desired width of groove is obtained.

Crosscutting Operations

For most crosscutting operations, the *miter gauge* is used. To cut stock to length with both ends squared, first check the miter gauge for accuracy. Hold a framing square against it and the side of the saw blade. Usually the miter gauge is operated in the left-hand groove. The right-hand groove is used only when it is more convenient.

Square one end of the stock by holding the work firmly against the miter gauge with one hand while pushing the miter gauge forward with the other hand. Measure the desired distance from the squared end. Mark on the front edge of the stock. Repeat the procedure, cutting to the layout line (*Figure 19–8*).

Figure 19–7 Rabbeting an edge using a single saw blade by making two passes.

Figure 19–6 Cutting a groove with a dado head installed on the table saw.

Figure 19–8 Using the miter gauge as a guide to crosscut (guard has been removed for clarity).

Cutting Identical Lengths

When a number of identical lengths need to be cut, first square one end of each piece. Clamp a stop to an *auxiliary wood fence* installed on the miter gauge. Place the square end of the stock against the stop block. Then make the cut. Slide the remaining stock across the table until its end comes in contact with the stop block. Then make another cut. Continue in this manner until the desired number of pieces is cut (*Figure 19–9*).

Another method for cutting identical lengths uses a block clamped to the rip fence in a location such that once the cut is made, there is clearance between the cut piece and the rip fence. The fence is adjusted so that the desired distance is between the face of the stop block and saw blade.

Square one end of the stock. Slide the squared end against the stop block. Then make a cut. Continue making cuts in this manner until the desired number of pieces is obtained (*Figure 19–10*).

Figure 19–9 Cutting identical lengths using a stop block on an auxiliary fence of the miter gauge (guard has been removed for clarity).

Figure 19–10 Using the rip fence and block as a stop to cut identical lengths (guard has been removed for clarity).

Mitering

Flat miters are cut in the same manner as square ends, except the miter gauge is turned to the desired angle. *End miters* are made by adjusting the miter gauge to a square position and making the cut with the blade tilted to the desired angle.

Compound miters are cut with the miter gauge turned and the blade tilted to the desired angles.

A *mitering jig* can be used with the table saw. **Procedure 19–A** shows the construction and use of the jig. Use of such a jig eliminates turning the miter gauge each time for left- and right-hand miters.

Dadoing

Dadoing is done in a similar manner as crosscutting except a dado set is used (*Figure 19–11*). The dado set is used only to cut *partway* through the stock thickness.

TABLE SAW AIDS

A very useful aid is an *auxiliary fence*. A straight piece of ¾-inch plywood about 12 inches wide and as long as the ripping fence is screwed or bolted to the metal fence. When cuts must be made close to the fence, the use of an auxiliary wood fence prevents the saw blade from cutting into the metal fence. Also, the additional height provided by the fence gives a broader surface to steady wide work when its edge is being cut. The auxiliary wood fence also provides a surface on which to clamp feather boards. An auxiliary fence is also useful when attached to the miter gauge.

StepbyStep Procedures

Procedure 19–A Making a Table Saw Jig to Cut 45-Degree Miters

STEP 1 FASTEN WOOD STRIPS, SIZED, SPACED, AND LOCATED TO FIT IN MITER GAUGE GROOVES, TO UNDERSIDE OF PLYWOOD. RUB WAX ON PLYWOOD AND WOOD STRIPS.

STEP 2 WITH SAW BLADE BELOW TABLE SURFACE PLACE JIG SO WOOD STRIPS FIT IN GROOVES AND SLIDE BACK AND FORTH. JIG SHOULD MOVE EASILY, BUT WITH NO PLAY. MAKE A SAW CUT ABOUT HALF-WAY ACROSS THE WIDTH.

STEP 3 HOLD FRAMING SQUARE AT 12 AND 12 ON IN-FEED EDGE WITH HEEL CENTERED ON THE SAW CUT. MARK LAYOUT LINES ON OUTSIDE EDGES OF SQUARE.

STEP 4 ATTACH GUIDE STRIPS SO EDGES ARE TO LAYOUT LINES.

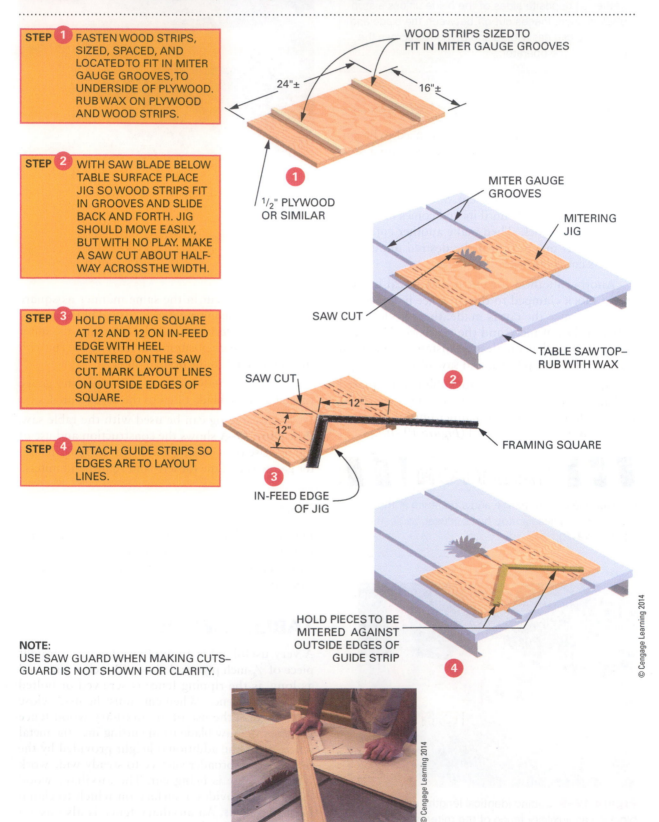

WOOD STRIPS SIZED TO FIT IN MITER GAUGE GROOVES

24"±   16"±

1

1/2" PLYWOOD OR SIMILAR

MITER GAUGE GROOVES

MITERING JIG

SAW CUT

TABLE SAW TOP– RUB WITH WAX

2

SAW CUT

12"

12"

FRAMING SQUARE

3

IN-FEED EDGE OF JIG

HOLD PIECES TO BE MITERED AGAINST OUTSIDE EDGES OF GUIDE STRIP

4

NOTE:
USE SAW GUARD WHEN MAKING CUTS– GUARD IS NOT SHOWN FOR CLARITY.

© Cengage Learning 2014

© Cengage Learning 2014

Figure 19–11 Cutting dadoes using a dado head.

Feather boards are useful aids to hold work against the fence as well as down on the table surface during ripping operations (*Figure 19–12*). Feather boards may be made of 1 × 6 nominal lumber, with one end cut at a 45-degree angle.

Saw cuts are made in this end about $\frac{1}{4}$ inch apart. This gives the end some spring and allows pressure to be applied to the piece being ripped while also allowing the ripped piece to move.

On the Job

Feather boards need only be installed to touch the stock. If it is too tight, the cutting operation becomes unsafe due to the increased force needed to feed the stock.

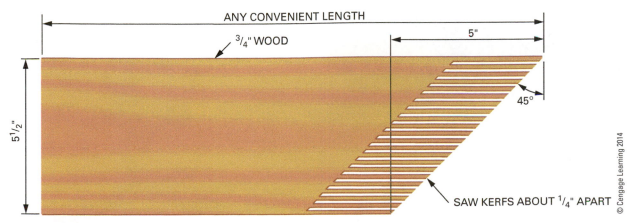

ANY CONVENIENT LENGTH

$\frac{3}{4}$" WOOD

5"

45°

$5\frac{1}{2}$"

SAW KERFS ABOUT $\frac{1}{4}$" APART

Figure 19–12 (A) Feather boards are useful aids to hold work during table saw operations (guard has been removed for clarity). Note that the locations of the feather boards do not cause the saw blade to bind. (B) Typical feather board design.

Band Saws

The band saw is a tool designed to cut irregular shapes in various thicknesses of material (*Figure 19–13*). It is usually operated freehand, moving the stock to follow curved lines. It can also be set up to use a miter gauge for straight cuts. In addition, the band saw is well suited to resaw thick boards into thinner ones (*Figure 19–14*).

The saw is named for its cutting blade, which is a continuous thin steel band with teeth. Band saws have two large diameter wheels, one above the table and one below (*Figure 19–15*). These wheels rotate the band into a cutting action that is always down, toward the table. The band is held straight for cutting by guide blocks and wheels. A set of guides is located above and below the table. The upper guide may be adjusted in height and should be positioned about $\frac{1}{4}$ inch above the stock being cut (*Figure 19–16*). This will keep the blade straight during the cut.

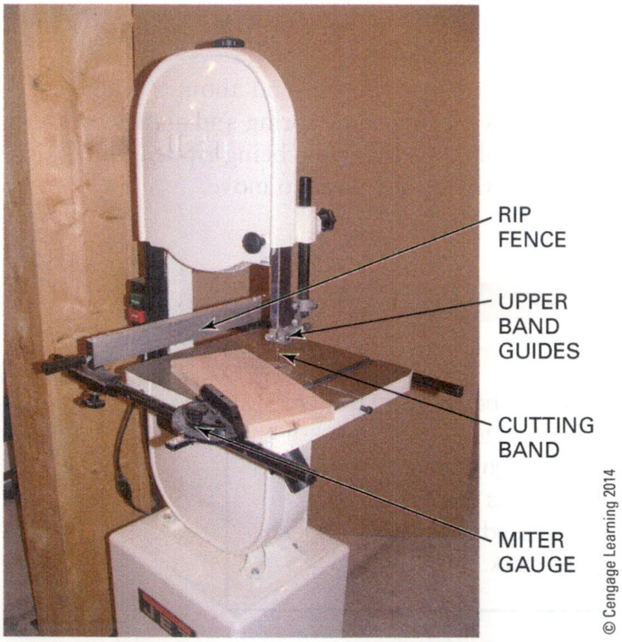

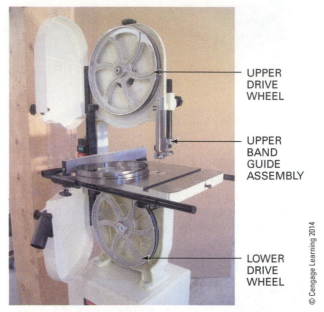

RIP FENCE

UPPER BAND GUIDES

CUTTING BAND

MITER GAUGE

© Cengage Learning 2014

Figure 19–13 The band saw can saw curves, irregular shapes as well as straight cuts.

UPPER DRIVE WHEEL

UPPER BAND GUIDE ASSEMBLY

LOWER DRIVE WHEEL

© Cengage Learning 2014

Figure 19–15 The band saw has a continuous cutting band spun on two large wheels.

© Cengage Learning 2014

Figure 19–14 The band saw may be used to resaw boards into thinner pieces.

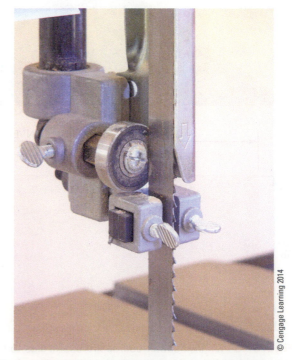

© Cengage Learning 2014

Figure 19–16 The guides help keep the band in place during cutting operations.

Band saw sizes are determined by the diameter of the drive wheels. They typically range from 10 inches to 18 inches in diameter. Band saw blades have varying numbers of teeth per inch and band widths ranging from $\frac{1}{8}$ inch to $\frac{3}{4}$ inch. This allows the bands to cut differently for the work needed. Fast rough cutting is done with wide coarse teeth while fine work requires a narrow, fine-toothed blade. Metal may also be cut with a band saw when set up with a metal cutting blade spinning at a slower rpm.

CAUTION

Take care to avoid backing out of long cuts. Backing out could cause the band to come off the drive wheels.

The width of the band also affects the turning radius on the cut (*Figure 19–17*). A narrow band on $\frac{1}{8}$ inch will allow a $\frac{1}{4}$ inch radius cut, where the $\frac{3}{4}$-inch width will cut only along a 2 $\frac{1}{2}$-inch radius.

Jointers

A *jointer* is used to smooth stock surfaces to remove the marks made by sawing (*Figure 19–18*). Its size is determined by the length of the cutting knives or roughly the width of the table, typically 4, 6, or 8 inches. A spring loaded safety shield swings in place to cover the portion of the cutter head not being used.

The jointer is constructed with two tables: infeed and outfeed. The two tables are always kept parallel to each other. The offset height of the tables determines the depth of the cut (*Figure 19–19*).

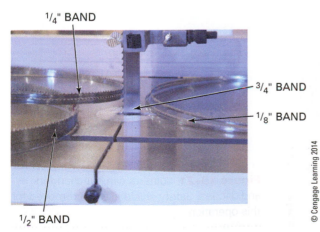

$\frac{1}{4}$" BAND
$\frac{3}{4}$" BAND
$\frac{1}{8}$" BAND
$\frac{1}{2}$" BAND

© Cengage Learning 2014

Figure 19–17 Bands come in various widths.

© Cengage Learning 2014

Figure 19–18 Jointers are used to smooth board surfaces.

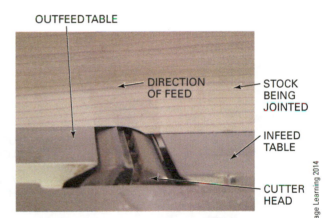

OUTFEED TABLE
DIRECTION OF FEED
STOCK BEING JOINTED
INFEED TABLE
CUTTER HEAD

© Cengage Learning 2014

NOTE: STOCK TOUCHES BOTH TABLES DURING THE CUTTER OPERATION

Figure 19–19 The outfeed table is set to the height of the cutters and the infeed table is adjustable for depth of cut.

The infeed table is adjustable in height by turning a wheel under the table. The outfeed table is stationary and set at a height equal to the extreme cutting height of the cutting head. This allows the jointed stock to have full support after it has passed over the cutting head.

CAUTION

Check the manufacturers recommendations for the maximum allowable depth of cut.

Safe operation requires that the operator secure the stock with two hands. Begin with the stock resting on the infeed table and fence.

Concentrate on keeping the stock against the table and fence during the entire pass, preventing it from wobbling. The slowly ease the stock toward the safety shield and into the cutter head (*Figure 19–20*). At some point halfway through the cutting pass, the operator's hands must shift to secure the stock to the outfeed table and fence. Take care to keep the stock moving and firmly held against the fence.

> **CAUTION**
>
> Avoid using your fingers to hold the stock directly over the cutters. Lift your hands over the area of the cutter head from infeed side to outfeed side. Accidents have happened in which cutters have kicked the stock out of the operator's hands. Always think ahead.

Other operations of the jointer include chamfers and rabbets (*Figure 19–21*). The fence may be tilted to joint bevels of various angles. Rabbets are performed by adjusting the fence over the cutter head such that only a small portion of it is available for cutting. Be sure to replace the safety shield after performing rabbet cuts.

Drill Press

A *drill press* in a precision drilling tool consisting of a base, column, table, and head (*Figure 19–22*). The base and column support the table and head. Power and control of the drill press are done from the head while the table supports the material being worked.

Bit speed is adjustable by typically changing the belt location on the cone-shaped pulleys. The smaller pulley on the motor will cause the bit to rotate slower. Wood is best bored by using a high

PUSH BLOCK

SAFETY SHIELD

© Cengage Learning 2014

Figure 19–20 Safe operation requires full attention to the machine and slow steady feeding of the stock.

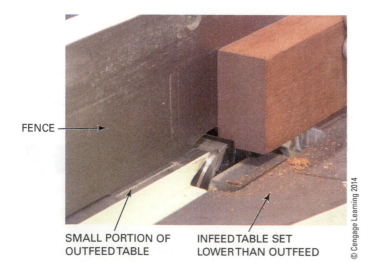

FENCE

SMALL PORTION OF OUTFEED TABLE

INFEED TABLE SET LOWER THAN OUTFEED

© Cengage Learning 2014

Figure 19–21 Jointers can cut chamfers and dadoes. Safety shield is often removed for this operation.

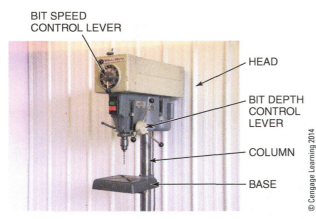

BIT SPEED CONTROL LEVER

HEAD

BIT DEPTH CONTROL LEVER

COLUMN

BASE

© Cengage Learning 2014

Figure 19–22 A drill press is used to make repetitive and precise holes.

speed with light pressure on the bit. Metal is best drilled using a very slow speed with higher pressure on the bit. Special cutting oil is recommended when drilling metal to keep the bit cool and speed up the removal of material in the hole.

> **CAUTION**
>
> Take care to secure the work to the table before drilling. If the bit binds in the hole the entire piece may spin out of control.

Shaper

A shaper is used to make molding and decorative edges in irregular surfaces (*Figure 19–23*). It is a heavy-duty router attached upside down to a base. It has a removable rip fence and vertical shaft

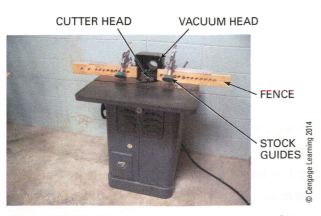

CUTTER HEAD   VACUUM HEAD

FENCE

STOCK GUIDES

© Cengage Learning 2014

Figure 19–23 Shapers are used to make decorative edges in wood.

projecting up through the center of the table that adjusts to the desired height. The spindle shaft is typically $\frac{1}{2}$ inch in diameter that spins about 9,000–18,000 RPM. This high speed allows for a smooth finished product. There are literally hundreds of shapes and styles of bits available.

The stock may be tooled by using the fence as a guide or by using a collar on the bit. The fence provides a quick set-up for straight edges such as raised panel doors or trim. A collar may be located on the bit itself, which rides along any irregular shape. This allows a decorative edge by cutting only the amount designed in a bit.

> **CAUTION**
>
> Great care must be taken when setting up the shaper. Check the cutters carefully to be sure they are sharp and in proper working condition. If a tip should break off while the cutter is spinning, the result could cause serious injury.

Bench Sander

Bench sanders in the form of a drum, disk, and belt are designed to smooth most any shaped piece of work (*Figure 19–24*). The material is moved over the sander to smooth the remover material fast.

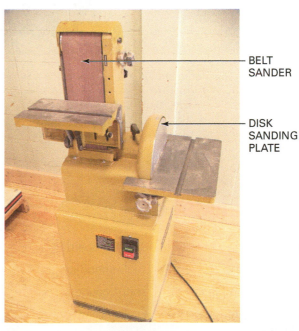

BELT SANDER

DISK SANDING PLATE

© Cengage Learning 2014

Figure 19–24 Bench sanders are used to sand material quickly and easily.

DECONSTRUCT THIS

Carefully study *Figure 19–25* and think about what is wrong and/or what is right. Consider all possibilities. What construction practice or method is different in your area of the country?

Figure 19–25 This photo shows a carpenter using a table saw.

© Cengage Learning 2014

KEY TERMS

bevel ripping

carbide blade

combination blade

crosscut circular saw blade

feather boards

high-speed steel blade

miter gauge

power miter saw

radial arm saw

rip fence

ripsaw blade

table saw

taper ripping jig

tungsten carbide-tipped blade

REVIEW QUESTIONS

Select the most appropriate answer.

1. To use a power tool properly, the operator should always

   a. wear eye protection.
   b. wear ear protection.
   c. understand the manufacturer's recommended instructions.
   d. all of the above.

2. The most frequently used circular saw blade in general carpentry is the

   a. combination planer blade.
   b. combination carbide-tipped blade.
   c. square-grind carbide-tipped blade.
   d. ripsaw blade.

3. The alternate top bevel grind, carbide-tipped circular saw blade is designed to

   a. rip solid lumber.
   b. crosscut solid lumber.
   c. cut in either direction of the grain.
   d. cut green lumber of heavy dimension.

4. A device used with stationary saws to make repeated cuts of the same length is called a

   a. rabbet.
   b. feather board.
   c. stop block.
   d. all of the above.

5. The saw best suited to rip material is the

   a. power miter box.
   b. radial arm saw.
   c. table saw.
   d. all of the above.

6. When using a table saw, it is a good idea to

   a. stand away from the line of the blade.
   b. know the location of the shut-off switch.
   c. keep wood scraps away from the blade.
   d. all of the above.

7. The guide of a table saw used for cutting material with the grain is called a

   a. rip fence.
   b. miter gauge.
   c. tilting arbor.
   d. ripping jig.

8. The tool designed to safely hold a piece of stock being cut with a table saw is a

   a. miter gauge.
   b. push stick.
   c. feather board.
   d. all of the above.

9. A dado is a wide cut partway through the thickness of the material and

   a. across the grain.
   b. with the grain.
   c. in either direction of the grain.
   d. close to the edge.

10. The table saw tool that should not be used at the same time as a rip fence is the

    a. push stick.
    b. miter gauge.
    c. dado head.
    d. none of the above.

11. Band saw sizes are determined by

    a. drive wheel diameters.
    b. depth of cut.
    c. width of blade.
    d. length of the band cutting blade.

12. Band saws are best suited to

    a. rip boards.
    b. crosscut stock.
    c. cut a compound miter.
    d. cut irregular shapes.

13. To safely operate a jointer it is important to

    a. keep stock moving smoothly along table.
    b. maintain pressure of stock to table and fence.
    c. avoid holding stock directly above cutters.
    d. all of the above.

14. The portion of a jointer that is always adjusted level with the cutter is the

    a. in-feed table.
    b. out-feed table.
    c. fence.
    d. safety shield.

15. The stationary power tool best suited to make a panel for a raised panel door is the

    a. jointer.
    b. bench sander.
    c. drill press.
    d. shaper.

UNIT 8
Jobsite Safety and Scaffolds

CHAPTER 20 Jobsite Safety and Construction Aids
CHAPTER 21 Scaffolds

The essential tools for jobsite safety involve attitude, knowledge, and skill. Workers must have a good attitude, be ready to learn and use their knowledge, and sharpen their skills every day in order for a jobsite to be safe. This affects all people on a jobsite.

Scaffolding and staging are terms that describe temporary working platforms. They are constructed at convenient heights above the floor or ground. They help workers perform their jobs quickly and safely. This unit includes information on the safe erection of various kinds of working platforms and the building of several construction aids.

OBJECTIVES

After completing this unit, the student should be able to:

- describe the characteristics of jobsite safety.
- identify personal safety devices use in construction.
- describe the function of various tools and equipment commonly found on a jobsite.
- describe the safe use of ladders, ladder jacks, and sawhorses.
- identify safety concerns with jobsite electricity.
- build a ladder and sawhorse.
- erect and dismantle metal scaffolding in accordance with recommended, safe procedures.
- safely set up, use, and dismantle pump jack scaffolding.
- name the parts of wood single-pole and double-pole scaffolding.
- build safe staging using roof brackets.

SAFETY REMINDER

Scaffolds are involved with most job site accidents. Know how to safely erect, use, and dismantle scaffolding.

Always remember where you are when working at elevated levels.

CHAPTER 20

Jobsite Safety and Construction Aids

Proper jobsite safety hinges on attitude, skill, and knowledge (A.S.K.). Never be afraid to ask questions when on the construction site. Your knowledge of safety rules and your safety skills as a tradesperson are required for your health and the safety of your coworkers. Everyone has the right to go home at the end of a shift as when they arrived.

JOBSITE ATTITUDE

Your attitude affects your work and the people around you. Jobsite attitude is created by every person on the job. Each person brings with them their viewpoint and approach to solving problems as they come to the job. All of these perspectives are blended into a "vibe" that pulses throughout the jobsite. But what does this mean really?

Attitude includes something as simple as using the correct tool for the job. It often seems quicker to use a tool because it is nearby, even if it is the wrong tool. The proper tool for a task is always safer and easier to use, even if it requires a few minutes to retrieve. For example, using a screwdriver from your pocket to pry apart materials may save a step, but the prybar in your tool box works better, more easily and more safely—and the material often comes apart with less damage.

Many times when a piece of material needs a little persuasion to go into place, the desire is to force it. Forcing is rarely a good idea. Recutting the material is good practice for doing a finer job. Also when tapping material into position, use a hammer. Not only do hammers work better than your hand, they do not get arthritis after a lifetime of hitting things.

Sometimes things do not go as planned. Often the same mistake is repeated two or three times in a row. Wasted material may start to accumulate. Do not give in to the alluring temptation of losing your temper. Throwing things will not help and is dangerous. Take the time to stop, breathe, and reflect on a different path. If you are able to do this, you are a jobsite leader.

Cell phones are an excellent way to communicate on the job. Information is easy to transmit to other workers. Smart phones offer a fast solution to get technical information on a new material, for example. Unfortunately they can also be a jobsite hazard. Many states have banned the use of cell phones while driving because they can distract the driver from the road. Jobsites are a worse place to be distracted. Do not let anything keep you from paying attention to safe behaviors and actions.

The safety of all people on the job depends on the attitude of all people on the job. Safety is like silence. Silence will not happen unless everybody involved is quiet. A safe jobsite requires everyone to pay attention and contribute to proper jobsite protocol. Every person on a jobsite has responsibilities.

Know where you are at all times. Visualize your surroundings, be aware of the tools and people nearby. For example, never work so fast that you forget you are on a scaffold (*Figure 20–1*). Place tools down thoughtfully.

Noise on the jobsite is normal (*Figure 20–2*). Jobsite sounds have a pattern and a rhythm.

Figure 20–1 It is important to remember where you are at all times.

FIGURE 20–2 Jobsite noise is to be expected. It is important to recognize changes from normal.

Almost every tool used in construction today makes noise. A safe worker listens to the noise tools make. Tools will often make a different sound when they are failing to perform as designed. Being alert to the tool's sounds is a first step in being safe.

Safety Responsibilities

The Occupational Safety and Health Administration (OSHA) was created in 1970 to assure safe and healthful working conditions for all workers. They set and enforce standards and provide training, outreach, education, and assistance. OSHA representatives can be found on any jobsite. Check the OSHA website, www.osha.gov and click the "regulations" tab for the exact safety standards. Many OSHA regulations refer to ANSI (American National Standards Institute) to describe the specific details for safety devices.

OSHA sees safety as the responsibility of all personnel on the jobsite, employer and employees alike. Some of the responsibilities of the employers are as follows:

- Provide a workplace free from serious recognized hazards, and comply with standards, rules, and regulations issued under the OSHA Act.

- Examine workplace conditions to make sure they conform to applicable OSHA standards.

- Make sure employees have and use safe tools and equipment and properly maintain this equipment.

- Establish or update operating procedures and communicate them so that employees follow safety and health requirements.

- Not to discriminate against employees who exercise their rights under the OSHA act to seek a safer work place.

- Correct cited violations by the deadline set in the OSHA citation, and submit required abatement verification documentation.

Some of the responsibilities of the employees are as follows:

- Comply with the standards, rules, regulations, and orders issued by his/her employer.

- Use safety equipment, personal protective equipment, and other devices and procedures provided or directed by the employer.

- Report unsafe and unhealthful working conditions to appropriate officials.

Falls are the number one cause of injuries on a jobsite. It should come as no surprise then that scaffolding and fall protection are most often cited by OSHA for problems.

Figure 20-3 Teamwork makes big jobs easier.

Teamwork

There is no replacement for teamwork on thejob site. People working together can outperform individuals working alone (*Figure 20–3*). An example of teamwork is when one person holds material to be fastened or cut for another. The difference between a team and a group of individuals is more dramatic when each member of the team anticipates the next move of the other. For example, while holding a board for another, a good team member watches out for safety and asks questions like the following: Can this be done more safely? Does a horse or ladder have to be moved into a better position? Will the tool being used cut only the desired material? Are the actions of the team putting others at risk? Safety must be continuous and ongoing.

Ever-changing Worksite

Jobsites are dangerous places to work. One of the reasons for this is that they change. From week to week, one can see how one phase or trade will finish and another trade will begin (*Figure 20–4*).

Figure 20-4 Changes happen fast on the jobsite.

The masons finish up and move out allowing the framers to move in, or finish stair carpenters begin work on stairs which creates a fall hazard that did not exist before. Change is part of the jobsite.

Changes also happen from moment to moment. A saw horse or ladder may be removed from the room, or an unsafe situation is created while a worker's back is turned. Workers must always be aware of what is going on around them. Communication between workers is essential on the job. Let others know when something changes, no matter how small, no matter who caused the change.

PERSONAL PROTECTIVE EQUIPMENT

Everyone on a jobsite must work and preform their tasks in a safe manner. This means operating tools and equipment appropriately and according to manufacturer's instructions. But safety begins with personal protective equipment.

Eye Protection

Wearing eye protection prevents injuries from flying particles and chemical splashes. They should be worn anytime work will cause the possibility of flying foreign objects, such as cutting, grinding, and nailing (*Figure 20–5*).

Safety glasses and goggles are made of impact-resistant plastic, side shields, and reinforced frames. In some instances, a face shield will cover the entire face.

Laser light from laser leveling devices can cause permanent eye damage. Special safety glasses designed for the particular laser light frequency should be worn, as normal jobsite safety glasses will not protect your eyes from the laser light. Never look directly into a laser beam.

Welders and metal workers must shield themselves with special face and eye protection. These eye shields are very dark to protect from extreme bright radiation. When on a jobsite where welders are working, do not look directly at their work. It does not matter how far away you are, unprotected viewing of welding light will cause eye injury (*Figure 20–6*).

Ear Protection

Jobsite workers are exposed to varying levels of noise. If you have to raise your voice for someone 3 feet away to hear you, the site is noisy, and ear protection should be worn. Excessive levels and/or extended periods of noise will result in permanent hearing loss. Hearing protection must be worn when noise levels or durations cannot be reduced.

Ear plugs made of moldable rubber, foam, or plastic are inserted into the ear. Ear muffs surround the ear and have soft cups that surround the ears. Some ear muffs are electronic, designed to cancel out impulse noises while allowing normal sounds to be heard.

OSHA has set limits of sound levels and the permissible exposure time limit. The perceived loudness doubles with an increase in 10 dB (decibels). Circular saws for example seem twice as loud

Figure 20–5 Safety glasses should be worn at all times.

Figure 20–6 Never look at welding arcs, even from a distance.

Hours per day	Sound level (dB)	Examples
8.0	90	Backhoes
6.0	92	Diesel Trucks
4.0	95	Hand Power tools
3.0	97	Bulldozers
2.0	100	Circular saws
1.5	102	Concrete saws
1.0	105	Jack Hammers,
0.25	115	Sandblasting, Chainsaws, Rock concerts

Figure 20–7 Ear protection requirements depend on the amount of time exposed to various jobsite noises.

as a backhoe. The louder the sound the less time a worker should be exposed to the sound without wearing ear protection (*Figure 20–7*).

Head Protection

Employees must wear head protection if any objects might fall from above. Employees should also wear head protection if they might bump their heads against fixed objects, such as exposed pipes or beams, or if there is a possibility of accidental head contact with electrical hazards.

Hard hats are usually made of light weight plastic outer shell with shock absorbing interior head straps (*Figure 20–8*). They come in different classes depending on the potential risk of the wearer, Class A, B, and C. Class A and B hats are both designed for falling object protection but Class B hats are better at electrical hazard protection. Class C hats are designed for minimum impact

protection, but no electrical protection. Bump hats are designed to protect in low headroom areas, not to protect from falling objects or electrical. It is important to check which hat you are wearing.

Hard hats must be cared for to make sure they perform as expected. They should be adjusted to fit properly. The hat should not bind, slip, or fall off, or irritate the skin. Hard hats may be worn with the bill backwards if indicated on the manufacturers label. Although hard hats should never be painted they may be purchased with a decorative overlay to the helmet. Ear protection may also be attached via slots in the rim of the hat (*Figure 20–9*).

Inspect them daily looking for cracks, holes, or defects. Do not drill holes, paint, or place labels on the hat. Do not store the hat in direct sun light since heat and sunlight can damage them. If it sustains an impact, it should be replaced.

Figure 20–8 Hard hats protect from head injuries and electrical contact.

Figure 20–9 Hard hats are available with accessories.

Respiratory Protection

Air quality on a jobsite varies according to the work being done. A respirator is required to protect the workers when air borne hazards exist (*Figure 20–10*). Dry cutting of dust from masonry work produces silica dust that may cause lung diseases years later. Some solvents can produce vapors that are known to cause cancer.

Figure 20–11 Jobsite clothing should fit well and be made of durable materials.

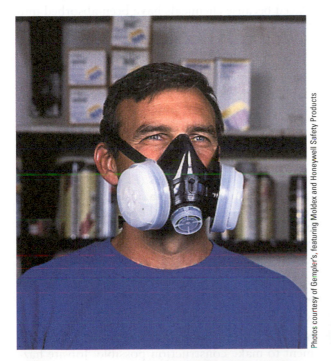

Employers are expected to provide respirators that are applicable and suitable for the work situation. The best type of respirator depends on the conditions. Read the packaging to determine if the respirator will protect under the present work conditions.

Protective Clothing

There are many varieties of protective clothing to suit the work conditions (*Figure 20–11*). In general jeans or denim are well suited for construction work. They offer protection from dust, abrasions, changing temperatures, and rough surfaces. Leather works well against heat and flames such as welding and metal cutting.

Clothes should be snug fitting to the body, not loose or hanging off, particularly at the waist. Loose, they will catch on material and equipment potentially causing injury. Jewelry of any kind, including rings and watches, should not be worn while working. They can catch and direct sharp objects into the body and they often conduct electricity.

When working in hot regions it is important to protect from the sun. Covering the skin with long sleeves and pants is better than working with bare skin (*Figure 20–12*). Use sun blocks, dark safety glasses and wide brimmed hard hats. In extreme situations avoid working during the times from 10 a.m. to 4 p.m. which is the most intense time of the day.

Hand and Foot Protection

A wide assortment of gloves is available today to protect hands from jobsite hazards. The type of

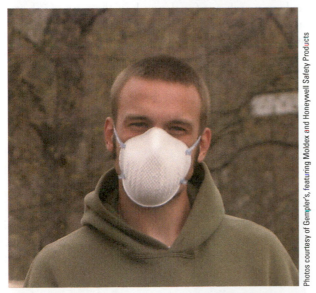

Photos courtesy of Gempler's, featuring Moldex and Honeywell Safety Products

Figure 20–10 Respirators are designed for different airborne hazards.

Figure 20–12 Clothes protect the skin from the sun.

Figure 20–13 Gloves offer protection from jobsite hazards to the hands.

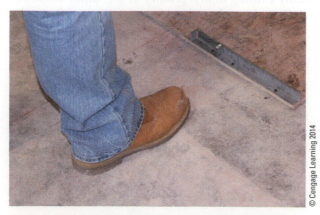

Figure 20–14 Work boots should have stiff soles and tread appropriate for the conditions.

work being done will affect the best choice of gloves to use (*Figure 20–13*). Leather gloves, for example, are good for general rough work and moderate heat. When working with diesel, nitrile gloves are recommended, not rubber gloves. It is important to check the glove manufacturer's recommendations against any particular jobsite hazard.

For gloves to provide maximum protection they should be in good condition. Gloves that are stiff or discolored may indicate they need to be replaced because chemicals have been absorbed into the glove material. Small pin holes leaks can allow electrical current to pass through the glove.

Work boots should be worn at all times on a construction site. They must meet minimum ANSI standards for compression and impact. Soles should be stiff to resist penetrations from nails and sharp objects. Steel toes offer good impact protection.

Work boots are not made for all situations. Some offer good traction in dirty environments while poor in oily situations (*Figure 20–14*). Some insulate from electrical hazards while some offer a grounding to reduce the risk of static charge buildup. It is important to check the manufacturer's recommendations before purchasing.

JOBSITE ACTIVITY

Activity around a jobsite is usually busy and often hectic. Each site requires different types of equipment to make construction possible. Jobsite hazards to watch out for depend on the type of jobsite activity (*Figure 20–15*). To be safe, everyone must watch where they are and where they are going.

Figure 20–15 Jobsites are often very busy with many activities.

Excavation Equipment

Some sites require deep excavations, others only shallow trenches. Many types of equipment can be found on the site at some time. These machines make a lot of noise and are operated by a person. These operators often back up with little or no warning. While there is a beeping sound emitted when equipment backing up occurs, it can be quick. When working or walking near these operators, it is important to make eye contact with them (*Figure 20–16*). This way you know that they know you are nearby. Trust is one thing, but being safely sure is another.

Bulldozers provide the bulk of excavation work on residential sites. Their rolling tracks give them superior traction to push material out of the way. They carve layers of soil and rock pushing them into piles or thicker layers. Topsoil is separated to be put back after construction is completed. Subsoil is loosen and removed as needed. They are limited to the depth of excavation by how large of an area they can operate. Excavations are much larger than the building.

When the project requires a deep hole, excavators are often used. They have a long arm with a bucket that curls under to loosen and remove material (*Figure 20–17*). The digging cycle involves pivoting in an axis. They can remove material from one direction and place it into a truck in the other direction. Excavators are also very handy for moving new material back into position. A good operator can make work on a jobsite much easier by reducing the amount of hand work necessary.

Backhoe-loaders are designed to be a universal tool. They can move loose material around the job and dig trenches (*Figure 20–18*). Often when pipe or wire is buried in trenches it must first be covered with a sandy material to protect it. The loader can bring new material from a remote location to the trench before it is filled in.

Backhoes and excavators operate on similar principles. They have many hydraulic pistons to move the arm as it digs. The smoothness and efficiency of operation can be seen when operators activate three or four valves at once. The arm behaves as though it were an extension of a human arm.

Skid steers are highly maneuverable machines. They perform many labor-saving tasks on the job (*Figure 20–19*). The machine can spin a full circle inside the space it takes up on the ground. The lift arms can accept many attachments to alter the skid steers function. They include buckets or forks to transport material, posthole diggers, pneumatic hammers, and road sweepers.

Figure 20–17 Excavators provide superior deep digging.

Figure 20–16 Equipment operators need room to maneuver and nearby workers must be wary.

Figure 20–18 Backhoes can transport material and dig trenches.

Figure 20–19 Skid steers are highly maneuverable machines performing many labor saving tasks.

A jobsite safety risk is the limited visibility of the operator. This machine allows for excellent view of the work being done in the front, but the view to the sides and back is poor. Whenever working around this machine keep good eye contact with the operator. Do not get too close behind the machine. This problem is made worse by the fact that the machine can instantly change direction.

Fork lifts or tellehandlers can move pallets of material from delivery trucks to where it will be installed by workers. They provide excellent mobility in off-road, rough terrain situations (*Figure 20–20*). They are four-wheel driven and can vary the way they steer from normal two-wheel steer to four-wheel, side-to-side steer. They can lift tons of materials to heights of 40 to 50 feet.

Self-propelled boom lifts are often used instead of scaffolding. They allow faster and more efficientwork to heights from 10 to 150 feet (*Figure 20–21*). The operators can move and adjust the boom height from the end of the lift.Boom lifts come in many style and sizes and are either electrically or engine propelled.

Fall Protection

Current OSHA standards require fall protection when workers are working at heights above 6 feet. These regulations allow the employer the option of a guardrail system or a personal fall protection system.

A guardrail system has a top-rail from 38 to 45 inches (0.97–1.2 m) above the work deck, with a mid-rail installed midway between the top rail and the platform. The work deck must also be equipped with a toeboard. These requirements are for all open sides of the scaffold, except for those sides of the scaffold that are within 14 inches of the face of the building or open to a ladder.

A typical personal fall protection system consists of five related parts: the harness, lanyard, lifeline, rope grab, and anchor (*Figure 20–22*). This system is designed to stop a worker after a fall has begun. The lanyard absorbs the shock of the fall. Failure of any one part means failure of the system. Therefore, constant monitoring of a lifeline system is a critical responsibility. It is easy for a system to lose its integrity, even on first use. Do not use the

Figure 20–20 Tellehandlers can move material across varying terrain and to great heights.

Figure 20–21 Boom lifts offer workers fast, safe access to varying working heights.

Figure 20–23 Expected ladder capacities and uses can be found on the ladder rails.

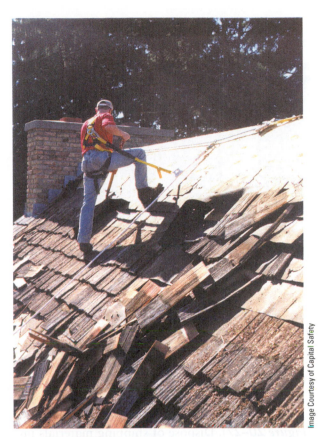

Figure 20–22 Components of a personal fall protection system.

scaffold components as an anchor point of the fall protection harness.

Falls are the leading construction work site injury, not only from heights but also from holes/excavations, ladders, stairways, walkways, and wall openings. Do not let water accumulate on floors. Ladders should be inspected for structural defects, and safety feet spreaders and rungs must be free of oil, dirt, or grease. Note the load capacity of the ladder before carrying anything up (*Figure 20–23*).

Excavations are not only an area of fall risk but also a risk of cave-ins. Excavations will require adequate cave-in protection if they are more than 5 feet deep. This protection is to be determined by a competent person. A competent person is one who has the authority to supervise safety conditions and make corrections if necessary. The soil type affects the type of cave in protection. Loose soil, such as sand, will require excavation side-wall shoring or sloped excavation walls.

JOBSITE HOUSEKEEPING

Jobsite conditions vary from site to site. Some sites are small and congested with material where the project seems too large for the site. Other sites are spacious and easy to get around. Most sites are busy with workers and materials (*Figure 20–24*). On all jobsites, the condition of the site affects potential risks to personal safety.

Cleanliness

Cleanliness refers to general conditions that allow for a safe working environment. A general rule for jobsite safety is to keep the site clean and orderly.

- Gather up and remove debris continually and daily to keep the work site orderly.
- Plan for the adequate disposal of scrap, waste, and surplus materials.
- Keep emergency exits, stairways, passageways, ladders, and scaffold free of material and supplies.
- Put tools and equipment away when they are not being used.
- Remove or bend over nails protruding from lumber.
- Keep hoses and power cords, etc. from laying in heavily travelled walkways or areas.
- Ensure structural openings are covered or adequately protected.
- Keep oily rags and debris in close metal containers.
- Keep materials at least 5 feet from openings, roof edges, excavations, or trenches.
- Secure material that is stored on roofs or on open floors.
- Do not throw tools or other materials.

Figure 20–24 Jobsites are often very busy and congested.

Figure 20–26 Materials are often delivered as they are needed.

- Do not raise or lower any tool or equipment by its own cable or supply hose
- Maintain good air quality by not using compressed air to clean.
- Store toxic, flammable, and hazardous materials in a clearly marked, secure location.
- Stack and store material such that it will not fall over or be damaged.
- Report any and all potential safety hazards to superiors.

A clean jobsite allows potential hazards to be more easily seen before they can cause injury. Scrap material located in a central location can be reviewed occasionally to see if it could be reused. Lowering the amount of waste is a green and more sustainable approach to construction (*Figure 20–25*).

Material Handling

Materials are continually delivered to the jobsite. They often are unloaded and stored nearby until they are needed (*Figure 20–26*). When they arrive

care should be taken to keep them in good condition, out of harm's way. Even sturdy and durable materials like concrete blocks (CMUs) should be handled appropriately. They are much easier to install if they are dry.

Materials should be kept in their shipping containers until needed and protected from moisture (*Figure 20–27*). Dispose of shipping materials immediately. Place materials on stickers or scraps to protect them from ground moisture.

Monitoring the jobsite inventory is easier when the materials are organized. A neat stack of boards clearly reveals the various lengths in the pile allowing the carpenter to select the most appropriate length. This helps reduces waste.

Properly stacked and stored materials reduce risk of injury of workers and protect the materials from damage. Materials that fall over can injure everything and everyone nearby.

Figure 20–25 An organized building site makes construction more efficient.

Figure 20–27 Packing covers serve to protect material until it is needed.

Lifting. When lifting materials and equipment it is important not to exceed your limits. Back injuries often take a long time to heal. Most back injuries are caused by improper lifting techniques. It is easy to get into the correct habit for lifting. The general rule is to use your legs not your back. The back should remain in a fixed, straight-line position and your legs should raise the object. Bend your knees, grasp the item to be lifted securely and then straighten the legs keeping your back as straight as possible (*Figure 20–28*). Carry smaller loads in two hands to stay balanced. This reduces twisting stress on the back. Keep the load close to your body as you stand up straight.

Often materials and equipment is raised into position by a fork lift or crane. Proper lifting these objects is important to maintain a safe site. Loads should be strongly secured with robes, cables, and slings (*Figure 20–29*). They should be properly balanced. Do not walk or work under loads being lifted. Accidents do happen, take steps to avoid disastrous consequences.

JOBSITE ELECTRICITY

Electricity is a universal source of power. It must be handled properly to protect everyone on the job. Shocks from a power tool can be a minor inconvenience or could cause death. It is difficult to determine ahead of time which effect will result, so precautions should always be taken. Power tools should always be operated according to manufacturer's recommendations. If the cord is frayed where wires are exposed the tool should be replaced. Repairs must be done as indicated by OSHA 29 standards.

Cords

Extension cords are used to provide greater reach of the tool from the source of power. They must be three-wire type and designed for hard or extra-hard usage. If they become damaged or frayed they should be replaced. Taping damaged cords is not allowed.

Keep cords away from high traffic areas to protect them. This also allows the work area to be safe from tripping hazards. Never use cords to move a tool from one work level to another.

Figure 20–28 Keep the back as straight as possible when lifting.

Figure 20–29 Never work or walk under hoisted material.

© Cengage Learning 2014

GFCI

GFIC (ground fault interrupter circuits) are designed to trip the power off if the electricity leaks out of the cord or tool. These devices are required in work areas where moisture is present (*Figure 20–30*). They trip fast enough that shocks are virtually eliminated. It is a good idea to use them for all situations.

Generators

Portable generators are internal combustion engines used to generate electricity. They are useful on jobsites for temporary power when local power is not available. Generators bring a variety of hazards to the site with them. They should be grounded to insure the GFIC works properly. Refer to the manufacturer's recommendations.

Portable generators should only be used in well-ventilated areas. They produce carbon monoxide (CO), a deadly gas. A carbon monoxide detector should be installed if there is any doubt about air quality. Symptoms of CO poisoning include dizziness, headaches, nausea, and tiredness. If you experience any of these symptoms, seek fresh air and medical attention. Keep them away and downwind from the building. (*Figure 20–31*).

The fuel used to run the generator is also a fire hazard. Generators have hot engine parts that could ignite the fuel during refueling. Fuel should be stored in approved containers and handled as a hazardous material.

HAZARDOUS MATERIALS

Hazardous materials and wastes come in many forms and varieties. Understanding what risks you are potentially being exposed to depends largely on

Figure 20–31 Portable gas-powered generators should be set up away from working areas.

the type of work you are doing. Hazardous materials for framing crews may be few while there are many remodeling and demolition hazards. Learning the problems associated with all construction material concerns is a lifetime endeavor.

MSDS

The MSDS (Material Safety and Data Sheet) is designed to keep workers informed of potential risks. Employers are required to have these sheets available for all employees. Construction workers should be well educated in the hazards they face. The MSDS provides a variety of information on all materials used (*Figure 20–32*). Product websites are an excellent place to find MSDS information for any building material.

Jobsite Waste

Jobsite waste is becoming more and more of a concern. Improving the way we handle jobsite waste is seen as a way of making the building process green. Cutting and fitting material usually produces waste (*Figure 20–33*). The issue is how can we reduce waste and make construction more efficient and sustainable.

Figure 20–30 GFCI devices should be used on the jobsite to protect from potentially fatal electrical shocks.

1. Product and manufacturer information
2. List of any hazardous ingredients in the product
3. Physical characteristics of the product
4. First Aid Measures
5. Fire and explosive nature of product
6. Spill cleanup procedures
7. Safe handling and storing information
8. Personal exposure concerns
9. Physical and chemical properties
10. Reactivity and stability of product
11. Health hazards
12. Ecological and environmental information
13. Disposal concerns
14. Transport information
15. U.S. Regulatory information
16. Special Precautions

© Cengage Learning 2014

Figure 20–32 Material safety data sheets contain safety information on any particular material.

© Cengage Learning 2014

Figure 20–33 Cutting and fitting usually produces jobsite waste.

What to do with the waste depends on its nature. Some waste is hazardous and must be discarded in a particular way to protect the environment. Follow the MSDS requirements for each material or concern.

Other waste materials are reusable. Whole pieces should be sent back to the supplier, sold to a third party, or given away. If there was a significant amount of leftover material, the estimating math should be adjusted to ensure that the correct amount of each material is delivered to the site.

Scraps from normal cutting and fitting are sometimes useable. Reuse these items on site whenever possible. Also protect them from damage and donate or sell them. Advertise reusable items in the newspaper; conduct a yard sale at the jobsite. Take recyclable scrap to dealers for recycling

(*Figure 20–34*).Simply sending useable scraps to the land fill is not a sustainable solution.

Packing materials used in shipping have a short lifespan. They quickly become waste. Request that vendors deliver materials in returnable containers. Wood pallets, for example, may be heavily constructed and reused by the material supplier. Choose materials that are delivered with minimal or no packaging. It is not hard to be green and sustainable. It only takes some thought.

FIRE PREVENTION

Construction by its very nature is a hazardous place where fires can happen easily. Discarded oily solvent rags left in an open pile may look neat and safe, but they can heat up over a few hours to the point they self-ignite. Always store these materials in closed metal containers.

Two words are used to describe materials' ability to burn—flammable and inflammable. These words both mean the same thing—the material will burn easily. Any material labeled with these words should be used thoughtfully. Protect them from heat sources that will cause them to ignite.

When utilizing equipment that produces heat, be sure the surrounding area is free of all potential fire hazards. Do not use heating devices like a salamander or other open-flamed heaters in an enclosed area. Make sure they vent to the outside air and are an appropriate distance from walls, ceilings, and floor. Fire extinguishers should be available whenever working with heat-producing equipment.

© Cengage Learning 2014

Figure 20–34 Some materials such as metal scraps are easily recycled.

Steel fabrication often involves cutting and fastening with lots of heat and sparks (*Figure 20–35*). Flammable materials should be protected. Do not allow combustible materials and rubbish to accumulate on a jobsite. Compressed gas cylinders, full or empty, should be stored and kept away from excessive heat.

Flammable liquids such as gas, oil, paint and paint thinner, solvents, and grease should be kept in plugged or capped containers and stored in non-combustible areas. Never smoke near flammable or volatile materials.

LADDERS

Carpenters must often use *ladders* to work from or to reach working platforms above the ground. The most commonly used ladders are the stepladder and the extension ladder. They are usually made of wood, aluminum, or fiberglass. Make sure all ladders are in good condition before using them.

Extension Ladders

To raise an extension ladder, the bottom of the ladder must be secured. This is done either with the help of another person or by placing the bottom against a solid object, such as the base of the wall. Pick up the other end. Walk forward under the ladder, pushing upward on each rung until the ladder is upright (*Figure 20–36*). With the ladder vertical, and leaning toward the wall, extend the ladder by pulling on the rope with one hand while holding the ladder upright with the other. Raise the ladder to the desired height. Make sure the spring-loaded hooks are over the rungs on both sides.

Lean the top of the ladder against the wall. Move the base out until the distance from the wall is about one-quarter of the vertical height. This will give the proper angle to the ladder. The proper angle for climbing the ladder can also be determined, as shown in *Figure 20–37*.

If the ladder is used to reach a roof or working platform, it must extend above the top support by at least 3 feet.

> ## CAUTION
>
> Be careful of overhead power lines when using ladders. Metal ladders conduct electricity. Contact with power lines could result in electrocution.

When the ladder is in position, shim one leg, if necessary, to prevent wobbling, and secure the top of the ladder to the building. Check that the feet of the ladder are secure and will not slip. Then face

Figure 20–36 Raising an extension ladder using the base of the building to secure the bottom of the ladder.

Figure 20–35 Some cutting and fitting involves heat. Care should be taken to prevent fires.

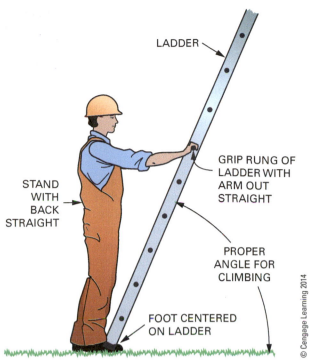

Figure 20–37 Techniques for finding the proper ladder angle before climbing.

In the figure, labels read: LADDER; GRIP RUNG OF LADDER WITH ARM OUT STRAIGHT; STAND WITH BACK STRAIGHT; PROPER ANGLE FOR CLIMBING; FOOT CENTERED ON LADDER. © Cengage Learning 2014

Figure 20–38 Face the ladder when climbing. Hold on with both hands.

© Cengage Learning 2014

the ladder to climb, grasping the side rails or rungs with both hands (*Figure 20–38*).

Stepladders

When using a stepladder, open the legs fully so the brackets are straight and locked. Make sure the ladder does not wobble. If necessary, place a shim under the leg to steady the ladder. Never work above the recommended top step indicated by the manufacturer. This is usually the second step down from the top, not including the top as a step. Do not use the ledge in back of the ladder as a step. The ledge is used to hold tools and materials only. Move the ladder as necessary to avoid overreaching. Make sure all materials and tools are removed from the ladder before moving it.

Ladder Jacks

Ladder jacks are metal brackets installed on ladders to hold scaffold planks (*Figure 20–39*). At least two ladders and two jacks are necessary for

a section. Ladders should be heavy-duty, free from defects, and placed no more than 8 feet apart. They should have devices to keep them from slipping.

The ladder jack should bear on the side rails in addition to the ladder rungs. If bearing on the rungs only, the bearing area should be at least 10 inches on each rung. No more than two persons should occupy any 8 feet of ladder jack scaffold at any one time. The platform width must not be less than 18 inches. Planks must overlap the bearing surface by at least 10 inches.

CONSTRUCTION AIDS

Sawhorses, work stools, ladders, and other construction aids are sometimes custom built by the carpenter on the job or in the shop.

© Cengage Learning 2014

Figure 20–39 Ladder jacks are used to support scaffold plank for short-term, light repair work.

Sawhorses

Sawhorses are used on practically every construction job. They support material that is being laid out or cut to size. They may be built of various materials depending on the desired strength of the sawhorse. Light-duty horses may be made with a 2 × 4 top and 1 × 4 legs, while heavy-duty horses are made with a 2 × 6 top and 2 × 4 legs. Both use plywood for leg braces.

Sawhorses are constructed in a number of ways according to the preference of the individual (*Procedure 20–A*). However, they should be of sufficient width, a comfortable working height, and light enough to be moved easily from one place to another. A typical sawhorse is 36 inches wide with 24-inch legs. A tall person may wish to make the leg 26 inches long.

Job-Made Ladders

At times it is necessary to build a ladder on the job. These are usually straight ladders no more than 24 feet in length. The side rails are made of clear, straight-grained 2 × 4 or 2 × 6 stock spaced 15 to 20 inches apart. *Cleats* or *rungs* are cut from 2 × 4 stock and inset into the edges of the side rails not more than $3/4$-inch. Filler blocks are sometimes used on the rails between the cleats. Cleats are uniformly spaced at 12 inches top to top (*Procedure 20–B*).

SAFETY SUMMARY

When you arrive at the jobsite, run a pre-work safety check. As you put on your hard hat and personal safety equipment, look for anything on the job that has changed from the previous day. Check to see your tools and equipment are in good working order. Keep your work site orderly and clean. Watch for anything that interferes with your safe work environment. Safety is an activity that never stops.

StepbyStep Procedures

Procedure 20–A Constructing a Typical Sawhorse

STEP 1 CUT 2" X 6" SAWHORSE TOP 36" LONG AND BEVEL BOTH EDGES OF EACH END AS SHOWN.

STEP 2 CUT FOUR LEGS TO 24" LENGTH OR AS DESIRED, WITH BEVEL ON EACH END AT SAME ANGLE AS TOP.

STEP 3 FASTEN ALL FOUR LEGS TO SAWHORSE TOP.

STEP 4 HOLD PLYWOOD BRACE AS SHOWN AND MARK ITS LENGTH AT THE TOP EDGE.

FROM EACH MARK, LAY OUT SAME ANGLE AS TOP AND LEGS. CUT, MAKE DUPLI-CATE, AND FASTEN ONE ON EACH END OF HORSE FLUSH WITH OUTSIDE FACE OF LEGS. MOVE LEGS TO FIT THE PIECE.

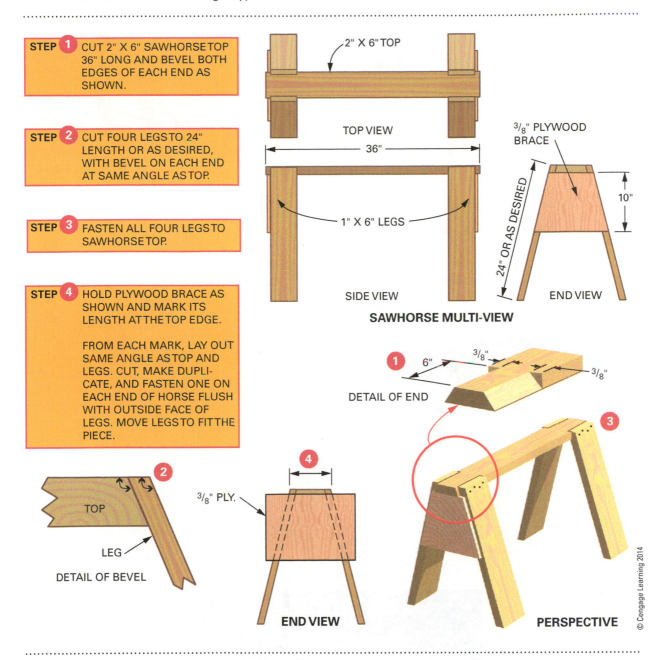

2" X 6" TOP

TOP VIEW

36"

1" X 6" LEGS

SIDE VIEW

3/8" PLYWOOD BRACE

10"

24" OR AS DESIRED

END VIEW

SAWHORSE MULTI-VIEW

6"   3/8"

3/8"

DETAIL OF END

TOP

LEG

DETAIL OF BEVEL

3/8" PLY.

END VIEW

PERSPECTIVE

© Cengage Learning 2014

StepbyStep Procedures

Procedure 20–B Constructing A Job-Built Ladder

STEP 1 TEMPORARILY FASTEN RAILS OF LADDER SIDE BY SIDE AND LAY OUT DADOES 12" OC AS SHOWN BELOW.

STEP 2 CUT DADOES, CUTTING ONLY 3/8" DEEP. ANY DEEPER WILL WEAKEN THE RAIL.

STEP 3 CUT RUNGS 15–20" LONG. FASTEN RUNGS IN DADOES WITH 16d NAILS KEEPING ENDS OF RUNGS FLUSH WITH OUTSIDE FACE OF RAILS.

STEP 4 CUT CLEATS TO FIT BETWEEN RUNGS.

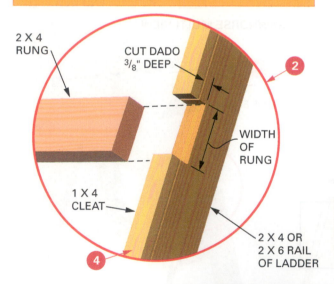

2 X 4 RUNG

CUT DADO 3/8" DEEP

WIDTH OF RUNG

1 X 4 CLEAT

2 X 4 OR 2 X 6 RAIL OF LADDER

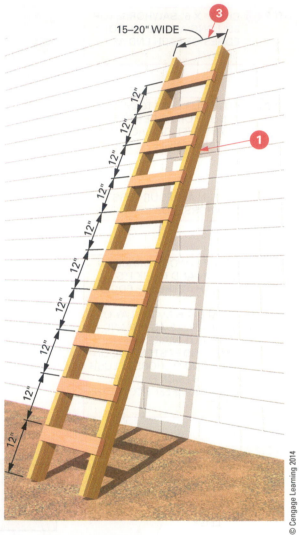

15–20" WIDE

12"

© Cengage Learning 2014

© Cengage Learning 2014

CHAPTER 21

Scaffolds

Scaffolds are an essential component of construction, because they allow work to be performed at various elevations. However, they also can create one of the most dangerous working environments. The U.S. Occupational Safety and Health Administration (OSHA) reports that in construction, falls are the number one killer, and 40 percent of those injured in falls had been on the job less than one year. A survey of scaffold accidents summarizes the problem (*Figure 21–1*). A scaffold fatality and catastrophe investigation conducted by OSHA revealed that the largest percentage, 47 percent, was due to equipment failure. In most instances, OSHA found the equipment did not just break; it was broken due to improper use and erection. Failures at the anchor points, allowing either the scaffold parts or its anchor points to break away, were often involved in these types of accidents. Other factors were improper, inadequate, and improvised construction, and inadequate fall protection. The point of this investigation is that accidents do not just happen; they are caused.

OSHA regulations on the fabrication of frame scaffolds are found in the Code of Federal Regulations 1926.450, 451, 452. These regulations should be thoroughly understood before any scaffold is erected and used. Furthermore, safety codes that are more restrictive than OSHA, such as those in Canada, California, Michigan, and Washington, should be consulted.

Scaffolds must be strong enough to support workers, tools, and materials. They must also provide an extra safety margin. The standard safety margin requirement is that all scaffolds must be capable of supporting at least four times the maximum intended load.

Those who erect scaffolding must be familiar with the different types and construction methods of scaffolding to provide a safe working platform for all workers. The type of scaffolding depends on its location, the kind of work being performed, the distance above the ground, and the load it is required to support. No job is so important as to justify risking one's safety and life. All workers deserve to be able to return to their families without injury.

The regulations on scaffolding enforced by OSHA make it clear that before erecting or using a scaffold, the worker must be trained about the hazards surrounding the use of such equipment. OSHA has not determined the length of training that should be required. Certainly that would depend on the expertise of the student in training.

Employers are responsible for ensuring that workers are trained to erect and use scaffolding. One level of training is required for workers, such as painters, to work from the scaffold. A higher level of training is required for workers involved in erecting, disassembling, moving, operating, repairing, maintaining, or inspecting scaffolds.

The employer is required to have a competent person to supervise and direct the scaffold erection. This individual must be able to identify existing and predictable hazards in the surroundings or working conditions that are unsanitary, hazardous, or dangerous to employees. This person also has authorization to take prompt corrective measures or eliminate such hazards. A competent person has the authority to take corrective measures and stop work if need be to ensure that scaffolding is safe to use.

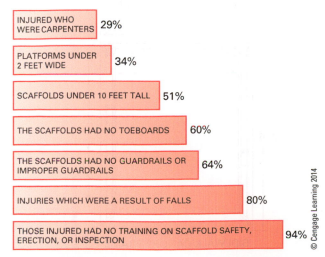

INJURED WHO WERE CARPENTERS	29%
PLATFORMS UNDER 2 FEET WIDE	34%
SCAFFOLDS UNDER 10 FEET TALL	51%
THE SCAFFOLDS HAD NO TOEBOARDS	60%
THE SCAFFOLDS HAD NO GUARDRAILS OR IMPROPER GUARDRAILS	64%
INJURIES WHICH WERE A RESULT OF FALLS	80%
THOSE INJURED HAD NO TRAINING ON SCAFFOLD SAFETY, ERECTION, OR INSPECTION	94%

© Cengage Learning 2014

Figure 21–1 Accident statistics involving scaffolding.

METAL SCAFFOLDING

Metal Tubular Frame Scaffold

Metal tubular frame scaffolding consists of manufactured *end frames* with folding *cross braces, adjustable screw legs, baseplates, platforms,* and *guardrail hardware* (***Figure 21–2***). They are erected in sections that consist of two end frames and two cross braces, and typically come in 5 × 7 modules. Frame scaffolds are easy to assemble, which can lead to carelessness. Because untrained erectors may think scaffolds are just stacked up, serious injury and death can result from a lack of training.

End frames consist of posts, horizontal bearers, and intermediate members. End frames come in a number of styles depending on the manufacturer. Frames can be wide or narrow, and some are designed for rolling tower scaffolds, while other frames have an access ladder built into the end frame (***Figure 21–3***).

© Cengage Learning 2014

Figure 21–2 A typical metal tubular frame scaffold.

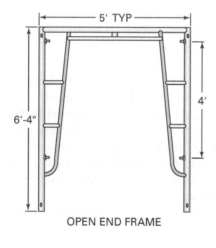

OPEN END FRAME

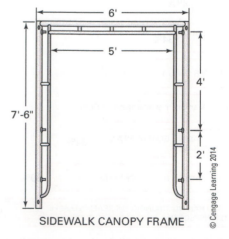

WALKTHROUGH FRAME
WITH BUILT IN LADDER

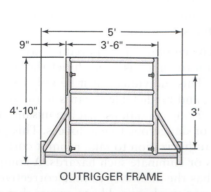

OUTRIGGER FRAME

SIDEWALK CANOPY FRAME

© Cengage Learning 2014

Figure 21–3 Examples of typical metal tubular end frames.

Cross braces rigidly connect one scaffold member to another member. Cross braces connect the bottoms and tops of frames. This diagonal bracing keeps the end frames plumb and provides the rigidity that allows them to attain their designed strength. The braces are connected to the end frames using a variety of locking devices (*Figure 21–4*).

OSHA regulations require the use of baseplates on all supported or ground-based scaffolds (*Figure 21–5*) in order to transfer the load of scaffolding, material, and workers to the supporting surface. It is extremely important to distribute this load over an area large enough to reduce the pounds per square foot load on the ground. If the scaffold sinks into the ground when it is being used, accidents could occur. Therefore, baseplates should sit on and be nailed to a mud sill. A mud sill is typically a 2 × 10 (5 × 25 cm) board approximately 18

to 24 inches (45–60 cm) long. On soft soil it may need to be longer and/or thicker (*Figure 21–6*).

To level an end frame while erecting a frame scaffold, screwjacks may be used. At least one-third of the screwjack must be inserted in the scaffold leg. Lumber may be used to crib up the legs of the scaffold (*Figure 21–7*). Cribbing height is restricted to equal the length of the mud sill. Therefore, using a 19-inch-long (48 cm), 2 × 10-inch (5 × 25 cm) mud sill, the crib height is limited to 19 inches (48 cm). OSHA also prohibits the use of concrete blocks to level scaffolding.

A guardrail system is a vertical fall protection barrier consisting of, but not limited to, top-rails, mid-rails, toe-boards, and posts (*Figure 21–8*). It prevents employees from falling off a scaffold platform or walkway. A guardrail system is required when the working height is 10 feet or more. Guardrail systems must have a top-rail capacity

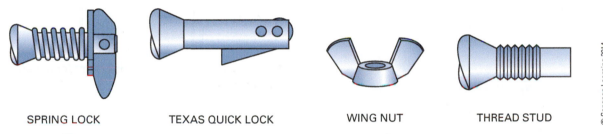

SPRING LOCK   TEXAS QUICK LOCK   WING NUT   THREAD STUD

© Cengage Learning 2014

Figure 21–4 Typical locking devices used to connect cross braces to end frames.

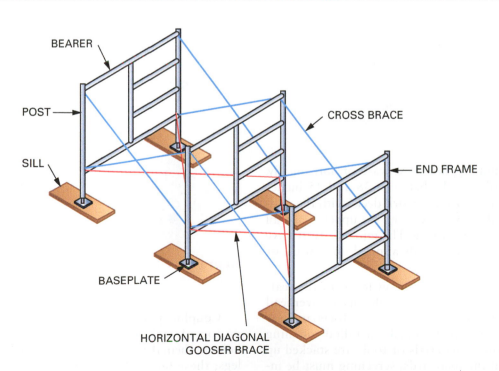

BEARER

POST

SILL

CROSS BRACE

END FRAME

BASEPLATE

HORIZONTAL DIAGONAL GOOSER BRACE

© Cengage Learning 2014

Figure 21–5 Typical baseplate setup for a metal tubular frame scaffold.

Figure 21–6 Baseplates and jackscrews should rest on mud sills.

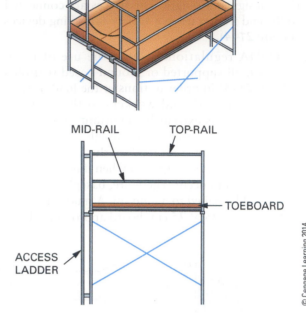

MID-RAIL   TOP-RAIL

TOEBOARD

ACCESS
LADDER

Figure 21–8 Typical guardrail system for a metal tubular frame scaffold.

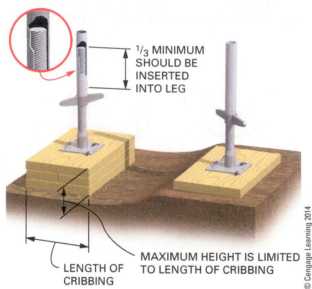

¹/₃ MINIMUM
SHOULD BE
INSERTED
INTO LEG

MAXIMUM HEIGHT IS LIMITED
TO LENGTH OF CRIBBING

LENGTH OF
CRIBBING

Figure 21–7 Cribbing, interlocked blocking may be used to level the ground under the scaffold.

Figure 21–9 Coupler pins to join end frames.

of 200 pounds applied downward or horizontally. The top-rail must be between 38 and 45 inches (0.97–1.2 m) above the work deck, with the mid-rail installed midway between the upper guardrail and the platform surface. The mid-rail must have a capacity of 150 pounds applied downward or horizontally.

If workers are on different levels of the scaffold, toeboards must be installed as an overhead protection for lower level workers. Toeboards are typically 1 × 4-inch boards installed touching the platform. If materials or tools are stacked up higher than the toeboards, screening must be installed. Moreover, all workers on the scaffold must wear hard hats.

Coupling pins are used to stack the end frames on top of each other (*Figure 21–9*). They have holes in them that match the holes in the end frame legs; these holes allow locking devices to be installed. Workers must ensure that the coupling pins are designed for the scaffold frames in use.

Figure 21–10 Coupler locking devices to prevent scaffold uplift.

© Cengage Learning 2014

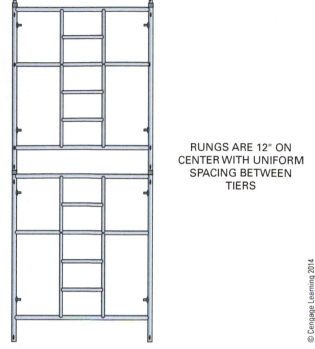

RUNGS ARE 12" ON CENTER WITH UNIFORM SPACING BETWEEN TIERS

© Cengage Learning 2014

Figure 21–11 The rungs of an end frame must be spaced no more than 16¾ inches (43 cm) apart if users are going to use it as an access ladder.

The scaffold end frames and platforms must have uplift protection installed when a potential for uplift exists. Installing locking devices through the legs of the scaffold and the coupling pins provides this protection (*Figure 21–10*). If the platforms are not equipped with uplift protection devices, they can be tied down to the frames with number nine steel-tie wire.

OSHA requires safe access onto the scaffold for both erectors and users of the scaffolds. Workers can climb end frames only if they meet OSHA regulations. Frames may be used as a ladder only if they are designed as such. Frames meeting such design guidelines must have level horizontal members that are parallel and are not more than 16¾ inches (43 cm) apart vertically (*Figure 21–11*). Scaffold erectors may climb end frame rungs that are spaced up to 22 inches (46 cm). Platform planks should not extend over the end frames where end frame rungs are used as a ladder access point. The cross braces should never be used as a means of access or egress. Attached ladders and stair units may be used (*Figure 21–12*). A rest platform is required for every 35 feet (10.7 m) of ladder.

Manually propelled mobile scaffolds use wheels or casters in the place of baseplates (*Figure 21–13*). Casters have a designed load capacity that should never be exceeded. Mobile scaffolds made from tubular end frames must use diagonal horizontal braces, or gooser braces, to keep the mobile tower frame square.

Side brackets are light-duty (35 pounds per square foot maximum) extension pieces used to increase the working platform (*Figure 21–14*). They are designed to hold personnel only and are not to be used for material storage. When side brackets are used, the scaffold must have tie-ins, braces, or outriggers to prevent the scaffold from tipping.

Hoist arms and wheel wells are sometimes attached to the top of the scaffold to hoist scaffold parts to the erector or material to the user of the scaffold (*Figure 21–15*). The load rating of these hoist arms and wheel wells is typically no more than 100 pounds. The scaffold must be secured from overturning at the level of the hoist arm, and workers should never stand directly under the hoist arm when hoisting a load. They should stand a slight distance away, but not too far to the side, because this will increase the lateral or side-loading force on the scaffold.

Scaffold Inspection

Almost half of all scaffold accidents, according to the U.S. Bureau of Labor Statistics, involve defective scaffolds or defective scaffold parts. This statistic means ongoing visual inspection of scaffold parts must play a major role in safe scaffold erection and use. OSHA requires that a competent person inspect all scaffolds at the beginning of every work shift.

Visual inspection of scaffold parts should take place at least five times: before erection, during erection, during scaffold use, during dismantling, and before scaffold parts are put back in storage. All damaged parts should be red-tagged and removed from service and then repaired or destroyed

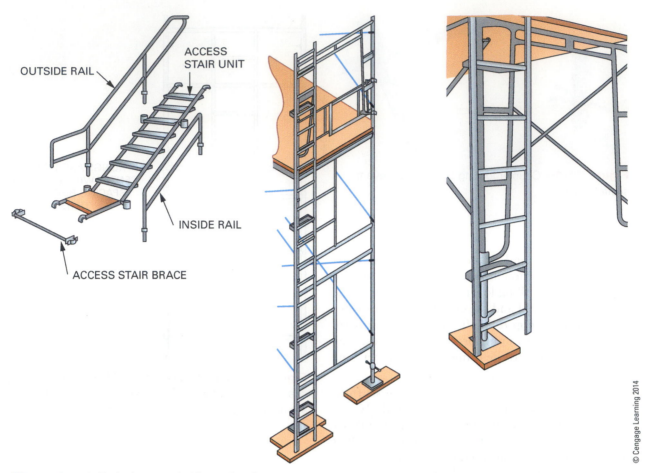

Figure 21–12 Typical access ladder and stairway.

Figure 21–13 Casters replace a baseplate to transform a metal tubular frame scaffold into a mobile scaffold.

as required. Things to look for during the inspection process include the following:

Broken and excessively rusted welds

Split, bent, or crushed tubes

Cracks in the tube circumference

Distorted members

Excessive rust

Damaged brace locks

Lack of straightness

Excessively worn rivets or bolts on braces

Split ends on cross braces

Bent or broken clamp parts

Damaged threads on screwjacks

Damaged caster brakes

Damaged swivels on casters

Corrosion of parts

Metal fatigue caused by temperature extremes

Leg ends filled with dirt or concrete

Scaffold Erection Procedure

The first thing to be done during the erection procedure is to inspect all scaffold components delivered to the job site. Defective parts must not be used.

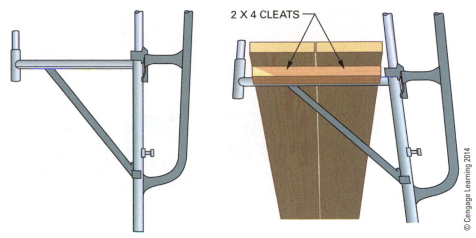

2 X 4 CLEATS

Figure 21–14 Side brackets are used to extend a scaffold work platform. These brackets should only be used for workers and never for material storage.

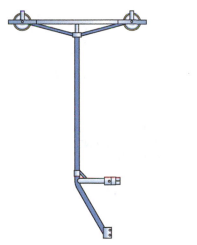

Figure 21–15 A hoist that attaches to the top of a scaffold is used to raise material and equipment.

The foundation of the scaffold must be stable and sound, able to support the scaffold and four times the maximum intended load without settling or displacement.

Always start erecting the scaffold at the highest elevation, which will allow the scaffold to be leveled without any excavating by installing cribbing, screwjacks, or shorter frames under the regular frames. The scaffold must always be level and plumb. Lay out the baseplates and screwjacks on mud sills so the guardrails and end frames with cross braces can be properly installed (*Figure 21–16*).

Stand one of the end frames up and attach the cross braces to each side, making sure the correct length cross braces have been selected for the job. Connect the other end of the braces to the second

end frame. All scaffold legs must be braced to at least one other leg (*Figure 21–17*). Make sure that all brace connections are secure. If any of these mechanisms are not in good working order, replace the frame with one that has properly functioning locks.

Use a level to plumb and level each frame (*Figure 21–18*). Remember that OSHA requires that all tubular welded frame scaffolds be plumb and level. Adjust screwjacks or cribbing to level the scaffold. As each frame is added, keep the scaffold bays square with each other. Repeat this procedure until the scaffold run is erected. Remember, if the first level of scaffolding is plumb and level, the remaining levels will be more easily assembled.

The next step is to place the planks on top of the end frames. All planking must meet OSHA requirements and be in good condition. If planks that do not have hooks are used, they must extend over their end supports by at least 6 inches (15 cm) and not more than 12 inches (30 cm). A cleat should be nailed to both ends of wood planks to prevent plank movement (*Figure 21–19*). Platform laps must be at least 12 inches, and all platforms must be secured from movement. Hooks on planks also have uplift protection installed on the ends.

It is a good practice to plank each layer fully as the scaffold is erected. If the deck is to be used for erecting, then a minimum of two planks can be used. However, full decking is preferred, because it is a safer method for the erector.

Before the second lift is erected, the erector must provide an access ladder. Access may be on the end frame, if it is so designed, or on attached ladder. If the ladder is bolted to a horizontal

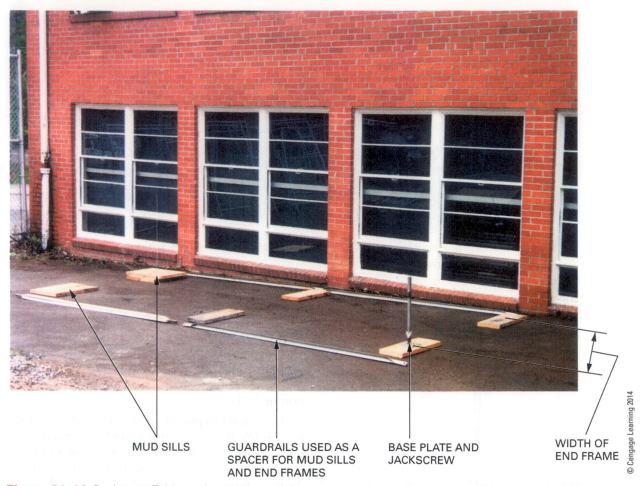

MUD SILLS   GUARDRAILS USED AS A SPACER FOR MUD SILLS AND END FRAMES   BASE PLATE AND JACKSCREW   WIDTH OF END FRAME

© Cengage Learning 2014

Figure 21–16 During scaffold erection, the baseplates are spaced according to guardrail braces and width of end frames.

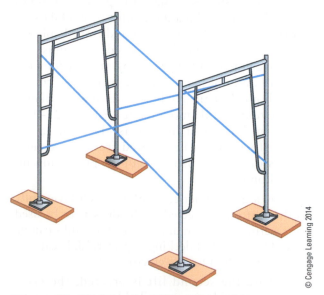

© Cengage Learning 2014

Figure 21–17 Cross braces connect end frames, keeping them rigid and plumb.

© Cengage Learning 2014

Figure 21–18 End frames and platform planks must be level. Level and plumb begins on the first row of scaffolding.

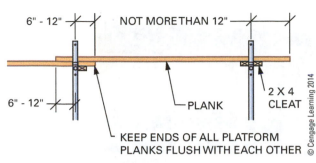

Figure 21–19 Recommended placement for scaffold planks.

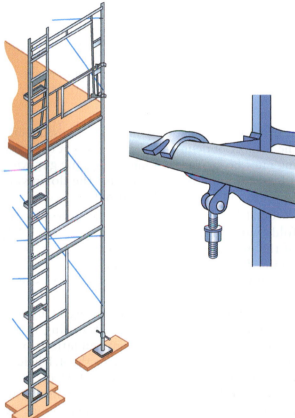

Figure 21–20 Attachable ladders are connected so that the bolt attaches from the bottom.

member, the bolt must face downward (*Figure 21–20*). Next, the second level of frames may be hung temporarily over the ends of the first frames and then installed onto the coupling pins of the first-level frames (*Figure 21–21*). Special care must be taken to ensure proper footing and balance when lifting and placing frames. OSHA requires erector fall protection—a full-body harness attached to a proper anchor point on the structure—

when it is feasible and not a greater hazard to do so. Never attach fall protection to the scaffold.

After the end frames have been set in place and braced, they should have uplift protection pins installed through the legs and coupling pins. Wind, side brackets, and wheel wells can cause uplift so it is a good practice to pin all scaffold legs together.

The remaining scaffolding is erected in the same manner as the first. Remember, all work platforms must be fully decked and have a guardrail system or personal fall-arrest system installed before it can be turned over to the scaffold users. If the scaffold is higher than four times its minimum base dimension, it must be restrained from tipping by guying, tying, bracing, or equivalent means (*Figure 21–22*). The scaffold is not allowed to tip into or away from the structure.

After the scaffold is complete, it is inspected again to make sure that it is plumb, level, and square before turning it over for workers to use. The inspection should also include checking that all legs are on baseplates or screwjacks and mud sills (if required), ensuring the scaffolding is properly braced with all brace connections secured, and making sure all tie-ins are properly placed and secured, both at the scaffold and at the structure. All platforms must be fully planked with proper decking and in good shape. Toeboards and/or screening should be installed as needed. Check that end and/or side brackets are fully secured and that any overturning forces are compensated for. All access units are inspected to ensure they are correctly installed and ladders and stairs are secured. Again, workers on the scaffold must wear hard hats.

After the scaffolding passes all inspections, it is ready to be turned over to the workers. Remember that this scaffolding must be inspected by a competent person at the beginning of each work shift and after any occurrence, such as a high wind or a rainstorm that could affect its structural integrity.

Scaffold Capacity

All scaffolds and their components must be capable of supporting, without failure, their own weight and at least four times the maximum intended load applied or transmitted to them. Erectors and users of scaffolding must never exceed this safety factor.

Erectors and users of the scaffold must know the maximum intended load and the load-carrying capacities of the scaffold they are using. The erector must also know the design criteria, maximum intended load-carrying capacity, and intended use of the scaffold.

Figure 21–21 The next lift of end frames may be preloaded by hanging them on the previous end frame.

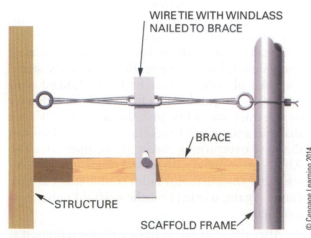

Figure 21–22 A wire tie and a windlass may be used to secure the scaffold tightly to the building.

When erecting a frame scaffold, the erector should know the load-carrying capacities of its components. The rated leg capacity of a frame may never be exceeded on any leg of the scaffold. Also, the capacity of the top horizontal member of the end frame, called the bearer, may never be exceeded. Remember, it is possible to overload the bottom legs of the scaffold without overloading the bearer or top horizontal member of any frame. It is also possible to overload the bearer or top horizontal member of the frame scaffold and not overload the leg of that same scaffold. Erectors must pay careful attention to the load capacities of all scaffold components.

If the scaffold is covered with weatherproofing plastic or tarps, the lateral pressure applied to the scaffold will dramatically increase. Consequently, the number of tie-ins attached to prevent overturning must be increased. Additionally, any guy wires added for support will increase the downward pressure and weight of the scaffold.

OSHA regulations state that supported scaffolds with a ratio larger than four-to-one (4:1) of the height to narrow base width must be restrained from tipping by guying, tying, bracing, or equivalent means. Guys, ties, and braces must be installed at locations where horizontal members support both inner and outer legs. Guy, ties, and braces must be installed according to the scaffold manufacturer's recommendations or at the closest horizontal member to the 4:1 height. For scaffolds greater than 3 feet (0.91 m) wide, the vertical locations of horizontal members are repeated every 26 feet (7.9 m). The top guy, tie, or brace of completed scaffolds must be placed no farther than the 4:1 height from the top. Such guys, ties, and braces must be installed at each end of the scaffold and at horizontal intervals not to exceed 30 feet (9.1 m). The tie or standoff should be able to take pushing and pulling forces so the scaffold does not fall into or away from the structure.

The supported scaffold poles, legs, post, frames, and uprights should bear on baseplates, mud sills, or other adequate, firm foundation. Because the mud sills have more surface area than baseplates, sills distribute loads over a larger area of the foundation. Sills are typically wood and come in many sizes. Erectors should choose a size according to

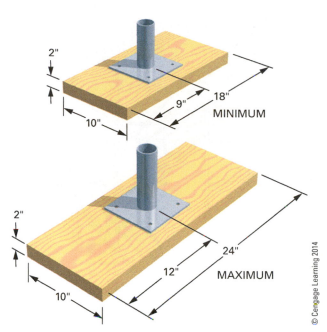

Figure 21–23 Baseplates should be centered on the mud sills.

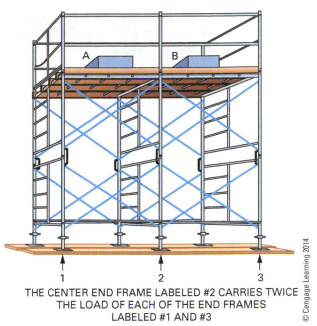

THE CENTER END FRAME LABELED #2 CARRIES TWICE THE LOAD OF EACH OF THE END FRAMES LABELED #1 AND #3

Figure 21–24 The inner frames, such as #2 shown here, often carry twice the load of the frames located at the end of the scaffold.

the load and the foundation strength required. Mud sills made of 2 × 10-inch (5 × 25 cm) full-thickness or nominal lumber should be 18 to 24 inches (46–60 cm) long and centered under each leg (*Figure 21–23*).

The loads exerted onto the legs of a scaffold are not equal. Consider a scaffold with two loads on two adjacent platforms (*Figure 21–24*). Half of load A is carried by end frame #1 and the other half is carried by #2. Half of load B is carried by end frames #2 and #3. End frame #2 carries two half loads or one full load, which is twice the load of end frames #1 and #3. At no time should the manufacturer's load rating for its scaffolding be exceeded.

Scaffold Platforms

The scaffolding's work area must be fully planked between the front uprights and the guardrail supports in order for the user to work from the scaffold. The planks should not have more than a 1-inch gap between them unless it is necessary to fit around uprights such as a scaffold leg. If the platform is planked as fully as possible, the remaining gap between the last plank and the uprights of the guardrail system must not exceed $9\frac{1}{2}$ inches.

Scaffold platforms must be at least 18 inches wide with a guardrail system in place. In areas where they cannot be 18 inches wide, they will be as wide as is feasible. The platform is allowed to be as much as 14 inches away from the face of the work.

Planking for the platforms, unless cleated or otherwise restrained by hooks or equivalent means, should extend over the centerline of their support at least 6 inches (15 cm) and no more than 12 inches (30 cm). If the platform is overlapped to create a long platform, the overlap shall occur only over supports and should not be less than 12 inches unless the platforms are nailed together or otherwise restrained to prevent movement.

When fully loaded with personnel, tools, and/or material, the wood plank used to make the platform must never deflect more than one-sixtieth ($\frac{1}{60}$) of its span. In other words, a 2 × 10-inch (5 × 25 cm) plank that is 12 feet (3.7 m) long and is sitting on two end frames spaced 10 feet (3 m) apart should not deflect more than 2 inches (5 cm) or one-sixtieth of the span.

Any solid sawn wood planks should be scaffold-grade lumber as set out by the grading rules for the species of lumber being used. A recognized lumber-grading association, such as the Western Wood Products Association (WWPA) or the National Lumber Grades Authority (NLGA), establishes these grading rules. A grade should be stamped on the scaffold-grade plank, indicating that it meets OSHA and industry requirements for scaffold planks. Two of the most common wood species used for scaffold planks are southern yellow pine and Douglas fir. OSHA does not require wood scaffold planks to bear grade stamps.

Maximum Permissible Plank Span

Maximum Intended Load	Rough 2× Full Thickness		2× Nominal Thickness	
Lbs/sq ft	Feet	Meters	Feet	Meters
25	10	3	8	2.4
50	8	2.4	6	1.8
75	6	1.8	—	—

© Cengage Learning 2014

Figure 21–25 Maximum spacing of planks based on the load rating of the scaffold.

The erector may use "equivalent" planks, which are determined to be equivalent by visually inspecting or test loading the wood plank in accordance with grading rules.

Scaffold platforms are usually rated for the intended load. Light-duty scaffolds are designed at 25 pounds per square foot, medium-duty scaffolds are rated at 50 pounds per square foot, and heavy-duty scaffolds at 75 pounds per square foot. The maximum span of a plank is tabulated in *Figure 21–25*. Using this chart, the maximum load that could be put on a nominal thickness plank (1½-inch or 3.8 cm) with a span of 7 feet (2.1 m) is 25 pounds per square foot. Note that a load of 50 pounds per square foot would require a span of no more than 6 feet (1.8 m).

Fabricated planks and platforms are often used in lieu of solid sawn wood planks. These planks and platforms include fabricated wood planks that use a pin to secure the lumber sideways, oriented strand board planks, fiberglass composite planks, aluminum-wood decked planks, and high-strength galvanized steel planks. The loading of fabricated planks or platforms should be obtained from the manufacturer and never exceeded. Scaffold platforms must be inspected for damage before each use.

Scaffold Access

A means of access must be provided to any scaffold platform that is 2 feet above or below a point of access. Such means include a hook-on or attachable ladder, a ramp, or a stair tower and are determined by the competent person on the job.

If a ladder is used, it should extend 3 feet above the platform and be secured at both the top and bottom. Hook-on and attachable ladders should be specifically designed for use with the type of scaffold used, have a minimum rung length of 11½ inches, and have uniformly spaced rungs with a maximum spacing between rung length of 16¾ inches. Sometimes a stair tower can be used for

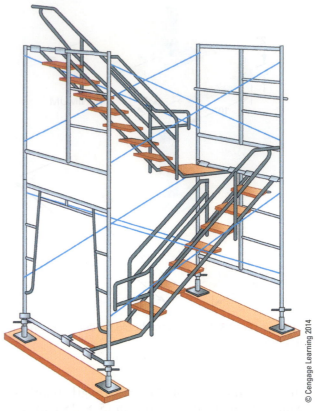

© Cengage Learning 2014

Figure 21–26 Scaffold access may be provided by a stair tower.

access to the work platform, usually on larger jobs (*Figure 21–26*). A ramp can also be used as access to the scaffold or the work platform. When using a ramp, it is important to remember that a guardrail system or fall protection is required at 6 feet above a lower level.

The worker using the scaffold can sometimes access the work platform using the end frames of the scaffold itself. According to regulations, the end frame must be specifically designed and constructed for use as ladder rungs. The rungs can run up the center or to one side of the end frame; some have the rungs all the way across the end frame. Scaffold users should never climb any end frame unless the manufacturer of that frame designated it to be used for access.

Scaffold Use

Scaffolds must not be loaded in excess of their maximum intended load or rated capacities, whichever is less. Workers must know the capacity of scaffolds they are erecting and/or using. Before the beginning of each work shift, or after any occurrence that could affect a scaffold's structural integrity, the competent person must inspect all scaffolds on the job.

Employees must not work on scaffolds covered with snow or ice except to remove the snow or ice. Generally, work on or from scaffolds is prohibited during storms or high winds. Debris must not be allowed to accumulate on the platforms. Makeshift scaffold devices, such as boxes or barrels, must not be used on the scaffold to increase workers' working height. Step ladders should not be used on the scaffold platform unless they are secured according to OSHA regulations.

Dismantling Scaffolds

Many guidelines and rules for erection also apply to scaffold dismantling. However, dismantling requires additional precautions to ensure the scaffold will come down in a controlled, safe, and logical manner. Important factors to consider include the following:

1. Check every scaffold before dismantling. Any loose or missing ties or bracing must be corrected.
2. If a hoist is to be used to lower the material, the scaffold must be tied to the structure at the level of the hoist arm to dispel any overturning effect of the wheel and rope.
3. The erector should be tied off for fall protection, as required by the regulations, unless it is infeasible or a greater hazard to do so.
4. Start at the top and work in reverse order, following the step-by-step procedures for erection. Leave the work platforms in place as long as possible.
5. Do not throw planks or material from the scaffold. This practice will damage the material and presents overhead hazards for workers below.

6. Building tie-ins and bracing can be removed only when the dismantling process has reached that level or location on the scaffold. An improperly removed tie can cause the entire scaffold to overturn.
7. Remove the ladders or the stairs only as the dismantling process reaches that level. Never climb or access the scaffold by using the cross braces.
8. As the scaffold parts come off the scaffold, they should be inspected for any wear or damage. If a defective part is found, it should be tagged for repair and not used again until inspected by the competent person.
9. Dismantled parts and materials should be organized, stacked, and placed in bins or racks out of the weather.
10. Secure the disassembled scaffold equipment to ensure that no unauthorized, untrained employees use it. All erectors must be trained, experienced, and under the supervision and direction of a competent person.
11. Always treat the scaffold components as if a life depended on them, because the next time the scaffold is erected, someone's life will indeed be depending on it being sound.

MOBILE SCAFFOLDS

The rolling tower, or mobile scaffold, is widely used for small jobs, generally not more than 20 feet in height (*Figure 21–27*). The components of the mobile scaffold are the same as those for the

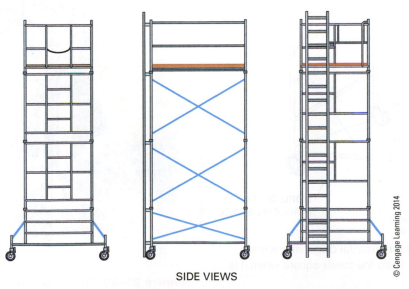

SIDE VIEWS

Figure 21–27 Typical setup for a mobile scaffold.

© Cengage Learning 2014

stationary frame scaffold, with the addition of casters and horizontal diagonal bracing. There are additional restrictions on rolling towers as well.

The height of a rolling tower must never exceed four times the minimum base dimension. For example, if the frame sections are 5 × 7, the rolling tower can be only 20 feet high. If the tower exceeds this height-to-base ratio, it must be secured to prevent overturning. When outriggers are used on a mobile tower, they must be used on both sides.

Casters on mobile towers must be locked with positive wheel swivel locks or the equivalent to prevent movement of the scaffold while it is stationary. Casters typically have a load capacity of 600 pounds each, and the legs of a frame scaffold can hold 2,000 to 3,000 pounds each. Care must be taken not to overload the casters.

Never put a cantilevered work platform, side bracket, or well wheels on the side or end of a mobile tower. Mobile towers can tip over if used incorrectly. Mobile towers must have horizontal, diagonal, or gooser braces at the base to prevent racking of the tower during movement (*Figure 21–28*). Metal hook planks also help prevent racking if they are secured to the frames.

The force to move the scaffold should be applied as close to the base as practicable, but not more than 5 feet (1.5 m) above the supporting surface. The casters must be locked after each movement before beginning work again. Employees are not allowed to ride on rolling tower scaffolds during movement unless the height-to-base width ratio is two-to-one or less. Before the scaffold is moved, each employee on the scaffold must be made aware of the move. Caster and wheel stems shall be pinned or otherwise secured in scaffold legs or adjustment screws. The surface that the mobile tower rolls on must be free of holes, pits, and obstructions and must be within 3 degrees of level. Only use a mobile scaffold on firm floors.

PUMP JACK SCAFFOLDS

Pump jack scaffolds consist of 4 × 4 poles, a pump jack mechanism, and metal braces for each pole (*Figure 21–29*). The braces are attached to the pole at intervals and near the top. The arms of the bracket extend from both sides of the pole at 45-degree angles. The arms are attached to the sidewall or roof to hold the pole steady.

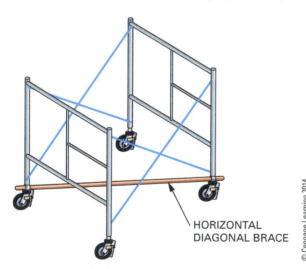

Figure 21–28 The horizontal diagonal brace (or gooser) is used to keep the tower square when it is rolled.

HORIZONTAL DIAGONAL BRACE

© Cengage Learning 2014

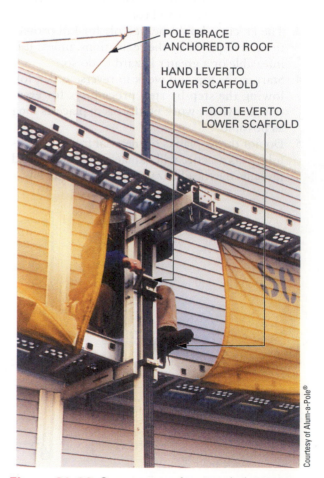

POLE BRACE ANCHORED TO ROOF

HAND LEVER TO LOWER SCAFFOLD

FOOT LEVER TO LOWER SCAFFOLD

Courtesy of Alum-a-Pole®

Figure 21–29 Components of a pump jack system.

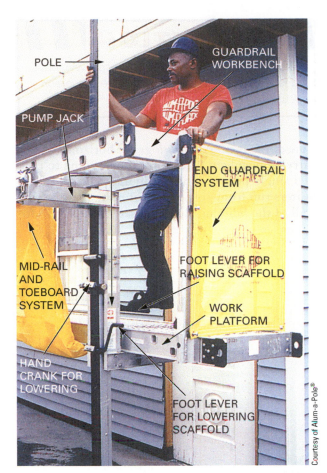

POLE

GUARDRAIL WORKBENCH

PUMP JACK

END GUARDRAIL SYSTEM

MID-RAIL AND TOEBOARD SYSTEM

FOOT LEVER FOR RAISING SCAFFOLD

WORK PLATFORM

HAND CRANK FOR LOWERING

FOOT LEVER FOR LOWERING SCAFFOLD

Courtesy of Alum-a-Pole®

Figure 21–30 Pump jacks are raised by pressing the foot lever.

The scaffold is raised by pressing on the foot pedal of the pump jack (*Figure 21–30*). The mechanism has brackets on which to place the scaffold plank. Other brackets hold a guardrail or platform. Reversing a lever allows the staging to be pumped downward.

Pump jack scaffolds are used widely for siding, where staging must be kept away from the walls, and when a steady working height is desired. However, pump jack scaffolds have their limitations. They should not be used when the working load exceeds 500 pounds. No more than two persons are permitted at one time between any two supports. Wood poles must not exceed 30 feet in height. Braces must be installed at a maximum vertical spacing of not more than 10 feet.

To pump the scaffold past a brace location, temporary braces are used. The temporary bracing is installed about 4 feet above the original bracing. Once the scaffold is past the location of the original brace, it can be reinstalled. The temporary brace is then removed.

Wood pump jack poles are constructed of two 2 × 4s nailed together. The nails should be 3-inch or 10d, and no less than 12 inches apart, staggered uniformly from opposite outside edges.

WOOD SCAFFOLDS

Wood scaffolds are single-pole or double-pole. They are used when working on walls. The single-pole scaffold is used when it can be attached to the wall and does not interfere with the work (*Figure 21–31*). The double-pole scaffold is used when the scaffolding must be kept clear of the wall for the application of materials or for other reasons (*Figure 21–32*). Wood scaffolds are designated as light-, medium-, or heavy-duty scaffolds, according to the loads they are required to support.

Scaffolding Terms

Poles. The vertical members of a scaffold are called poles. All poles should be set plumb. They should bear on a footing of sufficient size and strength to spread the load. This prevents the poles from settling (*Figure 21–33*). If wood poles need to be spliced for additional height,

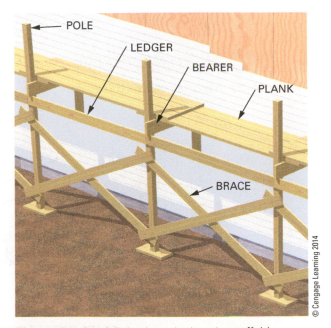

POLE

LEDGER

BEARER

PLANK

BRACE

© Cengage Learning 2014

Figure 21–31 A light-duty single-pole scaffold. Guardrails are required when the scaffolding is over 10 feet in height.

the ends are squared so the upper pole rests squarely on the lower pole. The joint is scabbed on at least two adjacent sides. Scabs should be at least 4 feet long. The scabs are fastened to the poles so they overlap the butted ends equally (*Figure 21–34*).

Bearers. Bearers or *putlogs* are horizontal load-carrying members. They run from building to pole in a single-pole staging. In double-pole scaffolds, bearers run from pole to pole at right angles to the wall of the building. They are set with their width

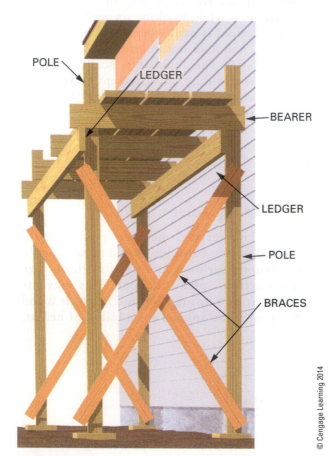

Figure 21–32 A double-pole wood scaffold.

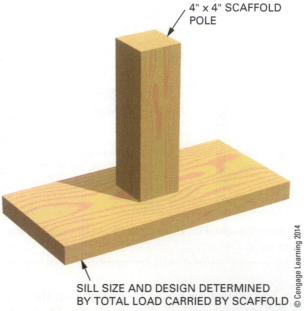

4" x 4" SCAFFOLD POLE

SILL SIZE AND DESIGN DETERMINED BY TOTAL LOAD CARRIED BY SCAFFOLD

Figure 21–33 The bottom ends of scaffold poles are set on footings or pads to prevent them from sinking into the ground.

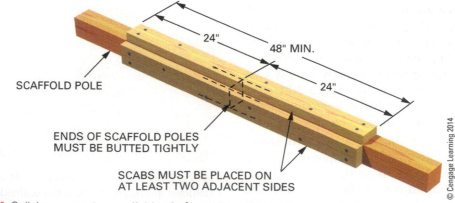

SCAFFOLD POLE

24"

48" MIN.

24"

ENDS OF SCAFFOLD POLES MUST BE BUTTED TIGHTLY

SCABS MUST BE PLACED ON AT LEAST TWO ADJACENT SIDES

Figure 21–34 Splicing a wooden scaffold pole for additional height.

oriented vertically. They must be long enough to project a few inches outside the staging pole.

When placed against the side of a building, bearers must be fastened to a notched *wall ledger*. At each end of the wall, bearers are fastened to the corners of the building (*Figure 21–35*).

Ledgers. Ledgers run horizontally from pole to pole. They are parallel with the building and support the bearers. Ledgers must be long enough to extend over two pole spaces. They must be overlapped at the pole and not spliced between them (*Figure 21–36*).

Braces. Braces are diagonal members. They stiffen the scaffolding and prevent the poles from moving or buckling. Full diagonal face bracing is applied across the entire face of the scaffold in both directions. On medium- and heavy-duty double-pole scaffolds, the inner row of poles is braced in the same manner. Cross bracing is also provided between the inner and outer sets of poles on all double-pole scaffolds. All braces are spliced on the poles (*Figure 21–37*).

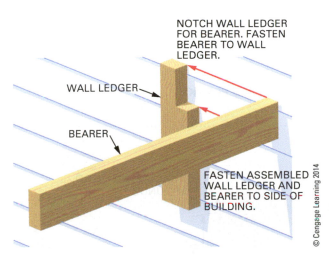

Figure 21–35 Bearers must be fastened in a notch of a wall ledge for placement against the side of a building.

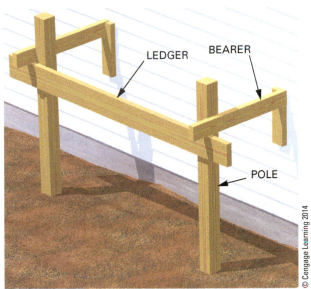

Figure 21–36 Ledgers run horizontally from pole to pole and support the bearers.

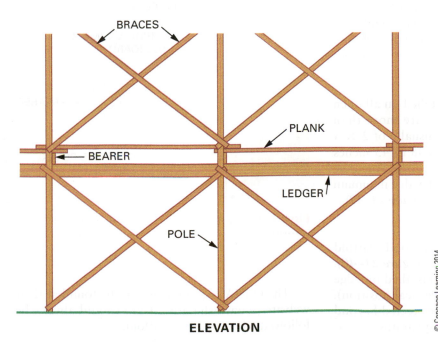

ELEVATION

Figure 21–37 Diagonal bracing is applied across the entire face of the scaffold.

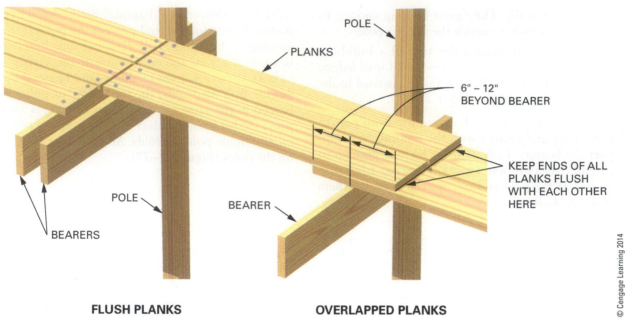

FLUSH PLANKS   **OVERLAPPED PLANKS**

© Cengage Learning 2014

Figure 21–38 Recommended placement of scaffold plank.

Plank. Staging planks rest on the bearers. They are laid with the edges close together so the platform is tight. There should be no spaces through which tools or materials can fall. All planking should be scaffold grade or its equivalent. Planking should have the ends banded with steel to prevent excessive checking.

Overlapped planks should extend at least 6 inches beyond the bearer. Where the ends of planks butt each other to form a flush floor, the butt joint is placed at the centerline of the pole. Each end rests on separate bearers. The planks are secured to prevent movement. End planks should not overhang the bearer by more than 12 inches (*Figure 21–38*).

Guardrails. Guardrails are installed on all open sides and ends of scaffolds that are more than 10 feet in height. The top-rail is usually of 2 × 4 lumber. It is fastened to the poles 38 to 45 inches (0.97–1.1 m) above the working platform. A middle rail of 1 × 6 lumber and a toeboard with a minimum height of 4 inches are also installed (*Figure 21–39*).

Size and Spacing of Scaffold Members. OSHA requires each member of a wood scaffold to be of a certain size and spacing (*Figure 21–40*). All members, except planks, are installed on edge (that is, with their width in the vertical position). Also, no wood scaffold should be erected beyond the reach of the local firefighting apparatus.

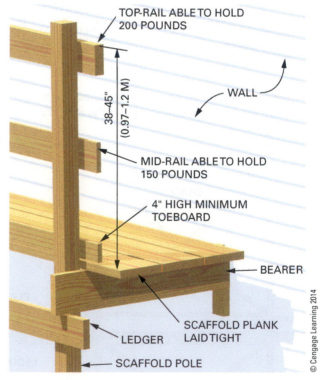

© Cengage Learning 2014

Figure 21–39 Typical wood guardrail specifications.

The OSHA requirements can be found at their website (www.osha.gov). Care should always be followed when erecting scaffold.

Single-Pole Wood Scaffolding				
	Light-Duty		Medium-Duty	Heavy-Duty
Uniformly distributed load	Not to exceed 25 pounds/square foot	Not to exceed 25 pounds/square foot	Not to exceed 50 pounds/square foot	Not to exceed 75 pounds/square foot
Maximum height of scaffold	20 ft	60 ft	60 ft	60 ft
Poles or uprights	2 × 4 in.	4 × 4 in.	4 × 4 in.	4 × 4 in.
Pole spacing (longitudinal)	6 ft 0 in.	10 ft 0 in.	8 ft 0 in.	6 ft 0 in.
Maximum width of scaffold	5 ft 0 in.	5 ft 0 in.	5 ft 0 in.	5 ft 0 in.
Bearers or putlogs	2 × 6 in. or 3 × 4 in.	2 × 6 in. or 3 × 4 in.	2 × 9 in. or 3 × 4 in.	2 × 9 in. or 2 × 5 in. (rough)
Spacing of bearers or putlogs	6 ft 0 in.	10 ft. 0 in.	8 ft 0 in.	6 ft 0 in.
Ledgers	1 × 4 in.	1¼ × 9 in.	2 × 9 in.	2 × 9 in.
Vertical spacing of horizontal members	7 ft 0 in.	7 ft 0 in.	9 ft 0 in.	6 ft 6 in.
Bracing, horizontal and diagonal	1 × 4 in.	1 × 4 in.	1 × 6 in. or 1¼ × 4 in.	2 × 4 in.
Planking	1¼ × 9 in.	2 × 9 in.	2 × 9 in.	2 × 9 in.
Toeboards	4 in. high (min)	4 in. high (min)	4 in. high (min)	4 in. high (min)
Guardrail	2 × 4 in.	2 × 4 in.	2 × 4 in.	2 × 4 in.

Independent (Double-Pole) Wood Scaffolding				
	Light-Duty		Medium-Duty	Heavy-Duty
Uniformly distributed load	Not to exceed 25 pounds/square foot	Not to exceed 25 pounds/square foot	Not to exceed 50 pounds/square foot	Not to exceed 75 pounds/square foot
Maximum height of scaffold	20 ft	60 ft	60 ft	60 ft
Poles or uprights	2 × 4 in.	4 × 4 in.	4 × 4 in.	4 × 4 in.
Pole spacing (longitudinal)	6 ft 0 in.	10 ft 0 in.	8 ft 0 in.	6 ft 0 in.
Pole spacing (transverse)	6 ft 0 in.	10 ft 0 in.	8 ft 0 in.	8 ft 0 in.
Ledgers	1¼ × 4 in.	1¼ × 9 in.	2 × 9 in.	2 × 9 in.
Vertical spacing of horizontal members	7 ft 0 in.	7 ft 0 in.	6 ft 0 in.	4 ft 6 in.
Spacing of bearers	6 ft 0 in.	10 ft 0 in.	8 ft 0 in.	8 ft 0 in.
Bearers	2 × 5 in. or 3 × 4 in.	2 × 9 in. (rough) or 3 × 8 in.	2 × 9 in. (rough)	2 × 9 in. (rough)
Bracing, horizontal	1 × 4 in.	1 × 4 in.	1 × 6 in. or 1¼ × 4 in.	2 × 4 in.
Bracing, diagonal	1 × 4 in.	1 × 4 in.	1 × 4 in.	2 × 4 in.
Planking	1¼ × 9 in.	2 × 9 in.	2 × 9 in.	2 × 9 in.
Toeboards	4 in. high (min)	4 in. high (min)	4 in. high (min)	4 in. high (min)
Guardrail	2 × 4 in.	2 × 4 in.	2 × 4 in.	2 × 4 in.

Figure 21–40 Sizes and spacing requirements of wood scaffold members.

© Cengage Learning 2014

ROOFING BRACKETS

Roofing brackets are used when the slope of the roof is too steep for carpenters to work on without slipping (*Figure 21–41*). Usually any roof with more than 4 on 12 slope requires roof brackets. They keep the worker from slipping and also hold the roofing materials. Roofing brackets are made of metal. Some are adjustable for roofs of different slopes.

A metal plate at the top of the bracket has slots in which to drive nails to fasten the bracket to the roof (*Figure 21–42*). The slots are open angled or *key holed*. This enables removal of the bracket from the fasteners without pulling the nails. The bracket is simply tapped upward from the bottom, and then lifted over the nail heads. The nails that remain are then driven all the way to the surface or flush.

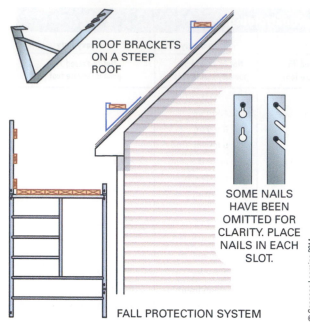

Figure 21–41 Roof brackets are used when the roof pitch is too steep for carpenters to work on without slipping.

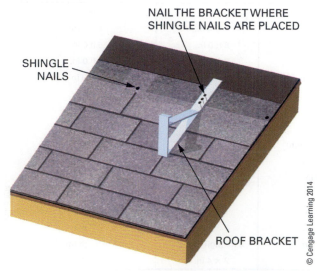

Figure 21–42 Roof brackets are nailed where a shingle nail is located.

Applying Roof Brackets

Roof brackets are usually used when the roof is being shingled. Apply roof brackets in rows. Space them out so that they can be reached without climbing off the roof bracket staging below.

On asphalt-shingled roofs, place the brackets at about 6- to 8-foot (1.8 to 2.4 m) horizontal intervals. The nails for the brackets should be installed so they land where a shingle nail is located (see Figure 21–42). The bottom nail of the bracket should

be at or slightly above the nailing line of the shingle. (See Chapter 51.) This will place the bracket so it will be in centered under the next shingle tab, ensuring that the bracket nails are covered and won't cause leaks. No joint or cutout in the course above should fall in line with the nails holding the bracket. Otherwise, the roof will leak. Use three $3\frac{1}{4}$-inch or 12d common nails driven flush. Try to get at least one nail into a rafter.

Open adjustable brackets so the top member is approximately level or slightly leaning toward the roof. Place a staging plank on the top of the brackets. Overlap them as in wall scaffolds. Keep the inner edges against the roof for greater support. A toeboard made of 1×6 or 1×8 lumber is usually placed flat on the roof with its bottom edge on top of the brackets. This protects the new roofing from damage by the workers' toes during construction (*Figure 21–43*). After the shingles are applied, tap the bottom of the bracket upward along the slope of the roof to release it from the nails. Raise the shingle tab and drive the nails flush so they do not stick up and damage the shingles.

Trestles

Trestle jacks are used when a low scaffold on a level surface is desired. A trestle is a low working platform supported by a bearer with spreading legs at each end. This type of scaffold may be used in the building interior for working on ceilings. They are available for frequent and prolonged use.

Trestle jacks are adjustable in height at about 3-inch intervals. They are clamped to a 2-inch

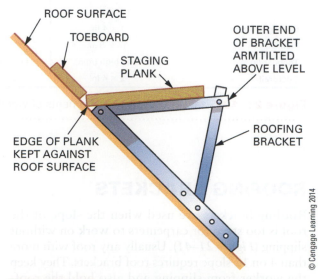

Figure 21–43 A toeboard protects the roofing material from damage from workers' toes.

© Cengage Learning 2014

Figure 21–44 Adjustable metal trestle jacks are manufactured for prolonged and repetitive use.

SAWHORSE

PLANKS

© Cengage Learning 2014

Figure 21–45 Sawhorses are sometimes used as supports for a trestle scaffold.

wood support on which the scaffold planks are placed. The size of the support depends on how far apart the trestle jacks are placed. Metal braces hold the trestle rigid (*Figure 21–44*).

Sawhorses are used for trestle supports when occasional use is required and their height meets the requirements of the job (*Figure 21–45*). For light-duty work, horses and trestle jacks should not be spaced more than 8 feet apart. Do not use horses that have become weak or defective. The horse should be strong enough to support four times the intended load.

If a horse scaffold is arranged in tiers, no more than two tiers should be used. The legs of the horses in the upper tier should be nailed to the planks. Each tier should be cross braced.

SCAFFOLD SAFETY

The safety of those working at a height depends on properly constructed scaffolds and proper use of scaffolds. Those who have the responsibility of constructing scaffolds must be thoroughly familiar with the sizes, spacing, and fastening of scaffold members and other scaffold construction techniques. Those who use the scaffold must be aware of where they are working at all times. They must watch their step and the material they are using to prevent accidents. They should also make it a habit to visually inspect the scaffold before each use.

DECONSTRUCT THIS

Carefully study *Figure 21–46* and think about what is wrong and/or what is right. Consider all possibilities. What construction practice or method is different in your area of the country?

© Cengage Learning 2014

Figure 21–46 This photo shows a carpenter using an extension ladder.

KEY TERMS

bearers	guardrails	pump jack	stepladder
braces	ladder jacks	roofing brackets	trestles
competent person	ledgers	sawhorses	toeboard
double-pole	mobile scaffold	scaffolds	
extension ladder	mud sill	single-pole	
fall protection	poles	staging planks	

REVIEW QUESTIONS

Select the most appropriate answer.

1. Which of the following demonstrates a good job site attitude?

   a. Using any tool that happens to be nearby for a particular job.
   b. Asking for more information about how to work safely.
   c. Using your fist to force a board into place.
   d. Shouting your anger when something goes wrong.

2. Which of the following is not recommended by OSHA?

   a. Workers are required to work in hazardous conditions.
   b. Employers must provide safety instruction.
   c. Workers are to use tools and equipment safely.
   d. Employers must maintain safe conditions on the job site.

3. The loudest job site equipment below is

   a. chainsaw
   b. concrete saw
   c. circular saw
   d. bulldozer

4. A versatile machine with interchangeable attachments to the lift arms is called

   a. excavator
   b. tellehandlers
   c. skid steer
   d. backhoe

5. Personal fall protection system includes

   a. work boots, harness, lanyards, and anchor
   b. rope grab, life line, gloves, and lanyards
   c. harness, life line, rope grab, lanyard, and anchor
   d. anchor, hard hat, harness, and life line

6. To lower a tool from a roof it is best to use

   a. the electrical cord
   b. the pneumatic hose
   c. tellehandler
   d. all of the above

7. Waste materials from a jobsite should be

   a. recycled
   b. reused
   c. reduced
   d. all of the above

8. The vertical members of a pump jack scaffold are called

   a. columns.
   b. piers.
   c. poles.
   d. uprights.

9. Bearers of a wood scaffold support

   a. ledgers.
   b. planks.
   c. rails.
   d. braces.

10. Scaffold planks should be at least

    a. 2 × 6.
    b. 2 × 8.
    c. 2 × 10.
    d. 2 × 12.

11. Overlapped planks should extend beyond the bearer at least _____ inches and no more than _____ inches.

    a. 3, 6
    b. 3, 8
    c. 6, 8
    d. 6, 12

12. Metal tubular frame scaffolding is held rigidly plumb by

    a. end frames.
    b. goosers.
    c. cross braces.
    d. cribbing.

13. The workers allowed to climb an access ladder for a metal tubular scaffold that has its rungs spaced 18 inches apart are the scaffold

    a. users only.
    b. erectors only.
    c. erector and dismantlers only.
    d. all of the above.

14. The height of a mobile scaffold must not exceed the minimum base dimension by

    a. three times.
    b. four times.
    c. five times.
    d. six times.

15. OSHA requires that guardrails be installed on all scaffolds more than

    a. 6 feet in height.
    b. 10 feet in height.
    c. 16 feet in height.
    d. 20 feet in height.

16. Tubular scaffold end frames should be installed level and plumb to sit on top of

    a. base plates.
    b. mud sills.
    c. cribbing.
    d. all of the above.

17. To access the work area of a scaffold, the user should use an approved ladder or

    a. the ladder built into the end frame.
    b. cross braces.
    c. the horizontal bearing points of the scaffold.
    d. all of the above.

18. All scaffolding should be able to support _____ times the intended load.

    a. four
    b. five
    c. ten
    d. twenty

19. The responsibility for safety on a job site relies on

    a. the local OSHA inspector.
    b. the general contractor.
    c. scaffold erectors.
    d. every worker on the job.

20. Pump jacks differ from metal tubular scaffolding in that metal tubular scaffolding

    a. often requires toeboards.
    b. is usually installed on a mud sill.
    c. has braces.
    d. none of the above.

UNIT 9
Architectural Plans and Building Codes

The ability to interpret architectural plans and to understand building codes, zoning ordinances, building permits, and inspection procedures is necessary for promotion on the job. With this ability, the carpenter has the competence needed to handle jobs of a supervisory nature, including those of foreman, superintendent, and contractor.

OBJECTIVES

After completing this unit, the student should be able to:

- recognize the difference between residential and commercial drawings.

- describe and explain the function of the various kinds of drawings contained in a set of architectural plans.

- demonstrate how specifications are used.

- read and use an architect's scale.

- identify various types of lines and read dimensions.

- identify and explain the meaning of symbols and abbreviations used on architectural plans.

- read and interpret plot, foundation, floor, and framing plans.

- locate and explain information found in exterior and interior elevations.

- identify and utilize information from sections and details.

- use schedules to identify and determine location of windows, doors, and interior finish.

- define and explain the purpose of building codes and zoning laws.

- explain the requirements for obtaining a building permit and the duties of a building inspector.

SAFETY REMINDER

A set of building prints is the vehicle used to communicate to the trades person the desired structure. Failing to read and understand prints properly can have costly and potentially disastrous results.

CHAPTER 22
Understanding Architectural Plans

RESIDENTIAL AND COMMERCIAL PRINTS

Plans and prints are a collection of large paper sheets designed to communicate the owner's desires to the builder. They consist of different views and close-up details of the construction process. In order to build from prints or plans the carpenter must know how to read and interpret these drawing.

Residential and commercial plans are written in similar formats, using similar lines and symbols. They have written tables and attached documents that explain what materials are to be selected and provide notes on how they are to be used. The scope of residential plans is smaller than for commercial prints. Commercial plan sets have more pages and a more complicated level of detail.

Commercial plans are more complicated for several reasons. Commercial projects are simply a lot bigger: Commercial buildings have more and different types of materials in them, and each material must be itemized and specified with explanations on how it is to be installed. Codes for commercial building are more restrictive and more encompassing than residential projects. A complete set of commercial prints and paper work is considered a legal document.

ARCHITECTURAL PLANS

Blueprinting is an early method of creating drawings and making copies. It uses a water wash to produce prints with white lines against a dark blue background. Although these are true blueprints, the process is seldom used today. The word *blueprint* remains in use, however, to mean any copy of the original drawing.

A later method of making prints uses the original black line drawings of graphite or ink on translucent paper, cloth, or polyester film. Copies were then made with a *diazo* printer. The copying process utilizes an ultraviolet light that passes through the paper, but not the lines. The light causes a chemical reaction on the copy paper, except where the lines are drawn. The copy paper is then exposed to ammonia vapor. This vapor develops the remaining chemical into blue, black, or brown lines against a white background. The diazo printing process is faster and less expensive than the early blueprinting methods.

Today most architectural plans, also called architectural drawings or construction drawings, are done using a CADD (computer-aided drafting and design) program. CADD drawings are produced faster and more easily than drawings done by hand. Changes to the plan can be made faster, allowing them to be easily customized to the desires of the customer. Once the plans are finished, they are sent to a plotter or copier, which prints as many copies as needed.

Prints are drawn using symbols and a standardized language. To adequately read and interpret these prints, the builder must be able to understand the language of architectural plans. While this is not necessarily difficult, it can be confusing.

TYPES OF VIEWS

The architect can choose from any of several types of views to describe a building with clarity, and often multiple views are prepared. Because many people and many different trades are needed to build a house, different views are required that contain different information.

Pictorial View

Pictorial drawings are usually three-dimensional (3D) *perspective* or isometric views (*Figure 22–1*). The lines in a perspective view diminish in size as they converge toward vanishing points located on a line called the *horizon*. In an isometric drawing, the horizontal lines are drawn at 30-degree angles. All lines are drawn to actual scale. They do not diminish or converge as in perspective drawings.

A *presentation drawing* is usually a perspective view. It shows the building from a desirable vantage point to display its most interesting features (*Figure 22–2*). Walks, streets, shrubs, trees, vehicles, and even people may be drawn. Presentation drawings are often colored for greater eye appeal. Presentation drawings provide little information for construction purposes. They are usually used as a marketing tool to show the client how the completed building will look.

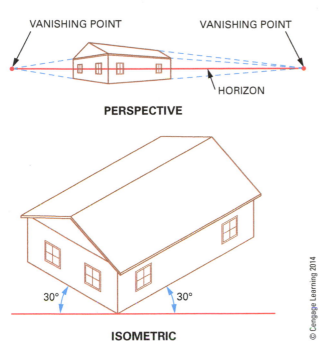

PERSPECTIVE

VANISHING POINT   VANISHING POINT

HORIZON

30°   30°

ISOMETRIC

Figure 22–1 Pictorial views used in architectural drawings.

Figure 22–2 A presentation drawing is usually a perspective view.

MULTIVIEWS

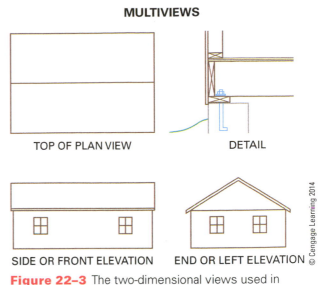

TOP OF PLAN VIEW   DETAIL

SIDE OR FRONT ELEVATION   END OR LEFT ELEVATION

Figure 22–3 The two-dimensional views used in architectural drawings.

© Cengage Learning 2014

Elevations show the building as seen from the street. They can show front, back, right side, and left side. *Section views* show a cross section as if the building were sliced open to reveal its skeleton (*Figure 22–5*). *Details* are blown up, with zoomed-in views of various items to show a closer view.

Plan Views

Plot Plan. The *plot plan* shows information about the lot, such as property lines and directions and measurements for the location of the building, walks, and driveways (*Figure 22–6*). It shows elevation heights and direction of the sloping ground. The drawing simulates a view looking down from a considerable height. It is made at a small scale because of the relatively large area it represents. This view helps the builder locate the building on the site and helps local officials to estimate the impact of the project on the lot.

Multiview Drawings

Different kinds of *multiview* drawings, also called orthographics, are required for the construction of a building. Multiview drawings are two-dimensional (2D) and offer separate views of the building (*Figure 22–3*). The bottom view is never used in architectural drawings.

These different views are called plan view, elevation, section view, and details. When together, they constitute a set of architectural plans. *Plan views* show the building from above, looking down. There are many kinds of plan views for different stages of construction (*Figure 22–4*).

Green Tip

Orienting a building to address the sun will affect a lifetime of heating and cooling dollars. Warm climates should include natural shading and northerly exposures. Cooler areas should orient longer building sides and larger glass toward southerly exposures.

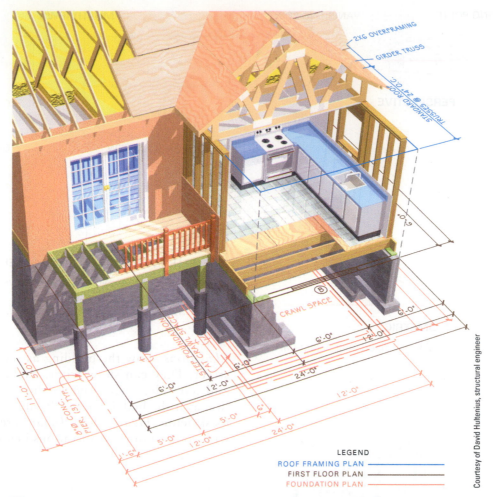

LEGEND
ROOF FRAMING PLAN ———————
FIRST FLOOR PLAN ———————
FOUNDATION PLAN ———————

Courtesy of David Hultenius, structural engineer

Figure 22–4 Plan views are horizontal cut views through a building.

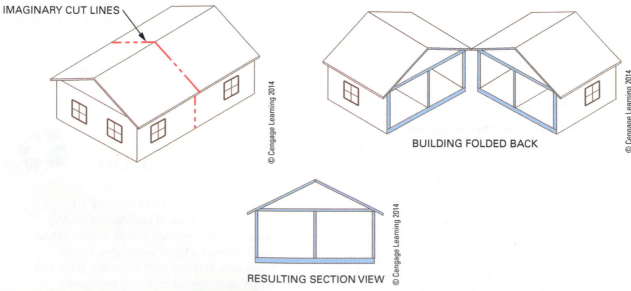

IMAGINARY CUT LINES

© Cengage Learning 2014

BUILDING FOLDED BACK

© Cengage Learning 2014

RESULTING SECTION VIEW

© Cengage Learning 2014

Figure 22–5 Section views are cutaway views.

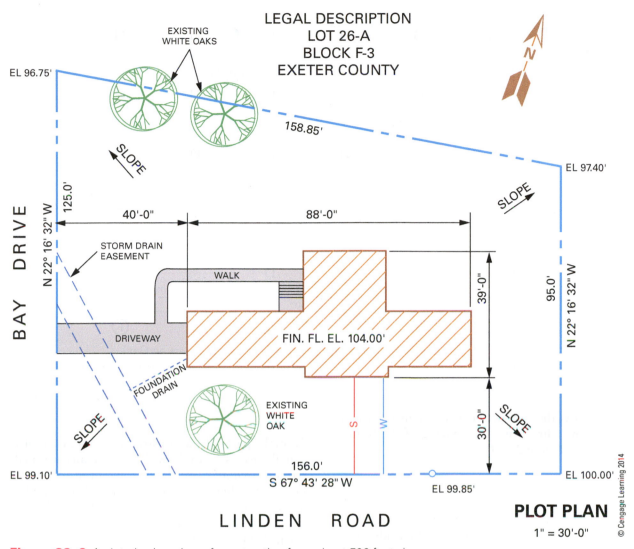

LEGAL DESCRIPTION
LOT 26-A
BLOCK F-3
EXETER COUNTY

Figure 22–6 A plot plan is a view of construction from about 500 feet above.

Structural Plans. Structural plans of commercial prints show the structural parts of the building, such as reinforced concrete, steel, and heavy timbers. The materials used and notes to communicate details are also shown on the structural plans.

Commercial floor plans are laid out within a grid or coordinate system. The grid lines are drawn where there is a structural load point or foundation. Structural columns and bearing points are located at the intersection of these lines (*Figure 22–7*). This is done so all columns and footings may be referenced to the grid by the numbering system. Grid lines going in one direction are numbered, and the opposite lines are lettered. Everything inside the building is measured and dimensioned from these grid lines.

Foundation Plan. The *foundation plan* (*Figure 22–8*) shows a horizontal cut through the foundation walls. It shows the shape and dimensions of the foundation walls and footings. Windows, doors, and stairs located in this level are included. First-floor framing material size, spacing, and direction are sometimes included.

Floor Plan. The floor plan is view of the horizontal cut made about 4 to 5 feet above the floor. It shows the locations of walls and partitions, windows, and doors and fixtures appropriate for each room (*Figure 22–9*). Dimensions are included.

Framing Plan. *Framing plans* are not always found in a set of architectural drawings. When used, they may be of the floor or roof framing.

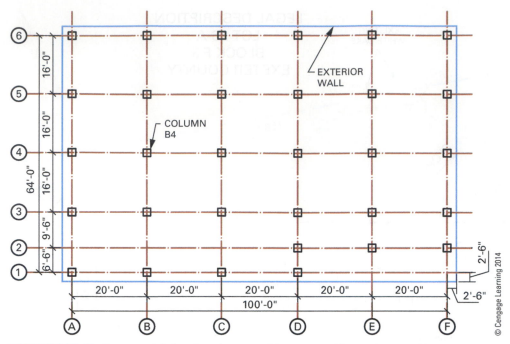

FIGURE 22–7 Commercial drawings have a grid system with numbers and letters.

They show the support beams and girders as well as the size, direction, and spacing of the framing members (*Figure 22–10*).

Elevations

Elevations are a group of drawings (*Figure 22–11*) that show the shape and finishes of all sides of the exterior of a building. *Interior elevations* are drawings of certain interior walls. The most common are kitchen and bathroom wall elevations. They show the design and size of cabinets built on the wall (*Figure 22–12*). Other walls that have special features, such as a fireplace, may require an elevation drawing. Occasionally found in some sets of plans are *framing elevations*. Similar to framing plans, they show the spacing, location, and sizes of wall framing members. No further description of framing drawings is required to be able to interpret them.

Section Views

A set of architectural plans may have many *section* views. Each is designed to reveal the structure or skeleton view of a particular part of the building (*Figure 22–13*). A section reference line is found on the plans or elevations to identify the section being viewed.

Details

To make parts of the construction more clear, it is usually necessary to draw *details*. Details are small parts drawn at a very large scale, even full size if necessary. Their existence is revealed on the plan and elevation views using a symbol (*Figure 22–14*). The location of the symbol shows the location of the vertical cut through the building. This symbol may have different shapes, yet all have numbers or letters that refer to the page where the detail is shown.

Page Numbering of Prints

A set of drawings contains a lot of information. To reduce confusion while reading them, the drawings are made limiting the information on each page. Particular pages can relate to electrical work, plumbing, heating, and ventilating, for example. They are usually printed on separate sheets and labeled according to the type of construction being drawn. Labels begin with a letter that helps identify the type of construction, and a number that follows indicates the page number. These are the labels:

A – Architectural
S – Structural
M – Mechanical
C – Civil (site work)
P – Plumbing
E – Electrical.

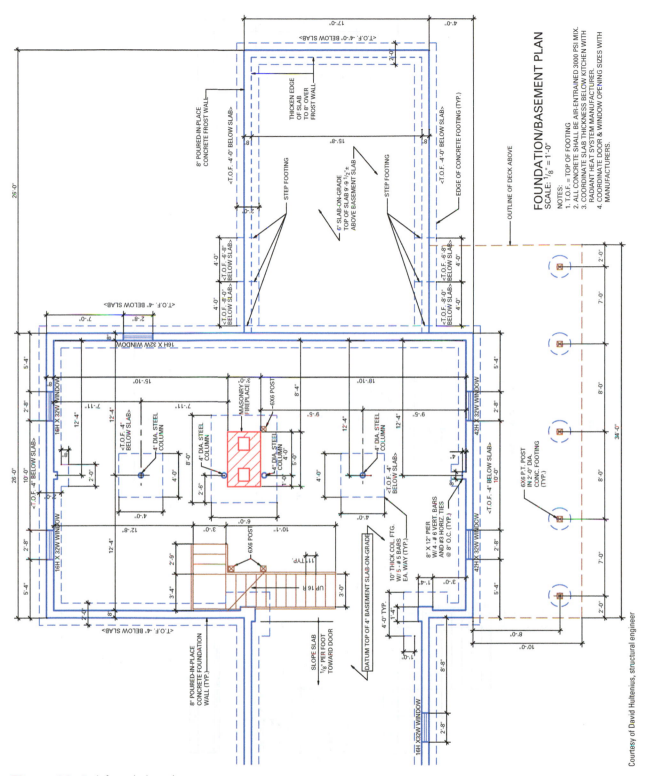

Figure 22–8 A foundation plan.

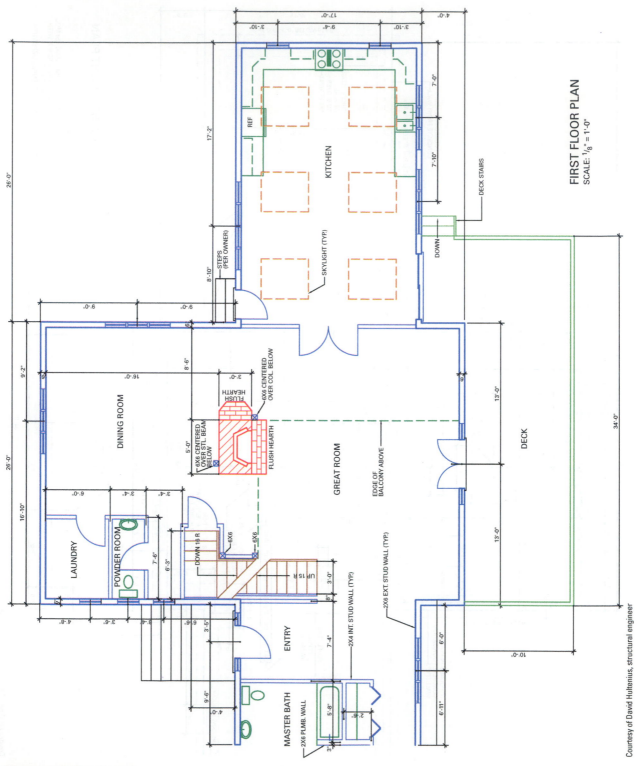

FIRST FLOOR PLAN
SCALE: 1/8" = 1'-0"

Figure 22–9 A floor plan.

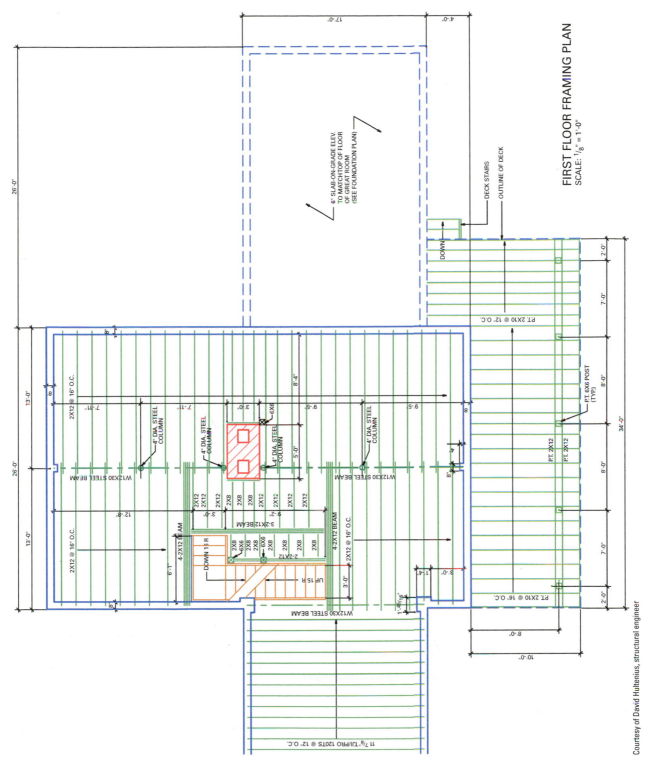

Figure 22–10 A first-floor framing plan.

Figure 22–11 Elevations.

For example the third page in the structural set would be S3.

The carpenter is responsible for building the structure to accommodate wiring, pipes, and ducts. He or she must be able to interpret these plans proficiently to understand the work involved.

SCHEDULES

Besides drawings, printed instructions are included in a set of drawings. Window schedules and door schedules (*Figure 22–15*) give information about the location, size, and kind of windows and doors

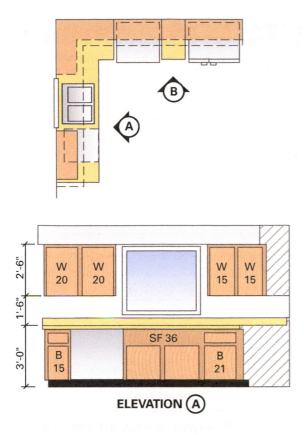

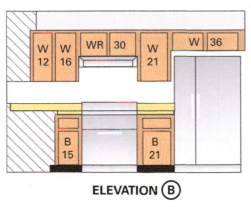

Figure 22–12 Interior wall elevations.

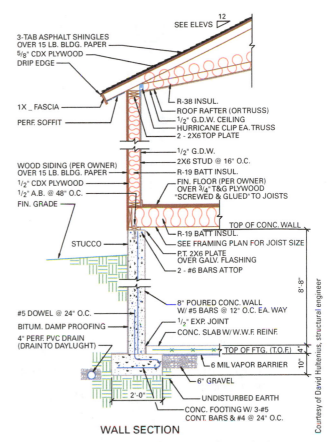

Figure 22–13 A section is a view of a vertical cut through part of the construction.

Window, door, and finish schedules are used for easy understanding and to conserve space on floor plans.

SPECIFICATIONS

Specifications, commonly called *specs,* are written to give information that cannot be completely provided in the drawings or schedules. They supplement the working drawings with more complete descriptions of the methods, materials, and quality of construction. If there is a conflict between any information in a set of prints, the specifications take precedence over the drawings. Any conflict should be pointed out to the architect so corrections can be made.

The amount of detail contained in the specs will vary, depending on the size of the project. On small jobs, they may be written by the architect. In large commercial jobs, a specifications writer, trained in the construction process, may be required. For commercial projects, a specifications guide, used by spec

to be installed in the building. Each of the different units is given a number or letter. A corresponding number or letter is found on the floor plan to show the location of the unit. Windows may be identified by letters and doors by numbers. The letters and numbers may be framed with various geometric figures, such as circles and triangles.

A finish schedule (*Figure 22–16*) may also be included in a set of plans. This schedule gives information on the kind of finish material to be used on the floors, walls, and ceilings of the individual rooms.

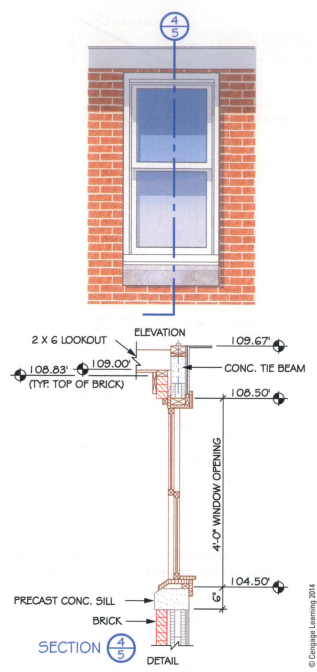

Figure 22–14 Detail of a window.

Window Schedule				
Sym	Size	Model	Rough Open	Quan.
A	1' × 5'	Job Built	Verify	2
B	8' × 5'	W 4 N 5 CSM.	8'-0¾ × 5'-0⅞	1
C	4' × 5'	W 2 N 5 CSM.	4'-0¾ × 5'-0⅞	2
D	4' × 3⁶	W 2 N 3 CSM.	4'-0¾ × 3'-6½	2
E	3⁶ × 3⁶	2 N 3 CSM.	3'-6½ × 3'-6½	2
F	6' × 4'	G 64 Sldg.	6'-0½ × 4'-0½	1
G	5' × 3⁶	G 536 Sldg.	5'-0½ × 3'-6½	4
H	4' × 3⁶	G 436 Sldg.	4'-0½ × 3'-6½	1
J	4' × 2'	A 41 Awn.	4'-0½ × 2'-0⅞	3

Door Schedule			
Sym	Size	Type	Quan.
1	3' × 6⁸	S.C. RP. Metal Insulated	1
2	3' × 6⁸	S.C. Flush Metal Insulated	2
3	2⁸ × 6⁸	S.C. Self Closing	2
4	2⁸ × 6⁸	Hollow Core	5
5	2⁶ × 6⁸	Hollow Core	5
6	2⁶ × 6⁸	Pocket Sldg.	2

Figure 22–15 A typical window and door schedule.

Interior Finish Schedule												
Room	Floor					Walls				Ceil.		
	Vinyl	Carpet	Tile	Hardwood	Concrete	Paint	Paper	Texture	Spray	Smooth	Brocade	Paint
Entry					●							
Foyer		●				●			●	●		●
Kitchen		●						●		●		●
Dining				●		●				●		●
Family		●				●				●		●
Living		●				●				●		●
Mstr. Bath			●				●			●		●
Bath #2			●			●			●	●	●	●
Mstr. Bed		●				●		●		●		●
Bed #2		●				●			●	●		●
Bed #3		●				●				●		●
Utility	●					●			●	●		●

Figure 22–16 A typical finish schedule.

writers, has been developed by the Construction Specifications Institute (CSI). The guide is the Master Format that allows for descriptions of various construction elements to follow a logical process. This is to keep information in a logical form and reduce confusion.

Two versions of the Master Format are currently in use: a 1995 version and a 2004 version. It does not matter which format is used because both are designed so that needed information can be easily found.

The 1995 guide has sixteen major divisions, each containing a number of subdivisions (*Figure 22–17*).

DIVISION 1—GENERAL REQUIREMENTS
01010 Summary of Work
01100 Alternatives
01150 Measurement & Payment
01200 Project Meetings
01300 Submittals
01400 Quality Control
01500 Temporary Facilities & Controls
01600 Material & Equipment
01700 Project Closeout

DIVISION 2—SITE WORK
02010 Subsurface Exploration
02100 Clearing
02110 Demolition
02200 Earthwork
02250 Soil Treatment
02300 Pile Foundations
02350 Caissons
02400 Shoring
02500 Site Drainage
02550 Site Utilities
02600 Paving & Surfacing
02700 Site Improvements
02800 Landscaping
02850 Railroad Work
02900 Marine Work
02950 Tunneling

DIVISION 3—CONCRETE
03100 Concrete Formwork
03150 Forms
03200 Concrete Reinforcement
03250 Concrete Accessories
03300 Cast-in-Place Concrete
03350 Specially Finished (Architectural) Concrete
03360 Specially Placed Concrete
03400 Precast Concrete
03500 Cementitious Decks
03600 Grout

DIVISION 4—MASONRY
04100 Mortar
04150 Masonry Accessories
04200 Unit Masonry
04400 Stone
04500 Masonry Restoration & Cleaning
04550 Refractories

DIVISION 5—METALS
05100 Structural Metal Framing
05200 Metal Joists
05300 Metal Decking
05400 Lightgage Metal Framing
05500 Metal Fabrications
05700 Ornamental Metal
05800 Expansion Control

DIVISION 6—WOOD & PLASTICS
06100 Rough Carpentry
06130 Heavy Timber Construction
06150 Trestles
06170 Prefabricated Structural Wood
06200 Finish Carpentry
06300 Wood Treatment
06400 Architectural Woodwork
06500 Prefabricated Structural Plastics
06600 Plastic Fabrications

DIVISION 7—THERMAL & MOISTURE PROTECTION
07100 Waterproofing
07150 Dampproofing
07200 Insulation

07300 Shingles & Roofing Tiles
07400 Preformed Roofing & Siding
07500 Membrane Roofing
07570 Traffic Topping
07600 Flashing & Sheet Metal
07800 Roof Accessories
07900 Sealants

DIVISION 8—DOOR & WINDOWS
08100 Metal Doors & Frames
08200 Wood & Plastic Doors
08300 Special Doors
08400 Entrances & Storefronts
08500 Metal Windows
08600 Wood & Plastic Windows
08650 Special Windows
08700 Hardware & Specialties
08800 Glazing
08900 Window Walls/Curtain Walls

DIVISION 9—FINISHES
09100 Lath & Plaster
09250 Gypsum Wallboard
09300 Tile
09400 Terrazzo
09500 Acoustical Treatment
09540 Ceiling Suspension Systems
09550 Wood Flooring
09650 Resilient Flooring
09680 Carpeting
09700 Special Flooring
09760 Floor Treatment
09800 Special Coatings
09900 Painting
09950 Wall Covering

DIVISION 10—SPECIALTIES
10100 Chalkboards & Tackboards
10150 Compartments & Cubicles
10200 Louvers & Vents
10240 Grilles & Screens
10260 Wall & Corner Guards
10270 Access Flooring
10280 Specialty Modules
10290 Pest Control
10300 Fireplaces
10350 Flagpoles
10400 Identifying Devices
10450 Pedestrian Control Devices
10500 Lockers
10530 Protective Covers
10550 Postal Specialties
10600 Partitions
10650 Scales
10670 Storage Shelving
10700 Sun Control Devices (Exterior)
10750 Telephone Enclosures
10800 Toilet & Bath Accessories
10900 Wardrobe Specialties

DIVISION 11—EQUIPMENT
11050 Built-in Maintenance Equipment
11100 Bank & Vault Equipment
11150 Commercial Equipment
11170 Checkroom Equipment
11180 Darkroom Equipment
11200 Ecclesiastical Equipment
11300 Educational Equipment
11400 Food Service Equipment
11480 Vending Equipment
11500 Athletic Equipment
11500 Industrial Equipment
11600 Laboratory Equipment
11630 Laundry Equipment
11650 Library Equipment
11700 Medical Equipment

11800 Mortuary Equipment
11830 Musical Equipment
11850 Parking Equipment
11860 Waste Handling Equipment
11870 Loading Dock Equipment
11880 Detention Equipment
11900 Residential Equipment
11970 Theater & Stage Equipment
11990 Registration Equipment

DIVISION 12—FURNISHINGS
12100 Artwork
12300 Cabinets & Storage
12500 Window Treatment
12550 Fabrics
12600 Furniture
12670 Rugs & Mats
12700 Seating
12800 Furnishing Accessories

DIVISION 13—SPECIAL CONSTRUCTION
13010 Air Supported Structures
13050 Integrated Assemblies
13100 Audiometric Room
13250 Clean Room
13350 Hyperbaric Room
13400 Incinerators
13440 Instrumentation
13450 Insulated Room
13500 Integrated Ceiling
13540 Nuclear Reactors
13550 Observatory
13600 Prefabricated Structures
13700 Special Purpose Rooms & Buildings
13750 Radiation Protection
13770 Sound & Vibration Control
13800 Vaults
13850 Swimming Pools

DIVISION 14—CONVEYING SYSTEMS
14100 Dumbwaiters
14200 Elevators
14300 Hoists & Cranes
14400 Lifts
14500 Material Handling Systems
14570 Turntables
14600 Moving Stairs & Walks
14700 Tube Systems
14800 Powered Scaffolding

DIVISION 15—MECHANICAL
15010 General Provisions
15050 Basic Materials & Methods
15180 Insulation
15200 Water Supply & Treatment
15300 Waste Water Disposal & Treatment
15400 Plumbing
15500 Fire Protection
15600 Power or Heat Generation
15650 Refrigeration
15700 Liquid Heat Transfer
15800 Air Distribution
15900 Controls & Instrumentation

DIVISION 16—ELECTRICAL
16010 General Provisions
16100 Basic Materials & Methods
16200 Power Generation
16300 Power Transmission
16400 Service & Distribution
16500 Lighting
16600 Special Systems
16700 Communications
16850 Heating & Cooling
16900 Controls & Instrumentation

© Cengage Learning 2014

Figure 22–17 The 1995 version of CSI format for specifications.

Division 6-Wood & Plastics

Section 06200—Finish Carpentry

General: This section covers all finish woodwork and related items not covered elsewhere in these specifications. The contractor shall furnish all materials, labor, and equipment necessary to complete the work, including rough hardware, finish hardware, and specialty items.

Protection of Materials: All millwork (finish woodwork*) and trim is to be delivered in a clean and dry condition and shall be stored to insure proper ventilation and protection from dampness. Do not install finish woodwork until concrete, masonry, plaster, and related work is dry.

Materials: All materials are to be the best of their respective kind. Lumber shall bear the mark and grade of the association under whose rules it is produced. All millwork shall be kiln dried to a maximum moisture content of 12%.

   1. Exterior trim shall be select grade white pine, S4S.

   2. Interior trim and millwork shall be select grade white pine, thoroughly sanded at the time of installation.

Installation: All millwork and trim shall be installed with tight fitting joints and formed to conceal future shrinkage due to drying. Interior woodwork shall be mitered or coped at corners (cut in a special way to form neat joints*). All nails are to be set below the surface of the wood and concealed with an approved putty or filler.

*(explanations in parentheses have been added to aid the student.)

© Cengage Learning 2014

Figure 22–18 Sample specifications following the CSI format.

An example of the content and the manner in which specifications are written, using the specification guide, under Division 6—WOOD & PLASTICS, Section 06200, is shown in *Figure 22–18*.

For some light commercial and residential construction, many sections of the spec guide would not apply. A shortened version is then used, eliminating divisions 12,000, 13,000, and 14,000. On simpler plans, notations made on the same sheets as the drawings may take the place of specifications. The ability to read notations and specifications accurately is essential in order to conform to the architect's design.

Both the 1995 version and the 2004 version are the same for the first 14 divisions. The 2004 version, which is designed for large commercial projects, is more detailed than the earlier version. Also, the 2004 version reserves some numbers for future expansion (*Figure 22–19*).

ARCHITECTURAL PLAN LANGUAGE

Carpenters must be able to read and understand the combination of lines, dimensions, symbols, and notations on the architectural drawings. Only then

can they build exactly as the architect has designed the construction. No deviation from the plans may be made without the approval of the architect.

Scales

It would be inconvenient and impractical to make full-size drawings of a building. Therefore, they are drawn to scale. This means that each line in the drawing is reduced proportionally to a size that clearly shows the information and can be handled conveniently. Not all drawings in a set of plans, or even on the same page, are drawn at the same scale. The scale of a drawing is stated in the title block of the page. It can also be listed directly below the drawing.

Architect's Scale. The architect's scale is commonly used to scale lines when making drawings (*Figure 22–20*). It has a triangular cross section giving space for two scales on each face. Six faces are produced, allowing room for a potential of six scales. But it doesn't stop there. The main scale is simply a ruler divided into 1-inch increments and fractions in $\frac{1}{16}$-inch increments. The other five scales actually show two scales in each. Each scale is doubled up with another scale. The desired scale used depends on which end of the scale is read.

Each of the five scales is divided into fractional increments depending on the number labeled at the end scale. For example, the $\frac{1}{4}$ scale is read from right to left. At the other end, the $\frac{1}{8}$ scale is read from left to right (*Figure 22–21*). One scale is twice as big (or half as big, depending) as its counterpart. To draw a line in $\frac{1}{4}$ scale, each foot of actual size is drawn as $\frac{1}{4}$ inch. Read the scales in one direction from the starting end, being careful not to confuse it with the scale running in the opposite direction.

The scales are paired as follows:

$1\frac{1}{2}'' = 1'\text{-}0''$	and	$3'' = 1'\text{-}0''$
$1'' = 1'\text{-}0''$	and	$\frac{1}{2}'' = 1'\text{-}0''$
$\frac{3}{8}'' = 1'\text{-}0''$	and	$\frac{3}{4}'' = 1'\text{-}0''$
$\frac{1}{8}'' = 1'\text{-}0''$	and	$\frac{1}{4}'' = 1'\text{-}0''$
$\frac{3}{32}'' = 1'\text{-}0''$	and	$\frac{3}{16}'' = 1'\text{-}0''$

On the end of each scale, a space representing 1 foot at that scale is divided up. This space is used when scaling off dimensions that include fractions of a foot. What these lines represent depends on the scale. In some cases, they are divided into fractions of an inch and in others they are divided into whole inches. For example, in *Figure 22–21*, the $1\frac{1}{2}$ scale's smallest line represents an actual

Pre-2004 CSI Version	2004 Version of CSI Divisions
DIVISION 1 GENERAL REQUIREMENTS	Division 0
DIVISION 2 SITE CONSTRUCTION	Division 01 General Requirements
DIVISION 3 CONCRETE	Division 02 Existing Conditions
DIVISION 4 MASONRY	Division 03 Concrete
DIVISION 5 METALS	Division 04 Masonry
DIVISION 6 WOOD AND PLASTICS	Division 05 Metals
DIVISION 7 THERMAL AND MOISTURE PROTECTION	Division 06 Wood, Plastics, and Composites
DIVISION 8 DOORS AND WINDOWS	Division 07 Thermal and Moisture Protection
DIVISION 9 FINISHES	Division 08 Openings
DIVISION 10 SPECIALTIES	Division 09 Finishes
DIVISION 11 EQUIPMENT	Division 10 Specialties
DIVISION 12 FURNISHINGS	Division 11 Equipment
DIVISION 13 SPECIAL CONSTRUCTION	Division 12 Furnishings
DIVISION 14 CONVEYING SYSTEMS	Division 13 Special Construction
DIVISION 15 MECHANICAL	Division 14 Conveying Equipment
DIVISION 16 ELECTRICAL	Division 21 Fire Suppression
	Division 22 Plumbing
	Division 23 Heating, Ventilating, and Air Conditioning
	Division 25 Integrated Automation
	Division 26 Electrical
	Division 27 Communications
	Division 28 Electronic Safety and
	Division 31 Earthwork
	Division 32 Exterior Improvements
	Division 33 Utilities
	Division 34 Transportation
	Division 40 Process Integration
	Division 41 Material Processing and Handling Equipment
	Division 42 Process Heating, Cooling, and Drying Equipment
	Division 43 Process Gas & Liquid Handling, Purification, & Storage Equipment
	Division 44 Pollution Control Equipment
	Division 45 Industry-Specific Manufacturing Equipment
	Division 48 Electrical Power Generation

© Cengage Learning 2014

Figure 22–19 The 2004 version of CSI formats has more detail and room for future expansion.

$\frac{1}{4}$ inch. In the $\frac{3}{8}$ scale, the smallest line represents 1 actual inch.

Commonly Used Scales. Probably the most commonly used scale found on architectural plans is $\frac{1}{4}$ inch equals 1 foot. This is indicated as $\frac{1}{4}'' = 1'\text{-}0''$. It is often referred to as a "quarter-inch scale." This means that every $\frac{1}{4}$ inch on the drawing will equal 1 foot in the building. Floor plans and exterior elevations for most residential buildings are drawn at this scale.

To show the location of a building on a lot and other details of the site, the architect may use a scale of $\frac{1}{16}'' = 1'\text{-}0''$. This reduces the size of the drawing to fit it on the paper. To show certain details more clearly, larger scales of $1\frac{1}{2}'' = 1'\text{-}0''$ or $3'' = 1'\text{-}0''$ are used. Complicated details may be drawn full size or half size. Other scales are used when appropriate. Section views showing the elevation of interior walls are often drawn at $\frac{1}{2}'' = 1'\text{-}0''$ or $\frac{3}{4}'' = 1'\text{-}0''$.

Drawing plans to scale is important. The building and its parts are shown in true proportion, making it easier for the builder to visualize the construction. However, the use of a scale rule to determine a dimension should be a last resort.

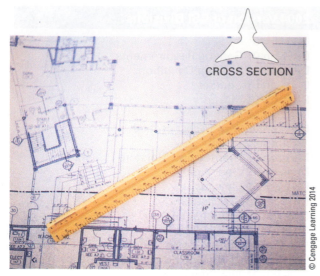

CROSS SECTION

© Cengage Learning 2014

Figure 22–20 The architect's scale is used to draw plans.

Dimensions on plans should be determined by either reading the dimension or adding and subtracting other dimensions to determine it. The use of a scale rule to determine a dimension will result in inaccuracies.

Types of Lines

Some lines in an architectural drawing look darker than others. They are broader so they stand out clearly from other lines. This variation in width is called *line contrast*. This technique, like all architectural drafting standards, is used to make the drawing easier to read and understand (*Figure 22–22*).

Object Line. Lines that outline the object being viewed are broad, solid lines called object lines. These lines represent the portion of the building visible in this view.

Hidden Line. To indicate an important object not visible in the view, a hidden line consisting of short, fine, uniform dashes is used. Hidden lines are used only when necessary. Otherwise the drawing becomes confusing to read.

Centerline. Centerlines are indicated by a fine, long dash, then a short dash, then a long dash, and so on. They show the centers of doors, windows, partitions, and similar parts of the construction.

Section Line. A section reference or cutting-plane line is, sometimes, a broad line consisting of a long dash followed by two short dashes. At its ends are arrows. The arrows show the direction in which the cross section is viewed. Letters identify

the cross-sectional view of that specific part of the building. More elaborate methods of labeling section reference lines are used in larger, more complicated sets of plans (*Figure 22–23*). The sectional drawings may be on the same page as the reference line or on other pages.

Break Line. A break line is used in a drawing to terminate part of an object that, in actuality, continues. It can be used only when there is no change in the drawing at the break. Its purpose is to shorten the drawing to utilize space.

Dimension Line. A dimension line is a fine, solid line used to indicate the location, length, width, or thickness of an object. It is terminated with arrowheads, dots, or slashes (*Figure 22–24*).

Extension Line. Extension lines are fine, solid lines projecting from an object to show the extent of a dimension.

Leader Line. A leader line is a fine solid line. It terminates with an arrowhead, and points to an object from a notation.

Dimensions

Dimension lines on a blueprint are generally drawn as continuous lines. The dimension appears above and near the center of the line. All dimensions on vertical lines should appear above the line when the print is rotated one-quarter turn clockwise. Extension lines are drawn from the object so that the end point of the dimension is clearly defined. When the space is too small to permit dimensions to be shown clearly, they may be drawn as shown in *Figure 22–25*.

Kinds of Dimensions. Dimensions on architectural blueprints are given in feet and inches, such as 3'-6", 4'-8", and 13'-7". A dash is always used to separate the foot measurement from the inch measurement. When the dimension is a whole number of feet with no inches, the dimension is written with zero inches, as 14'-0". The use of the dash prevents mistakes in reading dimensions.

Dimensions of 1 foot and under are given in inches, as 10", 8", and so on. Dimensions involving fractions of an inch are shown, for example, as $1'-0\frac{1}{2}''$, $2'-3\frac{3}{4}''$, or $6\frac{1}{2}''$. If there is a difference between a written dimension and a scaled dimension of the same distance, the written dimension should be followed.

Modular Measure

In recent years, **modular measurement** has been used extensively. A grid with a unit of 4 inches is used in designing buildings (*Figure 22–26*). The

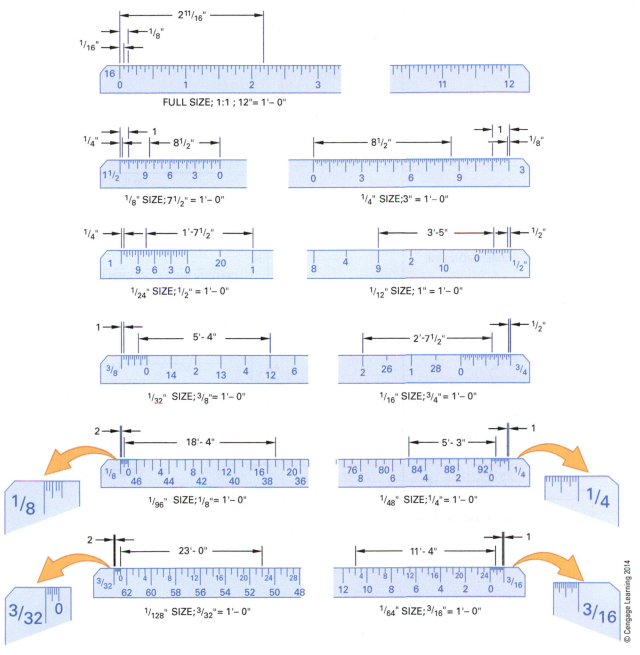

Figure 22–21 The eleven scales found on an architect's scale.

idea is to draw the plans to use material manufactured to fit the grid spaces. Drawing plans to a modular measure enables the builder to use manufactured component parts with less waste, such as 4 × 8 sheet materials and manufactured wall, floor, and roof sections that fit together with greater precision.

The spacing of framing members, and the location and size of windows and doors adhering to the concept of modular measurement cut down cost and conserve materials.

Symbols

Symbols are used on drawings to represent objects in the building, such as doors, windows, cabinets, plumbing, and electrical fixtures. Others are used in regard to the construction, such as for walls, stairs, fireplaces, and electrical circuits. They may be used for identification purposes, such as those used for section reference lines. The symbols for various construction materials, such as lumber, concrete, sand, and earth (*Figure 22–27*), are used when they make the drawing easier to read. (More

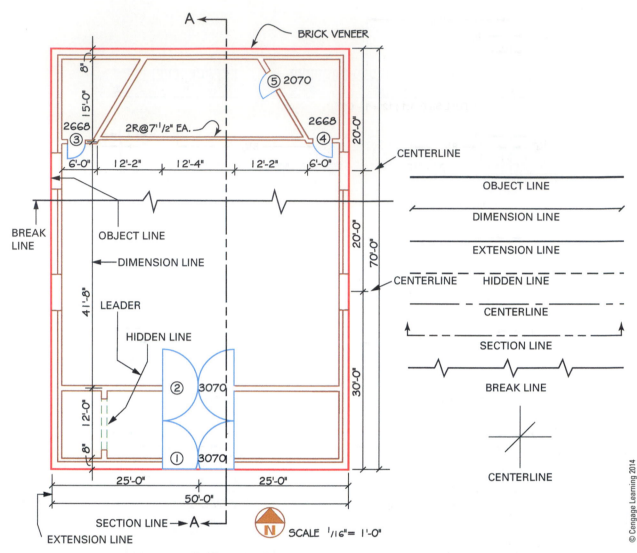

Figure 22–22 Types of lines on architectural drawings.

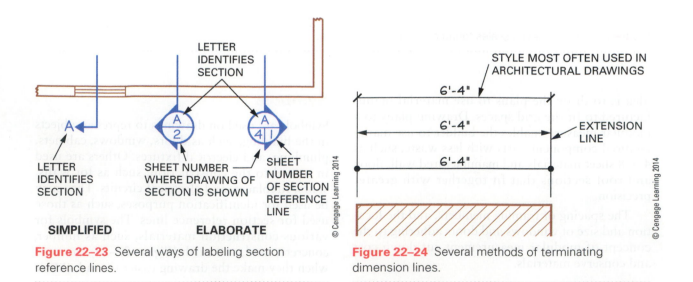

Figure 22–23 Several ways of labeling section reference lines.

Figure 22–24 Several methods of terminating dimension lines.

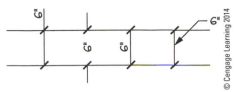

Figure 22–25 Methods of dimensioning small spaces.

detailed illustration, description, and use of architectural symbols are presented in later units where appropriate.)

Abbreviations

Architects find it necessary to use abbreviations on drawings to conserve space. Only capital

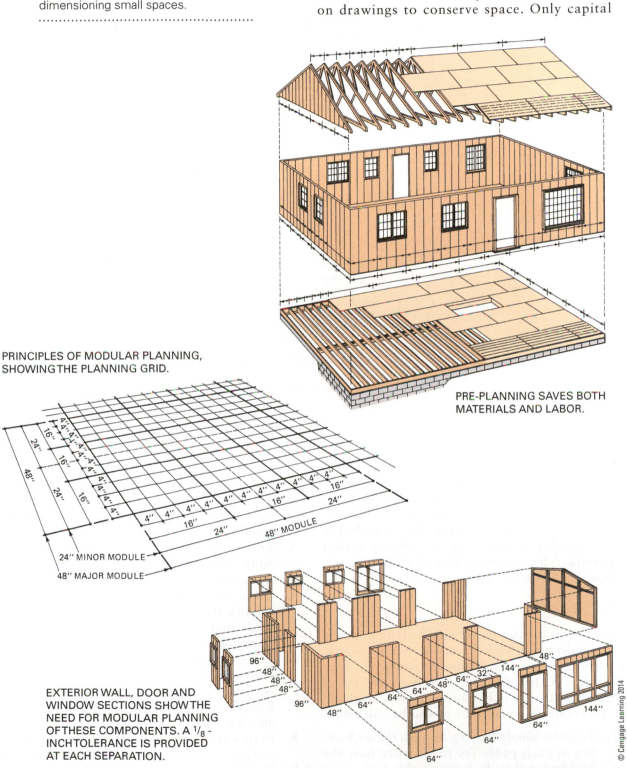

PRINCIPLES OF MODULAR PLANNING, SHOWING THE PLANNING GRID.

PRE-PLANNING SAVES BOTH MATERIALS AND LABOR.

24" MINOR MODULE
48" MAJOR MODULE

EXTERIOR WALL, DOOR AND WINDOW SECTIONS SHOW THE NEED FOR MODULAR PLANNING OF THESE COMPONENTS. A ¹/₈ - INCH TOLERANCE IS PROVIDED AT EACH SEPARATION.

Figure 22–26 Modular measurement uses a grid of 4 inches.

© Cengage Learning 2014

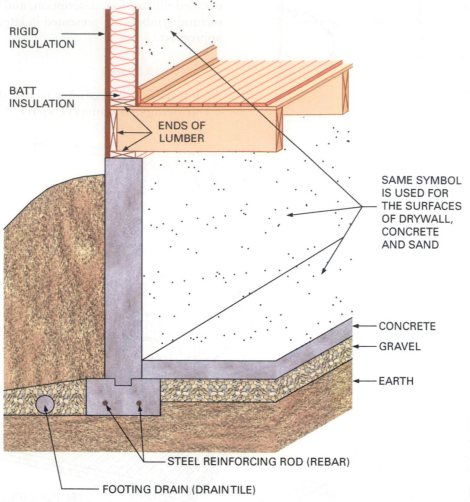

RIGID
INSULATION

BATT
INSULATION

ENDS OF
LUMBER

SAME SYMBOL
IS USED FOR
THE SURFACES
OF DRYWALL,
CONCRETE
AND SAND

CONCRETE

GRAVEL

EARTH

STEEL REINFORCING ROD (REBAR)

FOOTING DRAIN (DRAINTILE)

© Cengage Learning 2014

Figure 22–27 Symbols for commonly used construction materials.

letters, such as DR for door, are used. Abbreviations that make an actual word, such as FIN for finish, are followed by a period (for example, FIN.). Several words may use the same abbreviation, such as W for west, width, or with. The location of these abbreviations is the key to their meaning. A list of commonly used abbreviations is shown in *Figure 22–28*.

Reading Prints

A set of prints can be very intimidating to read and understand. There is an enormous quantity of information to absorb and comprehend. They are often put together with many pages and lots of notes. The trick is to be patient and read methodically. Following a procedure makes the task easier.

1. Begin by simply flipping through the set looking at each page. Try to visualize how the building will look. Note roughly how many pages of each type of the print, A or E or S, etc.

2. Compare the pages that refer to the same region of the building. Some pages show the building as it will look when completed, others show information for specific areas of the building.

3. Review the pages pertaining to the structure. Note what materials and types of framing styles are being used.

4. Check the interior walls and rooms, and try to visualize them completed.

5. Look carefully at the details and cross section views associated with each part of the building.

6. Go back and read all the general notes on drawings. These give detail descriptions and guidance for the builder.

7. Read the list of materials specified in the tables and charts of the set of prints.

8. Review the sheets pertaining to the electrical and plumbing components of the building.

9. Read the written documents and specifications for the project.

Access Panel . . . . AP	Dressed and Matched . . . . D & M	Plate . . . . PL
Acoustic . . . . ACST	Dryer . . . . D	Plate Glass . . . . PL GL
Acoustical Tile . . . . AT	Electric Panel . . . . EP	Platform . . . . PLAT
Aggregate . . . . AGGR	End to End . . . . E to E	Plumbing . . . . PLBG
Air Conditioning . . . . AIR COND	Excavate . . . . EXC	Plywood . . . . PLY
Aluminum . . . . AL	Expansion Joint . . . . EXP JT	Porch . . . . P
Anchor Bolt . . . . AB	Exterior . . . . EXT	Precast . . . . PRCST
Angle . . . . ∠	Finish . . . . FIN	Prefabricated . . . . PREFAB
Apartment . . . . APT	Finished Floor . . . . FIN FL	Pull Switch . . . . PS
Approximate . . . . APPROX	Firebrick . . . . FBRK	Quarry Tile Floor . . . . QTF
Architectural . . . . ARCH	Fireplace . . . . FP	Radiator . . . . RAD
Area . . . . A	Fireproof . . . . FPRF	Random . . . . RDM
Area Drain . . . . AD	Fixture . . . . FIX	Range . . . . R
Asbestos . . . . ASB	Flashing . . . . FL	Recessed . . . . REC
Asbestos Board . . . . AB	Floor . . . . FL	Refrigerator . . . . REF
Asphalt . . . . ASPH	Floor Drain . . . . FD	Register . . . . REG
Asphalt Tile . . . . AT	Flooring . . . . FLG	Reinforce or Reinforcing . . . REINF
Basement . . . . BSMT	Fluorescent . . . . FLUOR	Revision . . . . REV
Bathroom . . . . B	Flush . . . . FL	Riser . . . . R
Bathtub . . . . BT	Footing . . . . FTG	Roof . . . . RF
Beam . . . . BM	Foundation . . . . FND	Roof Drain . . . . RD
Bearing Plate . . . . BRG PL	Frame . . . . FR	Room . . . . RM or R
Bedroom . . . . BR	Full Size . . . . FS	Rough . . . . RGH
Blocking . . . . BLKG	Furring . . . . FUR	Rough Opening . . . . RO
Blueprint . . . . BP	Galvanized Iron . . . . GI	Rubber Tile . . . . R TILE
Boiler . . . . BLR	Garage . . . . GAR	Scale . . . . SC
Book Shelves . . . . BK SH	Gas . . . . G	Schedule . . . . SCH
Brass . . . . BRS	Glass . . . . GL	Screen . . . . SCR
Brick . . . . BRK	Glass Block . . . . GL BL	Scuttle . . . . S
Bronze . . . . BRZ	Grille . . . . G	Section . . . . SECT
Broom Closet . . . . BC	Gypsum . . . . GYP	Select . . . . SEL
Building . . . . BLDG	Hardware . . . . HDW	Service . . . . SERV
Building Line . . . . BL	Hollow Metal Door . . . . HMD	Sewer . . . . SEW
Cabinet . . . . CAB	Hose Bib . . . . HB	Sheathing . . . . SHTHG
Calking . . . . CLKG	Hot Air . . . . HA	Sheet . . . . SH
Casing . . . . CSG	Hot Water . . . . HW	Shelf and Rod . . . . SH & RD
Cast Iron CI	Hot Water Heater . . . . HWH	Shelving . . . . SHELV
Cast Stone . . . . CS	I Beam I	Shower . . . . SH
Catch Basin . . . . CB	Inside Diameter . . . . ID	Sill Cock . . . . SC
Cellar . . . . CEL	Insulation . . . . INS	Single Strength Glass . . . . SSG
Cement . . . . CEM	Interior . . . . INT	Sink . . . . SK or S
Cement Asbestos Board . . . . . CEM AB	Iron I	Soil Pipe . . . . SP
Cement Floor . . . . CEM FL	Jamb . . . . JB	Specification . . . . SPEC
Cement Mortar . . . . CEM MORT	Kitchen . . . . K	Square Feet . . . . SQ FT
Center . . . . CTR	Landing . . . . LDG	Stained . . . . STN
Center to Center . . . . C TO C	Lath . . . . LTH	Stairs . . . . ST
Center Line . . . . or CL	Laundry . . . . LAU	Stairway . . . . STWY
Center Matched . . . . CM	Laundry Tray . . . . LT	Standard . . . . STD
Ceramic . . . . CER	Lavatory . . . . LAV	Steel . . . . ST or STL
Channel . . . . CHAN	Leader L	Steel Sash . . . . SS
Cinder Block . . . . CIN BL	Length . . . . L, LG or LNG	Storage . . . . STG
Circuit Breaker . . . . CIR BKR	Library . . . . LIB	Switch . . . . SW or S
Cleanout . . . . CO	Light . . . . LT	Telephone . . . . TEL
Cleanout Door . . . . COD	Limestone . . . . LS	Terra Cotta . . . . TC
Clear Glass . . . . CL GL	Linen Closet . . . . L CL	Terrazzo . . . . TER
Closet . . . . C, CL or CLO	Lining . . . . LN	Thermostat . . . . THERMO
Cold Air . . . . CA	Living Room . . . . LR	Threshold . . . . TH
Cold Water . . . . CW	Louver LV	Toilet . . . . T
Collar Beam . . . . COL B	Main . . . . MN	Tongue and Groove . . . . T & G
Concrete . . . . CONC	Marble . . . . MR	Tread . . . . TR or T
Concrete Block . . . . CONC B	Masonry Opening . . . . MO	Typical . . . . TYP
Concrete Floor . . . . CONC FL	Material . . . . MATL	Unfinished . . . . UNF
Conduit . . . . CND	Maximum . . . . MAX	Unexcavated . . . . UNEXC
Construction . . . . CONST	Medicine Cabinet . . . . MC	Utility Room . . . . URM
Contract . . . . CONT	Minimum . . . . MIN	Vent V
Copper . . . . COP	Miscellaneous . . . . MISC	Vent Stack . . . . VS
Counter . . . . CTR	Mixture . . . . MIX	Vinyl Tile . . . . V TILE
Cubic Feet . . . . CU FT	Modular . . . . MOD	Warm Air . . . . WA
Cut Out . . . . CO	Mortar . . . . MOR	Washing Machine . . . . WM
Detail . . . . DET	Moulding . . . . MLDG	Water . . . . W
Diagram . . . . DIAG	Nosing . . . . NOS	Water Closet . . . . WC
Dimension . . . . DIM	Obscure Glass . . . . OBSC sL	Water Heater . . . . WH
Dining Room . . . . DR	On Center . . . . OC	Waterproof . . . . WP
Dishwasher . . . . DW	Opening . . . . OPNG	Weather Stripping . . . . WS
Ditto . . . . DO	Outlet . . . . OUT	Weephole . . . . WH
Double-Acting . . . . DA	Overall . . . . OA	White Pine . . . . WP
Double Strength Glass . . . . DSG	Overhead . . . . OVHD	Wide Flange . . . . WF
Down . . . . DN	Pantry . . . . PAN	Wood . . . . WD
Downspout . . . . DS	Partition . . . . PTN	Wood Frame . . . . WF
Drain . . . . D or DR	Plaster . . . . PL or PLAS	Yellow Pine . . . . YP
Drawing . . . . DWG	Plastered Opening . . . . PO	

Figure 22–28 Commonly used abbreviations found in construction drawings.

© Cengage Learning 2014

CHAPTER 23

Floor Plans

A house starts out as an idea drawn on paper. The 3D vision of the structure is converted to many 2D views (*Figure 23–1*). These views must be then read and interpreted by the builder, who makes the idea come alive.

FLOOR PLANS

Floor plans (*Figure 23–2*) contain a substantial amount of information. They are used more than any other kind of drawing. After consideration of many factors that determine the size and shape of the building, floor plans are drawn first. Others, such as the foundation plan and elevations, are derived from it. They are generally drawn at a scale of $\frac{1}{4}''$ = 1'-0'' or 1:48. A separate plan is made for each floor of buildings with more than one story. Commercial floor plans contain even more information. The grid markers help a builder reference specific areas of the plan (*Figure 23–3*). To keep the plan less cluttered, notes, abbreviations, and symbols are included on the print.

Figure 23–1 The 3D vision of a building is described on 2D paper.

© Cengage Learning 2014

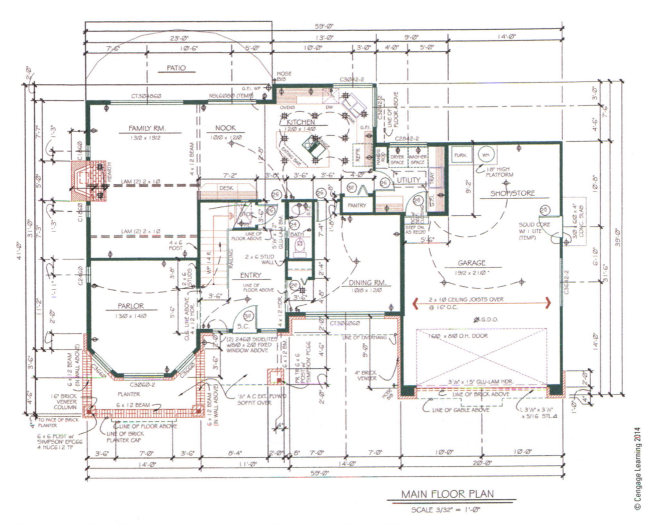

Figure 23–2 First-floor plans contain a substantial amount of information.

Floor Plan Symbols

To make the plan as uncluttered as possible, numerous *symbols* are used. Recognition of commonly used symbols makes it easier to read the floor plan as well as other plans that use the same symbols. Symbols used in elevation and section drawings are different from plan symbols. They are described in later chapters.

Green Tip

Building dimensions in 4-foot modules reduces material waste since standard building panels are 4′ × 8′.

Door Symbols. Symbols for exterior doors are drawn with a line representing the outside edge of the sill. Interior door symbols show no sill line. The symbols in *Figure 23–4* identify the swing, and show on which side of the opening to hang the door.

Similarly, exterior sliding door symbols show the sill line. The symbols for interior sliding doors, called bypass doors, show none. Pocket doors slide inside the wall (*Figure 23–5*).

Bifold doors open to almost the full width of a closet opening. They are used when complete access to the closet is desired. The sections or panels of the doors are clearly seen in the symbols (*Figure 23–6*).

Window Symbols. The inside and outside lines of window symbols represent the edges of the window sill. In between, other lines are drawn for the panes of glass. A window with a fixed, single sash

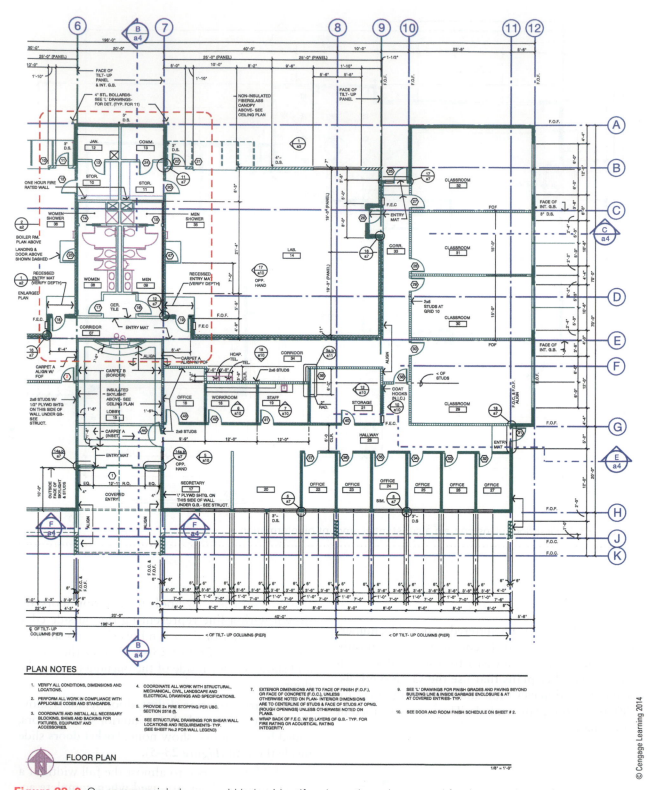

PLAN NOTES

1. VERIFY ALL CONDITIONS, DIMENSIONS AND LOCATIONS.

2. PERFORM ALL WORK IN COMPLIANCE WITH APPLICABLE CODES AND STANDARDS.

3. COORDINATE AND INSTALL ALL NECESSARY BLOCKING, SHIMS AND BACKING FOR FIXTURES, EQUIPMENT AND ACCESSORIES.

4. COORDINATE ALL WORK WITH STRUCTURAL, MECHANICAL, CIVIL, LANDSCAPE AND ELECTRICAL DRAWINGS AND SPECIFICATIONS.

5. PROVIDE 2x FIRE STOPPING PER UBC. SECTION 2516 (f).

6. SEE STRUCTURAL DRAWINGS FOR SHEAR WALL LOCATIONS AND REQUIREMENTS- TYP. (SEE SHEET No.2 FOR WALL LEGEND)

7. EXTERIOR DIMENSIONS ARE TO FACE OF FINISH (F.O.F.), OR FACE OF CONCRETE (F.O.C.), UNLESS OTHERWISE NOTED ON PLAN- INTERIOR DIMENSIONS ARE TO CENTERLINE OF STUDS & FACE OF STUDS AT OPNG. (ROUGH OPENINGS) UNLESS OTHERWISE NOTED ON PLANS.

8. WRAP BACK OF F.E.C. W/ (2) LAYERS OF G.B.- TYP. FOR FIRE RATING OR ACOUSTICAL RATING INTEGRITY.

9. SEE 'L' DRAWINGS FOR FINISH GRADES AND PAVING BEYOND BUILDING LINE & INSIDE GARBAGE ENCLOSURE & AT AT COVERED ENTRIES- TYP.

10. SEE DOOR AND ROOM FINISH SCHEDULE ON SHEET # 2.

FLOOR PLAN

1/8" = 1'-0"

Figure 23–3 On commercial plans, a grid helps identify where the columns and footings are located.

is indicated by one line. Because the **double-hung window** has two sashes that slide vertically, its symbol shows two lines (*Figure 23–7*).

The **casement window,** which swings outward, is depicted by a symbol similar to that for a swinging door. They may be shown having two or more units in each window (*Figure 23–8*). An **awning window** is similar to a casement except it swings outward from the top. Its open position is indicated by dashed lines (*Figure 23–9*).

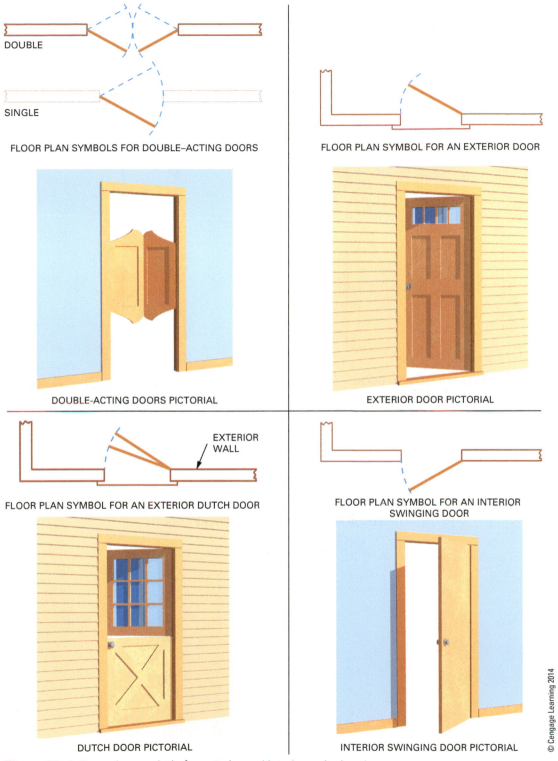

DOUBLE

SINGLE

FLOOR PLAN SYMBOLS FOR DOUBLE–ACTING DOORS

FLOOR PLAN SYMBOL FOR AN EXTERIOR DOOR

DOUBLE-ACTING DOORS PICTORIAL

EXTERIOR DOOR PICTORIAL

EXTERIOR WALL

FLOOR PLAN SYMBOL FOR AN EXTERIOR DUTCH DOOR

FLOOR PLAN SYMBOL FOR AN INTERIOR SWINGING DOOR

DUTCH DOOR PICTORIAL

INTERIOR SWINGING DOOR PICTORIAL

© Cengage Learning 2014

Figure 23–4 Floor plan symbols for exterior and interior swinging doors.

The symbol for a *sliding window* is similar to that for a sliding door except for a line indicating the inside edge of the sill (*Figure 23–10*).

Different kinds of windows may be used in combination. *Figure 23–11* shows a window with a fixed sash, with casements on both sides. Main entrances may consist of a door with a sidelight on one or both sides (*Figure 23–12*).

Close to the window and door symbols are letters and numbers that identify the units in the window and door schedules.

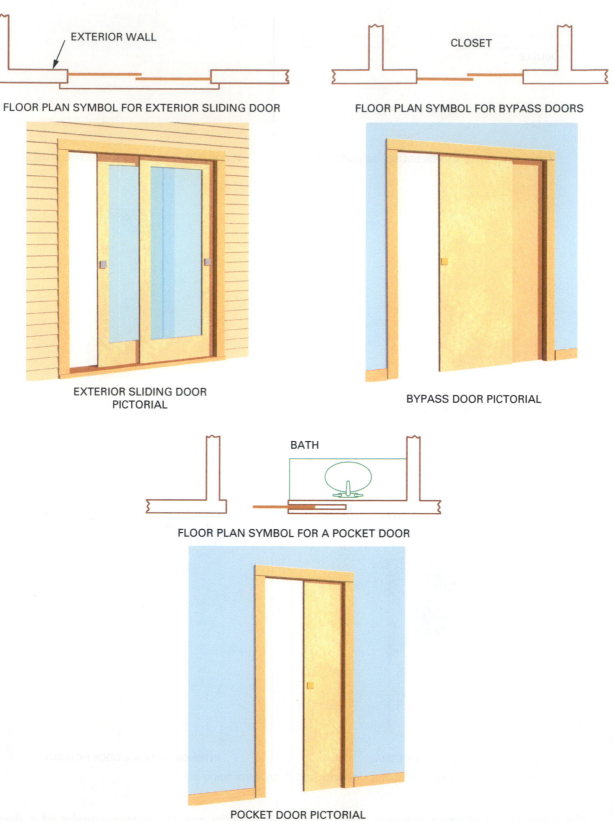

FLOOR PLAN SYMBOL FOR EXTERIOR SLIDING DOOR

EXTERIOR WALL

CLOSET

FLOOR PLAN SYMBOL FOR BYPASS DOORS

EXTERIOR SLIDING DOOR
PICTORIAL

BYPASS DOOR PICTORIAL

BATH

FLOOR PLAN SYMBOL FOR A POCKET DOOR

POCKET DOOR PICTORIAL

Figure 23–5 Symbols for exterior and interior sliding door.

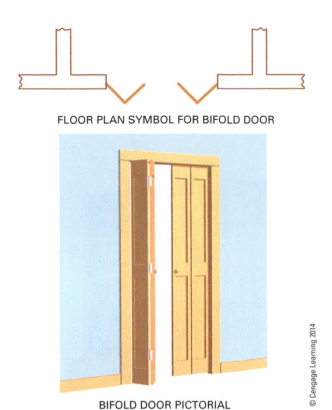

FLOOR PLAN SYMBOL FOR BIFOLD DOOR

BIFOLD DOOR PICTORIAL

Figure 23–6 Bifold doors are sometimes used on closets and wardrobes.

Structural Members. Openings without doors in interior walls for passage from one area to another are indicated by dashed lines. Notations are given if the opening is to be *cased* (trimmed with molding) and if the top is to be *arched* or treated in any other manner (*Figure 23–13*).

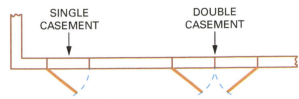

SINGLE
CASEMENT

DOUBLE
CASEMENT

FLOOR PLAN SYMBOLS

DOUBLE CASEMENT PICTORIAL

Figure 23–8 Symbols for casement windows.

WINDOW WITH
FIXED SASH

DOUBLE-HUNG
WINDOW

FLOOR PLAN SYMBOLS

WINDOW WITH FIXED SASH   DOUBLE-HUNG WINDOW
PICTORIAL

Figure 23–7 Fixed sash and double-hung window symbols.

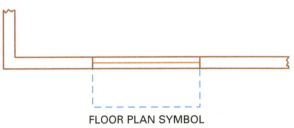

FLOOR PLAN SYMBOL

AWNING WINDOW PICTORIAL

Figure 23–9 The open position of an awning window is shown with dashed lines.

© Cengage Learning 2014

FLOOR PLAN SYMBOL

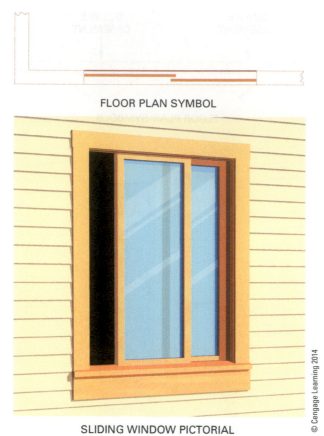

SLIDING WINDOW PICTORIAL

Figure 23–10 Sliding window floor plan symbol.

FLOOR PLAN SYMBOL

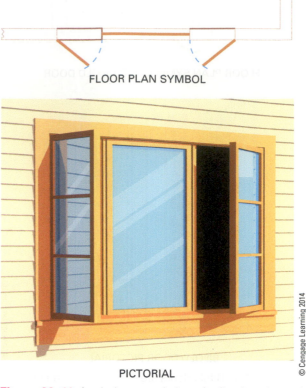

PICTORIAL

Figure 23–11 A window consisting of a fixed sash and casement units.

FLOOR PLAN SYMBOL

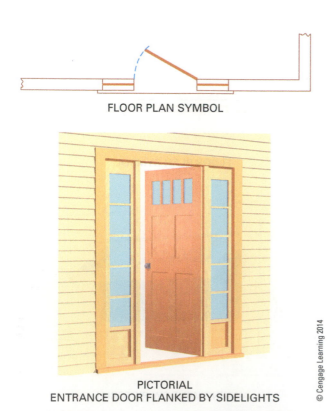

PICTORIAL
ENTRANCE DOOR FLANKED BY SIDELIGHTS

Figure 23–12 An entrance door flanked by sidelights.

CASED OPENING

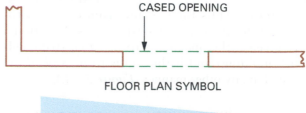

FLOOR PLAN SYMBOL

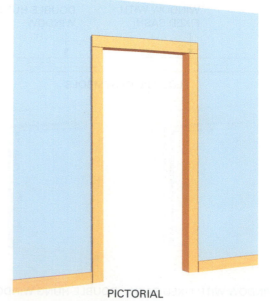

PICTORIAL

Figure 23–13 Dashed lines indicate an interior wall opening without a door.

© Cengage Learning 2014

The location of garage door *headers* may be shown by a series of dashes. Their size is usually indicated with a notation (*Figure 23–14*). Window and other headers are shown in the same manner.

Ceiling beams above the cutting plane of the floor, which support the ceiling joists, are also represented by a series of dashes. *Ceiling joists* or trusses are identified in the floor plan with a double-ended arrow showing the direction in which they run. Their size and spacing are noted alongside the arrow (*Figure 23–15*).

Kitchen, Bath, and Utility Room. The location of *bathroom* and *kitchen fixtures*, such as sinks, tubs, refrigerators, stoves, washers, and dryers, are shown by obvious symbols, abbreviations, and notations. The extent of the *base cabinets* is indicated by a line indicating the edge of the counter-top. Objects such as dishwashers, trash compactors, and lazy Susans are shown by a dashed line and notations or abbreviations. The *upper cabinets* are symbolized by dashed lines (*Figure 23–16*).

Other Floor Plan Symbols. The floor plan also shows the location of *stairways*. Lines indicate the outside edges of the treads. Also shown are the direction of travel and the number of risers (vertical distance from tread to tread) in the staircase (*Figure 23–17*).

The location and style of *chimneys, fireplaces,* and hearths are shown by the use of appropriate symbols. Fireplace dimensions are generally not

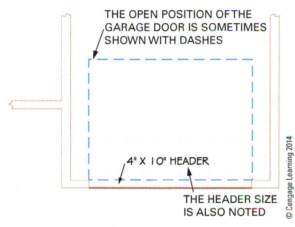

Figure 23–14 The symbol for a garage door header is a dashed line.

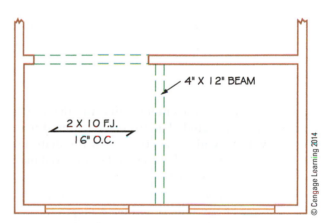

Figure 23–15 Symbols for ceiling beams and ceiling joists.

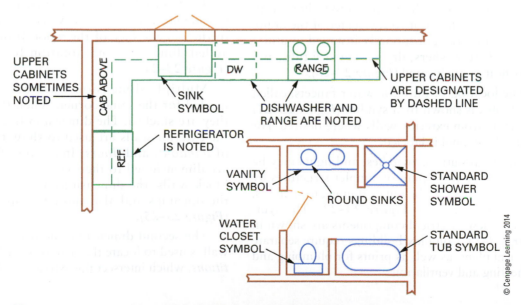

Figure 23–16 Floor plan symbols for kitchen and bath cabinets.

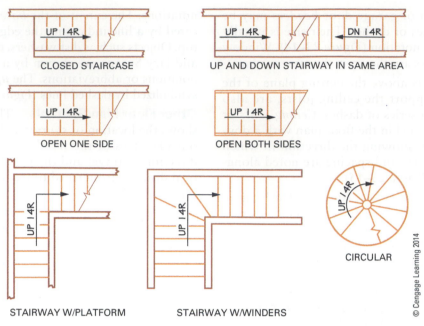

CLOSED STAIRCASE

UP AND DOWN STAIRWAY IN SAME AREA

OPEN ONE SIDE

OPEN BOTH SIDES

STAIRWAY W/PLATFORM

STAIRWAY W/WINDERS

CIRCULAR

© Cengage Learning 2014

Figure 23–17 Stair symbols vary according to the style of the staircase.

given. The sizes of the chimney flue and the hearth are usually indicated. The fireplace material may be shown by symbols according to the material specified (*Figure 23–18*). The kind of material may also be identified with a notation.

An *attic access,* also called a scuttle, is usually located in a closet, hall, or garage ceiling. It is outlined with dashed lines. It may also be identified by a notation (*Figure 23–19*).

The floor plan symbol for a *floor drain* is a small circle, square, or circle within a square. The slope of the floor is shown by straight lines from the corners of the floor to the center of the drain. Floor drains are appropriately installed in utility rooms where washers, dryers, laundry tubs, and water heaters are located (*Figure 23–20*).

The location of outdoor water faucets, called hose bibbs, is shown by a symbol (*Figure 23–21*) projecting from exterior walls where desired. For clarity, the symbol is labeled.

Electrical outlets, switches, and lights may be shown on the floor plan of simpler structures by the use of curved, dashed lines running to *switches, outlets,* and *fixtures* (*Figure 23–22*). The symbols for these electrical components are shown in *Figure 23–23.* Complex buildings require separate electrical plans, as well as prints for plumbing and for heating and ventilation.

Dimensions

Dimensions are placed and printed so that they are as easy to read as possible. Read dimensions carefully. A mistake in reading a dimension early in the construction process could have serious consequences later.

Exterior Dimensions.   On floor plans, the *overall* dimensions of the building are found on the outer dimension lines. In a *wood frame,* the dimensions are to the outside face of the frame. *Concrete block* walls are dimensioned to their outside face. *Brick veneer* walls are dimensioned to the outside face of the wood frame, with an added dimension and notation for the veneer (*Figure 23–24*).

Multiple dimension lines are sometimes needed for the same corner. When this happens they are stacked. The dimension lines closest to the exterior walls are used to show the location of *windows* and *doors.* In a wood frame, they are dimensioned to their centerline. In concrete block walls, the dimensions are to the edges of the openings and also show the opening width (*Figure 23–25*).

The second dimension line from the exterior wall is used to locate the centerline of *interior partitions,* which intersect the exterior wall.

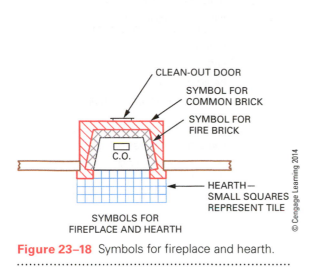

Figure 23–18 Symbols for fireplace and hearth.

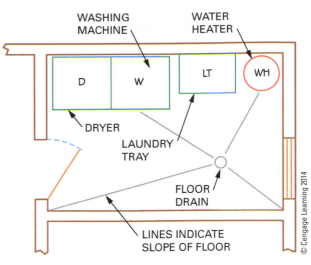

Figure 23–20 Some symbols found in a utility room.

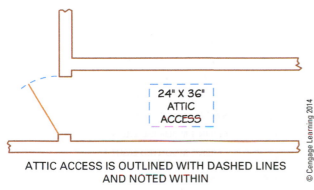

Figure 23–19 Attic access is outlined with dashed lines.

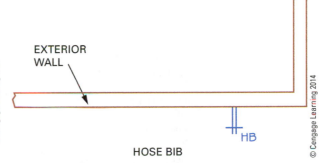

Figure 23–21 An exterior hose bibb.

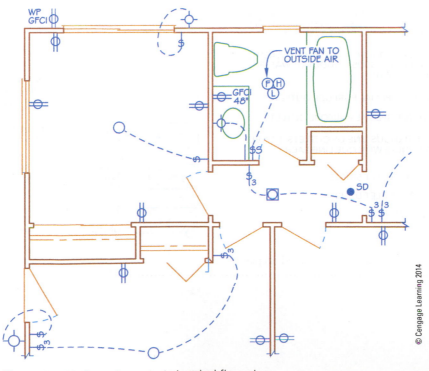

Figure 23–22 Part of a typical electrical floor plan.

D	ELECTRIC DOOR OPENER
	BELL
	BELL/BUZZER
	BUZZER
CH	CHIME
	CIRCUIT-BREAKER PANEL
CB	RECESSED FLUSH MOUNT
CB	WALL MOUNT
C	CLOCK
D	DROP CORD
	SWITCH LEG – ELECTRICAL CIRCUIT, HIGH VOLTAGE THIN LINES CURVED OR STRAIGHT 3/16" TO 1/4" DASHES
	SWITCH LEG – ELECTRICAL CIRCUIT, HIGH VOLTAGE THIN LINES CURVED OR STRAIGHT WITH LONG, 3/4" TO 1-1/2" AND TWO SHORT 3/16" TO 1/4" DASHES
F 8	FAN NOTE: VENT OUT IF REQUIRED
4	MULTIPLE DUPLEX CONVENIENCE OUTLET
	DUPLEX CONVENIENCE OUTLET (WALL OUTLET) 110 VOLTS
S	OUTLET WITH SWITCH 110 VOLTS
	SPLIT WIRED OUTLET, WIRE TO SWITCH
WP	WEATHERPROOF OUTLET 110 VOLTS
GFI	BATHROOM GROUND-FAULT CIRCUIT-INTERRUPTER 110 VOLTS (GFI OR GFCI)
R	RANGE OUTLET 220 VOLTS
220	220-VOLT OUTLET, USED FOR DOUBLE OVEN, FURNACE, ELECTRIC WATER HEATER, CLOTHES DRYER, SPA, WELDER; LABEL ITEM
DW	SPECIAL CONNECTION FOR DISHWASHER, GARBAGE DISPOSAL (GD), HOT WATER HEATER (WH)
	SINGLE SPECIAL PURPOSE OUTLET
	DOUBLE SPECIAL PURPOSE OUTLET
	CONTINUOUS LINE OF OUTLETS, DOUBLE OR SINGLE WITH ONE LINE EACH
	FLOOR OUTLET SINGLE
	FLOOR OUTLET DOUBLE
J J	JUNCTION BOX
G	GENERATOR

H F L	HEAT, LIGHT, FAN COMBINATION
I	SPECIAL INSTRUMENT SPECIFY TYPE, I.E., INTERCOM
M	MOTOR
P	PHONE
R	RADIO OUTLET
SD	SMOKE DETECTOR, CEILING MOUNT
SD	WALL MOUNT SPECIFY TYPE, I.E., INTERCOM
	WALL-MOUNTED LIGHT
S	SPECIAL PURPOSE COMMUNICATION OUTLET, SPECIFY TYPE
TV	ELECTRICAL CIRCUIT, HIGH VOLTAGE
TV	CABLE T.V. OUTLET
PC	PERSONAL COMPUTER CONNECTION
T	TRANSFORMER "T" IN CIRCLE = THERMOSTAT
S	STEREO SPEAKERS
	VACUUM
	YARD LIGHTS
	SINGLE-POLE SWITCH
2	DOUBLE-POLE SWITCH
3	THREE-WAY SWITCH
T	SWITCH WITH TIMER
	CEILING-MOUNTED LIGHT
PS	CEILING-MOUNTED LIGHT WITH PULL SWITCH (PS)
	CAN CEILING LIGHT FIXTURE
	SURFACE MOUNT FIXTURE. OUTLINE DENOTES FIXTURE SHAPE, USE DASH OUTLINES FOR RECESSED FIXTURES.
PS	WALL LIGHT OPTION WITH PULL SWITCH (PS)
	FLOOD LIGHTS
	FLUORESCENT LIGHT
	RECESSED FLUORESCENT LIGHT. THE FIXTURE OUTLINE DENOTES THE SIZE, PROVIDE NOTE SUCH AS 24" X 48"
	CEILING SURFACE-MOUNT FLUORESCENT LIGHT
	SIMPLIFIED FLUORESCENT LIGHT. SOLID SURFACE MOUNT DASHED RECESSED.
T	TRACK LIGHT
S	UNDER-CABINET LIGHT WITH SWITCH

Figure 23–23 Electrical symbols used on plans.

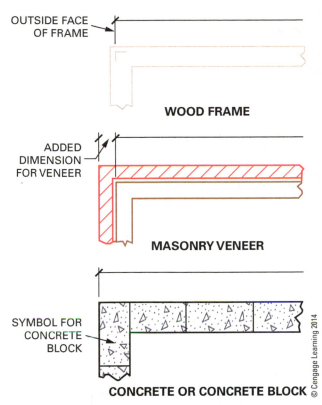

WOOD FRAME

OUTSIDE FACE OF FRAME

ADDED DIMENSION FOR VENEER

MASONRY VENEER

SYMBOL FOR CONCRETE BLOCK

CONCRETE OR CONCRETE BLOCK

Figure 23–24 Overall dimensions are made to the structural portion of a building.

Interior Dimensions. Dimensions are given from the outside of a wood frame or from the inside of concrete block walls, to the centerlines or edges of **partitions**. Interior *doors* and other openings are dimensioned to their centerline similar to exterior walls.

Not all interior dimensions are required. Some may be assumed. For instance, it can be clearly seen if a door is centered between two walls of a closet or hallway (*Figure 23–26*).

The minimum distance from the corner is typically determined by framing the jack and king studs starting at the corner of the room (*Figure 23–27*). This allows sufficient room for the door and **casing** to be applied. If wider custom casing is used, more room may be needed. The goal is to place the door in the corner with room to finish the corner, but not so close that the casing must be scribed to the wall.

Dimensions are also useful to calculate the area of a room from the plans for estimating purposes. Area is room length times room width. It is not necessary to make deductions for the thickness of the walls.

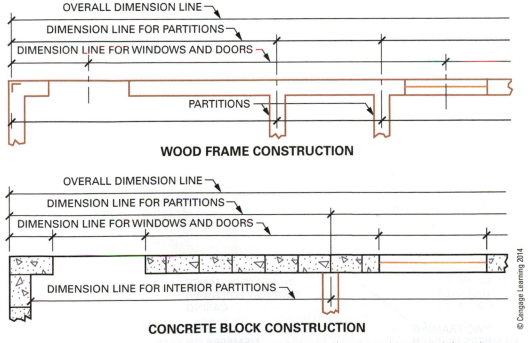

OVERALL DIMENSION LINE

DIMENSION LINE FOR PARTITIONS

DIMENSION LINE FOR WINDOWS AND DOORS

PARTITIONS

WOOD FRAME CONSTRUCTION

OVERALL DIMENSION LINE

DIMENSION LINE FOR PARTITIONS

DIMENSION LINE FOR WINDOWS AND DOORS

DIMENSION LINE FOR INTERIOR PARTITIONS

CONCRETE BLOCK CONSTRUCTION

Figure 23–25 Standard practice for dimensioning windows, doors, partitions, and then the overall dimension.

© Cengage Learning 2014

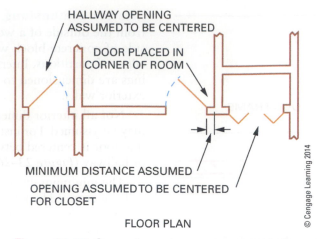

HALLWAY OPENING
ASSUMED TO BE CENTERED

DOOR PLACED IN
CORNER OF ROOM

MINIMUM DISTANCE ASSUMED

OPENING ASSUMED TO BE CENTERED
FOR CLOSET

FLOOR PLAN

© Cengage Learning 2014

Figure 23–26 Some dimensions are assumed to be centered or as close to the corner as possible.

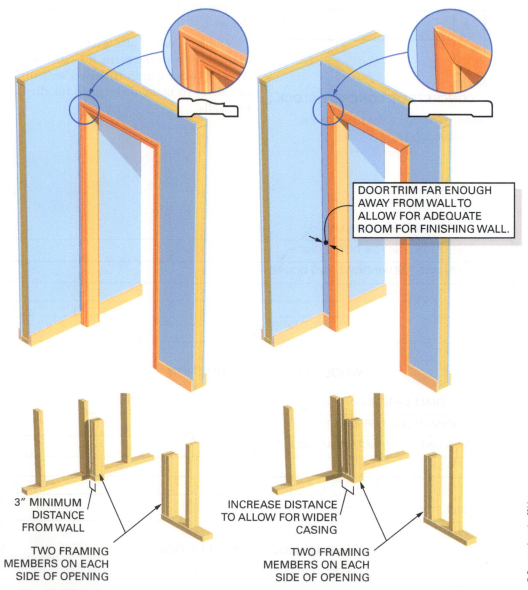

DOOR TRIM FAR ENOUGH
AWAY FROM WALL TO
ALLOW FOR ADEQUATE
ROOM FOR FINISHING WALL.

3" MINIMUM
DISTANCE
FROM WALL

TWO FRAMING
MEMBERS ON EACH
SIDE OF OPENING

INCREASE DISTANCE
TO ALLOW FOR WIDER
CASING

TWO FRAMING
MEMBERS ON EACH
SIDE OF OPENING

© Cengage Learning 2014

Figure 23–27 Locating a door to allow room for finish.

CHAPTER 24

Sections and Elevations

SECTIONS

Floor plans are views of a horizontal cut. Sections show *vertical* cuts called for on the floor, framing, and foundation plans (*Figure 24–1*). Sections provide information not shown on other drawings. The number and type of section drawings in a set of prints depend on what is required for a complete understanding of the construction. They are usually drawn at a larger scale.

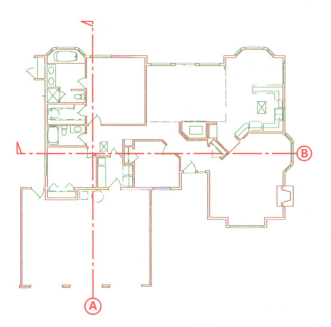

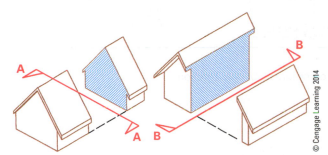

Figure 24–1 Sections are views of vertical cutting planes across the width or through the length of a building. Section reference lines identify the location of the section and the direction from which it is being viewed.

© Cengage Learning 2014

Kinds of Sections

Full sections cut across the width or through the length of the entire building (*Figure 24–2*). For a small residence, only one full section may be required to fully understand the construction. Commercial structures may require several full and many partial sections for complete understanding. Commercial sections also serve as a reference map for the building.

A partial section shows the vertical relationship of the parts of a small portion of the building. Partial sections through exterior walls are often used to give information about materials from foundation to roof (*Figure 24–3*). They are drawn at a larger scale. The information given in one wall section does not necessarily apply for all walls. It may not apply, in fact, for all parts of the same wall. The wall section, or any other section being viewed, applies only to that part of the construction located by the section reference lines. Because the construction changes throughout the building, many section views are needed to provide clear and accurate information.

An enlargement of part of a section is required when enough information cannot be given in the space of a smaller-scale drawing. These large-scale drawings are called *details* (*Figure 24–4*).

Reading Sections

Section views and details are densely packed with information and offer much guidance to the builder. Measurements for the height, thickness, and spacing of the building components are often given. Material types and specifications are indicated. Fastening instructions for normal and special situations are also labeled.

The section view may be referenced for information during all phases of construction. The requirements for the footings and foundation, floor and walls, ceiling and roofing are all drawn. Also included are the finish materials and any special installation instructions. The location and size of all building materials such as steel, wood, masonry, and any other material used in the building are shown.

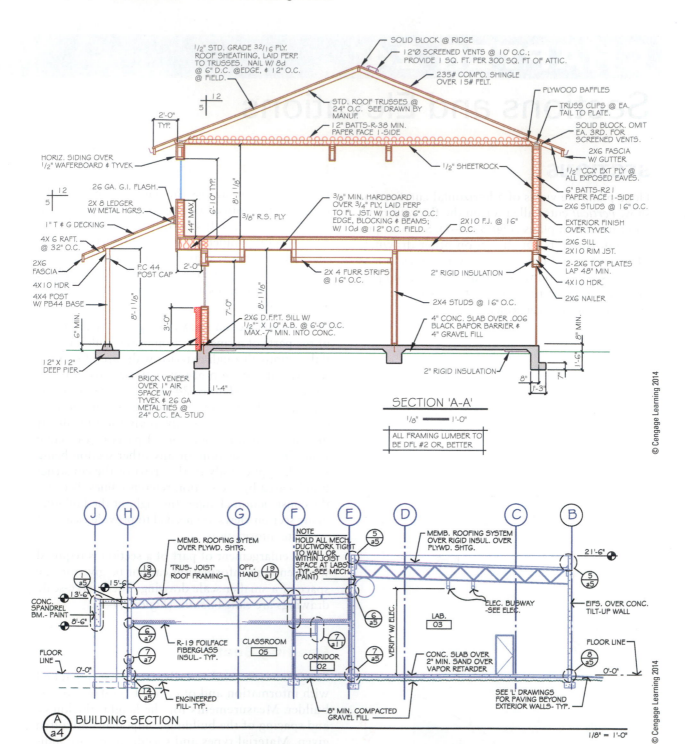

Figure 24–2 Typical section views for residential (a) and commercial (b) construction.

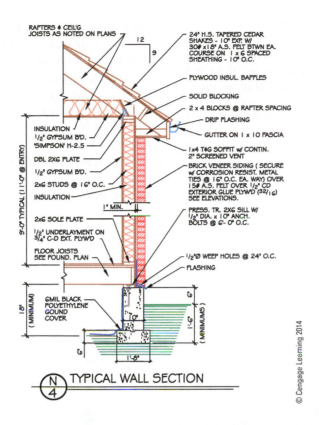

RAFTERS & CEIL'G
JOISTS AS NOTED ON PLANS

24" H.S. TAPERED CEDAR
SHAKES - 10" EXP. W/
30# x 18" A.S. FELT BTWN EA.
COURSE ON 1 x 6 SPACED
SHEATHING - 10" O.C.

PLYWOOD INSUL. BAFFLES

SOLID BLOCKING

2 x 4 BLOCKS @ RAFTER SPACING

DRIP FLASHING

GUTTER ON 1 x 10 FASCIA

1x4 T&G SOFFIT w/ CONTIN.
2" SCREENED VENT

BRICK VENEER SIDING (SECURE
w/ CORROSION RESIST. METAL
TIES @ 16" O.C. EA. WAY) OVER
15# A.S. FELT OVER 1/2" CD
EXTERIOR GLUE PLYW'D (32/16)
SEE ELEVATIONS.

PRESS. TR. 2X6 SILL W/
1/2" DIA. x 10" ANCH.
BOLTS @ 6'- 0" O.C.

1/2"Ø WEEP HOLES @ 24" O.C.

FLASHING

INSULATION
1/2" GYPSUM B'D.
'SIMPSON' H-2.5
DBL 2X6 PLATE
1/2" GYPSUM B'D.
2x6 STUDS @ 16" O.C.
INSULATION

2x6 SOLE PLATE
1/2" UNDERLAYMENT ON
3/4" C-D EXT. PLYW'D
FLOOR JOISTS
SEE FOUND. PLAN

6MIL BLACK
POLYETHYLENE
GOUND
COVER

1" MIN.

9'-0" TYPICAL (11'-0" @ ENTRY)

18" (MINIMUM)

6"

1'-6" (MINIMUMS)

6"

10"

1'-8"

TYPICAL WALL SECTION ⊙ N/4

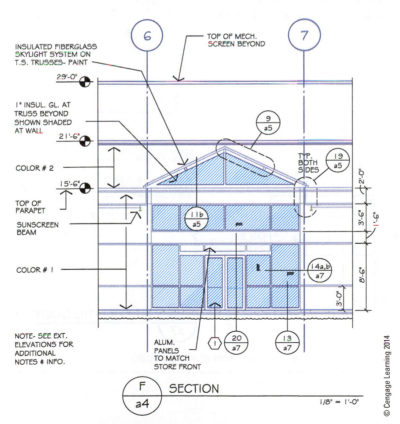

INSULATED FIBERGLASS
SKYLIGHT SYSTEM ON
T.S. TRUSSES- PAINT

TOP OF MECH.
SCREEN BEYOND

6   7

29'-0"

1" INSUL. GL. AT
TRUSS BEYOND
SHOWN SHADED
AT WALL

21'-6"

COLOR # 2

15'-6"

TOP OF
PARAPET

SUNSCREEN
BEAM

COLOR # 1

9 / a5

19 / a5

TYP.
BOTH
SIDES

11b / a5

14a,b / a7

2'-0"

3'-6"

1'-6"

8'-6"

3'-0"

NOTE- SEE EXT.
ELEVATIONS FOR
ADDITIONAL
NOTES & INFO.

ALUM.
PANELS
TO MATCH
STORE FRONT

1   20 / a7   13 / a7

SECTION ⊙ F/a4   1/8" = 1'-0"

Figure 24–3 Typical partial section views in residential (a) and commercial (b) construction.

ELEVATIONS

Elevations are orthographic drawings. They are usually drawn at the same scale as the floor plan. They show each side of the building as viewed from outside at a distance of about 100 feet. Generally four elevations, one for each side, are included in a set of drawings. They are titled Front, Rear, Left Side, and Right Side. They may also be titled according to the compass direction that they face, for instance, North, South, East, and West. From the exterior elevations, the general shape and design of the building can be determined (*Figure 24–5*).

Symbols

Elevation symbols are different from floor plan symbols for the same object. In elevation drawings, the symbols represent, as closely as possible, the actual object as it would appear to the eye. To make the drawing more clear, the symbols are usually identified with a notation.

The location of any steps, porches, dormers, skylights, and chimneys, although not dimensioned, can be seen in elevations. Foundation footings and walls below the grade level may be shown with hidden lines. The kind and size of exterior siding, railings, entrances, and special treatment around doors and windows are shown (*Figure 24–6*).

The elevations show the windows and doors in their exact location. Other openings, such as louvers, are shown in place. Their style and size are identified by appropriate symbols and notations (*Figure 24–7*).

The type of roofing material, the roof pitch, and the cornice style may also be determined from the exterior elevations (*Figure 24–8*).

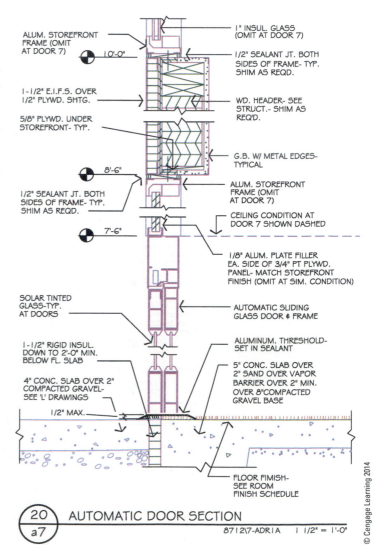

Figure 24–4 Details are a small part of a section drawn at a large scale.

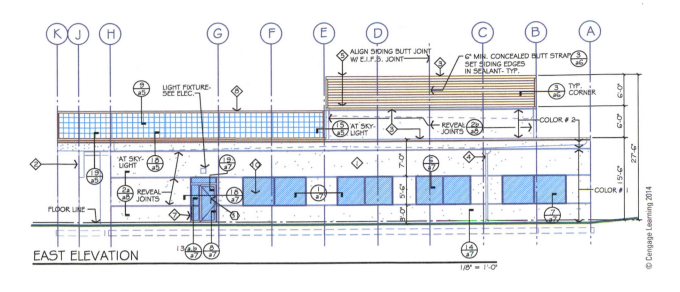

EAST ELEVATION

1/8" = 1'-0"

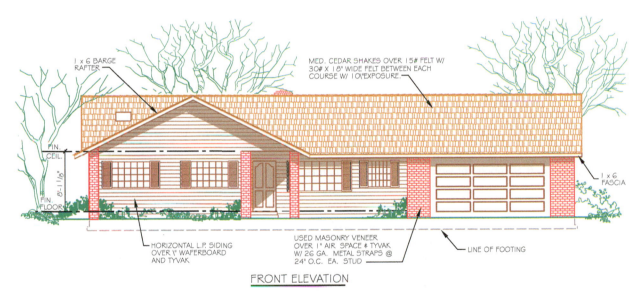

FRONT ELEVATION

1 x 6 BARGE RAFTER

MED. CEDAR SHAKES OVER 15# FELT W/ 30# X 18" WIDE FELT BETWEEN EACH COURSE W/ 10"EXPOSURE.

1 x 6 FASCIA

FIN. CEIL.

FIN. FLOOR

HORIZONTAL L.P. SIDING OVER 1" WAFERBOARD AND TYVAK

USED MASONRY VENEER OVER 1" AIR SPACE & TYVAK W/ 26 GA. METAL STRAPS @ 24" O.C. EA. STUD

LINE OF FOOTING

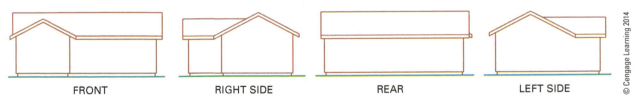

FRONT   RIGHT SIDE   REAR   LEFT SIDE

Figure 24–5 Elevations show the exterior of a building in commercial (a) and residential (b) construction.

Dimensions

In relation to other drawings, elevations have few dimensions. Some dimensions usually given are floor-to-floor heights, distance from grade level to finished floor, height of window openings from the finished floor, and distance from the ridge to the top of the chimney.

A number of other things may be shown on exterior elevations, depending on the complexity of the structure. Little information is given in elevations that cannot be seen in more detail in plans and sections. However, elevations serve an important purpose in making the total construction easier to visualize.

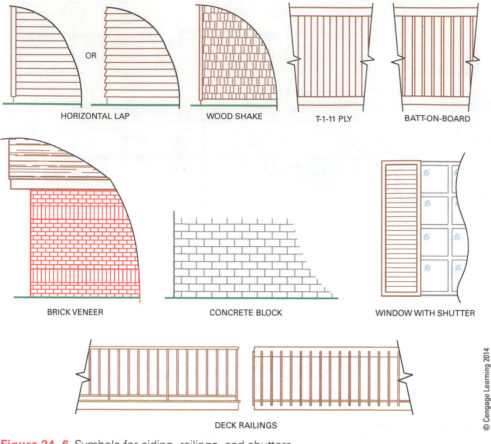

Figure 24–6 Symbols for siding, railings, and shutters.

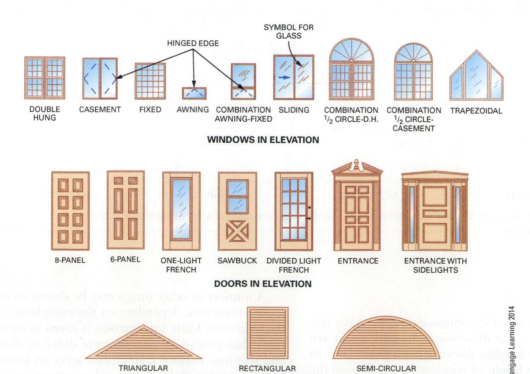

Figure 24–7 Symbols for windows, doors, and louvers.

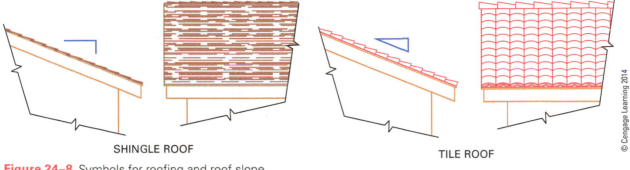

SHINGLE ROOF

TILE ROOF

© Cengage Learning 2014

Figure 24–8 Symbols for roofing and roof slope.

CHAPTER 25

Plot and Foundation Plans

PLOT PLANS

A plot plan is a map of a section of land used to show the proposed construction (*Figure 25–1*). Depending on the size, the scale of plot plans may vary from 1″ = 10′ to 1′ = 200′ or 1:100 to 1:250. It is a required drawing when applying for a permit to build in practically every community. It is a necessary drawing to plan construction that may be affected by various features of the land. The plan must show compliance with zoning and health regulations. Although plot plan requirements may vary with localities, certain items in the plan are standard.

Property Lines

The property line *measurements* and *bearings*, known as metes and bounds, show the shape and size of the parcel. They are standard in every plot plan.

Measurements. The boundary lines are measured in *feet, yards, rods, chains,* or *meters*. Typically, in the United States, the foot is the most commonly used measurement, while most of the world and Canada use meters.

> 3 feet equals 1 yard.
> $16\frac{1}{2}$ feet or $5\frac{1}{2}$ yards equal 1 rod.
> 66 feet or 22 yards or 4 rods equal 1 chain.

Parts of measurement units are expressed as decimals. For instance, a boundary line dimension is expressed as 100.50 feet, not 100 feet, 6 inches.

The measurement is shown centered on, close to, and inside the line.

North. The North compass direction is clearly marked on every plot plan. In a clear space, an arrow of any style labeled with the letter *N* is pointed in the north direction (*Figure 25–2*).

Bearings. In addition to the length of the boundary line, its bearing is shown. The bearing is a compass direction given in relation to a *quadrant* of a circle. There are 360 degrees in a circle and 90 degrees in each quadrant. Degrees are divided into *minutes* and *seconds*.

> 1 degree equals 60 minutes (60′).
> 1 minute equals 60 seconds (60″).

The boundary line bearing is expressed as a certain number of degrees clockwise or counterclockwise from either North or South. For instance, a bearing may be shown as N 30° W, N 45° E, S 60° E, S 30° W (*Figure 25–3*). No bearings begin with East or West as a direction. The bearing is shown centered close to and outside the boundary line opposite its length.

Point of Beginning. An object that is unlikely to be moved easily, such as a large rock, tree, or iron rod driven into the ground, is used for a point of beginning. To denote this point on the plot plan, one corner of the lot may be marked with the abbreviation POB. It is from this point that the lot is laid out and drawn (*Figure 25–4*).

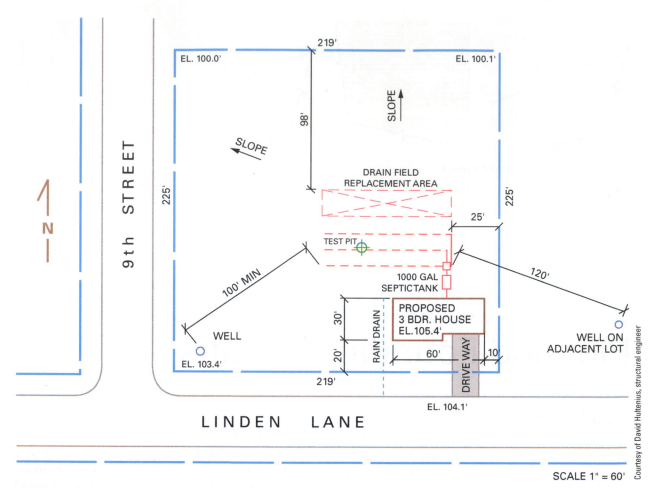

SCALE 1" = 60'

Courtesy of David Hultenius, structural engineer

Figure 25–1 A typical plot plan.

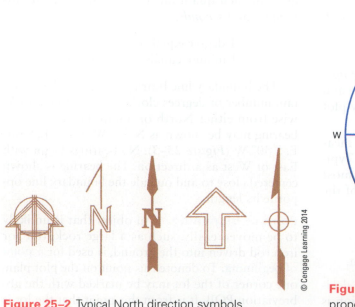

© Cengage Learning 2014

Figure 25–2 Typical North direction symbols.

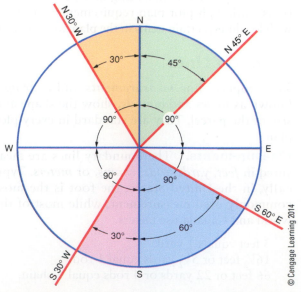

© Cengage Learning 2014

Figure 25–3 Method of indicating bearings for property lines.

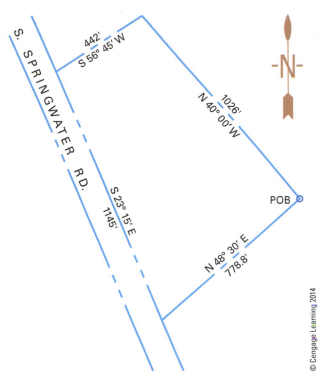

Figure 25–4 Measurements, bearings, and legal description of a parcel of land.

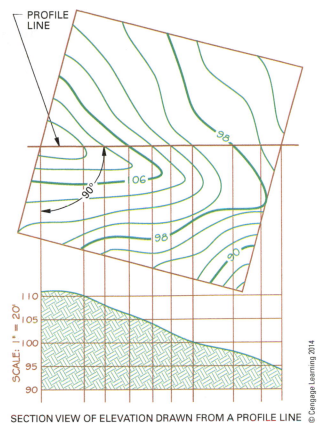

SECTION VIEW OF ELEVATION DRAWN FROM A PROFILE LINE

Figure 25–5 Contour lines show the elevation and slope of the land.

Topography

Topography is the detailed description of the land surface. It includes any outstanding physical features and differences in *elevation* of the building site. Elevation is the height of a surface above sea level. It is expressed in decimal feet and $\frac{1}{100}$ of a foot or meters and decimals.

Contour Lines. Contour lines are irregular, curved lines connecting points of the same elevation of the land. The vertical distance between contour lines is called the contour interval. It may vary depending on how specifically the contour of the land needs to be shown on the plot plan.

When contour lines are close together, the slope is steep. Widely spaced contour lines indicate a gradual slope. At intervals, the contour lines are broken and the elevation of the line is inserted in the space (*Figure 25–5*). On some plans, dashed contour lines indicate the existing grade, and solid lines depict the new grade. Topography is not always a requirement on plot plans. This is especially true for sites where there is little or no difference in elevation of the land surface. The slope of the finished grade may be shown by arrows instead of contour lines (*Figure 25–6*).

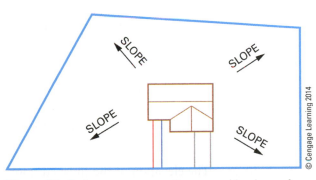

Figure 25–6 Arrows are sometimes used in place of contour lines to show the slope of the land.

Elevations

The height of several parts of the site and the construction are indicated on the plot plan. It is necessary to know these elevations for grading the lot and construction of the building and accessories.

Benchmark. Before construction begins, a reference point, called a **benchmark**, is established on or close to the site. It is used for conveniently

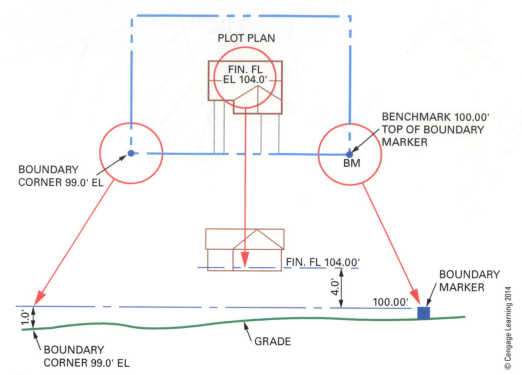

PLOT PLAN

FIN. FL
EL 104.0'

BENCHMARK 100.00'
TOP OF BOUNDARY
MARKER

BM

BOUNDARY
CORNER 99.0' EL

FIN. FL 104.00'

4.0'

BOUNDARY
MARKER

100.00'

1.0'

BOUNDARY
CORNER 99.0' EL

GRADE

© Cengage Learning 2014

Figure 25–7 A benchmark is a reference point used for determining differences in elevation.

determining differences in elevation of various land and building surfaces.

The benchmark is established on some permanent object, which will not be moved or destroyed, at least until the construction is complete. It may be the actual elevation in relation to sea level. It may also be given an arbitrary elevation of 100.00 feet. All points on the lot, therefore, would be either above, level with, or below the benchmark (*Figure 25–7*). The location of the benchmark is clearly shown on the plot plan with the abbreviation BM.

Finish Floor. The elevation of the finished floor or the top of the foundation levels may be shown and noted on the plot plan. This elevation helps the contractor to determine the bottom of the excavation. The bottom of excavation elevation must be calculated. To do this, subtract the building component heights from the finish floor height given. For example, what is the bottom of excavation elevation for the building shown in *Figure 25–8*. ✱

✱IRC
R403.1.7.3 To solve, add up the components:

$$10'' + 8'\text{-}0'' + 1'\text{-}2\tfrac{3}{4}'' = 10'\text{-}\tfrac{3}{4}''$$

Convert $\tfrac{3}{4}''$ to decimal feet:

$$\tfrac{3}{4} \div 12 = 0.0625$$

Add 0.0625 to 10 feet:

0.0625 + 10 = 10.0625 (rounded off to 10.06 feet)

Next, subtract:

$$104.50' - 10.06' = 94.44'$$

The bottom of excavation elevation is 94.44 feet.

Converting Decimals to Fractions. Calculations, especially those dealing with elevations and roof framing, require the carpenter to convert decimals of a foot to feet, inches, and sixteenths of an inch as found on the rule or tape. To convert, use the following method:

- Multiply a decimal of a foot by 12 (the number of inches in a foot) to get inches.
- Multiply any remainder decimal of an inch by 16 (the number of sixteenths in an inch) to get sixteenths of an inch. Round off any remainder to the nearest sixteenth of an inch.
- Combine whole feet, whole inches, and sixteenths of an inch to make the conversion.

For example, convert the finished floor elevation of 104.65 feet to feet, inches, and sixteenths of an inch.

1. Multiply 0.65 feet × 12 = 7.80 inches.
2. Multiply 0.80 inches × 16 = 12.8 sixteenths of an inch.

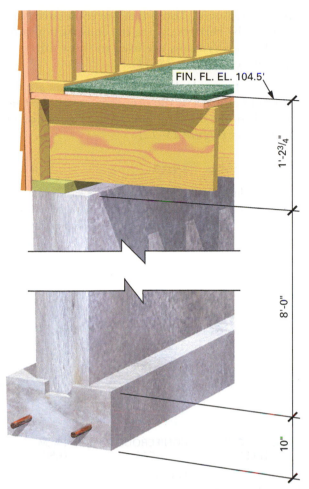

ADD COMPONENTS: 10" + 8'-0" + 1'-2³/₄" = 10'-0³/₄".

CONVERT TO DECIMAL FEET: ³/₄" ÷ 12 = 0.0625 FEET.

ADD TO 10 FEET = 10.0625 OR ROUNDED TO 10.06 FEET.

SUBTRACT 104.50' − 10.06' = 94.44 FEET IS THE EXCAVATION ELEVATION .

Figure 25–8 The distance from the finished floor elevation to the bottom of the footing may need to be calculated.

3. Round off 12.8 sixteenths of an inch to 13 sixteenths of an inch.
4. Combine feet, inches, and sixteenths = 104'–7³/₁₆".

It is more desirable to remember the method of conversion rather than use conversion tables. Reliance on conversion tables requires access to the tables and knowledge of their use, encourages dependence on them, and results in helplessness without them.

Other Elevations. In addition to contour lines, the elevation of each corner of the property is noted on the plot plan (*Figure 25–9*). The top of one of the boundary corner markers makes an excellent benchmark.

Figure 25–9 The elevations of property line corners are indicated on the plot line.

Existing and proposed roads adjacent to the property are shown as well as any easements. Easements are right-of-way strips running through the property. They are granted for various purposes, such as access to other property, storm drains, or utilities. Elevations of a street at a driveway and at its centerline are usually required. One of these is sometimes used as the benchmark.

The Structure

The shape and location of the building are shown. Distances, called *setbacks*, are dimensioned from the boundary lines to the building.

The shape, width, and location of patios, walks, driveways, and parking areas may also be shown on plot plans. Details of their construction are found in another drawing.

A plot plan may also show any *retaining walls*. These walls are used to hold back earth to make more level surfaces instead of steep slopes.

Utilities. The water supply and public sewer connections are shown by noted lines from the structure to the appropriate boundary line. If a private sewer disposal system is planned, it is shown on the plot plan (*Figure 25–10*). Although there are many kinds of sewer disposal systems, those most commonly used consist of a septic tank and leach or drain field, which are usually subject to strict regulations with regard to location and construction.

The location of gas lines is shown, if applicable. Sometimes the location of the nearest utility pole is given. It is shown using a small solid circle as a symbol and is noted. Foundation drain lines leading to a storm drain, drywell, or other drainage may be shown and labeled.

Landscaping. The location and kind of existing and proposed trees are shown on the plot plan. Existing trees are noted, whether they are to be

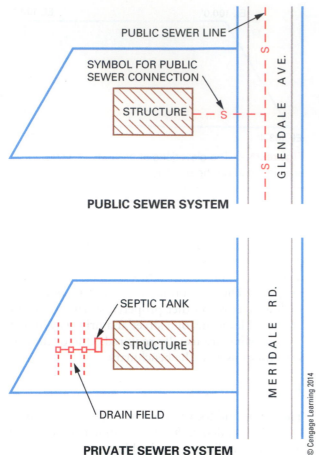

PUBLIC SEWER SYSTEM

PRIVATE SEWER SYSTEM

Figure 25–10 The style of sewage disposal system is shown on the plot plan.

saved or removed. Those that are to be saved are protected by barriers during the construction process. Various symbols are used to show different kinds of trees (*Figure 25–11*).

Identification.   Included on the plot plan are the name and address of the property owner, the title and scale of the drawing, and a legal description of the property (*Figure 25–12*).

Foundation Plans

The foundation plan is drawn at the same scale as the floor plan. It is a view from above of a horizontal cut through the foundation. Great care must

DECIDUOUS TREES   CONIFEROUS TREES   PALM TREES

Figure 25–11 Various symbols are used to indicate different kinds of trees.

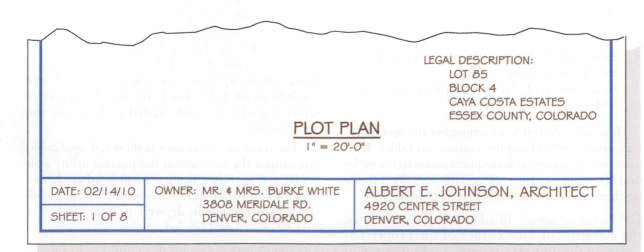

LEGAL DESCRIPTION:
LOT 85
BLOCK 4
CAYA COSTA ESTATES
ESSEX COUNTY, COLORADO

PLOT PLAN
1" = 20'-0"

DATE: 02/14/10

SHEET: 1 OF 8

OWNER: MR. & MRS. BURKE WHITE
3808 MERIDALE RD.
DENVER, COLORADO

ALBERT E. JOHNSON, ARCHITECT
4920 CENTER STREET
DENVER, COLORADO

Figure 25–12 Certain identification items are needed on the plot plan.

be taken when reading the foundation plan so no mistakes are made. A mistake in the foundation affects the whole structure, and generally requires adjustments throughout the construction process.

Two commonly used types of foundations are those having a crawl space, or basement below grade, and those with a concrete slab floor at grade level (*Figure 25–13*).

Crawl Space and Basement Foundations

The crawl space is the area enclosed by the foundation between the ground and the floor above. A minimum distance of 18 inches from the ground to the floor and 12 inches from the ground to the bottom of any beam is required. The ground is covered with a plastic sheet, called a vapor barrier, to prevent moisture rising from the ground from penetrating into the floor frame above.*

＊IRC R317.1

A foundation enclosing a basement is similar to that of a crawl space except the walls are higher, windows and doors may be installed, and a concrete floor is provided below grade. The basement may be used for additional living area, a garage, a utility room, or a workshop.

Reading Plans.   Whether the foundation supports a floor using closely spaced floor joists or more widely spaced post-and-beam construction, the information given in the foundation plan is similar. A typical foundation plan is shown in *Figure 25–14*.

The inside and outside of the foundation wall are clearly outlined. Dashed lines on both sides of the wall show the location of the foundation footing. The type, size, and spacing of anchor bolts are shown by a notation. Wall openings for windows, doors, or crawl space access and vents are shown with appropriate symbols and noted. Small retaining walls of concrete or metal, called areaways, that hold earth away from windows that are below grade may be shown (*Figure 25–15*).

Walls for *stoops* or platforms for entrances are shown. A notation is made in regard to the material with which to fill the enclosed area and

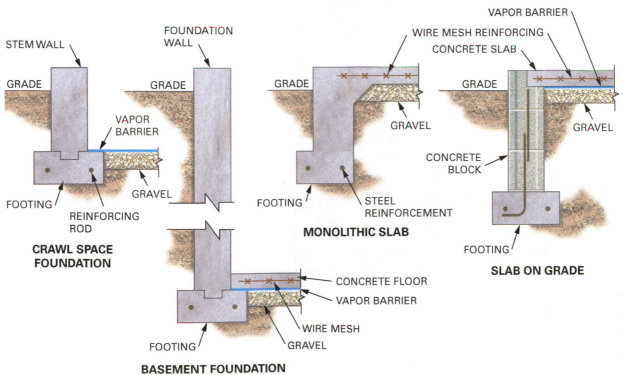

Figure 25–13 Commonly used foundation styles.

© Cengage Learning 2014

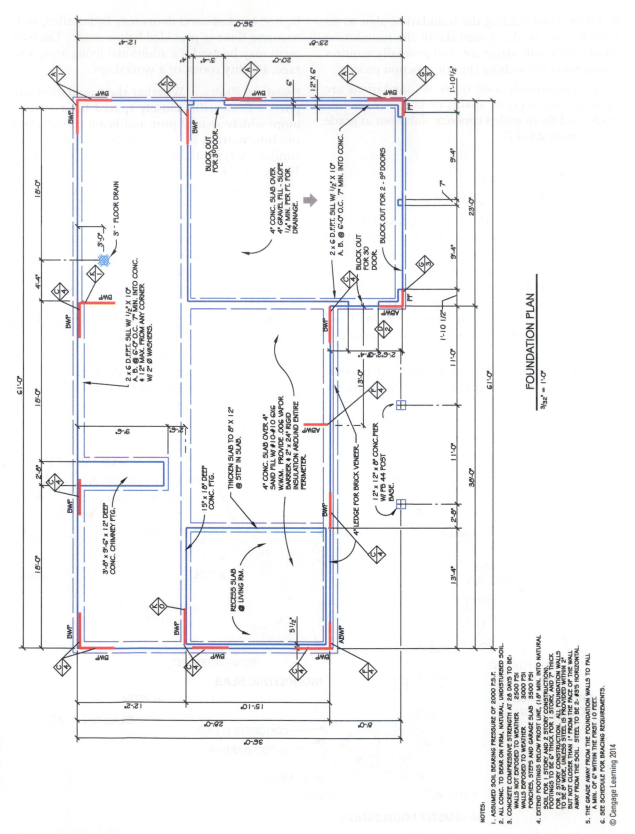

Figure 25–14 A foundation plan for a partial basement.

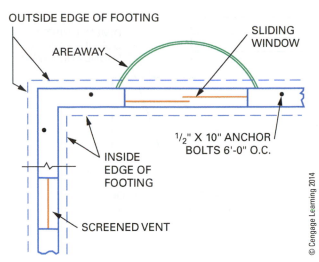

Figure 25–15 Partial plan of a foundation wall with various items indicated.

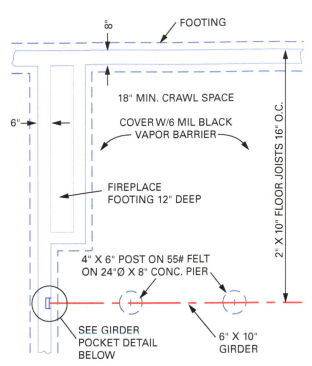

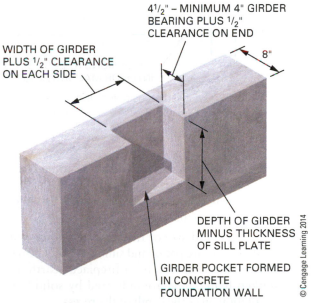

Figure 25–16 Girder pockets and column details are indicated on foundation plans.

cap the surface. Other footings shown by dashed lines include those for chimneys, fireplaces, and columns or posts. Columns or posts support girders shown by a series of long and short dashes directly over the center of the columns. A recess in the foundation wall, called a **beam pocket**, that is used to support the ends of the girder is shown. Notations are made to identify all of these items (*Figure 25–16*).

Floor joist direction installed above the plan view is shown by a line with arrows on both ends similar to those shown in the floor plan for ceiling joists. A notation gives the size and spacing of the joists.

The composition, thickness, and underlying material of the basement floor or crawl space surface are noted. The location of a stairway to the basement is shown with the same symbol as used in the floor plan.

Although it may be stated in the specifications, the strength of concrete used for various parts of the foundation, and the wood type and grade, in addition to size, may be specified by a notation. Plans for foundations with basements may show the location of furnaces and other items generally found in the floor plan if the basement is used as part of the living area.

Dimensions. It is important to understand how parts of the foundation are dimensioned. Foundation walls are dimensioned face-to-face. Interior footings, columns, posts, girders, and beams are dimensioned to their centerline (*Figure 25–17*).

SLAB-ON-GRADE FOUNDATION

The **slab-on-grade foundation** is used in many residential and commercial buildings. It takes less labor and material than foundations to support beam and joist floor framing. Often the concrete

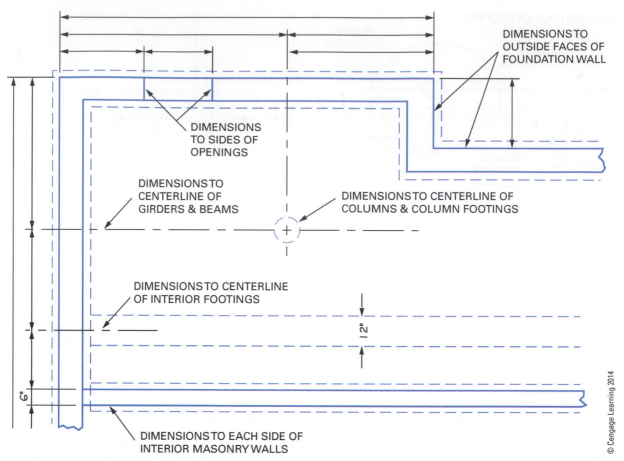

Figure 25–17 Typical dimensioning of crawl space and basement foundations.

for the footing, foundation, and slab can be placed at the same time. There are several kinds, but slab-on-grade foundation plans (*Figure 25–18*) show common components:

- The shape and size of the slab are shown with solid lines, as are patios and similar areas. Changes in floor level, such as for a fireplace hearth or a sunken living area, are indicated by solid lines. A notation gives the depth of the recess.

- Exterior and interior footing locations ordinarily below grade are indicated with dashed lines. Footings outside the slab and ordinarily above grade are shown with solid lines.

- Appropriate symbols and notations are used for fireplaces, floor drains, and ductwork

for heating and ventilation. Blueprints for larger buildings usually have separate drawings for electrical, plumbing, and mechanical work.

- Notations are written for the slab thickness, wire mesh reinforcing, fill material, and vapor barrier under the slab. Interior footing, mudsill, reinforcing steel, and anchor bolt size, location, and spacing are also noted. A typical commercial slab foundation is shown in *Figure 25–19*.

Dimensions. Overall dimensions are to the outside of the slab. Interior piers are located to their centerline. Door openings are dimensioned to their sides.

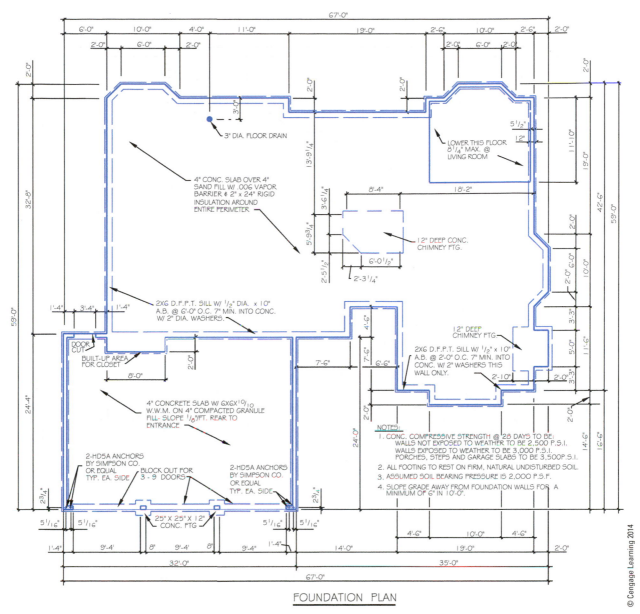

Figure 25–18 Slab-on-grade foundation plan.

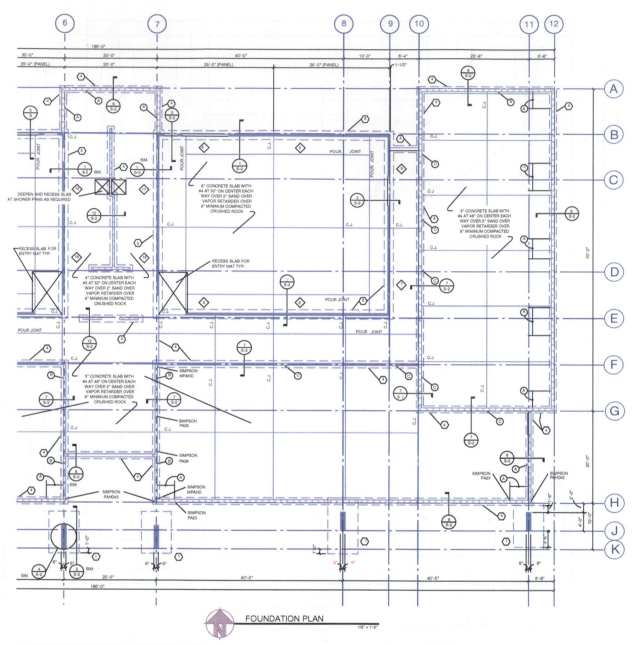

FOUNDATION PLAN
1/8" = 1'-0"

SHEAR WALL SCHEDULE		
MK	SHEATHING	NAILING
A	1/2" PLYWOOD ONE SIDE	8d AT 6" ON CENTER EDGES, 12" ON CENTER IN FIELD
B	1/2" PLYWOOD ONE SIDE	8d AT 4" ON CENTER EDGES, 12" ON CENTER IN FIELD
C	5/8" GYP. WALLBOARD ONE SIDE	6d COOLER OR WALLBOARD NAILS AT 7" ON CENTER TO ALL SUPPORTS (UNBLOCKED)

NOTES
1. USE COMMON NAILS U.O.N.
2. PROVIDE BLOCKING AT ALL UNSUPPORTED PLYWOOD EDGES.
3. WALLS NOTED ON PLAN ARE TYPE NOTED FULL LENGTH OF WALL (OR LENGTH SHOWN BY DIM. LINES)

FOOTING SCHEDULE		
MK	SIZE	REINFORCING
1	6'-6" x 9'-0" x 20" THICK	LONGIT. : (6) #4 TOP, (7) #5 BTM. TRANS. : (8) #4 TOP, (8) #5 BTM.
2	6'-6" x 10'-0" x 20" THICK	LONGIT. : (6) #4 TOP, (7) #5 BTM. TRANS. : (9) #4 TOP, (9) #5 BTM.
3	6'-6" x 12'-0" x 20" THICK	LONGIT. : (6) #4 TOP, (7) #7 BTM. TRANS. : (11) #4 TOP, (11) #5 BTM.
4	18" WIDE x 12" THICK x CONTINUOUS	(2) #4 CONTINUOUS BOTTOM
5	24" WIDE x 16" THICK	(2) #5 CONTINUOUS TOP AND BOTTOM
6	2'-0" X 2'-0" X 18" THICKENED SLAB	UNREINFORCED

Figure 25–19 A typical slab foundation in commercial construction.

CHAPTER 26

Building Codes and Zoning Regulations

Cities and towns have laws governing many aspects of new construction and remodeling. These laws protect the consumer and the community. Codes and regulations provide for safe, properly designed buildings in a planned environment. Contractors and carpenters should have knowledge of local zoning regulations and building codes.

ZONING REGULATIONS

Zoning regulations, generally speaking, deal with keeping buildings of similar size and purpose in areas for which they have been planned. They can also regulate the space in each of the areas. The community is divided into areas called zones, shown on *zoning maps*.

Zones

The names given to different zones vary from community to community. The zones are usually abbreviated with letters or a combination of letters and numbers. A large city may have thirty or more zoning districts.

There may be several *single-family residential zones*. Some zones have less strict requirements than others. Other areas may be zoned as *multi-family residential*. They may be further subdivided into areas according to the number of apartments. Other residential zones may be set aside for *mobile home parks,* and those that allow a combination of *residences, retail stores, and offices.*

Other zones may be designated for the *central business district,* or various kinds of *commercial districts* and different *industrial* zones.

Lots

Zoning laws regulate buildings and building sites. Most cities specify a *minimum lot size* for each zone and a *maximum ground coverage* by the structure. The *maximum height* of the building

for each zoning district is stipulated. A *minimum lot width* is usually specified as well as *minimum yards.*

Minimum yard refers to the distance buildings must be kept from property lines. These distances are called setbacks. They are usually different for front, rear, and side.

Some communities require a certain amount of landscaped area, called green space, to enhance the site. In some residential zones, as much as half the lot must be reserved for green space. In a central business area, only 5 to 10 percent may be required.

In most zones, off-street parking is required. For instance, in single-family residential zones, room for two parking spaces on the lot is required.

Nonconforming Buildings

Because some cities were in existence before the advent of zoning laws, many buildings and businesses may not be in their proper zone. They are called nonconforming. It would be unfair to require that buildings be torn down, or to stop businesses, in order to meet the requirements of zoning regulations.

Nonconforming businesses or buildings are allowed to remain. However, restrictions are placed on rebuilding. If partially destroyed, they may be allowed to rebuild, depending on the amount of destruction. If 75 percent or more is destroyed, they are not usually allowed to rebuild in the same manner or for the same purpose in the same zone.

Any hardships imposed by zoning regulations may be relieved by a variance. Variances are granted by a Zoning Board of Appeals within each community. A public hearing is held after a certain period of time. The general public, and, in particular, those abutting the property are notified. The petitioner must prove certain types of hardship specified in the zoning laws before the zoning variance can be granted.

BUILDING CODES

Building codes regulate the design and construction of buildings by establishing minimum safety standards. They prevent such things as roofs being ripped off by high winds, floors collapsing from inadequate support, buildings settling because of a poor foundation, and tragic deaths from fire due to lack of sufficient exits from buildings. In addition to building codes, other codes govern the mechanical, electrical, and plumbing trades.

Some communities have no building codes. Some write their own. Some have codes, but exempt residential construction. Some have adopted one of three national model building codes. Some use one of the national codes supplemented with their own. Some states have building codes that supersede national ones. There are literally hundreds of building codes.

It is important to have a general knowledge of the building code used by a particular community. Construction superintendents and contractors must have extensive knowledge of the codes.

National Building Codes

Many national building codes are in existence today. Many are being updated almost every year. These codes are written by various organizations whose purpose it is to standardize construction over a broad area. These organizations include the International Conference of Building Officials (ICBO), Southern Building Code Congress International, Inc. (SBCCI), and Building Officials and Code Administrators International, Inc. (BOCA). The codes produced by each organization are as follows:

> ICBO—*Uniform Building Code*
>
> SBCCI—*Standard Building Code*
>
> BOCA—*National Building Code.*

Recently another code was created that merges these three codes and includes input from the International Code Council (ICC). This code is called the *International Residential Code* (IRC). A goal of this code is to reduce the confusion of so many different codes. Many states have adopted this code or created another one using it as a model. It is currently in the 2012 edition.

In Canada, the *National Building Code* sets the minimum standard. Some provinces augment this code with more stringent requirements and publish the combination as a *Provincial Building Code.*

A few cities have charters, which allow them to publish their own building codes.

Use of Residential Codes

In addition to structural requirements, major topics covered by residential codes include:

- exit facilities, such as doors, halls, stairs, and windows as emergency exits, and smoke detectors.
- room dimensions, such as ceiling height and minimum area.
- light, ventilation, and sanitation, such as window size and placement, maximum limits of glass area, fans vented to the outside, requirements for baths, kitchens, and hot and cold water.

Use of Commercial Codes

Codes for commercial work are much more complicated than those for residential work. The structure must first be defined for code purposes. To define the structure, six classifications must be used.

1. The *occupancy group* classifies the structure by how and whom it will be used. The classification is designated by a letter, such as *R,* which includes not only single-family homes but also apartments and hotels.
2. The size and location of the building.
3. The type of construction. Five general types are given numbers 1 through 5. Types 1 and 2 require that all structural parts be noncombustible. Construction in types 3, 4, or 5 can be made of either masonry, steel, or wood.
4. The floor area of the building.
5. The height of the building. Zoning regulations may also affect the height.
6. The number of people who will use the building, called the *occupant load,* determines such things as the number and location of exits.

Once the structure is defined, the code requirements may be studied.

BUILDING PERMITS

A building permit is needed before construction can begin. Application is made to the office of the local building official. The building permit application form (*Figure 26–1*) requires a general description of the construction, legal description

CITY OF ANYWHERE, USA
APPLICATION FOR BUILDING PERMIT

RADON GAS FEE _____

FOR OFFICE USE ONLY

Permit Type _____   Permit # _____

Permit Class of Work _____   Log # _____

Permit Use Code _____   Issue Date _____

Lot _____ Block _____ Sub _____   Permit Cost _____

Fire Zone: IN _____ OUT _____   Zone: _____   T.I.F. Due (Y/N or NA) _____

Utility Notification   1. FL Power _____   B of A (Y/N) _____ Case No.
                       2. Peoples Gas _____   E.D.C. (Y/N) _____ Case No.
NOTE: Items with *   3. Water Dept. _____   C.R.A. (Y/N) _____ Case No.
must be entered in computer.   H.P.C. (Y/N) _____ Case No.

*Plat Page _____   *Sec _____   *Township _____   *Range _____   Zone _____

*Dept of Commerce Code _____   *Const. Type _____ Protected _____ Unprotected _____

*Additional Permits Required:
    Building _____   Plumbing _____   No. of W.C. _____   No. of Meters _____

    Electrical _____   Mechanical _____   Gas _____   Fire Sprk. _____   Landscape _____

    Park/Paving _____   Total Spaces _____   Handicap _____

*Flood Zone _____   *Setbacks: Front _____   Left Side _____   Right Side _____
                                    Rear _____   Other Requirements _____
Threshold Building YES _____ NO _____   _____

Special Notes/Comments to Inspector: _____
_____
_____

APPLICANT PLEASE FILL OUT THIS SECTION

JOB ADDRESS _____   Suite or Apt. No. _____

CONTRACTOR _____   Cert./Reg. No. _____   Telephone _____

PROPERTY OWNER'S Name _____   Address _____

        City _____   State _____ Zip _____   Telephone _____

Building Description: Total Sq. Ft. _____   Estimated Job Value _____
    LF-SF or Dimensions _____   Building Use _____
    Valuation of Work _____   Former Use _____
    No. of Units _____   No. of Suites _____   No. of Stories _____

Special Notes or Comments: _____
_____

PHONE 555-1234 FOR ALL INSPECTIONS

HCS-12 Rev. 6-1-88

(OVER)

© Cengage Learning 2014

Figure 26–1 A typical form used to apply for a building permit.

and location of the property, estimated cost of construction, and information about the applicant.

Drawings of the proposed construction are submitted with the application. The type and kind of drawings required depend on the complexity of the building. For commercial work, usually five sets of plot plans and two sets of other drawings are required. The drawings are reviewed by the building inspection department. If all is in order, a permit (*Figure 26–2*) is granted upon payment of a fee. The fee is usually based on the estimated cost of the construction. Electrical, mechanical, plumbing, water, and sewer permits are usually obtained by subcontractors. The permit card must be displayed on the site in a conspicuous place until the construction is completed.

Inspections

Building inspectors visit the job site to perform code inspections at various intervals. These inspections may include:

1. A *foundation inspection* takes place after the trenches have been excavated and forms erected and ready for the placement of concrete. No reinforcing steel or structural framework of any part of any building may be covered without an inspection and a release.
2. A *frame inspection* takes place after the roof, framing, fire blocking, and bracing are in place, and all concealed wiring, pipes, chimneys, ducts, and vents are complete.

3. The *final inspection* occurs when the building is finished. A Certificate of Occupancy or Completion is then granted.

Some communities require many more inspections. These are designed to verify that the building meets the code of that area. For example, in southern Florida a separate inspection is made of windows after installation to ensure that they are installed to withstand severe wind loads from hurricanes. In California, anchor bolts and metal shear walls require a separate inspection to ensure that they will withstand seismic loads from earthquakes.

It is the responsibility of the contractor to notify the building official when the construction is ready for a scheduled inspection. If all is in order, the inspector signs the permit card in the appropriate space and construction continues. If the inspector finds a code violation, it is brought to the attention of the contractor or architect for compliance.

These inspections ensure that construction is proceeding according to approved plans. They also make sure construction is meeting code requirements. This protects the future occupants of the building and the general public. In most cases, a good rapport exists between inspectors and builders, enabling construction to proceed smoothly and on schedule.

ANYWHERE, USA
DEPARTMENT OF HOUSING & CONSTRUCTION SERVICES
BUILDING PERMIT

THIS PERMIT BECOMES INVALID IF NO INSPECTIONS HAVE BEEN MADE DURING ANY 3 MONTH PERIOD.

Flood Elevation - _____ Lowest Floor Minimum Required

☐ New Construction   ☐ Moving   ☐ Siding
☐ Grounds Improvements   ☐ Fences   ☐ Walls
☐ Utility Building   ☐ Pool
☐ Reroofing   ☐ Other _____

Permit No. _____ (ZONE)

Job Address _____

Lot _____   Blk. _____   Sub. _____

Date _____   This permit covers building construction only. Additional permits are required for electric, plumbing, gas and/or mechanical installations.

BUILDING			ELECTRICAL			PLUMBING			MECHANICAL-GAS		
Type of Inspection	Date	Inspector	Type of Inspection	Date	Inspector	Type of Inspection	Date	Inspector	Type of Inspection	Date	Inspector

NOTE: Building, Electrical, Plumbing and Mech/Gas Inspections shall be dated and initialed by inspectors before walls and ceilings are covered.

BUILDING OK TO COVER		ELECTRICAL OK TO COVER		PLUMBING OK TO COVER		MECH/GAS OK TO COVER	
Date	Inspector	Date	Inspector	Date	Inspector	Date	Inspector

NOTE: This card shall remain posted at the job site until all final inspections have been dated and initialed by inspectors.

BUILDING FINAL OK		ELECTRICAL FINAL OK		PLUMBING FINAL OK		MECH/GAS FINAL OK	
Date	Inspector	Date	Inspector	Date	Inspector	Date	Inspector

THIS CARD MUST BE POSTED IN AN EASILY SEEN LOCATION.

For INSPECTIONS, call 555-1234. ■ For other information, call 555-6789.

© Cengage Learning 2014

Figure 26–2 A typical building permit.

DECONSTRUCT THIS

Carefully study **Figure 26–3** and think about what is wrong and/or what is right. Consider all possibilities. What construction practice or method is different in your area of the country?

Courtesy of Capital Safety

Figure 26–3 This photo shows worker using a personal safety harness.

KEY TERMS

architect's scale	building codes	details	full sections
areaways	building permit	door schedules	green space
awning window	bypass doors	dormers	hearth
beam pocket	casement window	double-hung window	hose bibbs
bearing	casing	easement	isometric
benchmark	contour interval	elevation	louvers
bifold doors	contour lines	finish schedules	metes and bounds
blueprinting	crawl space	floor plans	modular measurement

national building codes

nonconforming orthographics

partial section elevations

partition

plan view

plot plan

pocket doors

point of beginning

risers

scale

scuttle

section view

setbacks

sidelight

skylight

slab-on-grade foundation

sliding

specifications

specifications guide

specifications writer

swing

topography

treads

trusses

vapor barrier

variance

window schedules

zones

zoning regulations

REVIEW QUESTIONS

Select the most appropriate answer.

1. A drawing view looking from the top downward is called a(n)

   a. elevation.
   b. perspective.
   c. plan.
   d. section.

2. Grid lines on a commercial set of plans

   a. show the location of columns and footings.
   b. as designated by numbers and letters.
   c. do not have to be evenly spaced.
   d. all of the above.

3. A drawing view showing a vertical cut through the construction is called a(n)

   a. elevation.
   b. perspective.
   c. plan.
   d. section.

4. The more commonly used scale for floor plans is

   a. $\frac{1}{4}'' = 1'\text{-}0''$.
   b. $\frac{3}{4}'' = 1'\text{-}0''$.
   c. $1\frac{1}{2}'' = 1'\text{-}0''$.
   d. $3'' = 1'\text{-}0''$.

5. The length of a line on a $\frac{1}{4}'' = 1'\text{-}0''$ scaled print that represents an actual distance of 14'-0" is

   a. $3\frac{1}{2}''$.
   b. $4\frac{1}{2}''$.
   c. 14".
   d. 56".

6. The symbol "FL" written on a set of prints means

   a. floor.
   b. flush.
   c. flashing.
   d. all of the above.

7. Each page of a set of prints is labeled with a letter and number. This is done to

   a. show where the columns are located.
   b. identify the page number.
   c. type of construction being drawn.
   d. answers b and c.

8. To determine a dimension that is not written on a set of plans it is best to

   a. use an architect's scale to measure.
   b. calculate it from existing dimensions.
   c. read the specifications.
   d. use the plot plan.

9. Centerlines are indicated by a

   a. series of short, uniform dashes.
   b. series of long then short dashes.
   c. long dash followed by two short dashes.
   d. solid, broad, dark line.

10. Which of the style of dimensioning below would most likely be found on a set of prints?

    a. 3'.
    b. 3 ft.
    c. 3'-0".
    d. all of the above.

11. The setback of a building from the property lines would be found on a

    a. floor plan.
    b. plot plan.
    c. elevation drawing.
    d. foundation plan.

12. The bearing of S 20° E indicates the direction of

    a. 20 degrees from south towards east.
    b. 20 degrees form east towards south.
    c. 20 degrees from east toward west.
    d. all of the above.

13. To find out which edge of a door is to be hinged, look on the

    a. elevations.
    b. specifications.
    c. floor plan.
    d. wall section.

14. The direction, size, and spacing of the first-floor joists is found on the

    a. foundation plan.
    b. first-floor plan.
    c. second-floor plan.
    d. all of the above.

15. The finished floor height is usually found on the

    a. plot plan.
    b. floor plan.
    c. foundation plan.
    d. framing plan.

16. Topographical lines indicate

    a. where construction is to take place.
    b. earth that is to be removed during construction.
    c. the top view of the floor plan.
    d. where certain elevations exist on a plot plan.

17. An exterior wall stud height can best be determined from the

    a. floor plan.
    b. framing elevation.
    c. wall section.
    d. specifications.

18. The view of a set of prints most helpful in determining the material installed behind a brick veneer is a(n)

    a. section view.
    b. elevation.
    c. foundation plan.
    d. all of the above.

19. If there is a conflict of information in a set of plans the information that takes precedence is found in (on)

    a. floor plan
    b. specifications
    c. structural plans
    d. schedule

20. The laws that guide what type of building may be built in a particular area are called

    a. Zoning Regulations.
    b. National Building Codes.
    c. Residential Codes.
    d. Building Permits.

SECTION TWO

Rough Carpentry

Courtesy Korie A. Bishop

Korie A. Bishop
Title: Construction Project Manager
Company: Stop & Shop Supermarket Company

© iStockphoto.com/mattjeacock

EDUCATION

Korie Bishop wanted to apply her strength in math and science to a college major. She considered architecture and engineering, and decided that civil engineering would offer her the most opportunities. She earned an associate's degree in civil engineering technology from Vermont Technical College, and graduated in 1995 with a bachelor's degree in civil engineering at Norwich University in Vermont.

In college, Korie learned that she enjoyed structural aspects of civil engineering more than the environmental aspects. "I liked the steel and concrete aspects," she said. "I wanted a job in this field, but in the mid-'90s there weren't structural jobs in my region."

HISTORY

Korie was offered a job with The Whiting-Turner Contracting Company, a Baltimore-based company that had a project in Burlington, Vermont. "It was a good opportunity to get into the building fields." As project engineer, Korie managed multimillion-dollar industrial construction projects—manufacturing facilities called "clean rooms" where microchips are manufactured. "Where the people in the white suits work!" she explained.

"From day one I was given a lot of responsibility." Korie explained how building a clean room requires building all the systems. "I learned a lot on the job. It was a very hands-on position. As a design engineer, I would have been in the office, but I was in the field. Every day was different. I coordinated a lot of people—architects, engineers, customer—and I met a lot of people."

Korie's position required a lot of travel, which she liked initially. After five years, she was promoted to project manager. She rotated through job sites in Vermont, New York, and Pennsylvania. She explained, "With Whiting-Turner, you go where the work is. This was great in my twenties, but then I wanted to put down some roots." After about seven years, she bought a house and moved to Connecticut, a central location to the different job sites.

In 2005, after ten years with Whiting-Turner, Korie began looking for a new opportunity in the construction field that required less travel. "I was lucky and found a position with Stop & Shop as a construction project manager," which she explained was one step up from her previous position. In this role, as the project "owner," she is the one who hires the general contractor in the construction or remodel of stores. Her territory is Connecticut, upstate New York, and western Massachusetts, and she can be home every night.

ON THE JOB

Korie manages about three or four projects at a time. In a typical day, Korie visits a job site for meetings, respond to e-mails, and updates budgets. "I head out early to one of my projects and meet with the general contractor and subcontractors; the architect will also attend. I'm a go-between. There's a weekly status meeting for each job. I'll answer questions and go over paperwork." After the meetings Korie will work from the job site, return to her office, or work from home.

Korie might also attend a preplanning meeting for a project that is starting soon. She explained that remodeling a grocery store is a whole different type of construction from building a new store, and requires more operations meetings. "If it's a remodel, I'll review the plans with the store operations team. I'll work with the store manager to discuss how to work with limited impact to the customers." Korie explained the importance of doing the work at night and making sure that in the morning the store looks like a store and not a construction zone.

BEST ASPECTS

"I stumbled into construction by accident, and I love it . . . I seriously enjoy the people I work with and what I do. My job gives me the flexibility to have a challenging career that fits perfectly into the rest of my life."

CHALLENGES

Korie described the challenges of working with a lot of different people and that being educated as an engineer meant she did not necessarily learn people skills. "This is something that doesn't come naturally to a lot of engineers. Some people are easier to get along with than others. Each new project brings a whole different set of faces." Korie explained that learning how to read people and how to communicate is a "side skill" that she had to develop.

Korie also described the organizational challenges of her job. "I have to keep good files and take good notes. I have to keep track of all the things I have going on. I always have lots of irons in the fire. Things move quickly, and I'm usually working out of my car or a job trailer."

IMPORTANCE OF EDUCATION

Korie described how education benefits those in this business. "Education provides a foundation for the thought processes involved in any business. In school, you have to solve problems and be organized. These are skills you'll use going forward. If you're able to get through school successfully, you can do any job successfully." She added, "Right out of school I was given a lot of responsibility, and I met the challenge."

FUTURE OPPORTUNITIES

Korie said that opportunities exist at Stop & Shop Supermarket Company to move up within construction management department, and while she aspires to moving up, she is happy where she is. "Not one morning do I feel like I don't want to go to work," she said.

WORDS OF ADVICE

Korie recalls how shy she used to be and how much she learned when she started asking questions. "Know your strengths and have confidence. Don't be afraid to venture out. Don't be afraid to ask questions about things you don't know. I have learned so much from people in the field. These tradesmen, a plumber or electrician, have been doing this for decades, and they love to be in the role of an educator. People will respect you for asking questions, and as a boss, when people ask me questions it shows me they are interested and want to do a good job."

Interview courtesy of Michael E.C. Surguy

UNIT 10
Building Layout

Before construction begins, lines must be laid out showing the location and elevation of the building foundation. Accuracy in laying out these lines is essential in order to comply with local zoning ordinances. In addition, accurate layout lines provide for a foundation that is level and to specified dimensions. Accuracy in the beginning makes the work of the carpenter and other construction workers easier later. Layout for the location of a building and its component parts must be done properly. Failure to do so can be costly and time-consuming.

OBJECTIVES

After completing this unit, the student should be able to:

- establish level points across a building area using a water level and using a carpenter's hand spirit level in combination with a straightedge.

- accurately set up and use the builder's level, transit-level, and laser level for leveling, determining and establishing elevations, and laying out angles.

- lay out building lines by using the Pythagorean theorem method for squaring corners, and check the layout for accuracy.

- build batter boards and accurately establish layout lines for building using building layout instruments.

- Understand the advantages and operation of a total station.

CHAPTER 27
Leveling and Layout Tools

Building layout requires leveling and plumbing lines as well as laying out various angles over the length and width of the structure. It is interesting to realize that no matter where you go on the earth, plumb points toward the center of the earth.

Level is always perpendicular (at a right-angle) to plumb (*Figure 27–1*). The construction industry uses many tools to achieve level and plumb. The carpenter should be able to set up, adjust, and use these tools.

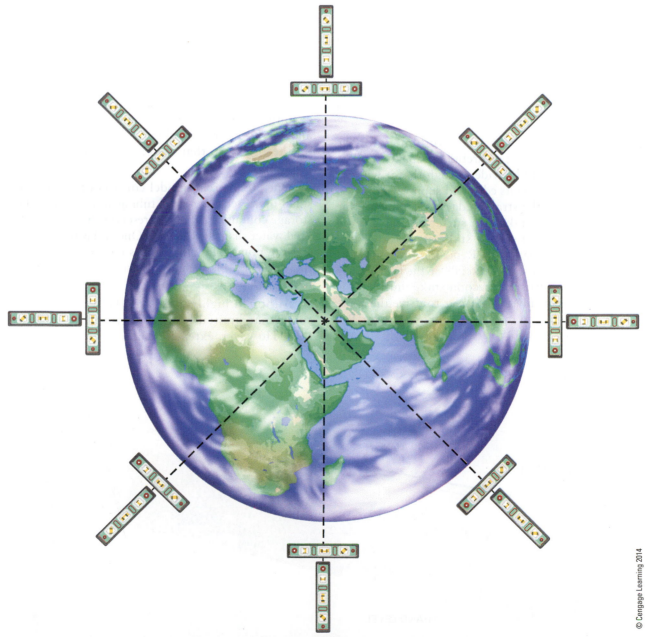

Figure 27–1 Plumb always points toward the center of the earth and level is a local reference perpendicular to plumb.

LEVELING TOOLS

Several tools, ranging from simple to state-of-the-art tools, are used to level the layout. More sophisticated leveling and layout tools, although preferred, are not always available.

Levels and Straightedges

If sophisticated leveling tools are not available, simple leveling tools may be used to level a building area. Such tools include a carpenter's hand level and a long straightedge. This leveling process begins at some point or stake that is at the desired elevation. Stakes are then placed across the building area to the desired distance from the starting point. This method can be an accurate, although time-consuming, method of leveling over a long distance. It can also be done by one person. Care should be taken to be sure each step is performed properly because slight errors, multiplied by each succeeding step, can grow into large ones.

Begin by selecting a length of lumber for the straightedge, sighting it carefully to make sure it is straight. It should be wide enough that it will not sag when placed on its edge and supported only on its ends. Place the straightedge on edge with one end on the first stake or a surface at the desired elevation. Drive a second stake at the other end slightly higher than level. Reposition the straightedge to the top of each stake. Place the level on top and carefully drive the second stake until level is achieved. Remember it is easier to drive the stake farther than it is to raise it, so don't go too far. Recheck the levelness of the two stakes.

Continue across the building area to the desired distance by moving the straightedge one stake at a time. Use the last stake driven as the new starting stake (*Figure 27–2*). Place the other end on another stake and drive it until the straightedge is again level.

If the ground is so hard that stakes are difficult to drive, a crow or shale bar may be used (*Figure 27–3*). This will also help to keep the stakes straight with relative ease. Continue moving the straightedge from stake to stake until the desired distance is leveled.

If you want to level to the corners of building layouts, start by driving a stake near the center so its top is to the desired height. Level from the center stake to each corner in the manner described in *Procedure 27–A*.

Water Levels

A *water level* is a very accurate tool, dating back centuries. It is used for leveling from one point to another. Its accuracy, within a pencil point, is based on the principle that water seeks its own level (*Figure 27–4*).

One commercial model consists of 50 feet of small-diameter, clear vinyl tubing and a small tube storage container. A built-in reservoir holds the colored water that fills the tube. One end is held to the starting point. The other end is moved down until the water level is seen and marked on the surface to be leveled (*Figure 27–5*).

Although highly accurate, the water level is somewhat limited by the length of the plastic tube. However, extension tubings are available.

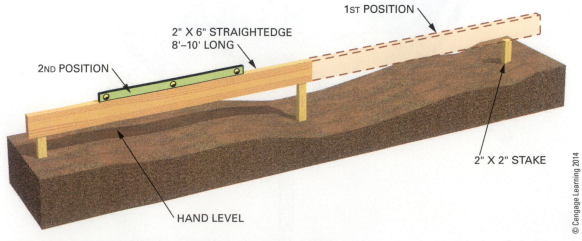

Figure 27–2 Leveling with a straightedge from stake to stake.

On the Job

Make a pilot hole in the earth for straight and easy driving of stakes.

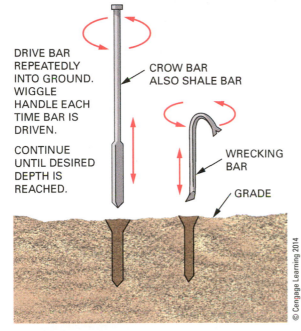

DRIVE BAR REPEATEDLY INTO GROUND. WIGGLE HANDLE EACH TIME BAR IS DRIVEN.

CONTINUE UNTIL DESIRED DEPTH IS REACHED.

CROW BAR ALSO SHALE BAR

WRECKING BAR

GRADE

Figure 27–3 Methods of starting a stake in hard ground.

Also, though slightly inconvenient, the water level may be moved from point to point.

A water level is accurate only if both ends of the tube are open to the air and there are no air bubbles in the length of the tube. Because both ends must be open, there may occasionally be some loss of liquid. However, this is replenished by the reservoir. Any air bubbles can be easily seen with the use of colored water.

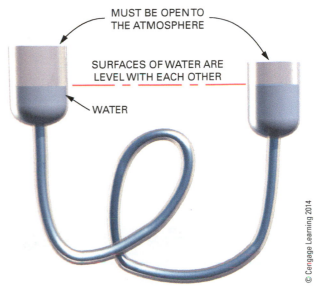

MUST BE OPEN TO THE ATMOSPHERE

SURFACES OF WATER ARE LEVEL WITH EACH OTHER

WATER

Figure 27–4 Water seeks its own level. Both ends of the water level must be open to the atmosphere.

StepbyStep Procedures

Procedure 27-A Leveling Corners Using a Level

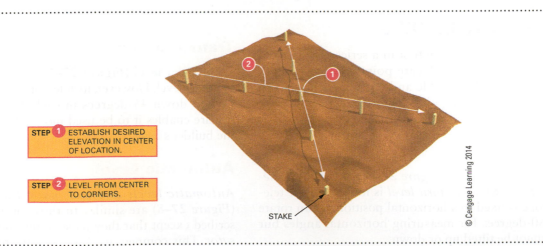

STEP **1** ESTABLISH DESIRED ELEVATION IN CENTER OF LOCATION.

STEP **2** LEVEL FROM CENTER TO CORNERS.

STAKE

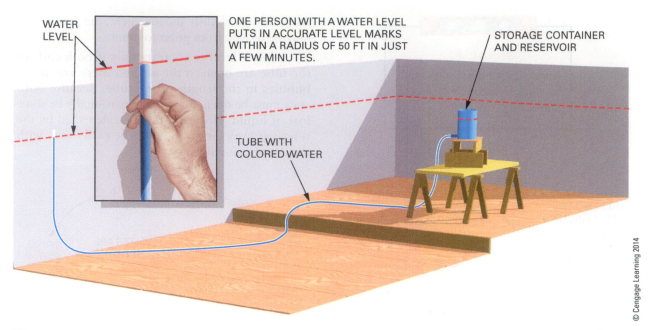

WATER LEVEL

ONE PERSON WITH A WATER LEVEL PUTS IN ACCURATE LEVEL MARKS WITHIN A RADIUS OF 50 FT IN JUST A FEW MINUTES.

STORAGE CONTAINER AND RESERVOIR

TUBE WITH COLORED WATER

© Cengage Learning 2014

Figure 27–5 The water level is a simple yet effective leveling tool.

Courtesy of David White

Figure 27–6 The builder's level.

Courtesy of David White

Figure 27–7 The telescope of the transit-level can be moved up and down 45-degrees each way.

In spite of these drawbacks, the water level is an extremely useful, inexpensive, simple tool for leveling from room to room, where walls obstruct views, down in a hole, or around obstructions. Another advantage is that leveling can be done by one person.

OPTICAL LEVELS

Optical levels are the first in a series of surveying equipment used to locate points on the ground. They are still available and useful as an inexpensive way to layout points. Optical levels include the builder's level and transit-level.

Builder's Levels

The builder's level (*Figure 27–6*) consists of a *telescope* to which a *spirit level* is mounted. The telescope is fixed in a horizontal position. It can rotate 360-degrees for measuring horizontal angles but cannot be tilted up or down.

Transit-Levels

The transit-level (*Figure 27–7*) is similar to the builder's level. However, its telescope can be moved up and down 45-degrees in each direction. This feature enables it to be used more effectively than the builder's level.

Automatic Levels

Automatic levels and *automatic transit-levels* (*Figure 27–8*) are similar to those previously described except that they have an internal *compensator*. This compensator uses gravity to maintain

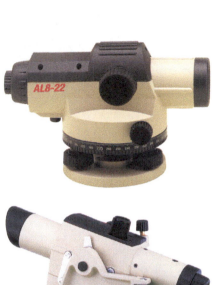

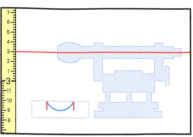

1) CONVENTIONAL INSTRUMENT CORRECTLY LEVELED. ROD READING IS 3'-3".

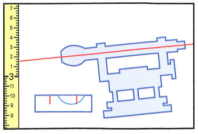

2) CONVENTIONAL INSTRUMENT SLIGHTLY OUT OF LEVEL. VIAL BUBBLE IS OFF CENTER AND INCORRECT ROD READING IS 3'-1½".

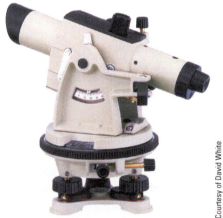

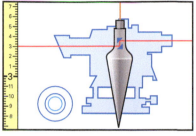

3) AUTOMATIC LEVEL-TRANSIT CORRECTLY LEVELED. ROD READING IS 3'-3".

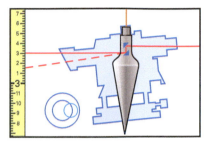

4) AUTOMATIC LEVEL-TRANSIT SLIGHTLY OUT OF LEVEL. CIRCULAR BUBBLE IS OFF CENTER, BUT THE COMPENSATOR CORRECTS FOR THE VARIATION FROM LEVEL AND MAINTAINS A CORRECT ROD READING OF 3'-3".

Courtesy of David White

Courtesy of David White

Figure 27–8 Automatic levels and automatic transit-levels level themselves when set up nearly level.

a true level line of sight. Even if the instrument is jarred, the line of sight stays true because gravity does not change.

Many models of leveling instruments are available. To become familiar with more sophisticated levels, study the manufacturers' literature. No matter what type of level is used, the basic procedures are the same.

Using Optical Levels

Before the level can be used, it must be placed on a *tripod* or some other solid support and leveled.

Setting Up and Adjusting the Level.
The telescope is adjusted to a level position by means of four *leveling screws* that rest on a *base leveling plate*. In higher quality levels, the base plate is part of the instrument. In less expensive models, the base plate is part of the tripod.

Open and adjust the legs of the tripod to a convenient height. Spread the legs of the tripod well apart, and firmly place its feet into the ground.

> **CAUTION**
>
> On a smooth surface, it is essential that the points on the feet hold without slipping. Make small holes or depressions for the tripod points to fit into. Or, attach wire or light chain to the lower ends of each leg (**Figure 27–9**).

When set up, the top of the tripod should be close to level. Sight by eye and tighten the tripod wing nuts. With the top of the tripod close to level, adjustment of the instrument is made easier.

Lift the instrument from its case by the frame. Note how it is stored so it can be replaced in the case in the same position. Make sure the horizontal clamp screw is loose so the telescope revolves freely. While holding onto the frame, secure the instrument to the tripod.

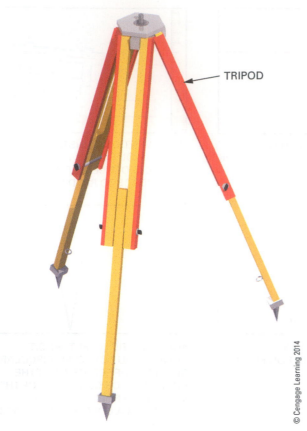

Figure 27–9 Make sure the feet of the tripod do not slip on smooth or hard surfaces.

CAUTION

Care must be taken not to damage the instrument. Never use force on any parts of the instrument. All moving parts turn freely and easily by hand. Excessive pressure on the leveling screws may damage the threads of the base plate. Unequal tension on the screws will cause the instrument to wobble on the base plate, resulting in leveling errors. Periodically use a toothbrush dipped in light instrument oil to clean and lubricate the threads of the adjusting screws.

Accurate leveling of the builder's level is important. Line up the telescope directly over two opposite leveling screws. Adjust these opposite screws so they are nearly snug. Back one screw off slightly to free both screws for leveling. Then turn the screws in opposite directions with forefingers and thumbs. Move the thumbs toward or away from each other, as the case may be, to center the bubble in the spirit level (*Figure 27–10*). The bubble will always move in the same direction as your left thumb is moving.

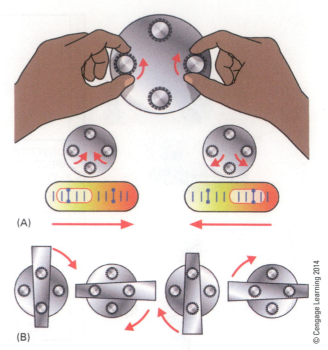

Figure 27–10 (A) Level the instrument by moving thumbs toward or away from each other. (B) The instrument is level when the bubble remains centered as the telescope is revolved in a complete circle.

Rotate the telescope 90-degrees over the other two opposite leveling screws and repeat the procedure. Make sure each of the screws has the same, but not too much, tension. Return to the original position, check, and make minor adjustments. Continue adjustments until the bubble remains exactly centered when the instrument is revolved in a complete circle.

CAUTION

Do not leave an instrument that has been set up unattended near moving equipment.

Sighting the Level. To sight an object, rotate the telescope and sight over its top, aiming it at the object. Look through the telescope. Focus it by turning the focusing knob one way or the other, until the object becomes clear. Keep both eyes open. This eliminates squinting, does not tire the eyes, and gives the best view through the telescope.

CAUTION

If the lenses need cleaning, dust them with a soft brush or rag. Do *not* rub the dirt off. Rubbing may scratch the lens coating.

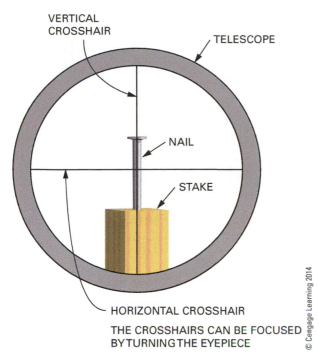

VERTICAL CROSSHAIR

TELESCOPE

NAIL

STAKE

HORIZONTAL CROSSHAIR

THE CROSSHAIRS CAN BE FOCUSED BY TURNING THE EYEPIECE

© Cengage Learning 2014

Figure 27–11 When looking into the telescope, vertical and horizontal crosshairs are seen.

When looking into the telescope, vertical and horizontal *crosshairs* are seen. They enable the target to be centered properly (*Figure 27–11*). The crosshairs themselves can be brought into focus by turning the eyepiece one way or the other. Center the crosshairs on the object by moving the telescope left or right. A fine adjustment can be made by tightening the horizontal clamp screw and turning the horizontal tangent screw one way or the other. The horizontal crosshair is used for reading elevations. The vertical crosshair is used when laying out angles and aligning vertical objects.

Leveling

When the instrument is leveled, a given point on the line of sight is exactly level with any other point. Any line whose points are the same distance below or above the line of sight is also level (*Figure 27–12*). To level one point with another, a helper must hold a *target* on the point to be leveled. A reading is taken. The target is then moved to selected points that are brought to the same elevation by moving those points up or down to get the same reading.

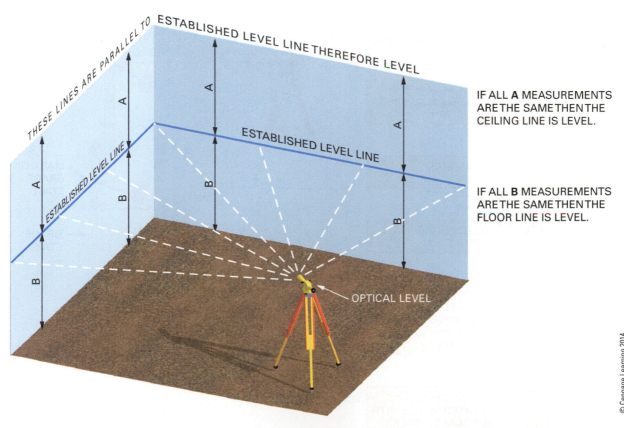

THESE LINES ARE PARALLEL TO ESTABLISHED LEVEL LINE THEREFORE LEVEL

A

A

A

A

ESTABLISHED LEVEL LINE

ESTABLISHED LEVEL LINE

B

B

B

B

B

OPTICAL LEVEL

IF ALL **A** MEASUREMENTS ARE THE SAME THEN THE CEILING LINE IS LEVEL.

IF ALL **B** MEASUREMENTS ARE THE SAME THEN THE FLOOR LINE IS LEVEL.

© Cengage Learning 2014

Figure 27–12 Any line parallel to the established level line is also level.

Targets. A measuring tape is often used as a target. The end of the tape is placed on the point to be leveled. The tape is then moved up or down until the same mark is read on the tape as was read at the starting point. Because of its flexibility, the tape may need to be backed up by a strip of wood to hold it rigid (*Figure 27–13*).

The simplest target is a plain 1 × 2 strip of wood. The end of the stick is held on the starting point of desired elevation. The line of sight is marked on the stick. The end of the stick is then placed on top of various points. They are moved up or down to bring the mark to the same height as the line of sight (*Procedure 27–B*). A stick of practically any length can be used.

Cut the stick to a length so that the mark to be sighted is at a noticeable distance off from the center of its length. It is then immediately noticeable if the stick is inadvertently turned upside down (*Figure 27–14*). If, for some reason, it is not desirable to cut the stick, clearly mark the top and bottom ends.

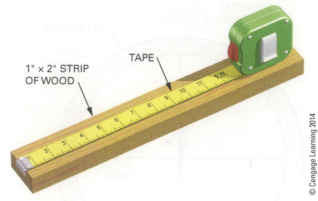

Figure 27–13 A tape can be backed up by a strip of wood to make it a stiff and steady target.

Leveling Rods. For longer sightings, the *leveling rod* is used because of its clearer graduations. A variety of rods are manufactured of wood or fiberglass for several leveling purposes. They are made with two or more sections that extend easily and

StepbyStep Procedures

Procedure 27-B Establishing Level Points

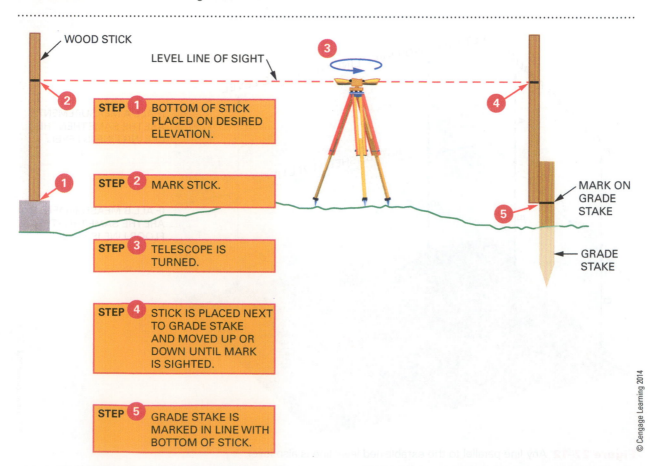

STEP **1** BOTTOM OF STICK PLACED ON DESIRED ELEVATION.

STEP **2** MARK STICK.

STEP **3** TELESCOPE IS TURNED.

STEP **4** STICK IS PLACED NEXT TO GRADE STAKE AND MOVED UP OR DOWN UNTIL MARK IS SIGHTED.

STEP **5** GRADE STAKE IS MARKED IN LINE WITH BOTTOM OF STICK.

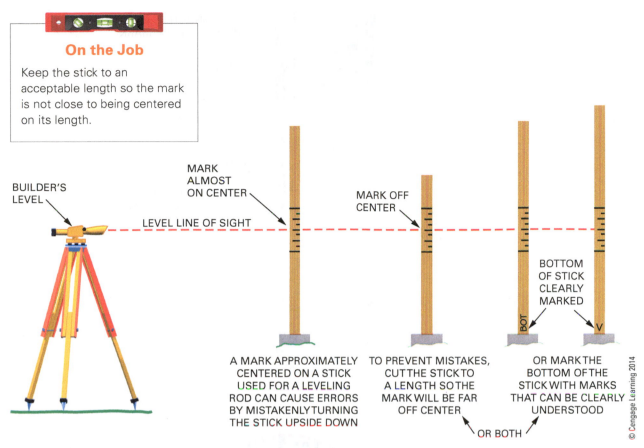

On the Job

Keep the stick to an acceptable length so the mark is not close to being centered on its length.

BUILDER'S LEVEL

MARK ALMOST ON CENTER

LEVEL LINE OF SIGHT

MARK OFF CENTER

BOTTOM OF STICK CLEARLY MARKED

BOT

V

A MARK APPROXIMATELY CENTERED ON A STICK USED FOR A LEVELING ROD CAN CAUSE ERRORS BY MISTAKENLY TURNING THE STICK UPSIDE DOWN

TO PREVENT MISTAKES, CUT THE STICK TO A LENGTH SO THE MARK WILL BE FAR OFF CENTER

OR MARK THE BOTTOM OF THE STICK WITH MARKS THAT CAN BE CLEARLY UNDERSTOOD

OR BOTH

Figure 27–14 Techniques for creating an easy-to-use marking stick.

lock into place. Rods vary in length—from two-section rods extending 9'-0" up to seven-section rods extending 25'-0".

The builder's rod has feet, inches, and eighths of an inch. The graduations are $\frac{1}{8}$ inch wide and $\frac{1}{8}$ inch apart. The engineer's rod is very similar yet the scale is slightly different. It is in feet, tenths, and hundredths of a foot. Instead of inches, the number markings represent a tenth of a foot. The smaller graduations are $\frac{1}{100}$ of a foot wide and $\frac{1}{100}$ foot apart, which is slightly smaller than $\frac{1}{8}$ inch. They are both designed for easy reading. An oval-shaped, red and white, movable target is available to fit on any rod for easy reading (*Figure 27–15*).

Communication. A responsible rod operator holds the rod vertical and faces the instrument so it can be read with ease and accuracy. Sighting distances are not usually over 100 to 150 feet, yet sometimes voice commands cannot be used. Hand signals are then given to the rod operator to move the target as desired by the instrument operator. Usually, appropriate hand signals are given even when distances are not great. Shouting on the job site is unnecessary, unprofessional, and creates confusion.

Establishing Elevations

Many points on the job site, such as the depth of excavations, the height of foundation footing and walls, and the elevation of finish floors, are required to be set at specified elevations or grades. These elevations are established by starting from the benchmark. The benchmark is a point of designated elevation. The instrument operator records elevations and rod readings in a notebook to make calculations.

Height of the Instrument (HI). When it is necessary to set a point at some definite elevation, first determine the **height of the instrument (HI).** To find HI, place the rod on the benchmark and add the reading to the elevation of the benchmark (*Figure 27–16*). For instance, if the benchmark has an elevation of 100.00 feet and the rod reads 5'-8", then the HI is 105'-8".

Grade Rod. What must be read on the rod when its base is at the desired elevation is called the **grade rod.** This is found by subtracting the desired elevation from the height of the instrument (HI). For instance, if the elevation to be established is

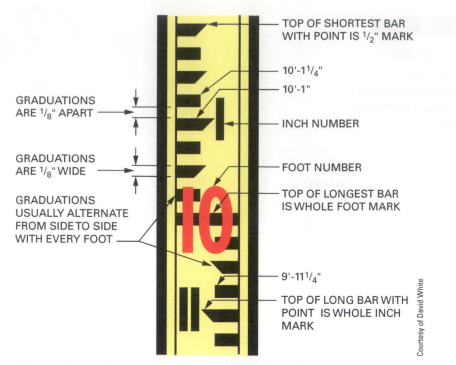

GRADUATIONS
ARE ¹/₈" APART

GRADUATIONS
ARE ¹/₈" WIDE

GRADUATIONS
USUALLY ALTERNATE
FROM SIDE TO SIDE
WITH EVERY FOOT

TOP OF SHORTEST BAR
WITH POINT IS ¹/₂" MARK

10'-1¹/₄"

10'-1"

INCH NUMBER

FOOT NUMBER

TOP OF LONGEST BAR
IS WHOLE FOOT MARK

9'-11¹/₄"

TOP OF LONG BAR WITH
POINT IS WHOLE INCH
MARK

Courtesy of David White

Figure 27–15 The builder's leveling rod is marked in feet, inches, and eighths of an inch.

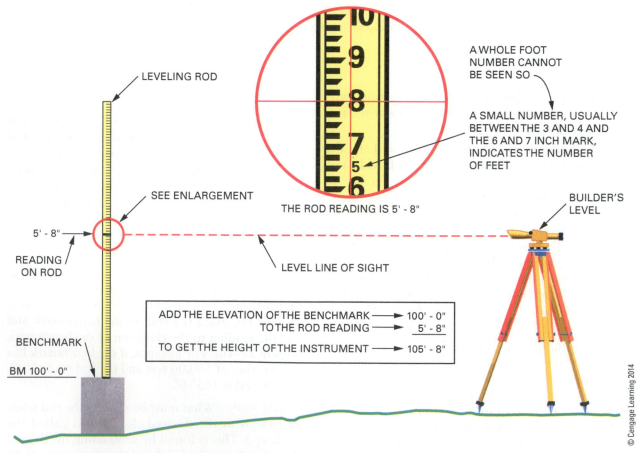

LEVELING ROD

SEE ENLARGEMENT

5' - 8"

READING
ON ROD

BENCHMARK

BM 100' - 0"

A WHOLE FOOT
NUMBER CANNOT
BE SEEN SO

A SMALL NUMBER, USUALLY
BETWEEN THE 3 AND 4 AND
THE 6 AND 7 INCH MARK,
INDICATES THE NUMBER
OF FEET

THE ROD READING IS 5' - 8"

BUILDER'S
LEVEL

LEVEL LINE OF SIGHT

ADD THE ELEVATION OF THE BENCHMARK	→	100' - 0"
TO THE ROD READING	→	5' - 8"
TO GET THE HEIGHT OF THE INSTRUMENT	→	105' - 8"

© Cengage Learning 2014

Figure 27–16 Determining the height of the instrument (HI).

102′-0″, subtract it from 105′-8″ (HI) to get 3′-8″ (the grade rod). The rod operator places the rod at the desired point. He or she then moves it up or down, at the direction of the instrument operator, until the grade rod of 3′-8″ is read on the rod. The base of the builder's rod is then at the desired elevation (*Figure 27–17*). A mark, drawn at the base of the rod on a stake or other object, establishes the elevation.

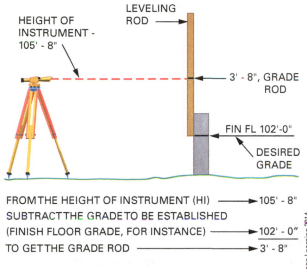

FROM THE HEIGHT OF INSTRUMENT (HI) ———▶ 105′- 8″

SUBTRACT THE GRADE TO BE ESTABLISHED
(FINISH FLOOR GRADE, FOR INSTANCE) ———▶ 102′- 0″

TO GET THE GRADE ROD ———▶ 3′- 8″

THE "GRADE ROD" IS WHAT THE ROD MUST READ WHEN ITS BASE IS AT THE DESIRED GRADE OR HEIGHT

Figure 27–17 Calculating the grade rod and establishing a desired elevation.

Determining Differences in Elevation

Differences in elevation need to be determined for such tasks as grading driveways, sidewalks, and parking areas, laying out drainage ditches, plotting contour lines, and estimating cut and fill requirements. The difference in elevation of two or more points is easily determined with the use of the builder's level or transit-level.

Single Setup. To find the difference in elevation between two points, set up the instrument about midway between them. Place the rod on the first point. Take a reading, and record it. Swing the level to the other point, take a reading, and record it. The difference in elevation is the difference between the recordings.

When making many readings, keeping track of the readings becomes more difficult. Using the surveying technique of tracking *backsight* and *foresight* makes this easier. Backsight is the reading from a level to a known or previously measured point. Foresight is a reading to a new location. All backsights are *plus* (+) *sights* and all foresights are *minus* (−) *sights*. These are recorded on a table for each setup of the transit or level (*Figure 27–18*).

Place the rod on point A and record the backsight reading as a plus (+) sight. Place the rod on point B, take the foresight reading, and record it as a minus (−) sight. Add the plus and minus sights to get the difference in elevation.

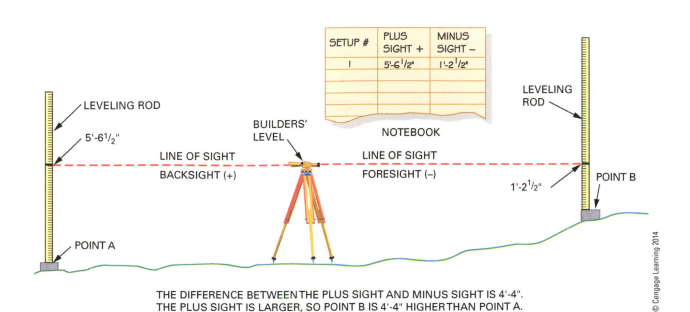

THE DIFFERENCE BETWEEN THE PLUS SIGHT AND MINUS SIGHT IS 4′-4″.
THE PLUS SIGHT IS LARGER, SO POINT B IS 4′-4″ HIGHER THAN POINT A.

Figure 27–18 Determining a difference of elevation between two points requiring only one setup.

Sometimes a reading is made above the level line of the transit. To keep this straight a rule is applied: for any backsight or foresight reading where the rod is flipped upside down, the plus or minus sign is reversed (*Figure 27–19*).

Multiple Setup. Sometimes the difference in elevation of two points is too great or the distance is too far apart. Then it is necessary to make more than one instrument setup to determine the difference. The procedure is similar to a series of one setup operations until the final point is reached.

> ## CAUTION
>
> When carrying a tripod-mounted instrument, handle it with care. Carry it in an upright position. Do not carry it over the shoulder or in a horizontal position. Be careful when going through buildings or close quarters not to bump the instrument.

Record all backsights as plus sights and all foresights as minus sights, unless the rod is upside down when read (*Figure 27–20*). Find the sum of all minus sights and all plus sights. The difference between them is the difference in elevation of the beginning and ending points. If the sum of the plus sights is larger, then the end point is higher than the starting point. If the sum of the minus sights is larger, then the end point is lower than the starting point.

Measuring and Laying Out Angles

To measure or lay out angles, the instrument must be set over a particular point on the ground. A hook, centered below the instrument, is provided for suspending a plumb bob. The plumb bob is used to place the level directly over this point. In more sophisticated instruments, a built-in *optical plumb* allows the operator to sight to a point below, exactly plumb with the center of the instrument. This enables quick and accurate setups over a point (*Figure 27–21*).

Setting Up over a Point. Suspend the plumb bob from the instrument. Secure it with a slip knot. Move the tripod and instrument so that the plumb bob appears to be over the point.

Press the legs of the tripod into the ground. Lower the plumb bob by moving the slip knot until it is about $\frac{1}{4}$ inch above the point on the ground. The final centering of the instrument can be made by loosening any two adjacent leveling screws and slowly shifting the instrument until the plumb bob is directly over the point (*Figure 27–22*). Retighten the same two leveling screws that were previously loosened, and level the instrument. Shift the instrument on the base plate until the plumb bob is directly over the point. Check the levelness of the instrument. Adjust, if necessary.

Circle Scale and Index. A horizontal circle scale (outside ring) is divided into 90-degree quadrants. A pointer, or *index,* turns with the telescope.

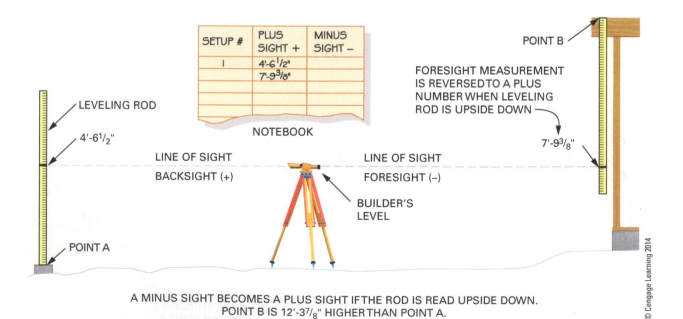

SETUP #	PLUS SIGHT +	MINUS SIGHT –
I	4'-6 1/2"	
		7'-9 3/8"

NOTEBOOK

LEVELING ROD

4'-6 1/2"

LINE OF SIGHT
BACKSIGHT (+)

POINT A

BUILDER'S LEVEL

POINT B

FORESIGHT MEASUREMENT IS REVERSED TO A PLUS NUMBER WHEN LEVELING ROD IS UPSIDE DOWN

7'-9 3/8"

LINE OF SIGHT
FORESIGHT (–)

A MINUS SIGHT BECOMES A PLUS SIGHT IF THE ROD IS READ UPSIDE DOWN.
POINT B IS 12'-3 7/8" HIGHER THAN POINT A.

© Cengage Learning 2014

Figure 27–19 For any readings taken with the rod upside down, the plus and minus signs of the sighting measurements are reversed.

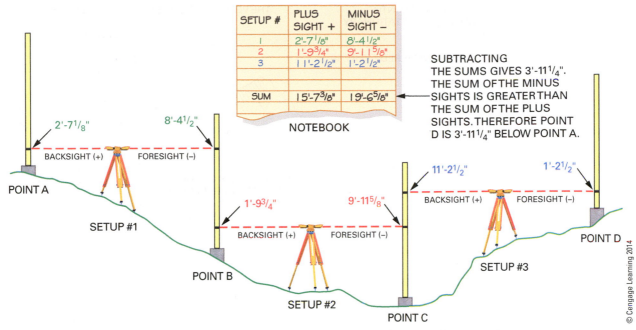

SETUP #	PLUS SIGHT +	MINUS SIGHT –
1	2'-7¹/₈"	8'-4¹/₂"
2	1'-9³/₄"	9'-11⁵/₈"
3	11'-2¹/₂"	1'-2¹/₂"
SUM	15'-7³/₈"	19'-6⁵/₈"

NOTEBOOK

SUBTRACTING THE SUMS GIVES 3'-11¹/₄". THE SUM OF THE MINUS SIGHTS IS GREATER THAN THE SUM OF THE PLUS SIGHTS. THEREFORE POINT D IS 3'-11¹/₄" BELOW POINT A.

Figure 27–20 Determining a difference of elevation between two points requiring more than one setup.

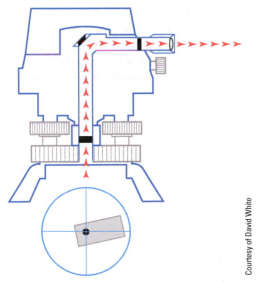

Figure 27–21 Some instruments have a device called an optical plumb for setting the instrument directly over a point.

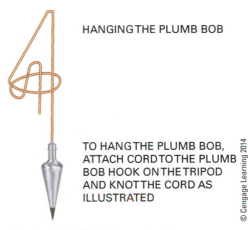

HANGING THE PLUMB BOB

TO HANG THE PLUMB BOB, ATTACH CORD TO THE PLUMB BOB HOOK ON THE TRIPOD AND KNOT THE CORD AS ILLUSTRATED

Figure 27–22 To locate the instrument directly over a point, a plumb bob is suspended from the level.

The circle scale remains stationary and indicates the number of degrees the telescope is turned. When desired, the horizontal circle may be turned by hand for setting to zero degrees, no matter which way the telescope is pointing. By starting at zero and rotating the telescope on it, any horizontal angle can be easily measured (*Figure 27–23*).

Reading the Horizontal Vernier. For more precise readings, the horizontal vernier is used to read minutes of a degree (*Figure 27–24*). A vernier is a smaller scale used to make more precise measurements when the zero index, the point where the reading is made, falls between the lines of the circle scale. The vernier is actually two verniers, one on each side of the vernier zero index. This makes it possible to read any angle, whether turned to the right or to the left.

The vernier scale turns with the telescope. If the eye end of the telescope is turned to the left (clockwise), the vernier scale on the left side of the zero index is used. If the eye end of the telescope is turned to the right (counterclockwise), the vernier

THE HORIZONTAL CIRCLE SCALE IS DIVIDED INTO QUADRANTS OF 90° EACH, AND REMAINS STATIONARY AS THE TELESCOPE IS TURNED.

IT IS GRADUATED IN DEGREES AND NUMBERED EVERY 10 DEGREES.

IT MAY BE ROTATED BY HAND TO ADJUST THE FIRST READING TO ZERO.

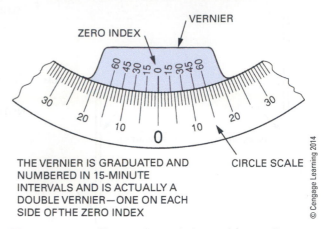

THE VERNIER ROTATES WITH THE TELESCOPE

THE VERNIER IS GRADUATED AND NUMBERED IN 15-MINUTE INTERVALS AND IS ACTUALLY A DOUBLE VERNIER—ONE ON EACH SIDE OF THE ZERO INDEX

Figure 27–24 The vernier scale is used for reading minutes of a degree.

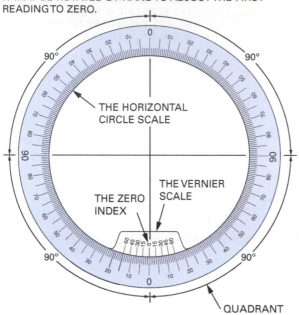

Figure 27–23 Reading the horizontal circle scale.

scale on the right side of the zero index is used. Use either the left or right vernier scale according to the direction in which the eye end of the telescope is turned when measuring or laying out angles (*Figure 27–25*).

To read the vernier scale, first lock the transit at the desired position. Read the degrees where the zero index lines up with the circle scale. Read the smaller of the two numbers (*Procedure 27–C*).

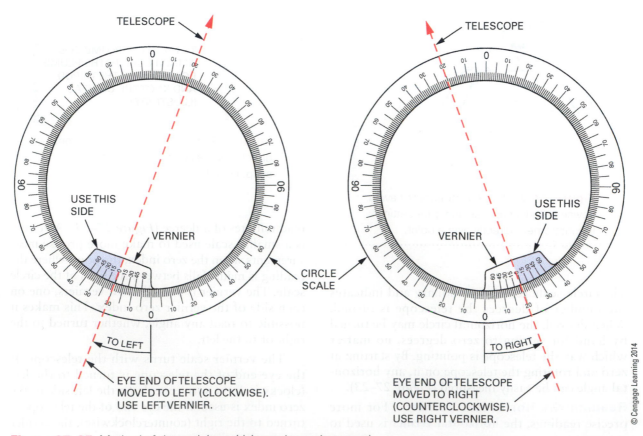

Figure 27–25 Method of determining which vernier scale to read.

StepbyStep Procedures

Procedure 27-C Reading a Vernier Scale

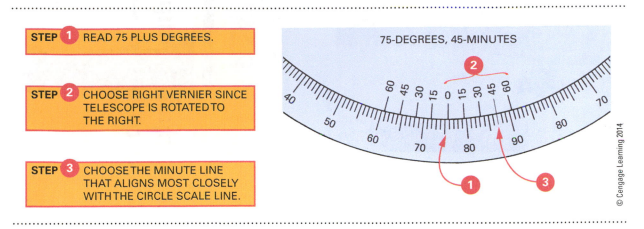

STEP 1 READ 75 PLUS DEGREES.

STEP 2 CHOOSE RIGHT VERNIER SINCE TELESCOPE IS ROTATED TO THE RIGHT.

STEP 3 CHOOSE THE MINUTE LINE THAT ALIGNS MOST CLOSELY WITH THE CIRCLE SCALE LINE.

75-DEGREES, 45-MINUTES

© Cengage Learning 2014

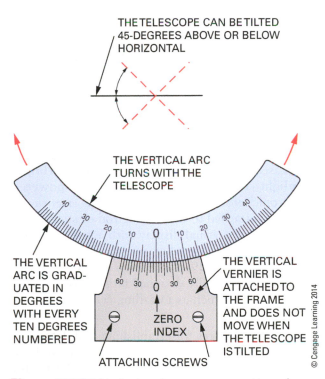

THE TELESCOPE CAN BE TILTED 45-DEGREES ABOVE OR BELOW HORIZONTAL

THE VERTICAL ARC TURNS WITH THE TELESCOPE

THE VERTICAL ARC IS GRADUATED IN DEGREES WITH EVERY TEN DEGREES NUMBERED

ZERO INDEX

THE VERTICAL VERNIER IS ATTACHED TO THE FRAME AND DOES NOT MOVE WHEN THE TELESCOPE IS TILTED

ATTACHING SCREWS

© Cengage Learning 2014

Figure 27–26 Vertical angles are measured by using the vertical arc and vernier scales.

In this case, the measurement is 75 plus degrees to the right. Therefore the right vernier is used. Now locate the vernier line that happens to line up best with the larger circle scale lines. In this case, the 45-minute line is best aligned with a degree line, thus the measurement is 75-degrees, 45-minutes.

Measuring Vertical Angles. The vertical arc scale is attached to the telescope. It measures vertical angles to 45-degrees above and below the horizontal. By tilting and rotating the telescope, set the horizontal crosshair on the points of the vertical angle being measured. Tighten the vertical clamp. Then turn the tangent screw for a fine adjustment to place the crosshair exactly on the point. Vertical angles are read by means of the vertical arc scale and the obvious vernier, similar to the reading of horizontal angles (*Figure 27–26*).

LASER LEVELS

A laser is a device in which light energy is released in a narrow beam. The light beam is absolutely straight. Unless interrupted by an obstruction or otherwise disturbed, the light beam can be seen for a long distance.

Lasers have many applications in space, medicine, agriculture, and engineering. The laser level has been developed for the construction industry to provide more efficient layout work (*Figure 27–27*). Laser levels are often used tool on the job site to determine level.

KINDS AND USES OF LASER LEVELS

Several manufacturers make laser levels in a number of different models. The least expensive models are the least sophisticated. A low-price unit is leveled and adjusted manually. More expensive ones are automatically adjusted to and maintained in level. Power sources include batteries, a rechargeable battery pack, or an AC/DC converter for 110 or 220 volts.

Establishing and Determining Elevations. A simple, easy-to-use model is mounted on a tripod or solid flat surface and leveled like manual or automatic optical instruments. The laser is turned

Figure 27–27 Laser levels have been developed for use in the construction industry.

Courtesy of Trimble

Courtesy of Leica Geosystems

HIGH .... FAST BEEPING

ON-GRADE .... SOLID TONE

LOW .... SLOW BEEPING

Figure 27–29 An electronic target senses the laser beam. An audio feature provides tones to match the visual display.

Figure 27–28 The laser beam rotates 360-degrees, creating a level plane of light.

© Cengage Learning 2014

on. It will emit a red beam. The beam rotates through a full 360-degrees, creating a level *plane* of light. As it rotates, it establishes equal points of elevation over the entire job site, similar to a line of sight being rotated by the telescope of an optical instrument (*Figure 27–28*).

Depending on the quality of the instrument, the laser head may rotate at various revolutions per second (RPS), up to 40 RPS. Its quality also determines its working range. This may vary from a 75- to 1,000-foot radius.

Laser beams may be restricted to only a small region of the job site. This is done to protect other workers from the light. These restrictions may be flaps that fold down to block the light. Some higher end lasers allow the instrument to be adjusted to limit the arc of light. Instead of rotating the head completely around, it simply moves back and forth.

Laser beams are difficult to see outdoors in bright sunlight. To detect the beam, a battery-powered electronic *sensor* target, also called a *receiver* or *detector,* is attached to the leveling rod or stick. Most sensors have a visual display with a selectable audio tone to indicate when it is close to or on the beam (*Figure 27–29*). In addition to electronic sensor targets, specially designed targets are used for interior work, such as installing ceiling grids and leveling floors.

The procedures for establishing and determining elevations with laser levels are similar to those with optical instruments. To establish elevations, the sensor is attached to the grade rod. The grade rod is moved up or down until the beam indicates that the base of the rod is at grade.

One of the great advantages of using laser levels is that, in most cases, only one person is needed to do the operations (*Figure 27–30*). Another advantage is that certain operations are accomplished more easily in less time.

Special Horizontal Operations. For leveling *suspended ceiling grids,* the laser level is mounted to an adjustable grid mount bracket. Once the first strip of angle trim has been installed, the laser and bracket can be attached easily and clamped into place.

Figure 27–30 When using laser levels, only one person is required for leveling operations.

Figure 27–32 Placed on its side, the laser level is used to layout and align partitions and walls.

Figure 27–31 Leveling suspended ceiling grids. The beam is viewed through the target.

Figure 27–33 Attaching a laser detector to a bulldozer allows operator to level the ground quickly.

The unit is leveled. The height of the laser beam is then adjusted for use with a special type of magnetic or clip-on target. Magnetic targets are used on steel grids. Clip-on targets are used on aluminum or plastic. Once the laser is set up, the ceiling grid is quickly and easily leveled using the rotating beam as a reference and viewing the beam through the target (*Figure 27–31*). Sprinkler heads, ceiling outlets, and similar objects can be set in a similar manner.

Special Vertical Operations. Most laser levels are designed to work laying on their side using special brackets or feet. The unit is placed on the floor and leveled. To lay out a partition, rotate the head of the laser downward. Position the laser beam over one end of the partition. Align the beam toward the far point by turning the head

toward and adjusting to the second point. Recheck the alignment and turn on to rotate the beam. The beam will be displayed as continuous straight and plumb lines on the floor, ceiling, and walls to both align and plumb the partition at the same time (*Figure 27–32*). Mounted on excavation equipment, lasers simplify and speed up construction work (*Figure 27–33*).

Layout Operations. Some laser levels emit a plumb line of light projecting upward from the top at a right-angle (90-degrees) to the plane of the rotating beam. The plumb reference beam allows one person to lay out 90-degree cross-walls or building lines. Set the unit on its side as for laying out a partition. The reference beam establishes a 90-degree corner with the rotating laser beam.

Plumbing Operations. The vertical beam provides a ready reference for plumbing posts, columns, elevator shafts, slip forming, and wherever a plumb reference beam is required.

Mount the laser unit on a tripod over a point of known offset from the work to be plumbed. Suspend a plumb bob from the center of the tripod directly over the point. Apply power to the unit. The vertical beam that is projected is ready for use as a reference. Move the top of the object until it is offset from the beam the same distance as the bottom point.

The heads on some laser units can be tilted so the rotating beam produces a plane of light at an angle to the horizontal. Such units are used for laying out slopes. Other units are manufactured for special purposes such as pipe laying and tunnel guidance. Marine laser units, with ranges up to 10 miles, are used in port and pier construction and offshore work.

Laser Safety

With a little common sense, the laser can be used safely. All laser instruments are required to have warning labels attached (*Figure 27–34*). The following are safety precautions for laser use:

- Only trained persons should set up and operate laser instruments.
- Never stare directly into the laser beam or view it with optical instruments.
- When possible, set the laser up so that it is above or below eye level.
- Turn the laser off when not in use.
- Do not point the laser at others.

TOTAL STATIONS

Total stations are instruments that measure angles and distances in an instant. They are a combination of laser transit and measuring tape. Angles are measured internally and displayed in an easy to read digital form (*Figure 27–35*). Distances are determined by a system called electronic distance measuring (EDM). The instrument emits an electromagnetic pulse, laser, or infrared beam. The beam is reflected back to the instrument. Some units

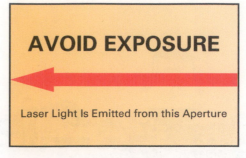

APERTURE LABEL

WARNING LABEL

Figure 27–34 Warning labels must be attached to every laser instrument.

Figure 27–35 Total station is a surveying tool designed to quickly and accurately measure distance and angles.

use a **prism** as a target (*Figure 27-36*). The prism is held plumb over a desired point on the ground as the measurement is taken. Plumb is determined by a **bull's eye bubble** located on the prism rod (*Figure 27-37*).

An on-board computer then turns this information into distance and angles. This information makes a 3D measurement of the prism's location from the instrument. As more points are measured,

Figure 27–36 A prism is used to reflect the light beam back to the total station.

Figure 27–37 Prism is held plumb by using a bull's eye bubble.

Figure 27–38 Survey data points are measured by pressing a button.

data is collected (*Figure 27–38*). Computerized data collectors, enable the measurements to be downloaded to the CAD file used to create a building.

The power and awe of the total station is in the blending of the survey and the structural drawing of the project. The entire site can be uploaded into the total station. It is then possible to locate the 3D position of any point of the jobsite. The information may also be converted to a topographical survey map.

Models vary in price and sophistication. Better models are able to measure distances of one kilometer (over 3000 feet) with an accuracy of 2–3 millimeters (1/8 inch). They can measure angles to within one arc-second (1/3600 of a degree). This equates to a pencil lead thickness at 100 meters (about 320 feet). They can cost $2,500 to $30,000 for the entire package.

Higher end models are robotic in nature. Only one person is needed to collect survey data

points. The instrument is set in a strategic place where the operator then walks the prism around the site. When the operator is ready to take a reading, a button is pushed emitting an infrared signal to the instrument. It then rotates to find the signal, positioning the laser on the prism towards the prism. When ready, the operator pushes another button and the data point distance and angles are recorded. As the operator moves to a new point, the instrument is then able to follow the prism as it moves and be instantly ready for a new reading.

It is important for the operator to read and follow the manufacturer's instructions. Because these tools are very expensive, using them is best done under the guidance of an experienced person. Small mistakes can be very costly. Understanding what to do with the data and using the available software requires learning surveying methods and techniques as well as computer programs.

CHAPTER 28

Laying Out Foundation Lines

Before any layout can be made, the builder must determine the dimensions of the building and its location on the site from the plot plan. This task often falls to a professional surveyor, but in many areas may be done by any qualified person. Care should be taken to position the building properly to avoid costly changes. It is usually the carpenter's responsibility to lay out building lines.

STAKING THE BUILDING

Proper layout begins with locating the property corners, then placing stakes at each corner of the building. Some lots are large enough that the building is measured from only one property line, while other lots are small enough to make it necessary to check all property lines with the building (*Figure 28–1*).

Begin by finding the survey rods that mark the corners of the property. Do not guess where the property lines are. Sometimes it is a good idea to stretch and secure lines between each corner, laying out all the property boundary lines.

Locate the front building line by measuring in from the front property line the specified front setback. Measure from both ends of the building line and drive a stake at each end. Stretch a line

between these stakes to better show the front edge of the building (*Figure 28–2*).

Along the front building line, measure in from the side property line the specified side setback. Drive a stake, stake A, firmly into the ground. Place a nail in the top of the stake to aid in making precise measurements (*Figure 28–3*). From this nail,

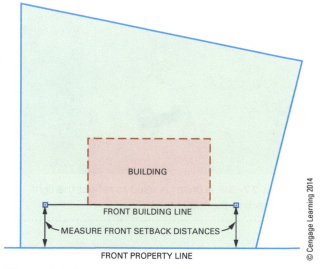

Figure 28–2 Locating the front building line.

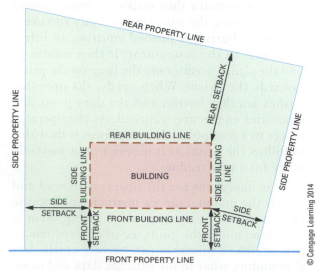

Figure 28–1 Locating the building on a lot from the dimensions found on the plot plan.

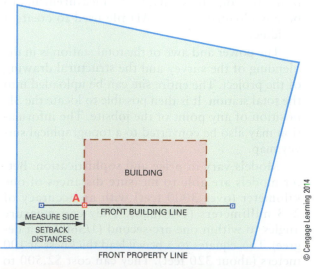

Figure 28–3 Measure from the side property line to locate the first building corner stake, Stake A.

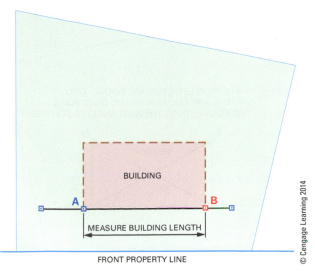

Figure 28–4 Measure along the building line to locate the second building corner stake, Stake B.

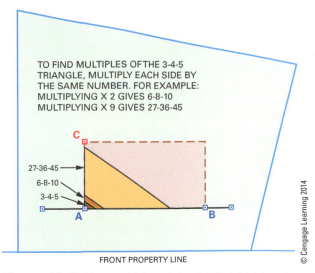

Figure 28–5 Use multiples of the 3-4-5 right triangle to place the third building corner stake, Stake C.

measure the front dimension of the building along the front building line. Drive stake B directly under the front building line string. Drive a nail in the top of the stake marking the exact length of the building (*Figure 28–4*).

The third stake, stake C, is placed to locate the back corner of the building. It must be square or at a right-angle to the front building line. This stake on the rear building line may be located using one of at least three methods. First, you could use an optical instrument such as a transit-level. It is set up directly over stake A and sighted to stake B, aligning the crosshairs. Then the telescope is rotated exactly 90-degrees. Stake C is located by using a tape measure and the crosshairs at the same time. The second method is to use the 3-4-5 method. This process uses multiples of 3-4-5 to create larger right triangles. Multiplying each side of a 3-4-5 triangle by the same number creates a larger triangle that also has a right angle. For example, 3-4-5 multiplied by 9 gives a 27-36-45 right triangle (*Figure 28–5*).

The third method is faster and more accurate. Two tapes are used to measure the building width from stake A and the **diagonal** of the building from stake B at the same time. To determine the diagonal of the building the **Pythagorean theorem,** $a^2 + b^2 = c^2$, is used. For example, consider a building whose length $a = 40'$ and width $b = 32'$. The diagonal equals c. Using the Pythagorean theorem, $c^2 = 32^2 + 40^2 = 2,624$. Taking the square root of the diagonal C gives us 51.2249939.

To convert to feet-inches to the nearest one-sixteenth, subtract 51 feet to leave the decimal.

PYTHAGOREAN THEOREM
$c^2 = a^2 + b^2$

FOR EXAMPLE, IF LENGTH = 40' AND WIDTH = 32', THEN DIAGONAL EQUALS c.

$c^2 = 32^2 + 40^2 = 2624$
c = 51.2249939' = 51' + 0.2249939'
0.2249939' × 12 = 2.6999268" = 2" + 0.6999268"
0.6999268" × 16 = 11.1988/16th WHICH ROUNDS OFF TO 11/16"
THUS THE DIAGONAL OF A 32' × 40' RECTANGLE EQUALS 51'-2 11/16".

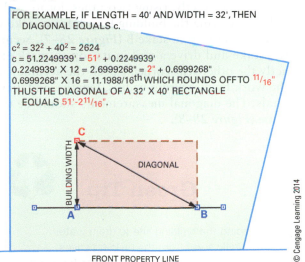

Figure 28–6 Use the Pythagorean theorem to place the third building corner stake, Stake C. Two tapes are used to measure the width and diagonal at the same time.

Convert the decimal 0.2249939 to inches by multiplying by 12: 0.2249939' × 12 = 2.6999268". Subtract 2 inches and write it down with 51 to make 51'-2". Convert 0.6999268" to a fraction by multiplying by 16, the desired denominator: 0.6999268" × 16 = 11.1988/16, which rounds off to 11/16". Thus, the diagonal of a 32' × 40' rectangle is 51'-2 11/16". Using two tapes, position stake C, which is located where the two tapes cross in *Figure 28–6*.

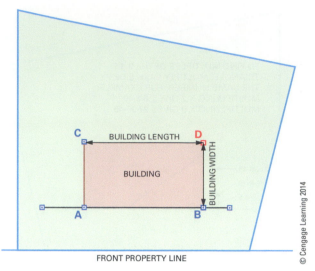

Figure 28–7 Measure from established front and rear corners (Stakes B and C) to locate the last building corner stake, Stake D.

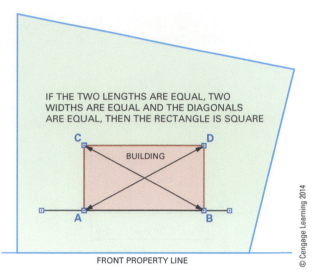

Figure 28–8 If the length and width measurements are accurate and the diagonal measurements are equal, then the corners are square.

When the locating of stake C is completed, drive a nail in the top of the stake marking the rear corner exactly. Using two tapes, locate stake D by measuring the building length from stake C and the width from stake B (*Figure 28–7*). Secure the stake and drive a nail in its top to mark exactly the other rear corner. Check the accuracy of the work by measuring widths, lengths, and diagonals. The diagonal measurements should be the same (*Figure 28–8*).

All measurements must be made on the level. If the land slopes, the tape is held level with a plumb bob suspended from it (*Figure 28–9*).

Layout of irregularly shaped buildings may seem complicated at first, but they are laid out using the same fundamental principles just outlined. The irregularly shaped building may be staked out from a large rectangle (*Procedure 28–A*). The large rectangle corners should align with as many of the building corners as possible. Stake the outermost corners for distance and square, then pull a string to make

the large rectangle. The intermediate corner stakes are then located by measuring from the four corners along the strings. Then the final (inside-corner) stakes are measured and located using two tapes.

Placing stakes accurately often requires the builder to move them slightly while installing them. This can sometimes be a nuisance. To speed the process, large nails may be driven into the ground through small squares of thin cardboard (*Figure 28–10*). The cardboard serves to make the nail head more visible. This process allows for faster erecting of **batter boards** later.

BATTER BOARDS

After the building is laid out the excavation process begins, which often removes the layout stakes in the process. Batter boards are often constructed to allow the layout stakes to be reinstalled after the excavation process is completed. If the stakes are

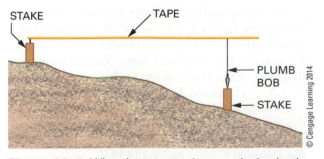

Figure 28–9 When layouts are done on sloping land, measurements must be taken on the level.

Step**by**Step Procedures

Procedure 28-A Establishing Multiple Corners for an Irregularly Shaped Building

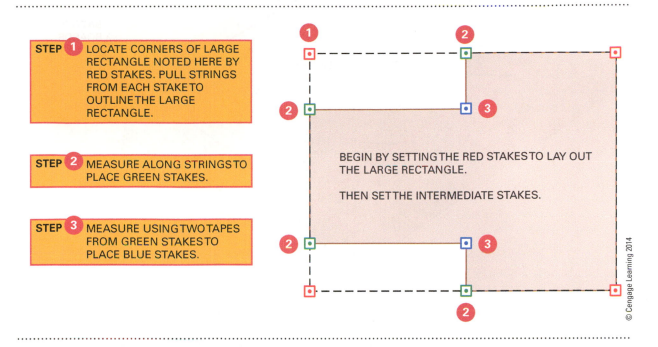

STEP 1 LOCATE CORNERS OF LARGE RECTANGLE NOTED HERE BY RED STAKES. PULL STRINGS FROM EACH STAKE TO OUTLINE THE LARGE RECTANGLE.

STEP 2 MEASURE ALONG STRINGS TO PLACE GREEN STAKES.

STEP 3 MEASURE USING TWO TAPES FROM GREEN STAKES TO PLACE BLUE STAKES.

BEGIN BY SETTING THE RED STAKES TO LAY OUT THE LARGE RECTANGLE.

THEN SET THE INTERMEDIATE STAKES.

© Cengage Learning 2014

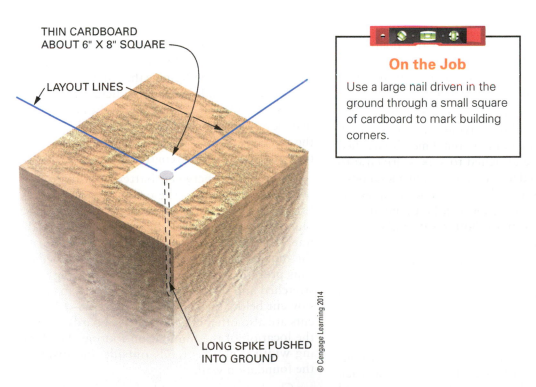

THIN CARDBOARD ABOUT 6" X 8" SQUARE

LAYOUT LINES

LONG SPIKE PUSHED INTO GROUND

© Cengage Learning 2014

On the Job

Use a large nail driven in the ground through a small square of cardboard to mark building corners.

Figure 28–10 Technique for marking the ground location of stakes.

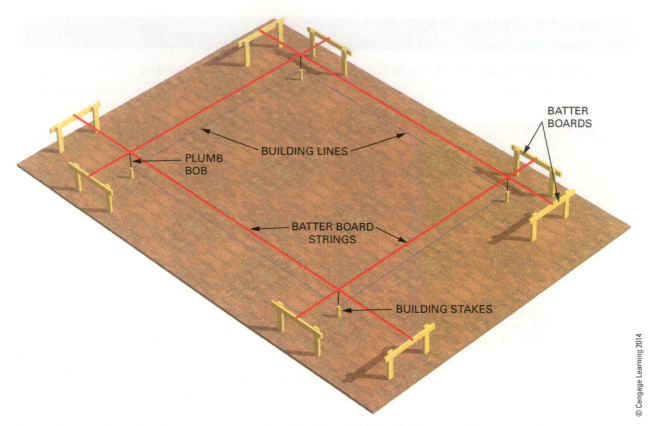

Figure 28–11 Batter boards are installed behind the building stakes and nearly level with the top of the foundation.

positioned with a total station, batter boards are not necessary. The information used to do the first layout may be reloaded into the instrument and a new layout done as quickly as before.

Batter boards are wood frames built behind the stakes to which building layout lines are secured. Batter boards consist of horizontal members, called ledgers. These are attached to stakes driven into the ground. The ledgers are fastened in a level position to the stakes, usually at the same height as the foundation wall (*Figure 28–11*). Batter boards are built-in the same way for both residential or commercial construction.

Batter boards are erected in such a manner that they will not be disturbed during excavation. Drive batter board stakes into the ground a minimum of 4 feet outside the building lines at each corner. When setting batter boards for large construction, increase this distance. This will allow room for heavy excavating equipment to operate without disturbing the batter boards. In loose soil or when stakes are higher than 3 feet, they must be braced (*Figure 28–12*).

Set up the builder's level about center on the building location. Sight to the benchmark, and record the sighting. Determine the difference between the benchmark sighting and the height of the ledgers. Sight and mark each corner stake at the specified elevation. Attach ledgers to the stakes so that the top edge of each ledger is on the mark. Brace the batter boards for strength, if necessary.

Stretch lines between batter boards directly over the nail heads in the original corner stakes. Locate the position of the lines by suspending a plumb bob directly over the nail heads. When the lines are accurately located, make a saw cut on the outside corner of the top edge of the ledger. This prevents the layout lines from moving when stretched and secured. Be careful not to make the saw cut below the top edge (*Figure 28–13*). Saw cuts are also often made on batter boards to mark the location of the foundation footing. The footing width usually extends outside and inside of the foundation wall.

Check the accuracy of the layout by again measuring the diagonals to see if they are equal. If not,

ON LARGE CONSTRUCTION, STRAIGHT BATTER BOARDS ARE USED IN ORDER TO
BE SET BACK FAR ENOUGH TO PROVIDE ROOM FOR HEAVY EXCAVATING EQUIPMENT

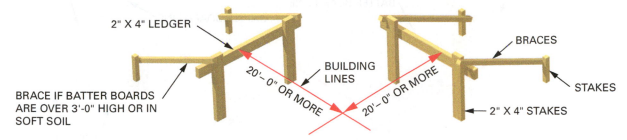

STRAIGHT BATTER BOARDS

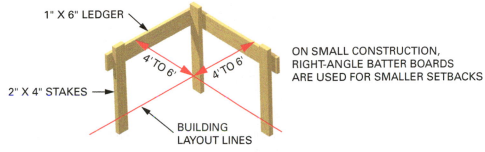

RIGHT-ANGLE BATTER BOARDS

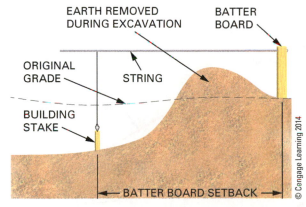

© Cengage Learning 2014

Figure 28–12 Batter boards are placed back far enough so they will not be disturbed during excavation operations.

make the necessary adjustment until they are equal (*Figure 28–14*). Once the excavation is completed, the building stakes may be relocated. Reattach the batter board strings in the appropriate saw kerfs. Use a plumb bob to determine the building stake location (*Figure 28–15*).

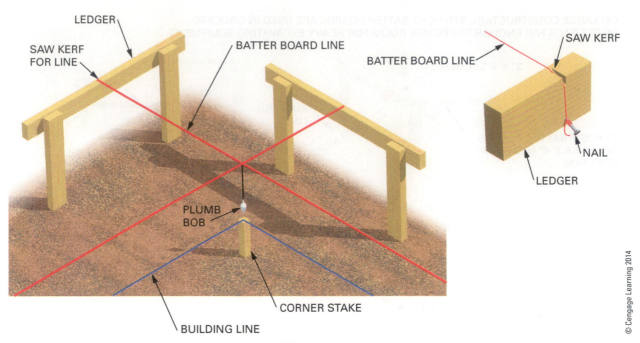

LEDGER

SAW KERF
FOR LINE

BATTER BOARD LINE

BATTER BOARD LINE

SAW KERF

NAIL

LEDGER

PLUMB
BOB

CORNER STAKE

BUILDING LINE

Figure 28–13 Saw kerfs are usually made in the ledger board to keep the strings from moving out of line.

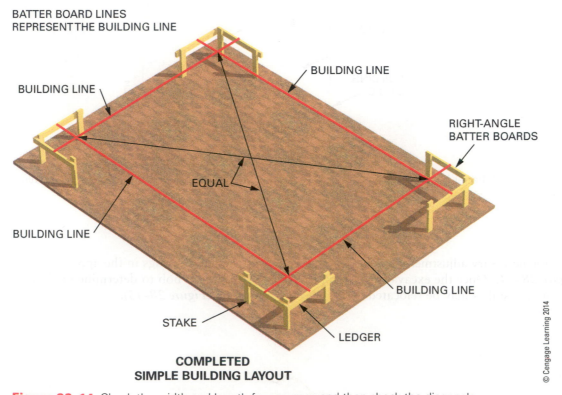

BATTER BOARD LINES
REPRESENT THE BUILDING LINE

BUILDING LINE

BUILDING LINE

BUILDING LINE

BUILDING LINE

RIGHT-ANGLE
BATTER BOARDS

EQUAL

BUILDING LINE

STAKE

LEDGER

**COMPLETED
SIMPLE BUILDING LAYOUT**

Figure 28–14 Check the width and length for accuracy and then check the diagonals.

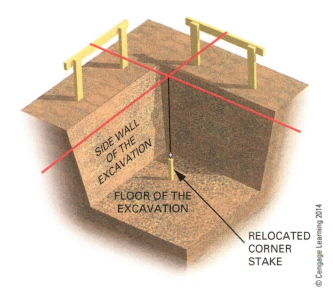

SIDE WALL OF THE EXCAVATION

FLOOR OF THE EXCAVATION

RELOCATED CORNER STAKE

© Cengage Learning 2014

Figure 28–15 Building corner stakes are easily repositioned later in the excavation using the batter board strings and a plumb bob.

DECONSTRUCT THIS

Carefully study *Figure 28–16* and think about what is wrong and/or what is right. Consider all possibilities. What construction practice or method is different in your area of the country?

Figure 28–16 This photo shows a construction job site dumpster.

KEY TERMS

batter boards	foundation wall	laser	Pythagorean theorem
builder's level	grade rod	laser level	total station
bulls eye bubble	height of the instrument	ledgers	transit-level
carpenter's hand level	horizontal circle scale	optical levels	vertical arc
diagonal	horizontal vernier	prism	

REVIEW QUESTIONS

Select the most appropriate answer.

1. The location of the building on the lot is determined from the

   a. foundation plan.
   b. floor plan.
   c. architect.
   d. plot plan.

2. The reference for establishing elevations on a construction site is called the

   a. starting point.
   b. reference point.
   c. benchmark.
   d. sight mark.

3. The builder's level is ordinarily used for

   a. laying out straight lines.
   b. reading elevations.
   c. reading vertical angles.
   d. all of these.

4. When leveling an optical level instrument with four leveling screws, turn the opposite leveling screws in

   a. opposite directions, noting the direction of your thumbs.
   b. same direction, noting the direction of your thumbs.
   c. opposite directions, noting the direction of your forefinger.
   d. same direction, noting the direction of your fingers.

5. The crosshairs of an optical leveling instrument are focused for individual users by rotating the

   a. focus knob on top of the telescope.
   b. eyepiece ring.
   c. tangent screw.
   d. base plate.

6. A tool used with a builder's level to check the level of objects is a

   a. pocket tape backed up by a strip of wood.
   b. grade rod.
   c. piece of wood with graduations marked on it.
   d. all of the above.

7. After a transit-level is moved and set up at a second location, a backsight reading of a normal rod is toward the

   a. previous point and is a plus measurement.
   b. unknown point and is a plus measurement.
   c. previous point and is a minus measurement.
   d. unknown point and is a minus measurement.

8. If the first reading of a normal rod at point A has an elevation of 6' and the second point B has an elevation of 2', then

   a. point A is higher than point B by 4'.
   b. point A is higher than point B by 8'.
   c. point B is higher than point A by 4'.
   d. point B is higher than point A by 8'.

9. The elevation difference between point A and point D where the backsight rod readings are 36", 48", and 59" and the foresight rod readings are 28", 42", and 60" is such that

   a. point A is 13" higher than point D.
   b. point A is 13" lower than point D.
   c. point A is 23" higher than point D.
   d. point A is 23" lower than point D.

10. When using a vernier scale to measure an angle, read the minutes at the

   a. zero index line.
   b. vernier line that best aligns with the circle scale line.
   c. circle scale.
   d. all of the above.

11. Laser leveling tools are used by

    a. carpenters.
    b. heavy equipment operators.
    c. plumbers.
    d. all of the above.

12. The diagonal of a rectangle whose dimensions are 30 feet × 40 feet is

    a. 45 feet.   c. 60 feet.
    b. 50 feet.   d. 70 feet.

13. The diagonal of a rectangle whose dimensions are 32 feet × 48 feet is

    a. 57′–8⅛″.   c. 57′–11″.
    b. 57′–8¼″.   d. 60′–0″.

14. A hanging plumb bob is actually pointing towards the

    a. ground.
    b. perpendicular level line.
    c. center of the earth.
    d. all of the above.

15. The instrument that uses a laser beam to measure distances and angles is called a

    a. laser level.
    b. theodolite.
    c. total station.
    d. builder's laser transit.

UNIT 11
Foundations, Concrete, and Formwork

©Shutterstock, Inc./Niki Crucillo

Concrete formwork is usually the responsibility of the carpenter, whether constructing wood forms or erecting a forming system. The forms must meet specified dimensions and be strong enough to withstand tremendous pressure. The carpenter must know that the forms will be strong enough before the concrete is placed. If a form fails to hold during concrete placement, costly labor and materials are lost. Formwork is unique in construction in that if it is built for a specific purpose, then it must be dismantled after the concrete cures. The carpenter must keep this in mind when designing the joints and seams in the formwork so the material can be taken apart easily and salvaged.

OBJECTIVES

After completing this unit, the student should be able to:

- describe the composition of concrete and factors affecting its strength, durability, and workability.

- explain the reasons for reinforcing concrete and describe the materials used.

- job-mix a batch of concrete and explain the method of and reasons for making a slump test.

- explain techniques used for the proper placement and curing of concrete.

- construct forms for footings, slabs, walks, and driveways.

- construct concrete forms for foundation walls.

- construct concrete column forms.

- lay out and build concrete forms for stairs.

- estimate quantities of concrete.

- describe methods of dampproofing a foundation

- describe how termites and fungi destroy wood, and state some construction techniques used to prevent destruction by wood pests.

SAFETY REMINDER

Concrete is a universal building material used for support of various structures. It must be handled and properly supported during placement. This is particularly important because concrete can be worked for only a short time before it sets up.

Avoid prolonged contact with fresh concrete or wet cement because of possible skin irritation. Wear protective clothing when working with newly mixed concrete. Wash skin areas that have been exposed to wet concrete as soon as possible.

CHAPTER 29

Foundations and Concrete

FOUNDATIONS

The foundation is a structural part of a building that touches the ground supporting the entire building. It is positioned in an excavation and protrudes from the ground surface at least 8 inches. It is made of materials that withstand the effects of water, heavy soil, rot, and decay. Masonry is the material most often used for this purpose (*Figure 29–1*).

Foundation Styles

Three basic styles of foundations are used for residential construction (*Figure 29–2*). The choice of which is used depends on architectural desires and soil conditions.

Full Basement Foundations. Full basement foundations are made when a basement below grade is desired. The sidewalls extend below grade to a point where full height basement of 7 to 9 feet is achieved. A slab is then placed as the floor. The first floor of the building is then built on top of the foundation walls.

Crawl Space Foundation. A crawl space foundation does not have a living area below the first floor. The sidewalls are short extending only to stable soil. Access to the bottom of the first floor allows for mechanical repairs and alterations that might be necessary. The bottom of floor system should be at least 18 inches above the interior ground level, and beams and girders should be at 12 inches minimum.

Slab-on-Grade Foundation. The first floor may be a concrete slab placed on the ground. This is commonly called slab-on-grade construction. This style is often used in warm climates, where frost penetration into the ground is not very deep. With improvements in the methods of construction, the need for lower construction costs, and the desire to give the structure a lower profile, slab-on-grade foundations are being used more often in all climates. Insulation is installed around the perimeter to prevent frost penetration. Electrical and plumbing below floor level must be done before concrete is placed.

Soil Conditions

Foundation styles vary with soil conditions, which are a local issue that should be considered before building begins. The soil must support the building without moving. Most soil is adequate to support residential buildings. Unstable soil or soil under large commercial buildings must be assessed by engineers to determine what type of foundation is required.

Foundations extend into the ground to rest on stable soil. Areas of cold weather require the foundation to begin at a level below the seasonal frost penetration. This is due to the fact that moist soil expands when it freezes. This can cause the foundation to heave and buckle, thereby failing. The foundation must rest on soil below this depth, called the **frost line**. In extreme northern climates, this boundary can be 8 feet below the surface (*Figure 29–3*). In tropical climates, footings only need to reach solid soil, with no consideration given to frost.

In regions where the soil is not stable, special foundations must be installed. For example, the soil in south central United States can be predominately clay. This soil will shrink and swell as much as 6 inches over the course of a year. This is caused by the natural wet-to-dry seasonal changes. A solution

Figure 29–1 Concrete is most often used as a foundation material.

© Cengage Learning 2014

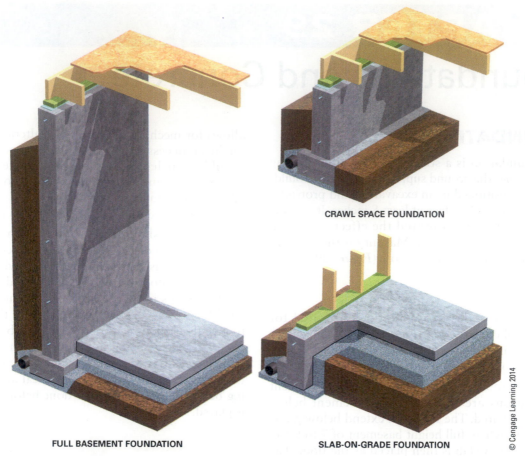

CRAWL SPACE FOUNDATION

FULL BASEMENT FOUNDATION

SLAB-ON-GRADE FOUNDATION

Figure 29–2 There three basic styles of foundations, full basement, crawl space and monolithic slab.

is a foundation made with trenches for concrete and steel bars to allow the building to rise and fall as one unit (*Figure 29–4*). This is generally referred to as a *reinforced concrete slab*. The steel bars may be encased by a plastic tube so to allow it to be pulled tight after the concrete cures. This process is called *post tensioning* and puts extreme stress on each bar.

FOUNDATION ALTERNATIVES FOUNDATION TYPES

Other types of foundations exist to meet various industry demands. They each provide the structural support necessary for the building while having different characteristics.

Concrete Block Foundation

The concrete block foundation is made by stacking together concrete masonry units (CMUs) in an array. CMUs are typically placed in a staggered bond where the vertical seams in the CMUs are not aligned one over the other. This provides maximum strength to the wall (*Figure 29–5*).

CMUs are sometimes called *concrete blocks* or *cinder blocks*. They come in regular and lightweight versions with varying widths. They all measure 8 inches high and typically 16 inches long; some are 18 inches long. Each measurement is slightly smaller to allow for a $\frac{3}{8}$-inch mortar joint. CMU widths range from 4 to 12 inches in 2-inch increments (*Figure 29–6*).

Most CMUs are called *stretchers*, which have the familiar shape of a CMU. Some are constructed with flush ends for building square openings and corners. These are called sash or corner blocks (*Figure 29–7*). They also have a slot midway for easy splitting to make a half block.

Installation of a CMU wall begins at the corners, on top of a slab or a footing. Several courses are laid up in a stepped arrangement. This is called a *lead*. Then a string is pulled tightly from one corner to the next on top of first course. CMUs are then installed along the line. This is continued until the wall height is achieved.

A variation to this system is the *insulated CMU*. It is made in two pieces that do not touch, separated by a layer of expanded polystyrene (*Figure 29–8*). This system creates a thermal break between the

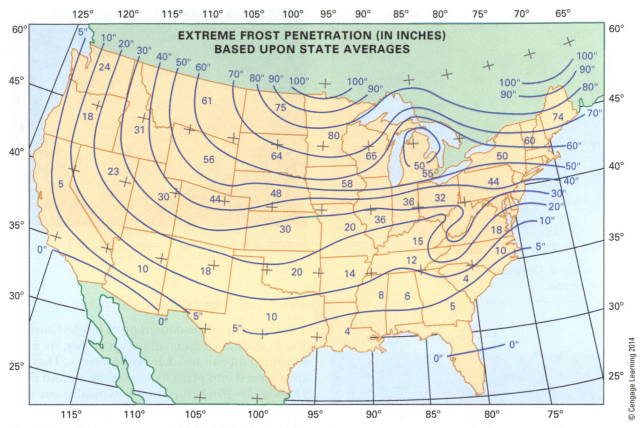

Figure 29-3 Frost line penetration in the United States.

Figure 29-4 A reinforced concrete slab may have steel rebar wrapped in plastic sleeves.

Figure 29-5 Concrete block walls are made of concrete masonry units (CMU).

Figure 29–6 Concrete masonry units are made in varying widths.

SLOT TO ALLOW FOR EASY CUTTING OF BLOCK IN HALF

STRETCHER BLOCK   CORNER BLOCK

Figure 29–7 Concrete masonry units are made in two styles.

inside and outside surface, thereby increasing the thermal resistance. They come in 10- and 12-inch widths and the standard height and length of 8 and 16 inches respectively.

All-Weather Wood Foundation

All-weather wood foundations(AWWFs)are made of pressure-treated lumber and plywood (*Figure 29–9*). This lumber is specially treated to resist decay and insect attack when buried below grade. It can be used to support light-frame buildings such as houses, schools, apartments, and office buildings.✱ Framing techniques are the same as for floors and walls (see Units 12 and 13).

✱ IRC
R404.2

Figure 29–8 A typical insulated concrete block.

Precast Concrete Foundations

Precast concrete foundations provide a fast alternative to traditional foundation types. They are manufactured off-site and delivered on truck. They are then craned into place by a specially trained crew (*Figure 29–10*). Setup time is typically one day. Sections are constructed with insulation and have a ribbed construction that allows for additional insulation. This makes the foundation energy efficient.

Each section is bolted together, and each seam is caulked to produce a watertight seal that is guaranteed by the manufacturer's installer (*Figure 29–11*). A slab is later placed as a floor to keep the bottom of the foundation from moving toward the inside during the backfilling of the foundation.

Insulated Concrete Forms

A rigid foam forming system called *insulated concrete forms,* often referred to as ICFs, uses 2-inch-thick expanded polystyrene (*Figure 29–12*). Units are manufactured into blocks with a metal or plastic rib system. Blocks size varies depending on manufacturer ranging from whatever wall thickness is needed × 16−18 inches high × 3−4 feet long (*Figure 29–13*). The blocks have fastening strips included on their surfaces to make it easy to screw on a wall finish later.

Assembly begins one course at a time with corner and straight blocks. The next course is set so the vertical seams are lapped (*Figure 29–14*). This interlock lends strength to the forms. Horizontal rebar is installed as each block layer is installed; vertical rebar may be inserted after the forms are completed. Special reinforced tape is sometimes used to help hold the course of block together. The success of this system is in the method of placing

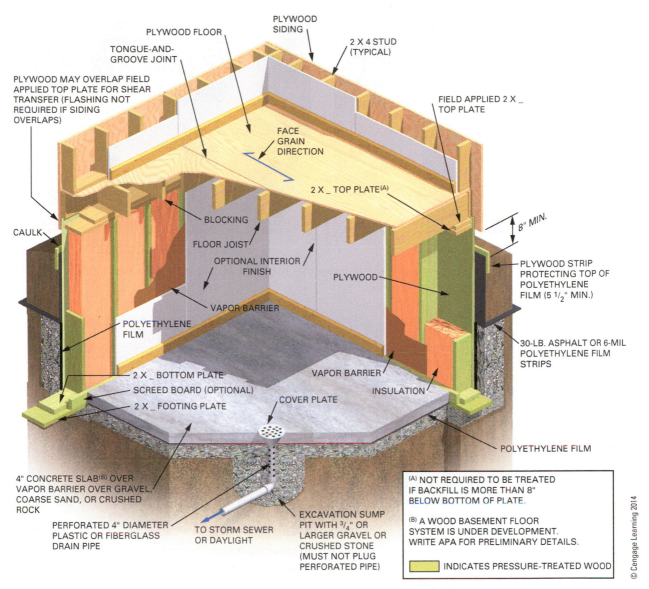

PLYWOOD FLOOR

PLYWOOD SIDING

TONGUE-AND-GROOVE JOINT

2 X 4 STUD (TYPICAL)

PLYWOOD MAY OVERLAP FIELD APPLIED TOP PLATE FOR SHEAR TRANSFER (FLASHING NOT REQUIRED IF SIDING OVERLAPS)

FIELD APPLIED 2 X _ TOP PLATE

FACE GRAIN DIRECTION

2 X _ TOP PLATE(A)

CAULK

8" MIN.

BLOCKING

FLOOR JOIST

OPTIONAL INTERIOR FINISH

PLYWOOD

PLYWOOD STRIP PROTECTING TOP OF POLYETHYLENE FILM (5 1/2" MIN.)

VAPOR BARRIER

POLYETHYLENE FILM

30-LB. ASPHALT OR 6-MIL POLYETHYLENE FILM STRIPS

2 X _ BOTTOM PLATE

SCREED BOARD (OPTIONAL)

2 X _ FOOTING PLATE

VAPOR BARRIER

COVER PLATE

INSULATION

POLYETHYLENE FILM

4" CONCRETE SLAB(B) OVER VAPOR BARRIER OVER GRAVEL, COARSE SAND, OR CRUSHED ROCK

PERFORATED 4" DIAMETER PLASTIC OR FIBERGLASS DRAIN PIPE

TO STORM SEWER OR DAYLIGHT

EXCAVATION SUMP PIT WITH 3/4" OR LARGER GRAVEL OR CRUSHED STONE (MUST NOT PLUG PERFORATED PIPE)

(A) NOT REQUIRED TO BE TREATED IF BACKFILL IS MORE THAN 8" BELOW BOTTOM OF PLATE.

(B) A WOOD BASEMENT FLOOR SYSTEM IS UNDER DEVELOPMENT. WRITE APA FOR PRELIMINARY DETAILS.

INDICATES PRESSURE-TREATED WOOD

© Cengage Learning 2014

Figure 29–9 All weather wood foundations are made with pressure treated lumber.

© Cengage Learning 2014

Figure 29–10 Precast concrete foundations are delivered on trucks and craned into place.

the concrete. Placement of the concrete is done in levels or lifts. Only one course is filled at a time, giving the concrete time to hydrate or harden slightly. Typically, by the time placement of the first lift is done for the entire perimeter, the concrete is ready to support another lift.

Crawl Space Foundations

In warmer climates like the southeastern United States, crawl space foundations are often used. They are place on undisturbed soil and only a foot or so below grade and project at least 8 inches above grade (*Figure 29–15*). A brick-block foundation is a combination of brick and block. It uses block below grade and brick for above grade where it shows. The pier and curtain wall foundation uses

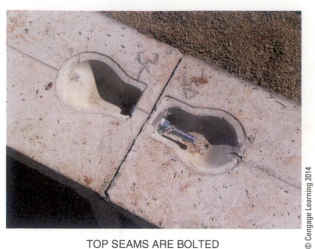

TOP SEAMS ARE BOLTED

ALL JOINTS ARE CAULKED

Figure 29–11 Precast concrete panels are fastened together to make them weather and watertight.

Figure 29–12 Insulated concrete forms are used for foundations and walls.

Figure 29–13 Insulated concrete forms are assembled from smaller units.

Figure 29–14 Insulated concrete form units are assembled to stagger the butt seams.

piers to carry the building load and then a curtain wall to fill in the space between piers. It can be made entirely of brick.

All crawl space foundations are placed on a properly sized spread footing. Proper size depends on the size and height of the building and is typically 8 inches wider than the wall itself. The foundation is centered on the footing. Steel reinforcement is placed in the footing to tie the building down through the piers and foundation. Internal piers to carry beams and girders can be made of brick, block, or precast concrete.

Ventilation placed in the top portion of the wall to aid in moisture control is often thought to be good practice. Experience has shown that it is not. It is best to place a moisture barrier on ground inside a crawl space. This will slow moisture from entering the building floor system. If ventilation is

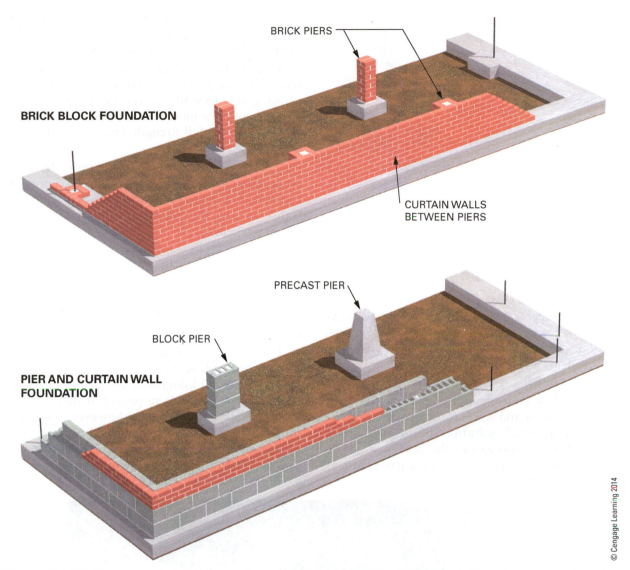

BRICK PIERS

BRICK BLOCK FOUNDATION

CURTAIN WALLS
BETWEEN PIERS

PRECAST PIER

BLOCK PIER

PIER AND CURTAIN WALL
FOUNDATION

© Cengage Learning 2014

Figure 29–15 Crawl space foundations may be made with a combination of CMU and brick or piers and curtain walls.

still desired it is best to close them off in the hot humid season. More on moisture issues for buildings may be found in Chapter 50.

CHARACTERISTICS OF CONCRETE

Concrete is a widely used, and it is the material of choice for foundations. Forms are temporarily constructed to contain the flowing concrete until it sets. Rebar and blockouts for openings such as windows are positioned first, then the concrete is placed.

Understanding the characteristics of concrete is essential for the construction of reliable concrete forms, the correct handling of freshly mixed material, and the final quality of hardened concrete.

PRECAUTIONS ABOUT USING CONCRETE

Avoid prolonged contact with fresh concrete or wet cement because of possible skin irritation. Wear protective clothing when working with newly mixed concrete. Wash skin areas that have been exposed to wet concrete as soon as possible. If any material containing cement gets into the eyes, flush immediately with water and get medical help.

CONCRETE

Concrete is a widely used building material. It can be formed into practically any shape for construction of buildings, bridges, dams, and roads. Improvements over the years have created a product that is strong, durable, and versatile (*Figure 29–16*).

Figure 29–16 Concrete is widely used in the construction industry.

Composition of Concrete

Concrete is a mixture of portland cement, fine and coarse *aggregates,* water, and various *admixtures.* Aggregates are fillers, usually sand, gravel, or stone (*Figure 29–17*). Admixtures are materials or chemicals added to the mix to achieve certain desired qualities.

When these concrete ingredients are mixed, the portland cement and water form a paste. A chemical reaction, called hydration, begins within this paste. This reaction causes the cement to set or harden, referred to as curing. As the paste hardens it binds the sand, which in turn binds the small and large aggregates. Together they form a strong, durable, and watertight mass called concrete.

Hydration (hardening) of the concrete is a relatively slow process. Initial setting, to the point of being able to walk on it, takes hours. It often takes a month or more for the concrete to achieve 90 percent of its full strength. Hydration continues for many years.

This process is somewhat fragile. If rapid evaporation or freezing occurs during the curing of concrete, the cement hydration process will stop. If this happens, the concrete will never achieve full design strength.

Portland Cement. In 1824, Joseph Aspdin, an Englishman, developed an improved type of cement. He named it Portland cement because it produced a concrete that resembled stone found on the Isle of Portland, England. Portland cement is a fine gray powder.

Portland cement is usually sold in plastic-lined paper bags that hold 1 cubic foot. Each bag weighs 94 pounds or 40 kilograms. The cement can still be used even after a long period of time as long as it remains dry. Eventually, however, it will absorb moisture from the air. It should not be used if it contains lumps that cannot be broken up easily. This condition is known as prehydration.

Figure 29–17 Concrete is made up of varying sizes of aggregates with cement binding them together.

Types of Portland Cement. New developments have produced several types of portland cement. Type I is the familiar gray kind most generally purchased and used. Type IA is an air-entrained cement that is more resistant to freezing and thawing. Types II through V have special uses, such as for low or high heat generation, high early strength, and resistance to severe frost. White portland cement, used for decorative purposes, differs only in color from gray cement. There are many other types, each with its specific purpose—ranging from underwater use to sealing oil wells.

Aggregates. Aggregates have no cementing value of their own. They serve only as a filler but are important ingredients. They constitute from 60 to 80 percent of the concrete volume.

Fine aggregate consists of particles $\frac{1}{4}$-inch or less in diameter. Sand is the most commonly used fine aggregate. *Coarse* aggregate, usually gravel or crushed stone, typically comes in sizes of $\frac{3}{8}$, $\frac{1}{2}$, $\frac{3}{4}$, 1, $1\frac{1}{2}$, and 2 inches. When strength is the only consideration, $\frac{3}{4}$-inch is the optimum size for most aggregates. Large-size aggregate uses less water and cement. Therefore, it is more economical. However, the maximum aggregate size must not exceed the following:

- one-fifth of the smallest dimension of the unreinforced concrete.
- one-third of the depth of unreinforced slabs on the ground (*Figure 29–18*).

MAXIMUM AGGREGATE SIZE

$\frac{1}{5}$ OF SMALLEST DIMENSION

SECTION THROUGH UNREINFORCED FOOTING

SLAB

$\frac{1}{3}$ OF DEPTH OF SLAB

SECTION THROUGH SLAB ON GROUND

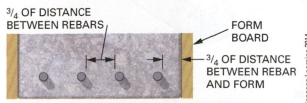

$\frac{3}{4}$ OF DISTANCE BETWEEN REBARS

FORM BOARD

$\frac{3}{4}$ OF DISTANCE BETWEEN REBAR AND FORM

SECTION THROUGH REINFORCED FOOTING

© Cengage Learning 2014

Figure 29–18 Aggregate size depends on the final dimensions of the concrete.

- three-quarters of the clear spacing between reinforcing bars or between the bars and the form.✱✱IRC R404.1.2.3.3

Green Tip

Some areas of the United States recycle concrete. Concrete is broken into sizes of large aggregate and used in new concrete, replacing crushed stone.

Fine and coarse particles must be in a proportion that allows the finer particles to fill the spaces between the larger particles. The aggregate should be clean and free of dust, loam, clay, or vegetable matter.

Admixtures. Admixtures are available to quicken or retard setting time, develop early strength, inhibit corrosion, retard moisture, control bacteria and fungus, improve pumping, and color concrete, among many other purposes.

Air. An important advance was made with the development of air-entrained concrete. It is produced by using air-entraining portland cement or an admixture. The intentionally made air bubbles are very small. Billions of them are contained in a cubic yard of concrete.

The introduction of air into concrete was designed to increase its resistance to freezing and thawing. It also has other benefits: Less water and sand are required, the workability is improved, separation of the water from the paste is reduced, and the concrete can be finished sooner and is more watertight than ordinary concrete. Air-entrained concrete is now recommended for almost all concrete projects.

Water. Water is the part of concrete mix that starts the hydration process. It also allows the concrete mixture to flow easily into place. Water used to make concrete must be clean. Other water may be used but should be tested first to make sure it is acceptable. The chemical reaction of hydration can be negatively affected by water impurities. A good general rule is, if it is safe to drink, it is safe to use in concrete.

The amount of water used can affect the quality and strength of the concrete. The amount of water in relation to the amount of portland cement is called the *water-to-cement ratio*. This ratio should be kept as small as possible while allowing for workability. Adding excessive water to a mix simply to increase the flow can weaken the concrete and must be avoided.

Concrete Strength.   Concrete strength is measured in pounds per square inch (psi). Test cylinders are compressed in large machines to determine how much stress they will withstand before breaking. Typical strength numbers with concrete are 2,500, 3,500, 4,000, and 4,500 psi (ranging from low to high).

To understand psi consider *Figure 29–19*. Cube A is 1-inch on all sides, referred to as a 1-inch cube. All columns and stacks are made up of the same 1-inch cubes. Assume cube A weighs 10 pounds. Then column B would be 20 pounds; columns C and D would be 40 pounds and 80 pounds, respectively. The floor under each column supports the weight of each column, which is measured in pounds. So because each column has the same footprint of 1 square inch, the psi for each is A = 10 psi (pounds per square inch), B = 20 psi, C = 40 psi, and D = 80 psi.

Stack E has sixty-four cubes and thus an overall weight of 640 pounds, but does not exert 640 psi on the floor. Notice that columns C and E are the same height. Stack E is made up of sixteen stacks of C-size columns. They cover an area of 4 × 4 = 16 square inches. But each of these columns exerts 40 psi on the floor; thus stack E can be see to exert 40 psi on the floor. Using math to solve

this, we take the overall weight of the stack and divide by the overall area, 640 pounds ÷ 16 square inches = 40 psi.

The overall strength of the concrete is determined by many things. It is affected by the relative amounts of its ingredients. Water is required for the reaction to begin, but too much water weakens the mix. Also, increasing the amount of portland cement increases strength. Aggregate sizes can also affect the strength.

Strength of concrete begins with the mix, but is also affected by how the concrete is placed and cured. Generally, a slow cure is best because the hydration process is slow. ✳   ✳IRC R402.2

Mixing Concrete

Concrete mixtures are designed to achieve the desired qualities of strength, durability, and workability in the most economical manner. The mix will vary according to the strength and other desired qualities, plus other factors such as the method and time of curing.

For very large jobs, such as bridges and dams, concrete mixing plants may be built on the job site. This is because freshly mixed concrete is needed on a round-the-clock basis. The construction engineer

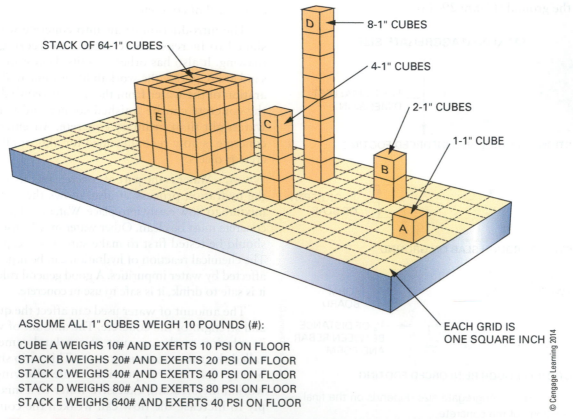

STACK OF 64-1" CUBES

D → 8-1" CUBES

4-1" CUBES

2-1" CUBES

1-1" CUBE

EACH GRID IS ONE SQUARE INCH

ASSUME ALL 1" CUBES WEIGH 10 POUNDS (#):

CUBE A WEIGHS 10# AND EXERTS 10 PSI ON FLOOR
STACK B WEIGHS 20# AND EXERTS 20 PSI ON FLOOR
STACK C WEIGHS 40# AND EXERTS 40 PSI ON FLOOR
STACK D WEIGHS 80# AND EXERTS 80 PSI ON FLOOR
STACK E WEIGHS 640# AND EXERTS 40 PSI ON FLOOR

© Cengage Learning 2014

Figure 29–19 Pounds per square inch (PSI) is affected by the weight and the area it rests upon.

calculates the proportions of the mix and supervises tests of the concrete.

Ready-Mixed Concrete. Most concrete used in construction is *ready-mixed* concrete. It is sold by the cubic yard or cubic meter. There are 27 cubic feet to the cubic yard. The purchaser usually specifies the amount and the desired strength of the concrete. The concrete supplier is then responsible for mixing the ingredients in the correct proportion to yield the desired strength.

Sometimes the purchasers specify the proportion of the ingredients. They then assume responsibility for the design of the mixture.

The ingredients are accurately measured at the plant with computerized equipment. The mixture is delivered to the jobsite in *transit-mix trucks* (*Figure 29–20*). The truck contains a large revolving drum, capable of holding from 1 to 10 cubic yards. There is a separate water tank with a water-measuring device. The drum rotates to mix the concrete as the truck is driven to the construction site.

Specifications require that each batch of concrete be delivered within one and a half hours after water has been added to the mix. If the job site is a short distance away, water is added to the cement and aggregates at the plant. If the job site is farther away, the proper amount of water is added to the dry mix from the water tank on the truck as it approaches or arrives at the job.

Job-Mixed Concrete. A small job may require that the concrete be mixed on the site either by hand or with a powered concrete mixer. All materials must be measured accurately. For measuring purposes, remember that a cubic foot of water weighs about $62\frac{1}{2}$ pounds and contains approximately $7\frac{1}{2}$ gallons.

All ingredients should be thoroughly mixed according to the proportions shown in *Figure 29–21*. A little water should be put in the mixer before the dry materials are added. The water is then added uniformly while the ingredients are mixed from one to three minutes.

Concrete Reinforcement

Concrete has high **compressive strength**. This means that it resists being crushed. It has, however, low **tensile strength**. It is not as resistant to bending or pulling apart. Steel bars, called **rebars**, are used in concrete to increase its tensile strength. Concrete is then called **reinforced concrete**. Fiberglass strands are sometimes used as reinforcement to add tensile strength. They are added to the concrete during the final mixing stage.

Rebars. Rebars used in construction are usually *deformed*. Their surface has ridges that increase the bond between the concrete and the steel. They come in standard sizes, identified by numbers that indicate the diameter in eighths. For instance, a #6 rebar has a diameter of $\frac{6}{8}$ or $\frac{3}{4}$-inch (*Figure 29–22*). Metric rebars are measured in millimeters.

The size, location, and spacing of rebars are determined by engineers and shown on the plans. The rebars are positioned inside the form before the concrete is placed (*Figure 29–23*). Rebars in bridges and roads that experience a salty environment, such as seawater or salting during the winter months, have an epoxy coating. This protects the rebar from rusting, and thus helps prevent premature concrete failure. The tasks of cutting, bending, placing, and tying rebars require workers who have been trained in that trade. * ✱ IRC R404.1.2(1)

Wire Mesh. Welded wire mesh is used to reinforce concrete floor slabs resting on the ground, driveways, and walks. It is identified by the gauge and spacing of the wire. Common gauges are #6, #8, and #10. The wire is usually spaced to make 6-inch squares (*Figure 29–24*).

Welded wire mesh is laid in the slab above the vapor barrier, if used, before the concrete is placed. It is spliced by lapping one full square plus 2 inches.

Figure 29–20 Ready-mixed concrete is delivered in trucks.

Water-to-Cement Ratio	Possible Strength (psi)	Water (gal)	Portland Cement (bags) (94 lb/bag)	Fine Aggregate (sand) (cubic feet)	Coarse Aggregate (size ¾" max) (cubic feet)
0.45	5500	5.6	1	1.4	2.4
0.50	4500	6.25	1	1.6	2.6
0.65	3000	8	1	2.3	3.4

Figure 29–21 Formulas for several concrete mixtures.

Bar #	Bar ∅ (inches)	Metric Sizes (mm)	Bar Weight (lbs per 100 lin ft)
2	¼	6	17
3	⅜	10	38
4	½	13	67
5	⅝	16	104
6	¾	19	150
7	⅞	22	204
8	1	26	267

Figure 29–22 Number and sizes of commonly used reinforcing steel bars.

Mesh Size (inches)	Mesh Gauge	Mesh Weight (lbs per 100 sq. ft.)
6 × 6	#6	42
6 × 6	#8	30
6 × 6	#10	21

Figure 29–24 Size, gauge, and weight of commonly used welded wire mesh.

Figure 29–23 Rebar is placed after the forms are created.

Placing Concrete

Prior to placing the concrete, sawdust, nails, and other debris should be removed from inside the forms. The inside surfaces are brushed or sprayed with oil to make form removal easier. No oil should be allowed to get on the steel reinforcement. This will reduce the steel/cement bond, thereby reducing the tensile strength provided by the steel. Also, before concrete is placed, the forms and subgrade are moistened with water. This is done to prevent rapid absorbing of water from the concrete.

Concrete is *placed*, not poured. Water should never be added so that concrete flows into forms without working it. Adding water alters the water-to-cement ratio on which the quality of the concrete depends.

Slump Test. Slump Tests are made by supervisors on the job to determine the consistency of the concrete. The concrete sample for a test should be taken just before the concrete is placed. The *slump cone* is first dampened. It is placed on a flat surface and filled to about one-third of its capacity with the fresh concrete. The concrete is then *rodded* by moving a metal rod up and down twenty-five times over the entire surface. Two more approximately equal layers are added to the cone. Each layer is rodded in a similar manner, with the rod penetrating the layer below. The excess is screeded from the top. The cone is then turned over onto a flat surface and removed by carefully lifting it vertically in three to seven seconds. The cone is gently placed beside the concrete, and the amount of slump measured between the top of the cone and the concrete (*Figure 29–25*). The test should be completed within two to three minutes.

Changes in slump should be corrected immediately. The table shown in *Figure 29–26* shows recommended slumps for various types of construction. Concrete with a slump greater than 6 inches should not be used unless a slump-increasing admixture has been added.

Figure 29–25 A slump test shows the wetness of a concrete mix.

Types of Construction	Slump in Inches	
	Maximum	Minimum
Reinforced foundation walls & footings	3	1
Plain footings, caissons & substructure walls	3	1
Beams & reinforced walls	4	1
Building columns	4	1
Pavement & slabs	3	1
Heavy mass concrete	2	1

Figure 29–26 Recommended slumps for different kinds of construction.

Placing in Forms. The concrete truck should get as close as possible. Concrete is placed by chutes where needed. It should not be pushed or dragged any more than necessary. It should not be dropped more than 4 to 6 feet, and should be dropped vertically and not angled. Drop chutes should be used in high forms to prevent the buildup of dry concrete on the side of the form or reinforcing bars above the level of the placement. Drop chutes also prevent separation caused by concrete striking and bouncing off the side of the form. Pumps are used on large jobs that need concrete continuously over long distances or to heights up to 500 feet.

Concrete must not be placed at a rapid rate, especially in high forms. The amount of pressure at any point on the form is determined by the height and weight of the concrete above it. Pressure is not affected by the thickness of the wall (*Figure 29–27*).

A slow rate of placement allows the concrete nearer the bottom to begin to harden. Once concrete hardens, it cannot exert more pressure on the forms even though liquid concrete continues to be placed above it (*Figure 29–28*). The use of stiff concrete with a low slump, which acts less like a liquid, will transmit less pressure. Rapid placing leaves the concrete in the bottom still in a fluid state. It will exert great lateral pressure on the forms at the bottom. This may cause the form ties to fail or the form to deflect excessively.

Concrete should be placed in forms in layers of not more than 12 to 18 inches thick. Each layer should be placed before setting occurs in the previous layer. The layers should be thoroughly consolidated (*Figure 29–29*). The rate of concrete placement in high forms should be carefully controlled.

Consolidation of Concrete. To eliminate voids or honeycombs in the concrete, it should be thoroughly worked by hand spading or vibrated after it goes into the form. Vibrators make it possible to use a stiff mixture that would be difficult to consolidate by hand.

An immersion-type vibrator, called a spud vibrator, has a metal tube on its end. This tube vibrates at a rapid rate. It is commonly used in construction to vibrate and consolidate concrete. Vibration makes the concrete more fluid and able to move, allowing trapped air to escape. This will prevent the formation of air pockets, honeycombs, and cold joints. The operator should be skilled in the use of the vibrator, keeping it moving up and down, uniformly vibrating the entire pour. Overvibrating should not be done because vibrating increases the lateral pressure on the form (*Figure 29–30*).✱

✱IRC R404.1.2.3.5

Curing Concrete

Concrete hardens and gains strength because of a chemical reaction called hydration. All the desirable properties of concrete are improved the longer this process takes place. A rapid loss of water from fresh concrete can stop the hydration process too soon, weakening the concrete. Curing prevents loss of moisture, allowing the process to continue so that the concrete can gain strength.

Concrete is *cured* either by keeping it moist or by preventing loss of its moisture for a period of time. For instance, if moist-cured for seven days, its strength is up to about 60 percent of full strength. A month later the strength is 95 percent, and up to full strength in about three months. Air-curing will reach only about 55 percent after three months and will never attain design strength. In addition, a rapid loss of moisture causes the concrete to shrink, resulting in cracks. Curing should be started as soon as the surface is hard enough to resist marring.

Methods of Curing. Flooding or constant sprinkling of the surface with water is the most effective

| RATE (FT/HR) | PRESSURES OF VIBRATED CONCRETE (PSF) | | | |
| | 50° F | | 70° F | |
	COLUMNS	WALLS	COLUMNS	WALLS
1	330	330	280	280
2	510	510	410	410
3	690	690	540	540
4	870	870	660	660
5	1050	1050	790	790
6	1230	1230	920	920
7	1410	1410	1050	1050
8	1590	1470	1180	1090
9	1770	1520	1310	1130
10	1950	1580	1440	1170

PRESSURE INCREASES AS PLACEMENT RATE INCREASES.
PRESSURE INCREASES AT LOWER TEMPERATURES.

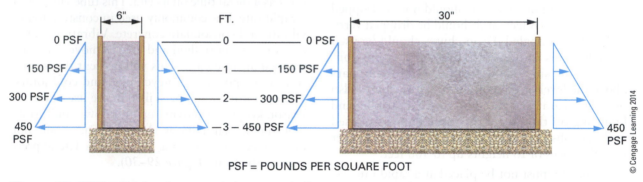

PSF = POUNDS PER SQUARE FOOT

© Cengage Learning 2014

Figure 29–27 The height of concrete being poured affects the amount of pressure against the forms. Pressure is not affected by the thickness of the wall.

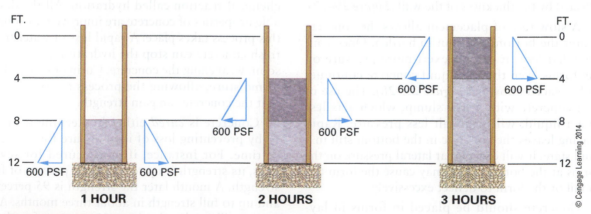

1 HOUR   **2 HOURS**   **3 HOURS**

© Cengage Learning 2014

Figure 29–28 As lower portions of concrete begin to hydrate, it begins to support the upper portions of concrete reducing the lateral pressure on the forms.

method of curing concrete. Curing can also be accomplished by keeping the forms in place, covering the concrete with burlap, straw, sand, or other material that retains water, and wetting it continuously.

In hot weather, the main concern is to prevent rapid evaporation of moisture. Sunshades or windbreaks may need to be erected. The formwork may be allowed to stay in place or the concrete surface may be covered with plastic film or other waterproof

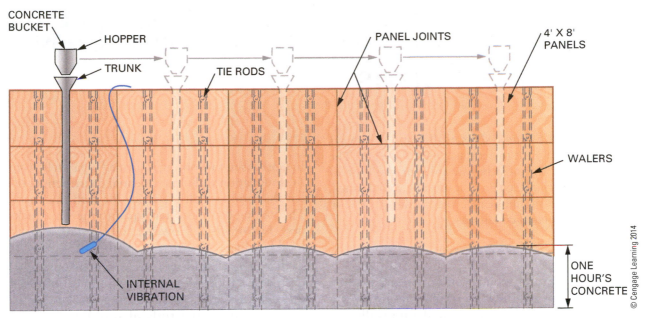

Figure 29–29 A safe, consistent pour rate is accomplished using drop chutes to place concrete in internally vibrated layers.

VIBRATE ONLY TO THE DEPTH OF THE FRESHLY PLACED CONCRETE. INSERTING THE VIBRATOR TOO FAR WILL CAUSE THE CONCRETE AT THE BOTTOM OF THE FORM TO REMAIN IN A LIQUID STATE LONGER THAN EXPECTED. THIS WILL RESULT IN HIGHER THAN EXPECTED LATERAL FORM PRESSURE AND MAY CAUSE THE FORM TO FAIL. THE DEPTH OF VIBRATION SHOULD JUST PENETRATE THE PREVIOUS LAYER OF CONCRETE BY A FEW INCHES.

Figure 29–30 Avoid excessive vibration of concrete.

sheets. The edges of the sheets are overlapped and sealed with tape or covered with planks. Liquid curing chemicals may be sprayed or mopped on to seal in moisture and prevent evaporation. However, manufacturers' directions for their use should be carefully followed.

Curing Time. Concrete should be cured for as long as practical. The curing time depends on the temperature. At or above 70 degrees, curing should take place for at least three days. At or above 50 degrees, concrete is cured at least five days. Near freezing, there is practically no hydration and concrete takes considerably longer to gain strength.

There is no strength gain while concrete is frozen. When thawed, hydration resumes with appropriate curing. If concrete is frozen within the first twenty-four hours after being placed, permanent damage to the concrete is almost certain. In cold weather, *accelerators* that shorten the setting time are sometimes used. Protect concrete from freezing for at least four days after being placed by providing insulation or artificial heat, if necessary.

Forms may be removed after the concrete has set and hardened enough to maintain its shape. This time will vary depending on the mix, temperature, humidity, and other factors.

CHAPTER 30

Flatwork Forms

Concrete for footings, slabs, walks, and driveways placed directly on the soil is called flatwork. The supporting soil must be suitable for the type of concrete work placed on it. It must be well drained. The concrete should be placed according to the local environmental conditions.

FOOTING FORMS

The footing for a foundation provides a base on which to spread the load of a structure over a wider area of the soil. For foundation walls, the most typical type is a *continuous* or *spread* footing. To provide support for columns and posts, pier footings of square, rectangular, circular, or tapered shape are used. Sometimes it is not practical to excavate deep enough to reach load-bearing soil. Then piles are driven and capped with a *grade beam* (*Figure 30–1*).

Continuous Wall Footings

In most cases, the footing is formed separately from the foundation wall. In residential construction, often the footing width is twice the wall thickness. The footing depth is often equal to the wall thickness (*Figure 30–2*). However, to be certain, consult local building codes.✳   ✳IRC R403.1.1

For larger buildings, architects or engineers design the footings to carry the load imposed on them. Usually, these footings are strengthened by reinforcing rods of specified size and spacing.

In areas where the soil is stiff, footings may be *trench poured* where no formwork is necessary. A trench is dug to the width and depth of the footing. The concrete is carefully placed in the trench (*Figure 30–3*). In other cases, forms need to be built for the footing.✳   ✳IRC R403.1.4

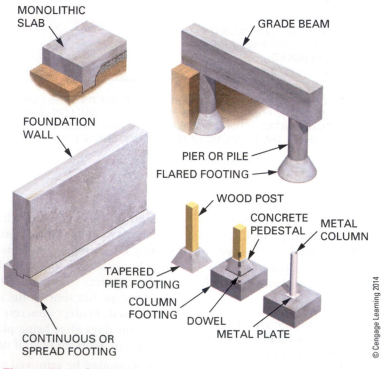

MONOLITHIC SLAB

GRADE BEAM

FOUNDATION WALL

PIER OR PILE

FLARED FOOTING

WOOD POST

CONCRETE PEDESTAL

METAL COLUMN

TAPERED PIER FOOTING

COLUMN FOOTING

DOWEL

METAL PLATE

CONTINUOUS OR SPREAD FOOTING

© Cengage Learning 2014

Figure 30–1 Several types of footings are constructed to support foundations.

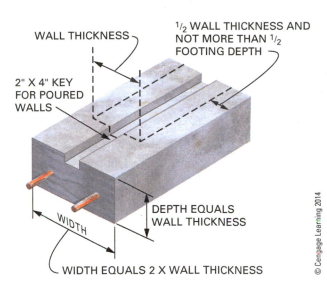

Figure 30-2 labels:
- WALL THICKNESS
- 1/2 WALL THICKNESS AND NOT MORE THAN 1/2 FOOTING DEPTH
- 2" X 4" KEY FOR POURED WALLS
- DEPTH EQUALS WALL THICKNESS
- WIDTH
- WIDTH EQUALS 2 X WALL THICKNESS

Figure 30–2 Typical footing for residential construction.

Figure 30–3 In stable soils, the soil acts as the footing form.

Locating Footings

To locate the footing forms, stretch lines on the batter boards in line with the outside of the footing. This is done by noting where the line should be on the batter board for the corner of the building. Then move the strings far enough away from the building to allow for the extra footing width (*Figure 30–4*). Suspend a plumb bob from the batter board lines at each corner. Drive stakes and attach lines to represent the outside surface of the footing.

Steel stakes are manufactured for use in building footing forms and other edge formwork. They come with prepunched nail holes for easy fastening (*Figure 30–5*). Spreaders and braces are typically made from 2 × 4s or other scrap material.

Building Wall Footing Forms

Stakes are used to hold the sides in position. Fasten the sides by driving nails through the stakes. Use duplex nails for easy removal. Various methods are used to build wall footing forms. One way is to erect the outside form around the perimeter of the building then return to assemble the inside form. This process will provide a strong, level form system (*Procedure 30–A*).

Set form stakes around the perimeter spaced 4 to 6 feet apart depending on the firmness of the soil. Use a gauge block that is the same thickness as the form board to set stakes. This will

ensure that the string does not touch any previous stakes while the current one is being adjusted. Snap a chalk line on the stakes at slightly above the proper elevation. Fasten the form board to the stakes and tap the stakes down until the form is level. Form the outside of the footing in this manner all around.

Before erecting the inside forms, cut a number of spreaders. Spreaders serve to tie the two sides together and keep them the correct distance apart. Nail one end to the top edges of the outside form. Erect the inside forms in a manner similar to that used in erecting the outside forms. Place stakes for the inside forms opposite those holding the outside form. Level across from the outside form to determine the height of the inside form. Fasten the spreaders as needed across the form at intervals to hold the form the correct distance apart.

Brace the stakes where necessary to hold the forms straight. In many cases, no bracing is necessary. Footing forms are sometimes braced by shoveling earth or placing large stones against the outside of the forms. (*Figure 30–6*) shows a typical setup for braced footing forms.

Keyways. A keyway is formed in the footing by pressing 2 × 4 lumber into the fresh concrete (*Figure 30–7*). The keyway form is beveled on both edges for easy removal after the concrete has set. The purpose of a keyway is to provide a lock between the footing and the foundation wall. This joint helps the foundation wall resist the pressure of the back-filled earth against it. It also helps to prevent seepage of water into the basement.

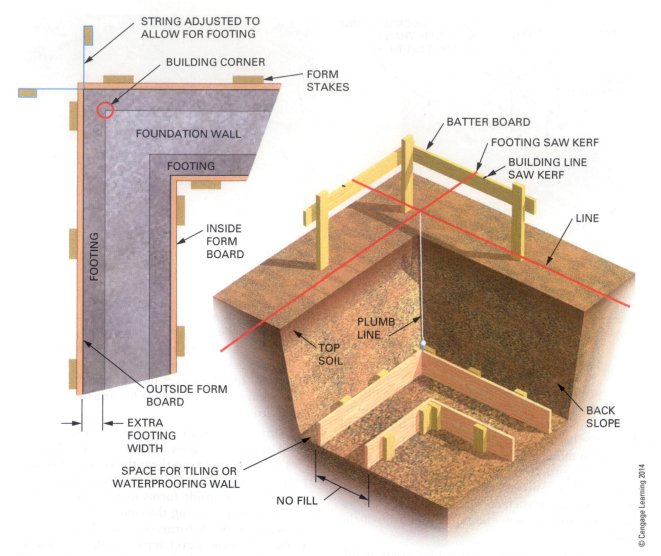

Figure 30–4 The footing is located by suspending a plumb bob from the batter board lines.

In some cases, where the design of the keyway is not so important, 2 × 4 pieces are not beveled on the edges, but are pressed into the fresh concrete at an angle.

Stepped Wall Footings

When the foundation is to be built on sloped land, it is sometimes necessary to *step* the footing. The footing is formed at different levels, to save material. In building stepped footing forms, the thickness of the footing must be maintained. The vertical and horizontal footing distances are adjusted so that a whole number of blocks or concrete forms can easily be placed into that section of the footing without cutting. The vertical part of each step should not exceed the footing thickness. The

horizontal part of the step must be at least twice the vertical part (*Figure 30–8*). Vertical boards are placed between the forms to retain the concrete at each step. ✳

✳**IRC R403.1.5**

Column Footings

Concrete for footings, supporting columns, posts, fireplaces, chimneys, and similar objects is usually placed at the same time as the wall footings. The size and shape of the column footing vary according to what it has to support. The dimensions are determined from the foundation plan.

In residential construction, these footing forms are usually built by nailing 2 × 8 pieces together in square, rectangular, or tapered shapes to the specified size (*Figure 30–9*).

Procedure 30–A Technique For Setting Footing Forms After Excavation

STEP 1 MOVE BATTER BOARD STRINGS TOWARD THE OUTSIDE TO ALLOW FOR EXTRA WIDTH OF FOOTING.

STEP 2 SUSPEND PLUMB BOB TO LOCATE OUTSIDE EDGES OF FOOTING. DRIVE TWO STAKES AT EACH CORNER AND STRETCH LINES FROM CORNER TO CORNER.

STEP 3 DRIVE INTERMEDIATE STAKES OUTSIDE OF THE STRETCH LINE USING A GAUGE BLOCK OF EQUAL THICKNESS TO THE FORM BOARD.

STEP 4 SNAP A LINE ON THE INSIDE FACE OF THE STAKES AT THE HEIGHT OF THE TOP OF THE FOOTING.

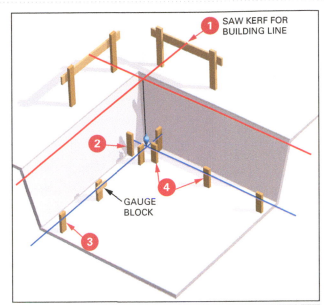

SAW KERF FOR BUILDING LINE

GAUGE BLOCK

STEP 5 NAIL THE FORM BOARD TO THE STAKE SLIGHTLY HIGHER THAN THE CHALK LINE.

STEP 6 LEVEL THE FORM BOARDS WITH A LASER OR OPTICAL TRANSIT BY TAPPING EACH STAKE DOWN AS NEEDED.

STEP 7 CUT AND ATTACH SPREADERS TO OUTSIDE FORM BOARD.

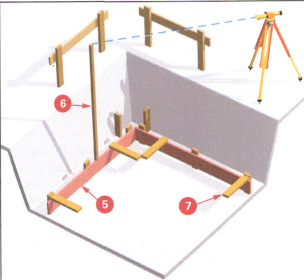

STEP 8 ATTACH THE INSIDE FORM BOARD TO THE SPREADERS AS INSIDE STAKES ARE PLACED.

STEP 9 USE A LEVEL TO POSITION THE HEIGHT OF THE INSIDE FORM.

STEP 10 BRACE AS NEEDED WITH STAKES DRIVEN AT AN ANGLE AND NAIL TO THE FORMS.

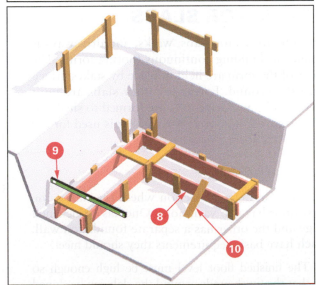

STEEL STAKES

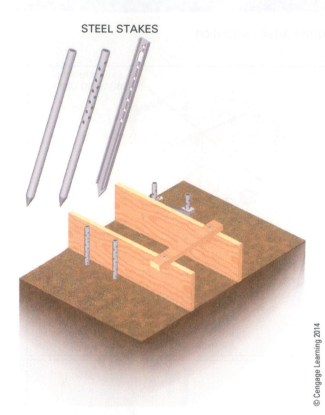

Figure 30–5 Manufactured steel stakes are often used with spreaders and braces when building concrete forms.

Figure 30–6 Forms must be braced as necessary.

2" X 4" PRESSED INTO FRESH CONCRETE AT AN ANGLE

2" X 4" WITH BEVELED EDGES PRESSED INTO FRESH CONCRETE

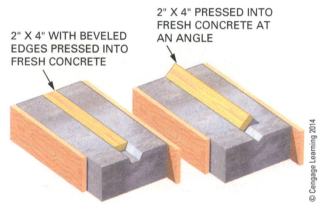

Figure 30–7 Methods of forming keyways in the footing.

Measurements are laid out on the wall footing forms to locate the column footings. Lines are stretched from opposite sides of the wall footing forms to locate the position of the forms. They are laid in position corresponding to the stretched lines (*Figure 30–10*). Stakes are driven. Forms are usually fastened in a position so that the top edges are level with the wall footing forms.

FORMS FOR SLABS

Building forms for slabs, walks, and driveways is similar to building continuous footing forms. The sides of the form are held in place by stakes driven into the ground. Forms for floor slabs are built level. Walks and driveways are formed to shed water. Usually 2 × 4 or 2 × 6 lumber is used for the sides of the form.

Slab-on-Grade

Slab-on-Grade construction where the slab is the first floor has two variations. One has a thickened edge and the other has a separate foundation wall. Each have basic requirements they should meet.

1. The finished floor level must be high enough so that the finish grade around the slab can be sloped away for good drainage. The top of the slab should be no less than 8 inches above the finish grade.✳ ✳IRC R401.3

2. All topsoil in the area in which the slab is to be placed must be removed. A base for the slab consisting of 4 to 6 inches of gravel, crushed stone, or other approved material must be well compacted in place.✳ ✳IRC R506.2.2

3. The soil under the slab may be treated with chemicals for control of termites, but caution is advised. Such treatment should be done only by those thoroughly trained in the use of these chemicals.✳ ✳IRC R318.2

4. All mechanicals (water and sewer lines, heating ducts, and other utilities) that are to run under the slab must be installed.

5. A vapor barrier should be placed under the concrete slab to prevent soil moisture from rising through the slab. The vapor barrier should be a heavy plastic film, such as 6-mil polyethylene or other material having equal or superior resistance to the passage of vapor. It should be strong enough to resist puncturing during

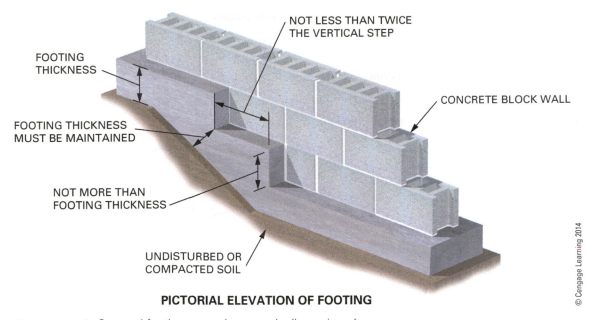

FOOTING THICKNESS

NOT LESS THAN TWICE THE VERTICAL STEP

FOOTING THICKNESS MUST BE MAINTAINED

CONCRETE BLOCK WALL

NOT MORE THAN FOOTING THICKNESS

UNDISTURBED OR COMPACTED SOIL

PICTORIAL ELEVATION OF FOOTING

Figure 30–8 Stepped footings must be properly dimensioned.

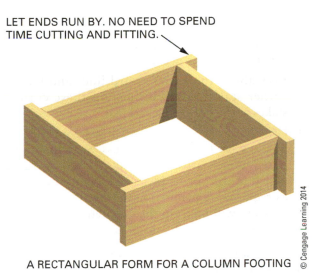

LET ENDS RUN BY. NO NEED TO SPEND TIME CUTTING AND FITTING.

A RECTANGULAR FORM FOR A COLUMN FOOTING

Figure 30–9 Construction of column footings.

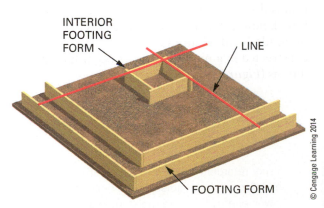

INTERIOR FOOTING FORM

LINE

FOOTING FORM

Figure 30–10 Locating interior footings using a string.

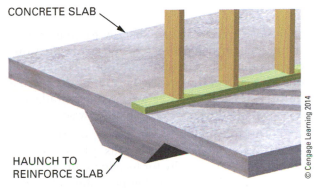

CONCRETE SLAB

HAUNCH TO REINFORCE SLAB

Figure 30–11 The slab is haunched under loadbearing walls.

the placing of the concrete. Joints in the vapor barrier must be lapped at least 4 inches and sealed. A layer of sand may be applied to protect the membrane during concrete placement.✱ ✱IRC R506.2.3

6. Where necessary, to prevent heat loss through the floor and foundation walls, waterproof, rigid insulation is installed around the perimeter of the slab.✱ ✱IRC Chapter 11 N1102.2.9

7. The slab should be reinforced with 6-inch by 6-inch, #10 welded wire mesh, or by other means to provide equal or superior reinforcing. The concrete slab must be at least 4 inches thick and haunched (made thicker) under loadbearing walls (*Figure 30–11*).

Monolithic Slabs A combined slab and foundation is called a monolithic slab (*Figure 30–12*). This type of slab is also referred to as a *thickened*

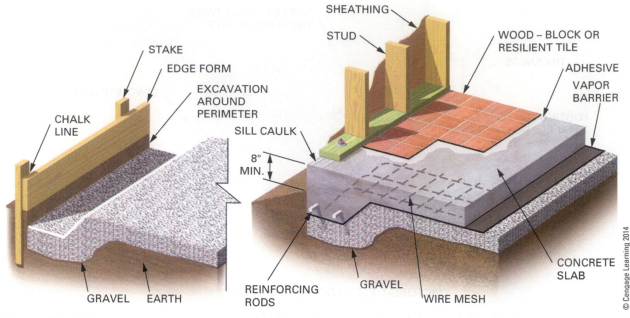

Figure 30–12 Typical monolithic slab-on-grade construction.

edge slab. It consists of a shallow footing around the perimeter. The perimeter is placed at the same time as the slab. The slab and footing make up a one-piece integral unit. The bottom of the footing must be at least 1 foot below the finish grade, unless local building codes dictate otherwise.

Forms for monolithic slabs are constructed using stakes and edge form boards, plank, or steel manufactured especially for forming slabs and similar objects. The construction procedure is similar to that for wall footing forms (see *Procedure 30–A*) except the inside form board is omitted (*Figure 30–13*).

*IRC R403.1.3.2

- From the batter board lines, plumb down to locate the building corner stakes. Stretch lines from these corner stakes and mark the soil to outline the perimeter of the building. Also mark

where mechanical trenches must be dug. Excavate the trench as needed.

- Reestablish the batter board lines and locate corner stakes. Stretch lines on these corner stakes at the desired elevation.

- Drive intermediate stakes using a gauge block to allow for the form thickness. Snap lines and fasten form boards to the stakes. Level the form with a transit- or builder's level. Brace the form as required, remembering that most of the side pressure will be from the concrete. Install reinforcement and mechanicals as required.

Independent Slabs In areas where the ground freezes to any appreciable depth during winter, the footing for the walls of the structure must extend below the frost line. If slab-on-grade construction is desired in these areas, the concrete slab and foundation wall may be separate. This type of slab-on-grade is called an **independent slab**. It may be constructed in a number of ways according to conditions (*Figure 30–14*).*

*IRC R506.1

Figure 30–13 Thickened edge slabs of two buildings.

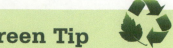

Green Tip

Vapor retarders under concrete footings and slabs reduce moisture in a building. This saves energy because moisture increases heating and cooling costs.

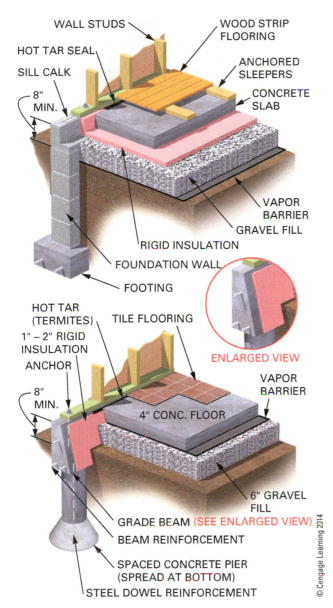

Figure 30-14 Independent slabs are constructed in a number of ways.

If the frost line is not too deep, the footing and wall may be an integral unit and placed at the same time. In colder climates, the foundation wall and footing are formed and placed separately. The wall may be set on piles or on a continuous footing, the forming of which has been described previously.

Slab Insulation

In colder climates, insulation is required under the slab. It is installed in a two or four foot width along the perimeter of the building. The thickness, amount, and location are governed by the energy code. The only insulation board suited for underground applications is extruded polystyrene. It is produced by various manufacturers in pink, green, blue, and gray colors. It has closed foam cells that keep water from getting into the insulation.*

Thin strips of extruded polystyrene may be used between slabs. These strips serve as isolation joints that allow the slabs to expand and contract with temperature and not interfere with each other.

*IRC
Chapter 11
N1102.2.9

FORMS FOR WALKS AND DRIVEWAYS

Forms for walks and driveways are usually built so water will drain from the surface of the concrete. In these cases, grade stakes must be established and grade lines carefully followed (*Figure 30–15*).

Establish the grade of the walk or driveway on stakes at both ends. Stretch lines tightly between the end stakes. Drive intermediate stakes. Fasten the edge pieces to the stakes following the line in a manner similar to making continuous footing forms.

Forms for Curved Walks and Driveways

In many instances, walks and driveways are curved. Special metal forms can be purchased to easily form curves, or wood forms can be constructed from $\frac{1}{4}$-inch plywood or hardboard. They may be used for small-radius curves. If using plywood, install it with the grain vertical for easier bending without breaking. Wetting the stock sometimes helps the bending process.

For curves of a large radius, 1 × 4 lumber can be used and satisfactorily bent. For smaller radius bends, dimension lumber forms can be curved by making saw kerfs uniformly spaced together. Soaking the kerf form with water before bending to aids in closing the saw kerfs (*Figure 30–16*). Be sure to select a good-quality board. Large knots and lumber defects will cause the bent board to break at the kerfs.

The spacing of the kerfs affects the radius of the curve. Closer kerfs yield a tighter bend. The number of kerfs affects the length of the arc—more kerfs equal more curve length.

To determine the kerf spacing, measure the radius distance from the end of the form (*Procedure 30–B*). Bend the form to close the kerf. The height of the board when the end is raised is the spacing of the kerfs.

The length of the form that has kerfs cut into it can be determined from the circumference of the

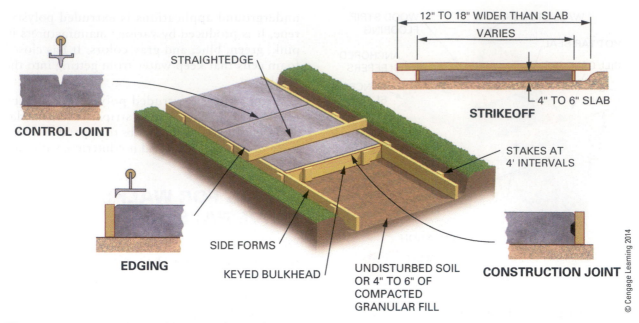

Figure 30–15 Typical way of forming of slabs for walkways.

© Cengage Learning 2014

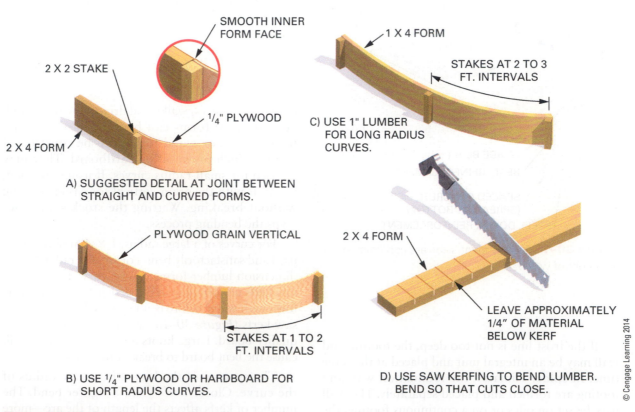

Figure 30–16 Forming curved edge slabs.

© Cengage Learning 2014

curve. First, calculate the circumference of the whole circle that the curve would make if it were stretched out to a full circle. Then determine what portion of a whole circle the curve occupies, that is, one-quarter, one-eighth, and so on. For example, what is the length of the form if it is a one-quarter bend with a

radius of 4 feet? The formula to determine circumference is $C = \pi d$, where π is found with a calculator and d = twice the radius. Thus, $C = \pi(8) = 25.13$ feet. One-quarter of $C = 25.13 \div 4 = 6.28$ feet. The length of form with kerfs cut into it is approximately $6'\text{-}3''$.

Step**by**Step Procedures

Procedure 30–B Making Radius Forms

STEP 1 DETERMINE THE RADIUS OF THE CURVE AND THE AMOUNT OF CURVE IN THE BEND, I.E. 4" RADIUS AND ONE-QUARTER BEND.

NOTE: SELECT GOOD QUALITY OR GRADE OF LUMBER FOR BENDING FORMS. KNOTS AND OTHER DEFECTS MIGHT CAUSE THE BEND TO BREAK.

STEP 2 FROM ONE END OF THE FORM, MEASURE A DISTANCE EQUAL TO THE RADIUS.

STEP 3 MAKE A SAW CUT TO WITHIN 1/4" OF THE BOTTOM OF THE FORM.

STEP 4 FASTEN OR HOLD THE STOCK AND BEND THE END UP TO CLOSE THE KERF.

STEP 5 MEASURE THE DISTANCE THE END HAS RISEN. THIS IS THE KERF SPACING.

STEP 6 CALCULATE THE CIRCUMFERENCE OF A WHOLE CIRCLE WITH THE DESIRED RADIUS.

STEP 7 DIVIDE THE WHOLE BY THE PORTION OF THE CURVE DESIRED.

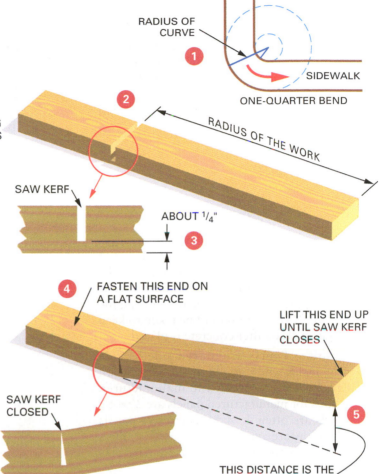

1 RADIUS OF CURVE

SIDEWALK

ONE-QUARTER BEND

2 RADIUS OF THE WORK

SAW KERF

ABOUT 1/4" **3**

4 FASTEN THIS END ON A FLAT SURFACE

LIFT THIS END UP UNTIL SAW KERF CLOSES

SAW KERF CLOSED

5

THIS DISTANCE IS THE SPACING OF SAW KERFS

6 EXAMPLE:
RADIUS = 4' AND THE BEND IS ONE-QUARTER.
THEN DIAMETER = 2(4) = 8'
$C = \pi d = \pi(8) = 25.13'$

7 THE ONE-QUARTER BEND HAS A CURVE LENGTH
$(25.13) \div 4 = 6.28'$
APPROXIMATELY 6'-3"

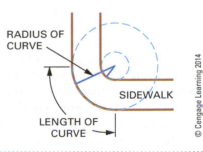

RADIUS OF CURVE

SIDEWALK

LENGTH OF CURVE

© Cengage Learning 2014

CHAPTER 31
Vertical Formwork

Foundation walls and columns are usually formed by using **panels** rather than building forms in place, piece by piece. Panel construction simplifies the erection and stripping of formwork. They also reduce the cost of forming by allowing the material to be used over again.

WALL FORMS

Wall Form Components

Various kinds of panels and panel systems are used. Some concrete panel systems are manufactured of steel, aluminum, or wood. They are designed to be used many times. Specially designed hardware is used for joining, spacing, aligning, and bracing the panels (*Figure 31–1*). Care should be taken to keep the form system clean after concrete is placed.

Form Panels. Panels are placed side by side to form the inside and outside of the foundation walls. Panel sizes vary with the manufacturer. Standard-sizes for manufactured panels are 2′ × 8′. Wood panels are often 4′ × 8′. Narrower panels of several widths are available to be used as fillers when the space is too narrow for a standard-size panel.

Snap Ties. Snap Ties hold the wall forms together at the desired distance apart. They support both sides against the lateral pressure of the concrete (*Figure 31–2*). They allow the side pressure on the outer panel to be carried or canceled out by the pressure on the inner panel. These ties reduce the need for external bracing and greatly simplify the erection of wall forms. The design of a form for a particular job, including the spacing of the ties and studs, is decided by a structural engineer.

These ties are called snap ties because after removal of the form, the projecting ends are snapped off slightly inside the concrete surface. A special snap tie wrench is used to break back the ties (*Procedure 31–A*). The small remaining holes are easily filled.

Because of the great variation in the size and shape of concrete forms, a large number of snap tie styles are used. For instance, *flat ties* of various styles are used with some manufactured panels.

Figure 31–1 Formwork for concrete walls may be manufactured or built on the jobsite.

Figure 31–2 A snap tie holds inner and outer forms from spreading when the concrete is placed.

For heavier formwork, *coil ties* and reusable *coil bolts* are used (*Figure 31–3*). For each kind of tie, there are also several sizes and styles. There are hundreds of kinds of form hardware. To become better acquainted with form hardware, study manufacturers' catalogs.

Walers. The snap ties run through and are wedged against form members called walers. Walers are doubled 2 × 4 pieces with space between them of about $\frac{1}{2}''$. They may be horizontal or vertical. Walers are spaced at right angles to the panel frame members. The number and spacing depend on the style of the form and the pressure exerted on the form (*Figure 31–4*).

The vertical spacing of the snap ties and walers depends on the height of the concrete wall. The vertical spacing is closer together near the bottom. This is because there is more lateral pressure from the concrete there than at the top (*Figure 31–5*).

For low wall forms less than 4 feet in height, the panel may be laid horizontally with vertical walers spaced as required (*Figure 31–6*).

Care of Forms. After the concrete placed in the footing has hardened sufficiently (sometimes three days), the forms are removed and cleaned. Any concrete clinging to the form should be scraped off. They are easier to clean at this stage rather than later when the concrete has reached full strength. Forms may be oiled to protect them and stored ready for reuse.

The forms last longer if they are handled properly. Place them gently and do not drop them. Damaged corners from dropping the panel will affect the surface appearance of the concrete. They

StepbyStep Procedures

Procedure 31–A Breaking Snap Ties

STEP 1 SLIDE THE SNAP TIE WRENCH UP AGAINST THE TIE SO THAT THE FRONT OF THE WRENCH IS TOUCHING THE CONCRETE.

STEP 2 KEEPING THE FRONT OF THE WRENCH TIGHT AGAINST THE CONCRETE, PUSH THE HANDLE END TOWARDS THE CONCRETE WALL SO THAT THE TIE IS BENT OVER AT APPROXIMATELY A 90° ANGLE.

STEP 3 ROTATE THE WRENCH AND TIE END 1/4 TO 1/2 TURN BREAKING OFF THE TIE END.

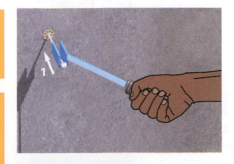

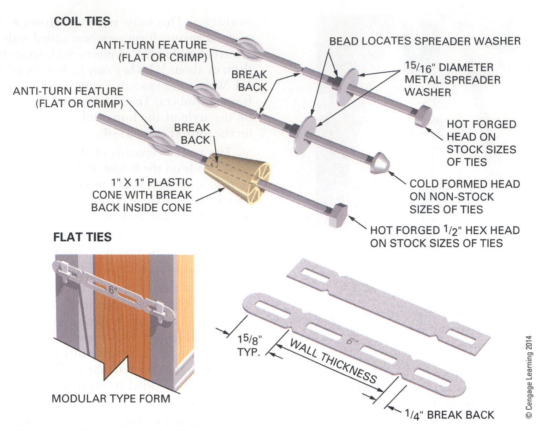

COIL TIES

ANTI-TURN FEATURE (FLAT OR CRIMP)

BEAD LOCATES SPREADER WASHER

BREAK BACK

$15/16$" DIAMETER METAL SPREADER WASHER

ANTI-TURN FEATURE (FLAT OR CRIMP)

BREAK BACK

HOT FORGED HEAD ON STOCK SIZES OF TIES

1" X 1" PLASTIC CONE WITH BREAK BACK INSIDE CONE

COLD FORMED HEAD ON NON-STOCK SIZES OF TIES

HOT FORGED $1/2$" HEX HEAD ON STOCK SIZES OF TIES

FLAT TIES

6"

MODULAR TYPE FORM

$15/8$" TYP.

WALL THICKNESS

6"

$1/4$" BREAK BACK

© Cengage Learning 2014

Figure 31–3 A large variety of snap ties are manufactured.

WALERS

SCAFFOLD

© Cengage Learning 2014

Figure 31–4 Walers are easily installed on forming systems when special hardware is used.

are also easier to use if they are not twisted or broken.

Preparing for Wall Form Assembly

Locating the Forms. Lines are stretched on the batter boards in line with the outside of the foundation wall. A plumb bob is suspended from the layout lines to the footing. Marks that are plumb

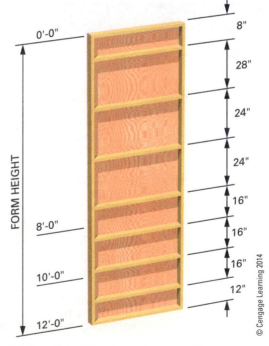

0'-0"

8"

28"

24"

24"

16"

FORM HEIGHT

8'-0"

16"

10'-0"

16"

16"

12"

12'-0"

© Cengage Learning 2014

Figure 31–5 Horizontal panel stiffeners are placed closer together near the bottom than they are at the top. Dimensions are for purposes of illustration only.

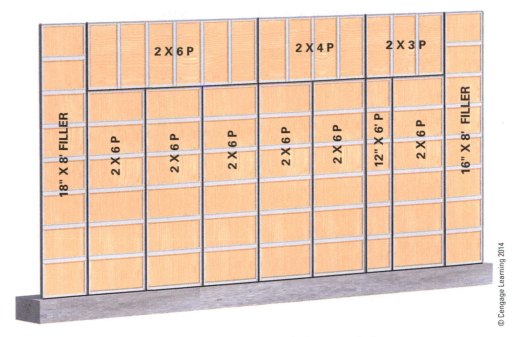

Figure 31–6 Forms can be laid horizontally or vertically as needed.

with the layout lines are placed on the footing at each corner. A chalk line is snapped on the top of the footing between the corner marks outlining the outside of the foundation wall.

Installing Plates. Sometimes panels are set on 2 × 4 or 2 × 6 lumber plates. Plates provide a positive online wall pattern. They can also level out rough areas on the footing. Plates function to locate the position and size of pilasters, and changes in wall thickness, corners, and other variations in the wall (*Figure 31–7*). The outer plate is fastened to the footing using masonry nails or pins driven by a powder-actuated tool. The inner plate is fastened only to the concrete form. This is done to allow the wall form thickness to swell slightly when the concrete is placed. This movement is due to the slack in the snap ties being taken up.

Manufactured Wall Form Assembly

Erecting Panels. Stack the number of panels necessary to form the inside of the wall in the center of the excavation. Lay the panels needed for the outside of the wall around the walls of the excavation. The face of all panels should be oiled or treated with a chemical releasing agent. This provides a smooth face to the hardened concrete and makes stripping of the forms easy.

Panels may be assembled in inside/outside pairs. But if rebar is required in the wall, then erect the outside wall forms first. This makes installing and tying rebar easier. Set panels in place at all corners first. Set base snap ties into slots while placing the panel, and attach the outside corners (*Figure 31–8*). Nail the panel into the plate with duplex nails. Make sure the corners are plumb by testing with a hand level and brace.

Fill in between the corners with panels, keeping the same width panels opposite each other.

Placing Snap Ties. Place snap ties in the dadoes between panels as work progresses. Tie panels together using the wedge bolts or the connection designed by the manufacturer (*Figure 31–9*). Snap ties must be positioned as each panel is placed. Be careful not to leave out any snap ties. Wedge bolts are not hammered tight, only drawn up snug (*Figure 31–10*).

Use filler panels as necessary to complete the outside wall section.✱ Brace the wall temporarily ✱**IRC** as needed. Install rebar as needed, then erect the **R611.5.4.2** panels for the inside of the wall. Keep joints between panels opposite to those for the outside of the wall (*Figure 31–11*). Insert the other ends of the snap ties between panels as they are erected.

Windows and other openings are sometimes made by inserting the window into the form before the concrete is placed. This saves a lot of time and

IT IS RECOMMENDED THAT THE INSIDE PLATE NOT BE FASTENED

FASTEN OUTSIDE PLATE INTO CONCRETE

THIS EDGE TO CHALK LINE

OUTSIDE EDGE OF FOOTING

CROSS SECTION A

BASE TIE

BASE TIE WEDGE BOLT

Figure 31–8 Base ties hold the panel bottoms from spreading.

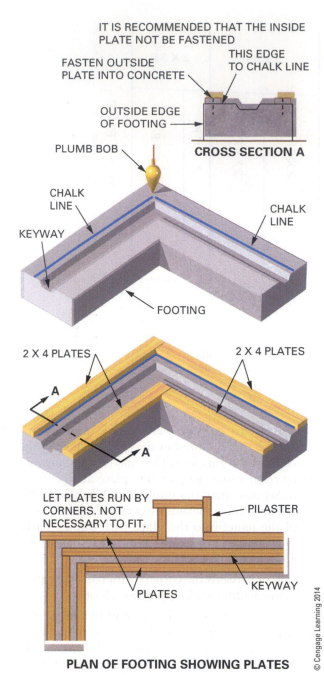

PLUMB BOB

CHALK LINE

CHALK LINE

KEYWAY

FOOTING

2 X 4 PLATES

2 X 4 PLATES

A

A

LET PLATES RUN BY CORNERS. NOT NECESSARY TO FIT.

PILASTER

PLATES

KEYWAY

PLAN OF FOOTING SHOWING PLATES

Figure 31–7 Attaching plates to the footing before forming walls.

CONNECTOR

SNAP TIE

WEDGE BOLTS

SNAP TIE

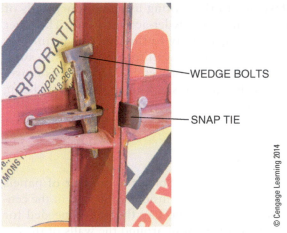

Figure 31–9 Form panels are connected by the snap ties, which are secured by special bolts and connectors.

makes a better tighter seal between the concrete and window (*Figure 31–12*).

Installing Walers. In a typical 8-foot-high wall, walers are used only to help keep the forms straight. Typically, the outside panels have one row of walers near the top edge. Special brackets, called waler ties, are used to attach walers to two rows of 2 × 4s (*Figure 31–13*). Let the bracket come between them and wedge into place.

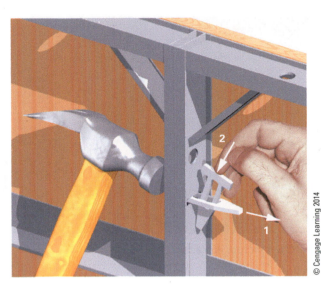

© Cengage Learning 2014

Figure 31–10 Tapping the side as the wedge is pushed down adequately tightens the form, yet leaves it loose enough to be easily removed later.

Forming Pilasters. The wall may be formed at intervals for the construction of pilasters. These are thickened portions of the wall that serve to give the wall more lateral (side-to-side) strength. They may also provide support for beams. They are formed with inside and outside corners in the usual manner. If the pilaster is large, longer snap ties are necessary (*Figure 31–14*).

Erecting Wood Wall Forms

Erecting Panels. Stack the number of panels necessary to form the inside of the wall in the center of the excavation. Lay the panels needed for the outside of the wall around the walls of the excavation. The face of all panels should be oiled or treated with a chemical releasing agent. This provides a smooth face to the hardened concrete and makes stripping of the forms easy.

Panels may be assembled in inside/outside pairs. But if rebar is required in the wall then erect the outside wall forms around the perimeter first. This makes installing and tying rebar easier. Set panels in place at all corners first. Nail the panel into the plate with duplex nails (*Figure 31–15*). Make sure the corners are plumb by testing with a hand level and brace.

Fill in between the corners with panels, keeping the same-width panels opposite each other. Place snap ties in the dadoes between panels as work progresses. Tie panels together by driving U-shaped clamps over the edge 2 × 4s or by nailing them together with duplex nails. Use filler panels as necessary to complete each wall section. Brace the wall temporarily as needed.

Placing Snap Ties. After the panels for the outside of the wall have been erected, place snap ties in the intermediate holes. Be careful not to leave

NOTICE PANELS HAVE ANOTHER PANEL OPPOSITE IT OF THE SAME SIZE. JOINTS BETWEEN INSIDE AND OUTSIDE PANELS SHOULD BE OPPOSITE EACH OTHER.

© Cengage Learning 2014

Figure 31–11 Panels should be assembled with opposing joints so snap ties can be easily installed.

Figure 31–12 Windows may be installed inside the form before concrete is placed.

Figure 31–13 Special brackets used to attach walers.

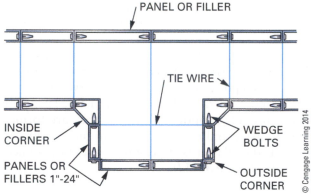

PANEL OR FILLER

TIE WIRE

INSIDE CORNER

WEDGE BOLTS

PANELS OR FILLERS 1"-24"

OUTSIDE CORNER

Figure 31–14 Pilasters may be formed like any other intersection. Note that inside the pilaster, tie wire is used to hold panel seams.

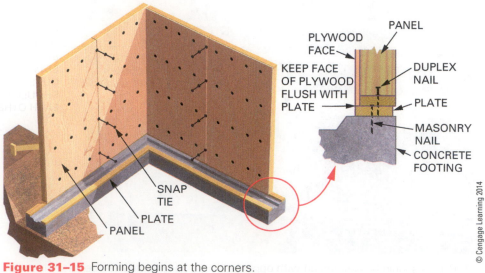

PANEL

PLYWOOD FACE

KEEP FACE OF PLYWOOD FLUSH WITH PLATE

DUPLEX NAIL

PLATE

MASONRY NAIL

CONCRETE FOOTING

SNAP TIE

PLATE

PANEL

Figure 31–15 Forming begins at the corners.

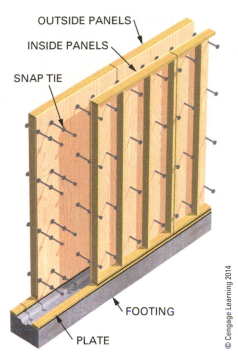

OUTSIDE PANELS

INSIDE PANELS

SNAP TIE

FOOTING

PLATE

© Cengage Learning 2014

Figure 31–16 Inside panels of the form are assembled later with snap ties.

out any snap ties. Erect the panels for the inside of the wall. Keep joints between panels opposite to those for the outside of the wall. Insert the other end of the snap ties between panels and in intermediate holes as panels are erected (*Figure 31–16*). If the concrete is to be reinforced, the rebars are tied in place before the inside panels are erected.

Installing Walers. When all panels are in place, install the walers. Let the snap ties come through them and wedge into place. Care must be taken when installing and driving snap tie wedges (*Figure 31–17*). Let the ends of the walers extend by the corners of the formwork. Reinforce the corners with vertical 2 × 4s (*Figure 31–18*). This is called yoking the corners.

Forming Pilasters. The wall may be formed at intervals for the construction of pilasters. These are thickened portions of the wall that serve to give the wall more lateral (side-to-side) strength. They may strengthen the wall or provide support for beams. They may be constructed on the inside or outside of the wall. In the pilaster area, longer snap ties are necessary (*Figure 31–19*).

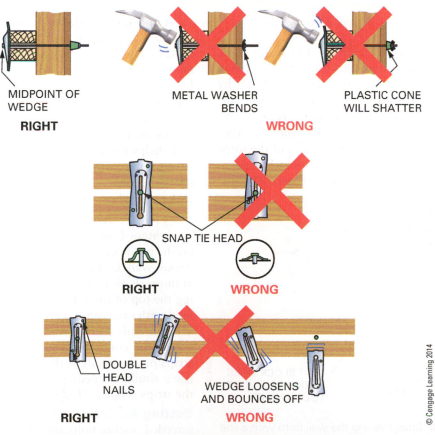

MIDPOINT OF WEDGE

RIGHT

METAL WASHER BENDS

WRONG

PLASTIC CONE WILL SHATTER

SNAP TIE HEAD

RIGHT **WRONG**

DOUBLE HEAD NAILS

WEDGE LOOSENS AND BOUNCES OFF

RIGHT **WRONG**

© Cengage Learning 2014

Figure 31–17 Care must be taken when installing snap tie wedges.

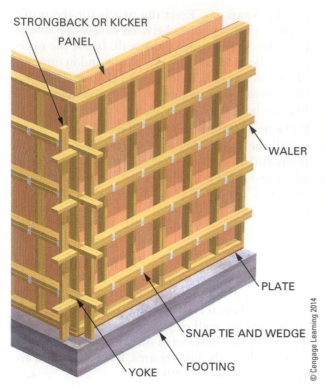

STRONGBACK OR KICKER

PANEL

WALER

PLATE

SNAP TIE AND WEDGE

YOKE

FOOTING

© Cengage Learning 2014

Figure 31–18 Yokes are vertical members that hold outside form corners together.

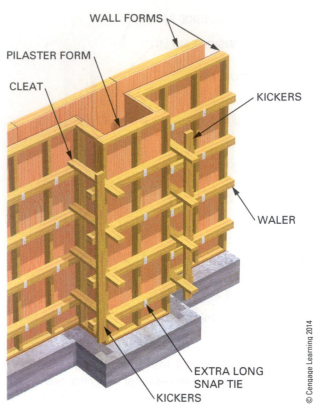

WALL FORMS

PILASTER FORM

CLEAT

KICKERS

WALER

EXTRA LONG SNAP TIE

KICKERS

© Cengage Learning 2014

Figure 31–19 Formwork for a pilaster.

Completing the Wall Forms

Straightening and Bracing. Brace the walls inside and outside as necessary to straighten them. Wall forms are easily straightened by sighting by eye along the top edge from corner to corner. Another method of straightening is to use line and gauge blocks, stretching a line from corner to corner at the top of the form over two blocks of the same thickness. Move the forms until a test block of equal thickness passes just under the line (*Figure 31–20*).

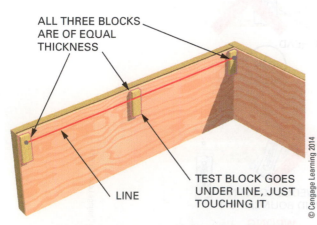

ALL THREE BLOCKS ARE OF EQUAL THICKNESS

LINE

TEST BLOCK GOES UNDER LINE, JUST TOUCHING IT

© Cengage Learning 2014

Figure 31–20 Straightening the wall form with a line and test block.

A special adjustable form brace and aligner are used to position and hold wall forms (*Figure 31–21*). This allows for easy prying when straightening the formwork simply by rotating the turnbuckle. Braces are cut with square ends and nailed into place against strongbacks. Strongbacks are placed across walers at right angles wherever braces are needed. The sharp corner of the square ends helps hold the braces in place. It also allows easy prying with a bar to tighten the braces and move the forms (*Figure 31–22*). There is no need to make a bevel cut on the ends of braces.

Leveling. After the wall forms have been straightened and braced, chalk lines are snapped on the inside of the form for the height of the foundation wall. Grade nails may be driven partway in at intervals along the chalk line as a guide for leveling the top of the wall. If the tops of the panels are level with each other, a short piece of stock notched at both ends can be run along the panel tops to screed the concrete. Another method is to fasten strips on the inside walls along the chalk line and use a similar screeding board, notched to go over the strips (*Figure 31–23*).

Setting Anchor Bolts. As soon as the wall is screeded, anchor bolts are set in the fresh concrete. A number of various styles and sizes are manufactured

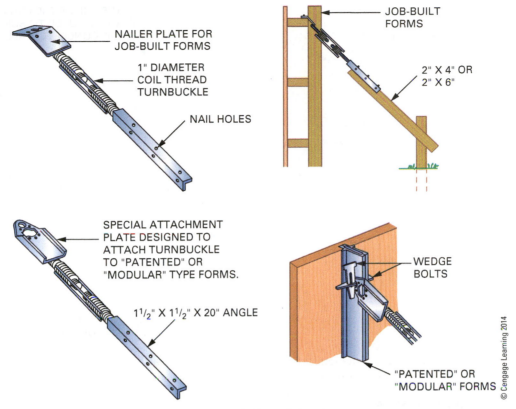

NAILER PLATE FOR JOB-BUILT FORMS

1" DIAMETER COIL THREAD TURNBUCKLE

NAIL HOLES

JOB-BUILT FORMS

2" X 4" OR 2" X 6"

SPECIAL ATTACHMENT PLATE DESIGNED TO ATTACH TURNBUCKLE TO "PATENTED" OR "MODULAR" TYPE FORMS.

1½" X 1½" X 20" ANGLE

WEDGE BOLTS

"PATENTED" OR "MODULAR" FORMS

© Cengage Learning 2014

Figure 31–21 Adjustable braces are available with forming systems.

Green Tip

Plan ahead to reduce waste of leftover concrete. Leftover concrete can be placed in premade forms for small pads, patio blocks, and splash blocks, or used as backfill for various posts.

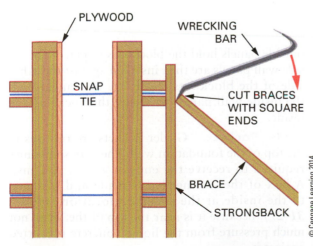

PLYWOOD

WRECKING BAR

SNAP TIE

CUT BRACES WITH SQUARE ENDS

BRACE

STRONGBACK

© Cengage Learning 2014

Figure 31–22 Cut braces with square ends. This allows for easy prying when straightening the formwork.

(*Figure 31–24*). The type is usually specified on the foundation plan. Care must be taken to set the anchor bolts at the correct height and at specified locations. Check local codes for anchor bolt spacing. An anchor bolt *template* is sometimes used to accurately place the bolts (*Figure 31–25*).✻

✻**IRC R403.1.6**

Openings in Concrete Walls

In many cases, openings must be formed in concrete foundation walls for such things as windows, doors, ducts, pipes, and beams. The forms used for providing the larger openings are called blockouts.

Constructing Blockouts. Blockouts are also called bucks. The blockout is usually made of 2-inch dimension lumber. Its width is the same as the thickness of the foundation wall (*Figure 31–26*). Nailing blocks or strips are often fastened to the outside of the bucks. These are beveled on both edges to lock them into the concrete when the form is stripped. They provide for the fastening of window and door frames in the openings. Intermediate pieces may be necessary in bucks for large openings to withstand the pressure of the concrete against them.

Large blockouts are made to the specified dimension. They are installed against the inside face of the outside panels. Duplex nails through the

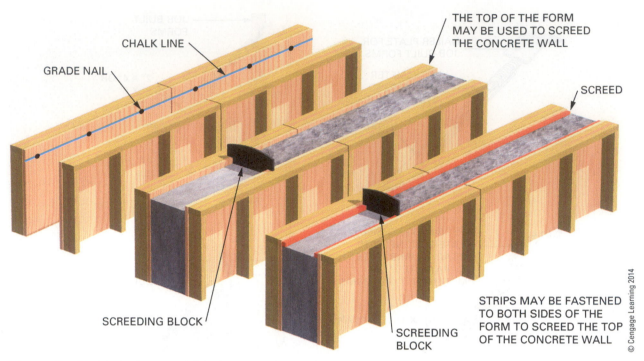

Figure 31–23 Methods to establish the top surface to the concrete in a wall form.

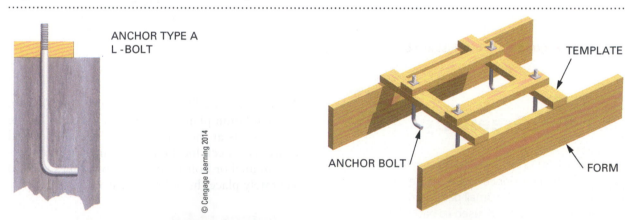

ANCHOR TYPE A
L -BOLT

Figure 31–24 Typical anchor bolts.

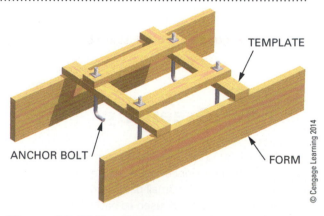

Figure 31–25 Templates are sometimes used to accurately place anchor bolts in fresh concrete.

Figure 31–26 Construction of a typical window and girder blockouts.

outside panels hold the blockouts in place. The inside wall panels are then installed against the other side of the blockouts. Nails are driven through the inside wall panels to secure the bucks on the inside.

Girder Pockets. Girder pockets are recesses in the top of the foundation wall. They are sometimes required to receive the ends of *girders* (beams). A box of the size needed is made and fastened to the inside at the specified location (*Figure 31–27*). Because it is near the top of the form, not much pressure from the liquid concrete is exerted against it.

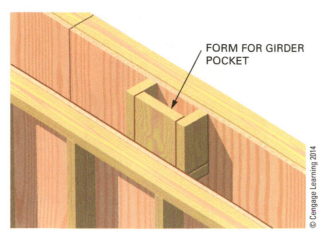

Figure 31–27 A small box is attached to the inside form for a girder pocket in a foundation wall.

COLUMN FORMS

To form columns, like all other kinds of form-work, as much use as possible is made of panels, manufactured or job built, to simplify erection and stripping. Columns may be formed in square, rectangular, circular, or a number of other shapes.

Erecting Wood Column Forms

Stretch lines and mark the location of the column on the footing. Fasten two 2 × 4 pieces to the footing on opposite sides of the column and outside

of the line by the panel thickness. Fasten the other two to the overlapping ends of the first pair in a similar location (*Figure 31–28*).

Build and erect two panels to form the thickness and height of the column. A cove molding of

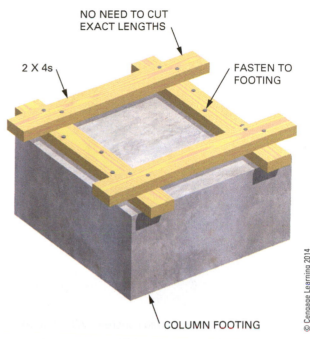

Figure 31–28 A yoke is constructed on the column footing to start the forming of a column.

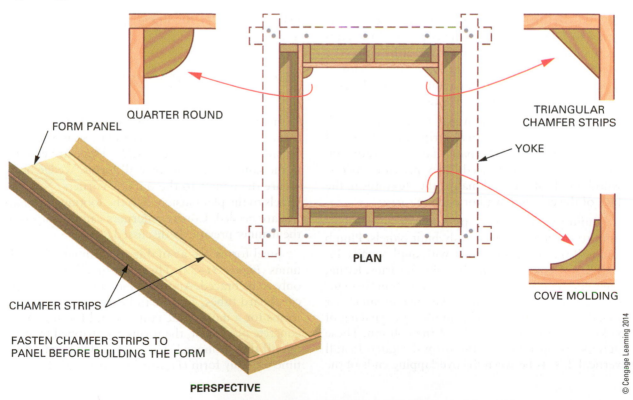

Figure 31–29 The corners of a concrete column can be formed in several ways.

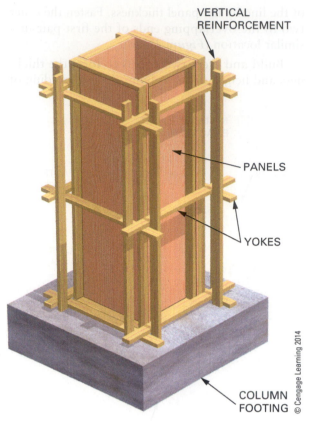

Figure 31–30 A completed concrete column form.

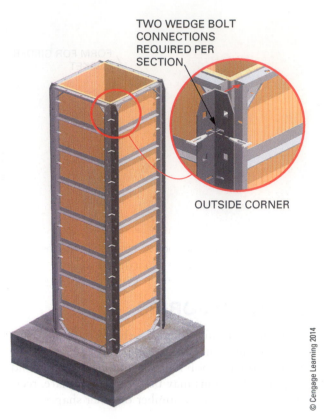

Figure 31–31 Yokes are not necessary to form columns when forming systems are used.

desired size may be fastened on the edges of this panel to form a radius to the corners of the concrete column. A quarter-round molding may be used for a cove shape. Triangular-shaped strips of wood can be used to form a chamfer on the corners of the column (*Figure 31–29*).

The face of the column may be decorated with **flutes** by fastening vertical strips of half-round molding spaced on the panel faces. In addition, *form liners* are often used. They provide various wood, brick, stone, and many other textures in the face of the concrete wall or column.

Build panels for the opposite sides of the column to overlap the previously built panels. Plumb and nail the corners together with duplex nails. Install 2 × 4 yokes around the column forms, letting their ends extend beyond the corners. Nail them together where they overlap. Yoke the column closer together at the bottom. The number and spacing of yokes depend on the height of the column. These details are specified by the form designer. Install vertical 2 × 4s between the overlapping ends of the

yokes (*Figure 31–30*). Brace the formwork securely to hold it plumb.

Erecting Manufactured Column Forms

Manufactured forms can be used instead of wood forms. Various-width forms can be assembled with outside corners (*Figure 31–31*). These are erected in the same fashion as for wall forms. Wedge bolts secure the panels to the outside corners. Nail the panels to the plate fastened to the footing. No snap ties are needed. Concrete corners can be shaped in the manner previously discussed.

Steel forms are available to form circular columns (*Figure 31–32*). They bolt together and need only to be braced. Another system uses heavy-duty cardboard tubes. These can be assembled on base forms for a monolithic column and footing. After concrete placement, the forms are stripped by peeling off the cardboard. This makes this form a one-time-use-only form (*Figure 31–33*).

COLUMN CAPITAL FORM

CIRCULAR COLUMN FORM

© Cengage Learning 2014

Figure 31–32 Manufactured circular column and column capital forms simplify forming.

© Cengage Learning 2014

Figure 31–33 Some forms are designed to be used only once.

CHAPTER 32
Concrete Stair Forms

STAIR FORMS

It may be necessary to refer to Chapters 47 and 48 on stair framing for definition of stair terms, types of stairs, stair layout, and methods of stair construction.

Concrete stairs may be suspended or supported by earth (*Figure 32–1*). Each type may be constructed between walls or have open ends.

Forms for Earth-Supported Stairs

Before placing concrete, stone, gravel, or other suitable fill is graded to provide proper thickness to the stairs. It should not be overly thick, or concrete will be wasted. It may be necessary to lay out the stairs before the supporting material is placed.

Forming between Existing Walls. When earth-supported stairs are formed between two existing walls, the **rise** and **run** of each step are laid out on the inside of the existing walls. *Rise* is the vertical distance that a step will rise. It is the height of each step, which is also called the *riser*. *Run* is the horizontal distance of a step. It is roughly the width of each step, which is called the tread. Boards are ripped to width to correspond to the height of each riser. The board is wedged and secured in place with its inside face aligned to the riser layout line. The top and bottom edges of the

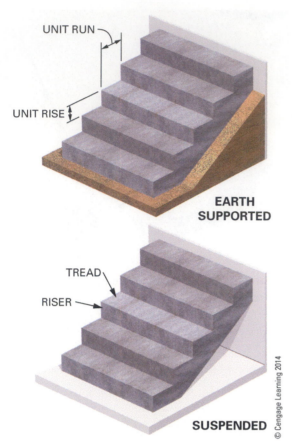

Figure 32–1 Concrete stairs may be formed on supporting soil or suspended.

board are aligned with tread layout lines. After the riser boards are secured in position, they are braced from top to bottom at midspan. This keeps them from bowing outward due to the pressure of the concrete (*Figure 32–2*).

The riser forms are beveled on the bottom edge. Beveling the bottom edge of the plank permits the mason to trowel the entire surface of the tread. Otherwise, the bottom edge of the riser form will leave its impression in the concrete tread.

Forming Stairs with Open Ends. In cases where the ends of the stairs are to be open, panels are erected on each end. It does no harm if the panels are larger than needed, the panels only need to be plumb. The distance between them is the desired width of the stairs. The end panels are then firmly fastened and braced in position.

The risers and treads are laid out on the inside surfaces of the panels. Cleats (short strips of wood) are fastened at each riser location. Allowances must be made for the thickness of the riser form board. Screws or duplex nails should be used to make stripping the form easier. The space should then be filled and compacted with the supporting material to the proper level. Any necessary reinforcing should be installed (*Figure 32–3*).

The boards used to form the risers are ripped to width and cut to length. The bottom edge is beveled to permit finishing of the total tread width. They are then fastened with duplex nails or screws through the side panels in a position against the cleats, with the top and bottom edges aligned to the tread layout lines. The riser forms are then braced from top to bottom at intervals between the two ends (*Figure 32–4*).

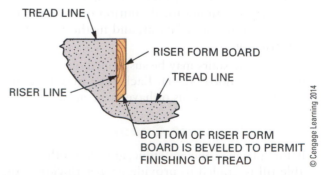

Figure 32–2 Concrete stairs may be formed between existing walls. Note that the bottom edge of the riser form may be beveled.

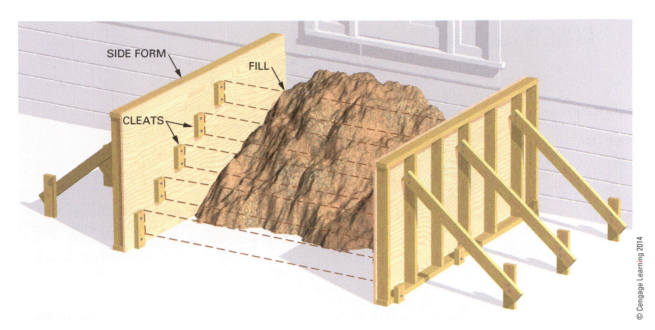

Figure 32–3 End panels are braced firmly in position. The stairs are laid out, cleats fastened to them, and the space filled with supporting material.

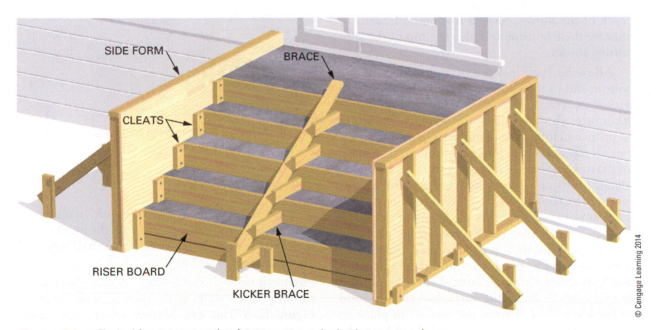

Figure 32–4 Typical form construction for concrete stairs having open ends.

Forms for Suspended Stairs

Forms for suspended stairs are more difficult to build. Instead of earth support, a form needs to be built on the bottom to support the stair slab. With proper design and reinforcement, the stairs are strong enough to support themselves in addition to the weight of the traffic. As with earth-supported stairs, suspended stairs may be formed between existing walls, with open ends, or both.

Forming between Existing Concrete Walls. A cross section of the formwork for suspended stairs between existing walls, with the form members identified, is shown in (*Figure 32–5*). First, lay out the treads, risers, and stair slab bottom on both of the walls between which the stairs run. This can

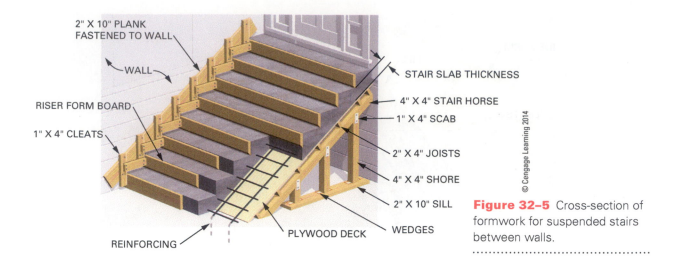

2" X 10" PLANK FASTENED TO WALL

WALL

RISER FORM BOARD

1" X 4" CLEATS

REINFORCING

PLYWOOD DECK

STAIR SLAB THICKNESS

4" X 4" STAIR HORSE

1" X 4" SCAB

2" X 4" JOISTS

4" X 4" SHORE

2" X 10" SILL

WEDGES

© Cengage Learning 2014

Figure 32–5 Cross-section of formwork for suspended stairs between walls.

be done by using a hand level, ruler, and chalk line (*Figure 32–6*).

The form for the bottom of the stair slab is then laid out. Allow for the thickness of the plywood deck, the width of the supporting joists, and the depth of the *horses*. Snap a line on both walls to indicate their bottom and also the top of the supporting *shores*.

Allowing for a plank sill and wedges at the bottom end of the shores, cut the shores to length with the correct angle at the top. They should be cut a little short to allow for wedging the shore to proper height. Install them at specified intervals on top of the sill. Brace in position, then place and fasten the horses on top with scabs or gusset plates. Scabs are short lengths of narrow boards fastened across a joint to strengthen it. Install joists at right angles across the horses at the specified spacing (*Figure 32–7*). Fasten the plywood form in position on top of the joists. Use only as many fasteners as needed to hold the plywood in place for easier stripping later. Wedge the shoring as necessary to bring the surface of the plywood to the layout line. Fasten the wedges in place so they will not move. The plywood should now be oiled to facilitate stripping. Install the reinforcing rods.

Rip the riser boards to width. Cut to length, allowing for wedges. Bevel the outside bottom edge of the treads to allow for easy leveling of the concrete at each tread. Install the riser form boards to the layout lines. Wedge in position as shown previously in Figure 32–2.

Another method, shown previously in Figure 32–5, secures the riser form with cleats fastened to a 2 × 10 attached to the wall. The 2 × 10 is

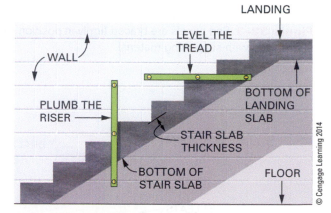

LANDING

WALL

PLUMB THE RISER

LEVEL THE TREAD

BOTTOM OF LANDING SLAB

STAIR SLAB THICKNESS

BOTTOM OF STAIR SLAB

FLOOR

© Cengage Learning 2014

Figure 32–6 Make the layout for suspended stairs on the inside of the walls.

attached about an inch above the stair layout. Project the riser layout lines upward on the planks. Fasten cleats to the plank. Then brace the cleats and fasten the riser board to them.

Forming with Open Ends. A completed form for suspended stairs with open ends is shown in *Figure 32–8*. This style of form sets the side panels for the stairs on the stair horse. The riser layout is made on the side panels and then set in place on top of the stair horse.

Laying Out the Side Forms. First, measure and snap a line on the side form up from the bottom edge a distance equal to the thickness of the slab. The inside corner intersection of the treads and risers lies along this line. Lay out the risers and treads above this line. The stair slab thickness is on the bottom edge of the form panels.

A **pitch board** speeds up the layout process (*Procedure 32–A*). It is a triangular scrap for which

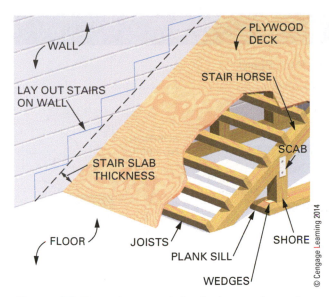

Figure 32–7 The framework for the bottom form of the suspended stairs.

the tread width and the riser height are the legs of a right triangle. To mark the tread and riser locations use the pitch board held to the previously snapped slab line.

Laying Out the Stair Horses. The length of the stair horse and shores can be determined from

the side form panels. Lay the side panel down on a floor surface. Snap a line that is parallel to the bottom edge of the side form. The distance away from the form bottom is equal to the thickness of the stair plywood deck and supporting joists (*Figure 32–9*).

From the side panel, extend the bottom floor line and the top plumb lines, making a large triangle on the floor. Lay out the thickness of the stair horse and shores including the sill. Measure and cut the length of the stair horse and shore. Cut one stair horse for each side to length at the angles indicated.

Installing Horses and Joists. Temporarily support and brace the horses in position. Install shores in a manner similar to that of closed stairs. Brace shores and horses firmly in position. Fasten joists at designated intervals to the horses, leaving an adequate and uniform overhang on both sides. Fasten decking to joists using a minimum number of nails.

Installing Side and Riser Forms. Snap lines on the deck for both sides of the stairs. Stand the side forms up with their inside face to the chalk line. Fasten them through the deck into the joists with duplex nails. Brace the side forms at intervals so they are plumb for their entire length (*Figure 32–10*). Apply form oil to the deck before reinforcing bars are installed.

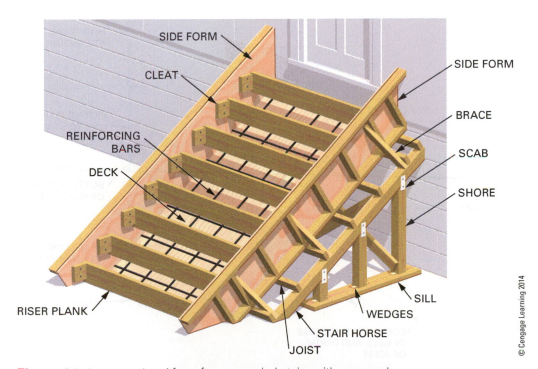

Figure 32–8 A completed form for suspended stairs with open ends.

StepbyStep Procedures

Procedure 32–A Laying Out a Set of Stairs Using a Pitch Board

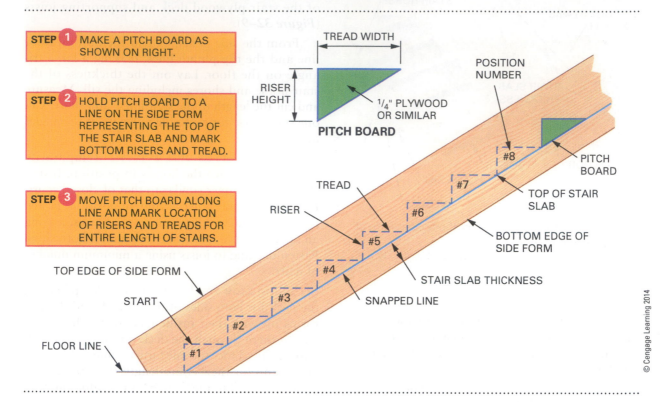

STEP 1 MAKE A PITCH BOARD AS SHOWN ON RIGHT.

STEP 2 HOLD PITCH BOARD TO A LINE ON THE SIDE FORM REPRESENTING THE TOP OF THE STAIR SLAB AND MARK BOTTOM RISERS AND TREAD.

STEP 3 MOVE PITCH BOARD ALONG LINE AND MARK LOCATION OF RISERS AND TREADS FOR ENTIRE LENGTH OF STAIRS.

TREAD WIDTH

RISER HEIGHT

1/4" PLYWOOD OR SIMILAR

PITCH BOARD

POSITION NUMBER

PITCH BOARD

TOP OF STAIR SLAB

BOTTOM EDGE OF SIDE FORM

TREAD

RISER

#8

#7

#6

#5

#4

#3

#2

#1

STAIR SLAB THICKNESS

SNAPPED LINE

TOP EDGE OF SIDE FORM

START

FLOOR LINE

© Cengage Learning 2014

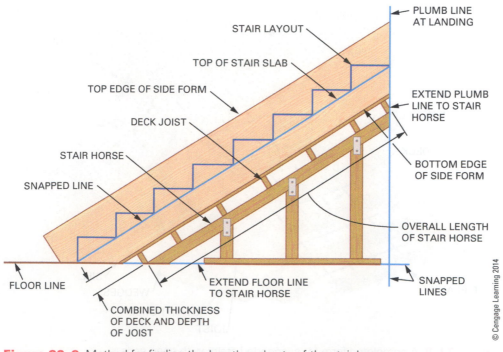

PLUMB LINE AT LANDING

STAIR LAYOUT

TOP OF STAIR SLAB

TOP EDGE OF SIDE FORM

DECK JOIST

STAIR HORSE

SNAPPED LINE

FLOOR LINE

COMBINED THICKNESS OF DECK AND DEPTH OF JOIST

EXTEND FLOOR LINE TO STAIR HORSE

EXTEND PLUMB LINE TO STAIR HORSE

BOTTOM EDGE OF SIDE FORM

OVERALL LENGTH OF STAIR HORSE

SNAPPED LINES

© Cengage Learning 2014

Figure 32–9 Method for finding the length and cuts of the stair horses.

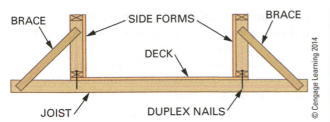

BRACE — SIDE FORMS — BRACE

DECK

JOIST — DUPLEX NAILS

Figure 32–10 Cross-section of the side form installed on the deck and braced.

Figure 32–11 Form boards may be left uncut to save time and make the forms more reusable.

Rip the riser boards to width and bevel the bottom edge. Install them on the riser layout line. Fasten them with duplex nails through the side forms into their ends. Install cleats against the riser boards on both ends for additional support. Install

CAUTION

Remove all protruding nails to eliminate the danger of stepping on or brushing against them.

intermediate braces to the riser form boards from top to bottom, if needed.

Economy and Conservation in Form Building

Economical concrete construction depends on the reuse of forms. Forms should be designed and built to facilitate stripping and reuse. Use panels to build forms whenever possible. Use only as many nails as necessary to make stripping forms easier.

Care must be taken when stripping forms to prevent damage to the panels so they can be reused. Stripped forms should be cleaned of all adhering concrete and stacked neatly.

Long lengths of lumber can often be used without trimming. Random-length boards can extend beyond the forms (*Figure 32–11*). There is no need to spend a lot of time cutting lumber to exact length. The important thing is to form the concrete to specified dimensions without spending too much time in unnecessary fitting.

CHAPTER 33

Foundation Dampproofing and Termite Control

DAMPPROOFING

Concrete improves in strength when it is kept moist. Wood and other building materials, on the other hand, do not. Water and dampness allow rot, decay, and attack from insects and fungi to occur. Indoor air quality is affected by high amounts of moisture in the air because it leads to mold and mildew. For these reasons keeping basements and crawl spaces dry is very important. Damp and water proofing is a process of stopping the flow of water to keep moisture from passing through the foundation wall. Water movement happens in several pathways and directions.

Water Movement

Rivers and creeks are proof that gravity causes water to run downhill. Drain pipes in plumbing systems are placed with a slight downward slope to allow water to yield to the force of gravity. Water can also travel upward, against gravity. In building materials such as masonry, water is drawn toward the drier region (*Figure 33–1*). This is called capillary action or wicking. This phenomenon involves liquids and small spaces, where molecular attractive forces between the liquid and the material itself are stronger than gravity.

Concrete is porous, having a vast array of small air pockets. These tiny voids and capillaries are created during the concrete hydration process. The water used to make concrete eventually bleeds and evaporates out. The space water once occupied is now open for other liquids and soil gases to pass. Soil gases tend to be toxic: they include methane, pesticides, and Radon.

Water can enter the building at any point along the entire foundation surface. Rain and melting snow can put liquid water against the building. Any cracks in the foundation will allow water flow inside. Water seeps into the ground, keeping it moist, and this moisture can move through solid, unbroken concrete by capillary action. Groundwater from below the building can enter through cracks and seams in the building materials. This may involve a small amount of dampness or great quantities of liquid water.

Liquid water that builds up against the foundation wall can exert a tremendous side pressure, called *hydrostatic pressure*. This pressure can easily be severe enough to cause the foundation wall to give way.

Water Drainage

Dry basements and crawl spaces begin once the excavation is completed. A layer of gravel or crushed stone should be placed under the footing, foundation, and slab (*Figure 33–2*). This allows

Figure 33–1 Water is drawn up into masonry by capillary action. These pieces were placed in 1/2″ of water for less than 15 minutes.

PIECE OF A CONCRETE SLAB

BRICK

WATER LEVEL IS LESS THAN 1/2″DEEP

CONCRETE BLOCK

© Cengage Learning 2014

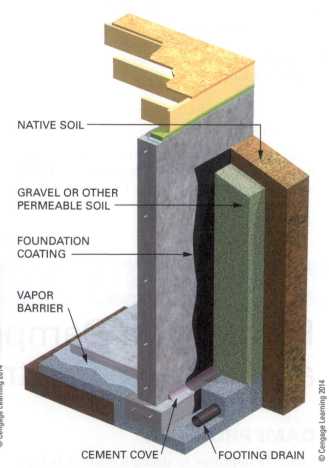

NATIVE SOIL

GRAVEL OR OTHER PERMEABLE SOIL

FOUNDATION COATING

VAPOR BARRIER

CEMENT COVE

FOOTING DRAIN

© Cengage Learning 2014

Figure 33–2 Dampproofing involves the entire foundation.

groundwater a pathway around the structure. The gravel layer keeps concrete above the groundwater and out of contact with it, breaking the potential for capillary action.

A perimeter footing drain allows any accumulated water to flow away from the building. The drain should be sloped over its entire length. The highest point of the perimeter drain should be below the floor level, and the drain pipe should be surrounded by gravel. The gravel layer should also be covered by a geotextile fabric. This material allows only water to pass through it, keeping dirt and soil from clogging the gravel and pipe. The drain should slope to a point of safe discharge.

Water discharge is best accomplished by simply extending the drain to the surface along a slightly downward slope. Determining the length of the drain away from the building depends on the site, often 50 feet or more. This drain system does not need electricity to operate and requires no maintenance.

If the site will not allow this type of the drain to extend to daylight, a sump pump may be required. A sump is a void or space built below the floor level sufficient to allow water to accumulate. An electric pump placed inside switches on when the water level rises, pumping it to an appropriate drain.

A good construction practice is to place cement in a coved shape on the footing against the foundation wall. This creates an angle and helps seal the construction seam between footing and foundation. Any water that flows down the wall is encouraged to find the footing drain and not enter the building.

Foundation Coatings

Foundation coatings help seal the wall from moisture penetration. They include liquid applications and sheet membranes.

Liquid Applications. Foundation coating that is liquid may be applied by spray, roller, or brush. These coating should be applied thick enough to fill all voids and cracks. Asphalt coatings, though least expensive, are least effective because over time they can dry out and crack.

Polymer-modified asphalt compounds perform better. The polymers are compounds of rubber or polyurethane added to the asphalt. These additives give the coating an ability to cover the small foundation cracks that may occur. They can adapt to minor movements caused by settling or seasonal temperature and moisture changes in building materials.

Sheet Membranes. Sheet membranes are applied to the entire surface of the foundation wall. There are two types. First is a thick modified asphalt product with a self-stick backing. They are elastomeric membranes that are applied by peeling off the release paper and pressing to the wall.

Other membranes are made of high density polyethylene attached to the wall with fasteners. They are dimpled and deformed to allow water a pathway to the footing drains (*Figure 33–3*). They keep liquid water from pooling against the wall, reducing capillary action into the concrete.

Slab Protection

Slabs also need to be protected from the effects of moisture. Sub-slab fill should be a gravel layer of about 6 inches thick. This will create a path of

Figure 33–3 Dampproofing may be a sheet membrane attached to the foundation.

Courtesy of Ewald Doerken AG

least resistance under the slab to the footing drain. A 6 mil polyethylene sheet is often installed under the slab. It should be installed with care and with the seams overlapped 12 inches. A polyurethane adhesive should be used to seal all seams and penetrations, then a layer of sand on top of the polyethylene will help protect it during concrete placement.

Backfill

Backfill of the foundation must be done carefully, because it is possible to crack the foundation wall with excessive side pressure. Installing the floor system before backfilling ties the top of the foundation together. This allows pressure from opposing sides of the building to help keep the wall where it belongs.

All materials that touch the foundation wall should allow liquid water to fall to the footing drain. Only gravel or other highly permeable soil should be used. A substitute for gravel is the sheet material mentioned previously. The dimpled, high-density polyethylene maintains a small space between the wall and the backfill. This creates a drainage plane for water to pass down to the footing drains and reducing the possibility of capillary action.

The slope of grade around the building must pitch downward, away from the building. This simple technique keeps water from pooling up against the building.

TERMITE CONTROL

Of all the destructive wood pests, termites are the most common. They cause tremendous economic loss annually. They attack wood throughout most of the country, but they are more prevalent in the warmer sections (*Figure 33–4*). Buildings should be designed and constructed to minimize termite attack. Chemicals should be applied only by trained technicians.

Termites play a beneficial role in their natural habitat. They break down dead or dying plant material to enrich the soil. However, when termites feed on wood structures, they become pests.

Kinds of Termites

There are three kinds of termites: *drywood, dampwood,* and *subterranean.*

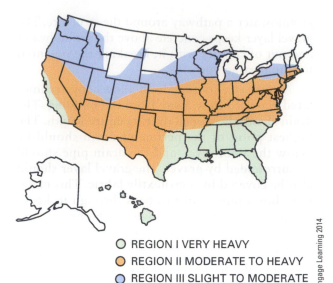

○ REGION I VERY HEAVY
● REGION II MODERATE TO HEAVY
● REGION III SLIGHT TO MODERATE
○ REGION IV NONE TO SLIGHT

© Cengage Learning 2014

Figure 33–4 Degree of subterranean termite hazard in the United States.

Drywood Termites. Drywood termites enter a building usually through attic or foundation vents. They attack sheathing and structural members. They also infest wood door and window frames and furniture. However, the colonies are small. Unchecked, they cause less damage to buildings than other kinds of termites. Treatment is usually tent fumigation of the entire building with a toxic gas.

Dampwood Termites. Dampwood termites infest wood kept wet by lawn sprinklers, leaking toilets, showers, pipes, roofs, and other places where wood is damp. They are capable of doing great damage to a structure if undetected. Entrance into the building is usually where a continual source of moisture keeps wood wet. Discovery of a leak sometimes reveals a dampwood termite infestation. Once the wood is dry, dampwood termites will leave.

Subterranean Termites. Subterranean termites live in the ground. They are the most destructive species because they have such large underground colonies (*Figure 33–5*). For protection against drying out, they must stay in close contact with the soil and its moisture. Above ground, they build earthen *shelter tubes* to protect themselves from the drying effects of the air. These tubes are usually built over a surface for support. Termites can build

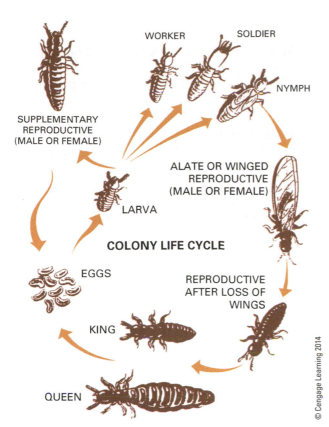

Figure 33–5 Typical subterranean termite life cycle.

© Cengage Learning 2014

practices could lead to termite infestation after completion of the building. Preventive efforts in the planning stage and during construction may save the future owner much anxiety and expense. All of the techniques used for the prevention of termite attack are based on keeping the wood in the structure dry (equilibrium moisture content) and making it as difficult as possible for termites to get to the wood. In lumber, a moisture content below 20 percent also prevents the growth of fungi, which cause wood to rot.

The Site

All tree stumps, roots, branches, and other wood debris should be removed from the building site. Do not bury it on the site. Footing and wall form planks, boards, stakes, spreaders, and scraps of lumber should be removed from the area before

unsupported tubes as high as 12 inches in their effort to reach wood (*Figure 33–6*).

Treatment for subterranean termites generally consists of correcting conditions favorable to infestation, installation of ground-to-wood termite barriers, and chemical treatment of the foundation and soil.

Green Tip

Insects tend not to consume or destroy dry wood. Buildings last longer if they are kept dry.

TECHNIQUES TO PREVENT TERMITES

Protection against subterranean termites should be considered during planning and construction of a building. Improper design and poor construction

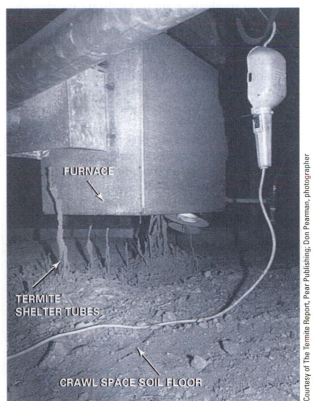

Figure 33–6 Subterranean termites can build unsupported shelter tubes as high as 12 inches.

Courtesy of The Termite Report, Pear Publishing; Don Pearman, photographer

backfilling around the foundation. Lumber scraps should not be buried anywhere on the building site. None should be left on the ground beneath or around the building after construction is completed.

The site should be graded to slope away from the building on all sides. The outside finished grade should always be equal to or below the level of the soil in crawl spaces. This ensures that water is not trapped underneath the building (*Figure 33–7*).

Chemical treatment of the soil before construction is one of the best methods of preventing termite attack. This should be a supplement to, and not a substitute for, proper building practices.

Perforated drain pipe should be placed around the foundation, alongside the footing. This will drain water away from the foundation (*Figure 33–8*). The foundation drain pipe should be sloped so water can drain to a lower elevation or a dry well some distance from the building. A dry well is a pit in the ground filled with stone to absorb the water from the drain pipe. Gutters and downspouts should be installed to lead roof water away from the foundation. Downspouts should be connected to a separate drain pipe to facilitate moving the water quickly.

Crawl Spaces

Solid concrete foundation walls should be properly reinforced and cured to prevent the formation of cracks. Cracks as little as $\frac{1}{32}$-inch wide permit the passage of termites.

Either concrete block walls should be capped with a minimum of 4 inches of reinforced concrete or the top course should be filled completely with concrete (*Figure 33–9*).

Air should be circulated in crawl spaces by means of ventilators placed to leave no pockets of stagnant air. In general, the total area of ventilation openings should be equal to $\frac{1}{150}$th of the ground area of the crawl space. Shrubbery should be kept away from openings to permit free circulation of air. There should be access to the crawl space for inspection of inner wall surfaces for termite tubes.

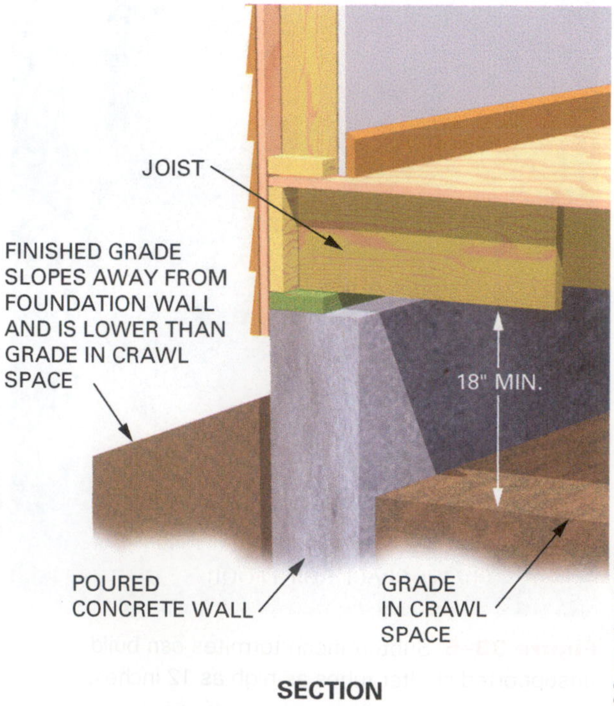

JOIST

FINISHED GRADE SLOPES AWAY FROM FOUNDATION WALL AND IS LOWER THAN GRADE IN CRAWL SPACE

18" MIN.

POURED CONCRETE WALL

GRADE IN CRAWL SPACE

SECTION

© Cengage Learning 2014

Figure 33–7 The finished grade should slope away from and be lower than the crawl space floor level.

Courtesy of Boccia, Inc.

Figure 33–8 Perforated drain pipe is placed alongside the foundation footing to drain water away from the building.

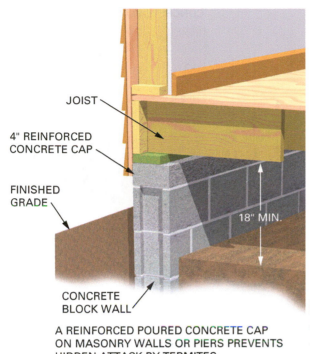

JOIST

4" REINFORCED
CONCRETE CAP

FINISHED
GRADE

18" MIN.

CONCRETE
BLOCK WALL

A REINFORCED POURED CONCRETE CAP
ON MASONRY WALLS OR PIERS PREVENTS
HIDDEN ATTACK BY TERMITES

SECTION

© Cengage Learning 2014

Figure 33–9 The top course of concrete block walls should be capped or filled completely with concrete.

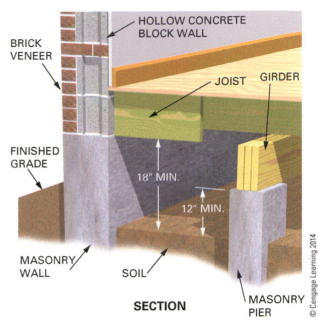

HOLLOW CONCRETE
BLOCK WALL

BRICK
VENEER

JOIST   GIRDER

FINISHED
GRADE

18" MIN.

12" MIN.

MASONRY
WALL   SOIL

SECTION

MASONRY
PIER

© Cengage Learning 2014

Figure 33–10 Provide adequate clearance between wood and soil in crawl spaces.

In crawl spaces and other concealed areas, clearance between the bottom of floor joists and the ground should be at least 18 inches and at least 12 inches for beams and girders (*Figure 33–10*).

Keep all plumbing and electrical conduits clear of the ground in crawl spaces. Suspend them from girders or joists. Do not support them with wood blocks or stakes in the ground. The soil around pipes extending from the ground to the wood above should be treated with chemicals.

Slab-on-Grade

Slab-on-grade is one of the most susceptible types of construction to termite attack. Termites gain access to the building over the edge of the slab, or through isolation joints, openings around plumbing, and cracks in the slab. Termite infestations in this type of construction are difficult to detect and control. For slab-on-grade construction, it is important to have the soil treated with chemicals before placing the concrete slab.

The monolithic slab provides the best protection against termites. The floor and footing are placed in one continuous operation, eliminating joints that permit hidden termite entry (*Figure 33–11*). Proper curing of the slab helps eliminate the

development of cracks through which termites can gain access to the wood above.

One type of independent slab extends completely across the top of the foundation. This prevents hidden termite entry. The lower edge of the slab should be open to view from the outside.

The top of the slab should be at least 8 inches above the grade (*Figure 33–12*).

Independent slabs that rest either partway on or against the side of the foundation wall are the least reliable. Termites may gain hidden access to the wood through expansion joints (*Figure 33–13*). Fill the spaces around expansion joints, pipes, conduit, ducts, or steel columns with hot roofing-grade coaltar pitch.

Exterior Slabs. Spaces beneath concrete slabs for porch floors, entrance platforms, and similar units against the foundation should not be filled. Leave them open with access doors for inspection. If this cannot be done and spaces must be filled, have the soil treated for termites by a professional.

Exterior Woodwork

Wall siding usually extends no more than 2 inches below the top of foundation walls. It should be at least 6 inches above the finished grade.

Porch supports should be placed not closer than 2 inches from the building to prevent hidden

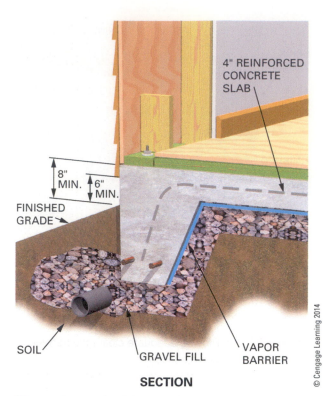

SECTION

Figure 33–11 In slab-on-grade construction, the monolithic slab provides the best protection against termites.

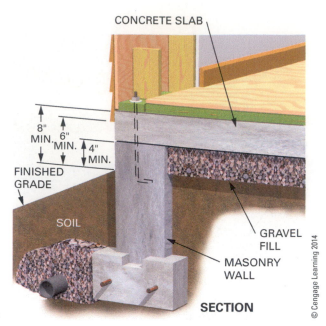

SECTION

Figure 33–12 An independent slab that extends across the top of the foundation wall prevents hidden termite attack.

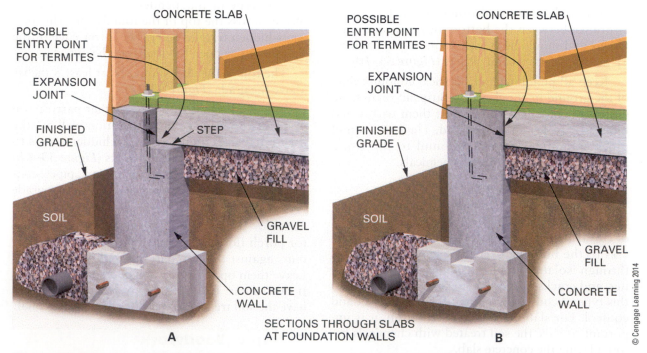

SECTIONS THROUGH SLABS AT FOUNDATION WALLS

Figure 33–13 Independent slab construction may allow a path for termite attack through isolation joints. (A) Edge of slab rests on ledge of the foundation wall. (B) Slab rests entirely on the ground (floating).

access by termites. Wood steps should rest on a concrete base that extends at least 6 inches above the ground.

Doorjambs, posts, and similar wood parts should never extend into or through concrete floors.

Termite Shields

If termite shields are properly designed, constructed, installed, and maintained, they will force termites into the open. This will reveal any tubes constructed around the edge and over the upper surface of the shield (*Figure 33–14*). However, research has shown that termite shields have not been effective in preventing termite infestations. This seems to be due to poor installation, inadvertent damage by the home owner, and infrequent

inspections. Check local building codes that may mandate their use.

Use of Pressure-Treated Lumber

All wood decays naturally when not kept dry. To prevent or slow this process, some wood species, for example, southern yellow pine, are sawed, kiln-dried, and then treated under pressure with preservatives (*Figure 32–15*). These preservatives enter the wood cells, virtually poisoning the wood for bacteria, fungi, and insects. This process creates pressure-treated lumber, which is used for foundation sills and structures that touch the ground.

Although other grades are available, two are generally used. *Above ground* is used for sill plates,

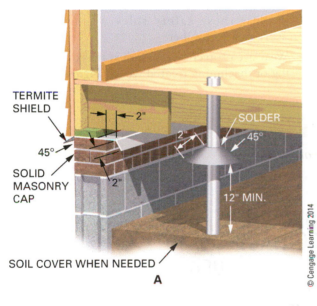

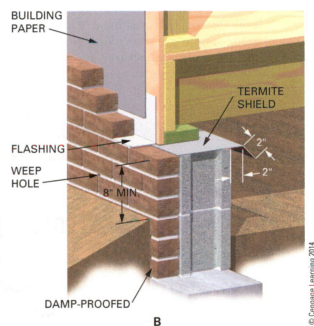

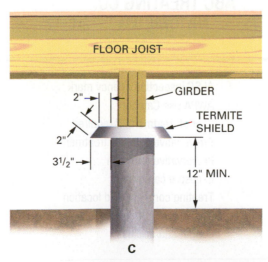

Figure 33–14 Typical installation of termite shields: (A) Exterior wall with wood siding, (B) exterior wall with brick veneer, and (C) over interior pier.

Figure 33–15 Preservatives are forced into lumber under pressure in large cylindrical tanks.

joists, girders, decks, and similar members. *Ground contact* is suitable for contact with soil or freshwater. Typical grade stamps are shown in *Figure 33–16*. Special grades are manufactured for saltwater immersion and wood foundations.

Check building codes for requirements concerning the use of pressure-treated lumber. Generally, building codes require the use of pressure-treated lumber for the following structural members:

- Wood joists or the bottom of structural floors without joists that are located closer than 18 inches to exposed soil.
- Wood girders that are closer than 12 inches to exposed soil in crawl spaces or unexcavated areas.
- Sleeper, sill, and foundation plates on a concrete or masonry slab that is in direct contact with the soil.

Termites generally will not eat treated lumber. They will tunnel over it to reach untreated wood. Their shelter tubes then may be exposed to view and their presence easily detected upon inspection.

Follow these safety rules when handling pressure-treated lumber:

- Wear eye protection and a dust mask when sawing or machining treated wood (*Figure 33–17*).

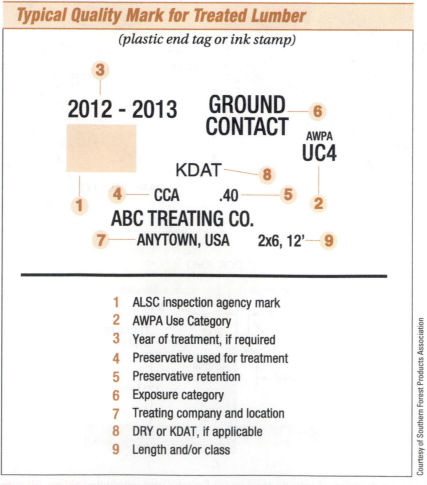

Typical Quality Mark for Treated Lumber

(plastic end tag or ink stamp)

2012 - 2013 ③

GROUND CONTACT ⑥

AWPA UC4 ②

KDAT ⑧

① ④ CCA .40 ⑤

ABC TREATING CO.

⑦ ANYTOWN, USA 2x6, 12' ⑨

1	ALSC inspection agency mark
2	AWPA Use Category
3	Year of treatment, if required
4	Preservative used for treatment
5	Preservative retention
6	Exposure category
7	Treating company and location
8	DRY or KDAT, if applicable
9	Length and/or class

Figure 33–16 Typical grade stamps for pressure-treated lumber.

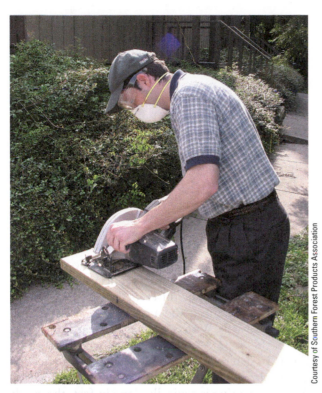

Figure 33–17 Use safety precautions when handling pressure-treated lumber.

- When the work is completed, wash areas of skin contact thoroughly before eating or drinking.
- Clothing that accumulates sawdust should be laundered separately from other clothing and before reuse.

- Dispose the treated wood by ordinary trash collection or burial. Do not burn treated wood. The chemical retained in the ash could pose a health hazard.

ESTIMATING CONCRETE CONSTRUCTION QUANTITIES

Estimating materials for construction often involves calculating for area and volume. Area calculations produce numbers with units that are in terms of square measure, volume in cubic measure.

The area for squares and rectangles is found by multiplying length times width. Circle area is π times the radius squared. Triangle area is ½ times the base times the height of the triangle (*Figure 33–18*). The height of a triangle is always measured perpendicular to the base. The example in this figure shows a right triangle which makes the height of the triangle also the side measurement of 6 inches.

Volume of many shapes is simply the cross sectional area (base shape) times the height of the figure. For example a slab has a rectangular surface area and a depth or height. Volume is found by multiplying the slab area times the depth of the slab.

The material quantities in concrete construction involve form boards and panels, rebar, and volume of concrete. Remember that if the number of pieces does not come out to a whole number, round them up to the next whole number of pieces or number of pieces in a package.

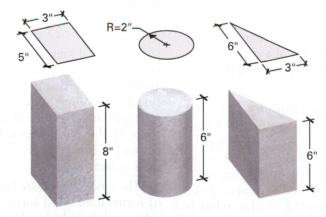

SHAPE	AREA	EXAMPLE	VOLUME	EXAMPLE
Rectangle	L × W	$3 \times 5 = 15$ in^2	Area × Height	$15 \times 8 = 120$ in^3
Circle	π r^2	$π \times 2^2 = 12.6$ in^2	Area × Height	$12.6 \times 7 = 88$ in^3
Triangle	½ BH	$½ \times 3 \times 6 = 9$ in^2	Area × Height	$9 \times 6 = 54$ in^3

Figure 33–18 Volumes of some shapes are easily calculated from the area of its base.

Forms Board

Form boards are sold by the piece of a certain size, which is determined from the architectural plans. Slab-on-grade form boards are estimated by dividing the perimeter by the board length. Footing form boards are estimated by multiplying the perimeter by 2 then dividing that by the board length. For example if the perimeter is 144 feet and the boards are 12 feet long, then the number of boards is $144 \times 2 \div 12 = 24$ boards. A minimal amount of waste is added for overlap and cuts. This extra is the waste factor.

Waste Factor

The waste factor changes depending on the material, but the process of including it in the estimate is the same. Multiply the quantity needed times (one + the waste percentage). This will increase the quantity by the additional material needed. For example, if the quantity is 100 pieces and the waste factor is 15 percent, then multiply 100 by (1 + 0.15 or 1.15) = 115 pieces.

Foundation Wall Forms

To find the foundation wall forms, multiply the perimeter by 2, then divide by the width of the panel. This will give the number of panels. For example, if the perimeter is 144 feet, then the number of 2-foot panels is $144 \times 2 \div 2 = 144$ panels.

Rebar

Rebar is placed in concrete to provide tensile strength. The quantity depends on where it is installed. Two rows of rebar are typically placed in footings. Poured foundation wall rebar is installed in a grid, typically $2' \times 2'$ OC (on center). Rebar is purchased by pieces 20 feet long or by a specific length.

Footing Rebar. The number of pieces of rebar may be found by doubling the perimeter or length of footings. Extra is needed since rebar must be overlapped and tied together at their ends. This extra should be about 10 percent more. For example, if the perimeter is 144 feet and the rebar piece length is 20 feet, the estimated footing rebar is $144 \times 2 \div 20 \times 1.10 = 15.8 = 16$ pieces.

Poured walls require rebar installed in a 2×2 grid pattern comprised of vertical and horizontal rows of bar.

Wall Horizontal Rebar. Horizontal rebar is the perimeter × number of rows of bar ÷ 20 feet per piece. The number of rows is the wall height ÷ grid spacing, then subtract 1 because a row is not needed at the top of the wall. Add 10 percent for overlap. For example, if the perimeter is 144 feet, grid spacing is 2 feet OC, and the wall height is 8 feet, then the horizontal rebar is $8 \div 2 - 1 = 3$ rows $\times 144 \div 20 \times 1.10 = 23.6 = 24$ bars.

Wall Vertical Rebar. Vertical rebar quantity is the number of verticals bar plus one per corner. The number of vertical bars is the perimeter ÷ grid spacing. For example, if the perimeter is 144 feet and the grid is 2 feet OC on a rectangular building, then the number of vertical rebars is $144 \div 2 = 72 + 4 = 76$ pieces.

Concrete

Ready-mix concrete is sold by the cubic yard (CY) or cubic meter. These amounts reflect the volume of concrete needed. The differences in sizes for a cubic yard, cubic foot, and a cubic inch are dramatic (*Figure 33–19*). With 36 inches in a yard, there are $36 \times 36 \times 36 = 46,656$ cubic inches in a cubic yard. To determine the volume of concrete needed for a job, multiply the width × length × thickness. Since measuring is typically done in feet, the result of this multiplication comes out as cubic feet. Cubic feet are converted to cubic yards by dividing cubic feet by 27 (the number of cubic feet in 1 cubic yard).

For example, how many cubic yards will be needed for a wall that is 8-inches-thick, 8-feet-high and 36 feet long. First, convert the 8 inches of thickness to feet: $8 \div 12 = 0.6666667$ feet. Note here that the 6 is repeated forever, but the calculator rounds the last displayed digit to 7. Then calculate the volume by multiplying thickness × width × length or $0.6666667 \times 8 \times 36 = 192$ cubic feet. Dividing by 27 yields 7.1 CY.

The actual amount of concrete needed is often not the same as the amount estimated. Slight variations in the forms may cause large errors in the calculated volume. The quantities from the ready-mix company are often close, but not perfect. Also, some spillage will occur. For these reasons, the amount of concrete ordered should be a little more than estimated.

The amount of waste factor depends on the type of forms used. Wall forms have smooth, uniform sides, and the calculated concrete quantities can be very close to actual quantities. Slabs, on the other hand, have a bottom surface that is irregular, making the actual thickness measurement difficult to determine. Therefore, take several thickness measurements and average them. Also, add a higher waste factor than you would for wall forms. In general, the waste factor amount is 5 to 10 percent. *Figure 30–20* shows a table of the short hand versions of these estimating techniques with an example.

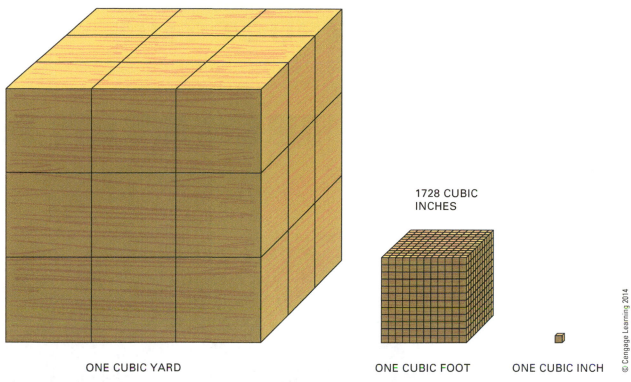

27 CUBIC FEET

1728 CUBIC INCHES

ONE CUBIC YARD

ONE CUBIC FOOT   ONE CUBIC INCH

Figure 33–19 Relative sizes of cubic measurements.

Estimate the materials for a foundation of a rectangular 30′ × 56″ building. Walls 8′ tall and 10″ thick, the footing is 2′ wide by 12″ thick, and slab is 5″ thick.			
Item	**Formula**	**Waste factor**	**Example**
Footing form boards	footing PERM × 2 sides ÷ 12′ = NUM of 12′ boards		172′ × 2 ÷ 12′ = 28.6 + 29 boards
Slab form boards	slab PERM ÷ board LEN = NUM of boards		172′ ÷ 12′ = 14.3 + 15 boards
Wall forms	wall PERM × 2 sides ÷ 2′ WID of form = NUM of form panels		172 × 2 ÷ 2′ = 172 panels
Rebar footing	PERM × 2 ÷ 20′ rebar LEN × waste = NUM of bars.	*Add 10% for bar overlap*	172′ × 2 ÷ 20′ × 1.10 518.9 + 19 rebars
Rebar walls 2′ × 2′ grid horizontal	PERM × [wall HGT ÷ 2 FT grid − 1] ÷ 20′ rebar LEN × waste = NUM of 20′ HOR bars	*Add 10% for bar overlap*	[8′ ÷ 2 − 1] × 172′ ÷ 20 × 1.10 5 28.3 + 29 rebars
Rebar walls 2′ × 2′ grid vertical	PERM ÷ 2 ft grid + 1 per corner = NUM of VERT bars		172 ÷ 2 + 4 = 90 vertical rebars
Concrete footing *(Be sure all measurements are in terms of feet)*	footing width × footing depth × PERM ÷ 27 = CY	*Add 5%*	2′ × 1′ × 172′ ÷ 27 × 1.05 = 5.6 + 5 3/4 CY
Concrete slab	slab WID × slab LEN × slab thickness ÷ 27 = CY concrete	*Add 10%*	30′ × 56′ × 5″/12 ÷ 27 × 1.10 5 28.5 + 28 3/4 CY
Concrete wall	PERM × wall HGT × wall thickness ÷ 27 = CY	*Add 5%*	172′ × 8 × 10″/12 ÷ 27 × 1.05 5 44.5 + 44 3/4 CY

Figure 33–20 Example of estimating form and concrete materials with formulas.

© Cengage Learning 2014

DECONSTRUCT THIS

Carefully study *Figure 33–21* and think about what is wrong and/or what is right. Consider all possibilities. What construction practice or method is different in your area of the country?

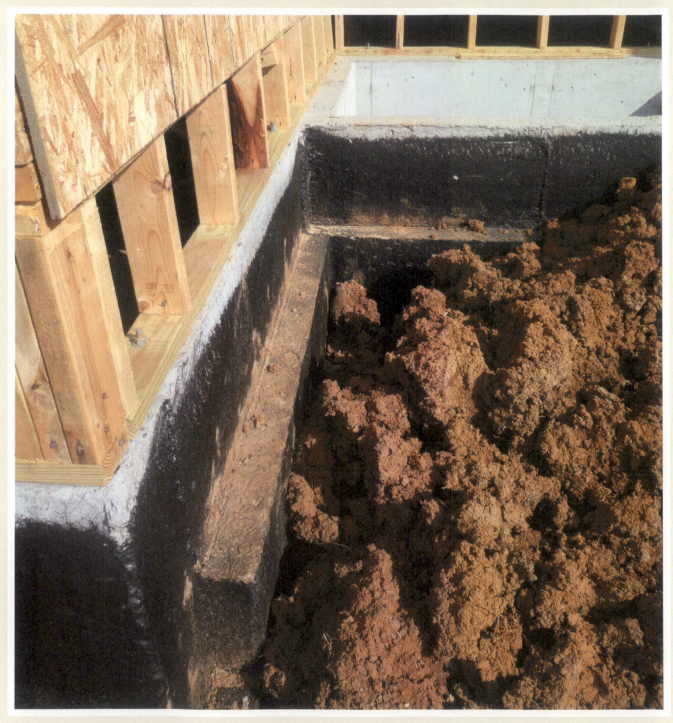

Figure 33–21 This photo shows a back-filled foundation and footing.

KEY TERMS

admixture	dry well elastomeric membranes	panels	screed
aggregates		pier	slump test
air-entrained cement	flutes	pilasters	snap ties
blockouts	footing	piles	spreaders
bucks	frost line	pitch board	spud vibrator
capillary action	girders	portland cement	strongbacks
cleats	gusset plates	pressure-treated	tensile strength
compressive strength	hydration	rebars	termites
concrete	independent slab	reinforced concrete	termite shields
course	kerfs	rise	walers
cove	keyway	run	waste factor
curing	monolithic slab	scabs	yoking

REVIEW QUESTIONS

Select the most appropriate answer.

1. Concrete is made of

   a. portland cement.
   b. large and small aggregates.
   c. water.
   d. all of the above.

2. The steel rods placed in concrete to increase its tensile strength are called

   a. reinforcing bars.
   b. aggregates.
   c. reinforcing nails.
   d. duplex nails.

3. The inside surfaces of forms are oiled to

   a. protect the forms from moisture.
   b. prevent the loss of moisture from concrete.
   c. strip the forms more easily.
   d. prevent honeycombs in the concrete.

4. Keyways are often put in spread footings to

   a. unlock the forms for easy removal.
   b. increase the compressive strength of concrete.
   c. keep the form boards from spreading.
   d. provide a stronger joint between footing and foundation.

5. Rapid placing of concrete

   a. omits the need for vibrating.
   b. may burst the forms.
   c. keeps the aggregate from separating.
   d. reduces voids and honeycombs.

6. Unless footings are placed below the frost line,

   a. the foundation will settle.
   b. the foundation may heave and crack.
   c. excavation will be difficult in winter.
   d. problems with form construction will result.

7. Spreaders for footing forms are used

   a. to allow easy placement of the concrete.
   b. to keep the forms straight.
   c. because they are easier to fasten.
   d. because they maintain the proper footing width.

8. A step in a footing should be dimensioned and sized to match the

   a. concrete block used in the foundation.
   b. form used to pour the foundation wall.
   c. building code requirements.
   d. all of the above.

9. The horizontal surface length of a stepped footing must be at least

   a. 4 feet.
   b. twice the vertical distance.
   c. the vertical distance.
   d. the thickness of the footing.

10. The typical order of installation of a manufactured forming system is

   a. inside forms, outside forms, reinforcing bars, then snap ties.
   b. inside forms, reinforcing bars, snap ties, then outside forms.

c. outside forms, snap ties, inside forms, then reinforcing bars.
d. outside forms, snap ties, reinforcing bars, then inside forms.

11. Walers are used on concrete wall forms to ____ the forms.

   a. stiffen
   b. straighten
   c. strengthen
   d. all of the above.

12. A concrete slab should be protected from

   a. freezing before it cures.
   b. curing too fast with a sealer.
   c. moisture after it cures by a subslab vapor retarder.
   d. all of the above.

13. While concrete sets, it should be protected from

   a. overheating.
   b. freezing.
   c. excessive vibrations.
   d. all of the above.

14. The minimum volume of concrete that should be ordered for a 6″ slab that measures 24′ × 36′ is

   a. 16 cubic yards.
   b. 36 cubic yards.
   c. 192 cubic yards.
   d. 432 cubic yards.

15. The procedure for erecting footing forms that follows locating the corner stakes with a plumb bob and batter board strings is installation of the

   a. spreaders.
   b. outside form boards.
   c. inside form boards.
   d. reinforcing bars.

16. Overvibrating concrete while placing it in wall forms causes

   a. voids and honeycombs.
   b. the aggregate to rise to the top.
   c. extra side pressure on the forms that could cause form failure.
   d. all of the above.

17. Accessories placed in a foundation wall form to create spaces are called

   a. blockouts.
   b. bucks.
   c. girder pocket forms.
   d. all of the above.

18. To aid in placing concete level at the top of a foundation wall,

   a. space nails along a chalk line.
   b. vibrate the top surface so it will flow level.
   c. use a chalk line only.
   d. add enough water to the concrete so that it will flow level.

19. The foundation type that has two variations is called

   a. full basement
   b. crawl space
   c. slab-on-grade
   d. all of the above

20. Foundation type that uses polystyrene as the concrete form is called

   a. CMU
   b. ICF
   c. AWWF
   d. crawl space

21. Water can enter masonry material by a process called

   a. capillary action
   b. crack location
   c. dampproofing
   d. all of the above

22. A dampproofed building includes

   a. a drainage plane.
   b. gravel around the foundation.
   c. a footing drain.
   d. all of the above.

23. When backfilling a foundation it is best to

   a. place native soil against the foundation.
   b. place a footing drain above the slab.
   c. coat the outside of the foundation.
   d. slope the surface of soil toward the building.

24. To protect against termites, keep wood in crawl spaces and other concealed areas above the ground at least

   a. 8 inches.
   b. 12 inches.
   c. 18 inches.
   d. 24 inches.

25. Pressure treatment is done on lumber to improve its

   a. decay resistance.
   b. pressure resistance
   c. nail-holding strength.
   d. all of the above.

UNIT 12
Floor Framing

Wood frame construction is used for residential and light commercial construction for important reasons of economy, durability, and variety. The cost for wood frame construction is generally less than for other types of construction. Fuel and air-conditioning expenses are reduced because wood frame construction provides better insulation.

Wood frame homes are very durable. If properly maintained, a wood frame building will last indefinitely. Many existing wood frame structures are hundreds of years old.

Because of the ease with which wood can be cut, shaped, fitted, and fastened, many different architectural styles are possible. In addition to single-family homes, wood frame construction is used for all kinds of low-rise buildings, such as apartments, condominiums, offices, motels, warehouses, and manufacturing plants.

OBJECTIVES

After completing this unit, the student should be able to:

- describe platform, balloon, and post-and-beam framing, and identify framing members of each.
- describe several energy and material conservation framing methods.
- read and understand member sizing tables found in building codes.
- build and install girders, erect columns, and lay out sills.
- lay out and install floor joists.
- frame openings in floors.
- lay out, cut, and install bridging.
- apply subflooring.

SAFETY REMINDER

Floor framing creates an elevated horizontal plane. Construction of the floor must proceed in a logical and thoughtful manner to reduce the risk of falling.

CHAPTER 34

Wood Frame Construction

There are several methods of framing a building. Some types are used less often today but still exist, so knowledge of them is necessary when remodeling. Other types are relatively new, and knowledge about them is not widespread. Some wood frames are built using a combination of types. New designs utilizing engineered lumber are increasing the height and width to which wood frame structures can be built. Sizing the members of the frame must be done to ensure the building preforms as expected.

PLATFORM FRAME CONSTRUCTION

The platform frame, sometimes called the *western* frame, is most commonly used in residential construction (*Figure 34–1*). In this type of construction, the floor is built and the walls are erected on top of it. When more than one story is built, the second-floor platform is erected on top of the walls of the first story.

A platform frame is easier to erect than a balloon frame. At each floor level, a flat surface is provided on which to work. A common practice is to assemble wall-framing units on the floor and then tilt the units up into place (*Figure 34–2*).

Effects of Shrinkage

Lumber shrinks mostly across width and thickness. A disadvantage of the platform frame is the relatively large amount of settling caused by the shrinkage of the large number of horizontal load-bearing frame members. However, because of the equal amount of horizontal lumber, the shrinkage is more or less equal throughout the building. To reduce shrinkage, only framing lumber with the proper moisture content should be used.

BALLOON FRAME CONSTRUCTION

In ballon frame construction, the wall *studs* and first-floor *joists* rest on the *sill*. The second-floor joists rest on a 1 × 4 ribbon that is cut in flush with the inside edges of the studs (*Figure 34–3*). This type of construction is used less often today, but a substantial number of structures built with this type of frame are still in use.

Effects of Shrinkage

Shrinkage of lumber along the length of a board is insignificant compared to shrinkage that can occur across the width of the board. Therefore, in the balloon frame, settling caused by shrinkage of lumber is held to a minimum in the exterior walls. This is because the studs are continuous from sill to top plate. To prevent unequal settling of the frame due to shrinkage, the studs of bearing partitions rest directly on the *girder*.

Fire-Stops

Fire-stop blocking is material installed to slow the movement of fire and smoke within smaller cavities of the building frame during a fire. It is sometimes called draft-stop blocking and *fire-blocking*. Should the building catch fire, this allows the occupants more time to get out and can reduce the overall damage to the building once the fire is put out. They must be installed in many places.

In a wood frame, a fire-stop in a wall might consist of dimension lumber blocking between studs. In the platform frame, the wall plates act as fire-stops.

Fire-stops must be installed in the following locations:

- in all stud walls, partitions, and furred spaces at ceiling and floor levels.
- between stair *stringers* at the top and bottom. (Stringers are stair framing members. They are sometimes called stair *horses*.)
- around chimneys, fireplaces, vents, pipes, and at ceiling and floor levels with noncombustible material.
- in the space between floor joists at the sill and girder.
- in all other locations as required by building codes (*Figure 34–4*).*

*IRC
R302.11

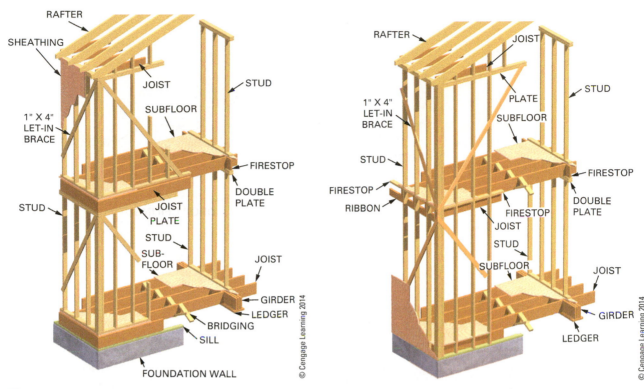

Figure 34–1 Platform frame construction.

Figure 34–3 Balloon frame construction.

Figure 34–2 Platform frame buildings are built one floor at a time.

POST-AND-BEAM FRAME CONSTRUCTION

The *post-and-beam* frame uses fewer but larger pieces. Large timbers, widely spaced, are used for joists, posts, and rafters. Matched boards (tongue and grooved) are often used for floors and roof sheathing (*Figure 34–5*).

Floors

APA-rated Sturd-I-Floor 48 on center (OC), which is $1\frac{3}{32}$ inches thick, may be used on floor joists that are spaced 4 feet OC instead of matched boards (*Figure 34–6*). In addition to being nailed, the plywood panels are glued to the floor beams with construction adhesive applied with caulking guns.

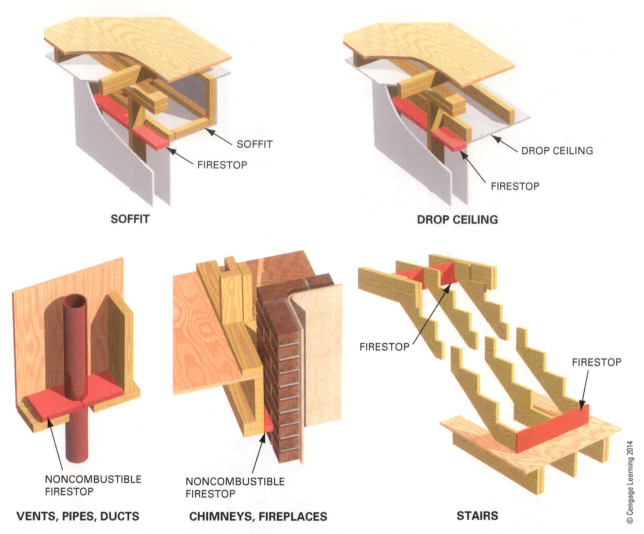

Figure 34–4 Some locations of firestops that help to slow the spread of fire.

The use of matched planks allows the floor beams to be more widely spaced.

Walls

Exterior walls of a post-and-beam frame may be constructed with widely spaced posts. This allows wide expanses of glass to be used from floor to ceiling. Usually some sections between posts in the wall are studded at close intervals, as in platform framing. This provides for door openings, fastening for finish, and wall sheathing. In addition, close spacing of the studs permits the wall to be adequately braced (*Figure 34–7*).

Roofs

The post-and-beam frame roof is widely used. The exposed roof beams and sheathing on the underside are attractive. Usually, the bottom surface of the roof planks is left exposed to serve as the finished ceiling. Roof planks come in 2-, 3-, and 4-inch nominal thicknesses. Some are end matched as well as edge matched. Some buildings may have a post-and-beam roof, while the walls and floors may be conventionally framed.

The post-and-beam roof may be constructed with a *longitudinal* frame. The beams run parallel to the ridge beam. Or they may have a *transverse* frame. The beams run at right angles to the ridge beam similar to roof rafters.

The ridge beam and longitudinal beams, if used, are supported at each end by posts in the end walls. They must also be supported at intervals along their length. This prevents the side walls from spreading and the roof from sagging (*Figure 34–8*). One of the disadvantages of a post-and-beam roof is that interior partitions and other

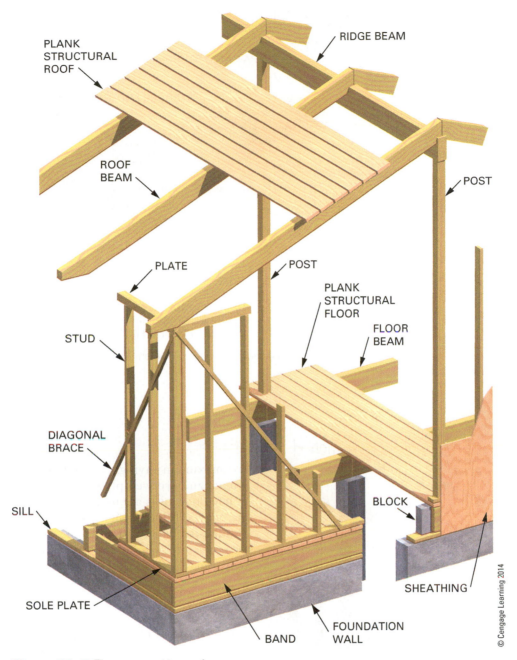

Figure 34–5 The post-and-beam frame.

interior features must be planned around the supporting roof beam posts.

Because of the fewer number of pieces used, a well-planned post-and-beam frame saves material and labor costs. Care must be taken when erecting the frame to protect the surfaces and make well-fitting joints on exposed posts and beams. Glulam beams are well suited for and frequently used in post-and-beam construction. A number of metal connectors are used to join members of the frame (*Figure 34–9*).

GREEN FRAMING METHODS

There has been much concern and thought about conserving energy and materials in building construction. Several systems have been devised that differ from conventional framing methods. They conserve energy and use less material and labor. Check state and local building codes for limitations.

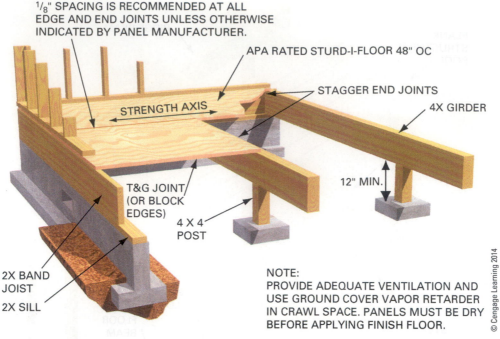

¹/₈" SPACING IS RECOMMENDED AT ALL EDGE AND END JOINTS UNLESS OTHERWISE INDICATED BY PANEL MANUFACTURER.

APA RATED STURD-I-FLOOR 48" OC

STAGGER END JOINTS

4X GIRDER

STRENGTH AXIS

12" MIN.

T&G JOINT (OR BLOCK EDGES)

4 X 4 POST

2X BAND JOIST

2X SILL

NOTE: PROVIDE ADEQUATE VENTILATION AND USE GROUND COVER VAPOR RETARDER IN CRAWL SPACE. PANELS MUST BE DRY BEFORE APPLYING FINISH FLOOR.

© Cengage Learning 2014

Figure 34–6 Floor beams may be spaced 4 feet OC when 1³⁄₃₂-inch-thick panels are used for a floor.

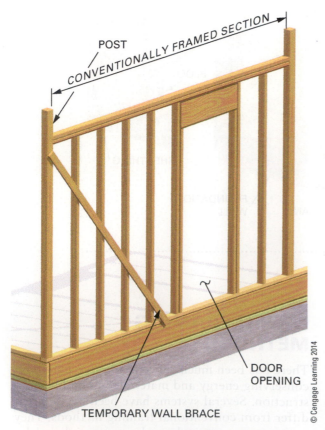

CONVENTIONALLY FRAMED SECTION

POST

DOOR OPENING

TEMPORARY WALL BRACE

© Cengage Learning 2014

Figure 34–7 Sections of the exterior walls of a post-and-beam wall may need to be conventionally framed.

Floors

For maximum savings, a single layer of ³/₄-inch tongue-and-grooved plywood is used over joists. In-line floor joists are used to make installation of the plywood floor easier (*Figure 34–10*). The use of adhesive when fastening the plywood floor is recommended. Gluing increases stiffness and prevents squeaky floors (*Figure 34–11*).

Walls

A single layer of plywood may act as both sheathing and exterior siding. In this case, the plywood must be at least ¹/₂ inch thick (*Figure 34–12*). If two-layer construction is used, ³/₈-inch plywood is acceptable.

Wall openings are planned to be located so that at least one side of the opening falls on an OC stud. Whenever possible, window and door sizes are selected so that the rough opening width is a multiple of the module (*Figure 34–13*). Also, locate partitions at OC wall stud positions if possible.

Roofs

Roof systems can be modified to improve the insulation over the exterior walls. The raised heel is built into the truss by the manufacturer

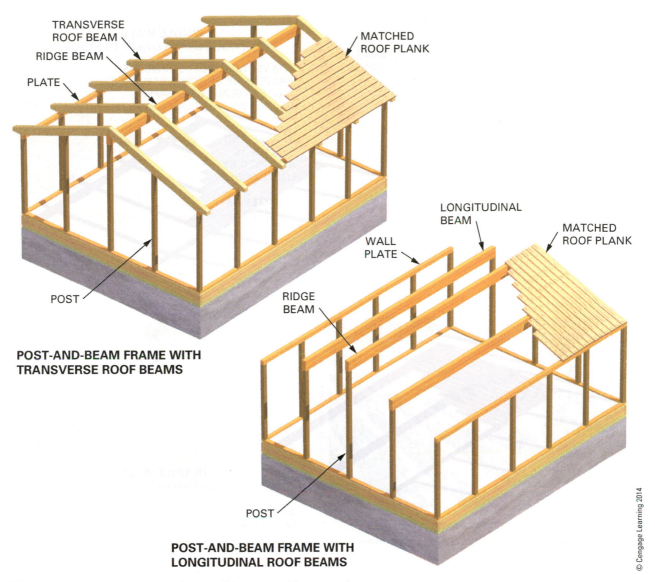

POST-AND-BEAM FRAME WITH TRANSVERSE ROOF BEAMS

POST-AND-BEAM FRAME WITH LONGITUDINAL ROOF BEAMS

© Cengage Learning 2014

Figure 34–8 Longitudinal and transverse post-and-beam roofs.

© Cengage Learning 2014

Figure 34–9 Metal connectors are specially made to join glulam beams.

(*Figure 34–14*). It raises the roof slightly to allow for full-thickness insulation at the **eaves**. Some areas of the country also use 2 × 6 wall studs to increase the wall insulation.

House Depths

House depths that are not evenly divisible by four may waste floor framing and sheathing. Lumber for floor joists is produced in increments of 2 feet. Assuming the girder remains in the center of the building, a house 25 feet wide would require 14-foot-long floor joists. These joists could be used uncut to make a building 28 feet wide. If the girder was installed offset from the center, 12- and 14-foot joists could be used to span 25 feet. Either way material is wasted.

IN-LINE FLOOR JOISTS

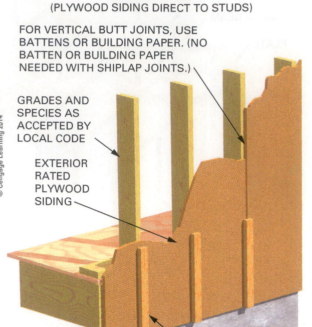

SCAB JOINTS

FLOOR PANELS SHOULD LAP BUTT JOINT TO PROVIDE TIE

Figure 34–10 In-line floor joists make installation of plywood subflooring simpler.

© Cengage Learning 2014

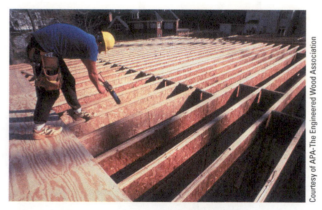

Courtesy of APA-The Engineered Wood Association

Figure 34–11 Using adhesive when fastening subflooring makes the floor frame stiffer, stronger, and quieter.

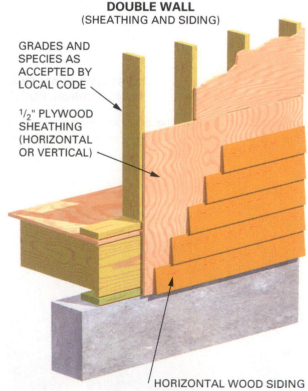

SINGLE WALL
(PLYWOOD SIDING DIRECT TO STUDS)

FOR VERTICAL BUTT JOINTS, USE BATTENS OR BUILDING PAPER. (NO BATTEN OR BUILDING PAPER NEEDED WITH SHIPLAP JOINTS.)

GRADES AND SPECIES AS ACCEPTED BY LOCAL CODE

EXTERIOR RATED PLYWOOD SIDING

BATTEN (OPTIONAL)

DOUBLE WALL
(SHEATHING AND SIDING)

GRADES AND SPECIES AS ACCEPTED BY LOCAL CODE

1/2" PLYWOOD SHEATHING (HORIZONTAL OR VERTICAL)

HORIZONTAL WOOD SIDING

© Cengage Learning 2014

Figure 34–12 Single-layer and double-layer exterior wall covering.

Full-width subfloor panels can be used without cutting on buildings that are 24-, 28-, and 32-feet wide. This decreases construction time and saves money.

SIZING STRUCTURAL MEMBERS

The size of structural members of a building is typically determined by professional engineers or ✱**IRC** architects.✱ Structural member and sizes are selected **R502.5** after consulting charts and tables. Carpenters are more successful when they understand how to communicate using engineering terms.

Girders are the main structural member that supports the inner end of floor joists. They are made of solid lumber, built-up materials, engineered

WINDOW ON MODULE

4' WALL SECTION

STUDS: 3 X 8' = 24 L.F.
JACKS: 2 X 7' = 14 L.F.
CRIPPLES: 3 X 3' = 9 L.F.
47 L.F.

O.C. O.C. O.C.

WINDOW OFF MODULE

4' WALL SECTION

STUDS: 4 X 8' = 32 L.F.
JACKS: 2 X 7' = 14 L.F.
CRIPPLES: 4 X 3' = 12 L.F.
58 L.F.

(23% MORE FRAMING REQUIRED)

O.C. O.C. O.C.

© Cengage Learning 2014

Figure 34–13 To conserve materials, locate wall openings so they fall on the OC studs.

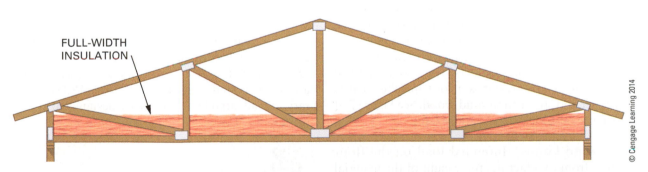

FULL-WIDTH INSULATION

© Cengage Learning 2014

Figure 34–14 Modified truss design accommodates thick ceiling insulation without compressing at eaves.

lumber, or steel. Each of these types is able to carry different amounts of load. Sizing girders also takes into account the width of the building, number of floors or stories in the building, and the distance between the columns that support them. These numbers are created by testing and experimentation.

The size and type of joist materials used in a building may be determined from building codes such as the 2012 International Residential Code, IRC. They are selected after considering the span charts. Joists are made of lumber, steel, or wood composite materials such as wood I-joists. Each material has a different strength and maximum span.

Structural Terms

Codes defining the span limits of joists use terms and numbers to describe how the material is to be used. These terms must be understood before reading the span charts. The overall strength of the floor system depends on many factors like, the grade and type of material used, the joist spacing, and the intended load on the floor. Stiffness of the floor must also be considered because a joist may be strong enough, but may move too much when under a load.

Joist Spacing. The space between joists affects the strength of the floor. The smaller the space between joists or the more joists used, make for a stronger floor. This space distance is normally 16 inches and is stated as 16 inches-on-center (OC). Other spacing includes 12, 19.2 and 24" OC.

The joist spacing is designed so an 8-foot panel (96 inches) of plywood is fully supported along its ends or shorter edges. Dividing 96 by a whole number of spaces gives the OC measurement (*Figure 34–15*). For example, six joists in eight feet have five open spaces. The joist spacing is then 96 ÷ 5 = 19.2 OC.

Material Strength. Wood species of dimension lumber are grouped according to their strengths. For example, lumber from the species of spruce, pine, and many fir trees have similar load characteristics and they are grouped in the species called spruce-pine-fir, (SPF). The grade of the wood and number of knots and defects changes the strength of lumber as well. Joist span charts give allowable spans for joists with these variations in mind.

Engineered lumber, such as I-joists, are made of manufactured wood products. They are assembled in an I-shape. The result is a joist member that is stronger and stiffer than solid wood. See Figure 7–4 for more on I-joist shapes and sizes.

Intended Load. Intended load on the floor comes from two factors, the weight of the material itself and the anticipated load of the everything in the building. Load is figured as pound per square foot or PSF.

Dead load is the term used to describe the weight of the lumber, floor material on top, and the ceiling material, if present, on the underside. This is the weight of all materials fastened to the joist to create the building. For residential floor framing, these numbers are either 10 PSF or 20 PSF. The difference between them on the charts is that 20 PSF has a ceiling attached and the 10 PSF does not.

Live load includes all possible increases to the load caused by people, furniture, and appliances. The main living floors in a house are normally

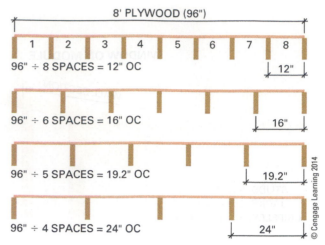

Figure 34–15 Spacing of joists relates to the number of joists under a plywood length.

40 PSF live load and second floor bedroom areas may be engineered to be 30 PSF. Note, for roof members, the snow load must be considered as well.

Stiffness. Stiffness of the floor is also a large factor in the desired strength of the floor system. Joists must be not only strong enough to support the load, but also stiff enough so not to sag or vibrate. This sag is referred to as deflection.

The amount of allowable deflection is governed by how the floor is to be used. Living floors should not allow dishes in a cabinet to vibrate when someone walks through the room. Roof members do not need this restriction and thus can be less stiff.

The deflection amount can be calculated. It is the joist span in inches divided by a number 180, 240, or 360. It is often written as L/180, L/240, or L/360 where L is joist span in inches. I-joists charts are often created to comply with an L/480 stiffness ratio. Residential living floors are designed at L/360 (*Figure 34–16*).

EXAMPLE

For example if a joist span is 13-6 feet or 162 inches, it should not deflect more than 162/360 = 0.45 inches (about 7/16"). If a joist is sized from an L/240 chart, the deflection limit would be 162/240 = 0.675 inches (about 11/16"). This minor increase in deflection makes the floor significantly less stiff.

Joist Span. Joist span is not the actual length of the joist. It is measured between the foundation sill plates. If a girder is used the span is from the plate to the center of the girder (*Figure 34–17*).

Typical Design Conditions for Residential Construction	
Region of Building	Deflection and Live Load
Living (non-sleeping) floors	L/360 & 40 PSF
Bedrooms and habitable attic floors	L/360 & 30 PSF
Attic floors with limited storage	L/240 & 10 PSF
Rafters	L/180 & 30,50, or 70 PSF for regional snow load

© Cengage Learning 2014

Figure 34–16 Typical residential strength and stiffness limits.

Joist Span Chart

Span charts have a lot of information packed into them (*Figure 34–18*). Reading them takes a little practice. Begin at the top by reading the title and second line of the chart. It is for residential floor joist in areas of 40 PSF live load with the stiffer, L/360 ratio.

The third line of the chart breaks the lower rows into two groups of columns. The left group is the Dead Load = 10 PSF for floors that will not have any ceiling material attached to the underside. The right group is Dead Load = 20 PSF for those that will have the ceiling attached. Each group has three sizes of joist materials, 2×8, 2×10, and 2×12.

The left most column of the chart breaks rows into four groups. Each group changes the joist spacing, 12, 16, 19.2, or 24" OC. Within each group is a choice of different species, grades, and materials. Each wood species is listed here in two different grades, #1 and #2. I-joists are listed in three sizes according to the flange size. It should be noted that charts from other sources will include more or less information, but reading the chart uses the same skill.

The numbers inside the bulk of the chart are given in feet and inches. These numbers refer to the maximum distance a material will span while providing the strength and stiffness indicated in the second and third line of the chart.

Sizing Joists. Deciding what size joists to use depends on information from the set of prints. This information is then brought to a span chart. Let us consider the building in the previous Figure 33–17 where no girder is used. The joist span is 17'−1". Assume the live load to be 40 PSF, the dead load is 10 PSF, and the joist spacing is 16" OC. Consider which joists could be used in this building.

To solve this, read the span chart beginning only with 16" OC section and only within the group of columns under the 10 PSF dead load. After reviewing, the maximum spans for all 2×8s and most of the 2 × 10s are not large enough to work in the 17' − 1" situation. But both I-joists listed will span this distance; TJI 110 spans 17' − 2" and TJI 230 spans 18' − 6". All the 2 × 12s will also span more than 17 − 1".

The best choice of material is often selected because it is the least expensive. Depending on joist material availability, it might be less expensive to reduce the joist spacing to 12". Looking in the 12" OC section reveals that all but one species of 2 × 10 will span 17' − 1". Thus a cost comparison should be done to determine the best choice of joist material.

EXAMPLE

Consider now making changes to the building design to effect the size of joists. If a ceiling was installed under the joist it would then have a dead load of 20 PSF. If a girder was also installed as in previous Figure 33−17, the span would be reduced to 9'−4". If 24" spacing was desired, a new section of the chart would have to be read. Using all this information, consider whether a #2 Hem-Fir 2×8 would work.

Solution. The maximum span of this joist would only be 9'−3", thus the answer is no, yet all the other 2×8s would function properly.

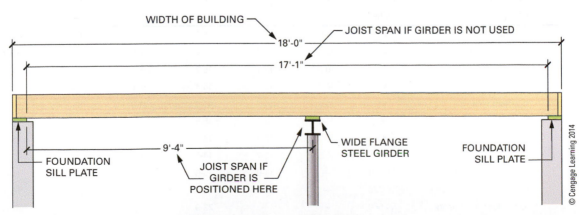

Figure 34–17 Determining joist span depends on how it is framed.

FLOOR JOIST SPACING FOR COMMON AND ENGINEERED LUMBER							
Residential living areas, live load = 40 psf, L/360							
		Dead Load = 10 psf			Dead Load = 20 psf		
		2 × 8	2 × 10	2 × 12	2 × 8	2 × 10	2 × 12
Joist Spacing	Joist Species or Material Type	Maximum Floor Joist Spans					
		(ft-in)	(ft-in)	(ft-in)	(ft-in)	(ft-in)	(ft-in)
12"	Douglas fir-larch #1	14-5	18-5	22-0	14-2	17-4	20-1
	Douglas fir-larch #2	14-2	17-9	20-7	13-3	16-3	18-10
	Hem-fir #1	13-10	17-8	21-6	13-10	16-11	19-7
	Hem-fir #2	13-2	16-10	20-4	13-1	16-0	18-6
	Southern pine #1	14-5	18-5	22-5	14-5	18-5	22-5
	Southern pine #2	14-2	18-0	21-9	14-2	16-11	19-10
	Spruce-pine-fir #1	13-6	17-3	20-7	13-3	16-3	18-10
	Spruce-pine-fir #2	13-6	17-3	20-7	13-3	16-3	18-10
	I-Joist 1 ¾" flange TJI 110	—	18-9	22-3	—	18-1	20-5
	I-Joist Joist 1 ⁵⁄₁₆" flange TJI 230	—	20-3	24-0	—	20-3	23-7
	I-Joist Joist 3 ½" flange TJI 560	—	—	28-10	—	—	28-10
16"	Douglas fir-larch #1	13-1	16-5	19-1	12-4	15-0	17-5
	Douglas fir-larch #2	12-7	15-5	17-10	11-6	14-1	16-3
	Hem-fir #1	12-7	16-0	18-7	12-0	14-8	17-0
	Hem-fir #2	12-0	15-2	17-7	11-4	13-10	16-1
	Southern pine #1	13-1	16-9	20-4	13-1	16-4	19-6
	Southern pine #2	12-10	16-1	18-10	12-4	14-8	17-2
	Spruce-pine-fir #1	12-3	15-5	17-10	11-6	14-1	16-3
	Spruce-pine-fir #2	12-3	15-5	17-10	11-6	14-1	16-3
	I-Joist Joist 1 ¾" flange TJI 110	—	17-2	19-4	—	15-8	17-8
	I-Joist Joist 1 ⁵⁄₁₆" flange TJI 230	—	18-6	21-11	—	18-1	20-5
	I-Joist Joist 3 ½" flange TJI 560	—	—	26-3	—	—	26-3
19.2"	Douglas fir-larch #1	12-4	15-0	17-5	11-3	13-8	15-11
	Douglas fir-larch #2	11-6	14-1	16-3	10-6	12-10	14-10
	Hem-fir #1	11-10	14-8	17-0	10-11	13-4	15-6
	Hem-fir #2	11-3	13-10	16-1	10-4	12-8	14-8
	Southern pine #1	12-4	15-9	19-2	12-4	14-11	17-9
	Southern pine #2	12-1	14-8	17-2	11-3	13-5	15-8
	Spruce-pine-fir #1	11-6	14-1	16-3	10-6	12-10	14-10
	Spruce-pine-fir #2	11-6	14-1	16-3	10-6	12-10	14-10
	I-Joist Joist 1 ¾" flange TJI 110	—	15-8	17-8	—	14-3	16-1
	I-Joist Joist 1 ⁵⁄₁₆" flange TJI 230	—	17-5	20-5	—	16-6	18-7
	I-Joist Joist 3 ½" flange TJI 560	—	—	24-9	—	—	24-9
24"	Douglas fir-larch #1	11-0	13-5	15-7	10-0	12-3	14-3
	Douglas fir-larch #2	10-3	12-7	14-7	9-5	11-6	13-4
	Hem-fir #1	10-9	13-1	15-2	9-9	11-11	13-10
	Hem-fir #2	10-2	12-5	14-4	9-3	11-4	13-1
	Southern pine #1	11-5	14-7	17-5	11-3	13-4	15-11
	Southern pine #2	11-0	13-1	15-5	10-0	12-0	14-0
	Spruce-pine-fir #1	10-3	12-7	14-7	9-5	11-6	13-4
	Spruce-pine-fir #2	10-3	12-7	14-7	9-5	11-6	13-4
	I-Joist Joist 1 ¾" flange TJI 110	—	14-0	15-9	—	12-9	14-4
	I-Joist Joist 1 ⁵⁄₁₆" flange TJI 230	—	16-2	18-3	—	14-9	16-7
	I-Joist Joist 3 ½" flange TJI 560	—	—	23-0	—	—	20-11

Figure 34–18 Typical building code span chart for joists.

Girder and Header Span Charts

Girders and headers are sized depending on the load they will experience. Headers are simply small girders placed over an opening in a wall. This load depends on where it is located within the house and what is being supported by them. Interior headers and girders support the floors and typically do not support any roof load. Exterior walls on the other hand support floors, ceilings and the roof (*Figure 34–19*).

Typical roof load changes from one region to another. It all depends on the local weather or the amount of snow for a particular region. Roofs with no concern for snow are designed with a 30 PSF live or snow load. Other regions will build with a design of 50 PSF snow load where heavy snow regions will design with 70 SPF snow load.

Another factor affecting the load on a girder is the length of the joists they support. They carry one-half the span of the joists, where the other half is supported by another structural member. Joists may span the building or to an interior support. This support is often installed at center-span of the building.

Charts and tables are made with information to select the proper sized girder or header (*Figure 34–20*). Reading them can be confusing at first. Begin reading the title of the chart to find it is for girders and headers of many wood species. Next, notice the first column has the seven major types of girders and headers described in the previous figure with the same corresponding letters A-G.

The second column of the chart shows the configurations of the headers and girders. These are made of dimension lumber of varying sizes and assembled in double, triple or quadruple layers. It should be noted that this chart is simplified for demonstration purposes. Building codes include more lumber sizes and variations.

The second and third lines of the chart show the three major groupings of snow load categories.

The next two lines show each live load group, 30, 50 or 70, with three variations governed by the width of the building. The sixth line of the chart has the titles of each column of numbers. One column is the span of the header or girder and the other column is the number of jacks studs required at each end of the girder. Jack studs are wall members cut short enough to support the header or girder and attached to full length wall studs.

Sizing Girders and Headers. Deciding what size header to use depends on information from the set of prints. This information is then brought to a span chart. Let us consider a building that is 26' × 48' built in region of moderate snow accumulation. Will a doubled 2 × 6 header adequately provide support over a 3' − 2" wide window in a first floor exterior 2 × 4 wall? The second floor is clear span joists?

The solution begins with choosing the header and girder grouping of C; the roof, ceiling and one clear span floor. This identifies the line of the chart, 2 − 2 × 6 in the C group. Next, we will assume a moderate snow is 50 PSF. Building width column must be 28 feet. This is because 20 feet is too small for the 26 feet. The intersection of 2 − 2 × 6 line and 50 PSF, 28 feet column is 3 − 4". The answer then is yes, a header made of 2 − 2 × 6's will span the required 3' − 2" opening. Two jacks will be needed at each end.

EXAMPLE

Consider what is the maximum span of a tripled 2 × 10 girder in an exterior wall supporting a roof and ceiling only. Building is 32' × 64' in Louisiana?

Solution. The line of the chart is in the A grouping of roof and ceiling only. Column choice is the 36 feet under the 30 PSF grouping. Answer is 8' − 2" with two jacks at either end.

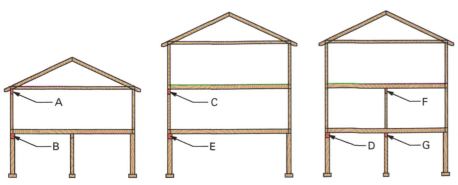

HEADER TYPES SUPPORTING

A - ROOF AND CEILING ONLY
B - ROOF, CEILING AND ONE CENTER-BEARING FLOOR
C - ROOF, CEILING AND ONE CLEAR SPAN FLOOR
D - ROOF, CEILING AND TWO CENTER-BEARING FLOORS
E - ROOF, CEILING AND TWO CLEAR SPAN FLOORS
F - ONE FLOOR ONLY
G - TWO FLOORS

© Cengage Learning 2014

Figure 34–19 Header and girder sizes depend on where in the building they are located.

GIRDER AND HEADER SPANS FOR <u>EXTERIOR</u> LOAD-BEARING WALLS
Douglas fir-larch, hem-fir, southern pine, and spruce-pine-fir

Girder and Header Supporting	Size	GROUND SNOW LOAD											
		30 PSF				50 PSF				70 PSF			
		Building Width				Building Width				Building Width			
		28 ft		36 ft		28 ft		36 ft		28 ft		36 ft	
		Span	NJ	Span	NJ	Span	NJ	Span	NJ	Span	NJ	Span	NJ
A Roof and ceiling only	2- 2 × 6	4-8	1	4-2	1	4-1	1	3-8	2	3-8	2	3-3	2
	2- 2 × 10	7-3	2	6-6	2	6-3	2	5-7	2	5-7	2	5-0	2
	3- 2 × 8	7-5	1	6-8	1	6-5	2	5-9	2	5-9	2	5-2	2
	3- 2 × 10	9-1	2	8-2	2	7-10	2	7-0	2	7-0	2	6-4	2
	4- 2 × 12	12-2	2	10-11	2	10-7	2	9-5	2	9-5	2	8-5	2
B Roof, ceiling and one center-bearing floor	2- 2 × 6	4-0	1	3-7	2	3-7	2	3-3	2	3-3	2	2-11	2
	2- 2 × 10	6-2	2	5-6	2	5-6	2	5-0	2	5-1	2	4-7	3
	3- 2 × 6	6-3	2	5-8	2	5-8	2	5-1	2	5-2	2	4-8	2
	3- 2 × 10	7-8	2	6-11	2	6-11	2	6-3	2	6-4	2	5-8	2
	4- 2 × 12	10-3	2	9-3	2	9-3	2	8-4	2	8-6	2	7-7	2
C Roof, ceiling and one clear span floor	2- 2 × 6	3-5	2	3-0	2	3-4	2	3-0	2	3-1	2	2-9	2
	2- 2 × 10	5-3	2	4-8	2	5-1	2	4-7	3	4-9	2	4-3	3
	3- 2 × 8	5-5	2	4-10	2	5-3	2	4-8	2	4-11	2	4-5	2
	3- 2 × 10	6-7	2	5-11	2	6-5	2	5-9	2	6-0	2	5-4	2
	4- 2 × 12	8-10	2	7-11	2	8-7	2	7-8	2	8-0	2	7-2	2
D Roof, ceiling and two center-bearing floors	2- 2 × 6	3-3	2	2-11	2	3-2	2	2-10	2	3-0	2	2-8	2
	2- 2 × 10	5-1	2	4-7	2	4-11	2	4-5	3	4-7	3	4-2	3
	3- 2 × 6	5-2	2	4-8	2	5-1	2	4-7	2	4-9	2	4-3	2
	3- 2 × 10	6-4	2	5-8	2	6-2	2	5-7	2	5-9	2	5-3	2
	4- 2 × 12	8-6	2	7-8	2	8-3	2	7-5	2	7-9	2	7-0	2
E Roof, ceiling and two clear span floors	2- 2 × 6	2-8	1	2-4	2	2-7	2	2-3	2	2-7	2	2-3	2
	2- 2 × 10	4-1	2	3-8	3	4-0	3	3-7	3	4-0	3	3-6	3
	3- 2 × 8	4-2	1	3-9	2	4-1	2	3-8	2	4-1	2	3-8	2
	3- 2 × 10	5-1	1	4-7	3	5-0	2	4-6	3	4-11	2	4-5	3
	4- 2 × 12	6-10	1	6-2	3	6-9	2	6-0	3	6-8	2	5-11	3

GIRDER AND HEADER SPANS FOR <u>INTERIOR</u> LOAD-BEARING WALLS
Douglas fir-larch, hem-fir, southern pine, and spruce-pine-fir

Girder and Header Supporting	Size	Span	NJ	Span	NJ								
F One floor only	2- 2 × 6	3-11	2	3-11	1								
	2- 2 × 10	6-1	2	6-1	2								
	3- 2 × 6	6-3	2	6-3	2								
	3- 2 × 10	7-7	2	7-7	2								
	4- 2 × 12	10-2	2	10-2	2								
G Two floors	2- 2 × 6	2-9	2	2-9	2								
	2- 2 × 10	4-3	2	4-3	3								
	3- 2 × 8	4-5	2	4-5	2								
	3- 2 × 10	5-4	2	5-4	2								
	4- 2 × 12	7-2	2	7-2	2								

Figure 34–20 Typical span chart for girder and headers from building codes.

CHAPTER 35

Layout and Construction of the Floor Frame

A floor frame consists of members fastened together to support the loads a floor is expected to bear. The floor frame is started after the foundation has been placed and has hardened. A straight and level floor frame makes it easier to frame and finish the rest of the building.

Because platform framing is used more than any other type, this chapter describes how to lay out and construct its floor frame. The knowledge gained in this chapter can be used to lay out and construct any type of floor frame.

DESCRIPTION AND INSTALLATION OF FLOOR FRAME MEMBERS

In the usual order of installation, the floor frame consists of *girders, posts* or *columns, sill plates, joists,* bridging, and *subflooring* (*Figure 35–1*).

Description of Girders

Girders are heavy beams that support the inner ends of the floor joists. Several types are commonly used.

Kinds of Girders. Girders may be made of solid wood or built up of three or more 2-inch planks. Laminated veneer lumber or glulam beams may also be used as girders (*Figure 35–2*). Sometimes, wide flange, I-shaped steel beams are used. A light steel is also available called Litesteel. It is cut and fitted with similar tools as used for wood.

Built-Up Girders. If built-up girders of dimension lumber are used, a minimum of three members are fastened together with three $3\frac{1}{2}$-inch or 16d nails at each end. The other nails are staggered not farther than 32 inches apart from end to end. Sometimes $\frac{1}{2}$-inch bolts are required. Applying glue between the pieces makes the bond stronger.

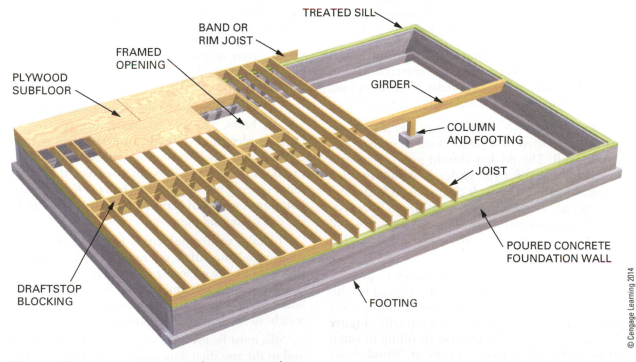

Figure 35–1 A floor frame of platform construction.

PLYWOOD SUBFLOOR

FRAMED OPENING

BAND OR RIM JOIST

TREATED SILL

GIRDER

COLUMN AND FOOTING

JOIST

POURED CONCRETE FOUNDATION WALL

FOOTING

DRAFTSTOP BLOCKING

© Cengage Learning 2014

© Cengage Learning 2014

Figure 35–2 Large glulam beams are often used as girders and headers.

Laminated veneer lumber of two layers may also be built up for use as girders.

Some engineers design built-up girders with butt splices that overhang from the support posts (*Figure 35–3*). The end joints are cantilevered over a distance that is within a range of ¼ or 1/6 for the post spacing. This design is similar to bridge girders.

> For Example if the post spacing is 8 feet or 96 inches, then the seam separation at the post should be less than ¼ × 96 = 24" and 1/6 × 96 = 16".

Some state codes require all girder butt seams to be supported over a post. The material used to make the girder in this case must be twice as long as the post spacing (*Figure 35–4*).

Girder Location. The ends of the girder are usually supported by a pocket formed in the foundation wall. The pocket should provide at least a 3-inch bearing for the girder. It should be wide enough to provide ½-inch clearance on both sides and the end. This allows any moisture to be evaporated by circulation of air. Thus no moisture will get into the girder, which would cause decay of the wood. ✳

✳**IRC R606.14**

The pocket is formed deep enough to provide for shimming the girder to its designated height. A steel bearing plate may be *grouted* in under the girder while it is supported temporarily (*Figure 35–5*). Grouting is the process of filling in small space with a thick paste of cement. Wood shims

are not usually suitable for use under girders. The weight imposed on them compresses the wood, causing the girder to sink below its designated level.

Installing Girders

Steel girders usually come in one piece and are set in place with a crane. A solid wood girder is often installed in a similar manner but is pieced together with half-lap joints. Joints are made near the posts or columns.

Built-up girders are erected in sections. This process begins by building one section at a time. One end is then set in the pocket in the foundation wall. The other end is placed and fastened on a braced temporary support. Continue building and erecting sections on posts until the girder is completed into the opposite pocket (*Figure 35–5*).

Sight the girder by eye from one end to the other. Place wedges under the temporary supports to straighten the girder. Permanent posts or columns are usually installed after the girder has some weight imposed on it by the *floor joists*. Temporary posts should be strong enough to support the weight imposed on them until permanent ones are installed.

SILLS

Sills, also called *mudsills* or *sill plates,* are horizontal members of a floor frame. They lie directly on the foundation wall and provide a bearing for *floor joists*. It is required that the sill be made with a decay-resistant material such as redwood, black locust, cedar, or pressure-treated lumber. Sills may consist of single 2 × 6, or doubled 2 × 6 lumber (*Figure 35–6*). ✳

✳**IRC R404**

The sill is attached to the foundation wall with anchor bolts. The size, number, and spacing of bolts are determined by local weather and seismic conditions. This information will be clearly indicated on the set of prints. In any case, the maximum spacing between anchor bolts is 6 feet, and a bolt must be located between 6 and 12 inches from the end of any sill piece. ✳

✳**IRC R403.1.6**

To take up irregularities between the foundation and the sill, a *sill sealer* is used. The sill sealer should be an insulating material used to seal against drafts, dirt, and insects. It comes 6 inches wide and in rolls of 50 feet. It compresses when the weight of the structure is upon it.

Sills must be installed so they are straight, level, and to the specified dimension of the building. The

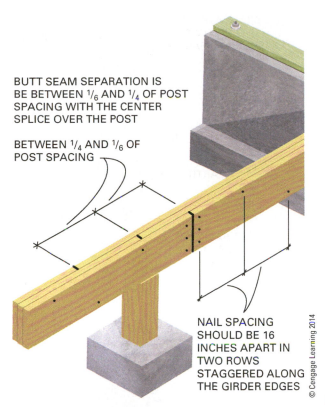

BUTT SEAM SEPARATION IS
BE BETWEEN 1/6 AND 1/4 OF POST
SPACING WITH THE CENTER
SPLICE OVER THE POST

BETWEEN 1/4 AND 1/6 OF
POST SPACING

NAIL SPACING
SHOULD BE 16
INCHES APART IN
TWO ROWS
STAGGERED ALONG
THE GIRDER EDGES

© Cengage Learning 2014

Figure 35–3 Spacing of fasteners and seams of a
built-up girder made with dimension lumber.

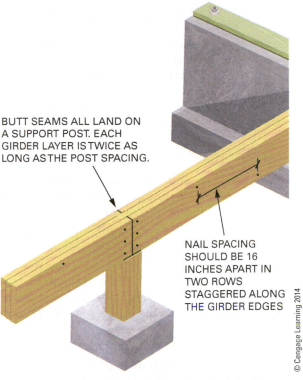

BUTT SEAMS ALL LAND ON
A SUPPORT POST. EACH
GIRDER LAYER IS TWICE AS
LONG AS THE POST SPACING.

NAIL SPACING
SHOULD BE 16
INCHES APART IN
TWO ROWS
STAGGERED ALONG
THE GIRDER EDGES

© Cengage Learning 2014

Figure 35–4 Spacing of fasteners with girder butt
seams positioned over the posts.

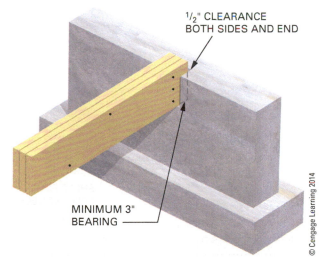

1/2" CLEARANCE
BOTH SIDES AND END

MINIMUM 3"
BEARING

© Cengage Learning 2014

Figure 35–5 A girder pocket of a foundation wall
should be large enough to provide air space around the
end and sides of the girder.

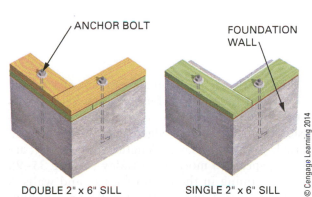

ANCHOR BOLT

FOUNDATION
WALL

DOUBLE 2" x 6" SILL

SINGLE 2" x 6" SILL

© Cengage Learning 2014

Figure 35–6 An anchor bolt should be located
between 6 and 12 inches from the end of each sill.

level of all other framing members depends on the care taken with the installation of the sill.

Sometimes the outside edge of the sill is flush with the outside of the foundation wall. Sometimes it is set in the thickness of the wall sheathing, depending on custom or design. In the case of brick-veneered exterior walls, the sill plate may be set back even farther (*Figure 35–7*).

Installing Sills

Remove washers and nuts from the anchor bolts. Snap a chalk line on the top of the foundation wall in line with the inside edge of the sill.

Cut the sill sections to length. Hold the sill in place against the anchor bolts. Square lines across the sill on each side of the bolts. Measure the distance from the center of each bolt to the chalk line. Transfer this distance at each bolt location to the sill by measuring from the inside edge (*Procedure 35–A*).

Bore holes in the sill for each anchor bolt. Bore the holes at least $\frac{1}{8}$ inch oversize to allow for adjustments. Place the sill sections in position over the anchor bolts after installing the sill sealer. The inside edges of the sill sections should be on the chalk line. Replace the nuts and washers. Be careful not to overtighten the nuts, especially if the concrete wall is still green. This may crack the wall.

If the inside edge of the sill plate comes inside the girder pocket, notch the sill plate around the end of the girder. Raise the ends of the girder so it is flush with the top of the sill plate. If steel girders are used, a dimension lumber sill is bolted to the top flanges. This allows for easy fastening of floor joists (*Figure 35–8*).

Floor Joists

Floor joists are horizontal members of a frame. They rest on and transfer the floor load to sills and girders. In residential construction, dimension lumber placed on edge has traditionally been used. Wood I-joists, with lengths up to 80 feet, are being specified more often today (*Figure 35–9*). Steel framing in the form of joists and walls is sometimes used. In general, when the price of lumber increases significantly or when termites are a problem, steel framing may be a solution.

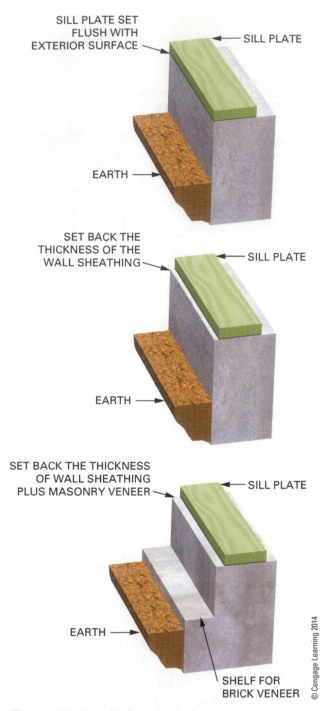

SILL PLATE SET FLUSH WITH EXTERIOR SURFACE — SILL PLATE

EARTH

SET BACK THE THICKNESS OF THE WALL SHEATHING — SILL PLATE

EARTH

SET BACK THE THICKNESS OF WALL SHEATHING PLUS MASONRY VENEER — SILL PLATE

EARTH

SHELF FOR BRICK VENEER

© Cengage Learning 2014

Figure 35–7 A sill plate may be located with different setbacks from the foundation edge.

StepbyStep Procedures

Procedure 35–A Installing a Sill Plate on Foundation with Anchor Bolts

STEP 1 SNAP CHALK LINE ON FOUNDATION WALL.

STEP 2 ALIGN SILL PLATE AGAINST ANCHOR BOLTS AND PARALLEL TO CHALK LINE.

STEP 3 SQUARE LINES ON SILL FROM BOTH SIDES OF EACH ANCHOR BOLT.

STEP 4 MEASURE EACH BOLT DISTANCE FROM CHALK LINE AND TRANSFER TO SILL.

STEP 5 DRILL HOLES APPROXIMATELY 1/8" LARGER THAN BOLT DIAMETER.

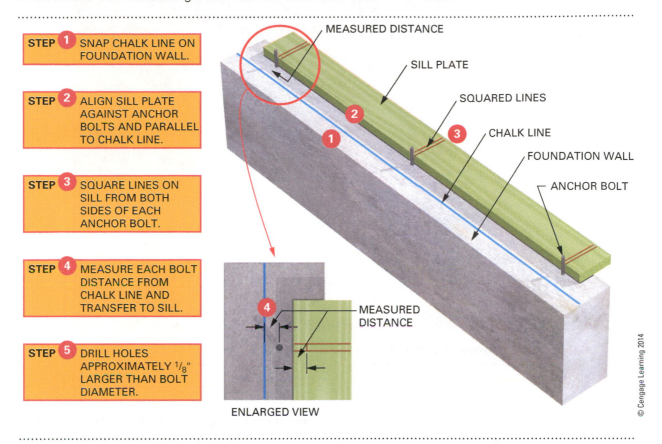

MEASURED DISTANCE

SILL PLATE

SQUARED LINES

CHALK LINE

FOUNDATION WALL

ANCHOR BOLT

MEASURED DISTANCE

ENLARGED VIEW

© Cengage Learning 2014

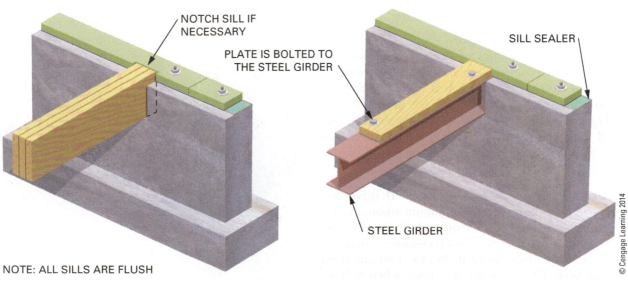

NOTCH SILL IF NECESSARY

PLATE IS BOLTED TO THE STEEL GIRDER

SILL SEALER

STEEL GIRDER

NOTE: ALL SILLS ARE FLUSH

Figure 35–8 Variations in girder and sill installations.

© Cengage Learning 2014

Figure 35–9 Engineered lumber makes a strong floor system.

Joists are generally spaced 16 inches OC in conventional framing. They may be spaced 12, 19.2, or 24 inches OC, depending on the type of construction and the load. The size of floor joists should be determined from the construction drawings.✳

✳IRC R502.3

Joist Framing at the Sill. Joists should rest on at least $1\frac{1}{2}$ inches of bearing on wood and 3 inches on masonry. In platform construction, the ends of floor joists are capped with a *band joist*, also called a *rim joist*, box header, or joist header (*Figure 35–10*).✳ The use of wood I-joists requires sill construction as recommended by the manufacturer for satisfactory performance of the frame (*Figure 35–11*).

✳IRC R502.6

Joist Framing at the Girder. If joists are lapped over the girder, the minimum amount of lap is 3 inches to allow for adequate nailing surface between lapped joists.✳ The maximum overhang of joists at the girder is 12 inches to eliminate floor squeaking. These squeaks are caused when walking on the floor at joist midspan; the overhung joist end raises up rubbing against the lapped joist. There is no need to lap wood I-joists. They come in lengths

✳IRC R502.6.1

long enough to span the building. However, they may need to be supported by girders depending on

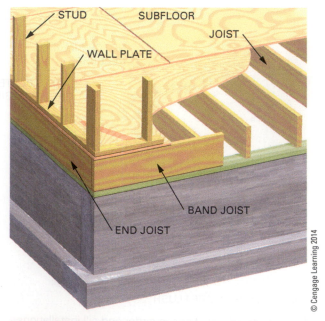

Figure 35–10 Typical framing near the sill using dimension lumber.

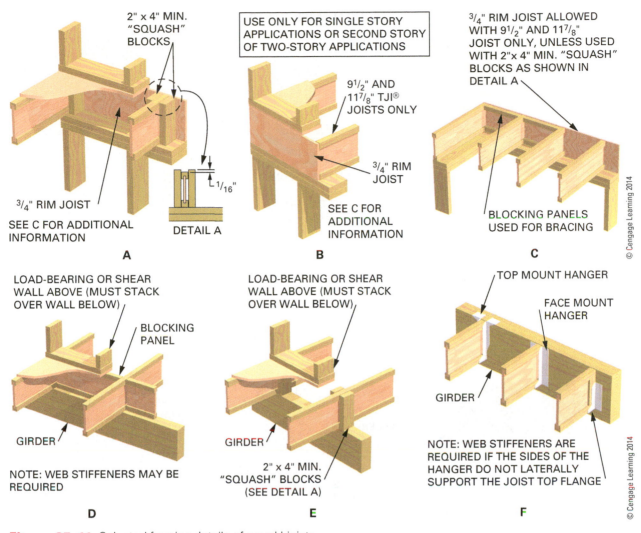

Figure 35–11 Selected framing details of wood I-joists.

the span and size of the wood I-joists. No matter how the joists are framed over the girder, drafted stop blocking is required. It should be installed using full-width framing lumber between joists on top of the girder (*Figure 35–12*).✳

✳IRC R502.13

Sometimes, to gain more headroom, joists may be framed into the side of the girder. There are a number of ways to do this. Joist hangers must be used to support wood I-beams. Web stiffeners should be applied to the beam ends if the hanger does not reach the top flange of the beam.✳

✳IRC R502.6.2

Holes bored in joists for piping or wiring should not be larger than one-third of the joist depth. They should not be closer than 2 inches to the top or bottom of the joist (*Figure 35–13*).✳

✳IRC R502.8

Some wood I-joists are manufactured with perforated knockouts in the web along its length. This allows for easy installation of wiring and pipes. To cut other-size holes in the web, consult the manufacturer's specifications guide. Do not cut or notch the flanges of wood I-joists.✳

✳IRC R502.8.2

Notching and Boring of Joists

Notches in the bottom or top of sawn lumber floor joists should not exceed one-sixth of the joist depth. Notches should not be located in the middle one-third of the joist span. Notches on the ends should not exceed one-fourth of the joist depth.

Laying Out Floor Joists

The locations of floor joists are marked on the sill plate. A squared line marks the side of the joist. An *X* to one side of the line indicates on which side of the line the joist is to be placed (*Figure 35–14*).

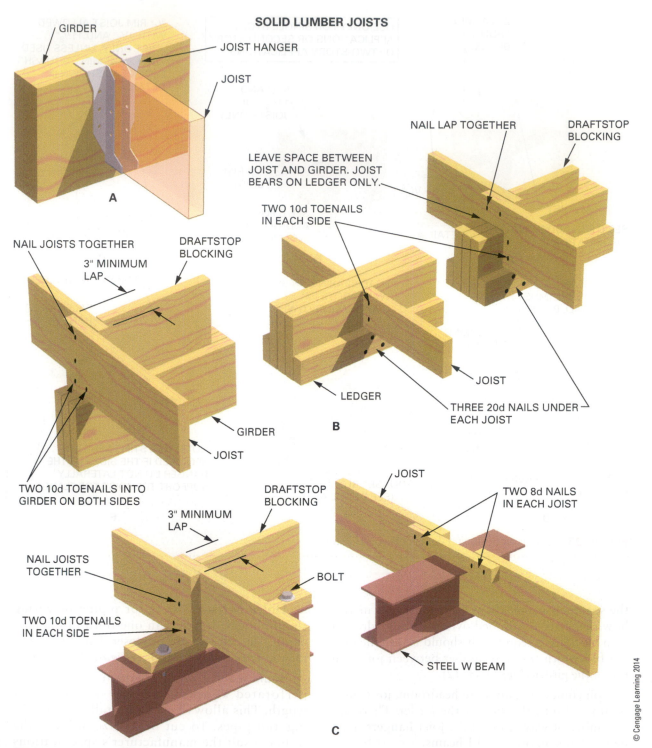

Figure 35–12 Various possible framing details at a girder.

Floor joists must be laid out so that the ends of *plywood subfloor* sheets fall directly on the center of floor joists. Start the joist layout by measuring the joist spacing from the end of the sill. Measure back one-half the thickness of the joist. Square a line across the sill. This line indicates the side of the joist closest to the corner. Place an X on the side of the line on which the joist is to be placed.

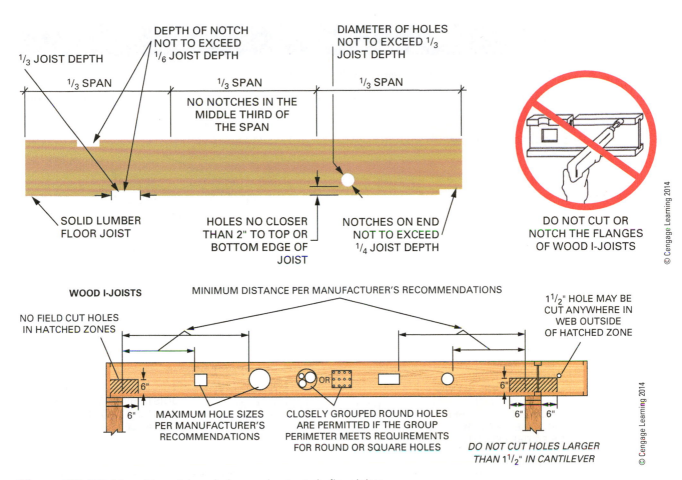

DEPTH OF NOTCH
NOT TO EXCEED
¹/₆ JOIST DEPTH

DIAMETER OF HOLES
NOT TO EXCEED ¹/₃
JOIST DEPTH

¹/₃ JOIST DEPTH

¹/₃ SPAN   ¹/₃ SPAN   ¹/₃ SPAN

NO NOTCHES IN THE
MIDDLE THIRD OF
THE SPAN

SOLID LUMBER
FLOOR JOIST

HOLES NO CLOSER
THAN 2" TO TOP OR
BOTTOM EDGE OF
JOIST

NOTCHES ON END
NOT TO EXCEED
¹/₄ JOIST DEPTH

DO NOT CUT OR
NOTCH THE FLANGES
OF WOOD I-JOISTS

WOOD I-JOISTS

MINIMUM DISTANCE PER MANUFACTURER'S RECOMMENDATIONS

NO FIELD CUT HOLES
IN HATCHED ZONES

1¹/₂" HOLE MAY BE
CUT ANYWHERE IN
WEB OUTSIDE
OF HATCHED ZONE

6"

6"   6"   6"

MAXIMUM HOLE SIZES
PER MANUFACTURER'S
RECOMMENDATIONS

CLOSELY GROUPED ROUND HOLES
ARE PERMITTED IF THE GROUP
PERIMETER MEETS REQUIREMENTS
FOR ROUND OR SQUARE HOLES

*DO NOT CUT HOLES LARGER
THAN 1¹/₂" IN CANTILEVER*

© Cengage Learning 2014

Figure 35–13 Allowable notches, holes, and cutouts in floor joists.

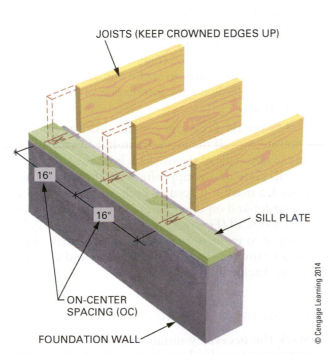

JOISTS (KEEP CROWNED EDGES UP)

16"

16"

SILL PLATE

ON-CENTER
SPACING (OC)

FOUNDATION WALL

Figure 35–14 A line is drawn to mark the edge of a joist; an *X* is marked to indicate on which side of the line the joist is placed.

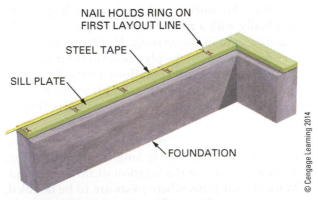

On the Job

Use a steel tape when laying out floor joists. An inaccurate layout leads to cutting back subfloor panels so their ends will fall on the centers of the floor joists. A steel tape prevents gaining on the spacing.

NAIL HOLDS RING ON
FIRST LAYOUT LINE

STEEL TAPE

SILL PLATE

FOUNDATION

© Cengage Learning 2014

Figure 35–15 Using a steel tape for layout reduces the possibility of step-off errors.

StepbyStep Procedures

Procedure 35–B Laying Out the Sill Plate for Floor Joists

STEP 1 MEASURE THE JOIST SPACING IN FROM THE CORNER.

STEP 2 MEASURE BACK ½ THE JOIST THICKNESS.

STEP 3 SQUARE A LINE ACROSS THE SILL PLATE AND PLACE AN X ON THE SIDE OF THE LINE WHERE THE JOIST IS TO BE PLACED.

STEP 4 CONTINUE THE ON-CENTER SPACING ALONG THE LENGTH OF THE BUILDING. USE A STEEL TAPE STRETCHED OVER THE ENTIRE LENGTH.

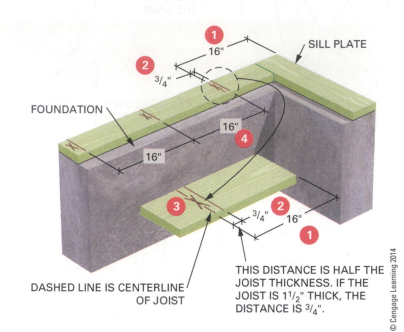

SILL PLATE

FOUNDATION

DASHED LINE IS CENTERLINE OF JOIST

THIS DISTANCE IS HALF THE JOIST THICKNESS. IF THE JOIST IS 1½" THICK, THE DISTANCE IS ¾".

© Cengage Learning 2014

From the squared line, measure and mark the spacing of the joists along the length of the building. Place an X on the same side of each line as for the first joist location (*Procedure 35–B*).

When measuring for the spacing of the joists, use a tape stretched along the length of the building. Most professional tapes have predominate markings for repetitive layout—black rectangles for 12- and 24-inch layouts, red rectangles for 16-inch, and small black diamonds for 19.2-inch layouts. Using a tape in this manner is more accurate. Measuring and marking each space individually with a rule or framing square generally causes a gain in the spacing. If the spacing is not laid out accurately, the plywood subfloor may not fall in the center of some floor joists. Time will then be lost either cutting the plywood back or adding strips of lumber to the floor joists (*Figure 35–15*).

Laying Out Floor Openings. After marking floor joists for the whole length of the building, study the plans for the location of floor openings. Mark the sill plate, where joists are to be doubled, on each side of large floor openings. Identify the layout marks that are not for full-length floor joists.

Shortened floor joists at the ends of floor openings are called tail joists. They are usually identified by changing the X to a T or a C for cripple joist. Lay out for partition supports, or wherever doubled floor joists are required (*Figure 35–16*). Check the mechanical drawings to make adjustments in the framing to allow for the installation of mechanical equipment.✱

✱IRC R502.10

Lay out the floor joists on the girder and on the opposite wall. If the joists are in-line, Xs are made on the same side of the mark on both the girder and the sill plate on the opposite wall. If the joists are lapped, a mark is placed on both sides of the line at the girder and on the opposite side of the mark on the other wall (*Figure 35–17*). These marks may be changed to make it easier to tell which mark is for the joist toward the front of the house and which is for the back.

Installing Floor Joists

Stack the necessary number of full-length floor joists at intervals along both walls. Each joist is carefully sighted along its length by eye. Any joist with a severe crook or other warp should

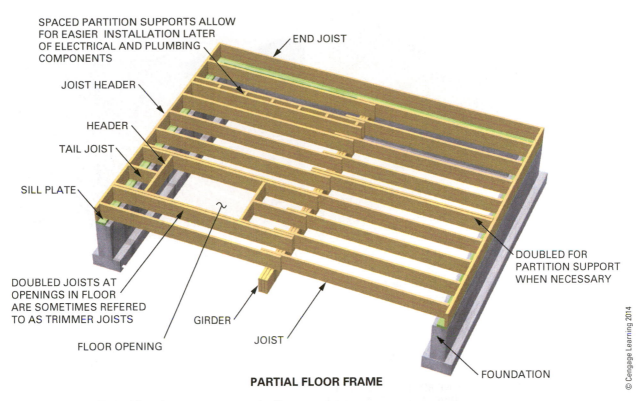

SPACED PARTITION SUPPORTS ALLOW FOR EASIER INSTALLATION LATER OF ELECTRICAL AND PLUMBING COMPONENTS

END JOIST

JOIST HEADER

HEADER

TAIL JOIST

SILL PLATE

DOUBLED JOISTS AT OPENINGS IN FLOOR ARE SOMETIMES REFERED TO AS TRIMMER JOISTS

FLOOR OPENING

GIRDER

JOIST

FOUNDATION

DOUBLED FOR PARTITION SUPPORT WHEN NECESSARY

PARTIAL FLOOR FRAME

© Cengage Learning 2014

Figure 35–16 Typical framing components of a floor system.

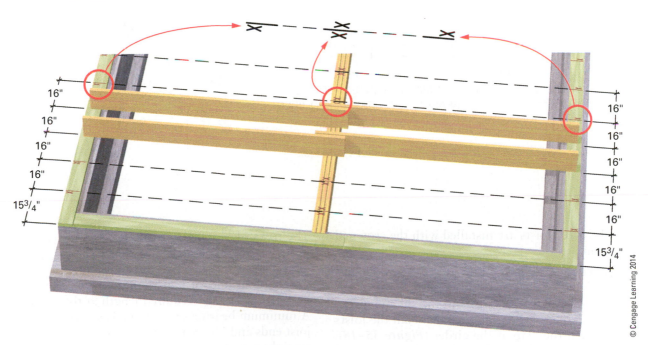

16"
16"
16"
16"
15³⁄₄"

16"
16"
16"
16"
16"
15³⁄₄"

© Cengage Learning 2014

Figure 35–17 Floor joist layout lines span the entire width of the building. If the joists are lapped at the girder, then the Xs are marked on different sides of the line.

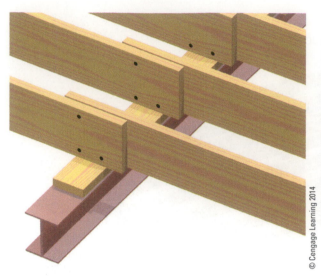

© Cengage Learning 2014

Figure 35–18 Lapped joists should be nailed together and toe-nailed to the sill plate.

Courtesy of Boise Cascade

Figure 35–19 Typical wood I-joist installation.

I-JOIST NAILING REQUIREMENTS.

SQUASH BLOCKS ARE REQUIRED IF A LOAD BEARING WALL IS TO BE INSTALLED ABOVE.

$1\frac{1}{4}$" TIMBERSTRAND® LSL OR $1\frac{1}{8}$" TJ® RIM BOARD

ONE 8D NAIL EACH SIDE. DRIVE NAILS AT AN ANGLE AT LEAST $1\frac{1}{2}$" FROM END.

ONE 10D NAIL INTO EACH FLANGE

$1\frac{3}{4}$" MINIMUM BEARING AT END SUPPORT

$3\frac{1}{2}$" MINIMUM INTERMEDIATE BEARING; $5\frac{1}{4}$" MAY BE REQUIRED FOR MAXIMUM CAPACITY

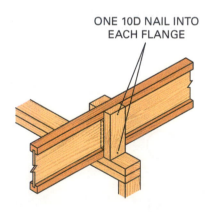

Courtesy of Boise Cascade

Fig 35-20 Nailing and fastening specifications for wood I-joist.

not be used. Joists are installed with the crowned edge up.

Keep the end of the floor joist in from the outside edge of the sill plate by the thickness of the band joist. Toenail the joists to the sill and girder with 10d or 3-inch common nails. Nail the joists together if they lap at the girder (*Figure 35–18*). When all floor joists are in position, they are sighted by eye from end to end and straightened. I-joists are unstable and will tip over until the sheathing is fastened. To make the work place safe a 1 × 4 is nailed to the top edges of each joist (*Figure 35–19*).

The rows of bracing must be less than 6 feet for the 1 3/4" flanged joist and 8 feet for the larger ones. Nails should be two 8ds every joist. Wood I-joists are installed using standard tools. They can be easily cut to any required length at the job site. A minimum bearing of 1 1/2 inches is required at joist ends and $3\frac{1}{2}$ inches over the girder. The wide, straight wood flanges on the joist make nailing easier, especially with pneumatic framing nailers (*Figure 35–20*). Nail joists at each bearing with one 8d or 10d nail on each side. Keep nails at least $1\frac{1}{2}$ inches from the ends to avoid splitting. ✱

✱IRC TABLE R602.3(1)

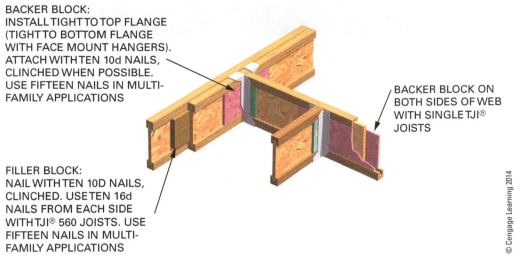

BACKER BLOCK:
INSTALL TIGHT TO TOP FLANGE
(TIGHT TO BOTTOM FLANGE
WITH FACE MOUNT HANGERS).
ATTACH WITH TEN 10d NAILS,
CLINCHED WHEN POSSIBLE.
USE FIFTEEN NAILS IN MULTI-
FAMILY APPLICATIONS

BACKER BLOCK ON
BOTH SIDES OF WEB
WITH SINGLE TJI®
JOISTS

FILLER BLOCK:
NAIL WITH TEN 10D NAILS,
CLINCHED. USE TEN 16d
NAILS FROM EACH SIDE
WITH TJI® 560 JOISTS. USE
FIFTEEN NAILS IN MULTI-
FAMILY APPLICATIONS

© Cengage Learning 2014

Fig 35-21 Doubled wood I-joist require blocking between the webs.

DOUBLING FLOOR JOISTS

For added strength, doubled floor joists must be securely fastened together. Their top edges must be even. In most cases, the top edges do not lie flush with each other. They must be brought even before they can be nailed together.

To bring them flush, toenail down through the top edge of the higher one, at about the center of their length. At the same time squeeze both together tightly by hand. Use as many toenails as necessary, spaced where needed, to bring the top edges flush (*Procedure 35–C*). Usually no more than two or three nails are needed. Then, fasten the two pieces securely together. Drive nails from both sides, staggered from top to bottom, about 2 feet apart. Angle nails slightly so they do not protrude.

Doubling up I-joists require the use of blocking (*Figure 34–21*). The thickness space between the webs of doubled I-joist must have filler blocks installed. Narrow flanged I-joists may be single 2 × 6 or 2 × 8. The wider flanged I-joists need an additional layer of plywood or two layers of 2× blocker blocks to fill this space. The space behind joist hangers must have backer blocks. These range from plywood pieces to 2× blocks. If the supporting I-joist is a single member, the blocking must be doubled, one for each side. Nailing requirements are found in the manufacturer's literature.

Framing Floor Openings

Large openings in floors should be framed before floor joists are installed. This is because room is needed for end nailing. To frame an opening in a floor, first fasten the trimmer joists in place. Trimmer joists are full-length joists that run along the inside of the opening. Mark the location of the

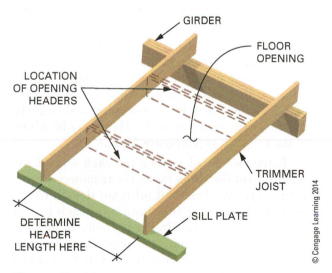

GIRDER

FLOOR
OPENING

LOCATION
OF OPENING
HEADERS

TRIMMER
JOIST

DETERMINE
HEADER
LENGTH HERE

SILL PLATE

© Cengage Learning 2014

Figure 35–22 Length of the header for a floor opening should be measured at the sill, not at the midspan.

StepbyStep Procedures

Procedure 35-C Aligning the top Edges of Joists

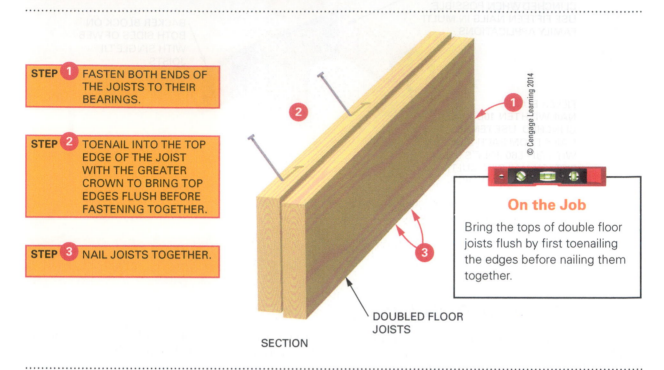

STEP 1 FASTEN BOTH ENDS OF THE JOISTS TO THEIR BEARINGS.

STEP 2 TOENAIL INTO THE TOP EDGE OF THE JOIST WITH THE GREATER CROWN TO BRING TOP EDGES FLUSH BEFORE FASTENING TOGETHER.

STEP 3 NAIL JOISTS TOGETHER.

© Cengage Learning 2014

On the Job

Bring the tops of double floor joists flush by first toenailing the edges before nailing them together.

DOUBLED FLOOR JOISTS

SECTION

---

headers on the trimmers. Headers are members of the opening that run at right angles to the floor joists. They should be doubled if they are more than 4 feet long.

Cut *headers* to length by taking the measurement at the sill between the trimmers. Taking the measurement at the sill where the trimmers are fastened, rather than at the opening, is standard practice. A measurement between trimmers taken at the opening may not be accurate. There may be a bow in the trimmer joists (*Figure 35–22*).

Place two headers, one for each end of the opening, on the sill between the trimmers. Transfer the layout of the tail joists on the sill to the headers. Fasten the first header on each end of the opening in position by driving nails through the side of the trimmer into the ends of the headers. Be sure the first header is the header that is farthest from the floor opening. Fasten the tail joists in position. Double up the headers. Finally, double up the trimmer joists. *Procedure 35–D* shows the sequence of operations used to frame a floor opening. This particular sequence allows you to end nail

the members rather than toenailing them. Use joist hangers as required by local codes. Joist hangers must be installed with the proper amount, size, and type of nails. Roofing nails are not acceptable for joist hangers.

Installing the Band Joist. After all the openings have been framed and all floor joists are fastened, install the band joist. This closes in the ends of the floor joists. Band joists may be lumber of the same size as the floor joists. They also may be a single or double layer of laminated strand lumber when wood I-joists are used.✱

Fasten the band joist into the end of each floor joist. If wood I-joists are used as floor joists, drive one nail into the top and bottom flange. The band joist is also toenailed to the sill plate at about 6-inch intervals.

✱**IRC R502.7**

BRIDGING

Bridging is added to a floor system make it stiffer and stronger. It transfers floor load from one joist to its neighbor. Properly installed bridging may

StepbyStep Procedures

Procedure 35-D Installing Framing Members Around a Floor Opening

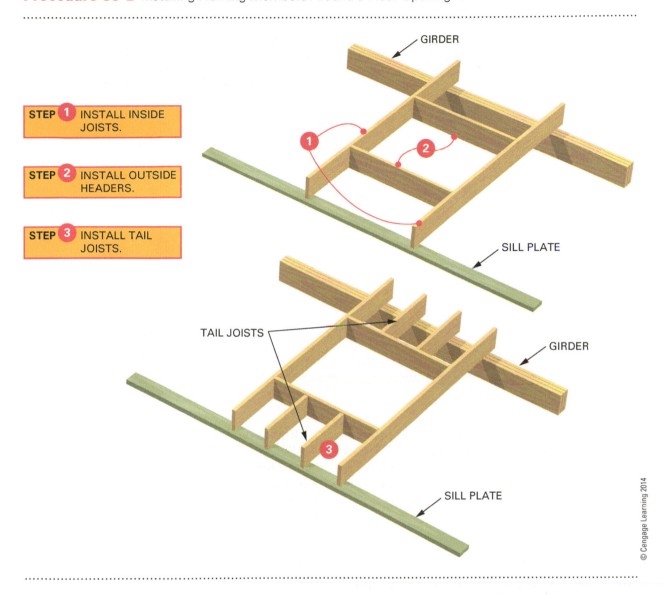

STEP 1 INSTALL INSIDE JOISTS.

STEP 2 INSTALL OUTSIDE HEADERS.

STEP 3 INSTALL TAIL JOISTS.

GIRDER

SILL PLATE

TAIL JOISTS

GIRDER

SILL PLATE

© Cengage Learning 2014

help prevent squeaky floors, improperly installed bridging will cause squeaks in floors. 2012 IR Code requires bridging only on joists larger than 2 × 12. Wood I-joist industry states that bridging is not necessary in most residential floor systems if the subfloor sheathing is adhered to the joists.

*IRC R502.7.1

Where bridging is used, it is installed in rows between floor joists. Row intervals not to exceed 8 feet. For instance, floor joists with spans 8 to 16 feet need one row of bridging near the center of the span.

Bridging may be solid wood, wood cross-bridging, metal cross-bridging, or 1 × 3 nailed to the bottom of joists (*Figure 35–23*). Usually solid wood bridging is the same size as the floor joists. It is installed in an offset fashion to permit end nailing.

Wood cross-bridging should be at least nominal 1 × 3 lumber with two 6d nails at each end. It is placed in double rows that cross each other in the joist space.

StepbyStep Procedures

Procedure 35-D *Continued*

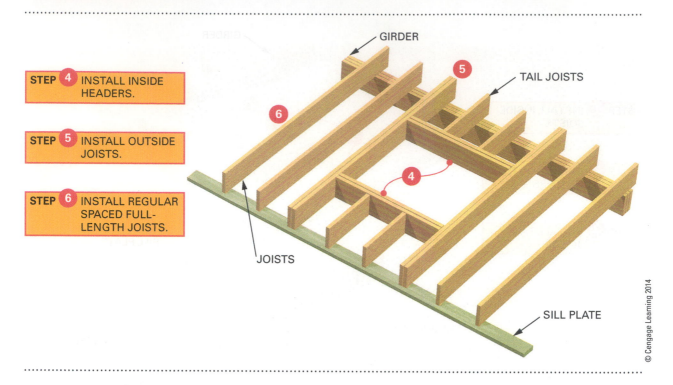

STEP 4 INSTALL INSIDE HEADERS.

STEP 5 INSTALL OUTSIDE JOISTS.

STEP 6 INSTALL REGULAR SPACED FULL-LENGTH JOISTS.

GIRDER

TAIL JOISTS

JOISTS

SILL PLATE

© Cengage Learning 2014

Metal cross-bridging is available in different lengths for particular joist size and spacing. It is usually made of 18-gauge steel, and is $\frac{3}{4}$ inch wide. It comes in a variety of styles. It is applied in a way similar to that used for wood cross-bridging.

Laying Out and Cutting Wood Cross-Bridging

Wood cross-bridging may be laid out using a framing square. Determine the actual distance between floor joists and the actual depth of the joist. For example, 2×10 floor joists 24 inches OC measure $22\frac{1}{2}$ inches between them. The actual depth of the joist is $9\frac{1}{4}$ inches.

Hold the framing square on the edge of a piece of bridging stock. Make sure the $9\frac{1}{4}$-inch mark of the tongue lines up with the upper edge of the stock. Also make sure the $22\frac{1}{2}$-inch mark of the blade lines up with the lower edge of the stock. Mark lines along the tongue and blade across the stock.

Rotate the square, keeping the same face up. Realign the square to the previous marks, then mark along the tongue (*Procedure 35–E*). The bridging may then be cut using a power miter box. Tilt the blade and use a stop set to cut duplicate lengths.

Installing Bridging

Determine the centerline of the bridging. Snap a chalk line across the tops of the floor joist from one end to the other. Square down from the chalk line to the bottom edge of the floor joists on both sides.

Solid Wood Bridging. To install solid wood bridging, cut the pieces to length. Install pieces in every other joist space on one side of the chalk line. Fasten the pieces by nailing through the joists into their ends. Keep the top edges flush with the floor joists. Install pieces in the remaining spaces on the opposite side of the line in a similar manner.

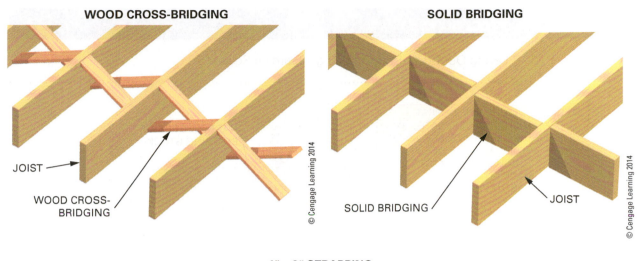

WOOD CROSS-BRIDGING

JOIST

WOOD CROSS-BRIDGING

© Cengage Learning 2014

SOLID BRIDGING

SOLID BRIDGING

JOIST

© Cengage Learning 2014

1" x 3" STRAPPING

1" x 3" STRAPPING

JOIST

© Cengage Learning 2014

Figure 35–23 Types of bridging.

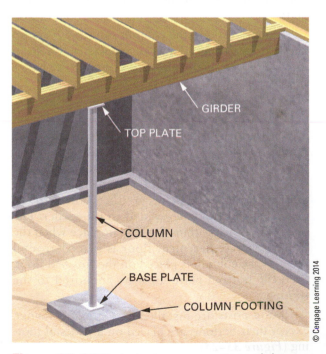

GIRDER

TOP PLATE

COLUMN

BASE PLATE

COLUMN FOOTING

© Cengage Learning 2014

Figure 35–24 Typical column supporting a girder.

Wood Cross-Bridging.   To install wood cross-bridging, start two 6d nails in one end of the bridging. Fasten it flush with the top of the joist on one side of the line. Nail only the top end. The bottom ends are not nailed until the subfloor is fastened.

Within the same joist cavity or space, fasten another piece of bridging to the other joist. Make sure it is flush with the top of the joist and positioned on the other side of the chalk line. Continue installing bridging in the other spaces, but alternate so that the top ends of the bridging pieces are opposite to each other where they are fastened to same joist. Nail the bottom of the bridging only after the subfloor has been installed and when the building has had a chance to dry from normal construction dampness. Be sure to leave a space between the bridging pieces where they cross forming the X. This will minimize floor squeaks.

StepbyStep Procedures

Procedure 35-E Laying Out Cross-Bridging Using a Framing Square

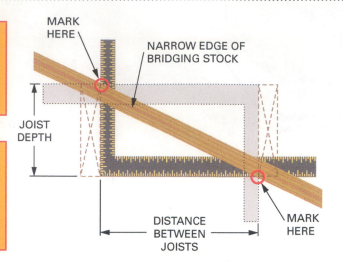

STEP 1 HOLD SQUARE IN FIRST POSITION AS INDICATED BY THE DARKER SQUARE. MARK STOCK ALONG BLADE AND TONGUE. THESE MARKS ARE MADE ON OPPOSITE EDGES OF BRIDGING EDGE.

STEP 2 ROTATE THE SQUARE TO THE POSITION AS INDICATED BY THE LIGHTER SQUARE. MARK ALONG THE TONGUE. THE ACTUAL BRIDGING LENGTH SHOULD BE ABOUT 1/4" SHORTER FOR EASE OF INSTALLATION.

MARK HERE

NARROW EDGE OF BRIDGING STOCK

JOIST DEPTH

DISTANCE BETWEEN JOISTS

MARK HERE

© Cengage Learning 2014

Metal Cross-Bridging. Metal cross-bridging is fastened in a manner similar to that used for wood cross-bridging. The method of fastening may differ according to the style of the bridging. Usually the bridging is fastened to the top of the joists through predrilled holes in the bridging. Because the metal is thin, nailing to the top of the joists does not interfere with the subfloor.

Some types of metal cross-bridging have steel prongs that are driven into the side of the floor joists. This bridging can be installed from below after the installation of the subfloor.

COLUMNS

Girders may be supported by framed walls, wood posts, or steel columns (*Figure 35–24*). Metal plates are used at the top and bottom of the columns to distribute the load over a wider area. The plates have predrilled holes so that they may be fastened to the girder. Notched sections prevent the columns from slipping off the plates. Column *IRC R407.3 size should be determined from the blueprints. *

Installing Columns

After the floor joists are installed and before any more weight is placed on the floor, the temporary posts supporting the girder are replaced with permanent posts or columns. Straighten the girder by stretching a line from end to end. Measure accurately from the column footing to the bottom of the girder. Hold a strip of lumber on the column footing. Mark it at the bottom of the girder. Transfer this mark to the column. Deduct the thickness of the top and bottom column plate.

To mark around the column so it has a square end, wrap a sheet of paper around it. Keeping the edges even, mark along the edge of the paper (*Procedure* 35-F).

Cut through the metal along the line using a hacksaw, reciprocating saw, or circular saw with a metal cutting blade. Install the columns in a plumb position under the girder and centered on the footing. Fasten the top plates to the girder with lag screws. If the girder is steel, then holes must be drilled. The plates are then bolted to the girder, or they may be welded to the girder. The bottoms of the columns are held in place when the finish concrete basement floor is placed around them. If the column is placed on the finished floor, then the bottom plate must be anchored to the footing. *

Wood posts are installed in a similar manner, *IRC R407.2* except their bottoms are placed on a pedestal footing (*Figure 35–25*). *

*IRC R407.1

StepbyStep Procedures

Procedure 35-F Marking a Square Line on a Column

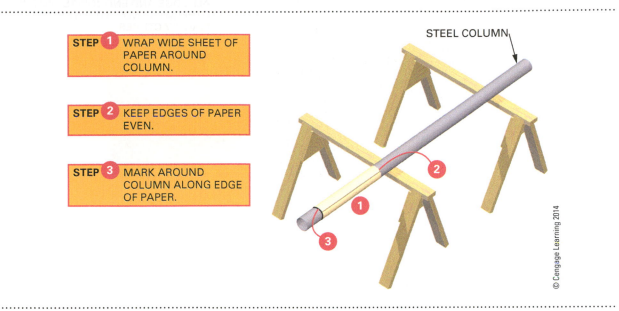

STEP 1 WRAP WIDE SHEET OF PAPER AROUND COLUMN.

STEP 2 KEEP EDGES OF PAPER EVEN.

STEP 3 MARK AROUND COLUMN ALONG EDGE OF PAPER.

STEEL COLUMN

© Cengage Learning 2014

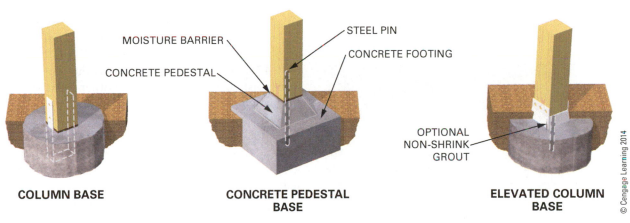

MOISTURE BARRIER

CONCRETE PEDESTAL

STEEL PIN

CONCRETE FOOTING

OPTIONAL NON-SHRINK GROUT

COLUMN BASE

CONCRETE PEDESTAL BASE

ELEVATED COLUMN BASE

© Cengage Learning 2014

Figure 35–25 The bottoms of wood posts sometimes rest on pedestal-type footings.

SUBFLOORING

Subflooring is used over joists to form a working platform. This is also a base for finish flooring, such as hardwood flooring, or underlayment for carpet or resilient tiles. Underlayment is a sheet material installed to level and smooth the floor before the floor finish. APA-rated Sheathing Exposure 1 is generally used for subflooring in a two-layer floor system. APA-rated Sturd-I-Floor panels are used when a single-layer subfloor and underlayment system is desired. Blocking is re-

quired under the joints of these panels unless tongue-and-groove edges are used. ***Figure 35–26** is a selection and fastening guide for APA Sturd-I-Floor panels. *****

*IRC
R503.2.1.1

*IRC
R502.1.2

APPLYING PLYWOOD SUBFLOORING

Starting at the corner from which the floor joists were laid out, measure in 4 feet. Note that tongue-and-groove (T&G) plywood subfloor is

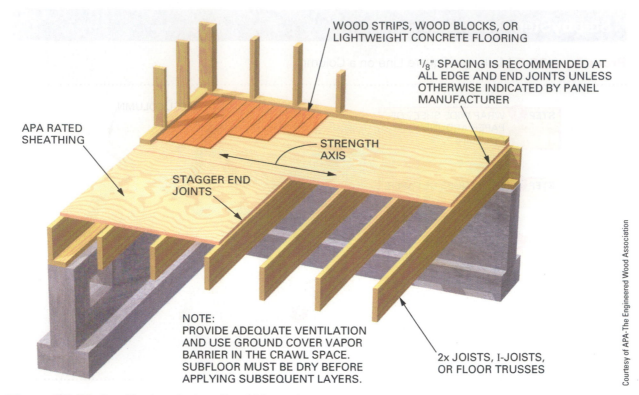

WOOD STRIPS, WOOD BLOCKS, OR
LIGHTWEIGHT CONCRETE FLOORING

$\frac{1}{8}$" SPACING IS RECOMMENDED AT
ALL EDGE AND END JOINTS UNLESS
OTHERWISE INDICATED BY PANEL
MANUFACTURER

APA RATED
SHEATHING

STRENGTH
AXIS

STAGGER END
JOINTS

NOTE:
PROVIDE ADEQUATE VENTILATION
AND USE GROUND COVER VAPOR
BARRIER IN THE CRAWL SPACE.
SUBFLOOR MUST BE DRY BEFORE
APPLYING SUBSEQUENT LAYERS.

2x JOISTS, I-JOISTS,
OR FLOOR TRUSSES

Courtesy of APA-The Engineered Wood Association

Figure 35–26 Specifications for installing APA panels.

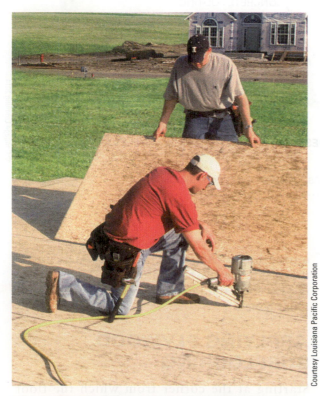

Courtesy Louisiana Pacific Corporation

Figure 35–27 Typical installation of APA-rated panel subfloor.

only $47\frac{1}{2}$" wide. Snap a line across the tops of the floor joists from one end to the other. Start with a full panel.

Position the groove of the sheet near the line with the tongue over the rim joist. This way the tongue and grooves of the remaining sheets that need to be encouraged into place with a hammer will not damage the tongues.

Fasten the first row to the chalk line (*Figure 35–27*) and align the joists to the correct spacing before nailing the panel. Leave a $\frac{1}{16}$-inch space at all panel end joints to allow for expansion.

Start the second row with a half-sheet to stagger the end joints. Continue with full panels to finish the row. Leave a $\frac{1}{8}$-inch space between panel edges. All end joints are made over joists.✳ ✳IRC R503.2.3

Continue laying and fastening plywood sheets in this manner until the entire floor is covered (*Procedure 35–G*). Leave out sheets where there are to be openings in the floor. Snap chalk lines across the edges and ends of the building. Trim overhanging plywood with a circular saw.

Procedure 35-G Layout Procedure for Installing Plywood Subfloor

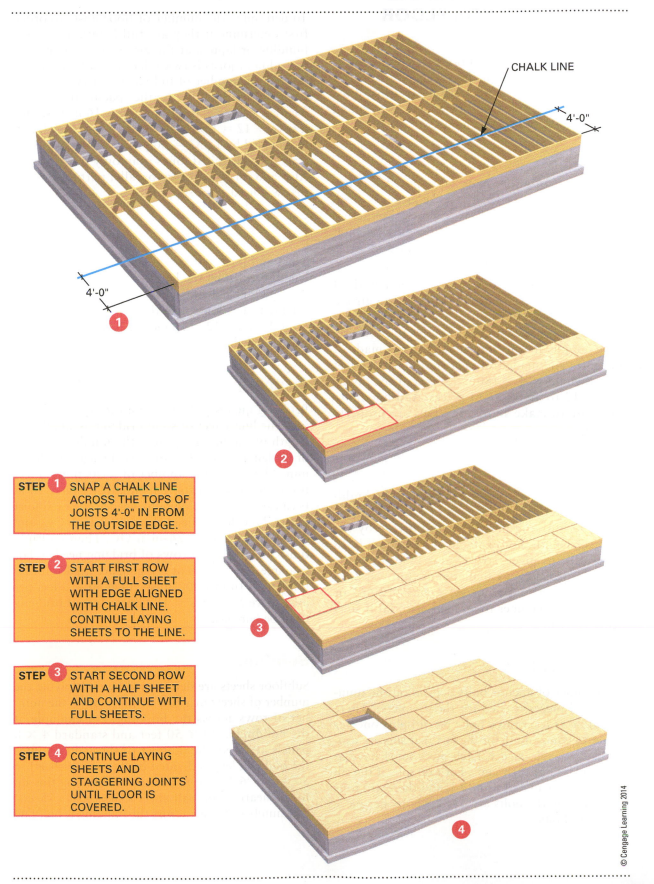

CHALK LINE

4'-0"

4'-0"

STEP 1 SNAP A CHALK LINE ACROSS THE TOPS OF JOISTS 4'-0" IN FROM THE OUTSIDE EDGE.

STEP 2 START FIRST ROW WITH A FULL SHEET WITH EDGE ALIGNED WITH CHALK LINE. CONTINUE LAYING SHEETS TO THE LINE.

STEP 3 START SECOND ROW WITH A HALF SHEET AND CONTINUE WITH FULL SHEETS.

STEP 4 CONTINUE LAYING SHEETS AND STAGGERING JOINTS UNTIL FLOOR IS COVERED.

ESTIMATING MATERIAL QUANTITIES FOR FLOOR FRAMING

Estimating the material in floor framing involves girder material, sills, anchor bolts, OC and rim joists, bridging, and subfloor. Remember that if the number of pieces does not come out even, round them up to the next whole number of pieces.

Girder

The amount of girder material required is determined by the girder length. If the girder is steel, then the quantity required is merely the actual length of the girder. If, however, the girder is built up, then a calculation must be performed to arrive at the amount of girder material required.

For a built-up girder, multiply the length of the girder by the number of plies in the girder. Then divide by the length of each piece of girder material to get the number of pieces to purchase. For example, if a 45-foot-long girder is made of four plies and made from material that is 12 feet long, then $45 \times 4 = 180$ linear feet. Then divide 180 by 12 to arrive at 15 pieces. Add one more for waste to make sure butt seams are properly aligned.

Sills

The number of sill pieces depends on the building perimeter and the length of each sill piece. Take the building perimeter and divide by the sill piece length. For example, if a building is 28 × 48 and sill pieces are 12 feet long, then the number of pieces is calculated as follows: $2(28 + 48) = 152$ feet (perimeter) $\div 12 = 12.667$ or 13 pieces. Remember to add sill material if steel girder is used.

Anchor Bolts

The number of anchor bolts depend on the number of sill pieces needed for the perimeter. There should be at least three bolts per sill plus one per corner to compensate for the fact that sills are usually cut at each corner. Each piece must have at least one bolt at their ends. For example, if the sill pieces are calculated to be 13, and there are four corners, then the bolts are 13×3 bolts per piece $+ 4 = 43$ bolts.

Floor Joists

To determine the number of floor joists to order, first determine if they are full length across the building or lapped at the girder. The quantity of lapped floor joists is twice that of full-length joists. To find the number of full-length joists, divide the length of the building by the spacing in terms of feet. For example, if the spacing is 16 inches, then it is $16 \div 12 = 1.333$ feet. So a 40-foot-long building divided by $1.333 = 30$ pieces. Add one joist to start. Double this if joists span only to the girder. Also add joists for every parallel partition above that will need extra support.

Band Joists

The band joist material is the twice the building length divided by the length of material used. For example, if the length is 60 feet and the material used is 12 feet long, then the band joist material is 60×2 sides $\div 12 = 10$ pieces. Add extra for headers as needed.

Bridging

Bridging quantity depends on the style of bridging. The linear feet of solid bridging is simply the length of the building times the number of rows of bridging. Linear feet of cross-bridging is determined by taking the number of joists times 3 feet for 16-inch OC joists. This number is arrived at because 3 feet of bridging is needed for each joist cavity. Four feet is needed for 19.2-inch OC, and 5 feet is needed for 24-inch OC. Then multiply times the number of rows of bridging needed. For example, if two rows of bridging are needed for 30 full-length 16-inch OC joists, then 30×3 feet per joist space $\times 2$ rows $= 180$ linear feet of bridging must be purchased.

Subfloor

Subfloor sheets are determined by multiplying the number of sheets in a row of subfloor by the number of rows across the building. For example, if the building is 30 × 50 feet and standard 4 × 8 panels are used, the number of sheets in a row is the building length divided by 8 (feet per sheet) $= 50 \div 8 = 6.25$ sheets per row. Rounding this up to the nearest one-half sheet makes it 6.5 sheets. The number of rows across the building is building width divided by 4 (feet per sheet) = 30 ÷ 4 = 7.5 rows. Leave this at the nearest one-half sheet. The number of pieces then is 6.5 × 7.5 = 48.75 or 49

sheets. See *Figure 35–28* for a listing of the short-hand versions of these estimating techniques with another example.

Estimate the materials for a floor system of a rectangular 30′ × 56′ building. A two-ply LVL full length girder, 16″ OC joists lapped at the girder, 12′ sill and band joists, 5 parallel partitions and one row of bridging each side of girder.

Item	Formula	Waste factor	Example
Girder Built-up	girder LEN × NUM of plys ÷ LEN of each ply = NUM of boards		30′ × 2 ÷ 30′ = 2 LVL boards
Sill	PERM ÷ LEN of each piece = NUM of boards	Add steel girder sill	172′ ÷ 12 = 14.3 = 15 boards
Anchor Bolts	NUM Sills × 3 + one per corner = NUM of bolts		15 × 3 + 4 = 49 bolts
Joists- Band joist	building LEN × 2 sides ÷ length of each board = NUM of boards		56′ × 2 ÷ 12 = 9.3 = 10 boards
Joists- OC (On Center) Joists	(building LEN ÷ OC spacing in ft + 1) × 2 if lapped + parallel partition supports = NUM joists		(56′ ÷ $^{16}/_{12}$ + 1) × 2 + 5 = 91 joists
Bridging- Solid	building LEN × NUM of rows ÷ LEN joists = NUM of boards		56′ × 2 ÷ 16 = 7 boards
Bridging- Wood Cross	building LEN ÷ OC spacing in ft × ft per OC spacing) × NUM rows = lineal ft bridging		56′ ÷ $^{16}/_{12}$ × 3 × 2 = 252 lineal ft bridging
Bridging-Metal	building LEN ÷ OC spacing in ft × 2 × NUM rows = NUM of pieces		56′ ÷ $^{16}/_{12}$ × 2 × 2 = 168 NUM of pieces
Subfloor- sheets	building LEN ÷ 8 (ft per sheet) = round up to nearest ½ sheet = NUM pieces per row Building WID ÷ 4 (ft per sheet) = round up to nearest ½ sheet = NUM of rows NUM of pieces per row × NUM of rows = NUM of pieces		56′ ÷ 8 = 7 pieces per row 30′ ÷ 4 = 7.5 rows 7 = 7.5 = 52.5 = 53 pieces

Courtesy Louisiana Pacific Corporation

Figure 35–28 Example of estimating floor framing materials with formulas.

DECONSTRUCT THIS

Carefully study **Figure 35–29** and think about what is wrong and/or what is right. Consider all possibilities. What construction practice or method is different in your area of the country?

Figure 35–29 This photo shows wood I-joist supported by a glulam girder.

© Cengage Learning 2014

KEY TERMS

balloon frame	edge	masonry	sheathing
bearing partitions	end	matched boards	shims
box header	fire-stop blocking	on center (OC)	sills
bridging	floor joists	plate	story
course	green	platform frame	tail joists
dead load	grouting	pocket	trimmer joists
deflection	joist header	rafters	underlayment
draft-stop blocking	linear feet	ribbon	web stiffener
eaves	live load	ridge	

REVIEW QUESTIONS

Select the most appropriate answer.

1. A platform frame is easy to erect because

    a. only one-story buildings are constructed with this type of frame.
    b. each platform may be constructed on the ground.
    c. at each level a flat surface is provided on which to work.
    d. fewer framing members are required.

2. A heavy beam that supports the inner ends of floor joists is often called a

    a. pier.
    b. girder.
    c. stud.
    d. sill.

3. The member of a floor frame that is fastened directly on the foundation wall is called a

    a. pier.
    b. girder.
    c. stud.
    d. sill.

4. Draftstop blocking is used to

    a. slow the spread of fire.
    b. reduce drafts and improve insulation.
    c. serve as nailing for finish material.
    d. all of the above.

5. To quickly and accurately mark a square end on a round column,

    a. use a square.
    b. measure down from the other end several times.
    c. use a pair of dividers.
    d. wrap a piece of paper around it.

6. When the ends of floor joists rest on a supporting member, they should have a bearing of at least

    a. 3 inches.
    b. $3\frac{1}{2}$ inches.
    c. $2\frac{1}{2}$ inches.
    d. $1\frac{1}{2}$ inches.

7. If floor joists lap over a girder, they should have a minimum lap of

    a. 2 inches.
    b. 3 inches.
    c. 6 inches.
    d. 12 inches.

8. It is important when installing floor joists to

    a. toenail them to the sill with at least two 8d nails.
    b. have the crowned edges up.
    c. face nail them to a band joist with at least three 8d nails.
    d. all of the above.

9. In stating the best order of installation for the members of a floor opening, the next member installed after the inside trimmer would be the

    a. outside trimmer.
    b. tail joists.
    c. inside header.
    d. outside header.

10. Holes bored in a floor joist should be no

    a. closer than 2 inches to the edge of the joist.
    b. larger than one-third of the joist width.
    c. larger than necessary.
    d. all of the above.

11. The bearing points of a girder on a foundation wall should be at least

    a. as long as the girder is deep (wide).
    b. 3 inches.
    c. 5 inches.
    d. 6 inches.

12. The maximum nail spacing for an engineered panel subfloor on 16 OC floor joists is

    a. 6 inches on the edge and 6 inches on intermediate supports.
    b. 6 inches on the edge and 8 inches on intermediate supports.
    c. 6 inches on the edge and 12 inches on intermediate supports.
    d. 8 inches on the edge and 8 inches on intermediate supports.

13. Numbers such as 16 and 19.2 in floor framing refer to

    a. joist spacing
    b. joist length
    c. blocking spacing
    d. blocking widths

14. To lay out the OC joists, the first one is set back a distance equal to

    a. one-half the joist thickness.
    b. the width of a joist.
    c. the thickness of a joist.
    d. 3/4 inch, always.

15. The dimension lumber species able to span the greatest distance is

    a. Douglas fir-larch
    b. Hem-fir
    c. Southern pine
    d. Spruce-pine-fir

16. The floor strength in a living area of residential construction is typically designed at

    a. 30 PSF L/240
    b. 30 PSF L/360
    c. 40 PSF L/240
    d. 40 PSF L/360

17. The max allowable deflection of a L/240 joist that spans 12'-6" is

    a. ¼ inch
    b. ⅝ inch
    c. ¾ inch
    d. ⅞ inch

18. The maximum span for a #2 Hem-Fir, 2×12 floor joist that is 19.2 OC with ceiling material attached to it is

    a. 14'–8"
    b. 16'–1"
    c. 17'–7"
    d. 18'–6"

19. The term used to refer to the weight of people a floor system will carry is

    a. dead load
    b. live load
    c. deflection
    d. span limit

20. The number of 19.2 OC wood I-joists needed that will span full width of a 32' × 56' building is

    a. 36
    b. 37
    c. 70
    d. 72

UNIT 13

Exterior Wall Framing

CHAPTER 36   Exterior Wall Frame Parts
CHAPTER 37   Framing the Exterior Wall

Wall-framing methods vary across the country and are affected by regional characteristics. These variations are easy to adjust for when the carpenter has an understanding of basic framing methods and practices. Exterior walls must be constructed to the correct height, corners braced plumb, walls straightened from corner to corner, and window and door openings framed to specified size. The techniques described in this unit will enable the apprentice carpenter to frame exterior walls with competence.

Setting the exterior wall frame is the first step in defining the outline of the house. Stack material close enough to the work area, yet not in the way of future work.

OBJECTIVES

After completing this unit, the student should be able to:

- identify and describe the function of each part of the wall frame.
- determine the length of exterior wall studs and the size of rough openings, and lay out a story pole.
- build corner posts and partition intersections, and describe several methods of forming them.
- lay out the wall plates.
- construct and erect wall sections to form the exterior wall frame.
- plumb, brace, and straighten the exterior wall frame.
- apply wall sheathing.

CHAPTER 36

Exterior Wall Frame Parts

The wall frame consists of a number of different parts. The student should know the name, function, location, and usual size of each member. Sometimes the names given to certain parts of a structure may differ according to the geographical area. For that reason, some members may be identified with more than one term.

PARTS OF AN EXTERIOR WALL FRAME

An exterior wall frame consists of *plates, studs, headers, sills, jack studs, trimmers, corner posts, partition intersections, ribbons,* and *braces* (*Figure 36–1*).

Plates

The top and bottom horizontal members of a wall frame are called *plates*. The bottom member is called a sole plate. It is also referred to as the *bottom plate* or *shoe*. The top members are called top plates. They usually consist of doubled 2-inch stock. In a balloon frame, the sole plate is not used. Instead, the studs rest directly on the sill plate.

Studs

Studs are vertical members of the wall frame. They run full-length between plates. *Jack studs* or *trimmers* are shortened studs that line the sides of an opening. They extend from the bottom plate up to the top of the opening. *Cripple* studs are shorter members above and below an opening, which extend from the top or the bottom plates to the opening.

Studs are usually 2 × 4s, but 2 × 6s are used when 6-inch insulation is desired in exterior walls. Studs are usually spaced 16 inches OC. **✴** ✴IRC R602.3.1

Headers

Headers or lintels run at right angles to studs. They form the top of window, door, and other wall openings, such as fireplaces. Headers must be strong

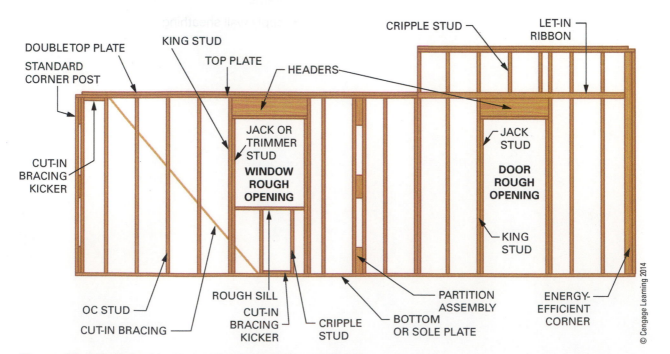

Figure 36–1 Typical component parts of an exterior wall frame.

DOUBLE TOP PLATE
STANDARD CORNER POST
KING STUD
TOP PLATE
HEADERS
CRIPPLE STUD
LET-IN RIBBON
JACK OR TRIMMER STUD
WINDOW ROUGH OPENING
JACK STUD
DOOR ROUGH OPENING
CUT-IN BRACING KICKER
KING STUD
OC STUD
CUT-IN BRACING
ROUGH SILL
CUT-IN BRACING KICKER
CRIPPLE STUD
PARTITION ASSEMBLY
BOTTOM OR SOLE PLATE
ENERGY-EFFICIENT CORNER

© Cengage Learning 2014

enough to support the load above the opening. The depth of the header depends on the width of the opening. As the width of the opening increases, so must the strength of the header. Check drawings, specifications, codes, or manufacturers' literature *IRC for header size.*

IRC R602.7

Kinds of Headers. Solid or built-up lumber may be used for headers. For 4-inch walls, two pieces of 2-inch lumber with ½-inch plywood or strand board sandwiched in between them gives the header the full 3½-inch thickness of the wall. In 6-inch walls, three pieces of 2-inch lumber with two pieces of ½-inch plywood or strand board in between makes up the 5½-inch wall thickness (*Figure 36–2*).

Green Tip

Properly size window and door headers; do not overbuild them. This practice saves wood and increases the amount of insulation in the wall.

Much engineered lumber is now being used for window and door opening headers. Figures 5–6, 6–1, and 8–1 show the use of laminated veneer lumber, parallel strand lumber, and glulam beams as opening headers (*Figure 36–3*).

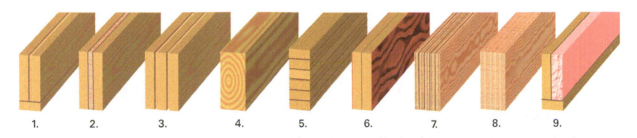

1. A BUILT-UP HEADER WITH A 2 X 4 OR 2 X 6 LAID FLAT ON THE BOTTOM.
2. A BUILT-UP HEADER WITH A ½" SPACER SANDWICHED IN BETWEEN.
3. A BUILT-UP HEADER FOR A 6" WALL.
4. A HEADER OF SOLID SAWN LUMBER.
5. GLULAM BEAMS ARE OFTEN USED FOR HEADERS.
6. A BUILT-UP HEADER OF LAMINATED VENEER LUMBER.
7. PARALLEL STRAND LUMBER MAKES EXCELLENT HEADERS.
8. LAMINATED STRAND LUMBER IS USED FOR LIGHT-DUTY HEADERS.
9. ENERGY-EFFICIENT HEADER WITH RIGID FOAM INSULATION.

© Cengage Learning 2014

Figure 36–2 Types and styles of headers.

GLULAM

© Cengage Learning 2014

PSL (PARALLEL STRAND LUMBER)

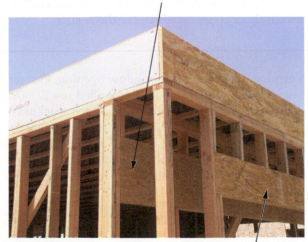

LSL (LAMINATED STRAND LUMBER)

© Cengage Learning 2014

Figure 36–3 Headers may be made of glulam, PSL, and LSL.

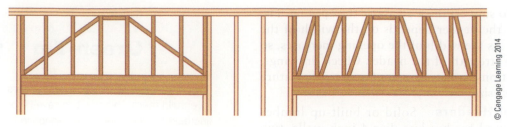

Figure 36–4 Two methods of trussing a large opening.

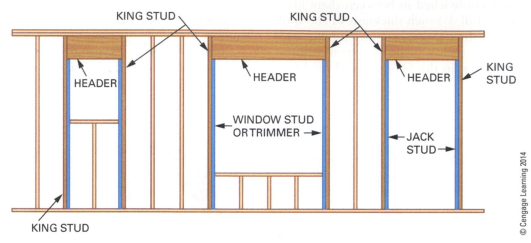

Figure 36–5 It is common practice to use the same header height for all wall openings.

In many buildings, when the opening must be supported without increasing the header size, the top of a wall opening may be trussed to provide support (*Figure 36–4*). However, when the opening is fairly close to the top plate, the depth of the header is increased. This completely fills the space between the plate and the top of the opening. In this case, the same size header is usually used for all wall openings, regardless of the width of the opening (*Figure 36–5*). This eliminates the need to install short cripple studs above the header.

Rough Sills

Forming the bottom of a window opening, at right angles to the studs, are members called rough sills. They usually consist of a single 2-inch thickness. They carry little load. However, many carpenters prefer to use a double 2-inch thickness rough sill. This provides more surface on which to fasten window trim in the later stages of construction.

Jack Studs and Trimmers

Jack studs and trimmers (also called *liners*) are shortened studs that support the headers. They are fastened to the king studs on each side of the opening. In window openings, the trimmer should be installed full-length from header to bottom plate (*Figure 36–6*). They should not be cut to allow

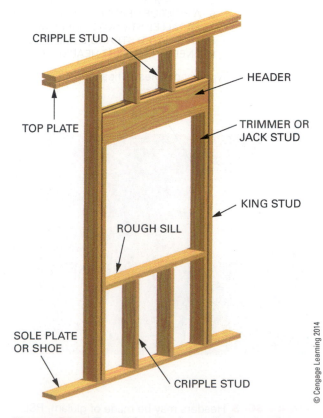

Figure 36–6 Typical framing for a window opening.

the sill to fit to the king stud. Door jack studs are installed the same as window trimmers (*Figure 36–7*).

Corner Posts

Corner posts are the same length as studs. They are constructed in a manner that provides an outside and an inside corner on which to fasten the exterior and interior wall coverings. They may be constructed in several ways.

Corner posts may be full-size solid lumber such as a 4 × 6 coupled with the 2 × dimension (aka 2×) lumber from the intersecting wall (*Figure 36–8*). This may also be parallel strand lumber or three 2× lumber nailed together. A second method uses three blocks in the space between two full pieces of 2× lumber. This method saves material by replacing the middle 2× lumber of the previous method with scraps. The third method is designed

to increase the amount of insulation in the corner. One 2× lumber is rotated. This method uses the same amount of full-length material as the second method, yet allows for a warmer corner. *
*IRC R602.3(2)

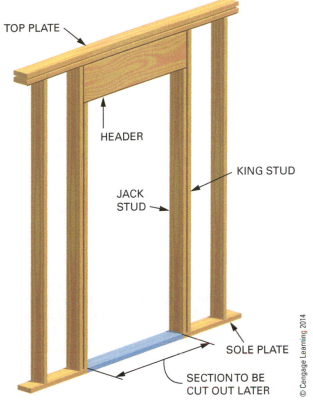

Figure 36–7 Typical framing for a door opening.

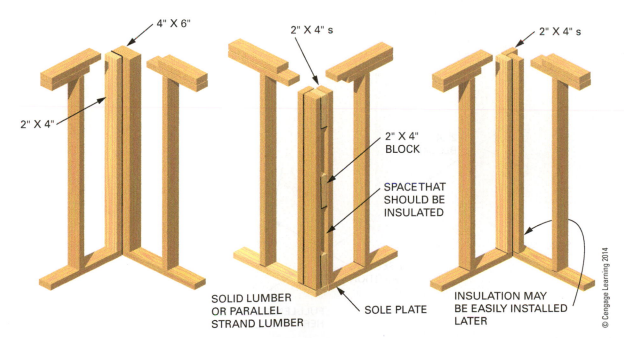

Figure 36–8 Methods of fabricating corner posts.

Partition Intersections

Wherever interior partitions meet an exterior wall, extra studs need to be put in the exterior wall. This provides wood for fastening the interior wall covering in the corner. In most cases, the *partition intersection* is made of two studs nailed to the edge of 2 × 4 blocks about a foot long. One block is placed at the bottom, one at the top, and one about center on the studs (*Figure 36–9*).

Another method is to maintain the regular spacing of the studs. Blocking is then installed between them wherever partitions occur. The block is set back from the inside edge of the stud the thickness of a board. A 1 × 6 board is then fastened vertically on the inside of the wall so that it is centered on the partition.

Another method is to nail a continuous 2 × 6 backer to a full-length stud. The edges of the backer project an equal distance beyond the edges of the stud.

Ribbons

Ribbons are horizontal members of the exterior wall frame in balloon construction. They are used to support the second-floor joists. The inside edge of the wall studs is notched so that the ribbon lays flush with the edge (*Figure 36–10*). Ribbons are usually made of 1 × 4 stock. Notches in the stud

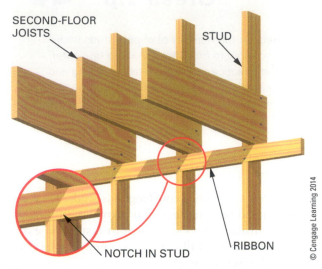

Figure 36–10 Ribbons are used to support floor joists in balloon frame walls.

© Cengage Learning 2014

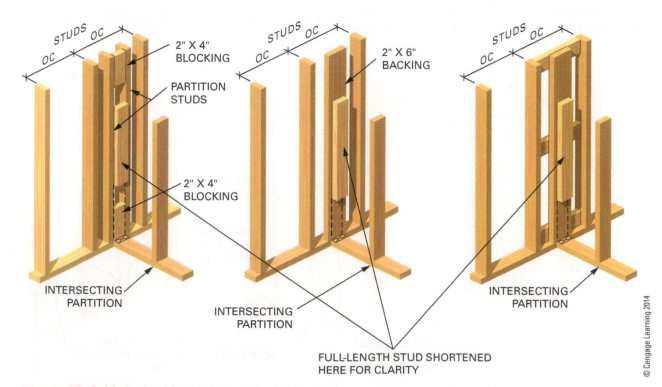

Figure 36–9 Methods of fabricating partition intersections.

© Cengage Learning 2014

should be made carefully so the ribbon fits snugly in the notch. This prevents the floor joists from settling. If the notch is cut too deep, the stud will be unnecessarily weakened. ✳

✳IRC R502.6

Corner Braces

Generally, no wall bracing is required if rated panel wall sheathing is used. In other cases, such as when insulating board sheathing is used, walls are braced with metal wall bracing. They come in gauges of 22 to 16 in flat, T-, or L-shapes. They are about 1½ inches wide and run diagonally from the top to the bottom plates. They are nailed to the stud edges before the sheathing is applied. The T- and L-shapes require a saw kerf in the stud to allow them to lay flat when installed. ✳

Another corner bracing technique is to use 1 × 4s called *let-in bracing* (*Figure 36–11*). A 1 × 4 is installed into notches cut out of the inside surface of the studs and plates. This allows the inside surface of all of the wall components to be continuously flush. The last method, *cut-in bracing,* uses a series of 2× lumber blocks cut between the studs. Kickers are nailed to the plates at the ends of the bracing system to spread out the racking load. Both of these methods require tight fits to achieve maximum stiffness.

✳IRC TABLE R602.10.4 B

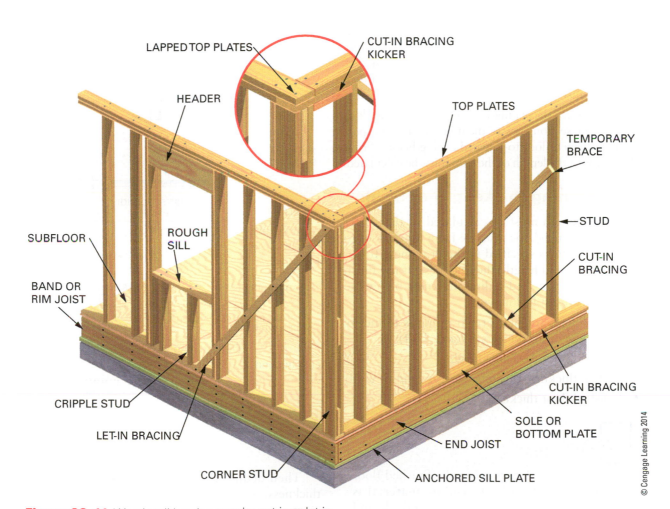

Figure 36–11 Wood wall bracing may be cut-in or let-in.

© Cengage Learning 2014

CHAPTER 37

Framing the Exterior Wall

Careful construction of the wall frame makes application of the exterior and interior finish easier. It also reduces problems for those who apply it later.

EXTERIOR WALL FRAMING

The standard height of a rough ceiling is usually 8'-1". Subtracting three plates of a total thickness of 4½ inches, the stud length is 92½ inches. Studs can be purchased precut to length, called precut studs, to save the carpenter time and wasted material. Note, however, that precut studs are usually 92⅝ inches long. The extra ⅛-inch length provides insurance that the finished material will be easily installed later.

Sometimes the ceiling height is not the standard height. The section view of the house plans will specify the finished-floor-to-finished-ceiling height. From this number, the length of the stud must be calculated.

Determining the Length of Studs

The stud length must be calculated so that, after the wall is framed, the distance from finish floor to ceiling will be as specified in the drawings. To determine the stud length, the thickness of the finish floor and the finished ceiling thickness below the ceiling joist must be known.

Stud length for platform framing is found by adding these measurements (*Figure 37–1*):

- Finished floor-to-ceiling height
- Ceiling thickness (includes *furring strips* if used)
- Finished floor thickness

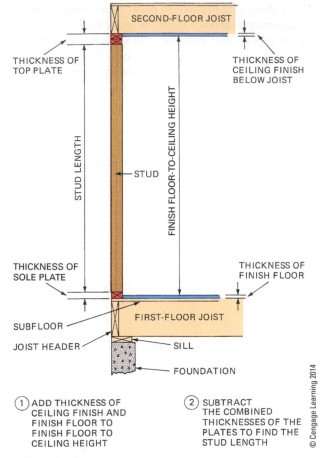

① ADD THICKNESS OF CEILING FINISH AND FINISH FLOOR TO FINISH FLOOR TO CEILING HEIGHT

② SUBTRACT THE COMBINED THICKNESSES OF THE PLATES TO FIND THE STUD LENGTH

© Cengage Learning 2014

Figure 37–1 Determining stud length in platform construction.

EXAMPLE

What is the stud length when the finished floor to ceiling height is 7'-9", ceiling material is ½-inch thick, finished floor is ¾-inch thick, and it is framed with three plates?

7'-9"	Finished height	7'-10¼"	Rough ceiling height
½"	Ceiling thickness	− 4½"	Total plate thickness
+ ¾"	Floor thickness	7'-5¾"	Stud length
7'-10¼"	Rough ceiling height		

Then deduct the top and bottom plate total thickness.

The stud length for balloon frame construction is found by adding these measurements (*Figure 37–2*):

- Finished floor-to-ceiling height of both stories
- Ceiling thickness of both stories (includes *furring strips* if used)
- Finished floor thickness of both stories
- Subfloor thickness of both stories
- Width of floor joists of both stories

Then deduct the top-plate total thickness.

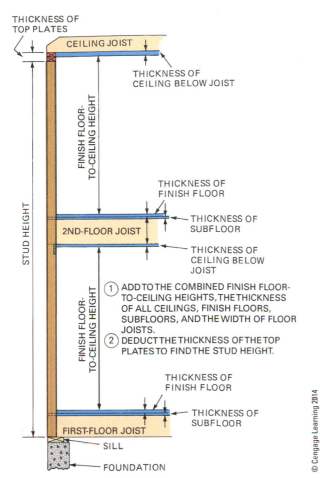

Figure 37–2 Determining stud length in balloon frame construction.

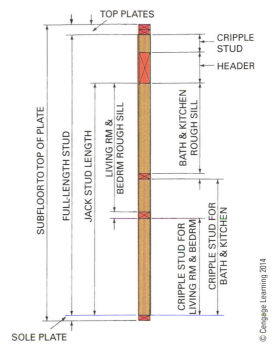

Figure 37–3 Layout of a typical story pole.

Figure 37–4 Door rough openings are made larger to accommodate the door and jamb. This is also true for windows.

Ribbon height to support second-floor joists is found by adding:

- Finished floor-to-ceiling height of first floor
- First-floor ceiling thickness (includes *furring strips* if used)
- Finished floor thickness of first floor
- Subfloor thickness of first floor
- Width of floor joists of first floor.

STORY POLE

The story pole is usually a straight strip of 1 × 2 or 1 × 3 lumber. Its length is the distance from the subfloor to the bottom of the joist above. Lines are squared across its width, indicating the height of the sole plates and top plates, headers, and rough sills for all openings.

The story pole is used to determine the length of studs, jack studs, trimmers, and cripple studs even if opening heights differ. The thickness and height of rough sills and the location of headers are also indicated and identified on the story pole (*Figure 37–3*).

A different story pole is made for each floor of the building. As framing progresses, the story pole is used to test the accurate location of framing members in the wall.

Determining Rough Opening Size

A rough opening is an opening framed in the wall in which to install doors and windows. The width and height of rough openings are made larger to accommodate the door or window unit (*Figure 37–4*). Rough opening dimensions are not indicated in the plans. It is the carpenter's responsibility to determine the rough opening size for the particular unit from the information given in the door and window schedule. The door and window schedule

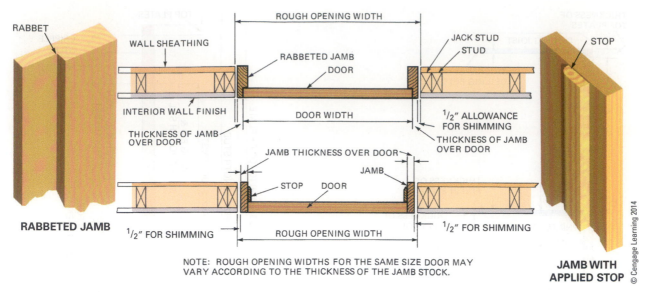

NOTE: ROUGH OPENING WIDTHS FOR THE SAME SIZE DOOR MAY VARY ACCORDING TO THE THICKNESS OF THE JAMB STOCK.

Figure 37–5 Determining the rough opening width of a door opening.

contains the kind, style, manufacturer's model number, size of each unit, and rough opening dimensions.

Rough Openings for Doors

The rough opening for a door must be large enough to accommodate the door, door frame, and space for shimming the frame to a level and plumb position. Usually ½ inch is allowed for shimming, at the top and both sides, between the door frame and the rough opening. The amount allowed for the door frame itself depends on the thickness of the door frame beyond the door.

Care must be taken not to make the rough opening oversize. If the opening is made too large, the window or door finish may not cover it.

The sides and top of a door frame are called jambs. Jambs may vary in thickness. Sometimes *rabbeted* wood jambs are used. The rabbet is that part of the jamb that the door stops up against. At other times nominal 1-inch lumber is used for the jamb. A separate stop is applied (*Figure 37–5*). Steel jambs have the door stop built-in similar to a rabbeted wood jamb.

The bottom member of the door frame is called a sill. Sills may be hardwood, metal, or a combination of wood and metal. The type of sill and its thickness must be known in order to figure the rough opening height.

Door Rough Opening Height. Rough opening height is determined by adding five dimensions (*Figure 37–6*). The rough opening heights for all openings in a house are usually the same, so only one rough opening height needs to be calculated.

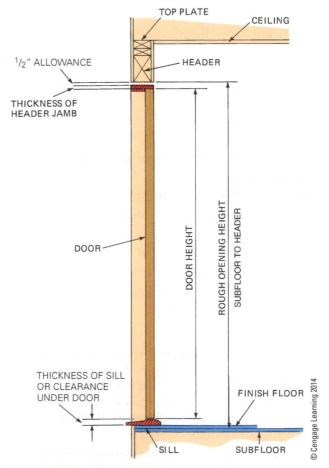

Figure 37–6 Determining the rough opening height of a door opening.

Because the wall rests on the subfloor, the subfloor is the starting point. To begin calculations add:

- Finished floor thickness
- Door sill (or threshold) thickness (if none, then add 1 inch for swing clearance under the door)
- Door height
- Head (top) jamb thickness
- Shim space (usually ½ inch)

EXAMPLE

What is the rough opening height for a 6'-8" (80") door with a ¾-inch finished floor, no threshold, and a ¾-inch jamb?

¾"	Finished floor
1"	Clearance under door
80"	Door height (this includes the small space between the door and jambs that allows opening of the door)
¾"	Jamb thickness
+ ½"	Shim space
83"	Total rough opening height

Jack stud length can be determined by subtracting bottom plate thickness, which is typically 1½ inches. In this case the jack stud length is 83" − 1½" = 81½".

Door Rough Opening Width. Refer back to *Figure 37–4* to see that the rough opening width is also found by adding five dimensions. Because door widths vary from room to room, a shorthand method is used. The rough opening width is found by adding 2½ inches to the door width. This number comes from adding ½ inch of shim space and ¾ inch of jamb thickness to both sides of the door; or ½" + ¾" = 1¼" times two = 2½". Note that if the jamb thickness is not ¾ inch, then the 2½" number must be adjusted accordingly. Thus, if the door is 2'-6" (30") wide, then the rough opening width for this door is 30" + 2½" = 32½".

Rough Openings for Windows

Many kinds of windows are manufactured by a number of firms. Because of the number of styles, sizes, and variety of construction methods, it is best to consult the manufacturer's catalog to obtain the rough opening sizes. These catalogs show the style and size of the window unit. They also give the rough opening (RO) for each unit (*Figure 37–7*). Catalogs are available from the lumber company that sells the windows.

Green Tip

Buying lumber in precut lengths reduces waste on the job and speeds the construction process.

WALL LAYOUT

Wall construction begins with careful layout of all wall components. This is usually done on the top and bottom plates. The layout of the walls usually begins at the same corner of the building where the floor joists began. This will help ensure that the load-bearing studs are directly over the joist below.✱ Before layout can begin, the carpenter must first determine if the wall to be laid out is load-bearing or non-load bearing. Note that exterior walls are referred to as *walls* and interior walls are referred to as *partitions*.

✱**IRC R601.2**

Determining Wall Type

The load-bearing walls (LBWs) are usually built first. They support the ceiling joist and rafters and typically run the length of the building. Non-load-bearing walls (NLBWs) are end walls and run parallel with the joists. Interior partitions are also load- and nonload bearing. They are load-bearing partitions (LBPs) if they run perpendicular to the joists and support the ends of the joists above. All other partitions are considered non-load-bearing partitions (NLBPs) (*Figure 37–8*).

Each type of wall has a slightly different layout characteristic. To reduce confusion, remember that all centerline dimensions for openings are measured from the building line, which is on the outside edge of the exterior framing. Layout must take this fact into account (*Figure 37–9*). *Figure 37–10* notes the similarities and differences of laying out walls and partitions.

Laying Out the Plates

To lay out the plates, measure in on the subfloor, at the corners, the thickness of the wall. Snap lines on the subfloor between the marks. This is done so the wall can be erected to a straight line later. *Tack the plates in position so their inside edges are to the chalk line. Do not drive nails home (they have to be pulled later). Use only as many as are needed to hold the pieces in place. Plan plate lengths so that joints between them fall in the center of a

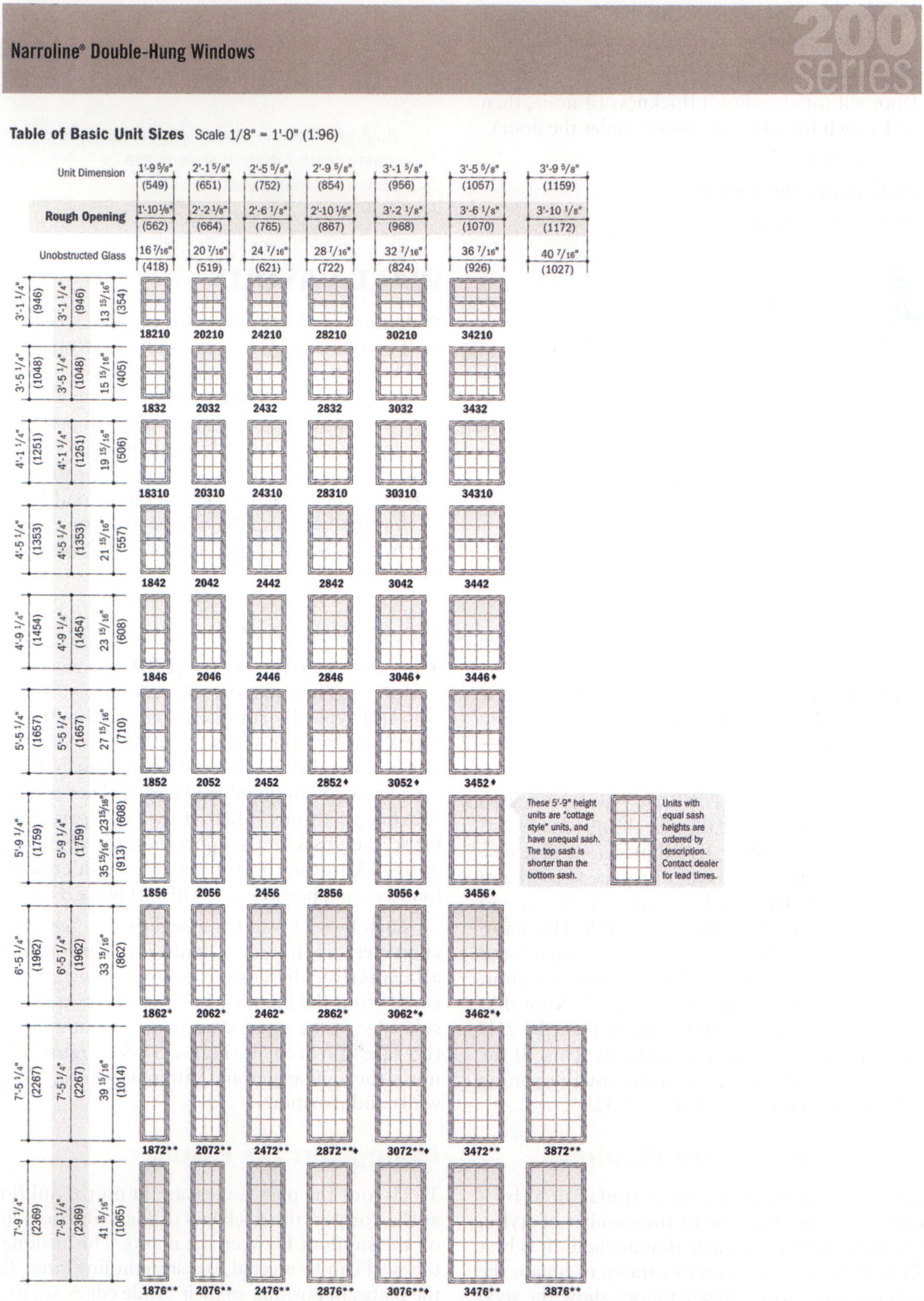

Narroline® Double-Hung Windows

200 series

Table of Basic Unit Sizes   Scale 1/8" = 1'-0" (1:96)

	1'-9 5/8" (549)	2'-1 5/8" (651)	2'-5 5/8" (752)	2'-9 5/8" (854)	3'-1 5/8" (956)	3'-5 5/8" (1057)	3'-9 5/8" (1159)
Unit Dimension	1'-9 5/8" (549)	2'-1 5/8" (651)	2'-5 5/8" (752)	2'-9 5/8" (854)	3'-1 5/8" (956)	3'-5 5/8" (1057)	3'-9 5/8" (1159)
Rough Opening	1'-10 1/8" (562)	2'-2 1/8" (664)	2'-6 1/8" (765)	2'-10 1/8" (867)	3'-2 1/8" (968)	3'-6 1/8" (1070)	3'-10 1/8" (1172)
Unobstructed Glass	16 7/16" (418)	20 7/16" (519)	24 7/16" (621)	28 7/16" (722)	32 7/16" (824)	36 7/16" (926)	40 7/16" (1027)

Height							
3'-1 1/4" (946) / 3'-1 1/4" (946) / 13 15/16" (354)	18210	20210	24210	28210	30210	34210	
3'-5 1/4" (1048) / 3'-5 1/4" (1048) / 15 15/16" (405)	1832	2032	2432	2832	3032	3432	
4'-1 1/4" (1251) / 4'-1 1/4" (1251) / 19 15/16" (506)	18310	20310	24310	28310	30310	34310	
4'-5 1/4" (1353) / 4'-5 1/4" (1353) / 21 15/16" (557)	1842	2042	2442	2842	3042	3442	
4'-9 1/4" (1454) / 4'-9 1/4" (1454) / 23 15/16" (608)	1846	2046	2446	2846	3046♦	3446♦	
5'-5 1/4" (1657) / 5'-5 1/4" (1657) / 27 15/16" (710)	1852	2052	2452	2852♦	3052♦	3452♦	
5'-9 1/4" (1759) / 5'-9 1/4" (1759) / 23 15/16" (608) / 35 15/16" (913)	1856	2056	2456	2856	3056♦	3456♦	
6'-5 1/4" (1962) / 6'-5 1/4" (1962) / 33 15/16" (862)	1862*	2062*	2462*	2862*	3062*♦	3462*♦	
7'-5 1/4" (2267) / 7'-5 1/4" (2267) / 39 15/16" (1014)	1872**	2072**	2472**	2872**♦	3072**♦	3472**	3872**
7'-9 1/4" (2369) / 7'-9 1/4" (2369) / 41 15/16" (1065)	1876**	2076**	2476**	2876**	3076**♦	3476**	3876**

These 5'-9" height units are "cottage style" units, and have unequal sash. The top sash is shorter than the bottom sash.

Units with equal sash heights are ordered by description. Contact dealer for lead times.

Courtesy of Andersen Windows

Figure 37–7 Sample of a manufacturer's catalog showing rough opening sizes for window units.

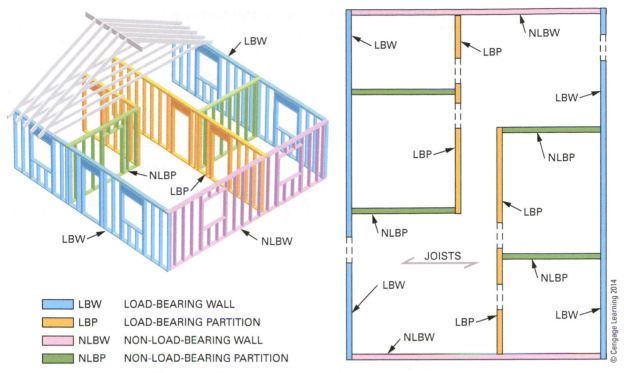

	LBW	LOAD-BEARING WALL
	LBP	LOAD-BEARING PARTITION
	NLBW	NON-LOAD-BEARING WALL
	NLBP	NON-LOAD-BEARING PARTITION

Figure 37–8 Walls in a building have different functions and characteristics.

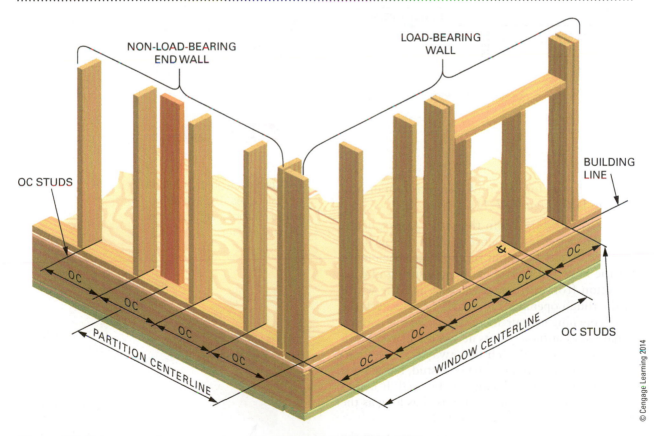

Figure 37–9 Layout wall components are measured from the building line.

LAYOUT VARIATIONS FOR WALLS AND PARTITIONS

	MEASURE TO OC STUDS	MEASURE TO CENTERLINES OF OPENINGS
LOAD-BEARING WALL (LBW)	FROM END OF PLATE	FROM END OF PLATE
NON-LOAD-BEARING WALL (NLBW)	INCLUDE WIDTH OF ABUTTING WALL AND SHEATHING THICKNESS	INCLUDE WIDTH OF ABUTTING WALL
LOAD-BEARING PARTITION (LBP)	INCLUDE WIDTH OF ABUTTING WALL	INCLUDE WIDTH OF ABUTTING WALL
NON-LOAD-BEARING PARTITION (NLBP)	FROM END OF PLATE	INCLUDE WIDTH OF ABUTTING WALL

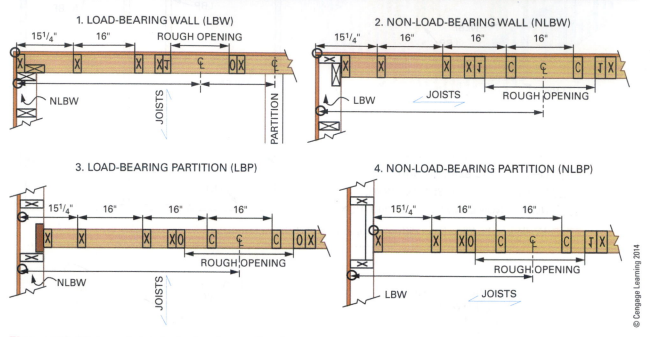

Figure 37–10 Layout details for four types of walls.

full-length stud for the convenient erection of wall sections later (*Figure 37–11*).

Wall Openings. From the set of prints, determine the centerline dimension of all the openings in the wall. Lay these out on the plate. Then mark for the king and jack studs by measuring in each direction from the centerline one-half the width of the rough opening (*Figure 37–12*). Recheck the rough opening measurement to be sure it is correct. Square lines at these points across the plates. Mark an *O* on the side of each line away from the centerline. The *O* represents the jack stud, but a *T* for trimmer or a *J* for jack can also be used. It makes little difference what marks are used as long as the builder understands what they mean.

From the squared lines, measure away from the centerline for the jack stud thickness. Square lines across. Mark *X*s on the side of the line away from center for the king studs on each side of the openings (see *Figure 37–12*).

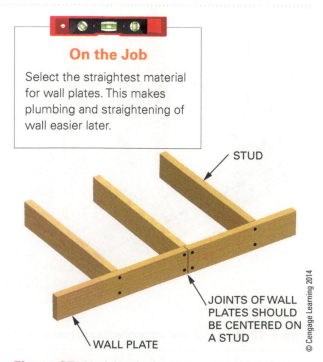

On the Job

Select the straightest material for wall plates. This makes plumbing and straightening of wall easier later.

Figure 37–11 Joints in the plates should fall at the center of a stud.

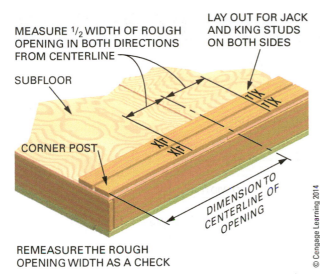

MEASURE ½ WIDTH OF ROUGH OPENING IN BOTH DIRECTIONS FROM CENTERLINE

LAY OUT FOR JACK AND KING STUDS ON BOTH SIDES

SUBFLOOR

CORNER POST

DIMENSION TO CENTERLINE OF OPENING

REMEASURE THE ROUGH OPENING WIDTH AS A CHECK

© Cengage Learning 2014

Figure 37–12 Laying out a rough opening width.

Partition Intersections

On architectural prints, interior partitions usually are dimensioned to their centerline. Mark on the plates the centerline of all partitions intersecting the wall (*Figure 37–13*). From the centerlines, measure in each direction one-half the partition stud thickness. Square lines across the plates. Mark Xs on the side of the lines away from center for the location of partition intersection studs.

Studs and Cripple Studs

After all openings and partitions have been laid out, start laying out all full-length studs and cripple studs. Proceed in the same manner and from the same end as laying out floor joists. This keeps studs directly in line with the joists below.

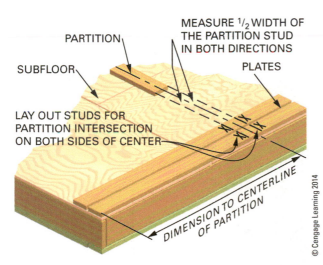

PARTITION

SUBFLOOR

MEASURE ½ WIDTH OF THE PARTITION STUD IN BOTH DIRECTIONS

PLATES

LAY OUT STUDS FOR PARTITION INTERSECTION ON BOTH SIDES OF CENTER

DIMENSION TO CENTERLINE OF PARTITION

© Cengage Learning 2014

Figure 37–13 Laying out a partition intersection.

Measure in from the outside corner the regular stud spacing. From this mark measure, in one direction or the other, one-half the stud thickness. Square a line across the plates. Place an X on the side of the line where the stud will be located.

Stretch a tape along the length of the plates from this first stud location. Square lines across the plates at each specified stud spacing. Place Xs on the same side of the line as the first line.

Where openings occur, mark the OC studs with a *C* instead of an *X*. This will indicate the location of cripple studs (*Procedure 37–A*). All regular and cripple studs should line up with the floor joists below. When laying out the opposite wall, start from the same end as the first wall to keep all framing lined up with joists below.

ASSEMBLING AND ERECTING WALL SECTIONS

The usual method of framing the exterior wall is to precut the wall frame members, assemble the wall frame on the subfloor, and erect the frame. With a small crew and without special equipment, the walls are raised section by section. When the frame is erected, the corners are plumbed and braced. Then the walls are straightened between corners. They are also braced securely in position. To prevent problems with the installation of the finish work later, it is important to keep the edges of the frame members flush wherever they join each other.

Precutting Wall Frame Members

Full-length Studs are often purchased precut to length for a standard 8 feet 1-inch wall height. Some builders buy in such high volume that lumber suppliers will also precut headers and jack and cripple studs. Most of the time framing members, other than studs, are cut to length on the job. A power miter saw is an effective tool for cutting studs and other framing to length. Set a stop the desired distance from the saw blade to cut duplicate lengths (*Figure 37–14*). Reject any studs that are severely warped. They may be cut into shorter lengths for blocking.

Corner Posts and Partition Intersections. Corner posts and partition intersections are often made up ahead of time to speed the assembly. Corners may be made by nailing two full-length studs together where one stud is rotated at a right angle. The detail in *Figure 37–15* allows for more insulation in the corner, thereby making the building more energy efficient. Corners may also be made by nailing

Step**by**Step Procedures

Procedure 37–A Laying out a typical load-bearing wall section

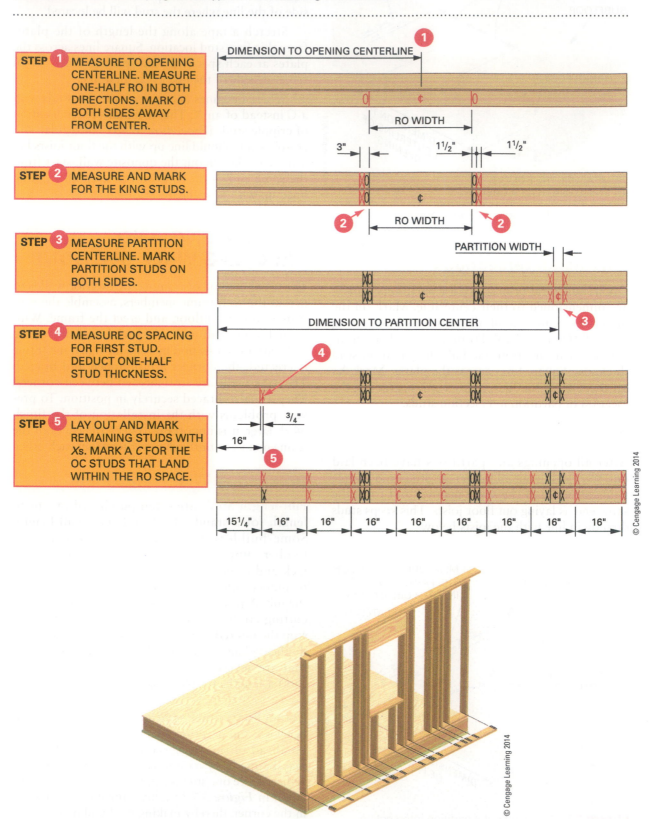

STEP 1 MEASURE TO OPENING CENTERLINE. MEASURE ONE-HALF RO IN BOTH DIRECTIONS. MARK *O* BOTH SIDES AWAY FROM CENTER.

STEP 2 MEASURE AND MARK FOR THE KING STUDS.

STEP 3 MEASURE PARTITION CENTERLINE. MARK PARTITION STUDS ON BOTH SIDES.

STEP 4 MEASURE OC SPACING FOR FIRST STUD. DEDUCT ONE-HALF STUD THICKNESS.

STEP 5 LAY OUT AND MARK REMAINING STUDS WITH *X*s. MARK A *C* FOR THE OC STUDS THAT LAND WITHIN THE RO SPACE.

© Cengage Learning 2014

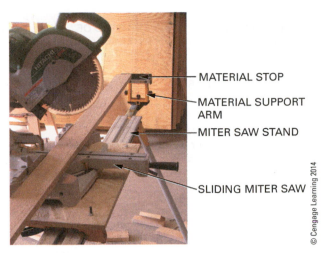

Figure 37–14 Techniques for making a cutoff jig for a circular saw.

together two full-length studs with short blocks of the same material laid flat between them.

Partition intersections may be made using ladder blocking, which again allows for more insulation in exterior walls. In this case, no extra framing layout is needed because the ladder blocking is installed between on-center studs. Another method also allows for insulation to be easily installed later. A 2 × 6 is nailed at right angles to a stud. A third method is similar to that used for corner posts except for the way the blocks are placed.

They are placed in a similar location between two full-length studs, yet with blocks on their edge (*Figure 37–16*).

Headers, Rough Sills, Jack Studs, and Trimmers. Cut all headers and rough sills. Their length can be determined from the layout on the plates.

Make a story pole. From it determine the length of all trimmers, jack studs, and cripple studs. Cut them accordingly. It may be necessary to place identifying marks on headers, rough sills, jacks, and trimmers if rough openings are different sizes. This will assist in locating the window or door unit to be placed in each rough opening.

Assembling Wall Sections

A variety of wall assembly procedures are used to build a wall frame. The following technique will quickly create a strong wall (*Procedure 37–B*). Separate the top plate from the bottom plate. Stand them on edge. To avoid a mistake, be careful not to turn any of the plates around. Be certain that the layout lines on top and bottom plates line up. Place all full-length studs, corner posts, and partition intersections in between them.

Very few studs are absolutely straight from end to end, so each stud crown will be faced the same way. Sight each full-length stud. It will be difficult for those who apply the interior finish if no attention is paid to the manner in which studs are

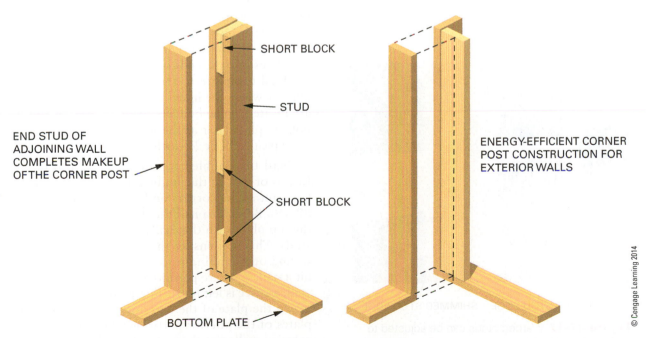

Figure 37–15 Construction of corner posts.

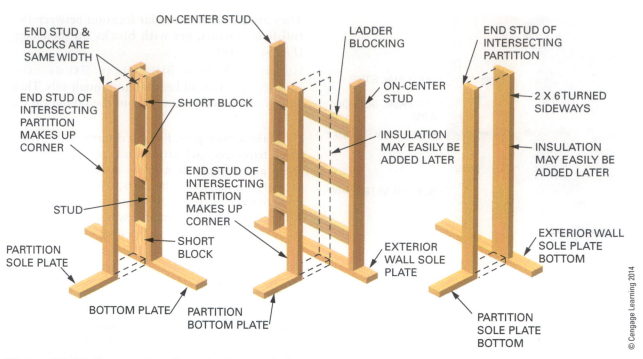

ON-CENTER STUD

END STUD & BLOCKS ARE SAME WIDTH

END STUD OF INTERSECTING PARTITION MAKES UP CORNER

SHORT BLOCK

STUD

END STUD OF INTERSECTING PARTITION MAKES UP CORNER

PARTITION SOLE PLATE

SHORT BLOCK

BOTTOM PLATE

PARTITION BOTTOM PLATE

LADDER BLOCKING

ON-CENTER STUD

INSULATION MAY EASILY BE ADDED LATER

EXTERIOR WALL SOLE PLATE

END STUD OF INTERSECTING PARTITION

2 X 6 TURNED SIDEWAYS

INSULATION MAY EASILY BE ADDED LATER

EXTERIOR WALL SOLE PLATE BOTTOM

PARTITION SOLE PLATE BOTTOM

© Cengage Learning 2014

Figure 37–16 Construction of partition intersections.

installed in the wall. A stud that is installed with its crowned edge out next to one with its crowned edge in will certainly present problems later. For this reason, some builders come back after the wall is built to flatten and adjust the crowns of the wall studs. Studs bowing inward are shaved with a power plane, and studs bowing outward are shimmed with strips of heavy cardboard (*Figure 37–17*).

Assemble the window and door openings first to ensure easy nailing. Fasten the headers, jacks, rough sills, and cripples in position, then fasten the king studs. If headers meet the top plate, nail through the plate into the header. Also fasten the jacks to the king studs by driving the nails at an angle. This is a stronger nailing technique, and eliminates any protruding nails. A pneumatic framing nailer makes the work easier (*Figure 37–18*). Fasten each stud, corner post, and partition intersection in the proper position by driving two 16d nails through the plates into the ends of each member.＊Nail the doubled top plate to the other top plate. Leave notches where the intersecting walls and partitions are located. Some builders toenail the studs to the bottom plate later as the wall is erected. Toenails are typically 8d or 2½-inch nails.

＊**IRC R602.3(1)**

Nail the doubled top plate to the top plate. Recess or extend the doubled top plate to make a lap joint at the corners and intersections (*Figure 37–19*). Be sure to nail the doubled top plate into the top plate such that nails are located above the studs. This will ensure that any holes drilled for wiring or plumbing into the top plates will not hit a nail.＊Where partitions intersect the exterior wall, a space is left for the top plate of the partition to lap the plate of the exterior wall. Lapping the plates of the interior partitions with those of the exterior walls ties them together. This results in a more rigid structure.＊

＊**IRC R602.6.1**

＊**IRC R602.3.2**

PLANED STUD EDGE   SHIMMED STUD EDGE

© Cengage Learning 2014

Figure 37–17 Warped studs can be adjusted to make a flat, straight wall.

StepbyStep Procedures

Procedure 37–B Assembling a Wall Section

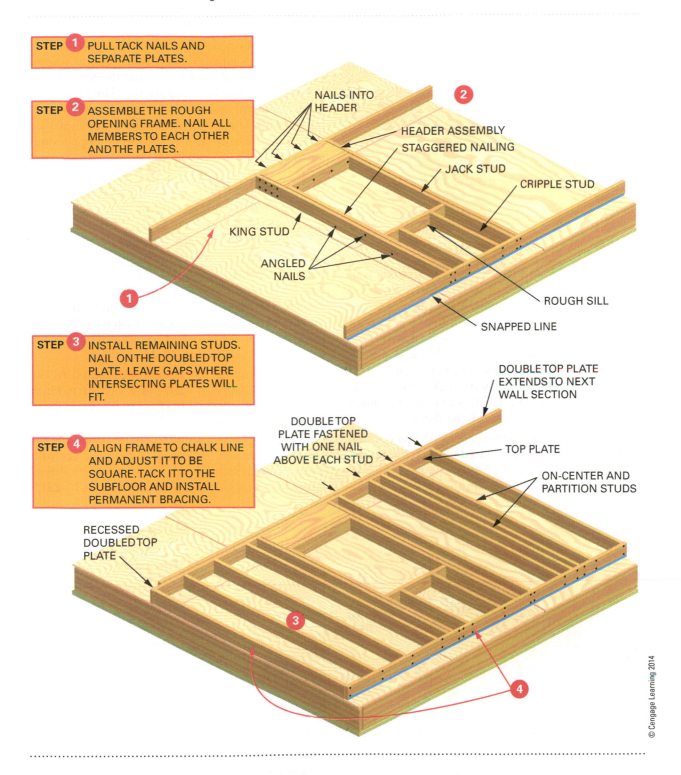

STEP 1 PULL TACK NAILS AND SEPARATE PLATES.

STEP 2 ASSEMBLE THE ROUGH OPENING FRAME. NAIL ALL MEMBERS TO EACH OTHER AND THE PLATES.

STEP 3 INSTALL REMAINING STUDS. NAIL ON THE DOUBLE TOP PLATE. LEAVE GAPS WHERE INTERSECTING PLATES WILL FIT.

STEP 4 ALIGN FRAME TO CHALK LINE AND ADJUST IT TO BE SQUARE. TACK IT TO THE SUBFLOOR AND INSTALL PERMANENT BRACING.

NAILS INTO HEADER

HEADER ASSEMBLY STAGGERED NAILING

JACK STUD

CRIPPLE STUD

KING STUD

ANGLED NAILS

ROUGH SILL

SNAPPED LINE

DOUBLE TOP PLATE EXTENDS TO NEXT WALL SECTION

DOUBLE TOP PLATE FASTENED WITH ONE NAIL ABOVE EACH STUD

TOP PLATE

ON-CENTER AND PARTITION STUDS

RECESSED DOUBLED TOP PLATE

© Cengage Learning 2014

Figure 37–18 Pneumatic nailers are industry standard tools for framing.

Bracing Walls

There are several methods of creating a strong wall section that will withstand racking loads. Plywood is the most popular material, but let-in and cut-in bracing are also sometimes used. In any case, wall bracing may be applied before the wall is erected. When plywood is used, some builders install it after the wall section is standing. Let-in and cut-in bracing must be installed with tight-fitting seams for maximum strength. For this reason, it is much **＊IRC R602.3.5** easier to do when the wall is lying down.**＊**

Before installing permanent wall bracing, the wall section should be squared while lying on the deck. To do this, align the bottom plate to the previously snapped chalk line on the subfloor where the inside edge of the wall plate will rest. Adjust the ends of the bottom plate into their proper position lengthwise (*Figure 37–20*). Toenail the sole plate to the subfloor with 16d nails spaced about every 6 to 8 feet along, what will be the top side of the bottom plate when the wall is in its final position.

Measure both the diagonals from corner to corner. If they are equal, the section is square. Toenail the top plate to the subfloor from the top side using one or two nails. These are used simply to keep the wall square and will be removed before standing the wall. The section is now ready for sheathing or bracing. Sheathing is applied in a manner similar to that used for a subfloor.

If let-in bracing is used, place the brace in position on top of the studs and plates at about a 45-degree angle. Make sure the top and bottom plates are covered by the brace. Mark along the sides of the brace at each stud and plate. Remove the brace. Using a circular saw with the blade set for the depth of the notch, make multiple saw cuts between the layout lines. With the claw of a straight-claw hammer, knock out and trim the remaining waste from the notch. Fasten the brace in the notches using two 10d common nails in each framing member (*Figure 37–21*).

Cut-in bracing has two blocks nailed to the plates called kickers at each end of the brace. These serve to transfer the racking load to the plates. First, snap a line at about a 45-degree angle on top of the stud edges. Use a speed square to determine the angle to cut the pieces. Fasten them into place with 16d nails, two at each end. The kicker nails should be angled so they do not protrude (*Figure 37–22*).**＊** **＊IRC TABLE R602.10.4**

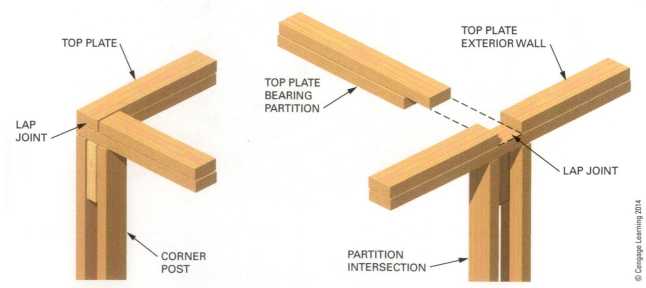

Figure 37–19 Doubled top plates are lapped at all intersections of wall sections.

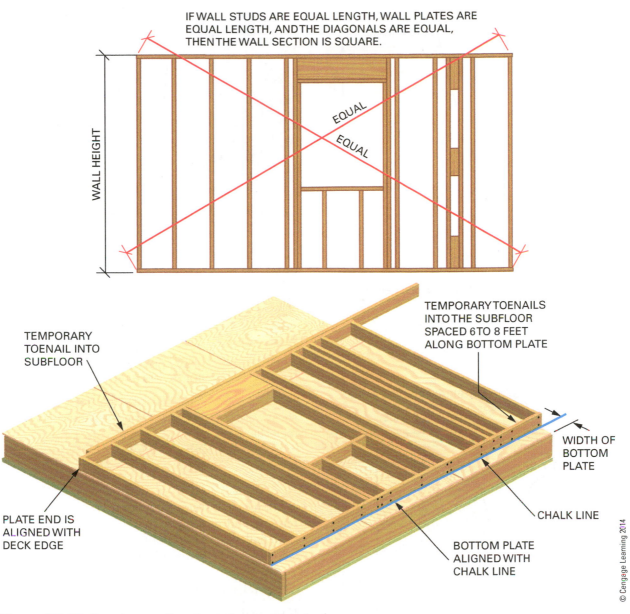

IF WALL STUDS ARE EQUAL LENGTH, WALL PLATES ARE EQUAL LENGTH, AND THE DIAGONALS ARE EQUAL, THEN THE WALL SECTION IS SQUARE.

EQUAL

EQUAL

WALL HEIGHT

TEMPORARY TOENAILS INTO THE SUBFLOOR SPACED 6 TO 8 FEET ALONG BOTTOM PLATE

TEMPORARY TOENAIL INTO SUBFLOOR

WIDTH OF BOTTOM PLATE

PLATE END IS ALIGNED WITH DECK EDGE

CHALK LINE

BOTTOM PLATE ALIGNED WITH CHALK LINE

© Cengage Learning 2014

Figure 37–20 Squaring a wall section before erecting the frame.

In regions where seismic activity is severe, such as California, wall bracing takes on a new meaning. Earthquakes occur with such severity that engineers must design the buildings to protect the occupants of the house long enough for them to escape the shaking building. Shear walls are built **IRC R602.10.3** into the building (*Figure 37–23*).* These are framed of 2 × 6s anchored to the slab and OSB sheathing heavily nailed. Metal strapping is nailed to the building, anchoring the building to the shear wall. The large wood/metal anchors are bolted to long $^{15}/_{16}$-inch-diameter bolts with $^{5}/_{8}$-inch anchor bolts spaced close together. The shear wall bottom plates are sometimes 3 inches thick (*Figure 37–24*).

To provide shear resistance in walls that have large openings such as a garage door, metal frames are used. They are bolted to large 1-inch-diameter bolts anchored 3 feet into concrete. These frames are made with $^{1}/_{8}$-inch-thick steel that is bent and folded into the wall thickness (*Figure 37–25*).* **IRC R602.10.6.3**

Erecting Wall Sections

To erect the wall, remove the toenails from the top plate while leaving the toenails in the bottom plate. The bottom toenails will remain until after the section is erected; they will serve as hinges to keep the bottom plate in position while the frame is raised.

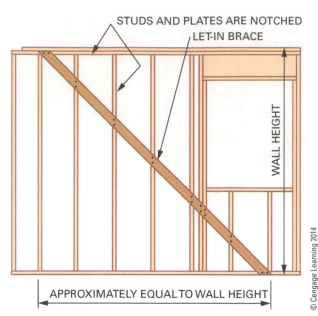

STUDS AND PLATES ARE NOTCHED
LET-IN BRACE

WALL HEIGHT

APPROXIMATELY EQUAL TO WALL HEIGHT

© Cengage Learning 2014

Figure 37–21 A let-in corner brace.

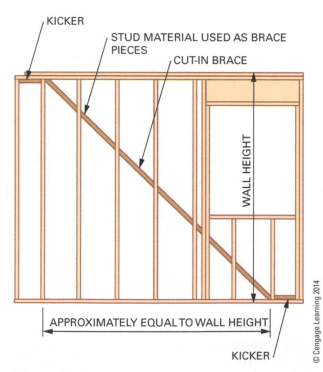

KICKER

STUD MATERIAL USED AS BRACE PIECES

CUT-IN BRACE

WALL HEIGHT

APPROXIMATELY EQUAL TO WALL HEIGHT

KICKER

© Cengage Learning 2014

Figure 37–22 A cut-in brace.

Lift the wall section into place, plumb, and temporarily brace. After checking to be sure that the bottom plate is on the chalk line, nail the bottom plate to the band and floor joists or about every 16 inches along the length. In corners, fasten end studs together to complete the construction of the corner post. A completed corner post provides surfaces to fasten both exterior and interior wall finish.

METAL STRAP TIES

SECTION LEFT UNSHEATHED UNTIL INSPECTION

HEAVY DUTY ANCHOR BOLTS AND HARDWARE

© Cengage Learning 2014

Figure 37–23 In earthquake-prone areas, seismic shear walls are constructed using heavy anchors and metal ties.

Brace each section temporarily as erected. Fasten one end of a brace to a 2 × 4 block that has been nailed to the subfloor and the other end to the side of a stud (*Figure 37–26*).

Bracing Walls Temporarily

If the walls have not been previously braced while being framed on the subfloor, all corners must be plumbed and temporarily braced. Install braces for both sides of each corner. Fasten the top end of the brace to the top plate near the corner post on the inside of the wall.

Temporary braces are fastened to the inside of the wall. They can remain in position until the exterior wall sheathing is applied. Sometimes they remain until it is absolutely necessary to remove them for the application of the interior wall finish. Care must be taken not to let the ends of the braces extend beyond the corner post or top plate. This might interfere with the application of wall sheathing or subsequent ceiling or roof framing.

Methods of Plumbing. Plumb the corner post, and fasten the bottom end of the brace. For accurate plumbing of the corner posts, use a 6-foot level

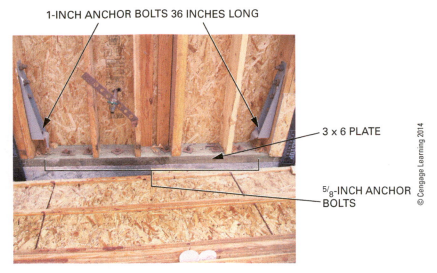

1-INCH ANCHOR BOLTS 36 INCHES LONG

3 x 6 PLATE

⅝-INCH ANCHOR BOLTS

Figure 37–24 The wall frame can be anchored to the foundation for seismic resistance.

Figure 37–25 Shear strength for walls with large openings can be supplied by metal panels bolted to the foundation.

Figure 37–26 Temporary braces hold the frame erect during construction.

with accessory aluminum blocks attached to each end. The blocks keep the level from resting against the entire surface of the corner post. This prevents any bow or irregularity in the surface from affecting the accurate reading of the level.

An alternate method is to use the carpenter's hand level in combination with a long straightedge. On the ends of the straightedge, two blocks of equal thickness are fastened to keep the edge of the straightedge away from the surface of the corner post (*Figure 37–27*). A plumb bob, transit-level, or laser level may also be used to plumb corner posts.

Straightening the Walls

After the corner posts have been plumbed and the top plates doubled, the tops of the walls must be straightened and braced. This can be done with

a line and gauge blocks (see *Figure 31–20*). Nail 2 × 4s at about a 45-degree angle. This will require a length of at least 12 feet for an 8-foot wall. Nail the brace into each plate and twice more into the studs at midspan. Another method is to straighten the brace by eye. After a little practice, eyeing for straightness is fast and surprisingly accurate over distances of less than 40 feet. As one person sights the top plate for straightness by getting the eye as close as possible to the plate, another person nails the brace.

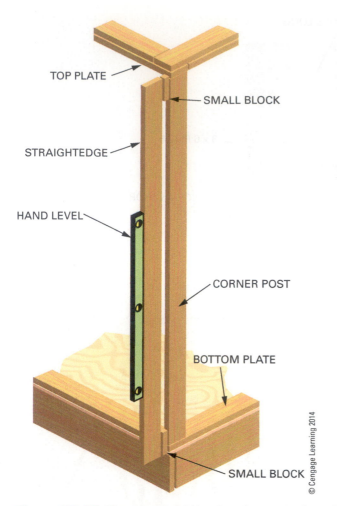

TOP PLATE

SMALL BLOCK

STRAIGHTEDGE

HAND LEVEL

CORNER POST

BOTTOM PLATE

SMALL BLOCK

© Cengage Learning 2014

© Cengage Learning 2014

Figure 37–27 Plumbing should be done between plates. This requires a special level or a straightedge with blocks.

For particularly stubborn wall sections that are difficult to move into plumb, a *spring brace* can be used. Variations of this brace can be used to move the wall top inward or outward.

To create an outward thrust, set a 12- to 16-foot-long 2 × 4 or 2 × 6 against the top plate over a stud (*Procedure 37–C*). The width of the board should be facing up. Nail a block to the subfloor behind the brace. Push down on the brace to make it arch. This will push the wall out. If more outward push is needed, nail the top of the brace to the stud while it is arched. Then lift the middle of the brace to straighten it. This will create a tremendous outward force. Care should be taken not to break anything.

To bring the wall inward, nail a 16-foot-long 1 × 6 to the top plate and the subfloor. These nails should be set firmly in a floor joist. Lift the brace to an arch. Block the arch with a short piece when

the wall is plumb. Care should be taken to watch the brace does not break.

Wall Sheathing

Wall sheathing covers the exterior walls. It usually consists of rated panels or rigid insulation board. Sometimes fiberboard or gypsum board are used. Older homes were built with boards.

Boards. Before plywood was created, boards were used predominantly but are seldom used today. Many buildings in existence today have board sheathing. The boards were applied diagonally or horizontally. The diagonally sheathed walls require no other bracing. They made the frame stiffer and stronger than boards applied horizontally. If used today, nail with two 2½-inch or 8d common nails at each stud for 6- and 8-inch boards, or three nails at each stud for 10- and 12-inch boards. End joints

StepbyStep Procedures

Procedure 37–C Plumbing Stubborn Wall Sections with Spring Braces

OUTWARD THRUST SPRING BRACE

STEP 1 SET A 2" X 4" -12' OR 16' AGAINST WALL WITH TOP AGAINST UPPER TOP PLATE.

STEP 2 NAIL BLOCK TO THE SUBFLOOR BEHIND THE BRACE.

STEP 3 PUSH BRACE DOWN AT MIDSPAN SO TOP END SLIDES DOWN AND DIGS INTO STUD.

STEP 4 IF MORE WALL MOVEMENT IS NEEDED, NAIL TOP WHILE BRACE IS BENT, THEN PICK UP ON BRACE TO STRAIGHTEN IT.

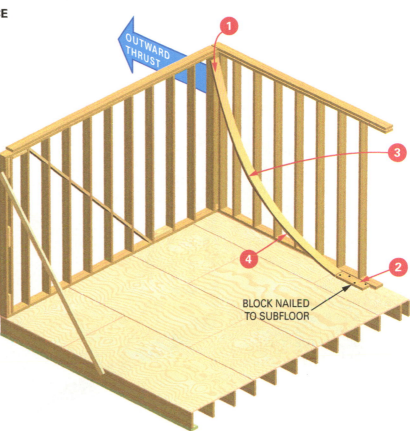

OUTWARD THRUST

BLOCK NAILED TO SUBFLOOR

© Cengage Learning 2014

CAUTION

NOTE: Do not exceed material strength where something might break causing personal injury.

---

must fall over the center of studs. They must be staggered so no two successive end joints fall on the same stud. *

*IRC R602.3(1)

Rated Panels. APA-rated wall sheathing panels are used most often today. They may be applied horizontally or vertically. There is no need for corner braces. A minimum ½-inch thickness is recommended when the sheathing is to be covered by exterior siding. Greater thicknesses are recommended when the sheathing also acts as the exterior finish siding. *

*IRC R602.12.2

Use 2-inch or 6d nails spaced 6 inches apart on the edges and 12 inches apart on intermediate studs for panels ½-inch thick or less. Use 2½-inch or 8d nails for thicker sheathing panels (*Figure 37–28*). *

*IRC R602.10.10

Green Tip

Installing fasteners to the recommended spacing makes the building strong enough to survive its intended load. Recommended spacing also saves material by not overbuilding when too many fasteners are used.

Rated sheathing panels are sometimes used in combination with rigid foam or gypsum board. When panels are applied vertically on both sides

StepbyStep Procedures

Procedure 37–C (*continued*)

INWARD THRUST SPRING BRACE

STEP 1 NAIL THE UPPER END OF A 1" X 6" - 16' BRACE TO THE UNDERSIDE OF THE TOP PLATE.

STEP 2 NAIL LOWER END OF BRACE TO FLOOR JOIST WITH SEVERAL 16D NAILS. CAUTION, THIS END MUST BE SECURELY FASTENED SO IT WILL NOT COME LOOSE WHEN UNDER STRESS.

STEP 3 USE A SHORT 2" X 4" AS A POST, ARC THE BRACE AT MIDSPAN. SECURE THE POST WITH A NAIL.

INWARD THRUST

NAILED SECURELY TO JOIST

© Cengage Learning 2014

---

of the corner, no other corner bracing may be necessary.

Rigid Foam Sheathing. *Rigid foam sheathing* is used to comply with the 2012 International Energy Conservation Code when greater wall insulation is desired (*Figure 37–29*). Rated panels are used in the corners to give the building adequate stiffness. Foam sheathing may be applied in thicknesses of ½" or 1". It may be foil faced to increase the thermal resistance. Special fasteners with plastic capped heads are used secure the sheets (*Figure 37–30*).

To comply with the 2012 International Energy Conservation Code rigid foam sheathing may be used. This applies to the colder regions of the country, such as the upper Midwest and Northeast.

One disadvantage of using these panels as sheathing is that they cannot be used as a nail base for siding. Any siding material applied must be fastened to the studs. The added insulation value to the building is significant making this a minor disadvantage.

Other Sheathing Panels. *Fiberboard* sheathing panels are available as wall sheathing (*Figure 37–31*). They are about ⅛"-thick panels with widths and lengths similar to those of plywood. They are made of specially treated, long wood fibers from recycled products. These fibers are pressed into plies that are pressure laminated with a water-resistant adhesive.

They come in three grades of strength. Green panels are for nonstructural applications, red is for permanent wall bracing in the corners, and blue is used where stronger bracing is required. Nails must be 1¼-inch roofing nails or 1-inch crown staples. Nail spacing for the green nonstructural panels is the same as for APA-rated panels, 6 inches on edges and 12 inches in the field. Structural panels are nailed 3 inches on edges and 6 inches in the field.

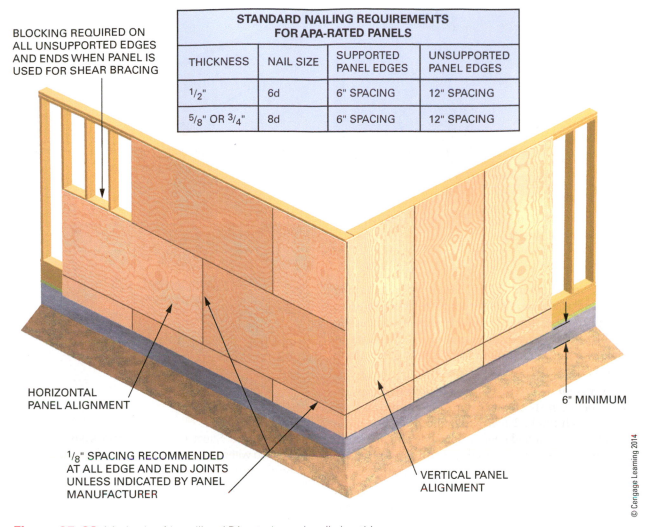

BLOCKING REQUIRED ON ALL UNSUPPORTED EDGES AND ENDS WHEN PANEL IS USED FOR SHEAR BRACING

STANDARD NAILING REQUIREMENTS FOR APA-RATED PANELS			
THICKNESS	NAIL SIZE	SUPPORTED PANEL EDGES	UNSUPPORTED PANEL EDGES
$\frac{1}{2}$"	6d	6" SPACING	12" SPACING
$\frac{5}{8}$" OR $\frac{3}{4}$"	8d	6" SPACING	12" SPACING

HORIZONTAL PANEL ALIGNMENT

$\frac{1}{8}$" SPACING RECOMMENDED AT ALL EDGE AND END JOINTS UNLESS INDICATED BY PANEL MANUFACTURER

VERTICAL PANEL ALIGNMENT

6" MINIMUM

© Cengage Learning 2014

Figure 37–28 Methods of installing APA-rated panel wall sheathing.

© Cengage Learning 2014

Figure 37–29 Rigid foam insulation may be used for exterior wall sheathing.

© Cengage Learning 2014

Figure 37–30 Rigid foam requires special fasteners with large plastic capped heads.

Figure 37–31 Thin sheets of ⅛″ fiberboard may be used as wall sheathing.

Figure 37–32 Structural insulated panels are made from thin panel skins and a foam core.

It is recommended that panels be nailed along one stud at a time. This serves to help keep ripples out of the sheathing. Nailing begins at one corner, nailing the first stud. Then nail along the plates to the next stud, which is in turn nailed off. Nailing four corners first should be avoided.

Gypsum sheathing consists of a treated gypsum filler between sheets of water-resistant paper. Usually ½ inch thick, 2 feet wide, and 8 feet long, the sheets have matched edges to provide a tighter wall. Because of the soft material, galvanized wall board nails must be used to fasten gypsum sheathing.

Figure 37–33 Structural insulated panels are assembled without the traditional wood framework.

Space the nails about 4 inches around the edges and 8 inches in the center. Gypsum board sheathing is used when a more fire-resistant sheathing is required.

Application of Sheathing Panels.   Sheathing panels are installed in a manner similar to that used for installing subfloors. If needed, snap lines to keep the plates and the edges of panels aligned. Panels should be installed with as few seams as possible. Each seam should be as tight as possible. This will make the building more airtight and make it easier to heat or cool the indoor air.

STRUCTURAL INSULATED PANELS

Another method of constructing a wall or roof section is to use **structural insulated panels (SIPs)**. They consist of two outer skin panels and an inner core of an insulating material (*Figure 37–32*). This sandwich of materials forms a rigid unit. Most panels use oriented strand board (OSB) for their facings. OSB makes panels available in large sizes up to sheets 12 feet by 36 feet. The cores of SIPs are made of foam insulation often molded from

expanded polystyrene, but sometimes from extruded polystyrene or urethane foam.

The insulating core and the two skins of a SIP are not structurally strong by themselves. But when molded together under pressure, these materials create a structural unit that is stiff and strong. The result is that no frame is necessary to support the panels. Panel manufacturers supply splines, connectors, adhesives, and fasteners to erect their systems (*Figure 37–33*).

SIPs are produced in thicknesses from 4½ inches to 12¼ inches. Panel sizes range from 4 feet by 8 feet up to 9 feet by 24 feet. Their R-values depend on the foam core and the thickness of the panel. For example, a 4½-inch panel has an R-value of 16, and a 12¼-inch thick panel has an R-value of 53.

Blank panels are available, but SIPs are typically prefabricated at the manufacturing plant with all windows and doors precut. They are delivered to the job site ready to install, with a shop drawing of how to assemble the panels using numbers written on the panels. This feature allows for faster assembly than that for wood-framed walls and roofs. SIPs have electrical chases precut in the core to accept electrical wiring.

Large panels require equipment to set them into position from the truck. The equipment typically used is a fork lift, boom truck, or crane depending on the structure being built and site conditions (*Figure 37–34*).

The fire-rated performance of SIPs is similar to other wood frame structures. Residential structures are typically required to meet a fifteen-minute standard by applying ½-inch common drywall. Thicker and multiple layers of drywall provide additional fire rating for light-commercial and multifamily requirements.

Building with SIPs offers several advantages and benefits over stick framing. It is easy to achieve an airtight home with SIPs (*Figure 37–35*), which increases the energy performance of the building. Overall, SIPs use less wood than does stick framing. The OSB panels are made from small, fast-growing trees, and so use wood more efficiently than does sawn lumber. They also reduce the amount of waste produced during home construction.

ESTIMATING MATERIALS FOR EXTERIOR WALLS

Estimating the amount of material needed for exterior walls involves studs, wall plates, headers, rough sills, wall sheathing, and temporary bracing. Remember that if the number of pieces does not come out even, round them up to the next whole number of pieces.

Figure 37–34 Structural insulated panels are typically assembled using heavy equipment.

Figure 37–35 Structural insulated panels are sealed at each seam.

Wall Studs

Wall studs are estimated from the total linear feet of exterior wall. If they are spaced 16 inches OC, then figure one stud for every linear foot of wall. This allows for the extras needed for corner posts, partition intersections, trimmers, door jacks, and blocking.

Wall Plates

For wall plates, multiply the total linear feet of wall by three (one sole plate and two top plates). Then divide by the desired length of the plate material, typically 12 feet. Add 5 percent for waste. For example, if the perimeter is 144 feet and the plate material is to be 12 feet long, the plates needed will be $144 \times 3 \div 12' \times 1.05 = 37.8 = 39$ plates.

Headers

For headers, add up the width of each opening. Add ½ foot per header to allow for extra length of header to sit on jack studs. Divide the total by the length of material being ordered. Add 10 percent for waste.

Rough Sills

Lineal feet of rough sill material would be the same number as the header material needed for windows.

Wall Sheathing

For wall sheathing, first find the total area to be covered. This is found by multiplying the total linear feet of wall by the wall height. Make sure the wall height allows for the area covering the box headers. Deduct the area of any large openings, like garage doors, but retain the area of normal-size openings of windows and doors.

To find the number of sheathing panels, divide the total wall area to be covered by the number of square feet in each sheet. A 4 × 8 panel sheet has 32 square feet. Add about 5 percent for waste. For example, if the perimeter of a rectangular building is 144 and the box header is 1 foot wide, the wall sheathing is 144 × (8 + 1) = 1,296 ÷ 32 × 1.05 = 42.5 = 43 sheets.

Temporary Wall Bracing

Temporary bracing is needed to hold wall sections plumb and secure. Material used is often a 2 × 4−12′. The number of pieces may be estimated as one piece per 20 feet of wall plus one per corner. For example, if the perimeter of a rectangular building is 144, the quantity of temporary bracing is 144 ÷ 20 + 4 = 11.2 = 12 pieces.

See *Figure 37–36* for a listing of the shorthand versions of these estimating techniques with another example.

Estimate the materials for an exterior wall of a rectangular building that measures 30′ × 62′. The box header is one foot wide, six 3246 windows and three 3′-0″ × 6′-8″ doors.			
Item	**Formula**	**Waste factor**	**Example**
Wall OC Studs	wall LEN in ft = NUM of studs		2 × (30 + 62) = 184′ = 184 PC
Wall plates	wall LEN × 3 ÷ 12′ = NUM 12′ plates	5%	184′ × 3 ÷ 12′ × 1.05 = 48.3 = 49 PC
Headers *win & door*	total of window & door WID plus 6″ per unit	5%	[6 (3′-2″ + 6″) + 3 (3′-0″ +6″)] × 1.05 = 32.5 = 33 LF
Rough sills	total of window WID plus 6″ per unit	5%	6 (3′-2″ + 6″) × 1.05 = 23.1 = 24 LF
Wall sheathing	wall area ÷ 32	5%	184′ × (8′ + 1′) ÷ 32 × 1.05 = 54.3 = 55 PC
Temporary braces	one per 20 feet of wall plus one per corner		184 ÷ 20 + 4 = 13.2 = 14 PC

Figure 37–36 Example of estimating wall framing materials with formulas.

© Cengage Learning 2014

DECONSTRUCT THIS

Carefully study **Figure 37–37** and think about what is wrong and/or what is right. Consider all possibilities. What construction practice or method is different in your area of the country?

Figure 37–37 This photo shows a wall being framed on a deck.

© Cengage Learning 2014

KEY TERMS

corner posts	lintel	structural insulated panels (SIPs)	threshold
door stop	racking	sole plate	top plates
gypsum board	rough opening	stop	trimmers
jack studs	rough sills	story pole	wall sheathing
jambs	shear walls	studs	
kickers	sill		

REVIEW QUESTIONS

Select the most appropriate answer.

1. The top and bottom horizontal members of a wall frame are called

   a. headers.
   b. plates.
   c. trimmers.
   d. sills.

2. The horizontal wall member supporting the load over an opening is called a

   a. header.
   b. rough sill.
   c. plate.
   d. truss.

3. Shortened studs above and below an opening are called

   a. shorts.
   b. lame.
   c. cripples.
   d. stubs.

4. Diagonal cut-in bracing requires the installation of

   a. kickers.
   b. backing.
   c. blocking.
   d. 1 × 4s.

5. The finish floor-to-ceiling height in a platform frame is specified to be 7'-10". The finish floor is ¾ inch thick and the ceiling material is ½ inch thick. A single bottom plate and a double top plate are used, each of which has an actual thickness of 1½ inches. What is the stud length?

   a. 7'-5¾"
   b. 7'-6¾"
   c. 7'-8¼"
   d. 7'-10½"

6. A doorjamb is ¾ inch thick. Adding ½ inch on each side for shimming the frame, what is the rough opening width for a door that is 2'-8" wide?

   a. 2'-9½"
   b. 2'-10½"
   c. 2'-11½"
   d. 3'-0½"

7. A story pole typically shows

   a. the length of headers.
   b. the length of rough sills.
   c. the length of jack studs.
   d. the width of the rough opening.

8. When laying out plates for walls and partitions, measurements for centerlines of openings start from the

   a. end of the plate.
   b. outside edge of the abutting wall.
   c. building line.
   d. nearest intersecting wall.

9. When laying out plates for any OC wall or partition stud, the measurement begins from the

   a. end of the plate.
   b. abutting wall or partition.
   c. opening centerlines.
   d. depends on the type of wall or partition.

10. The first OC wall stud is set back

    a. a distance that is usually ¾ inch.
    b. one-half stud thickness.
    c. to allow the first sheathing piece to be installed flush with the first stud.
    d. all of the above.

11. A corner stud that allows for ample room for insulation in the corner uses

    a. three small blocks.
    b. a stud that is rotated from the others in the wall.
    c. three full studs nailed as a post.
    d. all of the above.

12. Exterior walls are usually straightened before ceiling joists are installed by

    a. using only a carpenter's level.
    b. using a line stretched between two blocks and testing with a gauge block.
    c. using a plumb bob dropped to the bottom plate at intervals along the wall.
    d. by sighting along the length of the wall using a builder's level.

13. Bearing partitions

    a. have a single top plate.
    b. carry no load.
    c. are constructed like bearing walls.
    d. are erected after the roof sheathing is installed.

14. The top plate of the bearing partition

    a. laps the plate of the exterior wall.
    b. butts the headers.
    c. butts the stud of the exterior wall.
    d. is applied after the ceiling joists are installed.

15. Spring braces are typically used

    a. as temporary braces.
    b. as permanent bracing.
    c. only during certain times of the year.
    d. all of the above.

16. What is the rough opening height of a door opening for a 6'-8" door if the finish floor is ¾ inch thick, ½-inch clearance is allowed under the door, and the jam thickness is ¾ inch?

    a. 6'-9"
    b. 6'-9½"
    c. 6'-10"
    d. 6'-10½"

17. The type of plywood typically used for wall sheathing is

    a. CDX.
    b. AC.
    c. BC.
    d. hardwood.

18. The type of permanent wall bracing used most often in construction today is

    a. APA-rated sheathing.
    b. cut-in.
    c. let-in.
    d. all of the above.

19. Estimate the number of 16-inch OC exterior wall studs needed for a rectangular house that measures 28 × 48 by 8 feet high.

    a. 76
    b. 152
    c. 1,344
    d. 10,752

20. Estimate the number of pieces of wall sheathing needed for a rectangular house that measures 28 × 48 by 8 feet high. Figure an extra foot of material to cover the box header. Neglect the openings and gable end. Add 5 percent for cutting waste.

    a. 42
    b. 43
    c. 44
    d. 45

UNIT 14
Interior Rough Work

Interior rough work is constructed in the inside of a structure and later covered by some type of finish work. The interior rough work described in this unit includes the installation of *partitions*, *ceiling joists*, *furring strips*, and *backing* and *blocking*. The term *rough work* does not imply that the work is crude. It is a kind of work that will eventually be covered by other material. Careful construction of the rough frame makes application of the finish work easier and less complicated.

OBJECTIVES

After completing this unit, the student should be able to:

- assemble, erect, brace, and straighten bearing partitions.
- determine and make rough openings for doors.
- lay out, cut, and install ceiling joists.
- lay out and erect nonbearing partitions, and install backing in walls for fixtures.
- lay out and install furring for walls and ceilings.
- describe various components of light-gauge steel framing.
- lay out and frame light-gauge steel load-bearing and non-load-bearing walls.

SAFETY REMINDER

As tools and materials are gathered for interior wall framing, job site organization becomes very important for safety and efficiency. Keep waste material outside the work area in organized piles for removal.

CHAPTER 38

Interior Partitions and Ceiling Joists

Partitions and *ceiling joists* constitute some of the interior framing. Ceiling joists tie the exterior side walls together and support the ceiling finish.

Partitions supporting a load are called *bearing partitions*. Partitions that merely divide the area into rooms are called *nonbearing partitions* (*Figure 38–1*).

PARTITIONS

Load-bearing partitions (LBPs) support the inner ends of ceiling or floor joists. They are placed directly over the girder or the bearing partition in the lower level. If several bearing partitions are used on the same floor, supported girders or walls are placed directly under each.

Non-load-bearing partitions (NLBPs) are built to divide the space into rooms of varying size. They carry no structural load from the rest of the building, only the load of the partition material itself. They may be placed anywhere on the subfloor where joist reinforcement has been provided. They often run parallel to the joists and are nailed to blocking between joists (*Figure 38–2*).

Partitions are erected in a manner similar to that used for exterior walls. A double top plate is used to tie the wall and partition intersections together. If *roof trusses* are used, partitions have

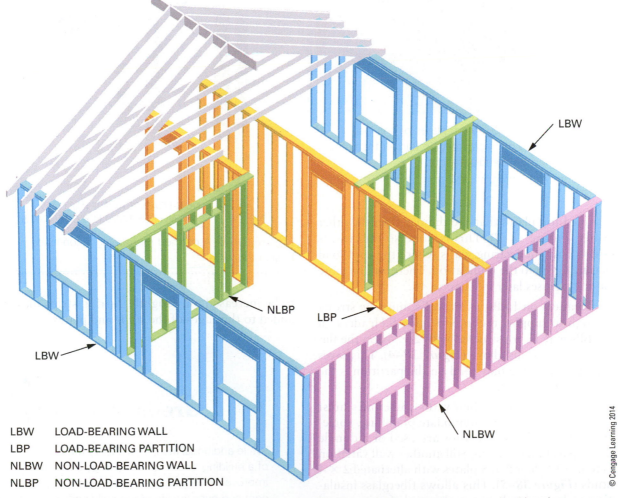

LBW	LOAD-BEARING WALL
LBP	LOAD-BEARING PARTITION
NLBW	NON-LOAD-BEARING WALL
NLBP	NON-LOAD-BEARING PARTITION

Figure 38–1 Walls are exterior and partitions are interior. Either may be load bearing or non-load bearing.

© Cengage Learning 2014

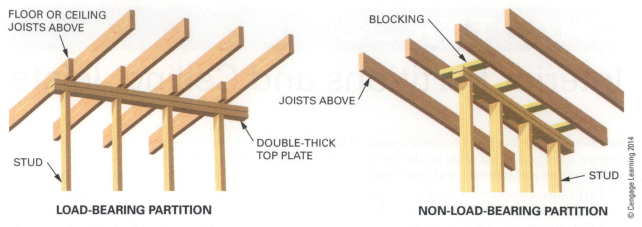

FLOOR OR CEILING JOISTS ABOVE

STUD

DOUBLE-THICK TOP PLATE

LOAD-BEARING PARTITION

BLOCKING

JOISTS ABOVE

STUD

NON-LOAD-BEARING PARTITION

© Cengage Learning 2014

Figure 38–2 Non-load-bearing partitions merely divide an area into rooms. Load-bearing partitions support the weight of the floor or ceiling above.

© Cengage Learning 2014

Figure 38–3 The double top plates of non-load-bearing walls and partitions may be thinner than those of load-bearing walls and partitions.

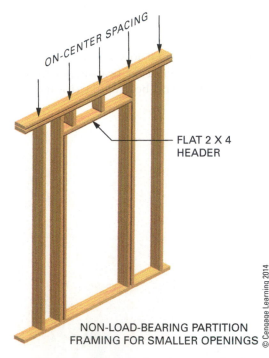

ON-CENTER SPACING

FLAT 2 X 4 HEADER

NON-LOAD-BEARING PARTITION FRAMING FOR SMALLER OPENINGS

© Cengage Learning 2014

Figure 38–4 A method for framing a non-load-bearing header.

doubled top plates that are thinner than the other walls (*Figure 38–3*). This is done so the trusses touch only the bearing walls where the roof load is transferred to the foundation. There will be more on roof trusses later in Chapter 46.

Headers on LBPs, as in walls, must be strong enough to support the intended load. Headers on NLBPs may be constructed with 2 × 4s on the flat with cripple studs above (*Figure 38–4*). This saves material since the only load on the partition is the wall finish material. ✻

✻IRC R602.7.3

Bathroom and kitchen walls sometimes must be made thicker to accommodate plumbing. Sometimes 2 × 6 plates and studs are used or a double 2 × 4 partition is erected. Still another wall variation is to use 2 × 6 or 2 × 8 plates with alternated 2 × 4 studs (*Figure 38–5*). This allows fiberglass insulation to be woven between the studs for increased soundproofing. Also, if the wall thickness needs

to be increased only slightly, furring strips may be added to the edges of the studs and plates.

Green Tip

While additional insulation affects the cost of a building, reductions in energy costs and improvements to the living environment continue over the life of the building.

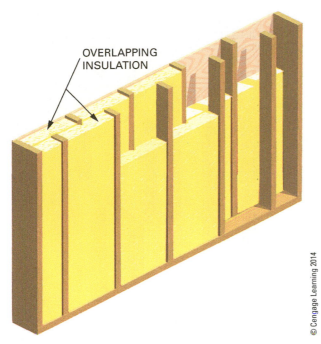

Figure 38–5 Sometimes a wide plate is used to build stagger-stud walls for increased soundproofing.

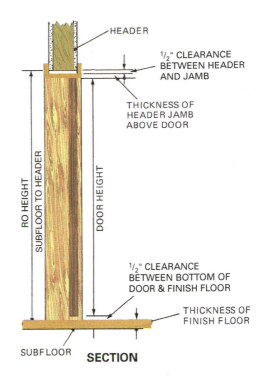

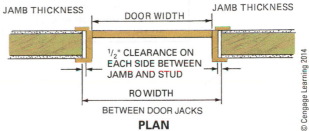

Figure 38–6 Figuring the rough opening size for an interior door.

Layout and Framing of Partitions

Partition layout is similar to wall layout. From the floor plan, determine the location of the partitions. Note that dimension lines are usually given to the centerlines of the partitions and sometimes to their edges. To locate all partitions, measure and snap chalk lines on the subfloor.

Lay the edge of the bottom plate to the chalk line, tacking it in position. Lay out the openings and partition intersections, then all on-center (OC) studdings including cripple studs. LBP OC studs should line up with floor and ceiling joists above and below. Lay the top plate next to the bottom plate and transfer the layout of the bottom plate to the top plate. Remember that all joints in the plates should end in the center of a stud.

Rough Opening Door Sizes. The rough opening sizes are found in the same way they are found for exterior walls (*Figure 38–6*). The width for a door opening is found by adding to the door width, twice the thickness of the jamb stock, and twice the $\frac{1}{2}$-inch shim space (one for each side). For example, a 2'-6" door rough opening width is 2'-8½" or 32½".

The rough opening height is found by adding five measurements from the door cross section. They include the following from the subfloor up:

- The thickness of the finish floor
- The 1-inch clearance between the finish floor and the bottom of the door
- The height of the door

- The thickness of the head jamb
- The $\frac{1}{2}$-inch shim space between the head jamb and the rough header

For example a 7' (84") door with $\frac{3}{4}$-inch jambs and $\frac{3}{4}$-inch finished floor needs a rough opening height of $\frac{3}{4}$-inch floor + 1-inch clearance + 84-inch door + $\frac{3}{4}$-inch jamb + $\frac{1}{2}$-inch shim space = 87".

Usually, no threshold is used under interior doors. However, if a threshold is used it will fit within the space allowed for by the 1-inch clearance.

Framing Partitions. For ease of erecting, construct the longer partitions first. Then construct shorter cross partitions, such as for closets, later. There is no hard-and-fast rule for constructing partitions; experience will allow the best process to emerge.

Pull the tack nails and separate the top and bottom plates. Place all full-length studs, corner posts, and partition intersections on the floor. Make sure their crowned edges run in the same direction. Nail the framing members around an opening first, then the OC studs to allow fast and easy assembly. Raise the section into position, locking the lapping top plates into position. Plumb and brace the wall section and nail the bottom plate to the floor about 16 inches apart or into floor joists where possible.

The end stud that butts against another wall can be straightened by toenailing through its edge into the center block of the partition intersection (*Figure 38–7*).

Other Openings.   Besides door openings, the carpenter must frame openings for heating and air-conditioning ducts, medicine cabinets, electrical panels, and other similar items. If the items do not fit in a stud space, the stud must be cut and a header installed. When ducts run in a wall through

On the Job

Straighten the edge of end studs in interior partitions that intersect with another wall.

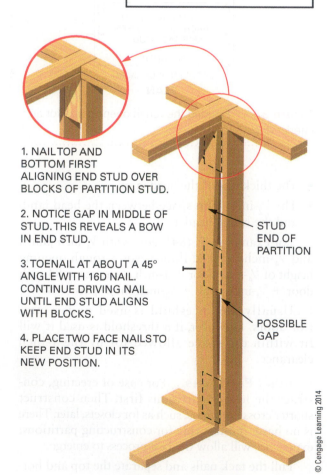

1. NAIL TOP AND BOTTOM FIRST ALIGNING END STUD OVER BLOCKS OF PARTITION STUD.

2. NOTICE GAP IN MIDDLE OF STUD. THIS REVEALS A BOW IN END STUD.

3. TOENAIL AT ABOUT A 45° ANGLE WITH 16D NAIL. CONTINUE DRIVING NAIL UNTIL END STUD ALIGNS WITH BLOCKS.

4. PLACE TWO FACE NAILS TO KEEP END STUD IN ITS NEW POSITION.

STUD END OF PARTITION

POSSIBLE GAP

© Cengage Learning 2014

Figure 38–7 Technique for straightening a crowned stud.

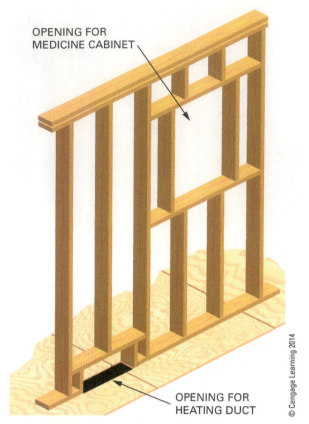

OPENING FOR MEDICINE CABINET

OPENING FOR HEATING DUCT

© Cengage Learning 2014

Figure 38–8 Miscellaneous openings in interior partitions are framed with non-load-bearing headers.

the floor, the bottom plate and subfloor must be cut out (*Figure 38–8*). The reciprocating saw is a useful tool for making these cuts.

CEILING JOISTS

Ceiling joists generally run from the exterior walls to the bearing partition across the width of the building. Construction design varies according to geographic location, traditional practices, and the size and style of the building. The size of ceiling joists is based on the span, spacing, load, and the kind and grade of lumber used. Determine the size and spacing from the plans or from local building codes. ✳ **✳IRC R802.4**

Methods of Installing Joists

In a conventionally framed roof, the *rafters* and the ceiling joists form a triangle. Framing a triangle is a common method of creating a strong and rigid building. The weight of the roof and weather is transferred from the roof to the exterior walls (*Figure 38–9*). The rafters are located over the studs and the ceiling joists are fastened to the sides of the rafters. This binds the rafters and ceiling joists together into a rigid triangle and keeps the walls from spreading outward due to the weight of the roof. The entire roof load is transferred to the foundation through vertically aligned framing members (*Figure 38–10*).✱

✱IRC R802.3.1

Ceiling joists may be made from engineered lumber and purchased in long lengths so that the rafter-ceiling triangle is easily formed. Typically, though, the ceiling joist lengths are half of the building width and therefore must be joined over a beam or bearing partition.

Sometimes the ceiling joists are installed in-line. Their ends butt each other at the centerline of the bearing partition. The joint must be *scabbed* to tie the joints together (*Figure 38–11*). Scabs are short boards fastened to the side of the joist and centered on the joint. They should be a minimum of 24 inches long. In-line ceiling joists are attached to the same side of each rafter pair (front and back rafter).

Another method of joining ceiling joists is to lap them over a bearing partition in the same manner as for floor joists (see Figure 34–18).✱ This puts a stagger in the alignment of the ceiling joist and consequently in the rafters as well (*Figure 38–12*). This stagger is visible at the ridgeboard. The layout lines for rafters and ceiling joists are measured from the outside end wall onto the top plate (*Figure 38–13*). This measurement is exactly the same, with the only difference being the side of the line on which the ceiling joists and rafters are placed.

✱IRC R802.3.2

Cutting the Ends of Ceiling Joists

The ends of ceiling joists on the exterior walls usually project above the top of the rafter. This is especially true when the roof has a low slope. These ends may be cut to the slope of the roof, flush with or slightly below the top edge of the rafter.

Lay out the cut, using a framing square. Cut one joist for a pattern. Use the pattern to mark the rest.

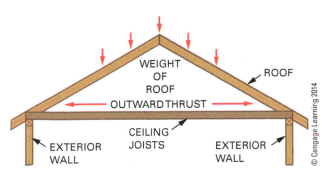

Figure 38–9 Ceiling joists tie the roof frame together into a triangle, which resists the outward thrust caused by the rafters.

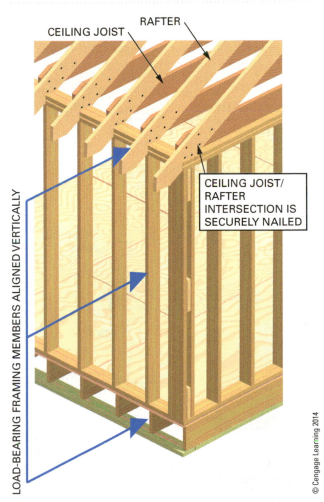

Figure 38–10 Ceiling joists are located so they can be fastened to the side of rafters and vertically aligned over the structural frame.

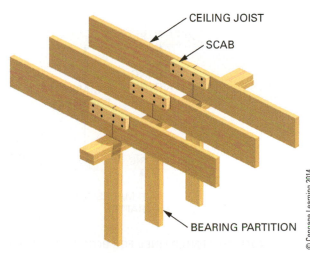

Figure 38–11 The joint of in-line ceiling joists must be scabbed at the interior bearing partition.

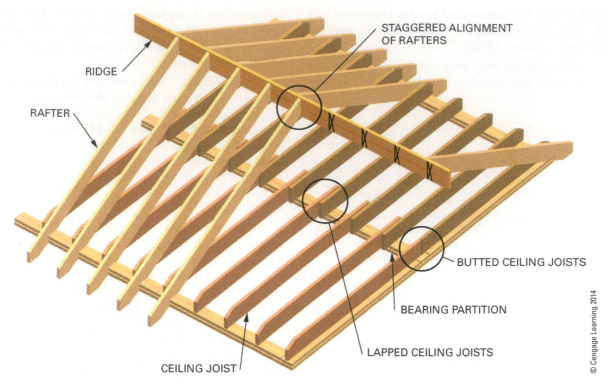

RIDGE

RAFTER

STAGGERED ALIGNMENT OF RAFTERS

BUTTED CEILING JOISTS

BEARING PARTITION

LAPPED CEILING JOISTS

CEILING JOIST

© Cengage Learning 2014

Figure 38–12 When ceiling joists are lapped, it causes a stagger in the rafters, which is visible at the ridge.

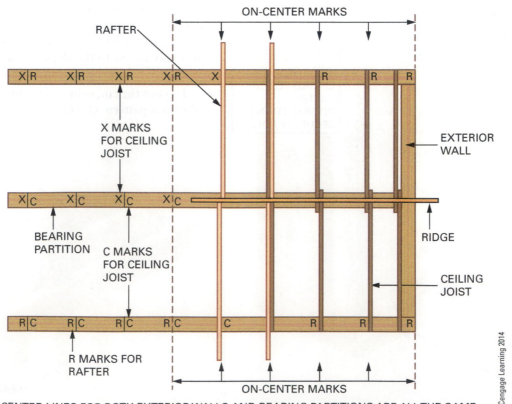

ON-CENTER MARKS

RAFTER

X MARKS FOR CEILING JOIST

EXTERIOR WALL

BEARING PARTITION

C MARKS FOR CEILING JOIST

RIDGE

CEILING JOIST

R MARKS FOR RAFTER

ON-CENTER MARKS

NOTE: ON-CENTER LINES FOR BOTH EXTERIOR WALLS AND BEARING PARTITIONS ARE ALL THE SAME.

© Cengage Learning 2014

Figure 38–13 Layout lines on all plates are the same measurements; only the positions of the rafters and ceiling joists vary.

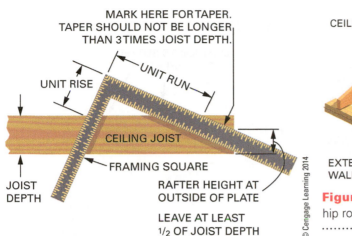

Figure 38–14 A framing square is used to mark the slope of a tapered cut on a ceiling joist.

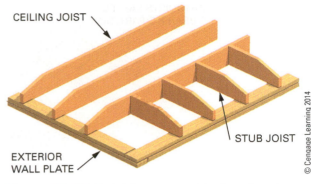

Figure 38–15 Stub joists are used for low-sloped hip roofs.

Make sure when laying out the joists that you sight each for a crown. Make the cut on the crowned edge so that edge is up when the joists are installed. Cut the taper on the ends of all ceiling joists before installation. Make sure the length of the taper cut does not exceed three times the depth of the member. Also make sure that the end of the joist remaining after cutting is at least half the member's **IRC** width (*Figure 38–14*).

R802.7.1.2

Stub Joists

Usually, ceiling joists run parallel to the end walls and in the same direction as the roof rafters. When low-sloped hip roofs are built, stub joists are used. These are joists that are shorter and run perpendicular to the normal joists. The use of stub joists allows clearance for the bottom edge of the rafters that run at right angles to the end wall (*Figure 38–15*).

Framing Ceiling Joists to a Beam

In many cases, the bearing partition does not run the length of the building because of large room areas. Some type of beam is then needed to support the inner ends of the ceiling joists in place of the supporting wall. Similar in purpose and design to a girder, the beam may be built-up, solid lumber or engineered lumber.

If the beam is to project below the ceiling, it is installed in the same manner as a header for an opening. The joists are then installed over the beam in the same manner as over the bearing partition (*Figure 38–16*).

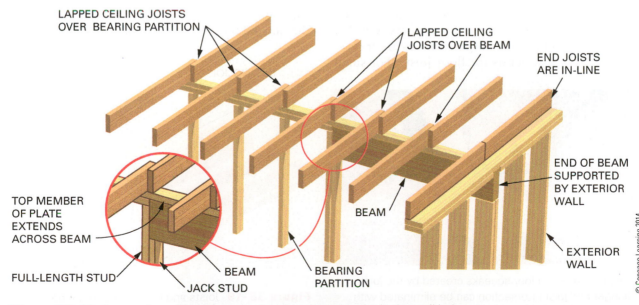

Figure 38–16 Support for ceiling joists may be placed as a header in the wall below.

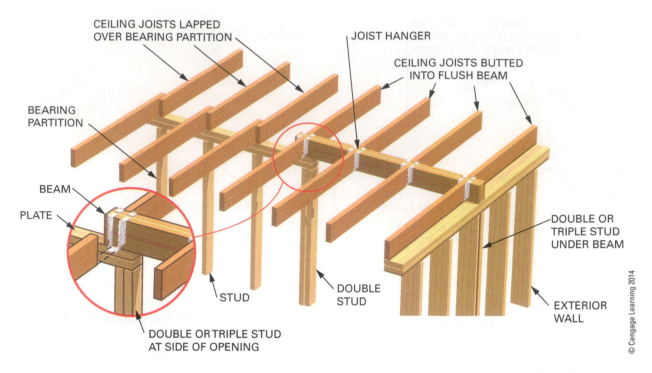

CEILING JOISTS LAPPED OVER BEARING PARTITION

JOIST HANGER

CEILING JOISTS BUTTED INTO FLUSH BEAM

BEARING PARTITION

BEAM

PLATE

DOUBLE OR TRIPLE STUD UNDER BEAM

STUD

DOUBLE STUD

EXTERIOR WALL

DOUBLE OR TRIPLE STUD AT SIDE OF OPENING

© Cengage Learning 2014

Figure 38–17 Support for ceiling joists may be a flush girder. This creates a flush ceiling through the partition below.

If the ceiling is to be flush and continuous through the partition, then the ceiling joists are cut and fitted to a flush beam (*Figure 38–17*). The beam is supported by the bearing partition and exterior wall. The joists are usually set in joist hangers. Adhesive may be used in the joist hanger to eliminate any squeaks that might occur between the joist and ceiling joist (*Figure 38–18*).

Openings

Openings in ceiling joists may need to be made for such things as chimneys, attic access (scuttle), or disappearing stairs. Large openings are framed in the same manner as for floor joists. For small openings, there is no need to double the joists or headers (*Figure 38–19*). ✶

✶IRC R802.9

Ribbands and Strongbacks

Ceilings are made stiffer by installing *ribbands* and *strongbacks*. Ribbands are 2 × 4s installed flat on top of the top of ceiling joists. They are placed at midspan to stiffen the joists as well as to keep the spacing uniform. They should be fastened with 16d nails and long enough to be attached to the end walls. With the addition of a 2 × 6 installed on edge, the ribband becomes a strongback. A strongback is

Figure 38–18 Floor squeaks created by the joist hanger and joist intersection can be eliminated with caulk applied when joists are installed.

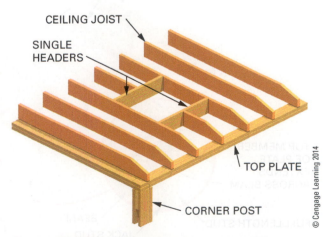

CEILING JOIST

SINGLE HEADERS

TOP PLATE

CORNER POST

© Cengage Learning 2014

Figure 38–19 Joists and headers need not be doubled for small ceiling openings.

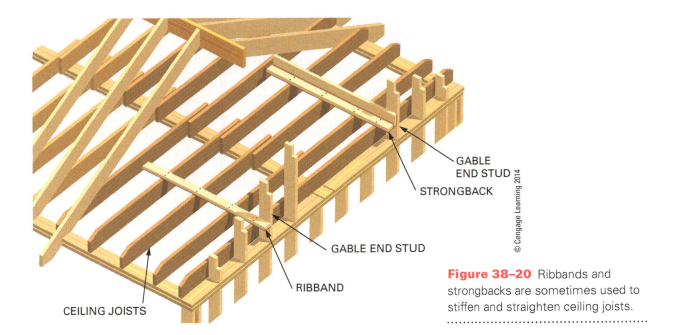

GABLE
END STUD

STRONGBACK

GABLE END STUD

RIBBAND

CEILING JOISTS

© Cengage Learning 2014

Figure 38–20 Ribbands and strongbacks are sometimes used to stiffen and straighten ceiling joists.

used when extra support and stiffness are required on the ceiling joists (*Figure 38–20*).

Layout and Spacing of Ceiling Joists

Roof rafters rest on the plate directly over the regularly spaced studs in the exterior wall. Ceiling joists are installed against the side of the rafters and fastened to them. Spacing of the ceiling joists and rafters should be the same so they can be tied together at the plate line.

- Start the ceiling joist layout from the same corner of the building where the floor joists and wall stud were laid out. Square up from the same side of each regularly spaced stud or cripple stud in the exterior wall and across the top of the plate. Mark an *R* on the side of the line over top of the stud for the rafter and an *X* or a *C* on the other side of the line for the location of the ceiling joists.

- Layout lines on the bearing partition and on the opposite exterior wall are on the same layout as on the first wall. This is similar to floor joist layout. These layout lines should all be the same distance from the end wall. The only difference is the location of the marks for rafters and ceiling joists. They vary depending on whether the ceiling joists are continuous or lapped at the load-bearing partition (*Figure 38–21*).

- Continuous and butted joists are placed on the same side of the layout line. The layout marks for both exterior walls are exactly the same. The load-bearing partition has only the joist layout line. This places the rafters on the same side of the joist.

- Lapped joist layout marks are reversed, similar to floor joists. Because the joists lap at the load-bearing partition, the layout on the opposite exterior wall is reversed from the first wall. Place on the opposite exterior wall an *R* and a *C* or an *X* on either side of the line but opposite from the first exterior wall. This allows for the lapped joist and creates staggered rafters. The layout for the bearing partition shows where front and back joists lap.

Installing Ceiling Joists

The ceiling joists on each end of the building are placed to allow for installation of gable end studs. The last ceiling joist is actually nailed to the gable studs that sit on the exterior wall. It is nailed from the inside of the building. These end joists are installed butted in-line regardless of how the other joists are laid out (*Figure 38–22*).

In addition to other functions, these end joists provide fastening for the ends of the ceiling finish. All other joists are fastened in position with their sides to the layout lines. If the outside ends of the joists have not been tapered, sight each joist for a crown. Install each one with the crowned edge up. If the outside ends have been tapered, install the ceiling joists with the cut edge up. Reject any badly warped joists.

Nailing of the rafter, ceiling joist, and wall plate is critical. This connection establishes the rigid triangle that supports the roof load. The number and size of nails depends on the slope of the roof, intended roof load, and the width of the building. Check local codes for nailing specs.✱

✱IRC
R802.3.1

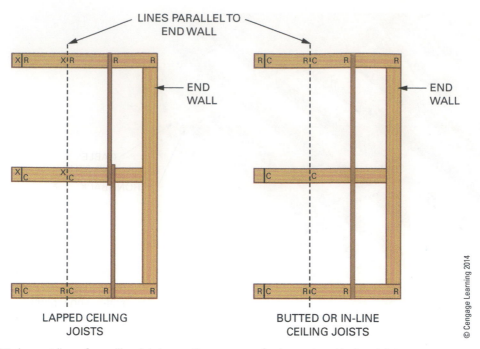

Figure 38–21 Layout lines for ceiling joists are the same as for lapped and in-line joists.

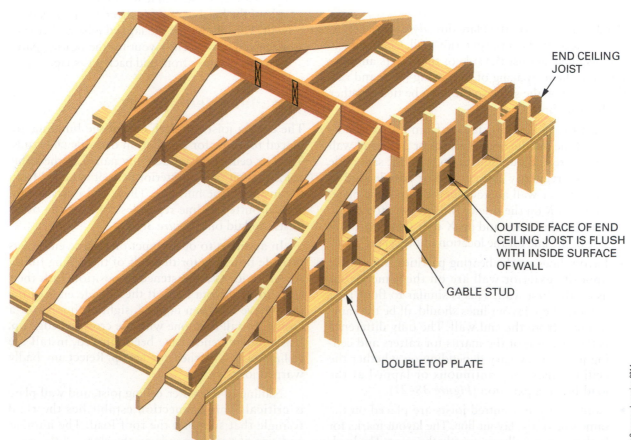

Figure 38–22 The end ceiling joist is attached to the gable end studs.

CHAPTER 39

Backing, Blocking, and Furring

Backing is ordinarily, a short block of lumber placed in floor, wall, and ceiling cavities to provide fastening for various parts and fixtures.

Blocking is installed for different purposes, such as providing support for parts of the structure, weather tightness, and fire-stopping. Sometimes blocking serves as backing.

Furring is attached to the structure, making the frame thicker. It is installed for various reasons but usually to ease the fastening of a wall or ceiling finish.

BACKING AND BLOCKING

There are many places in the structure where blocking and backing should be placed. It would be unusual to find directions for the placement of backing and blocking in a set of plans. It is the wise builder who installs them, much to the delight of those who must install fixtures and finish of all types in later stages of construction.

Backing

It is recommended to install short blocks with their ends against the sole plate and their sides against the studs on inside corners and on both sides of door openings. This provides more fastening surface for the ends of baseboard (**Figure 39–1**).

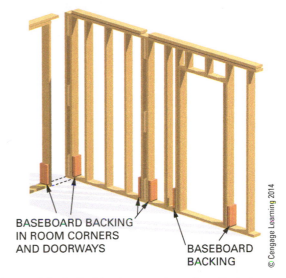

BASEBOARD BACKING IN ROOM CORNERS AND DOORWAYS

BASEBOARD BACKING

© Cengage Learning 2014

Figure 39–1 Backing is sometimes installed at corners and door openings for baseboard.

Much backing is needed in bathrooms. Plumbing rough-in work varies with the make and style of plumbing fixtures. The experienced carpenter will obtain the rough-in schedule from the plumber. He or she then installs backing in the proper location for such things as bathtub faucets, showerheads, lavatories, and water closets (*Figure 39–2*).

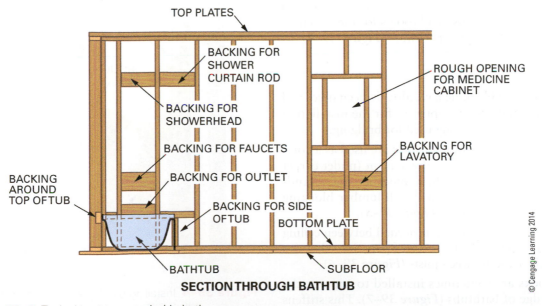

TOP PLATES

BACKING FOR SHOWER CURTAIN ROD

ROUGH OPENING FOR MEDICINE CABINET

BACKING FOR SHOWERHEAD

BACKING FOR FAUCETS

BACKING FOR LAVATORY

BACKING FOR OUTLET

BACKING AROUND TOP OF TUB

BACKING FOR SIDE OF TUB

BOTTOM PLATE

BATHTUB

SUBFLOOR

SECTION THROUGH BATHTUB

© Cengage Learning 2014

Figure 39–2 Typical backing needed in bathrooms.

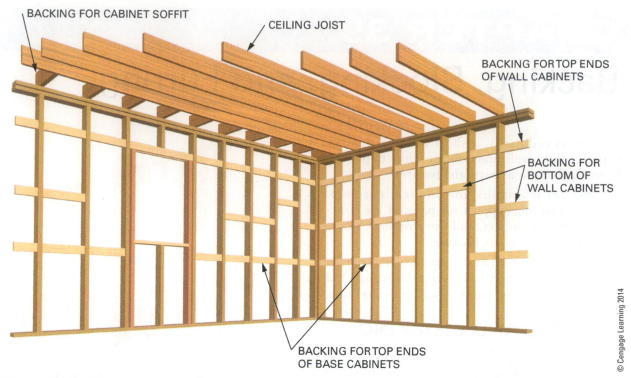

Figure 39–3 Considerable backing is needed in kitchens.

Backing should also be installed around the top of the bathtub.

In the kitchen, backing should be provided for the tops and bottoms of wall cabinets and for the tops of base cabinets. If the ceiling is to be built down to form a *soffit* at the tops of wall cabinets, backing should be installed to provide fastening for the soffit (*Figure 39–3*).

A home owner will appreciate the thoughtfulness of the builder who provides backing in appropriate locations in all rooms for the fastening of curtain and drapery hardware (*Figure 39–4*).

Blocking

Some types of blocking have already been described in earlier units. See Chapter 32 for the installation of solid lumber blocking used for bridging.

When floor sheathing panels are used as a combination subfloor and underlayment (under carpet and pad), the panel edges must be tongue-and-grooved or supported on 2-inch lumber blocking installed between joists (*Figure 39–5*).

Ladder-type blocking is needed between ceiling joists to support the top ends of partitions that run parallel to and between joists (*Figure 39–6*).

Blocks are sometimes installed to support the back edge of bathtubs (*Figure 39–7*). This stiffens the joint between the tub and the wall finish.

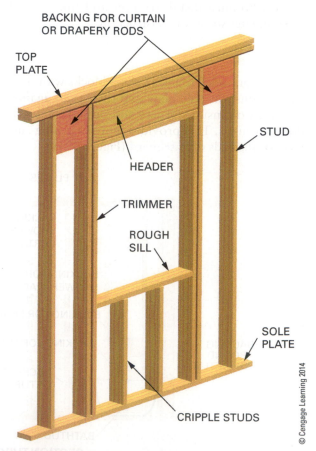

Figure 39–4 Installing backing around windows allows for easy installation of curtain and drapery hardware.

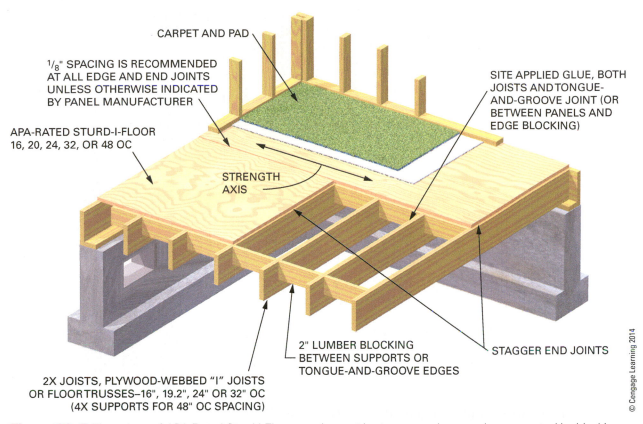

CARPET AND PAD

⅛" SPACING IS RECOMMENDED AT ALL EDGE AND END JOINTS UNLESS OTHERWISE INDICATED BY PANEL MANUFACTURER

SITE APPLIED GLUE, BOTH JOISTS AND TONGUE-AND-GROOVE JOINT (OR BETWEEN PANELS AND EDGE BLOCKING)

APA-RATED STURD-I-FLOOR 16, 20, 24, 32, OR 48 OC

STRENGTH AXIS

2" LUMBER BLOCKING BETWEEN SUPPORTS OR TONGUE-AND-GROOVE EDGES

STAGGER END JOINTS

2X JOISTS, PLYWOOD-WEBBED "I" JOISTS OR FLOOR TRUSSES–16", 19.2", 24" OR 32" OC (4X SUPPORTS FOR 48" OC SPACING)

© Cengage Learning 2014

Figure 39–5 The edges of APA-Rated Sturd-I-Floor panels must be tongue-and-grooved or supported by blocking.

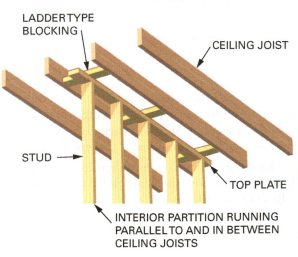

LADDER TYPE BLOCKING

CEILING JOIST

STUD

TOP PLATE

INTERIOR PARTITION RUNNING PARALLEL TO AND IN BETWEEN CEILING JOISTS

© Cengage Learning 2014

Figure 39–6 Ladder-type blocking provides support for the top plates of interior partitions.

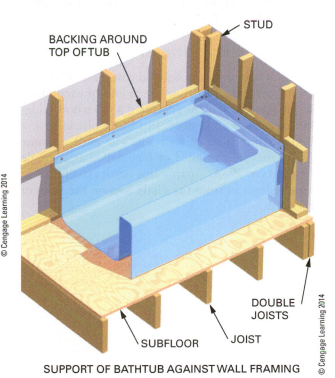

BACKING AROUND TOP OF TUB

STUD

SUBFLOOR

JOIST

DOUBLE JOISTS

© Cengage Learning 2014

SUPPORT OF BATHTUB AGAINST WALL FRAMING

Figure 39–7 Backing provides increased support between tubs and finish material.

Wall blocking is required to support the edge of rated panels used in structural shear walls. It is also used in wood foundations and panels permanently exposed to weather for weather tightness. When used for these purposes, blocking must be installed *IRC in a straight line (*Figure 39–8*).*

IRC R602.3

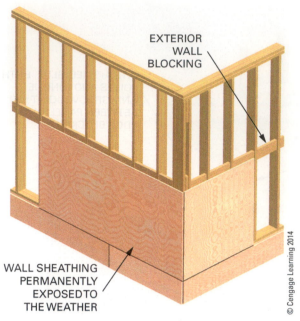

Figure 39–8 Straight-line blocking is used for structural or exterior wall sheathing.

Blocking between studs is required in walls over 8'-1" high. The purpose is to stiffen the studs and strengthen the structure. The required blocking also functions as a fire-stop in stud spaces. Blocking for these purposes may be installed in staggered *IRC fashion (*Figure 39–9*).*

R302.11

Installing Backing and Blocking

Install blocking in a staggered row. Fasten by nailing through the studs into each end of each block in the same manner as staggered solid wood bridging described previously. Installing blocking in a straight line is more difficult. The ends of some pieces may be toenailed or face nailed at an angle (*Procedure 39–A*). Snap a line across the framing. Square lines in from the chalk line on the sides of the studs.

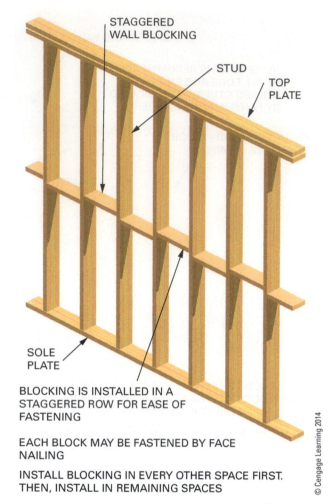

BLOCKING IS INSTALLED IN A STAGGERED ROW FOR EASE OF FASTENING

EACH BLOCK MAY BE FASTENED BY FACE NAILING

INSTALL BLOCKING IN EVERY OTHER SPACE FIRST. THEN, INSTALL IN REMAINING SPACES

Figure 39–9 Blocking used for fire-stopping may be installed in a staggered fashion for easier nailing.

It may be helpful to use a short post on one side of each stud to support the blocking while the end is being toenailed. Start the toenails in the end of the block before positioning it (*Figure 39–10*).

Backing may also be installed in a continuous length by notching the studs and fastening into its edges (*Figure 39–11*).

StepbyStep Procedures

Procedure 39-A Installing Straight, In-Line Blocking

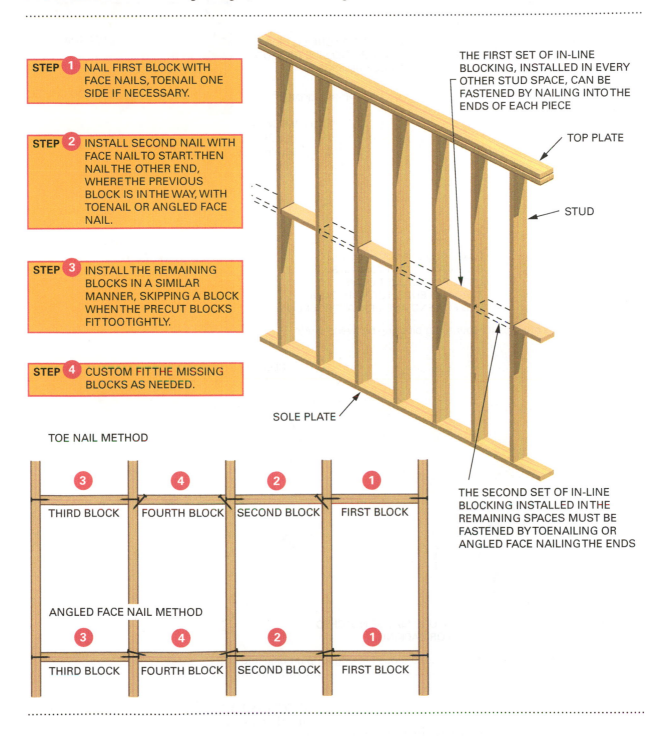

STEP 1 NAIL FIRST BLOCK WITH FACE NAILS, TOENAIL ONE SIDE IF NECESSARY.

STEP 2 INSTALL SECOND NAIL WITH FACE NAIL TO START. THEN NAIL THE OTHER END, WHERE THE PREVIOUS BLOCK IS IN THE WAY, WITH TOENAIL OR ANGLED FACE NAIL.

STEP 3 INSTALL THE REMAINING BLOCKS IN A SIMILAR MANNER, SKIPPING A BLOCK WHEN THE PRECUT BLOCKS FIT TOO TIGHTLY.

STEP 4 CUSTOM FIT THE MISSING BLOCKS AS NEEDED.

THE FIRST SET OF IN-LINE BLOCKING, INSTALLED IN EVERY OTHER STUD SPACE, CAN BE FASTENED BY NAILING INTO THE ENDS OF EACH PIECE

TOP PLATE

STUD

SOLE PLATE

THE SECOND SET OF IN-LINE BLOCKING INSTALLED IN THE REMAINING SPACES MUST BE FASTENED BY TOENAILING OR ANGLED FACE NAILING THE ENDS

TOE NAIL METHOD

3 THIRD BLOCK 4 FOURTH BLOCK 2 SECOND BLOCK 1 FIRST BLOCK

ANGLED FACE NAIL METHOD

3 THIRD BLOCK 4 FOURTH BLOCK 2 SECOND BLOCK 1 FIRST BLOCK

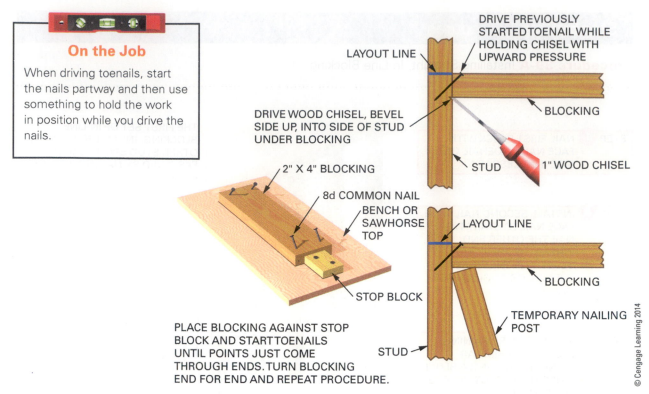

On the Job

When driving toenails, start the nails partway and then use something to hold the work in position while you drive the nails.

DRIVE PREVIOUSLY STARTED TOENAIL WHILE HOLDING CHISEL WITH UPWARD PRESSURE

LAYOUT LINE

BLOCKING

DRIVE WOOD CHISEL, BEVEL SIDE UP, INTO SIDE OF STUD UNDER BLOCKING

STUD   1" WOOD CHISEL

2" X 4" BLOCKING

8d COMMON NAIL

BENCH OR SAWHORSE TOP

LAYOUT LINE

BLOCKING

STOP BLOCK

TEMPORARY NAILING POST

STUD

PLACE BLOCKING AGAINST STOP BLOCK AND START TOENAILS UNTIL POINTS JUST COME THROUGH ENDS. TURN BLOCKING END FOR END AND REPEAT PROCEDURE.

Figure 39–10 Techniques for toenailing blocking between studs.

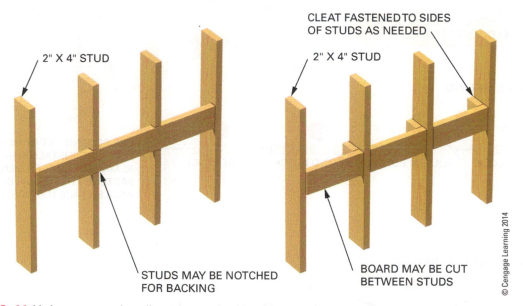

CLEAT FASTENED TO SIDES OF STUDS AS NEEDED

2" X 4" STUD

2" X 4" STUD

STUDS MAY BE NOTCHED FOR BACKING

BOARD MAY BE CUT BETWEEN STUDS

Figure 39–11 Various ways to install continuous backing for plumbing and other fixtures.

FURRING

Wall finish such as drywall is best installed on framing that is 16 inches on center. It is designed to easily handle this span. If installed on 24 inches on center it can develop an undesirable wavy surface or break with modest side pressure. Sometimes ceiling framing is installed 24 inches on center or more. To make the framing adequate for drywall, furring is often installed.

Furring is a general term used to describe wood or metal strips installed to framing or structural material that adds thickness to the section. Wood furring is 1x material that is from 2 to 6 inches wide. Wood attached to masonry walls should be treated to resist decay. Wood furring may be attached with nails or screws.

Furring attached to framing is usually installed perpendicular to the framing. If the framing surface is irregular, furring strips offer an opportunity to level and straighten the base framing for smooth flat drywall installation. If the finish material is to be installed vertically such as siding or paneling, furring is used to change the direction of the base framing (*Figure 39–12*).

Metal Furring

Metal furring is called channel or hat track. It gets its name from a hat-shaped cross section (*Figure 39–13*). It is attached using one #8 screw into each stud or joist. Metal furring may be used on walls or ceilings applied at right angles to joists. They may be applied vertically or horizontally to masonry walls.

Installing Ceiling Furring. Metal furring channels may be attached directly to structural ceiling members or suspended from them. More on the specifics on steel framing components will be addressed in the next chapter. Metal furring channels may be spliced. Overlap the ends by at least 8 inches, tie each end with wire (*Figure 39–14*).

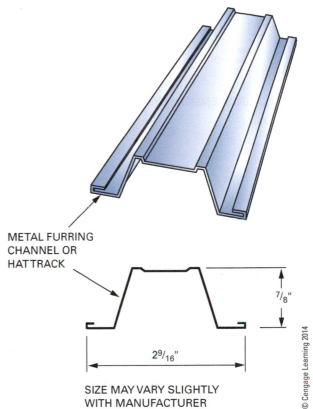

METAL FURRING
CHANNEL OR
HAT TRACK

7/8"

2⁹/₁₆"

SIZE MAY VARY SLIGHTLY
WITH MANUFACTURER

© Cengage Learning 2014

Figure 39–13 Metal furring has a hat-shaped cross section.

FURRING CHANNEL
OF HAT TRACK

VERTICAL EXTERIOR
SIDING

TREATED WOOD
FURRING

© Cengage Learning 2014

Figure 39–12 Vertical siding often requires furring to be installed on the studs.

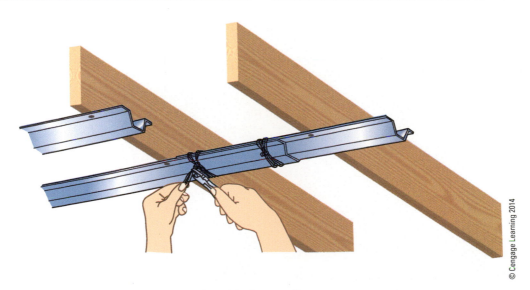

© Cengage Learning 2014

Figure 39–14 Metal furring may be spliced for added strength.

METAL FURRING CHANNEL

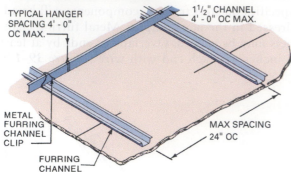

TYPICAL HANGER SPACING 4' - 0" OC MAX.

1½" CHANNEL 4' - 0" OC MAX.

METAL FURRING CHANNEL CLIP

FURRING CHANNEL

MAX SPACING 24" OC

DIRECT SUSPENSION SYSTEM

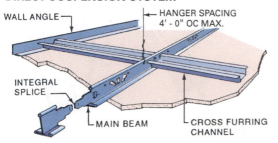

WALL ANGLE

HANGER SPACING 4' - 0" OC MAX.

INTEGRAL SPLICE

MAIN BEAM

CROSS FURRING CHANNEL

© Cengage Learning 2014

Figure 39–15 Metal channels are also used in suspended ceiling applications.

A 1-inch clearance between ends of furring and walls is typically done. Several methods of utilizing metal furring channels or steel studs in suspended ceiling applications are shown in *Figure 39–15*.

Masonry Wall Furring

When masonry walls require a wall finish, furring is installed directly to the masonry wall. (*Figure 39–16*) Fasteners must be long enough to penetrate into the the masonry about an inch. Masonry nails are harder and more brittle than common nails and must be driven with care. They can easily break and fly out while being driven. Masonry screws may also be used, but must be predrilled. This is best done by drilling through the furring and masonry as it lays against the wall. The screws are then easily fastened. Fasteners should be spaced about 24 inches apart.

Installing Wood Furring. First the perimeter furring is fastened to the wall. The perimeter furring gives ample support to the drywall. The bottom offers sufficient nailing surface for base molding. Next a 1x6 furring is fastened behind any intersecting walls. This gives adequate nailing surface for the drywall on either side of the intersecting wall. A second method places two furring strips behind the intersecting wall. A small space between the strips is acceptable so long as their edges make a bearing about 1 inch wider on each side for drywall. Remaining furring strips are installed in a similar fashion 16 inches on center.

Installing Wall Furring. Vertical application of steel furring channels is preferred. Secure the channels by staggering the fasteners from one side to the other not more than 24 inches on center (*Figure 39–17*). For horizontal application on walls, attach furring channels not more than 4 inches from the floor and ceiling. Fasten in the same manner as vertical furring.

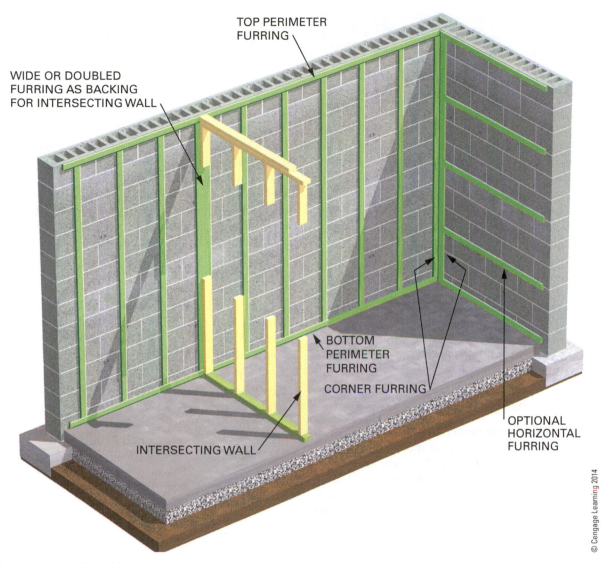

TOP PERIMETER FURRING

WIDE OR DOUBLED FURRING AS BACKING FOR INTERSECTING WALL

BOTTOM PERIMETER FURRING

CORNER FURRING

INTERSECTING WALL

OPTIONAL HORIZONTAL FURRING

© Cengage Learning 2014

Figure 39–16 Wood furring may be attached to masonry walls.

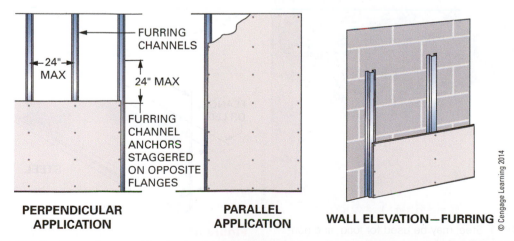

FURRING CHANNELS

24" MAX

24" MAX

FURRING CHANNEL ANCHORS STAGGERED ON OPPOSITE FLANGES

PERPENDICULAR APPLICATION

PARALLEL APPLICATION

WALL ELEVATION—FURRING

© Cengage Learning 2014

Figure 39–17 Fasteners for metal furring are installed 24 inches apart on alternate sides of the channel.

CHAPTER 40
Steel Framing

Light steel framing is used for structural framing and non-load-bearing partitions (*Figure 40–1*). Carpenters often frame these walls and apply *furring channels* of steel.

The strength of steel framing members of the same design and size may vary with the manufacturer. The size and spacing of steel framing members should be determined from the drawings or by a structural engineer.

STEEL FRAMING

The framing of steel stud walls is quite similar to framing with wood. Different kinds of fasteners and special tools are used.

Steel Framing Components

All steel frame components are cold-formed steel, which is made by one of two methods. The first method is to *press-brake* a steel section into shape. Most shapes are made using this process unless otherwise noted. The other method is to continuously *roll-form* the shape from a coil of steel. This is known as cold-rolled (CR) steel.

Before forming, the steel material is coated with zinc. Steel with this coating is commonly called gal-vanized steel. This coating is designed to protect ✳IRC the steel from corrosion.✳
R603.2.3

Figure 40–1 Steel may be used for load- and non-load-bearing building frames.

The major components of an steel frame are *studs, tracks,* and *channels*. Fasteners and accessories are needed to complete the system. The names used to refer to the dimensions of steel framing vary slightly from those used for wood (*Figure 40–2*). The length is the same for both wood and steel framing. The thickness of wood is typically $1\frac{1}{2}$ inches, whereas the term thickness for steel is the thickness of the steel sheet used to make the stud. The steel term that is similar to the *thickness* of a piece of wood is *leg* or flange size. The edge of the flange is bent over. This is referred to as the lip or return. A wood stud *width* is similar to steel stud *depth,* which is also referred to as the web.

The thicknesses of steel framing components vary according to the desired strength. Nonstructural members are 18 to 30 mil and structural

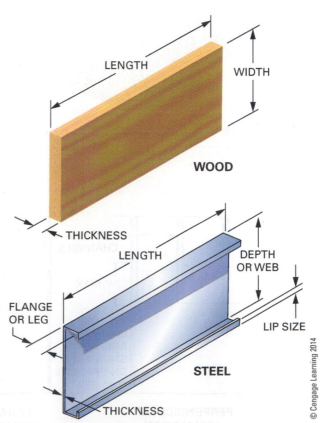

Figure 40–2 The names for steel stud material are somewhat different from those of wood.

Thickness (mills)	Thickness (in)	Thickness (mm)	Gauge No.
18	0.0188	0.457	25
27	0.0283	0.686	22
30	0.0312	0.762	20 – Drywall
33	0.0346	0.838	20 – Structural
43	0.0451	1.092	18
54	0.0566	1.372	16
68	0.0713	1.727	14
97	0.1017	2.464	12
118	0.1242	2.997	10

© Cengage Learning 2014

Figure 40–3 Steel framing is manufactured in various thicknesses.

Flange Width	Lip Length (in)
1¼"	³⁄₁₆
1³⁄₈"	³⁄₈
1⁵⁄₈"	½
2"	⁵⁄₈
2½"	⁵⁄₈
3"	⁵⁄₈
3½"	1

© Cengage Learning 2014

Figure 40–4 Flange width and lip length of steel studs are related.

framing is 33 to 118 mil. A mil is 1/1000th of an inch. (*Figure 40–3*). A gauge number is sometimes used to refer to thickness where the larger number denotes the smaller thickness. This system is obsolete for actual measurements.

Studs. Steel studs are C-shaped. They come in a variety of widths, including 1⁵⁄₈, 2½, 3½, 3⁵⁄₈, 4, 5½, 6 inches, and even numbered widths up to 16 inches. The flange widths also vary from 1¼ to 3½". This affects the lip length which ranges from ³⁄₈ inch to 1 inch (*Figure 40–4*). In general, the more steel a member contains, the stronger the member.

IRC R603.2.5.1 The stud web has holes made at intervals called knockouts or punch outs.* These allow bracing to be installed and wires to run through (*Figure 40–5*). They are spaced 24 inches apart along the stud length. Knockouts are not to be any closer to the end of the stud than 12 inches. Stock lengths of steel studs include 8, 9, 10, 12, 16, 20, and 24 feet. *IRC R603.3.2* Custom lengths are also available.*

Track. The top and bottom horizontal members of a steel-framed wall are U-shaped and called *track* or *runners*. They are installed on floors and ceilings to receive the studs. They are manufactured by thickness, widths, and leg size to match studs (*Figure 40–6*). Track is available in standard lengths of 10 feet.

Channels. Steel cold-rolled channels (CRCs) are formed from 54-mil (16-gauge) steel. They are available in several widths. They come in lengths of 10, 16, and 20 feet. Channels are used in suspended ceilings and through wall studs. When used for lateral bracing of walls, the channel is inserted through the stud punchouts. It is fastened with welds or clip angles to the studs (*Figure 40–7*).

Furring Channels. Furring channels (or *hat track*) are hat-shaped pieces made of 18- and 33-mil (25- and 20-gauge) steel. Their overall cross section size is ⁷⁄₈ inch by 2⁹⁄₁₆ inches. They are available in lengths of 12 feet (*Figure 40–8*). Furring channels are applied to walls and ceilings for the screw attachment of gypsum panels. Framing members may exceed spacing limits for various coverings. Furring can then be installed to meet spacing requirements and provide necessary support for the surfacing material.

Nomenclature. The Steel Stud Manufactures Association has standardized the nomenclature used to refer to steel components. It has a number, a letter, and then two more numbers. The first number denotes the width, and the letter indicates whether it is a stud, track, or furring. The next number is the flange thickness, and the last is the steel thickness (*Figure 40–9*).

Fasteners. Steel framing members and components are most commonly fastened with screws. Screws come in a variety of head styles and driving slots. Self-piercing points, like those of drywall

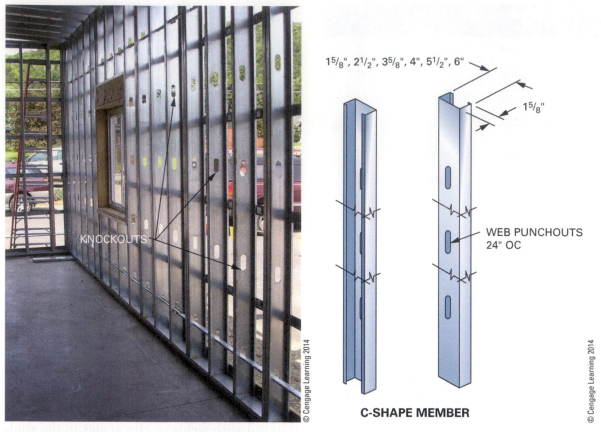

1⁵⁄₈", 2¹⁄₂", 3⁵⁄₈", 4", 5¹⁄₂", 6"

1⁵⁄₈"

WEB PUNCHOUTS
24" OC

C-SHAPE MEMBER

© Cengage Learning 2014

KNOCKOUTS

© Cengage Learning 2014

Figure 40–5 Knockouts allow for bracing and wiring.

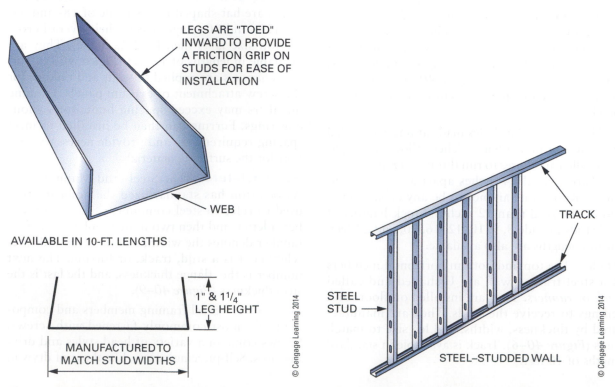

LEGS ARE "TOED"
INWARD TO PROVIDE
A FRICTION GRIP ON
STUDS FOR EASE OF
INSTALLATION

WEB

AVAILABLE IN 10-FT. LENGTHS

1" & 1¹⁄₄"
LEG HEIGHT

MANUFACTURED TO
MATCH STUD WIDTHS

TRACK

STEEL
STUD

STEEL–STUDDED WALL

© Cengage Learning 2014

© Cengage Learning 2014

Figure 40–6 Cold-formed channels called *track* are the top and bottom plates of steel-stud walls.

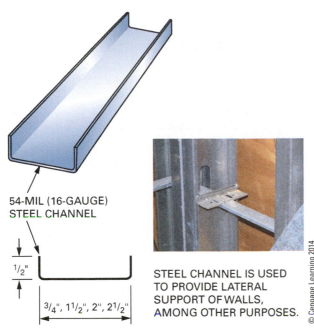

54-MIL (16-GAUGE)
STEEL CHANNEL

1/2"

3/4", 1 1/2", 2", 2 1/2"

STEEL CHANNEL IS USED
TO PROVIDE LATERAL
SUPPORT OF WALLS,
AMONG OTHER PURPOSES.

© Cengage Learning 2014

Figure 40–7 Cold-rolled channels are used to stiffen the framing members of walls and ceilings.

materials being fastened. A minimum of three threads should penetrate the steel.✳

✳IRC R603.2.4

Plywood may be attached using specially designed pneumatic nails. Powder-actuated fasteners are often used to attach the framed wall to other support material such as steel or masonry.

STRUCTURAL STEEL FRAMING

Layout and framing of steel-framed walls are similar to working with wood-framed walls. Concepts of measurement, square, and plumb all apply. The screw gun is used extensively. The assembly techniques between structural and nonstructural steel walls vary slightly.

Structural Wall Layout

Structural walls are constructed in the same fashion as wood walls. Layout begins by snapping lines on the subfloor where walls are to be constructed. Next, the top and bottom tracks are cut for length. Walls should be no longer than what the work crew can lift into place.

Splices in the track is done between studs. The splicing block must be at least 6 inches long with

screws, may be used on lighter-gauge studs. Heavy-gauge steel requires self-drilling points (*Figure 40–10*). Screws should be about 1/2 inch longer than the

© Cengage Learning 2014

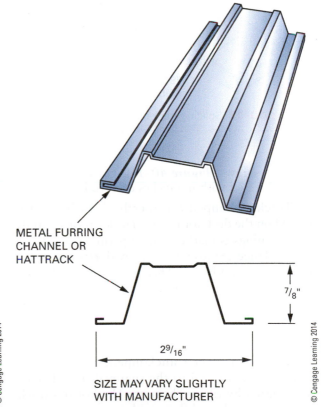

METAL FURRING
CHANNEL OR
HAT TRACK

7/8"

2 9/16"

SIZE MAY VARY SLIGHTLY
WITH MANUFACTURER

© Cengage Learning 2014

Figure 40–8 Furring channels (hat track) are used in both ceiling and wall installations.

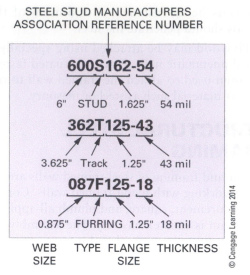

STEEL STUD MANUFACTURERS
ASSOCIATION REFERENCE NUMBER

600S162-54

6"   STUD   1.625"   54 mil

362T125-43

3.625"   Track   1.25"   43 mil

087F125-18

0.875"   FURRING   1.25"   18 mil

WEB   TYPE   FLANGE   THICKNESS
SIZE   SIZE

Figure 40–9 Manufacturer nomenclature for steel studs.

Figure 40–10 Heavy gauge steel studs require self-drilling screws. Non-structural studs may use self-piercing.

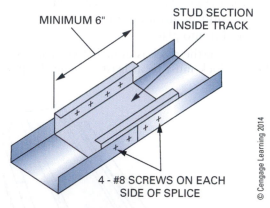

MINIMUM 6"

STUD SECTION
INSIDE TRACK

4 - #8 SCREWS ON EACH
SIDE OF SPLICE

Figure 40–11 Splicing track in steel framing.

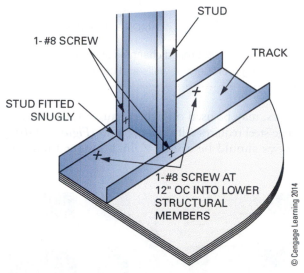

STUD

1- #8 SCREW

TRACK

STUD FITTED
SNUGLY

1- #8 SCREW AT
12" OC INTO LOWER
STRUCTURAL
MEMBERS

Figure 40–12 Load-bearing steel studs require four screws.

eight #8 screws (*Figure 40–11*). Thus splices cannot be located within 3 inches of a stud.

Tracks, clamped together back-to-back, are placed on the deck for layout. The location of studs and openings is marked with a permanent marker on the flanges. Studs must align with framing members above and below. This allows the load to be positively transferred to the foundation. The corner studs and other wall intersections are also marked.

Structural Wall Assembly

The tracks are then unclamped, and the bottom track is moved to the snapped line. Tracks are rotated by tilting them up. The bottom track is secured to the subfloor with temporary screws through the flanges. Studs are then placed into the bottom and top tracks. They should all be facing the same direction and have the knockouts aligned. Placing the colored ends of the studs into the bottom track will ensure that this happens.

The studs must touch the web of the track. This is done by tapping the tracks tight to the stud before fastening. The stud and track flanges must be temporarily clamped together to hold them tight as the screw is driven. Otherwise an undesirable space may result between the flanges. Load-bearing studs require four screws, two at each end (*Figure 40–12*).

The wall is lying on its side at this point, so only half of the screws can be driven. After lifting the wall into position, fasten the second pair of screws in the stud is fastened. Also, fasten screws through the bottom track into the framing below at 12″ on center.

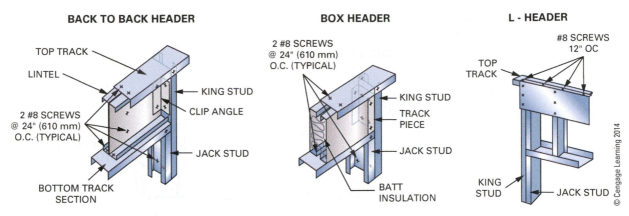

Figure 40–13 Supports over openings in structural steel framing.

King studs are positioned with the web side facing the center of the opening. Jack studs are reversed. Thus the jack studs and king studs are placed back-to-back to allow for their webs to be screwed together with #8 screws at about 24″ on center vertically. An additional track may be installed to cover the C-shape of the jack. This simplifies attaching the jamb of the door to the jack stud. High wind or seismic areas may require two king studs each side.

Headers. Headers, or *lintels*, are designed to carry loads from above to around the opening. This may be done by using methods called *box, back-to-back,* and *L-headers* (**Figure 40–13**). Box and back-to-back use the same material: only the header pieces are reversed. L-headers use one or two L-shaped pieces applied over the cripple studs over the opening.

Back-to-back headers are fastened through the webs with two #8 screws at 24 inches on center. The ends of the header are attached to the web of the king stud with a clip angle and four to eight screws, depending header span. For example, if the header is less than 8 feet then four screws may be use.

Box headers are installed with tracks attached to the bottom and top. These are attached with two #8 screws at 24″ on center. Ends of the headers are attached with clip angles with four to eight screws depending on header span. For insulated walls, the box header must be installed as it is assembled.

Framing at Intersections. Framing intersections and corners follows similar wood framing techniques (*Figure 40–14*). Corners may be assembled in two fashions; both provide adequate support of interior finish. Style A in must have a 2 inch space as noted in figure. This allows insulation, if necessary, to be installed later. There is also sufficient room to run wire around the

corner. Style B is better if the wall must be insulated, but makes running wires a little difficult. Style C shows the partition intersections made with a 2×6 attached. This allows for excellent support of interior finish, room for insulation, and running wires.

Intersecting walls must be tied together. This may be done with gusset plate of same thickness as the track with four #8 screws into each wall track. Alternatively the tracks may be overlapped and secured with four #8 screws.

Window Framing. Framing around a window rough opening may be done with an addition of track material wrapping the opening (*Figure 40–15*). Dimension lumber may also be installed to wrap the rough opening. A standard 2× lumber is screwed into to the track lining the rough opening or directly to the jack stud. Both of these framing methods simplify window installation and finish later by allowing residential wood carpentry techniques.

Horizontal Wall Bracing. Walls eight-feet high or less require one row of horizontal bracing at mid-span to resist stud twist when under load. This is similar to floor bridging and may be accomplished by strapping. Walls between 8 and 12 feet require two rows evenly spaced vertically. Strapping is installed on two sides of the stud and must be at least 1½ inch wide and 35 mil thick. It must be pulled taunt and fastened with one #8 screw in each stud on each side.

Both ends of wall bracing must be carefully secured to ensure the twisting force is adequately resisted. One method is by two screws into a corner stud assembly (*Figure 40–16*). Another method uses a block of track material: Track is cut, bent, and fastened between two studs. Strapping is then attached with #8 screws. If structural sheathing is used, the exterior side strap may be omitted, but interior side bracing is still needed.

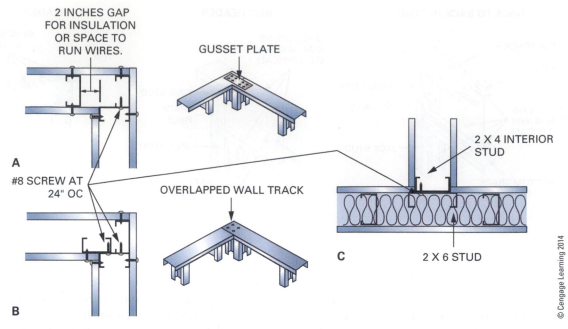

Figure 40–14 Details for wall framing at corners and intersections.

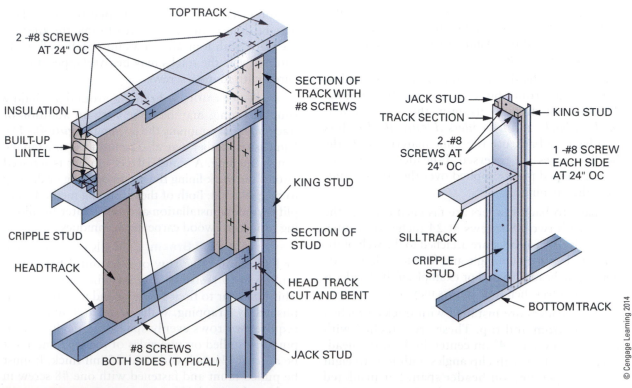

Figure 40–15 Track may be used to wrap a window rough opening.

Wind Bracing. Wall sections may be wind braced using metal straps (*Figure 40–17*). Bracing must be at least 45 mil and 3 inches wide. Attachment at each end must be with eleven #12 screws into the tracks and into the doubled studs installed at each end of bracing (*Figure 40–18*). The doubled stud webs should be installed back to back to allow insulation full access to the wall cavity.

Structural sheathing may be substituted for wind bracing. Its longer length axis should be positioned vertically with #8 screws or pneumatic pins. Care must be taken to ensure the panels are tight to the studs.

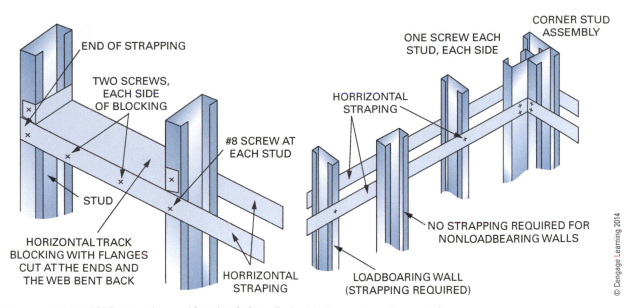

Figure 40–16 Wall strapping and bracing is installed with the ends well secured.

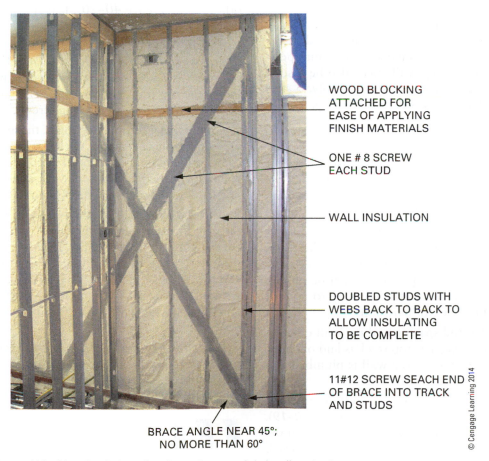

Figure 40–17 Wind bracing is installed from the top of the wall to the bottom.

© Cengage Learning 2014

Figure 40–18 Each end of wind bracing is attached with eleven screws.

© Cengage Learning 2014

Figure 40–19 Bottom track is fastened to the floor next to chalk lines.

NONSTRUCTURAL STEEL FRAMING

Constructing a nonstructural steel stud wall may be done the same as previously discussed in the structural section. Assembled and lifted into position. Non-load-bearing walls may also be assembled in place. Door partitions may involve a steel door bucks, chase walls, and metal furring.

Nonstructural Wall Layout

Sometimes interior walls must be constructed after the ceiling framing. The top and bottom track are fastened to the floor and ceiling before attaching the studs. This ensures that the wall is exactly the correct height.

Begin layout by snapping lines on the floor. Plumb up from partition ends. Snap lines on the ceiling. Make sure that partitions will be plumb. Using a laser level is an efficient way to lay out floor and ceiling lines for partitions.

Lay out the stud spacing and the wall opening on the bottom track. The top track is laid out after the first stud away from the wall is plumbed and fastened.

Installing Track. Fasten track to floor and ceiling so one edge is to the chalk line (*Figure 40–19*). Make sure both floor and ceiling track are on the same side of the line. Leave openings in floor track for door frames. Allow for the width of the door and thickness of the door frame. Tracks are usually fastened into concrete with powder-driven fasteners. Stub concrete nails or masonry screws may also be used. Fasten into wood with 1¼-inch oval head *IRC screws.*

R603.3.1

Attach the track with two fasteners about 2 inches from each end and a maximum of 24 inches on center in between. At corners, extend one track to the end. Then butt or overlap the other track (*Figure 40–20*). It is not desirable or necessary to make mitered joints.

Install backing, if necessary, between joists or *trusses* where the top track will be attached. Plumb up from the bottom track to the ceiling backing to locate the top track (*Figure 40–21*). Snap lines as needed and fasten with framing screws.

To cut metal framing to length, tin snips may be used on 18-mil (25-gauge) steel. Using tin snips becomes difficult on thicker metal. A power miter box, commonly called a *chop saw,* with a metal-cutting saw blade is the preferred tool (*Figure 40–22*).

Installing Studs. Cut the necessary number of full-length studs needed. For ease of installation, cut them about ¼ inch short. Install studs at partition intersections and corners. Fasten to bottom and top track. Use ³⁄₈-inch self-drilling pan head screws. If moisture may be present where a stud butts an exterior wall, place a strip of asphalt felt between the stud and the wall.

Place the first stud in from the corner between track. Fasten the bottom in position at the layout line. Plumb the stud. Using a magnetic level can be very helpful. Clamp the top end when plumb, and

CAUTION

The sharp ends of cut metal can cause serious injury. A pointed end presents an even greater danger. Avoid miter cuts on thin metal. Do not leave short ends of cut metal scattered around the job site. Dispose of them in a container as you cut them.

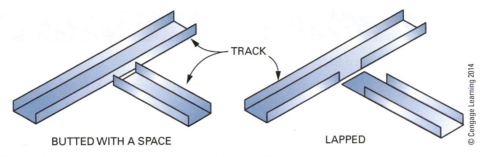

BUTTED WITH A SPACE   TRACK   LAPPED

© Cengage Learning 2014

Figure 40–20 Track may be butted or overlapped at intersections.

© Cengage Learning 2014

Figure 40–21 Top track may be located on ceiling members or blocking using a level.

© Cengage Learning 2014

Figure 40–22 Steel studs may be cut using a chop saw.

fasten to the top track. Lay out the stud spacing on the top track from this stud (*Procedure 40–A*).

Place all full-length studs in position between track with the open side facing in the same direction. The web punchouts should be aligned vertically. This provides for lateral bracing of the wall and the running of plumbing and wiring (*Figure 40–23*). Fasten all studs except those on each side of door openings securely to top and bottom track.

Wall Openings

Several methods are used to frame around door and window openings. One method involves installing wood jack studs and sills (*Figure 40–24*). This allows for conventional installation of interior and exterior finishes. A second method requires wood door frames to be installed with screws to the steel framing (*Figure 40–25*). A third method uses all steel studs with a metal door frame called a steel buck. There are two styles of bucks.

A one-piece metal door frame must be installed before the gypsum board is applied. A three-piece, knocked-down frame is set in place after the wall covering is applied (*Figure 40–26*).

Framing for a Three-Piece Frame. First, place full-length studs on each side of the opening in a plumb position. Fasten securely to the bottom and top plates. Cut a length of track for use as a header. Cut up to 6 inches longer than the width of the opening. Cut the flanges in appropriate places. Bend the web to fit over the studs on each side of the opening. Fasten the fabricated header to the

StepbyStep Procedures

Procedure 40-A Assembling a Steel Stud Wall

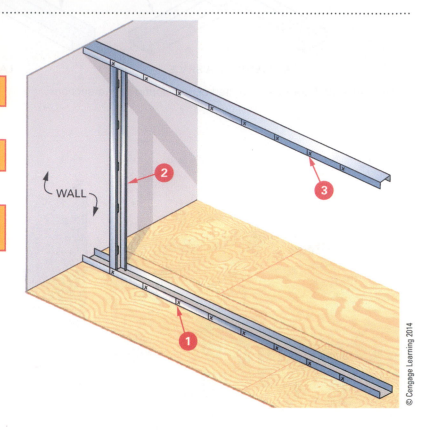

STEP 1 LAY OUT STUD LOCATION ON BOTTOM TRACK.

STEP 2 PLUMB UP FIRST STUD AWAY FROM WALL.

STEP 3 LAY OUT STUD SPACING ON TOP TRACK FROM PLUMBED STUD.

WALL

© Cengage Learning 2014

Figure 40-23 Punchouts should line up for bracing and mechanicals.

© Cengage Learning 2014

studs at the proper height. Install jack studs over the opening in positions that continue the regular stud spacing (*Figure 40-27*). Window openings are framed in the same manner. However, a rough sill is installed at the bottom of the opening.

Cripples are placed both above and below (*Figure 40-28*). ✱

Framing for a One-Piece Frame. Place the studs on each side of the opening. Do not fasten to the track. Set the one-piece door frame in place.

✱IRC
R603.6
and
R603.7

Figure 40-24 Window and door openings may be lined with wood for easier fastening of finish materials.

Level the door frame header by shimming under a jamb, if necessary. Fasten the bottom ends of the doorjambs to the floor in the proper location. Fasten the studs to the doorjambs. Then, fasten the studs to the bottom track. Plumb the door frame. Clamp the stud to the top track, and fasten with screws. Install header and jack studs in the same manner as described previously (**Procedure 40–B**).

The steel framing just described is suitable for average-weight doors up to 2′-8″ wide. For wider and heavier doors, the framing should be strengthened by using 33-mil or greater steel framing. Also installing more king studs strengthens the wall.

Chase Walls

A **chase wall** is made by constructing two closely spaced, parallel walls for the running of plumbing, heating and cooling ducts, and similar items. They are constructed in the same manner as described previously. However, the spacing between the outside edges of the wall frames must not exceed 24 inches.

The studs in each wall should be installed with the flanges running in the same direction. They should be directly across from each other. The walls should be tied to each other either with pieces

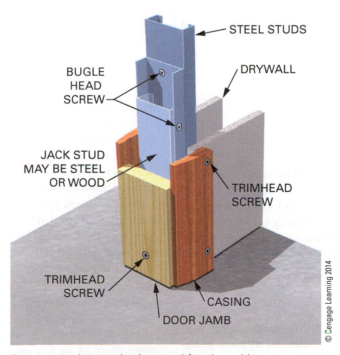

STEEL STUDS

DRYWALL

BUGLE HEAD SCREW

JACK STUD MAY BE STEEL OR WOOD

TRIMHEAD SCREW

TRIMHEAD SCREW

CASING

DOOR JAMB

Figure 40-25 Wood door frames may be attached to steel framing with screws.

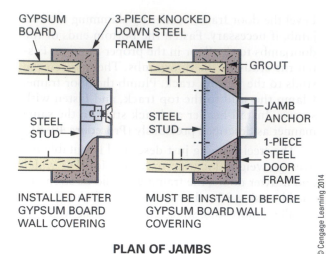

PLAN OF JAMBS

Figure 40–26 A one-piece or three-piece knocked-down metal door frame may be used in steel wall framing.

Figure 40–28 Rough sills are installed similarly to non-load-bearing headers.

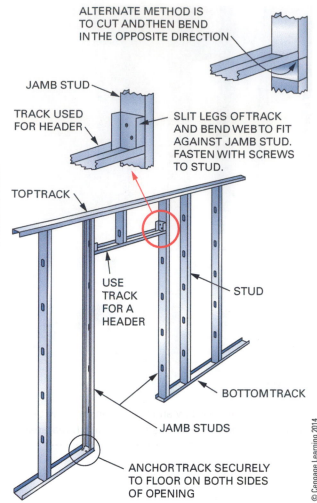

Figure 40–27 An opening framed for a three-piece, knocked-down, metal door frame.

of 12-inch-wide gypsum board or short lengths of steel stud. If the wall studs are not opposite to each other, install lengths of steel stud horizontally inside both walls. Tie together with shorter lengths of stud material spaced 24 inches on center (*Figure 40–29*). Wall ties should be spaced 48 inches on center vertically.

Installing Metal Furring

Metal furring may be used on ceilings applied at right angles to joists. They may be applied vertically or horizontally to framed or masonry walls. Space metal furring channels a maximum of 24 inches on center.

Ceiling Furring. Metal furring channels may be attached directly to structural ceiling members or suspended from them. For direct attachment, saddle tie with double-strand 43-mil (18-gauge) wire to each member (*Figure 40–30*). Leave a 1-inch clearance between ends of furring and walls. Metal furring channels may be spliced. Overlap the ends by at least 8 inches. Tie each end with wire. Steel studs may be used with their open side up for furring when support framing is widely spaced. Several methods of utilizing metal furring channels or steel studs in suspended ceiling applications are shown in *Figure 40–31*.

Wall Furring. Vertical application of steel furring channels is preferred. Secure the channels by staggering the fasteners from one side to the other not more than 24 inches on center. For horizontal application on walls, attach furring channels not more than 4 inches from the floor and ceiling. Fasten in the same manner as vertical furring.

StepbyStep Procedures

Procedure 40–B Assembling a One-Piece Door Frame

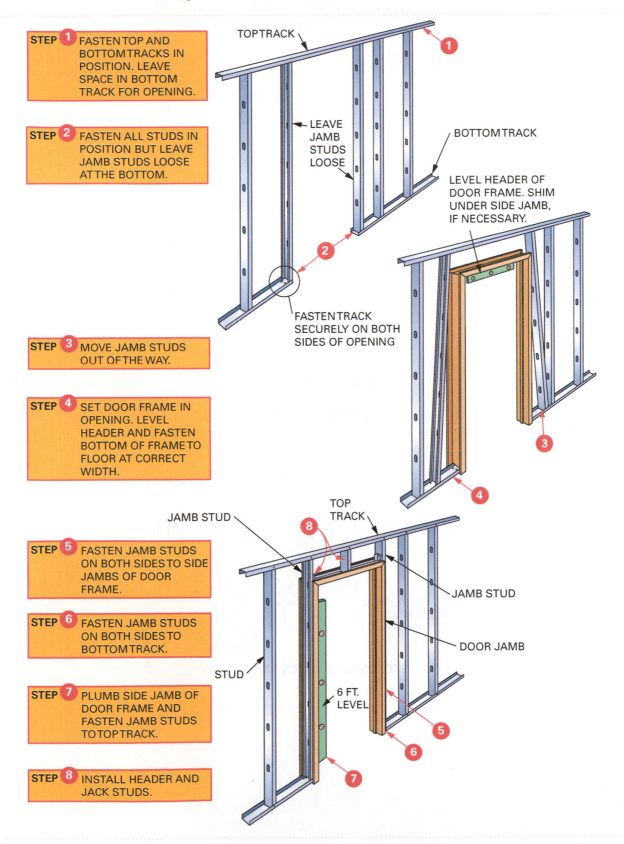

STEP 1 FASTEN TOP AND BOTTOM TRACKS IN POSITION. LEAVE SPACE IN BOTTOM TRACK FOR OPENING.

STEP 2 FASTEN ALL STUDS IN POSITION BUT LEAVE JAMB STUDS LOOSE AT THE BOTTOM.

STEP 3 MOVE JAMB STUDS OUT OF THE WAY.

STEP 4 SET DOOR FRAME IN OPENING. LEVEL HEADER AND FASTEN BOTTOM OF FRAME TO FLOOR AT CORRECT WIDTH.

STEP 5 FASTEN JAMB STUDS ON BOTH SIDES TO SIDE JAMBS OF DOOR FRAME.

STEP 6 FASTEN JAMB STUDS ON BOTH SIDES TO BOTTOM TRACK.

STEP 7 PLUMB SIDE JAMB OF DOOR FRAME AND FASTEN JAMB STUDS TO TOP TRACK.

STEP 8 INSTALL HEADER AND JACK STUDS.

TOP TRACK

LEAVE JAMB STUDS LOOSE

BOTTOM TRACK

FASTEN TRACK SECURELY ON BOTH SIDES OF OPENING

LEVEL HEADER OF DOOR FRAME. SHIM UNDER SIDE JAMB, IF NECESSARY.

JAMB STUD

TOP TRACK

JAMB STUD

DOOR JAMB

STUD

6 FT. LEVEL

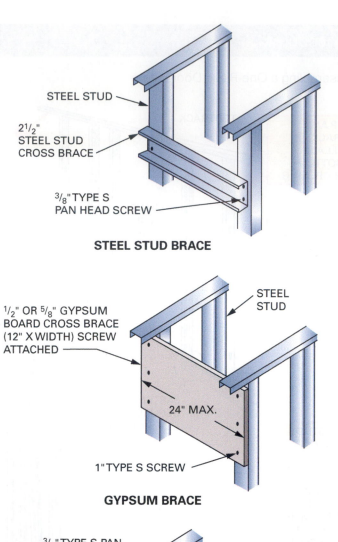

STEEL STUD BRACE

STEEL STUD

2¹⁄₂"
STEEL STUD
CROSS BRACE

³⁄₈" TYPE S
PAN HEAD SCREW

¹⁄₂" OR ⁵⁄₈" GYPSUM
BOARD CROSS BRACE
(12" X WIDTH) SCREW
ATTACHED

STEEL
STUD

24" MAX.

1" TYPE S SCREW

GYPSUM BRACE

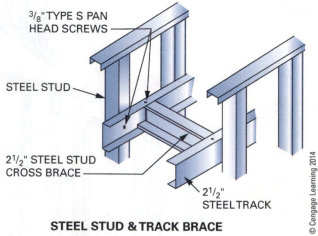

³⁄₈" TYPE S PAN
HEAD SCREWS

STEEL STUD

2¹⁄₂" STEEL STUD
CROSS BRACE

2¹⁄₂"
STEEL TRACK

STEEL STUD & TRACK BRACE

© Cengage Learning 2014

Figure 40–29 Chase wall construction details.

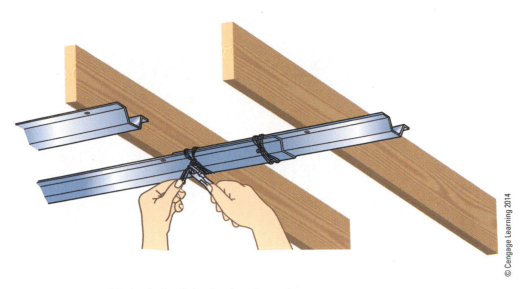

Figure 40–30 Method of splicing furring channels.

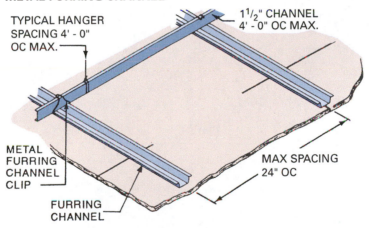

METAL FURRING CHANNEL

TYPICAL HANGER SPACING 4'-0" OC MAX.

1¹/₂" CHANNEL 4'-0" OC MAX.

METAL FURRING CHANNEL CLIP

FURRING CHANNEL

MAX SPACING 24" OC

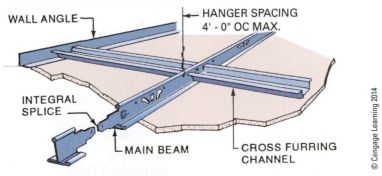

DIRECT SUSPENSION SYSTEM

WALL ANGLE

HANGER SPACING 4'-0" OC MAX.

INTEGRAL SPLICE

MAIN BEAM

CROSS FURRING CHANNEL

Figure 40–31 Metal channels are also used in suspended ceiling applications.

© Cengage Learning 2014

DECONSTRUCT THIS

Carefully study *Figure 40–-32* and think about what is wrong and/or what is right. Consider all possibilities. What construction practice or method is different in your area of the country?

Figure 40–32 This photo shows a wall built with steel studs.

KEY TERMS

backing	flange size	joist hangers	steel buck
baseboard	furring channels	load-bearing partitions (LBPs)	stub joists
blocking	gable end		underlayment
chase wall	galvanized steel	non-load-bearing partitions (NLBPs)	web
cold-rolled channels (CRCs)	hip roof		

REVIEW QUESTIONS

Select the most appropriate answer.

1. Wood-framed load-bearing partitions
   a. have a single top plate.
   b. carry no load from the roof.
   c. are constructed like exterior walls.
   d. are erected after the roof is tight.

2. The doubled top plate of a load-bearing partition
   a. is a 1 × member.
   b. laps the plate of the exterior wall.
   c. butts the top plate of the exterior wall.
   d. is applied after the ceiling joists are installed.

3. What is the rough opening height for a 6'-8" door if the finish floor is ¾ inch thick, a ½-inch clearance is allowed between the door and the finish floor, and the jamb thickness is ¾ inch?
   a. 6'-9" c. 6'-10"
   b. 6'-9½" d. 6'-10½"

4. The end stud of partitions that butt against another wall
   a. must be straightened as it is nailed to the intersecting wall.
   b. must be fastened to the intersecting wall with screws.
   c. is usually left out until the wall is erected.
   d. is usually not fastened near its center.

5. A roof load is supported by
   a. the strong triangle frame formed by the ceiling joists and rafters.
   b. framing members aligned vertically.
   c. ceiling joists resisting the outward thrust of rafters.
   d. all of the above.

6. Ceiling joists are installed
   a. with their end joints lapped at the bearing partition.
   b. full length along the building width.
   c. by being fastened to the rafter.
   d. all of the above.

7. The ends of ceiling joists are cut to the slope of the roof
   a. for easy application of the wall sheathing.
   b. so they will not project above the rafters.
   c. to mark the crowned edges up.
   d. all of the above.

8. Stub joists
   a. run at right angles to regular ceiling joists.
   b. are used on high-sloped roofs.
   c. are also called blocking and are installed between regular joists.
   d. span from the bearing partition to the exterior wall.

9. Openings in the ceiling may be framed with
   a. single headers and no trimmers.
   b. doubled headers and trimmers.
   c. joist hangers.
   d. all of the above.

10. Non-load-bearing partition headers are usually
    a. stud width material installed on the flat with cripples above.
    b. a doubled 2 × 6 with a plywood spacer.
    c. a doubled 2 × 10 with a plywood spacer.
    d. designed for size by a structural engineer.

11. Blocking and backing are installed
    a. using up scraps of lumber first.
    b. as a nail base for cabinets.
    c. to secure parallel partitions to ceiling joists.
    d. all of the above.

12. The measurement system for sizing steel studs for strength and thickness that uses numbers in a seemingly reverse order is
    a. metric.
    b. mil or 1/1000th.
    c. gauge.
    d. none of the above.

13. The depth of a steel stud
    a. is referred to as a web.
    b. is similar to width in a wood stud.
    c. is sized similar to wood studs.
    d. all of the above.

14. When working with steel framing, note that
    a. special self-tapping screws are needed.
    b. top plates are usually doubled.
    c. studs are also called track.
    d. all of the above.

15. The top track of non-load-bearing steel partitions is located by

    a. a long level.
    b. by plumbing with a laser transit from the sole plate.
    c. by hanging a plumb bob from the ceiling to the sole plate.
    d. all of the above

16. Furring is typically installed

    a. perpendicular to the framing members.
    b. on framing that is more than 16" on center.
    c. made of wood or metal.
    d. all of the above

17. Structural steel framing is made from steel that is

    a. more than 33 mil.
    b. less than 33 mil.
    c. more than 33 gauge.
    d. less than 33 gauge.

18. Top and bottom tracks of steel framing are

    a. spliced in structural framing with a 6" piece.
    b. butted with no splice piece in nonstructural framing.
    c. shaped similar to cold rolled channel.
    d. all of the above.

19. Structural headers in steel framing are made with

    a. LVL and PSL members.
    b. box, back-to-back, and L-shaped designs.
    c. I- and W-shaped steel beams.
    d. all of the above.

20. Structural wall assemblies are braced with

    a. structural sheathing.
    b. metal strapping.
    c. X-bracing spanning from top to bottom track.
    d. all of the above.

UNIT 15
Roof Framing

The ability to lay out rafters and frame all types of roofs is an indication of an experienced carpenter. On most jobs, the lead carpenter lays out the different rafters, while workers make duplicates of them. Those persons aspiring to supervisory positions in the construction field must know how to frame various kinds of roofs.

OBJECTIVES

After completing this unit, the student should be able to:

- describe roof types and define roof framing terms.
- describe the members of gable, shed, and flat roofs.
- lay out common rafters and erect gable, shed, and flat roofs.
- lay out and install gable studs.
- erect a trussed roof.
- apply roof sheathing.

SAFETY REMINDER

Roof framing involves sloped and often slippery surfaces while also requiring greater mental energy to perform more complicated framing details. Maintain a constant awareness of your surroundings and what your feet are doing as you focus on roofing details.

CHAPTER 41

Roof Types and Terms

Roofs may be framed in a stick-built fashion using rafters and ridgeboards, or they may be framed with trusses. Trusses will be discussed later in Chapter 43. Careful thought and patience are required for a carpenter who wants to become proficient at roof framing. Knowledge of diverse roof types and the associated terms is essential to framing roofs. Carpenters demonstrate their craftsmanship when constructing a roof frame.

ROOF TYPES

Several roof styles are in common use. These roofs are described in the following material and are illustrated in *Figure 41–1*.

Gable Roof. The gable roof is the most common roof style. Two sloping roof surfaces meet at the top. They form triangular shapes at each end of the building called *gables*.

Shed Roof. The shed roof slopes in one direction. It is sometimes referred to as a *lean-to*. It is commonly used on additions to existing structures. It is also used extensively on contemporary homes.

Hip Roof. The hip roof slopes upward to the ridge from all walls of the building. This style is used when the same overhang is desired all around the building. The hip roof eliminates maintenance of gable ends.

Intersecting Roof. An intersecting roof is required on buildings that have wings. Where two roofs intersect, valleys are formed. This requires several types of rafters.

Gambrel Roof. The gambrel roof is a variation of the gable roof. It has two slopes on each side instead of one. The lower slope is much steeper than the upper slope. It is framed somewhat like two separate gable roofs.

Mansard Roof. The mansard roof is a variation of the hip roof. It has two slopes on each of the four sides. It is framed somewhat like two separate hip roofs.

Butterfly Roof. The butterfly roof is an inverted gable roof. It is used on many modern homes. It resembles two shed roofs with their low ends placed against each other.

Other Roofs. Other roof styles are a combination of the styles just mentioned. The shape of the roof can be one of the most distinctive features of a building.

ROOF FRAME MEMBERS

A roof frame may consist of a ridgeboard and common, hip, or valley rafters. It may also have hip jacks, valley jacks, cripple jack rafters, collar ties, and gable end studs (*Figure 41–2*). Each of these components may be laid out and cut using similar mathematical principles and theory. Rafters are the sloping members of the roof that support the roof covering.

Ridgeboard. The ridgeboard is the uppermost member of a roof. Although not absolutely necessary, the ridgeboard simplifies the erection of the roof. It provides a place for the upper ends of rafters to be secured before the sheathing is applied.

Common Rafters. Common rafters are so named because they are the most common rafter. They make up the major portion of the roof, spanning from the ridgeboard to the wall. They extend at right angles from the plate to the ridge. They are used as a basis or starting point for all other rafters.

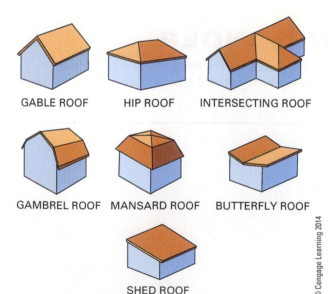

GABLE ROOF   HIP ROOF   INTERSECTING ROOF

GAMBREL ROOF   MANSARD ROOF   BUTTERFLY ROOF

SHED ROOF

© Cengage Learning 2014

Figure 41–1 Many roof styles can be used for residential buildings.

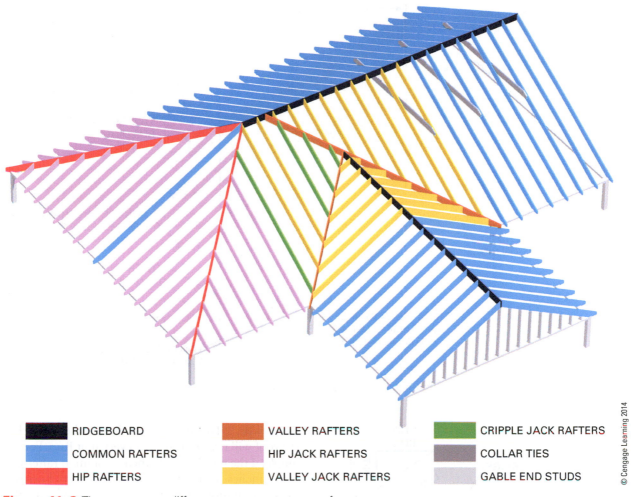

RIDGEBOARD

COMMON RAFTERS

HIP RAFTERS

VALLEY RAFTERS

HIP JACK RAFTERS

VALLEY JACK RAFTERS

CRIPPLE JACK RAFTERS

COLLAR TIES

GABLE END STUDS

© Cengage Learning 2014

Figure 41–2 There are many different components to a roof system.

Hip Rafters.   Hip rafters form an intersection of two roof sections. They project out from the roof plane, forming an outside corner. They usually run at a 45-degree angle to the plates.

Valley Rafters.   Valley rafters also create the intersection of two roof sections but project inward, forming an inside corner. Valley and hip rafters are similar in theory and layout.

Jack Rafters.   Jack rafters come in three types: hip jack rafters, valley jack rafters, and cripple jack rafters. They are essentially common rafters that are cut shorter to land on a hip, valley, or both.

Collar Ties.   Collar ties are horizontal members that add strength to the common rafter. They are raised and shortened ceiling joists.

Gable End Studs.   Gable end studs form the wall that closes in the triangular wall area under a gable roof.

Understanding rafters can be confusing at first, and it may take some time to become comfortable

with rafters. The various types have similarities and differences that make them unique (*Figure 41–3*). All rafters except the valley jack start from the wall plate. All rafters except the hip jack land on the ridge. Hip and valley rafters have longer lengths and tails because they are not parallel to the other rafters. They run at an angle from the plate to the ridge.

ROOF-FRAMING TERMS

Roof-framing theory has many terms. Successful roof construction begins with understanding these terms. They are defined as follows and most are illustrated in *Figure 41–4*. The unit Run is always a fixed number. A carpenter can always expect the total span and unit rise to be given. The other terms and measurements are marked with a square or calculated. These terms will also be discussed in more detail in later Chapters.

Unit Run.   The unit Run is a standardized horizontal distance. It is the number that is used as

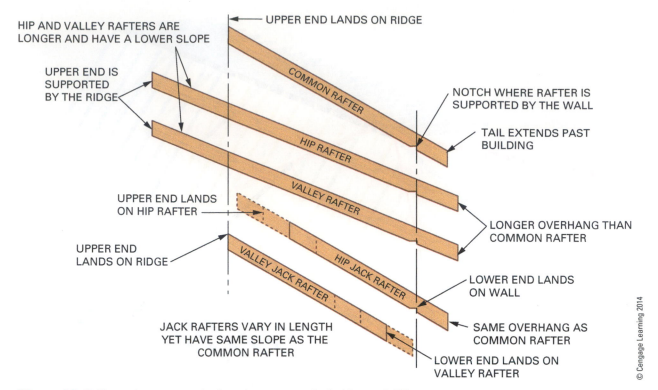

UPPER END LANDS ON RIDGE

HIP AND VALLEY RAFTERS ARE LONGER AND HAVE A LOWER SLOPE

UPPER END IS SUPPORTED BY THE RIDGE

COMMON RAFTER

NOTCH WHERE RAFTER IS SUPPORTED BY THE WALL

TAIL EXTENDS PAST BUILDING

HIP RAFTER

VALLEY RAFTER

UPPER END LANDS ON HIP RAFTER

LONGER OVERHANG THAN COMMON RAFTER

UPPER END LANDS ON RIDGE

VALLEY JACK RAFTER

HIP JACK RAFTER

LOWER END LANDS ON WALL

JACK RAFTERS VARY IN LENGTH YET HAVE SAME SLOPE AS THE COMMON RAFTER

SAME OVERHANG AS COMMON RAFTER

LOWER END LANDS ON VALLEY RAFTER

© Cengage Learning 2014

Figure 41–3 The various types of rafters have many similarities and differences.

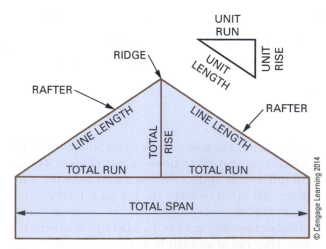

UNIT RUN

UNIT LENGTH

UNIT RISE

RIDGE

RAFTER

RAFTER

LINE LENGTH

LINE LENGTH

TOTAL RISE

TOTAL RUN

TOTAL RUN

TOTAL SPAN

© Cengage Learning 2014

Figure 41–4 The terms associated with roof theory.

a base for the roof angle, and it is a horizontal distance under a rafter. This distance is always 12 inches (or 200 mm in metric measure) for a common rafter. It is 16.97 inches (or 283 mm) for hip and valley rafters.

Total Span. The total span of a roof is the horizontal distance covered by the roof. This is usually the width of the building measured from the outer faces of the framing.

Unit span. The unit Span is twice the unit Run for a common rafter. Since unit Run is 12 inches by definition, then unit Span is always 24 inches.

Unit Rise. The unit Rise is the number of inches the roof will rise vertically for every unit Run. For example, if the unit Rise is 6, then a common rafter will rise 6 inches for every 12 inches it covers horizontally. This number is typically shown on a triangular symbol that is found on elevations and section views.

Total Run. The total Run of a rafter is the total horizontal distance over which the rafter slopes. This is usually one-half the span of the roof.

Total Rise. The total Rise is the vertical distance that the roof rises from plate to ridge. Total Rise may be calculated by multiplying the unit Rise by the total Run of the rafter. For example, if the unit Rise is 6 and the Run is 13, then the total Rise is $6 \times 13 = 78$ inches. This measurement is not to the top of the rafter, but rather to some point inside the rafter.

Theoretical Line Length. The line length of a rafter is the length of the rafter from the plate to the ridge. It is the hypotenuse (longest side) of the right triangle formed by the total Run as the base and the total Rise as the vertical leg. Line length gives no consideration to the thickness or width of the framing stock.

Unit Length. Unit Length is the length of rafter needed to cover a horizontal distance of one unit Run. It is the hypotenuse of a right triangle formed by the unit Run and unit Rise.

Bird's Mouth. Bird's mouth is the term used to refer to the notch cut in a rafter. This is done

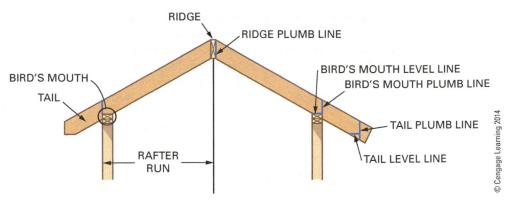

Figure 41–5 A bird's mouth is a notch for the rafter to sit on the wall.

to make the rafter sit securely on the wall plate so that it can be adequately fastened (*Figure 41–5*).

Pitch. The pitch of a roof is a fraction indicating the amount of incline of a roof. The pitch is found by dividing the Rise by the span. This fraction is the total Rise over total Span or the unit Rise over unit Span. For example, if a building is 32 feet and has a total Rise of 8 feet, it also has a unit Rise of 6 inches. Pitch will be total Rise over total Span = $^8/_{32}$ = ¼ pitch or unit Rise over unit Span = $^6/_{24}$ = ¼ (*Figure 41–6*).

Slope. Slope is the common term used to express the steepness of a roof. It is stated using the unit Rise and the unit Run. For example, if the unit Rise is 4 inches and the unit Run is always 12 inches, then the slope is said to be 4 on 12.

Plumb Line. A plumb line is any line on the rafter that is vertical when the rafter is in position. There is a plumb line at the ridge, at the wall plate, and usually at the end of the tail. They are marked using a square. A framing square is marked along the tongue. A Speed Square is marked along the edge of the square where the inch ruler is located.

Level Line. A level line is any line on the rafter that is horizontal when the rafter is in position. It is marked along the blade of the framing square

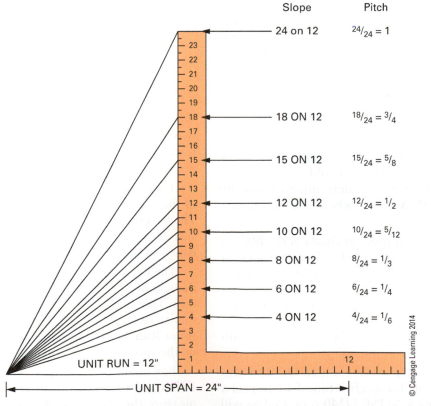

Figure 41–6 Pitch is a fraction or rise over span, slope relates unit rise to unit run.

when laying out level cuts on rafters. A Speed Square is marked along the long edge of the square where the degree scale is located after lining up the alignment guide with the plumb line.

SIZING RAFTER MEMBERS

To begin sizing rafters, information about the roof from the set of prints is consulted. The geographic location of the building is also considered with regard to snow. Charts are then read following a similar process as for joists.

Structural Terms

Structural terms for rafters are the same as for joists. These terms include on-center spacing, material strength, intended load (i.e., live and dead load PSF [pounds per square foot]), and deflection limits. For more information or a review of this procedure refer the Sizing Structural Members section in Chapter 34.

The major difference between joists and rafters is the span. Both spans are measured along a horizontal distance. For rafters this distance is the horizontal projection of the rafter, not the length of the rafter. The distance is the total run of the rafter, usually one-half the width of the building.

Intended load for rafters is more broadly considered than for joists. Dead load designs are the same. Dead load of 10 PSF is used where only the rafters are to be installed, and dead load of 20 PSF is used when a ceiling will be attached to the bottom of the rafters. Deflection limits are less strict for rafters than for joists. Rafters' stiffnesses are L/180 and L/240.

The International Residential Code (IRC) uses live load ranges of 20, 30, 50, and 70 PSF. This range takes into account the extra stress on a roof from snow loads of colder regions. The live load number used is determined from the region of North America where the building is located.

There are sixteen pages of charts in the IRC for rafters and four for ceiling joists. These charts adjust the factors of live load, dead load, and deflection limits. *Figure 41–7* is a list of selected information from the IRC rafter charts all for 16" OC. The sections include information for 30 PSF with L/180, 30 PSF with L/240 and 70 PSf with L/240.

As an exercise, determine if a 16 OC #2 Hem-fir 2×8 is useable for a rafter run of 14 feet on a building needing a 30 PSF, L/240 roof. Ceiling will

be attached to the rafters. The solutionis to find the intersection of the row and column according to the given information. The row is Hem-fir #2 in the second section of data because of 30 PSF and L/240 requirement. The column is under the 2×8 in the right-most section under the 20 PSF dead load because of the ceiling attached to rafters. The cell at the intersection reveals the maximum span is 13′−4″. Thus the answer is no, but a #1 grade of the same species will work.

EXAMPLE

What species and grades will work for a 2×6 rafter with no attached ceiling in a 30 PSF, L/180 to span 12 feet? Answer is in the first section under the 30 PSF, L/180. Secondly the left section, 10 PSF dead load, is used because of no ceiling is attached.

Solution: Looking at all species in the 2×6 column we find only #1 Douglas fir-larch at 12′−9″, #1 Hem-fir at 12′−5″, and #2 southern pine at 12′−6″ will work.

DETERMINING RAFTER LENGTHS

Three methods are available for finding the length of a rafter: the estimated, step-off, and calculated methods. They are used for different reasons and have varying degrees of speed and accuracy.

Estimated Rafter Length

With the estimated method, the carpenter may use two framing squares to scale the length of the rafter. It is only used as an estimate for ordering the proper length material. (*Figure 41–8*).

The blade of the square represents the total Run of the rafter. The tongue represents the total rise. The backside of the framing square, along the outside edge, is laid out in twelfths. The blade inside edge is in sixteenths and the tongue edge is in tenths.

The twelfths scale allows each inch mark to be thought of as a foot and each incremental line as an inch. This way no conversion of fractions is necessary to read the scale. To use this method read the rafter Run and Rise on one square and the length on the another.

For example, a Run of 12 feet becomes 12 inches and a Rise of 5 feet becomes 5 inches. Then measure the shortest distance between the blade

Rafter Spans for Common Lumber Species

Spacing	Species and Grade		Ground Snow load = 30 PSF, ceiling not attached to rafters, L/180									
			Dead Load = 10 PSF					Dead Load = 20 PSF				
			2×4	2×6	2×8	2×10	2×12	2×4	2×6	2×8	2×10	2×12
			Maximum Rafter Spans									
			ft-in	ft-in	ft-in	ft-in	ft-in	ft-in	ft-in	ft-in	ft-in	ft-in
16	Douglas fir-larch	#1	8-9	12-9	16-2	19-9	22-10	7-10	11-5	14-5	17-8	20-5
	Douglas fir-larch	#2	8-2	11-11	15-1	18-5	21-5	7-3	10-8	13-6	16-6	
	Hem-fir	#1	8-5	12-5	15-9	19-3	22-3	7-7	11-1	14-1	17-2	19-11
	Hem-fir	#2	8-0	11-9	14-11	18-2	21-1	7-2	10-6	13-4	16-3	18-10
	Southern pine	#1	8-9	13-9	18-1	21-5	25-7	8-8	12-10	16-2	19-2	22-10
	Southern pine	#2	8-7	12-6	16-2	19-3	22-7	7-10	11-2	14-5	17-3	20-2
	Spruce-pine-fir	#1	8-2	11-11	15-1	18-5	21-5	7-3	10-8	13-6	16-6	19-2
	Spruce-pine-fir	#2	8-2	11-11	15-1	18-5	21-5	7-3	10-8	13-6	16-6	19-2
	Ground Snow load = 30 PSF, ceiling attached to rafters, L/240											
16	Douglas fir-larch	#1	8-0	12-6	16-2	19-9	22-10	7-10	11-5	14-5	17-8	20-5
	Douglas fir-larch	#2	7-10	11-11	15-1	18-5	21-5	7-3	10-8	13-6	16-6	19-2
	Hem-fir	#1	7-8	12-0	15-9	19-3	22-3	7-7	11-1	14-1	17-2	19-11
	Hem-fir	#2	7-3	11-5	14-11	18-2	21-1	7-2	10-6	13-4	16-3	18-10
	Southern pine	#1	8-0	12-6	16-6	21-1	25-7	8-0	12-6	16-2	19-2	22-10
	Southern pine	#2	7-10	12-3	16-2	19-3	22-7	7-10	11-2	14-5	17-3	20-2
	Spruce-pine-fir	#1	7-6	11-9	15-1	18-5	21-5	7-3	10-8	13-6	16-6	19-2
	Spruce-pine-fir	#2	7-6	11-9	15-1	18-5	21-5	7-3	10-8	13-6	16-6	19-2
	Ground Snow load = 70 PSF, ceiling attached to rafters, L/240											
16	Douglas fir-larch	#1	6-0	9-0	11-5	13-11	16-2	5-10	8-6	10-9	13-2	15-3
	Douglas fir-larch	#2	5-9	8-5	10-8	13-1	15-2	5-5	7-11	10-1	12-4	14-3
	Hem-fir	#1	5-9	8-9	11-2	13-7	15-9	5-8	8-3	10-6	12-10	14-10
	Hem-fir	#2	5-6	8-4	10-6	12-10	14-11	5-4	7-10	9-11	12-1	14-1
	Southern pine	#1	6-0	9-5	12-5	15-2	18-1	6-0	9-5	12-0	14-4	17-1
	Southern pine	#2	5-11	8-10	11-5	13-7	16-0	5-10	8-4	10-9	12-10	15-1
	Spruce-pine-fir	#1	5-8	8-5	10-8	13-1	15-2	5-5	7-11	10-1	12-4	14-3
	Spruce-pine-fir	#2	5-8	8-5	10-8	13-1	15-2	5-5	7-11	10-1	12-4	14-3

© Cengage Learning 2014

Figure 41–7 Sample of rafter span tables.

and tongue edges to find 13 inches (feet). Add extra for the rafter overhang.

Two more accurate methods of rafter layout are required when building the roof. They are the step-off and calculated methods.

Step-Off Rafter Length Method

The step-off method uses a framing square to step-off the length of a rafter. It can be used for most types of rafters. The step-off method is based on the unit of Run (12 inches for the common rafter). The rafter stock is stepped off for each unit of Run until the desired number of units or parts of units are stepped off (*Figure 41–9*). If the rafter has a total Run of 16 feet, for example, the square is moved 16 times. This method works well in most situations, but it can cause errors in the length because small incremental errors that occur during each step add up.

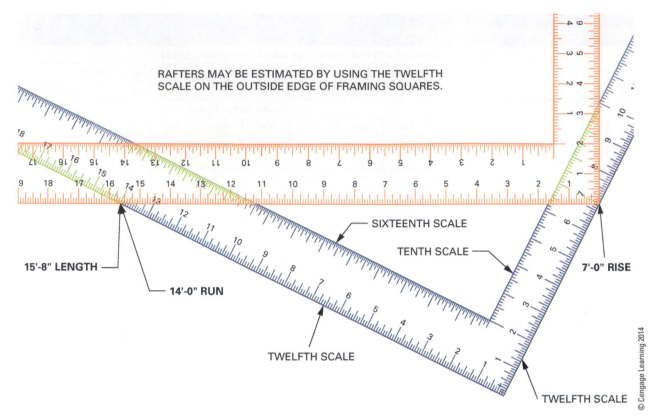

RAFTERS MAY BE ESTIMATED BY USING THE TWELFTH SCALE ON THE OUTSIDE EDGE OF FRAMING SQUARES.

15'-8" LENGTH

14'-0" RUN

SIXTEENTH SCALE

TENTH SCALE

7'-0" RISE

TWELFTH SCALE

TWELFTH SCALE

Figure 41–8 Rafter length may be estimated using two framing squares.

Calculated Rafter Length

The calculated rafter length method uses the rafter tables located on a framing square and a calculator. It is the most accurate way of determining rafter length. The procedure can be confusing at first, but once it is understood, any and all rafter lengths can be determined. This includes hip, valley, and jack rafters. The following Chapters of this book will embrace the calculated method.

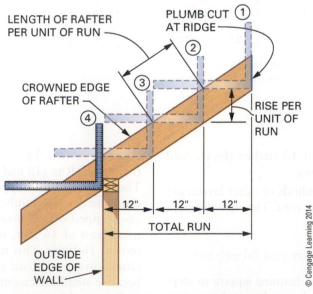

LENGTH OF RAFTER PER UNIT OF RUN

PLUMB CUT ① AT RIDGE

②

CROWNED EDGE OF RAFTER

③

④

RISE PER UNIT OF RUN

12"   12"   12"

TOTAL RUN

OUTSIDE EDGE OF WALL

Figure 41–9 Step-off method of determining rafter length.

CHAPTER 42

Gable, Shed, and Flat Roofs

The equal-pitched gable roof is the most commonly used roof style. Gable roofs have an equal slope on both sides intersecting the ridge in the center of the span (*Figure 42–1*). This roof is the simplest to frame. Only one type of rafter, the common rafter, needs to be laid out.

The saltbox is a gable roof with rafters having different slopes or different Runs on each side of the ridge (*Figure 42–2*). One rafter may slope faster to reach to the ridge from a lower floor.

COMPONENTS OF A COMMON RAFTER

The common rafter requires several cuts, which have to be laid out. The cuts and lines of a rafter have several names (*Figure 42–3*). The cut at the top is called the *plumb cut* or *ridge cut*. It fits against the ridge-board. The notch cut at the plate is called the *bird's mouth*, *seat cut*, or *heel*. It fits against the top and outside edge of the wall plate. It consists of a plumb line and a level line layout. These lines are named according to the notch name used. For example, if the notch is referred to as a seat cut, then it is laid out with a seat plumb line and a seat level line.

The distance between the ridge cut and the seat cut is referred to as the *rafter length*. It is also called the line length or theoretical line length of the rafter. This distance must be determined by the carpenter. At the bottom end of the rafter is the tail or overhang, which extends beyond the building. It supports the fascia and soffit. The plumb line is called the *tail plumb line* or *fascia cut*. The *tail level line* is also referred to as the *soffit cut*.

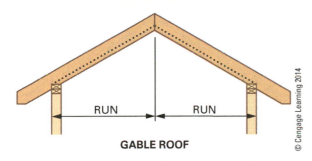

GABLE ROOF

© Cengage Learning 2014

Figure 42–1 The gable roof has opposing rafters that are symmetrical.

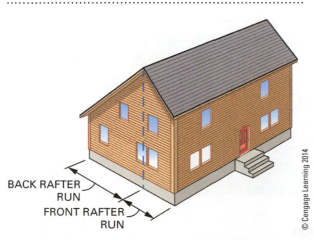

BACK RAFTER RUN
FRONT RAFTER RUN

© Cengage Learning 2014

Figure 42–2 The ridge of the saltbox roof is off-center.

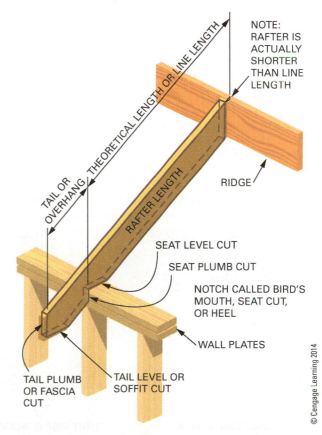

NOTE: RAFTER IS ACTUALLY SHORTER THAN LINE LENGTH

TAIL OR OVERHANG THEORETICAL LENGTH OR LINE LENGTH

RAFTER LENGTH

RIDGE

SEAT LEVEL CUT

SEAT PLUMB CUT

NOTCH CALLED BIRD'S MOUTH, SEAT CUT, OR HEEL

WALL PLATES

TAIL PLUMB OR FASCIA CUT

TAIL LEVEL OR SOFFIT CUT

© Cengage Learning 2014

Figure 42–3 Names of the cuts and lines of a common rafter.

COMMON RAFTER LAYOUT

Even though rafters may look different, they are laid out using a similar process. The principles for framing rafters, whether equally or unequally sloped, common or hip, are the same. Layout follows a logical flow (*Figure 42–4*). The ridge cut is marked, and the rafter length is adjusted for the ridge thickness. The line length is determined, and a plumb line is made for the seat cut. A level line is drawn to complete the bird's mouth. The tail length is determined, and a fascia line is drawn. The layout is finished with a soffit cut.

COMMON RAFTER PATTERN

Because common rafters are all cut the same way, only one layout is required. Once a rafter is laid out and cut, it becomes the pattern for the rest of the rafters. It should be tested and verified to fit before cutting additional rafters.

Select the straightest piece of stock for the pattern. Lay it across two sawhorses. Sight the stock along the edge for straightness and mark the crowned edge.

This edge will become the top side where the roof sheathing is attached. Further discussion of rafters will assume the carpenter is standing near the rafter's lower edge, opposite of the crowned edge.

Laying Out the Ridge Cut

Place the square down on the side of the stock at its left end. Hold the tongue of the square with the left hand and the blade with the right hand. Move the square until the outside edge of the tongue and the edge of the stock line up with the specified rise in inches. Make sure the blade of the square and the edge of the stock line up with the unit Run or 12 inches. Slide the square to the left until it reaches the end of the rafter stock. Recheck the alignment and mark along the outside edge of the tongue for the plumb cut at the ridge (*Figure 42–5*).

When using a Speed Square, place the pivot point of the square on the top edge of the rafter. Rotate the square with the pivot point touching the rafter. Looking in the rafter scale window, align the edge of the rafter with the number that corresponds to the rise per unit of Run desired. Note that there are two scales, one for a common rafter and one for hips and valleys. Mark the plumb line along the edge of the square that has the inch ruler marked on it.

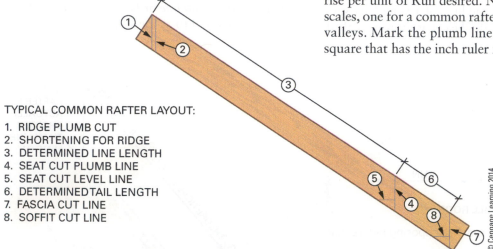

TYPICAL COMMON RAFTER LAYOUT:

1. RIDGE PLUMB CUT
2. SHORTENING FOR RIDGE
3. DETERMINED LINE LENGTH
4. SEAT CUT PLUMB LINE
5. SEAT CUT LEVEL LINE
6. DETERMINED TAIL LENGTH
7. FASCIA CUT LINE
8. SOFFIT CUT LINE

Figure 42–4 Rafters are laid out in a stepwise fashion.

UNIT RISE 6" FOR COMMON RAFTER

UNIT RISE 6" ALIGNED WITH EDGE OF THE BOARD

UNIT RUN 12" FOR COMMON RAFTER

Figure 42–5 Laying out the plumb cut of the common rafter at the ridge.

EXAMPLE

Assume the slope of the roof is 6 on 12. Hold the square so the 6-inch mark on the tongue and the 12-inch mark on the blade line up with the top edge of the rafter stock. Mark along the outside edge of the tongue of the square. Make a second plumb line for practice using a Speed Square. Place the Speed Square on the rafter stock and rotate the square, keeping the pivot point touching the edge of the rafter. Align the edge of the rafter with the number 6 on the common scale in the rafter scale window. Mark the plumb line along the edge of the square that has the ruler marked on it. Both layouts should look like that shown in Figure 42–5.

Rafter Shortening Due to Ridge

The length of the rafter must be shortened when a ridgeboard is inserted between abutting rafters (*Procedure 42–A*). The total shortening is equal to the thickness of the ridgeboard, thus each rafter will be shortened one-half the ridgeboard's thickness.

To shorten the rafter, measure at a right angle from the ridge plumb line a distance equal to one-half the thickness of the ridgeboard. Lay out another plumb line at this point. This will be the cut line. Note that shortening is always measured at right angles to the ridge cut, regardless of the slope of the roof.

StepbyStep Procedures

Procedure 42–A Shortening a Rafter That Abuts a Ridgeboard

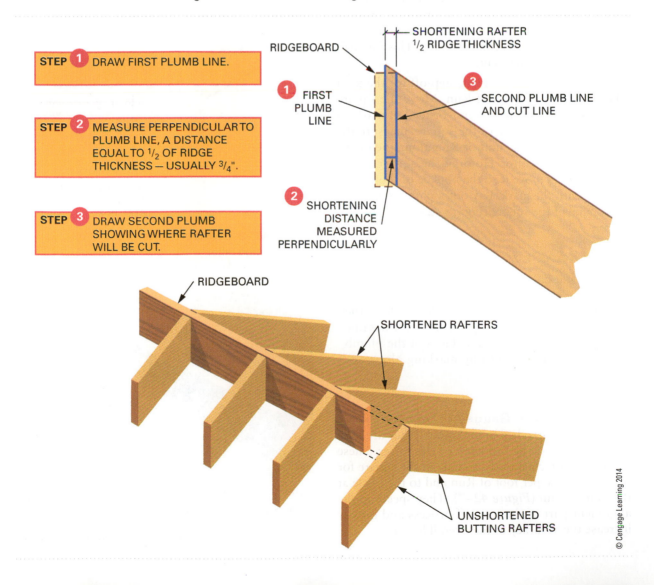

STEP 1 DRAW FIRST PLUMB LINE.

STEP 2 MEASURE PERPENDICULAR TO PLUMB LINE, A DISTANCE EQUAL TO 1/2 OF RIDGE THICKNESS — USUALLY 3/4".

STEP 3 DRAW SECOND PLUMB SHOWING WHERE RAFTER WILL BE CUT.

SHORTENING RAFTER 1/2 RIDGE THICKNESS

RIDGEBOARD

1 FIRST PLUMB LINE

2 SHORTENING DISTANCE MEASURED PERPENDICULARLY

3 SECOND PLUMB LINE AND CUT LINE

RIDGEBOARD

SHORTENED RAFTERS

UNSHORTENED BUTTING RAFTERS

© Cengage Learning 2014

COMMON RAFTER LENGTH

The length of a common rafter can be determined in two ways, via the step-off or calculation method. The step-off method requires close and fine measuring as each increment is made. The calculation method uses a calculator and the rafter tables found on a rafter square. Both methods can be used to determine the length of any kind of a rafter, but the calculation method is faster and more accurate.

Step-Off Method

The step-off method is based on the unit of Run (12 inches or 200 mm for the common rafter). The rafter stock is stepped off for each unit of Run until the desired number of units or parts of units have been stepped off.

First, lay out the ridge cut. Keeping the square in place, mark where the blade intersects with the top edge of the rafter. Hold the square at the same angle and slide it to the right until the tongue lines up with this mark. Move the square in a similar manner until the total Run of the rafter is laid out (*Procedure 42–B*). Mark a plumb line along the tongue of the square at the last step. This line will be parallel to the ridge cut.

Fractional Step-Off.   First, step-off for the total whole number of units of Run. Move the square as if to step off one more time. Instead of marking 12 inches on the blade of the square, measure the fractional part of the unit of Run along the blade. Mark it on the rafter. Holding the square in the same position, move it so the outside of the tongue lines up with the mark. Lay out a plumb line along the tongue of the square (*Figure 42–6*).

 EXAMPLE

If the rafter has a total Run of 16 feet, 7 inches, step-off sixteen times. Hold the square for the seventeenth step. Mark along the blade at the 7-inch mark. With the square held at the same angle, move it to the mark. Lay out the plumb cut of the bird's mouth by marking along the tongue.

Framing Square Gauges.   *Framing square* gauges can be attached to the square. They are used as stops against the top edge of the rafter. These gauges are attached to the tongue of the square for the desired rise per foot of Run and to the blade at the unit of Run (*Figure 42–7*). They speed up the alignment part of the step-off process and greatly increase the accuracy of the overall layout.

Calculation Method

The calculation method uses the idea that each step of the step-off method has a unit Length. Unit Length is the hypotenuse of the right triangle included in each step-off (*Figure 42–8*). Because the total Run is the number of step-offs, the rafter length is found by multiplying total Run by unit Length. For example, if the unit Length is 13 inches and the number of step-offs is 4, then the rafter length is $13 \times 4 = 52$ inches.

This may seem strange to multiply inches times feet. Normally in algebra, this will give a result that has no meaning. But in this case the Run is 4, a number of units, not 4 feet. Total Run must always look like feet, but it is always simply the number of unit Runs (step-offs) under the rafter.

Rafter tables provide rafter information in six rows for slopes of 2 through 18 inches of unit Rise. The table is stamped on the blade of a rafter (framing) square or sometimes printed in small booklets

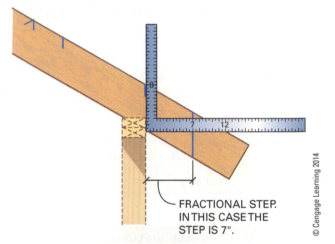

Figure 42–6 Laying out a fractional part of a Run for a common rafter.

Figure 42–7 Framing square gauges are attached to the square to hold it in the same position for easy alignment.

StepbyStep Procedures

Procedure 42–B Using the Step-Off Method for Rafter Layout of an 8 on 12 Rafter

STEP 1 ALIGN 8" AND 12" OF THE SQUARE ON THE TOP EDGE OF THE RAFTER.

STEP 2 SLIDE SQUARE LEFT TO THE END OF THE BOARD AND RECHECK 8" AND 12" ALIGNMENT.

STEP 3 HOLD AND MARK THE PLUMB CUT ALONG THE TONGUE.

STEP 4 STILL HOLDING SQUARE, MAKE A VERTICAL LINE UNDER 12" OF THE BLADE.

STEP 5 SLIDE SQUARE TO THE RIGHT UNTIL THE PREVIOUS MARK ALIGNS WITH THE OUTSIDE EDGE TONGUE. RECHECK THE 8" AND 12" ALIGNMENTS.

STEP 6 HOLD SQUARE AND MARK A VERTICAL LINE UNDER 12" OF THE BLADE.

STEP 7 REPEAT AS NEEDED.

STEP 8 LAST STEP IS MARKED WITH A FULL PLUMB LINE.

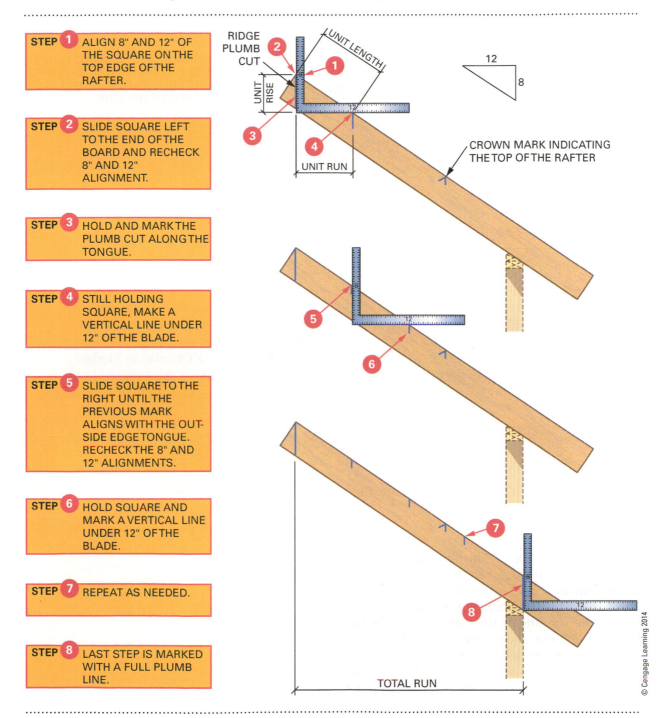

© Cengage Learning 2014

that come with the square (*Figure 42–9*). The inch marks 2 through 18 represent the unit Rises available with the information directly below each number.

The first row of the rafter table is the common rafter length per unit Run or, simply stated, unit Length. This number comes from using the Pythagorean theorem, $a^2 + b^2 = c^2$ on the unit triangle. The unit Rise and unit Run are the a and b with the unit Length being c. All unit Lengths are calculated this way. Knowing this, the unit Lengths of usual unit Rises, such as $6\frac{3}{4}$ inches, can be calculated.

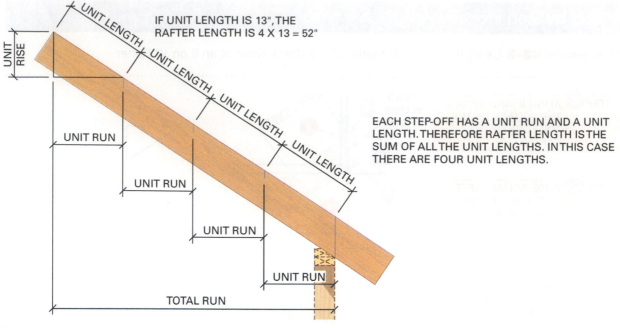

Figure 42–8 Unit length is the hypotenuse of the right triangle included in each step-off.

EXAMPLE

Suppose the unit Rise of a rafter is 17 inches. What is the unit Length? Unit Run is always 12 inches for a common rafter.

$$a^2 + b^2 = c^2$$
$$17^2 + 12^2 = c^2$$
$$289 + 144 = 433 = c^2$$
$$c = \sqrt{433} = 20.80865205 \text{ inches}$$

Unit length = 20.81 inches rounded to the nearest hundredths.

Using Rafter Tables. The unit length for normal rafters may be found on the rafter tables of a framing square. First, locate which line of the table will have the unit Lengths for a common rafter. This is the top line of the rafter table. Moving along the first row and stopping under the 17-inch mark shows the number 20 81 or 20.81 inches. The space represents a decimal point. Check to see that the unit length for a hip or valley rafter with a unit Rise of 8 is 18.76 inches.

Note that the calculated line length must be measured from the first ridge plumb line. If this line is drawn at the end of the rafter, a tape may be hooked on the end of the board to measure length. Measure from the first ridge cut along the top edge of the rafter, the length of the rafter (*Figure 42–10*). The measurement mark should be on the edge of the rafter. Mark the plumb line for the seat.

EXAMPLE

Rafter Length Using Calculation Method

Find the line length of a common rafter with a unit Rise of 8 inches for a 28-foot-wide building.

Step 1: Read below the 8-inch mark in the first row to find 14 42 or 14.42 inches unit length.

Step 2: Divide 28 by 2 to determine the Run.

28 ÷ 2 = 14 Run.

Step 3: Multiply unit Length times total Run.

14.42 × 14 = 201.88 inches.

That is 201 inches plus a fraction.

Step 4: Convert decimal to sixteenths.

0.88 × 16 = 14.08.

Step 5: Round off to nearest whole sixteenths.

14.08 = $^{14}/_{16}$, which reduces to $^7/_8$ inches.

Step 6: Add numbers to determine rafter length.

201 + $^7/_8''$ = 201$^7/_8''$.

To summarize, mark a plumb cut at the ridge. From the ridge cut, measure along the top edge of the rafter, the length of the rafter as determined by calculations. Mark the length and make a plumb line at the seat. At this point, the rafter length is theoretical because the rafter must be shortened due to the thickness of the ridge (see *Figure 42–10*).

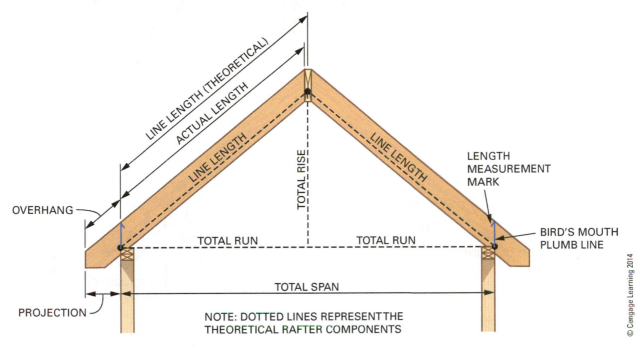

Figure 42–9 Rafter tables are found on the framing square.

Figure 42–10 The line length is theoretical, measured before rafter is shortened to allow for the ridge.

Seat Cut Layout

The seat cut or *bird's mouth* of the rafter is a combination of a level cut and a plumb cut. The level cut rests on top of the wall plate. The plumb cut fits snugly against the outside edge of the wall.

The seat cut level line is not a precise measurement. The main concern is that it be perpendicular to the plumb line and not cut too deeply. Position the level line so there is sufficient stock, about two-thirds of the rafter remaining above the seat cut, to ensure there is enough strength to support any overhang. Typically, the seat level line is 2 to 4 inches long. This line should not be longer than the width of the wall plate. ✱

IRC R802.6

To mark the seat level line, hold the framing square in the same manner as for marking plumb lines. Slide it along the stock until the blade lines up with the plumb line. Mark the appropriate level line along the blade.

With a Speed Square, hold the alignment guide of the square in line with the plumb line previously drawn. Mark along the long edge of the square to achieve level lines for seat cuts (*Figure 42–11*). On roofs with moderate slopes, the length of the level cut of the seat is sometimes the width of the wall plate. For steep roofs, the level cut is shorter.

Sometimes it is desirable to cut the seat cut to allow for wall sheathing (*Figure 42–12*). Extend the level line a distance equal to the wall sheathing thickness. Draw a new plumb line down from the end of the line.

Common Rafter Tails

The tail cut is the cut at the end of the rafter tail. It may be a plumb cut, a combination of cuts, or a square cut (*Figure 42–13*). Sometimes the rafter tails are left to run wild. This means they are slightly longer than needed. They are cut off later after the roof frame is erected.

On most plans the projection is given, not the overhang. The projection is a level measurement, whereas the overhang is a sloping measurement along the rafter. ✱

IRC R802.7.1.1

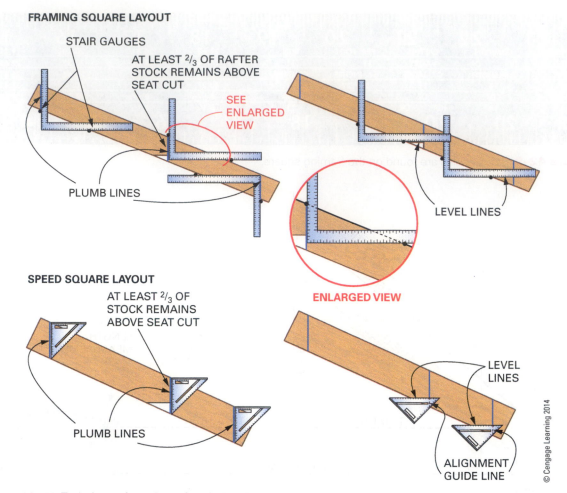

FRAMING SQUARE LAYOUT

STAIR GAUGES

AT LEAST ⅔ OF RAFTER STOCK REMAINS ABOVE SEAT CUT

SEE ENLARGED VIEW

PLUMB LINES

LEVEL LINES

ENLARGED VIEW

SPEED SQUARE LAYOUT

AT LEAST ⅔ OF STOCK REMAINS ABOVE SEAT CUT

PLUMB LINES

LEVEL LINES

ALIGNMENT GUIDE LINE

© Cengage Learning 2014

Figure 42–11 Techniques for using a framing and speed square to lay out a rafter.

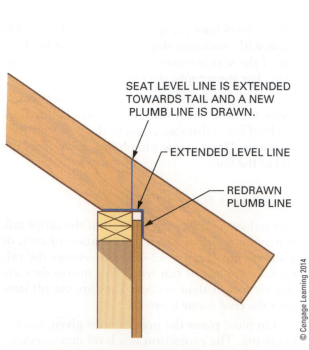

SEAT LEVEL LINE IS EXTENDED TOWARDS TAIL AND A NEW PLUMB LINE IS DRAWN.

EXTENDED LEVEL LINE

REDRAWN PLUMB LINE

© Cengage Learning 2014

Figure 42–12 Sometimes seat cuts are made deeper to allow for wall sheathing.

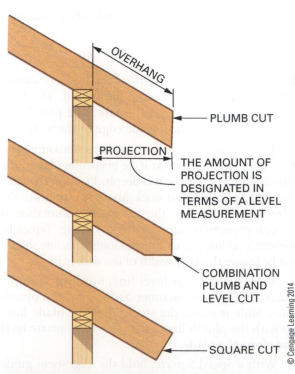

OVERHANG

PLUMB CUT

PROJECTION

THE AMOUNT OF PROJECTION IS DESIGNATED IN TERMS OF A LEVEL MEASUREMENT

COMBINATION PLUMB AND LEVEL CUT

SQUARE CUT

© Cengage Learning 2014

Figure 42–13 Various tail cut styles for the common rafter.

OVERHANG VS. PROJECTION

STEP-OFF METHOD

8 AND 12 ALIGNED ON THE LOWER EDGE OF RAFTER

OVERHANG

OVERHANG

21⅝"

PROJECTION

18"

PROJECTION

CALCULATION METHOD

OVERHANG LENGTH
STEP 1 18" ÷ 2 = 1.5 UNITS OF RUN
STEP 2 8" UNIT RISE HAS UNIT LENGTH OF 14.42"
STEP 3 UNIT LENGTH X RUN = RAFTER LENGTH
STEP 4 1.5 X 14.42" = 21.63"
STEP 5 0.63 X 16 = 10.08 SIXTEENTHS = ¹⁰⁄₁₆ = ⁵⁄₈
STEP 6 OVERHANG = 21⅝"

© Cengage Learning 2014

Figure 42–14 Laying out the tail of the common rafter using the projection or overhang calculation.

Stepping Off Projection. To mark the fascia cut, place a framing square on the rafter to mark a plumb cut. Align the square using the unit Rise on the tongue and unit Run on the blade. Larger projections will require the square to be moved to the lower edge of the rafter. *Figure 42–14* shows a fascia cut example when the unit Rise is 8 inches and the projection is 18 inches.

Calculating Overhang. The overhang may also be calculated and measured using the same formula for determining rafter length: unit length times Run. Convert the projection to Run by changing inches to feet. In this case, 18 inches is 1.5 units of Run and the unit length is 14.42 inches (see *Figure 42–14*). Thus the overhang measurement is 1.5 × 14.42 = 21.63 inches or 21⅝ inches. This number is measured along the rafter edge.

Soffit Cut. The soffit cut line varies according to the finish material. It should be made so the plumb line is long enough to adequately support the fascia (*Figure 42–15*).

The entire process of laying out a common rafter is displayed in *Procedure 42–C*

Wide Rafter Tails. On rafters made from wide boards such as TJI's or 2×10's, it is often desirable to cut the tail narrower. The tail is trimmed to look like a 2×6 (*Figure 42–16*). A parallel line to the top edge of the rafter is drawn and the seat cut plumb line is extended.

Cutting Wild Rafter Tails. To cut wild rafter tails after the rafters are installed, measure and mark the two rafters on either end for the amount of overhang. This is usually a level measurement from the outside of the wall studs to the tail plumb line (*Procedure 42–D*). Plumb the marks up to the top edge of the rafters with a level. Stretch a line across the top edges of all the rafters. Using a T-bevel, plumb down on the side of each rafter from the chalk line. Use a circular saw to cut each rafter. Cutting down along the line is the easiest way to make the cut.

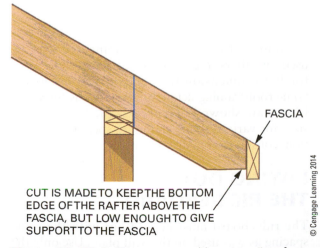

CUT IS MADE TO KEEP THE BOTTOM EDGE OF THE RAFTER ABOVE THE FASCIA, BUT LOW ENOUGH TO GIVE SUPPORT TO THE FASCIA

FASCIA

© Cengage Learning 2014

Figure 42–15 Soffit cuts are positioned to allow for adequate support of fascia.

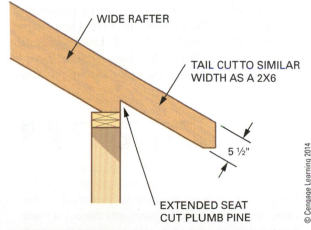

WIDE RAFTER

TAIL CUT TO SIMILAR WIDTH AS A 2X6

5½"

EXTENDED SEAT CUT PLUMB PINE

© Cengage Learning 2014

Figure 42–16 Wide rafter tails are ofen cut narrower.

StepbyStep Procedures

Procedure 42–C Using the Calculation Method for Common Rafter Layout

PROCEDURE 42-C COMMON RAFTER LAYOUT
CALCULATION METHOD

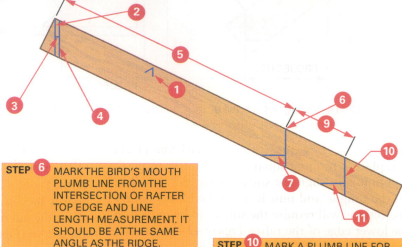

STEP 1 SELECT A STRAIGHT BOARD OR ONE WITH ONLY A SLIGHT CROWN.

STEP 2 DRAW A PLUMB LINE AT ONE END FOR THE RIDGE. MARK IT FROM THE END OF THE BOARD TO MAKE MEASURING THE LENGTH EASIER LATER.

STEP 3 MEASURE PERPENDICULAR TO PLUMB LINE A DISTANCE EQUAL TO ONE-HALF OF RIDGE THICKNESS. USUALLY 3/4".

STEP 4 DRAW SECOND PLUMB LINE PASSING THROUGH THE END OF THE HORIZONTAL MEASUREMENT. THIS WILL BE THE CUT LINE FOR THE RIDGE PLUMB CUT.

STEP 5 CALCULATE THE LINE LENGTH OF RAFTER. (SEE PREVIOUS EXAMPLE) MEASURE THE LINE LENGTH FROM THE FIRST PLUMB LINE (STEP 2) ALONG THE TOP EDGE OF RAFTER.

STEP 6 MARK THE BIRD'S MOUTH PLUMB LINE FROM THE INTERSECTION OF RAFTER TOP EDGE AND LINE LENGTH MEASUREMENT. IT SHOULD BE AT THE SAME ANGLE AS THE RIDGE.

STEP 7 MARK A SEAT CUT LINE THAT IS 2–4 INCHES LONG. IT SHOULD NOT BE LONGER THAN THE WALL PLATE IS WIDE.

STEP 8 CALCULATE THE TAIL LENGTH USING THE PROCEDURE SHOWN IN FIGURE 42–14.

STEP 9 MEASURE THE TAIL ALONG THE TOP EDGE OF RAFTER FROM THE BIRD'S MOUTH PLUMB LINE.

STEP 10 MARK A PLUMB LINE FOR THE FASCIA CUT FROM THE INTERSECTION OF RAFTER TOP EDGE AND TAIL LENGTH MEASUREMENT. IT SHOULD BE AT THE SAME ANGLE AS THE RIDGE.

STEP 11 MARK A SOFFIT CUT BACK TOWARD BIRD'S MOUTH. MAKE SURE THERE IS SUFFICIENT LENGTH ALONG THE PLUMB LINE TO ALLOW FOR THE FASCIA TO ATTACH.

STEP 12 CUT RAFTER. USE IT AS A PATTERN FOR THE REMAINING RAFTERS.

© Cengage Learning 2014

If the tail cut is a combination of plumb and level cuts, make the plumb cuts first. Then snap a line across the cut ends as desired. Level each soffit cut in from the chalk line. Working from the outside toward the wall is the easiest way to accomplish this.

CAUTION

On some cuts that are at a sharp angle with the edge of the stock (that is, the level cut of the seat), the guard of the circular saw may not retract. In this case, retract the guard by hand until the cut is made a few inches into the stock. Never wedge the guard in an open position to overcome this difficulty.

Wood I-Joist Roof Details

In addition to solid lumber, engineered may be used for rafters (*Figure 42–17*). Layout is the same for both dimension lumber and wood I-joists. Some roof-framing details and variations for wood I-joists are shown in *Figure 42–18*. To determine sizes for various spans and more specific information, consult the manufacturer's literature.

LAYING OUT THE RIDGEBOARD

The ridgeboard must be laid out with the same spacing as was used on the wall plate. Use only the straightest lumber for the ridge. The total length of the ridgeboard should be the same as the length

StepbyStep Procedures

Procedure 42–D Cutting Rafter Tails after Rafters are Installed

STEP 1 ON BOTH ENDS OF BUILDING, LAY OUT A LEVEL MEASUREMENT FROM THE WALL, AND MARK THE BOTTOM EDGE OF THE RAFTER.

STEP 2 ON BOTH ENDS, FROM THE BOTTOM MARK, LAY OUT A PLUMB LINE USING A CARPENTER'S HAND LEVEL.

STEP 3 STRETCH A LINE FROM END TO END. DO NOT SNAP A CHALK LINE. USE LINE AS A GUIDE.

IF TOP EDGES OF RAFTERS ARE NOT EXACTLY THE SAME HEIGHT, A SNAPPED LINE WILL PRODUCE END CUTS NOT IN LINE.

STEP 4 USING LINE AS A GUIDE, PLUMB DOWN ALONG ONE SIDE OF ALL RAFTERS USING A HAND LEVEL.

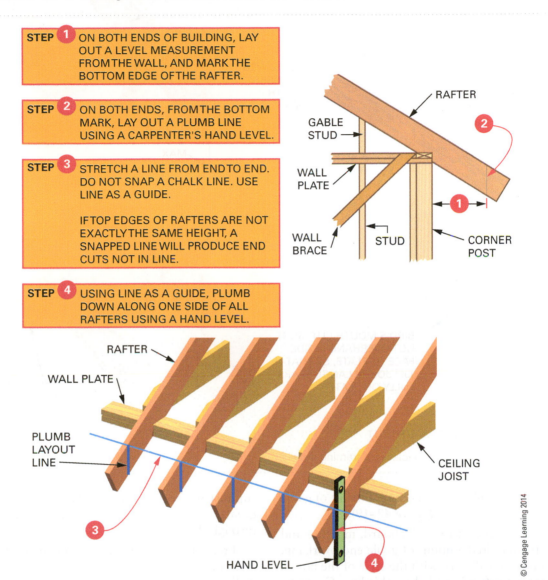

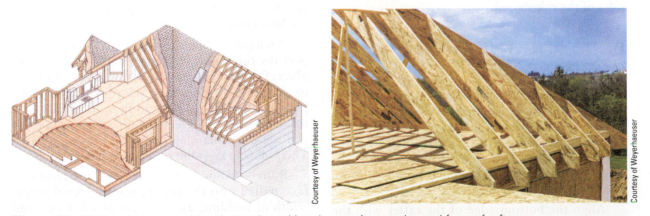

Figure 42–17 Wood I-joists and Laminated Strand Lumber are frequently used for roof rafters.

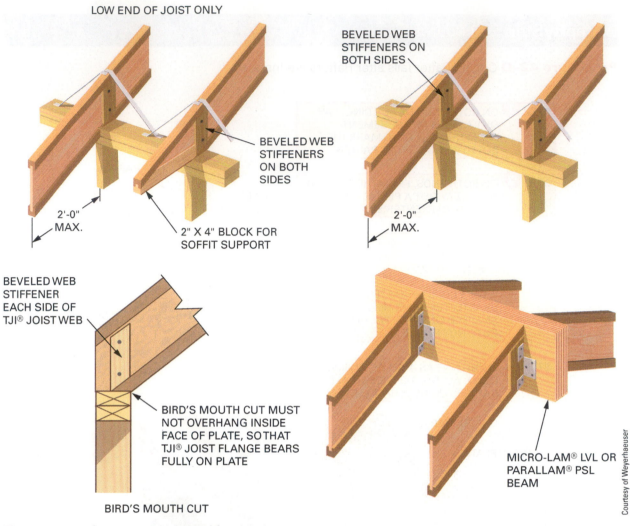

LOW END OF JOIST ONLY

BEVELED WEB STIFFENERS ON BOTH SIDES

BEVELED WEB STIFFENERS ON BOTH SIDES

2'-0" MAX.

2' X 4" BLOCK FOR SOFFIT SUPPORT

2'-0" MAX.

BEVELED WEB STIFFENER EACH SIDE OF TJI® JOIST WEB

BIRD'S MOUTH CUT MUST NOT OVERHANG INSIDE FACE OF PLATE, SO THAT TJI® JOIST FLANGE BEARS FULLY ON PLATE

BIRD'S MOUTH CUT

MICRO-LAM® LVL OR PARALLAM® PSL BEAM

Courtesy of Weyerhaeuser

Figure 42–18 Some wood I-joist roof framing details.

of the building plus the necessary amount of gable overhang on both ends (*Figure 42–19*).✳

✳IRC R802.3

From the end of the ridgeboard, measure and mark the required amount of gable end overhang. This distance will vary with the style of the house. The second rafter is nailed on this line. Continue the layout of remaining rafters, marking on both sides of the ridgeboard.

ERECTING THE GABLE ROOF FRAME

Prepare to erect the roof by placing plywood on top of the ceiling joists to serve as a safe work surface (*Procedure 42–E*). Tack them in place for safety. Place the ridgeboard on top in the direction it will be installed. Take care not to reverse the ridgeboard because the layout is usually not the same from one end to the other.

Align the bottom edge of the rafter with the bottom of the ridgeboard. This will allow greater

support of the entire rafter. This also allows for greater airflow over the top of the ridgeboard through the ridge vent.

Erection is most efficiently done by three workers, one at the ridge and one each at the bearing walls. Nail two rafters to the ridge at each end of the ridge. Raise and hold the ridge with rafters attached in position as the lower ends are nailed at the bird's mouth.

A temporary brace sometimes is helpful to support the ridge and two rafters while nailing takes place. Position and nail opposing pairs of rafters to complete the outline of the roof frame. Next, plumb and brace the ridge to the end wall. Attach the brace to the ridge and load-bearing wall or end wall. Fill in remaining rafters in opposing pairs. This will help keep the ridgeboard straight.

Rafters should be toenailed into the plate and face nailed into the ceiling joist.✳ Roof slope, width of building, and amount of snow or wind load will affect the number and size of nails needed

✳IRC R802.3.1

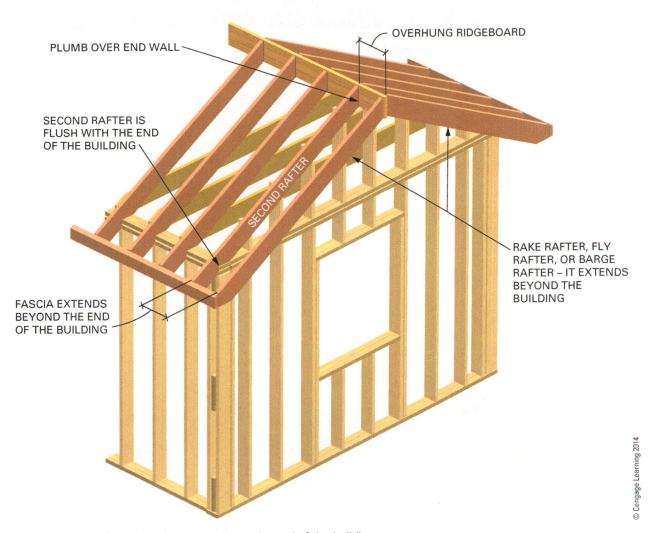

PLUMB OVER END WALL

OVERHUNG RIDGEBOARD

SECOND RAFTER IS FLUSH WITH THE END OF THE BUILDING

SECOND RAFTER

RAKE RAFTER, FLY RAFTER, OR BARGE RAFTER – IT EXTENDS BEYOND THE BUILDING

FASCIA EXTENDS BEYOND THE END OF THE BUILDING

© Cengage Learning 2014

Figure 42–19 Ridgeboards may overhang the end of the building.

to fasten the ceiling joist/rafter connection. Check local code requirements.

Collar Ties. Collar ties are horizontal members fastened to opposing pairs of rafters, which effectively reduce the span of the rafter. As a load is placed on one side of the roof, it is transferred to the other side through the collar tie. In effect, the load on one rafter supports the load on the other rafter and vice versa. ✳

✳**IRC TABLE R602.3(1)**

✳**IRC R802.3.1**

Install collar ties to every third rafter pair or as required by drawings or codes. The length of a collar tie varies, but they are usually about one-third to one-half of the building span. ✳

Constructing a Rake Overhang

The first and last rafters of a gable roof are called rake rafters, *fly rafters*, or *barge rafters*. They support the finish material. They may be constructed using several methods, depending on the amount of gable overhang and rafter length.

They should always be made using straight rafter pieces since they will form the roof edge. This is easily visible from the ground. They do not need to have the bird's mouth cut out.

Rake rafters are often supported by lookouts. Lookouts are framed in before the roof is sheathed. One method uses 2 × 4s on the flat notched into the second rafter (*Figure 42–20*). The other lookout method uses rafter width material and is strongest. This method looks similar to a cantilevered joist system. Lookouts may be spaced 16 to 32 inches on center (OC) according to the desired strength.

INSTALLING GABLE STUDS

The triangular areas formed by the rake rafters and the wall plate at the ends of the building are called gables. They must be framed with studs. These studs are called gable studs. The bottom ends are cut square. They fit against the top of the wall plate. The

StepbyStep Procedures

Procedure 42–E Erecting a Gable Roof

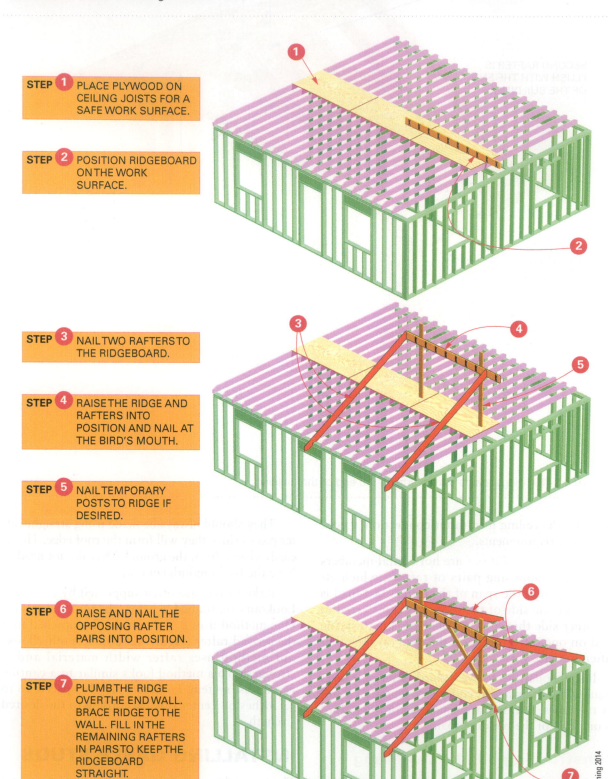

STEP 1 PLACE PLYWOOD ON CEILING JOISTS FOR A SAFE WORK SURFACE.

STEP 2 POSITION RIDGEBOARD ON THE WORK SURFACE.

STEP 3 NAIL TWO RAFTERS TO THE RIDGEBOARD.

STEP 4 RAISE THE RIDGE AND RAFTERS INTO POSITION AND NAIL AT THE BIRD'S MOUTH.

STEP 5 NAIL TEMPORARY POSTS TO RIDGE IF DESIRED.

STEP 6 RAISE AND NAIL THE OPPOSING RAFTER PAIRS INTO POSITION.

STEP 7 PLUMB THE RIDGE OVER THE END WALL. BRACE RIDGE TO THE WALL. FILL IN THE REMAINING RAFTERS IN PAIRS TO KEEP THE RIDGEBOARD STRAIGHT.

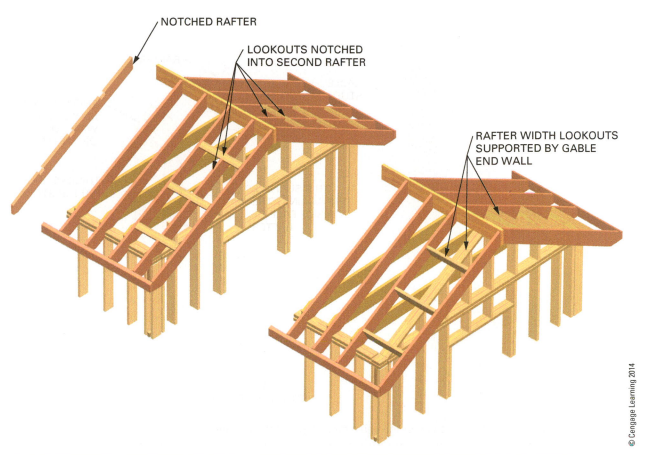

NOTCHED RAFTER

LOOKOUTS NOTCHED INTO SECOND RAFTER

RAFTER WIDTH LOOKOUTS SUPPORTED BY GABLE END WALL

© Cengage Learning 2014

Figure 42–20 Two styles of lookouts can support the rake overhang.

top ends fit snugly against the bottom edge and inside face of the end rafter. They are cut off flush with or below the top edge of the rafter (*Figure 42–21*).

Laying Out Gable Studs

The end wall plate is laid out for the location of the gable studs. Studs should be positioned directly above the end wall studs. This allows for easier installation of the wall sheathing. Square a line up from the wall studs over to the top of the wall plate.

Finding Gable Stud Length

Gable end stud lengths can be found using either of two methods.

Cut-and-Fit Method.   To use this method, stand each stud plumb in place on the layout line and then mark it along the bottom and top edge of the rafter (*Figure 42–22*). Remove the stud. Use a scrap piece of rafter stock to mark the depth of cut on the stud. Mark and cut all studs in a similar manner.

The studs are fastened by toenailing to the plate and by nailing through the rafter into the edge of the stud. Care must be taken when installing gable studs not to force a bow in the end rafters. This

creates a crown in them. Sight the top edge of the end rafters for straightness as gable studs are installed. After all gable studs are installed, the end ceiling joist is nailed to the inside edges of the studs.

Common Difference Method.   Gable studs that are spaced equally have a common difference in length. Each stud is shorter than the next one by the same amount (*Figure 42–23*). Once the length of the first stud and the common difference are known, gable studs can be easily laid out and gang cut on the ground.

To find the common difference in the length of gable studs, multiply the spacing, in terms of unit run, by the unit rise of the roof. For example, if the stud spacing is 16 inches OC and the unit Rise is 6, what is the common difference in length?

Step 1:   Convert OC spacing to unit Run.
16 ÷ 12 = $1\frac{1}{3}$ feet or units of Run.

Step 2:   Multiply unit Rise by unit Run.
6 × $1\frac{1}{3}$ = 8 inches common difference.

Next, determine the length of a stud. A framing square works well to measure the distance plumb and accurately. Mark it with the angle for the notch as described previously using a framing square. Assemble the stud material necessary for

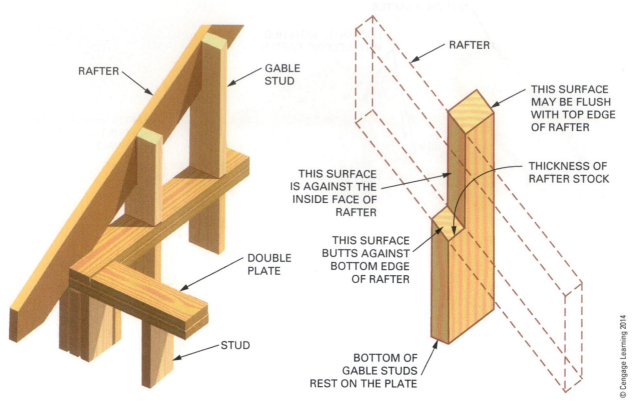

RAFTER

GABLE STUD

THIS SURFACE MAY BE FLUSH WITH TOP EDGE OF RAFTER

THICKNESS OF RAFTER STOCK

THIS SURFACE IS AGAINST THE INSIDE FACE OF RAFTER

THIS SURFACE BUTTS AGAINST BOTTOM EDGE OF RAFTER

DOUBLE PLATE

STUD

BOTTOM OF GABLE STUDS REST ON THE PLATE

© Cengage Learning 2014

Figure 42–21 The cut at the top end of the gable stud slopes with the rafter.

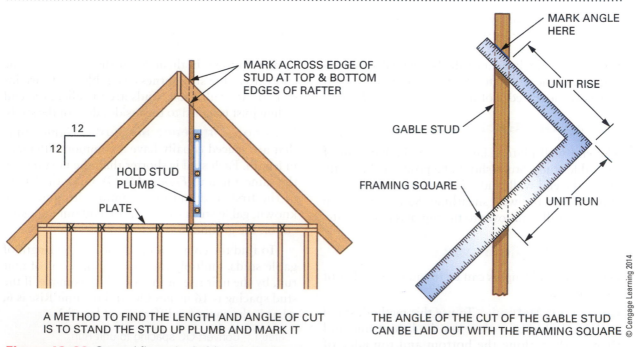

MARK ACROSS EDGE OF STUD AT TOP & BOTTOM EDGES OF RAFTER

12
12

HOLD STUD PLUMB

PLATE

A METHOD TO FIND THE LENGTH AND ANGLE OF CUT IS TO STAND THE STUD UP PLUMB AND MARK IT

MARK ANGLE HERE

UNIT RISE

GABLE STUD

FRAMING SQUARE

UNIT RUN

THE ANGLE OF THE CUT OF THE GABLE STUD CAN BE LAID OUT WITH THE FRAMING SQUARE

© Cengage Learning 2014

Figure 42–22 Cut-and-fit method of finding the length and cut of a gable stud.

one side of the gable. Lay them on horses next to each other where the bottom of each stud is separated from its neighbor by one common difference in length (*Procedure 42–F*).

Using the first measured stud as a reference, square a line across the ganged studs. Mark the

angle for each stud the same as the first. Be sure to measure and mark to the same point for each, that is, to the long point or short point of each notch.

Toenail each stud to the plate and face nail into the rafter. Check to be sure the rafter does not bow upward as each stud is installed.

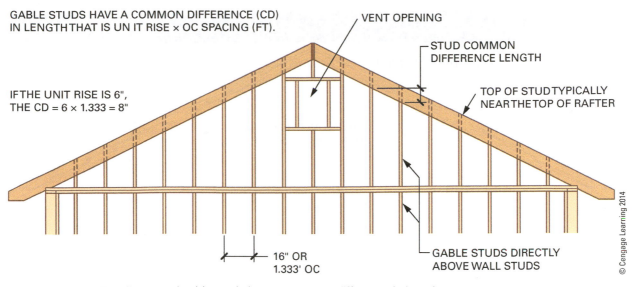

GABLE STUDS HAVE A COMMON DIFFERENCE (CD)
IN LENGTH THAT IS UN IT RISE × OC SPACING (FT).

VENT OPENING

STUD COMMON
DIFFERENCE LENGTH

IF THE UNIT RISE IS 6",
THE CD = 6 × 1.333 = 8"

TOP OF STUD TYPICALLY
NEAR THE TOP OF RAFTER

16" OR
1.333' OC

GABLE STUDS DIRECTLY
ABOVE WALL STUDS

© Cengage Learning 2014

Figure 42–23 Equally spaced gable studs have a common difference in length.

EXAMPLE

Gable studs are spaced 16 inches OC. The roof rises 8 inches per foot of run. Change 16 inches to $1\frac{1}{3}$ feet. Multiply $1\frac{1}{3}$ by 8 to get 10.666. The common difference in length for gable end studs here is $10\frac{2}{3}$ inches or roughly $10\frac{11}{16}$ inches.

SHED ROOF FRAMING

The shed roof slopes in only one direction. It is relatively easy to frame. A shed roof may be freestanding or one edge may rest on an exterior wall while the other edge butts against an existing wall (*Figure 42–24*). Flat roofs are shed roofs with little or not slope.

A shed roof is made of rafters that are all common rafters. A shed rafter Run is at a right angle to the plate line, and has a unit of Run is 12 inches. How the Run of the rafter is determined is similar to that of other common rafters, from where it

bears on other supporting members. Shed rafter Run is measured from the same side of the supporting walls (*Figure 42-25*). This makes the rafter Run

FREESTANDING SHED ROOF

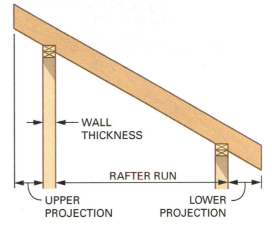

WALL
THICKNESS

RAFTER RUN

UPPER
PROJECTION

LOWER
PROJECTION

ATTACHED SHED ROOF

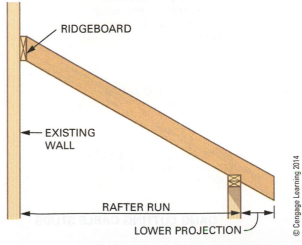

RIDGEBOARD

EXISTING
WALL

RAFTER RUN

LOWER PROJECTION

© Cengage Learning 2014

Figure 42-25 Styles of shed roofs.

Figure 42-24 A building with shed roofs butting each other.

© Cengage Learning 2014

StepbyStep Procedures

Procedure 42–F Laying Out Gable Studs

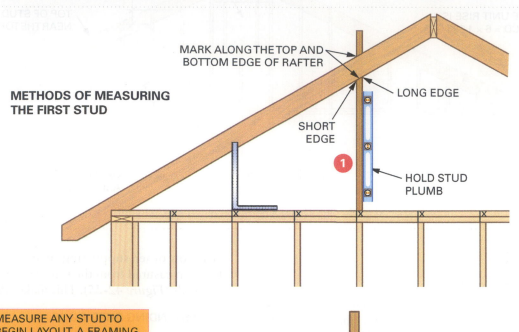

MARK ALONG THE TOP AND
BOTTOM EDGE OF RAFTER

LONG EDGE

SHORT EDGE

HOLD STUD PLUMB

METHODS OF MEASURING THE FIRST STUD

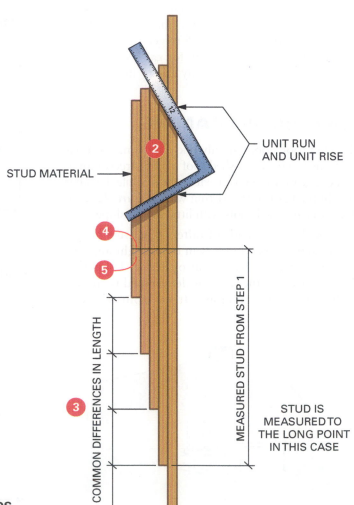

UNIT RUN AND UNIT RISE

STUD MATERIAL

COMMON DIFFERENCES IN LENGTH

MEASURED STUD FROM STEP 1

STUD IS MEASURED TO THE LONG POINT IN THIS CASE

STEP 1 MEASURE ANY STUD TO BEGIN LAYOUT. A FRAMING SQUARE MAY BE USED TO MEASURE A SMALL STUD.

STEP 2 ASSEMBLE STUD MATERIAL.

STEP 3 SPACE OUT STUDS SO THE BOTTOMS ARE SEPARATED BY THE COMMON DIFFERENCE.

STEP 4 SQUARE A LINE ACROSS STUD MATERIAL FROM THE STUD PREVIOUSLY MEASURED.

STEP 5 MARK ANGLED CUT ON EACH STUD.

STEP 6 MARK AND LAY OUT EXTRA FOR TOP TO FASTEN TO RAFTER.

GANG CUTTING GABLE STUDS

StepbyStep Procedures

Procedure 42–G Laying Out a Freestanding Shed Rafter (Double Overhang)

STEP 1 MARK A PLUMB LINE AT THE UPPER END OF THE BOARD.

STEP 2 STEP-OFF THE UPPER PROJECTION AND MARK THE SECOND PLUMB LINE.

STEP 3 MARK A REASONABLE SEAT CUT LINE FOR THE UPPER BIRD'S MOUTH AND NOTE THE HEIGHT ABOVE THE BIRD'S MOUTH.

STEP 4 MEASURE LINE LENGTH ALONG THE RAFTER FROM THE SECOND PLUMB LINE.

STEP 5 DRAW THE THIRD PLUMB LINE.

STEP 6 DRAW SEAT CUT AFTER MEASURING DOWN ALONG PLUMB LINE THE HEIGHT ABOVE THE BIRD'S MOUTH.

STEP 7 STEP-OFF THE LOWER PROJECTION.

STEP 8 DRAW THE FINAL PLUMB LINE.

© Cengage Learning 2014

the width of the building minus the thickness of the taller wall. If the upper end of the rafter is supported by a vertical surface such as a wall, the Run is measured from the same side of the wall surfaces. Rafters are then laid out similar to common rafters.

Laying Out a Shed Roof Rafter

Layout for a shed rafter begins at one end, such as the upper one, and progresses to the other end. The process is similar to that for any rafter.

Freestanding Shed Rafter.   To lay out a rafter for a freestanding shed roof, mark a plumb line at the upper end of the board (*Procedure 42–G*). This will be the fascia cut line. Deduct for fascia thickness, if appropriate. Step-off the upper projection and mark the plumb line. Mark a reasonable seat cut line for the upper bird's mouth and note the height above the bird's mouth.

Calculate the line length and measure it along the rafter from the upper bird's mouth plumb

Step**by**Step Procedures

Procedure 42-H Laying Out an Attached Shed Rafter (Single Overhang)

STEP **1** MARK A PLUMB LINE AT THE UPPER END OF THE BOARD.

STEP **2** MEASURE PERPENDICULAR THE FULL THICKNESS OF RIDGEBOARD.

STEP **3** MARK THE SECOND PLUMB LINE.

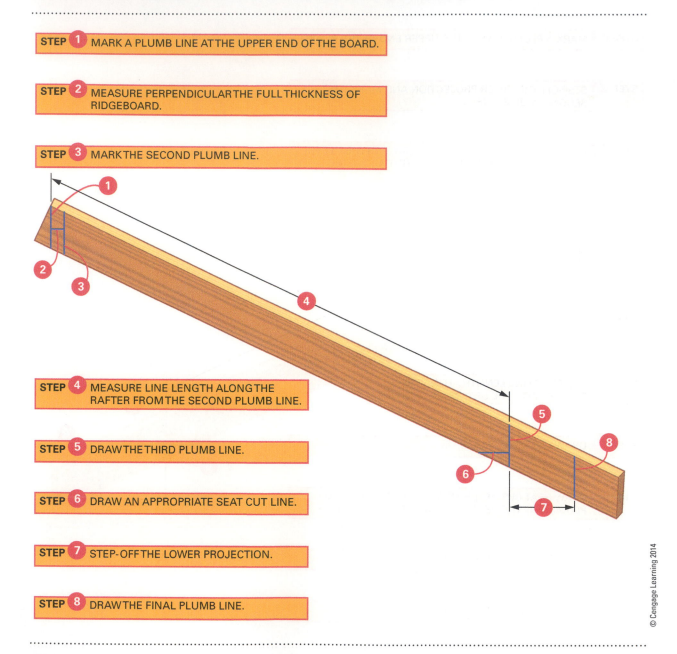

STEP **4** MEASURE LINE LENGTH ALONG THE RAFTER FROM THE SECOND PLUMB LINE.

STEP **5** DRAW THE THIRD PLUMB LINE.

STEP **6** DRAW AN APPROPRIATE SEAT CUT LINE.

STEP **7** STEP-OFF THE LOWER PROJECTION.

STEP **8** DRAW THE FINAL PLUMB LINE.

line. Draw the third plumb line for the lower bird's mouth. Measure down along this plumb line the height above the bird's mouth. Draw another seat cut. Step-off the lower projection and draw the final fascia plumb line. Level cuts may be added to each fascia cut as is done for common rafters.

Attached Shed Rafter.   Attached shed rafters are laid out in the same way as for common rafters (*Procedure 42–H*). The only difference is the

shortening for the ridge. A common rafter shortens by deducting one-half the ridge thickness. For an attached shed rafter, however, the shortening is a full ridge thickness.

Shed Rafter Erection.   Shed rafters are toenailed to the plates or ridge at the designated spacing. It is important to keep the bird's mouth snug up against the walls. In addition to nailing the rafters, metal framing connectors may be required. This includes all types of framed roofs.

FLAT ROOF FRAMING

Flat roofs are simply shed roofs with little or no slope. They are framed in a similar fashion as floors, yet typically cantilever all walls. This allows an overhang for shading on the windows and increasing weather resistance. The corners of the roof create a framing situation that compromises the roof strength at the corner. To overcome this difficulty two styles of framing rafters are used, the staggered frame and the mitered frame (*Figure 42–26*).

Staggered Frame

The staggered pattern has shortened cantilevered joists that attach to doubled rafters. These doubled rafters provides the extra support for the inner ends of the shortened lookout rafters. The lookout rafters should be at least three times longer than the overhang distance. For example if the overhang is 2 feet then the rafter must be 6 feet long. Thus the distance the doubled rafters are from the corner is twice the overhang distance.

Lookout rafters are installed along the entire width of the building. At the corners the shortened rafters are alternately cut shorter. They are attached to the previous lookout rafter. This makes only one rafter free floating, too short to extend into the building. The subfascia is then used to support the last rafter. This technique requires the fascia to be a structural member and long enough to land on the ends of at least four full length or cantilever rafters of proper length.

Mitered Frame

The mitered frame offers the greatest strength at the outside corner of the roof. It begins with the same doubled rafters as the previous method and is similarly set back from the corner a distance equal to twice the roof overhang length. Lookout rafters are then installed along the width of the building.

The doubled diagonal member is installed at a 45-degree angle with the walls. This member lands on the closest inside double rafter to the corner. The landing point is measured along the doubled rafter a distance equal to the twice the overhang.

Once the diagonal member is installed the mitered lookouts can be installed. They are each shortened by the same distance as the on-center spacing. The ends are cut at a 45-degree angle. The shortest lookout rafters are supported by the fascia.

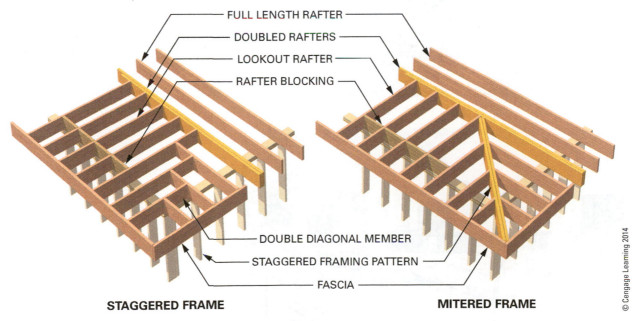

STAGGERED FRAME　　　　**MITERED FRAME**

Figure 42–26 Flat roofs may be framed in two styles.

CHAPTER 43
Trussed Roofs

ROOF TRUSSES

Roof trusses are used extensively in residential and commercial construction (*Figure 43–1*). They are designed to support the roof over a wide span, up to 100 feet, and eliminate the need for load-bearing partitions. They require less wood to manufacture, less time to install, and can reduce overall construction costs. Some styles, however, allow for little or no usable attic space.

Truss Construction

A roof truss consists of upper and lower *chords* and diagonal members called *webs*. The upper chords act as rafters, the lower chords serve as ceiling joists, and the webs are braces and stiffeners that replace collar ties. All truss members are fastened and joined securely with metal gusset plates (*Figure 43–2*). Plywood gussets may be used on jobsite–built trusses.

Figure 43–1 Trusses are used extensively in residential construction.

Figure 43–2 The members of roof trusses are securely fastened with metal gussets.

Truss Design

Most trusses are made in fabricating plants. They are transported to the job site. Trusses are designed by engineers to support prescribed loads. Trusses may also be built on the job, but approved designs must be used. The carpenter should not attempt to design a truss. Approved designs and instructions for job-built trusses are available from the American Plywood Association and the Truss Plate Institute.

The most common truss design for residential construction is the *Fink* truss (***Figure 43–3***). Other truss shapes are designed to meet special requirements (***Figure 43–4***).

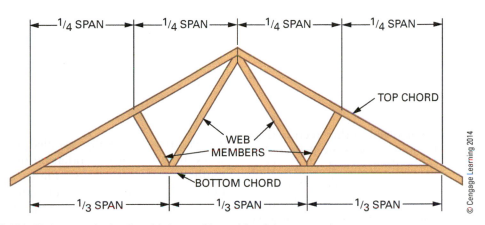

Figure 43–3 The Fink truss design is widely used in residential construction.

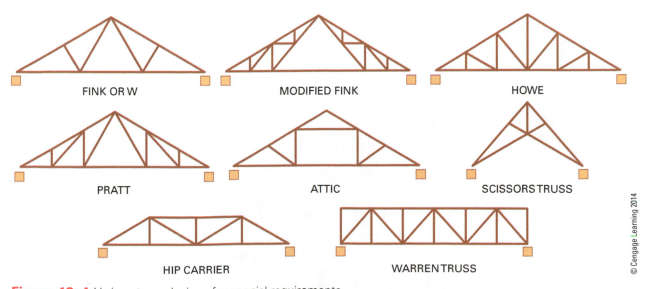

Figure 43–4 Various truss designs for special requirements.

Green Tip

Framing with trusses uses less material and labor than conventional framing.

CAUTION

Trusses are designed with smaller member sizes than are rafter ceiling joist systems. This causes higher stresses in the roof system members. Never cut any webs or chords of a truss unless directed by an engineer. Also, installing trusses can be very dangerous. Lives have been lost while installing trusses improperly. For these reasons, care must be employed by engineers in their designs and carpenters in their installation practices.

Erecting a Trussed Roof

Carpenters are more involved in the erection than the construction of trusses. Trusses are delivered to the job site using specially designed trucks. They should be unloaded and stored on a flat dry surface.

The erection and bracing of a trussed roof is a critical stage of construction. Failure to observe recommendations for erection and bracing could cause a collapse of the structure. This could result in loss of life or serious injury, not to mention the loss of time and material.

A print is provided showing the location of all trusses. A drawing of each truss is also provided that outlines important installation points (*Figure 43–5*).

The recommendations contained herein are technically sound. However, they are not the only method of bracing a roof system. They serve only as a guide. The builder must take necessary precautions during handling and erection to ensure that trusses are not damaged, which might reduce their strength.

Small trusses, which can be handled by hand, are placed upside down, hanging on the wall plates, toward one end of the building. Trusses for wide spans require the use of a crane to lift them into position. One at a time the truss is lifted, fastened, and braced in place. The end truss is installed first by swinging it up into place and bracing it securely in a plumb position (*Figure 43–5*).

A print is provided showing the location of all trusses. A drawing of each truss is also provided that outlines important installation points (*Figure 43–6*).

First Truss Installation

The end truss is usually installed first, but not always. In either case installation of the first truss requires great care. Trusses must be adequately braced in position, held secure, and plumb. They must be temporarily braced with enough strength to support the tip-over force of the trusses to follow. This may be achieved by bracing to securely anchored stakes driven into the ground or by bracing to the inside floor under the truss (*Figure 43–7*). These braces should be located directly in line with

Figure 43–5 Trusses are often lifted into place with a crane for speed and safety.

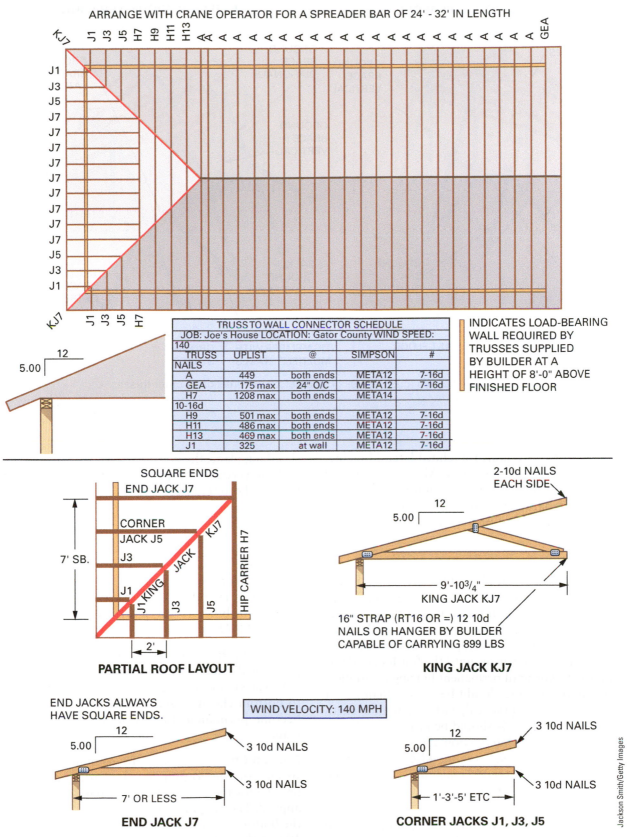

ARRANGE WITH CRANE OPERATOR FOR A SPREADER BAR OF 24' - 32' IN LENGTH

TRUSS TO WALL CONNECTOR SCHEDULE

JOB: Joe's House LOCATION: Gator County WIND SPEED: 140

TRUSS	UPLIST	@	SIMPSON	#
NAILS				
A	449	both ends	META12	7-16d
GEA	175 max	24" O/C	META12	7-16d
H7	1208 max	both ends	META14	
10-16d				
H9	501 max	both ends	META12	7-16d
H11	486 max	both ends	META12	7-16d
H13	469 max	both ends	META12	7-16d
J1	325	at wall	META12	7-16d

INDICATES LOAD-BEARING WALL REQUIRED BY TRUSSES SUPPLIED BY BUILDER AT A HEIGHT OF 8'-0" ABOVE FINISHED FLOOR

SQUARE ENDS
END JACK J7
CORNER JACK J5
KJ7
J3
HIP CARRIER H7
J1 KING JACK
J3
J5
7' SB.
2'

PARTIAL ROOF LAYOUT

2-10d NAILS EACH SIDE
12
5.00
9'-10³/₄"
KING JACK KJ7
16" STRAP (RT16 OR =) 12 10d NAILS OR HANGER BY BUILDER CAPABLE OF CARRYING 899 LBS

KING JACK KJ7

END JACKS ALWAYS HAVE SQUARE ENDS.

WIND VELOCITY: 140 MPH

12
5.00
3 10d NAILS
3 10d NAILS
7' OR LESS

END JACK J7

12
5.00
3 10d NAILS
3 10d NAILS
1'-3'-5' ETC

CORNER JACKS J1, J3, J5

Jackson Smith/Getty Images

Figure 43–6 Trusses are often delivered with an engineered set of prints showing truss labels and locations.

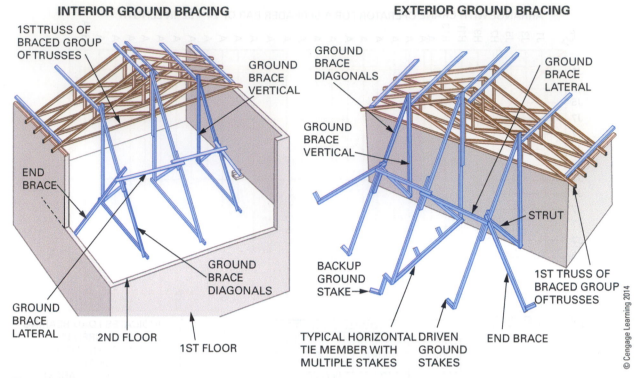

INTERIOR GROUND BRACING

1ST TRUSS OF BRACED GROUP OF TRUSSES

GROUND BRACE VERTICAL

END BRACE

GROUND BRACE DIAGONALS

GROUND BRACE LATERAL

2ND FLOOR

1ST FLOOR

EXTERIOR GROUND BRACING

GROUND BRACE DIAGONALS

GROUND BRACE LATERAL

GROUND BRACE VERTICAL

STRUT

BACKUP GROUND STAKE

TYPICAL HORIZONTAL TIE MEMBER WITH MULTIPLE STAKES

DRIVEN GROUND STAKES

1ST TRUSS OF BRACED GROUP OF TRUSSES

END BRACE

© Cengage Learning 2014

Figure 43–7 The first set of trusses must be well braced before the erection of other trusses.

all rows of continuous top chord **lateral bracing,** which will be installed later. All bracing should be securely fastened with the appropriate size and quantity of nails, remembering that lives depend on *IRC R802.11 the bracing doing its intended job.* *

Temporary Bracing of Trusses

As trusses are set in place, they are nailed to the plate. Metal framing connectors are usually applied (*Figure 43–8*). Information on these connectors may be found on the connector schedule of the set of prints. Sufficient temporary bracing must be applied to secure trusses until the finish material is applied and/or until permanent bracing is installed. Temporary bracing should be no less than 2 × 4 lumber as long as practical, with a minimum length of 8 feet. The 2 × 4s should be fastened with two 16d nails at every intersection and should be overlapped by two trusses (*Figure 43–9*). Exact spacing of the trusses should be maintained. Adjusting trusses later, while possible, is time consuming and *IRC R802.10.3 risky.* *

Temporary bracing must be applied to three planes of the truss assembly: the top chord or

sheathing plane, the bottom chord or ceiling plane, and the vertical web plane at right angles to the bottom chord (*Figure 43–10*).

Top Chord Bracing. Continuous lateral bracing should be installed within 6 inches of the ridge and at about 8- to 10-foot intervals between the ridge and wall plate. Diagonal bracing should be set at approximately 45-degree angles between the rows of lateral bracing. It forms triangles that provide stability to the plane of the top chord (*Figure 43–11*).

Web Plane Bracing. Temporary bracing in the plane of the web members is made up of diagonals placed at right angles to the trusses from top to bottom chords (*Figure 43–12*). They usually become permanent braces of the web member plane.

Bottom Chord Bracing. To maintain the proper spacing on the bottom chord, continuous lateral bracing for the full length of the building must be applied. The bracing should be nailed to the top of the bottom chord at intervals no greater than 8 to 10 feet along the width of the building.

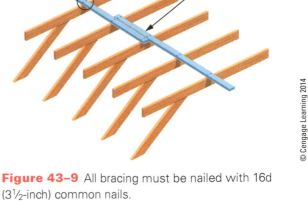

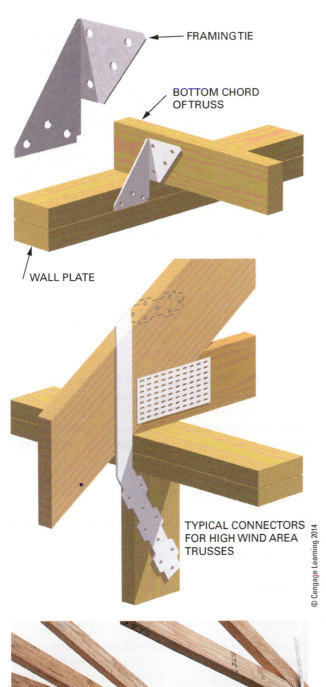

FRAMING TIE

BOTTOM CHORD OF TRUSS

WALL PLATE

TYPICAL CONNECTORS FOR HIGH WIND AREA TRUSSES

TWO 16d NAILS EACH TRUSS

TOP CHORD

OVERLAPPED LATERAL BRACING

Figure 43–9 All bracing must be nailed with 16d (3½-inch) common nails.

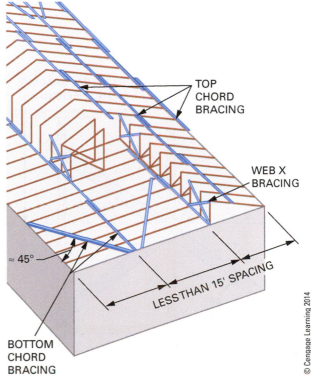

TOP CHORD BRACING

WEB X BRACING

≈ 45°

LESS THAN 15' SPACING

BOTTOM CHORD BRACING

Figure 43–10 Temporary bracing secures three planes of trusses.

Diagonal bracing should be installed, at least, at each end of the building (*Figure 43–13*). In most cases, temporary bracing of the plane of the bottom chord is left in place as permanent bracing.

Permanent Bracing

Permanent bracing is designed by the structural engineer when designing the truss, and all bracing designs should be carefully followed. The top chord

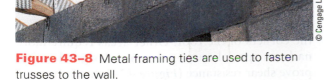

Figure 43–8 Metal framing ties are used to fasten trusses to the wall.

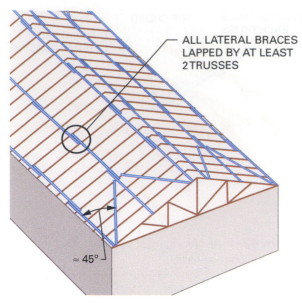

Figure 43–11 Typical temporary bracing of the top chord plane.

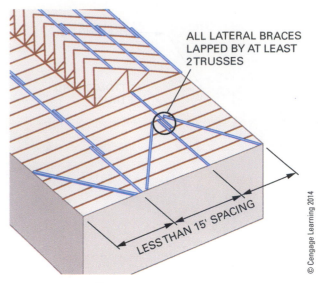

Figure 43–13 Typical bracing of the top chord plane.

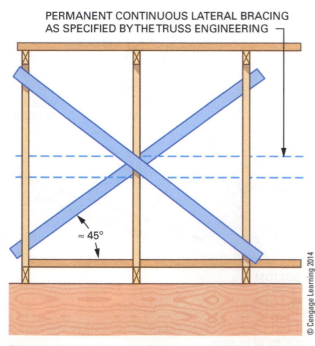

Figure 43–12 Bracing of the web member plane prevents lateral movement of the trusses.

permanent bracing is often provided by the roof sheathing. Web bracing may have X braces as well as lateral bracing and web stiffeners (**Figure 43–14**). These are usually installed after the trusses, but before the sheathing. *

*IRC R802.10.3

Framing Openings

Openings in the roof or ceiling for skylights or access ways must be framed within or between the trusses. The chords, braces, and webs of a truss system should never be cut or removed unless directed to do so by an engineer. Simply installing headers between trusses will create an opening. The sheathing or ceiling finish is then applied around the opening. *

*IRC R802.10.4

ROOF SHEATHING

Roof sheathing is applied after the roof frame is complete. Sheathing gives rigidity to the roof frame. It also provides a nailing base for the roof covering. Rated panels of plywood and strand board are commonly used to sheath roofs. *

*IRC R803.2.1

Panel Sheathing

Plywood and other rated panel roof sheathing are laid with the face grain running perpendicular to the rafter or top chord for greater strength (**Figure 43–15**). End joints are made on the framing member and staggered, similar to subfloor. Nails are typically spaced 6 inches apart on the ends and 12 inches apart on intermediate supports. *

*IRC R803.2.3

Additional nails are added according to local codes for increased strength. Some areas require extra nails in the sheathing to protect from uplift caused by high winds. These nailing zones put more nails along the perimeter of the roof and at the corners of the roof. Other areas require more nails on special trusses that are designed to improve shear resistance (*Figure 43–16*).

Adequate blocking, tongue-and-groove edges, or other suitable edge support such as *panel clips* must

"T" BRACING IS USED TO REPLACE LATERAL BRACING,
ONE "T" BRACE REPLACES ANY NUMBER OF LATERAL BRACES ON SAME WEB.

BRACE TO BE
APPLIED AS
"L" BRACE

"T" BRACE OR "L" BRACE MUST BE
90% THE LENGTH OF BRACED WEB.
SEE CHART BELOW FOR LUMBER
SIZE AND NAILING REQUIREMENTS.

TRUSS WEB

TRUSS WEB

"T" BRACE REPLACES LATERAL BRACE
TRUSS TO TRUSS. IF "T" BRACE USED,
DO NOT LATERAL BRACE TRUSS TO TRUSS

"T" BRACE ON ONE TRUSS

T OR L BRACE NAILING PATTERN		
T–BRACE SIZE	NAIL SIZE	NAIL SPACING
1 x 4	10D	8" OC
2 x 4	16D	8" OC
NAIL ALONG ENTIRE LENGTH OF T–BRACE ON EACH PLY.			

T-BRACE SAME LUMBER AS WEB OR #2 SPRUCE	T–BRACE OR L–BRACE SIZE			
	NUMBER OF PLYS OF TRUSS			
	1 PLY TRUSS		2 + PLY TRUSS	
	ROWS OF BRACING SHOWN ON ENGINEERING			
WEB SIZE	1	2	1	2
2 x 4	1 x 4	2 x 4	2 x 4	2 x 4
2 x 6	1 x 6	2 x 6	2 x 6	2 x 6
2 x 8	2 x 8	2 x 8	2 x 8	2 x 8

© Cengage Learning 2014

Figure 43–14 Some webs require stiffeners to be added by the carpenter.

Courtesy of APA–The Engineered Wood Association

Figure 43–15 Sheathing a trussed roof with plywood.

© Cengage Learning 2014

Figure 43–16 Sheathing in seismic-prone areas has extra nails indicated here by red painted lines.

be used when spans exceed the indicated value of the plywood roof sheathing. Panel clips are small metal pieces shaped like a capital *H*. They are used between adjoining edges of the plywood between rafters (*Figure 43–17*). One clip is typically used for 24-inch spans and two panel clips are used for 32- and 48-inch spans. If OSB (oriented strand board) is used as sheathing, two clips are required for 24-inch spans.✳

✳**IRC R803.2.2**

CAUTION

Sawdust from a circular saw on roof sheathing can be very slippery. Make sure of your footing as you cut.

The ends of sheathing may be allowed to slightly and randomly overhang at the *rakes* until sheathing is completed. A chalk line is then snapped between the ridge and plate at each end of the roof. The sheathing ends are trimmed straight to the line with a circular saw.

Plank Sheathing

Plank decking is used in post-and-beam construction where the roof supports are spaced farther apart. Plank roof sheathing may be of 2-inch nominal thickness or greater, depending on the span. Usually both edges and ends are matched. The plank roof often serves as the finish ceiling for the rooms below.✳

✳**IRC R803.1**

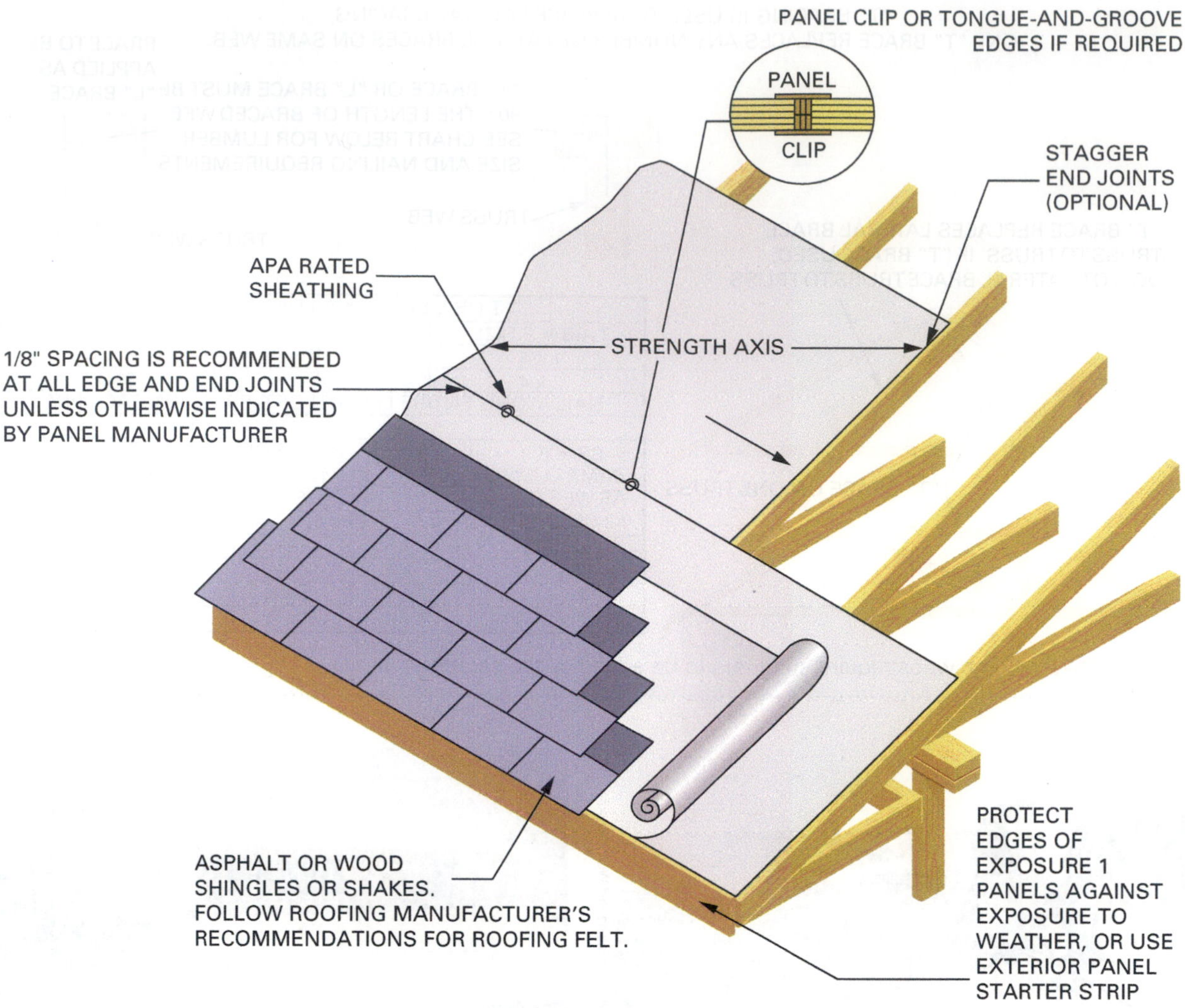

PANEL CLIP OR TONGUE-AND-GROOVE
EDGES IF REQUIRED

PANEL

CLIP

STAGGER
END JOINTS
(OPTIONAL)

APA RATED
SHEATHING

STRENGTH AXIS

1/8" SPACING IS RECOMMENDED
AT ALL EDGE AND END JOINTS
UNLESS OTHERWISE INDICATED
BY PANEL MANUFACTURER

ASPHALT OR WOOD
SHINGLES OR SHAKES.
FOLLOW ROOFING MANUFACTURER'S
RECOMMENDATIONS FOR ROOFING FELT.

PROTECT
EDGES OF
EXPOSURE 1
PANELS AGAINST
EXPOSURE TO
WEATHER, OR USE
EXTERIOR PANEL
STARTER STRIP

Courtesy of APA-The Engineered Wood Association

RECOMMENDED MINIMUM FASTENING SCHEDULE FOR APA PANEL ROOF SHEATHING (INCREASED NAIL SCHEDULES MAY BE REQUIRED IN HIGH WIND ZONES AND WHERE ROOF IS ENGINEERED AS A DIAPHRAGM.)

PANEL THICKNESS[B] (IN.)	NAILING[C][D]		
		MAXIMUM SPACING (IN.)	
	SIZE[F]	SUPPORTED PANEL EDGES[E]	INTERMEDIATE
$5/16 - 1$	8D	6	12[A]
$1-1/8$	8D OR 10D	6	12[A]

(A) FOR SPANS 48 INCHES OR GREATER, SPACE NAILS 6 INCHES AT ALL SUPPORTS.

(B) FOR STAPLING ASPHALT SHINGLES TO 5/16-INCH AND THICKER PANELS, USE STAPLES WITH A 15/16-INCH MINIMUM CROWN WIDTH AND A 1-INCH LEG LENGTH. SPACE ACCORDING TO SHINGLE MANUFACTURER'S RECOMMENDATIONS.

(C) USE COMMON SMOOTH OR DEFORMED SHANK NAILS WITH PANELS TO 1 INCH THICK. FOR 1-1/18-INCH PANELS, USE 8D RING-OR SCREW-SHANK OR 10D COMMON SMOOTH-SHANK NAILS.

(D) OTHER CODE-APPROVED FASTENERS MAY BE USED.

(E) SUPPORTED PANEL JOINTS SHALL OCCUR APPROXIMATELY ALONG THE CENTERLINE OF FRAMING WITH A MINIMUM BEARING OF 1/2". FASTENERS SHALL BE LOCATED 3/8 INCH FROM PANEL EDGES.

(F) SEE TABLE 5, PAGE 13, FOR NAIL DIMENSIONS.

NOTE: GLUING OF ROOF SHEATHING TO FRAMING IS NOT RECOMMENDED, EXCEPT WHEN RECOMMENDED BY THE ADHESIVE MANUFACTURER FOR ROOF SHEATHING THAT ALREADY HAS BEEN PERMANENTLY PROTECTED BY ROOFING.

Courtesy of APA-The Engineered Wood Association

Figure 43–17 Recommendations for the application of APA panel roof sheathing.

ESTIMATING MATERIALS FOR ROOF FRAMING

Estimating the amount of materials for roof framing involves rafters, ridgeboards, trusses and temporary bracing, gable end studs, and sheathing. Remember that if the number of pieces does not come out even, round them up to the next whole number of pieces.

Rafters

The number of rafters needed for a shed roof is half the number for a gable roof. The number of flat roof rafters is similar to the number of floor joists.

Common Rafters for Gable Roof. Divide the length of the building by the spacing of the rafters. Round up to the nearest whole number and add one, as a starter. Then multiply the total by two to include the other side. Add four, one for each rake rafter. For example, if the building is 50 feet, the number of 16-inch OC gable rafters is $50 \div {}^{16}/_{12} + 1 = 38.5 \Rightarrow 39 \times 2 + 4 = 82$ rafters.

Shed Roof Rafters. The number of shed roof rafters is half the number if a gable roof were installed. To the length add two times the overhang distance (in feet) and then divide by the on-center spacing (in feet). Add one to start. For example, if the length is 16 feet with 24 inches on-center and overhang is 12 inches, the number of rafters is $16 + (2 \times {}^{12}/_{12}) = 18 \div {}^{24}/_{12} = 9 + 1 \Rightarrow 10$ rafters.

Flat Roof Main Rafters. The portion of the lookouts that extend into the building (twice the overhang in feet) is subtracted twice from the length of the building. The remaining distance is divided by the on-center spacing (in feet). To this result, add one more to start and one for each double rafter. This is the number of rafters to run full width of the building. If they lap in the middle of the building, double this number for the other side. For example, if the length is 42 feet, overhang is 18 inches, the on-center spacing is 16 inches, and the rafters lap, then the number of main roof rafters is $42 - (2 \times {}^{18}/_{12}) = 36$ feet $\div {}^{16}/_{12} = 27 + 1 + 2$ (doubled rafters) $= 30 \times 2 = 60$ rafters.

Flat Roof Lookout Rafters. To the width of the building add two overhang distances (in feet). Divide this sum by the rafter on-center spacing (in feet), round up if necessary, and double it for the other side. Add two for each corner for the staggered or mitered lookouts. For example, if a rectangular building has a width of 30 feet, 16 inch on-center, and an overhang of 20 inches, the number of lookouts is 30 feet $+ (2 \times {}^{20}/_{12}) = 33.33$ feet. Dividing this by OC gives $33.33 \div {}^{16}/_{12} = 25$ pieces. Doubling it makes 50, and add $4 \times 2 \Rightarrow 58$ lookouts. The length of the lookout is three times the roof overhang. Divide this length by the desired board length. For example if the lookouts are 4 feet long three will come out 12 feet long.

Fascia. The subfascia material is best made from material 16 feet long. Divide the roof perimeter by 16. The roof perimeter must include the overhang distances. Add twice the overhang (in feet) to the width and the length. Find the perimeter from these numbers. For example, find the 16-foot fascia boards for a rectangular building that is 32 × 52 with an overhang of 2 feet. Length is 52 feet $+ (2 \times 2$ feet$) = 56$ feet long. The width is 32 $+ (2 \times 2$ feet$) = 36$ feet. Perimeter is $2 \times (56 + 36) = 184$, Fascia boards are 184 feet $\div 16 = 11.5 \Rightarrow 12$ fascia boards.

Flat Roof Diagonal Member. The length of the double diagonal member of a mitered frame is found with the Pythagorean Theorem using the lookout length in a^2 and b^2. Add two per corner. For example if the lookouts are 6 feet long, the diagonal member is square root of $(6^2 + 6^2) = 8.48$ feet $\Rightarrow 10$ foot board.

Ridgeboards

The ridgeboard for gable and shed roofs are found using the same process. The gable end overhang is added to the length of the building.

Ridgeboard for Gable and Shed Roof. Take the length of the building plus both rake overhangs. Divide this sum by the length of the material to be used for the ridge. Typically 12- or 16-foot boards are used because there is minimal cutting waste with the various OCs. For example, if the building is 50 feet and the gable end overhang is 12 inches, or one foot, then the number of 12-foot ridgeboards is $50 + 2 = 52 \div 12 = 4.3 \Rightarrow 5$ ridgeboards.

Gable End Studs

Divide the width of the building by the OC spacing. Add two extra studs. This number of studs will be enough to do both sides of the building. For example, if the width is 32 feet, the number of 16-inch OC studs needed will be $32 \div {}^{16}/_{12} + 2 = 26$ studs. Shed roofs need half this many.

Trusses

Divide the building length by the OC spacing, usually 2 feet. Subtract one. This is the number of common trusses. Two additional trusses need to be specified and ordered as end trusses for the ends of the building. *Note:* the number and sizes of hip trusses are determined by the manufacturer. For example, if the building is 50 feet, the number of common trusses is $50 \div 2 - 1 = 24$ trusses and two end trusses.

Temporary Truss Bracing Material

Divide the building width by six to get the number of rows of top, bottom, and web bracing. Round this number up to the nearest whole number. Multiply times length of building. Divide by the bracing length, typically 16 feet. Round to the nearest whole number. Add 20 percent for brace overlap and waste and any extra for ground bracing of the first truss. For example, if the building is 32 feet $\times$ 50 feet, the number of 16-feet $-$ 2 $\times$ 4 braces is 32 $\div$ 6 = 5.3 $\Rightarrow$ 6 rows. 50 $\div$ 16 = 3.1 $\Rightarrow$ 4 pieces per row. Thus, 6 rows $\times$ 4 pieces per row $\times$ 120% = 28.8 $\Rightarrow$ 29 braces.

Sheathing

Estimating sheathing involves finding the area to be covered by the sheathing and then dividing by the area of one sheet.

Sheathing Gable Roof.   Find the number of plywood rows by dividing the actual rafter length by 4 feet (width of plywood sheet). Round this number up to nearest one-half row. Since the waste of the other half is still usable. Next, find the number of sheets across the building by dividing the ridgeboard length by 8 feet (length of plywood sheet). Round this up to the nearest one-half sheet. Find the number of sheets by multiplying these numbers together and then doubling the result to account for the other side. Round up to the nearest whole number. Waste will already be allowed for in the rounding-up of rafter and ridgeboard lengths. For example, sheathing for a roof with 20′-6″ rafters on a building with a 52′ ridgeboard will have 20.5 $\div$ 4 = 5.1 $\Rightarrow$ 5.5 rows. It will have 52 $\div$ 8 = 6.5

sheets per row. Thus, we have 5.5 $\times$ 6.5 $\times$ 2 sides = 71.5 $\Rightarrow$ 72 sheets.

Sheathing Shed Roof.   The shed roof is one-half of a gable roof. To find the number of rafters, calculate as if it were a gable, except do not double for the other side.

Flat Roof Sheathing.   The roof area takes into account the overhangs on all sides. Add twice the overhang to the length and the width. Find the number of sheet to make a row along the length and multiply by the number of rows of sheets along the width (each rounded to the nearest ½ sheet). Then multiply sheets per row times the number of rows. For example find the number of 4×8 sheets of plywood for a rectangular building that is 30×52 with an overhang of 2 feet. Number of sheets along length is 54′ + 4′ = 58′ $\div$ 8 feet = 7.25 $\Rightarrow$ 7.5 sheets per row. Number of rows of sheets along the width is 30′ + 4′ = 34′ $\div$ 4 feet = 8.5 rows. Total number of sheets is 7.5 $\times$ 8.5 = 62.25 $\Rightarrow$ 63 sheets.

Gable End Sheathing.   Calculate the total Rise of the roof by multiplying the unit Rise times the rafter Run, then divide by 12 to change the answer to feet. Find the number of rows of sheathing by dividing this number by 4 feet, and round up to the nearest one-half sheet. Next, find the number of sheets across the building by dividing the building width by 8 feet. Round this number up to the nearest one-half sheet. Multiply these numbers, number rows, and sheets together, and round up to the nearest whole number. Waste material should be sufficient to sheath the opposite gable end. For example, the sheathing for the gable ends of a 32′-wide building with an 8″ unit Rise has 32′ $\div$ 8 = 4 pieces per row. There are 32 $\div$ 2 = 16 Run $\times$ 8 $\div$ 12 = 10.6′ total Rise $\div$ 4 = 2.6 $\Rightarrow$ 3 rows. Sheathing, then, is 3 $\times$ 4 = 12 pieces.

See *Figure 43–18* a listing of the shorthand version of these estimating techniques with another example.

Estimate the materials for a roof of a rectangular 28′ × 42′ building. Framing 16″ OC, 24″ OC trusses, 9″ gable end overhang, and a rafter length of 17′-4″. Shed roof on a building that is 12′ × 30′ with 9 inch gable end overhang, rafter length of 15′-7″, and 16″ OC. Flat roof is for a 28′ × 42′ with a mitered frame, an overhang on all sides of 18 inches, lapped rafters, and 16″ OC.

Item	Formula	Waste factor	Example
Common Rafters	(building LEN (in ft) ÷ OC (ft) + 1) × 2 + 1 per each rake rafter.		$42 ÷ {}^{16}/_{12} = 31.5 ⇒ 32 + 1$ $= 33 × 2 = 66 + 4 = 70\ PC$
Gable Ridgeboard	building LEN + (2 × gable end over hang)		$42 + 2 × {}^{9}/_{12} = 435 FT$
Gable end studs	building WID ÷ OC (ft) + 2		$28 ÷ {}^{16}/_{12} + 21 + 2 = 23 PC$
Trusses	building LEN ÷ OC (ft) − 1, plus two gable end trusses		$42 ÷ 2 − 1 = 20 PC$
Temp Truss Bracing	building WID ÷ 4 × building LEN ÷ 16 × 20% waste	20	$28 ÷ 4 × 42 ÷ 16 × 1.20 = 22.05 ⇒ 23 PC$
Gable Roof Sheathing	rafter LEN ÷ 4 (rounded up to nearest ½) × ridge LEN ÷ 8 × 2 (rounded up to nearest ½)		$(17.\overline{333}) ÷ 4 = 4.33 ⇒ 4.5 × (43.5 ÷ 8$ $= 5.4375 ⇒ 5.5) = 24.75 × 25 PC$
Gable End Sheathing	gable rise ÷ 4 (rounded up to nearest ½) × building WID ÷ 8 (rounded up to nearest ½)		$14 × {}^{7}/_{12} ÷ 4 = 2.04 ⇒ 2.5 × (28 ÷ 8$ $= 3.5 ⇒ 3.5) = 8.75 ⇒ 9 PC$
Shed Roof Rafters	(Building LEN + twice overhang) ÷ OC		$30 + (2 × {}^{9}/_{12}) = 31.5 ÷ {}^{16}/_{12} = 23.6 ⇒$ 24 rafters
Flat Roof Main Rafters	((Building LEN − 4 × overhang) ÷ OC +1 + 2) doubled		$((42 − 4 × {}^{18}/_{12}) ÷ {}^{16}/_{12} +1 + 2) × 2 = 60$ rafters
Flat Roof Lookouts	((Building WID +2 × Overhang) ÷ OC + 1 + # of corners x 2) doubled		$((28 + 2 × 18/12) ÷ {}^{16}/_{12} +1 + 2 × 4) × 2$ $= 64.5 ⇒ 65$ Lookouts
Flat Roof Fascia	(Building LEN + twice overhang + (Building WID + twice overhang) × 2 = LF		$(42 + 2 × {}^{18}/_{12} + 28 + 2 × {}^{18}/_{12}) × 2 =$ 152 LF
Shed Roof Sheathing	(Building LEN + 2 × Overhang (rounded up to nearest ½) ÷ 8′ × rafter length ÷ 4)(rounded up to nearest ½)		$30 + 2 × {}^{18}/_{12} ÷ 8 = 4.1 ⇒ 4.5$ sheets per row $15.583 ÷ 4 = 3.9 ⇒ 4$ rows × 4.5 sheets/row = 18 sheets
Flat Roof Sheathing	(Building LEN + twice overhang) ÷ sheet LEN = Sheets/row (Building WID + twice overhang) ÷ sheet WID = NUM Rows Rows × sheets per row = sheets.		$(42 + 2 × {}^{18}/_{12}) ÷ 8 = 5.6 ⇒ 6$ PC/row $(28 + 2 × {}^{18}/_{12}) ÷ 4$ rows. $6 × 4 = 24$ sheets

Figure 43–18 Example of estimating roof framing materials with formulas.

DECONSTRUCT THIS

Carefully study **Figure 43–19** and think about what is wrong and/or what is right. Consider all possibilities. What construction practice or method is different in your area of the country?

Figure 43–19 This photo shows a construction site.

KEY TERMS

bird's mouth	gables	plumb line	tail cut
butterfly roof	gambrel roof	rafter tables	total Rise
collar ties	line length	rake rafter	total Run
common rafter	lateral bracing	ridgeboard	total span
fascia	level line	seat cut	unit length
gable end stud	lookouts	shed roof	unit Rise
gable roof	mansard roof	slope	unit Run
gable studs	pitch	soffit	

© Cengage Learning 2014

REVIEW QUESTIONS

Select the most appropriate answer.

1. The type of roof that has the fewest number of different-size members is a
   a. gable.   c. gambrel.
   b. hip.   d. mansard.

2. The term used to represent horizontal distance under a rafter is
   a. run.   c. line length.
   b. span.   d. pitch.

3. The rafter that spans from a hip rafter to the wall plate is a
   a. hip jack.   c. valley.
   b. valley jack. d. cripple jack.

4. The rafter that spans from a wall plate to the ridge and whose Run is perpendicular to the ridge is a
   a. hip.   c. valley jack.
   b. valley.   d. common rafter.

5. The minimum amount of stock left above the seat cut of the common rafter to ensure enough strength to support the overhang is usually _____ of the rafter stock.
   a. one-quarter c. two-thirds
   b. one-half   d. three-quarters

6. What is the line length of a common rafter from the centerline of the ridge to the plate with a unit Rise of 5 inches, if the building is 28'-0" wide?
   a. 70 inches   c. 182 inches
   b. 140 inches   d. 364 inches

7. The common difference in the length of gable studs spaced 24 inches OC for a roof with a slope of 8 on 12 is
   a. 8 inches.   c. 16 inches.
   b. 12 inches.   d. 20 inches.

8. Unit Rise is similar to total Rise as unit Span is to _____.
   a. total Length
   b. total Span
   c. Run
   d. total Run

9. The twelfth scale on a framing square is best used to _____.
   a. layout rafters.
   b. calculate rafter length.
   c. estimate rafter length.
   d. measure twelve-sided objects.

10. Shortening a rafter at the ridge requires the measurement be
    a. parallel to plumb line.
    b. perpendicular to plumb line.
    c. the thickness of the rafter.
    d. all of the above.

11. The length of any rafter in a roof of specified slope may be found if its _____ is known.
    a. total run   c. unit length
    b. unit rise   d. all of the above

12. The run of a shed rafter for a freestanding roof is measured from the
    a. width of the building.
    b. length of the building.
    c. opposite sides of supporting walls.
    d. width minus thickness of taller supporting wall.

13. Flat roofs frames typically have
    a. two or four doubled rafters.
    b. lookout rafters.
    c. mitered or stagger framed corners.
    d. all of the above.

14. The gable ridgeboard is similar to a flat roof in that both
    a. lengths include two overhangs.
    b. have same OC spacing.
    c. have lookouts attached.
    d. all of the above.

15. The overhang length of the tail of a common rafter with a projection of 14 inches and a slope of 7 on 12 is
    a. 1.167 feet.
    b. 13.89 inches.
    c. 16.21 inches.
    d. 19.45 inches.

16. The member that provides the most resistance of trusses to tipping over is called a
    a. diagonal brace.   c. web.
    b. lateral brace.   d. gusset.

17. The part of a truss that may be cut if necessary is the
    a. web.
    b. chord.
    c. gusset.
    d. none of the above.

18. Temporary lateral bracing of a trussed roof is needed to
    a. provide the resistance to tipping over.
    b. keep the trusses evenly spaced.
    c. give a place to attach the roof sheathing.
    d. all of the above.

19. Including the rake rafters, the estimated number of 16″ OC gable rafters for a rectangular building measuring 28 × 48 feet is
    a. 72.   c. 78.
    b. 74.   d. 84.

20. The estimated number of pieces of panel sheathing needed for a gable roof with a rafter length of 14′–7″ and a ridgeboard length of 34′–9″ is
    a. 16.   c. 32.
    b. 18.   d. 36.

UNIT 16
Advanced Roof Framing

Complex roof framing situations offer the carpenter an opportunity to demonstrate a superior understanding of framing. Cutting rafter members accurately the first time saves time and material. Roof framing theory on how to cut and fit members takes a little time to understand, but the math to do the job is straightforward and doable. The precision and speed by which roof framing can be performed reveals an elegance in the process.

OBJECTIVES

After completing this unit, the student should be able to:

- describe and lay out the framing members of hip roof.

- describe and lay out the framing members of an equal-pitch intersecting roofs.

- layout and cut gable end triangular sheathing.

- layout and cut sheathing along hip and valley rafters.

- describe and lay out framing members of dormers.

- describe and lay out members of a gambrel roof system.

- describe and lay out a knee wall given its height or its run.

- describe and adjust for intersecting rafters with different heights above the bird's mouth.

- lay out and cut an unconventional hip rafter.

SAFETY REMINDER

Roof framing involves sloped and often slippery surfaces while also requiring greater mental energy to perform more complicated framing details. Maintain a constant awareness of your surroundings and what your feet are doing as you focus on roofing details.

CHAPTER 44

Hip Roofs

The hip roof is a little more complicated to frame than the gable roof. Two additional kinds of rafters need to be laid out.

To frame the hip roof, it is necessary to lay out not only common rafters and a ridge, but also hip rafters and hip jack rafters (*Figure 44–1*). Hip rafters form the outside corners where two sloping roof planes meet. They are longer than common rafters, and their slope is at a lower angle. Hip jacks are common rafters that must be cut shorter to land on a hip rafter. They all have a common difference in length.

HIP RAFTER THEORY

The theory behind fashioning a hip rafter is quite similar to that used for a common rafter, yet slight differences allow the hip rafter to take a different shape in the roof. The length of the hip rafter can be estimated using a framing square.

Unit Run

Both hip and common rafters have a total Run that always runs horizontal under the rafters. A common rafter Run forms a 90-degree angle with the plates, and the hip Run forms a 45-degree angle. Therefore, a hip rafter Run covers more distance than does a common run, but the number of units of Run is the same.

The hip unit Run forms the diagonal of a square created by the common unit Run (*Figure 44–2*). Using the Pythagorean theorem with 12 as a and b, c turns out to be 16.97 inches. This amount is rounded off to 17 inches when using a framing square to lay out the rafter.

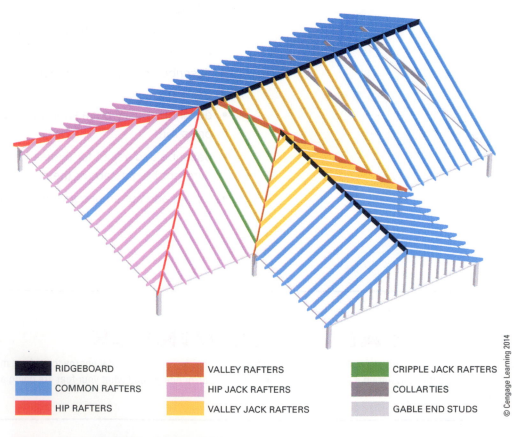

	RIDGEBOARD		VALLEY RAFTERS		CRIPPLE JACK RAFTERS
	COMMON RAFTERS		HIP JACK RAFTERS		COLLAR TIES
	HIP RAFTERS		VALLEY JACK RAFTERS		GABLE END STUDS

© Cengage Learning 2014

Figure 44–1 Members of a hip roof frame.

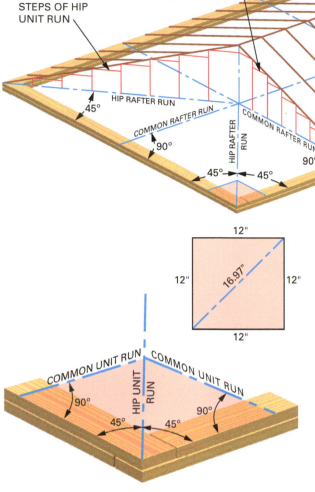

STEPS OF COMMON
UNIT RUN

STEPS OF HIP
UNIT RUN

45°
HIP RAFTER RUN
COMMON RAFTER RUN
COMMON RAFTER RUN
HIP RAFTER RUN
90°
90°
45° 45°

12"
12" 16.97" 12"
12"

COMMON UNIT RUN   COMMON UNIT RUN
HIP UNIT RUN
90° 90°
45° 45°

NOTES:

1. COMMON RAFTER RUN FORMS A 90-DEGREE ANGLE WITH WALL PLATES.

2. HIP RUN FORMS A 45-DEGREE ANGLE WITH WALL PLATES.

3. HIP RUN IS LONGER IN DISTANCE THAN THE COMMON RUN.

4. THE NUMBER OF UNITS OF RUN IS THE SAME FOR THE COMMON AND HIP RAFTERS. IN THIS CASE, THERE ARE EIGHT UNITS OF RUN IN EACH RAFTER.

5. THEREFORE, HIP UNIT RUN IS LARGER THAN 12".

6. HIP UNIT RUN IS THE DIAGONAL OF 12" SQUARE.

7. $a^2 + b^2 = c^2$
$12^2 + 12^2 = c^2 = 288$
$c = \sqrt{288} = 16.97$

8. AMOUNT OF RISE IN EACH UNIT RUN IS THE SAME FOR HIP AND COMMON RAFTERS.

© Cengage Learning 2014

Figure 44–2 The unit Run of the hip rafter is 16.97 inches, which is often rounded to 17 inches for layout purposes.

Unit Rise

Both hip and common rafters start at the wall plate and slope up to meet at the ridge, which is at the same height. Each rafter has the same number of unit runs under it. Therefore, the rise for each unit Run must be the same.

A framing square can be used to lay out a hip rafter. The square is held the same as for a common rafter. The unit Rise is held on the tongue and the unit Run on the blade. The difference is the unit Run is now 16.97 or 17 inches, not 12.

Estimating the Rough Length of the Hip Rafter

The rough length of the hip rafter can be found in the manner previously described for the common rafter. Let the blade of the square represent the length of common rafter. Let the tongue of the square represent its total Run. Measure across the square the distance between these two measurement points. A scale of 1 inch equals 1 foot gives the length of the hip rafter from plate to ridge (*Figure 44–3*).

EXAMPLE

The common rafter length is 15 feet. Its total Run is 12 feet. A measurement across these points scales off to 19'-2½". It will be necessary to order 20-foot lengths for the hip rafters unless extra is needed for overhang.

Laying Out the Hip Rafter

The steps to lay out a hip rafter are similar to those for a common rafter with some differences. The differences are caused by the fact that the hip runs at a 45-degree angle. There are at least eleven lines to draw in order to lay out a hip rafter (*Figure 44-4*).

Ridge Cut of the Hip

The ridge plumb line on a hip is drawn similar to that for a common rafter. The squares are held in the same manner except different numbers or scales are used (*Figure 44-5*). The hip ridge cut is also a compound angle called a **cheek cut** or *side cut*.

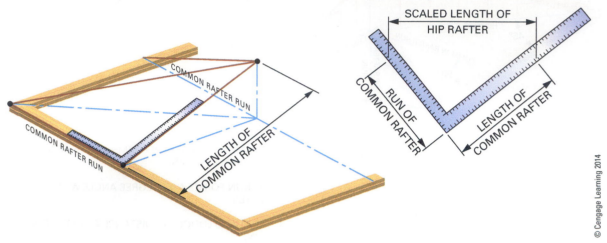

Figure 44-3 An estimated length of the hip rafter by scaling across the framing square.

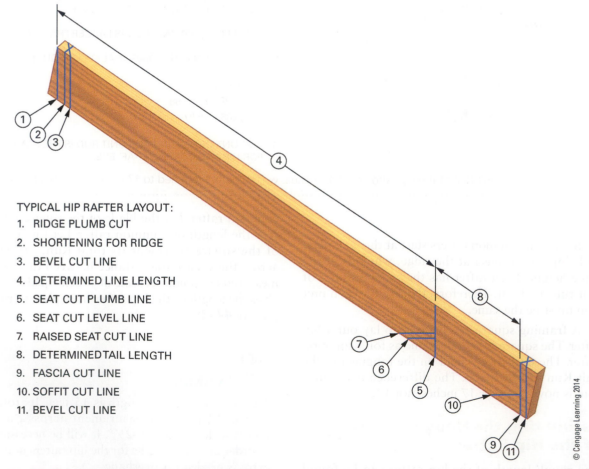

TYPICAL HIP RAFTER LAYOUT:

1. RIDGE PLUMB CUT
2. SHORTENING FOR RIDGE
3. BEVEL CUT LINE
4. DETERMINED LINE LENGTH
5. SEAT CUT PLUMB LINE
6. SEAT CUT LEVEL LINE
7. RAISED SEAT CUT LINE
8. DETERMINED TAIL LENGTH
9. FASCIA CUT LINE
10. SOFFIT CUT LINE
11. BEVEL CUT LINE

Figure 44-4 Layout for a hip rafter is done in steps.

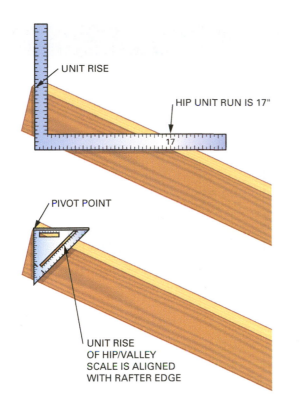

UNIT RISE

HIP UNIT RUN IS 17"

17

PIVOT POINT

UNIT RISE
OF HIP/VALLEY
SCALE IS ALIGNED
WITH RAFTER EDGE

© Cengage Learning 2014

Figure 44–5 Hip plumb lines are drawn in a fashion similar to that for common rafters.

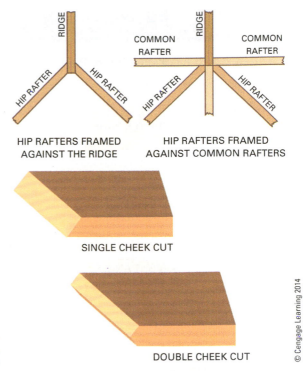

RIDGE

RIDGE

COMMON
RAFTER

COMMON
RAFTER

HIP RAFTER

HIP RAFTER

HIP RAFTER

HIP RAFTER

HIP RAFTERS FRAMED
AGAINST THE RIDGE

HIP RAFTERS FRAMED
AGAINST COMMON RAFTERS

SINGLE CHEEK CUT

DOUBLE CHEEK CUT

© Cengage Learning 2014

Figure 44–6 Single or double cheek cuts are used depending on the method of framing at the ridge.

A *single cheek* cut or a *double cheek* cut may be made on the hip rafter according to the way it is framed at the ridge (*Figure 44–6*).

Shortening the Hip

The rafter must be shortened before the cheek cuts are laid out. The hip rafter is shortened by one-half the 45-degree thickness of the ridge board. This shortening is the same whether it is a single or double cheek cut (*Figure 44–7*). This is due to the fact that the hip Run is at a 45-degree angle.

To lay out the ridge cut, select a straight length of stock for a pattern. Lay it across two sawhorses. Mark a plumb line at the left end. Hold the tongue of the square at the rise and the blade of the square

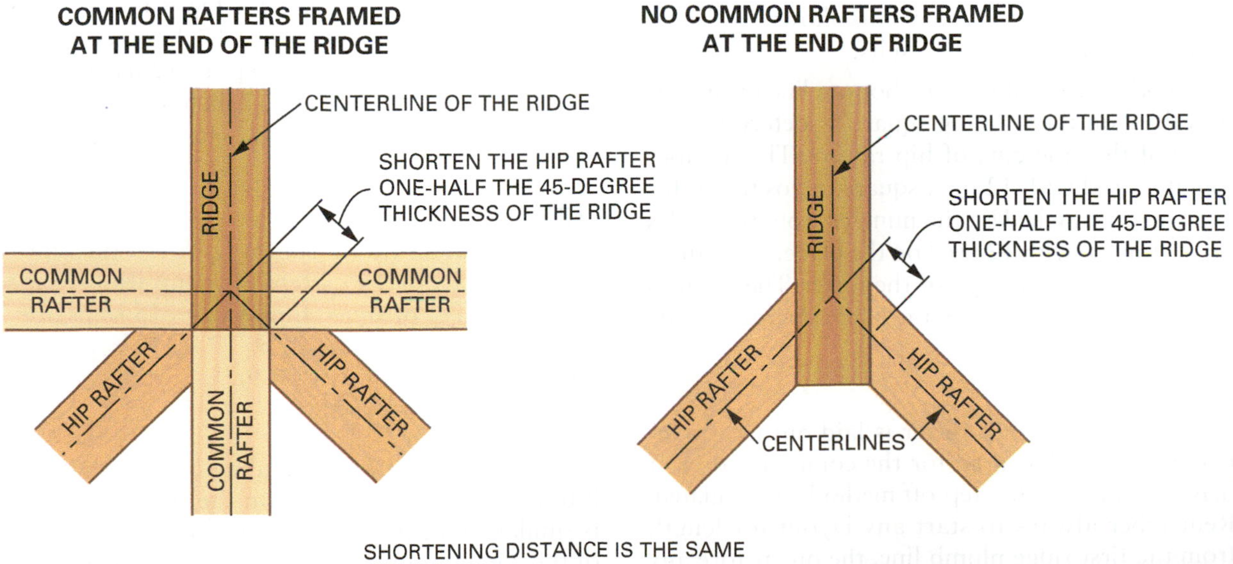

**COMMON RAFTERS FRAMED
AT THE END OF THE RIDGE**

CENTERLINE OF THE RIDGE

SHORTEN THE HIP RAFTER
ONE-HALF THE 45-DEGREE
THICKNESS OF THE RIDGE

RIDGE

COMMON
RAFTER

COMMON
RAFTER

HIP RAFTER

HIP RAFTER

COMMON
RAFTER

**NO COMMON RAFTERS FRAMED
AT THE END OF RIDGE**

CENTERLINE OF THE RIDGE

SHORTEN THE HIP RAFTER
ONE-HALF THE 45-DEGREE
THICKNESS OF THE RIDGE

RIDGE

HIP RAFTER

HIP RAFTER

CENTERLINES

© Cengage Learning 2014

SHORTENING DISTANCE IS THE SAME

Figure 44–7 Amount to shorten the hip rafter for either method of framing at the ridge.

StepbyStep Procedures

Procedure 44–A Marking a Rafter Plumb Line

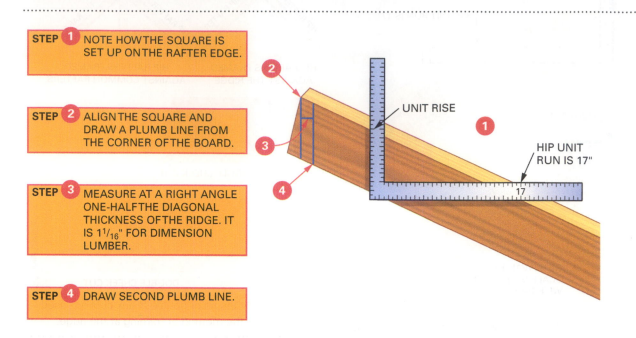

STEP 1 NOTE HOW THE SQUARE IS SET UP ON THE RAFTER EDGE.

STEP 2 ALIGN THE SQUARE AND DRAW A PLUMB LINE FROM THE CORNER OF THE BOARD.

STEP 3 MEASURE AT A RIGHT ANGLE ONE-HALF THE DIAGONAL THICKNESS OF THE RIDGE. IT IS 1¹¹⁄₁₆" FOR DIMENSION LUMBER.

STEP 4 DRAW SECOND PLUMB LINE.

UNIT RISE

HIP UNIT RUN IS 17"

© Cengage Learning 2014

at 17 inches, the unit of Run for the hip rafter. Shorten the rafter by measuring at right angles to the plumb line. This measurement is $^{11}/_{16}$ inches for dimension lumber. Lay out another plumb line at that point (*Procedure 44–A*).

Laying Out Cheek Cuts

To complete the layout of the ridge cut, mark lines for a single or double cheek cut as required. The method of laying out cheek cuts shown in *Procedure 44–B* gives accurate results.

Another method uses the bottom line of the rafter tables on the framing square to determine the angle of the side cuts of hip rafters. The number found is used with 12 on a square to position it for marking the angle. Use the number from the tables on the tongue with 12 used on the blade. Align these numbers along the edge of the rafter. The angle is laid out by marking along the blade (*Figure 44–8*).

Laying Out the Hip-Rafter Length

The length of the hip rafter is laid out in a manner similar to that used for the common rafter. It may be found by the step-off method or calculated. Remember always to start any layout for length from the first ridge plumb line, the one before any shortening.

Stepping Off the Hip. In the step-off method, the number of steps for the hip rafter is the same as for the common rafter in the same roof. The same rise is used, but the unit of Run for the hip is 17, not 12. For example, for a roof with a rise of 6 inches per foot of run, the square is held at 6 and 12 for the common rafter, and 6 and 17 for the hip rafter of the same roof.

If the total Run of the common rafter contains a fractional part of 12, then the hip Run must contain the same fractional part of 17. In other words, if the common Run is $12^1/_2$ steps of 12 inches, then the hip Run is $12^1/_2$ steps of 17 inches.

EXAMPLE

If the common Run distance is 15′-9″, then the Run is $15^9/_{12}$ or $15^3/_4$ steps. The hip Run is the same number of steps. In this case, 15 steps are made. The last step is $^3/_4$ of 17 or $12^3/_4$ inches along the blade (*Figure 44–9*).

Using Rafter Tables. Finding the length of the hip using the tables found on the framing square is similar to the process used for finding the length of the common rafter. However, the numbers from the second line are used instead of the first line.

Procedure 44–B Making Cheek Cuts Using the Measurement Method

CHEEK CUTS USING MEASUREMENT METHOD

SINGLE CHEEK LINE

STEP 1 SQUARE A LINE ACROSS THE TOP EDGE FROM THE SECOND PLUMB LINE.

STEP 2 MARK THE CENTER OF THE TOP EDGE.

STEP 3 FROM THE SECOND LINE, MEASURE AT A RIGHT ANGLE ONE-HALF THE THICKNESS OF THE HIP RAFTER. IT IS 3/4" FOR DIMENSION LUMBER.

STEP 4 DRAW THIRD PLUMB LINE.

STEP 5 DRAW A DIAGONAL LINE ACROSS THE TOP FROM THE THIRD LINE THROUGH THE CENTERLINE.

DOUBLE CHEEK LINES

STEP 1 SQUARE ACROSS FROM THE THIRD PLUMB LINE.

STEP 2 DRAW A DIAGONAL LINE FROM SQUARED LINE THROUGH THE CENTERLINE.

STEP 3 A FOURTH PLUMB LINE IS DRAWN AS A CUT LINE FOR THE DOUBLE CHEEK CUT. RAFTER MAY ALSO BE TURNED OVER TO REDRAW THE CUT PLUMB LINE.

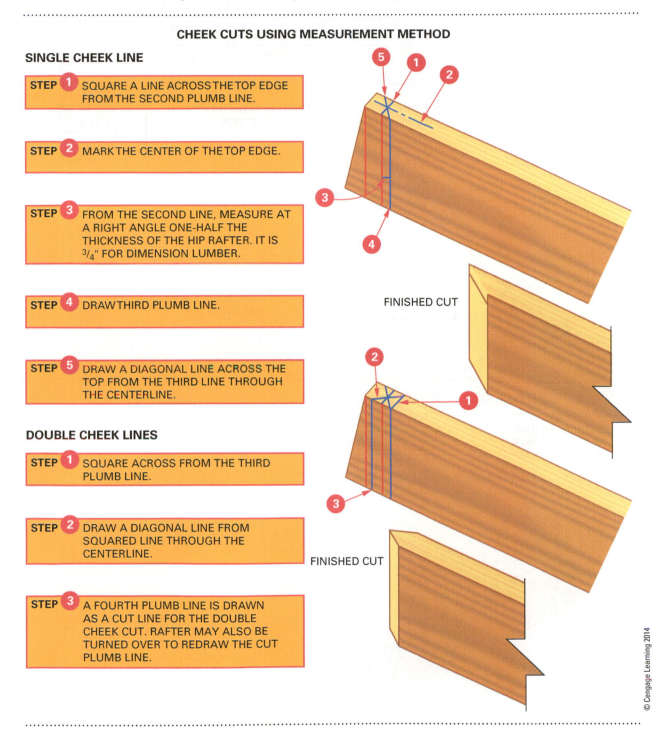

FINISHED CUT

FINISHED CUT

© Cengage Learning 2014

These numbers are the unit Length of the hip rafter in inches or length per unit Run.

As with common rafters, the numbers in the rafter tables can be calculated using the Pythagorean theorem. The unit Rise and unit Run are used as *a* and *b* to find *c*, unit Length. A common rafter uses unit Rise and 12 inches, where a hip uses unit Rise and 16.97 inches. It is important not to use the rounded-off number of 17 in these calculations.

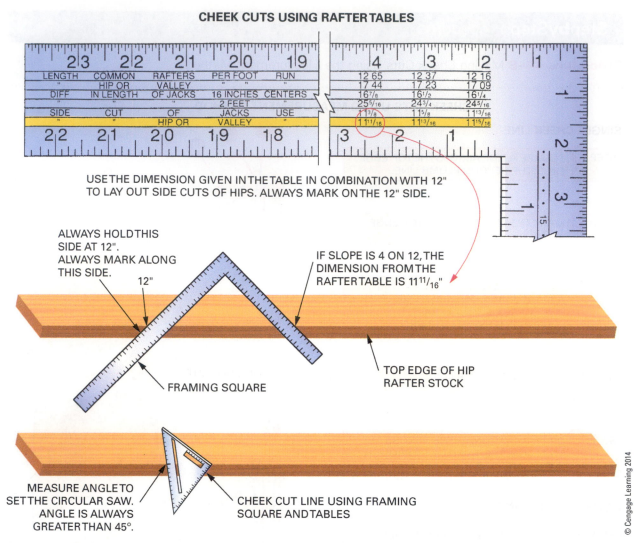

CHEEK CUTS USING RAFTER TABLES

					4	3	2
LENGTH	COMMON	RAFTERS	PER FOOT	RUN	12 65	12 37	12 16
"	HIP OR	VALLEY	"	"	17 44	17 23	17 09
DIFF	IN LENGTH	OF JACKS	16 INCHES	CENTERS	16⅞	16½	16¼
"	"	"	2 FEET	"	25⁵⁄₁₆	24¾	24⁵⁄₁₆
SIDE	CUT	OF	JACKS	USE	1³⁄₈	11⁵⁄₈	11¹³⁄₁₆
"	"	HIP OR	VALLEY	"	11¹¹⁄₁₆	11¹³⁄₁₆	11¹⁵⁄₁₆

USE THE DIMENSION GIVEN IN THE TABLE IN COMBINATION WITH 12" TO LAY OUT SIDE CUTS OF HIPS. ALWAYS MARK ON THE 12" SIDE.

ALWAYS HOLD THIS SIDE AT 12".
ALWAYS MARK ALONG THIS SIDE.

12"

IF SLOPE IS 4 ON 12, THE DIMENSION FROM THE RAFTER TABLE IS 11¹¹⁄₁₆"

FRAMING SQUARE

TOP EDGE OF HIP RAFTER STOCK

MEASURE ANGLE TO SET THE CIRCULAR SAW. ANGLE IS ALWAYS GREATER THAN 45°.

CHEEK CUT LINE USING FRAMING SQUARE AND TABLES

Figure 44–8 Cheek cuts using the rafter table method.

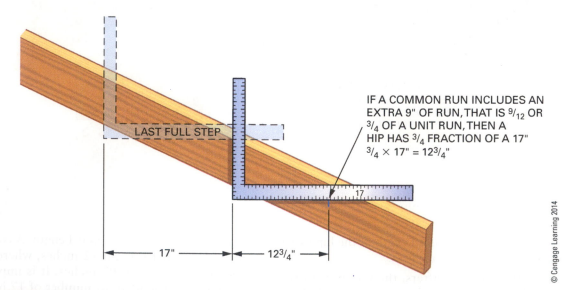

LAST FULL STEP

IF A COMMON RUN INCLUDES AN EXTRA 9" OF RUN, THAT IS ⁹⁄₁₂ OR ¾ OF A UNIT RUN, THEN A HIP HAS ¾ FRACTION OF A 17"
¾ × 17" = 12¾"

17

17"

12¾"

Figure 44–9 Stepping off a fractional unit of run on the hip rafter.

© Cengage Learning 2014

EXAMPLE

What is the unit Length for a hip rafter with a unit rise of 7 inches?

$a^2 + b^2 = c^2$

$7^2 + 16.97^2 = c^2$

$49 + \underline{287.9809} = 336.9809 = c^2$

$c = \sqrt{336.9809} = 18.35704$ inches

Unit Length = 18.36 inches rounded to the nearest hundredth. This number is also found in the rafter table: on the second line under the 7 is 18 36, which is 18.36.

As with the common rafter, the hip length is then found by multiplying the unit Length by the total Run of a common rafter. This is because the number of units of Run under a common rafter is the same as for a hip.

EXAMPLE

Find the line length of a hip rafter with a unit Rise of 8 inches for a 28-foot-wide building.

STEP 1: Read below the 8-inch mark in the first row to find 18 76 or 18.76 inches unit Length.

STEP 2: Divide 28 by 2 to determine the run. 28 ÷ 2 = 14 Run.

STEP 3: Multiply unit Length times total run. 18.76 × 14 = 262.64 inches. That is 262 inches plus a fraction.

STEP 4: Convert decimal to sixteenths. 0.64 × 16 = 10.24 sixteenths.

STEP 5: Round off to nearest whole sixteenths. 10.24 ⇒ $^{10}/_{16}$, which reduces to $^5/_8$ inches.

STEP 6: Collect information into rafter length. 262 + $^5/_8$″ = 262$^5/_8$″

If the total Run of the rafter contains a fractional part of a foot, multiply the figure in the tables by the whole number of feet plus the fractional part changed to a decimal. For instance, if the total Run of the rafter is 15′-6″, multiply the figure in the tables by 15$^1/_2$ changed to 15.5.

Lay out the length obtained along the top edge of the hip rafter stock. This must be done from the first plumb line before shortening and cheek cuts are made (*Figure 44–10*).

Laying Out the Seat Cut of the Hip Rafter

Like the common rafter, the seat cut of the hip rafter is a combination of plumb and level cuts. The height above the bird's mouth on a hip rafter is the same distance as the common rafter. To locate where the seat level line should be, first measure down along the plumb line from the top of the common rafter. Then mark that same measurement on the hip. No consideration needs to be given to fitting it around the corner of the wall.

Backing and Dropping the Hip. When making the level cut of the seat, special consideration must be given. The hip rafter is at the intersection of the two roof slopes. The center of the top edge is in perfect alignment with the other rafters, but the top outside corners of the top edge surface stick above the sheathing. To remedy this situation, the seat level line is cut deeper, a process called dropping the hip rafter, or the rafter top edge is bevel planed, a process called backing the hip (*Figure 44–11*).

Dropping the hip is much easier to do, so it is done more frequently than backing the hip. The amount of dropping must be determined for each roof, because this amount changes with the slope of the roof. As steepness increases, the amount of drop increases. The process for determining the amount of drop is shown in *Procedure 44–C*.

To back the rafter, the measurement of the drop is used to determine the bevel. This amount is the distance the top outside corners are beveled.

LAYING OUT THE TAIL OF THE HIP RAFTER

The hip tail can be found in one of two ways: Calculate the overhang or calculate the projection. Hip tail projection is determined with the fractional step-off method as shown earlier in *Figure 44–9*. For example, if the common projection is 15 inches, then the Run is 15 ÷ 12 = 1.25 or 1$^1/_4$ units of Run. The hip tail has the same run. To step this off, mark one full step plus a quarter step or $^1/_4$ × 16.97 = 4.25 inches, which is 4$^1/_4$ inches. The total hip projection here is 16.97 or 17 + 4$^1/_4$ = 21$^1/_4$ inches.

The overhang calculation uses the same formula as rafter length: unit Length times Run (tail) = line length (tail). For example, if the unit rise is 6, then the unit length of a hip is 18 inches. If the common projection has a Run of 1.25, then the overhang is 18 × 1.25 = 22$^1/_2$ inches (*Procedure 44–D*).

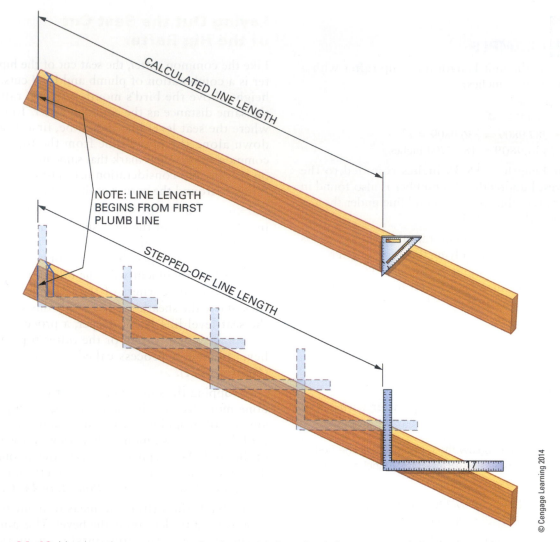

Figure 44–10 Line length measurement begins at the first plumb line.

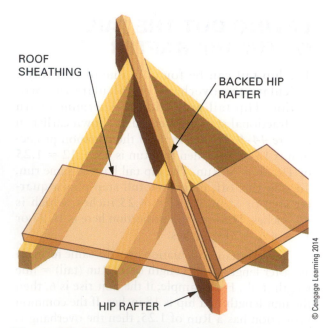

Figure 44–11 The hip rafter must be adjusted to allow for the roof sheathing.

HIP JACK RAFTERS

The hip jack rafter is a shortened common rafter. It is parallel to the common rafter, but is simply cut shorter to meet the hip. The seat cut and the tail are exactly like those of a common rafter, only the length varies as shown earlier in *Figure 44–1*.

Hip Jack Length

Each jack is shorter or longer than the next set by the same amount. This is called the *common difference* (*Figure 44–12*). The common difference is found in the rafter tables on the framing square for jacks 16 and 24 inches OC.

Once the length of one jack is determined, the length of all others can be found by making each set shorter or longer by the common difference. To find the length of any jack, its total Run must be known.

StepbyStep Procedures

Procedure 44–C Dropping or Backing a Hip Rafter

STEP 1 MEASURE HEIGHT ABOVE BIRD'S MOUTH ON COMMON RAFTER. MEASURE AND MARK THE SAME DISTANCE DOWN ON THE PLUMB LINE.

STEP 2 DRAW SEAT LEVEL LINE THROUGH THIS MEASUREMENT.

STEP 3 FROM LINE 1 MEASURE PERPENDICULAR TO PLUMB LINE TOWARD RIDGE CUT, A DISTANCE EQUAL TO ONE-HALF THE HIP RAFTER THICKNESS. THIS IS USUALLY 3/4" FOR DIMENSION LUMBER.

STEP 4 DRAW ANOTHER PLUMB LINE OF THE SAME LENGTH AS THE HEIGHT ABOVE THE BIRD'S MOUTH FOR COMMON RAFTER.

STEP 5 DRAW THE DROPPED SEAT CUT LINE.

STEP 6 NOTE THE BIRD'S MOUTH IS CUT ALONG THE FIRST PLUMB LINE AND SECOND LEVEL LINE.

STEP 7 DROP DISTANCE IS ALSO AMOUNT OF BEVEL FOR BACKING HIP RAFTER.

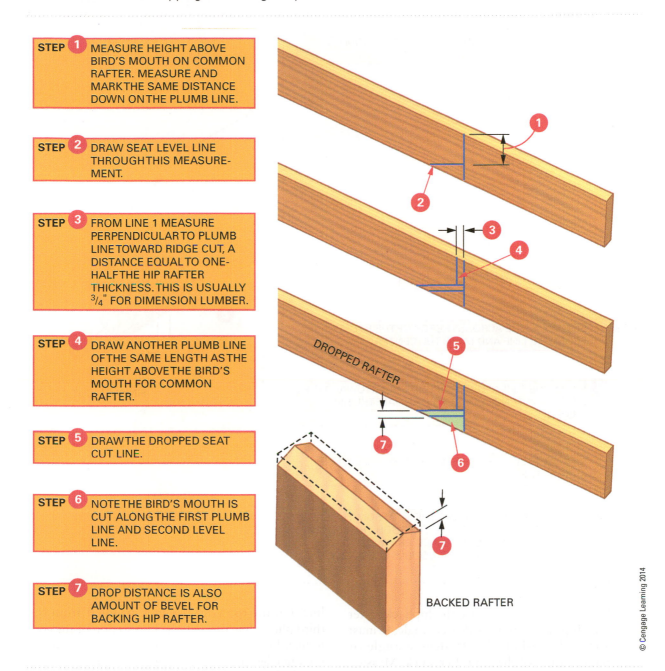

DROPPED RAFTER

BACKED RAFTER

© Cengage Learning 2014

Finding the Total Run. The total Run of any jack rafter and its distance from the corner along the outside edge of the wall plate form a square. Since all sides of a square are equal, the total Run of any hip jack rafter is equal to its distance from the corner of the building. To determine the jack Run, measure the distance from the corner to the center of where the jack will sit on the wall plate (*Figure 44–13*).

Find the length of the hip jack rafter by the step-off method, or use the rafter tables in the same way as for common rafters. The only difference is that the total Run of the jack found above is used. Use the common rafter pattern to lay out the jack tail.

StepbyStep Procedures

Procedure 44–D Laying Out a Hip Rafter Tail

STEP 1 UNIT RISE = 6, THEN UNIT LENGTH IS 18 FROM THE RAFTER TABLES.

STEP 2 PROJECTION IS 15".
15 ÷ 12 = 1.25 UNITS OF RUN.

STEP 3 OVERHANG LENGTH:
18 X 1.25 = 22½".

STEP 4 MEASURE THE OVERHANG LENGTH.

STEP 5 DRAW FASCIA PLUMB LINE.

STEP 6 SQUARE A LINE ACROSS THE TOP OF THE RAFTER FROM THIS LINE AND MARK THE CENTER.

STEP 7 FROM LINE 5 MEASURE PERPENDICULAR TOWARD THE RIDGE A DISTANCE EQUAL TO ONE-HALF THE HIP THICKNESS.

STEP 8 DRAW A SECOND PLUMB LINE AND SQUARE A LINE ACROSS THE TOP.

STEP 9 DRAW DIAGONALS ACROSS THE TOP FROM THE ENDS OF THE SECOND PLUMB LINE THROUGH THE CENTER MARK. TURN RAFTER OVER AND DRAW CUT PLUMB LINE ON THE BACK FOR THE DOUBLE CHEEK CUT.

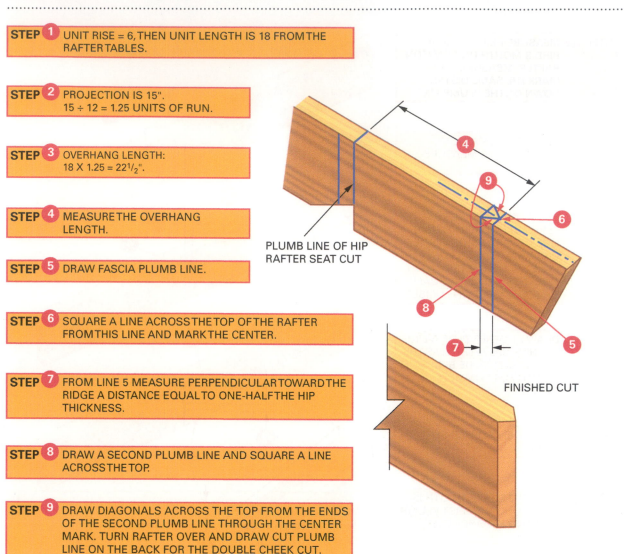

PLUMB LINE OF HIP RAFTER SEAT CUT

FINISHED CUT

© Cengage Learning 2014

Hip Jack Shortening. Since the hip jack rafter meets the hip rafter at a 45-degree angle, it must be shortened one-half the 45-degree angle or diagonal thickness of the hip rafter stock. Measure the distance at right angles to the original plumb line toward the tail. Draw another plumb line (**Procedure 44–E**).

Hip Jack Cheek Cuts. The hip jack rafter has a single cheek cut where it lands on the hip rafter. Square the shortened plumb line across the top of the rafter. Draw an intersecting centerline along the top edge. Measure back toward the tail at right angles from the second plumb line a distance equal to one-half the jack rafter stock. Draw another plumb line. On the top edge, draw a diagonal from the third plumb line through the intersection of the centerline. Take care when drawing the diagonal line on the top edge. The direction of the diagonal depends on which side of the hip the jack rafter is framed.

Determine the common difference of hip jack rafters from the rafter tables under the inch mark that coincides with the slope of the roof.

The common difference may also be calculated. For rafters spaced 24 inches OC, the common difference is equal to the common rafter length for two units of Run. For rafters spaced 16 inches OC, the common difference is equal to the common rafter length for 1⅓ units of Run (**Procedure 44–F**).

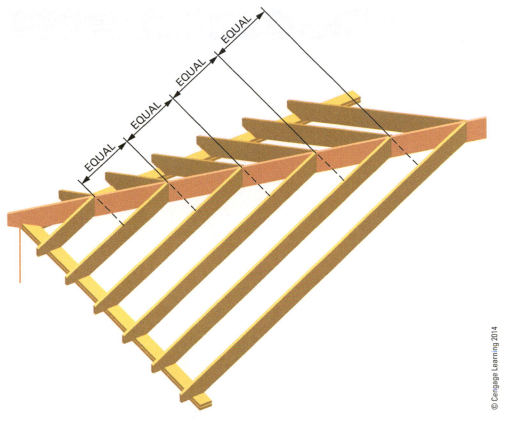

© Cengage Learning 2014

Figure 44–12 Hip jack rafters, like gable end studs, have a common difference in length.

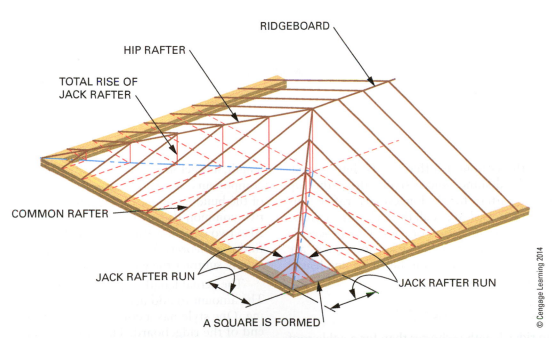

© Cengage Learning 2014

Figure 44–13 The total Run of any hip jack is equal to its distance from the corner of the building.

StepbyStep Procedures

Procedure 44–E Laying Out a Hip Jack Cheek Cut

STEP 1 LAY OUT PLUMB LINE.

STEP 2 MEASURE AT RIGHT ANGLE TO PLUMB LINE ONE-HALF 45-DEGREE THICKNESS OF HIP RAFTER AND DRAW SHORTENED PLUMB LINE.

STEP 3 SQUARE SHORTENED PLUMB LINE ACROSS TOP EDGE OF RAFTER STOCK.

STEP 4 MEASURE AT RIGHT ANGLE FROM SHORTENED PLUMB LINE ONE-HALF THE THICKNESS OF THE JACK RAFTER STOCK AND DRAW ANOTHER PLUMB LINE.

STEP 5 DRAW CENTERLINE ALONG TOP EDGE OF RAFTER.

STEP 6 DRAW DIAGONAL FROM LAST PLUMB LINE THROUGH CENTERLINE.

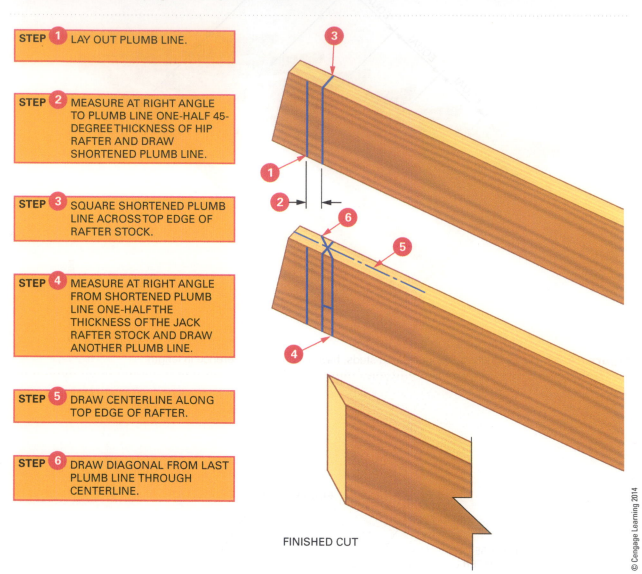

FINISHED CUT

© Cengage Learning 2014

Once the common difference is determined, measure the distance successively along the top edge of the pattern for the longest jack rafter. This pattern is then used to cut all the jack rafters necessary to frame that section of roof. This process is repeated for each section on jacks. Some jacks are framed in pairs depending on the width and length dimensions of the house.

Hip Roof Ridge Length

The hip ridge length is shorter than for a gable roof on the same sized building. To find the ridgeboard length, it is helpful to visualize the Run of the major components of a roof (*Figure 44–14*). The length of the building under the ridgeboard is made up of two common Runs and the hip ridgeboard length. Also two common Runs are equal to the width. To find the theoretical line length of the ridgeboard, simply subtract the width from the length.

The actual length is longer than the line length. The amount to add depends on two styles of framing. One style has a common rafter framed at the end of the ridgeboard. The other has no common rafter at the end (*Figure 44–15*).

StepbyStep Procedures

Procedure 44–F Calculating Jack Rafter Common Difference

THE JACK RAFTER SPACING AND RUN OF ITS COMMON DIFFERENCE IN LENGTH FORM THE SIDES OF A SQUARE AND ARE THEREFORE EQUAL TO EACH OTHER.

EXAMPLE: IF THE UNIT RISE IS 6" AND THE RAFTER SPACING IS 16", CALCULATE THE COMMON DIFFERENCE IN LENGTH FOR A JACK.

STEP 1 UNIT RISE = 6" THEN THE UNIT LENGTH IS 13.42" FOR A COMMON RAFTER.

STEP 2 CONVERT 16" OC SPACING TO UNITS OF RUN: $16 \div 12 = 1\frac{1}{3}$ UNITS.

STEP 3 MULTIPLY UNIT LENGTH X RUN: $13.42 \times 1\frac{1}{3} = 17.89"$.

STEP 4 COMMON DIFFERENCE IS $17\frac{7}{8}"$.

COMMON DIFFERENCE IN LENGTH OF THE JACK RAFTER

A SQUARE IS FORMED

HIP RAFTER

WALL PLATE

RUN OF THE JACK RAFTER COMMON DIFFERENCE

JACK RAFTERS

JACK RAFTER SPACING

RUN OF JACK RAFTER

© Cengage Learning 2014

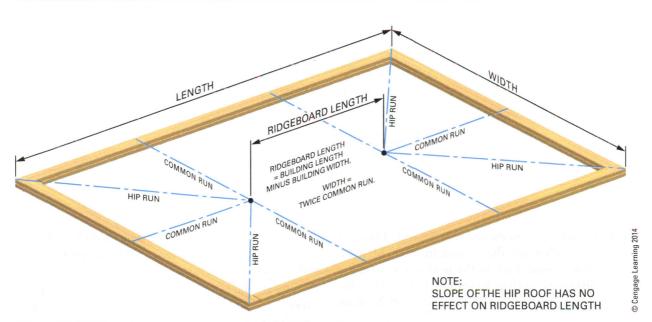

Figure 44–14 Determining the line length of the hip roof ridgeboard.

RIDGEBOARD LENGTH = BUILDING LENGTH MINUS BUILDING WIDTH.

WIDTH = TWICE COMMON RUN.

NOTE: SLOPE OF THE HIP ROOF HAS NO EFFECT ON RIDGEBOARD LENGTH

© Cengage Learning 2014

The amount to add in the first case is one-half the thickness of the common rafter at each end. This is $\frac{3}{4}$ inch for dimension lumber. The amount to add when no common rafter is framed at the end is one-half the thickness of the common rafter plus one-half the diagonal thickness of the hip. For dimension lumber, this is $\frac{3}{4} + 1\frac{1}{16} = 1\frac{13}{16}$ inches added to each end.

Raising the Hip Roof

Before raising the hip roof, the ridgeboard must first be laid out for the location of the common rafters (*Procedure 44–G*). Measure from the outside corner the common Run distance. Mark this length on the plate. Deduct the distance that the

COMMON RAFTERS FRAMED AT THE END OF THE RIDGE

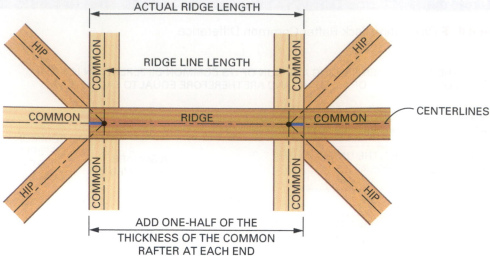

ACTUAL RIDGE LENGTH

RIDGE LINE LENGTH

CENTERLINES

COMMON · RIDGE · COMMON

HIP

COMMON

ADD ONE-HALF OF THE
THICKNESS OF THE COMMON
RAFTER AT EACH END

NO COMMON RAFTERS FRAMED AT THE END OF THE RIDGE

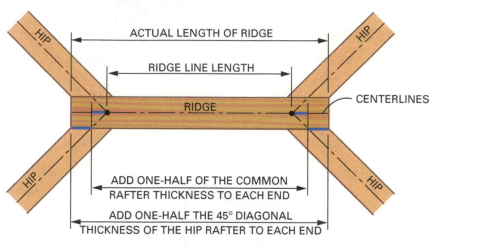

ACTUAL LENGTH OF RIDGE

RIDGE LINE LENGTH

RIDGE

CENTERLINES

HIP

ADD ONE-HALF OF THE COMMON
RAFTER THICKNESS TO EACH END

ADD ONE-HALF THE 45° DIAGONAL
THICKNESS OF THE HIP RAFTER TO EACH END

© Cengage Learning 2014

Figure 44–15 Determining the actual length of the hip roof ridge.

ridgeboard was lengthened on one end. This is $\frac{3}{4}$ or $1\frac{13}{16}$ inches for dimension lumber. Measure from this second mark to the next common rafter. This distance can then be transferred to the ridgeboard, measured from the end. Mark an X on the appropriate side.

Erect the common roof section in the same manner as described previously for a gable roof. If the hip rafters are framed against the ridge, install

them next. If they are framed against the common rafters, install the common rafters against the end of the ridge. Then install the hip rafters.

Fasten jack rafters to the plate and to hip rafters in pairs. As each pair of jacks is fastened, sight the hip rafter along its length. Keep it straight. Any *bow* in the hip is straightened by driving the jack a little tighter against the bowed-out side of the hip as the roof is framed (*Procedure 44–H*).

StepbyStep Procedures

Procedure 44–G Laying Out the Hip Ridgeboard

STEP 1 MEASURE ALONG WALL FROM THE CORNER THE DISTANCE EQUAL TO THE COMMON RUN. MARK ON THE WALL PLATE.

STEP 2 MEASURE ON PLATE, BACK TOWARDS THE CORNER, THE DISTANCE BEING ADDED TO THE RIDGEBOARD AS DETERMINED IN FIGURE 43—15.

STEP 3 FROM THE MARK IN STEP 2, MEASURE AWAY FROM THE CORNER TO THE NEXT COMMON RAFTER LAYOUT LINE.

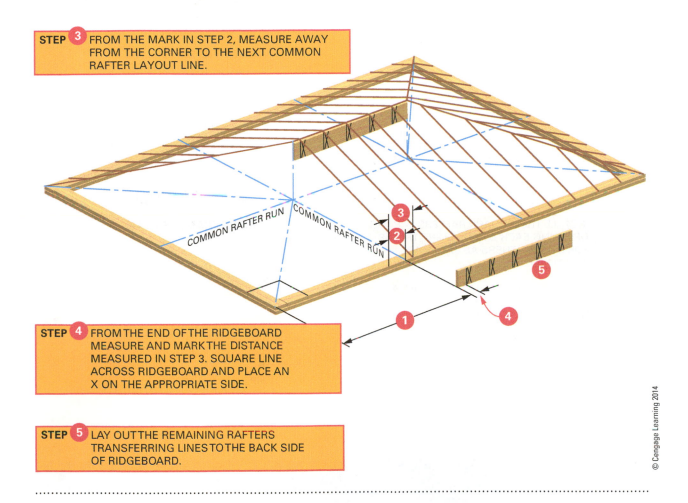

COMMON RAFTER RUN COMMON RAFTER RUN

STEP 4 FROM THE END OF THE RIDGEBOARD MEASURE AND MARK THE DISTANCE MEASURED IN STEP 3. SQUARE LINE ACROSS RIDGEBOARD AND PLACE AN X ON THE APPROPRIATE SIDE.

STEP 5 LAY OUT THE REMAINING RAFTERS TRANSFERRING LINES TO THE BACK SIDE OF RIDGEBOARD.

Procedure 44–H Erecting the Hip Roof Frame

STEP **1** ERECT THE RIDGE WITH ONLY AS MANY COMMON RAFTERS AS NEEDED.

STEP **2** INSTALL THE HIP RAFTERS. THE HIP RAFTERS EFFECTIVELY BRACE THE ASSEMBLY.

NOTE: THIS PROCEDURE SHOWS THE HIPS FRAMED TO THE RIDGE. IF HIPS ARE FRAMED TO THE COMMON RAFTERS, INSTALL THE COMMONS TO THE END OF THE RIDGE BEFORE THE HIPS.

STEP **3** INSTALL THE REMAINING COMMON RAFTERS IN PAIRS OPPOSING EACH OTHER. SIGHT THE TOP EDGE OF THE RIDGE FOR STRAIGHTNESS AS FRAMING PROGRESSES.

STEP **4** INSTALL THE HIP JACK RAFTERS IN PAIRS OPPOSING EACH OTHER. SIGHT THE HIP FOR STRAIGHTNESS AS JACKS ARE INSTALLED.

NOTE: IT IS BEST TO FIRST INSTALL A PAIR OF JACKS ABOUT HALFWAY UP THE HIP TO STRAIGHTEN IT.

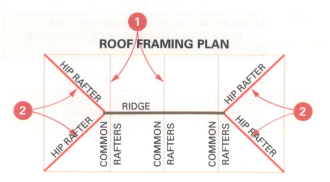

ROOF FRAMING PLAN

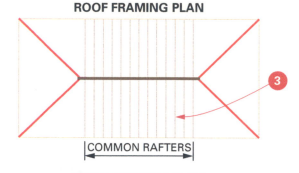

ROOF FRAMING PLAN

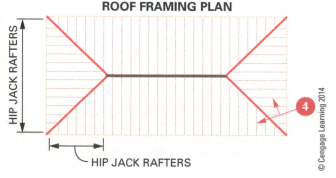

ROOF FRAMING PLAN

© Cengage Learning 2014

CHAPTER 45

Intersecting Roofs

Buildings of irregular shape, such as L-, H-, or T-shaped buildings, require a roof for each section. These roofs meet at an inside corner intersection, where valleys are formed. They are called *intersecting roofs*. The roof may be a gable, a hip, or a combination of types.

THE INTERSECTING ROOF

The intersecting roof requires the layout and installation of several kinds of rafters not previously described (*Figure 45–1*). Some buildings have sections

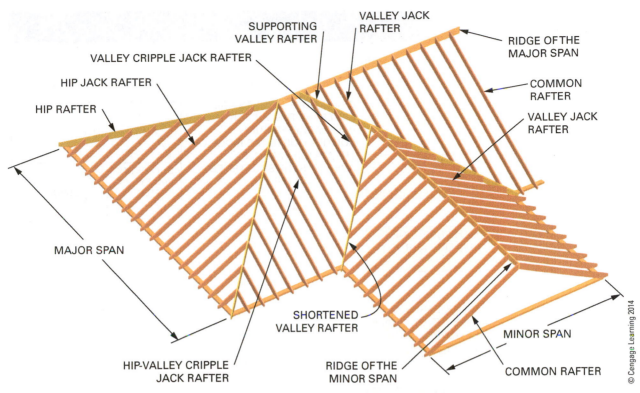

SUPPORTING
VALLEY RAFTER

VALLEY JACK
RAFTER

VALLEY CRIPPLE JACK RAFTER

RIDGE OF THE
MAJOR SPAN

HIP JACK RAFTER

COMMON
RAFTER

HIP RAFTER

VALLEY JACK
RAFTER

MAJOR SPAN

SHORTENED
VALLEY RAFTER

MINOR SPAN

HIP-VALLEY CRIPPLE
JACK RAFTER

RIDGE OF THE
MINOR SPAN

COMMON RAFTER

© Cengage Learning 2014

Figure 45–1 Members of the intersecting roof frame.

of different widths referred to as major spans and minor spans. A major span is the width of the main part of the building, whereas a minor span refers to the width of any extension to the building.

Valley rafters form the inside corner intersection of two roofs. If the heights of both roofs are different, two kinds of valley rafters are required.

Supporting valley rafters run from the plate to the ridge of the main roof. Their Run is one-half of the major span.

Shortened valley rafters run from the plate to the supporting valley rafter. Their Run is one-half of the minor span.

Valley jack rafters run from the ridge to the valley rafter.

Valley cripple jack rafters run between the supporting and shortened valley rafter.

Hip-valley cripple jack rafters run between a hip rafter and a valley rafter.

Confusion concerning the layout of so many different kinds of rafters can be eliminated by remembering the following.

- The length of any kind of rafter can be found using its Run.
- The amount of shortening is always measured at right angles to the plumb cut.

- Hip and valley rafters are similar. Common, jack, and cripple jack rafters are similar.
- The method previously described for laying out cheek cuts works on all rafters for any slope.

Supporting Valley Rafter Layout

The layout of a supporting valley rafter is more similar than not to that of a hip rafter. The unit of Run for both is 17 inches. The total Run for both is the same, one-half the major span. The valley rafter is also shortened by half the 45-degree or diagonal thickness of the ridge board (*Procedure 45–A*). Cheek cuts are made at the ridge depending on how the building is framed.

The line length of either valley is found by the step-off method or by the calculation method, using the formula unit Length × run.

Laying Out a Valley Rafter Seat Cut

Valley rafters are not dropped or backed like hip rafters. Instead, the seat cut line is made with side cuts or lengthened. This is done so the valley will clear the inside corner of the wall plate (*Procedure 45–B*).

To lay out a valley bird's mouth, first measure the line length and draw the seat plumb

StepbyStep Procedures

Procedure 45–A Laying Out a Valley Rafter

STEP 1 LAY OUT A PLUMB LINE.

STEP 2 MEASURE PERPENDICULARLY TO PLUMB LINE, ONE–HALF THE 45° THICKNESS OF THE RIDGE.

STEP 3 LAY OUT A SECOND PLUMB LINE. SQUARE THE PLUMB LINE ACROSS THE TOP EDGE.

STEP 4 MEASURE PERPENDICULARLY TO SECOND PLUMB LINE THE ONE-HALF THICKNESS OF THE SUPPORTING VALLEY RAFTER STOCK.

STEP 5 DRAW THIRD PLUMB LINE.

STEP 6 DRAW A SQUARE LINE AND CENTERLINE ACROSS THE TOP FROM SECOND PLUMB LINE.

STEP 7 DRAW A DIAGONAL FROM THIRD PLUMB LINE THROUGH THE CENTERLINE.

STEP 8 DRAW A SQUARE LINE ACROSS THE TOP FROM THIRD PLUMB LINE.

STEP 9 DRAW SECOND DIAGONAL.

STEP 10 DRAW A FOURTH PLUMB LINE ON BACK SIDE OF RAFTER.

WHEN RIDGE HEIGHTS ARE DIFFERENT, THE SUPPORTING VALLEY RAFTER RUNS CONTINUOUS TO THE RIDGE OF THE MAJOR SPAN AND SUPPORTS THE SHORTENED VALLEY RAFTER.

RIDGE HEIGHTS ARE THE SAME WHEN BOTH SPANS WITH THE SAME ROOF PITCH ARE EQUAL. BOTH VALLEY RAFTERS MEET AT THE RIDGE.

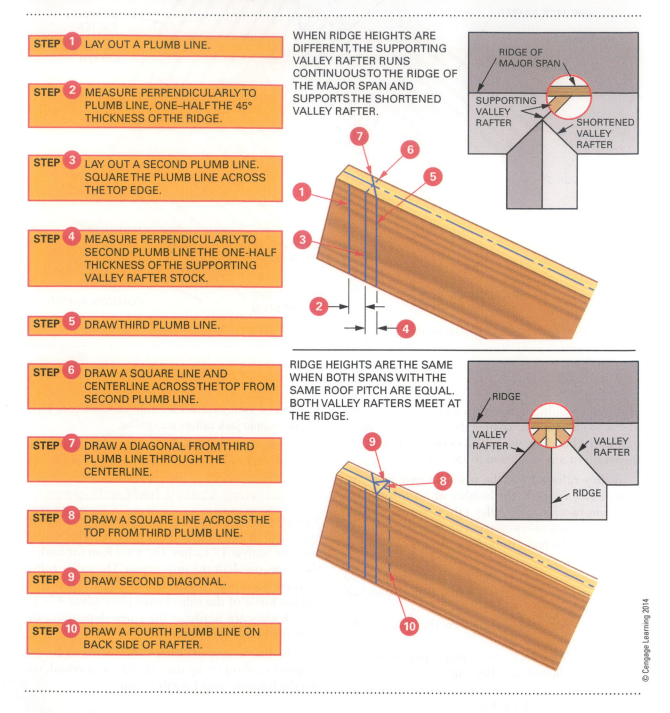

© Cengage Learning 2014

line. Then measure and mark down along this line, from the top edge of the rafter, a distance equal to the height above the bird's mouth on the common rafter. This makes the height above the bird's mouth the same for all rafters. Draw a level line at this point to make a bird's mouth. From the plumb line, measure at a right angle toward the tail a distance equal to one-half the valley rafter thickness. Draw a new plumb line and extend to it to the previously drawn seat cut line. On the bottom edge of the rafter, draw two squared lines from the two plumb lines. Find the center of the first line and draw the appropriate diagonals.

StepbyStep Procedures

Procedure 45–B Laying Out a Valley Seat Cut

STEP 1 MEASURE LINE LENGTH OF RAFTER AND DRAW PLUMB LINE.

STEP 2 MEASURE DOWN THE SAME DISTANCE AS HEIGHT ABOVE BIRD'S MOUTH OF COMMON RAFTER.

STEP 3 DRAW LEVEL LINE FOR THE SEAT CUT.

STEP 4 MEASURE PERPENDICULAR TO PLUMB LINE TOWARD THE TAIL, A DISTANCE THAT IS ONE-HALF THE VALLEY THICKNESS. THIS IS 3/4" FOR DIMENSION LUMBER.

STEP 5 DRAW SECOND PLUMB LINE.

STEP 6 DRAW TWO SQUARED LINES ACROSS BOTTOM FROM THE TWO PLUMB LINES.

STEP 7 FIND THE CENTER OF THE FIRST SQUARED LINE AND DRAW DIAGONALS FOR CHEEK CUTS.

NOTE: THE BIRD'S MOUTH MAY BE CUT SQUARE FROM THE SECOND PLUMB LINE.

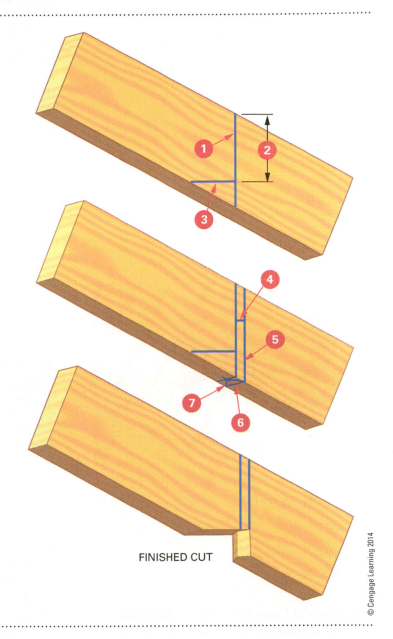

FINISHED CUT

© Cengage Learning 2014

There is no reason to drop or back a valley rafter as is done for hip rafters. The top corners of a valley are actually below the roof sheathing. The centerline of the top edge of the valley is aligned with the roof sheathing.

However, one edge of a supporting valley must be backed. This portion is located between the ridgeboards of the major and minor spans (*Figure 45–2*). Only the lower edge facing the wall plate is backed.

Laying Out a Valley Rafter Tail

The length of the valley rafter tail is found in the same manner as that used for the hip rafter. It may be stepped off as shown in *Figure 44–9* or calculated as shown in *Procedure 43–D*. Step-off the same number of times, from the plumb line of the seat cut, as stepped off for the common rafter. However, use a unit Run of 17 instead of 12.

The tail cut of the valley rafter is a double cheek cut. It is similar to that of the hip rafter, but

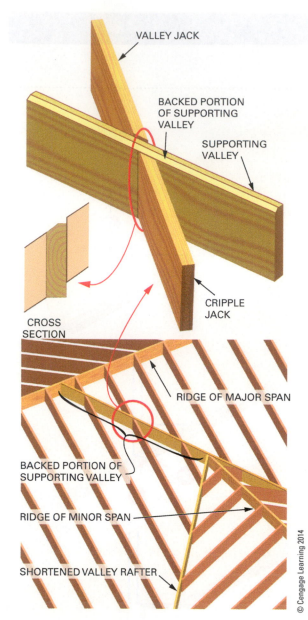

VALLEY JACK

BACKED PORTION
OF SUPPORTING
VALLEY

SUPPORTING
VALLEY

CRIPPLE
JACK

CROSS
SECTION

RIDGE OF MAJOR SPAN

BACKED PORTION OF
SUPPORTING VALLEY

RIDGE OF MINOR SPAN

SHORTENED VALLEY RAFTER

© Cengage Learning 2014

Figure 45–2 Backing the upper section of supporting valley.

angles outward instead. From the tail plumb line, measure at right angles to the plumb line toward the tail, one-half the thickness of the rafter stock. Lay out another plumb line. Square both plumb lines across the top of the rafter. Draw diagonals from the center (*Procedure 45–C*).

Shortening a Valley

The shortened valley differs from the supporting valley only in length and ridge cut. The length of

the shortened valley is found by using the Run of the common rafter of the minor span.

The plumb cut at the top end is different from that of the supporting valley. Since the two valley rafters meet at right angles, the cheek cut of the shortened valley is a squared miter, not a compound miter. It looks similar to the ridge plumb line of a common rafter. To shorten this rafter, measure back, at right angles from the plumb line at the top end, a distance that is half the thickness of the supporting valley rafter stock. Lay out another plumb line (*Procedure 45–D*).

Valley Jack Layout

The valley jack is a common rafter that is shortened at its bottom end where it meets the valley. The length of the valley jack is found, like any rafter, by multiplying unit Length × run. The total Run of any jack is the horizontal distance measured along the ridge to that rafter from the upper end of its valley rafter (*Figure 45–3*). Remember to use unit Run and unit length of a common rafter for the valley jack.

The Run of the largest valley jack may also be found by subtracting an amount from the Run of the common rafter (*Figure 45–4*). Begin by measuring from the center of the last OC rafter to the inside corner of the building. This is noted as measurement A. Subtract A from the OC spacing to find measurement B. Convert B to feet and then subtract it from the Run of the common rafter. Jack rafter length is then found by multiplying the valley jack Run times the common unit Length.

A valley jack has a common difference in length, similar to hip jacks. The layout for other valley jacks is made by adding or deducting the common difference from the jack length. This number is found on the fifth line of the rafter tables or may be calculated as shown earlier in *Procedure 45–F*.

The ridge cut of the valley jack is the same as a common rafter. It is also shortened in the same way. The cheek cut against the valley rafter is a single cheek cut that is shortened, like other rafters meeting at a diagonal, by deducting one-half the 45-degree thickness of the valley rafter.

Valley Cripple Jack Layout

As stated before, the length of any rafter can be found if its total Run is known. The Run of the valley cripple jack is always twice its horizontal

StepbyStep Procedures

Procedure 45–C Laying Out a Valley Rafter Tail

STEP 1 MEASURE TAIL LENGTH OR STEP-OFF PROJECTION.
NOTE: THIS IS DONE FROM THE FIRST BIRD'S MOUTH PLUMB LINE.

STEP 2 DRAW PLUMB LINE.

STEP 3 MEASURE PERPENDICULAR TO PLUMB LINE ONE-HALF THICKNESS OF VALLEY. THIS IS 3/4" FOR DIMENSION LUMBER.

STEP 4 DRAW TWO SQUARED LINES ACROSS TOP OF RAFTER AND MARK CENTERLINE.

STEP 5 DRAW DIAGONALS.

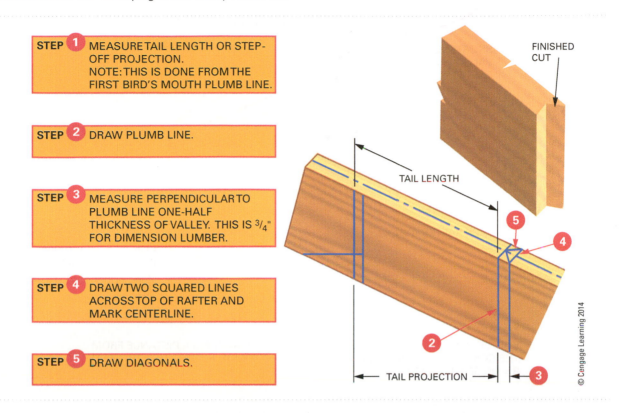

FINISHED CUT

TAIL LENGTH

TAIL PROJECTION

© Cengage Learning 2014

StepbyStep Procedures

Procedure 45–D Laying Out Shortened Valley Rafter Ridge Plumb Cut

STEP 1 DRAW PLUMB LINE USING SQUARE SETUP FOR A HIP OR VALLEY.

STEP 2 MEASURE PERPENDICULAR TO PLUMB LINE ONE-HALF THICKNESS OF SUPPORTING VALLEY. THIS IS 3/4" FOR DIMENSION LUMBER.

STEP 3 DRAW SECOND PLUMB LINE.

NOTE: CUT IS SQUARED ACROSS TOP OF RAFTER, SIMILAR TO COMMON RAFTERS.

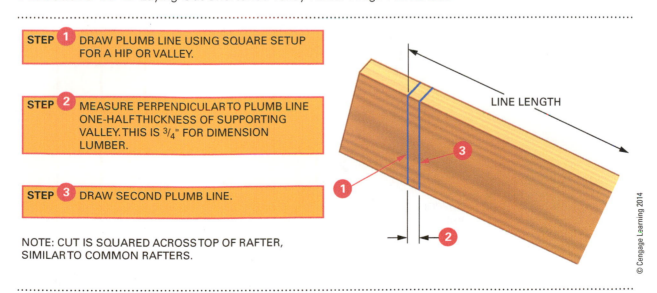

LINE LENGTH

© Cengage Learning 2014

distance from the intersection of the valley rafters (*Figure 45–5*). Use the common rafter tables or step-off in a manner similar to that used for common rafters to find its length. Shorten each end by one-half the 45-degree thickness of the valley rafter stock (*Procedure 45–E*). Noting that they angle in opposite directions, make a single cheek cut.

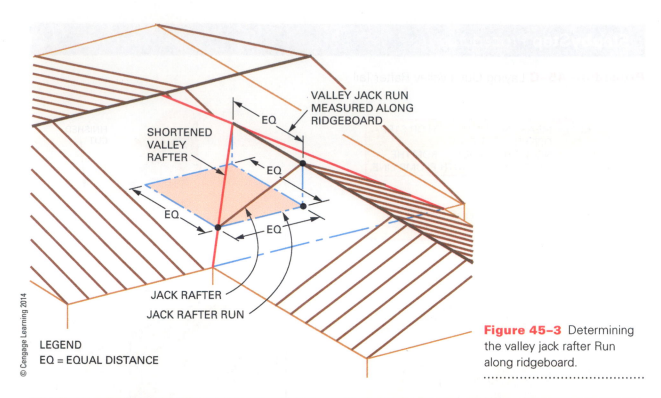

SHORTENED VALLEY RAFTER

VALLEY JACK RUN MEASURED ALONG RIDGEBOARD

EQ

EQ

EQ

EQ

EQ

JACK RAFTER

JACK RAFTER RUN

LEGEND
EQ = EQUAL DISTANCE

© Cengage Learning 2014

Figure 45–3 Determining the valley jack rafter Run along ridgeboard.

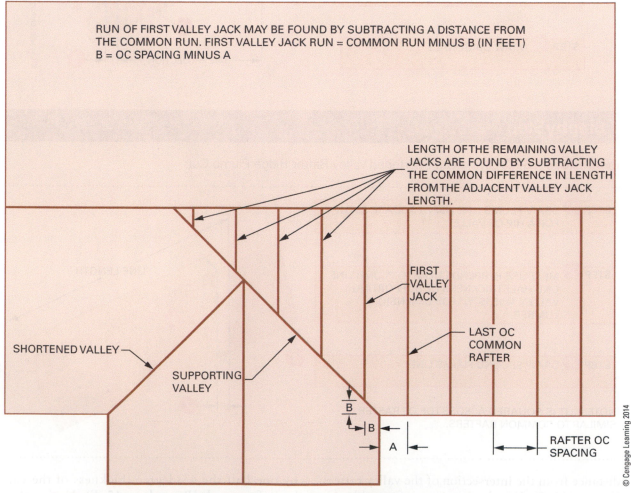

RUN OF FIRST VALLEY JACK MAY BE FOUND BY SUBTRACTING A DISTANCE FROM THE COMMON RUN. FIRST VALLEY JACK RUN = COMMON RUN MINUS B (IN FEET)
B = OC SPACING MINUS A

LENGTH OF THE REMAINING VALLEY JACKS ARE FOUND BY SUBTRACTING THE COMMON DIFFERENCE IN LENGTH FROM THE ADJACENT VALLEY JACK LENGTH.

FIRST VALLEY JACK

LAST OC COMMON RAFTER

SHORTENED VALLEY

SUPPORTING VALLEY

B

B

A

RAFTER OC SPACING

© Cengage Learning 2014

Figure 45–4 Determining a valley jack rafter Run by subtraction method.

StepbyStep Procedures

Procedure 45–E Laying Out a Valley Cripple Jack Rafter

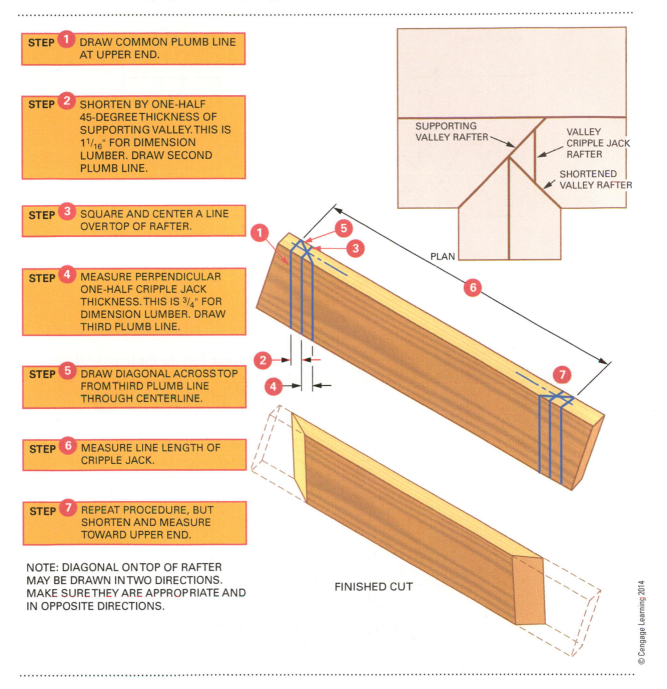

STEP 1 DRAW COMMON PLUMB LINE AT UPPER END.

STEP 2 SHORTEN BY ONE-HALF 45-DEGREE THICKNESS OF SUPPORTING VALLEY. THIS IS 1 1/16" FOR DIMENSION LUMBER. DRAW SECOND PLUMB LINE.

STEP 3 SQUARE AND CENTER A LINE OVER TOP OF RAFTER.

STEP 4 MEASURE PERPENDICULAR ONE-HALF CRIPPLE JACK THICKNESS. THIS IS 3/4" FOR DIMENSION LUMBER. DRAW THIRD PLUMB LINE.

STEP 5 DRAW DIAGONAL ACROSS TOP FROM THIRD PLUMB LINE THROUGH CENTERLINE.

STEP 6 MEASURE LINE LENGTH OF CRIPPLE JACK.

STEP 7 REPEAT PROCEDURE, BUT SHORTEN AND MEASURE TOWARD UPPER END.

NOTE: DIAGONAL ON TOP OF RAFTER MAY BE DRAWN IN TWO DIRECTIONS. MAKE SURE THEY ARE APPROPRIATE AND IN OPPOSITE DIRECTIONS.

SUPPORTING VALLEY RAFTER

VALLEY CRIPPLE JACK RAFTER

SHORTENED VALLEY RAFTER

PLAN

FINISHED CUT

© Cengage Learning 2014

Hip-Valley Cripple Jack Rafter Layout

All hip-valley cripple jacks cut between the same hip and valley rafters are the same. This is because the hip and valley rafters are parallel. To determine the length of the rafter, first find its total run. The Run of a hip-valley cripple jack rafter is equal to the plate line distance between the seat cuts of the hip and valley rafters (*Figure 45–6*). Remember to lay out all jack plumb lines the same as for common rafters.

Draw a plumb line at the top end. Shorten by measuring perpendicular toward the lower end,

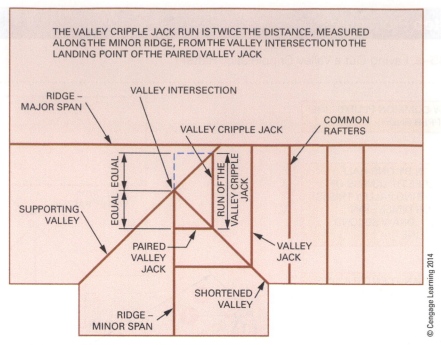

THE VALLEY CRIPPLE JACK RUN IS TWICE THE DISTANCE, MEASURED ALONG THE MINOR RIDGE, FROM THE VALLEY INTERSECTION TO THE LANDING POINT OF THE PAIRED VALLEY JACK

RIDGE – MAJOR SPAN

VALLEY INTERSECTION

VALLEY CRIPPLE JACK

COMMON RAFTERS

EQUAL EQUAL

RUN OF THE VALLEY CRIPPLE JACK

SUPPORTING VALLEY

PAIRED VALLEY JACK

VALLEY JACK

SHORTENED VALLEY

RIDGE – MINOR SPAN

© Cengage Learning 2014

Figure 45–5 Determining a valley cripple jack rafter Run.

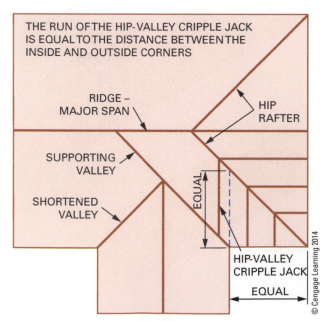

THE RUN OF THE HIP-VALLEY CRIPPLE JACK IS EQUAL TO THE DISTANCE BETWEEN THE INSIDE AND OUTSIDE CORNERS

RIDGE – MAJOR SPAN

HIP RAFTER

SUPPORTING VALLEY

SHORTENED VALLEY

EQUAL

HIP-VALLEY CRIPPLE JACK

EQUAL

© Cengage Learning 2014

Figure 45–6 Determining the Run of the hip-valley cripple jack rafter.

one-half 45-degree thickness of the hip rafter. This is $1\frac{1}{16}$ inches for dimension lumber (*Procedure 45–F*). Draw a second plumb line. Square a line over the top edge of the cripple also marking the center of the line. Measure and mark perpendicular toward

the lower end, one-half the thickness of the cripple rafter. This is $\frac{3}{4}$ inch for dimension lumber. Draw a third plumbline. Across the top edge, draw a diagonal from the top of the third line through the centerline.

Measure the line length of the cripple and repeat the exact layout of the upper end. This is done because the cuts are parallel. The only exception is that the shortening is one-half 45-degree thickness of the valley rafter which is $1\frac{1}{16}$ inches for dimension lumber.

RIDGE LENGTHS OF INTERSECTING ROOFS

Intersecting roofs have more than one ridgeboard. To reduce confusion, the width of the main roof is called the major span, while the width of the building wings is called the minor span.

The lengths of ridgeboards for intersecting roofs depend on a number of factors. These factors include the height differences of the major and minor ridges, the style of framing at the ridge intersection, and whether there are any hip rafters. The first step is to determine the overall or theoretical length, then modify it to fit depending on how the intersection is framed.

StepbyStep Procedures

Procedure 45–F Laying Out a Hip-Valley Cripple Jack Rafter

STEP 1 DRAW COMMON PLUMB LINE AT UPPER END.

STEP 2 SHORTEN ONE-HALF 45 DEGREE THICKNESS OF SUPPORTING VALLEY. THIS IS $1\frac{1}{16}$" FOR DIMENSION LUMBER. DRAW SECOND PLUMB LINE SQUARE AND CENTER A LINE OVER TOP OF RAFTER.

STEP 3 MEASURE PERPENDICULAR ONE-HALF CRIPPLE JACK THICKNESS. THIS IS $\frac{3}{4}$" FOR DIMENSION LUMBER. DRAW THIRD PLUMB LINE.

STEP 4 DRAW DIAGONAL ACROSS TOP FROM THIRD PLUMB LINE THROUGH CENTERLINE.

STEP 5 MEASURE LINE LENGTH OF HIP-VALLEY CRIPPLE JACK.

STEP 6 REPEAT PROCEDURE, BUT SHORTEN USING ONE-HALF 45-DEGREE THICKNESS OF VALLEY. THIS IS $1\frac{1}{16}$" FOR DIMENSION LUMBER.

NOTE: DIAGONALS ON TOP OF RAFTER MAY BE DRAWN IN TWO DIRECTIONS. MAKE SURE THEY ARE APPROPRIATE AND IN SAME DIRECTION.

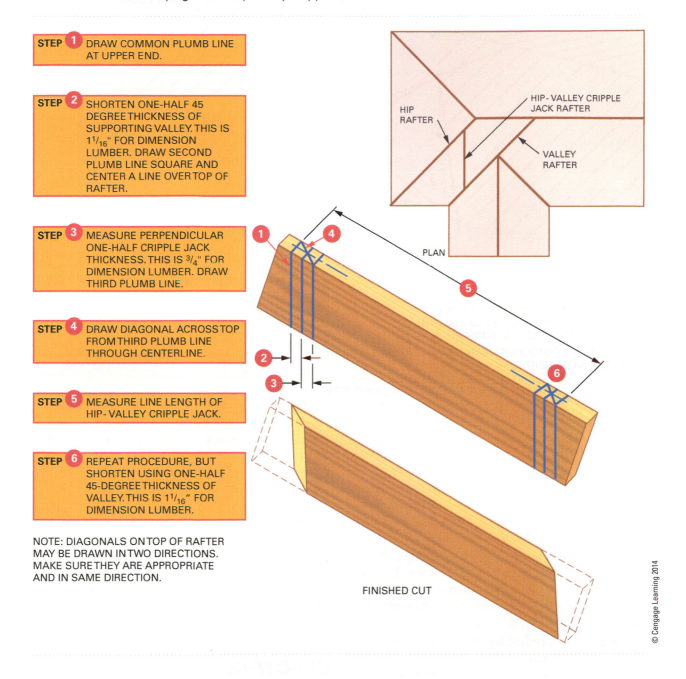

HIP-VALLEY CRIPPLE JACK RAFTER

HIP RAFTER

VALLEY RAFTER

PLAN

FINISHED CUT

© Cengage Learning 2014

Theoretical lengths of minor ridges come from adding two numbers. Ridge length equals the length of the extension or wing plus one-half the width of the minor span (*Figure 45–7*). The minor common rafter Run is one-half the minor span. The actual length of the minor ridge depends on the framing style. The typical adjustments to ridge length may be seen in *Figure 45–8.*

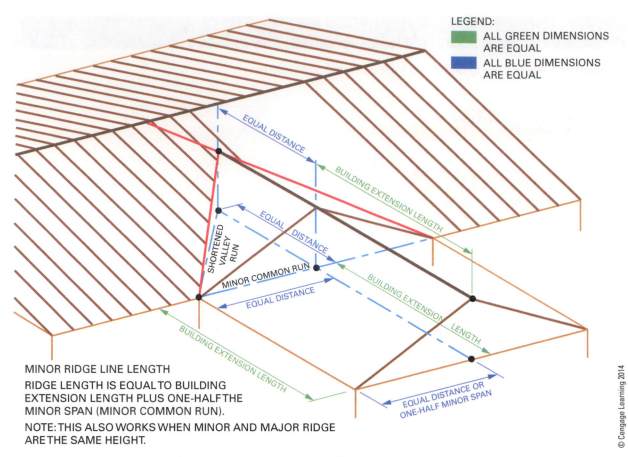

LEGEND:
- ALL GREEN DIMENSIONS ARE EQUAL
- ALL BLUE DIMENSIONS ARE EQUAL

EQUAL DISTANCE

BUILDING EXTENSION LENGTH

EQUAL DISTANCE

SHORTENED VALLEY RUN

MINOR COMMON RUN

BUILDING EXTENSION LENGTH

EQUAL DISTANCE

BUILDING EXTENSION LENGTH

EQUAL DISTANCE OR ONE-HALF MINOR SPAN

MINOR RIDGE LINE LENGTH

RIDGE LENGTH IS EQUAL TO BUILDING EXTENSION LENGTH PLUS ONE-HALF THE MINOR SPAN (MINOR COMMON RUN).

NOTE: THIS ALSO WORKS WHEN MINOR AND MAJOR RIDGE ARE THE SAME HEIGHT.

© Cengage Learning 2014

Figure 45–7 Determining the line length of the minor ridge.

EXAMPLE

Using the information in Figure 44–8 and dimension lumber framing, the theoretical length of **Ridge 1** is $48' - 24' - 24'$. Adjustments including adding $\frac{3}{4}''$ at both points A give an actual Ridge 1 (R1) length of $24'\text{-}1\frac{1}{2}''$.

Ridge 2 theoretical length is found by adding the wing length to one-half minor span. This is $9' + 12' = 21'$. The adjustment is a deduction at point B of one-half major ridge thickness or $21\text{-}\frac{3}{4}'' = 20'\text{-}11\frac{1}{4}''$.

Ridge 3 theoretical length is found similarly to Ridge 2. Thus $5' + 6.5'' = 11'\text{-}6''$. The necessary adjustment at point C is to deduct one-half the 45-degree thickness of the valley or $1\frac{1}{16}''$. The actual length of Ridge 3 is $11'\text{-}6''$ minus $1\frac{1}{16}'' = 11'\text{-}4\frac{15}{16}''$.

Ridge 4 length looks complicated but is not. Ridge 4 is a side of a parallelogram formed by the hip and valley on two sides and the ridge and wall plate on the others. Thus the theoretical length is the same as the wall plate measurement or simply 5 feet. Adjustments are made at either end. At point C, it is shortened by $1\frac{1}{16}''$ and at point D it is lengthened by $1\frac{13}{16}''$. The net result is to lengthen Ridge 4 by $\frac{3}{4}''$, giving an actual length of $5'\text{-}3\frac{3}{4}''$.

Sometimes we may want to measure on the ridge the location of a valley/ridge intersection. **Measurement 1** is noted in Figure 44–8 as M1. The theoretical distance of Measurement 1 is $6'\text{-}0''$ from the wall plate distance. Add $\frac{3}{4}''$ at point A and add $\frac{5}{16}''$ at point E. Thus the actual measurement line from the end of the ridge board is $6'$ plus $\frac{3}{4}''$ plus $\frac{5}{16}'' = 6'\text{-}1\frac{1}{16}''$.

ERECTING THE INTERSECTING ROOF

The intersecting roof is raised by framing opposing members of the main span first. Then, the valley rafters are installed. To prevent bowing the ridge of the main span, install rafters to oppose the valley rafters. Install common and jack rafters in sets opposing each other. Sight members of the roof as framing progresses, keeping all members in a straight line.

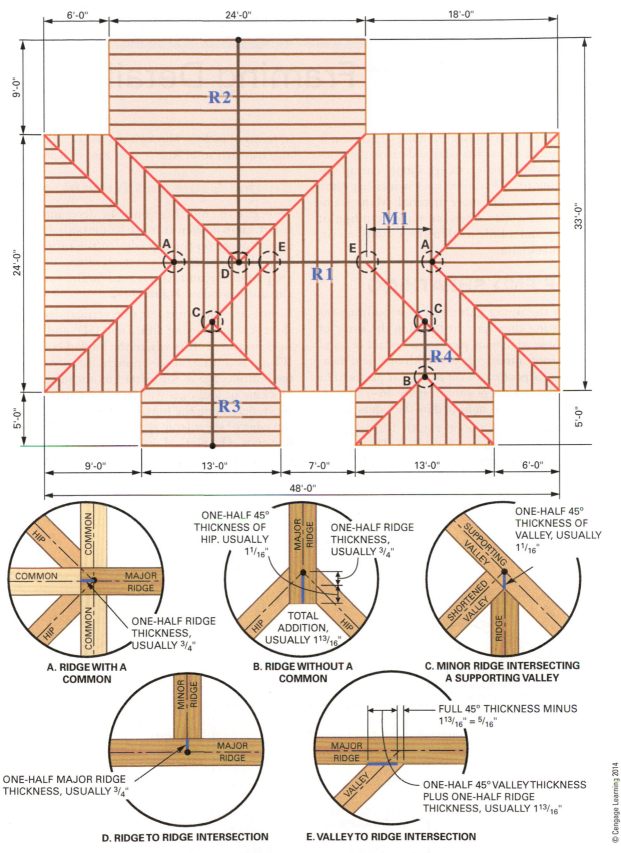

Figure 45–8 Ridgeboard line lengths are adjusted depending on the style of framing.

CHAPTER 46
Complex Roof Framing Details

When faced with an unusual framing situation it is often tempting to use the trial-and-error, cut-and-fit method where materials are simply cut up until something fits. This sometimes works but often takes longer and uses more material. Many framing situations occur where an understanding of roof theory and a little math make a solution fast, precise, and elegant.

GABLE END SHEATHING

Sheathing the upper gable end of a building involves cutting sheathing fit to along the slope of rafters. These triangular pieces can be laid out and cut using roof framing theory. With one field measurement a carpenter can cut triangular pieces to fit with as little waste as possible (*Figure 46–1*). Sheets are cut in sequence as indicated by the letters A – G.

This method uses the previously learned principle of run × unit Rise = total Rise. The run is a horizontal distance under a rafter and is always in terms of feet. Total rise is a vertical distance always in terms of inches. Through the use of algebra it is also true that total Rise ÷ unit Rise = Run.

Cutting Gable End Sheathing

To begin, the unit Rise must be determined. If it is not known, then it must be measured. Place a speed square on the top edge of the rafter with a level along the square's vertical edge. When they are plumb, the unit Rise is read on the scale in the alignment window *Figure 46–2*.

Sheet A. The short left side vertical distance of sheet A is the starting dimension (SD) and must be field measured. It is where the angled cut begins. When measuring, allow enough distance to cover the bottom edge of the rafter but not so much that blocking or lookouts will interfere with the sheet. Next, determine the horizontal distance along the top edge of sheet from its corner where the sloped line ends. The vertical distance and the horizontal distance provide enough information to cut the piece.

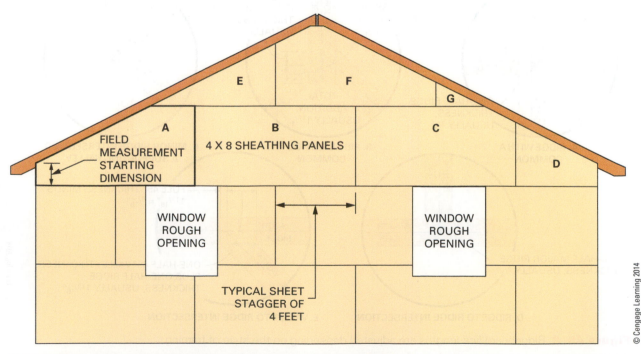

Figure 46–1 Gable end sheathing may be cut using roof framing theory.

© Cengage Learning 2014

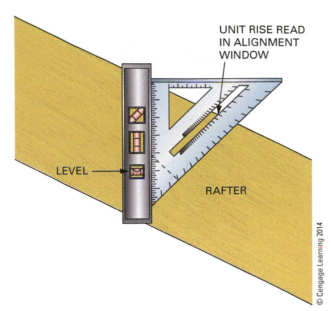

Figure 46–2 A speed square and a torpedo level may be used to determining the slope of an installed rafter.

Measure up on the vertical edge of a full sheet of sheathing and mark the SD. Measure the remaining vertical distance to the top of the sheet. This number, in inches, is a total Rise. If it is divided by unit Rise the result is a run or the needed horizontal distance in feet (*Figure 46–3*). Multiply this answer by 12 to convert to inches and convert the decimal to a fraction. For example if the unit Rise is 6 inches and the SD is 13½" then the remaining distance is 34½". The horizontal distance is 34½" ÷ 6 = 5.75 feet or 69 inches. Now the horizontal distance can be laid out and a sloped cut line drawn. The leftover piece is handy to keep around. It can be used as a template to layout another sloped lines or as a useable piece. Before sheet A is installed the remaining horizontal distance is measured and saved for future reference. This number is used to cut sheet E, the first piece in the next course.

Sheet B. This row, the third row of sheathing is continued to the other side of the building. Full sheets are often needed to complete the row as is sheet B.

Sheet C. Sheet C is cut in a similar manner as for sheet A. The starting dimension for this piece is measured vertically along the center of the stud. It also covers a similar amount of the rafter. Subtract the SD from 48 to get the total Rise of the cut off. Then divide by unit Rise, multiply times 12 to convert to inches, and convert the fraction. Save the starting dimension of this sheet for layout of sheet D. Mark and cut the slope on the sheet.

Sheet D. The last piece in the row, sheet D, is cut to length before laying out the sloped cut (*Figure 46–4*). The right side starting dimension should be the same as the left side of the building, but it is a good idea to check the measurement. The left-side vertical measurement of sheet D is the same as the saved measurement in sheet C. Measure, mark, and connect the marks for the slope cut.

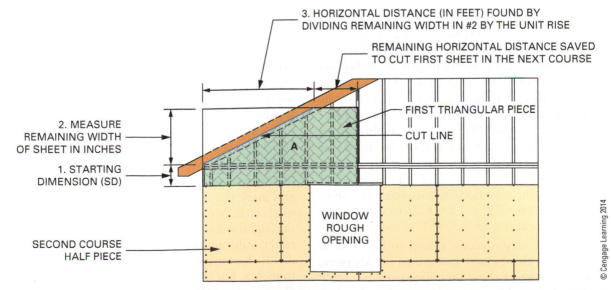

Figure 46–3 Sheathing may be cut by using the slope and one starting dimension.

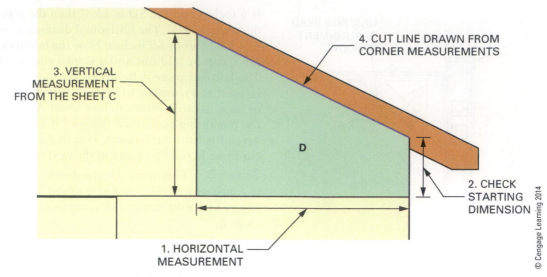

Figure 46–4 The last piece of sheathing in the row is cut to length and then cut on an angle.

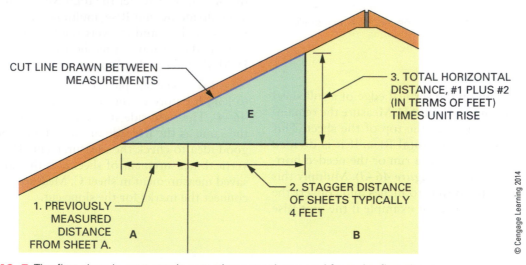

Figure 46–5 The first piece in next row is cut using a number saved from the first sheet.

Sheet E. The left side piece, sheet E is determined using the previously saved horizontal dimension from Sheet A. The entire length of the bottom of sheet E is found by adding the saved measurement from sheet A to the staggered distance of the sheets, typically 4 feet (*Figure 46–5*). This number is converted to decimal feet and then multiplied by the unit Rise to get the total rise of the triangular piece. For example if the save horizontal measurement is 27 inches and the sheet stagger is 4 feet or 48 inches, then the bottom horizontal measurement is $24 + 48 = 75'' \div 12 = 6.25$ feet. The vertical measurement is $6.25 \times 6 = 37\frac{1}{2}$ inches.

Check the waste pieces created so far to see if a piece exists that is large enough to fit. Measure, mark and connect these measurements to create the sloped piece. Measure and remember the right side vertical distance of sheet E. This number will be helpful in cutting sheet F.

Sheet F. Sheet F measurements come from a similar series of calculations (*Figure 46–6*). The left side vertical distance in inches is found by subtracting the saved number in sheet E from 48 inches. The left horizontal distance is calculated from dividing the vertical distance by the unit Rise and converting it to inches. For example if the save vertical dimension is $37\frac{1}{2}$ inches, then the horizontal measurement is $48 - 37\frac{1}{2}'' = 10\frac{1}{2}'' \div 6 = 1.75$ feet or 21 inches.

After sheet E is installed, measure horizontally and mark the center of the stud 8 feet to the right of sheet E. Measure vertically along the center of the stud the distance to include the overlap onto the rafter. Subtract this number from 48 inches. Then divide the remainder by the unit Rise and convert to inches. For example if the vertical measurement along stud is $14\frac{1}{2}$ inches, the remaining vertical distance is $48 - 14\frac{1}{2}'' = 37\frac{1}{2}$ inches. The horizontal measurement is $37\frac{1}{2} \div 6 = 6.25' = 75$ inches.

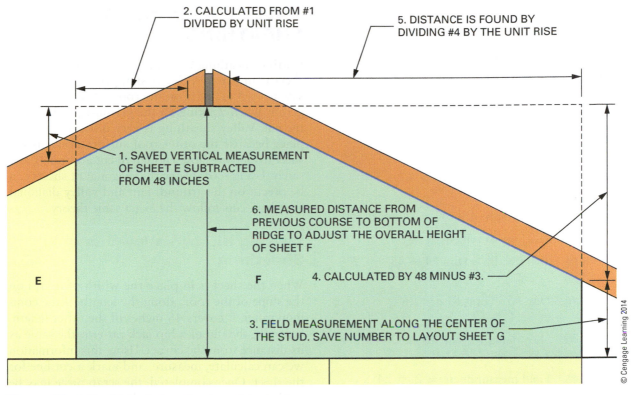

2. CALCULATED FROM #1 DIVIDED BY UNIT RISE

5. DISTANCE IS FOUND BY DIVIDING #4 BY THE UNIT RISE

1. SAVED VERTICAL MEASUREMENT OF SHEET E SUBTRACTED FROM 48 INCHES

6. MEASURED DISTANCE FROM PREVIOUS COURSE TO BOTTOM OF RIDGE TO ADJUST THE OVERALL HEIGHT OF SHEET F

4. CALCULATED BY 48 MINUS #3.

E

F

3. FIELD MEASUREMENT ALONG THE CENTER OF THE STUD. SAVE NUMBER TO LAYOUT SHEET G

© Cengage Learning 2014

Figure 46–6 The highest piece has two sloped cuts.

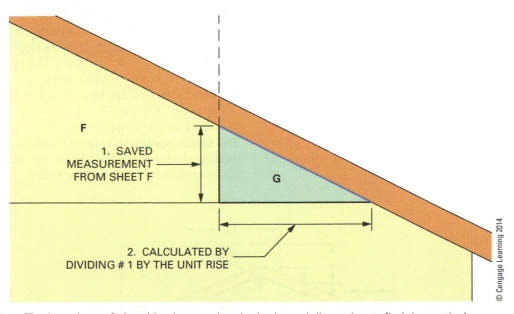

F

1. SAVED MEASUREMENT FROM SHEET F

G

2. CALCULATED BY DIVIDING # 1 BY THE UNIT RISE

© Cengage Learning 2014

Figure 46–7 The last piece of sheathing is cut using the horizontal dimension to find the vertical.

The overall height of the sheet must be cut along a horizontal line to fit under the ridge. Measuring from the lower course of sheathing to the bottom of the ridge gives the height limitations of the sheet. For example if the overheight is found to 47 inches, the sheet must have 48″ – 47″ = 1 inch cut off. This is done horizontally on the top of the sheet. Save the vertical distance that remains on the right side of the sheet F to aid in cutting sheet G.

Sheet G. Sheet G is laid out by using the saved right-side vertical measurement from sheet F. This number, in inches, is divided by the unit Rise to get the horizontal measurement (*Figure 46–7*). For example if the saved dimension is 14 ½ inches, the horizontal distance for sheet G is 14 ½″ ÷ 6 = 2.41′ = 29 inches. Layout and cut sheet G from a waste piece from the previous sheets to conserve material.

EXAMPLE

What are the sheet dimensions for the gable end sheathing for a house with a rafter unit Rise 5 inches, a starting dimension (SD) of 12 inches and is 26 feet wide (*Figure 46–8*)?

Solution:

Dim #1. Sheet width – SD $\Rightarrow$ 48″ – 12″ = 36″

Dim #2. Dim #1 ÷ unit Rise $\Rightarrow$ 36″ ÷ 5 = 7.2′ = 86.4″ = 86 $\frac{3}{8}$″

Dim #3. Sheet length – Dim #2 $\Rightarrow$ 8′ – 7.2′ = 0.8′

Dim #4. Field measurement $\Rightarrow$ 22″

Dim #5. Sheet width – Dim #4 $\Rightarrow$ 48″ – 22″ = 26″

Dim #6. Dim #5 ÷ unit Rise $\Rightarrow$ 26″ ÷ 5 = 5.2′ = 62.4″ = 62 $\frac{3}{8}$″

Dim #7. Sheet length – Dim #6 $\Rightarrow$ 8′ – 5.2′ = 2.8′ = 33.6″ = 33 $\frac{5}{8}$″

Dim #8. Field measurement $\Rightarrow$ 2′ = 24″

Dim #9. Dim #3 + Sheet stagger $\Rightarrow$ 4′ + 0.8 = 4.8′ = 57.6″ = 57 $\frac{5}{8}$″

Dim #10. Dim #9 × unit Rise $\Rightarrow$ 4.8 × 5 = 24″

Dim #11. Sheet width – Dim #10 $\Rightarrow$ 48″ – 24″ = 24″

Dim #12. Dim #11 ÷ unit Rise $\Rightarrow$ 24″ ÷ 5 = 4.8′ = 57.6″ = 57 $\frac{5}{8}$″

Dim #13. Sheet stagger + Dim #7 $\Rightarrow$ 4′ + 2.8 = 6.8′ = 81.6″ = 81 $\frac{5}{8}$″

Dim #14. Dim #13 × unit Rise $\Rightarrow$ 6.8 × 5 = 34″

Dim #15. Field measurement $\Rightarrow$ 27¼″

VALLEY AND HIP ROOF SHEATHING

Cutting sheathing to a hip is often done after the sheet is in place. A line is snapped and a circular saw is used to cut it on the roof. This can be dangerous when walking on the sawdust produced by the cut. Valley sheathing is very difficult to cut in place because the other roof surface prevents the sheet from lying close enough to cut accurately and safely. Using roof theory the diagonal of sheets can be precut on the ground. Hip and valley diagonal sheathing cuts follow the exact same theory.

Cutting Hip or Valley Diagonal Sheathing

When the sheet is in place the width of it lays on the slope of the roof, along the length of the common rafter. It covers 48 inches of the rafter length. Secondly, the run of a hip jack rafter is the same as its distance from the corner. Using this information we can calculate, measure, and mark a cut line for the sheet. Once completed, the scrap piece may be used a template to cut other sheets.

The run under the 48-inch-wide sheet is the same as the run of a hip jack that is 48 inches long. To find the run, divide the length by the unit length. For example if the common rafter slope is 5 on 12, then the unit length is 13 inches. The run under the sheet is 48 ÷ 13 = 3.69 feet = 44 $\frac{5}{16}$ inches. Measure and mark this distance from the upper corner of the sheet (*Figure 46–9*). Drawing a line from this mark to the lower corner gives the cut line. This theory applies to valley rafters as well. Simply rotating the hip sheet with the same face up makes it appropriate to fit to a valley rafter.

*IRC
R802.9

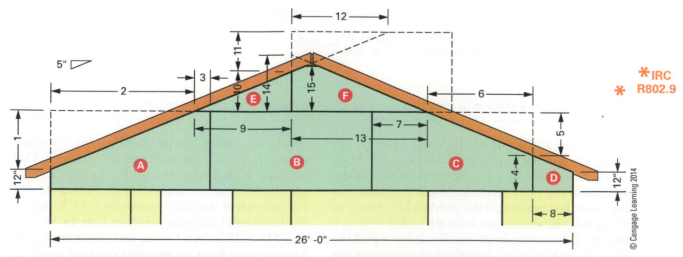

Figure 46–8 Gable end sheathing calculations may be done before actual installation.

© Cengage Learning 2014

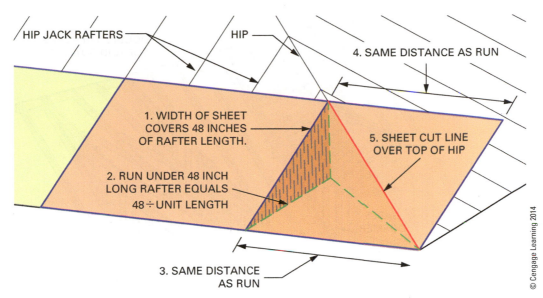

HIP JACK RAFTERS — HIP

4. SAME DISTANCE AS RUN

1. WIDTH OF SHEET COVERS 48 INCHES OF RAFTER LENGTH.

5. SHEET CUT LINE OVER TOP OF HIP

2. RUN UNDER 48 INCH LONG RAFTER EQUALS 48 ÷ UNIT LENGTH

3. SAME DISTANCE AS RUN

© Cengage Learning 2014

Figure 46–9 The run under the 48-inch width of sheathing is the same distance as the horizontal distance along the top edge of sheet.

DORMERS

Dormers typically have either a gable or a shed roof. A gable dormer roof is framed like an intersecting gable roof with valleys and jacks. A shed dormer is simpler in design and often framed to the ridge of the main roof. This is typically done to gain sufficient rafter slope (*Figure 46–10*).

When framing openings in the main roof for dormers, the rafters on both sides of the opening are doubled or tripled. Some dormers have their front walls directly over the exterior wall below. In this case, much of the load of the dormer is transferred into the load-bearing wall below. If dormers are placed such that the dormer front wall is recessed or if the dormers are partway up the main roof, the main roof or the floor joists must be strengthened. Double headers are also added at the top and bottom of the opening.

Dormer sidewall studs may be cut and fit with a common difference (CD) in length. Gable end studs of the dormer have the same CD as for the main roof gable end studs. The CD is unit Rise × OC spacing (ft).

Shed dormer sidewall studs CD is affected by the slopes of both the rafters, the main roof and the shed roof. The shed dormer sidewall stud CD is determined by subtracting the shed CD from the main roof CD (*Figure 46–11*).

 EXAMPLE

For example if the slope of the main roof is 11 on 12, the CD for 16″ OC studs is 11″ × 1.33 = 14.66 inches. If the shed dormer slope is 5 on 12, then the CD is 5″ × 1.33 = 6.66 inches. Subtracting gives the shed dormer sidewall studs CD of 14.66 − 6.66″ = 8 inches.

The top and bottom of the studs have angles cut on them. The bottom angle is determined from the main roof slope and the top from the shed roof slope.

A number of other types of dormer framing exist, but they are all related, in some way, to the framing theory previously described. To solve these various roof-framing situations, begin by returning to the basics of understanding rafter length as compared with Run and Rise.

Sometimes a dormer is framed to an already existing building. Dormer components are cut to fit

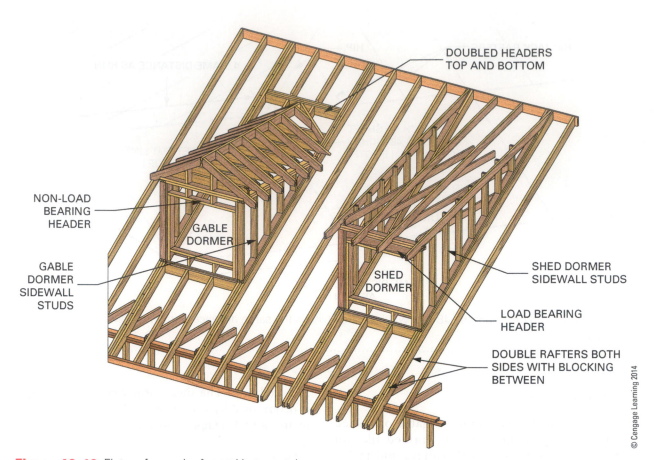

Figure 46–10 Flat roofs may be framed in two styles.

DOUBLED HEADERS
TOP AND BOTTOM

NON-LOAD
BEARING
HEADER

GABLE
DORMER

GABLE
DORMER
SIDEWALL
STUDS

SHED
DORMER

SHED DORMER
SIDEWALL STUDS

LOAD BEARING
HEADER

DOUBLE RAFTERS BOTH
SIDES WITH BLOCKING
BETWEEN

© Cengage Learning 2014

to the existing sheathing line and not to the main roof rafters.

Fitting a Gable Dormer to Roof Sheathing

The ridge must be fitted level into the existing roof. The simplest way to determine the ridge length is to first cut a slope on the end of the ridge, then install it with common rafters. The ridge is later cut for length after plumbing up from the end wall.

The ridge slope cut layout uses the unit rise for the existing roof (*Procedure 46–A*). Hold the unit Rise and 12 on a framing square and adjust to the end of the board. Mark the blade of the square.

Fitting a Valley Jack to Roof Sheathing After the ridge is placed against the existing roof, a 1 × 6 or 1 × 8 is nailed where a valley would be. It provides

a place for the new valley jacks to land. The upper end of the valley jack is laid out like a common rafter. The lower end is a compound miter.

Jack rafter Run is measured along the ridge from the center of the rafter where it lands on the ridge to the intersection of the ridge and the 1 × 6 or 1 × 8. No shortening need be done at the bottom miter because the Run is measured as actual.

The cross-sectional rise of the jack rafter must be determined to cut the compound miter at the proper angle. This amount of rise is related to the slope of the main roof and the thickness of the jack rafter. Layout is done from the center of the jack since the Run is measured from its center. Therefore the cross-sectional Run is one-half of the jack thickness, usually $3/4$″. Converting $3/4$″ to feet is $0.75″ ÷ 12 = 0.0625′$. The cross-sectional rise is unit rise times 0.0625 (*Procedure 46–B*).

SHED ROOF RAFTER

SHED ROOD COMMON DIFFERENCE

4"

5 5/16"

16" OC

TO BEGIN BY MEASURING THE FIRST STUD

MAIN ROOF COMMON DIFFERENCE

12"

8"

MAIN ROOF RAFTER

ANGLES COME FROM THE RAFTER SLOPE NEAREST THE STUD ENDS.

SHED DORMER SIDEWALL STUD COMMON DIFFERENCE (CD)
1. MEASURE THE LONGEST (FIRST) STUD.
2. MARK TOP ANGLE USING THE SLOPE OF SHED DORMER RAFTER AND BOTTOM ANGLE SAME AS THE SLOPE OF BOTTOM RAFTERS.
3. DETERMINE MAIN ROOF CD FOR STUDS $8 \times 1.33'$ (OC) = 12"
4. DETERMINE SHED ROOF CD FOR STUDS $4 \times 1.33' = 5\,5/16$"
5. SUBTRACT SHED ROOF CD FROM MAIN ROOF CD.
   $12" - 5\,5/16" = 6\,11/16"$
6. THE CD FOR SHED DORMER SIDEWALL STUDS IS $6\,11/16$"

© Cengage Learning 2014

Figure 46–11 Shed dormer side wall studs have a common difference (CD) in length.

Step**by**Step Procedures

Procedure 46–A Laying Out a Gable Dormer Ridge

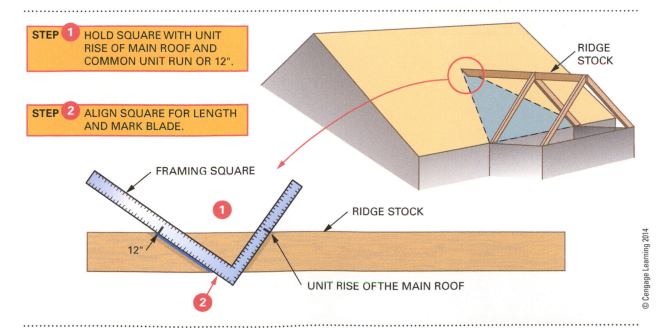

STEP 1 HOLD SQUARE WITH UNIT RISE OF MAIN ROOF AND COMMON UNIT RUN OR 12".

STEP 2 ALIGN SQUARE FOR LENGTH AND MARK BLADE.

RIDGE STOCK

FRAMING SQUARE

1

RIDGE STOCK

12"

2

UNIT RISE OF THE MAIN ROOF

© Cengage Learning 2014

StepbyStep Procedures

Procedure 46–B Laying Out a Valley Jack Framed to a Roof

STEP 1 MARK RIDGE PLUMB LINES AND SHORTEN AS FOR A COMMON RAFTER.

STEP 2 MEASURE RUN ALONG RIDGE OF MAIN ROOF FOR JACK RUN.

STEP 3 DETERMINE THE LENGTH, THEN MEASURE AND MARK ALONG TOP EDGE.

STEP 4 SQUARE A LINE ALONG THE TOP OF THE RAFTER AND MARK THE CENTER.

STEP 5 DETERMINE $1/2$ OF THE CROSS-SECTIONAL RISE OF THE RAFTER THICKNESS. FOR EXAMPLE: IF THE UNIT RISE OF THE MAIN ROOF IS 6, AND $1/2$ THE VALLEY JACK THICKNESS IS $3/4$" OR 0.0625', $1/2$ THE CROSS-SECTIONAL RISE IS 6 X 0.0625 = 0.375" OR $3/8$".

STEP 6 MARK FROM THE SQUARED LINE IN STEP 4, ALONG A PLUMB LINE, THE DISTANCE DETERMINED IN STEP 5 ($3/8$").

STEP 7 USING A FRAMING SQUARE, MARK A LEVEL LINE THAT JUST TOUCHES THE BOTTOM OF THE SHORT PLUMB LINE IN STEP 6.

STEP 8 DRAW A DIAGONAL LINE FROM THE END OF THE LEVEL LINE THROUGH PREVIOUSLY MARKED CENTER.

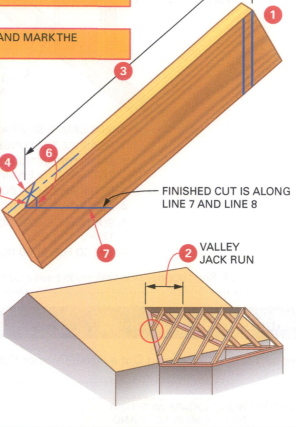

FINISHED CUT IS ALONG LINE 7 AND LINE 8

VALLEY JACK RUN

© Cengage Learning 2014

StepbyStep Procedures

Procedure 46–C Laying Out a Shed Dormer Rafter

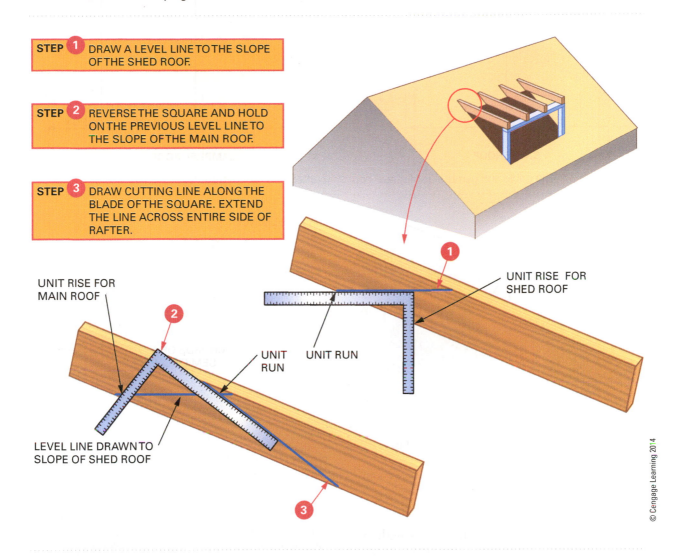

STEP 1 DRAW A LEVEL LINE TO THE SLOPE OF THE SHED ROOF.

STEP 2 REVERSE THE SQUARE AND HOLD ON THE PREVIOUS LEVEL LINE TO THE SLOPE OF THE MAIN ROOF.

STEP 3 DRAW CUTTING LINE ALONG THE BLADE OF THE SQUARE. EXTEND THE LINE ACROSS ENTIRE SIDE OF RAFTER.

UNIT RISE FOR MAIN ROOF

UNIT RISE FOR SHED ROOF

UNIT RUN

UNIT RUN

LEVEL LINE DRAWN TO SLOPE OF SHED ROOF

© Cengage Learning 2014

Fitting a Shed Roof Rafter to Roof Sheathing

The top ends of shed dormer rafters may be fit against the main roof ,which has a steeper slope. The cuts are laid out by using a framing square as outlined in *Procedure 46–C.*

GAMBREL ROOFS

Like the gable roof, the gambrel roof is symmetrical (see *Figure 46–12*). Each side of the building has rafters with two different slopes. They have different unit rises. Their Runs may or may not be the same depending on the desired roof style.

The saltbox is a gable roof with rafters having different slopes or different Runs on each side of the ridge (*Figure 46–13*). One rafter may slope faster to reach to the ridge from a lower floor.

FRAMING A GAMBREL ROOF

A gambrel roof is one where each side of the ridgeboard has two sloping rafters. The rafters of a true gambrel roof are chords of a semicircle whose diameter is the width of the building (*Figure 46–14*).

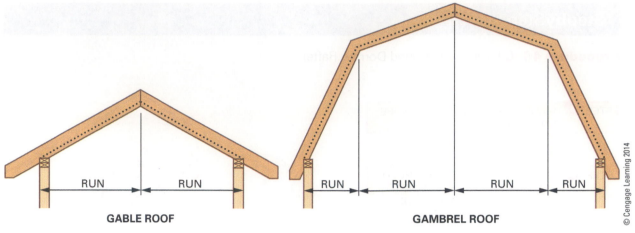

GABLE ROOF   GAMBREL ROOF

Figure 46–12 The gable and gambrel roofs have opposing rafters that are symmetrical.

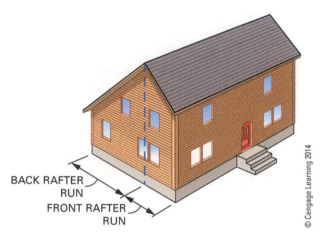

Figure 46–13 The ridge of the saltbox roof is off-center.

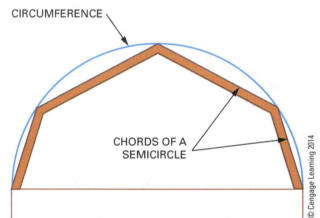

Figure 46–14 The slopes of a true gambrel roof form chords of a semicircle.

To frame a true gambrel roof, the rafter lengths are calculated using the Pythagorean theorem or determined from a full-scale layout on a large, flat surface, such as the subfloor. From the layout, rafter lengths and angles can be determined.

The calculated method involving the Pythagorean theorem may at first appear intimidating, but it really is only repetitive steps of the same process. Before calculations can begin, some information must be obtained from the plans or the architect. The measurements needed are the building span and the horizontal distances that the side rafter intersections are from the building's center and building line (*Figure 46–15*). Another method, which will not be described here, uses the ceiling height created at the rafter intersections. It is helpful to note that the total rise of the roof is equal to one-half the span. This happens because the height is also a radius of the semicircle. The distance from

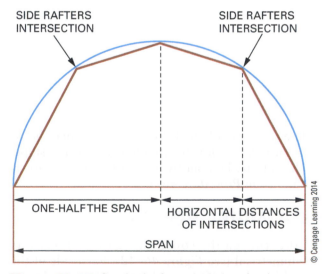

Figure 46–15 Gambrel rafter calculations begin with determining the horizontal distances of the intersections and the span.

the side rafters intersection to the center of the building is also a radius (*Figure 46–16*). Note there are only two different rafters because they come in pairs. This is because a gambrel roof is symmetrical. Bringing all of this information together show that right triangles are formed and the Pythagorean theorem can be used to solve for rafter lengths (*Figure 46–17*).

If the rafters are to be framed to a **purlin knee wall** as shown in *Figure 46–19*, the unit Rise for

each rafter may be found by dividing the total Rise of each rafter, in inches, by its total Run. For example, in Figure 42–27 U_1 is found by first converting (a) to inches, 10.954 feet times $12 = 131.448$ inches, then dividing it by 6, the Run, to get 21.908 or $21^{15}/_{16}$ inches unit Rise. U_2 is found similarly, (b), in inches, divided by 7. That is, 2.046 feet times $12 = 24.552$ inches, then divide by 7 to get 3.507 or $3^1/_2$ inches unit Rise.

Sometimes the slope of gambrel rafters is given in the drawings. The slopes, then, may not actually be chords of a semicircle (*Figure 46–20*). In this case, the rafter lengths and cuts can also be found

EXAMPLE

Find the lengths of the rafters labeled R_1 and R_2 for a gambrel roof system, given that the span is 26 feet and the horizontal distances of the side intersections are 6 and 7 feet (*Figure 46–18*).

1. $a^2 + b^2 = c^2$ therefore $a^2 = c^2 - b^2$

2. From triangle 1
$$a^2 = 13^2 - 7^2 = 169 - 49 = 120$$
$$a = \sqrt{120} = 10.954'$$

3. From triangle 2
$$R_1 = \sqrt{a^2 + 6^2} = \sqrt{10.952^2 + 6^2}$$
$$= \sqrt{120 + 36}$$
$$R_1 = \sqrt{156} = 12.490' = 12'\text{-}5^7/_8''$$

4. From triangle 3
$$b = \text{radius} - a = 13' - 10.954' = 2.046'$$

5. $R_2 = \sqrt{b^2 + 7^2} = \sqrt{2.046^2 + 7^2}$
$$= \sqrt{4.186 + 49}$$
$$R_2 = \sqrt{53.186} = 7.293' = 7'\text{-}3^1/_2''$$

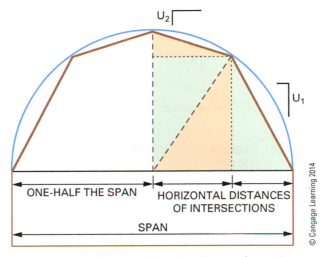

Figure 46–17 Various right triangles are formed by horizontal and vertical lines and radii of the semicircle.

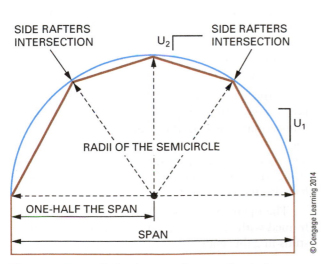

Figure 46–16 Rafter intersections are located a radius from the center of the span.

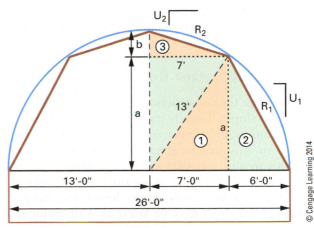

Figure 46–18 Starting information gathered for a gambrel calculation.

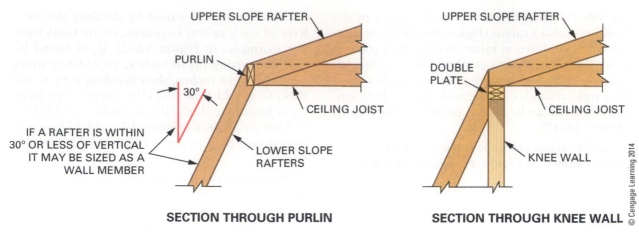

UPPER SLOPE RAFTER

PURLIN

30°

IF A RAFTER IS WITHIN 30° OR LESS OF VERTICAL IT MAY BE SIZED AS A WALL MEMBER

CEILING JOIST

LOWER SLOPE RAFTERS

SECTION THROUGH PURLIN

UPPER SLOPE RAFTER

DOUBLE PLATE

CEILING JOIST

KNEE WALL

SECTION THROUGH KNEE WALL

© Cengage Learning 2014

Figure 46–19 Gambrel roof rafters may intersect each other at a purlin or knee wall.

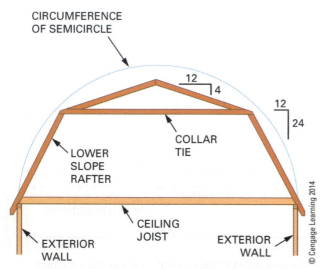

CIRCUMFERENCE OF SEMICIRCLE

12
4

12
24

COLLAR TIE

LOWER SLOPE RAFTER

CEILING JOIST

EXTERIOR WALL

EXTERIOR WALL

© Cengage Learning 2014

Figure 46–20 In some cases, the slope of the gambrel rafters is given in the drawings.

with a full-size layout. They are usually laid out with a framing square similar to the layout considerations of gable roof rafters.

Gambrel Rafter Layout

Determine the construction at the intersection of the two slopes. Usually gambrel roof rafters meet at a continuous member, similar to the ridge, called a *purlin*. Any structural member that runs at right angles to and supports rafters is called a *purlin*. Sometimes, the rafters of a gambrel roof meet at the top of a *knee wall*. In the lower and steeper slope, the rafters may be sized as wall members

rather than roof members if the rafter is within 30 degrees of vertical.

Determine the Run and Rise of the rafters for both slopes. Find their line length in the same way as for rafters in a gable roof. Lay out plumb lines at the top and bottom of rafters for each slope. Make the seat cut on the lower rafter. Notch all rafters where they intersect at the purlin (*Procedure 46–D*). Because of the steep slope of the lower roof, the level cut of the seat cannot be made the full width of the plate. At least two-thirds of the width of the rafter stock must remain.

Erecting the Gambrel Roof

If a purlin is used, fasten a rafter to each end of a section of purlin. Raise the assembly. Fasten the lower end of the rafters against ceiling joists and on the top plate of the wall. Brace the section under each end of the purlin. Continue framing sections until the other end of the building is reached. Place temporary and adequate bracing under the purlin where needed. If a knee wall is used, build, straighten, and brace the wall.

Frame the roof by installing both lower and upper slope rafters opposite each other. This maintains equal stress on the frame. Plumb and brace the roof after a few rafters have been installed. Sight along the ridge and purlin or knee wall for straightness as framing progresses. After all rafters have been erected, install ceiling joists.

The open ends formed by the gambrel roof are framed with studs in the same manner as installing studs in gable ends.

StepbyStep Procedures

Procedure 46–D Laying Out Gambrel Rafters

UPPER SLOPE RAFTER

STEP 1 LAY OUT PLUMB LINE.

STEP 2 SHORTEN ½ THICKNESS OF RIDGE.

STEP 3 LAY OUT LINE LENGTH.

STEP 4 MARK PLUMB LINE.

STEP 5 DRAW LEVEL LINE THE WIDTH OF KNEE WALL PLATE.

LOWER SLOPE RAFTER

STEP 1 LAY OUT PLUMB LINE— NO SHORTENING.

STEP 2 LAY OUT LINE LENGTH.

STEP 3 DRAW PLUMB LINE.

STEP 4 LAY OUT LEVEL LINE OF SEAT CUT. LEAVE ON MINIMUM OF ⅔ WIDTH OF RAFTER STOCK.

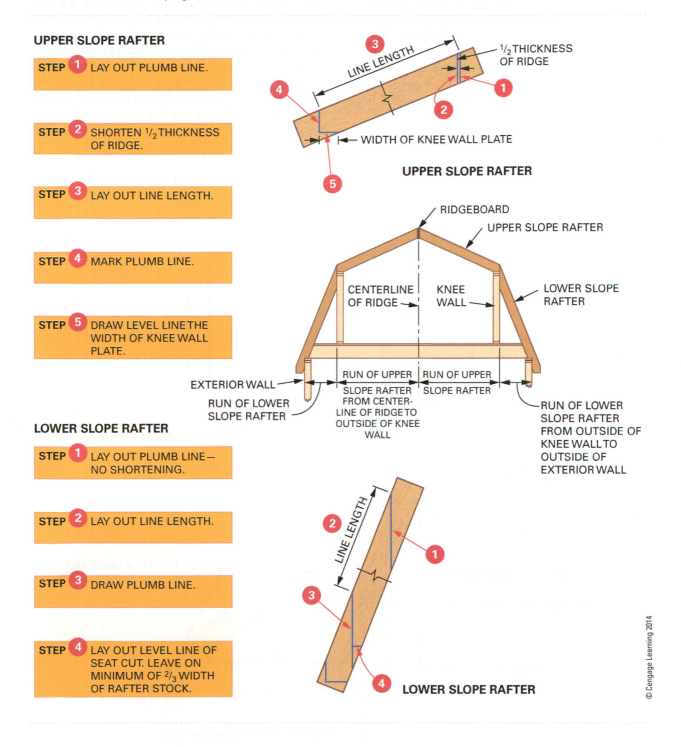

UPPER SLOPE RAFTER

LOWER SLOPE RAFTER

© Cengage Learning 2014

KNEE WALLS

Knee walls are shortened walls often found under rafters. They are built to turn an attic space in to useable space or provide support for long rafters. Sometimes long rafters sit on more than one wall where the long roof covers several floors (*Figure 46–21*). This situation requires the cutting of the rafter with multiple bird's mouths in coordination with determining the supporting wall heights. The knee wall and the mid span supporting wall follow the same theory.

Knee Wall Theory

Knee walls may be specified in two ways. The architect may want a certain wall height and is not concerned so much about the horizontal distance or floor space. Or he or she may want a certain room floor space dimension and not be too concerned with the actual knee wall height. Either situation uses the same rafter theory: run times unit Rise equals total Rise (*Figure 46–22*). Notice the knee wall run is measured from the same side of the two walls (right side in this case).

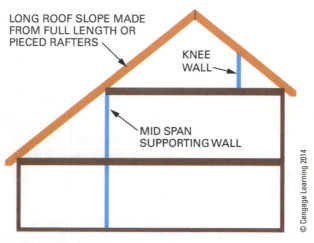

LONG ROOF SLOPE MADE FROM FULL LENGTH OR PIECED RAFTERS

KNEE WALL

MID SPAN SUPPORTING WALL

© Cengage Learning 2014

Figure 46–21 Knee walls and rafter supporting walls follow the same roof theory.

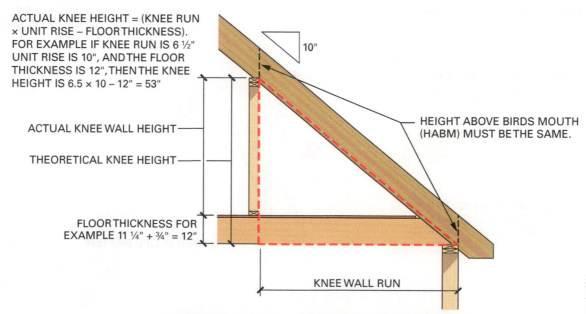

ACTUAL KNEE HEIGHT = (KNEE RUN × UNIT RISE – FLOOR THICKNESS). FOR EXAMPLE IF KNEE RUN IS 6 ½" UNIT RISE IS 10", AND THE FLOOR THICKNESS IS 12", THEN THE KNEE HEIGHT IS 6.5 × 10 – 12" = 53"

10"

ACTUAL KNEE WALL HEIGHT

THEORETICAL KNEE HEIGHT

HEIGHT ABOVE BIRDS MOUTH (HABM) MUST BE THE SAME.

FLOOR THICKNESS FOR EXAMPLE 11 ¼" + ¾" = 12"

KNEE WALL RUN

KNEE RUN = (ACTUAL KNEE HEIGHT + FLOOR THICKNESS) ÷ UNIT RISE. FOR EXAMPLE IF KNEE WALL IS TO BE 48", FLOOR THICKNESS IS 12" AND THE UNIT RISE IS 10", THEN KNEE RUN IS (48" + 12") ÷ 10 = 6.0 FEET

© Cengage Learning 2014

Figure 46–22 The floor thickness must be considered when calculating knee wall height.

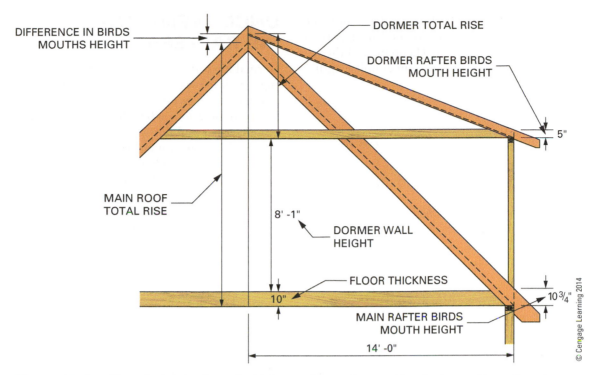

Figure 46–23 Often the height above the birds mouth must be considered when cutting rafters to meet at the ridge.

Determining Knee Height. The actual knee wall height depends on three factors: the knee wall run, the unit Rise of the rafter, and the thickness of the floor under it. These measurements are from the design of the building.

Actual knee wall height is found by multiplying the unit Rise times the knee wall run then deducting the floor thickness. For example if the unit Rise is 10 inches, the knee run is 6½ feet, and the floor thickness is 12 inches, the actual knee wall height is $10 \times 6.5 = 65'' - 12'' = 53$ inches. See Figure 46–22.

Determining the Knee Run. The knee wall run depends on three things: the actual height of the knee wall, the unit Rise of the rafter, and the thickness of the floor. The total rise of the knee wall is the actual knee height plus the floor thickness. The knee wall run is found by dividing the total knee rise by the unit Rise. For example, if the knee height is to be 48 inches, the unit rise is 10 inches and the floor thickness is 12 inches, the knee run is $48'' + 12'' = 60'' \div 10 = 6.0$ feet.

RAFTERS WITH DIFFERENT BIRD'S MOUTH HEIGHTS

Sometimes rafters of different widths meet at a ridgeboard, like a 2 × 6 shed dormer rafter meeting a 2 × 12 main roof rafter. The main roof rafter establishes the total rise and the dormer rafters are cut to fit it. The difference in bird's mouth heights of the two different rafters must be accounted for if the top edges of both rafters are to align at the ridge. Since the carpenter establishes the heights above the bird's mouth for both rafters, the carpenter must determine the unit Rise and total length of the dormer rafter (*Figure 46–23*).

This is accomplished by using some math and rafter theory. First determine the difference in bird's mouth heights of the dormer and main roof rafters and set it aside for a moment. Next determine the main roof total Rise by multiplying its unit Rise times its run. From this product subtract the floor thickness and the dormer wall height. Then add the difference in bird's mouth heights. The result is the total Rise of the dormer rafter. Divide this total Rise by the dormer rafter run to find the dormer unit Rise. Calculating the length of the dormer rafter is the same as normal rafters. Unit length is found using the Pythagorean Theorem, and the total length using the formula run times unit length.

EXAMPLE

For example, find the dormer rafter length if:
Unit Rise of the main roof is 12 inches

Building width is 28 feet, main and dormer rafter run = 14

Main floor thickness is 10 inches
Dormer wall height is 8' − 1" = 97"
Difference in bird's mouth heights (DBMH) is 4 ¼".

Solution:

Main roof total rise = run × unit Rise = 14 × 12 = 168"

Dormer total rise = main roof total rise − floor thickness − dormer wall height + DBMH

= 168 − 10 − 97" + 4 ¼" = 65 ¼"

Dormer unit rise = dormer total rise ÷ dormer run = 65 ¼" ÷ 14 = 4.66" = 4 ¹¹⁄₁₆".

Dormer rafter unit Length = square root (unit Rise² + unit Run²) = $\sqrt{(4.66^2 + 12^2)}$ = 12.87"

Dormer rafter length = dormer unit length × dormer run = 12.87" × 14 = 180.18" = 180 ⅛". Note if all the digits in the decimal of unit length are used the answer comes out as 180.23" = 180 ¼".

UNCONVENTIONAL HIP RAFTERS

Normal hip rafters have a run that is at a 45-degree angle with the wall plates. Also the common rafters of the two intersecting roofs have the same slope. There are situations where hip and valley rafters are installed on a building where the slopes of the two intersecting roofs are not the same. This occurs on L-shaped buildings where the ridgeboards are set at the same height and the widths of each section are not the same. In this situation the run of the hip and valley rafters will not be at 45 degrees with the wall plates (*Figure 46–24*). This type of problem is solved using the Pythagorean Theorem.

Lengths of Unconventional Hip Members

Normal hips and valleys have a unit Run of 16.97 or 17 feet. If we abandon this thought, thinking instead of the hip as just another common rafter, a solution is possible. Its unit Run will be 12 inches and its run will look like feet.

In the previous figure, the run of common rafter in section A is 11 and section B is 13. This is from the given building width information of 22 and 26 feet. The run of the hip is the diagonal of these two runs. In this case the diagonal is the square root of (run A² + Run B²) = $\sqrt{(11^2 + 13^2)}$ = 17.03 feet (*Figure 46–25*).

The total rise of all three rafters is the same since the ridgeboards are to be at the same height.

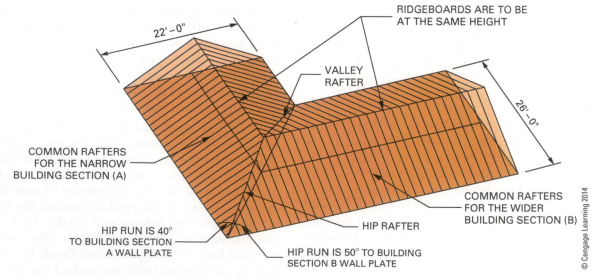

Figure 46–24 Hip and valley rafters of different width ell shaped buildings are calculated using the Pythagorean Theorem.

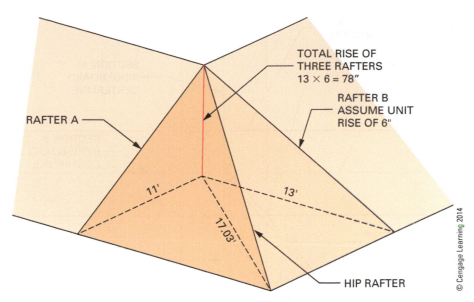

Figure 46–25 The run of the hip is the diagonal of a rectangle formed from the two rafter runs. The total rise of all rafters is the same.

To establish the total rise, one of rafter's unit Rise must be known. For example assume the unit Rise of rafter B is 6 inches. Then the total rise for all rafters equals rafter B run times rafter B unit Rise or $13 \times 6 = 78$ inches.

The lengths of the hip and common rafters may be determined by using the Pythagorean Theorem with total rise and run for each rafter. Remember to make the rise and run in the same units, such as inches. The unit rise or slope of the hip is needed to mark the hip plumb lines. This number is found by dividing the total rise by the hip run.

EXAMPLE

For example, what are the rafter lengths for the three rafters if

Run of Rafter A is 11 = 121″

Run of Rafter B is 13 = 169″

Unit Rise of Rafter B is 6″

Solution:

Run of Hip is $\sqrt{(\text{Run A}^2 + \text{Run B}^2)} =$
$\sqrt{(11^2 + 13^2)} = 17.03 = 204.35″$

Total rise is unit Rise B $\times$ Run B $= 6 \times 13 = 78″$

Length of Rafter A $= \sqrt{(\text{Run A}^2 + \text{Total Rise}^2)} =$
$\sqrt{(121^2 + 78^2)} = 143.96″ = 143\ ^{15}/_{16}″$

Length of Rafter B $= \sqrt{(\text{Run B}^2 + \text{Total Rise}^2)} =$
$\sqrt{(169^2 + 78^2)} = 186.13″ = 186\ ^{1}/_{8}″$

Length of Hip Rafter $= \sqrt{(\text{Run Hip}^2 + \text{Total Rise}^2)}$
$= \sqrt{(204.35^2 + 78^2)}$
$= 218.73″ = 218\ ^{3}/_{4}″$

Unit Rise of hip $=$ total Rise $\div$ hip run $=$
$78 \div 17.03 = 4.58″ = 4\ ^{9}/_{16}″$

Unconventional Side Cuts

Shortening and laying out side cuts for hips, valleys, and jacks are all done using the same principle. Layout begins with a plumb line at the ridge end. Secondly, the rafter is shorten by measuring perpendicular to the plumb line. Third, the side cut line is laid out by measuring perpendicular to the second plumb line. Normal shortening is $1\frac{1}{16}$ inches and normal side cut distances are ¾ inch. Shortening and side cuts on unconventional hips, jacks and valleys use a different numbers. These numbers are best determined from a full scale layout of the intersection of the ridgeboards, common and hip rafters. Once found they may be used for all unconventional members.

Begin by drawing a rectangle on a scrap of plywood with the dimensions of the building. Scale the feet to inches by using the twelfth scale of a framing square, marking inches to represent feet. For example 17 ½ feet would layout to 17 ½ inches.

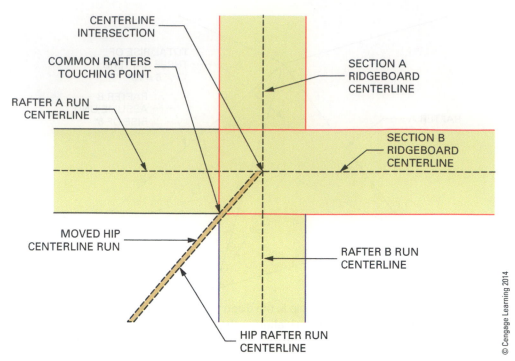

CENTERLINE INTERSECTION

COMMON RAFTERS TOUCHING POINT

RAFTER A RUN CENTERLINE

SECTION A RIDGEBOARD CENTERLINE

SECTION B RIDGEBOARD CENTERLINE

MOVED HIP CENTERLINE RUN

RAFTER B RUN CENTERLINE

HIP RAFTER RUN CENTERLINE

© Cengage Learning 2014

Figure 46–26 Drawing a scaled layout of the intersection of ridges and common rafters with centerlines makes it possible to determine side cut angles.

Only the runs of the building sections A and B need be drawn. Draw a diagonal from the corner to the intersection of the two rafter Runs. The sole purpose of this rectangle is to get an accurate angle at the intersection of the ridge and rafters centerlines. Notice, this diagonal is not at a 45 degree angle with the other sides of the rectangle.

Next is to draw full scale lines that will represent the edges of rafters and ridgeboards on either side of their centerlines. Center a scrap block of dimension lumber positioned on the center lines and mark both sides of the block. Mark the two ridges and two common rafters, leaving the hip for last because it will have to be moved slightly. These lines represent the actual thickness of the members.

Draw a line parallel to the hip centerline that passes through the corner where the two common rafters touch (*Figure 46–26*). Again center the

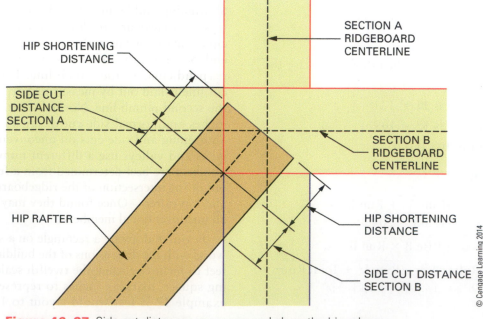

HIP SHORTENING DISTANCE

SIDE CUT DISTANCE SECTION A

SECTION A RIDGEBOARD CENTERLINE

SECTION B RIDGEBOARD CENTERLINE

HIP RAFTER

HIP SHORTENING DISTANCE

SIDE CUT DISTANCE SECTION B

© Cengage Learning 2014

Figure 46–27 Side cut distances are measured along the hip edges.

scrap block on the moved centerline and draw the sides of the hip rafter. Erase the first hip centerline to reduce confusion. Notice the hip centerline no longer passes through the intersection of the other centerlines (*Figure 46–27*). The drawing is now complete and may be used to measure the shortening and side cut distances. These distances will work to the layout line, i.e., the side cut lines for hips, valleys, and all jacks.

The shortening distance of the unconventional hip is measured from the end of the rafter to the touching point of the two common rafters. This is the perpendicular distance used in the layout shortening. Normal hip rafters are shortened 1¹⁄₁₆ inches, where this example would measure to be 1 inch. The side cut distances are normally ¾ inch, where in this example the A rafter would be ⅝ inch and the rafter B would be ⅞ inches (*Figure 46–28*).

Jack rafter side cut distances are the same as for the hip and valleys. Shortening distances are different than for the hips. These may be measured directly from the previous scaled drawing. The lines from the common rafter touching point to the intersections of the hip and common edges are the shortening distances. In the example used here the distances are 1 inch for the A section and 1³⁄₁₆ inches for the B Section jack rafters (*Figure 46–29*).

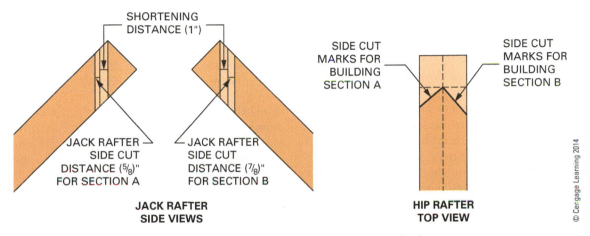

Figure 46–28 Measured side cut distances are laid out on the sides of rafters.

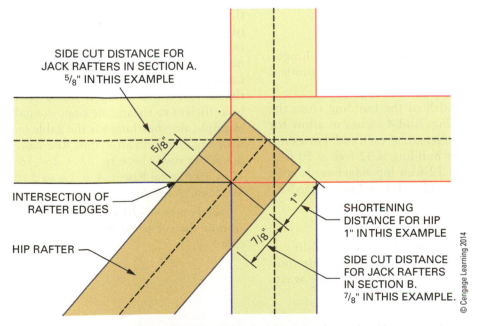

Figure 46–29 Shortening distances for jack rafters are determined by measuring on the layout jack rafters.

ESTIMATING MATERIALS FOR ROOF FRAMING

Estimating the amount of materials for roof framing involves rafters, ridgeboards, and sheathing. Remember that if the number of pieces does not come out even, round them up to the next whole number of pieces.

Rafters

There are many types of rafters in a hip roof. Estimating the material needed is easier than might be expected. The number of common rafters is almost the same whether for a gable roof or a hip roof.

Common Rafters for Hip or Valley Roofs.

The number of common rafters in a hip roof is similar to those in a gable roof. Divide the length of the building by the spacing of the rafters. Round up to the nearest whole number and add one, as a starter. Multiply the total by two to include the other side. Then add two rafters for each hip and valley rafter. This will allow material for the common and jack rafters. Two jack rafters may be cut from one board, one long and one short. For example, if a rectangular hip roofed building is 50 feet, the number of 16-inch OC gable rafters is $50 \div {}^{16}\!/_{12} + 1 = 38.5 \Rightarrow 39 \times 2 = 78 + 4 \times 2 = 86$ rafters.

Hip and Valley.

Count the number of hips and valleys. Remember that the width of the material used is not the same as the common rafter material.

Ridgeboards

The hip ridgeboard needs to be slightly longer, up to a possible ${}^{35}\!/_{8}$ inches than the theoretical calculation.

Subtract the width of the building from the length of the building. Add 4 inches to allow for possible additions in length at each end. For example, if a rectangular building is 32 feet $\times$ 50 feet, then the number of 12-foot hip ridgeboards is $50 - 32 = (18 + {}^{4}\!/_{12}) \div 12 = 1.5 \Rightarrow 2$ ridgeboards.

Sheathing

The amount of roof sheathing on a building is nearly the same whether it is a hip or gable roof. This is due to the fact that much of the waste pieces on a hip roof may be reused to do another roof section.

Sheathing Hip Roof.

Determine the sheathing as if it were a gable roof. Find the number of plywood rows by dividing the actual rafter length by 4 feet (width of plywood sheet). Round this number up to nearest one-half row. Next, find the number of sheets across the building by dividing the fascia board length by 8 feet (length of plywood sheet). Round this up to the nearest one-half sheet. Find the number of sheets by multiplying these numbers together and then doubling the result to account for the other side. Round up to the nearest whole number. Adding a waste factor of about 5 percent will cover the waste, which will be in triangular piece cuts. For example, sheathing for a roof with 20'-6" rafters on a building with a 52' fascia board will have $20.5 \div 4 = 5.1 \Rightarrow 5.5$ rows. It will have $52 \div 8 = 6.5$ sheets per row. Thus, we have $5.5 \times 6.5 \times 2$ sides $\times 1.05$ waster factor $= 75.05 \Rightarrow 76$ sheets.

Sheathing Intersecting Roof.

Calculate the number of panels for each building section as if the valleys and valley jacks were not present. Use the length of the each building section to the outside hip corner. The overlapped amount will compensate for waste.

Gable End Sheathing.

Calculate the total Rise of the roof by multiplying the unit Rise times the rafter Run, then divide by 12 to change the answer to feet. Find the number of rows of sheathing by dividing this number by 4 feet, and round up to the nearest one-half sheet. Next, find the number of sheets across the building by dividing the building width by 8 feet. Round this number up to the nearest one-half sheet. Multiply these numbers, number rows, and sheets together, and round up to the nearest whole number. Waste material should be sufficient to sheath the opposite gable end. For example, the sheathing for the gable ends of a 32'-wide building with an 8" unit Rise has $32' \div 8 = 4$ pieces per row. There are $32 \div 2 = 16$ Run $\times 8 \div 12 = 10.6'$ total Rise $\div 4 = 2.6 \Rightarrow 3$ rows. Sheathing, then, is $3 \times 4 = 12$ pieces.

See *Figure 46–30* a listing of the shorthand version of these estimating techniques with another example.

Estimate the materials for a roof of a rectangular 28′ × 42′ building. Framing 16″ OC, 24″ OC trusses, 9″ gable end overhang, and a rafter length of 17′-4″.

Item	Formula	Waste factor	Example
Commons for Hip Roof	common rafters + 2 × NUM hips		(70 − 4) + 2 (4) 74 PC
Hip Ridgeboard	building LEN − building WID + 4		42 − 28 + $^4/_{12}$ + 14.$\overline{333}$ FT
Hip Roof Sheathing	common rafter LEN ÷ 4 (rounded up to nearest ½) common ridge LEN ÷ 8 × 2 (rounded up to nearest ½) × waste	5%	(17.333) ÷ 4.33 ⇒ 4.5) × (43.5 ÷ 8 = 5.4375 ⇒ 5.5) * 1.05 = 25.8 ⇒ 26 PC

© Cengage Learning 2014

Figure 46–30 Example of estimating roof framing materials with formulas.

DECONSTRUCT THIS

Carefully study **Figure 46–31** and think about what is wrong and/or what is right. Consider all possibilities. What construction practice or method is different in your area of the country?

© Cengage Learning 2014

Figure 46–31 This photo shows the rafter farming of a gazebo.

KEY TEMS

backing the hip	hip jack rafter	jack rafter	shortened valley rafter
cheek cut	hip rafter	line length	supporting valley rafter
common rafter	hip roof	major span	valleys
cripple jack rafter	hip-valley cripple jack rafter	mansard roof	valley cripple jack rafter
dropping the hip		minor span	valley jack rafter
gambrel roof	intersecting roof	purlin knee wall	valley rafter

REVIEW QUESTIONS

Select the most appropriate answer.

1. The length of any rafter in a roof of specified slope may be found if its _____ is known.

   a. total run
   b. unit rise
   c. unit length
   d. all of the above

2. The rafter that spans from a valley rafter to a hip rafter is called

   a. valley jack.
   b. cripple jack.
   c. hip jack.
   d. hip or valley jack.

3. The line length of a hip rafter with a unit Rise of 6 inches and a total Run of 12 is

   a. 72 inches.
   b. 144 inches.
   c. 216 inches.
   d. 224 inches.

4. The jack rafter is most similar to a

   a. common rafter.
   b. hip rafter.
   c. valley rafter.
   d. gable end stud.

5. The total Run of any hip jack rafter is equal to its

   a. distance along the plate from the outside corner.
   b. distance along the plate from the inside corner.
   c. line length.
   d. common difference in length.

6. The total Run of the shortened valley rafter is equal to the total Run of the

   a. minor span.
   b. major span.

   c. common rafter of minor roof.
   d. hip rafter.

7. The amount that a dimension lumber hip rafter is shortened because of the ridge is

   a. ¾ inches.
   b. $1^{1}/_{16}$ inches.
   c. 1½ inches.
   d. none of the above.

8. The length of a ridgeboard for a hip roof installed on a rectangular building measuring 28 × 48 feet is slightly more than

   a. 20 feet.
   b. 28 feet.
   c. 48 feet.
   d. 76 feet.

9. The amount to add to the hip roof ridgeboard where there is not a common rafter butting the ridge is

   a. $1^{1}/_{16}$ inches
   b. 1 ½ inches
   c. 2 ⅛ inches
   d. 3 ⅝ inches

10. What is the hip projection, in inches, if the common rafter projection is 6 inches?

    a. 6 inches
    b. ⁶/₁₇ foot
    c. 8½ inches
    d. 17 inches minus 6 inches

11. Shed roof run of a building that is 16 × 24 built with 2 × 4 walls is most likely

    a. 7' – 8 ½"
    b. 8'
    c. 8' – 3 ½"
    d. 16'

12. What is the steepest rafter length of a gambrel rafter if its run is 7 feet and its total Rise is 8 feet?

    a. 9.45 feet
    b. 113 inches
    c. 10.63 feet
    d. 121.45 inches

13. What is the horizontal measurement of the first piece of gable end sheathing if the starting dimension is 16 ½ inches and the unit rise is 6 inches?

    a. 5.25 feet
    b. 31 ½ inches
    c. 2.75 feet
    d. 99 inches

14. Using the information in the previous problem, what is the vertical measurement of the first piece in the next course of gable end sheathing? Assume a normal sheet stagger of 4 feet.

    a. 2.75 feet
    b. 6.75 feet
    c. 16 ½ inches
    d. 40 ½ inches

15. For hip roof sheathing, what is the extra upper edge of sheet measurement (to the hip rafter) if the unit length is 15.62 inches?

    a. 11.12 inches
    b. 15.62 inches
    c. 36.88 inches
    d. 48 inches

16. What is the run of a knee wall if it is to be 52 inches tall? The unit Rise is 8 inches and the floor is 12 inches thick.

    a. 6.5 feet
    b. 8 feet
    c. 64 inches
    d. 96 inches

17. What numbers are needed to determine the total rise of narrow dormer rafters that are built to wide main roof rafters?

    a. run of main roof and unit Rise of main roof
    b. total Rise of main roof and dormer wall height
    c. floor thickness and difference in birds mouth heights.
    d. all of the above

18. What is unconventional hip length for an L-shaped building with a run of 10 on the narrower section and 12 on the wider section? The total Rise of the roof is 9 feet.

    a. 9 feet
    b. 15 feet
    c. 15.62 feet
    d. 18.03 feet

19. The estimated number of 16" OC common rafters for a hip roof of a rectangular building measuring 28 × 48 feet is

    a. 72.
    b. 74.
    c. 78.
    d. 82.

20. The estimated number of pieces of panel sheathing needed for a hip roof with a common rafter length of 14'–7" and a fascia board length of 34'–9" is

    a. 16.
    b. 18.
    c. 32.
    d. 36.

UNIT 17
Stair Framing

Chapter 47 Stairways and Stair Design
Chapter 48 Stair Layout and Construction

Staircases can be a showcase for carpenters to demonstrate their skill and talent. They are often intricate and ornate, requiring close cutting and fitting. Stairs must be carefully designed and laid out to ensure safe passage and ease of use. Also, many stair dimensions must comply with national and local codes. Building a set of stairs challenges the carpenter to work at his or her best while proving a place for great reward in pride of workmanship.

OBJECTIVES

After completing this unit, the student should be able to:

- describe several stairway designs.
- define terms used in stair framing.
- determine the rise and tread run of a stairway.
- determine the length of and frame a stairwell.
- lay out a stair carriage and frame a straight stairway.
- lay out and frame a stairway with a landing.
- lay out and frame a stairway with winders.
- lay out and frame service stairs.

CHAPTER 47

Stairways and Stair Design

A set of stairs or a staircase can be an outstanding feature of an entrance. A staircase provides beauty and grace to a room, generally affecting the character of the entire interior (*Figure 47–1*).

A set of stairs generally refers to one or more flights of steps leading from one level of a structure to another. *Staircase* is a term usually saved to refer to a finished set of stairs that has architectural appeal. Stairs are further defined as finish or service stairs. Finish stairs extend from one habitable level of a house to another. Service stairs extend from a habitable to a nonhabitable level, typically a basement.

Stairs are also governed closely by national and local building codes. Codes set limits on total stair height, step width, step rise, step depth, and the acceptable amount of variation between steps. These codes are designed to ensure a set of stairs will be safe for use by anyone.

Stairs, like rafters, are built by the most experienced carpenter on the job. Many design concepts and terms must be understood to successfully build a set of stairs. In addition to the fine carpentry of a staircase, consideration must be given to framing of the stairwell, or the opening in the floor through which a person must pass when climbing and going down the stairs (*Figure 47–2*). The stairwell is framed at the same time as the floor.

TYPES OF STAIRWAYS

A straight stairway is continuous from one floor to another. There are no turns or landings. Platform stairs have intermediate landings between floors. Platform-type stairs sometimes change direction at the landing. An L-type platform stairway changes direction 90 degrees. A U-type platform stairway changes direction 180 degrees.

Platform stairs are installed in buildings that have a high floor-to-floor level. They also provide a temporary resting place. They are a safety feature in case of falls. The landing is usually constructed at roughly the middle of the staircase.

Courtesy of L.J. Smith Stair Systems

Figure 47–1 Staircases are often the show piece of a buildings interior.

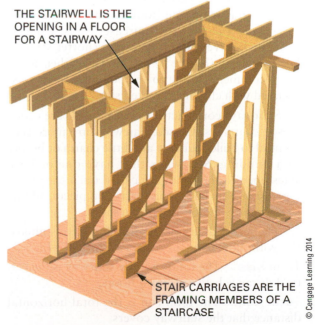

THE STAIRWELL IS THE OPENING IN A FLOOR FOR A STAIRWAY

STAIR CARRIAGES ARE THE FRAMING MEMBERS OF A STAIRCASE

© Cengage Learning 2014

Figure 47–2 Typical framing details for stairs.

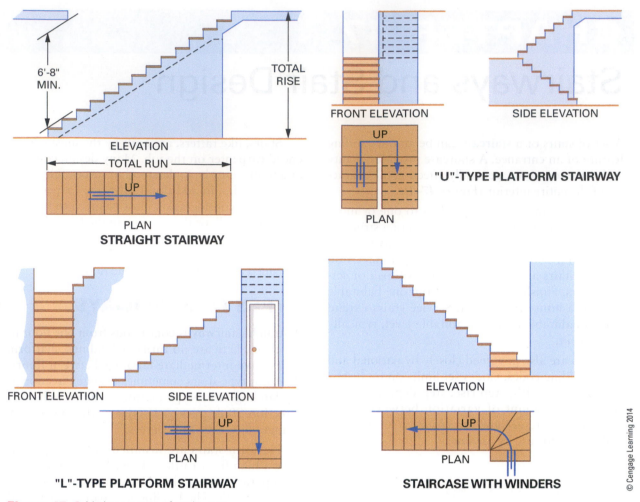

Figure 47–3 Various types of stairways.

A winding staircase gradually changes direction as it ascends from one floor to another. In many cases, only a part of the staircase winds. Winding stairs may solve the problem of a shorter straight horizontal run (*Figure 47–3*). However, their use is not recommended due to the potential danger associated with climbing tapered steps.

Stairways constructed between walls are called closed stairways. Closed stairways are more economical to build, but they add little charm or beauty to a building. Stairways that have one or both sides open to the room are called open stairways. Open stairways have more parts and pieces, adding to the charm and beauty of the stairs.

The terms used in stair framing have similarities with those used for rafters (*Figure 47–4*).

Total Rise. The total rise of a stairway is the vertical distance between finish floors.

Total Run. The total run is the total horizontal distance that the stairway covers.

Riser. The riser is the finish material used to cover the unit rise distance.

Unit Rise. The unit Rise is the vertical distance from one step to another.

Tread. The tread is the horizontal finish material used to make up the step on which the feet are placed when ascending or descending the stairs.

Unit Run. The unit Run is the horizontal distance between the faces of the risers.

Nosing. The nosing is that part of the tread that extends beyond the face of the riser. It is not part of the calculations for stairs, but rather an add-on to treads (*Figure 47–5*).

Stair Carriage. A stair carriage, sometimes called a stair horse, provides the main strength for the stairs. It is usually a nominal 2×10, 2×12, or 2×14 framing member cut to support the treads and risers.

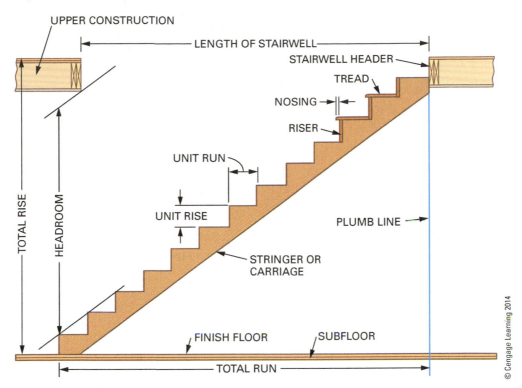

UPPER CONSTRUCTION

LENGTH OF STAIRWELL

STAIRWELL HEADER

TREAD

NOSING

RISER

UNIT RUN

UNIT RISE

PLUMB LINE

TOTAL RISE

HEADROOM

STRINGER OR CARRIAGE

FINISH FLOOR

SUBFLOOR

TOTAL RUN

© Cengage Learning 2014

Figure 47–4 The terms used in stair construction are similar to those used in roof framing.

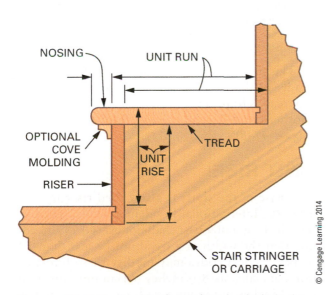

NOSING

UNIT RUN

OPTIONAL COVE MOLDING

TREAD

UNIT RISE

RISER

STAIR STRINGER OR CARRIAGE

© Cengage Learning 2014

Figure 47–5 The tread unit run does not include the nosing.

Stair Stringer. The stringer is the finish material applied to cover the stair carriage. It can also take the place of the stair carriage when side walls are used to support the stairs.

Stairwell. A stairwell is an opening in the floor for the stairway to pass through. It provides adequate headroom for persons using the stairs.

Headroom. The headroom of a set of stairs is the smallest vertical distance above the stairs from a line drawn from nosing to the upper construction.

STAIR DESIGN

Stairs in residential construction are at least 36 inches wide (**Figure 47–6**).✱ If more than fifty people occupy the building, the width must be 44 inches. The maximum height of a single flight of stairs is 12 feet, unless a platform is built in to break up the continuous run.✱ This platform must be as long as the stair is wide but need not be longer than 48 inches.✱

✱IRC R311.7.1

✱IRC R311.7.3

✱IRC R311.7.6

Staircases must be constructed at a proper angle for maximum ease in climbing and for safe descent. The relationship of the rise and run determines this angle (**Figure 47–7**). The preferred angle is between 30 and 38 degrees. Stairs with a slope of less than 20 degrees waste a lot of valuable space. Stairs with a slope that is excessively steep (50 degrees or over) are difficult to climb and dangerous to descend. Stair angles normally are not calculated, but rather are used only as a reference.

Every building presents a different situation for stair design. Each building will most likely have a

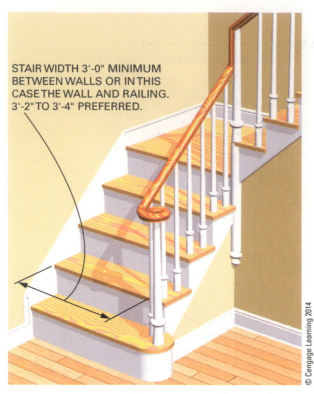

Figure 47–6 Stair width is measured from between the wall and railing.

STAIR WIDTH 3'-0" MINIMUM BETWEEN WALLS OR IN THIS CASE THE WALL AND RAILING. 3'-2" TO 3'-4" PREFERRED.

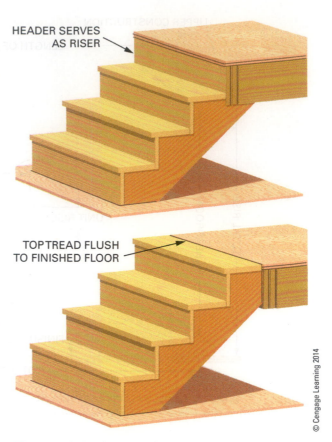

HEADER SERVES AS RISER

TOP TREAD FLUSH TO FINISHED FLOOR

Figure 47–8 Methods of building a staircase to the stairwell header.

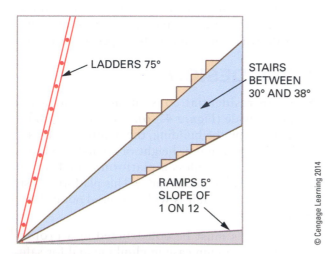

LADDERS 75°

STAIRS BETWEEN 30° AND 38°

RAMPS 5° SLOPE OF 1 ON 12

Figure 47–7 Appropriate angles for modes of egress from one level to another.

different total rise. To make matters more confusing, either the stairs may be designed to fit a stairwell, or the stairwell may be designed to fit the stairs. This makes building a set of stairs interesting with many design variations. The unit Rise and unit Run must be determined new for each building. The design process begins with determining the unit Rise.

This process assumes the stair is built where the header of the stairwell acts as the top riser. If the top tread is framed flush with the second floor, add another tread Run to the length of the stairwell (*Figure 47–8*).

UNIT RISE

Unit Rise of the stairs is also called riser height. The IRC, International Residential Code, specifies that the height of a riser shall not exceed $7\frac{3}{4}$ inches and that the width of a tread not be less than 10 inches. Local codes may modify these limits to numbers such as $8\frac{1}{4}$ inches maximum unit Rise and 9 inches minimum tread. Generally, smaller unit Rise and larger unit Run stairs are safer and easier to use.

✳ **IRC R311.7.5.1**

✳ **IRC R311.7.5.2**

To find the unit Rise first measure the total Rise of stairs. The measurement should be accurate and measured from the second floor above to the floor where the stairs will land. Next, the total Rise is divided by the maximum unit Rise allowed by local building codes. The IRC maximum is $7\frac{3}{4}$ inches. This is done to see how many risers will be needed. Rarely will this division work out to be a whole

✳ **IRC R311.7.5.1**

number. Round this number up to give the next whole number of risers that will fit in the stairs. This will make the unit Rise slightly smaller than the maximum unit rise. Now divide the total Rise by the whole number of risers to calculate the actual unit Rise.

EXAMPLE

A total rise of 9'-3$\frac{1}{4}$" (111$\frac{1}{4}$") is measured. Find the largest unit Rise for this stair that meets the IRC.

- Divide total Rise by maximum unit rise, 7$\frac{3}{4}$". 111$\frac{1}{4}$" ÷ 7.75 = 14.354 risers.
- Round up to the next whole number. 14.354 rounds up to 15 risers.
- Divide total Rise by number of risers. 111$\frac{1}{4}$" ÷ 15 = 7.417" or 7$\frac{7}{16}$ inches actual unit Rise.

The unit Rise just calculated is only one of several unit Rises that could be calculated. Adding more risers will give one or two more additional stair designs. Increasing the number of risers to sixteen, the unit Rise is 111.25 ÷ 16 = 6.953", which equals about 6$\frac{15}{16}$ inches. Increasing the number of risers to seventeen yields a unit Rise of 111.25 ÷ 14 = 6.544", which is about 6$\frac{9}{16}$ inches (*Figure 47-9*). Each design will produce an appropriate set of stairs with a unit Run.

UNIT RUN

The unit Run of a stair is also referred to as a tread. The unit Run may be determined in several ways. Simple designs have no headroom or total run constraints. Some stairwell openings are constructed to fit the stairs. Some stairs are built to fit a stairwell opening. For each situation, the unit Run is determined after the unit Rise is calculated.

Design	Total Rise	Number of Risers		Unit Rise
1	111$\frac{1}{4}$"	15	7.41"	7$\frac{7}{16}$"
2	111$\frac{1}{4}$"	16	6.92"	6$\frac{15}{16}$"
3	111$\frac{1}{4}$"	17	6.54"	6$\frac{9}{16}$"

© Cengage Learning 2014

Figure 47-9 Possible riser heights for a total rise of 111¼ inches.

Simple Unit Run Design

Simple designs may be used when the headroom is not close enough to cause concern. This allows the unit Run of the stairs to be any reasonable number. Increasing the unit Run decreases the steepness of the stairs. This makes the stairs easier to climb. It also uses up more space in the building. Decreasing the unit run increases the steepness. This makes the stairs more difficult to climb but uses less space.

There are two formulas that may be used to assist in making the selection of unit run. Each will produce an appropriate and safe stair angle.

17-18 Method. First, the sum of one riser and one tread should equal between 17 and 18. For example, if the unit rise is 7$\frac{1}{2}$ inches, then the minimum unit Run may be 17 inches minus 7$\frac{1}{2}$ inches. This equals 9$\frac{1}{2}$ inches. The maximum tread width may be 18 inches minus 7$\frac{1}{2}$ inches. This equals 10$\frac{1}{2}$ inches. Thus any tread width between 9$\frac{1}{2}$ and 10$\frac{1}{2}$ inches will produce an acceptable stair design.

24-25 Method. Another method for determining the tread width found in many building codes is the sum of two risers and one tread shall not be less than 24 inches nor more than 25 inches. For example, a unit Rise of 7$\frac{1}{2}$ inches calls for a minimum unit Run of 9 inches and a maximum of 10 inches (*Figure 47-10*).

These numbers are further restricted by the local building code in effect. The IRC requires a minimum run of 10 inches. Thus, the run to meet the IRC should be 10 wide. ✱

✱IRC R311.7.5.2

Unit Run to Find Stairwell Length

Framing a stairwell to fit a staircase is done when a particular stair design has been decided. The unit Rise and unit Run are determined as described earlier. Then the minimum stairwell length may be determined.

The length of the stairwell depends on the slope of the staircase. Stairs with a low angle require a longer stairwell to provide adequate headroom (*Figure 47-11*). Building codes require a minimum of 6'-8" for headroom. However, this is a minimum; more headroom may be preferred. ✱

✱IRC R311.7.2

To find the minimum length of the stairwell, add the desired headroom to the thickness of the upper-floor construction. The upper-floor construction includes subfloor and floor joists of the second floor as well as the first-floor ceiling thickness. Divide this sum by the unit Rise. The result is the number of risers in the stairwell that must clear the headroom. This number is rarely a whole

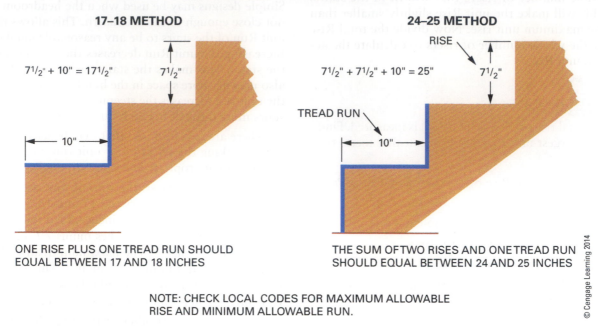

TWO METHODS TO FIND UNIT RUN FOR STAIRS
AFTER UNIT RISE HAS BEEN DETERMINED

17–18 METHOD

7¹/₂" + 10" = 17¹/₂"

7¹/₂"

10"

ONE RISE PLUS ONE TREAD RUN SHOULD
EQUAL BETWEEN 17 AND 18 INCHES

24–25 METHOD

RISE

7¹/₂" + 7¹/₂" + 10" = 25"

7¹/₂"

TREAD RUN

10"

THE SUM OF TWO RISES AND ONE TREAD RUN
SHOULD EQUAL BETWEEN 24 AND 25 INCHES

NOTE: CHECK LOCAL CODES FOR MAXIMUM ALLOWABLE
RISE AND MINIMUM ALLOWABLE RUN.

© Cengage Learning 2014

Figure 47–10 Two techniques for determining the unit run from a desired unit rise when the headroom is not a concern.

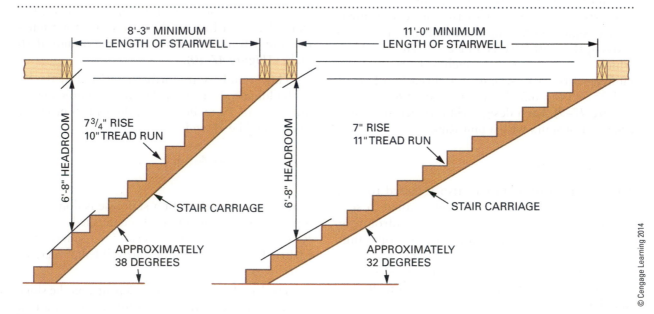

8'-3" MINIMUM
LENGTH OF STAIRWELL

6'-8" HEADROOM

7³/₄" RISE
10" TREAD RUN

STAIR CARRIAGE

APPROXIMATELY
38 DEGREES

11'-0" MINIMUM
LENGTH OF STAIRWELL

6'-8" HEADROOM

7" RISE
11" TREAD RUN

STAIR CARRIAGE

APPROXIMATELY
32 DEGREES

© Cengage Learning 2014

Figure 47–11 Angles formed by stairs and the need for headroom affect the length of the stairwell openings.

number, and that is okay. Multiply this number by the unit Run to find the length of the stairwell (*Procedure 47–A*).

Unit Run for Fixed-Length Stairwells

Often a staircase is built to fit a stairwell opening already constructed. The unit Run is calculated to fit the stairwell headroom constraints. This process is repeated several times with one or two fewer risers used in the calculations. This allows the carpenter to select the best stair design to fit the opening. Each design will fit into the same space (*Figure 47–12*).

The unit Rise is first calculated as described in Figure 47–7. This number serves as a starting point in the design. The unit Run that goes with the unit Rise is calculated from the headroom and stairwell length.

StepbyStep Procedures

Procedure 47–A Determining Stair Riser Height Using the Step-Off Method

FINDING STAIRWELL LENGTH USING UNIT RUN

NOTE:

IF THE TOP TREAD IS BUILT FLUSH WITH THE SECOND FLOOR THEN ADD ANOTHER RISER TO THE ANSWER IN STEP 2.

STAIRWELL LENGTH

UPPER CONSTRUCTION

TOTAL HEADROOM

DESIRED HEADROOM

EXAMPLE:

UNIT RISE IS 7.417" (7 $7/_{16}$" FROM PREVIOUS EXAMPLE)
UNIT RUN IS 10
DESIRED HEADROOM IS 80"
UPPER CONSTRUCTION IS 12 $1/_4$"

STEP 1 ADD THE DESIRED HEADROOM TO THE UPPER CONSTRUCTION THICKNESS. THIS BECOMES THE TOTAL HEADROOM.

12 $1/_4$ + 80 = 92 $1/_4$" TOTAL HEADROOM

STEP 2 DIVIDE TOTAL HEADROOM BY THE UNIT RISE. THIS GIVES TOTAL RISERS IN THE STAIRWELL.

92 $1/_4$ ÷ 7.417 = 12.44 RISERS

STEP 3 MULTIPLY THE NUMBER OF RISERS BY THE UNIT RUN. THIS GIVES STAIRWELL LENGTH.

12.44 x 10 = 124.4 = 124 $3/_8$" STAIRWELL LENGTH

STAIRWELL LENGTH SHOULD BE MORE THAN 124 $3/_8$".

© Cengage Learning 2014

TWO SETS OF STAIRS DESIGNED TO FIT THE SAME SPACE

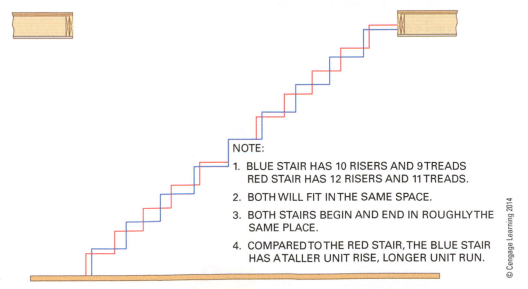

NOTE:

1. BLUE STAIR HAS 10 RISERS AND 9 TREADS RED STAIR HAS 12 RISERS AND 11 TREADS.

2. BOTH WILL FIT IN THE SAME SPACE.

3. BOTH STAIRS BEGIN AND END IN ROUGHLY THE SAME PLACE.

4. COMPARED TO THE RED STAIR, THE BLUE STAIR HAS A TALLER UNIT RISE, LONGER UNIT RUN.

© Cengage Learning 2014

Figure 47–12 Several stair designs can fit into the same space inside a stairwell opening.

To find the unit Run, add the desired headroom to the thickness of the upper-floor construction. The upper-floor construction includes subfloor and floor joists of the second floor as well as the first-floor ceiling thickness. Divide this number by the unit Rise. The result is the number of risers in the stairwell that must clear the headroom. This number is rarely a whole number and that is okay. Measure the stairwell length and divide it by the number of rises in the stairwell. The result is the unit Run (*Procedure 47–B*).

Now that a stair design has been calculated, additional unit Rise and unit Run calculations may be made to see if there is another design that might be better. Then the designs are compared to the local building code requirements to verify compliance (*Figure 47–13*). In this case, there is only one design that fits the IRC, design #1. Some local codes will allow all three. The decision here as to which design to use hinges on personal preferences.

STAIR DESIGN EXAMPLE

Calculate a set of three designs of unit Rise and unit Run for a stair with a total Rise of $102\frac{1}{2}$ inches and a stairwell length of 126 inches. The local code is to have limits of 8 inches unit Rise and 10 inches unit Run. The desired headroom is 82 inches and the upper construction is $10\frac{1}{4}$ inches. Assume the upper construction serves as a riser.

Number of risers

- $102.5 \div 8 = 12.8125 \rightarrow 13$ risers

Unit Rise

- $102.5 \div 13 = 7.885 \rightarrow 7\frac{7}{8}''$

Unit Run

- 17–18 method $17 - 7.885 = 9.12 \rightarrow 9\frac{1}{8}''$ min and $10\frac{1}{8}''$ max
- 24–25 method $24 - 2(7.885) = 8.23 \rightarrow 8\frac{1}{4}''$ min and $9\frac{1}{4}''$ max

Unit Run for fixed stairwell lengths
Total headroom

- $82 + 10.25 = 92.25''$ inches

Number of risers and treads in stairwell

- $92.25 \div 7.885 = 11.70$

Unit Run

- $126 \div 11.70 = 10.77 \rightarrow 10\frac{3}{4}''$

StepbyStep Procedures

Procedure 47–B Determining Stairwell Length

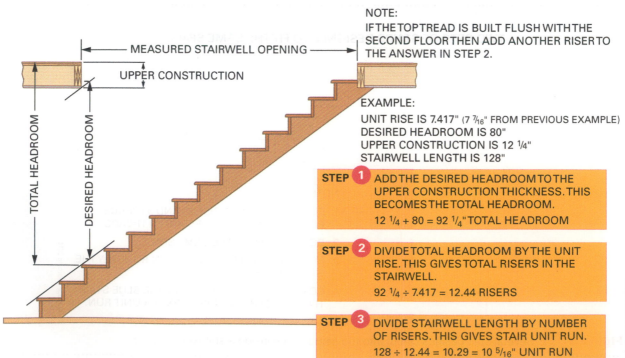

FINDING UNIT RUN FOR FIXED STAIRWELL LENGTH

MEASURED STAIRWELL OPENING

UPPER CONSTRUCTION

TOTAL HEADROOM

DESIRED HEADROOM

NOTE:
IF THE TOP TREAD IS BUILT FLUSH WITH THE SECOND FLOOR THEN ADD ANOTHER RISER TO THE ANSWER IN STEP 2.

EXAMPLE:
UNIT RISE IS 7.417" (7 $\frac{7}{16}$" FROM PREVIOUS EXAMPLE)
DESIRED HEADROOM IS 80"
UPPER CONSTRUCTION IS 12 $\frac{1}{4}$"
STAIRWELL LENGTH IS 128"

STEP 1 ADD THE DESIRED HEADROOM TO THE UPPER CONSTRUCTION THICKNESS. THIS BECOMES THE TOTAL HEADROOM.
12 $\frac{1}{4}$ + 80 = 92 $\frac{1}{4}$" TOTAL HEADROOM

STEP 2 DIVIDE TOTAL HEADROOM BY THE UNIT RISE. THIS GIVES TOTAL RISERS IN THE STAIRWELL.
92 $\frac{1}{4}$ ÷ 7.417 = 12.44 RISERS

STEP 3 DIVIDE STAIRWELL LENGTH BY NUMBER OF RISERS. THIS GIVES STAIR UNIT RUN.
128 ÷ 12.44 = 10.29 = 10 $\frac{5}{16}$" UNIT RUN

© Cengage Learning 2014

Repeating the procedure for two more unit Rises will yield the numbers in *Figure 47–14*. The process reveals there are two designs that will fit the code requirements.

A stairwell is framed in the same manner as any large floor opening, as discussed in Unit 11, Floor Framing. Several methods of framing stairwells are illustrated in *Figure 47–15*.✱

✱IRC R502.10

Total Rise	Number of Risers	Unit rise		Unit Run for Open Total Run Stairwells				Unit Run for Fixed Stairwell Length			
				17–18 Method		24–25 Method		Stairwell Length	Total Headroom	# of Risers in Stairwell	Unit Run
				Min	Max	Min	Max				
111.25	15	7.417	7 7/16	9.58	10.58	9.17	10.17	128	92.25	12.44	10.29
111.25	16	6.953	6 15/16	10.05	11.05	10.09	11.09	128	92.25	13.27	9.65
111.25	17	6.544	6 9/16	10.46	11.46	10.91	11.91	128	92.25	14.10	9.08

Note: In the example above, Unit Run Fixed Stairwell method above uses stairwell length of 128", the headroom of 80" and the upper construction 12¼.

Figure 47–13 Completed list of calculations for three sets of stairs that will fit into a particular stairwell.

Total Rise	Number of Risers	Unit rise		Unit Run for Open Total Run Stairwells				Unit Run for Fixed Stairwell Length			
				17–18 Method		24–25 Method		Stairwell Length	Total Headroom	# of Risers in Stairwell	Unit Run
				Min	Max	Min	Max				
102.5	13	7.885	7 7/8	9.12	10.12	8.23	9.23	126	92.25	11.70	10.77
102.5	14	7.321	7 5/16	9.68	10.68	9.36	10.36	126	92.25	12.60	10.00
102.5	15	6.833	6 13/16	10.17	11.17	10.33	11.33	126	92.25	13.50	9.33

Figure 47–14 Completed list of calculations for three sets of stairs that will fit into a particular stairwell.

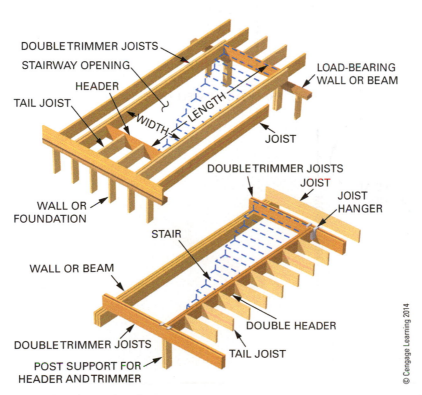

Figure 47–15 Methods of framing stairwells.

Variations in Width and Upper Construction

The width of the stairwell depends on the width of the staircase. The drawings show the finish width of the staircase. However, the stairwell must be made wider than the staircase to allow for wall and stair finish (*Figure 47–16*). Extra width will be required for a handrail and other finish parts of an open staircase that makes a U-turn on the landing above. The carpenter must be able to determine the width of the stairwell by studying the prints for size, type, and placement of the stair finish before framing the stairs.

The upper construction may be framed on a slope that matches the stairs. This creates a look that has two separate effects. The second floor may be extended into the opening to create more floor space on the second floor. Or the header may be backed up to improve the headroom in the stairwell. The framing is done in the same manner in either case (*Figure 47–17*).

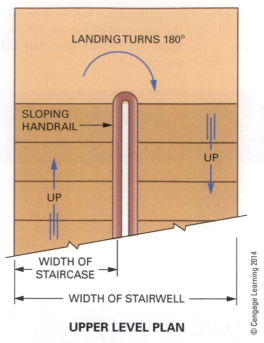

UPPER LEVEL PLAN

Figure 47–16 The stairwell must be wide enough to make room for the entire staircase.

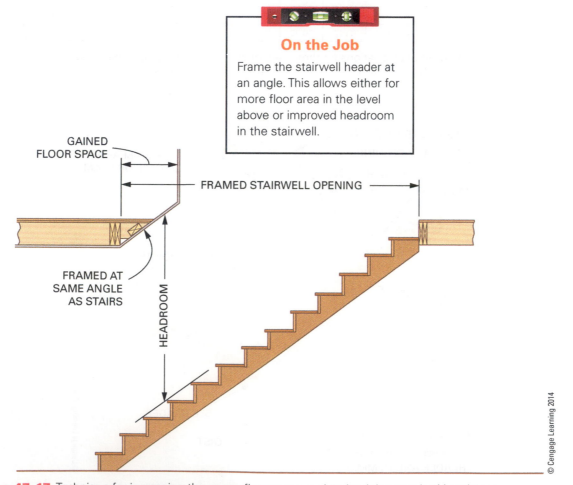

On the Job

Frame the stairwell header at an angle. This allows either for more floor area in the level above or improved headroom in the stairwell.

Figure 47–17 Technique for increasing the upper floor space and maintaining required headroom.

CHAPTER 48

Stair Layout and Construction

All stairs are laid out in approximately the same way. The use, location, and cost of stairs determine the way they are built. Regardless of the kind of stairs, where they are located, or how much they cost, care should be taken in their layout and construction.

METHODS OF STAIR CONSTRUCTION

There are two principal methods of stair construction. The housed stringer is laid out and cut to be a finished product. It is often fabricated off site in a shop and installed near the conclusion of the construction process. The job-built staircase uses stair carriages that are usually built on site. These are finished later as the construction process comes to an end.

Housed Stringer Staircase

For the housed stringer staircase, the framing crew frames only the stairwell. The staircase is installed when the house is ready for finishing. Dadoes are routed into the sides of the finish stringer. They *house* (make a place for) and support the risers and treads (*Figure 48–1*).

Occasionally, the finish carpenter builds a housed stringer staircase on the job site. A router and jig are used to dado the stringers. Then, the treads, risers, and other stair parts are cut to size. The staircase is then assembled either in place or as a unit and then installed into place. Stair carriages are not required when the housed stringer method of construction is used. Building of housed stringer stairs is described in more detail in Chapter 81.

Job-Built Staircase

The *job-built staircase* uses stair carriages that are installed when the structure is being framed. The carriage is laid out and cut with risers and treads fastened to the cutouts (*Figure 48–2*). This style uses temporary rough treads installed for easy access to upper levels during construction. Later the carriage is fitted with finish treads and risers with other stair trim.

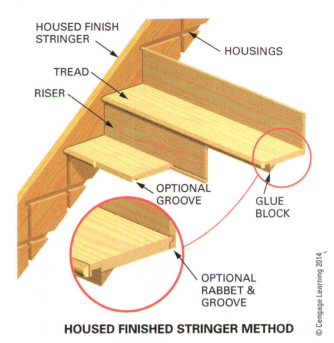

HOUSED FINISHED STRINGER METHOD

Figure 48–1 A housed finish stringer has dadoed sides to accept treads and risers.

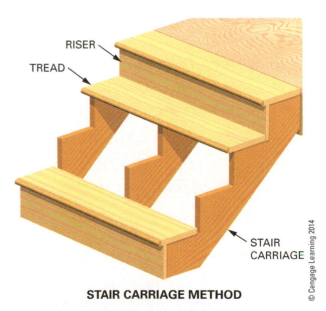

STAIR CARRIAGE METHOD

Figure 48–2 Stair carriages have notches cut to support treads and risers.

STAIR CARRIAGE LAYOUT

It is dangerous while using a flight of stairs to experience an unexpected variation in stair dimensions. A variation in riser height could cause someone to trip while ascending or fall while descending. Changes in tread width, narrower or wider, change the rhythm and pattern of a person's gait. This makes using the stairs more difficult. When laying out stairs, make sure that all riser heights are equal and all tread widths are equal.

This problem is addressed by building codes, which require all dimensions in stair layout to be accurate. The height from one riser to the next must be within $\frac{3}{16}$ inch. Total variation between the largest and smallest riser and largest and smallest tread is **✱IRC R311.7.5.1** not to exceed $\frac{3}{8}$ inch.✱ Note that a carpenter can **✱IRC R311.7.5.2** easily construct a set of stairs where measurements are within $\frac{1}{16}$ of an inch.✱

Scaling Rough Carriage Length

The length of lumber needed for the stair carriage is often determined using the Pythagorean theorem. It also can be found by scaling across the framing square. Use the edge of the square that is graduated in twelfths of an inch. Mark the total rise on the tongue. Then mark the total run on the blade. Scale off in between the marks (*Figure 48–3*).

 EXAMPLE

A stairway has a total rise of 8′-9″ and a total run of 12′-3″. What is the length of material needed to build the carriage?

Pythagorean Theorem

- Change dimensions to decimals.

$$8'\text{-}9'' = 8.75'; 12'\text{-}3'' = 12.25'$$

- Substitute into $a^2 + b^2$ and c^2 and solve.

$$8.75^2 + 12.25^2 = c^2 = 226.625$$

$$c = \sqrt{226.625} = 15.25'$$

- Round up to nearest even number.

$$15.05 \rightarrow 16 \text{ feet}$$

Scaling

Locate 8-$\frac{9}{12}$ths and 12-$\frac{3}{12}$ths on framing square. Measuring across the square between these dimensions results in a reading of a little over 15. At a scale of 1 inch = 1 foot, board length ordered should be 16 feet.

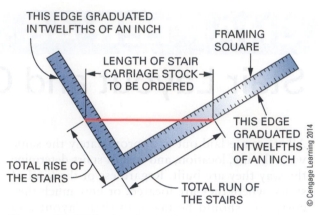

Figure 48–3 The framing square may be used to estimate the rough length of a stair carriage.

Stepping Off the Stair Carriage

Select the straightest material available for the stair carriage and place it on a pair of sawhorses. Sight the stock for a crowned edge. Set gauges on the framing square with the Rise on the tongue and the tread Run on the blade. Lines laid out along the tongue will be plumb lines. Those laid out along the blade will be level lines when the stair carriage is in its final position.

Using the framing square gauges against the top edge of the carriage, step off on the carriage the necessary number of times. Mark both Rise and Run along the outside of the tongue and blade. Lay out enough level and plumb lines to include the top tread and the lower finished floor line (*Figure 48–4*). The layout lines represent the bottom edge of the tread and back side of the riser.

Equalizing the Bottom Riser

The bottom of the carriage must be adjusted to allow for the finished floor. This amount depends on how the finish material is applied. This will make the bottom riser equal in height to all of the other risers when the staircase is finished. This process is known as dropping the stair carriage.

If the carriage rests on the finish floor, then the first riser is cut shorter by the thickness of the tread stock (*Figure 48–5*). If the carriage rests on the subfloor the riser height must be adjusted using the tread and finished floor thicknesses. To achieve the first riser height dimension, take the riser height minus the tread thickness plus finished floor thickness.

Sometimes the finish floor and tread stock are the same thickness. In this special case, nothing is cut off the bottom end of the stair carriage.

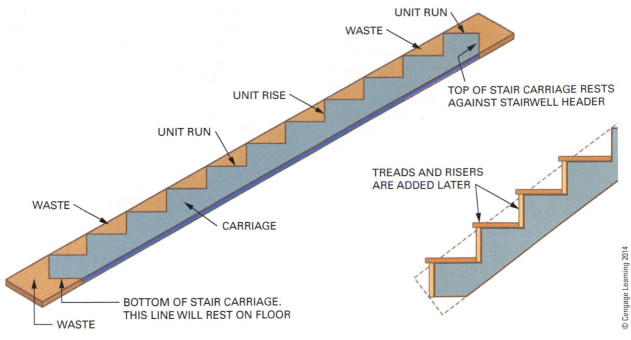

UNIT RUN

WASTE

UNIT RISE

UNIT RUN

WASTE

CARRIAGE

TOP OF STAIR CARRIAGE RESTS
AGAINST STAIRWELL HEADER

TREADS AND RISERS
ARE ADDED LATER

BOTTOM OF STAIR CARRIAGE.
THIS LINE WILL REST ON FLOOR

WASTE

Figure 48–4 A completed stair carriage layout.

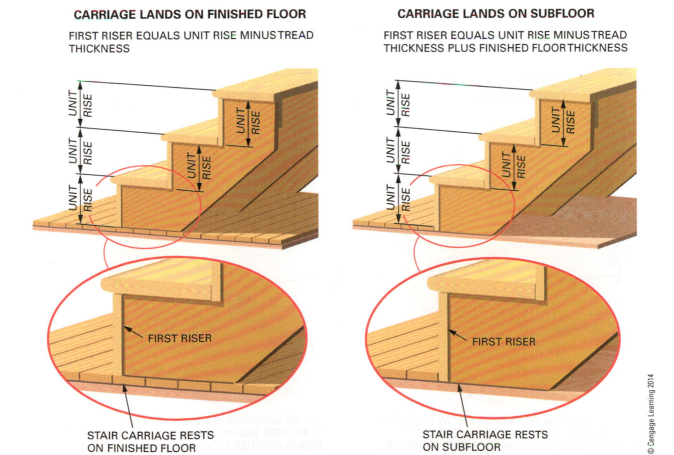

CARRIAGE LANDS ON FINISHED FLOOR

FIRST RISER EQUALS UNIT RISE MINUS TREAD
THICKNESS

UNIT RISE

UNIT RISE

UNIT RISE

UNIT RISE

UNIT RISE

FIRST RISER

STAIR CARRIAGE RESTS
ON FINISHED FLOOR

CARRIAGE LANDS ON SUBFLOOR

FIRST RISER EQUALS UNIT RISE MINUS TREAD
THICKNESS PLUS FINISHED FLOOR THICKNESS

UNIT RISE

UNIT RISE

UNIT RISE

UNIT RISE

UNIT RISE

FIRST RISER

STAIR CARRIAGE RESTS
ON SUBFLOOR

Figure 48–5 Adjusting the stair carriage bottom to equalize the first riser.

© Cengage Learning 2014

Equalizing the Top Riser

The top of the carriage must be properly located against the upper construction to maintain the unit Rise. This location depends on the style of the stairs. When the header of the upper construction acts as the top riser, the top of the carriage is roughly one riser height below the subfloor (*Figure 48–6*). This distance is equal to riser height plus tread thickness minus the finished floor thickness.

If the top tread is flush with the finished floor, the carriage elevation is only slightly below the subfloor. This distance is the tread thickness minus the finished floor thickness. If the finished floor thickness is larger than the tread, a negative number will result. In this case, raise the carriage this calculated distance above the subfloor.

Attaching the Stair Carriage to Upper Construction

When the upper construction is part of the top riser, most of the carriage is below the header. This offers poor support and room for fastening. Two methods of securing the carriage are shown in *Figure 48–7*. One method uses an extra header under the existing one. The carriage can then be attached in a variety of ways. The second method uses a wider riser on the top. It is long enough to fasten into the header and the carriage. It should be made with plywood for strength.

If no finished riser is desired, the carriage must be adjusted. With no riser board, the plumb line needs to be cut back a distance equal to one riser's thickness. This will allow the back edge of the top tread to rest against the header.

Cutting the Stair Carriages

After the first carriage is laid out, cut it. Follow the layout lines carefully because this will be a pattern for others.

> ### CAUTION
> When making a cut at a sharp angle to the edge, the guard of the saw may not retract. Retract the guard by hand until the cut is made a few inches into the stock. Then release the guard and continue the cut. Never wedge the guard in an open position.

Finish the cuts at the intersection of the riser and tread run with a handsaw. Use the first carriage as a pattern. Lay out and cut as many other carriages as needed. Three carriages are often used for

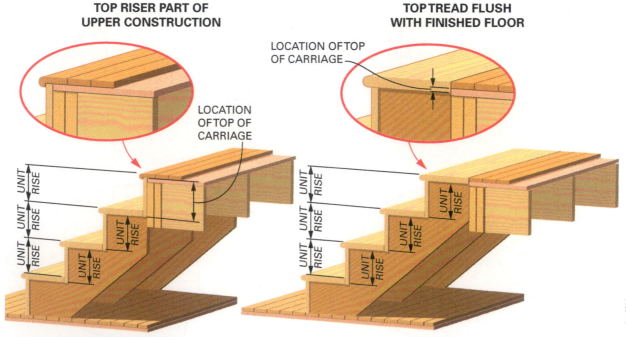

TOP RISER PART OF UPPER CONSTRUCTION

LOCATION OF TOP OF CARRIAGE

TOP TREAD FLUSH WITH FINISHED FLOOR

LOCATION OF TOP OF CARRIAGE

TOP OF CARRIAGE IS LOCATED BELOW SUBFLOOR. DISTANCE EQUALS UNIT RISE PLUS TREAD THICKNESS MINUS FINISHED FLOOR THICKNESS.

TOP OF CARRIAGE IS LOCATED BELOW SUBFLOOR. DISTANCE EQUALS TREAD THICKNESS MINUS FINISHED FLOOR THICKNESS.

© Cengage Learning 2014

Figure 48–6 Locating the top of the carriage on the header.

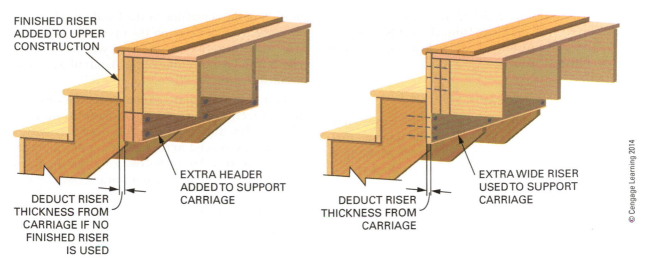

FINISHED RISER
ADDED TO UPPER
CONSTRUCTION

DEDUCT RISER
THICKNESS FROM
CARRIAGE IF NO
FINISHED RISER
IS USED

EXTRA HEADER
ADDED TO SUPPORT
CARRIAGE

DEDUCT RISER
THICKNESS FROM
CARRIAGE

EXTRA WIDE RISER
USED TO SUPPORT
CARRIAGE

© Cengage Learning 2014

Figure 48–7 Methods of framing the stair carriage to the stairwell header.

residential staircases of average width. For wider stairs, the number of carriages depends on such factors as whether or not risers are used and the thickness of the tread stock. Check the drawings or building code for the spacing of carriages for wider staircases.

FRAMING A STRAIGHT STAIRWAY

If the stairway is either completely closed or closed on one side, the walls must be prepared before the stair carriages are fastened in position.

Preparing the Walls of the Staircase

Gypsum board (drywall) is sometimes applied to walls before the stair carriages are installed against them. This procedure saves time. It eliminates the need to cut the drywall around the cutouts of the stair carriage. This method also requires no blocking between the studs to fasten the ends of the drywall panels.

However, blocking between the studs in back of the stair carriage provides backing for fastening the stair trim. Lack of it may cause difficulty for those who apply the finish.

If the drywall is to be applied after the stairs are framed, blocking is required between studs in back of the stair carriage to fasten the ends of the gypsum board. Snap a chalk line along the wall sloped at the same angle as the stairs. Be sure the top of the blocking is sufficiently above the stair

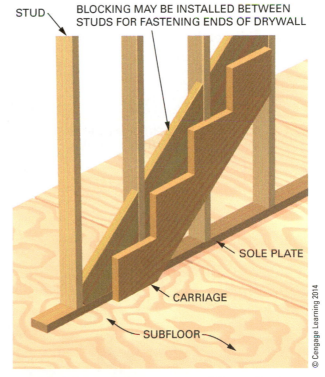

STUD

BLOCKING MAY BE INSTALLED BETWEEN
STUDS FOR FASTENING ENDS OF DRYWALL

SOLE PLATE

CARRIAGE

SUBFLOOR

© Cengage Learning 2014

Figure 48–8 Preparation of the wall for application of drywall before stair carriages are installed.

carriage to be useful. Install 2 × 6 or 2 × 8 blocking, on edge, between and flush with the edges of the studs. Their top edge should be to the chalk line (*Figure 48–8*).

Housed staircases may be installed after the walls are finished. Sometimes they are installed before the walls are finished. In such a case, they must be furred out away from the studs. This allows the wall finish to extend below the top edge of the finished stringer.

Installing the Stair Carriage

When installing the stair carriage, fasten the first carriage in position on one side of the stairway. Attach it at the top to the stairwell header. Make sure the distance from the second floor subfloor to the top tread Run is correct (see *Figure 48–6*).

Fasten the bottom end of the stair carriage to the sole plate of the wall and with intermediate fastenings into the studs. Drive nails near the bottom edge of the carriage, away from cutouts. This prevents splitting of the triangular sections.

Fasten a second carriage on the other wall in the same manner as the first. If the stairway is to be open on one side, fasten the carriage at the top and bottom of the staircase. The location of the stair carriage on the open end of a stairway is in relation to the position of the handrail. First, determine the

location of the centerline of the handrail. Then, position the stair carriage on the open side of a staircase. Make sure its outside face will be in a line plumb with the centerline of the handrail when it is installed (*Figure 48–9*).

Fasten intermediate carriages at the top into the stairwell header and at the bottom into the subfloor. Test the tread run and riser cuts with a straightedge placed across the outside carriages (*Figure 48–10*). About halfway up the flight, or where necessary, fasten a temporary riser board. This straightens and maintains the spacing of the carriages (*Figure 48–11*).

If a wall is to be framed under the stair carriage at the open side, fasten a bottom plate to the subfloor plumb with the outside face of the carriage. Lay out the studs on the plate. Cut and install studs under the carriage in a manner similar to that used to install gable studs. Be careful to keep the carriage straight. Do not crown it up in the center (*Figure 48–12*). Install rough lumber treads on the carriages until the stairway is ready for the finish treads.

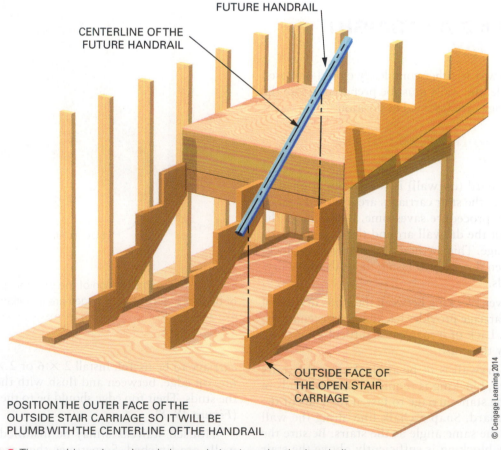

FUTURE HANDRAIL

CENTERLINE OF THE
FUTURE HANDRAIL

OUTSIDE FACE OF
THE OPEN STAIR
CARRIAGE

POSITION THE OUTER FACE OF THE
OUTSIDE STAIR CARRIAGE SO IT WILL BE
PLUMB WITH THE CENTERLINE OF THE HANDRAIL

© Cengage Learning 2014

Figure 48–9 The outside stair carriage is located plumb under the handrail.

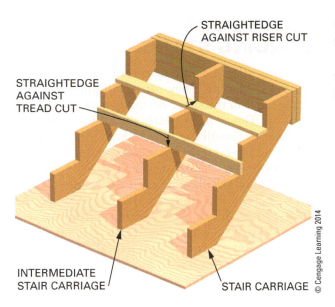

Figure 48–10 The alignment of tread and riser cuts on carriages must be checked with a straightedge.

STAIRWAY LANDINGS

A stair landing is an intermediate platform between two flights of stairs. A landing is designed for changing the direction of the stairs and as a resting place for long stair runs. The landing usually is floored with the same materials as the main floors of the structure. Some codes require that the minimum length of a landing be not less than 2'-6".

Other codes require the minimum dimension to be the width of the stairway.* 　　　　　*IRC R311.7.6

L-type stairs may have the landing near the bottom or the top. U-type stairs usually have the landing about midway on the flight. Many codes state

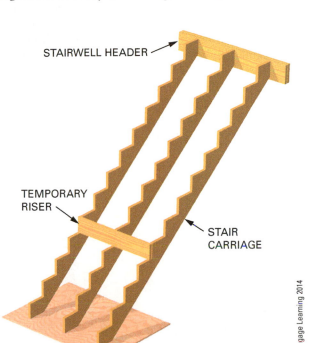

Figure 48–11 A temporary riser about halfway up the flight straightens and maintains the carriage spacing.

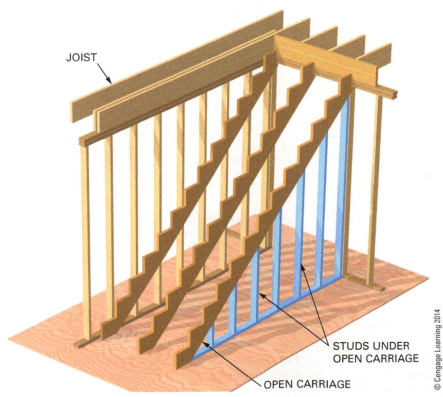

Figure 48–12 Studs are typically installed under the stair carriage on the open side of the stairway.

that no flight of stairs shall have a vertical rise of more than 12 feet. Therefore, any staircase running between floors with a vertical distance of more than 12 feet must have at least one intermediate landing or platform.✱

✱IRC
R311.7.3

Platform stairs are built by first erecting the platform. The finished floor of the platform may be thought of as an extra-wide tread. It should be the same height as if it were a finish tread in the staircase. This allows an equal riser height for both flights (*Figure 48–13*). The stairs are then framed to the platform as two straight flights (*Figure 48–14*). Either the stair carriage or the housed stringer method of construction may be used.

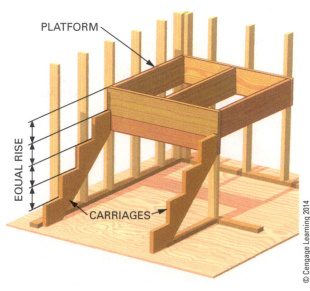

Figure 48–13 The top side of the stair platform is located as if it were a tread.

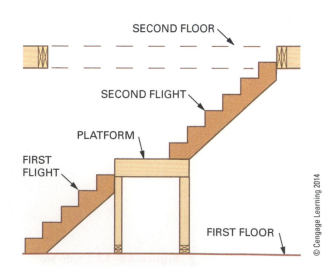

Figure 48–14 Stair carriages are framed to the platform as two straight flights of stairs.

LAYING OUT WINDING STAIRS

Winding stairs change direction without a conventional landing. They often will allow for two extra risers in the space normally occupied by the landing. This is particularly useful for stairwells that are too small for normal stair construction. Winders, however, are not recommended for safety reasons. Some codes state that they may be used in individual dwelling units, if the required tread width is provided along an arc. This is called the *line of travel*. It is a certain distance from the side of the stairway where the treads are narrower. The IRC (International Residential Code) states that the line of travel is 12 inches from the narrow edge of the tread. The minimum tread width at the line of travel is 11 inches and the minimum width at the narrow edge is 6 inches ✱ (*Figure 48–15*). Other codes may permit a narrower tread. Check the building code for the area in which the work is being performed.

✱IRC
R311.7.5.2.1

To lay out a winding turn of 90 degrees, draw a full-size winder layout on the floor directly below where the winders are to be installed (*Procedure 48–A*). On a closed staircase, the walls at the floor line represent its sides. For stairs open on one side, lay out lines on the floor representing the outside of the staircase. The wall represents the inside. Swing an arc, showing the line of travel. Use the outside corner of the wall under the outer carriage or layout lines as center. The radius of the arc may be 12 to 18 inches, as the codes permit.

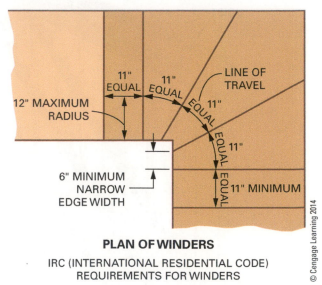

PLAN OF WINDERS
IRC (INTERNATIONAL RESIDENTIAL CODE)
REQUIREMENTS FOR WINDERS

Figure 48–15 International Residential Building Code specifications for winding stairs.

Step**by**Step Procedures

Procedure 48–A Laying Out Winding Stairs

STEP ❶ BEGIN WITH LAYOUT OF STAIR WIDTH AND THE LOWER TREADS. MARK THEM TOWARD THE PLATFORM.

STEP ❷ FROM POINT A OF LAST TREAD, MEASURE 6" TOWARD PLATFORM AND MARK IT AS POINT B.

STEP ❸ MEASURE FROM POINT A ALONG TREAD LINE 12" TO POINT C.

STEP ❹ SWING 11" ARC FROM POINT C TOWARD WHERE POINT D WILL BE. SWING 12" ARC FROM POINT B TO LOCATE POINT D.

STEP ❺ DRAW LINE TO WALL FROM POINT B THROUGH POINT D. THIS IS TREAD LINE FOR NEXT STEP.

STEP ❻ SWING 6" ARC FROM POINT B TO LOCATE POINT E ON THE STAIR WIDTH LINE.

STEP ❼ SWING 11" ARC FROM POINT D TOWARD WHERE POINT F WILL BE. SWING 12" ARC FROM POINT E TO LOCATE POINT F.

STEP ❽ DRAW LINE TO WALL FROM POINT E THROUGH POINT F.

STEP ❾ REPEAT THIS PROCEDURE FOR THE REMAINING WINDERS.

STEP ❿ FIRST NORMAL TREAD WILL BE 11" FROM LAST WINDER MEASURED 12" OUT AS SEEN BETWEEN G AND H.

STEP ⓫ LAY OUT REMAINING NORMAL TREADS.

STEP ⓬ DRAW PLUMB LINES UP THE WALL TO LOCATE RISER LINES FOR EACH STEP.

PLAN

PLAN

PICTORIAL

From the same center, lay out the width of the narrow end of the treads in both directions. Square lines from the end points to the opposite side of the staircase. Divide the arc into equal parts. Project lines from the narrow end of the tread, through the intersections at the arc, to the wide end at the wall. These lines represent the faces of the risers. Draw lines parallel to these to indicate the riser thickness. Plumb these lines up the wall to intersect with the tread run for each winder.

The cuts on the stair carriage for the winding steps are obtained from the full-size layout. Lay out and cut the carriage. Fasten it to the wall. Install rough treads until the stairs are ready for finishing.

If one side of the staircase is to be open, a **newel post** is installed. Then, the risers are mitered to or **mortised** into the post (*Figure 48–16*). A mortise is a rectangular cavity in which the riser is inserted. Newel posts are part of the stair finish. They are described in more detail in Chapters 81 and 83.

SERVICE STAIRS

Service stairs, typically used for basement stairs, are built as a quick set of stairs. They are not considered finish stairs and are often built without risers. If so, riser and railing spaces must be small enough to keep a 4 inch sphere from passing through. **✱** Two carriages are used with nominal 2 × 10 treads cut between them. The carriages are not always cut out, like those previously described. They may be dadoed to receive the treads. An alternative method is to fasten cleats to the carriages to support the treads (*Figure 48–17*).

✱ IRC R311.7.5.1

Lay out the carriages in the usual manner. Cut the bottoms on a level line to fit the floor. Cut the tops along a plumb line to fit against the header of the stairwell. "Drop" the carriages as necessary to provide a starting riser with a height equal to the rest of the risers.

Dadoed Carriages

If the treads of rough stairs are to be dadoed into the carriages, lay out the thickness of the tread on the stringer below the layout line. The top of the tread is to the original layout line. Mark the depth

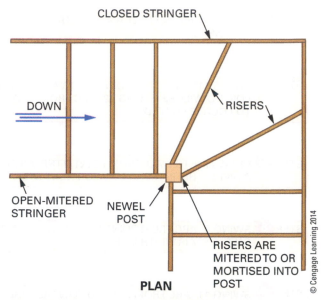

Figure 48–16 Risers of open-sided winders are mitered against or mortised into a newel post.

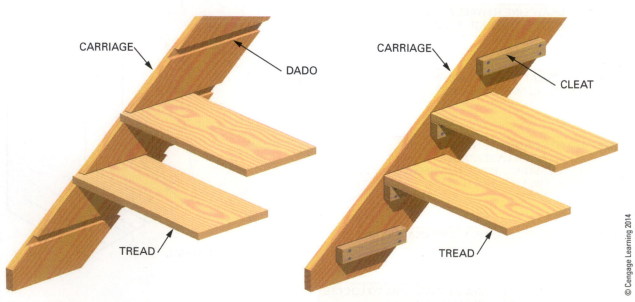

Figure 48–17 Service stair carriages may be dadoed or cleated to support the treads.

of the dado on both edges of the carriage. Set the circular saw to cut the depth of the dado. Make cuts along the layout lines for the top and bottom of each tread. Then, make a series of saw cuts between those just made.

Chisel from both edges toward the center, removing the excess to make the dado. Nail all treads into the dadoes. Assemble the staircase. Fasten the assembled staircase in position. Locate the top tread at a height to obtain an equal riser at the top. The lower end of a basement stairway is sometimes anchored by installing a kicker plate, which is fastened to the floor (*Figure 48–18*).

Cleated Carriages

If the treads are to be supported by cleats, measure down from the top of each tread a distance equal to its thickness. Draw another level line. Fasten 1 × 3 cleats to the carriages with screws. Make sure their top edges are to the bottom line. Fasten the carriages in position. Cut the treads to length. Install the treads between the carriages so the treads rest on the cleats. Fasten the treads by nailing the stair carriage into their ends.

Temporary Handrail

Jobsites often leave carriages and treads unfinished until the house is closer to completion. This allows construction to continue without undue damage to finish treads and risers. But the stairs must always be safe to use. Handrails and guardrails must be installed to protect the workers from falls (see *Figure 48–19*).

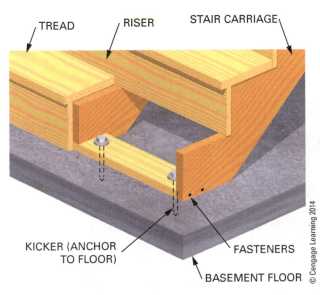

TREAD   RISER   STAIR CARRIAGE

KICKER (ANCHOR TO FLOOR)   FASTENERS

BASEMENT FLOOR

© Cengage Learning 2014

Figure 48–18 A kicker plate may be used to anchor a carriage to floor.

© Cengage Learning 2014

Figure 48–19 Temporary handrails are used to protect workers on a jobsite.

DECONSTRUCT THIS

Carefully study **Figure 48–20** and think about what is wrong and/or what is right. Consider all possibilities. What construction practice or method is different in your area of the country?

Figure 48–20 This photo shows a framing site.

© Cengage Learning 2014

KEY TERMS

closed stairways	mortise	stair carriage	stringer
handrail	newel post	stair horse	winders
housed stringer	nosing	staircase	
landing	open stairways	stairwell	

REVIEW QUESTIONS

Select the most appropriate answer.

1. The rounded outside edge of a tread that extends beyond the riser is called a
   a. housing.
   b. coving.
   c. turnout.
   d. nosing.

2. The unit Rise of a stair refers to the
   a. vertical part of a step.
   b. riser.
   c. result of dividing total rise by the number of risers.
   d. all of the above.

3. The unit Run of a stair refers to the
   a. tread width without nosing.
   b. horizontal portion of each step plus the nosing.
   c. stairwell opening.
   d. all of the above.

4. Stairways in residential construction should have a minimum width of
   a. 30 inches.
   b. 36 inches.
   c. 32 inches.
   d. 40 inches.

5. The IRC (International Residential Code) calls for the maximum riser height for a residence to be
   a. $8\frac{1}{4}$ inches.
   b. 7¾ inches.
   c. 7½ inches.
   d. 7¼ inches.

6. The largest riser height allowed by the IRC for a stairway with a total Rise of 8'-11" is
   a. 8 inches.
   b. 7¾ inches.
   c. 7⅝ inches.
   d. 7½ inches.

7. Using the 17–18 method, what is the range of unit Run for a stair with a unit Rise of 7¼ inches?
   a. 8¾ to 9¾ inches
   b. 9 to 10 inches
   c. 9¾ to 10¾ inches
   d. 10 to 11 inches

8. Using the 24–25 method, what is the range of unit Run for a stair with a unit Rise of $7\frac{1}{4}$ inches?
   a. 8½ to 9½ inches
   b. 9¾ to 10¾ inches
   c. 9½ to 10½ inches
   d. 10 to 11 inches

9. The IRC specifies a minimum unit Run of
   a. 8½ inches.
   b. 9½ inches.
   c. 9 inches.
   d. 10 inches.

10. The IRC specifies a minimum headroom clearance of
    a. 6'-6".
    b. 6'-8".
    c. 7'-0".
    d. 7'-6".

11. The smallest stairwell length for a set of stairs with a unit Rise of 7½ inches, a unit Run of 10 inches, a desired headroom of 82 inches, and an upper-construction dimension of 11 inches is
    a. 93¼".
    b. 110.75".
    c. 124".
    d. 130".

12. What is the unit Run for a stair to fit in a 10'-10" stairwell that has a unit Rise of 7½ inches, a desired headroom of 82 inches, and an upper-construction dimension of 11 inches?
    a. 10½ inches
    b. $10^{11}/_{16}$ inches
    c. $10^{15}/_{16}$ inches
    d. 12⅜ inches

13. The stair carriage with a unit Rise of $7\frac{1}{2}$ inches rests on the finish floor. What is the riser height of the first step if the tread thickness is $\frac{3}{4}$ inch?
    a. 6¾ inches
    b. 7¾ inches
    c. 7½ inches
    d. 8½ inches

14. A stairway has a unit Rise of 7½ inches and the stairwell header acts as the top riser. The tread stock thickness is $1^1/_{16}$ inches. The finish floor thickness of the upper floor is ¾ inch. What is the distance down from the top of the subfloor of the upper floor to the rough carriage?
    a. $7^3/_{16}$ inches
    b. 7½ inches
    c. 7¾ inches
    d. $7^{13}/_{16}$ inches

15. The line of travel for a set of winding stairs
    a. is the location of the minimum tread width for the stairs.
    b. is drawn for layout of treads.
    c. is at least 12 inches from the narrow edge.
    d. all of the above.

UNIT 18

Insulation and Ventilation

CHAPTER 49 Thermal and Acoustical Insulation
CHAPTER 50 Condensation and Ventilation

©Shutterstock, Inc./Niki Crucillo

Thermal insulation prevents the loss of heat in buildings during cold seasons. It also resists the passage of heat into air-conditioned buildings in hot seasons. Moisture in the air may cause severe problems for a building. Vapor retarders are essential to a long-lived building. Adequate ventilation must be provided in the living environment and within the building materials. Ventilation encourages evaporation of harmful moisture formed in living spaces and within the insulation. Acoustical insulation reduces the passage of sound from one area to another.

Insulation and ventilation should be thought of together. If a house has insulation, it must also have allowances for ventilation. This will ensure that the insulation performance is maximized while also making the house pleasant to live in.

OBJECTIVES

After completing this unit, the student should be able to:

- describe how insulation works and define insulating terms and requirements.
- describe the commonly used insulating materials, and state where insulation is placed.
- properly install various insulation materials.
- explain relative humidity and moisture migration.
- explain the need for ventilating a structure.
- explain the kinds and purpose of vapor retarders and how they are applied.
- describe various methods of construction to reduce the transmission of sound.

CHAPTER 49

Thermal and Acoustical Insulation

All materials used in construction have some insulating value. Insulation is a material whose purpose is to interrupt or slow the transfer of heat. Heat transfer is a complex process involving three mechanisms: conduction, convection, and radiation.

Conduction is the transfer of heat by contact. Warmer particles touch cooler ones, causing them to vibrate with more energy, thus transferring energy into the cooler material. Heat energy is thought to move from warmer materials into the cooler ones in an attempt to reach an equilibrium.

Convection involves a fluid, such as a gas or a liquid. The fluids affecting a house are usually air and water. When a portion of the fluid is warmed, it becomes less dense than the surrounding fluid. The warmer fluid requires more space than the cooler fluid. Gravity causes the cooler, more dense fluid to force the warmer, less dense fluid to rise. This action can be seen in the streams of fluid in a pan of hot oil on a cooking stove. A hot-air balloon encapsulates less dense hot air, which creates lift, causing the craft to rise.

Radiation is a general word referring to electromagnetic radiation, which includes microwaves, radio waves, infrared, visible light, ultraviolet, rays, and cosmic rays. These can be thought of as particles of energy with varying wavelengths and vibration frequencies that all travel at the same speed, the speed of light. Protection from radiation is best achieved by reflecting it with clean shiny surfaces such as aluminum foil.

Popcorn may be used to help remember these three forms of heat transfer. The three different methods of cooking popcorn use different modes of heat transfer. All three methods heat the moisture in the kernel to steam, which then ruptures the shell with the familiar popping sound. Stove-top popcorn absorbs heat by direct contact with burners. Hot-air poppers transfer heat from electric heating elements to the air that blows on to the kernels. Microwave packages use microwave radiation to heat the kernels (*Figure 49–1*).

Figure 49–1 Various poppers use different types of heat transfer to pop corn.

Figure 49–2 The R-value is broadly stamped on insulation packaging.

Slowing heat transfer is the goal of installing insulating materials. The best way to slow heat transfer within a house depends on where the house is located. In southern climates, protection from solar radiation is important, whereas in northern climates this importance is low. In both climates, protection from conduction and convection are important.

HOW INSULATION WORKS

Insulating materials create a space between two surfaces, thereby breaking contact and reducing conduction. They trap air and slow convection. They also can either reflect or absorb some radiation.

Air is an excellent insulator if confined to a small space and kept still. Insulation materials are designed to trap air into small unmoving pockets. Insulation effectiveness increases as the air spaces become smaller in size and greater in number. Many tiny air cells, trapped in its unique cellular structure, make wood a better insulator than concrete, steel, or aluminum.

All insulating materials are manufactured from materials that are themselves poor insulators. For example, fiberglass insulation is made from glass, which is a good conductor of heat. The improved insulation value comes from air trapped within the insulating material. Insulation also provides resistance to sound travel.

THERMAL INSULATION

Among the materials used for insulating are glass fibers, mineral fibers (rock), organic fibers (paper), and plastic. Aluminum foil is also used because it reflects heat.

Resistance Value of Insulation

The R-value of insulation is a number indicating its resistance to the flow of heat. These numbers are clearly indicated on all insulating materials

(*Figure 49–2*). The number for an R-value comes from the reciprocal of a materials measured thermal conductance. That is their ability to conduct heat. Most building materials have been laboratory tested to measure their conductivity.✱The lower the conductivity, the less heat that will pass through the material. Heat is measured with a Btu (British thermal unit). This is roughly equivalent to the energy in one kitchen match (*Figure 49–3*).

✱**IRC Chapter 11N1102.1.2**

For example, an inch of steel allows 312.5 Btus of heat to pass through it every hour. A softwood board allows 0.8 Btus to pass through it every hour. Taking the reciprocal of these numbers changes 312.5 to 0.0032 by using 1 ÷ 312.5 = 0.0032.

Thermal Conductivity (k-value)	a measure of a materials ability to conduct heat. Materials were tested at 1" thickness in terms of Btus per hour per square foot per 1° F difference.
Conductance (C-value)	similar to conductivity except materials were tested at their commonly used thicknesses, e.g., ½" plywood, ¾" hardwood floor. In terms of Btus per hour per square foot per 1° F difference.
U-Value	same as k-value except multiple layers of differing materials are calculated together, e.g., windows
Thermal Resistance (R-Value)	the reciprocal (inverse) of U-value or k-value. 1/u or 1/k = R-value. It is considered a measure of a materials resistance allowing heat to pass through it. i.e. if k = 0.5 then R = 1 ÷ 0.5 = 2

Figure 49–3 Definitions for heat loss terms relating to conductivity.

Foundation Materials	
8" concrete block (2-hole core)	1.11
12" concrete block (2-hole core)	1.28
8" lightweight aggregate block	2.18
12" lightweight aggregate block	2.48
Common brick	0.20/inch
Sand or stone	0.08/inch
Concrete	0.08/inch

Structural Materials	
Softwood	1.25/inch
Hardwood	0.91/inch
Steel	0.0032/inch
Aluminum	0.00070/inch

Sheathing Materials	
½" plywood	0.63
⅝" plywood	0.78
¾" plywood	0.94
½" aspenite, OSB	0.91
¾" aspenite, OSB	1.37

Insulating Materials	
Batts and Blankets	
3½" fiberglass	11
6" fiberglass	19
8" fiberglass	25
12" fiberglass	38
3½" high-density fiberglass	13
5½" high-density fiberglass	21
8½" high-density fiberglass	30
10" high-density fiberglass	38
fiberglass	3.17/inch

Loose Fill	
Fiberglass	3.17/inch
Cellulose	3.70/inch

Reflective Foil	
Foil-faced bubble pack	1.0
1-layer foil	0.22
4-layer foil	11

Rigid Foam	
Expanded polystyrene foam (bead board)	4.0/inch
Extruded polystyrene foam	5.0/inch
Polyisocyanurate/urethane foam	5.6/inch

Spray Foams	
Low-density polyurethane	3.60/inch
High-density polyurethane	6.5/inch

Finish Materials	
Wood shingles	0.87
Vinyl siding	Negligible
Aluminum siding	Negligible
Wood siding (½" × 8")	0.81
½" hardboard siding	0.36
Polyethylene film	Negligible
Builder's felt (15#)	0.06
25/32" hardwood flooring	0.68
Vinyl tile (1/8")	0.05
Carpet and pad	1.23
¼" ceramic tile	0.05
⅜" gypsum board	0.32
½" gypsum board	0.45
⅝" gypsum board	0.56
½" plaster	0.09
Asphalt shingles	0.27

Figure 49–4 Insulating R-values of various building materials.

Source: Professor Richard Harrington, SUNY College of Technology at Delhi.

The figure 0.8 changes to 1.25 by using 1 ÷ 0.8 = 1.25. Now the numbers represent not the amount of heat transferred but rather an indication of thermal resistance. Higher numbers mean better insulators (*Figure 49–4*). The additional benefit to this process is if you were to use two layers of an R 1.25 material, the new R-value would be 2.5. Adding conductance does not make sense.

Some R-values are given per inch of material. Some are given as per the normal thickness of the material. For example, the R-value of standard fiberglass is 3.17 per inch and a 3 ½-inch batt of high-density fiberglass is 13.

Building Section R-Value Calculations

Walls, ceilings, and floors are constructed with different materials. This makes the actual R-value of each building section dependent on the material used to construct it. Most building materials add to the overall R-value of the wall. The R-values for each material is found in the chart of the previous figure. Stud and joist R-values are calculated from the width of the member multiplied times the R-value 1.25 per inch for softwood. For example, a 2 × 6 stud R-value is 1.25 × 5.5" = 6.875.

An adjustment is necessary to compensate for the fact that not all materials are continuous throughout the entire wall. Sheathing and gypsum board cover the entire wall; studs are not continuous throughout the wall, but instead are spaced within the wall section. And since studs and insulation occupy the same cross-sectional region, fiberglass is also not continuous. To compensate for this we assume walls with 16" OC spaced studs have 25 percent wood

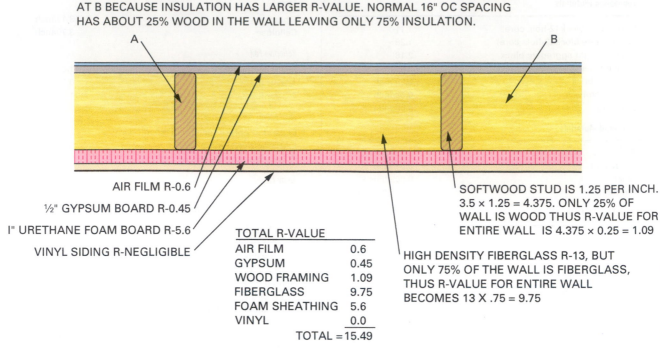

HEAT LOSS THROUGH THE STUD AT A IS HIGHER THAN THROUGH INSULATION AT B BECAUSE INSULATION HAS LARGER R-VALUE. NORMAL 16" OC SPACING HAS ABOUT 25% WOOD IN THE WALL LEAVING ONLY 75% INSULATION.

A

B

AIR FILM R-0.6
½" GYPSUM BOARD R-0.45
I" URETHANE FOAM BOARD R-5.6
VINYL SIDING R-NEGLIGIBLE

SOFTWOOD STUD IS 1.25 PER INCH. 3.5 × 1.25 = 4.375. ONLY 25% OF WALL IS WOOD THUS R-VALUE FOR ENTIRE WALL IS 4.375 × 0.25 = 1.09

HIGH DENSITY FIBERGLASS R-13, BUT ONLY 75% OF THE WALL IS FIBERGLASS, THUS R-VALUE FOR ENTIRE WALL BECOMES 13 X .75 = 9.75

TOTAL R-VALUE	
AIR FILM	0.6
GYPSUM	0.45
WOOD FRAMING	1.09
FIBERGLASS	9.75
FOAM SHEATHING	5.6
VINYL	0.0
TOTAL =	15.49

© Cengage Learning 2014

Figure 49–5 R-vlaues of a wall section vary according to the actual path of heat loss.

framing in the wall. This leaves 75 percent of the wall available for insulation. These numbers include the plates and headers. Multiplying the stud R-value by 0.25 and the fiberglass R-value by 0.75 adjusts the numbers to actual R-values.

Consider a wall section made with 2 × 4s, 1-inch urethane rigid foam sheathing, high density fiberglass, ½-inch gypsum board, and vinyl siding (*Figure 49–5*). Look up the R-value for each material. Also add an R-0.6 for the air film next to the wall surface.

Insulation Requirements

The rising costs of energy coupled with the ecological need to conserve have resulted in higher R-value recommendations for new construction than in previous years. ✱Average winter low-temperature zones of the United States are shown in *Figure 49–6*. This information is used to determine the R-value of insulation installed in walls, ceilings, and floors. Insulation requirements vary according to the average low temperature. ✱

In warmer climates, less insulation is needed to conserve energy and provide comfort in the cold season. However, air-conditioned homes should also receive more insulation in walls, ceilings, and floors. This ensures economy in the operation of air-conditioning equipment in hot climates.

The DOE (Department of Energy) rates each region of the US according to climate. These regions are referred to as *climate zones,* from 1 to 8 (*Figure 49–7*). Portions of Alaska are in zone 8. Some states like South Carolina and Nevada are entirely within one climate zone, and others like Colorado and Arizona have four zones. The IECC (International Energy Conservation Code) sets expectations for the R-values for many parts of a building according to climate zones. The website http://energycode.pnl.gov/EnergyCodeReqs/ allows the visitor to check the 2009 IECC zones for any state. As of this edition the 2012 version is not available for free.

In each climate zone is a list of 10 parts of a building and the code requirements (*Figure 49–8*). Some of these parts are self-explanatory and others need a description. Note the numbers given by the IECC are a minimum and builders are encouraged to exceed these requirements. Also the R-values

✱IRC Chapter 11N1102.2

✱IRC Chapter 11N1101.10

Figure 49–6 Average low-temperature zones of the United States.

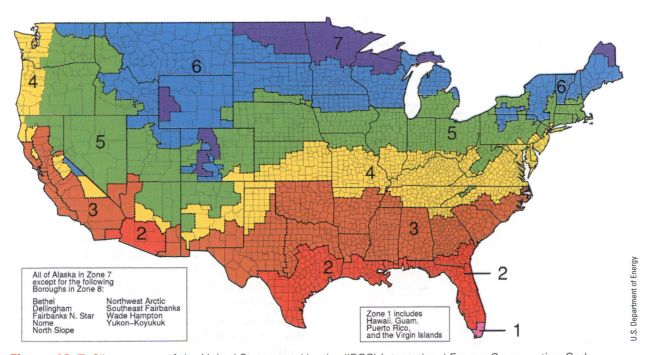

All of Alaska in Zone 7
except for the following
Boroughs in Zone 8:

Bethel	Northwest Arctic
Dellingham	Southeast Fairbanks
Fairbanks N. Star	Wade Hampton
Nome	Yukon–Koyukuk
North Slope	

Zone 1 includes
Hawaii, Guam,
Puerto Rico,
and the Virgin Islands

Figure 49–7 Climate zone of the United States used by the (IECC) International Energy Conservation Code.

listed refer only to the insulation, not the sum of all materials in the building section.

Comfort and operating economy are dual benefits. Insulating for maximum comfort automatically provides maximum economy of heating and cooling operations.

Where Heat Is Lost

The amount of heat lost from the average house varies with the type of construction. The principal areas and approximate amount of heat loss for a typical house with moderate insulation are shown in *Figure 49–9.*

International Energy Conservation Code (IECC)

		Excerpts from Climate Zone 5	
1	Ceiling R-value	38 (2009) 49 (2012)	Insulation R-value for ceilings is stated as R-38 in the 2009 code, but has changed to R-49 in the 2012 code.
2	Wood Frame Wall R-value	20 or 13+5	20 refers to a wall with R-20 cavity insulation. 13+5 refers to walls with R-13 cavity insulation and a continuous sheathing layer of R-5.
3	Mass Wall R-value	13/17	Mass walls are masonry walls built above ground such as concrete block. Insulation is R-13 unless more than half of the insulation is on the interior side of the wall, then R-17 is required.
4	Floor R-value	30	R-30 is required or sufficient insulation to fill the joist cavity with an R-19 as minimum.
5	Basement Wall R-value	10/13	R-10 is continuous insulated sheathing on the interior or exterior of the home or R-13 cavity insulation installed in an interior basement wall.
6	Slab R-value, Depth	10, 2 ft	R-10 is required along the building perimeter, installed vertically to a depth of 2 feet. Additional R-5 is required for heated slabs.
7	Crawlspace Wall R-value	10/13	R-10 is continuous insulated sheathing on the interior or exterior of the home or R-13 cavity insulation installed in an interior basement wall. This is the same as for basements.
8	Fenestration U-Value	0.35 (2009) 0.32 (2012)	Fenestration refers to windows, doors and skylights. Window U values have gotten stricter for 2012 IECC. The fenestration U-value excludes skylights.
9	Skylight U-Value	0.60 (2009) 0.55 (2012)	Code allows skylights to less energy efficient than windows in a wall.
10	Glazed fenestration SHGC	NR	Fenestration U-value considers the heat loss through the unit for heating season. SHGC (Solar Heat Gain Coefficient) considers the heat gain from the sun during air-conditioning season. Climate zone 5 has no SHGC requirement. The SHGC column applies to all glazed fenestration.

© Cengage Learning 2014

Figure 49–8 A portion of the IECC 2012 Clime zone 5 insulation requirements.

1%   THROUGH BASEMENT FLOOR
3%   THROUGH DOORS
5%   THROUGH CEILINGS
16% THROUGH WINDOWS
17% THROUGH FRAME WALLS
20% THROUGH BASEMENT WALLS
38% AIR LEAKAGE THROUGH CRACKS IN WALLS,
      AND AROUND WINDOWS AND DOORS

© Cengage Learning 2014

Figure 49–9 Typical heat loss for a house built with moderate insulation.

Houses of different architectural styles vary in their heat loss characteristics. A single-story house, for example, contains less exterior wall area than a two-story house, but has a proportionately greater ceiling area. Greater heat loss through floors is experienced in homes erected on concrete slabs or unheated crawl spaces unless these areas are well insulated.

The transfer of heat through uninsulated ceilings, walls, or floors can be reduced to almost any desired amount by installing insulation. Maximum quantities in these areas can cut heat losses. The use of 2 × 6 studs in exterior walls permits installation of 6-inch-thick insulation. This achieves an R-19, or, with higher density insulation, an R-21.

Windows and doors are generally sources of great heat loss. Weather stripping around windows and doors reduces heat loss. Heat loss through glass surfaces can be reduced 50 percent or more by installing double- or triple-glazed windows. This is referred to as insulated glass. Improved insulated glass design uses argon gas between two panes of glass. The trapped gas improves the R-value significantly. A low-emissivity coating on the inner glass surface also improves the window performance. This coating allows ultraviolet rays from the sun to pass but reflects infrared (radiant) rays. Windows having both of these features are called low E argon windows.

Adding a storm window or storm door is also effective. Adding additional layers of trapped air creates additional thermal resistance.

Constructing homes to be more airtight is an effective method of reducing energy needs. Since heat is lost through air leakage, anything that stops air movement will improve energy efficiency. An additional benefit of airtightness includes reducing moisture migration into building components. Moisture migration is covered further in Chapter 50. ✳

✳IRC
Chapter
11N1102.4

TYPES OF INSULATION

Insulation is manufactured in a variety of forms and types. These are commonly grouped as *flexible, loose-fill, rigid, reflective,* and *miscellaneous.* All insulations should be installed behind materials such as drywall. These materials provide air barriers to allow the insulation to perform well. They also provide a fire barrier, for those insulations that burn easily.

Flexible Insulation

Flexible insulation is manufactured in *blanket* or *batt* form. Blanket insulation (*Figure 49–10*) comes in rolls. Widths are suited to 16- and 24-inch stud and joist spacing. The usual thickness is from $3\frac{1}{2}$ inches and 6 inches. The body of the blanket is made of fluffy rockwool or glass fibers. Blanket insulation is unfaced or faced with asphalt-laminated Kraft paper or aluminum foil with flanges on both edges for fastening to studs or joists. These facings may be considered a vapor retarder, also called a vapor barrier, but installation of the flanges must be airtight to function as a vapor retarder. These facings should always face the warm side of the wall. ✳

✳R
702.7

Batt insulation (*Figure 49–11*) is made of the same material as blanket insulation. It comes in thicknesses up to 12 inches. Widths are for standard stud and joist spacing. Lengths are either 48 or 93 inches. Batts come faced or unfaced.

Green Tip

Fiberglass is made from 25 percent recycled materials.

Loose-Fill Insulation

Loose-fill insulation is usually composed of materials in bulk form. It is supplied in bags or bales. It is placed by pouring, blowing, or packing by hand. Materials include rockwool, fiberglass, and cellulose.

Loose-fill insulation is well suited for use between ceiling joists and in the sidewalls of existing houses that were not insulated during construction (*Figure 49–12*). There is an application process where loose-fill may be installed in stud walls of

Figure 49–11 Batt insulation is made up to 12 inches thick.

Figure 49–10 Blanket insulation comes in rolls.

Figure 49–12 Loose fill cellulose insulation being blown into place.

new construction. The insulation is damp with an adhesive that causes it to stick to itself and the wall sheathing (*Figure 49–13*).

Rigid Insulation

Rigid insulation is usually a foamed plastic material in sheet or board forms (*Figure 49–14*). The material is available in a wide variety of sizes, with widths up to 4 feet and lengths up to 12 feet. The most common types are made from polystyrene and polyurethane.

Expanded polystyrene or white bead board has good insulating qualities, R-value 4 per inch. It is sometimes installed on the inside of dry basement walls. If expanded polystyrene comes in contact with water, it will absorb moisture. The absorbed moisture will significantly reduce the R-value. Therefore, this foam should not be installed in damp areas.

Extruded polystyrene comes in a variety of colors depending on the manufacturer with an R-value 5 per inch. It is often installed as insulation for footings and basement walls. Extruded polystyrene is a closed-cell foam that will not absorb moisture, keeping its R-value while wet. It is the only insulation material suitable for wet locations such as under slabs. It is sometimes used for flotation billets under docks in lakes.

Polyurethane or polyisocyanurate foam has a superior R-value of 5.6 per inch. It is usually made with a facing of foil or building paper. In the foil face version, it is used as wall sheathing. The foil adds a reflective radiant barrier, making it perform well. The paper face version is used under roofing materials. Polyurethane foam should not be used in damp areas.

Reflective Insulation

Aluminum foil conducts heat well and does not have an R-value rating since R-values are based on conduction. But foil effectively reflects radiant energy. It is typically used in warm climates to protect the building from the sun and to reflect heat. Reflective insulation usually consists of foil bonded to a surface of some other material such as drywall or foam insulation. The reflective effectiveness depends on being installed with a minimum of ¾ inch of space.

Miscellaneous Insulation

Foamed-in-place insulation is sometimes used. Urethane foam is produced by mixing two chemicals together. It is sprayed into place and expands on contact with the surface. This product requires special equipment provided by an insulating contractor. Urethane comes in two densities, referred to as low- and high-density foams. Low-density is open celled, which means air fills the cells providing the insulation value. It has an R-value of 3.7 per inch. This foam expands one hundred times its original volume. The stud cavity is filled to overflowing, and the excess is trimmed flush (*Figure 49–15*).

High-density foam is a closed-cell system where a refrigerant gas fills the cells. This gas is a better insulator than air, thus it has a better R-value of 6.5 per inch (*Figure 49–16*). This foam does not expand as much as low-density foam and has a harder cured surface. Both low and high density foams offer superior insulating qualities.

Foamed-in-place insulation adds rigidity to the overall structure. The foam forms a structural bond

Figure 49–13 Cellulose may be applied into stud cavities.

Image courtesy of Pat Dundon, Dundon Insulation, Inc., www.insulationman.com.

Figure 49–14 Types of rigid insulation boards: (A) Foil-faced polyurethane. (B) Extruded polystyrene. (C) Expanded polystyrene.

© Cengage Learning 2014

Figure 49–15 Low density urethane insulation being applied to stud cavities.

Figure 49–16 High density urethane insulation applied between ceiling joists.

Figure 49–17 Urethane foam may be purchased in small cans.

between the sheathing and studs, giving a racking strength. It also seals the thermal envelope to air movement. The foam expands into cracks to fill and seal all seams in building materials. This also makes a good soundproofing material.

Spray foams are available in aerosol cans (*Figure 49–17*). They are used to seal and insulate gaps between doors and window units and the wall frame. They are also used to seal mechanical penetrations such as those made by electricians. As the foam is sprayed into gaps it reacts and expands, filling the space, creating an airtight seal.

There are two types of foam. One expands many times its volume and seals well in many situations. The other type does not expand as much. It is designed for the space between rough openings and window and doors. Since it does not expand much the risk of distorting the door or window jamb causing poor unit operations is reduced.

Other types of insulating materials include lightweight vermiculite and perlite aggregates. Vermiculite is made by superheating mica, which causes it to expand quickly, creating air spaces. It is able to withstand high temperatures and can be used around chimneys. Perlite is made from heating volcanic glass, causing it to expand. It is sometimes used in plaster and concrete to make it lighter and more thermal resistant.

WHERE AND HOW TO INSULATE

Any building surfaces that separate conditioned air, heated or cooled, from unconditioned air must be insulated. This will save energy and money over the life of the building. Insulation should be placed in all outside walls and top-floor ceilings. Houses with basements or crawl spaces should have their foundation walls insulated. Collectively these surfaces are called the thermal envelope.

Great care should be exercised when installing insulation. Insulation that is not properly installed can render the material useless and a waste of time and money. Pay attention to details around outlets, pipes, and any obstructions. Make the insulation conform to irregularities by cutting and piecing without bunching or squeezing. Keep the natural

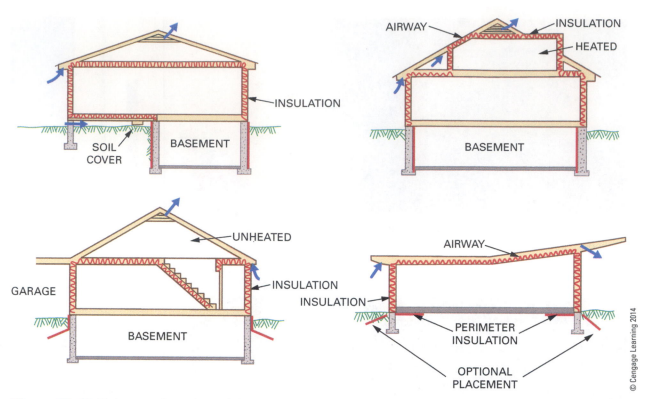

Figure 49–18 Various configurations of the thermal envelope.

fluffiness of the insulation intact at all times. Voids in insulation of only 5 percent can create overall efficiency reduction of 25 percent. This is caused by the process of the path of least resistance whereby heat moves move to the colder areas. Insulation should be installed neatly. Generally, if insulation looks neat, it will perform well.

In houses with flat or low-pitched roofs, insulation should be used in the ceiling area. Insulation is also used along the perimeter of houses built on slabs (*Figure 49–18*).

Installing Flexible Insulation

Cut the batts or blankets with a knife. Make lengths slightly oversize by about an inch at the top and bottom. Measure out from one wall a distance equal to the desired lengths of insulation. Roll out the material from the wall. Compress the insulation with a straightedge onto a cutting board to protect the floor, then cut with a knife (*Figure 49–19*). Cut the necessary number of lengths.

Figure 49–19 Flexible insulation is compressed with a straightedge as it is cut.

Place the batts or blankets between the studs. Staple the flanges of the facing to the stud thickness edge of the studs. Use a hammer tacker or a hand stapler to fasten the insulation in place (*Figure 49–20*).

Fill any spaces around windows and doors with spray-can foam of the proper type. Foam will fill the voids with an airtight seal and protect the

CAUTION

Always protect lungs, sinuses, eyes, and skin from insulation fibers. Long-term effects could be severe.

Figure 49–20 A hand stapler can be used to fasten wall insulation.

SPRAY A BEAD OF FOAM JUST LARGE ENOUGH TO SEAL DOOR OR WINDOW UNIT TO STUD FRAMING. DO NOT ATTEMPT TO FILL THE ENTIRE SPACE.

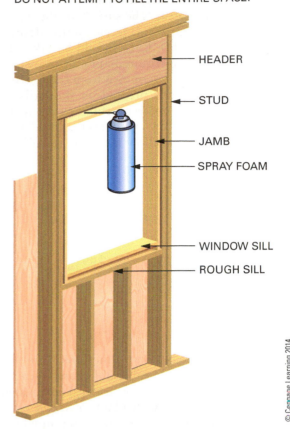

HEADER
STUD
JAMB
SPRAY FOAM
WINDOW SILL
ROUGH SILL

Figure 49–21 Spaces around doors and windows are filled with spray-foam insulation to seal them to the rough opening.

house from air leakage (*Figure 49–21*). After the foam cures, flexible insulation may be added to fill the remaining space, but this is not necessary.

Do not pack the insulation tightly. Squeezing or compressing it reduces its effectiveness. If the insulation has no covering or stapling tabs, it is friction-fitted between the framing members.

Ceiling Insulation

Ceiling insulation is installed by stapling it to the ceiling joists or by friction-fitting it between them. If furring strips have been applied to the ceiling joists, the insulation is simply laid on top of the furring strips. Extend the insulation across onto the top plate. In most cases, the use of unfaced insulation is recommended. This makes it easier to determine proper fit of the insulation as well as lowering the cost of materials.

In northern regions of North America, attics should be well ventilated. This ensures that any moisture and heat that escapes the building will not be trapped in the attic space. As warm/moist air is vented out, cool air from the outside replaces it. This has a drying effect on the house. This venting usually begins at the eaves and ends with ridge or attic vents. Therefore, the ceiling insulation must not block this venting.✱

✱**IRC R806.5 IRC Chapter 11 N1102.2.3**

It may be necessary to compress the insulation against the top of the wall plate to permit the free flow of air. An air-insulation dam should also be installed to protect the insulation from air movement inside the fibers. Otherwise, the insulation's R-value and performance will be reduced (*Figure 49–22*).

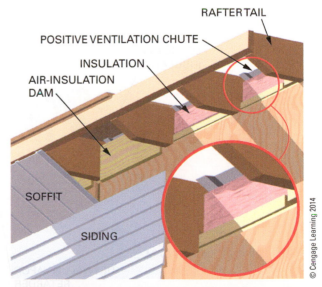

RAFTER TAIL
POSITIVE VENTILATION CHUTE
INSULATION
AIR-INSULATION DAM
SOFFIT
SIDING

Figure 49–22 Air-insulation dams protect the insulation from air infiltration.

Installing Loose-Fill Insulation

Loose-fill insulation is typically blown into place with special equipment. The surface of the material between ceiling joists is leveled to the desired depth (*Figure 49–23*). Care should be taken to evenly distribute the required amount of insulation. This amount is determined from the instructions and proportions printed on the bails of insulation. Take care not to overfill the eave areas. This will restrict air flow in regions that need attic ventilation and might allow insulation to fill the soffits.

Green Tip

Cellulose insulation is produced from 100 percent recycled newspaper.

Installing Rigid Insulation

Rigid insulation has many functions, some of which are not interchangeable. Take care when selecting the material if it will be installed near water. All foams are easily cut with a knife or saw. A table saw can also be used to cut rigid foam, in which case a respirator should be worn to protect from the dust.

CAUTION

Wear a respirator to protect lungs and sinuses from the fine airborne particles released from insulation when it is cut with a power saw.

Rigid foams may be attached with fasteners and adhesives or applied by friction-fitting between the framing members. Fasteners include plastic capped nails that are ring shanked with large plastic washers. They hold the foam in place without pulling through the foam. Adhesives are used to glue the sheet to a surface. Some adhesives will dissolve the plastic foam, so be sure to use the correct adhesive as listed on the adhesive tube.

Polyurethane rigid foam boards are sometimes used on roofs of cathedral ceilings (*Figure 49–24*). Various thicknesses may be used depending on the desired R-value. Roof sheathing should be installed on furring strips above the foam in northern regions.

Insulating Masonry Walls

Masonry walls can be insulated on the inside or outside of the building (*Figure 49–25*). Interior applications require the masonry wall to be furred for attaching the wall finish. Furring can be wood or metal hat track. Fasten furring strips 16 or 24 inches on center (OC). Dry walls may be insulated with expanded polystyrene, or polyurethane sheets.

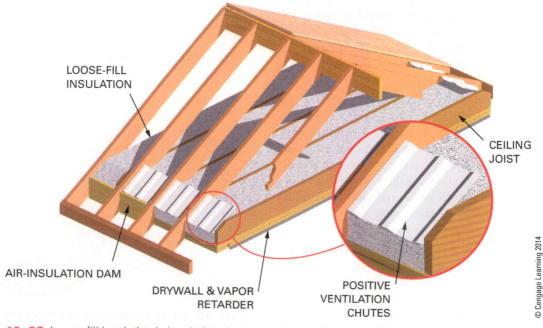

LOOSE-FILL INSULATION

CEILING JOIST

AIR-INSULATION DAM

DRYWALL & VAPOR RETARDER

POSITIVE VENTILATION CHUTES

© Cengage Learning 2014

Figure 49-23 Loose-fill insulation is leveled to the desired depth.

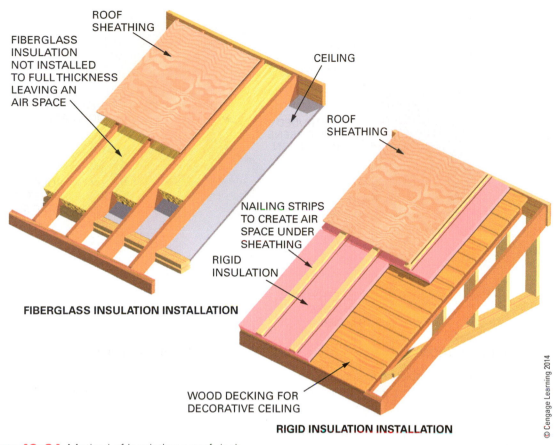

Figure 49–24 Method of insulating a roof deck.

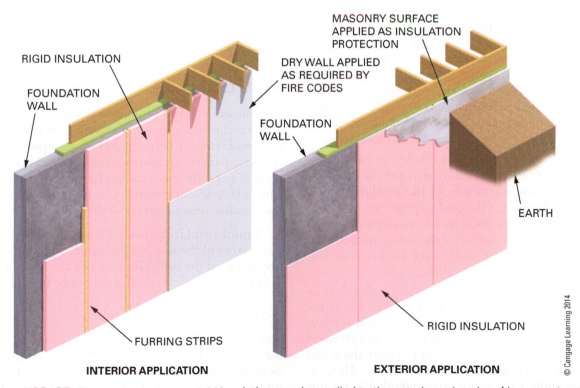

Figure 49–25 Extruded polystyrene rigid insulation may be applied to the exterior or interior of basement walls.

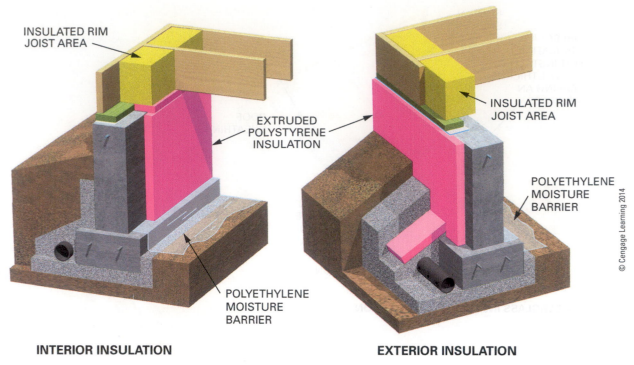

INSULATED RIM JOIST AREA

EXTRUDED POLYSTYRENE INSULATION

INSULATED RIM JOIST AREA

POLYETHYLENE MOISTURE BARRIER

POLYETHYLENE MOISTURE BARRIER

INTERIOR INSULATION

EXTERIOR INSULATION

© Cengage Learning 2014

Figure 49–26 Crawl space insulation methods.

For damp walls, only extruded polystyrene boards should be used.

Crawl space foundation walls may be insulated in the same fashion. Each method uses extruded polystyrene (*Figure 49–26*). The interior application does not need to be furred like basement walls. The band or rim joist area of the floor must be insulated. The entire ground area inside the building must be covered with a moisture barrier to keep ground moisture from entering the building. A layer of 6 mill polyethylene is typically used. Also the entire floor joist system should not be insulated. This will help keep the space warmer and dryer.

Exterior application of insulation has advantages over interior insulation. First the masonry wall is on the warm side of the insulation. This makes it harder for the cold to penetrate under the footing. Basement slabs are less susceptible to cold and frost heaves. Secondly the entire surface need not be covered with wall board for fire code reasons as for interior applications.

The disadvantage with exterior applications is that a protective layer must be installed over the foam where it extends above the ground surface. This layer should be a masonry or treated plywood layer to protect the foam from physical abuse and the sun, which will degrade it over time. Also, exterior application is not recommended in regions where termites are a problem. Termites are more difficult to detect behind the insulation layer. ✱

✱IRC Chapter11 N1101.13.1

Installing Reflective Insulation

Reflective insulation usually is installed between framing members in the same manner as blanket insulation. It is attached to either the face or side of the studs. However, an air space of at least $\frac{3}{4}$ inch must be maintained between its surface and the inside edge of the stud.

ACOUSTICAL INSULATION

Acoustical or *sound* insulation resists the passage of noise through a building section. The reduction of sound transfer between rooms is important in offices, apartments, motels, and homes. Excessive noise is not only annoying but harmful. It not only causes fatigue and irritability, but also can damage the sensitive hearing nerves of the inner ear.

Sound insulation between active areas and quiet areas of the home is desirable. Sound insulation between the bedroom area and the living area is important. Bathrooms also should be insulated.

Sound Transmission

Noises create sound waves. These waves radiate outward from the source until they strike a wall, floor, or ceiling. These surfaces then begin to vibrate by the pressure of the sound waves in the air. Because the wall vibrates, it conducts sound to the other side in varying degrees, depending on the wall construction.

25	Normal speech can be understood quite easily
30	Loud speech can be understood fairly well
35	Loud speech audible but not intelligible
42	Loud speech audible as a murmur
45	Must strain to hear loud speech
48	Some loud speech barely audible
50	Loud speech not audible

This chart from the Acoustical and Insulating Materials Association illustrates the degree of noise control achieved with barriers having different STC numbers.

© Cengage Learning 2014

Figure 49–27 Approximate effectiveness of sound reduction in walls with varying STC ratings.

Sound Transmission Class. The resistance of a building section, such as a wall, to the passage of sound is rated by its Sound Transmission Class (STC). The higher the number, the better the sound barrier. The approximate effectiveness of walls with varying STC numbers is shown in *Figure 49–27*.

Sound travels readily through the air and through some materials. When airborne sound strikes a wall, the studs act as conductors unless they are separated in some way from the covering material. Electrical outlet boxes placed back-to-back in a wall easily transmit sound. Faulty construction, such as poorly fitted doors, often allows sound to pass through. Therefore, good, airtight construction practices are the first line of defense in controlling sound.

Wall Construction

A wall that provides sufficient resistance to airborne sound should have an STC rating of 45 or greater. At one time, the resistance usually was provided only by double walls, which resulted in increased costs. However, a system of using sound-deadening insulating board with a gypsum board outer covering has been developed. This system provides good sound resistance. Resilient steel channels placed at right angles to the studs isolate the gypsum board from the stud. *Figure 49–28* shows various types of wall construction and their STC rating.

Floor and Ceiling Construction

Sound insulation between an upper floor and the ceiling below involves not only the resistance of airborne sounds, but also that of impact noise. Impact noise is caused by such things as dropped objects, footsteps, or moving furniture. The floor is vibrated by the impact. Sound is then radiated from

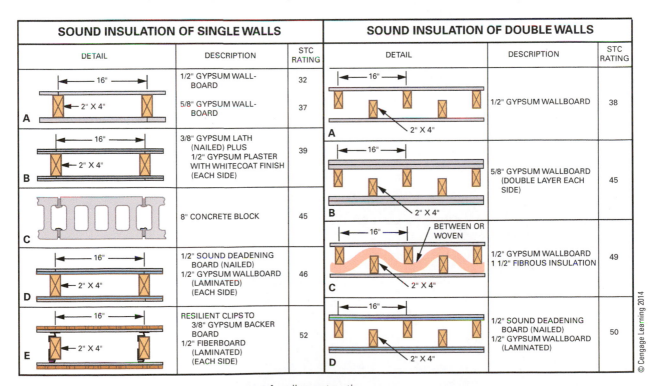

Figure 49–28 STC ratings of various types of wall construction.

both sides of the floor. Impact noise control must be considered as well as airborne sounds when constructing floor sections for sound insulation.

An **Impact Noise Rating (INR)** shows the resistance of various types of floor-ceiling construction to impact noises. The higher the positive value of the INR, the more resistant the insulation is to impact noise transfer. *Figure 49–29* shows various types of floor-ceiling construction with their STC and INR ratings.

Sound Absorption

The amount of noise in a room can be minimized by the use of *sound-absorbing materials*. Perhaps the most commonly used material is acoustical tile made of fiberboard. These tiles are most often used in the ceiling where they are not subjected to damage. The tiles are soft. The tile surface consists of small holes or fissures or a combination of both (*Figure 49–30*). These holes or fissures act as sound traps. The sound waves enter, bounce back and forth, and finally die out.

RELATIVE IMPACT AND SOUND TRANSFER IN FLOOR-CEILING COMBINATIONS (2" X 10" JOISTS)		ESTIMATED VALUE		RELATIVE IMPACT AND SOUND TRANSFER IN FLOOR-CEILING COMBINATIONS (2" X 8" JOISTS)		ESTIMATED VALUE	
DETAIL	DESCRIPTION	STC RATING	APPROX. INR	DETAIL	DESCRIPTION	STC RATING	APPROX. INR
A 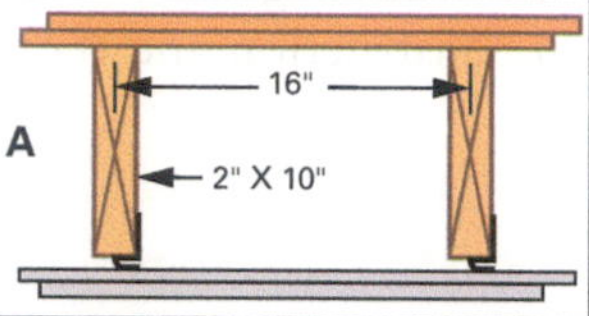	FLOOR 3/4" SUBFLOOR BUILDING PAPER 3/4" FINISH FLOOR CEILING GYPSUM LATH AND SPRING CLIPS 1/2" GYPSUM PLASTER	52	−2	D	FLOOR 7/8" T. & G. FLOORING CEILING 3/8" GYPSUM BOARD	30	−18
B 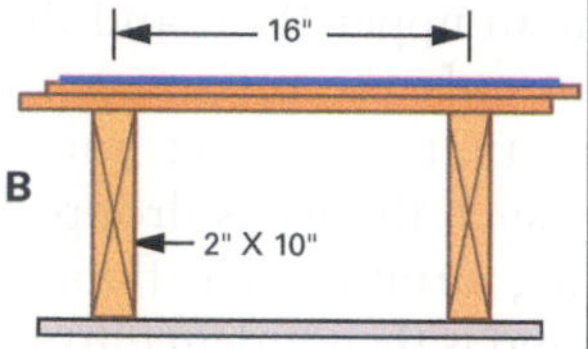	FLOOR 5/8" PLYWOOD SUBFLOOR 1/2" PLYWOOD UNDERLAYMENT 1/8" VINYL-ASBESTOS TILE CEILING 1/2" GYPSUM WALLBOARD	31	−17	E	FLOOR 3/4" SUBFLOOR 3/4" FINISH FLOOR CEILING 3/4" FIBERBOARD	42	−12
C 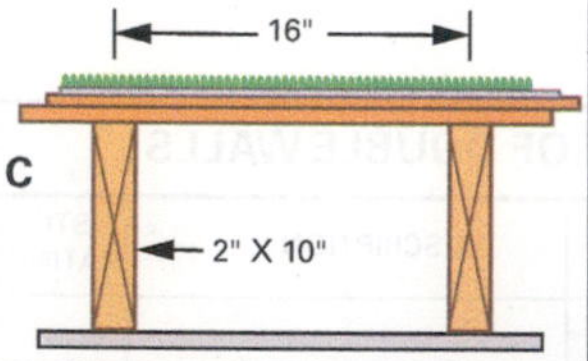	FLOOR 5/8" PLYWOOD SUBFLOOR 1/2" PLYWOOD UNDERLAYMENT FOAM RUBBER PAD 3/8" NYLON CARPET CEILING 1/2" GYPSUM WALLBOARD	45	+5	F 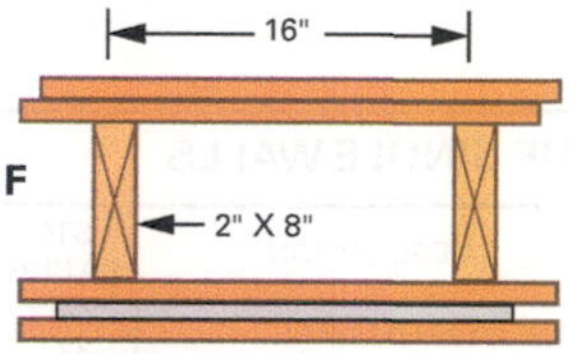	FLOOR 3/4" SUBFLOOR 3/4" FINISH FLOOR CEILING 1/2" FIBERBOARD LATH 1/2" GYPSUM PLASTER 3/4" FIBERBOARD	45	−4

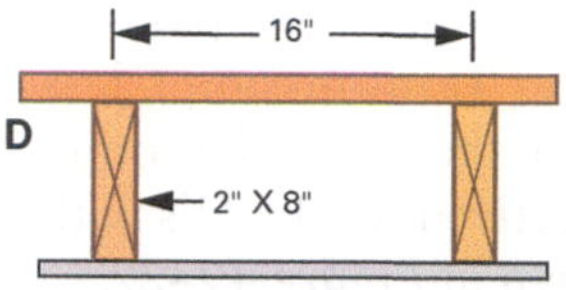

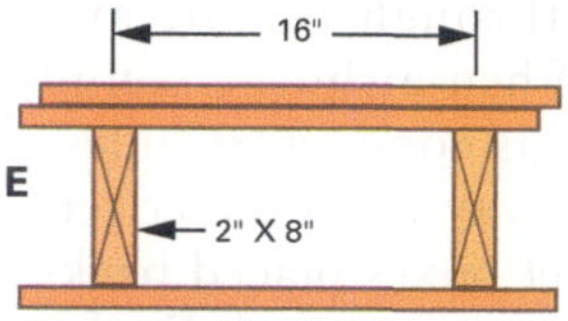

© Cengage Learning 2014

Figure 49–29 STC and INR for floor and ceiling constructions.

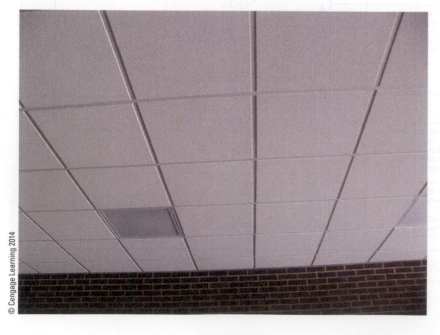

Figure 49–30 Sound-absorbing ceiling tiles.

© Cengage Learning 2014

CHAPTER 50
Condensation and Ventilation

Energy use and costs are reduced when buildings are insulated. Unfortunately, when a building is made more airtight and more energy efficient, a negative side effect occurs. This effect often involves water.

Water is the enemy of a building. Roofing and siding are installed to protect the building from water. Water can also cause the interior to be musty and moldy, the insulation to degrade and perform poorly, and even lead to structural failures. Moisture in and around a building must be understood and dealt with for the building to function properly and last a long time.

Energy-efficient construction is desired to reduce energy costs and make the building more comfortable. But the problems of excess moisture within a building must be addressed. Solutions include controlling air leakage, using vapor retarders, also called vapor barriers, and building drying potential into building systems.

CONDENSATION

Water exists in solid, liquid, and gaseous forms. Most familiar are the solid and liquid forms, which we all know as ice and water. Gaseous water is called steam when it is very hot and vapor when it is cool (*Figure 50–1*). Vapor is normally invisible. Condensation is when the vapor falls out of the air on to cooler surfaces. Dew point is a term used to identify the air temperature when condensation has occurred. Fog is an example of air that has reached its dew point.

Relative Humidity

The amount of moisture held in the air is referred to as relative humidity. This effect of this amount varies with the temperature of the air. Warm air can hold more moisture than cool air. Consider four identical 1-cubic-foot containers of air, each having about five drops (0.00047 pound) of water suspended in

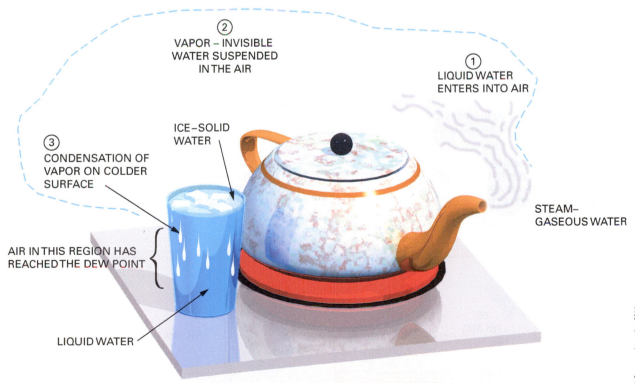

② VAPOR – INVISIBLE WATER SUSPENDED IN THE AIR

① LIQUID WATER ENTERS INTO AIR

ICE – SOLID WATER

③ CONDENSATION OF VAPOR ON COLDER SURFACE

STEAM – GASEOUS WATER

AIR IN THIS REGION HAS REACHED THE DEW POINT

LIQUID WATER

© Cengage Learning 2014

Figure 50–1 Water can exist in three forms or states. Moisture in warm air condenses when it comes in contact with a cold surface.

the air as vapor (*Figure 50–2*). The relative humidity (RH) in each container depends on the temperature of the air. At 70°, the RH would be 40 percent, or put another way, the air would be holding 40 percent of the moisture it could hold at that temperature.

As the temperature drops, the air is less able to hold moisture. When the temperature reaches 43°, the air can hold no more than the five drops; thus the relative humidity is 100 percent. The air is completely saturated with moisture. If we could feel that air it would feel moist and clammy yet it is the same amount of moisture as in the previous situations. A slight lowering of temperature will cause the air to be at the dew point and the container walls begin to feel moist.

Dew point is not always at 43°; it can be any temperature where the RH reaches 100 percent. The air RH is 100 percent just after a rain shower on an 80° day. The dew point occurs when the air can no longer hold any more moisture.

Moisture in Buildings

Building materials perform best and last longest when the relative humidity averages at or below 50 percent, but the RH in the living environment is often much higher.

Moisture in the form of water vapor inside a building comes from many activities. It is produced by cooking, bathing, washing, drying, and cleaning as well as many other sources. Reducing the moisture production within the building is one step in solving the problem, but can go only so far.

Older uninsulated homes did not have the interior moisture problems that we have today. They were drafty enough to dry the house during the heating season. Often, humidifiers were operated to add moisture to the air. Today, tighter homes might require dehumidifiers during the heating seasons (*Figure 50–3*).

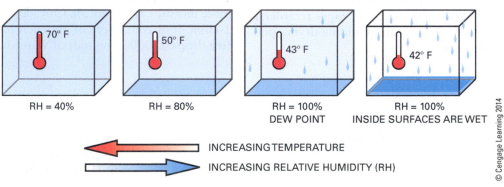

IDENTICAL BOXES, IDENTICAL AMOUNTS OF WATER VAPOR INSIDE.
TEMPERATURE DIFFERENCE CAUSES RELATIVE HUMIDITY TO CHANGE.

70° F   50° F   43° F   42° F

RH = 40%   RH = 80%   RH = 100%   RH = 100%
DEW POINT   INSIDE SURFACES ARE WET

INCREASING TEMPERATURE
INCREASING RELATIVE HUMIDITY (RH)

© Cengage Learning 2014

Figure 50–2 Relative humidity changes with air temperature.

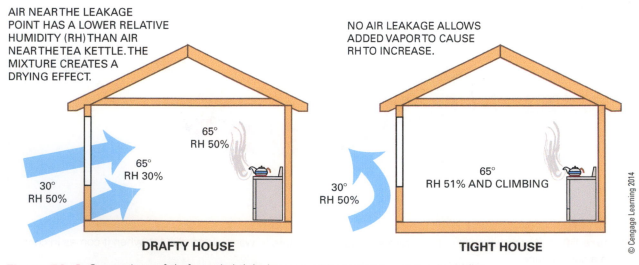

AIR NEAR THE LEAKAGE POINT HAS A LOWER RELATIVE HUMIDITY (RH) THAN AIR NEAR THE TEA KETTLE. THE MIXTURE CREATES A DRYING EFFECT.

NO AIR LEAKAGE ALLOWS ADDED VAPOR TO CAUSE RH TO INCREASE.

65°
RH 50%

65°
RH 30%

30°
RH 50%

DRAFTY HOUSE

30°
RH 50%

65°
RH 51% AND CLIMBING

TIGHT HOUSE

© Cengage Learning 2014

Figure 50–3 Comparison of drafty and airtight houses with respect to relative humidity.

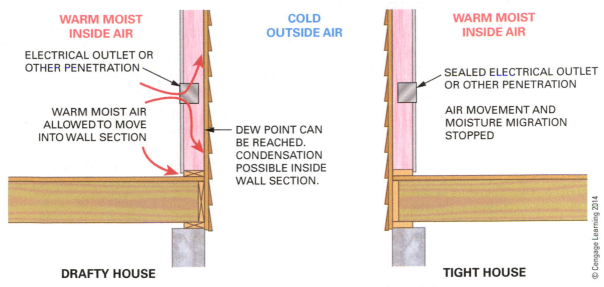

Figure 50–4 Moisture migration into building materials is caused mostly by air leakage.

If water vapor moves into the thermal envelope (insulated walls, ceilings, and floors), it will condense on cooler surfaces (*Figure 50–4*). Inside the wall, this contact point may be on the inside surface of the exterior wall sheathing. In a crawl space, condensation can form on the floor frame and subfloor. In attics, the ceiling joists and roof frame can become saturated with moisture.

Controlling moisture in the building is essential. Condensation of water vapor inside walls, attics, roofs, and floors could lead to serious problems.

Reducing the production of moisture within the house is ultimately the responsibility of the home owner. Home owners must be educated about the problems of moisture. Proper maintenance of exhaust piping for clothes dryers, and using bathroom and kitchen fans regularly are important. If they are defective, constricted, or unused, the moist air will not be removed.

Moisture Problems

- High relative humidity and a warm environment will allow mold and mildew to grow. This can be seen in bathrooms and basements.
- When insulation absorbs water, the dead air spaces in the insulation may become filled with water. Insulation may compress when it gets wet and will not return to its original shape. This significantly reduces the insulation R-value.
- Uncontrolled moisture can move through the wall. This may cause exterior paint to blister and peel.

- A warm attic will cause the formation of ice dams at the eaves. After a heavy snowfall, lost heat from the building causes the snow next to the roof to melt. Water then runs down the roof and freezes on the cold roof overhang, forming ice buildup. As this continues, an ice dam is formed. This causes water to back up on the roof, under the shingles, and into the walls and ceiling (*Figure 50–5*).

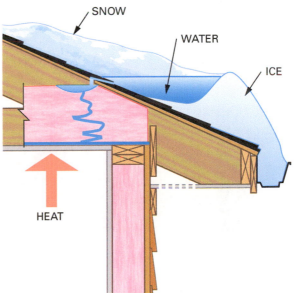

HEAT ESCAPING FROM CEILING MELTS SNOW. WATER FLOWS TO THE OVERHANG WHERE IT FREEZES INTO AN ICE DAM. WATER BUILDUP BEHIND DAM BACKS UP UNDER ROOFING MATERIAL.

Figure 50–5 A properly constructed and ventilated attic will keep ice dams from forming.

PREVENTION OF MOISTURE PROBLEMS

The goal is not to remove all moisture, because this would be virtually impossible and undesirable from the standpoint of human comfort. The goal is to reduce moisture migration into the building components and remove excess moisture through ventilation.

Vapor Retarder

A vapor retarder (barrier) is a material used to slow the flow of airborne moisture from passing through building materials. Polyethylene sheeting installed behind the drywall or interior finish serves as a vapor retarder. It is a transparent plastic sheet. It comes in rolls of usually 100 feet in length and 8, 10, 12, 14, 16, and 20 feet in width. The most commonly used thicknesses are 4 mils and 6 mils.

Blanket and batt insulations are manufactured with facings that also may serve as a vapor retarder. In order for these facings to perform as a vapor barrier, the stapling flanges must be installed in a tight manner. The flanges should be overlapped and stapled to the stud thickness, not the side (*Figure 50–6*). The drywall or finish material will help hold the flanges tight.

Figure 50–6 The vapor retarder material of faced insulation must be overlapped and stapled to the stud thickness to be effective.

© Cengage Learning 2014

Air Leakage

The major cause of moisture migration into building components is air leakage. If air is allowed to move into a wall section, moisture will be included. Thus, if air movement into the wall section is stopped, so is moisture migration.

Many methods and techniques are available to reduce air leakage. Some involve only a little extra time and money:

- Apply caulk under exterior wall plates before they are stood up to seal the plates to the subfloor.
- Install wall sheathing panels with adhesive applied to studs and plates to seal perimeter of each sheet.
- Apply sheathing tape to the seams of wall sheathing.
- Seal all penetrations interior and exterior framing. These include those created by plumbing, electrical, heating, air conditioning, and ventilating installations.
- Install foamed-in-place insulation.
- Install drywall panels with construction adhesive applied to studs and plates to seal the sheet perimeter.

It is not necessary to use every technique to reduce air leakage; each will make a difference. For example, an effective approach could be to seal the wall plates to the subfloor, seal holes and penetrations, and seal wall sheathing seams with sheathing tape.✴

✴IRC Chapter 11 N1102.4

Ventilation

Ventilation is the exchange of air to allow for drying and improved air quality. This often must take place both inside the living environment and within the building components.

Some believe buildings can be built too airtight. This is not an accurate statement. More to the point, buildings must control unwanted air leakage and ventilate unwanted moisture. With airtight construction techniques, the energy costs of the building are reduced. With proper ventilation, the building is comfortable and will last a long time.

Ventilation can be achieved either by allowing the natural flow of air or by using a fan. The interior living areas are vented with fans installed in rooms where air quality needs to be improved. These areas include bathrooms, kitchens, and laundry rooms. Using them daily exhausts the moisture-laden air to the outside.

Drying Potential

Materials of a building are exposed to severe weather. The building should be constructed such that building materials can dry easily. This is referred to as drying potential. Drying will allow condensed moisture and wind-driven water to be removed by evaporation. It can be achieved by constructing natural ventilation in the building.

Proper ventilation depends on where the building is built. The relative humidity of the region must be considered. In cool climes where the building will require heat, the inside air usually has a higher RH than the outside air. In warm moist climes requiring air-conditioning, the outside air usually has the higher RH.

When air leakage into the building takes place, the outside air brought in is adjusted to the same temperature as the inside air. This may be accomplished heating or cooling. When air is warmed, its ability to hold moisture increases and its RH is reduced. This has a drying effect. If the air must be cooled, its ability to hold moisture goes down and its RH goes up. This will possibly cause dampness.

In crawl spaces under floors where the air is usually cooler than the outside air, no ventilation should be installed. Here, warmer outside air entering the crawl space area would be cooled. This would raise the relative humidity, causing more moisture problems. The region in a crawl space should be kept dry and air tight. It should be insulated and considered inside the thermal envelope.

Drying would happen more effectively when the outside air was cooler. The best solution for crawl spaces is to install the vapor retarder on the ground. A sheet of polyethylene will inhibit moisture from leaving the ground and getting into the floor system above. It should have the seams sealed and a layer of sand over the sheet will protect it from accidental perforations.

Siding is designed to stop weather borne water from entering the building, but it is never perfect. Any wind driven rain that does penetrate the siding layer must be given a chance to leave. To accomplish this the siding should be furred off the wall sheathing. This is referred to as a rain screen detail. First the sheathing surface must be waterproof as with foil-faced foam sheathing or an air barrier material. Next furring is applied to the wall over the studs. Screening is applied at the bottom of the wall to allow air to enter behind the siding and keep out insects (*Figure 50–7*). Then siding is fastened over the furring into the studs.

Attic Ventilation

With a well-insulated ceiling and adequate ventilation, attic temperatures are lower and excess moisture is removed.

On roofs where the ceiling finish is attached to the roof rafters, insulation is usually installed between the rafters. An adequate air space of at least 2 inches should be maintained between the insulation and the roof sheathing. The air space must be connected to air inlets in the soffit and outlets at the ridge (*Figure 50–8*). Failure to do so will result in formation of ice dams at the eaves in cold climates. ✱ ✱R R806.1

Green Tip

Mold problems are a result of excess moisture, not necessarily poor housekeeping. Dry buildings are healthy buildings.

Types of Ventilators

There are many types of ventilators. Their location and size are factors in providing adequate ventilation. ✱ ✱R R806.4

Ventilating Gable Roofs. The best way to vent an attic is with a combination of ridge and soffit vents (*Figure 50–9*). In this way each rafter cavity is vented from soffit to ridge. The roof sheathing is cut back about 1 inch from the edge of the ridge board on each side of the ridge, and the vent material is nailed over this slot.

Cap shingles then can be nailed directly to the vent installed over the vent space. Perforated material or screen vents are installed in the soffits to provide the entry point for the ventilation. Positive-ventilation chutes should be installed to prevent any air obstructions caused by the ceiling

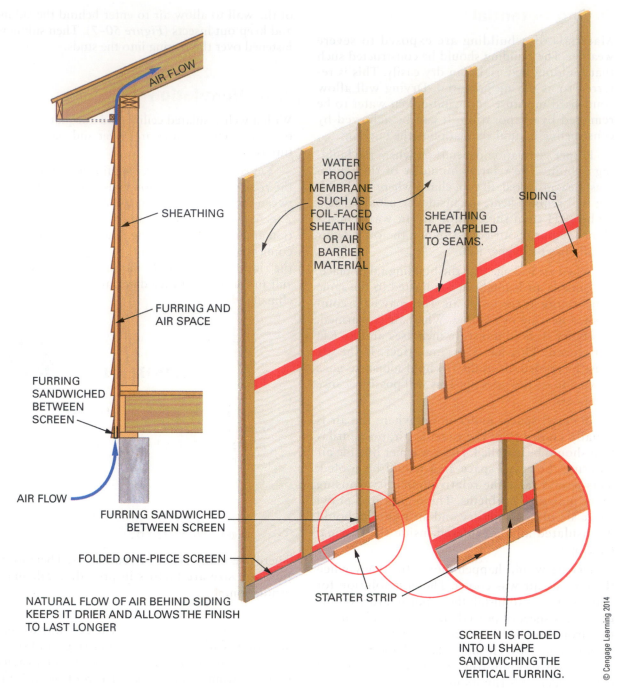

AIR FLOW

SHEATHING

WATER PROOF MEMBRANE SUCH AS FOIL-FACED SHEATHING OR AIR BARRIER MATERIAL

SHEATHING TAPE APPLIED TO SEAMS.

SIDING

FURRING AND AIR SPACE

FURRING SANDWICHED BETWEEN SCREEN

AIR FLOW

FURRING SANDWICHED BETWEEN SCREEN

FOLDED ONE-PIECE SCREEN

STARTER STRIP

NATURAL FLOW OF AIR BEHIND SIDING KEEPS IT DRIER AND ALLOWS THE FINISH TO LAST LONGER

SCREEN IS FOLDED INTO U SHAPE SANDWICHING THE VERTICAL FURRING.

© Cengage Learning 2014

Figure 50–7 Furring siding away from the sheathing allows for drying potential.

insulation near the eaves (*Figure 50–10*). This system can adequately vent the attic space of a house that is up to 50 feet wide. ✱

✱ **IRC R806.3**

Triangularly shaped louver vents are sometimes installed in both end walls of a gable roof. They come in various shapes and sizes and are installed as close to the roof peak as possible. Their effectiveness depends on the prevailing wind direct and are not necessary if continuous ridge and soffits vents are used.

The minimum free-air area for attic ventilators is based on the ceiling area of the rooms below. The free-air area for the openings should be $1/300$th of the ceiling area. For example, if the ceiling area is 1,200 square feet, the minimum total free-air area of the ventilators should be 1,200 divided by 300 = 4 square feet of open area. ✱ This area should be evenly divided between points of entry and exit of the air. Also, if no ceiling vapor retarder is installed then the free area opening space should be $1/150$.

✱ **IRC R806.2**

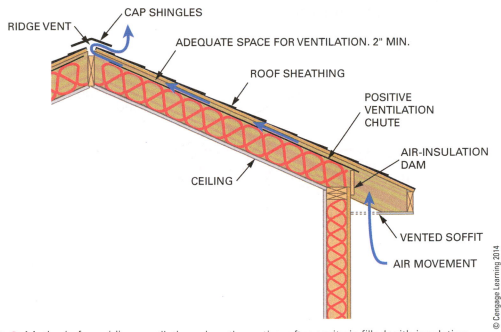

Figure 50–8 Method of providing ventilation when the entire rafter cavity is filled with insulation.

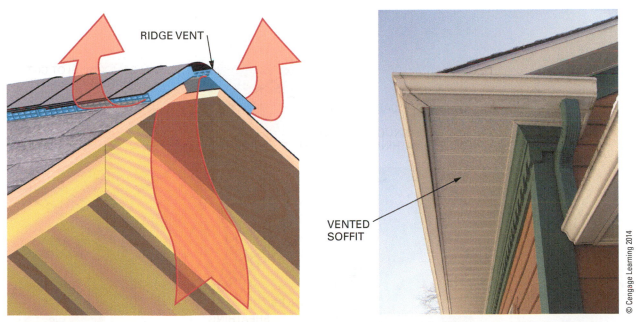

Figure 50–9 Ridge and soffit vents work together to provide adequate attic ventilation.

Ventilating Hip Roofs. Hip roofs should have additional continuous venting along each hip rafter. This allows each hip-jack rafter cavity to be vented. When cutting a 1-inch-wide slot on either side of the hip rafter for the vents, it is recommended to leave a 1-foot section of sheathing uncut between every 2 feet of slot section (*Figure 50–11*). This allows for adequate ventilation while maintaining the integrity sheathing for the hip roof.

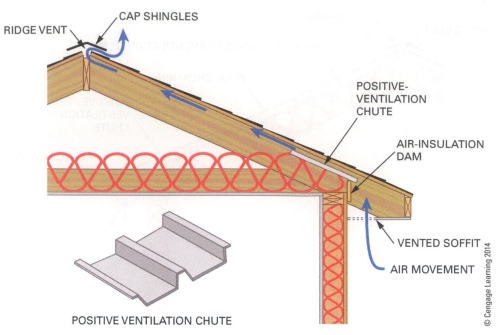

Figure 50–10 Positive-ventilation chutes maintain air space between compressed insulation and roof sheathing.

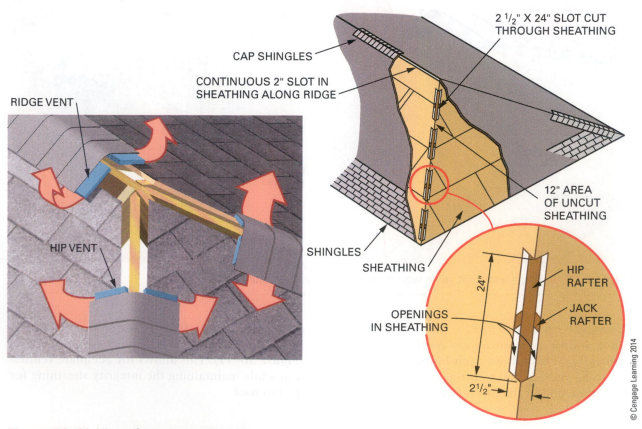

Figure 50–11 Hip roofs can be vented with continuous ridge and hip vents.

Reducing Air Leakage

Reducing air leakage is done by making the building more airtight. This may be done in several ways. Airtight sheathing is installed by adding construction adhesive. Apply a continuous bead on the studs and plates where the perimeter of the panel will fit seals the panel to the wall frame. Nail the panel as required by nailing codes. Another technique uses sheathing tape that is installed such that it covers every panel seam. Either technique is completed with a bead of adhesive applied under the bottom plate before the wall is raised into position.

Airtight drywall is installed in a similar fashion as airtight sheathing (*Figure 50–12*). Adhesive is applied continuously to the studs and plates along the sheet perimeter. The plate should also be sealed to the subfloor before the wall is erected.

Foamed-in-place insulation creates an airtight thermal envelope. It is installed by insulation contractors with special equipment (*Figure 50–13*).

All penetrations should be sealed before the interior finish is applied. No crack or hole is too small to be sealed (*Figure 50–14*). Interior and exterior walls are equally important. Any air movement in the structure can cause air leakage and moisture migration.

The 2012 IECC (International Energy Conservation Code) requires that all new buildings be inspected for airtightness. This involves a review of a check list followed by a blower door test. The check list is provides detailed instruction on proper installation of air barriers and insulation (*Figure 50–15*).

Air leakage in a building is measured in air changes per hour. This refers to the number of times the entire volume of inside air is changed without side air. The IECC 2012 requires all buildings to be tested for air leakage. The rate must not exceed 5 air changes per hour in Climate Zones 1 and 2. The rate must not exceed 3 air changes per hour in Climate Zones 3 through 8.

The testing must be done with a blower door. A blower door unit is designed to temporarily fit into a door opening. It has pressure gauges and a large variable speed fan (*Figure 50–16*). It is also very useful in locating where air leakage is occurring with the use of a smoke producing devise. The cloud of smoke follows the air movements.

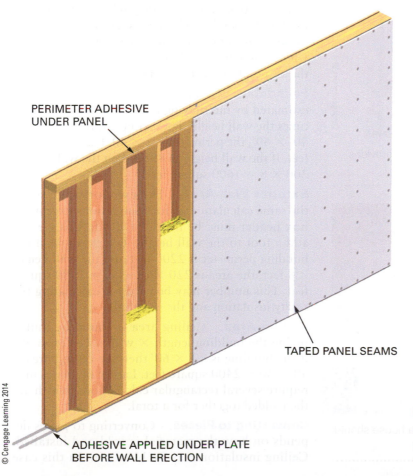

PERIMETER ADHESIVE UNDER PANEL

TAPED PANEL SEAMS

ADHESIVE APPLIED UNDER PLATE BEFORE WALL ERECTION

© Cengage Learning 2014

Figure 50–12 Airtight drywall may be achieved with adhesive placed on the perimeter of panel.

Installing a Vapor Retarder

Polyethylene sheeting is installed by unrolling a length long enough to cover the wall. Add extra length and width to ensure proper coverage. All seams should be along a nailing surface such as a plate or stud.

Partially unfold the section to expose the long edge. Staple it along the top plate about 6 to 12 inches apart, letting the rest drape down to the

Figure 50–13 Foamed-in-place insulation provides an excellent R-value and airtightness.

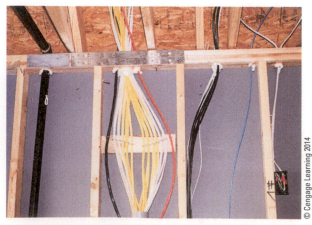

Figure 50–14 All penetrations within a house should be sealed.

floor. Next, begin along the stud in the middle of the sheeting. Smooth out wrinkles downward as it is stapled. The result is that the staples will form a large T shape in the sheeting. Have someone pull each unfastened corner snug to smooth wrinkles. Staple as needed (*Procedure 50–A*).

Carefully fit it around all openings. Lap all joints by several inches, keeping them on a surface nailing. Repair any tears with sheathing tape. Cut off the excess at the floor line.

Some carpenters cut the film out of openings after the drywall finish is applied. This assures a more positive seal. The retarder should be fitted tightly around outlet boxes. Add a ribbon of sealing compound around outlets and switch boxes.

ESTIMATING INSULATION MATERIALS

Estimating the amount of material needed for insulation materials involves finding the square footage of various sections of the thermal envelope. Remember that if the number of pieces does not come out even, round them up to the next whole number of pieces.

Thermal Envelope Area

Insulation types vary depending where they are to be used in the building. Ceiling insulation is a different thickness and often a different material than the walls, while foundation insulation is different yet. Finding the area of the insulation section is the first step.

Foundation Wall Area. Foundation wall area is estimated by multiplying the foundation perimeter times the wall height. For example if the building is $40' \times 60'$, the perimeter is $2 \times 40 + 2 \times 60 = 200$ feet. If the wall height is 8 feet then the wall area is $200 \times 8 = 1600$ square feet.

Exterior Wall Area. The exterior wall area is the same calculation as for foundations except the box header must be included. To do this we can add a foot to the wall height. For example, if the building perimeter is 2200 feet and the wall height is 8 feet, the area is $220 \times (8 + 1) = 1980$ square feet. This number may be used for calculating the cavity insulation and the sheathing.

Ceiling Area. Ceiling area is found by multiplying the building length $\times$ width. For example, if the building is $40' \times 60'$ then the ceiling area is $40 \times 60 = 2400$ square feet. Larger buildings may require several rectangular calculations, which are then added together for a total.

Converting to Pieces. Converting to pieces depends on how the material is sold and installed. Ceiling insulation may be blown-in. In this case

Checklist Excerpts from IECC Table 402.4.1.1 Air Barrier and Insulation Installation	
Air barrier and thermal barrier	A continuous air barrier shall be installed in the building envelope. Exterior thermal envelope contains a continuous air barrier. Breaks or joints in the air barrier shall be sealed. Air-permeable insulation shall not be used as a sealing material.
Ceiling/attic	The air barrier in any dropped ceiling/soffit shall be aligned with the insulation and any gaps in the air barrier sealed. Access openings, drop down stair or knee wall doors to unconditioned attic spaces shall be sealed.
Walls	Corners and headers shall be insulated and the junction of the foundation and sill plate shall be sealed. The junction of the top plate and top of exterior walls shall be sealed. Exterior thermal envelope insulation for framed walls shall be installed in substantial contact and continuous alignment with the air barrier. Knee walls shall be sealed.
Windows, skylights and doors	The space between window/door jambs and framing and skylights and framing shall be sealed.
Rim joists	Rim joists shall be insulated and include the air barrier.
Floors (including above-garage and cantilevered floors)	Insulation shall be installed to maintain permanent contact with underside of subfloor decking. The air barrier shall be installed at any exposed edge of insulation.
Crawl space walls	Where provided in lieu of floor insulation, insulation shall be permanently attached to the crawl space walls. Exposed earth in unvented crawl spaces shall be covered with a Class I vapor retarder with overlapping joints taped.
Shafts, penetration	Duct shafts, utility penetrations and flue shafts opening to exterior or unconditioned space shall be sealed.
Narrow cavities	Batts in narrow cavities shall be cut to fit, or narrow cavities shall be filled by insulation that on installation readily conforms to the available cavity space.
Garage separation	Air sealing shall be provided between the garage and conditioned spaces.
Recessed lighting	Recessed light fixtures installed in the building thermal envelope shall be air tight, IC rated, and sealed to the drywall.
Plumbing and wiring	Batt insulation shall be cut neatly to fit around wiring and plumbing in exterior walls, or insulation that on installation readily conforms to available space shall extend behind piping and wiring.
Shower/tub on exterior wall	Exterior walls adjacent to showers and tubs shall be insulated and the air barrier installed separating them from the showers and tubs.
Electrical/phone box on exterior walls	The air barrier shall be installed behind electrical or communication boxes or air sealed boxes shall be installed.
HVAC register boots	HVAC register boots that penetrate building thermal envelope shall be sealed to the subfloor or drywall.
Fireplace	An air barrier shall be installed on fireplace walls. Fireplaces shall have gasketed doors.

© Cengage Learning 2014

Figure 50-15 Excerpts of the check list of insulation details required by IECC 2012.

©Caro/Alamy

Figure 50–16 A blow door tests the air tightness of a building.

the installer may charge by the square foot. If the insulation is fiberglass batts, then the number of rolls is the area divided by the square feet (SF) per roll. For example, if each roll contains 98 SF then the number of rolls is 2400 SF ÷ 98 = 24.5 or 25 rolls. Since the cut pieces from installation are still useable no waste factor need to be considered. If sheets of rigid foam are used the area is divided by the area per sheet, usually 4 × 8 = 32 SF.

Miscellaneous Insulation Parts

Other insulation material types must also be considered. Spray can foam for windows and doors may be estimated at one can for two windows or doors. Sum up the number of windows and doors; then divide by two. Positive ventilation chutes are needed in each rafter cavity. To find this take the building length and divide by the OC spacing in feet, then double it for the other side. For example a rectangular gable roof building that is 40′ × 60′ with 16″ OC spaced rafters, then 60′ ÷ 16″/12″ = 45 × 2 = 90 pieces. Ridge vents may be ordered from the LF measurement of the ridge.

StepbyStep Procedures

Procedure 50–A Installing a Polyethylene Film Vapor Retarder

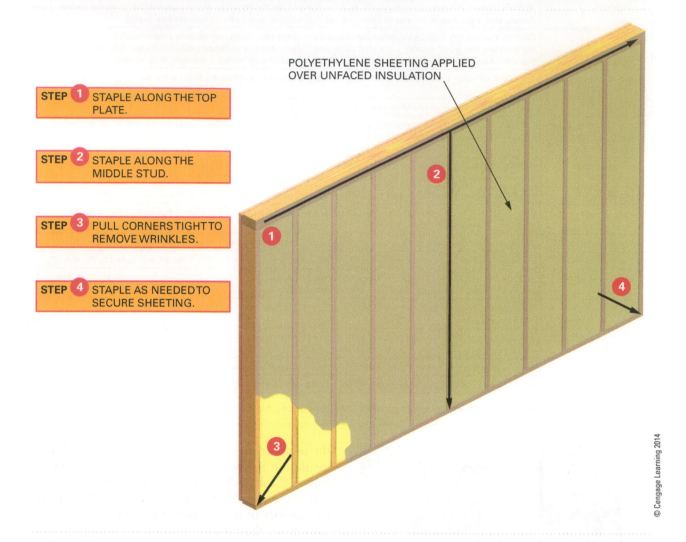

STEP 1 STAPLE ALONG THE TOP PLATE.

STEP 2 STAPLE ALONG THE MIDDLE STUD.

STEP 3 PULL CORNERS TIGHT TO REMOVE WRINKLES.

STEP 4 STAPLE AS NEEDED TO SECURE SHEETING.

POLYETHYLENE SHEETING APPLIED OVER UNFACED INSULATION

© Cengage Learning 2014

Estimate the insulation for a 48′ × 72′ rectangular building with 8 foot walls with 4 × 8 sheets of extruded rigid foam on foundation walls, fiberglass batts in exterior walls from 98 SF rolls and positive ventilation chutes between 16″ OC rafters.

Item	Formula	Waste Factor	Example
Foundation Wall	wall PERM × wall height		2(48 + 72) = 240 LF × 8 = 1920 SF ÷ 32 = 60 pieces
Exterior wall	wall PERM × (wall height + 1 foot)		2(48 + 72) = 240 LF × (8 + 1) = 2160 SF
Ceiling	ceiling area ÷ SF per roll		48 × 72 = 3456 SF ÷ 98 = 35.2 × 36 rolls
Posi-Vent Chutes	building LEN × OC (FT) × 2		72 ÷ 16″/12″ = 54 × 2 = 108 pieces

Figure 50-17 Example of estimating insulation with formulas.

© Cengage Learning 2014

DECONSTRUCT THIS

Carefully study **Figure 50–18** and think about what is wrong and/or what is right. Consider all possibilities. What construction practice or method is different in your area of the country?

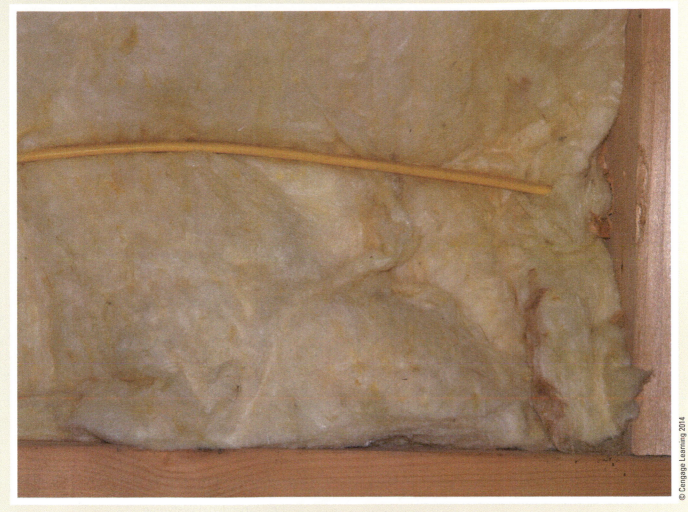

Figure 50–18 This photo shows a wall cavity with fiberglass insulation in it.

KEY TERMS

acoustical tile	dew point	flexible insulation	insulated glass insulation
Btu	drying potential	foamed-in-place	loose-fill insulation
condensation	equilibrium	ice dam	low-emissivity coating
conduction	expanded polystyrene	impact noise	perlite
convection	extruded polystyrene	Impact Noise Rating (INR)	

polyurethane	sound-deadening insulating board	spray foams	vapor retarder
radiation		storm window	vermiculite
R-value	Sound Transmission Class (STC)	thermal envelope	weather stripping
relative humidity		vapor	

REVIEW QUESTIONS

Select the most appropriate answer.

1. Which method of popping popcorn uses conduction as a mode of heat transfer?

   a. stove top
   b. hot air
   c. microwave
   d. solar

2. The structural building material with the greater R-value is

   a. concrete.
   b. steel.
   c. stone.
   d. wood.

3. The insulating term R-value is defined as the measure of

   a. resistance of a material to the flow of heat.
   b. relative amount of the heat lost through a building section.
   c. the conductivity of a material.
   d. the total heat transfer through a building section.

4. The boundary between conditioned and unconditioned air is called the thermal

   a. envelope.
   b. resistance.
   c. retarder.
   d. dam.

5. The material used to protect insulation from air infiltration into the Insulation layer at the eaves is called an air

   a. barrier.
   b. dam.
   c. retarder.
   d. stopper.

6. The term used to identify the situation when water droplets form on cooler surfaces is called

   a. evaporation point.
   b. vapor point.
   c. water point.
   d. dew point.

7. Moisture migration into the insulation layer can be reduced by

   a. installing a vapor retarder.
   b. placing sheathing tape on the seams of exterior sheathing.
   c. airtight construction techniques.
   d. all of the above.

8. A vapor retarder in a crawl space should be installed

   a. just below the subfloor.
   b. just under the joists.
   c. on top of the ground.
   d. all of the above.

9. If air temperature suddenly increases, the relative humidity of the air would

   a. increase.
   b. decrease.
   c. remain the same.
   d. all of the above.

10. The best choice of insulation to insulate under a concrete slab is

    a. urethane.
    b. polyisocyanurate.
    c. expanded polystyrene.
    d. extruded polystyrene.

11. The type of insulation that may be installed in a blown or sprayed in fashion is

    a. cellulose.
    b. fiberglass.
    c. urethane.
    d. all of the above.

12. Squeezing or compressing flexible insulation tightly into spaces

    a. reduces its effectiveness.
    b. increases its efficiency by creating more air spaces.
    c. is necessary to hold it in place.
    d. helps prevent air leakage by sealing cracks.

13. When insulation is placed between roof framing members, there should be an air space between the insulation and the roof sheathing of at least

    a. 3 inches.
    b. 2 inches.
    c. 1 inch.
    d. ½ inch.

14. Ice dams on top of the roofing can be eliminated by properly installed

    a. attic ventilation.
    b. roofing material.
    c. vapor retarder.
    d. air dam.

15. The best method of venting an attic space is with

    a. gable vents.
    b. hip vents.
    c. roof windows.
    d. continuous ridge and soffit vents.

16. What state has five climate zones?

    a. California
    b. Arizona
    c. Colorado
    d. all of the above

17. Fenestration refers to

    a. windows
    b. doors
    c. skylights
    d. all of the above

18. U-value is a term

    a. similar to k-value
    b. is the inverse of R-value
    c. used for building assemblies like windows
    d. all of the above

19. The term used to describe a method of maintaining an air space behind siding is called

    a. drying potential
    b. air barrier
    c. vapor barrier
    d. vapor retarder

20. The actual R-value of R-21 insulation installed in 16" OC cavities is

    a. 21
    b. 19
    c. 15.75
    d. 9.75

Exterior Finish

Courtesy John Slaughter

John Slaughter
Title: President and CEO
Company: Retroactive Home Construction, Inc.

© iStockphoto.com/mattjeacock

EDUCATION

Although John intended to pursue a degree in architecture upon entering Texas A&M, he earned a degree in building construction in 1988. "I changed my major right away. The construction group was more my style. I learned the basics of how to do everything."

HISTORY

After graduation, he got a job as a laborer for a massive home builder in the Sacramento, California, area. "My job was entry level—keeping houses clean and hauling lumber. I moved into reading plans and quality control. I did that for about two years."

John was eventually promoted to assistant superintendent, and after another three years he was promoted to superintendent. "Through my career, I climbed the ladder. I had about six jobs in seven and a half years. Each one was a promotion until I became vice president of construction."

In 2003, John changed companies. He left the public-corporation career to join a large local builder as vice president of construction and operations, which allowed him to work directly for the owner. "The support was down the hall, not in another city or state. I thought I would learn more by interacting with the owner," he explained. The company built communities from Yuba City to Bakersfield, an area extending about 350 miles. "It was a large company, a wonderful job, and I loved every minute of it," Unfortunately, the downturn in the economy halted the growth in home construction in California seen in recent decades and led to John's company going bankrupt from having too much debt in landholdings.

In 2008, John started his own home-building company. He started with one home that had been foreclosed upon before completion. "I bid on the completion of it, and they hired me to complete the project." John is currently working with banks, bidding on homes that have been left incomplete. "As much as this economy is sad and depressing, there is opportunity."

ON THE JOB

Being self-employed allows him to set his own schedule. In a typical day, he meets with subcontractors and walks homes. He also spends time learning about accounting processes, invoicing, lien releases, and liability insurance—all aspects of the home-building industry that personnel in other departments usually handle.

BEST ASPECTS

There are pros and cons to it all. John enjoys being responsible for his own company. "I'm bidding ten homes right now but on my own time. I don't answer to anyone. I can do things the way I want." John explained how on his first home, he earned 100 percent of the profit and got hands-on experience with all aspects of running a business. "It has been a great learning experience. Without the bankruptcy, I never would have done it. Being forced into it [by the economic downturn] was a blessing. I'm proud of what I have done. I'm very fortunate that it has worked out for me."

CHALLENGES

John enjoys the freedom that comes with having his own company, but admits the downsides. "There's no steady paycheck or benefits, and I'm always wondering about the next job—these are challenges." John also described the unpredictability of working with banks. "I think the banks in general are completely overwhelmed by what they have foreclosed upon."

IMPORTANCE OF EDUCATION

John described how fortunate he was to find his path through a technical college and earn his degree, an achievement that has come in handy over the years. "Like a lot of people at that age, I was feeling my way through life . . . The degree has really helped me in my career. When it came time for opportunities and promotions, the company I was at always noticed that I had my degree, and it made a difference."

John is required to have a contractor's license to do what he is doing now. "I have had the license for seventeen years, but this is the first time I have been using it. It's a good thing I had it. It has benefited me to have that education early in my career," he explained.

John also explained how an education can get one's foot in the door. "A lot of the huge corporations are involved with the NAHB [National Association of Home Builders] or the BIA [Building Industry Association]. To go through a program sponsored by such an association means getting a foot in the door."

FUTURE OPPORTUNITIES

In the short term, John is working to get more houses and grow his company. "By getting more business, I can hire help and get even more business. I can keep growing the company." John is also pursuing an MBA degree with the University of Phoenix. "I'm in my fourth class, and it's a lot more interesting than I thought it would be. More education is never a bad thing. Having my MBA will make me a little more credible in the future." He explained that he does not really need the master's degree but that if he ever wants to be a company president or to work for a big company again, the MBA will be asset. "I'm setting myself up for the future now. No matter when it is, education is good thing."

WORDS OF ADVICE

John shared pieces of advice he received in his life that he has never forgotten. "First, watch what everyone else around you is doing, and do something better. Ask yourself what is missing, what you can do better to be different." John explained how throughout his career, having his degree coupled with this awareness has given him an edge.

John also stressed that ethics and honesty and treating people with respect—while they sound cliché—are crucial to success. "You have got to really treat people right." Throughout his career John has experienced how ethics apply to his field. Specifically, he is always honest with his bids, which include a fair profit.

Finally, John advises, "Do the job to earn the job. I always took on a lot of work outside my job descriptions. I did a lot of work without the title. Ultimately you get the title, and the rewards will come."

Courtesy John Slaughter

UNIT 19
Roofing

©Shutterstock, Inc./Mau Horng

Materials used to cover a roof and make it weather tight are part of the exterior finish called *roofing*. Roofing adds beauty to the exterior and protects the interior. Before roofing is applied, the roof deck must be securely fastened. There must be no loose or protruding nails, and the deck must be clean of all debris. Properly applied roofing gives years of dependable service.

OBJECTIVES

After completing this unit, the student should be able to:

- define roofing terms.
- describe and apply roofing felt, organic or fiber-glass asphalt shingles, tile roofing, and roll roofing.
- describe various grades and sizes of wood shingles and shakes and apply them.
- flash valleys, sidewalls, chimneys, and other roof obstructions.
- estimate needed roofing materials.

SAFETY REMINDER

A variety of roofing materials is available to protect a building from weather. Application of materials is usually straightforward and designed to be installed with speed. Care must be taken to remember the dangers of falling.

CHAPTER 51

Asphalt Shingles and Tile Roofing

Asphalt shingles are the most commonly used roof covering for residential and light commercial construction. They are designed to provide protection from the weather for a period ranging from twenty to fifty years, and are available in many colors and styles. Tile roofing is used in warmer areas where the effects of the sun are substantial.

ROOF SAFETY

Safety, which is always an important consideration on the job, is never more important than when working on a roof. OSHA has rules for when working at heights. The slope of the roof has an impact on which safety method may be used. The slope designations are low slope roofs and steep roofs. Low slope roofs are those with a slope of 4 on 12 or less.

Both roof situations require protection when working 6 feet (1.8 meters) above a lower level. Protection may be in the form of a guardrail system, safety net system, and personal fall arrest system. These systems all involve hardware to be installed or worn. Steep roofs must have a guardrail system with toeboards, safety net system, or personal fall arrest system. More information can be found in Chapter 20.

Low slope roofs may also use a combination of protections that include a warning line system or safety monitoring system. Warning line is a barrier erected on a roof to warn employees they are approaching a distance of 15 feet from an unprotected roof side or edge. It is constructed with ropes, wires, or chains, with supporting stanchions. It must be flagged at not more than 6-foot (1.8 m) intervals with high-visibility material. It should act similar to a railing but need not be as strong; it is a warning device.

A safety monitoring system is a competent person whose job it is to constantly monitor the workers activities and who is authorized to take prompt corrective measures to eliminate hazardous conditions. He or she must be capable of identifying and correcting roofing hazards and must be trained and capable of identifying existing and predictable roofing conditions that are hazardous or dangerous to roofing employees. The monitor must be on

the same roof as and be able to visibly observe all other employees for whom he or she is responsible and must also be close enough to verbally communicate with them.

For more detailed information go to OSHA's website at www.osha.gov, then follow the links for "Regulations," "Construction," and "1926 Subpart M - Fall Protection."

ROOFING TERMS

An understanding of the terms most commonly used in connection with roofing is essential for efficient application of roofing material.

- A square is the amount of roofing required to cover 100 square feet of roof surface. There are three to five bundles of shingles per square (*Figure 51–1*). One square of shingles can weigh between 235 and 425 pounds depending on shingle quality.

- Deck is the wood roof surface to which roofing materials are applied.

- A shingle butt is the bottom exposed edge of a shingle.

- Courses are horizontal rows of shingles or roofing.

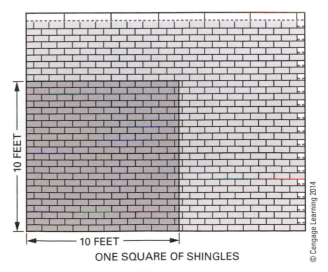

10 FEET

10 FEET

ONE SQUARE OF SHINGLES

© Cengage Learning 2014

Figure 51–1 One square of shingles will cover 100 square feet.

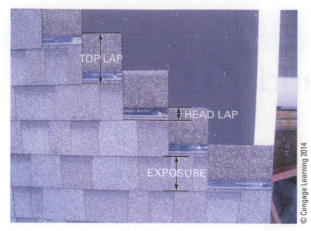

Figure 51–2 Asphalt strip shingle exposure and laps.

- **Exposure** is the distance between courses of roofing. It is the amount of roofing in each course exposed to the weather (*Figure 51–2*).

- The **top lap** is the height of the shingle or other roofing minus the exposure. In roll roofing, this is also known as the **selvage**.

- The **head lap** is the distance from the bottom edge of an overlapping shingle to the top of a shingle two courses under, measured up the slope.

- **End lap** is the horizontal distance that the ends of roofing in the same course overlap each other.

- **Flashing** is strips of thin roofing material. It is usually made of lead, zinc, copper, vinyl, or aluminum. It may also be strips of roofing material used to make watertight joints on a roof. Metal flashing comes in rolls of various widths that are cut to the desired length.

- *Asphalt cements* and *coatings* are manufactured to various consistencies depending on the purpose for which they are to be used. **Cements** are classified as *plastic, lap,* and *quick-setting*. They will not flow at summer temperatures. They are used as adhesives to bond asphalt roofing products and flashings. They are usually troweled on the surface. **Coatings** are usually thin enough to be applied with a brush. They are used to resurface old roofing or metal that has become weathered.

- Electrolysis is a reaction that occurs when unlike metals come in contact with water. This contact causes one of the metals to corrode. A simple way to prevent the disintegration caused by electrolysis is to secure metal roofing material with fasteners of the same metal.

PREPARING THE DECK

In preparation for installing roofing materials, the deck should be clean and clear of debris. This will prevent any damage to the roofing material and make a safer working surface. A metal **drip edge** is often installed along the roof edges. The metal drip edge is usually made of aluminum or galvanized steel (*Figure 51–3*). The drip edge is used to

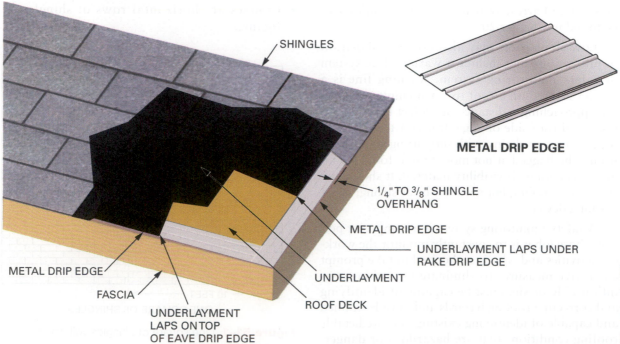

Figure 51–3 A metal drip edge may be used to support the shingle edge overhang.

support the asphalt shingle overhang. Otherwise the shingles would droop from the heat of the sun.

Install the metal drip edge by using roofing nails of the same metal spaced 16–24 inches along its inner edge. End joints may be butted or lapped by about 2 inches. *

✱IRC
R905.2.8.5

Underlayment

✱IRC
R905.2.7

The deck is next covered with an underlayment. It is applied as an extra level of protection for the house from wet weather particularly wind driven rain. The underlayment serves to protect sheathing from moisture before the roofing material is applied (*Figure 51–4*). Historically, underlayment was installed to protect the shingles from the pitch of the board sheathing. *

Asphalt Felts

Asphalt felts consist of heavy felt paper saturated with asphalt or coal tar. They are usually made in various weights of pounds per square (*Figure 51–5*). Asphalt felt comes in 36-inch-wide rolls. The rolls are 36, 72 or 144 feet long covering one, two or four squares.

Usually the lightest-weight felt is used as an underlayment for asphalt shingles.

Apply a layer of asphalt felt underlayment over the deck starting at the bottom. Lay each course of felt over the lower course at least 2 inches. Make any end laps at least 4 inches. Lap the felt 6 inches from both sides over all hips and ridges.

Nail or staple through each lap and through the center of each layer about 16 inches apart. Roofing nails driven through the center of metal discs or specially designed, large-head felt fasteners hold the underlayment securely in strong winds until shingles are applied. A metal drip edge is often installed along the rakes after the application of the underlayment (*Figure 51–6*).

Synthetic Underlayment

Synthetic underlayments are made of two layers of polypropylene. They come in 54-inch-wide 40-pound rolls that are over 200 feet long. They cover about ten squares per roll.

Figure 51–4 Asphalt felt underlayment is applied to the roof deck before shingling.

Courtesy of APA—The Engineered Wood Association

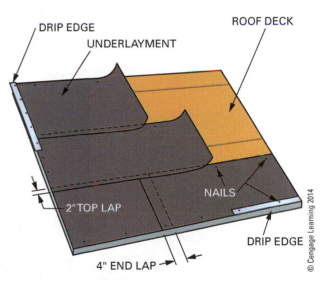

Figure 51–6 Application specifications for asphalt shingle underlayment.

© Cengage Learning 2014

	Approx. Weight Per Roll	Approx. Weight Per Square	Squares Per Roll	Roll Length
	60 #	15 #	4	144"
	60 #	30 #	2	72"
	60 #	60 #	1	36"

SATURATED FELT

© Cengage Learning 2014

Figure 51–5 Sizes and weights of asphalt-saturated felts.

The advantage with synthetic underlayments is they are more tear resistant. This makes them safer to walk on during installation. They also lay flatter than asphalt felts after they get wet (*Figure 51–7*).

Ice & Water Shield®

As an added measure of protection to roofs, an Ice & Water Shield® may be applied. An Ice & Water Shield® is a roofing membrane composed of two waterproofing materials bonded into one layer. It is composed of a rubberized asphalt adhesive backed by a layer of polyethylene. It comes in 3-foot × 75-foot rolls. The rubberized asphalt surface is backed with a release paper to protect the sticky side. During installation, the release paper is peeled off, allowing the asphalt to bond to the roof deck. ✳

✳IRC R905.2.7.1

This material is used in the trouble spots of a roof such as along the eaves, in valleys, and in unique areas where leaks are more likely (*Figure 51–8*). Low-slope roofs and roofs exposed to severe blowing weather are at increased risk of leaks.✳ An Ice & Water Shield® is often used during reroofing applications of houses that experience ice damming at the eaves. (See Chapter 50.)

✳IRC R905.2.7.2

Green Tip

Higher quality shingles last longer and save money in the end. Lesser quality shingles need to be replaced more often, which drives up labor and waste disposal costs.

The Ice & Water Shield® is installed directly on the roof deck. The deck should be clean and clear of debris. There should be no voids in the deck. Laps should be 6 inches at the ends and 4 inches at the sides. Cut pieces to the desired size. After positioning the sheet, fold back enough membrane to peel off some release paper. Reposition the membrane and press the membrane to the deck. Pull remaining release paper off and press the membrane to the deck. This will ensure proper adherence to the roof deck. This membrane is not recommended for use in the hot climes of intense sun such as the desert areas of the Southwest.

Figure 51–7 Synthetic underlayments are installed under the single layer.

Courtesy of GAF Materials Corporation

© W. R. Grace & Co. Conn (Used with permission)

Figure 51–8 Illustration of Ice & Water Shield® roofing underlayment invented by Grace Construction Products for protecting against water damage arising from ice dam formation in trouble spots of a roof such as valleys.

KINDS OF ASPHALT SHINGLES

Two types of **asphalt shingles** are manufactured. *Organic* shingles have a base made of heavy asphalt-saturated paper felt coated with additional asphalt. *Fiberglass* shingles have a base mat of glass fibers. The mat does not require the saturation process and requires only an asphalt coating. Both kinds of shingles are surfaced with selected mineral granules. The asphalt coating provides weatherproofing qualities. The granules protect the shingle from the sun, provide color, and protect against fire.

Fiberglass-based asphalt shingles have an Underwriters Laboratories Class A fire resistance rating. The Class A rating is the highest standard for resistance to fire. The Class C rating for organic shingles, while not as high, will meet most residential building codes.

Asphalt shingles come in a wide variety of colors, shapes, and weights (***Figure 51–9***). Three-tab

Product	Configuration	Per Square			Size			
		Approx. Shipping Weight	Shingles	Bundles	Width	Length	Exposure	Underwriters' Listing
Wood appearance strip shingle more than one thickness per strip 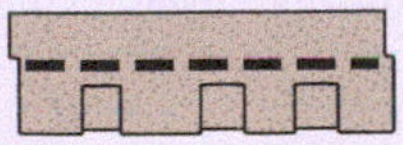 Laminated or job applied	Various edge, surface texture, & application treatments	285# to 390#	67 to 90	4 or 5	11½" to 15"	36" or 40"	4" to 6"	A or C—many wind resistant
Wood appearance strip shingle single thickness per strip 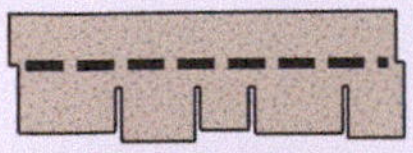	Various edge, surface texture, & application treatments	Various 250# to 350#	78 to 90	3 or 4	12" or 12¼"	30" or 40"	4" to 5⅝"	A or C—many wind resistant
Self-sealing strip shingle 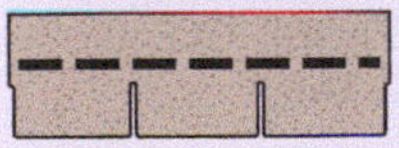	Conventional 3 tab	205#-240#	78 or 80	3	12" or 12¼"	36"	5" or 5⅛"	A or C—all wind resistant
	2 or 4 tab	Various 215# to 325#	78 or 80	3 or 4	12" or 12¼"	36"	5" or 5⅛a	
Self-sealing strip shingle 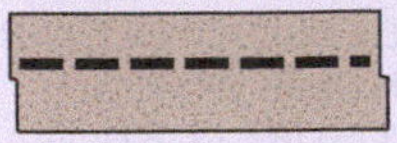 No cut out	Various edge and texture treatments	Various 215# to 290#	78 to 81	3 or 4	12" or 12¼"	36" or 36¼"	5"	A or C—all wind resistant
Individual lock down 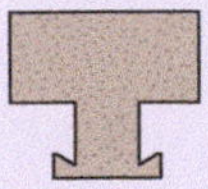 Basic design	Several design variations	180# to 250#	72 to 120	3 or 4	18" to 22¼"	20 to 22½"	—	C—many wind resistant

Courtesy of Asphalt Roofing Manufacturers Association

Figure 51–9 Asphalt shingles are available in a wide variety of sizes, shapes, and weights.

shingles or strip shingles are made with cut out slots every 12 inches. Laminated made are thicker and in patterns to look like expensive slate or wood shake roofs. High-quality varieties come with longer warrantees of fifty years or more.

Shingle quality and longevity are generally determined by the weight per square. Asphalt shingles can weigh anywhere from 235 to 425 pounds per square. Most asphalt shingles are manufactured with factory-applied adhesive. The adhesive sticks the layers together after the sun has a chance to heat up the installed shingles. This increases their resistance to the wind. Each shingle style is applied in a similar manner.

APPLYING ASPHALT SHINGLES

Before applying strip shingles, make sure that the roof deck is properly prepared. The underlayment and drip edge should be applied. Asphalt roofing products become soft in hot weather. Be careful not to damage them by digging in with heavily cleated shoes during application or by unnecessary walking on the surface after application. The slope of a roof should not be less than 4 inches per unit of run when conventional methods of asphalt shingle application are used.✱ Make sure the directions on the back of the bundle are followed. The warrantee will otherwise be voided.

✱IRC R905.2.2

Asphalt Shingle Layout

Layout may begin from either rake edge. Before any shingles are installed, it is best to precut the starter course shingles and the rake edge pieces that begin each course. This makes installation go more smoothly.

Eave Starter Course. The eave starter course backs up and fills in the spaces between tabs of the first regular course of shingles. It is made from shingles cut lengthwise. The full exposure portion of the shingle is removed (*Figure 51–10*). Apply these pieces so the factory-applied sealing strip is near the eave edge. This will seal the first shingle course tabs. Cut the first eave starter about 6 inches short so the butt seams do not align with the first full-tab course of shingle butt seams. Overhang the shingles past the drip edge $\frac{1}{4}$ to $\frac{3}{8}$ inches. Nail the eave starter shingle with four nails in the space just above the seal strip. Make sure there are no nails closer than 12 inches from the butt seams. This will make sure the next course will adequately cover the nails.

BUTT SEAM

FIRST STARTER PIECE IS CUT 6" SHORT TO KEEP BUTT SEAMS FROM ALIGNING.

FIRST SHINGLE IS FULL TAB

EAVE STARTER COURSE MADE FROM FULL TAB SHINGLES WITH THE EXPOSURE PORTION CUT OFF.

© Cengage Learning 2014

Figure 51–10 Eave starters are installed under the first course of shingles.

Some roofers install the eave starter strips up the rake edge as well, keeping the sealing strips closest to the edge of the roof. This gives the shingles greater wind resistance.

Rake Starter Pieces. The first course begins with a full-tab shingle. Each of the next courses begin with a portion of a full-tab shingle. Each of the following rake starter pieces is cut smaller for each course (*Figure 51–11*). The amount cut off depends on the length of each full-tab shingle and their style. If shingles are the three-tab strip shingle type, the amount cut off is important to make the slots align vertically. The amount cut off must be multiples of 6 inches or one-half tab. This makes the amount cut off 6 inches, 12 inches, 18 inches, 24 inches, and 30 inches.

If the shingles are the laminated type, the actual amount cut off is not as important. Typically multiples of 6½ inches is the cut off. But multiples of 4 or 5 inches may be used as well. Laminated shingles are typically 1 meter long or about 39 inches. If multiples of 6½ inches is removed for each course, the removed partition of the cuts are 6½, 13, 19½, 26, and 32½ inches (*Procedure 51–A*).

After the rake starter pieces are installed, full-tab shingles are applied across and up the roof. A full-tab shingle is then applied at the rake edge to cover the last rake starter. This makes room for another set of rake starter pieces to be installed (*Figure 51–12*).

> ## CAUTION
> No butt seams of any three consecutive courses should line up. This maximizes the life of the shingles. No rake tab should be less than 3 inches in width.

Shingles are cut by scoring them on the back side with a utility knife. This saves the blade by keeping it away from the grit of the granules. Use a square or straightedge as a guide for the knife. Bend the shingle back to break it along the scored line.

Fastening Asphalt Shingles

Keeping the shingle courses straight is important in all styles of shingles. If the underlayment is installed straight, the lines on it may used to see if the courses are straight (*Figure 51–13*). If the underlayment is not straight, snap lines periodically. Lines should be parallel to the eave. Keep the bottom edges of the each shingle flush with the one next to it.

Selecting suitable fasteners, using the recommended number, and putting them in the right

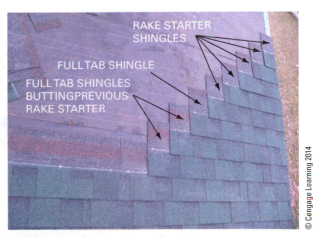

Figure 51–12 Full tab shingles are installed butting the rake starter pieces.

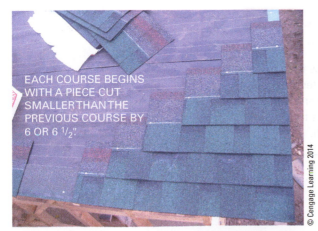

Figure 51–11 Rake starter pieces are installed in a stepped back fashion.

Figure 51–13 The lines on underlayment that is installed straight may be used as guide lines for shingles.

StepbyStep Procedures

Procedure 51–A Laying Out Rake Starters

STEP 1 BEGIN EAVE STARTER WITH A PIECE THAT HAS ABOUT 6" REMOVED. OVERHANG BOTH EDGES ABOUT ¼ OR ³/₈ INCH. INSTALL SEVERAL PIECES ALONG EAVE MAINTAINING A STRAIGHT EDGE.

STEP 2 START THE FIRST FULL TAB SHINGLE FULL LENGTH. INSTALL SEVERAL MORE ALONG EAVE.

STEP 3 BEGIN THE SECOND COURSE WITH A FULL TAB SHINGLE THAT HAS 6 OR 6 ½" REMOVED FROM THE RAKE EDGE END OF THE SHINGLE. INSTALL SEVERAL FULL TAB SHINGLES OF THE SECOND COURSE.

STEP 4 BEGIN THE THIRD COURSE WITH A SHINGLE THAT HAS TWICE THE 6 OR 6 ½" INCREMENT. INSTALL SEVERAL FULL LENGTH PIECES ON TOP OF FIRST COURSE. WATCH THAT EXPOSE IS UNIFORMLY SPACED.

STEP 5 FOURTH COURSE BEGINS WITH ONE-HALF A SHINGLE. REPEAT AS ABOVE CUTTING MORE OFF EACH STARTER UNTIL LAST COURSE HAS A PIECE THAT IS 6 OR 6 ½" LONG.

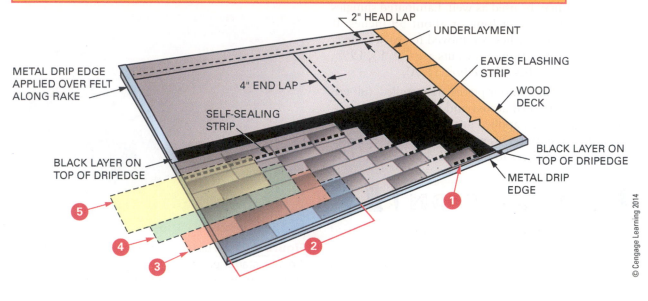

© Cengage Learning 2014

On the Job

Cut off pieces from making the rake starter strips are used on the other end of the roof. This amounts to very little waste overall.

places are important steps in the application of asphalt shingles (*Figure 51–14*). Use a minimum of four fasteners in each strip shingle. Do not nail into or above the factory-applied adhesive. ✱

✱**IRC**
R905.2.6

The fastener length should be sufficient to penetrate the sheathing at least $\frac{1}{2}$ inch, or through approved panel sheathing. Roofing nails should be 11- or 12-gauge galvanized steel or aluminum with barbed shanks. They should have $\frac{3}{8}$- to $\frac{7}{16}$-inch heads. Roofing nails may be driven by hand or with power nailers (*Figure 51–15*).✱

✱**IRC**
R905.2.5

Align each shingle of the first course carefully. Fasten each shingle from the end nearest the shingle just laid. This prevents buckling. Drive fasteners straight so that the nail heads will not cut into the shingles. The entire nail head should bear tightly against the shingle. It should not penetrate its surface (*Figure 51–16*).

Shingle Exposure

The maximum *exposure* of asphalt shingles to the weather depends on the type of shingle. Refer to manufacturer's recommendations on the packaging material of the bundle. Most commonly used asphalt shingles have an exposure of 5 inches or $5\frac{5}{8}$ inches. Less than the recommended exposure may be used, if desired.

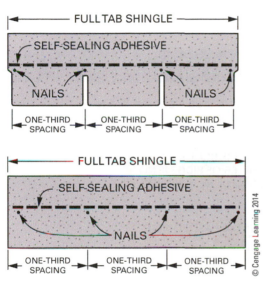

Figure 51–14 Recommended fastener locations for asphalt strip shingles.

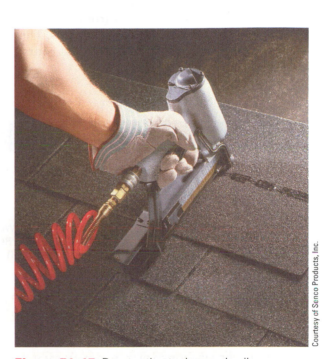

Figure 51–15 Pneumatic staplers and nailers are often used to fasten asphalt shingles.

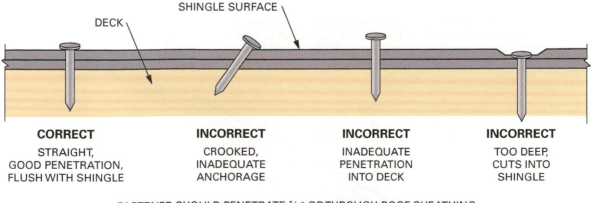

FASTENER SHOULD PENETRATE 1/2" OR THROUGH ROOF SHEATHING

Figure 51–16 It is important that fasteners are installed correctly.

LAYING OUT SHINGLE COURSES

The last course of shingles below the *ridge cap* or *ridge vent* should be exposed by about the same amount as all the other shingle courses. This can be achieved by changing slightly the exposure of the last ten or so courses near the ridge. First, make sure the plywood is sufficiently cut back on both sides of the roof for the ridge vent. The edge of the slot should be about $^3/_4$ inch from the ridge board or a total slot width of 3 inches. Next, measure down $3^1/_2$ inches from the edge of the plywood on both ends of the building. If no ridge vent is being used, measure down 5 inches from the center of the ridge. Snap a line between the marks. This line will be the top edge of the next to last course of shingles.

Measure the distance between the snapped line and the top edge of the last shingle course installed. Divide this distance by the normal exposure of the shingles. This is the number of courses remaining to the snapped line. Round up this number up to the nearest whole number. Now, divide this whole number into the previous distance measurement. The result is the slightly smaller exposure of the remaining courses that will allow the top edge of the next to last course to land on the snapped line (*Procedure 51–B*). Snap lines across the roof as needed, and shingle up to the ridge.

The line for the last course of shingles is snapped on the face of the course below. Lay out the exposure from the bottom edge of the course on both ends of the roof. Snap a line across the roof. Fasten the last course of shingles by placing

Step**by**Step Procedures

Procedure 51–B Evenly Spacing Upper Shingle Course Exposures Near the Ridge

STEP 1 MEASURE DOWN 3 $^1/_2$" FROM THE TOP OF PLYWOOD SLOT IF ROOF IS TO HAVE A RIDGE VENT. MEASURE DOWN 5" FROM CENTER OF RIDGE IF NO RIDGE VENT IS TO BE INSTALLED. DO THIS ON BOTH ENDS OF HE BUILDING.

STEP 2 MEASURE FROM MARKS IN STEP 1 TO THE TOP EDGE OF THE LAST COURSE OF SHINGLES INSTALLED 3 TO 4' FROM RIDGE.

STEP 3 DIVIDE MEASURED DISTANCE BY NORMAL SHINGLE EXPOSURE. ROUND RESULT UP TO NEXT WHOLE NUMBER OF COURSES. DIVIDE THE MEASURED DISTANCE AGAIN NOW BY THIS ROUNDED NUMBER. THIS RESULT IS THE SLIGHTLY SMALLER EXPOSURE THAT WILL MAKE COURSES WORK OUT TO A FULL EXPOSURE AT THE RIDGE.

STEP 4 INSTALL REMAINING COURSE WITH THIS NEW EXPOSURE. SNAP SEVERAL LINES IF NEED BE. LAST COURSE IS EITHER CUT TO THE PLYWOOD IF A RIDGE VENT IS INSTALLED OR LAPPED OVER TO CLOSE THE RIDGE.

5"

TOP EDGE OF LAST COURSE INSTALLED AT LEAST 3 TO 4 FEET FROM THE RIDGE.

© Cengage Learning 2014

their butts to the line. If the ridge is to be vented, cut the top edge of the shingle at the sheathing. If the ridge is not vented, bend the top shingle edges over the ridge. Fasten their top edges to the opposite slope.

Applying the Ridge Cap

The ridge cap is applied after both sides of the roof have been shingled. Cut ridge cap and hip shingles from shingle strips to make approximately 12-inch × 12-inch squares. Cut shingles from the top of the cutout to the top edge on a slight taper. The top edge should be narrower than the bottom edge (*Figure 51–17*). Cutting the shingles in this manner keeps the top half of the shingle from protruding when it is bent over the ridge.

Snap a line on shingles keep the vent or cap shingles on a straight line. To do this, center a vent strip or a cap shingle on the ridge at each end of the roof. Bend it over the ridge. Mark its bottom edge on the front slope or the one most visible. Snap a line between the marks. Install the ridge vent as recommended by manufacturer.

Beginning at the bottom of a hip or at one end of the ridge, apply the shingles over the hip or ridge. Each exposure should be the same as the shingles. In cold weather, ridge cap shingles may

Figure 51–18 Applying ridge cap shingles to a vented ridge.

have to be warmed in order to prevent cracking when bending them over the ridge. On the ridge, shingles are started from the end away from prevailing winds. The wind should blow over the shingle butts, not against them. Secure each shingle with one fastener on each side, $5\frac{1}{2}$ inches from the butt and 1 inch up from each edge (*Figure 51–18*). Apply the cap shingles across the ridge until 3 or 4 feet from the end. Then space the cap to the end in the same manner as spacing the shingle course to the ridge. The last ridge shingle is cut to size. It is applied with one fastener on each side of the ridge. The two fasteners are covered with asphalt cement to reduce corrosion and prevent leakage.

CEMENT ROOF TILES

Another roofing option is cement roof tiles (*Figure 51–19*), which are typically used in hot locales. They are designed for various architectural styles and will last thirty to sixty years. The lifetime range is determined by the amount of sun and windblown rain the roof receives.

Concrete and clay tiles are manufactured in a variety of styles and thicknesses. Consequently, their weight can range from 500 to 1,200 pounds per square. Regular tiles range between 700 and 800 pounds per square. This increased weight affects the strength required from the roof system. Many truss manufacturers will require the roof to be loaded with the tiles before any finishes are applied (*Figure 51–20*). Only weeks later will the interior framing surfaces be shimmed and trimmed flush. This time allows the trusses to shift and adjust to the load.

Underlayment requirements vary with local codes. Most areas use at least one layer of 30# felt fastened with nails and metal caps (*Figure 51–21*).

CUT ALONG DOTTED LINE, TAPERING TOP PORTION SLIGHTLY

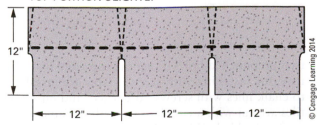

Figure 51–17 Ridge and hip cap shingles are cut from strip shingles.

Figure 51–19 Cement tile roofing is applied to many homes in warm climates.

Figure 51–20 Cement tile must be preloaded on roofs to allow the roof system to settle.

Figure 51–21 First step in tile roofing is felt applied with capped nails.

Figure 51–22 Areas of windblown rain have a second layer of hot-mopped felt.

Figure 51–23 Tiles screwed in place.

Coastal areas prone to windblown rain typically use another layer composed of 90# asphalt roll installed in hot, mopped-in-place, and bitumen (*Figure 51–22*). ✱

The tile pieces range in sizes from 7 to 13 inches wide by 15 to 18 inches long. They typically have two prepunched holes along the top edge that are used for fastening with screws. Tiles are installed to chalk lines with screws. At corners and hips, the tiles are cut with a circular saw with a diamond or composite blade (*Figure 51–23*). ✱

✱IRC
R905.3.3

✱IRC
R905.3.6

CHAPTER 52

Wood Shingles and Shakes

Wood shingles and shakes are common roof coverings (*Figure 52–1*). Most shingles and shakes are produced from western red cedar. Cedar logs are first cut into desired lengths. They are then split into sections from which shingles and shakes are sawn or split. All shingles are sawed. Most shakes are split. Shingles, therefore, have a relatively smooth surface. Most shakes have at least one highly textured, natural-grain surface. Most shakes also have thicker butts than shingles (*Figure 52–2*).

Most wood shingles and shakes are produced by mills that are members of the Cedar Shake and Shingle Bureau, an industry authority that educates

*IRC R905.7.7 the public and building code officials. *

Figure 52–2 All shingles are sawed from the log. Most shakes are split and have a rough surface.

DESCRIPTION OF WOOD SHINGLES AND SHAKES

Wood shingles are available for use on roofs in four standard grades. Shakes are manufactured by different methods to produce four types. Both shingles and shakes may be treated to resist fire or premature decay in areas of high humidity (*Figure 52–3*).

Sizes and Coverage

Shingles come in lengths of 16, 18, and 24 inches. The butt thickness increases with the length. Shakes are available in lengths of 15, 18, and 24 inches. Their butt thicknesses are from $\frac{3}{8}$ to $\frac{3}{4}$ inch. The 15-inch length is used for starter and finish courses. *Figure 52–4* shows the sizes and the

Figure 52–1 Wood shingles and shakes may be used as siding and roof covering.

SHINGLES

NO. 1 BLUE LABEL®   NO. 2 RED LABEL   NO. 3 BLACK LABEL   NO. 4 UNDERCOURSING

SHAKES

NO. 1 HANDSPLIT & RESAWN   NO. 1 CERTISAWN ®   NO. 1 TAPERSPLIT   NO. 1 STRAIGHTSPLIT

Figure 52–3 Example of grades and kinds of wood shingles and shakes.

Coverage and Exposure Tables

Shingle Coverage Table

Length and Thickness	Approximate Coverage of One Square (4 bundles) of Shingles Based on Following Weather Exposures								
	3½″	4″	4½″	5″	5½″	6″	6½″	7″	7½″
16″ × 5/2″	70	80	90	100*	—	—	—	—	—
18″ × 5/2¼″	—	72½	81½	90½	100*	—	—	—	—
24″ × 4/2″	—	—	—	—	73½	80	86½	93	100*

NOTE *Maximum exposure recommended for roofs.

Shake Coverage Table

Shake Type, Length and Thickness	Approximate Coverage (in sq. ft.) of One Square, When Shakes are Applied with an Average ½″ Spacing, at Following Weather Exposures, in Inches:				
	5	5½	7½	8½	10
18″ × ½″ Handsplit-and-Resawn Mediums(a)	—	75	100	—	—
18″ × ¾″ Handsplit-and-Resawn Heavies(a)	—	75	100	—	—
18″ × ⅝″ Tapersawn —		75			
24″ × ⅜″ Handsplit 50(e)	—	75	—	—	—
24″ × ½″ Handsplit-and-Resawn Mediums	—	—	75	85	100
24″ × ¾″ Handsplit-and-Resawn Heavies	—	—	75	85	100
24″ × ⅝″ Tapersawn —		—	75	85	100
24″ × ½″ Tapersplit —		—	75	85	100
18″ × ⅜″ Straight-Split	65		90(c)	—	—
24″ × ⅜″ Straight-Split	—	—	75(b)	85	100(c)
15″ Starter-Finish course	Use supplementary with shakes applied not over 10″ weather exposure				

Figure 52–4 Tables show the sizes and coverage of wood shingles and shakes.

Shingle Exposure Table

Slope	Maximum Exposure Recommended for Roofs								
	Length								
	No. 1 Blue Label			No. 2 Red Label			No. 3 Black Label		
	16″	18″	24″	16″	18″	24″	16″	18″	24″
3/12 to 4/12	3¼″	4¼″	5½″	3½″	4″	5½″	3″	3½″	5″
4/12 and steeper	5″	5½″	7½″	4″	4½″	6½″	3½″	4″	5½″

Shake Exposure Table

Slope	Maximum exposure recommended for roofs	
	Length	
	18″	24″
4/12 and steeper	7½″	10″

© Cengage Learning 2014

Figure 52–5 Maximum exposures of wood shingles and shakes for various roof pitches.

amount of roof area that one square, laid at various exposures, will cover.

Maximum Exposures

The area covered by one square of shingles or shakes depends on the amount exposed to the weather. The maximum amount of shingle exposure depends on the length and grade of the shingle or shake and the pitch of the roof. Shakes are not generally applied to roofs with slopes of less than 4 inches rise per foot.✱ Shingles, with reduced exposures, may be used on slopes down to 3 inches rise per foot. *Figure 52–5* shows the maximum recommended roof exposure for wood shingles and shakes.   ✱IRC R905.7.2

SHEATHING AND UNDERLAYMENT

Shingles and shakes may be applied over spaced or solid roof sheathing. Spaced sheathing or *skip sheathing* is usually 1 × 4 or 1 × 6 boards. Solid sheathing is usually APA-rated panels. Solid Sheathing may be required in regions subject to frequent earthquakes or under treated shingles and shakes. It is also recommended for use with shakes in areas where wind-driven weather is common. ✱ ✱IRC R905.7.1.1

Spaced Sheathing

Solid wood sheathing is applied from the eaves up to a point that is plumb with a line 12 to 24 inches inside the wall line. An eaves flashing is installed, if required. Spaced sheathing may then be used above the solid sheathing to the ridge.

For shingles, either 1 × 4 or 1 × 6 spaced sheathing may be used. Space 4-inch boards the same amount as the shingles are exposed to the weather. In this method of application, each course of shingles is nailed to the center of the board. If 6-inch boards are used, they are spaced two exposures. Two courses of shingles are nailed to the same board when courses are exposed up to, but not exceeding, 5½ inches. For shingles with greater exposures, the sheathing is spaced a distance of one exposure (*Figure 52–6*). ✱   ✱IRC R905.7.1

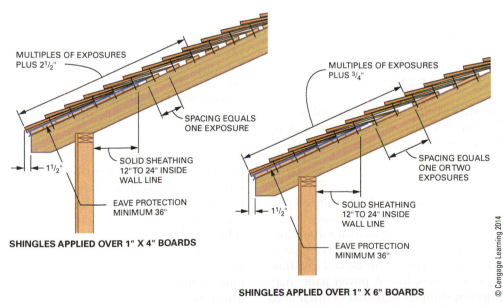

MULTIPLES OF EXPOSURES PLUS 2½″

SPACING EQUALS ONE EXPOSURE

SOLID SHEATHING 12″ TO 24″ INSIDE WALL LINE

EAVE PROTECTION MINIMUM 36″

1½″

SHINGLES APPLIED OVER 1″ X 4″ BOARDS

MULTIPLES OF EXPOSURES PLUS ¾″

SPACING EQUALS ONE OR TWO EXPOSURES

SOLID SHEATHING 12″ TO 24″ INSIDE WALL LINE

EAVE PROTECTION MINIMUM 36″

1½″

SHINGLES APPLIED OVER 1″ X 6″ BOARDS

© Cengage Learning 2014

Figure 52–6 Application of wood shingles on spaced sheathing.

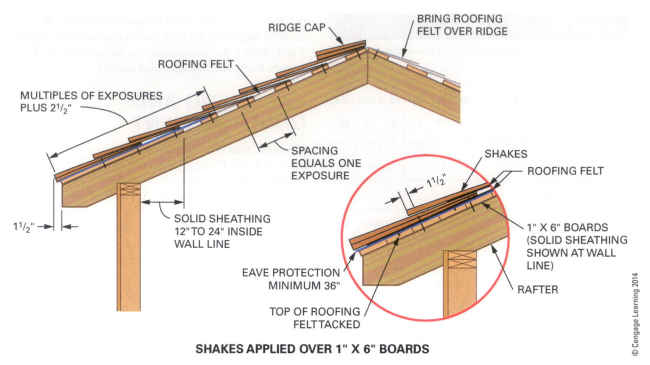

RIDGE CAP

BRING ROOFING FELT OVER RIDGE

ROOFING FELT

MULTIPLES OF EXPOSURES PLUS 2½"

SPACING EQUALS ONE EXPOSURE

SHAKES

ROOFING FELT

1½"

1½"

SOLID SHEATHING 12" TO 24" INSIDE WALL LINE

EAVE PROTECTION MINIMUM 36"

TOP OF ROOFING FELT TACKED

1" X 6" BOARDS (SOLID SHEATHING SHOWN AT WALL LINE)

RAFTER

SHAKES APPLIED OVER 1" X 6" BOARDS

© Cengage Learning 2014

Figure 52–7 Method of applying shakes on spaced sheathing.

In shake application, spaced sheathing is usually 1 × 6 boards spaced the same distance, on center (OC), as the shake exposure (*Figure 52–7*). The spacing should never be more than 7½ inches for 18-inch shakes and 10 inches for 24-inch shakes installed on roofs. ✱

✱IRC R905.8.1

Underlayment

No underlayment is required under wood shingles. A breather-type roofing felt may be used over solid or spaced sheathing. Underlayment is typically applied under shakes.

APPLICATION TOOLS AND FASTENERS

Shingles and shakes are usually applied with a shingling hatchet. Recommendations for the type and size of fasteners should be closely followed.

Shingling Hatchet

A shingling hatchet (*Figure 52–8*) should be lightweight. It should have both a sharp *blade* and a *heel*. A sliding gauge is sometimes used for fast and accurate checking of shingle exposure. The gauge permits several shingle courses to be laid at a time without snapping a chalk line. A power nailer may be used. However, shingles and shakes often need to be trimmed or split with the hatchet. More time may be lost than gained by using a power nailer.

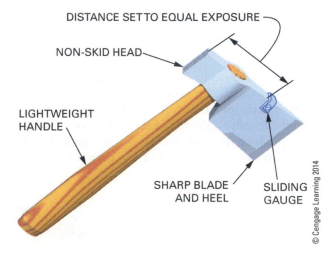

DISTANCE SET TO EQUAL EXPOSURE

NON-SKID HEAD

LIGHTWEIGHT HANDLE

SHARP BLADE AND HEEL

SLIDING GAUGE

© Cengage Learning 2014

Figure 52–8 A shingling hatchet is commonly used to apply wood shingles and shakes.

Fasteners

Apply each shingle with only two corrosion-resistant nails, such as stainless steel, hot-dipped galvanized, or aluminum nails. Box nails usually are used because their smaller gauge minimizes splitting. Minimum nail lengths for shingles and shakes are shown in *Figure 52–9*. Staples should be 16-gauge aluminum or stainless steel with a minimum crown of 7/16 inch. The staple should be driven with its crown across the grain of the shingle or shake. Staples should be long enough to penetrate the sheathing at least ½ inch.

Fasteners	
Type of Certi-label Shake or Shingle	**Nail Type and Minimum Length**
Certi-Split & Certi-Sawn Shakes	**Type (in)**
18" Straight-Split	5d Box 1 3/4
18" and 24" Handsplit and Resawn	6d Box 2
24" Tapersplit	5d Box 1 3/4
18" and 24" Tapersawn	6d Box 2
Certigrade Shingles	**Type (in)**
16" and 18" Shingles	3d Box 1 1/4
24" Shingles	4d Box 1 1/2

© Cengage Learning 2014

Figure 52–9 Recommended nail types and sizes for wood shingles and shakes.

APPLYING WOOD SHINGLES

With a string pulled along the bottom edge of shingles, install the first layer of a starter course of wood shingles. If a gutter is used, overhang the shingles plumb with the center of the gutter. If no gutter is installed, overhang the starter course $1\frac{1}{2}$ inches beyond the fascia. Space adjacent shingles between $\frac{1}{4}$ and $\frac{3}{8}$ inch apart. Place each fastener about $\frac{3}{4}$ inch in from the edge of the shingle and not more than 1 inch above the exposure line. Drive fasteners flush with the surface. Make sure that the head does not crush the surface.

Apply another layer of shingles on top of the first layer of the starter course. The starter course may be tripled, if desired, for appearance. This procedure is recommended in regions of heavy snowfall.

Lay succeeding courses across the roof. Apply several courses at a time. Stagger the joints in adjacent courses at least $1\frac{1}{2}$ inches. There should be no joint in any three adjacent courses in alignment. Joints should not line up with the centerline of the heart of wood or any knots or defects. Flat grain or *slash grain* shingles wider than 8 inches should be split in two before fastening. Trim shingle edges with the hatchet to keep their butts in line (*Figure 52–10*). ✱

✱IRC R905.7.5

After laying several courses, snap a chalk line to straighten the next course. Proceed shingling up the roof. When 3 or 4 feet from the ridge, check the distance on both ends of the roof. Divide the distance as close as possible to the exposure used. A full course should show below the ridge cap. Shingle tips are cut flush with the ridge. Tips can be easily cut across the grain with a hatchet if cut on an angle to the side of the shingle.

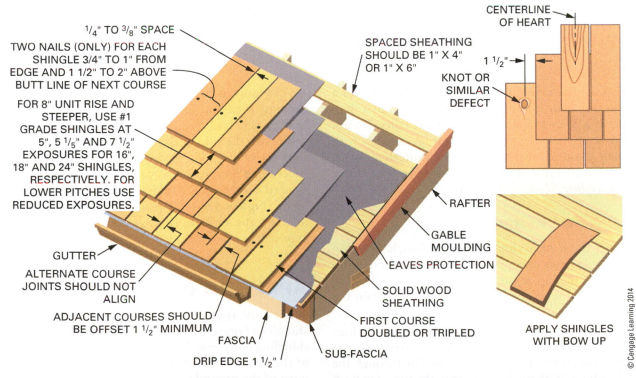

¼" TO ³⁄₈" SPACE

TWO NAILS (ONLY) FOR EACH SHINGLE 3/4" TO 1" FROM EDGE AND 1 1/2" TO 2" ABOVE BUTT LINE OF NEXT COURSE

FOR 8" UNIT RISE AND STEEPER, USE #1 GRADE SHINGLES AT 5", 5 1/5" AND 7 1/2" EXPOSURES FOR 16", 18" AND 24" SHINGLES, RESPECTIVELY. FOR LOWER PITCHES USE REDUCED EXPOSURES.

GUTTER

ALTERNATE COURSE JOINTS SHOULD NOT ALIGN

ADJACENT COURSES SHOULD BE OFFSET 1 1/2" MINIMUM

FASCIA

DRIP EDGE 1 1/2"

SUB-FASCIA

FIRST COURSE DOUBLED OR TRIPLED

SOLID WOOD SHEATHING

EAVES PROTECTION

GABLE MOULDING

RAFTER

SPACED SHEATHING SHOULD BE 1" X 4" OR 1" X 6"

CENTERLINE OF HEART

1 1/2"

KNOT OR SIMILAR DEFECT

APPLY SHINGLES WITH BOW UP

© Cengage Learning 2014

Figure 52–10 Details of wood shingle application.

Procedure 52–A Applying Shingles or Shakes Along an Open Valley

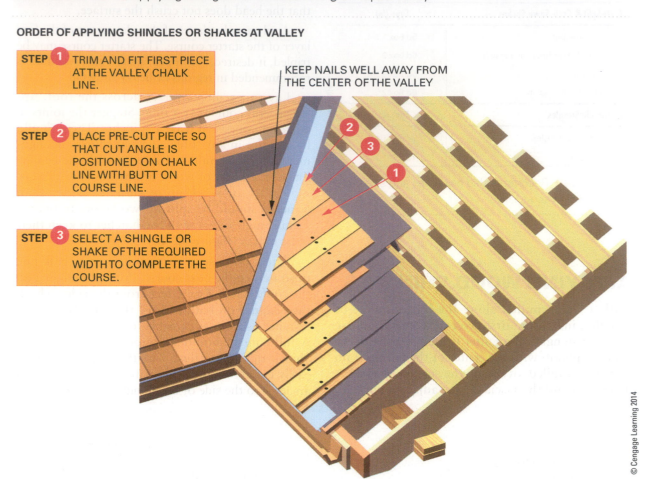

ORDER OF APPLYING SHINGLES OR SHAKES AT VALLEY

STEP 1 TRIM AND FIT FIRST PIECE AT THE VALLEY CHALK LINE.

KEEP NAILS WELL AWAY FROM THE CENTER OF THE VALLEY

STEP 2 PLACE PRE-CUT PIECE SO THAT CUT ANGLE IS POSITIONED ON CHALK LINE WITH BUTT ON COURSE LINE.

STEP 3 SELECT A SHINGLE OR SHAKE OF THE REQUIRED WIDTH TO COMPLETE THE COURSE.

© Cengage Learning 2014

On intersecting roofs, stop shingling the roof a few feet away from the valley. Select and cut a shingle at a proper taper. Apply it to the valley. Do not break joints in the valley. Do not lay shingles with their grain parallel to the valley centerline. Work back out. Fit a shingle to complete the course (*Procedure 52–A*). ✱

✱**IRC R905.7.6**

Hips and Ridges

After the roof is shingled, 4- to 5-inch-wide hip and ridge caps are applied. Measure down on both sides of the hip or ridge at each end for a distance equal to the exposure. Snap a line between the marks.

Lay a shingle so its bottom edge is to the line. Fasten with two nails. Use longer nails so that they penetrate at least $\frac{1}{2}$ inch into or through the sheathing. With the hatchet, trim the top edge flush with the opposite slope.

Lay another shingle on the other side with the butt even and its upper edge overlapping the first shingle laid. Trim its top edge at a bevel and flush with the side of the first shingle. Double this first set of shingles, alternating the joint. Apply succeeding layers of cap, with the same exposure as used on the roof. Alternate the overlap of each layer (*Figure 52–11*). Space the exposure when nearing the end so all caps are about equally exposed.

APPLYING WOOD SHAKES

Shakes are applied in much the same manner as shingles (*Figure 52–12*). Mark the handle of the shingling hatchet at $7\frac{1}{2}$ and 10 inches from the top of the head. These are the exposures that are used most of the time when applying wood shakes.

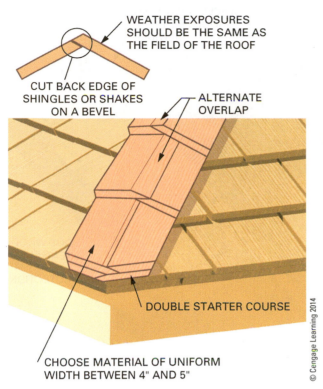

WEATHER EXPOSURES SHOULD BE THE SAME AS THE FIELD OF THE ROOF

CUT BACK EDGE OF SHINGLES OR SHAKES ON A BEVEL

ALTERNATE OVERLAP

DOUBLE STARTER COURSE

CHOOSE MATERIAL OF UNIFORM WIDTH BETWEEN 4" AND 5"

Figure 52–11 When applying hip and ridge shingles, alternate the overlap.

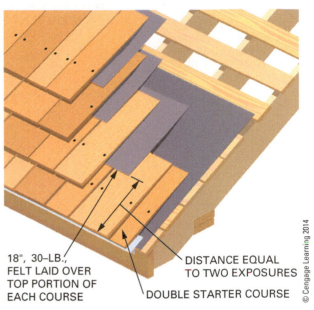

18", 30–LB., FELT LAID OVER TOP PORTION OF EACH COURSE

DISTANCE EQUAL TO TWO EXPOSURES

DOUBLE STARTER COURSE

Figure 52–13 An underlayment of felt is required when laying shakes.

Underlayment

*IRC R905.8.3

A full width of felt underlayment is first applied under the starter course. * Butts of the starter course should project 1½ inches beyond the fascia. Next, lay an 18-inch-wide strip of #30 roofing felt over the upper end of the starter course.

Its bottom edge should be positioned at a distance equal to twice the shake exposure above the butt line of the first exposed course of shakes (*Figure 52–13*). * For example, 24-inch shakes, laid with 10 inches of exposure, would have felt applied 20 inches above the butts of the shakes. The felt will cover the top 4 inches of the shakes. It will extend up 14 inches on the sheathing. The top edge of the felt must rest on the spaced sheathing, if used.

*IRC R905.8.7

Figure 52–12 Three or four shake courses at a time are carried across the roof.

© Cengage Learning 2014

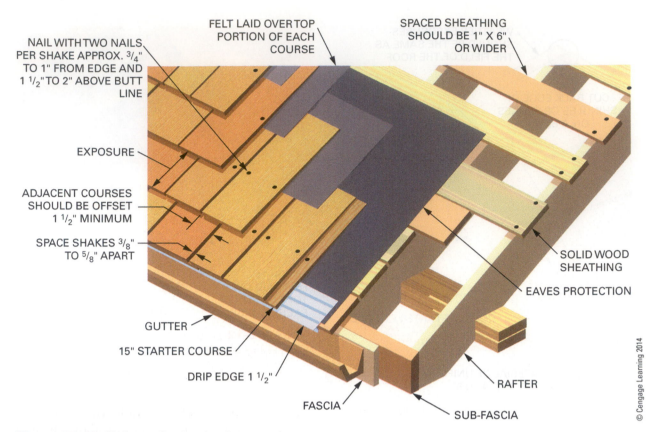

NAIL WITH TWO NAILS PER SHAKE APPROX. 3/4" TO 1" FROM EDGE AND 1 1/2" TO 2" ABOVE BUTT LINE

FELT LAID OVER TOP PORTION OF EACH COURSE

SPACED SHEATHING SHOULD BE 1" X 6" OR WIDER

EXPOSURE

ADJACENT COURSES SHOULD BE OFFSET 1 1/2" MINIMUM

SPACE SHAKES 3/8" TO 5/8" APART

SOLID WOOD SHEATHING

EAVES PROTECTION

GUTTER

15" STARTER COURSE

DRIP EDGE 1 1/2"

FASCIA

SUB-FASCIA

RAFTER

© Cengage Learning 2014

Figure 52–14 Shake application details.

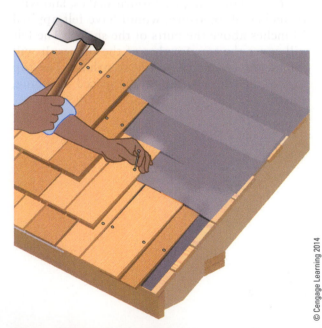

© Cengage Learning 2014

Figure 52–15 The tip of the shake is inserted between the layers of underlayment.

Nail only the top edge of the felt. Fasten successive strips on their top edge only. Their bottom edges should be one shake exposure from the bottom of the previous strip. It is important to lay the felt straight. It serves as a guide for applying shakes. After the roof is felted, the tips of the shakes are tucked under the felt. The bottom should be exposed by the distance of twice the exposure.

Apply the second and successive courses with joints staggered and fasteners placed the same as for shingles. The spacing between shakes should be at least 3/8 inch, but not more than 5/8 inch to allow wood to dry properly after rain. Lay straight-split shakes with their smooth end toward the ridge (*Figure 52–14*).✱

✱IRC
R905.8.6

Maintaining Shake Exposure

There is a tendency to angle toward the eave. Therefore, check the exposure regularly with the hatchet handle. An easy way to be sure of correct exposure is to look through the joint between the edges of the shakes in the course below the one being nailed. The bottom edge of the felt will be visible. The butt of the shake being nailed is positioned directly above it (*Figure 52–15*).

Ridges and Hips

Adjust the exposure so that tips of shakes in the next-to-last course just come to the ridge. Use economical 15-inch starter-finish shakes for the last course. They save time by eliminating the need to trim shake tips at the ridge. Cap ridges and hips with shakes in the same manner as that used for wood shingles.

CHAPTER 53

Flashing

Flashing is a material used in various locations that are susceptible to leaking. It prevents water from entering a building (*Figure 53–1*). The words *flash, flashed,* and *flashing* are also used as verbs to describe the installation of the material. Various kinds of flashing are applied at the eaves, valleys, chimneys, vents, and other roof projections. They prevent leakage at the intersections.

KINDS OF FLASHING

Flashing material may be sheet copper, zinc, aluminum, galvanized steel, vinyl, or mineral-surfaced asphalt roll roofing. Copper and zinc are high-quality flashing materials, but they are more expensive. Roll roofing is less expensive. Colors that match or contrast with the roof covering can be used. If properly applied, roll roofing used as a valley flashing will outlast the main roof covering. Sheet metal, especially copper, may last longer. However, it is good practice to replace all flashing when reroofing.

Eaves Flashing

Whenever there is a possibility of ice dams forming along the eaves and causing a backup of water,

Ice & Water Shield® flashing is needed. Apply the flashing such that it overhangs the drip edge by $\frac{1}{4}$ to $\frac{3}{8}$ inch. The flashing should extend up the roof far enough to cover a point at least 12 inches inside the wall line of the building. If the overhang of the eaves requires that the flashing be wider than 36 inches, the necessary horizontal lap joint is located on the portion of the roof that extends outside the wall line (*Figure 53–2*).

For a slope of at least 4 inches rise per foot of run, install a single course of 36-inch-wide flashing covering the underlayment and drip edge. On lower slopes greater protection against water leakage is gained by applying Ice & Water Shield® flashing over the entire roof.

Valley Flashing

Roof valleys are especially vulnerable to leakage. This is because of the great volume of water that flows down through them. Valleys must be carefully flashed according to recommended procedures. Valleys are flashed in two ways: *open* or *closed*. Open valleys are constructed with no shingle or roofing material installed within several inches of

Figure 53–1 Flashings are used to seal against leakage where the roofing butts against adjoining surfaces.

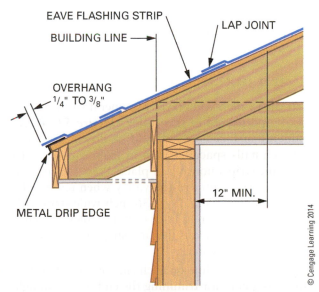

Figure 53–2 If there is danger of ice dams forming along the eaves, eave flashing is installed with seams away from the building line.

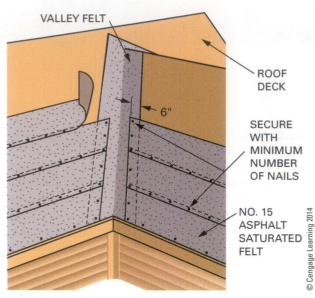

Figure 53–3 Felt underlayment is applied in the valley before roof underlayment.

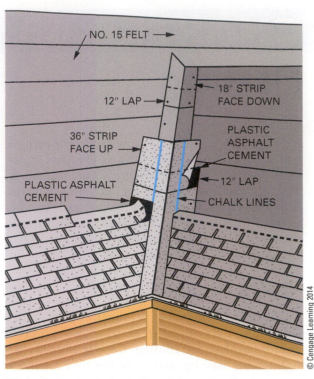

Figure 53–4 Method of applying roll roofing open valley flashing.

© Cengage Learning 2014

the valley center. **Closed valleys** have the roofing material covering the entire valley centerline. ✱

✱IRC R905.2.8.2

Open Valley Flashing

Begin valley flashing by applying a 36-inch-wide strip of asphalt felt centered in the valley (*Figure 53–3*). Fold or crease the roll along the length and seat it well into the valley. Be careful not to cause any break in the felt. Fasten it with only enough nails along its edges to hold it in place. Let the courses of felt underlayment applied to the roof overlap the valley underlayment by not less than 6 inches. The eave flashing, if required, is then applied.

Using Roll Roofing Flashing

Lay an 18-inch-wide layer of mineral-surfaced roll roofing centered in the valley (*Figure 53–4*). Its mineral-surfaced side should be down. Use only enough nails spaced 1 inch in from each edge to hold the strip smoothly in place. Press the roofing firmly in the center of the valley when nailing the opposite edge. Next, lay a 36-inch-wide strip with its surfaced side up. Center it in the valley. Fasten it in the same manner as the first strip pressing firmly into the valley center.

Snap a chalk line on each side of the valley. Use them as guides for trimming the ends of the shingle courses. These lines are spaced 6 inches apart at the ridge. Because more water may be present in the valley at the eave than at the ridge, these lines

are spread $\frac{1}{8}$ inch per foot as they approach the eave. Thus, a valley 16 feet long will be 8 inches wide at the eaves.

The upper corner of each end asphalt shingle is clipped. This helps keep water from entering between the courses. Each roof shingle is cemented to the valley flashing with plastic asphalt cement.

Metal Flashing

Prepare the valley with underlayment in the same manner as described previously. Next, lay a strip of sheet metal flashing centered in the valley (*Figure 53–5*). The metal should extend at least 10 inches on each side of the valley centerline for slopes with a 6-inch rise or less and 7 inches on each side for a steeper slope. Carefully press and form it into the valley. Fasten the metal with nails of similar material spaced close to its outside edges. Use only enough fasteners to hold it smoothly in place.

Snap lines on each side of the valley, as described previously. Use them as guides for cutting the ends of the shingle courses. Trim the last shingle of each course to fit on the chalk line. Clip 1 inch from its upper corner at a 45-degree angle. To form a tight seal, cement the shingle to the metal flashing with a 3-inch width of asphalt plastic cement.

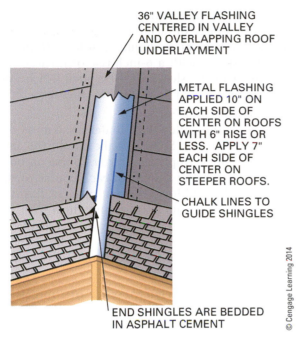

36" VALLEY FLASHING CENTERED IN VALLEY AND OVERLAPPING ROOF UNDERLAYMENT

METAL FLASHING APPLIED 10" ON EACH SIDE OF CENTER ON ROOFS WITH 6" RISE OR LESS. APPLY 7" EACH SIDE OF CENTER ON STEEPER ROOFS.

CHALK LINES TO GUIDE SHINGLES

END SHINGLES ARE BEDDED IN ASPHALT CEMENT

Figure 53–5 Method of applying metal open valley flashing.

Figure 53–6 Valleys are sometimes flashed by weaving shingles together.

EACH STRIP TO EXTEND AT LEAST 12" BEYOND CENTER OF VALLEY

36" ROLL ROOFING 50# OR HEAVIER

6" MIN

EXTRA NAIL IN END OF STRIP

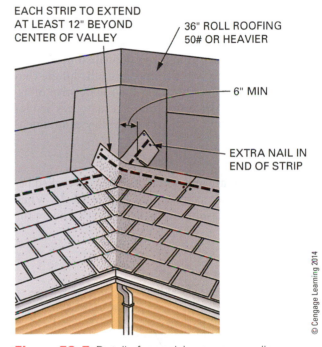

Figure 53–7 Details for applying a woven valley.

If a valley is formed by the intersection of a low-pitched roof and a much steeper one, a 1-inch-high, crimped standing seam should be made in the center of the metal flashing of an open valley. The seam will keep heavy rain water flowing down the steeper roof from overrunning the valley and possibly being forced under the shingles of the lower slope.

Closed Valley Flashing

A closed valley protects the valley flashing. It thus adds to the weather resistance at vulnerable points. Several methods are used to flash closed valleys.

The first step for any method is to apply the asphalt felt underlayment as previously described for open valleys. Then, center a 36-inch width of smooth or mineral surface roll roofing, 50-pound-per-square, or heavier, in the valley over the underlayment. Form it smoothly in the valley. Secure it with only as many nails as necessary. Another method uses a strip of wide metal flashing on top of the felt underlayment. *****

*IRC R905.2.8.2

Woven Valley Method

Valleys may be flashed by applying asphalt shingles on both sides of the valley and alternately weaving each course over and across the valley. This is called a woven valley (*Figure 53–6*).

Lay the first course of shingles along the edge of one roof up to and over the valley for a distance

of at least 12 inches. Lay the first course along the edge of the adjacent roof. Extend the shingles over the valley on top of the previously applied shingles (*Figure 53–7*).

Succeeding courses are then applied. Weave the valley shingles alternately, first on one roof and then on the other. When weaving the shingles, make sure they are pressed tightly into the valley. Also make sure no nail is closer than 6 inches to the valley centerline. Fasten the end of the woven shingle with two nails. Most carpenters prefer to cover each roof area with shingles to a point approximately 3 feet from the valley. They weave the valley shingles in place later.

No end joints should occur within 6 inches of the center of the valley. Therefore, it may be necessary to occasionally cut a strip short that would otherwise end near the center. Continue from this cut end with a full-length strip over the valley.

Closed Cut Valley Method

Apply the shingles to one roof area. Let the end shingle of every course overlap the valley by at least 12 inches (*Figure 53–8*). Make sure no end joints occur within 6 inches of the center of the valley. Place fasteners no closer than 6 inches from the center of the valley. Form the end shingle of each course snugly in the valley. Secure its end with two fasteners.

Snap a chalk line 2 inches short of and parallel to the center of the valley on top of the overlapping shingles. Apply shingles to the adjacent roof area. Cut the end shingle to the chalk line. Clip the upper corner of each shingle as described previously for open valleys. Bed the end of each shingle that lies in the valley in about a 3-inch-wide strip of asphalt cement. Make sure that no fasteners are located closer than 6 inches to the valley centerline.

Step Valley Flashing Method

Step flashings are individual metal pieces tucked between courses of shingles (see *Figure 53–1*). When applying step flashings in valleys, first estimate the

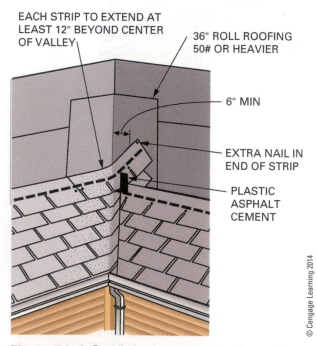

EACH STRIP TO EXTEND AT LEAST 12" BEYOND CENTER OF VALLEY

36" ROLL ROOFING 50# OR HEAVIER

6" MIN

EXTRA NAIL IN END OF STRIP

PLASTIC ASPHALT CEMENT

© Cengage Learning 2014

Figure 53–8 Details for applying a closed cut valley.

number of shingle courses required to reach the ridge. Cut a piece of metal flashing for each course of shingles. Each piece should be at least 18 inches wide for slopes with a 6-inch rise or greater and 24 inches wide for slopes with less pitch. The height of each piece should be at least 3 inches more than the shingle exposure.

Prepare the valley with underlayment as described previously. Snap a chalk line in the center of the valley. Apply the starter course on both roofs. Trim the end shingle of each course to the chalk line. Fit and form the first piece of flashing to the valley. Trim its bottom edge flush with the drip edge. Fasten it in the valley over the first layer of the starter course. Use fasteners of like material to prevent electrolysis. Fasten the upper corners of the flashing only.

Apply the first regular course of shingles to both roofs on each side of the valley. Trim the valley shingles so their ends lay on the chalk line. Bed them in plastic asphalt cement. Do not drive nails through the metal flashing.

Apply the next piece of flashing in the valley over the first course of shingles. Keep its bottom edge about $\frac{1}{2}$ inch above the butts of the next course of shingles. Apply the second course of shingles in the same manner as the first. Secure a flashing over the second course. Apply succeeding courses and flashings in this manner (*Figure 53–9*). Remember, flashing is placed over each course of shingles. Do not leave any flashings out. When the valley is completely flashed, no metal flashing surface is exposed. If the valley does not extend all the way to the ridge of the main roof, a *saddle* is applied over the ridge of the minor roof (*Figure 53–10*).

Flashing Against a Wall

When roof shingles butt up against a vertical wall, the joint must be made watertight. The usual method of making the joint tight is with the use of metal step flashings (*Figure 53–11*).

The flashings are purchased or cut about 8 inches in width. They are bent at right angles in the center so they will lay about 4 inches on the roof and extend about 4 inches up the sidewall. The length of the flashings is about 3 inches more than the exposure of the shingles. When used with shingles exposed 5 inches to the weather, they are made 8 inches in length. Cut and bend the necessary number of metal flashings. ✳

✳**IRC R905.2.8.3**

The roofing is applied and flashed before the siding is applied to the vertical wall. First, apply an underlayment of asphalt felt to the roof deck. Turn the ends up on the vertical wall by about 3 to 4 inches.

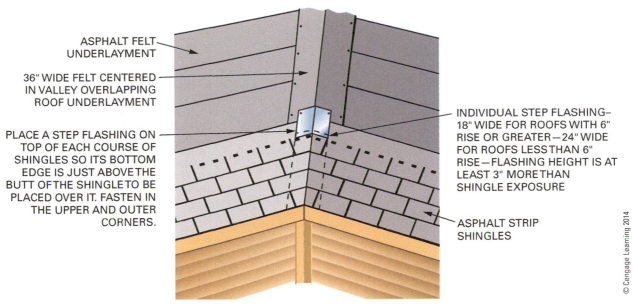

ASPHALT FELT UNDERLAYMENT

36" WIDE FELT CENTERED IN VALLEY OVERLAPPING ROOF UNDERLAYMENT

PLACE A STEP FLASHING ON TOP OF EACH COURSE OF SHINGLES SO ITS BOTTOM EDGE IS JUST ABOVE THE BUTT OF THE SHINGLE TO BE PLACED OVER IT. FASTEN IN THE UPPER AND OUTER CORNERS.

INDIVIDUAL STEP FLASHING— 18" WIDE FOR ROOFS WITH 6" RISE OR GREATER—24" WIDE FOR ROOFS LESS THAN 6" RISE—FLASHING HEIGHT IS AT LEAST 3" MORE THAN SHINGLE EXPOSURE

ASPHALT STRIP SHINGLES

© Cengage Learning 2014

Figure 53–9 Details for applying metal step flashings in a valley.

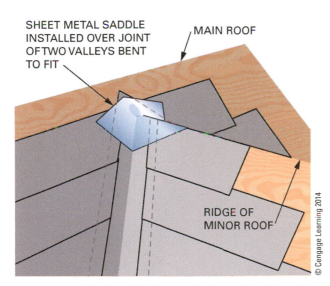

SHEET METAL SADDLE INSTALLED OVER JOINT OF TWO VALLEYS BENT TO FIT

MAIN ROOF

RIDGE OF MINOR ROOF

© Cengage Learning 2014

Figure 53–10 A saddle is installed over the ridge of a minor roof where it intersects with the main roof.

Apply the first layer of the starter course, working toward the vertical wall. Fasten a metal flashing on top of the starter course of shingles. Its bottom edge should be flush with the drip edge. Use one fastener in each top corner. Lay the first regular course with its end shingle over the flashing and against the sheathing of the sidewall. Do not drive any fasteners through the flashings. It is usually not necessary to bed the shingles to the flashings with asphalt cement. The step flashing holds down the end of the shingle below it.

Apply a flashing over the upper side of the first course and against the wall. Keep its bottom edge at a point that will be about $\frac{1}{2}$ inch above the butt of the next course of shingles. Continue applying shingles and flashings in this manner until the ridge is reached. Some carpenters prefer to cut the shingles back if a tab cutout occurs over a flashing. This prevents metal from being exposed to view.

Flashing a Chimney

Chimneys and other large protrusions through the roof, such as skylights, must be installed so water and weather do not enter the building. Chimneys are a greater challenge to weather proofing because their height or length changes over the course of a year due to the masonry expanding and contracting from normal operations. A hot chimney (e.g., from a running furnace) is taller than a cool one. The change in distance for tall chimneys can be measured in inches. To accomplish waterproofing a combination of flashing layers is used.

In many cases, especially on steep sloped roof and wide chimneys, a cricket or saddle is built behind the chimney between the upper side of chimney and roof deck. The cricket prevents water and debris from accumulating behind the chimney (*Figure 53–12*). Counter flashings are installed by *masons* who build the chimney in conjunction with the step flashing the carpenters apply. The top edge of counter flashing is bent at a right angle at about $\frac{1}{2}$ inch to be inserted into a mortar joint. This edge

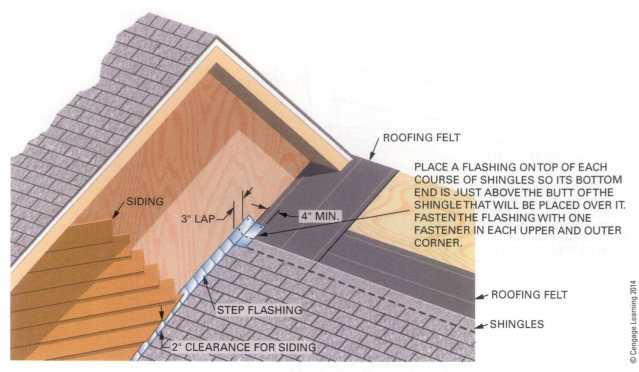

ROOFING FELT

PLACE A FLASHING ON TOP OF EACH COURSE OF SHINGLES SO ITS BOTTOM END IS JUST ABOVE THE BUTT OF THE SHINGLE THAT WILL BE PLACED OVER IT. FASTEN THE FLASHING WITH ONE FASTENER IN EACH UPPER AND OUTER CORNER.

SIDING

3" LAP

4" MIN.

ROOFING FELT

SHINGLES

STEP FLASHING

2" CLEARANCE FOR SIDING

Figure 53–11 Using metal step flashing where a roof butts a wall.

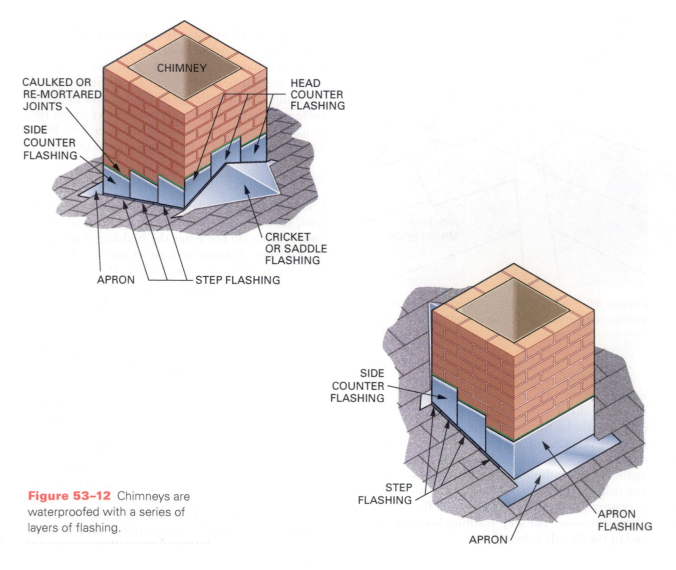

CHIMNEY

CAULKED OR RE-MORTARED JOINTS

SIDE COUNTER FLASHING

HEAD COUNTER FLASHING

CRICKET OR SADDLE FLASHING

APRON

STEP FLASHING

SIDE COUNTER FLASHING

STEP FLASHING

APRON FLASHING

APRON

Figure 53–12 Chimneys are waterproofed with a series of layers of flashing.

© Cengage Learning 2014

is then sealed in the joint with mortar or caulk. The lower edge of the counter flashing covers the step flashing that is installed on each course of shingles along the sides of the chimney. Flashing of a chimney may be installed while the roof is being shingled or after the roofing is complete. The mess of building a chimney is easier to clean up before shingles are installed. Bending metal flashing is best done on a brake which is discussed in Chapter 62.

Underlayment is applied to the roof deck and tucked under any existing flashings. The shingle courses are brought up to the chimney. They are applied under the flashing on the lower side of the chimney. The top edges of the shingles are cut as necessary, until the only a shingle exposure shows below (downhill from) the chimney. The apron flashing is applied on top of the lower course of shingles. Two cuts are made in it to allow a portion of the flashing to be bent vertically (*Procedure 53–A*). The lower edge of the apron flashing is pressed into a bed of asphalt or plastic cement on top of the shingles. The overall width should be about 12 inches and about 6 inches longer on both sides of the chimney. Where the bend is made is not critical; the vertical bend or the surface touching the roof should be at least 4 inches.

The first piece of step flashing is installed on top of the apron flashing with its end trimmed so about 1 inch wraps around to the chimney's front surface. Step flashings are then installed by tucking them in between each shingle course and against the apron flashing or chimney as they progress up the roof. Step flashings are typically 8 × 8 inch pieces folded in half at a right angle. This is a similar process as in flashing against a wall. Nails are carefully placed into the roof sheathing along the flashing edge as far away from the chimney as possible.

The apron is counter flashing bent to fit in the mortar joint. It is also cut long enough to wrap around and up the sides of the chimney about 4 inches. Side counter flashings, also installed into mortar joints, overlap the previous flashing about 3 inches. The mid and lower portions of counter flashings are bedded to the chimney with asphalt cement. It is very important not to attach the counter flashing to the step flashing. Remember, anything attached to the chimney will move up and down while anything attached to the roof will not move.

The last piece of sidewall step flashing is cut and fitted around the upper corners of the chimney. One more piece of step flashing is installed on the backside of the chimney and has a portion of it bent around to the side covering the last sidewall piece about 1 inch. Apply a layer of mastic to adhere the step flashings to themselves.

A cricket is typically built from plywood and lumber as needed for strength. Wide chimneys require more wood than smaller ones. The slope of the cricket is not critical; it only needs to slope enough to adequately shed water. The cricket subframe is installed on top of the last step flashing previously applied to the back side of the chimney. Large crickets may be covered with shingles creating a small valley section. These courses of shingles are step flashed to the backside of the chimney. Smaller crickets are totally covered with flashing material. This flashing is called head flashing.

Head flashing covers the cricket and is bent up the backside of the chimney. It extends into the small valley and up the main roof far enough for the roof main shingles to cover it by at least 6 inches. It also extends down to the valley past the cricket sub-frame by at least 6 inches. It is cemented to the cricket sub-frame and roof deck, but not to the chimney. The lower edge of the cricket must be on top of shingles. The upper course of underlayment lies on top of the valley portion of the cricket.

Attach a flashing piece to cover the V-shaped gap in the vertical surfaces of the head flashing where the brick is exposed. It should be attached with mastic to the flashing only, not the chimney.

The head-counter flashing is installed into the mortar joint and down to cover the cricket's vertical surfaces. It begins at the chimney corners with the last piece of side counter flashing that wraps the corner to the backside about 4 inches. The flashing process is completed when the center-counter flashing is installed at the cricket's ridge. Vertical surfaces are attached to the chimney with mastic.

Shingles are cemented to the head flashing and saddle flashing. A saddle flashing is installed over the upper-most portion of the cricket.

Working with metal flashing can be a little tricky to get right the first time; be patient. Gentle tapping with a hammer can help move and shape the metal. Chimneys may be flashed by methods and materials other than described previously, depending on the custom of certain geographical areas. The major thought to keep in mind when flashing is "water runs downhill." Every joint or seam should overlap towards the downhill side. Lapping water proofing materials always begins at the lowest portion of the roof.

StepbyStep Procedures

Procedure 53–A A Flashing a Chimney.

STEP 1 APPLY UNDERLAYMENT TO THE ROOF DECK AND TUCK UNDER ANY EXISTING FLASHINGS. INSTALL SHINGLES UP TO CHIMNEY AND CUT THE TOP EDGES OF THE SHINGLES AROUND CHIMNEY.

STEP 2 PRECUT AND BEND APRON FLASHING, THEN INSTALL IT BY ADHERING WITH MASTIC TO THE SHINGLES. PLACE ONE NAIL INTO THE ROOF DECK THROUGH EACH OF THE UPPER, OUTER MOST CORNERS OF THE FLASHING.

STEP 3 INSTALL FIRST STEP FLASHING WITH A BEND THAT WRAPS TO THE FRONT SURFACE OF CHIMNEY. APPLY A SMALL AMOUNT OF MASTIC TO THE FOLDED EDGE OF STEP FLASHING TO ADHERE IT TO THE APRON AND NOT THE CHIMNEY.

STEP 4 INSTALL SHINGLES AND STEP FLASHINGS ALONG THE SIDE OF THE CHIMNEY. NAIL STEP FLASHING TO ROOF SHEATHING WITH ONE NAIL IN THE UPPER CORNER.

STEP 5 PRECUT AND BEND APRON, CUTTING IT LONG ENOUGH TO WRAP THE SIDES OF THE CHIMNEY ABOUT 4 INCHES. ATTACH BY ADHERING THE VERTICAL SURFACES TO CHIMNEY BUT NOT THE STEP FLASHING.

STEP 6 INSTALL SIDE COUNTER FLASHING, BEGINNING WITH LOWER ONE, OVERLAPPING EACH PIECE ABOUT 3 INCHES. POSTPONE LAST (UPPER) PIECE OF SIDE COUNTER FLASHING.

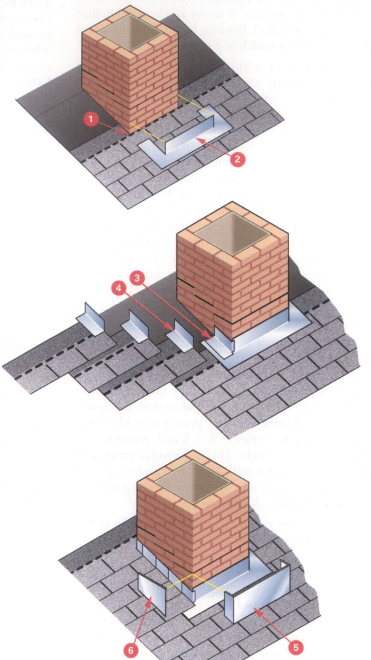

(continued)

Procedure 53–A (continued)

STEP **7** INSTALL LAST SIDEWALL STEP FLASHING BY FITTING AROUND THE CORNER OF THE CHIMNEY.

STEP **8** INSTALL ONE STEP FLASHING ON THE BACKSIDE OF THE CHIMNEY WITH A PORTION OF IT BENT AROUND THE CORNER ABOUT 1 INCH. APPLY A LAYER OF MASTIC TO ADHERE THE STEP FLASHINGS TO THE PREVIOUS ONE.

STEP **9** CUT AND FIT CRICKET FROM PLYWOOD AND LUMBER AS NEEDED.

STEP **10** CUT AND FIT HEAD FLASHING TO COVER CRICKET WOOD FRAME. BEND IT UP THE BACKSIDE OF THE CHIMNEY ABOUT 4 INCHES. ALSO ALLOW ABOUT 6 INCHES OR MORE TO RUN UP THE ROOF PAST THE VALLEY. ATTACH BY ADHERING FLASHING TO THE CRICKET WOOD-FRAME AND ROOF DECK, BUT NOT TO THE CHIMNEY. LOWER END SHOULD BE ON THE SHINGLES.

STEP **11** ATTACH A SMALL FLASHING PIECE TO THE HEAD FLASHING VERTICAL SURFACE TO COVER THE V-SHAPED GAP. ADHERE TO FLASHING NOT CHIMNEY.

STEP **12** CUT, FIT AND INSTALL LAST SIDE WALL COUNTER FLASHING TO WRAP CHIMNEY CORNER ABOUT 4 INCHES.

STEP **13** INSTALL FIRST HEAD COUNTER TO COVER PREVIOUS FLASHING. ATTACH BY ADHERING VERTICAL SURFACE TO CHIMNEY WITH MASTIC.

STEP **14** INSTALL THE CENTER-COUNTER FLASHING AT THE CRICKET RIDGE. MAKE SURE MASTIC ADHERES TO CHIMNEY ONLY.

STEP **15** INSTALL SHINGLES UP THE ROOF. INSTALL A SADDLE FLASHING ON TOP OF CRICKET AND UNDER NEAREST SHINGLE LAYER.

STEP **16** REPOINT THE MORTAR JOINTS.

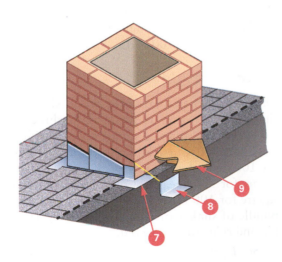

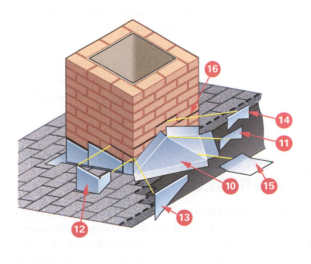

© Cengage Learning 2014

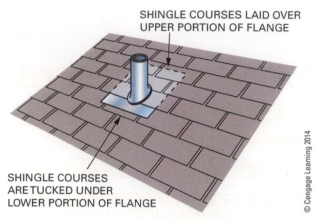

SHINGLE COURSES LAID OVER UPPER PORTION OF FLANGE

SHINGLE COURSES ARE TUCKED UNDER LOWER PORTION OF FLANGE

© Cengage Learning 2014

Figure 53–13 A vent stack flashing boot.

Flashing Vents

Flashings for round pipes, such as *stack vents* for plumbing systems and *roof ventilators,* usually come as *flashing collars* made for various roof pitches. They fit around the stack. They have a wide flange on the bottom that rests on the roof deck. The flashing is installed over the stack vent, with its flange on the roof sheathing. It is fastened in place with one fastener in each upper corner.

Shingle up to the lower end of the stack vent flashing. Lift the lower part of the flange. Apply shingle courses under it. Cut the top edge of the shingles, where necessary, until the shingle exposure, or less, shows below the lower edge of the flashing. Apply asphalt cement under the lower end of the flashing. Press it into place on top of the shingle courses.

Apply shingles around the stack and over the flange. Do not drive nails through the flashing. Bed shingles to the flashing with asphalt cement, where necessary (*Figure 53–13*).

ESTIMATING ROOFING MATERIALS

Estimating the amount of material needed for roofing involves underlayment and shingles. Remember that if the number of pieces does not come out even, round it up to the next whole number of pieces.

Roof Area

Roofing materials estimates begins with calculation roof area in square feet. Multiply the rafter length including overhang × ridge board length. Double this number for the other side and divide the total by 100 to determine the number of squares needed. For example, if the rafter is 15 feet and the ridge is 66 feet, the roof area is $15 \times 66 \times 2 \div 100 = 19.8$ squares. Hip roof area is the same as the gable roof area on the same size building.

Underlayment

Once roof area is determined, the number of rolls of underlayment is estimated by dividing roof area by the area each roll will cover. Rounding the estimated number up to a whole roll usually compensates for any waste. Add 10 percent waste if the roof has hips and valleys. For example, if the roof area for a roof with hips and valleys is 1,920, the number of rolls of $15\# - 4$ square per roll felt is $1,920 \times 110\% \div 100 \div 4 = 5.2 \Rightarrow 6$ rolls.

Asphalt Shingles

Extra shingles are needed for eave starters and cap shingles. Extra is also needed if the roof has dormers, hips, and valleys. The waste factor percentage added will vary depending on the complicated nature of the roof. Add 5 to 15 percent, for waste is typical. The number of bundles depends on the number of bundles per square. They range from three to five bundles per square. For example, if the area is 1,920 square feet and the waste is 7 percent for four bundles per square shingle, the bundles needed is $1,920 \times 107\% \div 100 \times 4 = 82.1 \Rightarrow 83$ bundles.

Wood Shingles

Calculating the squares for wood shingles and shakes is the same as for asphalt shingles. For wood shingles, add one square for every 240 linear feet of starter course and one extra square for every 100 linear feet of valleys. For shakes, add one square for 120 linear feet of starter course and one square for 100 lineal feet of valleys. Add an extra bundle of shakes or shingles for every 16 feet of hip and ridge to be covered.

See *Figure 53–14* for a listing of the shorthand versions of these estimating techniques with another example.

Item	Formula	Waste Factor	Example
Roof Area	gable rafter LEN × gable ridgeboard LEN × 2 ÷ 100 SF per square = squares of roof area		15.5 × 54' × 2 ÷ 100 = 16.74 squares
Underlayment	roof squares ÷ 4 square per roll = rolls		16.74 ÷ 4 = 4.1 ⇒ 5 rolls
Asphalt Shingles	roof squares × bundles per square × waste factor = bundles	5%	16.74 × 4 × 105% = 70.3 ⇒ 71 bundles
Wood Shingles	[roof squares + one per 240 ft eave starter] × 7 bundles per square + 1 bundle per 16 ft ridge = bundles		[16.74 + 54 × 2 ÷ 240] × 7 + 54 ÷ 16 = 123.6 ⇒ 121 bundles
Wood Shakes	[roof squares + one per 120 ft eave starter] × 7 bundles per square + 1 bundle per 16 ft ridge = bundles		[16.74 + 54 × 2 ÷ 120] × 7 + 54 ÷ 16 = 126.7 ⇒ 127 bundles

Estimate the materials for a roof with a 15'-6" gable rafter, 54' ridgeboard, and 4 bundles per square shingles.

© Cengage Learning 2014

Figure 53–14 Example of estimating roof materials with formulas.

DECONSTRUCT THIS

Carefully study *Figure 53–15* and think about what is wrong and/or what is right. Consider all possibilities. What construction practice or method is different in your area of the country?

© Cengage Learning 2014

Figure 53–15 This photo shows workers loading a roof with clay tiles.

KEY TERMS

apron	drip edge	mortared	solid sheathing
apron flashing	end lap	open valley	spaced sheathing
asphalt felts	exposure	saddle	square
asphalt shingles	flashing	selvage	starter course
cements	flat grain	shakes	top lap
closed valley	head flashing	shingle butt	wood shingles
coatings	head lap	side flashing	woven valley
counter flashing cricket	Ice & Water Shield®	sliding gauge	
deck			

REVIEW QUESTIONS

Select the most appropriate answer.

1. A square is the amount of roofing required to cover
   a. 1 square foot.
   b. 100 square feet.
   c. 150 square feet.
   d. 200 square feet.

2. One roll of #15 asphalt felt will cover about
   a. one square.
   b. two squares.
   c. three squares.
   d. four squares.

3. When applying asphalt felt on a roof deck as underlayment, lap each course over the lower course by at least
   a. 2 inches.   c. 6 inches.
   b. 4 inches.   d. 12 inches.

4. When cutting eave starter strips, it is important to
   a. cut the shingle lengthwise, removing the width of one exposure.
   b. cut the first eave starter shorter than a full-tab shingle.
   c. have the sealing tabs at the bottom of the strip.
   d. all of the above.

5. Laying out rake edge starters requires the starters be
   a. cut 6 or 6½ inches shorter for each course.
   b. installed so butt seams of nearby courses align.
   c. nailed so butt seams of next course align over nails.
   d. all of the above.

6. Nailing for asphalt shingles should be located the self-sealing strip.
   a. above.
   b. on.
   c. below.
   d. above or below.

7. Checking and adjusting shingle exposure should be done when the top shingle course is from ridge.
   a. 1–2 feet.
   b. 3–4 feet.
   c. 5–6 feet.
   d. 7–8 feet.

8. Most wood shingles and shakes are made from
   a. cypress.
   b. redwood.
   c. western red cedar.
   d. all of the above.

9. The longest available length of wood shingles and shakes is
   a. 16 inches.
   b. 18 inches.
   c. 24 inches.
   d. 28 inches.

10. Wood shingles normally overhang the fascia by
    a. ⅜ inch.
    b. 1 inch.
    c. 1½ inches.
    d. 2 inches.

11. Prior to applying shakes, it is important to lay underlayment straight

    a. so there are no wrinkles in the felt.
    b. to obtain the proper lap.
    c. to improve its looks after the roof is completed.
    d. because the felt serves as an installation guide for shingles.

12. The first step in installation of roll roof valley flashing is

    a. installing an 18-inch-wide flashing piece faceup.
    b. installing an 18-inch-wide flashing piece facedown.
    c. installing a full-width flashing piece faceup.
    d. installing a full-width underlayment.

13. When nailing shingles, do not locate any nails closer to the valley centerline than

    a. 6 inches.
    b. 8 inches.
    c. 10 inches.
    d. 12 inches.

14. Step flashings used against a vertical wall are cut about 8 inches wide. They should be bent so that

    a. 3 inches lay on the wall and 5 inches lay on the roof.
    b. 4 inches lay on the wall and 4 inches lay on the roof.
    c. 5 inches lay on the wall and 3 inches lay on the roof.
    d. 2 inches lay on the wall and 6 inches lay on the roof.

15. A built-up section between the roof and the upper side of chimney is called a

    a. cricket.
    b. dutchman.
    c. furring.
    d. counterflashing.

UNIT 20
Windows

©Shutterstock, Inc./Mau Horng

Windows are normally installed prior to the application of exterior *siding*. Care must be taken to provide easy-operating, weather-tight, attractive units. Quality workmanship results in a more comfortable interior, saves energy by reducing fuel costs, minimizes maintenance, gives longer life to the units, and makes application of the exterior siding easier.

O B J E C T I V E S

After completing this unit, the student should be able to:

- describe the most popular styles of windows and name their parts.

- select and specify desired sizes and styles of windows from manufacturers' catalogs.

- install various types of windows in an approved manner.

- cut glass and glaze a sash.

SAFETY REMINDER

Windows are made in many styles, sizes, and shapes. They are often a focal point of large rooms. Follow manufacturer's recommendations and local code requirements carefully when installing them to avoid injury.

CHAPTER 54

Window Terms and Types

Wood *windows* are one of many types of millwork (*Figure 54–1*). Millwork is a term used to describe products, such as windows, doors, and cabinets, fabricated in woodworking plants that are used in the construction of a building. Windows are usually fully assembled and ready for installation when delivered to the construction site. Windows are also made of aluminum and steel. Windows made with exposed wood parts encased in vinyl are called vinyl-clad windows. The names given to various parts of a window are the same, in most cases, regardless of the window type.

PARTS OF A WINDOW

When shipped from the factory, the window is complete except for the interior trim. It is important that the installer know the names, location, and functions of the parts of a window in order to understand, or to give, instructions concerning them.

The Sash

The sash is a frame in a window that holds the glass. The type of window is generally determined by the way the sash operates. The sash may be installed in a fixed position, move vertically or horizontally, or swing outward or inward.

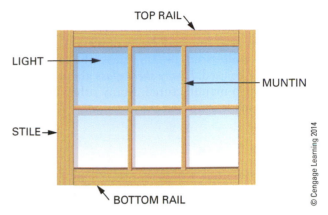

Figure 54–2 A window sash and its parts.

Sash Parts. Vertical edge members of the sash are called stiles. Top and bottom horizontal members are called rails. The pieces of glass in a sash are called lights. There may be more than one light of glass in a sash. Small strips of wood that divide the glass into smaller lights are called muntins. Muntins may divide the glass into rectangular, diamond, or other shapes (*Figure 54–2*).

Many windows come with false muntins called grilles. Grilles do not actually separate or support the glass. They are applied as an overlay to simulate small lights of glass. They are made of wood or plastic. They snap in and out of the sash for easy cleaning of the lights (*Figure 54–3*). They may also

Figure 54–1 Windows of many types and sizes are fully assembled in millwork plants and ready for installation.

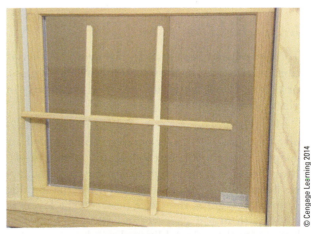

Figure 54–3 Removable grilles simulate true divided-light muntins.

be preinstalled between the layers of glass in double- or triple-glazed windows.

WINDOW GLASS

Several qualities and thicknesses of sheet glass are manufactured for glazing and other purposes. The installation of glass in a window sash is called glazing. Single-strength (SS) glass is about $3/32$ inch thick. It is used for small lights of glass. For larger lights, double-strength (DS) glass about $1/8$ inch thick may be used. *Heavy sheet glass* about $3/16$ inch thick is also manufactured. Many other kinds of glass are made for use in construction.

Safety Glass

Most residential windows are not glazed with safety glass, so if they break, they could fragment into large pieces and cause injury. Care must be taken to handle windows in a manner to prevent breaking the glass. Some codes, however, do require a type of safety glass in windows with low sill heights or located near doors. Skylights and roof windows are generally required to be glazed with safety glass.

Safety glass is constructed, treated, or combined with other materials to minimize the possibility of injuries resulting from contact with it. Several types of safety glass are manufactured.

Laminated glass consists of two or more layers of glass with inner layers of transparent plastic bonded together. Tempered glass is treated with heat or chemicals. When broken at any point, the entire piece immediately disintegrates into a multitude of small granular pieces. *Transparent plastic* is also used for safety glazing material.

Insulated Glass

To help prevent heat loss, and to avoid condensation of moisture on glass surfaces, insulated glass, or *thermal pane windows,* are used frequently in place of single-thickness glass.

Insulated glass consists of two or three (generally two) layers of glass separated by a sealed air space $3/16$ to 1 inch in thickness (*Figure 54–4*). Moisture is removed from the air between the layers before the edges are sealed. To raise the R-value of insulated glass, the space between the layers is filled with argon gas. Argon conducts heat at a lower rate than air. Additional window insulation may be provided with the use of *removable glass panels* or *combination storm sash.*

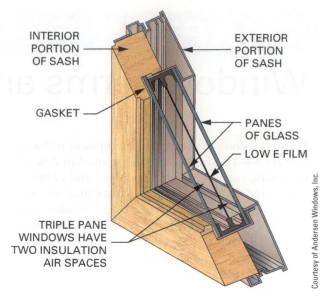

Figure 54–4 Crossectional view of energy efficient window.

Labels in figure: INTERIOR PORTION OF SASH — EXTERIOR PORTION OF SASH — GASKET — PANES OF GLASS — LOW E FILM — TRIPLE PANE WINDOWS HAVE TWO INSULATION AIR SPACES — Courtesy of Andersen Windows, Inc.

Solar Control Glass. The R-value of windows may also be increased by using special *solar-control insulated glass,* called *high performance* or *Low-E glass.* Low-E is an abbreviation for low emissivity. It is used to designate a type of glazing that reflects heat back into the room in winter and blocks heat from entering in the summer (*Figure 54–5*). An invisible, thin, metallic coating is bonded to the air space side of the inner glass. This lets light through, but reflects heat.

THE WINDOW FRAME

The sash is hinged to, slides on, or is fixed in a window frame. The frame usually comes with the exterior trim applied. It consists of several distinct parts (*Figure 54–6*).

The Sill

The bottom horizontal member of the window frame is called a sill. It is set or shaped at an angle to shed water. Its bottom side usually is grooved so a weather-tight joint can be made with the wall siding.

Jambs

The vertical sides of the window frame are called side jambs. The top horizontal member is called a head jamb.

Extension Jambs. The inside edge of the jamb should be flush with the finished interior wall surface when the window is installed. In some cases,

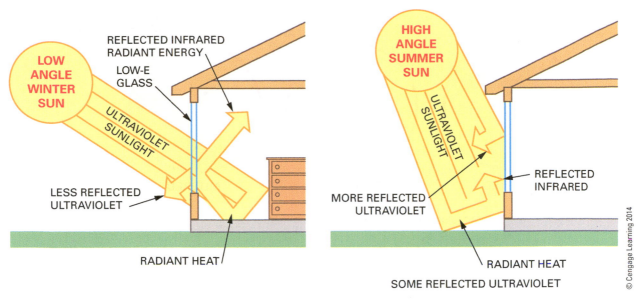

Figure 54–5 Low-E glass is used in windows to keep heat in during cold weather and out during hot weather.

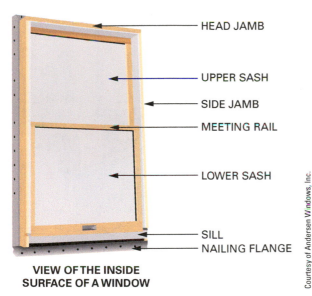

**VIEW OF THE INSIDE
SURFACE OF A WINDOW**

Figure 54–6 A window frame consists of parts with specific terms.

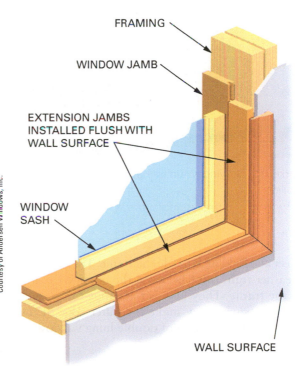

EXTENSION JAMBS MAY BE INSTALLED ON ALL FOUR SIDES AS SHOWN OR MAY EXCLUDE THE BOTTOM WHERE A SILL IS INSTALLED

Figure 54–7 To compensate for varying wall thicknesses, extension jambs are provided with some window units.

windows can be ordered with jamb widths for standard wall thicknesses. In other cases, jambs are made narrow. Extension jambs are then provided with the window unit. The extensions are cut to width to accommodate various wall thicknesses. They are applied to the inside edge of the jambs of the window frame (*Figure 54–7*). The extension jambs are installed at a later stage of construction when the interior trim is applied. They should be stored for safekeeping until needed.

Windows may also be purchased with extension jambs already installed. Care should always be taken to protect the jambs throughout the construction process.

Blind Stops

Blind stops are sometimes applied to certain types of window frames. They are strips of wood attached to the outside edges of the jambs. Their inside edges

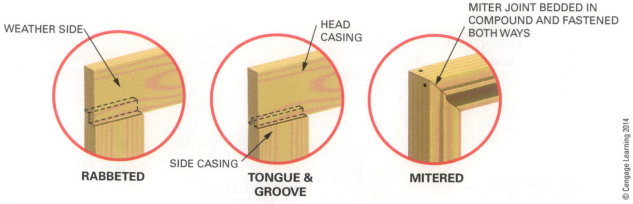

Figure 54–8 A weathertight joint is made between side and header casings.

Figure 54–9 Window units that are joined create a mullion.

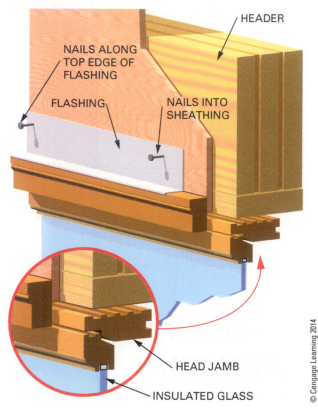

Figure 54–10 A window flashing covers the top edge of the header casing.

project about $\frac{1}{2}$ inch inside the frame. They provide a weather-tight joint between the outside casings and the frame. They also act as stops for screens and storm sash. They make the outer edge of the channel for the top sash of double-hung windows.

Casings

Window units usually come with exterior casings applied. The side members are called side casings. In most windows, the lower ends are cut at a bevel and rest on the sill. The top member is called the head casing. On flat casings, a weather-tight *rabbeted* or *tongue-and-grooved* joint is made between them. When molded casings are used, the mitered joints at the head are usually bedded in compound (*Figure 54–8*).

When windows are installed or manufactured, side by side, in multiple units, a mullion is formed where the two side jambs are joined together. The casing covering the joint is called a mullion casing (*Figure 54–9*).

Window Flashing

In some cases, a window flashing is also provided. This is a piece of metal as long as the head casing, which is also called a drip cap. It is bent to fit over the head casing and against the exterior wall and under the siding (*Figure 54–10*). The flashing prevents the entrance of water at this point. Flashings are usually made of aluminum. The vinyl flanges of vinyl-clad wood windows are usually formed as an integral part of the window. No additional head flashings are required.

Protective Coatings

Most window units with wood exterior casings are primed with a first coat of paint applied at the factory. Priming should be done before installation. Store the units under cover and protected from the weather until installed. Additional protective coats should be applied as soon as practical. Vinyl-clad wood windows are designed to eliminate painting.

TYPES OF WINDOWS

Common types of windows are fixed, single-or double-hung, casement, sliding, awning, and hopper windows.

Fixed Windows

Fixed windows consist of a frame in which a sash is fitted in a fixed position. They are manufactured in many shapes (*Figure 54–11*).

Oval and *circular* windows are usually installed as individual units. *Elliptical, half rounds,* and *quarter rounds* are widely used in combination with other types. In addition, fixed windows are manufactured in other *geometric* shapes (*squares, rectangles, triangles, parallelograms, diamonds, trapezoids, pentagons, hexagons,* and *octagons*). They may be assembled or combined with other types of windows in a great variety of shapes (*Figure 54–12*). In addition to factory-assembled units, lengths of the frame stock may be purchased for cutting and assembling odd-shape or-size units on the job.

Arch windows have a curved top or head that make them well suited to be joined in combination with a number of other types of windows or doors. All of the windows mentioned come in a variety of sizes. With so many shapes and sizes, hundreds of interesting and pleasing combinations can be made. Arched windows may be made as part of the sash (*Figure 54–13*).

Single- and Double-Hung Windows

The double-hung window consists of an upper and a lower sash that slide vertically by each other in separate channels of the side jambs (*Figure 54–14*).

Figure 54–12 Windows come in a variety of shapes and sizes.

Figure 54–11 Fixed windows are often used in combinations.

Figure 54–13 Arched windows can be part of a window sash.

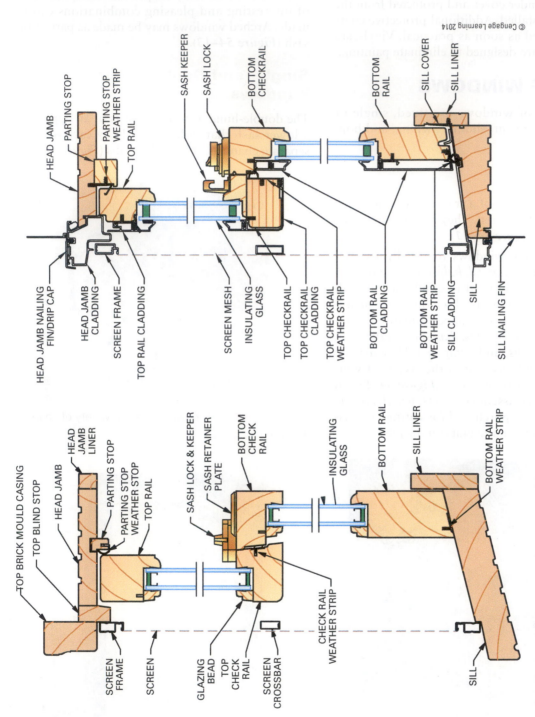

Figure 54–14 The double-hung window and its parts.

Labels (top diagram):
HEAD JAMB
PARTING STOP
PARTING STOP WEATHER STRIP
TOP RAIL
SASH KEEPER
SASH LOCK
BOTTOM CHECKRAIL
BOTTOM RAIL
SILL COVER
SILL LINER

HEAD JAMB NAILING FIN/DRIP CAP
HEAD JAMB CLADDING
SCREEN FRAME
TOP RAIL CLADDING
SCREEN MESH
INSULATING GLASS
TOP CHECKRAIL
TOP CHECKRAIL CLADDING
TOP CHECKRAIL WEATHER STRIP
BOTTOM RAIL CLADDING
BOTTOM RAIL WEATHER STRIP
SILL CLADDING
SILL
SILL NAILING FIN

Labels (bottom diagram):
TOP BRICK MOULD CASING
TOP BLIND STOP
HEAD JAMB
HEAD JAMB LINER
PARTING STOP
PARTING STOP WEATHER STOP
TOP RAIL
SASH LOCK & KEEPER
SASH RETAINER PLATE
BOTTOM CHECK RAIL
INSULATING GLASS
BOTTOM RAIL
SILL LINER
BOTTOM RAIL WEATHER STRIP

SCREEN FRAME
SCREEN
GLAZING BEAD
TOP CHECK RAIL
SCREEN CROSSBAR
CHECK RAIL WEATHER STRIP
SILL

© Cengage Learning 2014

The single-hung window is similar except the upper sash is fixed. A strip separating the sash is called a parting strip.

In most units, the sash slides in metal channels that are installed in the frames. Each sash is provided with *springs, sash balances,* or *compression weather stripping* to hold it in place in any position. Compression weather stripping prevents air infiltration, provides tension, and acts as a counterbalance. Some types provide for easy removal of the sash for painting, repair, and cleaning.

When the sashes are closed, specially shaped *check* or *meeting rails* come together to make a weather-tight joint. *Sash locks* located at this point not only lock the window, but draw the rails tightly together. Other hardware consists of *sash lifts* that are fastened to the bottom rail of the bottom sash. They provide an uplifting force to make raising the sash easier and help keep the sash in the position it is placed in.

Most double-hung windows are also designed to be removed via a tilt-in action (*Figure 54–15*). This makes cleaning both inside and outside surfaces easy. Double-hung windows can be arranged in a number of ways. They can be installed side by side in multiple units or in combination with other types. Bow window units project out from the building, often creating more floor space (*Figure 54–16*). The look is of a smooth curve. A bay window unit is similar to a bow except the sides are straight with corners at the window intersections.

Casement Windows

The casement window consists of a sash hinged at the side. It swings outward by means of a crank or lever. Most casements swing outward. The inswinging type is very difficult to make weathertight. An advantage of the casement type is that the entire sash can be opened for maximum ventilation. *Figure 54–17* shows the use of casement windows in a bow window unit.

Sliding Windows

Sliding windows have sashes that slide horizontally in separate tracks located on the header jamb and sill (*Figure 54–18*). When a window-wall effect is desired, many units can be placed side by side. Most units come with all necessary hardware applied.

Awning and Hopper Windows

An awning window unit consists of a frame in which a sash hinged at the top swings outward by means of a crank or lever. A similar type, called the hopper window, is hinged at the bottom and swings inward.

Each sash is provided with an individual frame so that many combinations of width and height can be used. These windows are often used in combination with other types (*Figure 54–19*).

Figure 54–15 Double-hung windows may tilt in for easy cleaning.

Figure 54–16 Bow window units are made by joining smaller units into a curved wall.

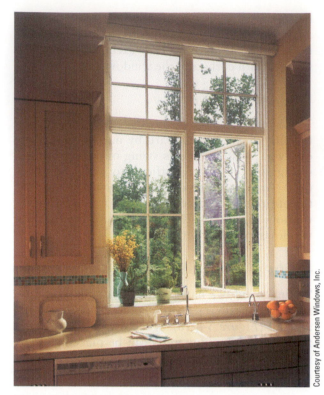

Courtesy of Andersen Windows, Inc.

Figure 54–17 Casement windows swing outward.

Courtesy of Marvin Windows and Doors

Figure 54–19 Awning windows are often used in stacks or in combination with other types of windows.

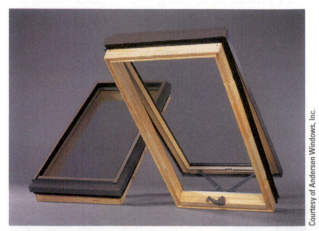

Courtesy of Andersen Windows, Inc.

Figure 54–20 Skylights and roof windows are made in a number of styles and sizes.

Courtesy of Andersen Windows, Inc.

Figure 54–18 The sashes in sliding windows move horizontally by each other.

Skylight and Roof Windows

Skylights provide light only. Roof windows contain operating sashes to provide light and ventilation (*Figure 54–20*). One type of roof window comes with a tilting sash that allows access to the outside surface for cleaning. Special flashings are used when multiple skylights or roof windows are ganged together.

CHAPTER 55

Window Installation and Glazing

There are numerous window manufacturers that produce hundreds of kinds, shapes, and sizes of windows. Because of the tremendous variety and design differences, follow the manufacturer's instructions closely to ensure a correct installation. Directions in this unit are basic to most window installations. They are intended as a guide to be supplemented by procedures recommended by the manufacturer.

SELECTING AND ORDERING WINDOWS

The builder must study the plans to find the type and location of the windows to be installed. The floor plan shows the location of each unit. Each unit is usually identified by a number or a letter next to the window symbol.

Those responsible for designing and drawing plans for building or selecting windows must be aware of, and comply with, building codes that set certain standards in regard to windows. ✳ Most codes require minimum areas of natural light. Codes also require minimum ventilation by windows unless provided by other means. Some codes stipulate minimum window sizes in certain rooms for use as emergency egress. ✳

✳**IRC 612.2**

✳**IRC 312.2.1**

Out-swinging windows, such as awning and casement windows, should not swing out over decks, patios, and similar areas unless they are high enough to permit persons to travel under them. When too low, the projecting sash could cause serious injury.

Window Schedule

The numbers or letters found in the floor plan identify the window in more detail in the window schedule. This is, usually, part of a set of plans (*Figure 55–1*). This schedule normally includes window style, size, manufacturer's name, and unit number. Rough opening sizes may or may not be shown.

Manufacturers' Catalogs

Sometimes a window schedule is not included. Units are identified only by the manufacturer's name and number on the floor plan. To get more

Window Schedule				
Ident.	**Quan.**	**Manufacturer**	**Size**	**Remarks**
A	6	Andersen	TW28310	D.H. Tiltwash
B	1	Andersen	WDH2442	Woodwright D.H.
C	2	Andersen	3062	Narrowline D.H.
D	1	Andersen	CW24	Casement Single
E	1	Andersen	C34	Casement Triple
F	1	Andersen	C23	Casement Double

© Cengage Learning 2014

Figure 55–1 Typical window schedule found in a set of plans.

information, the builder must refer to the window manufacturer's catalog.

The catalog usually includes a complete description of the manufactured units and optional accessories, such as insect screens, glazing panels, and grilles. For a particular window style, the catalog typically shows overall unit dimensions, rough opening widths and heights, and glass sizes of manufactured units. Large-scale, cross-section details of the window unit also usually are included so the builder can more clearly understand its construction (*Figure 55–2*).

Green Tip

Window performance is only as good as the installation. Follow manufacturer's recommendations. Correct installations will maintain manufacturers warranties and keep energy costs at their lowest.

Order window units giving the type and identification letters and/or numbers found in the window schedule or manufacturer's catalog. The size of all existing rough openings should be checked to make sure they correspond to the size given in the catalog before windows are ordered.

INSTALLING WINDOWS

All rough window openings should be prepared to ensure weather-tight window installations.

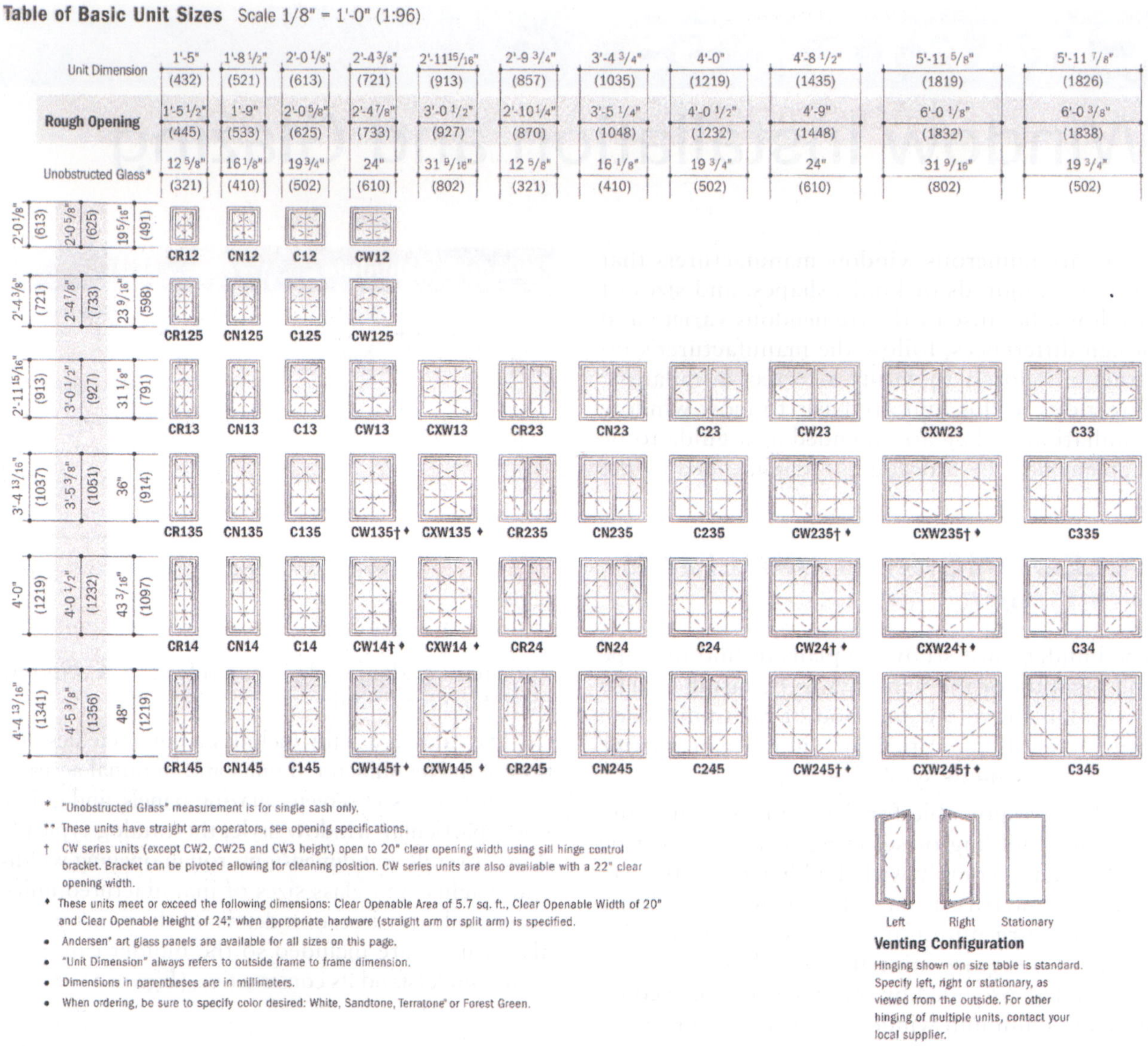

Table of Basic Unit Sizes Scale 1/8" = 1'-0" (1:96)

	1'-5" (432)	1'-8 1/2" (521)	2'-0 1/8" (613)	2'-4 3/8" (721)	2'-11 15/16" (913)	2'-9 3/4" (857)	3'-4 3/4" (1035)	4'-0" (1219)	4'-8 1/2" (1435)	5'-11 5/8" (1819)	5'-11 7/8" (1826)
Unit Dimension											
Rough Opening	1'-5 1/2" (445)	1'-9" (533)	2'-0 5/8" (625)	2'-4 7/8" (733)	3'-0 1/2" (927)	2'-10 1/4" (870)	3'-5 1/4" (1048)	4'-0 1/2" (1232)	4'-9" (1448)	6'-0 1/8" (1832)	6'-0 3/8" (1838)
Unobstructed Glass*	12 5/8" (321)	16 1/8" (410)	19 3/4" (502)	24" (610)	31 9/16" (802)	12 5/8" (321)	16 1/8" (410)	19 3/4" (502)	24" (610)	31 9/16" (802)	19 3/4" (502)

Row heights (left side):

- 2'-0 1/8" (613) / 2'-0 5/8" (625) / 19 5/16" (491): **CR12 CN12 C12 CW12**
- 2'-4 3/8" (721) / 2'-4 7/8" (733) / 23 9/16" (598): **CR125 CN125 C125 CW125**
- 2'-11 15/16" (913) / 3'-0 1/2" (927) / 31 1/8" (791): **CR13 CN13 C13 CW13 CXW13 CR23 CN23 C23 CW23 CXW23 C33**
- 3'-4 13/16" (1037) / 3'-5 3/8" (1051) / 36" (914): **CR135 CN135 C135 CW135† ◆ CXW135 ◆ CR235 CN235 C235 CW235† ◆ CXW235† ◆ C335**
- 4'-0" (1219) / 4'-0 1/2" (1232) / 43 3/16" (1097): **CR14 CN14 C14 CW14† ◆ CXW14 ◆ CR24 CN24 C24 CW24† ◆ CXW24† ◆ C34**
- 4'-4 13/16" (1341) / 4'-5 3/8" (1356) / 48" (1219): **CR145 CN145 C145 CW145† ◆ CXW145 ◆ CR245 CN245 C245 CW245† ◆ CXW245† ◆ C345**

* "Unobstructed Glass" measurement is for single sash only.

** These units have straight arm operators, see opening specifications.

† CW series units (except CW2, CW25 and CW3 height) open to 20" clear opening width using sill hinge control bracket. Bracket can be pivoted allowing for cleaning position. CW series units are also available with a 22" clear opening width.

◆ These units meet or exceed the following dimensions: Clear Openable Area of 5.7 sq. ft., Clear Openable Width of 20" and Clear Openable Height of 24", when appropriate hardware (straight arm or split arm) is specified.

• Andersen® art glass panels are available for all sizes on this page.

• "Unit Dimension" always refers to outside frame to frame dimension.

• Dimensions in parentheses are in millimeters.

• When ordering, be sure to specify color desired: White, Sandtone, Terratone® or Forest Green.

Left Right Stationary

Venting Configuration
Hinging shown on size table is standard. Specify left, right or stationary, as viewed from the outside. For other hinging of multiple units, contact your local supplier.

Courtesy of Andersen Windows, Inc.

Figure 55–2 Typical page from a window manufacturer's catalog.

Housewraps and Building Paper

Exterior walls are sometimes covered with a building paper, 15# asphalt felt, prior to the application of siding. It serves as a water barrier. In place of building paper, exterior walls are often covered with a type of air infiltration barrier commonly called housewrap. This material prevents the infiltration of air into the structure, yet it also allows the passage of water vapor to the outside.

Housewrap is a very thin, tough plastic material. It is used to cover the sheathing on exterior walls for the same purpose as building paper (*Figure 55–3*). Housewraps are commonly known by the brand names of Typar and Tyvek. They are more resistant than building paper to air leakage and are virtually tear-proof. Yet they are also

Courtesy of fiberweb

Figure 55–3 Housewrap is widely used as an air infiltration barrier on sidewalls.

breathable to allow water vapor to escape. Building paper comes in 36-inch-wide rolls. Housewrap rolls are 1.5, 3, 4.5, 5, 9, and 10 feet wide.

Housewrap gets its name because it is completely wrapped around the building. It covers corners, window and door openings, plates, and sills. Housewrap is designed to survive prolonged periods of exposure to the weather. It can and usually is applied immediately after framing is completed, but before doors and windows have been installed. The wrap, then, serves also as a flashing for the sides of windows and doors.

Applying Housewrap. Begin at the corner holding the roll vertically on the wall. Unroll it a short distance. Make sure the roll is plumb. Secure the sheet to the corners, leaving about a foot extending beyond the corner to overlap later. Continue to unroll (*Figure 55–4*). Make sure the sheet is straight, with no buckles. Fasten every 12 to 18 inches along the edges and 24 inches on center in the field.

Unroll directly over window and door openings and around the entire perimeter of the building. Overlap all joints by at least 3 inches. Secure them with sheathing tape. Horizontal joints must overlap the upper layer over the lower layer.

The cutting out of the window and door openings is done with waterproofing in mind. First, cut a rectangle out of housewrap in the opening that leaves about 1 inch at the top and enough material along sides and bottom to wrap the thickness of the wall. This is about 4 or 6 inches (*Figure 55–5*). Cut diagonally from the corners of openings. Fold theremaining edges into the opening and staple to the rough opening frame.

CAUTION

Housewraps are slippery. They should not be used in any application where they will be walked on.

Water-Resistant Sheathing

Water-resistant sheathing may be used for wall and roof sheathing. It is marketed under the name Zip System. It is made from $7/16$-inch rated panel with a coating adhered to the panel. The protective coating replaces the need for housewrap and shingle underlayment (*Figure 55–6*). After sheathing panels are installed the seams are taped with a special tape. This saves time and creates a fast, airtight, weather-resistant structure, even before the finishes are applied. It should be noted that foil-faced foam sheathing, when taped, also eliminates the need for housewrap.

Figure 55–4 Housewrap is unrolled and stapled to the building.

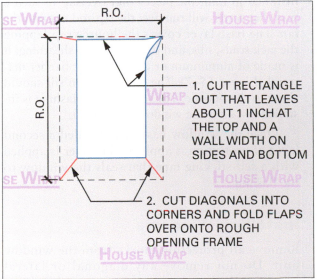

R.O. = WINDOW ROUGH OPENING

1. CUT RECTANGLE OUT THAT LEAVES ABOUT 1 INCH AT THE TOP AND A WALL WIDTH ON SIDES AND BOTTOM

2. CUT DIAGONALS INTO CORNERS AND FOLD FLAPS OVER ONTO ROUGH OPENING FRAME

Figure 55–5 Housewrap is cut to folded back into rough openings.

© Cengage Learning 2014

Figure 55–6 Weather-resistant sheathing eliminates the need for house wrap.

Window Base Flashing

A base layer provides an added layer of protection for the framing. Water penetration is more likely in areas of severe wind driven rain. The base layer is installed in multiple pieces that overlap in such a way that water will run *over* the lap and *not into* the lap. The base layer covers the housewrap lapping the jack studs, sills, and the edge of the sheathing. It is made of aluminium or ice and water barrier material (*Figure 55–7*). The base layer on sill should slope to the outside. The top piece of flashing is installed last after the window is installed.

Once the window has been installed a second layer of protection is applied. This layer is applied with a self-sticking tape that seals the window to the flashing.

Installing Windows

Remove all protection blocks from the window unit. Do not remove any diagonal or lateral braces applied at the factory. Close and lock the sash. If windows are stored inside, they can easily be moved through the openings and set in place while standing inside but the window should be opened to provide something to grasp. (*Figure 55–8*). Caulk may be applied to the backside of the nailing flanges before window is installed, but the bottom flange should not be caulked. This will allow any leaked in water to have an escape route. Place unit in the opening. It is important to center the unit in the opening from side-to-side.

> **CAUTION**
>
> Have sufficient help when setting large units. Handle them carefully to avoid damaging the unit or breaking the glass. Broken glass can cut through protective clothing and cause serious injury.

Place a level on the window sill. If not level, determine which side of the window is the highest. Check to see if the higher side of the unit is at the desired height. If not, bring it up by shimming with a wood shim between the rough sill and the bottom

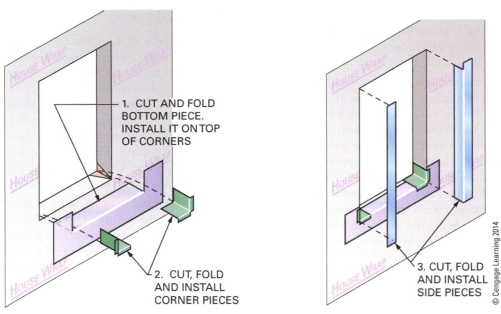

1. CUT AND FOLD BOTTOM PIECE. INSTALL IT ON TOP OF CORNERS

2. CUT, FOLD AND INSTALL CORNER PIECES

3. CUT, FOLD AND INSTALL SIDE PIECES

© Cengage Learning 2014

Figure 55–7 Base layer of flashing is installed in three pieces.

© Cengage Learning 2014

Figure 55–8 Windows can be installed from the inside, through the opening.

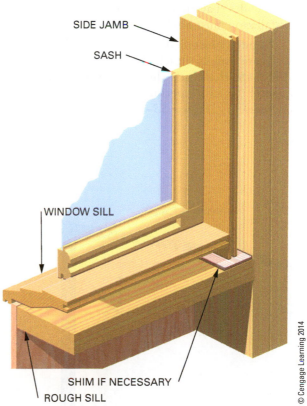

SIDE JAMB

SASH

WINDOW SILL

SHIM IF NECESSARY

ROUGH SILL

© Cengage Learning 2014

Figure 55–9 Use shims under the side jambs to level the window unit.

end of the window's side jamb (*Figure 55–9*). Level the window by shimming the low side of window. Tack the lower end of the casing on both sides.

On wide windows with long sills, shim at intermediate points so that the sill is perfectly straight and level with no sag. Use either a long level or a shorter level in combination with a straightedge. Also, sight the sill by eye from end to end to make sure it is straight. Place one nail at the lower ends of each side nailing flange or casing.

Plumb the ends of the side jambs with a level. Tack the top ends of the side nailing flanges or casings. Straighten the side jambs between sill and head jamb if necessary.

Check the joint between the upper and lower sash. It should be closely aligned. Make sure the sash operates properly. If not, make necessary adjustments. Then fasten the window permanently in place. On windows with wood casings use galvanized casing nails spaced about 16 inches apart. Keep nails about 2 inches back from the ends of the casings to avoid splitting. Nail length depends on the thickness of the casing. Nails should be long enough to penetrate the sheathing and into the framing members. Set the nails so they can be puttied over later.

Vinyl-clad windows have a vinyl nailing flange. Large-head roofing nails are driven through the flange instead of through the casing (*Figure 55–10*). Drive roofing nails in every nailing slot of flanges that penetrate the framing by 1½ inches.

It should be noted that areas of high wind have local codes that affect the methods of installation. Installation must also be inspected.

✻IRC 612.7.2 IRC 612.7.2.1 Windows installed in masonry and brick veneer walls are usually attached to a treated wood buck. Adequate clearance should be left for caulking around the entire perimeter between the window and masonry (*Figure 55–11*). ✻

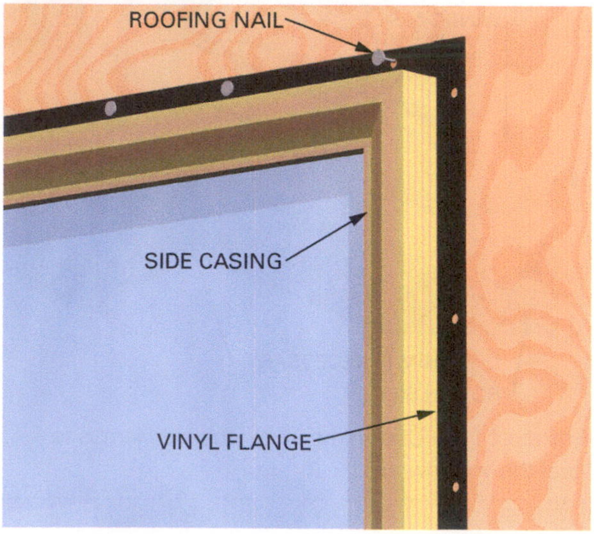

Sealing the Window

Sealing a window begins by making sure the house-wrap laps over on top of the nailing flange. The side and bottom flanges are nailed on top of the wrap. Flashing tape is then applied to the perimeter, covering the nailing flanges (*Figure 55–12*). Flashing material comes in rolls and is applied like tape. It is important to begin at the bottom and work up the sides, ending with the top piece. Each layer is over lapped onto the previous piece. This will keep the any water able to flow over the lap and not into the lap.

Additional head flashing may be applied over the window and under the siding. This will help direct away any water from behind the siding (*Figure 55–13*). Its length should be equal to the overall width of the window. Do not let the ends project beyond the side casings more than about ⅛ inch each side. This will make the application of siding difficult. Place the flashing firmly on top of the head casing. Secure with fasteners along its top edge and into the wall sheathing.

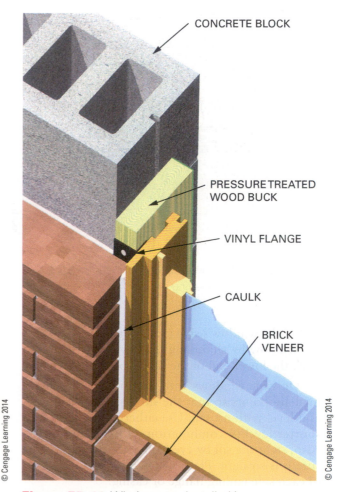

© Cengage Learning 2014

Figure 55–11 Windows are installed in masonry openings against wood bucks.

Figure 55–10 Roofing nails are used to fasten the flanges of vinyl-clad windows.

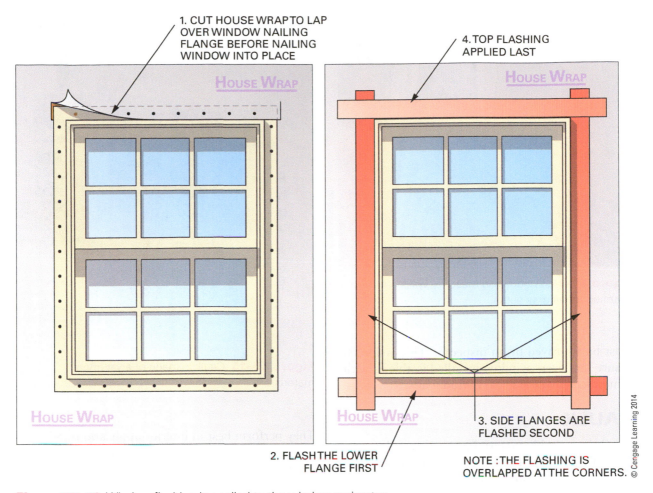

1. CUT HOUSE WRAP TO LAP OVER WINDOW NAILING FLANGE BEFORE NAILING WINDOW INTO PLACE

HOUSE WRAP

HOUSE WRAP

4. TOP FLASHING APPLIED LAST

HOUSE WRAP

HOUSE WRAP

2. FLASH THE LOWER FLANGE FIRST

3. SIDE FLANGES ARE FLASHED SECOND

NOTE : THE FLASHING IS OVERLAPPED AT THE CORNERS.

© Cengage Learning 2014

Figure 55–12 Window flashing is applied to the window perimeter.

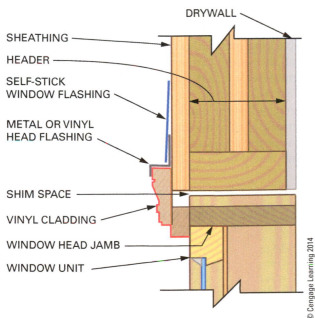

DRYWALL

SHEATHING

HEADER

SELF-STICK WINDOW FLASHING

METAL OR VINYL HEAD FLASHING

SHIM SPACE

VINYL CLADDING

WINDOW HEAD JAMB

WINDOW UNIT

© Cengage Learning 2014

Figure 55–13 A drip cap is often installed as an extra layer of flashing over the top of the head casing.

METAL WINDOWS

Metal windows are available in the same styles as wood windows. The shape and sizes of the parts vary with the manufacturer and the intended use.

In frame construction, if metal windows are used, they may be set in a wood frame. The frame is then installed in the same manner as for wood windows. They may also be set in the opening with their flanges overlapping the siding or sheathing. Caulking is applied under the flanges on sides and top, but not the bottom to allow any leaked water to escape. The unit is then screwed to the wall.

In masonry construction, wood bucks are fastened to the sides of the opening. Metal windows are installed against them. The flanges on the two sides and top are bedded in caulking (*Figure 55–14*). They are then screwed to the bucks.

Carefully follow the installation directions provided with the units, whether wood or metal. In areas prone to hurricanes, window installation

Figure 55–14 Windows in masonry walls are fastened to window bucks.

Figure 55–15 Caulk is made of a variety of materials.

***IRC
612.7.2
IRC
612.7.2.1** must be approved by an inspector. This is done to ensure that every window installed is able to withstand the anticipated heavy wind loads. *****

CAULK

Caulk is a pliable material applied to fill gaps in building materials. It bonds to the surrounding material during cure and remains flexible after cure. This flexibility allows building materials to expand and contract with heat and moisture, maintaining the seal.

Caulks are made with a variety of materials, each with a different function. The materials include acrylic, butyl rubber, latex, polyurethane, and silicone. Many caulks are made with a mixture of these ingredients to blend the desirable characteristics (*Figure 55–15*). Read the manufacturer's recommendations on the tube to determine the best usage.

Generally speaking, acrylic and latex caulks are used when expected material movement is small.

They perform best in interior applications. Silicone is used when resistance to mold and mildew are required such as in kitchens and bathrooms. Silicone can be used for interior and exterior applications and has very good flexibility. Polyurethane is designed for exterior applications where severe material movement is expected. It can fill large gaps and has superior bonding and flexibility characteristics.

Caulk performs best when it is installed with a backing material (*Figure 55–16*). Backing material allows the caulk to bond to the materials on opposite sides of the joint. When the building material moves, the caulk is compressed and stretched. If the caulk is allowed to bond on three surfaces, as the material moves, the bond begins to tear in the corners and eventually the bond can be completely broken.

TWO SURFACE BOND

CAULK

BACKER ROD

MATERIAL MOVEMENT

THREE SURFACE BOND

CAULK

SHEAR STRESS CAUSES CAULK BOND TO FAIL OVER TIME

MATERIAL MOVEMENT

CAULK

FAILED BOND

MATERIAL MOVEMENT

Figure 55–16 Beads of caulk should be installed with a backer rod.

Green Tip

Flashing is a vital, yet simple component of a building. Sealants are not a substitute for proper flashing.

GLAZING

The art of cutting and installing lights of glass in sash and doors is called *glazing*. Those who do such work are called glaziers. Sometimes the carpenter may have to replace a light of glass.

Glazing Materials

Sashes are made so that the glass is usually held in place with glazing points and glazing compound (*Figure 55–17*). Glazing points are small triangular or diamond-shaped pieces of thin metal. They are driven into the sash parts to hold the glass in place. Glazing compound is commonly called putty. It is used to cover and seal around the edges of the light.

A light of glass is installed with its convex side, or crown of its bow, up in a thin bed of compound against the rabbet of the opening. Glass set in this manner is not as apt to break when installed.

Installing Glass

To replace or glaze a light of glass, first remove the broken glass.

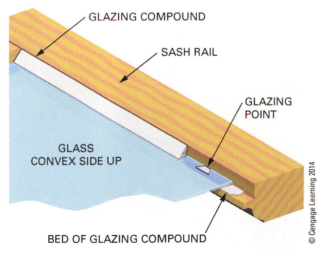

Figure 55–17 Panes of glass are secured with glazing points and glazing compound.

GLAZING COMPOUND

SASH RAIL

GLAZING POINT

GLASS CONVEX SIDE UP

BED OF GLAZING COMPOUND

© Cengage Learning 2014

CAUTION

Use heavy gloves to handle the broken glass. Broken glass edges are sharp and can cut easily.

Clean all compound and glazier points from the rabbeted section of the sash. Apply a thin bed of glazing compound to the surface of the rabbet on which the glass will lay. Lay the glass in the sash with its crowned side up. Carefully seat the glass in the bed by moving it back and forth slightly.

Fasten the glass in place with glazier points. Slide the driver along the glass. Do not lift the driver off the glass. Special glazier point-driving tools prevent glass breakage. If a driving tool is not available, drive the points with the side of an old chisel or a putty knife.

Lay a bead of compound on the glass along the rabbet on top of the light of glass. Trim the compound at a bevel by drawing the putty knife along. One edge should be flush with the inside edge of the glass opening. The other edge of the compound should feather to the outside edge of the opening. Prime the compound as soon as possible after glazing. Lap the paint about $1/16$ inch onto the glass to make a seal against the weather.

Cutting Glass

Sometimes it may be necessary to cut a light of glass to size so it will fit in the opening. Lay the glass on a clean, smooth surface. Brush some mineral spirits along the line of cut. Hold a straightedge on the line. Draw a glass cutter with firm pressure along the straightedge to make a clean, uniformly scored line (*Figure 55–18*).

Do not score over the line more than once. This will dull the glass cutter. The line must be scored along the whole length the first time with no skips. Otherwise, the glass may not break where desired. Practice on scrap pieces to become proficient in making clean breaks.

Grab the glass with both hands close to the line using the thumbs and index fingers (*Figure 55–19*). With outward pulling action and a twist of the wrist, jerk your hands up and down in a short, quick motion. The glass will break along the line. If it resists, move the glass so the scored line is even with the edge of the workbench. Apply downward pressure on the overhanging glass. If the glass is properly scored, it will break along the scored line.

Figure 55–18 The glass must be scored with a single stroke of the glass cutter.

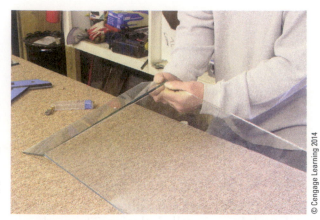

Figure 55–19 After glass is scored completely, it will break cleanly along the scored line.

DECONSTRUCT THIS

Carefully study *Figure 55–20* and think about what is wrong and/or what is right. Consider all possibilities. What construction practice or method is different in your area of the country?

Figure 55–20 This photo shows a window installed with flashing.

KEY TERMS

arch windows	fixed windows	lights	side jambs
argon gas	glaziers	low emissivity	sill
bay window	glazing	mullion	single-hung window
blind stops	glazing compound	muntins	single-strength (SS) glass
bow window	glazing points	parting strip	
bucks	grilles	putty	sliding windows
caulk	head casing	rails	stiles
double-strength (DS) glass	head jamb	roof windows	tempered glass
	hopper window	safety glass	vinyl-clad
drip cap	housewrap	sash	window flashing
extension jambs	laminated glass	side casings	window frame

REVIEW QUESTIONS

Select the most appropriate answer.

1. A frame holding a pane of glass is called a
   a. light.
   b. mullion.
   c. sash.
   d. stile.

2. Small strips that divide the glass into smaller panes are called
   a. mantels.
   b. margins.
   c. mullions.
   d. muntins.

3. When windows are installed in multiple units, the joining of the side jambs forms a
   a. mantel.
   b. margin.
   c. mullion.
   d. muntin.

4. A window that consists of an upper and a lower sash, both of which slide vertically, is called a
   a. casement window.
   b. double-hung window.
   c. hopper window.
   d. sliding window.

5. A window that has a sash hinged on one side and swings outward is called
   a. an awning window.
   b. a casement window.
   c. a double-hung window.
   d. a hopper window.

6. The difference between a hopper and an awning window is that the hopper window
   a. swings inward instead of outward.
   b. swings outward instead of inward.
   c. is hinged at the top rather than at the bottom.
   d. is hinged on the side rather than on the bottom.

7. The term *fixed window* refers to a
   a. window with an unmovable sash.
   b. repaired window.
   c. window unit properly flashed.
   d. all of the above.

8. Single-hung windows
   a. have one sash.
   b. have one sash that moves.
   c. are installed with no other windows nearby.
   d. slide from side to side.

9. Multiple window units fastened together to form a large, outward curved window are referred to as

   a. bay windows.
   b. bow windows.
   c. double casement windows.
   d. fixed double-awning windows.

10. Windows installed in areas prone to hurricanes should be

    a. flashed on all sides of the opening.
    b. caulked into place.
    c. inspected by an inspector.
    d. all of the above.

11. The best choice of caulk for interior bathroom applications is

    a. latex.
    b. silicone.
    c. urethane.
    d. all of the above.

12. The art of cutting and installing glass is called

    a. gauging.
    b. glazing.
    c. gouging.
    d. glassing.

13. A gas used to make windows more energy efficient is

    a. argon.
    b. oxygen.
    c. nitrogen.
    d. radon.

14. Beads of caulking placed in large gaps should be installed

    a. to fill the gap.
    b. with a backer rod.
    c. in multiple layers allowing each to dry first.
    d. all of the above.

15. The term "Low E" for windows refers to low

    a. emissivity.
    b. energy.
    c. egress.
    d. extension.

UNIT 21
Exterior Doors

Exterior doors, like windows, are manufactured in millwork plants in a wide range of styles and sizes. Many entrance doors come prehung in frames, complete with exterior casings applied, and ready for installation. In other cases, the door is fitted and hinged to a door frame.

OBJECTIVES

After completing this unit, the student should be able to:

- describe the standard designs of exterior doors.
- name the parts of an exterior door frame.
- fit and hang a prehung exterior door.
- build and set a door frame.
- install locksets in doors.

SAFETY REMINDER

Installing exterior doors may involve fastening large, heavy sections together. Be sure to lift and secure units properly to prevent injuries to people and damage to materials.

CHAPTER 56

Exterior Door Styles

Exterior doors are often the central focus of a house. As one approaches it from the street, the architectural style of the house is revealed. Entrances to houses have many variations, yet the names of the parts and pieces are similar.

DOOR STYLES AND SIZES

Exterior flush and panel doors are available in many styles. There are many choices when designing entrances, and only a select few are considered here. To get a broad view of what is available consult a manufacturer's literature.

Flush Doors

An exterior flush door has a smooth, flat surface of wood veneer or metal. It has a framed, *solid core* of staggered wood blocks or composition board. Wood *core blocks* are inserted in appropriate locations in composition cores. They serve as *backing* for door locks (*Figure 56–1*). Openings may be cut in flush doors either in the factory or on the job. Lights of various kinds and shapes are installed in them. Molding of various shapes may be applied in many designs to make the door more attractive.

Panel Doors

Panel doors are classified by one large manufacturer as *high-style, panel, sash, fire, insulated, French,* and *Dutch,* doors. Sidelights, although not actually doors, constitute part of some entrances. They are fixed in the door frame on one or both sides of the door (*Figure 56–2*). Transoms are similar to sidelights. When used, they are installed above the door.

Panel Door Styles.

High-style doors, as the name implies, are highly crafted designer doors. They may have a variety of cut-glass designs. Panel doors are made with raised panels of various shapes. Sash doors have panels of tempered or insulated glass that allow the passage of light (*Figure 56–3*). Fire doors are used where required by codes. These doors prevent the spread of fire for a certain period of time. Insulated doors have thicker panels with Low-E or argon-filled insulated glass. French doors may contain one, five, ten, twelve, fifteen, or eighteen lights of glass for the total width and length inside the frame. A Dutch door consists of top and bottom units, hinged independently of each other.

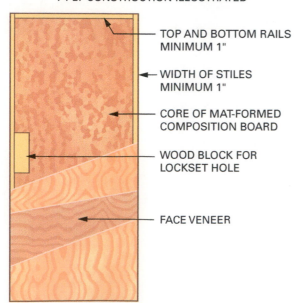

MAT-FORMED COMPOSITION BOARD CORE
7-PLY CONSTRUCTION ILLUSTRATED

- TOP AND BOTTOM RAILS MINIMUM 1"
- WIDTH OF STILES MINIMUM 1"
- CORE OF MAT-FORMED COMPOSITION BOARD
- WOOD BLOCK FOR LOCKSET HOLE
- FACE VENEER

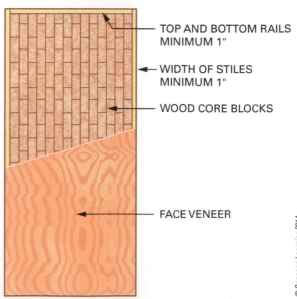

FRAMED BLOCK NON-GLUED CORE
5-PLY CONSTRUCTION ILLUSTRATED

- TOP AND BOTTOM RAILS MINIMUM 1"
- WIDTH OF STILES MINIMUM 1"
- WOOD CORE BLOCKS
- FACE VENEER

© Cengage Learning 2014

Figure 56–1 Composition or solid wood cores are possible in exterior flush doors.

Parts of a Panel Door. A panel door consists of a frame that surrounds panels of solid wood and glass, or louvers. Some door parts are given the same terms as a window sash.

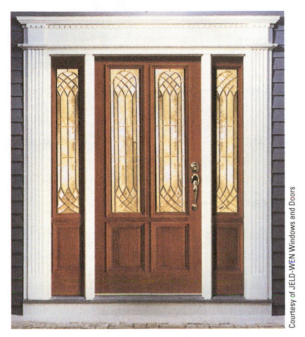

Figure 56–2 Sidelights are installed on one or both sides of the main entrance door.

The outside vertical members are called stiles. Horizontal frame members are called rails. The *top rail* is generally the same width as the stiles. The *bottom rail* is the widest of all rails. A rail situated at lockset height, usually 38 inches from the finish floor to its center, is called the lock rail. Almost all other rails are called *intermediate rails*. Mullions are vertical members between rails dividing panels in a door.

The molded shape on the edges of stiles, rails, mullions, and bars, adjacent to panels, is called the sticking. The name is derived from the molding machine, commonly called a sticker, used to shape the parts. Several shapes are used to *stick* frame members.

Bars are narrow horizontal or vertical rabbeted members. They extend the total length or width of a glass opening from rail to rail or from stile to stile. Door muntins are short members, similar to and extending from bars to a stile, rail, or another bar. Bars and muntins divide the overall length and width of the glass area into smaller lights.

Panels fit between and are usually thinner than the stiles, rails, and mullions. They may be raised on one side or on both sides for improved appearance (*Figure 56–4*).

EXTERIOR PANEL DOORS

EXTERIOR SASH DOORS

SIDELIGHTS

Figure 56–3 Several kinds of exterior doors are made in many designs.

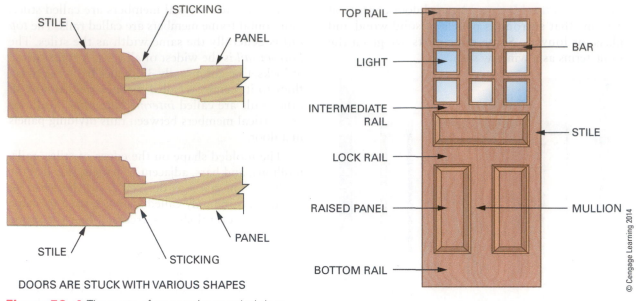

DOORS ARE STUCK WITH VARIOUS SHAPES

Figure 56–4 The parts of an exterior paneled door.

PARTS OF AN EXTERIOR DOOR FRAME

Terms given to members of an exterior door frame are the same as similar members of a window frame. The bottom member is called a *sill* or *stool*. The vertical side members are called *side jambs*. The top horizontal part is a *head jamb*. The exterior door trim may consist of many parts to make an elaborate and eye-appealing entrance or a few parts for a more simple doorway. The door casings, if not too complex, are usually applied to the door frame before it is set (*Figure 56–5*). When more intricate trim is specified, it is usually applied after the frame is set (*Figure 56–6*).

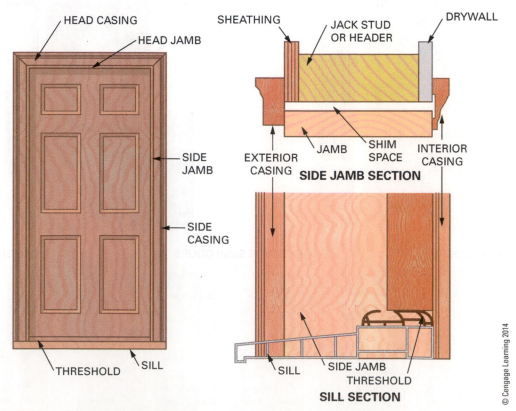

Figure 56–5 Parts of an exterior door frame.

Sills

In residential construction, door frames usually are designed and constructed for entrance doors that swing inward. Codes require that doors swing outward in buildings used by the general public. The shape of a wood door sill for an inswinging door is different from that for an outswinging door (*Figure 56–7*).

In addition to wood, extruded aluminum sills, also called thresholds, of many styles are manufactured for both inswinging and outswinging doors. They usually come with vinyl inserts to weatherstrip the bottom of the door. Some are adjustable for exact fitting between the sill and door (*Figure 56–8*).

Jambs

Side and header jambs are the same shape. Jambs may be square edge pieces of stock to which door stops are later applied. Or, they may be rabbeted jambs, with single or double rabbets. On double-rabbeted jambs, one rabbet is used as a stop for the main door. The other is used as a stop for

Figure 56–6 Elaborate entrance trim is available.

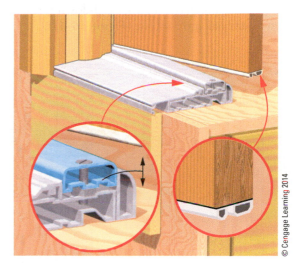

Figure 56–8 Some metal sills are adjustable for exact fitting at the bottom of the door.

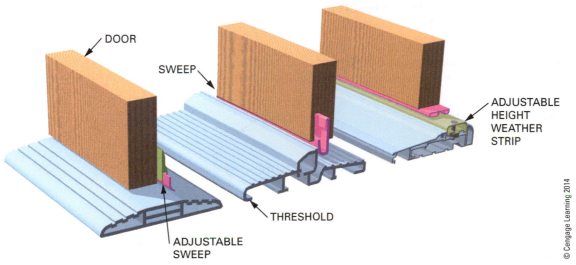

Figure 56–7 Wood sill shapes and styles vary according to the swing of the door.

storm and screen doors (*Figure 56–9*). Several jamb widths are available for different wall thicknesses. For walls of odd thicknesses, jambs, except double-rabbeted ones, may be ripped to any desired width.

Exterior Casings

Exterior casings may be plain square-edge stock. Moldings are sometimes applied around the outside edges of the casings. This is done to improve the appearance of the entrance. Because the main entrance is such a distinctive feature of a building, the exterior casing may be enhanced on the job with fluted, or otherwise shaped, pieces and appropriate caps and moldings applied (*Figure 56–10*). Flutes are narrow, closely spaced, concave grooves that run parallel to the edge of the trim. In addition, ornate main entrance trim may be purchased in knocked-down form. It is then assembled at the jobsite (*Figure 56–11*).

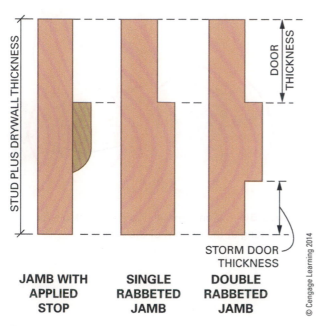

Figure 56–9 Door jamb cross-sections may be square edged, single rabbeted, or double rabbeted.

Weatherstripping

Exterior doors have weatherstripping to protect from air and weather infiltration. It is a soft pliable strip usually applied to the door jamb. When

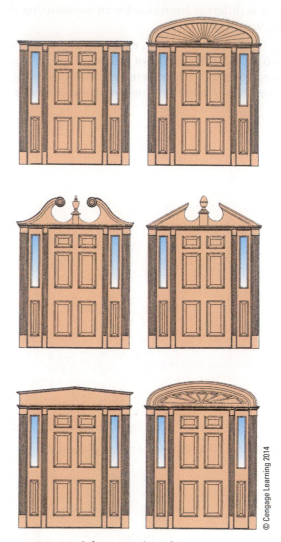

Figure 56–11 A few samples of the many manufactured entrance door styles.

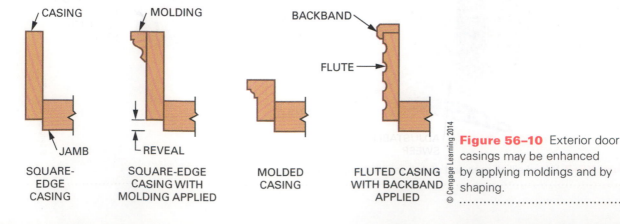

Figure 56–10 Exterior door casings may be enhanced by applying moldings and by shaping.

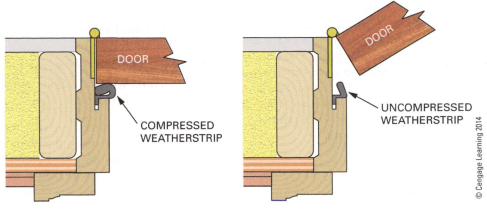

Figure 56–12 Plastic weather strips compress when the door closes to seal the opening.

the door closes it is compressed to seal the sides and top of the door to the jamb (*Figure 56–12*). Weatherstripping acts with the threshold to keep weather and air from passing through while the door is closed.

EXTERIOR DOOR SIZES

Residential exterior entrance doors are typically 3'-0" wide × 6'-8" high. The IRC states at least one entrance door must have a clear opening width of 32 inches as measured between the face of an open door and the doorstop on the opposite side. The height must be 78 inches as measured between the threshold and the head jamb doorstop. The smallest door to accomplish this is a 3'-0" × 6'-8" door. Entrance doors are normally manufactured in a thickness of 1 ¾ inches.

Some styles, such as French, panel, and sash doors, are available in narrower widths, such as 1'-6", 2'-0", and 2'-4", when they are installed as double doors. Sidelights are made in both 1⅜-inch and 1 ¾-inch thicknesses.

DETERMINING THE SWING OF THE DOOR

When ordering or making doors, the swing direction of a door is important to know. The set of prints for the building show all doors and the direction of the door swing. The door schedule will list the doors, their styles and swings. The carpenter should be able to independently verify the swing of a door. It can be confusing because there are several ways of referring to swing, depending on whom you are speaking. Exterior doors add more confusion because they have surfaces designed to only be the outside or the inside. Door swing is labeled either left-hand or right-hand.

Interior Door Swing

A straightforward method for determining the swing or handing of interior doors is to consider your forearm when approaching the door as it swings away. The hand of forearm that puts your elbow at the hinge and the hand toward the knob gives the handing (*Figure 56–13*).

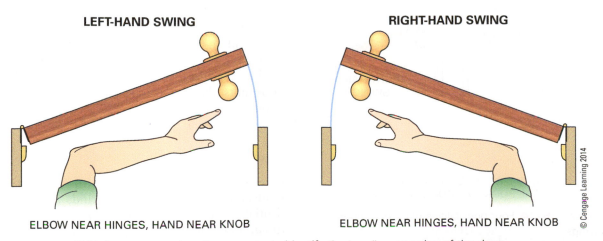

LEFT-HAND SWING

RIGHT-HAND SWING

ELBOW NEAR HINGES, HAND NEAR KNOB

ELBOW NEAR HINGES, HAND NEAR KNOB

Figure 56–13 With forearm out, the elbow serves to identify the handing or swing of the door.

Some people consider the door with the viewer on the side of the door where the hinges are visible when the door is closed; this is the side of the door where the door opens towards you. The handing of the door is the side the knob is. In other words, the hand easiest to pull open the door is the handing. This method is limiting when we consider exterior and commercial doors. We shall use the forearm method for exterior doors.

Exterior Door Swing

Exterior doors add a complication because they have an inside and outside surface. Door parts must be gasketed and caulked watertight on the outside. They may also swing inward or outward. Commercial doors that have pre-bored holes for crash bars and other side-specific hardware must be precisely described before ordering them.

The handing of exterior doors is considered from the side of the door as you approach to enter the room. This is also the keyed side or outside of the building. If it swings away, it is considered normal and may be thought of as for interior doors. If it swings towards you it is considered reversed. This makes for four handing possibilities; left-hand, right-hand, left-hand reverse, and right-hand reverse (*Figure 56–14*).

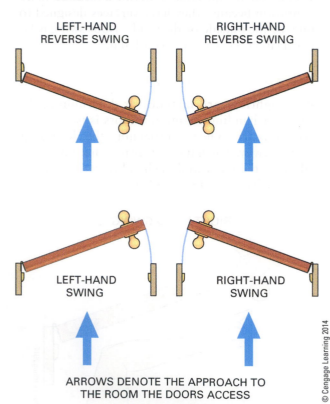

Figure 56–14 Four possibilities of exterior door swings.

- Right-Hand – door swings away, hinges on the right
- Right-Hand Reverse – door swings towards you, hinges on the right
- Left-Hand – door swings away, hinges on the left
- Left-Hand Reverse – door swings towards you, hinges on the left

It should be noted here that the handing of the door is simplified when doors have both surfaces alike. A left-hand door is the same as a right-hand reverse door. The only reason to make the different designations is to deal with inside and outside surfaces of an exterior door.

When communicating doors with another person, it is a good idea to spend time making sure the correct door is ordered or assembled. Confusion here can cause lots of lost time and money.

OTHER EXTERIOR DOORS

Other exterior doors, such as swinging double doors, are fitted and hung in a similar manner to single doors. Allowance must be made, when fitting swinging double doors, for an astragal between them for weathertightness (*Figure 56–15*). An astragal is a *molding* that is rabbeted on both edges. It is designed to cover the joint between double doors. One edge has a square rabbet. The other has a beveled rabbet to allow for the swing of one of the doors.

Patio Doors

Patio door units normally consist of two or three sliding or swinging glass doors completely assembled in the frame (*Figure 56–16*). Instructions for assembly are included with the unit. They should be followed carefully. Installation of patio door frames is similar to setting frames for swinging doors. After the frame is set, the doors are installed using special hardware supplied with the unit.

In sliding patio door units, one door is usually stationary. In two-door swinging units, either one or the other door is the swinging door. In three-door units, the center door is usually the swinging door.

Garage Doors

Overhead garage doors come in many styles, kinds, and sizes. Two popular kinds used in residential construction are the *one-piece* and the *sectional* door. The rigid one-piece unit swings out at

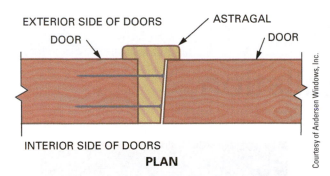

EXTERIOR SIDE OF DOORS

DOOR

ASTRAGAL

DOOR

INTERIOR SIDE OF DOORS
PLAN

Courtesy of Andersen Windows, Inc.

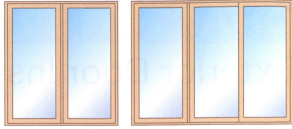

SLIDING DOORS

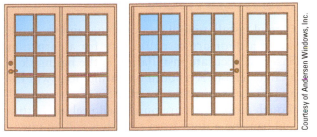

SWINGING DOORS

Courtesy of Andersen Windows, Inc.

Courtesy of Andersen Windows, Inc.

Figure 56–15 An astragal is required between double doors for weathertightness.

Courtesy of Andersen Windows, Inc.

Figure 56–16 Two or three doors usually are used in sliding- or swinging-type patio door units.

the bottom. It then slides up and overhead. The sectional type has hinged sections. These sections move upward on rollers and turn to a horizontal position overhead. A *rolling steel door,* used mostly in commercial construction, consists of narrow metal slats that roll up on a drum installed above the opening.

Special hardware, required for all types, is supplied with the door. Equipment is available for power operation of garage doors, including remote

control. Also supplied are the manufacturers' directions for installation. These should be followed carefully. There are differences in the door design and hardware of many manufacturers.

CHAPTER 57

Exterior Door Installation

EXTERIOR DOOR INSTALLATION

Careful installation of a door frame and the hanging of a door are essential jobs for a carpenter. Fast and accurate installation is the hallmark of a craftsperson. A smooth operating door will last and protect against the weather for many years. Although most doors are purchased prehung, fitting and hanging a door from scratch is still an important part of the carpentry trade. There are situations in remodeling work where doors are hung on preexisting and trimmed openings.

PREHUNG DOORS

The term *prehung door* is used to describe a door frame that is preassembled and ready to stand into the rough opening (**Figure 57–1**). It has side and head jambs fastened together with the sill. The door is bored or mortised for locksets and is attached to the frame via the hinges. The exterior casing and weatherstripping are also attached. The unit is ready for installation.

Installing a Prehung Door

Installation begins with unpacking the door unit. It comes with pads and blocks stapled to the frame, which are designed to protect the door frame during shipping. Remove all packing and be careful to remove all staples.

Verify the opening is ready for the door unit. The width and height dimensions of the rough opening should be large enough for the unit to fit inside and also allow space to level and plumb the frame. The sheathing and drywall must not project into the opening which will interfere with fitting the unit. The floor must be checked for level. If it is not level, the sill must be shimmed and fully supported by the shim pieces.

Base flashing should be installed on top of subfloor to protect the framing from water penetration. It should also slope to the outside slightly. This will encourage any water to exit to the outside. Housewrap must lap under door-side trim and over the head flashing. Sheathing surfaces must be sealed to the door in a similar fashion as for windows (see Chapter 55).

Figure 57–1 Doors can be prehung to speed up installation.

© Cengage Learning 2014

Applying caulk to the base flashing before the unit is installed helps make the unit more air- and weathertight. The door frame is then set in the opening, bedding the sill in caulk to the base flashing. The unit is centered in the opening. Only the bottom is of concern at first, the head will be adjusted later. Tack the lower end of both side casings to the wall. The unit can be checked for level at the sill or at head jamb.

The side jambs are then plumbed using a 6-foot level. Read the level as close to perfect as possible. If a long level is not available, a combination of a straight edge and a carpenter's level will work. A short carpenter's level is not suitable because any bows in the jambs are harder to realize and remove. If both jambs are bowed, small blocks of equal thickness may be placed at the top and bottom of the level. Once the unit is level and plumb it may be tacked into the upper corners of the casing.

At this point it is a good idea to check the operation of the door. Open the door and step to the inside.

Close the door again and look at the gap between the door and the jamb. The gap should be the same all the way around the door. This will give the best operation of the door and weatherstrip. Once the operation is satisfactory, fasten the casing to the wall. Set casing fasteners as needed for finishing.

Slice the house wrap about a one-half inch above the head casing. Insert an L-shaped piece of flashing material under wrap and lap over the head casing (*Figure 57–2*). Make the flashing about ½

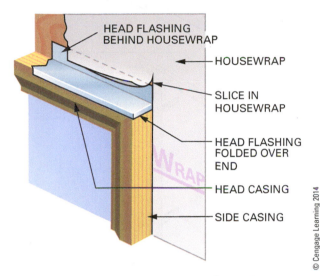

Figure 57–2 Shimming techniques for setting a door frame into a rough opening.

inch longer than the head casing, then snip and fold the ends over. Apply flashing tape over the cut. Note if flashing tape is applied to the sides of the door, it should lap under the head flashing tape.

Jambs must now be shimmed to the trimmer studs. Shims are about 1½ inch wide by 12 inch long tapered wood pieces. They are installed in pairs with their tapers reversed. This makes the shim set thickness adjustable, yet always the same thickness throughout the set. Shim sets are installed from the inside (*Figure 57–3*).

The first shim is trimmed as necessary such that the wider portion is as thick as possible and extends to the casing. The second piece is trimmed to length and slid into place such that it tightens up against the jamb and the first shim before touching the casing. A fastener through the jamb into the shim keeps the shim sets in place. Removing the weatherstrip to place the fastener under it will eliminate having to finish the holes later. Shims should be snug, not overly tight because they can create a bow in the jamb.

Wood jambs need at least six shim sets, three on each side. Other jambs, such those made of plastic, should have more—five on each side and one on the head jamb shim set is recommended. It is also desirable to have a shim set behind each hinge and the lockset strike plate. These are *minimum* standards; increasing the number of shims makes the door more stable and secure.

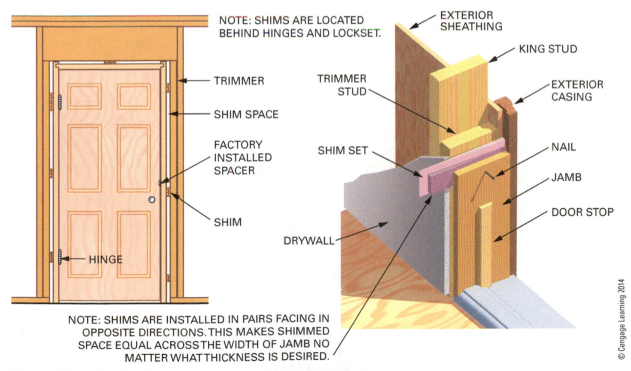

Figure 57–3 Head flashing laps under housewrap and over the head casing.

StepbyStep Procedures

Procedure 57–A Installing a Pre-hung Exterior Door.

STEP 1 REMOVE PACKING MATERIAL. CHECK OVER THE ROUGH OPENING FOR PROPER SIZE, DRYWALL AND SHEATHING CUT BACK, AND LEVELNESS OF FLOOR. LEVEL THE FLOOR IF NECESSARY WITH WIDE SHIMS.

STEP 2 INSTALL A BASE FLASHING. APPLY SEVERAL BEADS OF CHALK ON THE BASE FLASHING THAT WILL ADHERE TO THE SILL.

STEP 3 PLACE DOOR FRAME IN THE OPENING. CENTER THE BOTTOM CASING IN THE OPENING AND LEVEL THE FRAME.

STEP 4 TACK THE BOTTOM ENDS OF EACH SIDE CASING TO THE WALL.

STEP 5 PLUMB EACH SIDE JAMB AND TACK THE TOP ENDS OF CASING TO THE WALL.

STEP 6 OPEN DOOR TO CHECK OPERATION. STRAIGHTEN JAMBS AS NECESSARY. FASTEN THE CASING TO THE WALL.

STEP 7 SHIM AND ADJUST THE JAMB TO THE ROUGH OPENING.

STEP 8 REPLACE A TOP HINGE SCREW WITH THE LONGER ONE PROVIDED BY MANUFACTURER.

STEP 9 CUT HOUSEWRAP (BLUE LINE). INSERT HEAD FLASHING ON TOP OF THE HEAD CASING AND UNDER THE WRAP.

STEP 10 INSTALL FLASHING TAPE ALONG PERIMETER OF THE DOOR STARTING AT THE BOTTOM AND WORKING TO THE TOP. OMIT THE BOTTOM TAPE TO ALLOW A PLACE FOR ANY WATER THAT LEAKS IN TO ESCAPE.

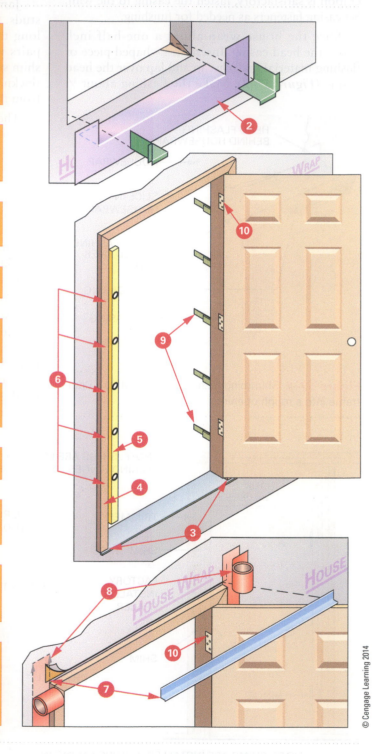

© Cengage Learning 2014

Exterior doors come with a long screw attached in the packaging. This is to replace a top hinge screw. The long screw is driven through a shim set into the trimmer stud. It secures the top hinge and jamb back to the rough opening. This prevents the door hinge from sagging away, causing the door to rub the jamb as it closes (*Procedure 57–A*).

SETTING DOOR FRAMES IN MASONRY WALLS

In commercial construction, exterior wood or metal door frames are sometimes set in place before masonry walls are built. The frames must be set and firmly braced in a level and plumb position. The head jamb is checked for level. The bottom ends of the side jambs are secured in place. It may be necessary to shim one or the other side jamb in order to level the head jamb. The side jambs are then plumbed in a sideways direction. They are braced in position. Then, the frame is plumbed and braced at a right angle to the wall (*Procedure 57–B*).

Finally, the frame is checked to see if it has a **wind.** The term wind is pronounced the same way it's pronounced in the phrase "wind a clock." A wind is a twist in the door frame caused when the side jambs do not line up vertically with each other. No matter how carefully the side jambs of a door frame are plumbed, it is always best to check the frame to see if it has a wind.

One method of checking the door frame for a wind is to stand to one side. Sight through the door frame to see if the outer edge of one side jamb lines up with the inner edge of the other side jamb. If they do not line up, the frame is in a wind. Make

StepbyStep Procedures

Procedure 57–B Installing an Exterior Door Frame in a Masonry Wall

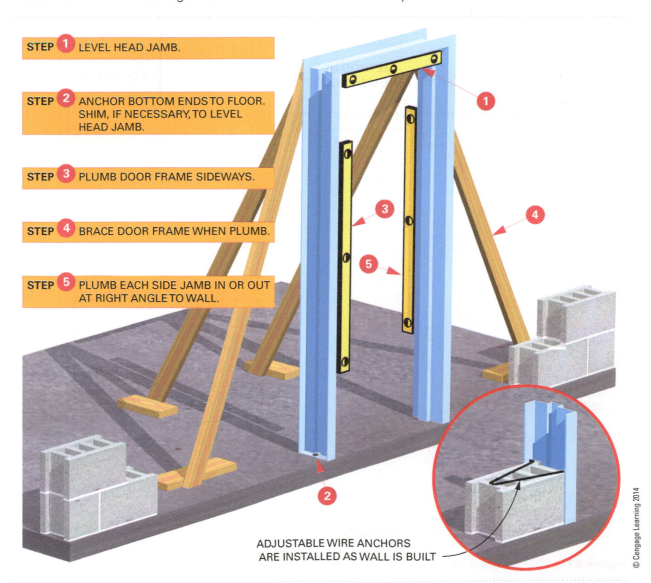

STEP 1 LEVEL HEAD JAMB.

STEP 2 ANCHOR BOTTOM ENDS TO FLOOR. SHIM, IF NECESSARY, TO LEVEL HEAD JAMB.

STEP 3 PLUMB DOOR FRAME SIDEWAYS.

STEP 4 BRACE DOOR FRAME WHEN PLUMB.

STEP 5 PLUMB EACH SIDE JAMB IN OR OUT AT RIGHT ANGLE TO WALL.

ADJUSTABLE WIRE ANCHORS ARE INSTALLED AS WALL IS BUILT

© Cengage Learning 2014

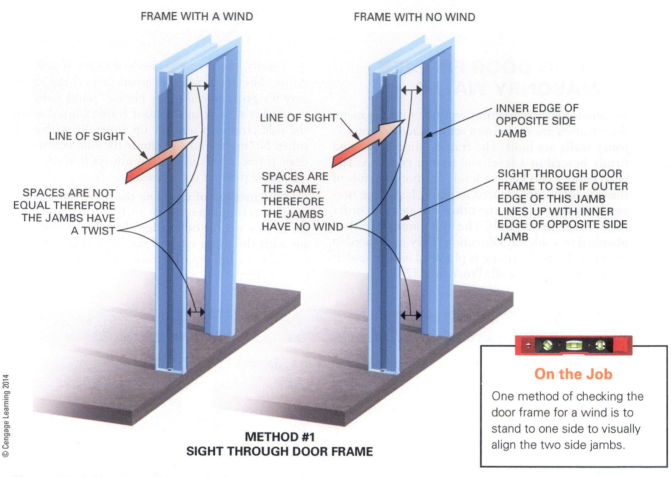

FRAME WITH A WIND

FRAME WITH NO WIND

LINE OF SIGHT

SPACES ARE NOT EQUAL THEREFORE THE JAMBS HAVE A TWIST

LINE OF SIGHT

SPACES ARE THE SAME, THEREFORE THE JAMBS HAVE NO WIND

INNER EDGE OF OPPOSITE SIDE JAMB

SIGHT THROUGH DOOR FRAME TO SEE IF OUTER EDGE OF THIS JAMB LINES UP WITH INNER EDGE OF OPPOSITE SIDE JAMB

On the Job

One method of checking the door frame for a wind is to stand to one side to visually align the two side jambs.

**METHOD #1
SIGHT THROUGH DOOR FRAME**

© Cengage Learning 2014

Figure 57–4 Visual inspection method for checking for a wind or twist in a door frame.

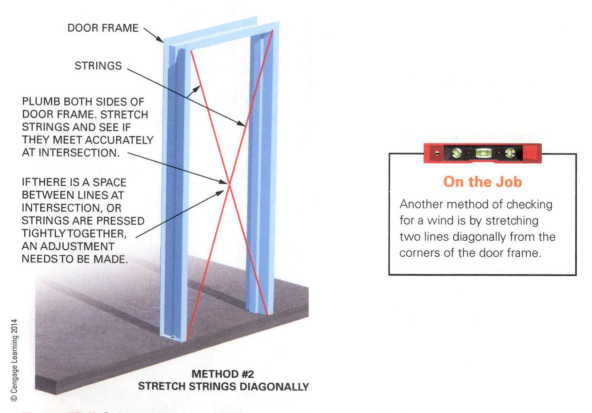

DOOR FRAME

STRINGS

PLUMB BOTH SIDES OF DOOR FRAME. STRETCH STRINGS AND SEE IF THEY MEET ACCURATELY AT INTERSECTION.

IF THERE IS A SPACE BETWEEN LINES AT INTERSECTION, OR STRINGS ARE PRESSED TIGHTLY TOGETHER, AN ADJUSTMENT NEEDS TO BE MADE.

On the Job

Another method of checking for a wind is by stretching two lines diagonally from the corners of the door frame.

**METHOD #2
STRETCH STRINGS DIAGONALLY**

© Cengage Learning 2014

Figure 57–5 String method for checking for a wind or twist in a door frame.

adjustments until they do. One way of making the adjustment is to plumb and brace one side at a right angle to the wall. Then sight, line up, and brace the other side jamb (*Figure 57–4*).

Some workers check for a wind by stretching two strings diagonally from the corners of the frame. If both strings meet accurately at their intersections, the frame does not have a wind (*Figure 57–5*).

MAKING A DOOR UNIT

The first step in making a door unit is to determine the side that will close against the stops on the door frame. If the design of the door permits that either side may be used toward the stops, sight along the door stiles from top to bottom to see if the door is bowed. Hardly any doors are perfectly straight. Most are bowed to some extent.

Determining Stop Side of Door

The door should be fitted to the opening so that the hollow side of the bow will be against the door stops. Hanging a door in this manner allows the top and bottom of the closed door to come up tight against stops. The center comes up tight when the door is latched. Also, the door will not rattle.

If no attention is paid to which side the door stops against, then the reverse may happen. The door will come up against the stop at the center and away from the stop at the bottom and top (*Figure 57–6*).

Determining Exposed Side of Sash Doors

It is important to hang exterior doors containing lights of glass with the proper side exposed to the weather. This prevents wind-driven rainwater from seeping through joints. Manufacturers clearly indicate, with warning labels glued to the door, which side should face the exterior. Any door warranty is voided if the door is improperly hung. Do not hang exterior doors with the removable glass bead facing outward. Glass bead is small molding used to hold lights of glass in the opening. The bead can be identified by holes made when fasteners of the bead were set (*Figure 57–7*).

Some doors are manufactured with compression glazing. This virtually eliminates the possibility of any water seeping through the joints. When Low-E insulating glass is used in a door, it is especially important to have the door facing in the direction indicated by the manufacturer.

Fitting the Door

Place the door on sawhorses, with its face side up. Measure carefully the width and height of the door frame. The frame should be level and plumb, but this may not be the case and should not be taken for granted.

The process of fitting a door into a frame is called jointing. The door must be carefully jointed. An even joint of approximately $\frac{3}{32}$ inch must be made between the door and the frame on all sides.

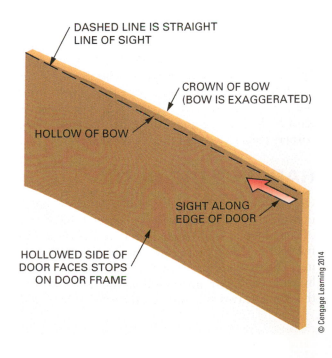

DASHED LINE IS STRAIGHT LINE OF SIGHT

CROWN OF BOW (BOW IS EXAGGERATED)

HOLLOW OF BOW

SIGHT ALONG EDGE OF DOOR

HOLLOWED SIDE OF DOOR FACES STOPS ON DOOR FRAME

© Cengage Learning 2014

Figure 57–6 Method of determining which side of the door will rest against the top when the door is closed.

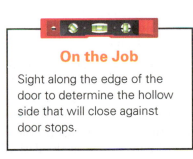

On the Job

Sight along the edge of the door to determine the hollow side that will close against door stops.

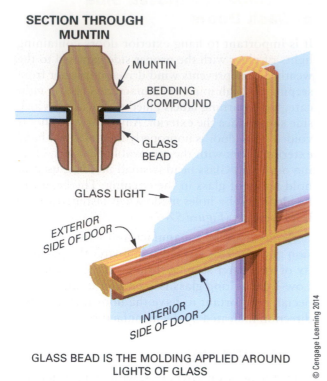

SECTION THROUGH MUNTIN

MUNTIN

BEDDING COMPOUND

GLASS BEAD

GLASS LIGHT

EXTERIOR SIDE OF DOOR

INTERIOR SIDE OF DOOR

GLASS BEAD IS THE MOLDING APPLIED AROUND LIGHTS OF GLASS

© Cengage Learning 2014

Figure 57–7 Doors containing glass lights should face in the direction recommended by the manufacturer.

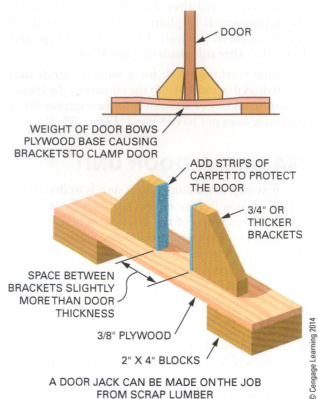

DOOR

WEIGHT OF DOOR BOWS PLYWOOD BASE CAUSING BRACKETS TO CLAMP DOOR

ADD STRIPS OF CARPET TO PROTECT THE DOOR

3/4" OR THICKER BRACKETS

SPACE BETWEEN BRACKETS SLIGHTLY MORE THAN DOOR THICKNESS

3/8" PLYWOOD

2" X 4" BLOCKS

A DOOR JACK CAN BE MADE ON THE JOB FROM SCRAP LUMBER

© Cengage Learning 2014

Figure 57–8 Door jacks make working on a door edge easier.

A wider joint of approximately $\frac{1}{8}$ inch must be made to allow for swelling of the door and frame in extremely damp weather.

Use a **door jack** to hold the door steady. A manufactured or job-made jack may be used (*Figure 57–8*). Use a jointer plane, either hand or power. Plane the edges and ends of the door so that it fits snugly in the door frame.

Fit the top end against the head jamb. Then fit the bottom against the sill so that a proper joint is obtained. Careful jointing may require moving the door in and out of the frame several times.

Next, joint the hinge edge against the side jamb. Finally, plane the lock edge so the desired joint is obtained on both sides. The lock edge must also be planed on a *bevel*. This is so the back edge does not strike the jamb when the door is opened. The bevel is determined by making the same joint between the back edge of the door and the side jamb when the door is slightly open, as when the door is closed (*Figure 57–9*).

Extreme care must be taken when fitting doors not to get them undersize. The door can always be jointed a little more. However, it cannot be made wider or longer (*Figure 57–10*). Do not cut more than $\frac{1}{2}$ inch total from the width of a door. Cut no more than 2 inches from its height. Cut equal amounts from ends and edges when approaching maximum amounts. Check the fit frequently by placing the door in the opening, even if this takes a little extra time. Most entrance doors are very expensive. Care should be taken not to ruin one. Speed will come with practice. Handle the door carefully to avoid marring it or other finish. After the door is fitted, *ease* all sharp corners slightly with a block plane and sandpaper. To ease sharp corners means to round them over slightly.

HANGING THE DOOR

On swinging doors, the loose-pin type **butt hinge** is ordinarily used. The pin is removed. Each *leaf* of the hinge is applied separately to the door and frame. The door is hung by placing it in the opening. The pins are inserted to rejoin the separated hinge leaves. Extreme care must be taken so that the hinge leaves line up exactly on the door and

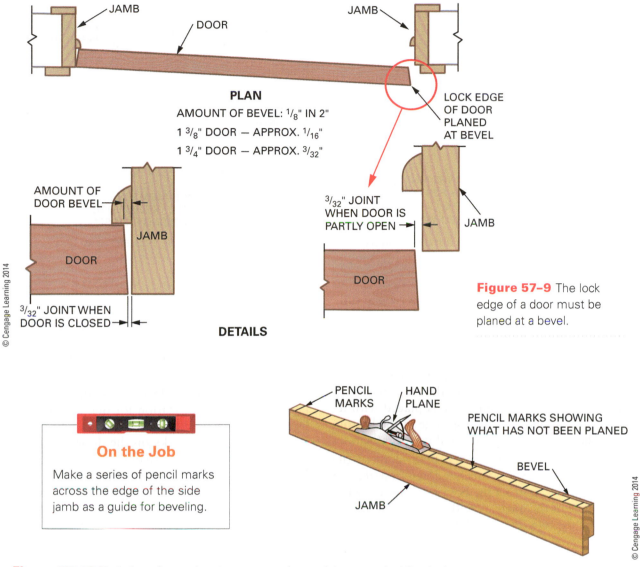

PLAN

AMOUNT OF BEVEL: 1/8" IN 2"

1 3/8" DOOR — APPROX. 1/16"

1 3/4" DOOR — APPROX. 3/32"

LOCK EDGE OF DOOR PLANED AT BEVEL

3/32" JOINT WHEN DOOR IS PARTLY OPEN →

JAMB

AMOUNT OF DOOR BEVEL

JAMB

DOOR

3/32" JOINT WHEN DOOR IS CLOSED →

DOOR

DETAILS

Figure 57–9 The lock edge of a door must be planed at a bevel.

On the Job

Make a series of pencil marks across the edge of the side jamb as a guide for beveling.

PENCIL MARKS

HAND PLANE

PENCIL MARKS SHOWING WHAT HAS NOT BEEN PLANED

BEVEL

JAMB

Figure 57–10 Technique for seeing the amount of material removed while planing.

frame. Three 4 × 4 hinges on 1¾-inch doors 7'-0", or less, in height are often used. Use four hinges on doors over 7'-0" in height.

The hinge leaves are recessed flush with the door edge and only partway across. The recess for the hinge is called a hinge gain or hinge mortise. Butt hinges come in two styles. They vary according to the method of making the mortise (*Figure 57–11*). Hinge gains are only made partway across the edge of the door. This is so that the edge of the hinge is not exposed when the door is opened. The remaining distance, from the edge of the hinge to the side of the door, is called the backset of the hinge. Butt hinges must be wide enough so that the

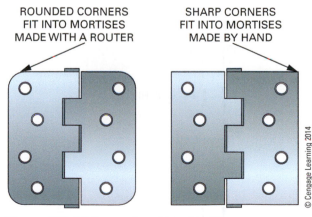

ROUNDED CORNERS FIT INTO MORTISES MADE WITH A ROUTER

SHARP CORNERS FIT INTO MORTISES MADE BY HAND

Figure 57–11 Two styles of butt hinges.

© Cengage Learning 2014

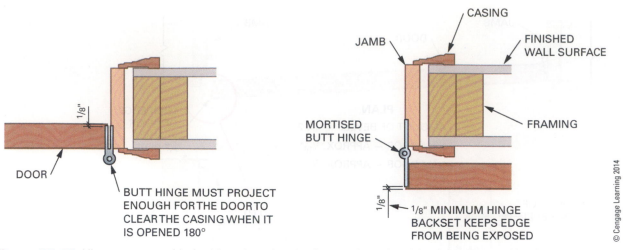

© Cengage Learning 2014

Figure 57–12 Hinges are set with the hinge barrel projecting out from the side of the door and jamb.

pin is located far enough beyond the door face to allow the door to clear the door trim when fully opened (*Figure 57–12*).

Location and Size of Door Hinges

On paneled doors, the top hinge is usually placed with its upper end in line with the bottom edge of the top rail. The bottom hinge is placed with its lower end in line with the top edge of the bottom rail. The middle hinge is centered between them.

On flush doors, the usual placement of the hinge is approximately 9 inches down from the top and 13 inches up from the bottom, as measured to the center of the hinge. A middle hinge is centered between the two (*Figure 57–13*).

Laying Out Hinge Locations

Place the door in the frame. Shim the top and bottom so the proper joint is obtained. Place shims between the lock edge of the door and side jamb of the frame. The hinge edge should be tightly against the door jamb.

Use a sharp knife. Mark across the door and jamb at the desired location for one end of each hinge. A knife is used because it makes a finer line than a pencil. The marks on the door and jamb should not be any longer than the hinge thickness. Place a small X, with a pencil, on both the door and the jamb. This indicates on which side of the knife mark to cut the gain for the hinge (*Figure 57–14*). Care must be taken to cut hinge gains on

Figure 57–13 Recommended placement of hinges on doors.

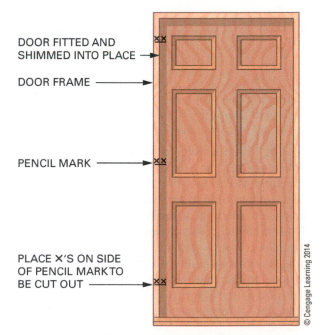

Figure 57–14 Mark the location of hinges on the door and frame with a sharp knife.

the same side of the layout line on both the door and the door frame. Remove the door from the frame. Place it in the door jack with its hinge edge up in order to lay out and cut the hinge gains.

Laying Out the Hinge Gain

The first step in laying out a hinge gain is to mark its ends. Place a hinge leaf on the door edge with its end on the knife mark previously made. With the *barrel* of the leaf against the side of the door, hold the leaf firmly. Score a line along one end with a sharp knife. Then, tap the other end, until the leaf just covers the line. Score a line along the other end (*Figure 57–15*). Score only partway across the door edge.

Cutting Hinge Gains

Use a sharp knife to deepen the scored backset line. It may be necessary to draw the knife along the line several times to score to the bottom of the gain. Take care if using a chisel for scoring. Using a chisel will easily split the edge of the door (*Figure 57–16*).

With the bevel of the chisel down, cut a small chip from each end of the gain. The chips will break off at the scored end marks (see *Procedure 12–B*). With the flat of the chisel against the shoulders of the gain, cut down to the bottom of the gain.

Make a series of small chisel cuts along the length of the gain. Brush off the chips. Then, with the flat of the chisel down, pare and smooth the excess down to the depth of the gain. Be careful not to slip and cut off the backset (*Figure 57–17*).

After the gain is made, the hinge leaf should press-fit into it. It should be flush with the door

edge. If the hinge leaf is above the surface, deepen the gain until the leaf lies flush. If the leaf is below the surface, it may be shimmed flush with thin pieces of cardboard from the hinge carton.

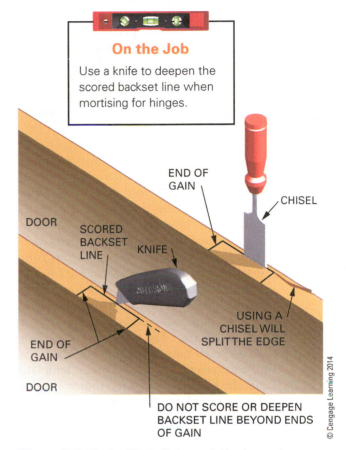

On the Job

Use a knife to deepen the scored backset line when mortising for hinges.

Figure 57–16 A utility knife is useful in deepening the cut that is parallel to the grain.

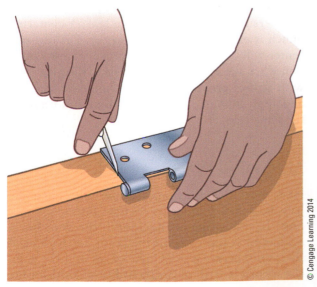

Figure 57–15 The ends of hinge gains may be laid out by scoring with a knife along a hinge leaf.

Figure 57–17 Chiseling out the hinge gain.

Using Butt Hinge Markers. Instead of laying out hinge gains with a knife, butt hinge markers of several sizes are often used (see *Figure 11–25*. With markers, the location of the hinge leaves must still be marked on the door and jamb, as described. The width and length of the hinge gain are outlined by simply placing the marker in the proper location on the door edge, then tapping it with a hammer (*Figure 57–18*). However, the depth of the gain must be scored with a butt gauge or some other gauging method. The gain is then chiseled out in the manner previously described.

Routing a Butt Hinge. A butt hinge template fixture or jig and portable electric router are usually used when many doors need to be hung. Most hinge routing jigs contain three adjustable templates secured on a long rod. The templates are temporarily attached into position. The jig is used on both the door and frame as a guide to rout hinge gains. The templates are adjustable for different size hinges, hinge locations, and door thicknesses. An attachment positions the hinge template jig to provide the required joint between the top of the door and the frame. The gains are cut using a router with a special hinge mortising bit. The bit is set to cut the gain to the required depth. A template guide is attached to the base of the router. It rides against the jig template when routing the hinge gain (*Figure 57–19*).

Figure 57–19 A butt hinge template can be used on both door and frame when routing hinge gains.

Courtesy of Porter Cable

CAUTION

Care must be taken not to clip the template with a rotating router bit when lifting the router from the template. It is best to let the router come to a stop before removing it. Both the template and bit could be damaged beyond repair.

Butt hinges with rounded corners are used in routed gains. Hinges with square corners require that the rounded corner of a routed gain be chiseled to a square corner.

Applying Hinges

Press the hinge leaf in the gain. Drill pilot holes for the screws. Center the pilot holes carefully on the countersunk holes of the hinge leaf. A centering punch is often used for this purpose. Drilling off-center will cause the hinge to move from its position when the screw is driven. Fasten the hinge leaf with screws provided with the hinges. Cut gains and apply all hinge leafs on the door and frame in the same manner.

Shimming Hinges

Hang the door in the frame by inserting a pin in the barrels of the top hinge leaves first. Insert the other pins. Try the swing of the door. If the door binds against or is too close to the jamb on the hinged side, shim between the hinge leaves and the gain. Use a narrow strip of cardboard on the side of the

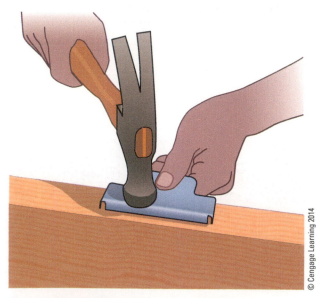

Figure 57–18 Sometimes butt hinge markers are used to lay out hinge gains.

© Cengage Learning 2014

screws nearest the pin. This will move the door toward the lock side of the door frame. If the door binds against or is too close to the jamb on the lock side, apply shims in the same manner. However, apply them on the opposite side of the hinge screws (*Figure 57–20*). Check the bevel on the lock edge of the door. Plane to the proper bevel, if necessary. Ease all sharp, exposed corners.

Butt hinges with rounded corners are used in routed gains. Hinges with square corners require that the rounded corner of a routed gain be chiseled to a square corner.

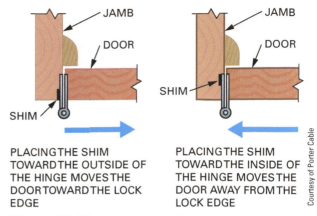

PLACING THE SHIM TOWARD THE OUTSIDE OF THE HINGE MOVES THE DOOR TOWARD THE LOCK EDGE

PLACING THE SHIM TOWARD THE INSIDE OF THE HINGE MOVES THE DOOR AWAY FROM THE LOCK EDGE

Courtesy of Porter Cable

Figure 57–20 Shimming the hinge edges will move the door toward or away from the lock edge.

CHAPTER 58

Installing Exterior Door Hardware

After the door has been fitted and hung in the frame, the *lockset* and other door hardware are installed. A large varity of locks are available from several large manufacturers in numerous styles and qualities, providing a wide range of choices. Doors must be prepared to accept the lockset (*Figure 58–1*).

CYLINDRICAL LOCKSET

Cylindrical locksets are often called *key-in-the-knob locksets* (*Figure 58–2*). They are the most commonly used type in both residential and commercial construction. This is primarily because of the ease and speed of installing them. They may be obtained in several groups, from light-duty residential to extra-heavy-duty commercial and industrial applications.

Green Tip

Hinged doors and windows are more airtight than sliding ones because they compress and seal to the gaskets better.

In place of knobs, lever handles are provided on locksets likely to be used by handicapped persons or in other situations where a lever is more suitable than a knob (*Figure 58–3*). Deadbolt locks are used for both primary and auxiliary locking of doors in residential and commercial buildings (*Figure 58–4*). They provide additional security. They also make an attractive design in combination with *grip-handle* locksets or latches **✻** (*Figure 58–5*). Mortised locksets are **✻ IRC 311.2** more often used in commercial doors than residential doors. The name comes from the mortised hole needed in the door to accept the lockset. Today these doors typically come precut with a mortised slot for the lockset to slip into. Mortised locksets are designed to be durable able to withstand constant use (*Figure 58–6*).

INSTALLING CYLINDRICAL LOCKSETS

To install a cylindrical lockset, first check the contents and read the manufacturer's directions carefully. There are so many kinds of locks manufactured that the mechanisms vary greatly. The directions included with the lockset must be followed carefully.

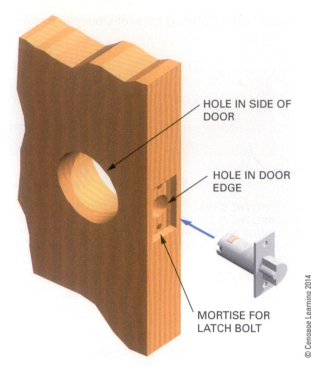

HOLE IN SIDE OF DOOR

HOLE IN DOOR EDGE

MORTISE FOR LATCH BOLT

© Cengage Learning 2014

Figure 58–1 The method of preparing the door for cylindrical locksets.

Courtesy of Schlage Lock Co.

Figure 58–2 Cylindrical locksets are the most commonly used type in both residential and commercial construction.

Courtesy of Schlage Lock Co.

Figure 58–3 Locksets with lever handles are used when difficulty turning knobs is expected.

Courtesy of Schlage Lock Co.

Figure 58–4 Deadbolt locks are used primarily as auxiliary locks for added security.

Courtesy of Schlage Lock Co.

Figure 58–5 A grip-handle lockset combines well with a deadbolt lock.

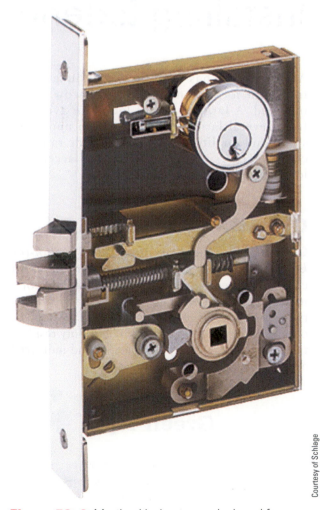

Courtesy of Schlage

Figure 58–6 Mortised locksets are designed for doors with heavy traffic and use.

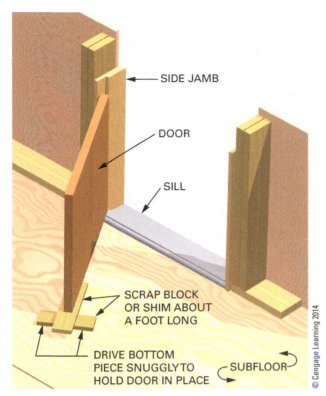

Figure 58-7 labels: SIDE JAMB, DOOR, SILL, SCRAP BLOCK OR SHIM ABOUT A FOOT LONG, DRIVE BOTTOM PIECE SNUGGLY TO HOLD DOOR IN PLACE, SUBFLOOR

© Cengage Learning 2014

Figure 58–7 A door may be shimmed from the floor to hold it plumb during installation.

However, there are certain basic procedures. Open the door to a convenient position. Wedge the bottom to hold it in place (*Figure 58–7*). Measure up, from the floor, the recommended distance to the centerline of the lock. This is usually 36 to 40 inches. At this height, square a light line across the edge and stile of the door.

Marking and Boring Holes

Position the center of the paper template supplied with the lock on the squared lines. Lay out the centers of the holes that need to be bored (*Figure 58–8*). It is important that the template be folded over the high corner of the beveled door edge. The distance from the door edge to the center of the hole through the side of the door is called the *backset* of the lock. Usual backsets are $2\frac{3}{8}$ inches for residential and $2\frac{3}{4}$ inches for commercial. Make sure the backset is marked correctly before boring the hole. One hole must be bored through the side and one into the edge of the door. The manufacturers' directions specify the hole sizes where a 1-inch hole for bolts and $2\frac{1}{8}$-inch hole for locksets are common.

© Cengage Learning 2014

Figure 58–8 Using a template to lay out the centers of the holes for a lockset.

The hole through the side of the door should be bored first. Stock for the center of the boring bit is lost if the hole in the edge of the door is bored first. It can be bored with hand tools, using an expansion bit in a bit brace. However, it is a difficult job. If using hand tools, bore from one side until only the point of the bit comes through. Then bore from the other side to avoid splintering the door.

Using a Boring Jig. A boring jig is frequently used. It is clamped to the door to guide power-driven multispur bits. With a boring jig, holes can be bored completely through the door from one side. The clamping action of the jig prevents splintering (*Figure 58–9*).

After the holes are bored, insert the latchbolt in the hole bored in the door edge. Hold it firmly and

© Cengage Learning 2014

Figure 58–9 Boring jigs are frequently used to guide bits when boring holes for locksets.

Figure 58–10 Using a faceplate marker.

© Cengage Learning 2014

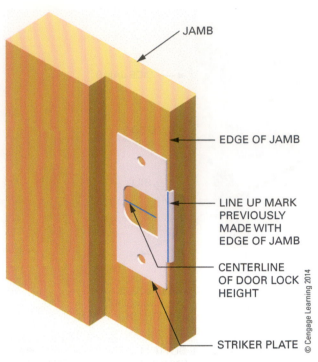

Figure 58–11 Installing the striker plate.

© Cengage Learning 2014

score around its faceplate with a sharp knife. Remove the latch unit. Deepen the vertical lines with the knife in the same manner as with hinges. Take great care when using a chisel along these lines. This may split out the edge of the door. Then, chisel out the recess so that the faceplate of the latch lays flush with the door edge.

Faceplate markers, if available, may be used to lay out the mortise for the latch faceplate. A marker of the appropriate size is held in the bored latch hole and tapped with a hammer (*Figure 58–10*). Complete the installation of the lockset by following specific manufacturers' directions.

Installing the Striker Plate

The **striker plate** is installed on the door jamb so when the door is closed it latches tightly with no play. If the plate is installed too far out, the door will not close tightly against the stop. It will then rattle. If the plate is installed too far in, the door will not latch.

To locate the striker plate in the correct position, place it over the latch in the door. Close the door snugly against the stops. Push the striker plate in against the latch. Draw a vertical line on the face of the plate flush with the outside face of the door (*Figure 58–11*).

Open the door. Place the striker plate on the jamb. The vertical line, previously drawn on it, should be in line with the edge of the jamb. Center the plate on the latch. Hold it firmly while scoring a line around the plate with a sharp knife. Chisel out the mortise so the plate lies flush with the jamb. Screw the plate in place. Chisel out the center to receive the latch.

DOOR HARDWARE

A variety of hardware for doors is available. The kind of trim and finish are factors, in addition to the kind and style, which determine the quality of hardware. **Escutcheons** are decorative plates of various shapes that are installed between the door and the lock handle or knob. Locksets and escutcheons are available in various metals and finishes. More expensive locksets and trim are made of brass, bronze, or stainless steel. Less expensive ones may be of steel that is plated or coated with a finish. It is important to consider conditions and usage when selecting locksets. This is especially the case in areas with a humid climate or near salt water.

Security hinges are designed to prevent a closed door from being opened from the hinge side of the door. Most butt hinges have pins that may be removed while the door is closed. This makes it possible to remove the door from the opening when it is closed and locked. Security hinges have pins that are not easily removed. Should the hinge pin be removed, the interlocking hinge leafs lock the door in the opening by resisting the leafs from sliding past each other (*Figure 58–12*).

Door closers are made to automatically close a door. They are attached to the door and the jamb. An adjustable spring mechanism allows variation to the speed and strength of the closer (*Figure 58–13*).

Labels on figure:
JAMB
EDGE OF JAMB
LINE UP MARK PREVIOUSLY MADE WITH EDGE OF JAMB
CENTERLINE OF DOOR LOCK HEIGHT
STRIKER PLATE

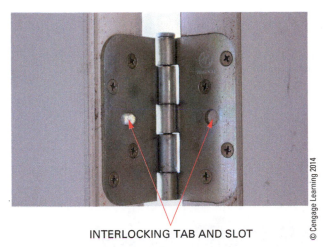

INTERLOCKING TAB AND SLOT

Figure 58–12 Security hinges are designed to keep hinge leaves from sliding apart if pin is removed.

Door stops or keepers restrict the travel of the door. They may be installed to the door, floor, or the wall. They protect the door and wall from damage if door is opened too far (*Figure 58–14*).

Door viewers or peep holes allow view through the door from one side to the other. They offer a 190° limited view while blurring and obscuring the view from the other side. For this reason it must be installed with the direction of the view in mind (*Figure 58–15*).

Figure 58–14 Stops are used to confine opening doors.

TC805

TC830

TC901

TC931

Figure 58–13 Door closers are designed to keep door closed after opening.

Figure 58–15 Peep holes allow secure side occupant to view the other side without opening the door.

DECONSTRUCT THIS

Carefully study *Figure 58–16* and think about what is wrong and/or what is right. Consider all possibilities. What construction practice or method is different in your area of the country?

Figure 58–16 This photo shows a carpenter using a drill to bore a hole in a door.

KEY TERMS

astragal	door casings	hinge mortise	prehung door
backset	door jack	jointing	rabbeted jamb
boring jig	escutcheons	lever handle	sticking
butt hinge	faceplate marker	lock rail	stool
butt hinge template	flush	lockset	striker plate
compression glazing	flush door	multispur bit	transom
cylindrical lockset	glass bead	panel door	wind
deadbolt	hinge gain	patio door	

REVIEW QUESTIONS

Select the most appropriate answer.

1. The standard thickness of exterior wood doors in residential construction is

   a. $1\frac{3}{8}$ inches.
   b. $1\frac{1}{2}$ inches.
   c. $1\frac{3}{4}$ inches.
   d. $2\frac{1}{4}$ inches.

2. The typical minimum width of exterior entrance doors is

   a. 3'-0".
   b. 2'-8".
   c. 2'-6".
   d. 2'-4".

3. The height of exterior entrance doors in residential construction is generally not less than

   a. 7'-0".
   b. 6'-10".
   c. 6'-8".
   d. 6'-6".

4. The term used to describe a door frame that is twisted is

   a. twist.
   b. warp.
   c. bowed.
   d. wind.

5. A narrow member dividing the glass in a door into smaller lights is called a

   a. bar.
   b. mullion.
   c. rail.
   d. all of the above.

6. Shims should be installed

   a. in pairs.
   b. with nails through them.
   c. behind hinges.
   d. all of the above.

7. A left-hand swinging door is one that

   a. is installed on the left side of the room.
   b. has hinges on the left when the door swings away.
   c. has the lockset on the left when the door swings away.
   d. requires a left hand to open it.

8. Before hanging a door, sight along its length for a bow. The hollow side of the bow should

   a. face the door stops.
   b. face the inside.
   c. be straightened with a plane.
   d. all of the above.

9. The joint between the door and door frame should be close to

   a. $\frac{3}{32}$ inch.
   b. $\frac{3}{64}$ inch.
   c. $\frac{1}{4}$ inch.
   d. $\frac{3}{16}$ inch.

10. The top hinges on a paneled door has its center placed

    a. in line with the top edge of the bottom rail.
    b. in line with the top edge of the intermediate rail.
    c. 9 inches down from the top of the door.
    d. 13 inches down from the top of the door.

11. The center of the hole for a lockset in a residential door is typically set back from the door edge

    a. 2 inches.
    b. $2\frac{1}{8}$ inches.
    c. $2\frac{3}{8}$ inches.
    d. $2\frac{3}{4}$ inches.

12. The lockset hole diameter in residential doors is typically

    a. 2 inches.
    b. $2\frac{1}{8}$ inches.
    c. $2\frac{3}{8}$ inches.
    d. $2\frac{3}{4}$ inches.

13. The term *sidelights* refers to

    a. outside lighting.
    b. horizontal lights.
    c. glass panels.
    d. all of the above.

14. The term used to refer to the part of a door frame located under a door is called a

    a. weatherstrip.
    b. sole.
    c. threshold.
    d. escutcheon.

15. Door hardware designed to prevent a closed door from being removed is called a

    a. security hinge.
    b. door keeper.
    c. door closer.
    d. mortised lockset.

UNIT 22

Siding and Cornice Construction

©Shutterstock, Inc./Mau Horng

The exterior finish work is the major visible part of the architectural design of a building. Because the exterior is so prominent, it is important that all finish parts be installed straight and true with well-fitted joints.

The portion of the finish that covers the vertical area of a building is the siding. Siding does not include masonry covering, such as stucco or brick veneer. Siding is used extensively in both residential and commercial construction.

That area where the lower portion of the roof, or eaves, overhangs the walls is called the cornice. Variations in cornice design and detail can set the appearance of one building apart from another.

Protecting a building from weather is the major function of exterior finish. Use care to install siding with the nature of rain and wind-blown water in mind. This will maximize the life of the finish and the building.

OBJECTIVES

After completing this unit, the student should be able to:

- describe the shapes, sizes, and grades of various siding products.
- install corner boards and prepare sidewalls for siding.
- apply horizontal and vertical lumber siding.
- apply plywood and hardboard panel and lap siding.
- apply wood shingles and shakes to sidewalls.
- apply aluminum and vinyl siding.
- estimate required amounts of siding.
- describe various types of cornices and name their parts.
- install gutters and downspouts.

CHAPTER 59
Siding Types and Sizes

Siding is manufactured from solid lumber, plywood, hardboard, aluminum, concrete, and vinyl. It comes in many different patterns. Prefinished types eliminate the need to refinish for many years, if at all. Siding may be applied horizontally, vertically, or in other directions, to make many interesting designs (*Figure 59–1*).

WOOD SIDING

The natural beauty and durability of solid wood have long made it an ideal material for siding. The *Western Wood Products Association (WWPA)* and the *California Redwood Association (CRA)* are two major organizations whose member mills manufacture siding and other wood products. They have to meet standards supervised by their associations. Grade stamps of the WWPA and CRA and other associations of lumber manufacturers are placed on siding produced by member mills. Grade stamps ensure the consumer of a quality product that complies with established standards (*Figure 59–2*). WWPA member mills produce wood siding from species such as fir, larch, hemlock, pine, spruce, and cedar. Most redwood siding is produced by mills that belong to the CRA.

Wood Siding Grades

Grain. Some siding is available in *vertical grain, flat grain,* or *mixed grain.* In vertical grain siding

Figure 59–2 Association trademarks ensure that the product on which they appear has met established standards of quality.

the annual growth rings, viewed in cross section, must form an angle of 45 degrees or more with the surface of the piece. All other lumber is classified as flat grain (*Figure 59–3*). Vertical grain siding is the highest quality. It warps less, takes and holds finishes better, has fewer defects, and is easier to work.

Surface Texture. Sidings are manufactured with *smooth, rough,* or *saw-textured* surfaces. Saw-textured surface finishes are obtained by resawing

Figure 59–1 Wood siding is used in both residential and commercial construction.

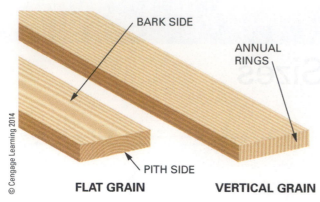

© Cengage Learning 2014

Figure 59–3 In some species, siding is available in vertical, flat, or mixed grain.

Grade	Description
Clear All Heart	A superior grade for fine sidings and architectural uses. It is all heartwood and the graded face of each piece is free of knots.
Clear	Similar in quality to clear all heart, except that it includes sapwood in varying amounts. Some boards may have one or two small, tight knots on the graded face.
Heart B	An economical all-heartwood grade containing a limited number of tight knots and characteristics not permitted in clear or clear all heart. It is graded on one face and one edge.
B Grade	An economical grade containing a limited number of tight knots with sapwood accenting the heartwood.

Redwood grades are established by the redwood inspection service

Courtesy of California Redwood Association

Figure 59–5 Some redwood siding grades.

General Categories (Note that there are additional grades for bevel pattern)	Grades		
	Western Species		Cedar
	Selects	Finish	Western & Canadian
All Patterns — Premium Grades	C Select	Superior	Clear Heart A Grade
	D Select	Prime	
			B Grade
Additional Grades for Bevel Patterns — Premium		Superior Bevel	Clear VG Heart A Bevel B Bevel Rustic C Bevel
Additional Grades for Bevel Patterns — Knotty		Prime Bevel	Select Knotty Quality Knotty
All Patterns — Knotty Grades	Commons	Alternate Boards	
	#2 Common	Select Merch.	Select Knotty
	#3 Common	Construction Standard	Quality Knotty

Courtesy of Western Wood Products Association

Figure 59–4 WWPA grade rules for siding products.

in the mill. They generally hold finishes longer than smooth surfaces.

WWPA Grades. Grades published by the WWPA for siding products are shown in *Figure 59–4.* Siding graded as *premium* has fewer defects such as knots and pitch pockets. The highest premium grade is produced from clear, all-heart lumber. *Knotty* grade siding is divided into #1, #2, and #3 *common.* The grade depends on the type and number of knots and other defects.

CRA Grades. There are over thirty grades of redwood lumber. The best grades are grouped in a category called *architectural.* They are used for high-quality exterior and interior uses, including siding (*Figure 59–5*).

Siding Patterns and Sizes

The names, descriptions, and sizes of siding patterns are shown in *Figure 59–6.* Some patterns can be used only for either horizontal or vertical applications. Others can be used for both. *Drop* and *tongue-and-grooved* sidings are manufactured in a variety of patterns other than those shown. *Bevel* siding, more commonly known as *clapboards,* is a widely used kind (*Figure 59–7*).

PANEL AND LAP SIDING

Most *panel* and *lap* siding is manufactured from plywood, hardboard, and fiber-cement boards. They come in a variety of sizes, patterns, and surface textures. Plywood siding manufactured by American Plywood Association (APA) member mills is known as APA303 siding. It is produced in a variety of surface textures and patterns (*Figure 59–8*).

Hardboard siding is made of high-density fiberboard with a hard tempered surface. It typically comes primed or prefinished in a variety of colors. It also comes in a variety of surface styles.

Fiber-cement siding is made of a fiber-reinforced cementitious material. It also comes in a variety of

		Nominal Sizes
Siding Patterns		**Thickness & Width**
TRIM **BOARD-ON-BOARD** **BOARD-AND-BATTEN** Boards are surfaced smooth, rough or saw-textured. Rustic ranch-style appearance. Provide horizontal nailing members. Do not nail through overlapping pieces. Vertical applications only.		1×2 1×4 1×6 1×8 1×10 1×12 $1\frac{1}{4} \times 6$ $1\frac{1}{4} \times 8$ $1\frac{1}{4} \times 10$ $1\frac{1}{4} \times 12$
BEVEL OR BUNGALOW Bungalow ("Colonial") is slightly thicker than Bevel. Either can be used with the smooth or saw-faced surface exposed. Patterns provide a traditional-style appearance. Recommend a 1" overlap. Do not nail through overlapping pieces. Horizontal applications only. Cedar Bevel is also available in $\frac{7}{8} \times 10$, 12.		$\frac{1}{2} \times 4$ $\frac{1}{2} \times 5$ $\frac{1}{2} \times 6$ $\frac{5}{8} \times 8$ $\frac{5}{8} \times 10$ $\frac{3}{4} \times 6$ $\frac{3}{4} \times 8$ $\frac{3}{4} \times 10$
DOLLY VARDEN Dolly Varden is thicker than bevel and has a rabbeted edge. Surfaced smooth or saw-textured. Provides traditional-style appearance. Allows for $\frac{1}{2}$" overlap, including an approximate $\frac{1}{8}$" gap. Do not nail through overlapping pieces. Horizontal applications only. Cedar Dolly Varden is also available $\frac{7}{8} \times 10$, 12.		Standard Dolly Varden $\frac{3}{4} \times 6$ $\frac{3}{4} \times 8$ $\frac{3}{4} \times 10$ Thick Dolly Varden 1×6 1×8 1×10 1×12
DROP Drop siding is available in 13 patterns, in smooth, rough and saw-textured surfaces. Some are T&G (as shown), others are shiplapped. A variety of looks can be achieved with the different patterns. Do not nail through overlapping pieces. Horizontal or vertical applications.		$\frac{3}{4} \times 6$ $\frac{3}{4} \times 8$ $\frac{3}{4} \times 10$

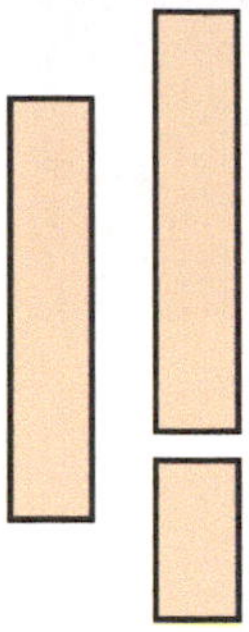

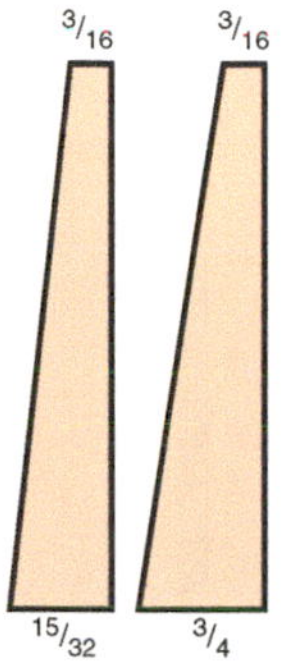

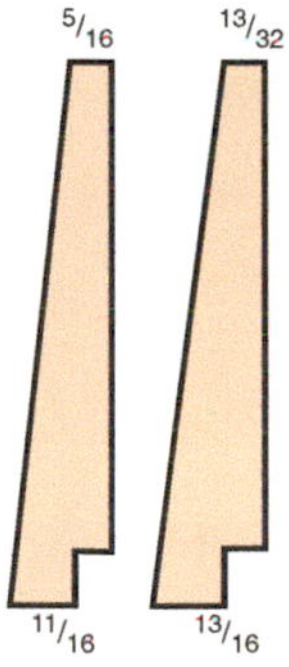

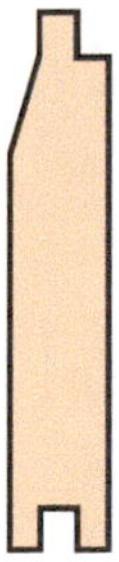

Figure 59–6 Names, descriptions, and sizes of natural wood siding patterns.

Figure 59–7 Bevel siding is commonly known as clapboards.

		Nominal Sizes
Siding Patterns		**Thickness & Width**
TONGUE AND GROOVE Tongue & groove siding is available in a variety of patterns. T&G lends itself to different effects aesthetically. Sizes given here are for Plain Tongue & Groove. Do not nail through overlapping pieces. Vertical or horizontal applications.		1×4 1×6 1×8 1×10 Note: T&G patterns may be ordered with $\frac{1}{4}$, $\frac{3}{8}$ or $\frac{7}{16}$" tongues. For wider widths, specify the longer tongue and pattern.
CHANNEL RUSTIC Channel Rustic has $\frac{1}{2}$" overlap (including an approximate $\frac{1}{8}$" gap) and a 1" to $1\frac{1}{4}$" channel when installed. The profile allows for maximum dimensional change without adversely affecting appearance in climates of highly variable moisture levels between seasons. Available smooth, rough or saw-textured. Do not nail through overlapping pieces. Horizontal or vertical applications.		$\frac{3}{4} \times 6$ $\frac{3}{4} \times 8$ $\frac{3}{4} \times 10$
LOG CABIN Log Cabin siding is $1\frac{1}{2}$" thick at the thickest point. Ideally suited to informal buildings in rustic settings. The pattern may be milled from appearance grades (Commons) or dimension grades ($2 \times$ material). Allows for $\frac{1}{2}$" overlap, including an approximately $\frac{1}{8}$" gap. Do not nail through overlapping pieces. Horizontal or vertical applications.		$1\frac{1}{2} \times 6$ $1\frac{1}{2} \times 8$ $1\frac{1}{2} \times 10$ $1\frac{1}{2} \times 12$

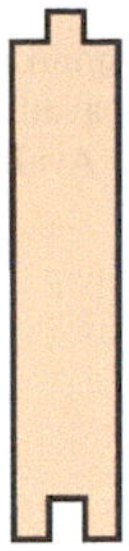

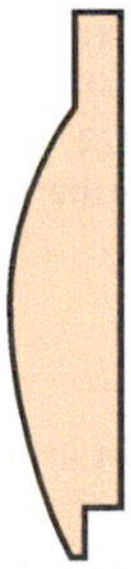

699

BRUSHED

Brushed or relief-grain surfaces accent the natural grain pattern to create striking textured surfaces. Generally available in 11/32", 3/8", 1/2", 19/32" and 5/8" thicknesses. Available in redwood, Douglas fir, cedar, and other species.

KERFED ROUGH-SAWN

Rough-sawn surface with narrow grooves providing a distinctive effect. Long edges shiplapped for continuous pattern. Grooves are typically 4" OC. Also available with grooves in multiples of 2" OC Generally available in 11/32", 3/8", 1/2", 19/32" and 5/8" thicknesses. Depth of kerfgroove varies with panel thickness.

APA TEXTURE 1-11

Special 303 Siding panel with shiplapped edges and parallel grooves 1/4" deep, 3/8" wide; grooves 4" or 8" OC are standard. Other spacings sometimes available are 2", 6" and 12" OC, check local availability. T 1-11 is generally available in 19/32" and 5/8" thicknesses. Also available with scratch-sanded, overlaid, rough-sawn, brushed and other surfaces. Available in Douglas fir, cedar, redwood, southern pine, other species.

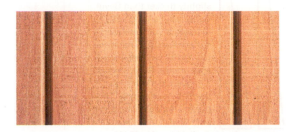

ROUGH-SAWN

Manufactured with a slight, rough-sawn texture running across panel. Available without grooves, or with grooves of various styles; in lap sidings, as well as in panel form. Generally available in 11/32", 3/8", 1/2", 19/32" and 5/8" thicknesses. Rough-sawn also available in Texture 1-11, reverse board-and-batten (5/8" thick), channel groove (3/8" thick), and V-groove (1/2" or 5/8" thick). Available in Douglas fir, redwood, cedar, southern pine, other species.

CHANNEL GROOVE

Shallow grooves typically 1/16" deep, 3/8" wide, cut into faces of 3/8" thick panels, 4" or 8" OC. Other groove spacings available. Shiplapped for continuous patterns. Generally available in surface patterns and textures similar to Texture 1-11 and in 11/32", 3/8" and 1/2" thicknesses. Available in redwood, Douglas fir, cedar, southern pine and other species.

REVERSE BOARD-AND-BATTEN

Deep, wide grooves cut into brushed, roughsawn, coarse sanded or other textured surfaces. Grooves about 1/4" deep, 1" to 1-1/2" wide, spaced 8", 12" or 16" OC with panel thickness of 19/32" and 5/8". Provides deep, sharp shadow lines. Long edges shiplapped for continuous pattern. Available in redwood, cedar, Douglas fir, southern pine and other species.

Courtesy of APA-The Engineered Wood Association

Figure 59–8 APA303 plywood panel siding is produced in a wide variety of sizes, surface textures, and patterns.

finishes and has excellent decay and termite resistance properties. Special considerations must be made when cutting this type of siding. Cutting of fiber-cement boards produces silica dust, which is known to cause cancer. All cutting should be done with a special dust-reducing circular saw blade or a set of shears. Cutting should be done outside and downwind of any other workers.

Panel Siding

Panel siding comes in 4-foot widths and lengths of 8 and 10 feet, and sometimes 9 feet. It is usually applied vertically, but may also be applied horizontally. Most panel siding is shaped with shiplapped edges for weather-tight joints.

Lap Siding

Lap siding is applied horizontally and is manufactured in styles with rough-sawn, weathered wood grain, or other embossed surface textures so that they look like wood. Some surfaces are grooved or beaded with square or beveled edges (*Figure 59–9*). They come in widths from 6 to 12 inches, and lengths of 12 or 16 feet.

Lap siding should overlap by at least $1\frac{1}{4}$ inches. It may be face nailed or blind nailed (*Figure 59–10*). Check local codes for feasibility. Blind nails are placed 1 inch down from the top. Face nails are placed 1 inch up from the bottom. Nails on hardboard siding should penetrate only one layer of siding. This will allow the material to expand and contract with moisture. Fiber-cement siding may have two nails per board in vertical alignment.

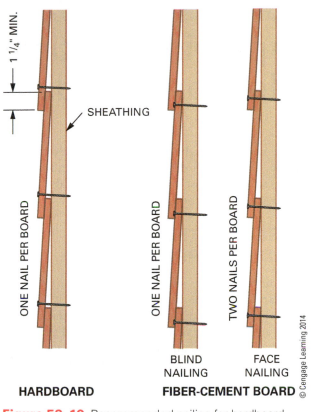

Figure 59–10 Recommended nailing for hardboard and fiber-cement board lap siding.

Nails are driven just snug or flush with the surface. If nails are driven too deep, they lose holding power. These should be caulked closed and another nail driven nearby.

Butt seams may be made either with moderate contact or with a gap. The gap is later caulked shut.

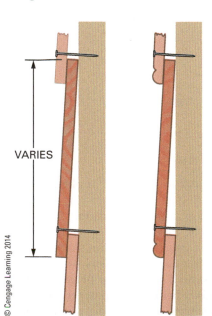

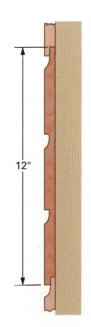

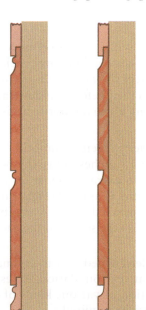

Figure 59–9 Styles of horizontal lap siding.

CHAPTER 60

Applying Vertical and Horizontal Siding

The method of siding application varies with the type. This chapter describes the application procedures for the most commonly used kinds of solid and engineered wood siding.

PREPARATION FOR SIDING APPLICATION

To maximize the siding paint longevity, drying potential must be built into the siding. This can be achieved by furring the siding off the sheathing and allowing air to circulate behind the siding (*Figure 60–1*). As air circulates behind the siding, it mixes with soffit air and eventually leaves at the ridge.

Green Tip

Furring wood siding away from sheathing allows painted siding surfaces to last two to three times longer.

Furring may be 1 × 3, 1 × 4, or 4-inch strips of plywood nailed to the sheathing over each stud. Screen is used to protect the airspace from insects. Furring should be installed before windows are installed. This maintains normal exterior finish details (*Figure 60–2*).

A 6- to 8-inch-wide screen is smoothly taped to the sheathing about 4 inches above the bottom edge of the siding. The extra screen is folded up and over after the furring is applied (*Figure 60–3*). Caulking or adhesive applied to the screen before the siding is applied seals the screen to the backside of the siding.

Next, it must be determined how the siding will be ended or treated at the foundation, eaves, and corners. The installation of various kinds of exterior wall trim may first be required.

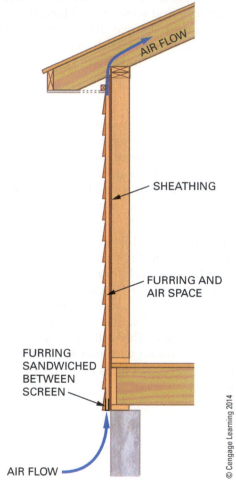

Figure 60–1 Siding lasts longer when it is furred off the sheathing.

Foundation Trim

In most cases, no additional trim is applied at the foundation. The siding is started so that it extends slightly below the top of the foundation. However, a water table may be installed for appearance. It sheds water a little farther away from the foundation. The water table usually consists of a board and a drip cap installed around the perimeter. Its bottom edge is slightly below the top of the foundation. The siding is started on top of the water table.

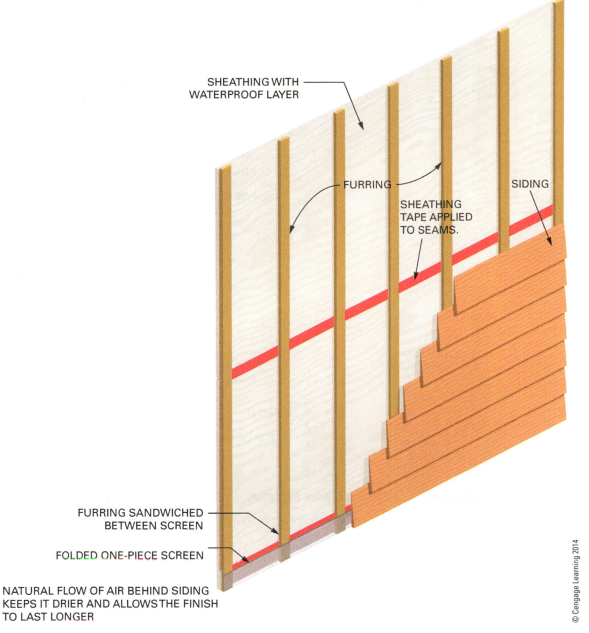

SHEATHING WITH WATERPROOF LAYER

FURRING

SIDING

SHEATHING TAPE APPLIED TO SEAMS.

FURRING SANDWICHED BETWEEN SCREEN

FOLDED ONE-PIECE SCREEN

NATURAL FLOW OF AIR BEHIND SIDING KEEPS IT DRIER AND ALLOWS THE FINISH TO LAST LONGER

© Cengage Learning 2014

Figure 60–2 Siding lasts longer when it is furred off the sheathing.

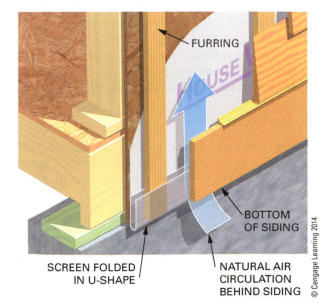

FURRING

BOTTOM OF SIDING

SCREEN FOLDED IN U-SHAPE

NATURAL AIR CIRCULATION BEHIND SIDING

© Cengage Learning 2014

Figure 60–3 Air flow behind siding begins through the screen.

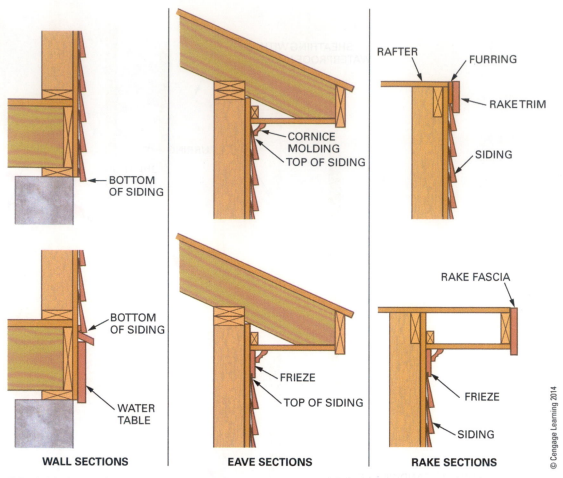

Figure 60–4 Methods of ending the siding at the foundation, eaves, and rakes.

Eaves Treatment

At the eaves, the siding may end against the bottom edge of the frieze. The width of the frieze may vary. It is necessary to know its width when laying out horizontal siding courses. The siding may also terminate against the soffit, if no frieze is used. The joint between is then covered by a cornice molding. The size of the molding must be known to plan exposures on courses of horizontal siding.

Rake Trim

At the rakes, the siding may be applied under a furred-out rake fascia. When the rake overhangs the sidewall, the siding may be fitted against the rake frieze. When fitted against the rake soffit, the joint is covered with a molding (*Figure 60–4*).

Gable End Treatment

Sometimes a different kind of siding is used on the gable ends than on the sidewalls below the plate. The joint between the two types must be weather tight. One method of making the joint is to use a drip cap and flashing between the two types of material.

In another method, the plate and studs of the gable end are extended out from the wall a short distance. This allows the gable end siding to overlap the siding below (*Figure 60–5*). Furring strips may also be used on the gable end framing in place of extending the gable plate and studs.

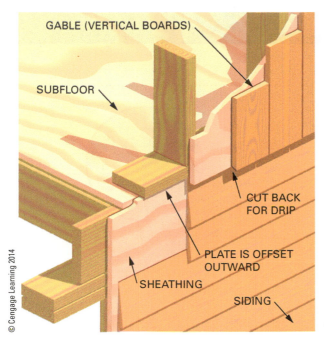

Figure 60-5 The upper gable end may be constructed to overhang the wall siding.

Treating Corners

One method of treating corners is with the use of corner boards. Horizontal siding may be mitered around exterior corners. Or, metal corners may be used on each course of siding. In interior corners, siding courses may butt against a square corner board or against each other (*Figure 60-6*). The thickness of corner boards depends on the type of siding used. The corner boards should be thick enough so that the siding does not project beyond the face of the corner board.

The width of the corner boards depends on the effect desired. However, one of the two pieces, making up an outside corner, should be narrower than the other by the thickness of the stock. Then, after the wider piece is fastened to the narrower one, the same width is exposed on both sides of the corner. The joint between the two pieces should be on the side of the building that is least viewed.

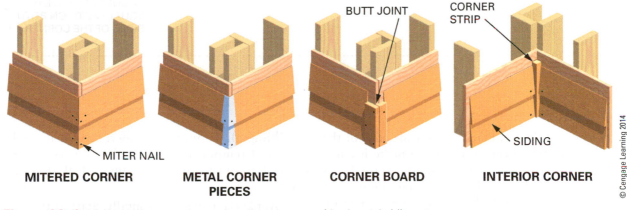

Figure 60-6 Methods for returning and ending courses of horizontal siding at corners.

INSTALLING CORNER BOARDS

Before installing corner boards, flash both sides of the corner. Install a strip flashing vertically on each side. One edge should extend beyond the edge of the corner board at least 2 inches. Tuck the top ends under any previously applied flashing.

With a sharp plane, slightly back-bevel one edge of the narrower of the two pieces that make up the outside corner board. This ensures a tight fit between the two boards. Cut, fit, and fasten the narrow piece. Start at one end and work toward the bottom. Keep the beveled edge flush with the corner. Fasten with galvanized or other noncorroding nails spaced about 16 inches apart along its inside edge.

Cut, fit, and fasten the wider piece to the corner in a similar manner. Make sure its outside edge is flush with the surface of the narrower piece. The outside row of nails is driven into the edge of

StepbyStep Procedures

Procedure 60–A Installing Corner Boards

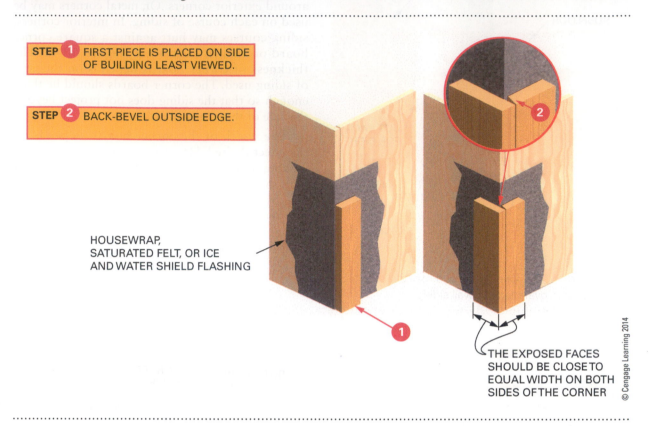

STEP 1 FIRST PIECE IS PLACED ON SIDE OF BUILDING LEAST VIEWED.

STEP 2 BACK-BEVEL OUTSIDE EDGE.

HOUSEWRAP, SATURATED FELT, OR ICE AND WATER SHIELD FLASHING

THE EXPOSED FACES SHOULD BE CLOSE TO EQUAL WIDTH ON BOTH SIDES OF THE CORNER

© Cengage Learning 2014

the narrower piece. Plane the outside edge of the wide piece wherever necessary to make it come flush. Slightly round over all sharp exposed corners by planing a small chamfer and sanding. Set all nails so they can be filled over later. Make sure a tight joint is obtained between the two pieces (*Procedure 60–A*).

Corner boards may also be applied by fastening the two pieces together first. Then install the assembly on the corner.

APPLYING HOUSEWRAP

A description of the kinds and purposes of housewrap has been previously given in Chapter 55. A *breathable* type of material should be applied to sidewalls. It is typically applied to the whole building all at once. This serves as an immediate and temporary weather barrier as well as a backup to shed water if a future leak develops in the siding.

Apply the paper horizontally. Start at the bottom of the wall. Make sure the paper lies flat with no wrinkles. Fasten it in position. If nailing, use large-head roofing nails. Fasten in rows near the bottom, the center, and the top about 16 inches apart. Each succeeding layer should lap the lower layer by about 4 inches.

The sheathing paper should lap over any flashing applied at the sides and tops of windows and doors and at corner boards. It should be tucked under any flashings applied under the bottoms of windows or frieze boards. In any case, all laps should be over the paper below. * **✳IRC 703.2**

INSTALLING HORIZONTAL WOOD SIDING

One of the important differences between bevel siding and other types with tongue-and-groove, shiplap, or rabbeted edges is that exposure of courses of bevel siding can be varied somewhat. With other types, the amount exposed to the weather is constant with every course and cannot vary.

The ability to vary the exposure is a decided advantage. It is desirable from the standpoint of appearance, weather tightness, and ease of application to have a full course of horizontal siding exposed above and below windows and over the tops of doors (*Figure 60–7*). The exposure of the siding may vary gradually up to a total of $\frac{1}{2}$ inch over the entire wall, but the width of each exposure should not vary more than $\frac{1}{4}$ inch from its neighbor. ✷

✷**IRC 703.3.2**

Determining Siding Exposure

To determine the siding exposure so that it is about equal both above and below the window sill, divide the overall height of the window frame by the amount of exposure. For example, consider a window that is 52 inches in height with the coverage required above and below the window being $12\frac{1}{2}$ inches and $40\frac{1}{2}$ inches, respectively (*Figure 60–8*). The exposure of each section may be adjusted to allow full laps above and below the window.

First, the number of courses for each section must be determined. This is done by dividing the coverage distance by the maximum exposure of the siding. Then, round that number up to the next whole number of courses.

Next, the number of courses is divided into the coverage distances to find the exposure for that area. Note that because windows vary in height around the house, this process does not always work out neatly. Sometimes all that can be done is to adjust the last exposure to the largest that it can be.

Layout lines are then transferred to a story pole. The story pole is used to lay out the courses all around the building (*Figure 60–9*).

Instead of calculating the number and height of siding courses mathematically, *dividers* may be used to space off the distances.

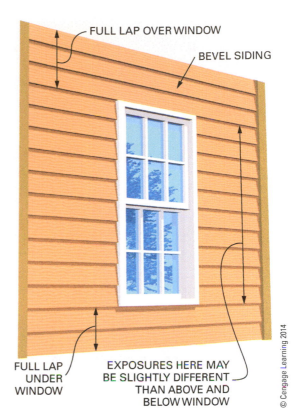

FULL LAP OVER WINDOW

BEVEL SIDING

FULL LAP UNDER WINDOW

EXPOSURES HERE MAY BE SLIGHTLY DIFFERENT THAN ABOVE AND BELOW WINDOW

© Cengage Learning 2014

Figure 60–7 Bevel siding exposure may be varied from top to bottom.

Starting Bevel Siding from the Top

Another advantage of using bevel siding is that application may be made starting at the top and working toward the bottom if more convenient. With this method, a number of chalk lines may be snapped without being covered by a previous course. This saves time. Any scaffolding already erected may be used and then dismantled as work progresses toward the bottom.

Starting Siding from the Bottom

Most horizontal siding, however, is usually started at the bottom. For bevel siding, a **furring strip**, called a **starter strip**, of the same thickness and width of the siding head lap is first fastened along the bottom edge of the sheathing (*Figure 60–10*). For the first course, a line is snapped on the wall at a height that is in line with the top edge of the first course of siding.

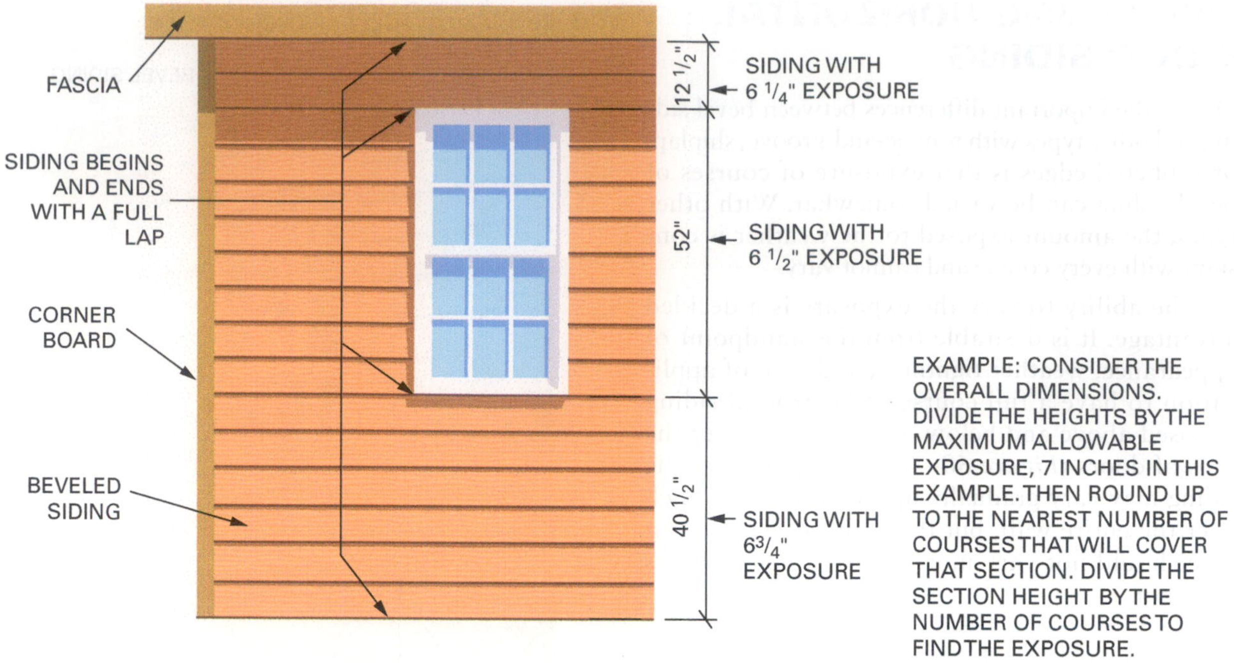

FASCIA

SIDING BEGINS AND ENDS WITH A FULL LAP

CORNER BOARD

BEVELED SIDING

12 1/2"

SIDING WITH 6 1/4" EXPOSURE

52"

SIDING WITH 6 1/2" EXPOSURE

40 1/2"

SIDING WITH 6³/4" EXPOSURE

EXAMPLE: CONSIDER THE OVERALL DIMENSIONS. DIVIDE THE HEIGHTS BY THE MAXIMUM ALLOWABLE EXPOSURE, 7 INCHES IN THIS EXAMPLE. THEN ROUND UP TO THE NEAREST NUMBER OF COURSES THAT WILL COVER THAT SECTION. DIVIDE THE SECTION HEIGHT BY THE NUMBER OF COURSES TO FIND THE EXPOSURE.

40 1/2" ÷ 7 = 5.8 ⇒ 6 COURSES   40 1/2" ÷ 6 = 6.75 OR 6³/4" EXPOSURE

52" ÷ 7 = 7.4 ⇒ 8 COURSES   52" ÷ 8 = 6.5 OR 6 1/2" EXPOSURE

12 1/2" ÷ 7 = 1.8 ⇒ 2 COURSES   12 1/2" ÷ 2 = 6.25 OR 6 1/4" EXPOSURE

© Cengage Learning 2014

Figure 60–8 Example of determining siding exposures around a window.

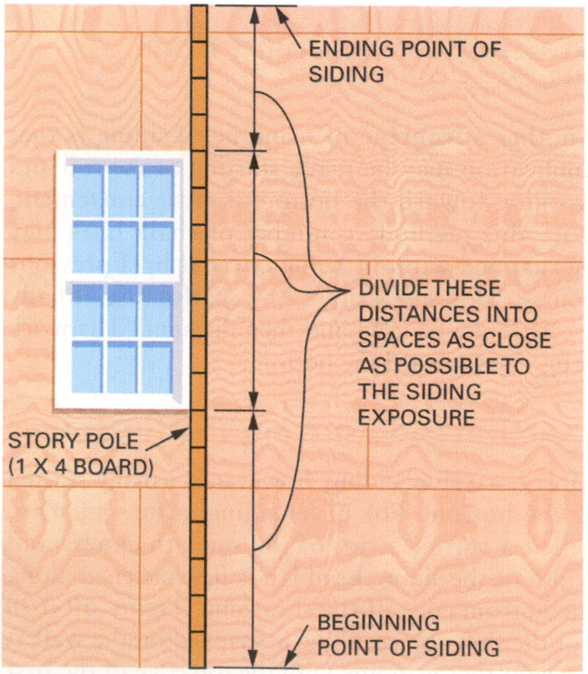

ENDING POINT OF SIDING

DIVIDE THESE DISTANCES INTO SPACES AS CLOSE AS POSSIBLE TO THE SIDING EXPOSURE

STORY POLE (1 X 4 BOARD)

BEGINNING POINT OF SIDING

© Cengage Learning 2014

Figure 60–9 Laying out a story pole for courses of horizontal siding.

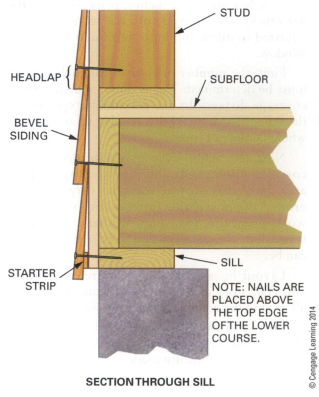

STUD

SUBFLOOR

HEADLAP

BEVEL SIDING

STARTER STRIP

SILL

NOTE: NAILS ARE PLACED ABOVE THE TOP EDGE OF THE LOWER COURSE.

© Cengage Learning 2014

SECTION THROUGH SILL

Figure 60–10 For bevel siding, a strip of wood the same thickness and width of the headlap is used as a starter strip.

For siding with constant exposure, the only other lines snapped are across the tops of wide entrances, windows, and similar objects to keep the courses in alignment. For lap siding, with exposures that may vary, lines are snapped for each successive course in line with their top edge. Stagger joints in adjacent courses as far apart as possible. A small piece of felt paper often is used behind the butt seams to ensure the weather tightness of the siding.

Fitting Siding

When applying a course of siding, start from one end and work toward the other end. With this procedure, only the last piece will have to be fitted. Tight-fitting butt joints must be made between pieces. If an end joint must be fitted, use a block plane to trim the end as needed. When a piece has to be fitted between other pieces, measure carefully. Cut it slightly long. Place one end in position. Bow the piece outward, position the other end, and snap into place (*Figure 60–11*).

A preacher is often used for accurate layout of siding where it butts against corner boards, casings, and similar trim. The siding is allowed to overlap the trim. The preacher is held against the trim. A line is then marked on the siding along the face of the preacher (*Figure 60–12*).

When fitting siding under windows, make sure the siding fits snugly in the groove on the underside of the window sill for weather tightness (*Figure 60–13*).

Fastening Siding

Siding is fastened to each bearing stud or about every 16 inches. On bevel siding, fasten through the butt edge just above the top edge of the course below. Do not fasten through the lap. This prevents splitting of the siding that might be caused by slight swelling or shrinking due to moisture changes.

Care must be taken to fasten as low as possible to avoid splitting the siding in the center. The location and number of fasteners recommended for siding are shown in *Figure 60–14*. ✱

✱**IRC 703.4**

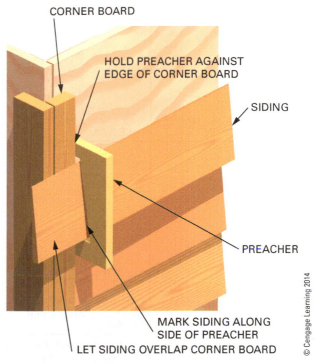

CORNER BOARD

HOLD PREACHER AGAINST EDGE OF CORNER BOARD

SIDING

PREACHER

MARK SIDING ALONG SIDE OF PREACHER

LET SIDING OVERLAP CORNER BOARD

© Cengage Learning 2014

Figure 60–12 A preacher may be used for accurately marking a siding piece for length.

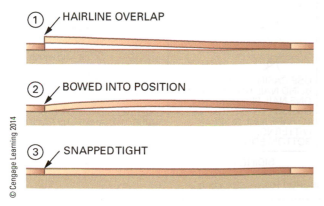

① HAIRLINE OVERLAP

② BOWED INTO POSITION

③ SNAPPED TIGHT

© Cengage Learning 2014

Figure 60–11 Method of cutting horizontal siding to fit snugly.

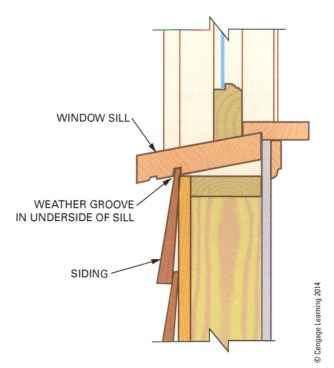

WINDOW SILL

WEATHER GROOVE IN UNDERSIDE OF SILL

SIDING

© Cengage Learning 2014

Figure 60–13 Fit siding into weather groove on the underside of the window sill.

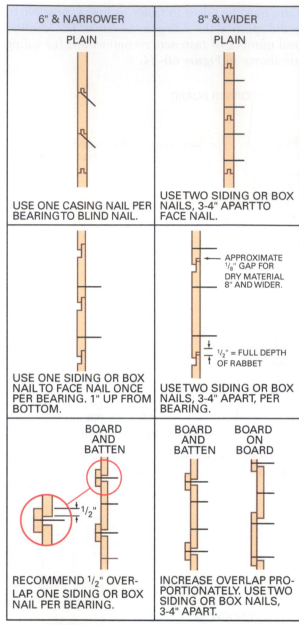

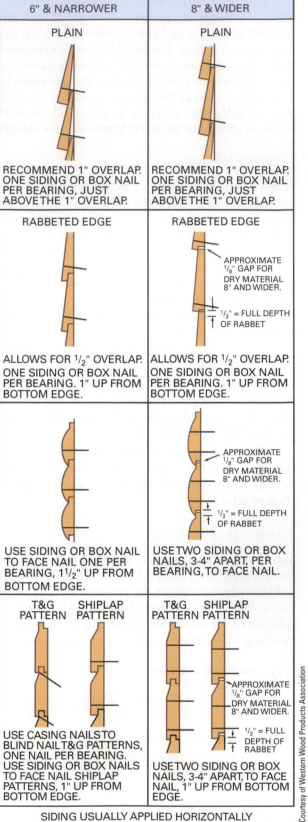

SIDING USUALLY APPLIED HORIZONTALLY

Courtesy of Western Wood Products Association

Figure 60–14 Location and number of fasteners recommended for wood siding.

VERTICAL APPLICATION OF WOOD SIDING

Bevel sidings are designed for horizontal applications only. Board on board and board and batten are applied only vertically. Almost all other patterns may be applied in either direction.

INSTALLING VERTICAL TONGUE-AND-GROOVE SIDING

Corner boards usually are not used when wood siding is applied vertically. The siding boards are fitted around the corner (*Figure 60–15*). Rip the grooved edge from the starting piece. Slightly back-bevel the ripped edge. Place it vertically on the wall with the beveled edge flush with the corner similar to making a corner board.

The tongue edge should be plumb; the bottom end should be about 1 inch below the sheathing (*Figure 60–16*). The top end should butt or be tucked under any trim above. Face nail the edge nearest the corner. Blind nail into the tongue edge. Nails should be placed from 16 to 24 inches apart. Blocking must be provided between studs if siding is applied directly to the frame.

Fasten a temporary piece on the other end of the wall projecting below the sheathing by the same amount. Stretch a line to keep the bottom

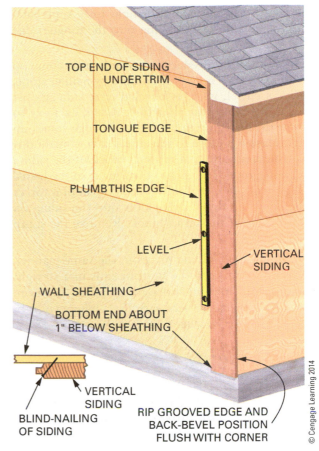

TOP END OF SIDING UNDER TRIM

TONGUE EDGE

PLUMB THIS EDGE

LEVEL

VERTICAL SIDING

WALL SHEATHING

BOTTOM END ABOUT 1" BELOW SHEATHING

VERTICAL SIDING

BLIND-NAILING OF SIDING

RIP GROOVED EDGE AND BACK-BEVEL POSITION FLUSH WITH CORNER

© Cengage Learning 2014

Figure 60–16 Starting the application of vertical board siding.

Courtesy of California Redwood Association

Figure 60–15 Vertical tongue-and-groove siding needs little accessory trim, such as corner boards.

ends of other pieces in a straight line. Apply succeeding pieces by toenailing into the tongue edge of each piece.

Make sure the edges between boards come up tight. If they do not come up tight by nailing alone, drive a chisel, with its beveled edge side against the tongue, into the sheathing. Use it as a pry to force the board up tight. When the edge comes up tight, fasten it close to the chisel. If this method is not successful, toenail a short block of the siding with its grooved edge into the tongue of the board (*Figure 60–17*). Drive the nail home until it forces the board up tight. Drive nails into the siding on both sides of the scrap block. Remove the scrap block. **✳**

✳IRC 703.4

Continue applying pieces in the same manner. Make sure to keep the bottom ends in a straight line. Avoid making horizontal joints between lengths. If joints are necessary, use a mitered or rabbeted joint for weather tightness (*Figure 60–18*).

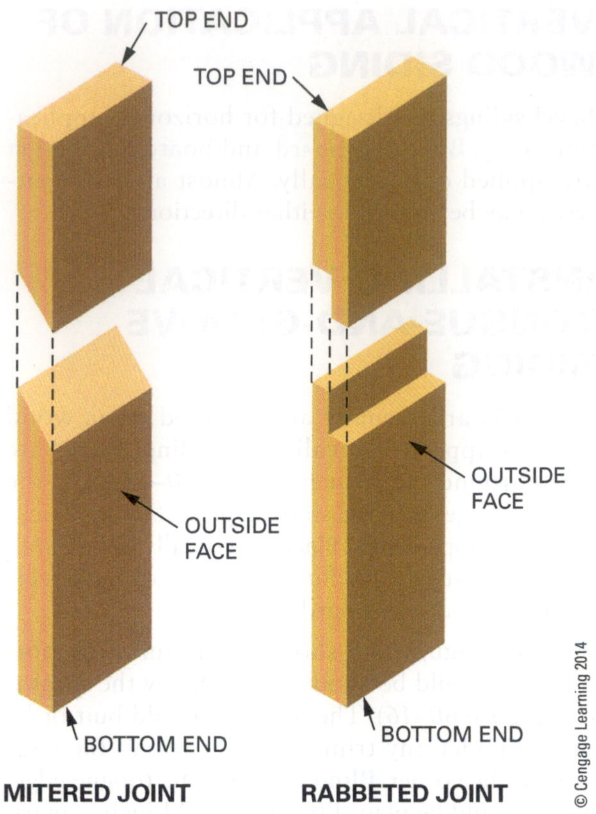

Fitting around Doors and Windows

Vertical siding is fitted tightly around window and door casings with different methods than those used for horizontal siding.

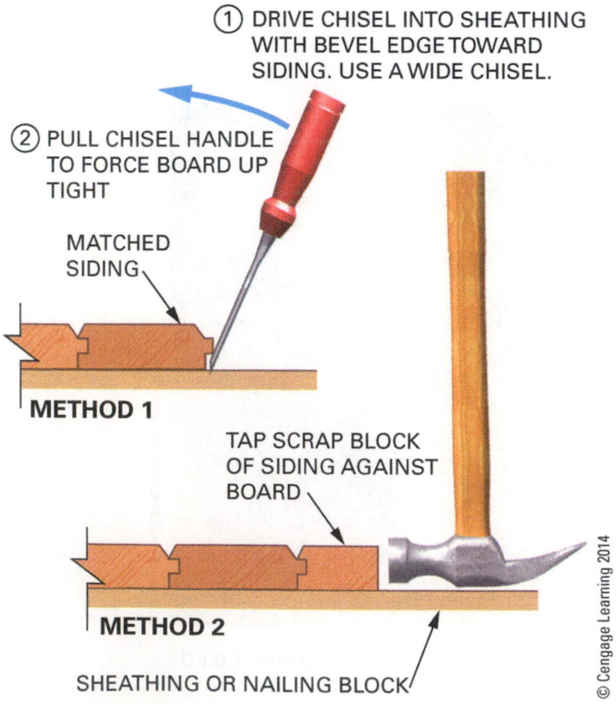

MITERED JOINT   **RABBETED JOINT**

Figure 60–18 Use mitered or rabbeted end joints between lengths of vertical board siding.

Figure 60–17 Techniques for tightening the joints of boards during installation.

Approaching a Wall Opening. When approaching a door or window, cut and fit the piece just before the one to be fitted against the casing. Then remove it. Set it aside for the time being. Cut, fit, and tack the piece to be fitted against the casing in place of the next to last piece. Level from the top of the window casing and the bottom of the sill. Mark the piece.

To lay out the piece so it will fit snugly against the side casing, first cut a scrap block of the siding material, about 6 inches long (*Procedure 60–B*). Remove the tongue from one edge. Be careful to remove all of the tongue, but no more. Hold the block so its grooved edge is against the side casing and the other edge is on top of the siding to be fitted. Mark the piece by holding a pencil against the outer edge of the block while moving the block along the length of the side casing.

Cut the piece, following the layout lines carefully. When laying out to fit against the bottom of the sill, make allowance to rabbet the siding to fit in the weather groove on the bottom side of the window sill. Place and fasten the pieces in position.

StepbyStep Procedures

Procedure 60–B Method for Fitting Vertical Board Siding When Approaching a Window Casing

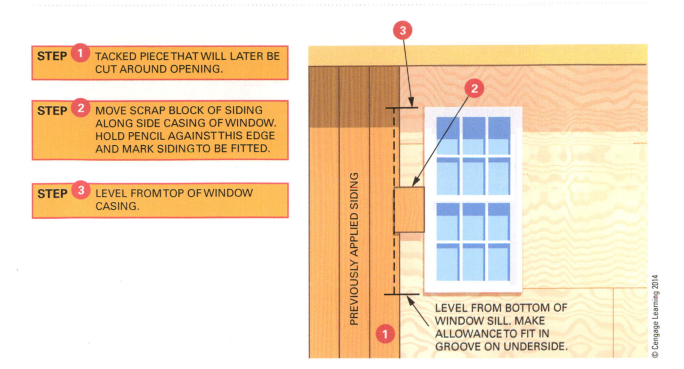

STEP 1 TACKED PIECE THAT WILL LATER BE CUT AROUND OPENING.

STEP 2 MOVE SCRAP BLOCK OF SIDING ALONG SIDE CASING OF WINDOW. HOLD PENCIL AGAINST THIS EDGE AND MARK SIDING TO BE FITTED.

STEP 3 LEVEL FROM TOP OF WINDOW CASING.

PREVIOUSLY APPLIED SIDING

LEVEL FROM BOTTOM OF WINDOW SILL. MAKE ALLOWANCE TO FIT IN GROOVE ON UNDERSIDE.

© Cengage Learning 2014

Continue to apply the siding with short lengths across the top and bottom of the window. Each length under a window must be rabbeted at the top end to fit in the weather groove at the sill.

Leaving a Wall Opening. A full length must also be fitted to the casing on the other side of the window. To mark the piece, first tack a short length of scrap siding above and below the window and against the last pieces of siding installed. Tack the length of siding to be fitted against these blocks. Level from the top and bottom of the window. Mark the piece for the horizontal cuts.

To lay out the piece for the vertical cut that fits against the side casing, use the same block with the tongue removed that was used previously. Hold the grooved edge against the side casing. With a pencil against the other edge, ride the block along the side casing while marking the piece to be fitted (*Procedure 60–C*).

Remove the piece and the scrap blocks from the wall. Carefully cut the piece to the layout lines. Then fasten in position. Continue applying the rest of the siding until you are almost to the other end of the wall.

Method of Ending Vertical Siding

The last piece of vertical siding should be close to the same width of previously installed pieces. If siding is installed in random widths, plan the application. The width of the last piece should be equal, at least, to the width of the narrowest piece. It is not good practice to allow vertical siding to end with a narrow sliver.

Stop several feet from the end. Space off, and determine the width of the last piece. If it will not be a satisfactory width, install narrower or wider pieces for the remainder, as required. It may be necessary to rip available siding to narrower widths and reshape the grooves (*Figure 60–19*).

When the corner is reached, the board is ripped to width along its tongue edge. It is slightly back-beveled for the first piece on the next wall to butt against. When the last corner is reached, the board is ripped in a similar manner. However, it is smoothed to a square edge to fit flush with the surface of the first piece installed. All exposed sharp corners should be eased or slightly rounded.

StepbyStep Procedures

Procedure 60–C Method of fitting vertical board siding when leaving a window casing

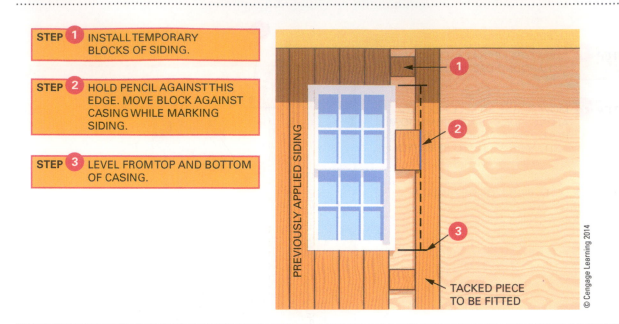

STEP 1 INSTALL TEMPORARY BLOCKS OF SIDING.

STEP 2 HOLD PENCIL AGAINST THIS EDGE. MOVE BLOCK AGAINST CASING WHILE MARKING SIDING.

STEP 3 LEVEL FROM TOP AND BOTTOM OF CASING.

PREVIOUSLY APPLIED SIDING

TACKED PIECE TO BE FITTED

© Cengage Learning 2014

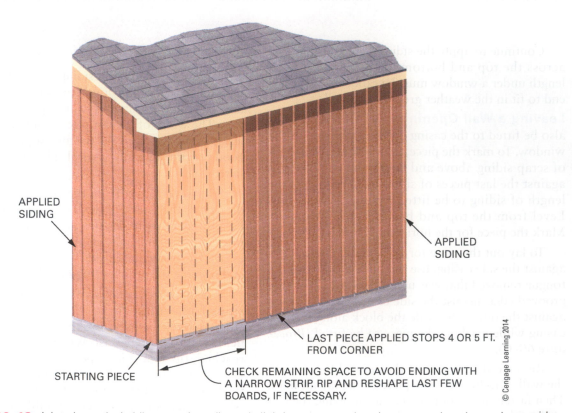

APPLIED SIDING

APPLIED SIDING

STARTING PIECE

LAST PIECE APPLIED STOPS 4 OR 5 FT. FROM CORNER

CHECK REMAINING SPACE TO AVOID ENDING WITH A NARROW STRIP. RIP AND RESHAPE LAST FEW BOARDS, IF NECESSARY.

© Cengage Learning 2014

Figure 60–19 Joints in vertical siding may be adjusted slightly to ensure that the corner piece is nearly as wide as the others.

INSTALLING PANEL SIDING

Plywood, hardboard, and other panel siding is usually installed vertically. It can be installed horizontally, if desired. Lap siding panels are ordinarily applied horizontally.

Installing Vertical Panel Siding

Start a vertical panel so it is plumb, with one edge squared and flush with the starting corner. The inner edge should fall on the center of a stud. Fasten panels, of ½-inch thickness or less, with 6d siding nails. Use 8d siding nails for thicker panels. Fasten panel edges about every 6 inches and about every 12 inches along intermediate studs. ✳

✳IRC 703.4

Apply successive sheets. Leave a ⅛-inch space between panels. Panels must be installed with their bottom ends in a straight line. There should be a minimum of 6 inches above the finished grade line.

Installing Horizontal Panel Siding

Mark the height of the first course of horizontal panel siding on both ends of the wall. Snap a chalk line between marks. Fasten a full-length panel with its top edge to the line, its inner end on the center of a stud, and its outer end flush with the corner. Fasten in the same way as for vertical panels.

Apply the remaining sheets in the first course in like manner. Trim the end of the last sheet flush with the corner. Start the next course so joints will line up with those in the course below.

Both vertical and horizontal panels may be applied to sheathing or directly to studs if backing is provided for all joints (*Figure 60–20*).

Carefully fit and caulk around doors and windows. It is important that horizontal butt joints be either offset and lapped, rabbeted, or flashed (*Figure 60–21*). Vertical joints are either shiplapped or covered with batterns. ✳

✳IRC 703.3.1

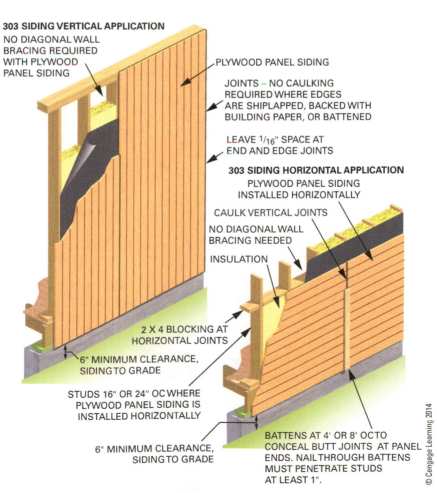

303 SIDING VERTICAL APPLICATION
NO DIAGONAL WALL BRACING REQUIRED WITH PLYWOOD PANEL SIDING

PLYWOOD PANEL SIDING

JOINTS – NO CAULKING REQUIRED WHERE EDGES ARE SHIPLAPPED, BACKED WITH BUILDING PAPER, OR BATTENED

LEAVE 1/16" SPACE AT END AND EDGE JOINTS

303 SIDING HORIZONTAL APPLICATION
PLYWOOD PANEL SIDING INSTALLED HORIZONTALLY

CAULK VERTICAL JOINTS

NO DIAGONAL WALL BRACING NEEDED

INSULATION

2 X 4 BLOCKING AT HORIZONTAL JOINTS

6" MINIMUM CLEARANCE, SIDING TO GRADE

STUDS 16" OR 24" OC WHERE PLYWOOD PANEL SIDING IS INSTALLED HORIZONTALLY

6" MINIMUM CLEARANCE, SIDING TO GRADE

BATTENS AT 4' OR 8' OC TO CONCEAL BUTT JOINTS AT PANEL ENDS. NAIL THROUGH BATTENS MUST PENETRATE STUDS AT LEAST 1".

© Cengage Learning 2014

Figure 60–20 Panel siding may be applied vertically or horizontally to sheathing or directly to studs.

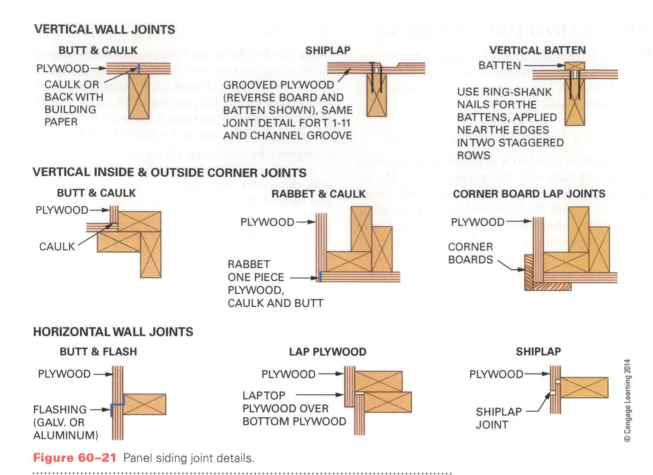

VERTICAL WALL JOINTS

BUTT & CAULK

PLYWOOD →
CAULK OR
BACK WITH
BUILDING
PAPER

SHIPLAP

GROOVED PLYWOOD
(REVERSE BOARD AND
BATTEN SHOWN), SAME
JOINT DETAIL FOR T 1-11
AND CHANNEL GROOVE

VERTICAL BATTEN

BATTEN →
USE RING-SHANK
NAILS FOR THE
BATTENS, APPLIED
NEAR THE EDGES
IN TWO STAGGERED
ROWS

VERTICAL INSIDE & OUTSIDE CORNER JOINTS

BUTT & CAULK

PLYWOOD →
CAULK

RABBET & CAULK

PLYWOOD →
RABBET
ONE PIECE
PLYWOOD,
CAULK AND BUTT

CORNER BOARD LAP JOINTS

PLYWOOD →
CORNER
BOARDS →

HORIZONTAL WALL JOINTS

BUTT & FLASH

PLYWOOD →
FLASHING →
(GALV. OR
ALUMINUM)

LAP PLYWOOD

PLYWOOD →
LAP TOP
PLYWOOD OVER
BOTTOM PLYWOOD →

SHIPLAP

PLYWOOD →
SHIPLAP
JOINT

© Cengage Learning 2014

Figure 60–21 Panel siding joint details.

Applying Lap Siding Panels

Panels of lap siding are applied in much the same manner as wood bevel siding with some exceptions. First, install a strip, of the same thickness and width of the siding headlap, along the bottom of the wall. Determine the height of the top edge of the first course. Snap a chalk line across the wall. Apply the first course with its top edge to the snapped line.

When applied over nailable sheathing, space nails 8 inches apart in a line about $\frac{3}{4}$ inch above the bottom edge of the siding. When applied directly to framing, fasten at each stud location. Joints between ends of siding should also be flashed with a narrow strip of #15 felt centered behind the joint and backed with a wood shingle wedge (*Figure 60–22*). *

*IRC
703.4

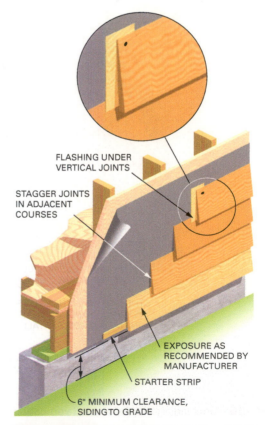

FLASHING UNDER
VERTICAL JOINTS

STAGGER JOINTS
IN ADJACENT
COURSES

EXPOSURE AS
RECOMMENDED BY
MANUFACTURER

STARTER STRIP

6" MINIMUM CLEARANCE,
SIDING TO GRADE

© Cengage Learning 2014

Figure 60–22 Lap siding application details.

CHAPTER 61

Wood Shingle and Shake Siding

Wood shingles and shakes may be used for siding, as well as roofing (*Figure 61–1*). Those previously described in Chapter 52 for roofing may also be applied to sidewalls.

Figure 61–1 Wood shingles and shakes may be used as siding.

SIDEWALL SHINGLES AND SHAKES

Some kinds of shingles and shakes are designed for sidewall use only (*Figure 61–2*). Rebutted and rejointed ones are machine trimmed with parallel edges and square butts for sidewall application. Rebutted and rejointed machine-grooved, sidewall shakes have striated faces.

Special fancy butt shingles are available in a variety of designs. They provide interesting patterns, in combination with square butts or other types of siding (*Figure 61–3*).

APPLYING WOOD SHINGLES AND SHAKES

Wood shingles and shakes may be applied to sidewalls in either single-layer or double-layer courses. In single coursing, shingles are applied to walls with a single layer in each course, in a way similar to roof application. However, greater exposures are allowed on sidewalls than on roofs (*Figure 61–4*).✳

✳IRC
703.5.1

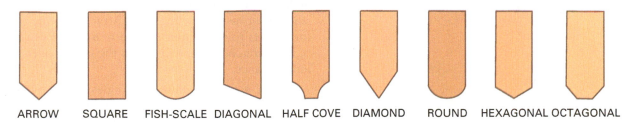

ARROW   SQUARE   FISH-SCALE   DIAGONAL   HALF COVE   DIAMOND   ROUND   HEXAGONAL   OCTAGONAL

FANCY BUTT RED CEDAR SHINGLES. NINE OF THE MOST POPULAR DESIGNS ARE SHOWN. FANCY BUTT SHINGLES CAN BE CUSTOM PRODUCED TO INDIVIDUAL ORDERS.

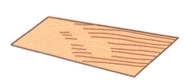

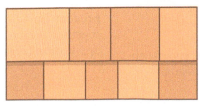

REBUTTED AND REJOINTED. MACHINE TRIMMED FOR PARALLEL EDGES WITH BUTTS SAWN AT RIGHT ANGLES. FOR SIDEWALL APPLICATION WHERE TIGHTLY FITTING JOINTS ARE DESIRED.

PANELS. WESTERN RED CEDAR SHINGLES ARE AVAILABLE IN 4- AND 8-FOOT PANELIZED FORM.

MACHINE GROOVED. MACHINE-GROOVED SHAKES ARE MANUFACTURED FROM SHINGLES AND HAVE STRIATED FACES AND PARALLEL EDGES. USED DOUBLE-COURSED ON EXTERIOR SIDEWALLS.

Figure 61–2 Some wood shingles and shakes are made for sidewall applications only.

ROUND FANCY BUTT SHINGLES

© Cengage Learning 2014

Figure 61–3 Fancy butt shingles are still used to accent sidewalls with distinctive designs.

SINGLE COURSING

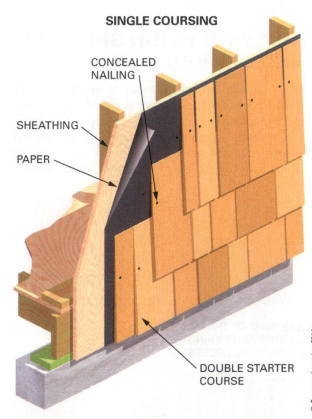

CONCEALED NAILING

SHEATHING

PAPER

DOUBLE STARTER COURSE

© Cengage Learning 2014

Figure 61–4 Single-coursed shingle wall application is similar to roof application with greater weather exposures allowed.

In **double coursing**, two layers are applied in one course. Consequently, even greater weather exposures are allowed. Double coursing is used when wide courses with deep, bold shadow lines are desired (*Figure 61–5*). ✴

✴IRC 703.5.2

Applying the Starter Course

The starter course of sidewall shingles and shakes is applied in much the same way as the starter course on roofs. A double layer is used for single-course applications. A triple layer is used for triple coursing. Less expensive **undercourse** shingles are used for underlayers. ✴

✴IRC 703.5.4

Fasten a shingle on both ends of the wall with its butt about 1 inch below the top of the foundation. Stretch a line between them at the butts. Sight the line for straightness. Fasten additional shingles at necessary intervals. Attach the line to their butts to straighten it (*Figure 61–6*). Even a tightly stretched line will sag in the center over a long distance.

Apply a single course of shingles so the butts are as close to the chalk line as possible without touching it. Remove the line. Apply another course on top of the first course. Offset the joints in the outer layer at least $1\frac{1}{2}$ inches from those in the bottom layer. Untreated shingles should be spaced $\frac{1}{8}$ to $\frac{1}{4}$ inch apart to allow for swelling and to prevent buckling. Shingles can be applied close together if factory primed or if treated soon after application (*Figure 61–7*).

Single Coursing

A story pole may be used to lay out shingle courses in the same manner as with horizontal wood siding. Snap a chalk line across the wall, at the shingle butt line, to apply the first course. Using only as many finish nails as necessary, tack 1 × 3 straightedges to the wall with their top edges to the line.

DOUBLE COURSING

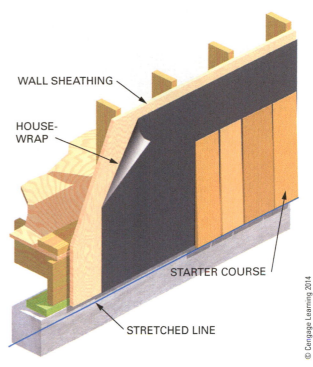

NO. 3 OR UNDERCOURSING GRADE SHINGLES

HOUSEWRAP

LUMBER OR PLYWOOD SHEATHING

OUTER COURSE 1/2" LOWER THAN UNDERCOURSE

JOINTS SHOULD BE OPEN FOR UNSTAINED SHINGLES AND MAY BE CLOSED FOR STAINED SHINGLES

DOUBLE UNDER-COURSE

APPLY NAILS IN STRAIGHT LINE 2" ABOVE SHINGLE BUTTS

© Cengage Learning 2014

WALL SHEATHING

HOUSE-WRAP

STARTER COURSE

STRETCHED LINE

© Cengage Learning 2014

Figure 61–6 Stretch a straight line as a guide for the butts of the starter course.

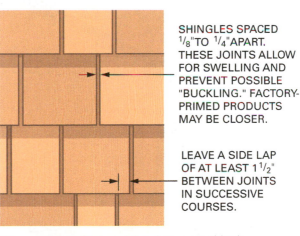

SHINGLES SPACED 1/8"TO 1/4"APART. THESE JOINTS ALLOW FOR SWELLING AND PREVENT POSSIBLE "BUCKLING." FACTORY-PRIMED PRODUCTS MAY BE CLOSER.

LEAVE A SIDE LAP OF AT LEAST 1 1/2" BETWEEN JOINTS IN SUCCESSIVE COURSES.

© Cengage Learning 2014

Figure 61–7 Stagger joints between shingle courses.

Max. Weather Exposure		
Shingle Length	Single Course	Double Course
16"	7"	12"
18"	8"	14"
24"	10½"	16"
Shake Length & Type	Single Course	Double Course
16" Centigroove	7"	12"
18" Centigroove	8"	14"
18" resawn	8"	14"
24" resawn	10½"	16"
18" taper-sawn	8"	14"
24" tapersplit	10½"	18"
24" taper-sawn	10½"	18"
18" straight-split	8"	16"

© Cengage Learning 2014

Figure 61–5 Double-coursed shingles, with two layers in each row, permit even greater weather exposures.

Lay individual shingles with their butts on the straightedge (*Figure 61–8*). Use a shingling hatchet to trim and fit the edges, if necessary. Butt ends are not trimmed. If rebutted and rejointed shingles are used, no trimming should be necessary.

At times it may be necessary to fit a shingle between others in the same course. Tack the next to last shingle in place with one nail. Slip the last shingle under it. Score along the overlapping edge with the hatchet. Cut along the scored line. Fasten both shingles in place (*Figure 61–9*).

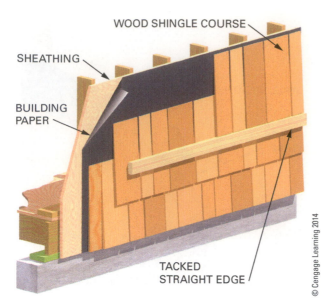

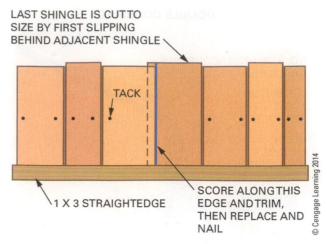

SHEATHING

BUILDING PAPER

WOOD SHINGLE COURSE

TACKED STRAIGHT EDGE

© Cengage Learning 2014

Figure 61–8 Rest shingle butts on a straightedge when single coursing on sidewalls.

LAST SHINGLE IS CUT TO SIZE BY FIRST SLIPPING BEHIND ADJACENT SHINGLE

TACK

1 X 3 STRAIGHTEDGE

SCORE ALONG THIS EDGE AND TRIM, THEN REPLACE AND NAIL

© Cengage Learning 2014

Figure 61–9 Technique for cutting last shingle to the proper size.

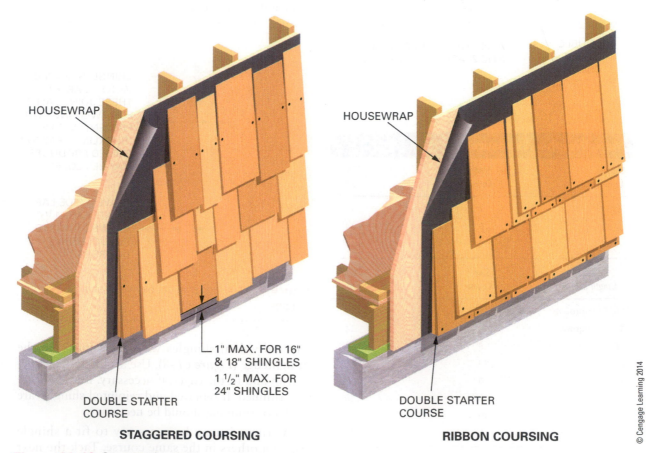

HOUSEWRAP

DOUBLE STARTER COURSE

1" MAX. FOR 16" & 18" SHINGLES
1 1/2" MAX. FOR 24" SHINGLES

STAGGERED COURSING

HOUSEWRAP

DOUBLE STARTER COURSE

RIBBON COURSING

© Cengage Learning 2014

Figure 61–10 Staggered and ribbon courses are alternatives to straight-line courses.

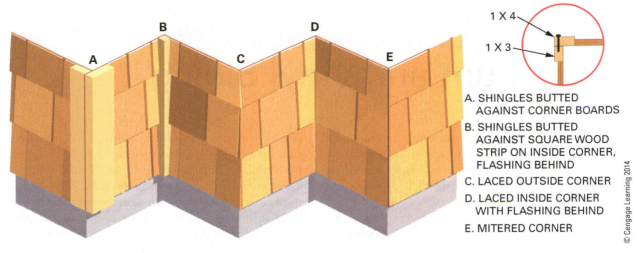

A. SHINGLES BUTTED AGAINST CORNER BOARDS

B. SHINGLES BUTTED AGAINST SQUARE WOOD STRIP ON INSIDE CORNER, FLASHING BEHIND

C. LACED OUTSIDE CORNER

D. LACED INSIDE CORNER WITH FLASHING BEHIND

E. MITERED CORNER

Figure 61–11 Corner details for wood shingle siding.

Fasteners. Each shingle, up to 8 inches wide, is fastened with two nails or staples about $\frac{3}{4}$ inch in from each edge. On shingles wider than 8 inches, drive two additional nails about 1 inch apart near the center. Fasteners should be hot-dipped galvanized, stainless steel, or aluminum. They should be driven about 1 inch above the butt line of the next course. Fasteners must be long enough to penetrate the sheathing by at least $\frac{3}{4}$ inch.✱

✱**IRC**
703.5.3
703.5.3.1

Staggered and Ribbon Coursing. An alternative to straight-line courses are staggered and ribbon coursing. In staggered coursing, the butt lines of the shingles are alternately offset below, but not above, the horizontal line. Maximum offsets are 1 inch for 16- and 18-inch shingles and $1\frac{1}{2}$ inches for 24-inch shingles.

In ribbon coursing, both layers are applied in straight lines. The outer course is raised about 1 inch above the inner course (*Figure 61–10*).

Corners

Shingles may be butted to corner boards like any horizontal wood siding. On outside corners, they may also be applied by alternately overlapping each course in the same manner as in applying a wood shingle ridge. Inside corners may be woven by alternating the corner shingle first on one wall and then the other (*Figure 61–11*).

Double Coursing

When double coursing, the first course is tripled. The outer layer of the course is applied $\frac{1}{2}$ inch lower than the inner layer. For ease in application, use a rabbeted straightedge or one composed of two pieces with offset edges (*Figure 61–12*).

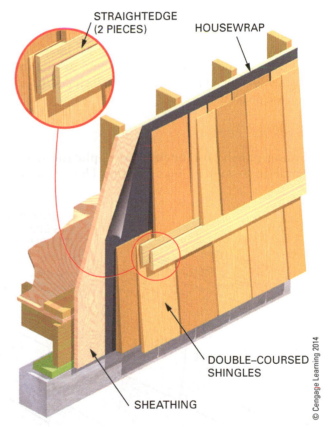

Figure 61–12 Use a straightedge made of two pieces with offset edges for double-coursed application.

Fastening. Each inner-layer shingle is applied with one fastener at the top center. Each outer course shingle is face-nailed with two 5d galvanized box or special 14-gauge shingle nails. The fasteners are driven about 2 inches above the butts, and about $\frac{3}{4}$ inch in from each edge.

CHAPTER 62

Aluminum and Vinyl Siding

Except for the material, aluminum and vinyl siding systems are similar. *Aluminum siding* is finished with a baked-on enamel. In *vinyl siding,* the color is embedded in the material itself. Both kinds are manufactured with interlocking edges, for horizontal and vertical applications. Descriptions and instructions are given here for vinyl siding systems, much of which can be applied to aluminum systems.✽

✽IRC
703.11.1

configurations to simulate one, two, or three courses of bevel or drop siding. Panels designed for vertical application come in 12-inch widths. They are shaped to resemble boards. They can be used in combination with horizontal siding. Vertical siding panels with solid surfaces may also be used for soffits. For ventilation, perforated soffit panels of the same configuration are used (*Figure 62–1*).

SIDING PANELS AND ACCESSORIES

Siding systems are composed of siding panels and several specially shaped moldings. Moldings are used on various parts of the building to trim the installation. In addition, the system includes shapes for use on soffits.

Siding Panels

Siding panels, for horizontal application, are made in 8- and 12-inch widths. They come in

Siding System Accessories

Siding systems require the use of several specially shaped accessories. *Inside* and *outside corner posts* are used to provide a weather-resistant joint to corners. Corner posts are available with channels of appropriate widths to accommodate various configurations of siding.

Some other accessories include **J-channels,** *starter strips,* and **undersill trim,** also known as *finish trim.* J-channels are made with several opening sizes. They are used in a number of places such

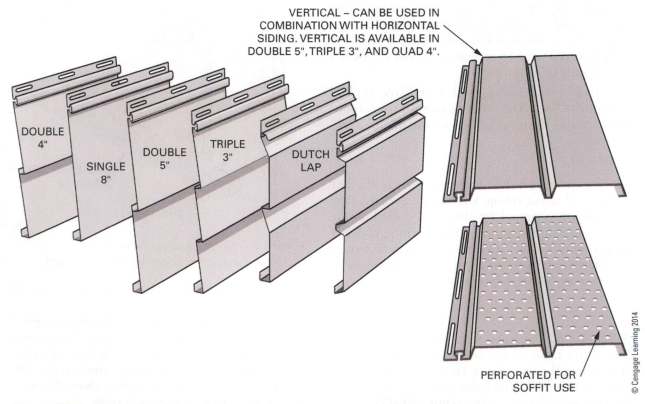

VERTICAL – CAN BE USED IN
COMBINATION WITH HORIZONTAL
SIDING. VERTICAL IS AVAILABLE IN
DOUBLE 5", TRIPLE 3", AND QUAD 4".

DOUBLE 4"

SINGLE 8"

DOUBLE 5"

TRIPLE 3"

DUTCH LAP

PERFORATED FOR SOFFIT USE

© Cengage Learning 2014

Figure 62–1 Commonly used configurations of horizontal and vertical vinyl siding.

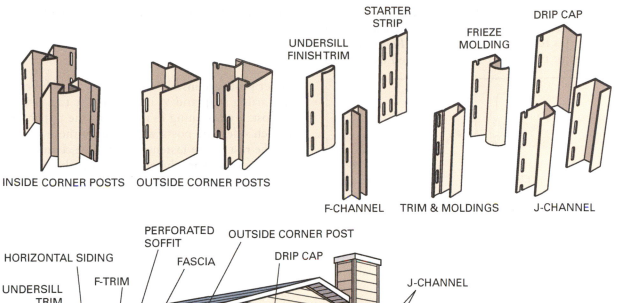

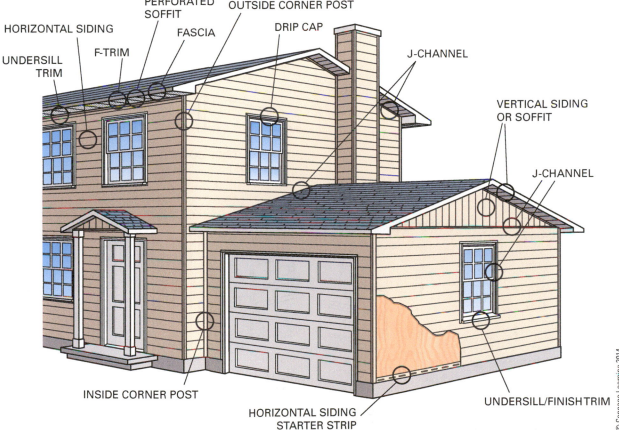

Figure 62–2 Various accessories are used to trim a vinyl siding installation.

as around doors and windows, at transition of materials, against soffits, and in many other places (*Figure 62–2*). The majority of vinyl siding panels and accessories are manufactured in 12′-6″ lengths. Trim accessories and molding are typically 12 feet long.

APPLYING HORIZONTAL SIDING

The siding may expand and contract as much as $\frac{1}{4}$-inch over a 12′-6″ length with changes in temperature. For this reason, it is important to center

fasteners in the slots. Do not drive them too tightly. There should be about $\frac{1}{32}$ inch between the head of the fastener, when driven, and the siding (*Figure 62–3*). After the panel is fastened, it should be easily moved from side to side in the nail slots. Space fasteners 16 inches apart for horizontal siding and 6 to 12 inches apart for accessories unless otherwise specified by the manufacturer.

Applying the Starter Strip

Snap a level line to the height of the starter strip all around the bottom of the building. Fasten the strips to the wall with their top edges to the chalk line. Leave a $\frac{1}{4}$-inch space between them and other accessories to allow for expansion (*Figure 62–4*). Make sure the starter strip is applied as straight as possible. It controls the straightness of entire installation.

Installing Corner Posts

Corner posts are installed in corners $\frac{1}{4}$ inch below the starting strip and $\frac{1}{4}$ inch from the top. Attach the posts by fastening in the top of the upper slot on each side. The posts will hang on these fasteners. The rest of the fasteners should be centered on the slots. Make sure the posts are straight and true from top to bottom (*Figure 62–5*).

Figure 62–3 Fasten siding so as to allow for expansion and contraction.

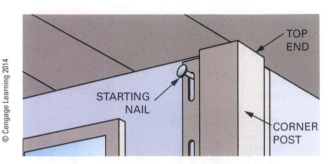

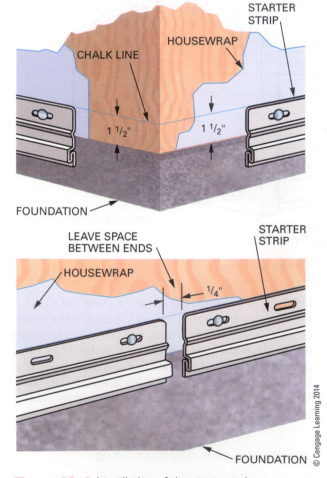

Figure 62–4 Installation of the starter strip.

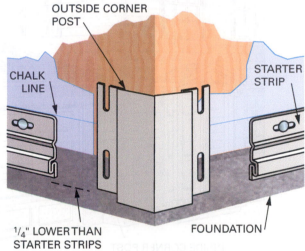

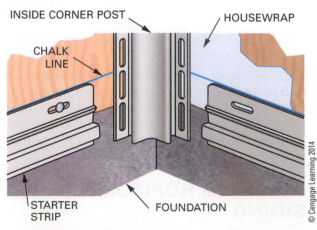

Figure 62–5 Inside and outside corner posts are installed in a similar way.

Installing J-Channel

Install J-channel pieces across the top and along the sides of window and door casings. They may also be installed under windows or doors with the undersill trim nailed inside of the channel. To miter the corners, cut all pieces to extend, on both ends, beyond the casings and sills a distance equal to the width of the channel face. On both ends of the side J-channels, cut a $\frac{3}{4}$-inch notch out of the bottom of the J-channel. Fasten in place. On both ends of the top and bottom channels, make $\frac{3}{4}$-inch cuts. Bend down the tabs and miter the faces. Install them so the mitered faces are in front of the faces of the side channels (*Figure 62–6*).

Installing Siding Panels

Snap the bottom of the first panel into the starter strip. Fasten it to the wall. Start from a back corner, leaving a $\frac{1}{4}$-inch space in the corner post channel. Work toward the front with other panels. Overlap each panel about 1 inch (*Figure 62–7*).

The seams of overlapped panels are more visible from one direction and less so from the other. Lap the panels so they are visible from the direction least traveled. This will put the best side of the siding toward the most often viewed side. If either direction is equally traveled, alternate the direction of the lap on each course. This will reduce the number of seams visible to almost a half.

Also keep vertical seams as far away from each other as possible. Do not let them align vertically.

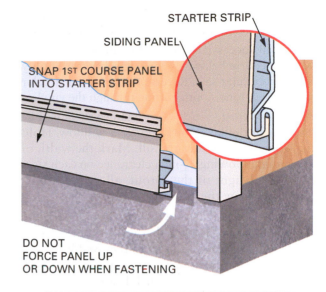

SNAP 1ST COURSE PANEL INTO STARTER STRIP

SIDING PANEL

STARTER STRIP

DO NOT FORCE PANEL UP OR DOWN WHEN FASTENING

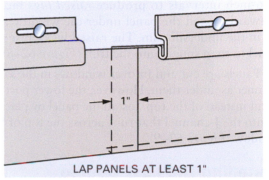

LAP PANELS AT LEAST 1"

© Cengage Learning 2014

Figure 62–7 Installing the first course of horizontal siding panels.

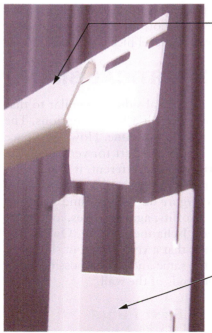

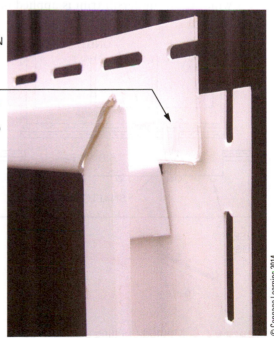

TOP J-CHANNEL WITH CUTS AND FOLDED DOWN TAB. THE FACE SIDE IS ALSO CUT ON A MITER.

TOP J-CHANNEL INSERTED WITH TAB INSIDE SIDE CHANNEL. MITERED FACE OVERLAPS SIDE CHANNEL FACE.

SIDE J-CHANNEL WITH CUTS AND TAB REMOVED.

© Cengage Learning 2014

Figure 62–6 Cutting J-channel to fit around windows and doors so water does not get behind siding.

A random pattern in the seams will be less noticeable to the eye than vertically aligned seams.

Install successive courses by interlocking them with the course below. Use tin snips, hacksaw, utility knife, or circular saw. Reverse the blade if a circular saw is used, for smooth cutting through the vinyl.

Fitting around Windows.   Plan so there will be no joint in the last course under a window. Hold the siding panel under the sill. Mark the width of the cutout, allowing $1/4$-inch clearance on each side. Mark the height of the cutout, allowing $1/4$-inch clearance below the sill.

Make vertical cuts with tin snips. Score the horizontal layout lines with a utility knife or scoring tool. Bend the section to be removed back and forth until it separates. Using a special snaplock punch, punch the panel $1/4$ inch below the cut edge at 6-inch intervals to produce *raised lugs* facing outward. Install the panel under the window and up in the undersill trim. The raised lugs cause the panel to snap snugly into the trim (*Figure 62–8*).

Panels are cut and fit over windows in the same manner as under them. However, the lower portion is cut instead of the top. Install the panel by placing it into the J-channel that runs across the top of the window (*Figure 62–9*).

Installing the Last Course under the Soffit

The last course of siding panels under the soffit is installed in a manner similar to that for fitting under a window. An undersill trim is applied on the

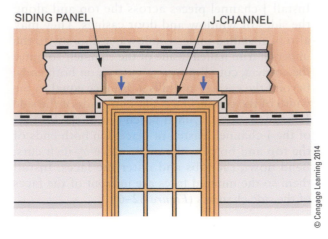

CUT EDGE OF PANEL FITS INTO J-CHANNEL OVER TOP OF WINDOW

SIDING PANEL   J-CHANNEL

© Cengage Learning 2014

Figure 62–9 Fitting a panel over a window.

wall, up against the soffit. Panels in the last course are cut to width. Lugs are punched along the cut edges. The panels are then snapped firmly into place into the undersill trim (*Procedure 62–A*).

Gable End Installation

The rakes of a gable end are first trimmed with J-channels. The panel ends are inserted into the channel with a $1/4$-inch expansion gap. Make a pattern for cutting gable end panels at an angle where they intersect with the rake. Use two scrap pieces of siding to make the pattern (*Figure 62–10*). Interlock one piece with an installed siding panel below. Hold the other piece on top of it and against the rake. Mark along the bottom edge of the slanted piece on the face of the level piece.

APPLYING VERTICAL SIDING

The installation of vertical siding is similar to that for horizontal siding with a few exceptions. The method of fastening is the same. However, space fasteners about 12 inches apart for vertical siding panels. The starter strip is different. It may be $1/2$-inch J-channel or drip cap flush with and fitted into the corner posts (*Figure 62–11*). Around windows and doors, under soffits, against rakes, and other locations, $1/2$-inch J-channel is used. One of the major differences is that a vertical layout should be planned so that the same and widest possible piece is exposed at both ends of the wall.

Installing the First Panel

To install the first panel, start by determining the widest possible width of the first and last panel.

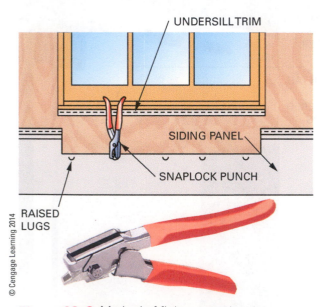

UNDERSILL TRIM

SIDING PANEL

SNAPLOCK PUNCH

RAISED LUGS

© Cengage Learning 2014

Figure 62–8 Method of fitting a panel under a window.

StepbyStep Procedures

Procedure 62–A Fitting the Last Course of Horizontal Siding under the Soffit

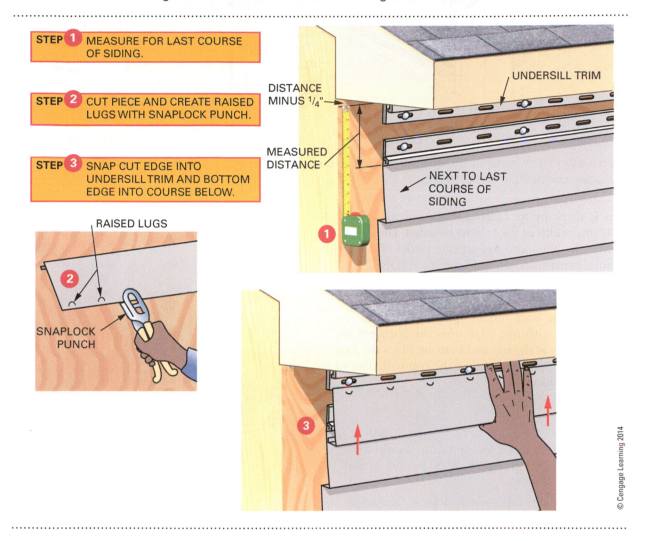

STEP 1 MEASURE FOR LAST COURSE OF SIDING.

STEP 2 CUT PIECE AND CREATE RAISED LUGS WITH SNAPLOCK PUNCH.

STEP 3 SNAP CUT EDGE INTO UNDERSILL TRIM AND BOTTOM EDGE INTO COURSE BELOW.

RAISED LUGS

SNAPLOCK PUNCH

DISTANCE MINUS ¼"

MEASURED DISTANCE

UNDERSILL TRIM

NEXT TO LAST COURSE OF SIDING

© Cengage Learning 2014

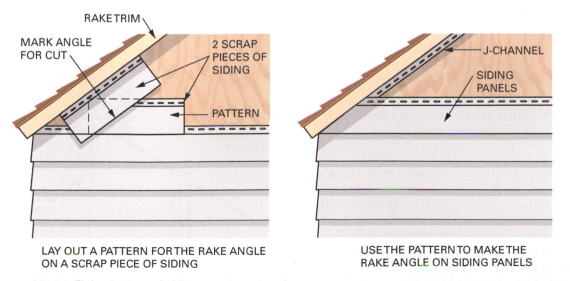

RAKE TRIM

MARK ANGLE FOR CUT

2 SCRAP PIECES OF SIDING

PATTERN

J-CHANNEL

SIDING PANELS

LAY OUT A PATTERN FOR THE RAKE ANGLE ON A SCRAP PIECE OF SIDING

USE THE PATTERN TO MAKE THE RAKE ANGLE ON SIDING PANELS

© Cengage Learning 2014

Figure 62–10 Fitting horizontal siding panels to the rakes.

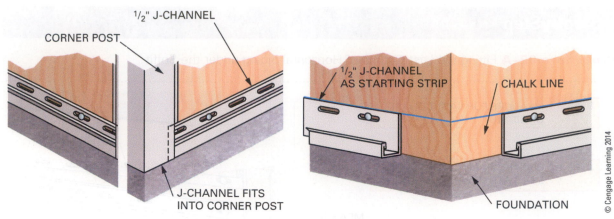

Figure 62–11 The starter strip shape and its intersection with corner posts is different for vertical application of vinyl siding compared to horizontal.

This is done by measuring, between the corner posts, the width of the face to be sided. Divide this number by the exposure of one panel. Take the decimal remainder and add one to it. Divide this number by two. This will be the size of the first and last panel (*Figure 62–12*).

Install the first vertical panel plumb on the starter strip with one edge into the corner post. Allow $\frac{1}{4}$ inch at the top and bottom. Place the first nails in the uppermost end of the top nail slots to hold it in position.

EXAMPLE

What are the starting and finishing widths for a wall section that measures 18'-9" for siding that is 12 inches wide?

Convert this measurement to a decimal by first dividing the inches portion by 12 and then adding it to the feet to get 18.75 feet.

Divide this by the siding exposure, in feet: $18.75 \div 1$ foot = 18.75 pieces.

Subtract the decimal portion along with one full piece giving 1.75 pieces. Next $1.75 \div 2 = 0.875$, multiplied by 12 giving $10\frac{1}{2}$ inches.

This is the size of the starting and finishing piece. Thus there are seventeen full-width pieces and two $10\frac{1}{2}$-inch-wide pieces.

The edge of the panel may need to be cut in order for equal widths to be exposed on both ends of the wall. If the panel is cut on the flat surface, place a piece of undersill trim backed by furring into the channel of the corner post. Punch lugs along the cut edge of the panel at 6-inch intervals. Snap the panel into the undersill trim. Edges of vertical panels cut to fit in J-channels around windows and doors are treated in the same way (*Figure 62–13*).

Figure 62–12 Example for finding first and last panel width of vertical siding.

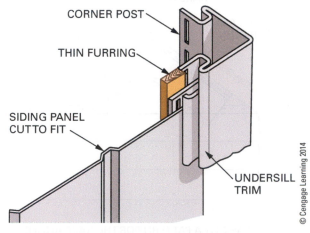

Figure 62–13 Undersill trim and furring are required when vertical siding is cut to fit into corner posts and J-channels.

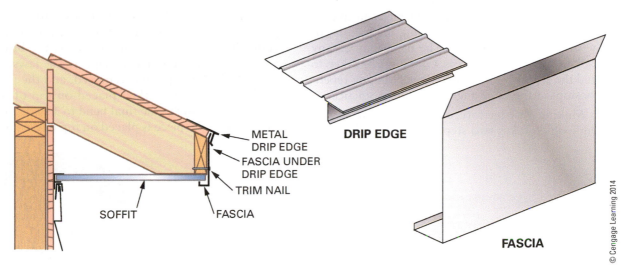

Figure 62–14 Metal fascia material is fitted under metal drip edge.

ALUMINUM ACCESSORIES

Fascia is completed by covering the wood subfascia with an L-shaped piece of siding. This piece may be made of either metal or vinyl. The top edge is installed by slipping it under the metal drip edge. The bottom is held in place with nails (*Figure 62–14*). Nails are aluminum or stainless steel painted to the same color as the trim.

Aluminum trim pieces are often fabricated on the job. This is done by using a tool, called a brake, to bend sheet metal (*Figure 62–15*). Light-duty brakes are designed for aluminum only, whereas others will bend heavy sheet metal.

Aluminum stock is sold in 50-foot-long rolls of various widths ranging from 12 to 24 inches. These rolls are referred to as coil stock. Each side of the sheet is colored with a baked-on enamel finish. One side is usually white and the other one of a variety of colors produced by the manufacturer.

Coil stock can be cut with a utility knife, but using a straightedge makes the cut look professional. Stock is cut with one score of the knife. The cut is then completed by bending the piece back and forth through the cut.

Stock can also be cut with a cutter designed to work with the brake (*Figure 62–16*). A coil handler makes unrolling the coil easier, and the jaws provide the rail for the cutter to ride. Coil stock is ripped to the desired width using the same cutter and the jaws of the brake. Brake jaws are clamped and unlocked using a clamping lever.

Once the pieces are cut to width, they are then bent to the desired configuration. Care must be taken to visualize the piece as it is bent. Mistakes

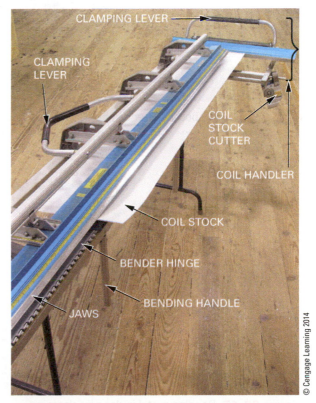

Figure 62–15 An aluminum brake is used to shape coil stock.

are easy to make. For example, the piece may have the correct shape but the wrong color facing outward.

Bending the stock begins with careful positioning of the piece in the jaws. The same amount of stock should be revealed from the jaws on both ends. This will ensure that the piece is not tapered.

Locking the jaws tight secures the piece during the bend. The bend is made by raising the handles of the brake (*Figure 62–17*). The bend may be any angle from 0 to 180 degrees. Care must be taken to bend the stock to the desired angle. Making repetitive stops at the same angle takes some practice.

Making an edge return uses a full 180-degree bend, which takes two steps. First, align the piece to as small a bend as possible, usually about $5/8$ inch. The piece should be flush with the brake bender. Then lock and bend as far as the brake will allow. This is about 150 degrees. The jaws are then unlocked, the piece is removed, and the jaws relocked. The piece is then placed on top of the jaws (*Figure 62–18*). The final bend to 180 degrees is made while the piece is placed on top of the jaws.

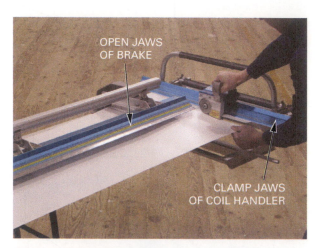

Figure 62–16 A special cutter can be used to cut stock to length and width.

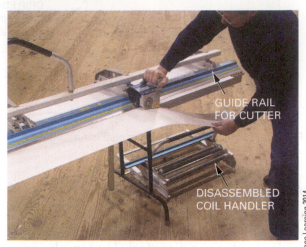

NOTE: CUTTER MUST BE KEPT TIGHT TO GUIDE RAIL AS CUT IS MADE.

© Cengage Learning 2014

Figure 62–17 Bending metal is done by clamping jaws and lifting handles.

© Cengage Learning 2014

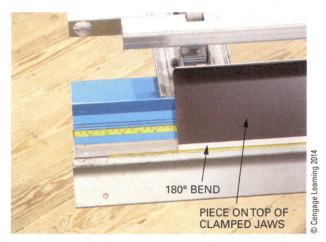

Figure 62–18 Full 180-degree bends are completed on top of the jaws.

© Cengage Learning 2014

CHAPTER 63
Cornice Terms and Design

Cornice terms from earlier times still remain in use. However, cornice design has changed considerably. In earlier times, cornices were very elaborate and required much time to build. Now, cornice design, in most cases, is much more simplified. Only occasionally is a building designed with an ornate cornice similar to those built in years gone by.

PARTS OF A CORNICE

Several finish parts are used to build the cornice. In some cases, additional framing members are required.

Subfascia

The subfascia is sometimes called the false fascia or rough fascia (*Figure 63–1*). It is a horizontal framing member fastened to the rafter tails. It provides an even, solid, and continuous surface for the attachment of other cornice members. When used, the subfascia is generally a nominal 1- or 2-inch-thick piece. Its width depends on the slope of the roof, the tail cut of the rafters, or the type of cornice construction.

Soffit

The finished member on the underside of the cornice is called a plancier and is often referred to as a soffit (*Figure 63–2*). Soffit material may include solid lumber, plywood, strand board, fiberboard, or corrugated aluminum and vinyl panels. Soffits may be perforated or constructed with screen openings to allow for ventilation of the rafter cavities. Soffits may be fastened to the bottom edge of the rafter tails to the slope of the roof. The soffit is an ideal location for the placement of attic ventilation.

Fascia

The fascia is fastened to the subfascia or to the ends of the rafter tails. It may be a piece of lumber grooved to receive the soffit. It also may be made from bent aluminum and vinyl material used to wrap the subfascia. Fascia provides a surface for the attachment of a gutter. The fascia may be built up from one or more members to enhance the beauty of the cornice.

The bottom edge of the fascia usually extends below the soffit by $\frac{1}{4}$ to $\frac{3}{8}$ inch. The portion of

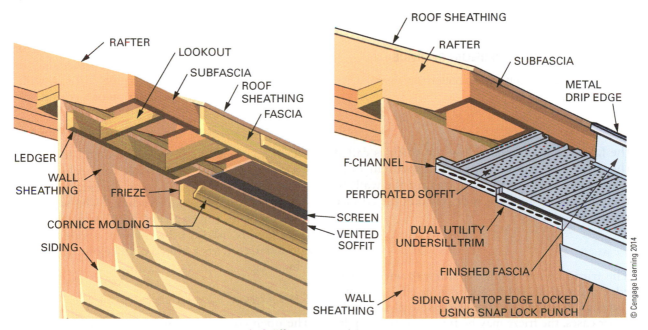

Figure 63–1 Cornices may be constructed of different materials.

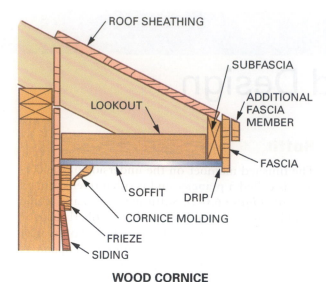

WOOD CORNICE

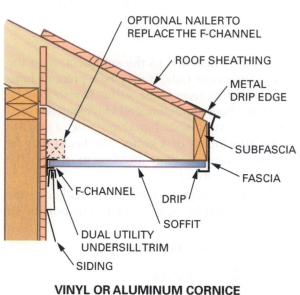

VINYL OR ALUMINUM CORNICE

Figure 63–2 The soffit is the bottom finish member of the cornice.

© Cengage Learning 2014

the fascia that extends below the soffit is called the drip. The drip is necessary to prevent rainwater from being swept back against the walls of the building. In addition, a drip makes the cornice more attractive.

Frieze

The frieze is fastened to the sidewall with its top edge against the soffit. Its bottom edge is sometimes rabbeted to receive the sidewall finish. In other cases, the frieze may be furred away from the sidewall to allow the siding to extend above and behind its bottom edge. However, the frieze is not always used. The sidewall finish may be allowed to come up to the soffit. The joint between the siding and the soffit is then covered by a molding.

Cornice Molding

The cornice molding is used to cover the joint between the frieze and the soffit. If the frieze is not used, the cornice molding covers the joint between the siding and the soffit.

Lookouts

Lookouts are framing members, usually 2 × 4 stock, that are used to provide a fastening surface for the soffit. They run horizontally from the end of the rafter to the wall, adding extra strength to larger overhangs. Lookouts may be installed at every rafter or spaced 48 inches on center (OC), depending on the material being used for the soffit (*Figure 63–2*).

CORNICE DESIGN

Cornices are generally classified into three main types: box, snub, and open (*Figure 63–3*).

The Box Cornice

The box cornice is probably most common. It gives a finished appearance to this section of the exterior. Because of its overhang, it helps protect the sidewalls from the weather. It also provides shade for windows.

Box cornices may be designated as narrow or wide. They may be constructed with level or sloping soffits. A *narrow box cornice* is one in which the cuts on the rafters serve as nailing surfaces for the cornice members. A *wide box cornice* may be constructed with a level or sloping soffit. A wide, level soffit requires the installation of lookouts.

The Snub Cornice

The snub cornice is the finished detail for a roof with no overhang. Sometimes for architectural appeal or economic concerns the rafters are installed with little or no overhang. The snub cornice completes this exterior finish. A snub cornice is frequently used on the rakes of a gable end in combination with a boxed cornice on the sides of a building.

The Open Cornice

The open cornice has an overhang but no soffit. It is used when it is desirable to expose the rafter tails. It is often used when the rafters are large,

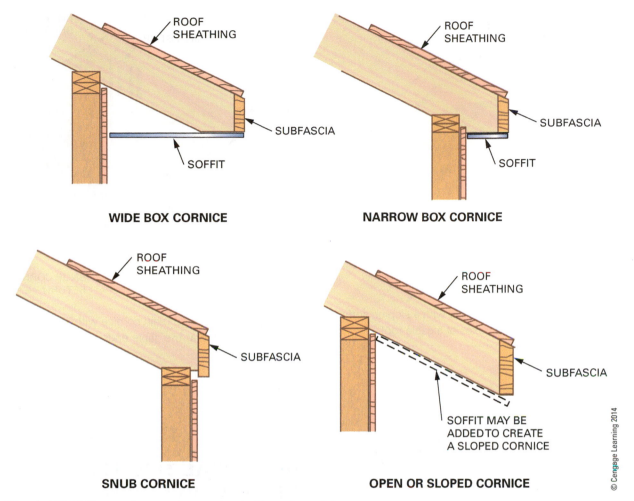

ROOF SHEATHING

SUBFASCIA

SOFFIT

WIDE BOX CORNICE

ROOF SHEATHING

SUBFASCIA

SOFFIT

NARROW BOX CORNICE

ROOF SHEATHING

SUBFASCIA

SNUB CORNICE

ROOF SHEATHING

SUBFASCIA

SOFFIT MAY BE ADDED TO CREATE A SLOPED CORNICE

OPEN OR SLOPED CORNICE

© Cengage Learning 2014

Figure 63–3 The cornice may be constructed in various styles.

laminated or solid beams with a wide overhang that exposes the roof decking on the bottom side. Open cornices give a contemporary or rustic design look to post-and-beam framing. They provide protection to sidewalls at low cost. This cornice might also be used for conventionally framed buildings for reasons of design and to reduce costs.

By adding a soffit, a *sloped cornice* is created. The soffit is installed directly to the underside of the rafter tails. This is sometimes done to simplify the cornice detail when there is also an overhang over the gable end of the building.

Rake Cornices

The main cornice is constructed on the rafter tails where they meet the walls of a building. On buildings with hip or mansard roofs, the main cornice, regardless of the type, extends around the entire building.

On buildings with gable roofs, a boxed main cornice with a sloping soffit, attached to the bottom edge of the rafter tails, may be returned up the rakes to the ridge. The cornice that runs up the rakes is called a rake cornice (*Figure 63–4*).

Cornice Returns. A main cornice with a horizontal soffit attached to level lookouts may, at times, be terminated at each end wall against a snub rake cornice (*Figure 63–5*). At other times, a cornice return must be constructed to change the level box cornice to the angle of the roof.

A main cornice with a level soffit may also be returned upon itself. That is, the main cornice is mitered at each end. It is turned 90 degrees toward and beyond the corner as much as it overhangs on the side of the building. This short section on each end of a gable roof is called a cornice return. The cornice return provides a stop for the rake cornice. It often adds to the architectural design and statement of the building (*Figure 63–6*).

Figure 63–4 A boxed cornice with a sloping soffit may be returned up the rakes of a gable roof.

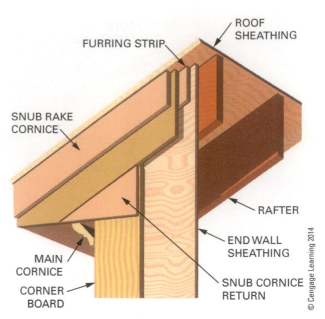

Figure 63–5 A boxed main cornice with a level soffit may be terminated at the gable ends against the rake cornice.

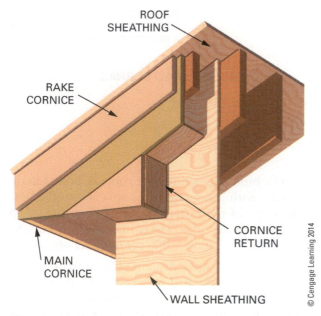

Figure 63–6 A boxed main cornice with a level soffit may be changed to the angle of the roof with a cornice return.

PRACTICAL TIPS FOR INSTALLING CORNICES

Install cornice members in a straight and true line with well-fitting and tight joints. Do not dismiss the use of hand tools for cutting, fitting, and fastening exterior trim.

All fasteners should be noncorrosive. Stainless steel fasteners offer the best protection against corrosion. Hot-dipped galvanized fasteners provide better protection than plated ones. Fasteners used in wood should be well set. They should be puttied to conceal the fastener. This also prevents the heads from corroding. Setting and concealing fasteners in exterior trim is a mark of quality.

Paint or otherwise seal and protect all exterior trim as soon as possible after installation. Properly installed and protected wood exterior trim will last indefinitely.

CHAPTER 64
Gutters and Downspouts

A gutter is a shallow trough or conduit set below the edge of the roof along the fascia. It catches and carries off rainwater from the roof. A downspout, also called a conductor, is a rectangular or round pipe. It carries water from the gutter downward and away from the foundation (*Figure 64–1*).

Green Tip

Removing rainwater away from the building in gutters and downspouts keeps the building drier. Drier buildings have lower energy needs and last longer.

GUTTERS

Gutters or *eaves troughs* may be made of wood, galvanized iron, aluminum, copper, or vinyl. Copper gutters require no finishing. Vinyl and aluminum gutters are prefinished and ready to install. Wood and galvanized metal gutters need several protective coats of finish after installation.

The size of a gutter is determined by area of the roof for which it handles the water runoff. Under ordinary conditions, 1 square inch of gutter cross section is required for every 100 square feet of roof area. For instance, a 4 × 5 inch gutter has a cross-section area of 20 square inches. It is, therefore, capable of handling the runoff from 2,000 square feet of roof surface.

Wood Gutters

Wood gutters usually are made of Douglas fir, western red cedar, or California redwood. They come in sizes of 3 × 4, 4 × 5, 4 × 6, and 5 × 7, and lengths of over 30 feet. Although wood gutters are not used as extensively as in the past, they enhance the cornice design. When properly installed and maintained, they will last as long as the building itself.

Metal Gutters

Metal gutters are made in rectangular, beveled, ogee, or semicircular shapes (*Figure 64–2*). They

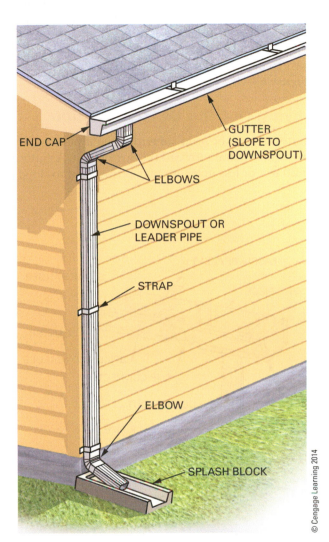

Figure 64–1 Gutters and downspouts form an important system for conducting water away from the building.

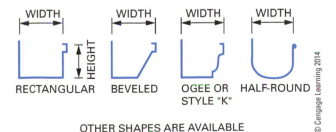

Figure 64–2 Metal gutters are available in several shapes.

come in a variety of sizes, from $2\frac{1}{2}$ inches to 6 inches in height and from 3 to 8 inches in width. Stock lengths run from 10 to 40 feet. Forming machines are often brought to the job site to form aluminum into gutters of practically any desired length.

Besides straight lengths, metal gutter systems have components comprised of *inside* and *outside corners, joint connectors, outlets, end caps,* and others. *Metal brackets* or *spikes and ferrules* are used to support the gutter sections on the fascia (*Figure 64–3*).

Laying Out the Gutter Position

Gutters should be installed with a slight pitch to allow water to drain toward the downspout. A gutter of 20 feet or less in length may be installed to slope in one direction with a downspout on one end. On longer buildings, the gutter is usually crowned in the center. This allows water to drain to both ends.

On both ends of the fascia, mark the location of the bottom side of the gutter. The top outside edge of the gutter should be in relation to a straight line projected from the top surface of the roof. The height of the gutter depends on the pitch of the roof (*Figure 64–4*).

Stretch a chalk line between the two marks. Move the center of the chalk line up enough to give the gutter the proper pitch. Hold the line down in the center. Snap it on both sides of the center. For a slope in one direction only, snap a straight line lower on one end than the other to obtain the proper pitch. It is important to install gutters on the chalk line. This avoids any dips that may prevent complete draining of the gutter.

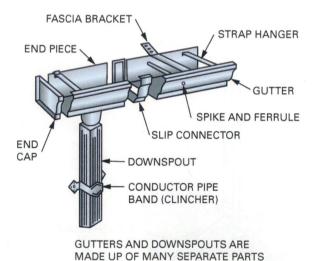

GUTTERS AND DOWNSPOUTS ARE MADE UP OF MANY SEPARATE PARTS

PIECE NEEDED	DESCRIPTION
	GUTTER COMES IN VARIOUS LENGTHS
	SLIP JOINT CONNECTOR USED TO CONNECT JOINTS OF GUTTER
	END CAPS – WITH OUTLET USED AT ENDS OF GUTTER RUNS
	END PIECE – WITH OUTLET USED WHERE DOWNSPOUT CONNECTS
	OUTSIDE MITER USED FOR OUTSIDE TURN IN GUTTER
	INSIDE MITER USED FOR INSIDE TURN IN GUTTER

PIECE NEEDED	DESCRIPTION	PIECE NEEDED	DESCRIPTION
	FASCIA BRACKET USED TO HOLD GUTTER TO FASCIA ON WALL		ELBOW – STYLE B FOR DIVERTING DOWNSPOUT TO LEFT OR RIGHT
	STRAP HANGER CONNECTS TO EAVE OF ROOF TO HOLD GUTTER		CONNECTOR PIPE BAND OR CLINCHER USED TO HOLD DOWNSPOUT TO LEFT OR RIGHT
	STRAINER CAP SLIPS OVER OUTLET IN END PIECE AS A STRAINER		SHOE USED TO LEAD WATER TO SPLASHER BLOCK
	DOWNSPOUT COMES IN 10' LENGTHS		MASTIC USED TO SEAL ALUMINUM GUTTERS AT JOINTS
	ELBOW – STYLE A FOR DIVERTING DOWNSPOUT IN OR OUT FROM WALL		SPIKE AND FERRULE USED TO HOLD GUTTER TO EAVES OF ROOF

Figure 64–3 Components of a metal gutter system.

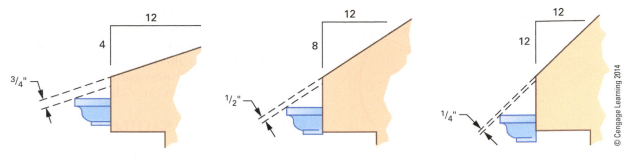

Figure 64-4 The height of the gutter on the fascia is in relation to the slope of the roof.

INSTALLING METAL AND VINYL GUTTERS

Fasten the gutter brackets to the chalk line on the fascia with screws. All screws should be made of stainless steel or other material that is corrosion resistant. Aluminum brackets may be spaced up to 30 inches on center (OC). Steel brackets may be spaced up to 48 inches OC. Install the gutter sections in the brackets. Use slip-joint connectors to join the sections. Apply the recommended gutter sealant to connectors before joining.

Locate the outlet tubes as required, keeping in mind that the downspout should be positioned plumb with the building corner and square with the building. Join with a connector. Add the end cap. Use either inside or outside corners where gutters make a turn.

Vinyl gutters and components are installed in a manner similar to metal ones.

INSTALLING DOWNSPOUTS

Metal or vinyl downspouts are fastened to the wall of the building in specified locations with aluminum straps. Downspouts should be fastened at the top and bottom and every 6 feet in between for long lengths.

The connection between the downspout and the gutter is made with 45-degree elbows and short straight lengths of downspout (*Figure 64–5*). The connection will depend on the offset of the gutter from the downspout. Because water runs downhill, care should be taken when putting the downspout pieces together. The downspout components are assembled where the upper piece is inserted into the lower one (*Figure 64–6*). This makes the joint lap such that the water cannot escape until it reaches the bottom-most piece. An elbow, called a *shoe*, should be used with a splash block at the bottom of the downspout. This leads water away from the

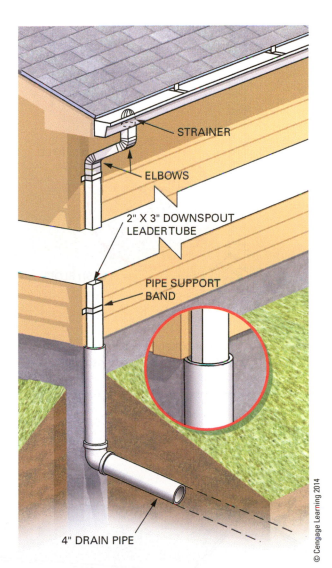

Figure 64-5 Typical downspout leader tubes fastened in place with support bands.

foundation. An alternate method is to connect the downspout with underground piping that carries the water away from the foundation. The piping can be connected to storm drains and drywells or piped to the surface elsewhere. Storm water as

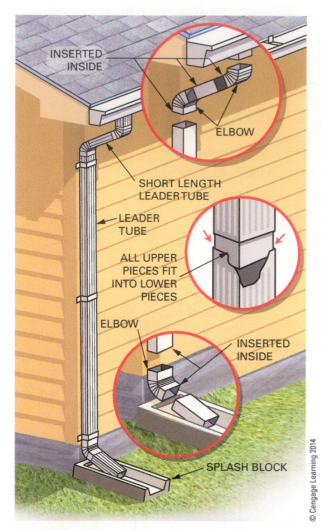

Figure 64–6 Upper gutter components are inserted into lower ones to ensure that the downspout does not leak.

found in gutter downspouts should never be connected to footing or foundation drains. Strainer caps should be placed over gutter outlets if water is conducted by this alternate method. Leaves and other debris that fall into gutters flow into the drainage system and can cause clogging problems.

ESTIMATING SIDING

Estimating the amount of material needed for some siding is similar to roofing as it is sold by the square, or 100 square feet. Others are sold by the lineal foot, and still others are sold by the board foot. All estimates begin with the building area to be covered by siding. Remember that if the number of pieces does not come out even, round it up to the next whole number of pieces.

Building Area

Building area calculations are done by adding together the smaller areas of the building. First, calculate the wall area by multiplying its length by its height. The length may be the perimeter of the building excluding the gable areas. Then, the triangular gable areas are calculated using the formula $A = \frac{1}{2}$ base $\times$ height. The base is the width of the building. This triangular area is doubled to add the second gable. Add all the square foot areas together that will be sided, such as dormers, bays, and porches.

Windows and doors will not be covered, so their total surface areas should be deducted. Multiply the width $\times$ height $\times$ number of windows that size. Subtract the total results from the total area to be covered by siding. (*Figure 64–7*).

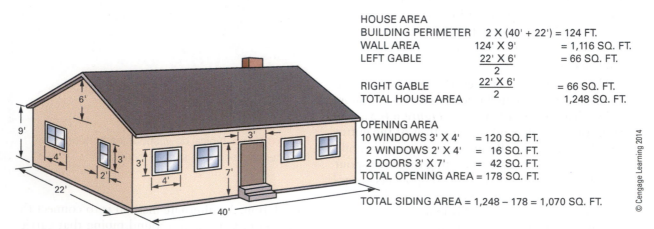

Figure 64–7 Estimating the area to be covered by siding.

Wood Siding

The amount of siding to order depends on the kind of siding used. Some siding is sold by the lineal foot and some by the board foot. Either way it comes, an allowance must be added to compensate for the overlap of each piece. The more pieces, the more overlap in a square foot. ***Figure 64–8*** shows *conversion factor* numbers used to convert square foot (SF) area to be covered to lineal or board feet. Multiply SF by these factors to convert to lineal feet or board feet. For example, if 1,177 square feet are to be covered with 6-inch Drop T&G siding, then 1,177 × 2.34 = 2,754.1 ⇒ 2,755 lineal feet is needed. Or 1,177 × 1.17 = 1,377.1 ⇒ 1,378 board feet. Add 10 percent for waste.

Wood Shingle Siding

The number of squares of shingles needed to cover a certain area depends on how much they are exposed to the weather. Since a 16-inch shingle could have an exposure from $3^1/_2$ inches to $9^1/_2$ inches, the area that one bundle covers will vary. Wood shingles are packaged so four bundles of 16-inch shingles exposed 5 inches will cover 100 square feet, one square. If it is desired to expose the shingle more, then the number of bundles needed will go down.

Conversion numbers shown in (***Figure 64–9***) are used to adjust squares needed as they relate to desired exposure. Divide the area to be covered by the conversion number, and the result is squares of shingles. For example, if 1,177 square feet is to be covered with 16-inch shingles exposed to $5^1/_2$ inches, then 1,177 ÷ 100 = 11.77 squares. But if the same 16-inch shingles will have a 7-inch exposure, then 1,177 SF ÷ 140 = 8.4 squares are needed. Add 10 percent for waste.

Aluminum Vinyl and Siding

Aluminum and vinyl siding panels are sold by the square. Exposure and coverage is predetermined because the exposure is not adjustable. Determine the wall area to be covered. Divide by 100. Add 10 percent of the area for waste. For example, an 1,177-square-foot building needs 1,177 ÷ 100 × 110% = 12.9 ⇒ 13 squares.

Pattern	Nominal Width	Width: Dressed	Width: Exposed Face	Factor for Converting SF to Lineal Feet	Factor for Converting SF to Board Feet
Bevel & Bungalow	4	3½	2½	4.8	1.60
	6	5½	4½	2.67	1.33
	8	7¼	6¼	1.92	1.28
	10	9¼	8¼	1.45	1.21
Dolly Varden	4	3½	3	4.0	1.33
	6	5½	5	2.4	1.2
	8	7¼	6¾	1.78	1.19
	10	9¼	8¾	1.37	1.14
	12	11¼	10¾	1.12	1.12
Drop T&G & Channel Rustic	4	3⅜	3⅛	3.84	1.28
	6	5⅜	5⅛	2.34	1.17
	8	7⅛	6⅞	1.75	1.16
	10	9⅛	8⅞	1.35	1.13
Log Cabin	6	$5\frac{7}{16}$	$4\frac{15}{16}$	2.43	2.43
	8	7⅛	6⅝	1.81	2.42
	10	9⅛	8⅝	1.39	2.32
Boards	2	1½	The exposed face width will vary depending on size selected and on how the boards-and-battens or boards-on-boards are applied.		
	4	3½			
	6	5½			
	8	7¼			
	10	9¼			

Coverage Estimator

Courtesy of Western Wood Products Association

Figure 64–8 Estimating information for converting area to dimensions required for material purchase.

The accessories are typically sold in 12-feet-long pieces. Take the lineal feet needed for each type of accessory, and divide by the length per piece. For example, if ten windows with a perimeter of 14 feet need J-channels that are 12 feet long, then $10 \times 14 \div 12 = 11.6 \Rightarrow 12$ pieces.

See *Figure 64–10* for a listing of the shorthand versions of these estimating techniques with another example.

	Approximate Coverage of One Square (4-bundle roof-pack) of Shingles at Indicated Weather Exposures:												
Length	3½"	4"	4½"	5"	5½"	6"	6½"	7"	7½"	8"	8½"	9"	9½"
16"	70	80	90	100	110	120	130	140	150	160	170	180	190
18"	—	72½	81½	90½	100	109	118	127	136	145½	154½	163½	172½
24"	—	—	—	—	73½	80	86½	93	100	106½	113	120	126½

Length	10"	10½"	11"	11½"	12"	12½"	13"	13½"	14"	14½"	15"	15½"	16"
16"	200	210	220	230	240	—	—	—	—	—	—	—	—
18"	181½	191	200	209	218	227	236	245½	254½	—	—	—	—
24"	133	140	146½	153	160	166½	173	180	186½	193	200	206½	213

© Cengage Learning 2014

Figure 64–9 Roof shingle coverage at various exposures.

Estimate the siding needed to cover a rectangular building the measures 30'×56' with 9' high walls, 9' gable height, three 3949 windows, four 4959 windows and two 3'×7' doors. Consider the lineal feet of 8" bevel siding, squares of 18 inch wood shingles exposed to 7 ½", and number of squares of vinyl siding.

Item	Formula	Waste factor	Example
Building Area	PERM × wall HGT + 2 × gable area − opening area = build area		172 × 9 + (2 × ½ × 30 × 9) − (3 × 3 × 4 + 4 × 4 × 5 + 2 × 3 ×7) = 1548 − 158 = 1390 SF
8" Bevel Siding	building area × conversion factor (Figure 64–8) × waste = lineal ft	10%	1390 × 1.92 × 110% = 2935.6 ⇒ 2936 lineal feet
18" Wood Shingles exposed to 7½"	building area ÷ conversion factor (Figure 64–9) × waste = squares of siding	10%	1390 ÷ 136 × 110% = 11.2 ⇒ 12 squares of siding
Vinyl Siding	building area ÷ 100 × waste = squares of siding	10%	1390 ÷ 100 × 110% = 15.2 ⇒ 16 squares of siding

© Cengage Learning 2014

Figure 64–10 Example of estimating siding materials with formulas.

DECONSTRUCT THIS

Carefully study *Figure 64–11* and think about what is wrong and/or what is right. Consider all possibilities. What construction practice or method is different in your area of the country?

Figure 64–11 This photo shows installed wood siding.

© Cengage Learning 2014

KEY TERMS

battens

blind nail

board and batten

board on board

box cornice

brake

coil stock

conductor

corner board

cornice

cornice molding

cornice return

double coursing

downspout

drip

face nail

fancy butt

frieze

furring strip

gutter

J-channel

open cornice

panel siding

plancier

preacher

rake

rake cornice

rake frieze

return

ribbon coursing

single coursing

snaplock punch

snub cornice

staggered
coursing

starter strip

striated

subfascia

undercourse

undersill trim

vertical grain

water table

REVIEW QUESTIONS

Select the most appropriate answer.

1. Bevel siding is applied

    a. horizontally.
    b. vertically.
    c. horizontally or vertically.
    d. horizontally, vertically, or diagonally.

2. Fiber-cement siding products are

    a. made to look like concrete blocks.
    b. installed similar to wood siding.
    c. usually cut to size indoors.
    d. only installed horizontally.

3. A particular advantage of bevel siding over other types of horizontal siding is that

    a. it comes in a variety of prefinished colors.
    b. it has a constant weather exposure.
    c. the weather exposure can be varied slightly.
    d. application can be made in any direction.

4. When applying horizontal siding, it is desirable to

    a. maintain exactly the same exposure with every course.
    b. apply full courses above and below windows.
    c. work from the top down.
    d. use a water table.

5. In order to lay out a story pole for courses of horizontal siding, which of the following must be known?

    a. the width of windows and doors
    b. the kind and size of finish at the eaves and foundation

    c. the location of windows, doors, and other openings
    d. the length of the wall to which siding is to be applied

6. With wood shingle siding

    a. the exposure may be varied.
    b. butt seams should line up vertically.
    c. each piece should be fastened with four nails.
    d. all of the above.

7. When installing aluminum or vinyl siding, drive nails

    a. tightly against the flange.
    b. loosely against the flange.
    c. with small heads.
    d. colored the same as the siding.

8. Vinyl siding is installed with

    a. vertical butt seams randomly placed.
    b. butt seam overlapping away from view.
    c. a loose fit to trim pieces.
    d. all of the above.

9. To allow for expansion when installing solid vinyl starter strips, leave a space between the ends of at least

    a. $\frac{1}{8}$ inch.
    b. $\frac{1}{4}$ inch.
    c. $\frac{3}{8}$ inch.
    d. $\frac{1}{2}$ inch.

10. The exterior trim that extends up the slope of the roof on a gable end is called the

    a. box finish.
    b. rake finish.
    c. return finish.
    d. snub finish.

11. A member of the cornice fastened with a vertical surface to the rafter tails is called the

    a. drip.
    b. fascia.
    c. soffit.
    d. frieze.

12. A soffit is the part of a cornice that

    a. may be horizontal under the rafter tails.
    b. is vertical and attached to the ends of rafters.
    c. serves as a drip cap.
    d. is often made of 2 × 4s.

13. Care should be taken when installing gutters so that

    a. they are level with the fascia.
    b. downspouts are in the center.
    c. downspout leader tubes are connected to the foundation drains.
    d. parts are installed with the idea that water runs downhill.

14. The square feet (SF) area of siding needed for a rectangular house with a hip roof measuring 30 × 40 feet where the wall height is 8 feet is

    a. unable to be determined because roof height is not known.
    b. 78 SF minus window and door areas.
    c. 1,120 SF minus window and door areas.
    d. 9,600 SF minus window and door areas.

UNIT 23
Decks, Porches, and Fences

©Shutterstock, Inc./Mau Horng

Among the final steps in finishing the exterior is the building of porches, decks, fences, and other accessory structures. Plans may not always show specific construction details. Therefore, it is important to know some of the techniques used to build these structures.

Accents to a building may come in the form of decks and porches. They serve as reminders to the overall workmanship of the house. Take care to install material in a neat and professional manner.

OBJECTIVES

After completing this unit, the student should be able to:

- describe the construction of and kinds of materials used in decks and porches.
- lay out and construct footings for decks and porches.
- install supporting posts, girders, and joists.
- apply decking in the recommended manner and install flashing, for an exposed deck, against a wall.
- construct deck stairs and railings.
- describe several basic fence styles.
- design and build a straight and sturdy fence.

CHAPTER 65
Deck and Porch Construction

Wood porches and decks are built to provide outdoor living areas for various reasons, in both residential and commercial construction (*Figure 65–1*). The construction of both is similar. However, a porch is covered by a roof. Its walls may be enclosed with wire mesh screens for protection against insects. With screen and storm window combinations, glass replaces screens to keep the porch comfortably warm in the cold months.

DECK MATERIALS

Decking materials must be chosen for strength and durability, as well as appearance and resistance to decay. Redwood, cedar, and pressure-treated southern yellow pine are often used as decking boards. Other decking materials available include Timber Tech® and Trex®. These decking products are made from a mixture of plastic and sawdust. They are cut, fit, and fastened in the same manner as wood and have the added benefit of being made mostly from recycled material.✱

✱IRC
507.3.1

If not specified, the kind, grade, and sizes of material must be selected before building a deck. Also, the size and kind of fasteners, connectors, anchors, and other hardware must be determined.

Figure 65–1 Wood decks are built in many styles. This multilevel deck blends well with the landscape.

Courtesy of California Redwood Association

Lumber

All lumber used to build decks should be pressure-treated with preservatives or be from a decay-resistant species, such as redwood or cedar.✱ Remember, it is the heartwood of these species that is resistant to decay, not the sapwood. Either *all-heart* or pressure-treated lumber should be used wherever there is a potential for decay. This is essential for posts that are close to the ground and other parts subject to constant moisture. (A description of pressure-treated lumber and its uses can be found in Chapter 33.)

✱IRC
317.1.2

Lumber Grades. For pressure-treated southern pine and western cedar, #2 grade is structurally adequate for most applications. Appearance can be a deciding factor when choosing a grade. If a better appearance is desired, higher grades should be considered.✱

✱IRC
317.2

A grade called **construction heart** is the most suitable and most economical grade of California redwood for deck posts, beams, and joists. For decking and rails, a grade called construction common is recommended. Better-appearing grades are available. However, they are more expensive.

Two grades of redwood, called redwood deck heart and redwood deck common, are manufactured especially for exterior walking surfaces. Two grades of decking, standard and premium, are also available in pressure-treated southern pine. Special decking grades of western cedar may also be obtained. The lumber grade or special purpose is shown in the grade stamp (*Figure 65–2*).

Lumber Sizes. Specific sizes of supporting posts, girders, and joists depend on the spacing and height of supporting posts and the spacing of girders and joists. In addition, the sizes of structural members depend on the type of wood used and the weight imposed on the members. Too many factors prohibit generalization about sizes of structural members. Check with local building officials or with a professional to determine the sizes of structural members for specific deck construction. Determining sizes with incomplete information may result in failure of undersize members. Unnecessary expense is incurred with the use of oversize members.

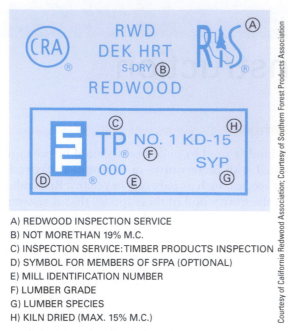

A) REDWOOD INSPECTION SERVICE
B) NOT MORE THAN 19% M.C.
C) INSPECTION SERVICE: TIMBER PRODUCTS INSPECTION
D) SYMBOL FOR MEMBERS OF SFPA (OPTIONAL)
E) MILL IDENTIFICATION NUMBER
F) LUMBER GRADE
G) LUMBER SPECIES
H) KILN DRIED (MAX. 15% M.C.)

Courtesy of California Redwood Association; Courtesy of Southern Forest Products Association

Figure 65–2 Grade stamps show the grade and special purpose of lumber used in deck construction.

Fasteners

All nails, fasteners, and hardware should be stainless steel, aluminum, or hot-dipped galvanized. Electroplated galvanizing is not acceptable because the coating is too thin. In addition to corroding and eventual failure, poor-quality fasteners will react with substances in decay-resistant woods and cause unsightly stains. ✱

✱**IRC 317.3.1**

BUILDING A DECK

Most wood decks consist of posts, set on footings, supporting a platform of *girders* and *joists* covered with **deck boards.** Posts, rails, **balusters,** and other special parts make up the railing (*Figure 65–3*). Other parts, such as shading devices, privacy screens, benches, and planters, lend finishing touches to the area.

Installing a Ledger

If the deck is to be built against a building, a **ledger,** usually the same size as the joists, is bolted to the wall for the entire length of the deck

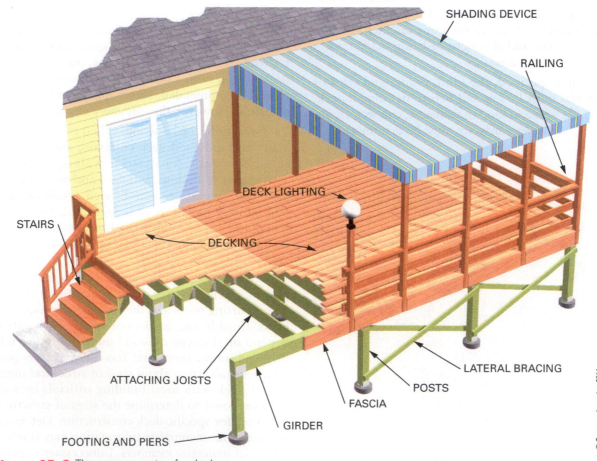

Figure 65–3 The components of a deck.

© Cengage Learning 2014

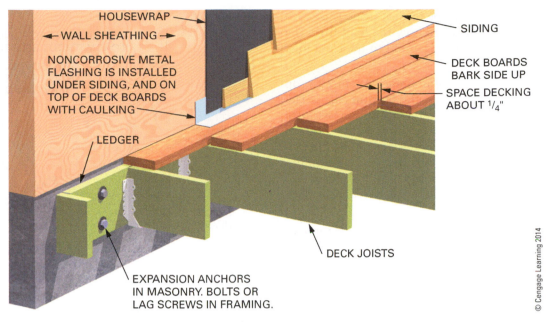

HOUSEWRAP
◄─ WALL SHEATHING ─►
NONCORROSIVE METAL
FLASHING IS INSTALLED
UNDER SIDING, AND ON
TOP OF DECK BOARDS
WITH CAULKING
LEDGER
EXPANSION ANCHORS
IN MASONRY. BOLTS OR
LAG SCREWS IN FRAMING.

SIDING
DECK BOARDS
BARK SIDE UP
SPACE DECKING
ABOUT ¼"
DECK JOISTS

© Cengage Learning 2014

Figure 65–4 A ledger is made weathertight with flashing.

*IRC
507.2.1
*IRC
507.2

(*Figure 65–4*). ✱ The ledger acts as a beam to support joists that run at right angles to the wall. It is installed to a level line. Its top edge is located to provide a comfortable step down from the building after the decking is applied. The ledger height may be used as a benchmark for establishing the elevations of supporting posts and girders. ✱

When fastening the ledger to the building, you must be careful and mind the details to ensure the deck will not fail if a large group of people assemble. The ledger must be bolted to the band joists made of 2 inch nominal lumber or engineered lumber. Anchors must be ½ inch bolts or lag screws installed with washers, and they must

be made with hot-dipped galvanized steel or stainless steel. The faces of the ledger and band joist should not be separated by more than $^{15}/_{32}$ inch sheathing. Only 1 inch separation is allowed if stacked washers are used on $^{15}/_{32}$ to make up the difference. The spacing of the fasteners depends on the length of the deck joists (*Figure 65–5*): The longer the joist, the greater the load on the ledger.

Ledger and band joists rarely will align their top and bottom edges. For this reason the position of each fastener must be carefully considered. The top edge of the band joist and the bottom edge of the ledger may have a bolts close to the edge. Bolts in the bottom edge of the band joist and the top

Joist Span	6' and less	6' 1" to 8'	8' 1" to 10'	10' 1" to 12'	12' 1" to 14'	14' 1" to 16'	16' 1" to 18'
Connection details	On-center spacing of fasteners (inch)						
½ inch diameter lag screw with $^{15}/_{32}$ inch maximum sheathing.	30	23	18	15	13	11	10
½ inch diameter bolt with $^{15}/_{32}$ inch maximum sheathing.	36	36	34	29	24	21	19
½ inch diameter bolt with $^{15}/_{32}$ inch maximum sheathing and ½ inch stacked washers.	36	36	29	24	21	18	16

© Cengage Learning 2014

Figure 65–5 Ledger anchors are spaced according to the length of the deck joists.

edge of the ledger must be further away. This is because of the difference in how the load from the deck is applied to the ledger and to the band joist (*Figure 65–6*).

The band joist has a minimum distance to top of ¾ inch and 2 inches to the bottom edges. The ledger is reversed where the minimum distance to the top is 2 inches and ¼ inch to the bottom (*Figure 65–7*). The on-center spacing of the fasteners is staggered into two rows. The minimum vertical distance between these rows is 1⅝ inches. Fasteners must be 2 inches or more from the ends of band joist or ledger pieces.

After the deck frame and surface is applied, flashing is installed under the siding and on top of the deck board. Caulking is applied between the deck and the flashing. The flashing is then fastened, close to and along its outside vertical edge. The outside horizontal edge of the flashing should extend beyond the ledger.

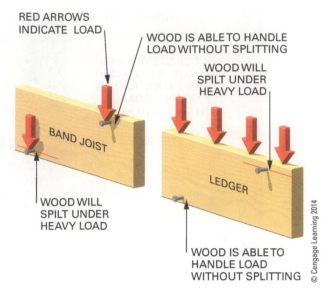

Figure 65–6 Closeness of anchors to board edges depnds on how the deck load is applied to the bolt.

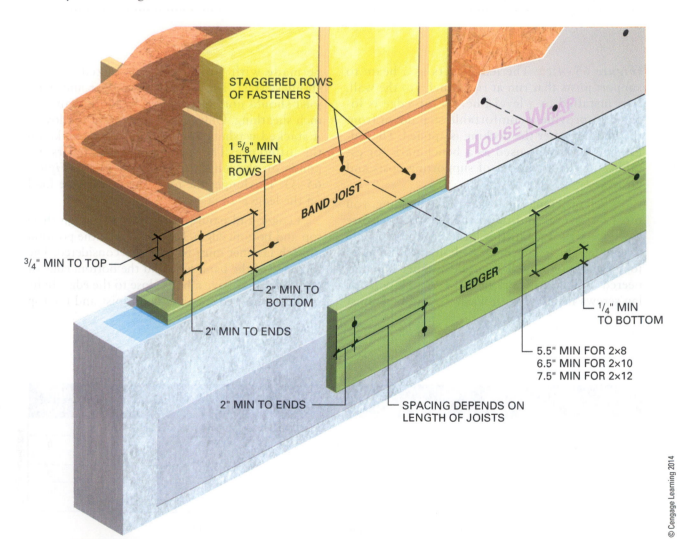

Figure 65–7 Fasteners in the band joist and ledger must follow careful spacing guidelines.

FOOTING AND POST LAYOUT AND EXCAVATION

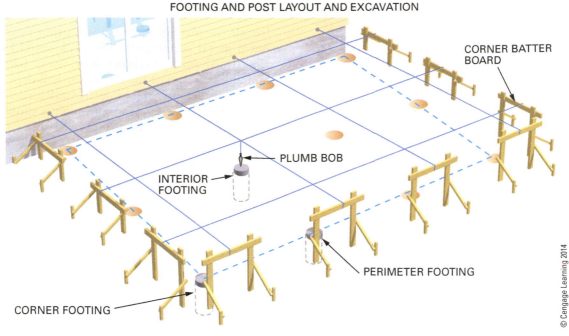

Figure 65–8 Footing and post layout and excavation.

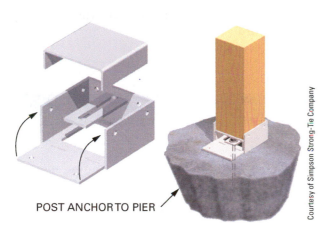

POST ANCHOR TO PIER

Figure 65–9 Post anchors may be fastened to concrete using an expansion bolt.

Footing Layout and Construction

Footings for the supporting posts must be accurately located. To determine their location, erect batter boards and stretch lines in a manner previously described in Chapter 26. All footings require digging a hole and filling it with concrete (*Figure 65–8*).

Footing Size and Style

In stable soil and temperate climate, the footing width is usually made twice the width of the post it is supporting. The footing depth reaches undisturbed soil, at least 12 inches below grade. In cold climates, the footing should extend below the frost line.

Several footing styles are commonly used. One method is to partially fill the footing hole with concrete within a few inches from the top. Set a precast pier 2 inches into the wet concrete. After the concrete has set, attach a post anchor to the top of the precast pier (*Figure 65–9*). Align piers and anchors with the layout lines.

The top of the footing may be brought above grade with the use of a wood box or fiber-tube form. Place concrete in the footing hole. Bring it to the top of the form. Set post anchors while the concrete is still wet (*Figure 65–10*).

Erecting Supporting Posts

All supporting posts are set on footings. They are then plumbed and braced. Cut posts, for each footing, a few inches longer than their final length. Tack the bottom of each post to the anchor. Brace them in a plumb position in both directions (*Procedure 65–A*).

When all posts are plumbed and braced, the tops must be cut level to the proper height. From the height of the deck, deduct the deck thickness and the depth of the girder. Mark on a corner post. Mark the other posts by leveling from the first post marked. Mark each post completely around using a square. Cut the tops with a portable circular saw (*Figure 65–11*).

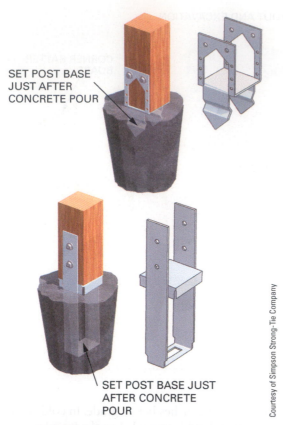

SET POST BASE
JUST AFTER
CONCRETE POUR

SET POST BASE JUST
AFTER CONCRETE
POUR

Courtesy of Simpson Strong-Tie Company

Figure 65–10 Post anchors may be set into concrete after it is placed.

Installing Girders

Install the girders on the posts using post-and-beam metal connectors. The deck should slope slightly, about $\frac{1}{8}$ inch per foot, away from the building. The size of the connector will depend on the size of the posts and girders. Install girders with the crowned edge up. Any splice joints should fall over the center of the post (*Figure 65–12*).

Installing Joists

Joists may be placed over the top or between the girders. When joists are hung between the girders, the overall depth of the deck is reduced. This provides more clearance between the frame and the ground. For decking run at right angles, joists may be spaced 24 inches on center. Joists should be spaced 16 inches OC (on center) for diagonal decking.

Lay out and install the joists in the same manner as described in earlier chapters. Use appropriate hangers if joists are installed between girders (*Figure 65–13*). When joists are installed over girders, use recommended framing anchors. Make sure all joists are installed with their crowned edges facing upward.

StepbyStep Procedures

Procedure 65–A Erecting and Bracing Supporting Posts

STEP 1 FASTEN POST TO ANCHOR WITH NAILS OR BOLTS.

STEP 2 NAIL BRACE TOPS TO POSTS.

STEP 3 DRIVE STAKE AT THE END OF EACH BRACE.

STEP 4 PLUMB POSTS EACH WAY.

STEP 5 FASTEN BRACES TO STAKES.

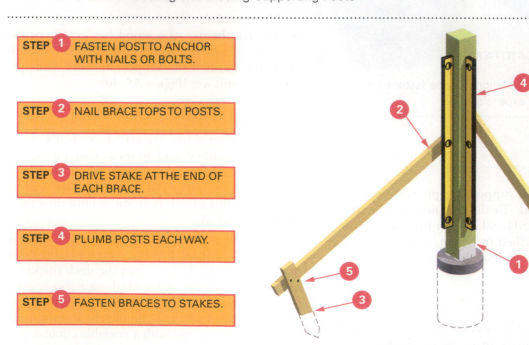

© Cengage Learning 2014

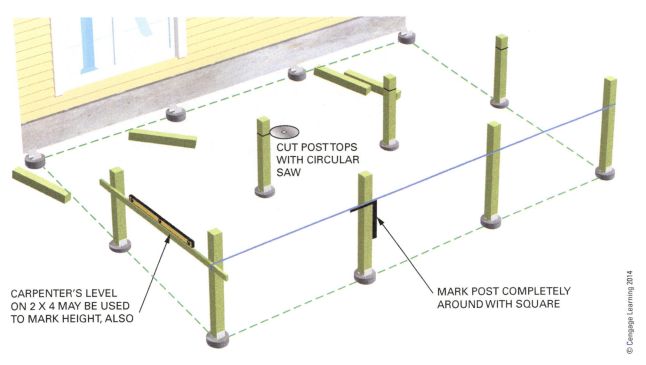

CUT POST TOPS
WITH CIRCULAR
SAW

CARPENTER'S LEVEL
ON 2 X 4 MAY BE USED
TO MARK HEIGHT, ALSO

MARK POST COMPLETELY
AROUND WITH SQUARE

© Cengage Learning 2014

Figure 65–11 The post height is determined and the tops cut level with each other.

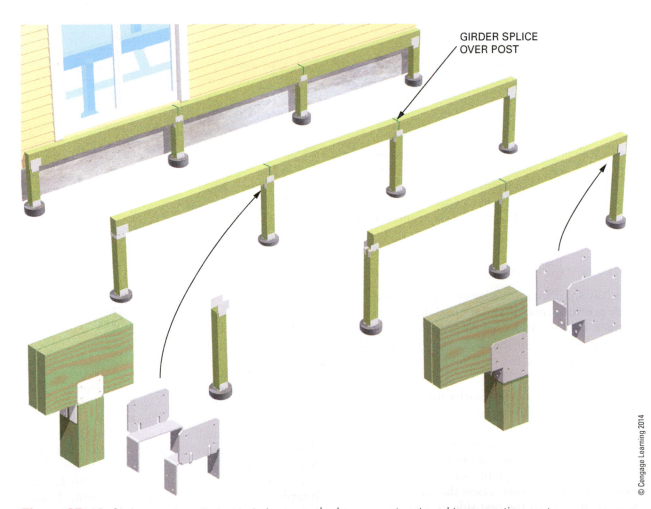

GIRDER SPLICE
OVER POST

© Cengage Learning 2014

Figure 65–12 Girders are installed with their crowned edges up and anchored to supporting posts.

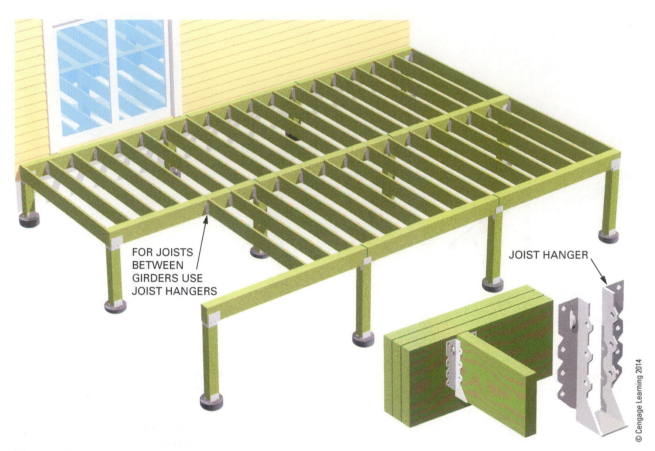

Figure 65–13 Deck joists are installed between girders with joist hangers.

Bracing Supporting Posts

If the deck is 4 feet or more above the ground, the supporting posts should be braced in a manner similar to that shown in *Figure 65–14*. Use minimum 1 × 6 braces for heights up to 8 feet. Use minimum 1 × 8 braces applied continuously around the perimeter for higher decks.

Applying Deck Boards

Specially shaped radius edge decking is available in both pressure-treated and natural decay-resistant lumber. It is usually used to provide the surface and walking area of the deck. Dimension lumber of 2-inch thickness and widths of 4 and 6 inches is also widely used.

Wood swells as it absorbs moisture, and it shrinks as it dries. As it does so, boards cup. It is best to install the board so the cup is down, leaving a shell to shed water. The USDA Forestry Service says that deck boards should be oriented with bark side up. Also poorer quality surfaces will produce more splinters. For this reason they recommend installing the boards the best side up and applying a good coat of water repellent. (*Figure 65–15*).

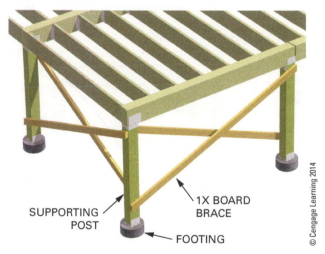

Figure 65–14 Supporting posts must be braced if the deck is 4 feet or more above the ground.

Boards are usually laid parallel with the long dimension of the deck (*Figure 65–16*). However, because deck boards usually do not come longer than 16 feet, it may be desirable to lay the boards parallel to the short dimension, if their length will span it, to eliminate end joints in

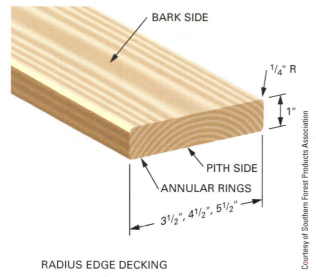

BARK SIDE

$\frac{1}{4}$" R

1"

PITH SIDE

ANNULAR RINGS

$3\frac{1}{2}$", $4\frac{1}{2}$", $5\frac{1}{2}$"

RADIUS EDGE DECKING

Courtesy of Southern Forest Products Association

Figure 65–15 Special radius edge decking is used as decking boards.

Courtesy of California Redwood Association

Figure 65–16 Deck boards are often installed parallel to the length of the deck.

the decking. Boards may also be laid in a variety of patterns including diagonal, herringbone, and parquet. Make sure the supporting structure has been designed and built to accommodate the design (*Figure 65–17*).

Much care should be taken with the application of deck boards. Snap a straight line as a guide to

apply the starting row. Start at the outside edge if the deck is built against a building. A ripped and narrower ending row of decking is not as noticeable against the building as it is on the outside edge of the deck (*Figure 65–18*).

Straighten the boards as they are installed. Maintain about a $\frac{1}{4}$-inch space between dry

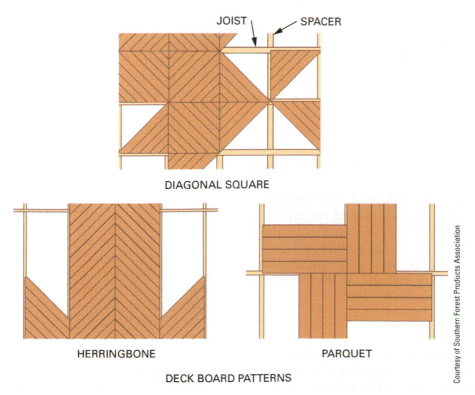

JOIST   SPACER

DIAGONAL SQUARE

HERRINGBONE   PARQUET

DECK BOARD PATTERNS

Courtesy of Southern Forest Products Association

Figure 65–17 Deck boards may be installed in various arrangements.

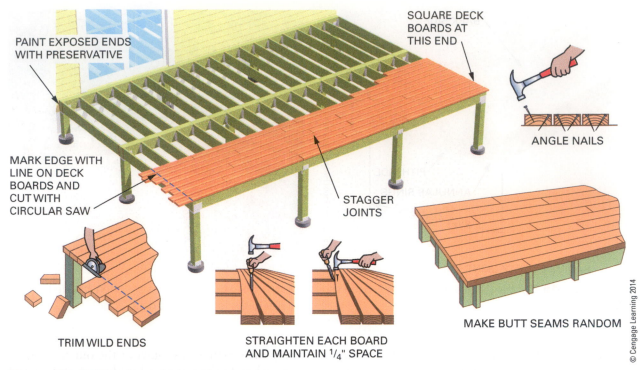

PAINT EXPOSED ENDS WITH PRESERVATIVE

SQUARE DECK BOARDS AT THIS END

ANGLE NAILS

MARK EDGE WITH LINE ON DECK BOARDS AND CUT WITH CIRCULAR SAW

STAGGER JOINTS

TRIM WILD ENDS

STRAIGHTEN EACH BOARD AND MAINTAIN ¼" SPACE

MAKE BUTT SEAMS RANDOM

© Cengage Learning 2014

Figure 65–18 Techniques for installing deck boards.

boards. If the decking boards are wet, as with most pressure-treated boards, they will shrink as they dry. Nailing them tight together is the preferred method as the ¼" space will appear when the lumber reaches equilibrium moisture content.

If the deck boards do not span the entire width of the deck, cut the boards so their ends are centered over joists. Make tight-fitting end joints. Stagger them between adjacent rows. Predrill holes for fasteners to prevent splitting at the ends. Let the end of each row overhang the deck. Use two screws or nails in each joist. Drive nails at an angle. Set the heads below the surface. This will keep the nails from working loose and the heads from staining the surface.

When approaching the end, it is better to increase or decrease the spacing of the last six or seven rows of decking. This way you will end with a row that is nearly equal in width to all the rest, rather than with a narrow strip. When all the deck boards are laid, snap lines and cut the overhanging ends. Apply a preservative to them.

Applying Trim

A fascia board may be fastened around the perimeter of the deck. Its top edge should be flush with the deck surface. The fascia board conceals the cut ends of the decking and the supporting members below. The fascia board is optional.

Stairs and Railings

Stairs. Most decks require at least one or two steps leading to the ground. To protect the bottom ends of the stair carriage, they should be treated with preservative and supported by an above-grade concrete pad (*Figure 65–19*). Stair layout and construction are described in Chapter 48. Stairs with more than two risers are generally required to have at least one handrail. The design and construction of the stair handrail should conform to that of the deck railing.

Rails. There are numerous designs for deck railings. All designs must conform to certain code requirements. Most codes require at least a 36-inch-high railing around the exposed sides, ✳ if the deck is more than 30 inches above the ground.✳ In addition, some codes specify that no openings in the railing should allow a 4-inch sphere to pass through it.✳ Each linear foot of railing must be strong enough to resist a pressure of 20 pounds per square foot applied horizontally at a right angle against the top rail. Check local building codes for deck stair and railing requirements.

✳IRC 312.1.2
✳IRC 312.1.1
✳IRC 312.1.3

TREAD

STAIR CARRIAGE

CONCRETE PAD

© Cengage Learning 2014

Figure 65–19 Stairs for decks are usually constructed with a simple basement or utility design.

Railings may consist of posts, top, bottom, and intermediate rails, and balusters. Sometimes lattice work is used to fill in the rail spaces above the deck. It is frequently used to close the space between the deck and the ground. Posts, rails, balusters, and other deck parts are manufactured in several shapes especially for use on decks (*Figure 65–20*).

Stanchions or posts are sometimes notched on their bottom ends to fit over the edge of the deck. They are usually spaced about 4 feet apart. They are fastened with lag screws or bolts. The top rail may go over the tops or be cut between the posts. The bottom rail is cut between the posts. It is kept a few inches (no more than 8)

above the deck. The remaining space may be filled with intermediate rails, balusters, lattice work, or other parts in designs as desired or as specified (*Figure 65–21*).

Deck Accessories

There are many details that can turn a plain deck into an attractive and more comfortable living area. *Shading structures* are built in many different designs. They may be completely closed in or spaced to provide filtered light and air circulation. Benches partially or entirely around the deck may double as a place to sit and act as a railing (*Figure 65–22*). Bench seats should be 18 inches from the deck. Make allowance for cushion thickness, if used. The depth of the seat should be from 18 to 24 inches.

PORCHES

The porch deck is constructed in a similar manner to an open deck. Members of the supporting structure may need to be increased in size and the spans decreased to support the weight of the walls and roof above. Work from plans drawn by professionals or check with building officials before starting. Porch walls and roofs are framed and finished as described in previous chapters (*Figure 65–23*).

SUMMARY

Decks and porches are designed in many different ways. There are other ways to construct them besides the procedures described in this chapter. However, making the layout, building the supporting structure, applying the deck, constructing the railing and stairs, and, in the case of a porch, building the walls and roof are basic steps that can be applied to the construction of practically any deck or porch.

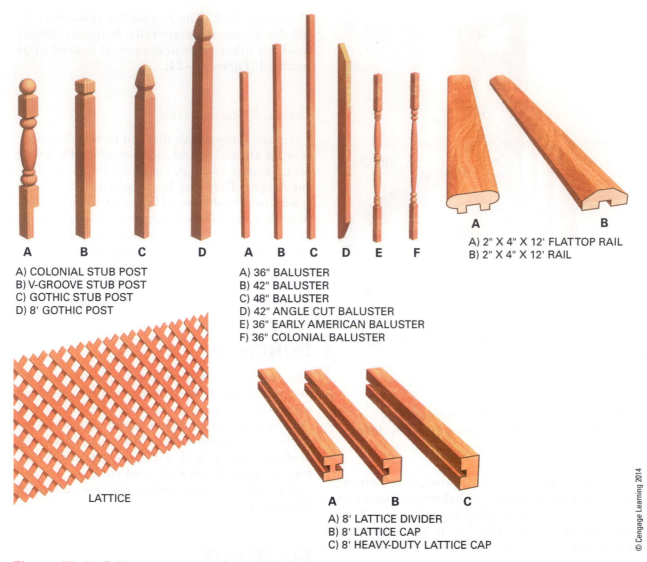

A) COLONIAL STUB POST
B) V-GROOVE STUB POST
C) GOTHIC STUB POST
D) 8' GOTHIC POST

A) 36" BALUSTER
B) 42" BALUSTER
C) 48" BALUSTER
D) 42" ANGLE CUT BALUSTER
E) 36" EARLY AMERICAN BALUSTER
F) 36" COLONIAL BALUSTER

A) 2" X 4" X 12' FLATTOP RAIL
B) 2" X 4" X 12' RAIL

LATTICE

A) 8' LATTICE DIVIDER
B) 8' LATTICE CAP
C) 8' HEAVY-DUTY LATTICE CAP

Figure 65–20 Railing parts are manufactured in many shapes.

© Cengage Learning 2014

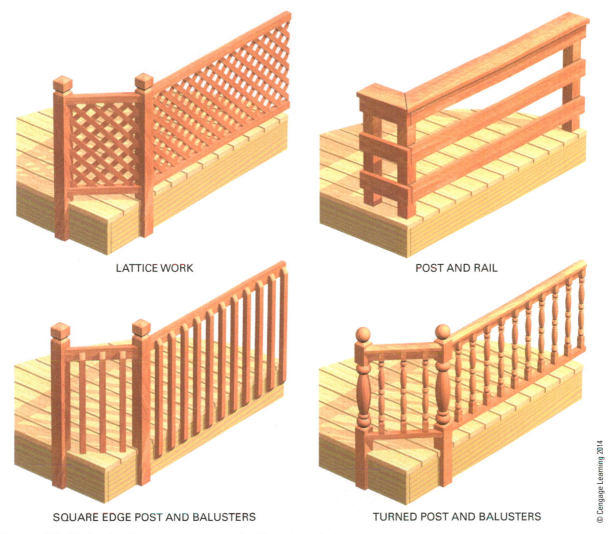

LATTICE WORK

POST AND RAIL

SQUARE EDGE POST AND BALUSTERS

TURNED POST AND BALUSTERS

© Cengage Learning 2014

Figure 65–21 Deck railings are constructed with various designs.

Courtesy of California Redwood Association

Figure 65–22 A plain deck can be made more attractive and comfortable with benches and shading devices.

Figure 65–23 A porch is composed of a deck enclosed by walls and a roof.

© Cengage Learning 2014

CHAPTER 66

Fence Design and Erection

A superior fence combines utility and beauty (*Figure 66–1*). It may define space, create privacy, provide shade or shelter, screen areas from view, or form required barriers around swimming pools and other areas for the protection of small children.

The design of a fence is often the responsibility of the builder. With creativity and imagination, he or she can construct an object of beauty and elegance that also fulfills its function. The objective of this chapter is to create an awareness of the importance of design by showing several styles of fences and the methods used to erect them.

FENCE MATERIAL

Fences are not supporting structures. Lower, knotty grades of lumber may be used to build them. This is not the case if appearance is a factor and you want to show the natural grain of the wood. Many kinds of softwood may be used as long as they are shielded from the weather by protective coatings of paint, stains, or preservatives. However, for wood posts that are set into the ground and other parts that may be subjected to constant moisture, pressure-treated or all-heart, decay-resistant lumber must be used.

The same types of fasteners that are used on exposed decks should be used to build fences. These include stainless steel, aluminum, and hot-dipped galvanized fasteners. Inferior hardware and fasteners will corrode and cause unsightly stains when in contact with moisture.

FENCE DESIGN

Fences consist of posts, rails, and fence boards. Fences may be constructed in almost limitless designs. Zoning regulations sometimes restrict their height or placement. Often the site will affect the design. For example, the fence may have to be stepped like a staircase on a steep slope (*Figure 66–2*). Fences on property lines can be designed to look attractive from either side. Fence designs may block wind and sunlight. Fence boards may be spaced in many attractive patterns. Most fences are constructed in several basic styles, each of which can be designed in numerous ways. Provisions should be made in the fence design to drain water from any area where it may otherwise be trapped.

Courtesy of California Redwood Association

Figure 66–1 A fence can serve its intended purpose and also enhance the surroundings.

© Cengage Learning 2014

Figure 66–2 Sometimes the site, such as a steep slope, may affect the design of a fence.

Courtesy of California Redwood Association

Courtesy of California Redwood Association

Figure 66–3 The picket fence is constructed in many styles.

Picket Fence

The picket fence is commonly used on boundary lines or as barriers for pets and small children. Usually not more than 4 feet high, the pickets are spaced to provide plenty of air and also to conserve material. The tops of the pickets may be shaped in various styles. Or, they may be cut square, with ends exposed or capped with a molding. The pickets may be applied with their tops in a straight line or in curves between posts (*Figure 66–3*). When pickets are applied with their edges tightly together, the assembly is called a stockade fence. Stockade fences are usually higher and are used when privacy is desired.

Board-on-Board Fence

The board-on-board fence is similar to a picket fence. However, the boards are alternated from side to side so that the fence looks the same from both sides (*Figure 66–4*). The boards may vary in height and spacing according to the degree of privacy and protection from wind and sun desired. The tops or edges of the boards may be shaped in many different designs.

Lattice Fence

This fence gets its name from narrow strips of wood called lattice. Strips are spaced by their own width. Two layers are applied at right angles to each other, either diagonally or in a horizontal and vertical fashion, to form a lattice work panel. Panels of various sizes can be prefabricated and installed between posts and rails, similar to a lattice work deck railing as shown in Figure 65–17.

Panel Fence

The panel fence creates a solid barrier with boards or panels fitted between top and bottom rails (*Figure 66–5*). Fence boards may be installed diagonally or in other appealing designs. Alternating the panel design provides variety and adds to the visual appeal of the fence. A small space should be left between panel boards to allow for swelling in periods of high humidity. In many cases, two or more basic styles may be combined to enhance the design. *Figure 66–6* shows lattice panels combined with board panels.

Louvered Fence

The louvered fence is a panel fence with vertical boards set at an angle (*Figure 66–7*). The fence

© Cengage Learning 2014

Figure 66–4 A board-on-board fence is similar to a picket fence. However, it looks the same on both sides.

Figure 66–5 The panel fence provides shade and privacy, and it also restricts air movement.

permits the flow of air through it and yet provides privacy. This fence is usually used around patios and pools.

Post-and-Rail Fence

The post-and-rail fence is a basic and inexpensive style normally used for long boundaries (*Figure*

66–8). Designs include two or more square edge or round rails, of various thicknesses, widths, and diameters, cut between the posts or fastened to their edges. Most post-and-rail fence designs have large openings. They are not intended as barriers.

FENCE POST DESIGN

Fence posts are usually wood or iron. Wood posts are usually 4 × 4. Larger sizes may be used, depending on the design. The post tops may be shaped in various ways to enhance the design of the fence (*Figure 66–9*). To conserve material and reduce expenses, a 4 × 4 post may be made to appear larger by applying furring and then boxing it in with 1-inch boards. The top may then be capped with shaped members and molding in various designs to make an attractive fence post (*Figure 66–10*).

Iron posts may be pipe or solid rod ranging in size from 1 inch to $1\frac{1}{4}$ inches in diameter for fences 3 to 4 feet tall. Larger diameters should be used for higher fences. Iron fence posts may be boxed in to simulate large wood posts. The tops are then usually capped with various shaped members similar to boxed wood posts (*Figure 66–11*). Iron posts should be galvanized or otherwise coated to resist corrosion.

BUILDING A FENCE

The first step in building a fence is to set the fence posts. Locate and stretch a line between the end posts. If it is not possible to stretch a line because

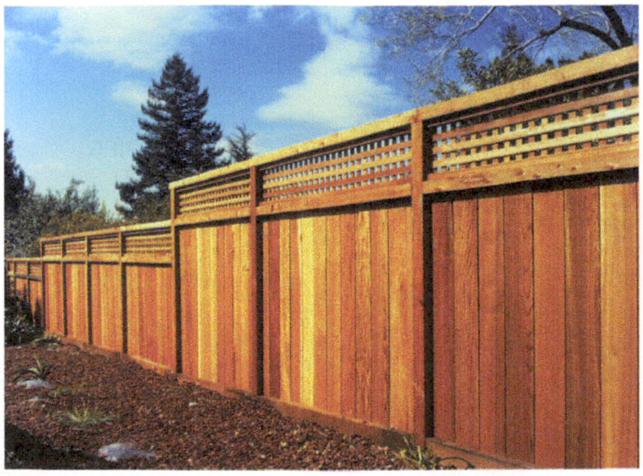

Figure 66–6 Solid panels are combined with lattice work panels for an attractive design.

Figure 66–7 The louvered fence provides privacy and lets air through.

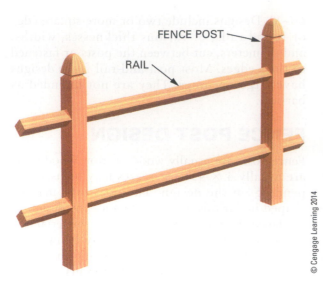

Figure 66–8 The post-and-rail fence is an inexpensive design for long boundary lines.

of steep sloping land, set up a transit-level to lay out a straight line. (See Chapter 26, "Laying Out Foundation Lines.") If the fence is to be built on a property line, make sure that the exact locations of the boundary markers are known.

Setting Posts

Posts are generally placed about 8 feet apart, to their centerlines. Mark the post locations with stakes along the fence line. Dig holes about

Figure 66–9 Fence post tops may be shaped in various ways to enhance the design of the fence.

10 inches in diameter with a post hole digger. The depth of the hole depends on the height of the fence. Higher fences require that posts be set deeper (*Figure 66–12*). The bottom of the hole should be filled with gravel or stone. This provides drainage and helps eliminate moisture to extend the life of the post.

Posts may be set in the earth. For the strongest fences, however, set the posts in concrete. Set the end posts first. Use a level to ensure that the posts are plumb in both directions. Brace them securely. The height of the post is determined by measuring from the ground, especially when the ground slopes.

If top fence rails run across the tops of the posts, then the posts are left long. The tops are cut later in a straight or contoured line, as necessary. Make sure the face edges of the posts are aligned with the length of the fence. After the posts are braced, stretch lines between the end posts at the top and at the bottom.

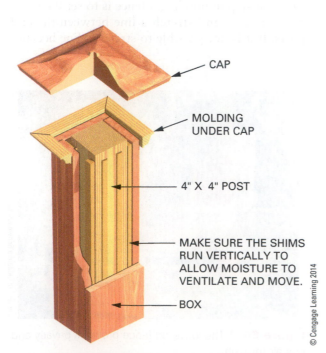

Figure 66–10 Wood posts may be boxed and capped to give a more solid look.

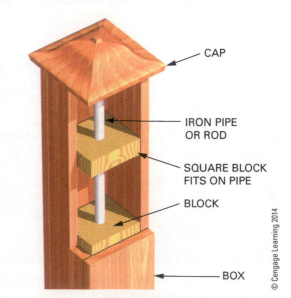

Figure 66–11 An iron fence post may be boxed and capped.

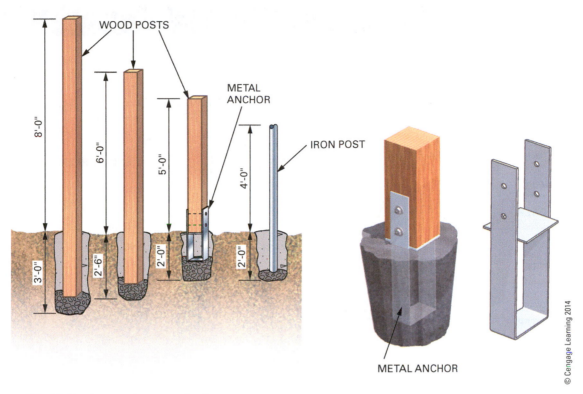

Figure 66–12 Design requirements for fence posts.

Set and brace intermediate posts. The face edges of the posts are kept to the stretched line at the top and bottom. They are plumbed in the other direction with a level. Lag screws or spikes partially driven into the bottom of the post strengthen the set of fence posts in concrete.

Place concrete around the posts. The concrete may be placed in the hole dry as the moisture from the ground will provide enough water to hydrate the cement. Tamp it well into the holes. Form the top so the surface pitches down from the posts. Instead of setting the posts in concrete, metal anchors may be embedded. The posts are installed on them after the concrete has hardened (*Procedure 66–A*).

Installing Fence Rails

Usually two or three horizontal rails are used on most fences. However, the height or the design may require more. Rails may run across the face or be cut in between the fence posts. If rails are cut between posts, they are installed to a snapped chalk line across the edges of the posts. They are secured by toenailing or with metal framing connectors. The bottom rail should be kept at least 6 inches above the ground.

Very often, the faces of wood posts are not in the same line as the rails. Therefore, rail ends must be cut with other than square ends to fit between posts. *Procedure 66–B* shows a method of laying out rail ends to fit between twisted posts.

When iron fence posts are used, holes are bored in the rails. The rails are then installed over the previously set iron posts. Splices are made on the ends to continue the rail in a straight, unbroken line. Special metal pipe grips are made to fasten fence rails to iron posts (*Figure 66–13*). If iron posts are boxed with wood, then the rails are installed in the same manner as for wood posts.

On some rustic-style post-and-rail fences, the rails are doweled into the posts. In this case, post and rails must be installed together, one section at a time.

Applying Spaced Fence Boards and Pickets

Fasten pickets in plumb positions with their tops to the correct height at the starting and ending points. Stretch a line tightly between the tops of the two pickets. If the fence is long, temporarily install intermediate pickets to support the line from sagging.

StepbyStep Procedures

Procedure 66–A Setting Fence Posts

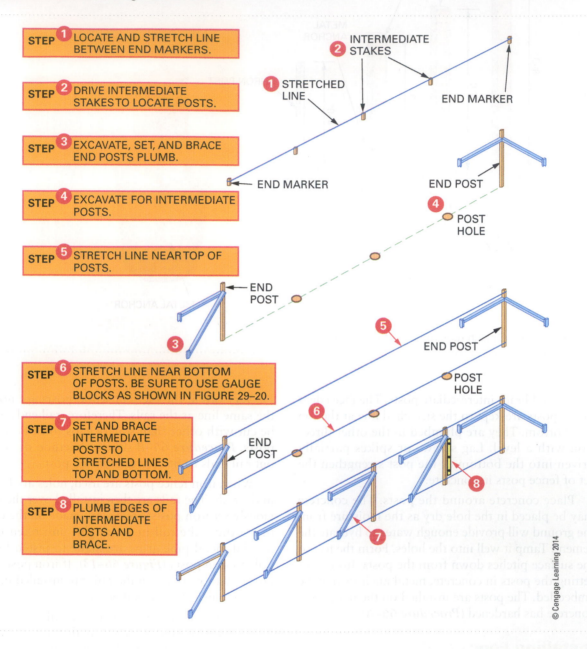

STEP 1 LOCATE AND STRETCH LINE BETWEEN END MARKERS.

STEP 2 DRIVE INTERMEDIATE STAKES TO LOCATE POSTS.

STEP 3 EXCAVATE, SET, AND BRACE END POSTS PLUMB.

STEP 4 EXCAVATE FOR INTERMEDIATE POSTS.

STEP 5 STRETCH LINE NEAR TOP OF POSTS.

STEP 6 STRETCH LINE NEAR BOTTOM OF POSTS. BE SURE TO USE GAUGE BLOCKS AS SHOWN IN FIGURE 29–20.

STEP 7 SET AND BRACE INTERMEDIATE POSTS TO STRETCHED LINES TOP AND BOTTOM.

STEP 8 PLUMB EDGES OF INTERMEDIATE POSTS AND BRACE.

2 INTERMEDIATE STAKES

1 STRETCHED LINE

END MARKER

END MARKER

END POST

4 POST HOLE

END POST

5

END POST

POST HOLE

6

END POST

7

8

© Cengage Learning 2014

Sight the line by eye to see if it is straight. If not, make adjustments and add more support pickets if necessary. Use a picket for a spacer, and fasten pickets to the rails. If the spacing is different, rip a piece of lumber for use as a spacer.

Cut only the bottom end of the pickets when trimming their height. The bottom of pickets should not touch the ground. Place a 2-inch block on the ground. Turn the picket upside down with its top end on the block. Mark it at the chalk line. Fasten the picket with its top end to the stretched line (*Procedure 66–C*).

Continue cutting and fastening pickets using the spacer. Keep their tops to the line. Check the pickets for plumb frequently. If not plumb, bring back into plumb gradually with the installation of three or four pickets.

Stop 3 or 4 feet from the end. Check to see if the spacing will come out even. Usually the spacing has to be either increased or decreased slightly. Set the dividers for the width of a picket plus a space, increased or decreased slightly, whichever is appropriate. Space off the remaining distance. Adjust the dividers until the spacing comes out even. Any

StepbyStep Procedures

Procedure 66–B Cutting Railing to Meet a Twisted Post

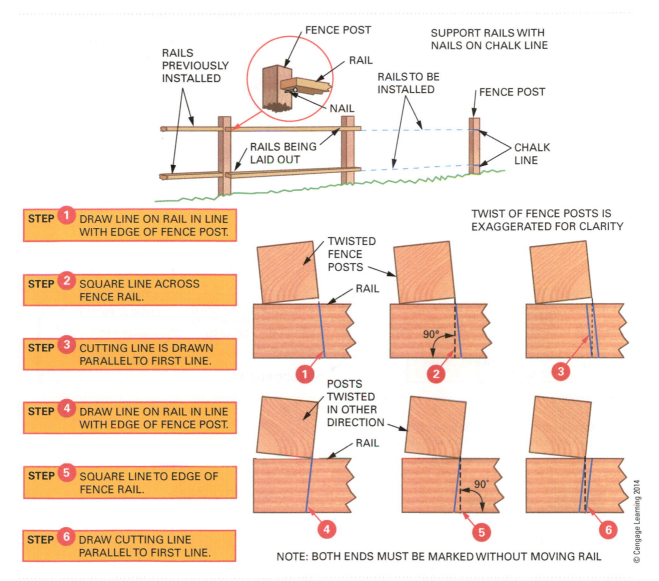

STEP ① DRAW LINE ON RAIL IN LINE WITH EDGE OF FENCE POST.

STEP ② SQUARE LINE ACROSS FENCE RAIL.

STEP ③ CUTTING LINE IS DRAWN PARALLEL TO FIRST LINE.

STEP ④ DRAW LINE ON RAIL IN LINE WITH EDGE OF FENCE POST.

STEP ⑤ SQUARE LINE TO EDGE OF FENCE RAIL.

STEP ⑥ DRAW CUTTING LINE PARALLEL TO FIRST LINE.

NOTE: BOTH ENDS MUST BE MARKED WITHOUT MOVING RAIL

© Cengage Learning 2014

slight difference in a few spaces is not noticeable. This is much better than ending up with one narrow, conspicuous space (*Figure 66–14*).

Installing Pickets in Concave Curves

In some cases, the pickets or other fence boards are installed with their tops in concave curved lines between fence posts. If the fence boards have shaped tops that cannot be cut, erect the fence using the following procedure.

Install a picket on each end at the high point of the curve. In the center, temporarily install a picket with its top to the low point of the curve. Fasten a flexible strip of wood in a curve to the top of the three pickets. Start from the center. Work both ways to install the remainder of the pickets with their tops to the curved strip. Space the pickets to each end (*Figure 66–15*). Other fence boards, such as board-on-board and louvers, are installed in a similar manner.

If fence board tops are to be cut in the shape of the curve, tack them in place with their tops above the curve. Bend the flexible strip to the curve. Mark all the fence board tops. Remove the fence boards, if necessary. Cut the tops and replace them.

StepbyStep Procedures

Procedure 66–C Installing Spaced Pickets

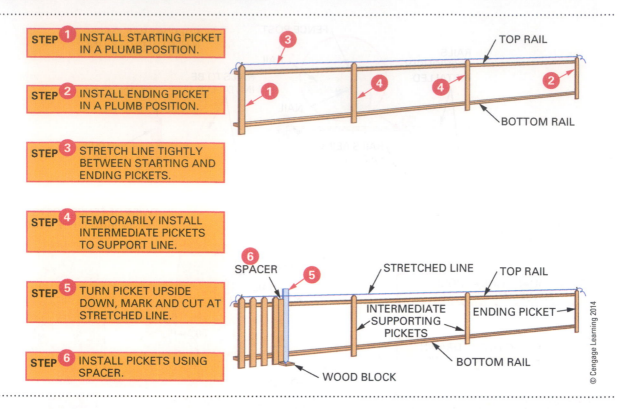

STEP 1 INSTALL STARTING PICKET IN A PLUMB POSITION.

STEP 2 INSTALL ENDING PICKET IN A PLUMB POSITION.

STEP 3 STRETCH LINE TIGHTLY BETWEEN STARTING AND ENDING PICKETS.

STEP 4 TEMPORARILY INSTALL INTERMEDIATE PICKETS TO SUPPORT LINE.

STEP 5 TURN PICKET UPSIDE DOWN, MARK AND CUT AT STRETCHED LINE.

STEP 6 INSTALL PICKETS USING SPACER.

TOP RAIL

BOTTOM RAIL

SPACER
STRETCHED LINE
TOP RAIL
INTERMEDIATE SUPPORTING PICKETS
ENDING PICKET
BOTTOM RAIL
WOOD BLOCK

© Cengage Learning 2014

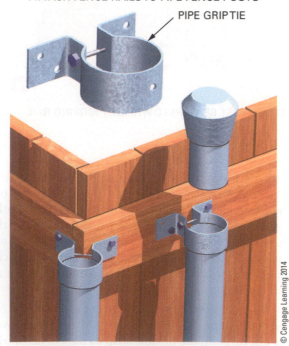

METAL GRIP TIES ARE MANUFACTURED TO ATTACH FENCE RAILS TO PIPE FENCE POSTS

PIPE GRIP TIE

© Cengage Learning 2014

Figure 66–13 Fence rails may be attached to iron fence posts.

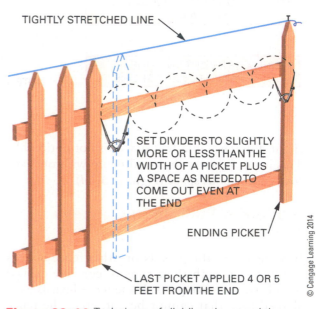

TIGHTLY STRETCHED LINE

SET DIVIDERS TO SLIGHTLY MORE OR LESS THAN THE WIDTH OF A PICKET PLUS A SPACE AS NEEDED TO COME OUT EVEN AT THE END

ENDING PICKET

LAST PICKET APPLIED 4 OR 5 FEET FROM THE END

© Cengage Learning 2014

Figure 66–14 Technique of dividing the remaining distance into equal spaces when ending a picket fence installation.

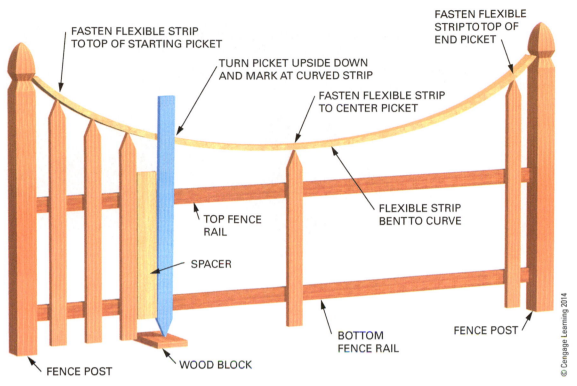

FASTEN FLEXIBLE STRIP
TO TOP OF STARTING PICKET

TURN PICKET UPSIDE DOWN
AND MARK AT CURVED STRIP

FASTEN FLEXIBLE
STRIP TO TOP OF
END PICKET

FASTEN FLEXIBLE STRIP
TO CENTER PICKET

FLEXIBLE STRIP
BENT TO CURVE

TOP FENCE
RAIL

SPACER

BOTTOM
FENCE RAIL

FENCE POST

FENCE POST

WOOD BLOCK

FENCE POST

© Cengage Learning 2014

Figure 66–15 A method for installing spaced fenceboards with their tops in a concave curve.

DECONSTRUCT THIS

Carefully study *Figure 66–16* and think about what is wrong and/or what is right. Consider all possibilities. What construction practice or method is different in your area of the country?

Figure 66–16 This photo shows a corner of a deck.

KEY TERMS

balusters	deck board	louvered fence	radius edge decking
board-on-board fence	lattice work	picket fence	stockade fence
construction heart	ledger	post-and-rail fence	

REVIEW QUESTIONS

Select the most appropriate answer.

1. Fasteners and hardware used on exposed decks and fences should not be

   a. aluminum.
   b. hot-dipped galvanized.
   c. electroplated.
   d. stainless steel.

2. A ledger is a beam

   a. attached to the side of a building.
   b. supported by a girder.
   c. used to support girders.
   d. installed on supporting posts.

3. A footing for supporting fence posts must extend

   a. at least 12 inches below grade.
   b. below the frost line.
   c. to stable soil.
   d. all of the above.

4. Deck joists must be installed

   a. between girders.
   b. over girders.
   c. crowned edge up.
   d. using adhesive or glue.

5. A railing is required on deck stairs with more than

   a. 30 inches total rise.
   b. two risers.
   c. 3 feet total rise.
   d. four risers.

6. A post that is to extend 6 feet above grade should be set _____ below grade for stability.

   a. 2'-0".
   b. 2'-6".
   c. 3'-0".
   d. 4'-0".

7. The usual height of a bench seat without a cushion is

   a. 14 inches.
   b. 16 inches.
   c. 18 inches.
   d. 20 inches.

8. A high fence with picket edges applied tightly together is called a

   a. board-on-board fence.
   b. panel fence.
   c. post-and-rail fence.
   d. stockade fence.

9. The bottom rail of any type of fence is installed above the ground by at least,

   a. the thickness of a 2 × 4 block.
   b. the width of a 2 × 4 block.
   c. 6 inches.
   d. 8 inches.

10. When applying fence pickets to rails

    a. level each one before fastening.
    b. plumb them frequently.
    c. cut the top ends to the line.
    d. fasten bottom ends flush with bottom rail.

SECTION FOUR

Interior Finish

Courtesy Jeffrey S. Rainforth

Jeffrey S. Rainforth
Vice President,
Phelps Construction Group,
Boonton, New Jersey

© iStockphoto.com/mattjeacock

EDUCATION

As a boy, Jeff Rainforth knew exactly what he wanted to do, to build and be a builder. He did carpentry work in the summers, framing houses and installing siding and roofing for home builders. Admitting that he didn't push himself in high school, Jeff and his parents thought State University of New York College of Technology at Delhi was a good option for him. His major was carpentry. "Coming out of high school, I figured I would take a couple years of classes and then go back to building houses." His plans changed, however. His Delhi instructors, Floyd Vogt and Richard Harrington, pushed him and showed him different perspectives on construction and advanced framing principles. Jeff buckled down with his studies and started taking more advanced math and science courses, which he discovered came easily to him.

Jeff graduated from Delhi in 1995 and transferred to the College of Environmental Science & Forestry at Syracuse University, graduating in 1997 with a degree in construction management and civil engineering. He went to work as a project manager for a large general contractor where he had interned—a firm that builds $1 million to $50 million projects. In this position, Jeff's responsibilities included planning site logistics, developing and maintaining project schedules, and coordinating subcontractors. He also became an OSHA trainer. "I continued my pursuit of continuing my education and took a few courses and jumped into the role of Director of Safety."

While in this role, Jeff was project manager on jobs that included the New Jersey Devils' marketing center, the New York football Giants' executive office expansion and locker room renovations, the restoration of buildings on Ellis Island, and likely his proudest achievement, the high-profile renovations of the Statue of Liberty after she was closed for security purposes following 9/11. The renovations included closing access into the copper statue while making the last pedestal level blast resistant, installing a viewing gallery and light show in the ceiling of the 6P level, and installing an emergency fire standpipe system up to the statue's crown. "It was a very intense, demanding project, but certainly the most rewarding project I will probably ever be involved with."

After eleven years, Jeff and several colleagues started their own company, which now employs sixteen people and builds projects ranging in value from $1 million $15 million. "We were awarded a few smaller-size projects shortly after opening our doors, which got the wheels moving forward. About four months in, we were awarded our first high-profile, challenging project. The New Jersey Devils hired us to build their executive offices in the Prudential Arena, which was under construction at the time. The project had to be completed prior to the first game in the new arena. We made the impossible happen, and completed their 18,000-square-foot offices in ten weeks. This project put our new company on the map."

As vice president, Jeff manages projects, estimates, schedules the work, and lines up more business. His company will tackle projects including building, renovating, and restoring schools, malls, hotels, churches, post offices, libraries, police headquarters, office buildings, and warehouses.

ON THE JOB

Jeff described how his company builds five to seven projects at a time while estimating several others, and explained what he does in a typical day. "I speak to subcontractors, estimate the work, manage building projects, and we're always trying to build new relationships for future work. Last night I attended a gathering with the Professional Women in Construction, The Construction Financial Management Association, and the American Subcontractors Association in an effort to network with others in our industry and build relationships with people I hope to do business with in the years to come."

BEST ASPECTS

Jeff appreciates the dynamic nature of his job. "Every day is something different, something new. Just when you think you've completely figured something out, you look at it from another angle and see something new." Jeff described two current projects: renovating an occupied twenty-three-story residential tower in Fort Lee, New Jersey, right next to the George Washington Bridge, and renovating a hundred-year-old brownstone church. He added, "Each project has its own unique challenges."

IMPORTANCE OF EDUCATION

Jeff admitted that he undervalued education at the start. "Through going to school and meeting people like Floyd, I learned the importance of education. I have since learned its importance if you want to move forward in this industry. I also learned that it doesn't end at your graduation ceremony."

FUTURE OPPORTUNITIES

Very satisfied about where he is in his career, Jeff said, "I was part of building my company from the ground up. I'm very proud of the team that we have put together. I look forward to the challenges and opportunities that lay ahead."

CHALLENGES

Jeff described the greatest challenge as keeping up with change: "From efficiently managing changes on the job, to keeping up with new technology and materials. There are so many management tools and software offered to plan and schedule construction projects. It still comes down to keeping yourself educated and knowledgeable about your profession."

WORDS OF ADVICE

"We only get to do this thing called life once; don't ever sell yourself short. Continue to challenge yourself to go further and achieve your goals. One of my favorite quotes, which could frame my career to date, was stated by Sir Isaac Newton: 'If I have been able to see further than others, it is because I have stood on the shoulders of giants.' Don't be afraid to ask questions and lean from your peers' experiences."

Courtesy Jeffrey S. Rainforth

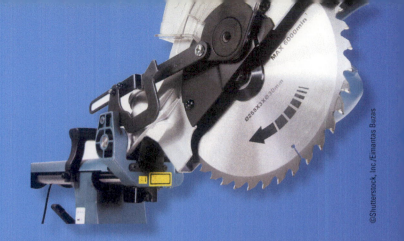

UNIT 24
Drywall Construction

The term *drywall construction* generally means the application of gypsum board. Drywall is used extensively as interior finish and is produced from gypsum, which is mined from the earth.

Installing the interior finish covers up other work performed. Check that all the work—framing, insulating, and mechanicals—is truly complete before proceeding.

OBJECTIVES

After completing this unit, the student should be able to:

- describe various kinds, sizes, and uses of gypsum panels.

- describe the kinds and sizes of nails, screws, and adhesives used to attach gypsum panels.

- make single-ply and multi-ply gypsum board applications to interior walls and ceilings.

- conceal gypsum board fasteners and corner beads.

- reinforce and conceal joints with tape and compound.

CHAPTER 67
Gypsum Board

Gypsum board is sometimes called *wallboard*, *plasterboard*, drywall, or *Sheetrock*. It is used extensively in construction (*Figure 67–1*). The term Sheetrock is a brand name for gypsum panels made by the U.S. Gypsum Company. However, the brand name is in such popular use, it has become a generic name for gypsum panels. Gypsum board makes a strong, high-quality, fire-resistant wall and ceiling covering. It is readily available, easy to apply, decorate, or repair, and relatively inexpensive.

GYPSUM BOARD

Many types of gypsum board are available for a variety of applications. The board or panel is composed of a gypsum core encased in a strong, smooth-finish paper on the face side and a natural-finish paper on the back side. The face paper is folded around the long edges. This reinforces and protects the core. The long edges are usually tapered. This allows the joints to be concealed with compound without any noticeable crown joint (*Figure 67–2*). A crowned joint is a buildup of the compound above the surface.

Types of Gypsum Panels

Most gypsum panels can be purchased, if desired, with an aluminum foil backing. The backing functions as a vapor retarder. It helps prevent

This image is a copyrighted work of USG Corporation.

Figure 67–1 The application of gypsum board to interior walls and ceilings is called drywall construction.

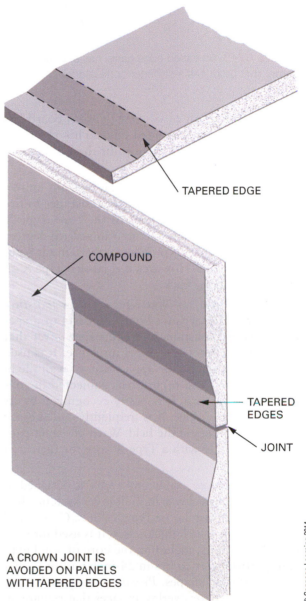

TAPERED EDGE

COMPOUND

TAPERED EDGES

JOINT

A CROWN JOINT IS AVOIDED ON PANELS WITH TAPERED EDGES

© Cengage Learning 2014

Figure 67–2 The long edges of gypsum panels usually are tapered for effective joint concealment.

the passage of interior water vapor into wall and ceiling spaces.

Regular. Regular gypsum panels are most commonly used for single- or multilayer application. They are applied to interior walls and ceilings in new construction and remodeling.

EASED EDGE

© Cengage Learning 2014

Figure 67–3 An eased edge panel has a rounded corner that produces a stronger concealed joint.

Eased Edge. An eased-edge gypsum board has a special tapered, rounded edge. This produces a much stronger concealed joint than a tapered, square edge (*Figure 67–3*).

Type X. Type X gypsum board is typically known as fire code board. It has greater resistance to fire because of special additives in the core. Type X gypsum board is manufactured in several degrees of resistance to fire. Type X looks the same as regular gypsum board. However, it is labeled Type X on the edge or on the back.

Water-Resistant. Water-resistant or *moisture resistant* (MR) gypsum board consists of a special moisture-resistant core and paper cover that is chemically treated to repel moisture. It is used frequently as a base for application of wall tile in bath, showers, and other areas subjected to considerable moisture. It is easily recognized by its distinctive green face. It is frequently called *green board* by workers in the field. Water-resistant panels are available with a Type X core for increased fire resistance.

Special Purpose. Backing board is designed to be used as a base layer in multilayer systems. It is available with regular or Type X cores. *Core-board* is available in 1-inch thicknesses. It is used for various applications, including the core of solid gypsum partitions. It comes in 24-inch widths with a variety of edge shapes. *Predecorated* panels have coated, printed, or overlay surfaces that require no further treatment. *Liner board* has a special fire-resistant core encased in a moisture-repellent paper. It is used to cover shaft walls, stairwells, chaseways, and similar areas.

Veneer Plaster Base. Veneer plaster bases are commonly called blue board. They are large, 4-foot-wide gypsum board panels faced with a specially treated blue paper. This paper is designed to receive applications of veneer plaster. Conventional plaster is applied about ⅜-inch thick and takes

considerable time to dry. In contrast, specially formulated veneer plaster is applied in one coat of about ¹⁄₁₆ inch, or two coats totaling about ⅛ inch. It takes only about forty-eight hours to dry.

Gypsum Lath. Gypsum lath is used as a base to receive conventional plaster. Other gypsum panels, such as soffit board and sheathing, are manufactured for exterior use.

Sizes of Gypsum Panels

Widths and Lengths. Coreboards and liner boards come in 2-foot widths and from 8 to 12 feet long. Other gypsum panels are manufactured 4 feet wide and in lengths of 8, 9, 10, 12, 14, or 16 feet. Gypsum board is made in a number of thicknesses. Not all lengths are available in every thickness.

Thicknesses.

- A ¼-inch thickness is used as a base layer in multilayer applications. It is also used to cover existing walls and ceilings in remodeling work. It can be applied in several layers for forming curved surfaces with short radii.

- A ⅜-inch thickness is usually applied as a face layer in repair and remodeling work over existing surfaces. It is also used in multilayer applications in new construction.

- Both ½ inch and ⅝ inch are commonly used thicknesses of gypsum panels for single-layer wall and ceiling application in residential and commercial construction. The ⅝-inch-thick panel is more rigid and has greater resistance to impacts and fire than does the ½-inch-thick panel.

- Coreboards and liner boards come in thicknesses of ¾ and 1 inch.

CEMENT BOARD

Like gypsum board, cement board and *wonder board* are panel products. However, they have a core of portland cement reinforced with a glass fiber mesh embedded in both sides (*Figure 67–4*). The core resists water penetration and will not deteriorate when wet. It is designed for use in areas that may be subjected to high-moisture conditions. It is used extensively in bathtub, shower, kitchen, and laundry areas as a base for ceramic tile. In fact, some building codes require its use in these areas.✳ ✳IRC 702.4.2

Panels are manufactured in sizes designed for easy installation in tub and shower areas with a minimum of cutting. Standard cement board panels come in a thickness of ½ inch, in widths of 32 or 36 inches, and in 5-foot lengths. Custom panels are available in a thickness of ⅝ inch, widths of 32 or 48 inches, and lengths from 32 to 96 inches.

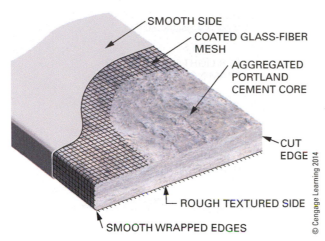

Figure 67–4 Composition of cement board.

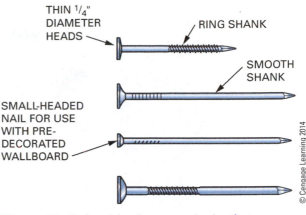

Figure 67–5 Special nails are required to fasten gypsum board.

Cement board is also manufactured in a $^5/_{16}$-inch thickness. It is used as an underlayment for floors and countertops. Exterior cement board is used primarily as a base for various finishes on building exteriors.

DRYWALL FASTENERS

Specially designed nails and screws are used to fasten drywall panels. Ordinary nails or screws are not recommended. The heads of common nails are too small in relation to the shank. They are likely to cut the paper surface when driven. Staples may be used only to fasten the base layer in multilayer applications. They must penetrate at least $^5/_8$ inch into supports. Using the correct fastener is extremely important for proper performance of the application. Fasteners with corrosion-resistant coatings must be used when applying water-resistant gypsum board or cement board. Care should be taken to drive the fasteners straight and at right angles to the wallboard to prevent the fastener head from breaking the face paper.

Nails

Gypsum board nails should have flat or concave heads that taper to thin edges at the rim. Nails should have relatively small-diameter shanks with heads at least $^1/_4$ inch, but no more than $^5/_{16}$ inch in diameter. For greater holding power, nails with annular ring shanks are used (**Figure 67–5**).

Smooth shank nails should penetrate at least $^7/_8$ inch into framing members. Only $^3/_4$-inch penetration is required when ring shank nails are used. Greater nail penetrations are required for fire-rated applications.

Nails should be driven with a drywall hammer that has a convex face. This hammer is designed to compress the gypsum panel face to form a dimple

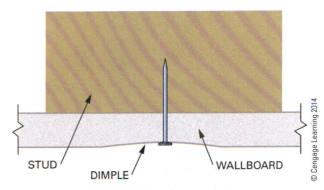

Figure 67–6 Fasteners are set with a dimple in the board for easier concealing with compound.

of not more than $^1/_{16}$ inch when the nail is driven home (**Figure 67–6**). The dimple is made so the nail head can later be covered with compound.

Screws

Special drywall screws are used to fasten gypsum panels to steel or wood framing or to other panels. They are made with Phillips heads designed to be driven with a drywall screwgun (**Figure 67–7**). A proper setting of the nosepiece on the power screwdriver ensures correct countersinking of the screw head. When driven correctly, the specially contoured bugle head makes a uniform depression in the panel surface without breaking the paper.

Different kinds of drywall screws are used for fastening into wood, metal, and gypsum panels. *Type W* screws are used for wood, *Type S* and *Type S-12* for metal framing, and *Type G* for fastening to gypsum backing boards (**Figure 67–8**).

Type W screws have sharp points. They have specially designed threads for easy penetration and

Figure 67–7 Drywall screws are driven with a screwgun to the desired depth.

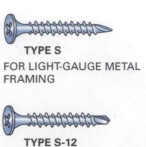

TYPE S
FOR LIGHT-GAUGE METAL FRAMING

TYPE S-12
FOR 20-GAUGE OR HEAVIER METAL FRAMING

TYPE G
FOR FASTENING INTO BASE LAYERS OF GYPSUM BOARD

TYPE W
FOR WOOD FRAMING

Figure 67–8 Several types of screws are used to fasten gypsum panels. Selection of the proper type is important.

excellent holding power. The screw should penetrate the supporting wood frame by at least $\frac{5}{8}$ inch.

Type S screws are self-drilling and self-tapping. The point is designed to penetrate sheet metal with little pressure. This is an important feature because thin metal studs have a tendency to bend away when driving screws. Type S-12 screws have a different drill point designed for heavier-gauge metal framing.

Type G screws have a deep, special thread design for effectively fastening gypsum panels together. These screws must penetrate into the supporting board by at least $\frac{1}{2}$ inch. If the supporting board is not thick enough, longer fasteners should be used. Make sure there is sufficient penetration into framing members.∗

✳IRC 702.3.6

Adhesives

Drywall adhesives are used to bond single layers directly to supports or to laminate gypsum board to base layers. Adhesives used to apply gypsum board are classified as stud adhesives and laminating adhesives.

For bonding gypsum board directly to supports, special drywall stud adhesive or approved *construction adhesive* is used. Supplemental fasteners must be used with stud adhesives. Stud adhesives are available in large cartridges. They are applied to framing members with hand or powered adhesive guns (*Figure 67–9*).

Figure 67–9 Applying drywall adhesive to studs.

CAUTION

Some types of drywall adhesives may contain a flammable solvent. Do not use these types near an open flame or in poorly ventilated areas.

For laminating gypsum boards to each other, joint compound adhesives and contact adhesives are used. Joint compound adhesives are applied over the entire board with a suitable spreader prior to lamination. Boards laminated with joint compound adhesive require supplemental fasteners.

When contact adhesives are used, no supplemental fasteners are necessary. However, the board cannot be moved after contact has been made.

The adhesive is applied to both surfaces by brush, roller, or spray gun. It is allowed to dry before laminating. A modified contact adhesive is also used. It permits an open time of up to thirty minutes during which the board can be repositioned, if necessary.

CHAPTER 68
Drywall Application

Single-layer gypsum board applications are widely used in light commercial and residential construction. They adequately meet building code requirements. Multilayer applications are more often used in commercial construction. They have increased resistance to fire and sound transmission. Both systems provide a smooth, unbroken, quality surface if recommended application procedures are followed.

DRYWALL APPLICATION

Drywall should not be delivered to the job site until shortly before installation begins. Boards stored on the job for long periods are subject to damage. The boards must be stored under cover and stacked flat on supports. Supports should be at least 4 inches wide and placed fairly close together (*Figure 68–1*).

Leaning boards against framing for long periods may cause the boards to warp. This makes application more difficult. To avoid damaging the edges, carry the boards. Do not drag them. Then, set the boards down gently. Be careful not to drop them.

Cutting and Fitting Gypsum Board

Take measurements accurately. Cut the board by first scoring the face side through the paper to the core. Use a utility knife. Cutting only the paper will produce the cleanest break. Pressing hard during the cut not only wastes energy, but it also produces a more ragged cut. Guide it with a *drywall T-square,* if cutting a square end (*Figure 68–2*). The board is then broken along the scored face. The back paper is scored along the fold. The sheet is then broken by snapping the board in the reverse direction (*Procedure 68–A*).

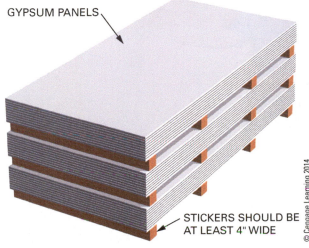

Figure 68–1 Correct method of stacking gypsum board.

GYPSUM PANELS

STICKERS SHOULD BE AT LEAST 4" WIDE

© Cengage Learning 2014

Figure 68–2 Using a drywall T-square as a guide when scoring across the width of a board.

This image is a copyrighted work of USG Corporation.

StepbyStep Procedures

Procedure 68–A Scoring and Breaking Drywall

STEP 1 SCORE THE PAPER AND BREAK THE SHEET BY LIFTING SLIGHTLY AND STEPPING BACK. AS YOU STEP BACK, DRAG THE SHEET SLIGHTLY ON THE FLOOR.

STEP 2 SCORE THE BACK OF THE SHEET BEHIND IN THE CREASE. LEAVE A PORTION OF THE PAPER INTACT AT THE TOP AND BOTTOM. THIS WILL SERVE AS A TEMPORARY HINGE.

STEP 3 RAISE THE SHORTER PIECE END OF THE SHEET AND SUPPORT IT ON YOUR TOE. KEEP YOUR TOE UNDER THE LARGER PIECE SIDE OF THE BREAK. WITH A QUICK SUDDEN MOVEMENT, CLOSE AND STRAIGHTEN THE SHEET TRYING TO OVER CLOSE IT. THE MOMENTUM OF THE PIECE WILL CAUSE THE REMAINING PAPER TO SNAP.

Figure 68-3 Technique for cutting gypsum board parallel to the edge.

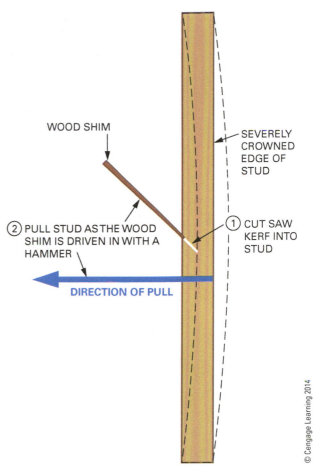

WOOD SHIM

SEVERELY CROWNED EDGE OF STUD

② PULL STUD AS THE WOOD SHIM IS DRIVEN IN WITH A HAMMER

① CUT SAW KERF INTO STUD

DIRECTION OF PULL

Figure 68-4 Technique for straightening a severely crowned stud.

To make cuts parallel to the long edges, the board is often gauged with a tape and scored with a utility knife. A *tape guide* and *tape tip* are sometimes used to aid the procedure. The tape guide permits more accurate gauging and protects the fingers. The tape tip contains a slot into which the knife is inserted. This prevents slipping off the end of the tape (*Figure 68-3*).

When making parallel cuts close to long edges, it is usually necessary to score both sides of the board to obtain a clean break.

Smooth ragged edges with a drywall rasp, coarse sanding block, or knife. A job-made drywall rasp can be made by fastening a piece of metal lath to a wood block. Cut panels should fit easily into place without being forced. Forcing the panel may cause it to break.

Aligning Framing Members

Before applying the gypsum board, check the framing members for alignment. Stud edges should not be out of alignment more than $1/8$ inch with adjacent studs. A wood stud that is out of alignment can be straightened by the procedure shown in *Figure 68-4*. Ceiling joists are sometimes brought into alignment with the installation of a strongback across the tops of the joists at about the center of the span (*Figure 68-5*).

Fastening Gypsum Panels

Drywall is fastened to framing members with nails or screws. Hand pressure should be applied on the panel next to the fastener being driven. This ensures that the panel is in tight contact with the framing member. The use of adhesives reduces the number of nails or screws required. A single or double method of nailing may be used.✴

✴IRC 702.3.2

Single-Nailing Method. With this method, nails are spaced a maximum of 7 inches on center (OC) on ceilings and 8 inches OC on walls into frame members. Nails should be first driven in the center of the board and then outward toward the edges. Perimeter fasteners should be at least 3/8 inch, but not more than 1 inch from the edge.✴

✴IRC 702.3.5

Double-Nailing Method. In double nailing, the perimeter fasteners are spaced as for single nailing. In the field of the panel, space a first set of nails 12 inches OC. Space a second set 2 to $2^{1}/_{2}$ inches from the first set. The first nail driven is reseated after driving the second nail of each set. This assures solid contact with framing members (*Figure 68-6*).

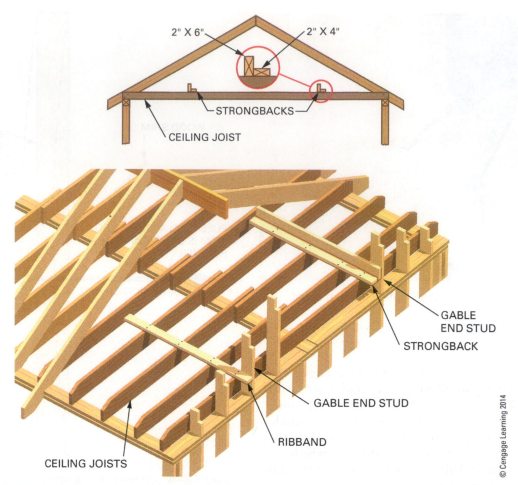

Figure 68–5 A strongback is sometimes used to align ceiling joists or the bottom chord of roof trusses.

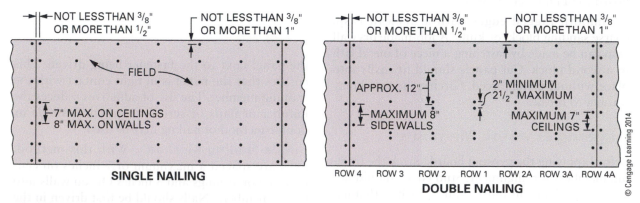

Figure 68–6 Spacing of single nailed or double nailed panels. Greater fastener spacing is used when panels are screwed.

Screw Attachment. Screws are spaced 12 inches OC on ceilings and 16 inches OC on walls when framing members are spaced 16 inches OC. If framing members are spaced 24 inches OC, then screws are spaced a maximum of 12 inches OC on both walls and ceilings.

Using Adhesives. Apply a straight bead about ¼ inch in diameter to the centerline of the stud edge. On studs where panels are joined, two parallel beads of adhesive are applied, one on each side of the centerline. Zigzag beads should be avoided to prevent the adhesive from squeezing out at the joint.

On wall applications, supplemental fasteners are used around the perimeter. Space them about 16 inches apart. On ceilings, in addition to perimeter fastening, the field is fastened at about 24-inch intervals (*Figure 68–7*).

Gypsum panels may be prebowed. This reduces the number of supplemental fasteners required. Prebow the panels by one of the methods shown in *Figure 68–8*. Make sure the finish side of the panel faces in the correct direction. Allow them to remain overnight or until the boards have a 2-inch permanent bow. Apply adhesive to the studs. Fasten the panel at top and bottom plates. The bow keeps the center of the board in tight contact with the adhesive until bonded.

Ceiling Application

Gypsum panels are applied first to ceilings and then to the walls. Panels may be applied parallel, or at right angles, to joists or furring. If applied parallel, edges and ends must bear completely on framing. If applied at right angles, the edges are fastened where they cross over each framing member. Ends must be fastened completely to joists or furring strips.

Carefully measure and cut the first board to width and length. Cut edges should be against the wall. Lay out lines on the panel face indicating the location of the framing in order to place fasteners accurately.

Gypsum board panels are heavy. At least two or more people are needed for ceiling application unless a drywall jack is available. Lift the panel overhead

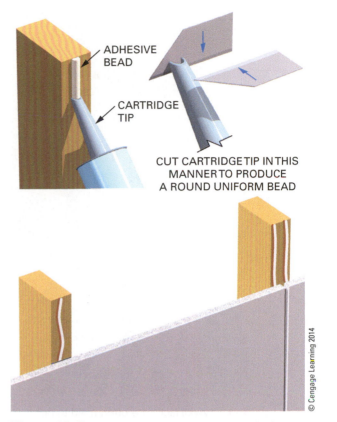

ADHESIVE BEAD

CARTRIDGE TIP

CUT CARTRIDGE TIP IN THIS MANNER TO PRODUCE A ROUND UNIFORM BEAD

© Cengage Learning 2014

Figure 68–7 Two beads of adhesive are applied under joints in the board.

and place it in position (*Figure 68–9*). Install two deadmen under the panel to hold it in position.

Deadmen are supports made in the form of a "T." They are easily made on the job using 1 × 3

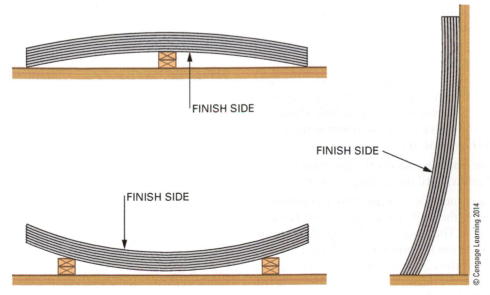

FINISH SIDE

FINISH SIDE

FINISH SIDE

© Cengage Learning 2014

Figure 68–8 Prebowing keeps the board in tight contact with the adhesive until bonded and reduces the number of fasteners required.

Figure 68–9 Drywall jacks are often used to raise drywall panels to the ceiling while fastening.

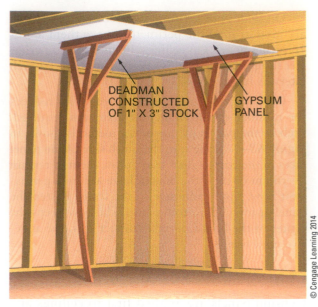

Figure 68–10 A deadman may be used to hold a board in place while fastening.

lumber with short braces from the vertical member to the horizontal member. The leg of the support is made about $\frac{1}{4}$ inch longer than the floor-to-ceiling height. The deadmen are wedged between the floor and the panel. They hold the panel in position while it is being fastened (*Figure 68–10*). Using deadmen is much easier than trying to hold the sheet in position and fasten it at the same time.

Fasten the sheet in one of the recommended manners. Hold the board firmly against framing to avoid nail pops or protrusions. Drive fasteners straight into the member. Fasteners that miss supports should be removed. The nail hole should be dimpled so that later it can be covered with joint compound (*Figure 68–11*).

Continue applying sheets in this manner, staggering end joints, until the ceiling is covered.

To cut a corner out of a panel to accommodate a protrusion in the wall, make the shortest cut with a drywall saw. Then, score and snap the sheet in the other direction (*Figure 68–12*). To cut a circular hole, mark the circle with pencil dividers, then twist and push the drywall saw through the board. Cut out the hole, following the circular line.

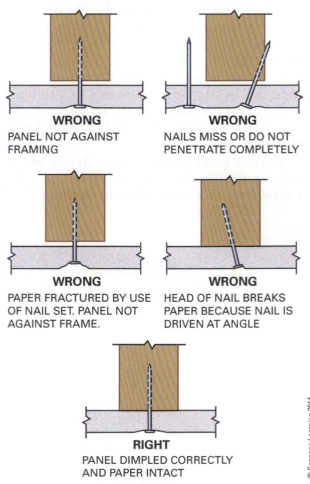

Figure 68–11 It is important to drive fasteners correctly for secure attachment of gypsum panels.

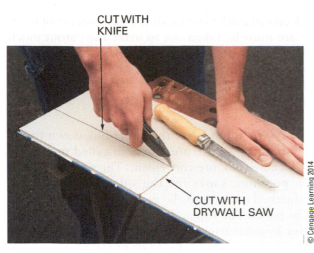

Figure 68–12 Use a knife and drywall saw to cut the corner out of a gypsum panel.

Horizontal Application on Walls

When walls are less than 8′-1″ high, wallboard is usually installed horizontally, at right angles to the studs. If possible, use a board of sufficient length to go from corner to corner. Otherwise, use as long a board as possible to minimize end joints because they are difficult to conceal. Stagger end joints or center them over and below window and door openings if possible. That way not so much of the joint is visible. End joints should not fall on the same stud as those on the opposite side of the partition. ✳

✳**IRC 702.3.5**

The top panel is installed first. Cut the board to length to fit easily into place without forcing it. Stand the board on edge against the wall. Start fasteners along the top edge opposite each stud. Raise the sheet so the top edge is firmly against the ceiling and drive nails. Fasten the rest of the sheet in the recommended manner (*Figure 68–13*).

Measure and cut the bottom panel to width and length. Cut the width about $\frac{1}{4}$ inch narrower than the distance measured. Lay the panel against the wall. Raise it with a drywall foot lifter against the bottom edge of the previously installed top panel. A drywall foot lifter is a tool especially designed for this purpose. However, one can be made on the job by tapering the end of a short piece of 1×3 or 1×4 lumber (*Figure 68–14*). Fasten the sheet as recommended. Install all others in a similar manner. Stagger any necessary end joints. Locate them as far from the center of the wall as possible so they will be less conspicuous. Avoid placing end joints over the ends of window and door headers. This will reduce the potential for wallboard cracks.

Where end joints occur on steel studs, attach the end of the first panel to the open or unsupported edge of the stud. This holds the stud flange in a rigid position for the attachment of the end of the adjoining panel. Making end joints in the

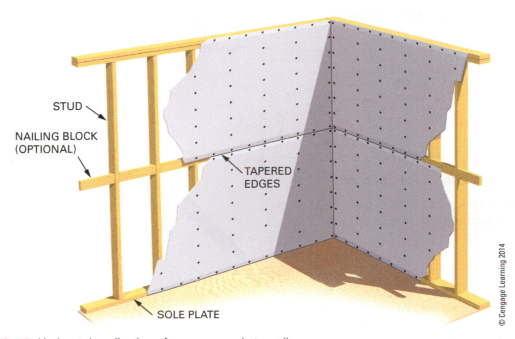

STUD

NAILING BLOCK (OPTIONAL)

TAPERED EDGES

SOLE PLATE

Figure 68–13 Horizontal application of gypsum panels to walls.

Figure 68–14 A drywall foot lifter is used to lift gypsum panels in the bottom course up against the top panels.

This image is a copyrighted work of USG Corporation.

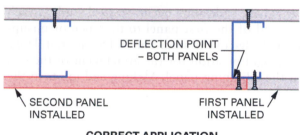

DEFLECTION POINT – BOTH PANELS

SECOND PANEL INSTALLED

FIRST PANEL INSTALLED

CORRECT APPLICATION

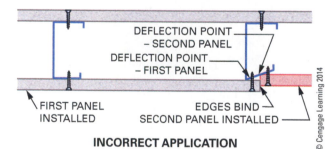

DEFLECTION POINT – SECOND PANEL

DEFLECTION POINT – FIRST PANEL

FIRST PANEL INSTALLED

EDGES BIND

SECOND PANEL INSTALLED

INCORRECT APPLICATION

© Cengage Learning 2014

Figure 68–15 Sequence of making end joints when attaching gypsum panels to steel studs.

opposite manner usually causes the stud edge to deflect. This results in an uneven surface at the joint (*Figure 68–15*).

Making Cutouts in Wall Panels. There are several ways of making cutouts in wall panels for

electrical outlet boxes, ducts, and similar objects. Care must be taken not to make the cutout much larger than the outlet. Most cover plates do not cover by much. If cut too large, much extra time has to be taken to patch up around the outlet, replace the panel, or install oversize outlet cover plates.

Plumb the sides of the outlet box down to the floor, or up to the previously installed top panel, whichever is more convenient. The panel is placed in position. Lines for the sides of the box are plumbed on it from the marks on the floor or on the panel. The top and bottom of the box are laid out by measuring down from the bottom edge of the top panel. With a saw or utility knife, cut the outline of the box. Take care not to damage the vapor retarder by pulling the lower end of the sheet away from the wall as you cut.

A fast, easy, and accurate way of making cutouts is with the use of a portable electric **drywall cutout tool** (*Figure 68–16*). The approximate location of the center of the outlet box is determined and marked on the panel. The panel is then installed over the box. Using the cutout tool, a hole is plunged through the panel in the approximate center of the outlet box. Care must be taken not to make contact with wiring. The tool is not recommended for use around live wires. The tool is moved in any direction until the bit hits a side of the box. It is then withdrawn slightly to ride over the edge to the outside of the box. The tool is then moved so the bit rides around the outside of the box to make the cutout. Usually cutouts are made for outlets after all the panels in a room have been installed.

Courtesy of Porter-Cable

Figure 68–16 A portable electric drywall cutout tool often is used to make cutouts for outlet boxes and similar objects.

To make cutouts around door openings, either mark and cut out the panel before it is applied, or make the cutout after it is applied. To make the cutout after the panel is applied, use a saw to cut in one direction. Then score it flush with the opening on the back side in the other direction. Bend and score it on the face to make the cutout. Another method uses the drywall cutout tool around framing.

Vertical Application on Walls

Vertical application of gypsum panels on walls, with long edges parallel to studs, is more practical if the ceiling height is more than 8'-1" or the wall is 4'-0" wide or less. Note that vertical application requires more lineal feet of drywall seams, and finishing the seams is more physically demanding than with horizontal application. ✱

✱IRC
702.3.5

To install vertical panels, cut the first board, in the corner, to length and width. Its length should be about ¼ inch shorter than the height from floor to ceiling. It should be cut to width so the edge away from the corner falls on the center of a stud. All cut edges must be in the corners. None should butt edges of adjacent panels.

With a foot lifter, raise the sheet so it is snug against the ceiling. The tapered edge should be plumb and centered on the stud. Fasten it in the specified manner. Continue applying sheets around the room with tapered uncut edges against each other. There should be no horizontal joints between floor and ceiling (*Figure 68–17*). Make any necessary cutouts as previously described.

Floating Interior Angle Construction

To help prevent nail popping and cracking due to structural stresses where walls and ceilings meet, the *floating angle* method of drywall application may be used. When joists or furring strips are at right angles to walls, fasteners in ceiling panels are located 7 inches from the wall for single nailing, and 11 to 12 inches for double nailing or screw attachment. When joists or furring are parallel to the wall, nailing is started at the intersection.

Gypsum panels applied on walls are fitted tightly against ceiling panels. This provides a firm support for the floating edges of the ceiling panels. The top fastener into each stud is located 8 inches down from the ceiling for single nailing and 11 to 12 inches down for double nailing or screw attachment.

At interior wall corners, the underlying wallboard is not fastened. The overlapping board is fitted snugly against the underlying board. This brings it in firm contact with the face of the stud. The overlapping panel is nailed or screwed into the interior corner stud (*Figure 68–18*).

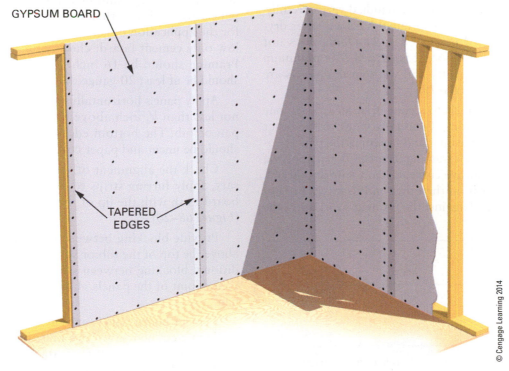

GYPSUM BOARD

TAPERED EDGES

© Cengage Learning 2014

Figure 68–17 Applying gypsum panels vertically.

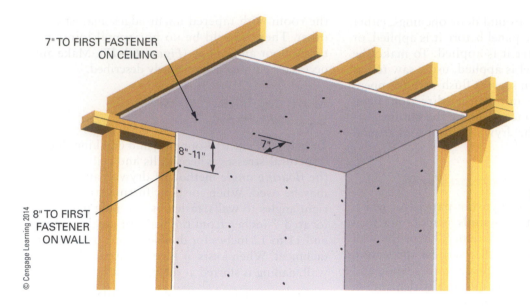

7" TO FIRST FASTENER
ON CEILING

8"–11"

7"

8" TO FIRST
FASTENER
ON WALL

© Cengage Learning 2014

Figure 68–18
Floating angle
method of applying
drywall has no
fasteners at the
corners of sheets.

APPLYING GYPSUM PANELS TO CURVED SURFACES

Gypsum panels may be applied to curved surfaces. However, closer spacing of the frame members may be required to prevent flat areas from occurring on the face of the panel. If the paper and core of gypsum panels are moistened, they may be bent to curves with shorter radii than when dry. After the boards are thoroughly moistened, they should be stacked on a flat surface. They should be allowed to stand for at least one hour before bending. Moistened boards must be handled very carefully. They will regain their original hardness after drying. Wallboard marketed as *bendable* does not need to be wet before it is shaped. The minimum bending radii for dry and wet gypsum panels are shown in *Figure 68–19*.

To apply panels to a convex surface, fasten one end to the framing. Gradually work to the other end by gently pushing and fastening the panel progressively to each framing member. When applying panels to a concave curve, fasten a stop at one end. Carefully push on the other end to force the center of the panel against the framing. Work from the

end against the stop. Fasten the panel successively to each framing member.

Gypsum board may be applied to the curved inner surfaces of arched openings. If the dry board cannot be bent to the desired curve, it may be moistened or parallel knife scores made about 1 inch apart across its width.

DRYWALL APPLICATION TO BATH AND SHOWER AREAS

Water-resistant gypsum board and cement board panels are used in bath and shower areas as bases for the application of ceramic tile. (Some areas allow only cement board; check your local codes.) Framing should be 16 inches OC. Steel framing should be at least 20-gauge thickness. ✷

✷ IRC
702.3.8
702.3.8.1

Apply panels horizontally with the bottom edge not less than $\frac{1}{4}$ inch above the lip of the shower pan or tub. The bottom edges of gypsum panels should be uncut and paper covered.

Check the alignment of the framing. If necessary, apply furring strips to bring the face of the board flush with the lip of the tub or shower pan (*Figure 68–20*).

Provide blocking between studs about 1 inch above the top of the tub or shower pan. Install additional blocking between studs behind the horizontal joint of the panels above the tub or shower pan.

Cement board panels are cut using a masonry blade in a circular saw. Care must be taken to reduce exposure to dust. Before attaching panels, apply thinned ceramic-tile adhesive to all cut edges around holes and other locations.

Board Thickness in Inches	Dry	Wet
¼	5 ft.	2 to 2½ ft.
⅜	7½ ft.	3 to 3½ ft.
½	20 ft.	4 to 4½ ft.

© Cengage Learning 2014

Figure 68–19 Minimum bending radii of gypsum panels.

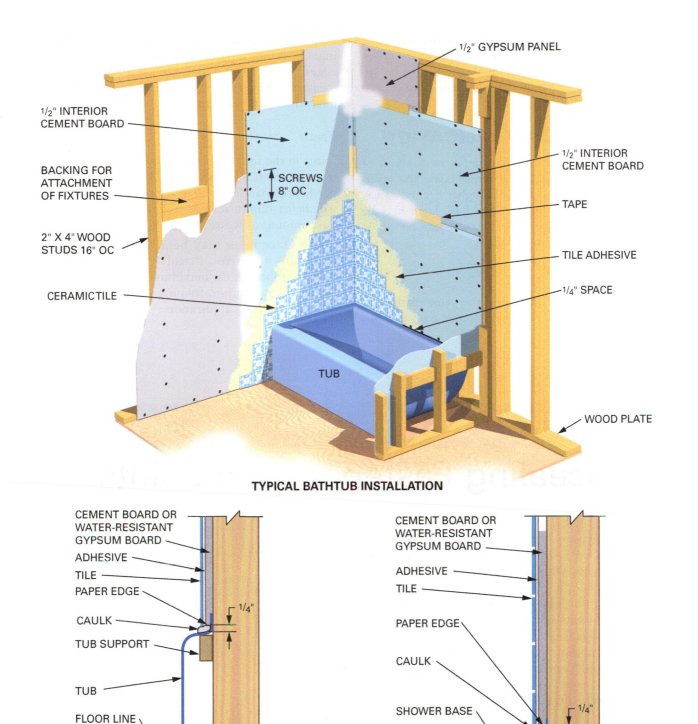

½" GYPSUM PANEL

½" INTERIOR CEMENT BOARD

BACKING FOR ATTACHMENT OF FIXTURES

2" X 4" WOOD STUDS 16" OC

CERAMIC TILE

SCREWS 8" OC

½" INTERIOR CEMENT BOARD

TAPE

TILE ADHESIVE

¼" SPACE

TUB

WOOD PLATE

TYPICAL BATHTUB INSTALLATION

CEMENT BOARD OR WATER-RESISTANT GYPSUM BOARD

ADHESIVE

TILE

PAPER EDGE

CAULK

TUB SUPPORT

TUB

FLOOR LINE

¼"

MULTILAYER APPLICATION

CEMENT BOARD OR WATER-RESISTANT GYPSUM BOARD

ADHESIVE

TILE

PAPER EDGE

CAULK

SHOWER BASE

¼"

TYPICAL SHOWER INSTALLATION

Figure 68–20 Installation details around bathtubs and showers.

© Cengage Learning 2014

Attach panels with corrosion-resistant screws or nails spaced not more than 8 inches apart. When ceramic tile more than $\frac{3}{8}$ inch thick is to be applied, the nail or screw spacing should not exceed 4 inches OC.

MULTILAYER APPLICATION

A multilayer application has one or more layers of gypsum board applied over a base layer. This layering provides greater strength, higher fire resistance, and better sound control. The base layer may be gypsum backing board, regular gypsum board, or other gypsum base material.

Base Layer

The base layer is fastened in the same manner as single-layer panels. However, double nailing is not necessary, and staples may be used in wood framing. On ceilings, panels are applied with the long edges either at right angles or parallel to framing members. On walls, the panels are applied with the long edges parallel to the studs.

Face Layer

Joints in the face layer are offset at least 10 inches from joints in the base layer. The face layer is applied either parallel to or at right angles to framing, whichever minimizes end joints and results in the least amount of waste.

The face layer may be attached with nails, screws, or adhesives. If nails or screws are used without adhesive, the maximum spacing and minimum penetration into framing should be the same as for single-layer application.

CHAPTER 69
Concealing Fasteners and Joints

After the gypsum board is installed, it is necessary to conceal the fasteners and to reinforce and conceal the joints. One of several levels of finish may be specified for a gypsum board surface. The lowest level of finish may simply require the taping of wallboard joints and *spotting* of fastener heads on surfaces. This is done in warehouses and other areas where appearance is normally not critical. The level of finish depends, among other things, on the number of coats of compound applied to joints and fasteners (*Figure 69–1*).

DRYWALL FINISHING TOOLS

Drywall finish requires special tools that involve a little practice to become proficient at using them. They include hand tools such as hawks and mudpans, knives, flat and curved trowels, and inside and outside corner trowels (*Figure 69–2*).

Mud pans are used to hold the material being applied to the wall. Hawks, knives, and trowels are used to apply, smooth, and shape the compound to the desired effect. They are held and drawn over the surface at a slight angle. The leading edge of

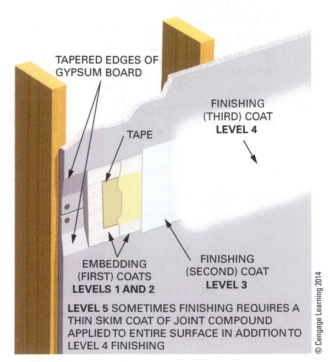

Figure 69–1 The level of finish varies with the type of final decoration to be applied to drywall panels.

TAPERED EDGES OF GYPSUM BOARD

TAPE

FINISHING (THIRD) COAT **LEVEL 4**

EMBEDDING (FIRST) COATS **LEVELS 1 AND 2**

FINISHING (SECOND) COAT **LEVEL 3**

LEVEL 5 SOMETIMES FINISHING REQUIRES A THIN SKIM COAT OF JOINT COMPOUND APPLIED TO ENTIRE SURFACE IN ADDITION TO LEVEL 4 FINISHING

© Cengage Learning 2014

Figure 69–2 Hand tools used to finish drywall.

© Cengage Learning 2014

the tool is raised off the surface, and the trailing edge lightly scrapes the surface smooth. The angle of the tool changes the amount of compound left behind. As the angle is reduced, more material is allowed to flow past the trowel. A shaper angle scrapes the surface more, pulling material away. To use these tools effectively, the angle held is constant and adjusted as needed during the entire pass of the tool.

Automatic or mechanical tools are designed to speed up the finishing process. The banjo is designed to apply joint compound and tape in one pass. The bazooka provides a similar junction, but allows for a longer reach to the ceiling (*Figure 69–3*). It is filled with compound with a specially designed hand pump.

The nail spotter applies compound over the fasteners, smoothing the surface of extra material. Flat finishers apply a wide, thin layer of compound over flat seams (*Figure 69–4*). These tools are used on the third layer (second coat). They have a slight crown built into them so to apply a layer slightly raised in the middle. When it dries, it is nearly flat. This makes a fourth layer often unnecessary.

DESCRIPTION OF MATERIALS

Fasteners are concealed with *joint compound*. Joints are reinforced with *joint tape* and covered with joint compound. Exterior corners are reinforced with *corner bead*. Other kinds of drywall trim may be used around doors, windows, and other openings (*Figure 69–5*).

Joint Compounds

Drying-type joint compounds for joint finishing and fastener spotting are made in both a dry

BAZOOKA

Courtesy of AMES Taping Tools

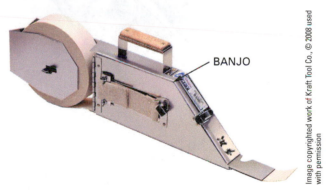

BANJO

Image copyrighted work of Kraft Tool Co., © 2008 used with permission

Figure 69–3 Bazooka and banjo are automatic drywall tools designed to apply tape and compound in one pass.

powder form and a ready-mixed form in three general types. Drying-type compounds provide smooth application and ample working time. A *taping compound* is used to embed and adhere tape to the board over the joint. A *topping compound* is used for second and third coats over taped joints. An *all-purpose compound* is used for both bedding the tape and finishing the joint. All-purpose compounds do not possess the strength or workability of two-step taping and topping compound systems. ✳

✳IRC

702.2.1

FLAT FINISHER

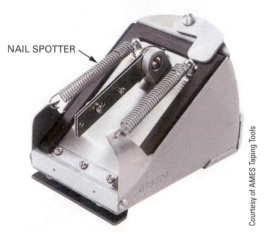

NAIL SPOTTER

Figure 69–4 Automatic drywall tools designed to apply joint compound and smooth the surface in one pass.

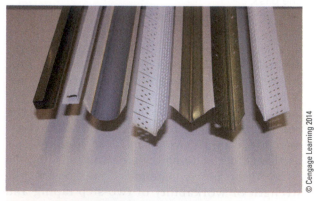

Figure 69–5 Marerials typicaly used to finish drywall.

Setting-type joint compounds are used when a faster setting time than that of drying types is desired. Drying-type compounds harden through the loss of water by evaporation. They usually cannot be recoated until the next day. Setting-type compounds harden through a chemical reaction when water is added to the dry powder. Therefore, they come only in a dry powder form and not ready-mixed. They are formulated in several different setting times. The fastest setting type will harden in as little as twenty to thirty minutes. The slowest type takes four to six hours to set up. Setting-type joint compounds permit finishing of drywall interiors in a single day.

Joint Reinforcing Tape

Joint reinforcing tape is used to cover, strengthen, and provide crack resistance to drywall joints. One type is made of *high-strength fiber paper*. It is designed for use with joint compounds on gypsum panels. It is creased along its center to simplify folding for application in corners (*Figure 69–6*).

Another type is made of *glass fiber mesh*. It is designed to reinforce joints on veneer plaster gypsum panels. It is not recommended for use with conventional compounds for general drywall joint finishing. It may be used with special high-strength setting compounds. Glass fiber mesh tape is available with a plain back or with an adhesive backing for quick application (*Figure 69–7*). Joint tape is normally available 2 and $2\frac{1}{2}$ inches wide in 300-foot rolls.

Corner Bead and Other Drywall Trim

Corner beads are applied to protect exterior corners of drywall construction from damage by impact.

Figure 69–6 Applying joint tape to an interior corner.

Figure 69-7 An adhesive-backed glass fiber mesh tape is quickly applied to drywall joints.

Figure 69-9 A clinching tool is sometimes used to fasten corner beads to exterior corners.

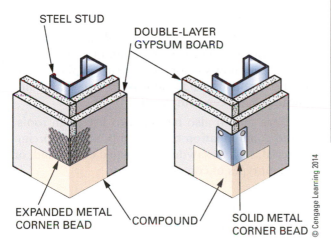

STEEL STUD

DOUBLE-LAYER GYPSUM BOARD

EXPANDED METAL CORNER BEAD

COMPOUND

SOLID METAL CORNER BEAD

Figure 69-8 Corner beads are used to finish and protect exterior corners of drywall panels.

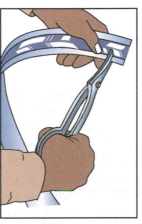

CUT TAPE WITH SNIPS

EMBED IN JOINT COMPOUND

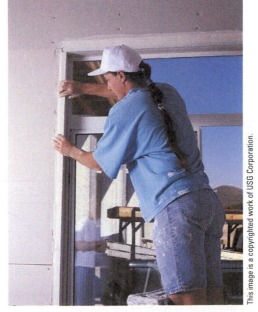

Figure 69-10 Flexible metal corner tape is applied to exterior corners by embedding in compound.

One type with solid metal flanges is widely used. Another type has flanges of expanded metal with a fine mesh. This provides excellent keying of the compound (*Figure 69-8*).

Corner bead is fastened through the drywall panel into the framing with nails or staples. Instead of using fasteners, a *clinching tool* is sometimes used. It crimps the solid flanges and locks the bead to the corner (*Figure 69-9*).

Metal corner tape is applied by embedding it in joint compound. It is used for corner protection on arches, windows with no trim, and other locations (*Figure 69-10*).

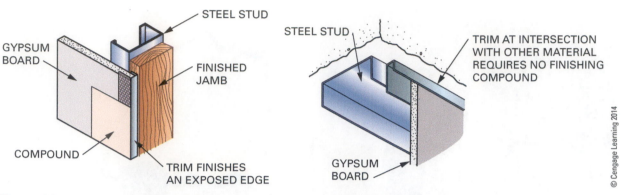

Figure 69–11 Various types of metal trim are used to provide finished edges to gypsum panels.

A variety of *metal trim* is used to provide protection and finished edges to drywall panels. Metal trim is used at windows, doors, inside corners, and intersections. Such trim is fastened through their flanges into the framing (*Figure 69–11*).

Control joints are metal strips with flanges on both sides of a $\frac{1}{4}$-inch, V-shaped slot. Control joints are placed in large drywall areas. They relieve stresses induced by expansion or contraction. They are used from floor to ceiling in long partitions and from wall to wall in large ceiling areas (*Figure 69–12*). The flanges are concealed with compound in a manner similar to corner beads and other trim.

A wide assortment of rigid vinyl drywall accessories is available (*Figure 69–13*), including the metal trim previously discussed. They are designed for easy installation and workability to reduce installation time. Most have edges to guide the drywall knife, which allows for an even application of joint compound. Some have edges that are later torn away when the painting is done. This allows the finish to be applied more quickly and at the same time more uniformly. Vinyl accessories make it possible to create smooth joints easily, whether they are curved or straight.

APPLYING JOINT COMPOUND AND TAPE

In cold weather, care should be taken to maintain the interior temperature at a minimum of 50°F for twenty-four hours before and during application of joint compound, and for at least four days after application has been completed.

Care should also be taken to use clean tools. Avoid contamination of the compound by foreign material, such as sawdust, hardened compound, or different types of compounds.

Checking and Prefilling Joints

Before applying compound to drywall panels, check the surface for fasteners that have not been sufficiently recessed. Also look for other conditions that may affect the finishing. Prefill any joints between panels of $\frac{1}{4}$ inch or more and all V-groove joints between eased-edged panels with compound. A twenty-four-hour drying period can be eliminated with the use of setting compounds for prefilling operations. Normally the flat, tapered seams are finished first before the corners.

Embedding Joint Tape

Fill the recess formed by the tapered edges of the sheets with the specified type of joint compound. Use a joint knife (*Figure 69–14*). Center the tape on the joint. Lightly press it into the compound. Draw the knife along the joint with sufficient pressure to *embed* the tape and remove excess compound (*Figure 69–15*).

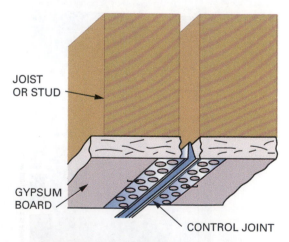

Figure 69–12 Control joints are used in large wall and ceiling areas subject to movement by expansion and contraction.

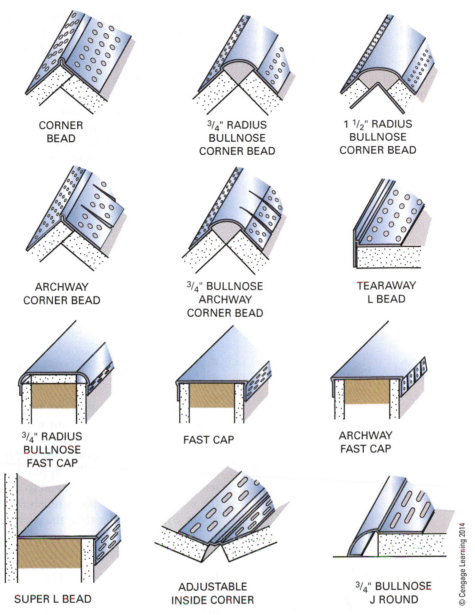

Figure 69–13 Many rigid vinyl drywall accessories are available.

Figure 69–14 Taping compound is first applied to the channel formed by the tapered edges between panels.

Figure 69–15 The tape is embedded into the compound.

There should be enough compound under the tape for a proper bond, but not over $\frac{1}{32}$ inch under the edges. Make sure there are no air bubbles under the tape. The tape edges should be well adhered to the compound. If not satisfactory, lift the portion. Add compound and embed the tape again. A *taping tool* sometimes is used. It applies the compound and embeds the tape at the same time (*Figure 69–16*).

A

B

Figure 69–16 (A) A taping tool applies tape and compound at the same time. (B) The corner is then smoothed and finished before compound sets.

Immediately after embedding, apply a thin coat of joint compound over the tape. This helps prevent the edges from wrinkling. It also makes easier concealment of the tape with following coats. Draw the knife to bring the coat to a feather edge on both sides of the joint. Make sure there is no excess compound left on the surface. After the compound has set up, but not completely hardened, wipe the surface with a damp sponge. This eliminates the need for sanding any excess after the compound has hardened.

Spotting Fasteners

Fasteners should be *spotted* immediately before or after embedding joint tape. Spotting is the application of compound to conceal fastener heads. Apply enough pressure on the taping knife to fill only the depression. Level the compound with the panel surface. Spotting is repeated each time additional coats of compound are applied to joints.

Applying Compound to Corner Beads and Other Trim

The first coat of compound is applied to corner beads and other metal trim when first coats are given to joints and fasteners. The nose of the bead or trim serves as a guide for applying the compound. The compound is applied about 6 inches wide from the nose of the bead to a feather edge on the wall. Each subsequent finishing coat is applied about 2 inches wider than the previous one.

Interior Corners

Interior corners are finished in a similar way. However, the tape is folded in the center to fit in the corner. After the tape is embedded, drywall finishers usually apply a setting compound to one side only of each interior corner. By the time they have finished all interior corners in a room, the compound has set enough to finish the other side of the corners.

Fill and Finishing Coats

Allow the first coat to dry thoroughly. This may take twenty-four hours or more depending on temperature and humidity unless a setting type compound has been used. It is common to use setting compounds for first coats and slower setting types for finishing coats. Feel the entire surface to see if any excess compound has hardened on the surface. Sand any excess, if necessary, to avoid interfering with the next coat of compound.

The Level 3 or second coat is sometimes called a fill coat (*Figure 69–17*). It is feathered out about

Figure 69–17 Applying a coat of compound to a drywall joint.

Figure 69–18 Some walls are finished with a skim coat of compound.

2 inches beyond the edges of all first coats, approximately 7 to 10 inches wide. Care must always be taken to remove all excess compound so that it does not harden on the surface. Professional drywall finishers rarely have to sand any excess in preparation for following coats. Remember, a damp sponge rubbed over the joint after the compound starts to set up will remove any small particles of excess. It will also help bring edges to a feather edge.

If the level of finish requires it, apply a third and finishing coat of compound over all fill coats. The edges of the finishing coat should be feathered out about 2 inches beyond the edges of the second coat. Some drywall finishers apply a skim coat of compound or plaster over the entire wall surface as a veneer (*Figure 69–18*).

This provides a more uniform surface where reflected light shows variations between the panels and compound.

Green Tip

New buildings should be dried and vented before occupation to reduce moisture problems such as mold and mildew.

Sanding

The final step in finishing is sanding. All surfaces should be smooth and clean before paint is applied. Sanding may be done by hand using a pole sander or, on larger jobs, with a portable electric sander

(*Figure 69–19*). Sandpaper with a grit of 100, 120, or 150 is appropriate to erase any lumps, bumps, or extra material.

Care should be taken not to raise a nap on the drywall paper with too much sanding. A damp sponge can also be used to smooth the surface. It does not remove material as fast as sandpaper or raise the a nap on the drywall paper. It also produces virtually no dust.

Textures and Coatings

Textures are sometimes added to a wall or ceiling surface to add character and charm. They are applied to a mostly finished wall surface; sanding is not usually necessary. The texture itself is joint compound applied in various thicknesses and methods to create a desired effect (*Figure 69–20*).

REMEDIES FOR DRYWALL PROBLEMS

In drywall finishing often times a situation arises that must be repaired. The surface may not yet acceptable for paint because the drywall finish requires an extra process.

Drywall Butt Edge Finish

The non-tapered ends of drywall often meet on the wall surface and create a more finishing difficulties. Since there is no taper to the drywall, a smooth

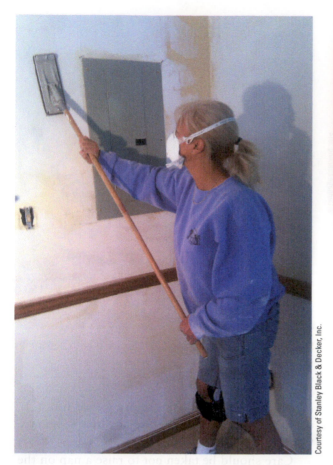

Courtesy of Stanley Black & Decker, Inc.

Courtesy of Linda Tubbs, Tubbspro.com

Figure 69–19 A pole and electric sander for finishing drywall.

© Cengage Learning 2014

Figure 69–20 Texture may be added to the wall surface.

it to dry. Embed a tape over the joint in the same manner as with the tapered seams. Next apply a full trowel width of compound to each side of the seam and tape (*Figure 69–21*). Leave a space about two-thirds the width of your trowel between these layers. Allow these to dry and then apply a closing coat to fill the space. Sand and touch up as needed.

Bubbles in Tape

Joint tape sometimes will bubble away from the drywall surface as it dries. This happens because the tape has not bonded to the drywall, which is usually because there is no joint compound between it and the drywall. Small bubbles may be sliced and opened with a utility knife, then more compound added under the flap. It is all then closed up and smoothed with a taping knife. Large long bubbles are best cut and removed. Scrape out dried compound enough to make a new compound and tape application.

Fastener Pops

Fastener or nail pops occur when fastener heads work themselves out from beneath the finished surface. This happens for several reasons, most often because the fastener was not installed correctly. Pops occur when the fastener does not penetrate solidly into the framing or when the fastener is driven too far, breaking through the surface paper. It also happens when there is a space between the drywall and framing. Either the drywall was not tight in the first place or the framing material shrank because it was not previously dry.

Repair of fastener pops begins by removing the old fastener. Drive screws nearby making sure they hit solid framing while pressing the drywall tight to the framing. Spot finish the heads as before making

level finished surface is not possible. The approach is to widen the joint finish making it harder to see. Each level must dry before the next is applied. A setting type compound allows the finish to proceed more quickly.

Begin by checking if the two surfaces are flush and even with each other. If not, apply joint compound to bring the lower surface up flush. Allow

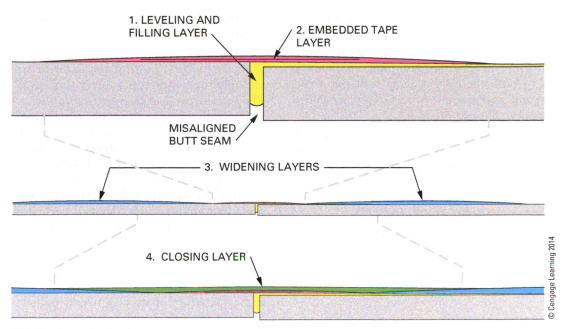

1. LEVELING AND FILLING LAYER

2. EMBEDDED TAPE LAYER

MISALIGNED BUTT SEAM

3. WIDENING LAYERS

4. CLOSING LAYER

© Cengage Learning 2014

Figure 69–21 Steps for finishing butt seams in drywall.

sure several coats are applied to compensate for normal compound shrinkage.

Patching Drywall

Damage often happens to drywall such that the surface is broken and holes are formed. Small holes about 2 inches or less may be repaired by applying joint compound. A small piece of tape embedded over the hole prevents the patch from cracking.

Medium-sized holes that occur between framing are patched with a new piece of drywall (*Figure 69–22*). The hole is cut larger to make a square hole. Then a new piece of drywall is cut to fit and finished. The new piece may be secured by first screwing pieces of plywood inside the hole to the existing drywall. The new piece is then secured to the portions of plywood that span over the hole.

Another method uses a larger repair piece to create bonding flaps. The piece is cut about 3 inches larger than the opening (*Figure 69–23*).

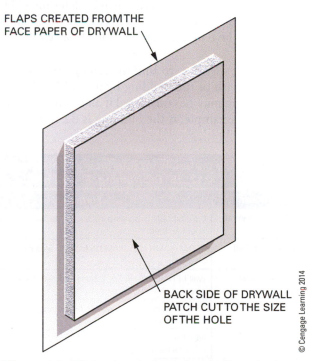

FLAPS CREATED FROM THE FACE PAPER OF DRYWALL

BACK SIDE OF DRYWALL PATCH CUT TO THE SIZE OF THE HOLE

© Cengage Learning 2014

HOLE TO BE PATCHED

PLYWOOD SCRAPS SCREWED TO WALLBOARD

© Cengage Learning 2014

Figure 69–22 Patch techniques for medium sized holes in drywall using plywood supports.

Figure 69–23 Patch techniques for medium sized holes in drywall using bonding flaps.

Dimensions of the hole size are scored on the backside of the piece. Breaking the edges and removing the gypsum while leaving the face paper intact creates the bonding flaps. Joint compound is applied to both the flaps and the hole perimeter. Pressing and smoothing the piece into place with a knife beds the flaps to the wall.

Large holes are repaired by cutting the hole large enough to expose the framing. Portions of the hole perimeter have one-half the thickness of framing member available to secure the new piece.

ESTIMATING DRYWALL

Estimating the amount of drywall material needed involves drywall, fasteners, and joint compound. Remember, if the number of pieces does not come out even, round it up to the next whole number of pieces.

Drywall

Drywall used on walls and ceilings are determined separately then added together. Many contractors use 12-foot sheets for both areas. They use up the scrap pieces in small areas after the larger areas are covered. To find the total number of square feet of drywall required, the areas of the ceiling, exterior walls, and interior walls are determined separately. Then the areas are added together. Only the large wall openings, such as double doors, are deducted since the cutouts for windows are mostly waste. The number of drywall sheets can then be determined by dividing the total area to be covered by the area of one panel. Five percent may be added for waste.

Ceiling Drywall. To find the ceiling area, multiply the length of the building by its width. The entire building may be considered if all ceilings are the same. For example, if the building is 32 feet by 50 feet, then the ceiling drywall is $32 \times 50 \div (4 \times 12) \times 105\% = 35$ sheets.

Wall Drywall. To find the exterior wall area, multiply the building perimeter times wall height. The interior wall area is the overall length of interior walls times wall height times two sides of walls. For example, the number of 4×12 sheets (48 square feet) for the exterior walls of a 32-foot by 50-foot building with wall height of 8 feet is $164 \times 8 = 1,12$ SF $\div 48 \times 105\% = 28.7 \Rightarrow 29$ sheets. If the interior wall length is 120 feet for the same building, then the drywall needed is $120 \times 8 \times 2 = 1,920$ SF $\div 48 \times 105\% = 42$ sheets.

Fasteners and Finish

Fasteners. About a 1½ fastener is needed for each square foot of drywall when applied to framing 16 inches OC, and 1 fastener per 1 square foot is needed for 24-inch OC framing. Screws may be purchased in boxes of 5,000 screws per box. For example, the fasteners for 4×12 sheets in the previous example attached to 16-inch OC framing is $35 + 31 + 42 = 108$ sheets $\times 48 = 5,184$ SF $\times 1.5 = 7,776 \div 5,000 = 2$ boxes of fasteners.

Joint Compound. Approximately 7½ gallons or 1½ five-gallon pails of joint compound will be needed to finish every 1,000 square feet of drywall. Also, one large 250-foot roll of joint tape will be needed for every 5-gallon pail of joint compound. For example, the number of 5-gallon pails of joint compound for a 5,184-SF drywall is $5,184 \div 1,000 \times 1.5 = 7.7 \Rightarrow 8$ pails = 8 large rolls of tape.

See *Figure 69–24* for a listing of the shorthand versions of these estimating techniques, along with another example.

Item	Formula	Waste factor	Example
Ceiling drywall	building LEN × building WID × waste factor ÷ area per sheet = sheets	5%	64 × 28 × 105% ÷ 48 = 39.2 ⇒ 40 sheets
Exterior Wall Drywall	PERM × wall HGT × waste factor ÷ area per sheet = sheets	5%	184 × 8′ × 105% ÷ 48 = 32.2 ⇒ 33 sheets
Interior Wall Drywall	wall LEN × wall HGT × 2 sides × waste factor ÷ area per sheet = sheets	5%	150 × 8′ × 2 × 105% ÷ 48 = 52.5 ⇒ 53 sheets
Fasteners	NUM sheets × sheet area × 1.5 fastener/SF ÷ fasteners per box = boxes		(40 + 33 + 53) × 48 × 1.5 ÷ 5000 = 1.8 ⇒ 2 boxes
Joint Compound	NUM sheets × sheet area ÷ 1000 × 1.5 pails per 1000 SF = NUM five gallon pails		126 × 48 ÷ 1000 × 1.5 = 9.1 ⇒ 10 pails
Joint Tape	1 – 250′ roll per 5 gallon pail		10 pails = 10 rolls

Estimate the materials for interior wall finish materials for a rectangular 28′ × 64′ building. Drywall sheets are 4 × 12, 16″ OC framing, 8′ wall height and 150′ of interior walls.

Figure 69–24 Example of estimating drywall materials with formulas.

© Cengage Learning 2014

DECONSTRUCT THIS

Carefully study *Figure 69–25* and think about what is wrong and/or what is right. Consider all possibilities. What construction practice or method is different in your area of the country?

© Cengage Learning 2014

Figure 69–25 This photo shows a drywall finisher working.

KEY TERMS

backing board	deadman	face	joint reinforcing tape
cement board	drywall	fill coat	metal corner tape
contact adhesives	drywall cutout tool	finishing coat	tile
control joints	drywall foot lifter	gypsum lath	Type X
corner beads	eased edge	joint compounds	

REVIEW QUESTIONS

Select the most appropriate answer.

1. Standard gypsum board width is

   a. 36 inches.
   b. 48 inches.
   c. 54 inches.
   d. 60 inches.

2. Standard gypsum board lengths are

   a. 8, 10, and 12 feet.
   b. 8, 10, 12, and 14 feet.
   c. 8, 9, 10, 12, and 14 feet.
   d. 8, 9, 10, 12, 14, and 16 feet.

3. When fastening drywall, minimum penetration of ring-shanked nails into the framing member is

   a. $\frac{1}{2}$ inch.
   b. $\frac{3}{4}$ inch.
   c. $\frac{7}{8}$ inch.
   d. 1 inch.

4. Gypsum board is usually installed vertically on walls when the wall height is greater than

   a. 8'-0".
   b. 8'-1".
   c. 8'-4".
   d. 8'-6".

5. Ceiling joists are sometimes aligned by the use of a

   a. deadman.
   b. dutchman.
   c. strongback.
   d. straightedge.

6. In the single-nailing method, nails are spaced a maximum of

   a. 8 inches OC on walls; 7 inches OC on ceilings.
   b. 10 inches OC on walls; 8 inches OC on ceilings.
   c. 12 inches OC on walls; 10 inches OC on ceilings.
   d. 12 inches OC on walls and ceilings.

7. Screws are spaced

   a. 12 inches OC on walls; 10 inches OC on ceilings.
   b. 12 inches OC on walls and ceilings.
   c. 16 inches OC on walls and ceilings.
   d. 16 inches OC on walls; 12 inches OC on ceilings.

8. Joints in the face layer of a multilayer application are offset from joints in the base layer by at least

   a. 6 inches.
   b. 8 inches.
   c. 10 inches.
   d. 12 inches.

9. The paper-covered edge of water-resistant gypsum board is applied above the lip of the tub or shower pan not less than

   a. $\frac{1}{4}$ inch.
   b. $\frac{3}{8}$ inch.
   c. $\frac{1}{2}$ inch.
   d. $\frac{3}{4}$ inch.

10. When ceramic tile more than $\frac{3}{8}$ inch thick is to be applied over water-resistant gypsum board, fasten the board with screws or nails spaced not more than

    a. 4 inches OC.
    b. 6 inches OC.
    c. 8 inches OC.
    d. 10 inches OC.

11. The drywall finishing tool used to hold joint compound during the finishing process is called a

    a. trowel.
    b. hawk.
    c. knife.
    d. corner trowel.

12. The automatic drywall tool that is designed to apply joint compound and tape to seams in one pass is called a

    a. banjo.
    b. bazooka.
    c. flat.
    d. both a and b.

13. The automatic drywall finishing tool designed to apply a wide thin coat of compound is called a

    a. bazooka.
    b. flat finisher.
    c. nail spotter.
    d. banjo.

14. Fastener pops are caused by

    a. improperly driven fasteners.
    b. using wet framing lumber.
    c. a space between drywall and framing.
    d. all of the above

15. Small-sized patch repairs in drywall (less than 2 inches wide) are most often done with

    a. joint compound and tape.
    b. squared out holes and plywood.
    c. Squared out holes that exposing framing.
    d. all of the above.

UNIT 25
Wall and Ceiling Finish

©Shutterstock, Inc./Eimantas Buzas

Interior finishes serve as accents to design and protection of the structure from water. Long-term use and function relies on proper workmanship.

Plans and specifications often call for the installation of *wall paneling* in certain rooms of both residential and commercial construction. *Ceramic tile* is widely used in restrooms, baths, showers, kitchens, and similar areas. Inexpensive and attractive ceilings may be created by installing suspended ceilings or ceiling tiles. They may be installed in new construction beneath exposed joists or when remodeling below existing ceilings.

Suspended or tile ceiling finish is installed in sections and pieces. The layout of the pattern and border tiles is important for the ceiling to look professionally installed.

OBJECTIVES

After completing this unit, the student should be able to:

- describe and apply several kinds of sheet wall paneling.

- describe and apply various patterns of solid lumber wall paneling.

- describe and install ceramic wall tile to bathroom walls.

- estimate quantities of wall paneling and ceramic wall tile.

- describe the sizes, kinds, and shapes of ceiling tile and suspended ceiling panels.

- lay out and install suspended ceilings.

- lay out and install ceiling tile.

- estimate quantities of ceiling finish materials.

CHAPTER 70

Types of Wall Paneling

Two basic kinds of wall paneling are sheets of various prefinished material and solid *wood boards.* Many compositions, colors, textures, and patterns are available in sheet form. Solid wood boards of many species and shapes are used for both rustic and elegant interiors (*Figure 70–1*).

DESCRIPTION OF SHEET PANELING

Sheets of prefinished plywood, hardboard, particleboard, plastic laminate, and other material are used to panel walls.

Plywood

Prefinished plywood is probably the most widely used sheet paneling. A tremendous variety is available in both hardwoods and softwoods. The more expensive types have a face veneer of real wood. The less expensive kinds of plywood paneling are prefinished with a printed wood grain or other design on a thin vinyl covering. Care must be taken not to scratch or scrape the surface when handling these types. Unfinished plywood panels are also available.

Some sheets are scored lengthwise at random intervals to imitate solid wood paneling. There is always a score 16, 24, and 32 inches from the edge. This facilitates fastening of the sheets and in case the sheet has to be ripped lengthwise to fit stud spacing.

Most commonly used panel thicknesses are $3/16$ and $1/4$ inch. Sheets are normally 4 feet wide and

7 to 10 feet long. An 8-foot length is most commonly used. Panels may be shaped with square, beveled, or shiplapped edges (*Figure 70–2*). Matching molding is available to cover edges, corners, and joints. Thin ring-shanked nails, called *color pins,* are available in colors to match panels. They are used when exposed fastening is necessary.

Hardboard

Hardboard is available in many surface colors, textures, and designs. Some of these designs simulate stone, brick, stucco, leather, weathered or smooth wood, and other materials. Unfinished hardboard is also used that has a smooth, dark brown surface suitable for painting and other decorating. Unfinished hardboard may be solid or perforated in a number of designs.

Tileboard is a hardboard panel with a baked-on plastic finish. It is embossed to simulate ceramic wall tile. The sheets come in a variety of solid colors, marble, floral, and other patterns. Tileboard is designed for use in bathrooms, kitchens, and similar areas.

Hardboard paneling comes in widths of 4 feet and in lengths of from 8 to 12 feet. Commonly used thicknesses are from $1/8$ to $1/4$ inch. Color-coordinated molding and trim are available for use with hardboard paneling.

Particleboard

Panels of particleboard come with wood grain or other designs applied to the surface, similar to plywood and hardboard. Sheets are usually $1/4$ inch thick, 4 feet wide, and 8 feet long. Prefinished particleboard is used when an inexpensive wall covering is desired. Because the sheets are brittle and break easily, care must be taken when handling them. They must be applied only on a solid wall backing.

Unfinished particleboard is not usually used as an interior wall finish. One exception, made from aromatic cedar chips, is used to cover walls in closets to repel moths.

Plastic Laminates

Plastic laminates are widely used for surfacing kitchen cabinets and countertops. They are also

Figure 70–1 Solid wood board paneling provides warmth and beauty to interiors of buildings.

Courtesy of California Redwood Association

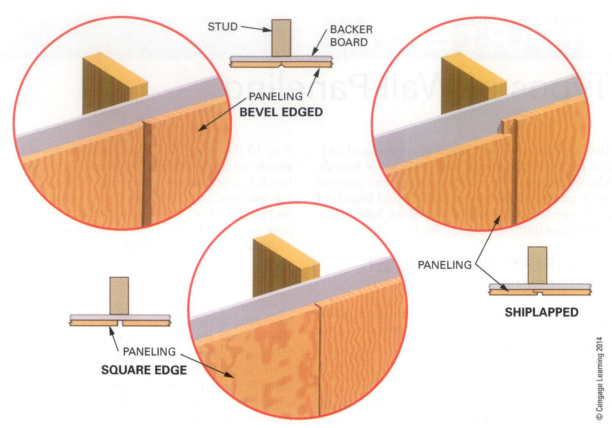

Figure 70–2 Sheet paneling comes with various edge shapes.

used to cover walls or parts of walls in kitchens, bathrooms, restrooms, and similar areas where a durable, easy-to-clean surface is desired. Laminates can be scorched by an open flame. However, they resist mild heat, alcohol, acids, and stains. They clean easily with a mild detergent.

Laminates are manufactured in many colors and designs, including wood grain patterns. Surfaces are available in gloss, satin, and textured finishes, among others.

Laminates are ordinarily used in two thicknesses. Vertical-type laminate is relatively thin (about $\frac{1}{32}$ inch). It is used for vertical surfaces, such as walls and cabinet sides. Vertical-type laminate is available only in widths of 4 feet or 8 feet. Regular or standard laminate is about $\frac{1}{16}$ inch thick. It comes in widths of 24, 36, 48, and 60 inches and in lengths of 5, 6, 8, 10, and 12 feet. It is generally used on horizontal surfaces, such as countertops. It can be used on walls, if desired, or if the size required is not available in vertical type. Sheets are usually manufactured 1 inch wider and longer than the nominal size.

Laminates are difficult to apply to wall surfaces because they are so thin and brittle. Also, because a *contact cement* is used, the sheet cannot be moved once it makes contact with the surface. Thus, prefabricated panels, with sheets of laminate already bonded

to a backer, are normally used to panel walls. See Chapter 85 for installation techniques for laminates.

DESCRIPTION OF BOARD PANELING

Board paneling is used on interior walls when the warmth and beauty of solid wood is desired. Wood paneling is available in softwoods and hardwoods of many species. Each has its own distinctive appearance, unique grain, and knot pattern.

Wood Species

Woods may be described as light, medium, and dark toned. Light tones include birch, maple, spruce, and white pine. Some medium tones are cherry, cypress, hemlock, oak, ponderosa pine, and fir. Among the darker-toned woods are cedar, mahogany, redwood, and walnut. For special effects, knotty pine, wormy chestnut, pecky cypress, and white-pocketed Douglas fir board paneling may be used.

Surface Textures and Patterns

Wood paneling is available in many shapes. It is either planed for smooth finishing, or rough-sawn for a rustic, informal effect. Square-edge boards may be joined edge to edge, spaced on a dark background, or applied in *board-and-batten* or *board-on-board* patterns. *Tongue-and-grooved* or *shiplapped* paneling comes in patterns, a few of which are illustrated in *Figure 70–3*.

Sizes

Most wood paneling comes in a $3/4$-inch thickness and in nominal widths of 4, 6, 8, 10, and 12 inches. A few patterns are manufactured in a $9/16$-inch thickness. Aromatic cedar paneling is used in clothes closets. It runs from $3/8$ to $5/16$ inch thick, depending on the mill. It is usually *edge-* and *end-matched* (tongue and-grooved) for application to a backing surface.

Moisture Content

To avoid shrinkage, wood paneling, like all interior finish, should be dried to a *equilibrium moisture content (EMC)*. That is, its moisture content should be about the same as the area in which it is to be used. Except for arid, desert areas and some coastal regions, the average moisture content of the air in the United States is about 8 percent (*Figure 70–4*). Interior finish applied with an excessive moisture content will eventually shrink, causing open joints, warping, loose fasteners, and many other problems.

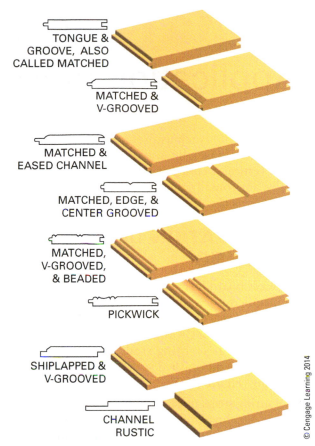

TONGUE & GROOVE, ALSO CALLED MATCHED

MATCHED & V-GROOVED

MATCHED & EASED CHANNEL

MATCHED, EDGE, & CENTER GROOVED

MATCHED, V-GROOVED, & BEADED

PICKWICK

SHIPLAPPED & V-GROOVED

CHANNEL RUSTIC

© Cengage Learning 2014

Figure 70–3 Solid wood paneling is available in a number of patterns.

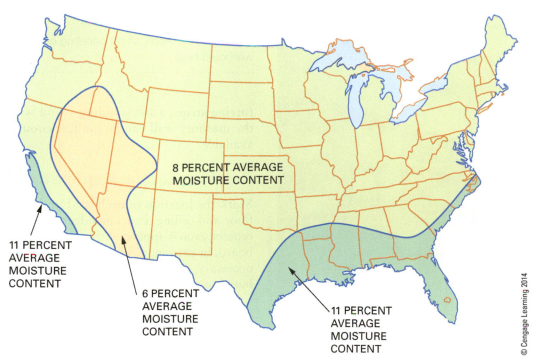

8 PERCENT AVERAGE MOISTURE CONTENT

11 PERCENT AVERAGE MOISTURE CONTENT

6 PERCENT AVERAGE MOISTURE CONTENT

11 PERCENT AVERAGE MOISTURE CONTENT

© Cengage Learning 2014

Figure 70–4 Recommended average moisture content for interior finish woodwork in different parts of the United States.

CHAPTER 71
Application of Wall Paneling

Sheet paneling, such as prefinished plywood and hardboard, is usually applied to walls with long edges vertical. *Board paneling* may be installed vertically, horizontally, diagonally, or in many interesting patterns (*Figure 71–1*).

INSTALLATION OF SHEET PANELING

A backer board layer, at least 3/8 inch thick, should be installed on walls prior to the application of sheet paneling. The backing makes a stronger and more fire-resistant wall, helps block sound transmission, tends to bring studs in alignment, and provides a rigid finished surface for application of paneling (*Figure 71–2*). **✱**

✱IRC R702.5

Furring strips must be applied to masonry walls before paneling is installed. The strips are usually applied by driving hardened nails. Instead of using

SHEET PANELING   GYPSUM BOARD AS BACKER

PARTITION STUD

© Cengage Learning 2014

Figure 71–2 Apply sheet paneling over a gypsum wallboard base.

furring strips, a freestanding wood wall close to the masonry wall can be built, if enough space is available.

Preparation for Application

Mark the location of each stud in the wall on the floor and ceiling. Paneling edges must fall on stud centers, even if applied with adhesive over a backer board, in case supplemental nailing of the edges is necessary. Panels are usually fastened with a combination of color pins and adhesive.

Apply narrow strips of color on the wall from floor to ceiling where joints between paneling will occur. Use paint or magic marker colored the same as the joint of the panel. If joints between sheets open slightly because of shrinkage during extended dry periods or heating seasons, it is not so noticeable with a similar color behind the joint.

Courtesy of California Redwood Association

Figure 71–1 Wood paneling can be applied in many directions.

Sometimes paneling does not extend to the ceiling but covers only the lower portion of the wall. This partial paneling is called **wainscoting**. It is usually installed about 3 feet above the floor (*Figure 71–3*).

If the wall is to be wainscoted, snap a horizontal line across the wall to indicate its height.

Stand panels on their long edge against the wall for at least forty-eight hours before installation. This allows them to adjust to room temperature and humidity. Otherwise sheets may buckle after installation, especially if they were very dry when installed.

Just prior to application, stand panels on end, side by side, around the room. Arrange them by matching the grain and color to obtain the most pleasing appearance.

Starting the Application

Start in a corner and continue installing consecutive sheets around the room. Select the starting corner, remembering that it will also be the ending point. This corner should be the least visible, such as behind an often-open door. Cut the first sheet to a length about $\frac{1}{4}$ inch less than the wall height. Place the sheet in the corner. Plumb the outside edge and tack it in the plumb position. Set the distance between the points of the dividers the same as the amount the sheet overlaps the center of the stud (*Figure 71–4*). **Scribe** this amount on the edge of the sheet butting the corner (*Figure 71–5*).

Remove the sheet from the wall. Place it on saw-horses on which a sheet of plywood, or two 8-foot lengths of 2 × 4s, have been placed for support. Cut to the scribed lines with a sharp, fine-toothed,

WALLBOARD

CHAIR RAIL

PANELING

BASEBOARD

© Cengage Learning 2014

Figure 71–3 Wainscoting is a wall finish applied to the lower portion of the wall that is different from the *upper portion*.

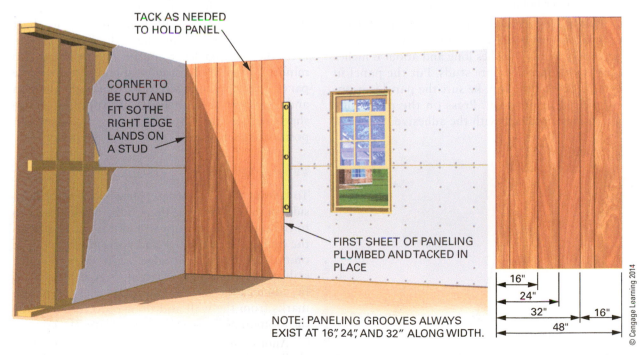

TACK AS NEEDED TO HOLD PANEL

CORNER TO BE CUT AND FIT SO THE RIGHT EDGE LANDS ON A STUD

FIRST SHEET OF PANELING PLUMBED AND TACKED IN PLACE

NOTE: PANELING GROOVES ALWAYS EXIST AT 16″, 24″, AND 32″ ALONG WIDTH.

16″
24″
32″
16″
48″

© Cengage Learning 2014

Figure 71–4 The first sheet must be set plumb in the corner.

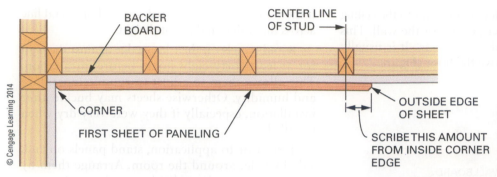

BACKER BOARD

CENTER LINE OF STUD

CORNER

FIRST SHEET OF PANELING

OUTSIDE EDGE OF SHEET

SCRIBE THIS AMOUNT FROM INSIDE CORNER EDGE

Figure 71–5
Set scribers equal to the largest space and scribe the edge of the first sheet to the corner.

hand crosscut saw. Handsaws may be used on the face side because they cut on the downstroke. This action will keep the splintering of the cut on the backside of the panel. Circular and jig saws should be used from the backside.

Replace the sheet with the cut edge fitting snugly in the corner. The joint at the ceiling need not be fitted if a molding is to be used. If a tight fit between the panel and ceiling is desired, set the dividers and scribe a small amount at the ceiling line. Remove the sheet again. Cut to the scribed line. Replace the sheet, and raise it snugly against the ceiling. The space at the bottom will be covered later by a baseboard.

Fastening

If only nails are used, fasten about 6 inches apart along edges and about 12 inches apart on intermediate studs for $\frac{1}{4}$-inch-thick paneling. Nails may be spaced farther apart on thicker paneling. Drive nails at a slight angle for better holding power (*Figure 71–6*).

If adhesives are used, apply a $\frac{1}{8}$-inch continuous bead where panel edges and ends make contact. Apply beads 3 inches long and about 6 inches apart on all intermediate studs. Put the panel in place. Tack it at the top. Be sure the panel is properly placed in position. Press on the panel surface to make contact with the adhesive. Use firm,

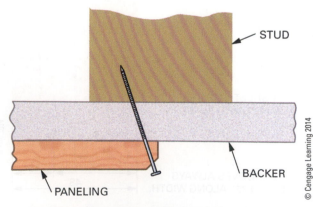

STUD

BACKER

PANELING

Figure 71–6 Drive nails or color pins at a slight angle for better holding power.

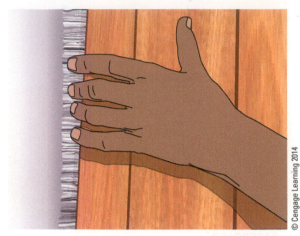

Figure 71–7 Press sheet into place against adhesive, then pull sheet away a short distance to speed up the setting of adhesive.

uniform pressure to spread the adhesive beads evenly between the wall and the panel. Then, grasp the panel and slowly pull the bottom of the sheet a few inches away from the wall (*Figure 71–7*). Press the sheet back into position after about two minutes. After about twenty minutes, recheck the panel. Apply pressure to ensure thorough adhesion and to smooth the panel surface. Apply successive sheets in the same manner. Do not force panels in position. Panels should touch lightly at joints.

Wall Outlets

To lay out for wall outlets, plumb and mark both sides of the outlet to the floor. If the opening is close to the ceiling, plumb upward and mark lightly on the ceiling. Level the top and bottom of the outlet on the wall beyond the edge of the sheet to be installed. Or, level on the adjacent sheet, if closer. Cut, fit, and tack the sheet in position. Level and plumb marks from the wall and floor onto the sheet for the location of the opening (*Procedure 71–A*).

Another method is to rub a cake of carpenter's chalk on the edges of the outlet box. Fit and tack the sheet in position. Tap on the sheet directly over the outlet to transfer the chalked edges to the back

StepbyStep Procedures

Procedure 71–A Cutting Outlet Holes in Paneling

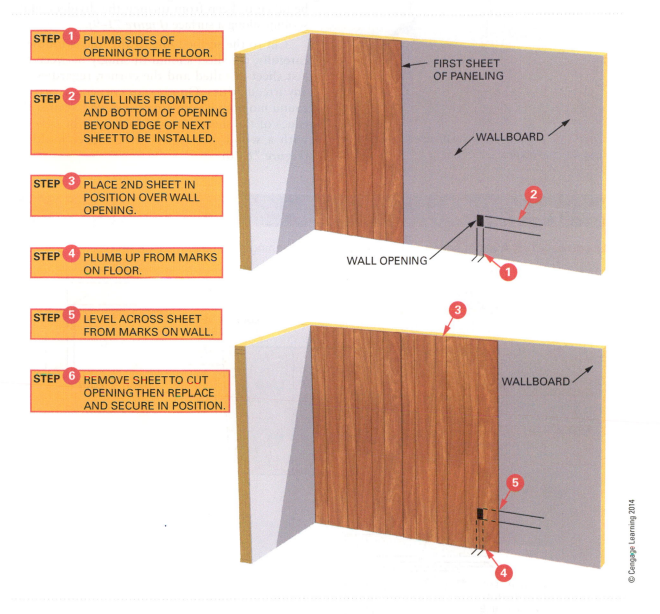

STEP 1 PLUMB SIDES OF OPENING TO THE FLOOR.

STEP 2 LEVEL LINES FROM TOP AND BOTTOM OF OPENING BEYOND EDGE OF NEXT SHEET TO BE INSTALLED.

STEP 3 PLACE 2ND SHEET IN POSITION OVER WALL OPENING.

STEP 4 PLUMB UP FROM MARKS ON FLOOR.

STEP 5 LEVEL ACROSS SHEET FROM MARKS ON WALL.

STEP 6 REMOVE SHEET TO CUT OPENING THEN REPLACE AND SECURE IN POSITION.

FIRST SHEET OF PANELING

WALLBOARD

WALL OPENING

WALLBOARD

© Cengage Learning 2014

of the sheet. Remove the sheet. Cut the opening for the outlet. Openings for wall outlets, such as electrical boxes, must be cut fairly close to the location and size. The cover plate may not cover if the cutout is not made accurately. This could require replacement of the sheet. A jig saw may be used to cut these openings. Remember to cut from the back of the panel to avoid splintering the face (*Figure 71–8*).

Ending the Application

The final sheet in the wall need not fit snugly in the corner if the adjacent wall is to be paneled or if interior corner molding is to be used. Take measurements at the top, center, and bottom. Cut the sheet to width, and install.

If the last sheet butts against a finished wall and no corner molding is used, the sheet must be cut to fit snugly in the corner. To mark the sheet accurately, first measure the remaining space at the top, bottom, and about the center. Rip the panel about $1/2$ inch wider than the greatest distance. Place the sheet with the cut edge in the corner and the other edge overlapping the last sheet installed. Tack the sheet in position. The amount of overlap should be exactly the same from top to bottom. Set the dividers for the amount of overlap. Scribe this amount

Figure 71–8 A saber saw may be used to cut an opening for an electrical outlet box.

on the edge in the corner (*Procedure 71–B*). Instead of dividers, it is sometimes more exact to use a small block of wood for scribing. The width is cut the same as the amount of overlap. Care must be taken to keep from turning the dividers while scribing along a surface (*Figure 71–9*).

Cut to the scribed line. If the line is followed carefully, the sheet should fit snugly between the last sheet installed and the corner, regardless of any irregularities. On exterior corners, a quarter-round molding is sometimes installed against the edges of the sheets. Or, the joint may be covered with a wood, metal, or vinyl corner molding (*Figure 71–10*).

Step**by**Step Procedures

Procedure 71–B A saber saw may be used to cut an opening for an electrical outlet box

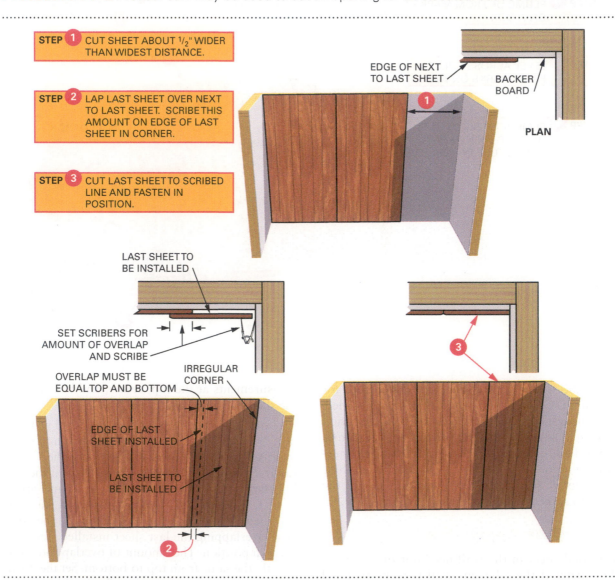

STEP 1 CUT SHEET ABOUT ½" WIDER THAN WIDEST DISTANCE.

STEP 2 LAP LAST SHEET OVER NEXT TO LAST SHEET. SCRIBE THIS AMOUNT ON EDGE OF LAST SHEET IN CORNER.

STEP 3 CUT LAST SHEET TO SCRIBED LINE AND FASTEN IN POSITION.

EDGE OF NEXT TO LAST SHEET

BACKER BOARD

PLAN

LAST SHEET TO BE INSTALLED

SET SCRIBERS FOR AMOUNT OF OVERLAP AND SCRIBE

IRREGULAR CORNER

OVERLAP MUST BE EQUAL TOP AND BOTTOM

EDGE OF LAST SHEET INSTALLED

LAST SHEET TO BE INSTALLED

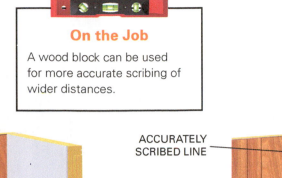

On the Job

A wood block can be used for more accurate scribing of wider distances.

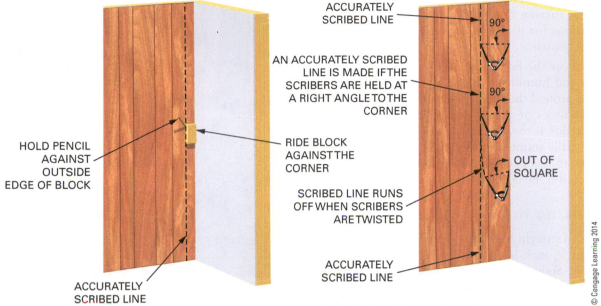

ACCURATELY SCRIBED LINE

90°

AN ACCURATELY SCRIBED LINE IS MADE IF THE SCRIBERS ARE HELD AT A RIGHT ANGLE TO THE CORNER

90°

HOLD PENCIL AGAINST OUTSIDE EDGE OF BLOCK

RIDE BLOCK AGAINST THE CORNER

OUT OF SQUARE

SCRIBED LINE RUNS OFF WHEN SCRIBERS ARE TWISTED

ACCURATELY SCRIBED LINE

ACCURATELY SCRIBED LINE

© Cengage Learning 2014

Figure 71–9 Accurate scribing requires that the marked line be made perpendicular to the corner.

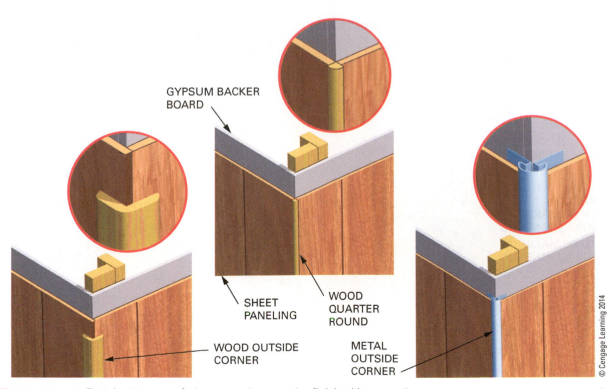

GYPSUM BACKER BOARD

SHEET PANELING

WOOD QUARTER ROUND

WOOD OUTSIDE CORNER

METAL OUTSIDE CORNER

© Cengage Learning 2014

Figure 71–10 Exterior corners of sheet paneling may be finished in several ways.

INSTALLING SOLID WOOD BOARD PANELING

Horizontal board paneling may be fastened to studs in new and existing walls (*Figure 71–11*). For vertical application of board paneling in a frame wall, blocking must be provided between studs (*Figure 71–12*). On existing and masonry walls, horizontal furring strips must be installed. Blocking or furring must be provided in appropriate locations for diagonal or pattern applications of board paneling.

Allow the boards to adjust to room temperature and humidity by standing them against the walls around the room. At the same time, put them in the order of application. Match them for grain and color. If tongue-and-grooved boards are to be eventually stained or painted, apply the same finish to the tongues so that an unfinished surface is not exposed if the boards shrink after installation.

Starting the Application

Select a straight board to start with. Cut it to length, about ¼ inch less than the height of the wall. If tongue-and-grooved stock is used, tack it in a plumb position with the grooved edge in the corner. If a tight fit is desired, adjust the dividers to scribe an amount a little more than the depth of the groove.

Courtesy of California Redwood Association

Figure 71–11 No wall blocking is required for horizontal application of board paneling.

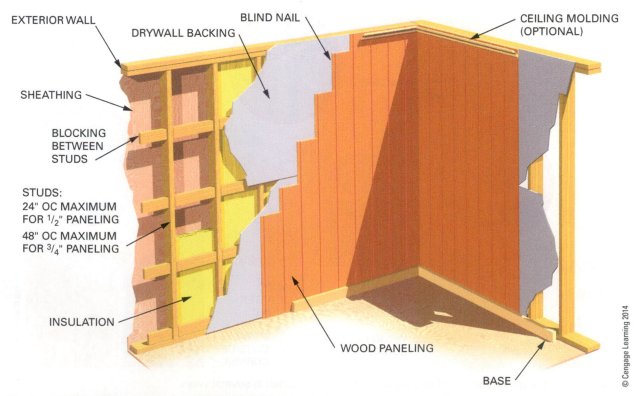

© Cengage Learning 2014

Figure 71–12 Blocking must be provided between studs for vertical board paneling.

Figure 71–13 Tongue-and-grooved paneling is blind nailed.
© Cengage Learning 2014

Rip to the scribed line. Face nail along the cut edge into the corner with finish nails about 16 inches apart. Blind nail the other edge through the tongue.

Continuing the Application

Apply succeeding boards by blind nailing into the tongue only (*Figure 71–13*). Make sure the joints between boards come up tightly. See Figure 60–15 for methods of bringing edge joints of matched boards up tightly if they are slightly crooked. Severely warped boards should not be used. As installation progresses, check the paneling for plumb. If out of plumb, gradually bring back by driving one end of several boards a little tighter than the other end. Cut out openings in the same manner as described for sheet paneling.

Applying the Last Board

If the last board in the installation must fit snugly in the corner without a molding, the layout should be planned so that the last board will be as wide as possible. If boards are a uniform width, the width of the starting board must be planned to avoid ending with a narrow strip. If random widths are used, they can be arranged when nearing the end.

Cut and fit the next to the last board. Then remove it. Tack the last board in the place of the next to the last board. Cut a scrap block about 6 inches long and equal in width to the finished face of the next to the last board. Use this block to scribe the last board by running one edge along the corner and holding a pencil against the other edge (*Figure 71–14*). Remove the board from the wall. Cut it to the scribed line. Fasten the next to the last board in position. Fasten the last board in position with the cut edge in the corner. Face nail the edge nearest the corner.

Horizontal application of wood paneling is done in a similar manner. However, blocking between studs on open walls or furring strips on existing walls are not necessary. On existing walls, locate and snap lines to indicate the position of stud centerlines. The thickness of wood paneling should be at least $\frac{3}{8}$ inch for 16-inch spacing of frame members and $\frac{5}{8}$ inch for 24-inch spacing. Diagonal and pattern application of board paneling is similar to vertical and horizontal applications. If wainscoting is applied to a wall, the joint between the different materials may treated in several ways (*Figure 71–15*).

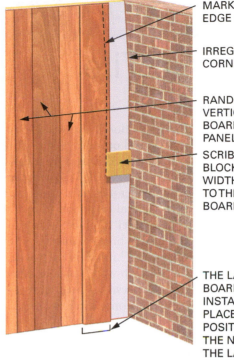

Figure 71–14 Laying out the last board to fit against a finished corner.

MARK INSIDE EDGE OF BLOCK

IRREGULAR CORNER

RANDOM WIDTH VERTICAL BOARD WALL PANELING

SCRIBING BLOCK OF SAME WIDTH AS NEXT TO THE LAST BOARD

THE LAST BOARD TO BE INSTALLED IS PLACED IN THE POSITION OF THE NEXT TO THE LAST BOARD

© Cengage Learning 2014

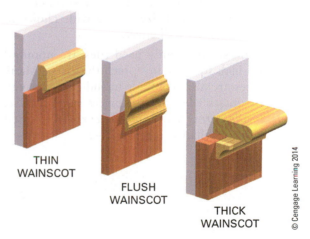

THIN WAINSCOT

FLUSH WAINSCOT

THICK WAINSCOT

Figure 71–15 Methods of finishing the joint at the top of wainscoting.

APPLYING PLASTIC LAMINATES

Plastic laminates are not usually applied to walls unless first prefabricated to sheets of plywood or similar material. They are then installed on walls in the same manner as sheet paneling. Special matching molding is used between sheets and on interior and exterior corners. (The application of plastic laminates is described in greater detail in chapter 85 on kitchen cabinets and countertops.)

CHAPTER 72
Ceramic Tile

Ceramic tile is used to cover floors and walls in restrooms, baths, showers, and other high-moisture areas that need to be cleaned easily and frequently (*Figure 72–1*). On large jobs, the tile is usually applied by specialists. On smaller jobs, it is sometimes more expedient for the general carpenter to install tile. Ceramic tile is usually set in place using thin set, a mortar-type adhesive, but is sometimes set using an organic adhesive. Thin set is used when tile is likely to get soaked with water, such as when used in showers. Organic adhesive is used when tile will not get soaked, such as on a kitchen counter backsplash.

DESCRIPTION OF CERAMIC TILE

Ceramic tiles are usually rectangular or square, but many geometric shapes, such as hexagons and octagons, are manufactured. Many solid colors, patterns, designs, and sizes give a wide choice to achieve the desired wall effect.

The most commonly used tiles are nominal 4- and 6-inch squares, in $\frac{1}{4}$-inch thickness. Many other sizes are also available including 1-through 12-inch-square tiles. Special pieces such as *base, caps, inside corners,* and *outside corners* are used to trim the installation (*Figure 72–2*).

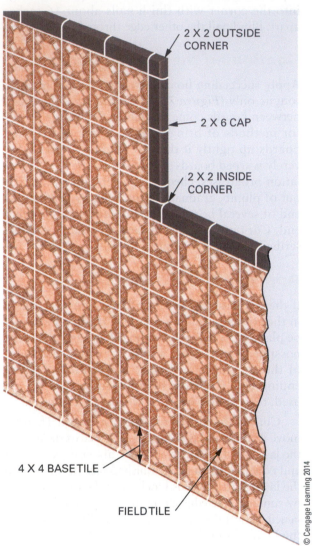

2 X 2 OUTSIDE CORNER

2 X 6 CAP

2 X 2 INSIDE CORNER

4 X 4 BASE TILE

FIELD TILE

© Cengage Learning 2014

Figure 72–2 Special pieces are used to trim a ceramic tile installation.

© Cengage Learning 2014

Figure 72–1 Ceramic tile is used extensively in high-moisture areas such as showers.

WALL PREPARATION FOR TILE APPLICATION

Cement board is the recommended backing for ceramic tile when tile will become wet daily. Otherwise, water-resistant gypsum board may be used. Installation instructions for these products are given in *Figure 68–20.*✳

✳IRC
R702.4.2

Minimum Backer and Tile Area in Baths and Showers

Tiles should overlap the lip and be applied down to the shower floor or top edge of the bathtub. On tubs without a showerhead, they should be installed to extend to a minimum of 6 inches above the rim. Around bathtubs with showerheads, tiles should extend a minimum of 5 feet above the rim or 6 inches above the showerhead, whichever is higher.

In shower stalls, tiles should be a minimum of 6 feet above the shower floor or 6 inches above the showerhead, whichever is higher. A 4-inch minimum extension of the full height is recommended beyond the outside face of the tub or shower (*Figure 72–3*).

Calculating Border Tiles

Before beginning the application of ceramic wall tile, the width of the *border tiles* must be calculated. Border tiles are those that fit to the corners and edges of the tiled area. The installation has a professional appearance if the border tiles are the same width and also as wide as possible (*Figure 72–4*).

Measure the width of the wall from corner to corner. Change the measurement to inches. Divide it by the width of a tile. Measure the tile accurately. Sometimes, the actual size of a tile is different than its nominal size. Add the width of a tile to the remainder. Divide by two to find the width of border tiles.

EXAMPLE

A wall section measures 8′-4″, or 100 inches from wall to wall. If $4\frac{1}{4}$-inch, actual size, tiles are used, dividing 100 by $4\frac{1}{4}$ equals 23 full tiles with 0.53 of a tile remainder. Add 1 to the decimal remainder and multiply it by the width of a tile, or $1.53 \times 4.25 = 6.5$ inches. Divide this by 2 to get $3\frac{1}{4}$ inches. This will give twenty-two full tiles and two $3\frac{1}{4}$-inch border tiles across the tiled area.

Tile Layout

A layout line needs to be placed as a guide for the individual tiles. Measure out from the corner and mark a point that will be the edge of a full tile.

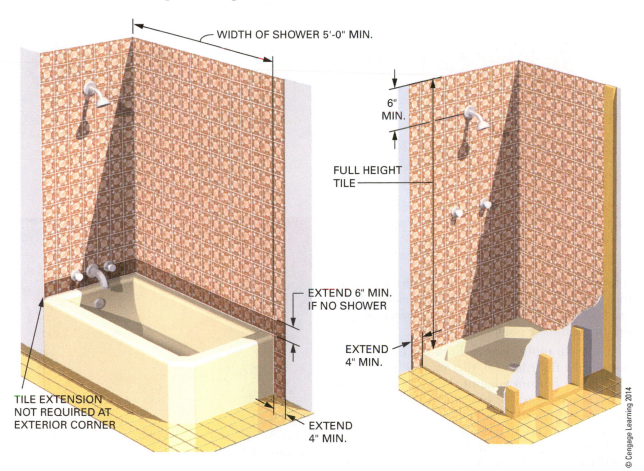

Figure 72–3 Minimum areas recommended for the installation of ceramic tile around bathtubs and shower stalls.

Figure 72–4 Border tiles should be as wide as possible and close to the same width on both sides of the wall.

BORDER TILES

© Cengage Learning 2014

If the section to be tiled is a wall, plumb a line from top to bottom. If the section is a floor, measure and mark another point from the other corner. Snap a line through these points.

For wall applications, check the level of the floor. If the floor is level, tiles may be applied by placing the bottom edge of the first row on the floor using plain tile or base tile. If the floor is not level, the bottom edge of the tiles must be cut to fit the floor while keeping the top edges level. Place a level on the floor and find the low point of the tile installation. From this point, measure up and mark on the wall the height of a tile. Draw a level line on the wall through the mark. Tack a straightedge on

the wall with its top edge to the line. Tiles are then laid to the straightedge. When tiling is completed above, the straightedge is removed. The tiles in the bottom row are then cut and fitted to the floor.

For floor applications of tile, check that the line is parallel to the opposite wall. Adjust the line to split any differences. This line is best placed where the tile will be most often viewed from a distance. This will ensure the tiles are as straight as possible where they are most visible.

TILE APPLICATION

When tiling walls such as a shower, it usually is best to install tile on the back wall first. This way, the joint in the corner is the least visible and the most water-tight. Apply the recommended adhesive to the wall or floor with a trowel. Use a flat trowel with grooved edges that allows the recommended amount of adhesive to remain on the surface (*Figure 72–5*). Too heavy a coat results in adhesive being squeezed out between the joints of applied tile, causing a mess. Too little adhesive results in failure of tiles to adhere to the wall. Follow the manufacturer's directions for the type of trowel to use and the amount of adhesive to be spread at any one time. Be careful not to spread adhesive beyond the area to be covered.

Green Tip

Masonry and ceramic tile surfaces can store solar energy when installed in areas of direct sunlight. This helps balance the daily high and low room temperatures.

Figure 72–5 Use a trowel that is properly grooved to allow the correct amount of adhesive to remain on the surface.

© Cengage Learning 2014

Apply the first tile to the guide line. Press the tile firmly into the adhesive. Apply other tiles in the same manner. Start from the center guide line and pyramid upward and outward. As tiles are applied, slight adjustments may need to be made to keep them lined up. Tile spacers may be used to help keep tile properly spaced (*Figure 72–6*). Tile spacers are rubber pieces with the same dimensions as the joints between tiles. Keep fingers clean and adhesive off the face of the tiles. Clean tiles with a damp cloth.

Cutting Border Tiles

After all *field tiles* are applied, it is necessary to cut and apply border tiles. Field tiles are whole tiles that are applied in the center of the wall. Ceramic tile may be cut in any of several ways. Tile saws can be used to cut all types of tile (*Figure 72–7*).

They are operated in a manner similar to that of a power miter box except that the material is eased into the blade on a sliding tray. Water is pumped from the reservoir below to the blade, keeping it cool during the cutting process. Thin ceramic tile is often cut using a hand cutter (*Figure 72–8*). First the tool scores the tile. The back edge of the cutter is then pressed against the tile to break it. This tool cannot make small or narrow cuts. A nibbler is used to make small or irregular cuts (*Figure 72–9*). Nibblers chip small pieces off in sometimes random directions. Care should be taken to cut only small pieces at a time. This will allow a successful cut to be made gradually.

To finish the edges and ends of the installation, *caps* are sometimes used. Caps may be 2 × 6 or 4 × 4 pieces with one rounded finished edge. Special trim pieces are used to finish interior and exterior corners (*Figure 72–10*).

Figure 72–6 Rubber tile spacers may be used to maintain a uniform grout spacing.

Figure 72–8 Hand cutter for thin ceramic tile.

Figure 72–7 Tile cutters are often used for cutting ceramic tile.

Figure 72–9 Nibblers are used to cut curves.

Figure 72–10 Special trim pieces are often used to add accents to a tiled surface.

Grouting Tile Joints

After all tile has been applied, the joints are filled with tile grout. Grout comes in a powder form. It is mixed with water to form a paste of the desired consistency. It is spread over the face of the tile with a *rubber trowel* to fill the joints. The grout is worked into the joints. Then, the surface is wiped as clean as possible with the trowel.

Wall tile grout is allowed to set up, but not harden. The joints are then *pointed*. Removing excess grout and smoothing joints is called *pointing*. A small hardwood stick with a rounded end can be used as a pointing tool. The entire surface is wiped clean with a dry cloth after the grout has dried.

After floor tile grout has set up slightly, the excess is removed. Wipe the tile across the grout lines at a 45-degree angle with a damp sponge. Wipe once, then turn the sponge over and wipe another area. Rinse and repeat. The key is to keep the sponge clean for one wiping at a time. Let it set up more and repeat. Finish the grouting by buffing with a dry clean rag.

After the grout has set and cured for several weeks, silicone grout sealer may be applied. This product seals tiny pores in the grout, making it more resistant to staining. Sealer is liberally brushed on the grout, allowing it to soak in. The excess is wiped clean before it dries.

CHAPTER 73

Ceiling Finish

Suspended ceilings are widely used in commercial and residential construction as a ceiling finish. They also provide space for recessed lighting, ducts, pipes, and other necessary conduits (*Figure 73–1*). There are different systems each providing a different look, but they are installed with similar layouts and structural systems. Some systems can provide another insulation layer to better conserve energy. All systems provide an ability better control sound transmissions.

In remodeling work, a suspended system can be easily installed beneath an existing ceiling. In residential basements, where overhead pipes and ducts may make other types of ceiling application difficult, a suspended type is easily installed. In addition, removable panels make pipes, ducts, and wiring accessible.

SUSPENDED CEILING COMPONENTS

Suspended ceiling can be a generic name for all ceiling systems where the ceiling hangs below the framing. It is also a term use to refer to a particular type ceiling, sometimes called a drop ceiling.

A suspended ceiling system consists of panels that are laid into a metal grid. The grid consists of main runners, cross tees, and wall angles. It is constructed in a 2 × 4 rectangular or 2 × 2 square pattern (*Figure 73–2*). Grid members come prefinished in white, black, brass, chrome, and wood grain patterns, among others.

Wall Angles

Wall angles are L-shaped pieces that are fastened to the wall to support the ends of main runners and cross tees. They come in 10- and 12-foot lengths. They provide a continuous finished edge around the perimeter of the ceiling, where it meets the wall.

Main Runners

Main runners or tees are shaped in the form of an upside-down T. They come in 12-foot lengths. End splices make it possible to join lengths of main runners together. Slots are punched in the side of the runner at 12-inch intervals to receive cross tees. Along the top edge, punched holes are spaced at

Figure 73–1 Suspended ceilings offer a variety of design styles.

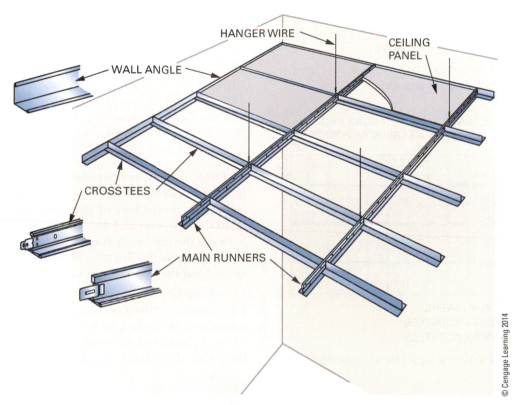

Figure 73–2 A suspended ceiling consists of grid members and ceiling panels.

intervals for suspending main runners with *hanger wire*. Main runners extend from wall to wall. They are the primary support of a ceiling's weight.

Cross Tees

Cross tees come in 2- and 4-foot lengths. A slot, of similar shape and size as those in main runners, is centered on the 4-foot cross tees for use when turning 2 × 4 grid into a 2 × 2 grid. They come with connecting tabs on each end. These tabs are inserted and locked into main runners and other cross tees.

Ceiling Panels

Ceiling panels are manufactured of many different kinds of material, such as gypsum, glass fibers, mineral fibers, and wood fibers. Panel selection is based on considerations such as fire resistance, sound control, thermal insulation, light reflectance, moisture resistance, maintenance, appearance, and cost. Panels are given a variety of surface textures, designs, and finishes. They are available in 2 × 2 and 2 × 4 sizes with square or rabbeted edges (*Figure 73–3*).

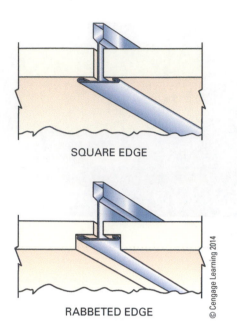

SQUARE EDGE

RABBETED EDGE

© Cengage Learning 2014

Figure 73–3 Suspended ceiling panels may have square or rabbeted edges and ends.

CORNER TILE DIMENSIONS SHOULD BE AS LARGE AS POSSIBLE

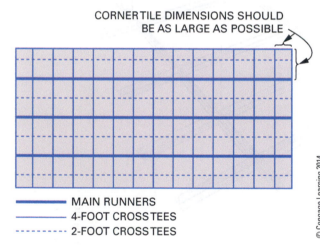

——— MAIN RUNNERS
——— 4-FOOT CROSS TEES
- - - - 2-FOOT CROSS TEES

© Cengage Learning 2014

Figure 73–4 A typical layout for a suspended ceiling grid.

SUSPENDED CEILING LAYOUT

Before the actual installation of a suspended ceiling, a scaled sketch of the ceiling grid should be made. The sketch should indicate the direction and location of the main runners, cross tees, light panels, and border panels.

Main runners usually are spaced 4 feet apart. They usually run parallel with the long dimension of the room. For a standard 2 × 4 pattern, 4-foot cross tees are then spaced 2 feet apart between main runners. If a 2 × 2 pattern is used, 2-foot cross tees are installed between the midpoints of the 4-foot cross tees. Main runners and cross tees should be located in such a way that *border panels*

on both sides of the room are equal and as large as possible (*Figure 73–4*). Sketching the ceiling layout also helps when estimating materials.

Sketching the Layout

Sketch a grid plan by first drawing the overall size of the ceiling to a convenient scale. Use special care in measuring around irregular walls.

Locating Main Runners.   To locate main runners, change the width of room to inches and divide by 48. Add 48 inches to any remainder. Divide the sum by 2 to find the distance from the wall to the first main runner. This distance is also the length of border panels.

 EXAMPLE

If the width of the room is 15'-8", changing to inches equals 188. Dividing 188 by 48 equals 3, with a remainder of 44 inches. Adding 48 to 44 equals 92 inches. Dividing 92 by 2 equals 46 inches. Thus there will be 3 (Figure 73-5) main tees and two border tiles 46 inches long.

Draw a main runner the calculated distance from, and parallel to, the long dimension of the ceiling. Draw the rest of the main runners parallel to the first, and at 4-foot intervals. The distance between the last main runner and the wall should be the same as the distance between the first main runner and the opposite wall (*Figure 73–5*).

Locating Cross Tees.   To locate 4-foot cross tees between main runners, first change the long dimension of the ceiling to inches. Divide by 24. Add 24 to the remainder. Divide the sum by 2 to find the width of the border panels on the other walls.

 EXAMPLE

If the long dimension of the room is 27'-10", changing it to inches equals 334. Dividing 334 by 24 equals 13, with a remainder of 22 inches. Adding 24 to 22 equals 46 inches. Dividing 46 by 2 equals 23 inches. Thus along the room length there will be 13 (Figure 73-6) cross tees and two border tiles 23 inches wide.

Draw the first row of cross tees the calculated distance from, and parallel to, the short wall. Draw the remaining rows of cross tees parallel to each other at 2-foot intervals. The distance from the last row of cross tees to the wall should be the same as the distance from the first row of cross tees to the opposite wall (*Figure 73–6*).

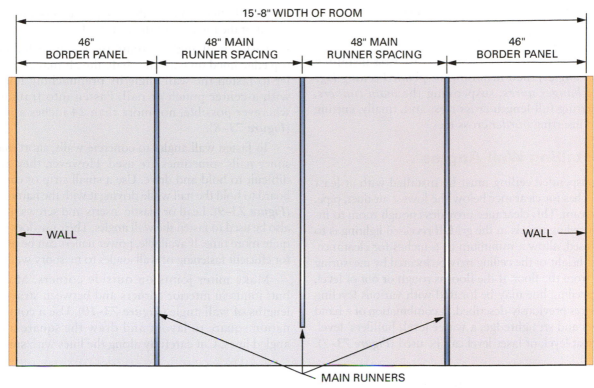

1. CHANGE ROOM WIDTH DIMENSION TO INCHES: 15 X 12 + 8 = 188"
2. DIVIDE 188 BY 48 = 3.91667 TILES
3. ADD 1 TO REMAINDER = 1.91667 TILES
4. DIVIDE THIS NUMBER BY 2: 1.91667 ÷ 2 = 0.95833
5. MULTIPLY FRACTION OF A TILE BY 48: 0.95833 X 48 = 46" BORDER TILE LENGTH

Figure 73–5 Method of determining the location of main runners.

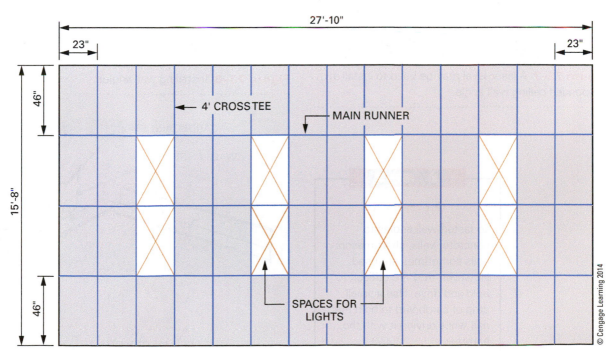

Figure 73–6 Completed sketch of a suspended ceiling layout.

CONSTRUCTING THE CEILING GRID

The ceiling grid is constructed by first installing *wall angles,* then installing *suspended ceiling lags* and *hanger wires,* suspending the *main runners,* inserting full-length *cross tees,* and, finally, cutting and inserting *border* cross tees.

Installing Wall Angles

A suspended ceiling must be installed with at least 3 inches for clearance below the lowest air duct, pipe, or beam. This clearance provides enough room to insert ceiling panels in the grid. If recessed lighting is to be used, allow a minimum of 6 inches for clearance. The height of the ceiling may be located by measuring up from the floor. If the floor is rough or out of level, the ceiling line may be located with various leveling devices previously described. A combination of a hand level and straightedge, a water level, builders' level, transit-level, or laser level can be used (*Figure 73–7*).

Snap chalk lines on all walls around the room to the height of the top edge of the wall angle.

Fasten wall angles around the room with their top edge lined up with the chalk line. It may be easier to fasten the wall angle by prepunching holes with a center punch or nail. Fasten into framing wherever possible, not more than 24 inches apart (*Figure 73–8*).

To fasten wall angles to concrete walls, short masonry nails sometimes are used. However, they are difficult to hold and drive. Use a small strip of cardboard to hold the nail while driving it with the hammer (*Figure 73–9*). Lead or plastic inserts and screws may also be used to fasten the wall angles. Their use does require more time. If available, power nailers can be used for efficient fastening of wall angles to masonry walls.

Make miter joints on outside corners. Make butt joints in interior corners and between straight lengths of wall angle (*Figure 73–10*). Use a combination square to layout and draw the square and angled lines. Cut carefully along the lines with snips.

Figure 73–7 A laser level may be used to install a suspended ceiling wall angle.

Figure 73–8 Installing wall angles.

On the Job

To fasten wall angles to concrete walls, short masonry nails sometimes are used. However, they are difficult to hold and drive. Use a small strip of cardboard to hold the nail while driving it with the hammer.

Figure 73–9 Technique for driving short nails.

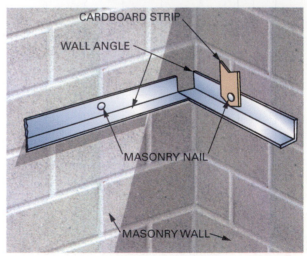

CARDBOARD STRIP

WALL ANGLE

MASONRY NAIL

MASONRY WALL

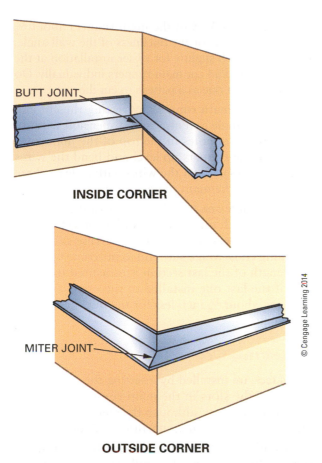

BUTT JOINT

INSIDE CORNER

MITER JOINT

OUTSIDE CORNER

© Cengage Learning 2014

Figure 73–10 Methods of joining a wall angle at corners.

CAUTION

Use care in handling the cut ends of wall angle and other grid members. Cut metal ends are very sharp and can cause serious injury.

Installing Hanger Lags

Suspended ceiling hanger lags are installed in a grid no larger than 4 feet by 4 feet. From the ceiling sketch, determine the position of the first main runner. Stretch a line at this location across the room from the top edges of the wall angle. Stretch the line tightly on nails inserted between the wall and wall angle (*Figure 73–11*). The line serves as a guide for installing *hanger lags* or *screw eyes* and *hanger wires* from which main runners are suspended.

Install hanger lags not more than 4 feet apart and directly over the stretched line. Hanger lags should be of the type commonly used for suspended ceilings. They must be long enough to penetrate wood joists a minimum of 1 inch to provide strong support. *Eye pins* are driven into concrete with a *powder-actuated* fastening tool (*Figure 16–13*). Hanger wires may also be attached directly around the lower chord of bar joists or trusses.

Installing Hanger Wire

Cut a number of hanger wires using wire cutters. The wires should be about 12 inches longer than the distance between the overhead construction and the stretched line. For residential work, 16-gauge wire is usually used. For commercial work, 12-gauge and heavier wire is used.

Attach the hanger wires to the hanger lags. Insert about 6 inches of the wire through the screw eye. Securely wrap the wire around itself three times. Pull on each wire to remove any kinks. Then make a 90-degree bend where it crosses the stretched line (*Figure 73–12*). Stretch lines, install hanger lags, and attach and bend hanger wires in

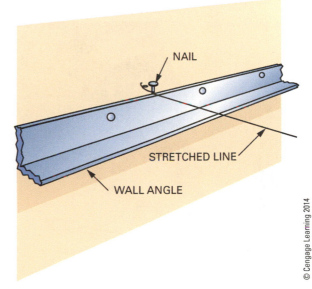

NAIL

STRETCHED LINE

WALL ANGLE

© Cengage Learning 2014

On the Job

Stretch lines for main runners on nails inserted between the wall and wall angle.

Figure 73–11 Technique for stretching a line between wall angles.

the same manner at each main runner location. Leave the last line stretched tightly in position.

Installing Main Runners

The ends of the main runners rest on the wall angles. They must be cut so that a cross-tee slot in the web of the runner lines up with the first row of cross tees. A cross-tee line must be stretched, at wall angle height, across the short dimension of the room to line up the slots in the main runners. The line must run exactly at right angles to the main runner line and at a distance from the wall equal to the width of the border panels. If the walls are at right angles to each other, the location of the cross-tee line can be determined by measuring out from both ends of the wall.

When the walls are not at right angles, the Pythagorean theorem is used to square the grid system (*Figure 73–13*). After the main runner line is installed, measure out from the short wall, along the stretched main runner line, a distance equal to the width of the border panel. Mark the line. Stretch the cross-tee line through this mark and at right angles to the main runner line.

At each main runner location, measure from the short wall to the stretched cross-tee line. Transfer the measurement to the main runner. Measure from the first cross-tee slot beyond the measurement, so as to cut as little as possible from the end of the main runner (*Procedure 73–A*). Cut the main runners about $\frac{1}{8}$ inch less to allow for the thickness of the wall angle. Backcut the web slightly for easier installation at the wall. Measure and cut main runners individually. Do not use the first one as a pattern to cut the rest.

Hang the main runners by resting the cut end on the wall angle and inserting suspension wires in the appropriate holes in the top of the main runner. Bring the runners up level and bend the wires (*Figure 73–14*). Twist the wires with at least three turns to hold the main runners securely.

More than one length of main runner may be needed to reach the opposite wall. Connect lengths of main runners together by inserting tabs into matching ends. Make sure end joints come up tight. The length of the last section is measured from the end of the last one installed to the opposite wall, allowing about $\frac{1}{8}$ inch less for the thickness of the wall angle.

Installing Cross Tees

Cross tees are installed by inserting the tabs on the ends into the slots in the main runners. These fit into position easily, although the method of attaching varies from one manufacturer to another.

Install all full-length cross tees between main runners first. Lay in a few full-size ceiling panels. This stabilizes the grid while installing cross tees for border panels. Cut and install cross tees along the border. Insert the connecting tab of one end in the main runner and rest the cut end on the wall angle (*Figure 73–15*). If the walls are not straight or square, it is necessary to cut cross tees for border tiles individually. For 2 × 2 panels, install 2-foot cross tees at the midpoints of the 4-foot cross tees.

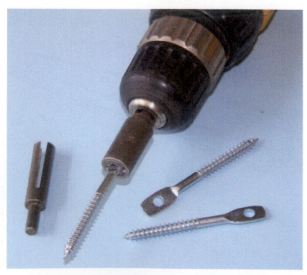

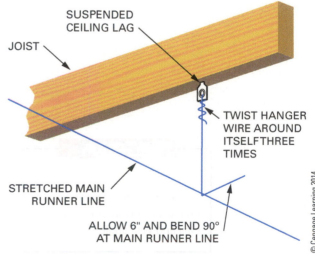

Figure 73–12 Suspended ceiling lags are used to support hanger wire. Wire may be prebent to accept main runners.

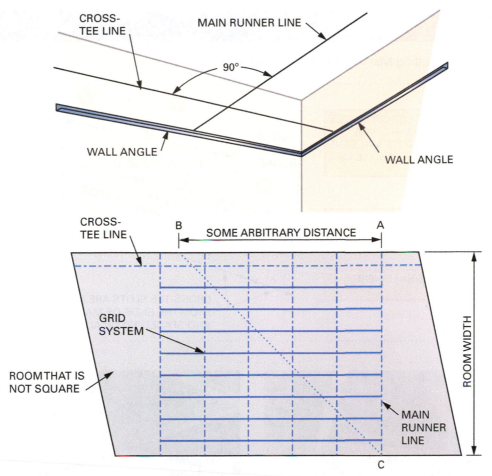

1) From A measure some distance to B. The actual distance does not matter; it merely needs to be large enough to make a big triangle.
2) Starting at point A, measure the room width keeping your tape as square as possible with the wall towards where point C will be.
3) Use these numbers in the Pythagorean theorem to determine the distance from B to C.
4) Measure and mark the distance from B to C and mark C. C is now square with point A.
5) Connect A and C with a string and measure each successive row of main runner from it.
6) **EXAMPLE** If AC, the room width, is 18'-9", then measure AB to be, say, 16'-0". Convert these dimensions to inches. 16'-0" becomes 192" and 18'-9" becomes 225" (18 × 12 + 9 = 225"). Put these dimensions in the Pythagorean theorem. $C^2 = A^2 + B^2$

$C = \sqrt{192^2 + 225^2} = \sqrt{36864 + 50625} = \sqrt{87489} = 295.7853952"$

To convert the decimal to sixteenths

$0.7853952 \times 16 = 12.566$ sixteenths $\Rightarrow {}^{13}/_{16}"$

Thus, the measurement from B to C is 295 $^{13}/_{16}$ inches.

Figure 73–13 Method of stretching two perpendicular lines using the Pythagorean theorem.

© Cengage Learning 2014

After the grid is complete, sight sections by eye. Straighten where necessary by making adjustments to border cross tees or hanger wires.

Installing Ceiling Panels

Ceiling panels are placed in position by tilting them slightly, lifting them above the grid, and letting them fall into place (*Figure 73–16*). When handling panels, be careful to avoid marring the finished surface. Install all border panels first. Then install full-size field panels. Measure each border panel individually, if necessary. Cut them slightly smaller than measured so they can drop into place easily. Cut the panels with a sharp utility knife using a straightedge as a guide. A scrap piece of cross-tee material can be used as a straightedge. Always cut with the finished side of the panel up.

StepbyStep Procedures

Procedure 73–A Cutting Main Runners so Cross-Tee Slots Align

STEP 1 LOCATE THE FIRST CROSS-TEE SLOT THAT WILL ALLOW A BORDER TILE TO FIT. MAKE THE CUTOFF PIECE AS SMALL AS POSSIBLE.

STEP 2 MEASURE BACK WIDTH OF BORDER PANEL.

STEP 3 CUT MAIN RUNNER HERE.

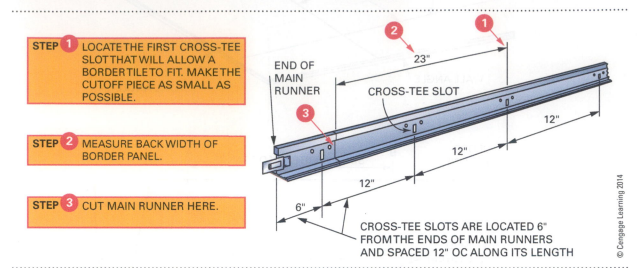

END OF MAIN RUNNER

CROSS-TEE SLOT

23"

12"

12"

12"

6"

CROSS-TEE SLOTS ARE LOCATED 6" FROM THE ENDS OF MAIN RUNNERS AND SPACED 12" OC ALONG ITS LENGTH

© Cengage Learning 2014

Figure 73–14 Method of adjusting the main runner level.

Courtesy of Trimble

Figure 73–16 Installing ceiling panels.

© Cengage Learning 2014

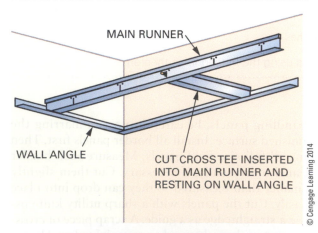

MAIN RUNNER

WALL ANGLE

CUT CROSS TEE INSERTED INTO MAIN RUNNER AND RESTING ON WALL ANGLE

© Cengage Learning 2014

Figure 73–15 Inserting a cross tee in the main runner.

Cutting Ceiling Panels Around Columns

When a column is near the center of a ceiling panel, cut the panel at the midpoint of the column. Cut semicircles from the cut edge to the size required for the panel pieces to fit snugly around the column. After the two pieces are rejoined around the column, glue scrap pieces of panel material to the back of the installed panel.

If the column is close to the edge or end of a panel, cut the panel from the nearest edge or end to fit around the column. The small piece is also fitted around the column and joined to the panel by gluing scrap pieces to its back side (*Figure 73–17*).

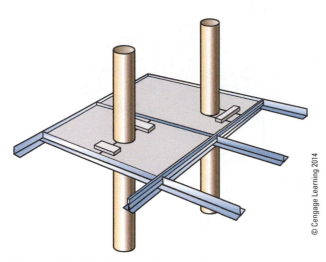

© Cengage Learning 2014

Figure 73–17 Fitting ceiling panels around columns.

LINEAR METAL CEILING SYSTEM

The linear metal ceiling system is used in commercial work to provide an architecturally attractive ceiling. It can be formed into flat or curved ceilings (*Figure 73–18*). Main tees are installed in a similar manner as the drop ceiling. The cross tees are placed 4 feet on center or closer as needed to keep the main tees secure.

The ceiling panels are 3 ¼ inches or 7 ¼ inches wide and 12 feet long metal pans. They are shaped to snap into place leaving a ¾ inch space between adjacent panels. They attach to the main tee that has special clips to hold each panel (*Figure 73–19*).

ACOUSTICAL CEILING TILE

Another form of suspended ceiling is the acoustical ceiling tile. It is usually stapled to furring strips that are fastened to drywall or exposed joists. If installing tiles on an existing ceiling that is not sound, furring strips should be installed and fastened through the ceiling into the joists above (*Figure 73–20*).

Description of Ceiling Tiles

Most ceiling tiles are made of wood fiber or mineral fiber. Wood fiber tiles are lowest in cost and are adequate for many installations. Mineral fiber tiles are used when a more fire-resistant type is required.

Manufacture. Wood fibers are pressed into large sheets that are ⁷⁄₁₆ to ¾ inch thick. Mineral fiber tiles are made of rock that is heated to a molten state. The fibers are then sprayed into a sheet form. The surfaces of some sheets are fissured or perforated for sound absorption. The surfaces of other sheets are embossed with different designs or left smooth. Then, they are given a factory finish and cut into individual tiles. Most tiles are cut with *chamfered, tongue-and-grooved* edges with two adjacent *stapling flanges* for concealed fastening (*Figure 73–21*).

Sizes. The most popular sizes of ceiling tile are a 12-inch square and a 12 × 24-inch rectangle in thicknesses of ½ inch. Tiles are also manufactured in squares of 16 inches and in rectangles of 16 × 32 inches.

Ceiling Tile Layout

Before installation begins, it is necessary to calculate the size of *border tiles* that run along the walls. It is desirable for border tiles to be as wide as possible and of equal widths on opposite walls.

Calculating Border Tile Sizes. To find the width of the border tiles along the long walls of a room, first determine the dimension of the short wall. In most cases, the measurement will be a number of full feet, plus a few inches. Each foot will be a full tile. Neglect the foot measurement and add 12 more inches to the remaining inch measurement. Divide this sum by 2 to find the width of border tiles for each long wall edge.

EXAMPLE

The distance between the long walls (width of the room) is 10'-6". If 12 × 12 tiles are used add 12 inches to the remainder, 12 + 6 = 18 inches. Divide this number by 2, 18 ÷ 2 = 9 inches. Thus there are 9 full tiles and two border tiles 9" wide.

The width of border tiles along the short walls is calculated in the same manner, only using the room length.

EXAMPLE

The distance between the short walls (length of the room) is 19'-8". If 12 × 12 tiles are used add 12 inches to the remainder, 12 + 8 = 20 inches. Divide this number by 2, 20 ÷ 2 = 10 inches. Thus there are 18 full tiles and two border tiles 10" long.

Preparation for Ceiling Tile Application

Unless an adhesive application to an existing ceiling is to be made, furring strips must be provided on which to fasten ceiling tiles. Furring strips are usually fastened directly to exposed joists. They are sometimes applied to an existing ceiling and fastened into the concealed joists above.

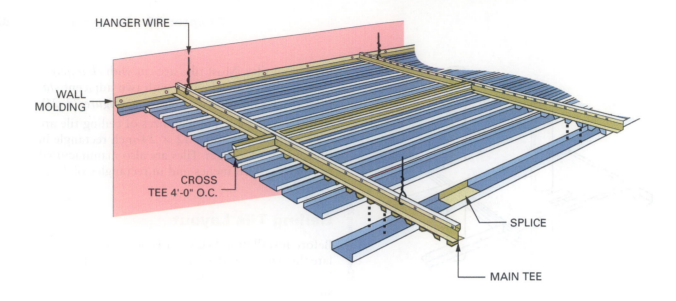

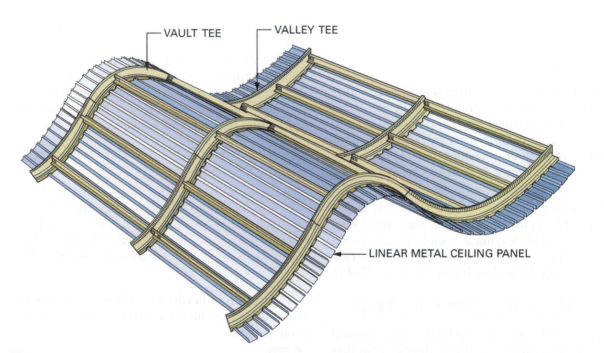

Figure 73–18 Linear metal ceilings may be flat or follow curved layouts.

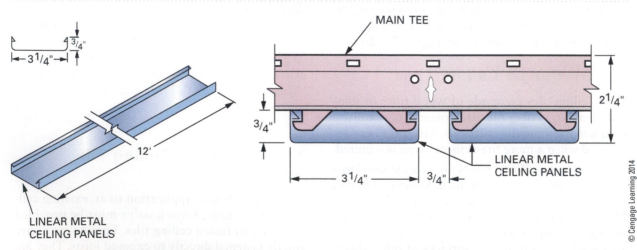

Figure 73–19 Linear panels are attached to the main tees by clips in the main tee.

Figure 73–20 Installing ceiling tile on wood furring strips.

Courtesy of Armstrong World Industries

Locating Concealed Joists. If the joists are hidden by an existing ceiling, tap on the ceiling with a hammer. Drive a nail into the spot where a dull thud is heard, to find a concealed joist. Locate other joists by the same method or by measuring from the first location. Usually, ceiling joists are spaced 16 inches on center (OC) and run parallel to the short dimension of the room. When all joists are located, snap lines on the existing ceiling directly below and in line with the concealed joists.

Laying Out and Applying Furring Strips

For fastening 12-inch tiles, furring strips must be installed 12 inches OC. From the corner, measure out the width of the border tiles. This measurement is the center of the first furring strip away from the wall. To mark the edge, measure from the center, in either direction, half the width of the furring strip. Mark an X on the side of the mark toward the center of the furring strip. From the edge of the first furring strip, measure and mark, every 12 inches, across the room. Place Xs on the same side of the mark as the first one (*Figure 73–22*).

Lay out the other end of the room in the same manner. Snap lines between the marks for the location of the furring strips. The strips are fastened by keeping one edge to the chalk line with the strip on the side of the line indicated by the X. Fasteners must penetrate at least 1 inch into the joist. Starting and ending furring strips are also installed against both walls.

Squaring the Room. First, snap a chalk line on a furring strip as a guide for installing border tiles against the long wall. The line is snapped parallel to, and the width of the border tiles, away from the wall.

A second chalk line must be snapped to guide the application of the short wall border tiles. The line must be snapped at exactly at 90 degrees to the first chalk line. Otherwise tiles will not line up properly. From the short wall, measure in along the first chalk line, the width of the short wall border tiles. From this point, use the Pythagorean theorem and snap another line at a right angle to the first line. This method of squaring lines has been previously described in *Figure 73–13* and *Figure 26–6*.

Installing Ceiling Tile

Ceiling tiles should be allowed to adjust to normal interior conditions for twenty-four hours before installation. Some carpenters sprinkle talcum powder or cornstarch on their hands to keep them dry. This prevents fingerprints and smudges on the finished ceiling.

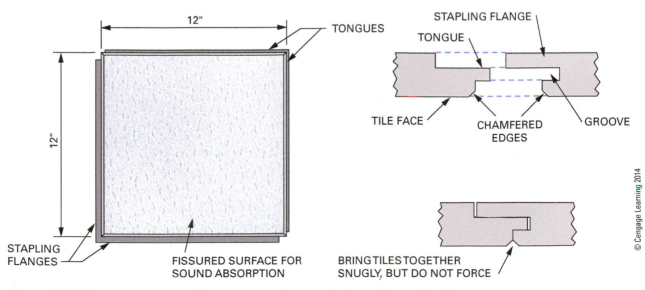

STAPLING FLANGE

TONGUE

12"

12"

TONGUES

TILE FACE

CHAMFERED EDGES

GROOVE

STAPLING FLANGES

FISSURED SURFACE FOR SOUND ABSORPTION

BRING TILES TOGETHER SNUGLY, BUT DO NOT FORCE

© Cengage Learning 2014

Figure 73–21 A typical ceiling tile has tongue-and-grooved edges with stapling flanges.

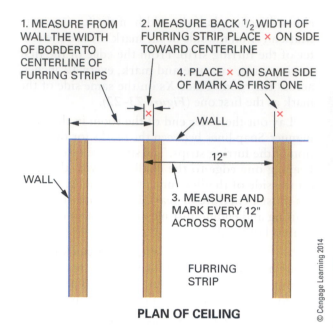

1. MEASURE FROM WALL THE WIDTH OF BORDER TO CENTERLINE OF FURRING STRIPS

2. MEASURE BACK ½ WIDTH OF FURRING STRIP, PLACE × ON SIDE TOWARD CENTERLINE

4. PLACE × ON SAME SIDE OF MARK AS FIRST ONE

WALL

WALL

12"

3. MEASURE AND MARK EVERY 12" ACROSS ROOM

FURRING STRIP

PLAN OF CEILING

Figure 73–22 Furring strip layout for ceiling tile.

Cut tiles face up with a sharp utility knife guided by a straightedge. All cut edges should be against a wall.

Starting the Installation.   To start the installation, cut a tile to fit in the corner. The outside edges of the tile should line up exactly with both chalk lines. Because this tile fits in the corner, it must be cut to the size of border tiles on both long and short sides of the room. For example, if the border

tiles on the long wall are 9 inches and those on the short wall are 10 inches, the corner tile should be cut twice to make it 9 × 10. Staple the tile in position. Be careful to line up the outside edges with both chalk lines (*Figure 73–23*).

Completing the Installation.   After the corner tile is in place, work across the ceiling. Install two or three border tiles at a time. Then fill in with full-size field tiles (*Figure 73–24*). Tiles are applied so they are snug to each other, but not jammed tightly.

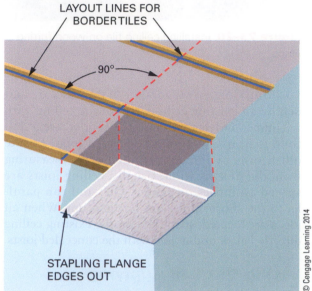

LAYOUT LINES FOR BORDER TILES

90°

STAPLING FLANGE EDGES OUT

Figure 73–23 Install the first tile in the corner with its outside edges to the layout lines.

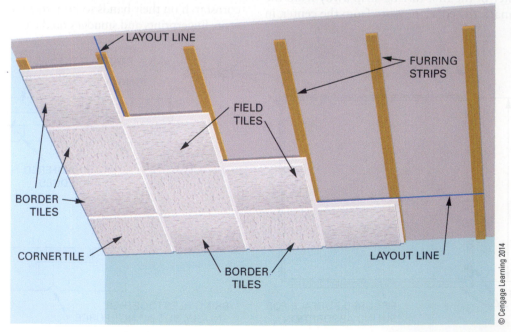

LAYOUT LINE

FURRING STRIPS

FIELD TILES

BORDER TILES

CORNER TILE

LAYOUT LINE

BORDER TILES

Figure 73–24 Install a few border tiles. Then fill in with full-sized field tiles.

Fasten each tile with four ½- or 9/16-inch staples. Place two in each flange, using a hand stapler. Use six staples in 12 × 24 tiles. Continue applying tiles in this manner until the last row is reached.

When the last row is reached, measure and cut each border tile individually. Cut the tiles slightly less than measured for easy installation. Do not force the tiles in place. Face nail the last tile in the corner near the wall where the nailhead will be covered by the wall molding. After all tiles are in place, the ceiling is finished by applying molding between the wall and ceiling. (The application of wall molding is discussed in a later unit.)

Adhesive Application of Ceiling Tile. Ceiling tile is sometimes cemented to existing plaster and drywall ceilings. These ceilings must be completely dry, solid, level, and free of dust and dirt. If the existing ceiling is in poor condition or has loose paint, adhesive application is not recommended.

Special tile adhesive is used to cement tiles to ceilings. Four daubs of cement are used on 12 × 12 tile. Six daubs are used on 12 × 24 tile. Before applying the daubs, prime each spot by using the putty knife to force a thin layer into the back surface of the tile. Apply the daubs, about the size of a walnut. Press the tile into position. Keep the adhesive away from the edges to allow for spreading when the tile is pressed into position. No staples are required to hold the tile in place.

ESTIMATING OF WALL AND CEILING FINISHES

Estimating the amount of materials for interior finishes involves paneling of various sorts, ceramic tile and suspended ceiling components. Remember that if the number of pieces does not come out even, round it up to the next whole number of pieces.

Sheet Paneling

To determine the number of sheets of paneling needed, measure the perimeter of the room. Divide the perimeter by the width of the panels to be used. Deduct two-thirds of a panel for each large opening such as doors or fireplaces. For example, if the perimeter is 62 feet with two doors, the number of 4-foot panels is 62 ÷ 4 − 2 ($^2/_3$) = 14.1 ⇒ 15 panels.

Board Paneling

Determine the square foot area to be covered by multiplying the perimeter by the height. Deduct the area of any large openings such as doors and large windows. Additional material is needed because of

Nominal Size	Width		Area Factor*
	Dress	Face	
Shiplap			
1 × 6	5½	5⅛	1.17
1 × 8	7¼	6⅞	1.16
1 × 10	9¼	8⅞	1.13
1 × 12	11¼	10⅞	1.10
Tongue-and-Groove			
1 × 4	3⅜	3⅛	1.28
1 × 6	5⅜	5⅛	1.17
1 × 8	7⅛	6⅞	1.16
1 × 10	9⅛	8⅞	1.13
1 × 12	11⅛	10⅞	1.10
S4S			
1 × 4	3½	3½	1.14
1 × 6	5½	5½	1.09
1 × 8	7¼	7¼	1.10
1 × 10	9¼	9¼	1.08
1 × 12	11¼	11¼	1.07
Paneling and Siding Patterns			
1 × 6	5⁷⁄₁₆	5¹⁄₁₆	1.19
1 × 8	7⅛	6¾	1.19
1 × 10	9⅛	8¾	1.14
1 × 12	11⅛	10¾	1.12

*Number multiplied by square feet to convert square feet to board feet

Figure 73–25 Factors used to estimate amounts of board paneling.

© Cengage Learning 2014

the difference in the nominal size of board paneling and its actual size. To make this allowance, multiply the area to be covered by area factor in table found in *Figure 73–25*. For example, a 1 × 6 S4S paneling actually has a face that covers 5½ inches, so multiply the area to be covered by 1.09 to find board feet. An additional 5 percent for waste in cutting is recommended. For example, the total area is 850 SF, and 1 × 8 T&G board paneling is to be used, so the board feet of material needed is 850 × 1.16 × 105% = 1035.3 ⇒ 1036 board feet. To further reduce waste in cutting, order suitable lengths to the area being covered.

Ceramic Tile

Styles and sizes of tiles vary between manufacturers as do the grout joint between the tiles. These factors affect the number of tiles that will cover a square foot. So the number of square feet coverage from a box of tile may be found from the supplier. Determine the square feet area to be covered. To find the number of boxes of tiles needed, divide the area by the number of square feet in a box. Add 5 percent for waste. For example, a 222-SF room to be covered by tile that comes in 10 SF per box will require $222 \div 10 \times 105\% = 23.3 \Rightarrow 24$ boxes. Adhesive, thin set, and grout are determined in the same fashion using conversion numbers from the manufacturers.

The number of straight pieces of cap is found by determining the total linear feet to be covered and dividing by the length of the cap piece. The number of interior and exterior corners is determined by counting from a layout of the installation.

Suspended Ceiling Components

The parts and pieces for a suspended ceiling are estimated using perimeter and area of the room. Consult the set of prints for rooms to have a suspended ceiling installed and the dimensions for each.

Wall Angle. Wall angle is installed on the perimeter of the room. To find the number needed, divide the perimeter of the room by 10, the length of a wall angle. For example, if the perimeter is 135 feet, then the number of wall angles is $135 \div 10 = 13.5 \Rightarrow 14$ pieces.

Main Runner. Main runners run the length of the room. To determine the number needed, first the number of rows is found by dividing the room width by 4 feet (runner spacing). Subtract one because it is not needed at the wall angles. Round off this number to the nearest whole number of rows. To find the number of pieces in each row, divide the room length by 12 feet (length of each runner). Round up this number to the nearest one-half a runner. This is because the waste end of a runner is usable if it is longer than one-half a piece. To find the number of main runners, multiply the rows by the number of pieces in a row. For example, if the room is 17'-3" × 28'-9", the number of rows is $17.25 \div 4 = 4.3 - 1 = 3.3 \Rightarrow 3$ rows. The number of pieces per row is $28.75 \div 12 = 2.3 \Rightarrow 2.5$. The number of main runners is $3 \times 2.5 = 7.5 \Rightarrow 8$ pieces.

Cross Tees. Determining the number of 4-foot cross tees is determined from the number of rows of main runners. Since there is one more row of cross tees than main runners, add one to the number of main runner rows. The number of tees along the length of the room is found by dividing the length by 2 feet (cross-tee spacing) and subtracting one. Finally, the number of 4-foot tees is the number of rows times pieces per row. For example, if the main runner rows is 3 and the room length is 28'-9", the number of rows of tees is $3 + 1 = 4$. The pieces per row is $28.75 \div 2 - 1 = 13.3 \Rightarrow 14$ pieces. The number of pieces for the room is $4 \times 14 = 56$.

There is the same number of 2-foot cross tees as 4-foot plus a few. One more tee is needed per row of 4-foot cross tees. For example, if there are 56 − 4-foot tees in four rows, then the number of 2-foot tees is $56 + 4 = 60$.

Hanger Lags and Wire. There are three hanger lags per main tee. Multiply main runners by three. The length of wire is the number of lags times the length of each wire. Measure the distance from the supporting surface for the main runners and add 2 feet for wire wrap. For example, the number of 50-foot rolls of wire needed for a ceiling suspended 6 feet with eight main runners is $8 \times (6 + 2) = 64$ feet $\div 50 = 1.2 \Rightarrow 2$ rolls.

Suspended Ceiling Panels. Each border panel requires a full-size ceiling panel. The number of 2 × 4 panels is the same number as would be for 2-foot cross tees. It is the number of 4-foot tees plus one per main runner rows plus one more. Deduct one panel per light fixture. For example, if the number of 4-foot tees is 56 and there are four rows of main runner and eight lights, then the number of boxes of panels (ten per box) is $56 + 4 + 1 - 8 = 53 \div 10 = 5.3 \Rightarrow 6$ boxes. If 2 × 2 panels are used, simply double the 2 × 4 panel count.

Acoustical Ceiling Panels. To estimate furring strips and ceiling tiles for acoustical ceilings, measure the width and length of the room, rounding each measurement to the next whole foot measurement. Multiply these numbers together to find the area of the ceiling in square feet. Divide the ceiling area by the number of square feet contained in one of the ceiling tiles being used to find the number of ceiling tiles needed. Two percent may be added for damage and waste. For example, if the ceiling is 17'-3" × 28'-9", then the width is 18 feet and length is 29 feet. The area is 522. The number of 12 × 12 tiles is $522 \times 102\% = 533$ tiles.

To estimate furring strips, the number of rows and length of each row must be found. To find the number of rows, divide the rounded-up width by the width of the tile. Add one row to start. Multiply the room's rounded-up length by the number of rows and divide by the length of furring purchased. Add 5 percent for waste. For example, if the room is 18 feet by 29 feet, the number of 12-foot pieces of furring for 12 × 12 tile ceiling is $(18 + 1) \times 29 = 19 \times 29 \div 12 \times 105\% = 48.2 \Rightarrow 49$ pieces.

See *Figure 73–26* a listing of the shorthand version of these estimating techniques, along with another example.

Estimate the materials for the wall and ceiling finish materials for the following 8' high rooms. One room 14' × 18' with three doors and 4 × 8 sheet paneling. One room 18' × 22' with 1" × 4" T&G paneling (*see Figure 73–23 for area factor*). One bathroom with 155 SF of tile, 15 SF per box. One room 24' × 38' with a suspended ceiling, 2 × 4 grid, and 12 lights. One room 12' × 14' with 12" × 12" acoustical ceiling tiles.

Item	Formula	Waste factor	Example
Sheet Paneling	PERM ÷ sheet WID − ²/₃ sheet per door = sheets		2 × 14 + 2 × 18 = 64' ÷ 4 − 2 (²/₃) = 14.6 ⇒ 15 sheets
Board Paneling	PERM × room HGT × area factor × waste = board ft	5%	2 × 18 + 2 × 22 = 80 × 8 = 640 × 1.28 × 105% = 860.1 ⇒ 861 board feet
Ceramic Tile	area to be covered ÷ SF per box × waste = boxes	5%	155 ÷ 15 × 105% = 10.8 ⇒ 11 boxes
Suspended Ceiling Wall Angle	PERM ÷ wall angle LEN = pieces		2 × 24 + 2 × 33 = 114 ÷ 10 = 11.4 ⇒ 12 pieces
Suspended Ceiling Main Runner	[room WID ÷ main tee spacing − 1] × [room LEN ÷ main tee LEN rounded up to nearest ½ piece] = pieces		[24 ÷ 4 − 1] × [38 ÷ 12] = 5 × 3.5 = 17.5 ⇒ 18 pieces
Suspended Ceiling 4 feet Cross Tees	[main runner rows +1] × [room LEN ÷ 2 − 1] = NUM of 4 ft cross tees		[5 +1] × [38 ÷ 2 − 1] = 6 × 18 = 108 cross tees
Suspended Ceiling Hanger Lags	NUM main tees × 3 = pieces		18 × 3 = 54 pieces
Suspended Ceiling Wire (50' roll)	NUM lags × [suspended distance +2'] ÷ 50 = rolls		54 × [2 +2] ÷ 50 = 4.3 ⇒ 5 rolls
Suspended Ceiling Panels	[NUM 4 feet cross tees + NUM main tee rows + 1 − one per light] ÷ NUM per box = boxes		[108 + 5 + 1 − 12] ÷ 10 = 10.2 ⇒ 11 boxes
Acoustical Ceiling Tiles	Room area ÷ SF per tile × waste = pieces	2%	12 × 14 ÷ 1 × 102% = 171.3 ⇒ 172 pieces
Acoustical Ceiling Furring	[room WID ÷ tile WID +1] × room LEN ÷ furring LEN × waste = pieces	5%	[12 ÷ 1 +1] × 14 ÷ 12 × 105% = 13.6 ⇒ 14 pieces

Figure 73–26 Example of estimating wall and ceiling finish materials with formulas.

DECONSTRUCT THIS

Carefully study *Figure 73–27* and think about what is wrong and/or what is right. Consider all possibilities. What construction practice or method is different in your area of the country?

© Cengage Learning 2014

Figure 73–27 This photo shows workers installing a deck support structure for an elevated concrete slab form tied off to a personal fall protection systems.

KEY TERMS

board paneling	furring strips	scribe	wainscoting
ceramic tile	main runners	sheet paneling	wall angles
cross tees	perforated	thin set	
fissured	plastic laminate	tile grout	

REVIEW QUESTIONS

Select the most appropriate answer.

1. On prefinished plywood paneling that is scored to simulate boards, some scores are always placed in from the edge

   a. 12 and 16 inches.
   b. 16 and 20 inches.
   c. 12 and 24 inches.
   d. 16 and 32 inches.

2. A wainscoting is a wall finish

   a. applied diagonally.
   b. applied partway up the wall.
   c. used as a coating on prefinished wall panels.
   d. used around tubs and showers.

3. For most parts of the country, wood used for interior finish should be dried to a moisture content of about

   a. 8 percent.
   b. 12 percent.
   c. 15 percent.
   d. 20 percent.

4. The thickest plastic laminate described in this unit is called

   a. vertical type.
   b. regular type.
   c. backer type.
   d. all-purpose type.

5. If $\frac{3}{4}$-inch-thick wood paneling is to be applied vertically over open studs, wood blocking must be provided between studs for nailing at intervals of

   a. 16 inches.
   b. 24 inches.
   c. 32 inches.
   d. 48 inches.

6. When applying board paneling vertically to an existing wall,

   a. nail paneling to existing studs.
   b. apply horizontal furring strips.
   c. remove the wall covering and install blocking between studs.
   d. apply vertical furring strips.

7. The recommended backing for use behind ceramic tile for showers is

   a. waterproof drywall.
   b. water-resistant gypsum board.
   c. cement board.
   d. all of the above.

8. Cutting ceramic tile is done with a

   a. nibbler.
   b. ceramic tile saw.
   c. hand cutter.
   d. all of the above.

9. Bathroom tile should extend over the tops of showerheads a minimum of

   a. 4 inches.
   b. 6 inches.
   c. 8 inches.
   d. 12 inches.

10. The number of ceramic tiles, nine tiles to a square foot, needed to cover 150 square feet of wall area is

    a. 1,152.
    b. 1,350.
    c. 1,500.
    d. 1,674.

11. The most common sizes in inches of suspended ceiling panels are

    a. 8 × 12 and 12 × 12.
    b. 12 × 12 and 16 × 16.
    c. 12 × 12 and 12 × 24.
    d. 24 × 24 and 24 × 48.

12. The dimension of a border panel for a 12'-6" room, when 24" × 24" suspended ceiling panels are used, is

    a. 3.
    b. 6.
    c. 15.
    d. 16.

13. The diagonal measurement of a 16" × 24" rectangle is

    a. 384".
    b. 832".
    c. 28$\frac{7}{8}$".
    d. 28"-10$\frac{1}{8}$".

14. The number of usable pieces that may normally be cut from one main runner is

    a. 2.
    b. 3.
    c. 4.
    d. more than 4.

15. In a suspended ceiling, hanger wire is used to suspend

    a. cross tees.
    b. main runners.
    c. wall angle.
    d. furring strips.

16. The first step in installing a suspended ceiling is to

    a. square the room.
    b. make a sketch of the planned ceiling.
    c. calculate border tiles.
    d. install the wall angle.

17. The number of 10-foot wall angle needed for a room that measures 17'-9" × 23'-9" is at least

    a. 8.
    b. 9.
    c. 83.
    d. 432.

18. Suspended ceiling lags should be installed in a grid that is

    a. 2' × 2'.
    b. 4' × 4'.
    c. 8' × 8'.
    d. 12' × 12'.

19. The number of 4-foot cross tees for a suspended ceiling with six rows of main runners that are 30 feet long is

    a. 84.
    b. 90.
    c. 98.
    c. 105.

20. The most common sizes in inches of stapled-place ceiling tile are

    a. 8 × 12 and 12 × 12.
    b. 12 × 12 and 16 × 16.
    c. 12 × 12 and 12 × 24.
    d. 14 × 24 and 24 × 48.

©Shutterstock, Inc./Eimantas Buzas

UNIT 26
Finish Floors

Many times a layer of underlayment is installed over the subfloor as part of the finished floor. A number of materials are then used for finish floors. Each may be applied by a flooring specialist, such as carpet by carpet installers. But in some areas carpenters are asked to install various flooring materials.

Resilient flooring sheets and tiles are often used in kitchens and bathrooms. Wood flooring comes as strips of solid wood or a wood-plastic combination. Solid wood flooring is a longtime favorite because of its durability, beauty, and warmth.

OBJECTIVES

After completing this unit, the student should be able to:

- describe the kinds, sizes, and grades of hardwood finish flooring.
- apply strip, plank, and parquet finish flooring.
- estimate quantities of wood finish flooring required for various installations.
- apply underlayment and resilient tile flooring.
- estimate required amounts of underlayment and resilient tile for various installations.

SAFETY REMINDER

Installing flooring requires the installer to bend or work on knees for long periods of time. Take care to protect your body from injury due to repetitive movements and contact with hard surfaces.

CHAPTER 74

Description of Wood Floors

Most hardwood finish flooring is made from white or red oak. Beech, birch, hard maple, and pecan finish flooring are also manufactured. For less expensive finish floors, some softwoods such as Douglas fir, hemlock, and southern yellow pine are used.

KINDS OF HARDWOOD FLOORING

The four basic types of solid wood finish flooring are strip, plank, parquet strip, and parquet block. Laminated strip wood flooring is a relatively new type that is gaining in popularity. *Laminated parquet blocks* are also manufactured.

Solid Wood Strip Flooring

Solid wood strip flooring is probably the most widely used type. Most strips are tongue-and-grooved on edges and ends to fit precisely together.

Unfinished strip flooring is milled with square, sharp corners at the intersections of the face and edges. After the floor is laid, any unevenness in the faces of adjoining pieces is removed by sanding the surface so strips are flush with each other.

Prefinished strips are sanded, finished, and waxed at the factory. They are not sanded after installation. A chamfer is machined between the face side and edges of the flooring prior to prefinishing. When installed, these chamfered edges form small V-grooves between adjoining pieces. This obscures any unevenness in the surface.

The most popular size of hardwood strip flooring is ¾ inch thick with a face width of 2¼ inches. The face width is the width of the exposed surface between adjoining strips. It does not include the tongue. Other thicknesses and widths are manufactured (*Figure 74–1*).

Laminated Wood Strip Flooring

Laminated strip flooring is a five-ply prefinished wood assembly. Each board is ⁹⁄₁₆ inch thick, 7½ inches wide, and 7 feet, 11½ inches long. The board consists of a bottom veneer, a three-ply cross-laminated core, and a face layer. The face layer consists of three rows of hardwood strips joined snugly edge to edge (*Figure 74–2*).

The uniqueness of this flooring is in the exact milling of edge and end tongue and grooves. This precision allows the boards to be joined with no noticeable unevenness of the prefinished surface. This eliminates the need to chamfer the edges. Without the chamfers, a smooth continuous surface without V-grooves results when the floor is laid (*Figure 74–3*).

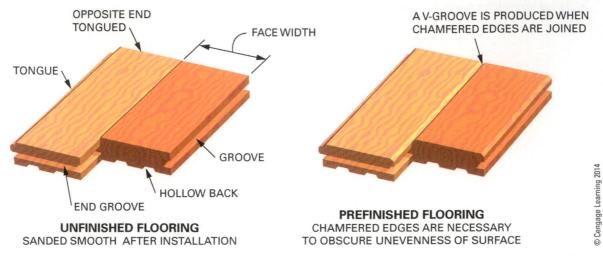

OPPOSITE END TONGUED

FACE WIDTH

TONGUE

GROOVE

END GROOVE

HOLLOW BACK

UNFINISHED FLOORING
SANDED SMOOTH AFTER INSTALLATION

A V-GROOVE IS PRODUCED WHEN CHAMFERED EDGES ARE JOINED

PREFINISHED FLOORING
CHAMFERED EDGES ARE NECESSARY TO OBSCURE UNEVENNESS OF SURFACE

© Cengage Learning 2014

Figure 74–1 Hardwood strip flooring is edge and end matched. The edges of prefinished flooring are chamfered.

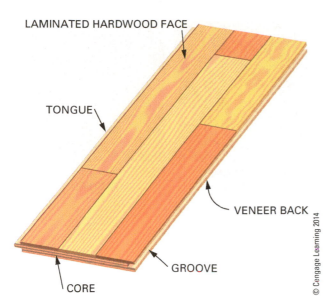

Figure 74–2 labels: LAMINATED HARDWOOD FACE, TONGUE, VENEER BACK, GROOVE, CORE

© Cengage Learning 2014

Figure 74–2 Cross-section of a board of laminated strip flooring.

© Cengage Learning 2014

Figure 74–3 Completed installation of a laminated strip floor.

Figure 74–4 label: FACTORY-INSTALLED WALNUT PLUGS

© Cengage Learning 2014

Figure 74–4 Plank flooring usually is applied in rows of alternating widths. Plugs of contrasting color may be added to simulate screw fastening.

Plank Flooring

Solid wood plank flooring is similar to strip flooring. However, it comes in various mixed combinations ranging from 3 to 8 inches in width. For instance, plank flooring may be laid with alternating widths of 3 and 4 inches, 3, 4, and 6 inches, 3, 5, and 7 inches, or any random-width combination.

Like strips, planks are available unfinished or prefinished. The edges of some prefinished planks have deeper chamfers to accentuate the plank widths. The surface of some prefinished plank flooring may have plugs of contrasting color already installed to simulate screw fastening. One or more plugs, depending on the width of the plank, are used across the face at each end (*Figure 74–4*).

Unfinished plank flooring comes with either square or chamfered edges and with or without plugs. The planks may be bored for plugs on the job, if desired.

Parquet Strips

Parquet strip flooring has short strips that are laid to form various mosaic designs. The original parquet floors were laid by using short strips. Some, at the present time, are laid in the same manner. This type is manufactured in precise, short lengths, which are multiples of its width. For instance, 2¼-inch parquet strips come in lengths of 9, 11¼, 13½, and 15¾ inches. Each piece is tongue-and-grooved on the edges and ends. Herringbone, basket weave, and other interesting patterns can be made using parquet strips (*Figure 74–5*).

Parquet Blocks

Parquet block flooring consists of square or rectangular blocks, sometimes installed in combination with strips, to form mosaic designs. The three basic types are the *unit, laminated,* and *slat* block (*Figure 74–6*).

Unit Blocks.   The highest-quality parquet block is made with ¾-inch-thick, tongue-and-groove solid hardwood, usually oak. The widely used 9 × 9 unit block is made with six strips, each 1½ inches wide, or with four strips, each 2¼ inches wide. Unit blocks are laid with the direction of the strips at right angles to adjacent blocks (*Figure 74–7*). Unit blocks are made in other sizes and used in combination with parquet strips. Several patterns have gained popularity.

FRENCH HERRINGBONE
PATTERN

BASKET WEAVE
PATTERN

HERRINGBONE
PATTERN

STONE
PATTERN

© Cengage Learning 2014

Figure 74–5 Parquet strips are made in lengths that are multiples of its width.

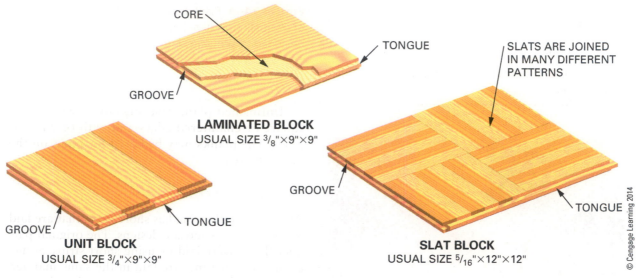

CORE

TONGUE

GROOVE

LAMINATED BLOCK
USUAL SIZE ³⁄₈"×9"×9"

SLATS ARE JOINED
IN MANY DIFFERENT
PATTERNS

GROOVE

TONGUE

UNIT BLOCK
USUAL SIZE ³⁄₄"×9"×9"

SLAT BLOCK
USUAL SIZE ⁵⁄₁₆"×12"×12"

© Cengage Learning 2014

Figure 74–6 Three basic types of parquet block.

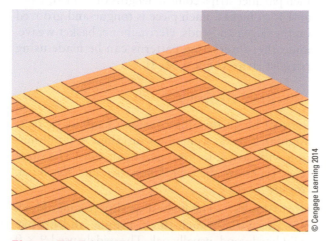

© Cengage Learning 2014

Figure 74-7 Unit blocks are widely used in an alternating pattern.

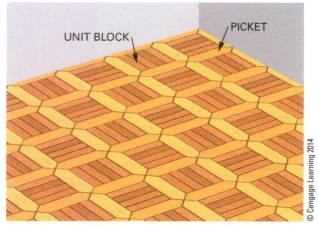

UNIT BLOCK

PICKET

© Cengage Learning 2014

Figure 74–8 The Monticello pattern is a famous parquet that uses square center blocks and picket-shaped strips.

Monticello is the name of a parquet originally designed by Thomas Jefferson, the third president of the United States. The pattern consists of a 6 × 6 center unit block surrounded by 2¼-inch-wide pointed *pickets*. Each center unit block is made

of four 1½-inch-wide strips (*Figure 74–8*). Each block comes with three pickets joined to it at the factory.

Another popular parquet, called the *Marie Antoinette*, is copied from part of the Versailles Palace

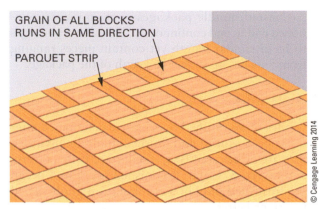

Figure 74–9 A popular parquet called the Marie Antoinette uses center blocks and parquet strips.

Figure 74–10 Rectangular parquet blocks may be used to make herringbone and similar patterns.

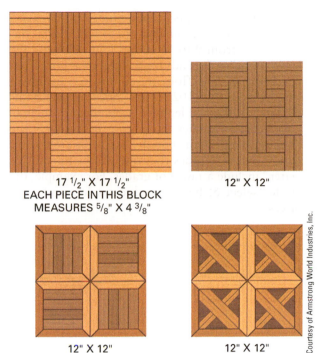

17 1/2" X 17 1/2"
EACH PIECE IN THIS BLOCK
MEASURES 5/8" X 4 3/8"

12" X 12"

12" X 12"

12" X 12"

Figure 74–11 Slat blocks are made of short, narrow strips joined in various patterns.

in France. Square center unit blocks are enclosed by strips applied in a basket weave design (*Figure 74–9*).

Rectangular parquet blocks are often used in a *herringbone* pattern. One commonly used block is 4½ × 9 and made of three strips each 1½ inches wide and 9 inches long (*Figure 74–10*).

Wood Blocks. Wood blocks are generally made of three-ply laminated oak in a ⅜-inch thickness. Most blocks come in 8" × 8" or 9" × 9" sizes. A 2" × 12" wood strip is manufactured for use in herringbone and similar patterns.

Slat Blocks. Slat blocks are also called *finger blocks*. They are made by joining many short, narrow strips together in small squares of various patterns. Some strips may be as narrow as ⅝ inch and as short as 2 inches or less. Several squares are assembled to make the block (*Figure 74–11*). The squares are held together with a mesh backing or with a paper on the face side that is removed after the block is laid.

Presanded Flooring

Some strip, plank, and parquet finish flooring can be obtained presanded at the factory, but without the finish. The presanded surface eliminates the necessity for all but a touch-up sanding after installation to remove surface marks before the finish is applied.

GRADES OF HARDWOOD FLOORING

Uniform grading rules have been established for unfinished and factory finished solid hardwood flooring by the National Wood Flooring Association (NWFA) (*Figure 74–12*). The certification trademark on flooring assures consumers that the flooring is manufactured and graded in compliance with established quality standards. Other types of wood finish flooring, such as parquet and laminated flooring, have no official grade rules.

Unfinished Flooring

Oak flooring is available quarter-sawed and plain-sawed. The grades for unfinished oak flooring, in declining order, are clear, select, No. 1 common, No. 2 common, and Shorts.

Birch, beech, and hard maple flooring are graded in declining order as first grade, second grade, second & better, third grade, and third & better. Grades of hickory and pecan flooring are first grade, second grade, second & better, third grade, and third & better.

In addition to appearance, grades are based on length. For instance, bundles of 1¼-foot shorts contain pieces from 9 to 18 inches long. The average length bundles depends on the grade ranging from 2 - 3½ feet. The flooring comes in bundles in lengths from 9 to 102 inches. Pieces in each bundle are not of equal lengths. A bundle may include pieces from 6 inches under to 6 inches over the nominal length of the bundle. No pieces are shorter than 9 inches. Grades that include the phrase "& better" include a range of grades. For example, the grade "select & better" includes select and clear pieces.

Nested bundle packages contain pieces positioned end to end continuously. This allows packages to be of similar length yet contain pieces ranging from 9 inches up to the the length of the package.

Prefinished Flooring

Grades of prefinished flooring are determined after it has been sanded and finished. In declining order, they are prime, standard, standard & better, tavern, and tavern & better. Prefinished beech and pecan are furnished only in a combination grade called tavern or better.

Unfinished Oak Flooring (Red & White Separated)	Unfinished Hard Maple, Beech & Birch	Unfinished Pecan Flooring	Prefinished Oak Flooring
Clear Plain or CLear Quartered* Best appearance, mostly heartwood. Best grade, most uniform color, limited small character marks. Bundles 1¼ ft. and up. Average length 3¾ ft.	**First Grade** Best appearance. Natural color variation, limited character marks. Bundles 1¼ ft. and up. Average length 3 ft.	**First Grade** Excellent appearance. Natural color variation, limited character marks, unlimited sapwood. Bundles 1¼ ft. & up. Average length 3 ft.	**Prime Prefinished Oak (Special Order)** Excellent appearance. Natural color variation, limited character marks, unlimited sapwood. Bundles 1¼ ft. & up. Average length 3¼ ft.
Select Plain or Select Quartered Excellent appearance. Limited character marks, unlimited sound sapwood. Bundles 1¼ ft. and up. Average length 3¼ ft. **Select & Better** A combination of Clear and Select grades.	**Second Grade** Color variations in appearance. Varying sound wood characteristics of species. Bundles 1¼ ft. and up. Average length 2¾ ft.	**Second Grade** Color variation in appearance. Varying sound wood characteristics of species. Bundles 1¼ ft. and up. Average length 2½ ft.	**Standard Prefinished Oak** Variegated appearance. Varying sound wood characteristics of species. Bundles 1¼ ft. & up. Average length 2½ ft. **Standard & Better Grade** Bundles 1¼ ft. & up. Average length 2½ ft.
No. 1 Common Variegated appearance. Light and dark colors; knots, flag worm holes and other character marks allowed to provide a variegated appearance after imperfections are filled and finished. Bundles 1¼ ft. and up. Average length 2½ ft.	**Second & Better Grade** A combination of First & Second Grades. Lengths equivalent to Second Grade. Average length 2¾ ft. **Third Grade** Rustic appearance. All wood characteristics of species. Serviceable economical floor after filling. Bundles 1¼ ft. and up. Average length 2¼ ft.	**Second & Better Grade** A combination of FIRST and SECOND GRADES. Bundles 1¼ ft. and up. Average length 2½ ft.	
No. 2 Common Rustic appearance. All wood characteristics of species. A serviceable economical floor after knot holes, worm holes, checks and other imperfections are filled and finished. Bundles 1¼ ft. and up. Average length 2¼ ft. Red and White may be mixed.	**Third & Better Grade** A combination of First, Second & Third Grades. Bundles 1¼ ft. and up. Average length 2¼ ft.	**Third Grade** Rustic appearance. All wood characteristics of species. A serviceable, economical floor after filling. Bundles 1¼ ft. and up. Average length 2 ft. **Third & Better Grade** A combination of FIRST, SECOND and THIRD GRADES. Bundles 1¼ ft. and up.	**Tavern Prefinished Oak** Rustic appearance. All wood characteristics of species. A serviceable, economical floor. Bundles 1¼ ft. & up. Average length 2 ft. **Tavern & Better Prefinished Oak** All wood characteristics of species. Bundles 1¼ ft. & up. Average length 2 ft.
1¼ Shorts Pieces 9 to 18 inches. Bundles average nominal 1¼ ft. **No. 1 Common & Better Shorts** A combination grade of CLEAR, SELECT, & NO. 1 COMMON **No. 2 Common Shorts** Same as No. 2 Common.			

Arrows: SELECT AND BETTER; SECOND AND BETTER; THIRD AND BETTER; STANDARD AND BETTER; TAVERN AND BETTER

Courtesy of National Wood Flooring Association

Figure 74–12 Guide to hardwood flooring grades.

CHAPTER 75

Installing Wood Floor

Lumber used in the manufacture of hardwood flooring has been air-dried, kiln-dried, cooled, and then accurately machined to exacting standards. It is a fine product that should receive proper care during handling and installation.

HANDLING AND STORAGE OF WOOD FLOOR

Maintain moisture content of the flooring by observing recommended procedures. Flooring should not be delivered to the job site until the building has been closed in. Outside windows and doors should be in place. Cement work, plastering, and other materials must be thoroughly dry. In warm seasons, the building should be well ventilated. During cold months, the building should be heated, not exceeding 72°F, for at least five days before delivery and until flooring is installed and finished.

Check with the supplier of the flooring as to how to handle the material. Some suppliers store the flooring in dry conditioned warehouses and recommend installing the floor immediately upon delivery. Some suppliers recommend acclimating the flooring to the atmosphere for weeks before installation.

CONCRETE SLAB PREPARATION

Wood finish floors can be installed on an on-grade or above-grade concrete slab. Floors should not be installed on below-grade slabs. New slabs should be at least thirty days old. Before testing for moisture and installing wood floor. Flooring should not be installed when tests indicate excessive moisture in the slab.

Testing for Moisture

Concrete can be tested for the moisture in a variety of ways. These tests require special tools, expertise, and many days to complete. The best testing methods reveal the actual amount or quantity of moisture that passes through it. The results will be one of three possible scenarios. One, the slab is dry enough to install wood directly to the slab, two, the slab is too wet for wood floor application, and three the moisture is a little high and a moisture barrier should be installed. However, in slab-on-grade and below-grade applications, adding a moisture barrier is always recommended.

Moisture Barriers

The NWFA allows the moisture barrier to have a perm rating 0.15 perm or below. Moisture barriers may be achieved using many products. They include a premium construction grade polyethylene film of 6 mil thickness minimum, a double layer of #15 asphalt saturated felt paper where each layer is adhered to the surface below in a skim coat of asphalt mastic type adhesive, a waterproof, dimple type membrane, providing a thermal air gap, and an attached elastomeric membrane with sealed seams.

Installing Plywood Subfloor

Subfloor of wood flooring also may be installed in many ways. Each method provides adequate nail base for wood flooring. A floated subfloor uses two layers of ⅜ plywood installed perpendicular to each other. There is a ¾ inch gab at the walls and ⅛ inch gab between all seams (*Figure 75–1*). The first layer is laid in place and the second layer is screw and glued to the first.

A second method adheres the plywood to the concrete. Panels are cut in 2'×8' of 4'×4' panels. The back side is then scored ½ the thickness in a 12"×12" grid. The sections are adhered in a staggered joint pattern with ⅛" spacing between panels and ¾" minimum expansion space at walls.

The third method uses direct fastening to the concrete. Fasteners may be powder-driven pins, pneumatic driven nails, or screws. Panels joints are staggered allowing ⅛" expansion space around all panels and a ¾" expansion space at all vertical obstructions. Nailing requirements have a minimum 32 fasteners per 4' × 8' panel.

Applying Sleepers

Finish flooring may also be fastened to sleepers installed on the slab. Sleepers are short lengths of lumber cemented to the slab. They must be pressure-treated and dried to a suitable moisture content. Usually, 2 × 4 lumber, from 18 to 48 inches long, is used.

Sleepers are laid on their side and cemented to the slab with mastic. They are staggered, with end

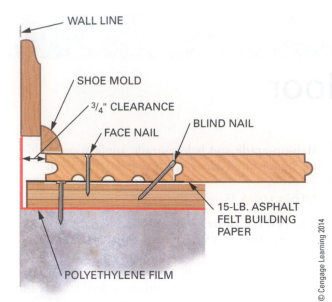

Figure 75–1 Installation details of a plywood subfloor over a concrete slab.

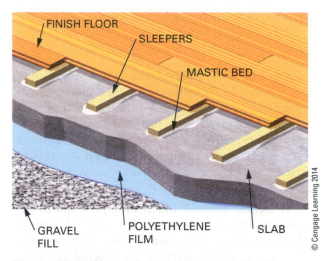

Figure 75–2 Sleepers are cemented to a concrete slab to provide fastening for strip or plank finish flooring.

laps of at least 4 inches, in rows 12 inches on center and at right angles to the direction of the finish floor. If the slab was not installed with a vapor retarder, a polyethylene vapor barrier is then placed over the sleepers. The edges are lapped over the rows (*Figure 75–2*). With end-matched flooring, end joints need not meet over the sleepers.

SUBFLOORS ON JOISTS

Exterior plywood or boards are recommended for use as subfloors on joists when wood finish floors are installed on them. If plywood is used on 16" spaced joists, a full ⅝-inch thickness is required. Thickness of ¾ inch is preferred for ¾-inch strip finish flooring. Use ¾-inch-thick subfloor for ½-inch strip flooring.

INSTALLING WOOD FLOOR

In new construction, the base or door casings are not usually applied yet for easier application of the finish floor. In remodeling, the base and base shoe must be removed. Use a scrap piece of finish flooring as a guide on which to lay a jambsaw. Cut the ends of any door casings that are extending below the finish floor surface.

Before installing any type of wood finish flooring, nail any loose areas. Sweep the subfloor clean. Scraping may be necessary to remove all plaster, taping compound, or other materials. Verify the subfloor is flat. Repair any raised areas.

Installing Strip Flooring

Strip flooring laid in the direction of the longest dimension of the room gives the best appearance. The flooring may be laid in either direction. Installing perpendicular to the joists is preferred. Flooring installed parallel to the joists should have an additional layer of ½" plywood underlayment (*Figure 75–3*).

When the subfloor is clean, cover it with an acceptable moisture barrier. This will help keep out dust, prevent squeaks in dry seasons, and minimize the impact of seasonal humidly changes.

Starting Strip.   The location and straight alignment of the first course is important. Place a strip of flooring on each end of the room, ¾ inch from the starter wall with the groove side toward the wall. Mark along the edge of the flooring tongue. Snap a chalk line between the two points. The gap between the flooring and the wall is needed for expansion. It will eventually be covered by the base.

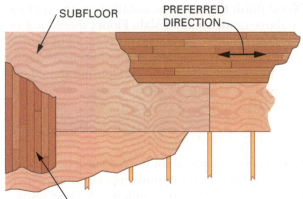

IF A PARALLEL DIRECTION OF FLOORING IS DESIRED AN ADDITION LAYER OF 1/2" UNDERLAYMENT IS NEEDED.

PANEL SUBFLOOR

Figure 75–3 Several factors determine the direction in which strip flooring is laid.

Hold the strip with its tongue edge to the chalk line (*Figure 75–4*). Face nail and blind nail as needed to install a strip along the entire wall. Face nail spacing should be 10-12" and blind nails should be 6'-8" apart. Work from left to right with the grooved end of the first piece toward the wall. If you are right handed, Left is determined by having the back of the person laying the floor to the wall where the starting strip is laid. This allows for easy use of the rubber mallet to keep the end joints between strips are driven up tight.

When necessary to cut a strip to fit to the right wall, use a strip long enough so that the cut-off piece is 8 inches or longer. Start the next course on the left wall with this piece.

Blind Nailing

Hardwood flooring must be installed over a proper subfloor using a fastener specifically designed for the installation of wood flooring. Tongue and groove flooring must be blind nailed using the appropriate fastener that is specifically made for the type of product being installed. Smooth fasteners (finish nails, etc.)

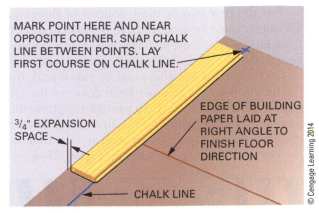

Figure 75–4 A chalk line is snapped on the floor for alignment of the starting row of strip flooring.

may only be used for the purpose of attaching the start and finish rows either by face or blind nailing.

Flooring is blind nailed by driving nails at about a 45-degree angle through the flooring. Start the nail in the corner at the top edge of the tongue. Recommendations for fastening are shown in *Figure 75–5*.

Wood Flooring Type	Fastener to be Used	Fastener Spacing
Solid strip T&G ¾" x less than 3"	1½" to 2" fastener, or 6d-8d casing or finish nails. On slab with ¾" underlayment, use 1½" fastener	Blind fastener spacing along the lengths of the strips, minimum two fasteners per piece near the ends (1"-3"). In addition, every 8"-10" apart for blind nailing, 10"-12" for face nailing.
Solid strip T&G ½" x 1½", ½" x 2"	1½" fastener	Blind fastener spacing along the lengths of the strips, minimum two fasteners per piece near the ends (1"-3"). In addition, every 10" apart. ½" flooring must be installed over a minimum ²³⁄₃₂" thick subfloor.
Solid strip T&G ⅜" x 1½", ⅜" x 2"	1¼" fastener	Blind fastener spacing along the lengths of the strips, minimum two fasteners per piece near the ends (1"-3"). In addition, every 8" apart.
Solid strip T&G ⁵⁄₁₆"	Narrow crowned (under ⅜") 1"-1½" staples or 1"- ½" hardwood flooring cleats	Space fasteners at 3"-4" intervals for staples, 4"-6" for cleats, and within 1"-2" of end joints, or as recommended by the flooring manufacturer.
Solid plank ¾" x 3" or wider	1½"-2" fastener, or 6d-8d casing or finish nails. On slab with ¾" underlayment, use 1½" fastener	Blind fastener spacing along the lengths of the strips, minimum two fasteners per piece near the ends (1"-3"). In addition, every 6"-8" apart for blind nailing, 10"-12" for face nailing. To assist the nailing schedule, option is to apply adhesive.
Engineered wood flooring	Narrow crowned (under ⅜") 1" to 1½" staples or 1"-1½" hardwood flooring cleats designed for engineered flooring	Space fasteners at 3"-4" intervals for staples, 4"-6" for cleats, and within 1"-2" of end joints, or as recommended by the flooring manufacturer.

Figure 75–5 Nailing guide for strip and plank finish flooring.

For the first two or three courses of flooring, a hammer must be used to drive the fasteners. For floor laying, care must be taken not to let the hammer glance off the nail. This may damage the edge of the flooring. Care also must be taken that, on the final blows, the hammer head does not hit the top corner of the flooring. To prevent this, raise the hammer handle slightly on the final blow so that the hammer head hits the nail head and the tongue, but not the corner of the flooring (*Figure 75–6*).

> ⚠️ **CAUTION**
>
> Eye protection should be worn when driving hardened steel nails with a hammer. A small piece of steel may break off the hammer or the nail, and fly out in any direction. This could cause serious injury to an unprotected eye.

After the nail is driven home, its head must be set slightly using a cleat set. This allows adjoining strips to come up tightly against each other. Nail heads may be used to set blind nails. After the nail is driven most the of the way, a nail head is place on the nail head to be set. With one sharp blow the nail is set (*Figure 75–7*).

Racking the Floor

After the second course of flooring is fastened, lay out seven or eight loose rows of flooring, end to end. Lay out in a staggered pattern. End joints should be at least 6 inches apart. In general the wider the board, the greater the stagger. Find or cut pieces to fit within ¾ inch of the end wall. Distribute long and short pieces evenly for the best appearance. Avoid clusters of short strips. Avoid butts seems from alternating rows from lining up creating H patterns. Laying out loose flooring in this manner is called racking the floor. Racking is done to save time and material (*Figure 75–8*).

Take note of where the most traffic will pass through the room. Use longer pieces near door openings to reduce the number of end joints. This will give a better look and reduce the possibility of floor squeaks.

Using the Power Nailer

At least two courses of flooring must be laid by hand to provide clearance from the wall before a power nailer can be used. The power nailer holds strips of special barbed fasteners. The fasteners are driven and set through the tongue of the flooring at the proper angle. Although it is called a power nailer, a heavy hammer is swung by the operator against a plunger to drive the fastener (*Figure 75–9*).

The hammer is double ended. One end is rubber and the other end is steel. The flooring strip is placed in position. The rubber end of the hammer is used to drive the edges and ends of the strips up

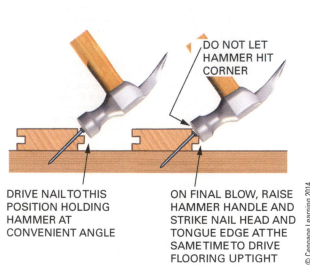

DRIVE NAIL TO THIS POSITION HOLDING HAMMER AT CONVENIENT ANGLE

DO NOT LET HAMMER HIT CORNER

ON FINAL BLOW, RAISE HAMMER HANDLE AND STRIKE NAIL HEAD AND TONGUE EDGE AT THE SAME TIME TO DRIVE FLOORING UP TIGHT

© Cengage Learning 2014

Figure 75–6 Technique for driving a blind nail.

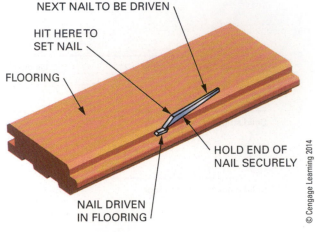

NEXT NAIL TO BE DRIVEN

HIT HERE TO SET NAIL

FLOORING

HOLD END OF NAIL SECURELY

NAIL DRIVEN IN FLOORING

© Cengage Learning 2014

Figure 75–7 Method to set nails driven by hand.

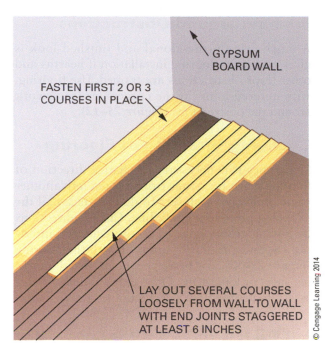

Figure 75–8 Racking the floor places the strips in position for efficient installation.

Figure 75–9 A power nailer is widely used to fasten strip flooring.

tight. The steel end is used against the plunger of the power nailer to drive the fasteners. Slide the power nailer along the tongue edge. Drive fasteners about 8 to 10 inches apart or as needed to bring the strip up tight against previously laid strips.

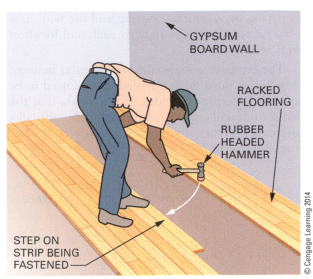

Figure 75–10 Technique of aligning boards together tightly before nailing.

Note: When using the power nailer, one heavy blow is used to drive the fastener. Some nailer models have a ratcheting drive shaft that allows for multiple blows to set the nail. In either case, after the nail is set, the shaft returns and another fastener drops into place ready to be driven. Make sure the wood strip is fit fairly tight before nailing.

Whether laying floor, the floor layer stands with heels on strips already fastened, and toes on the loose strip to be fastened. With weight applied to the joint, easier alignment of the tongue and groove is possible (*Figure 75–10*). The weight of the worker also prevents the loose strip from bouncing when it is driven to make the edge joint tight. Avoid using a power nailer, pneumatic nailer, and hammer-driven fasteners on the same strip of flooring. Each method of fastening places the strips together with varying degrees of tightness. This variation, compounded over multiple strips, will cause waves in the straightness of the flooring.

Ending the Flooring

Continue across the room. Rack seven or eight courses as work progresses. The last three or four courses from the opposite wall must be nailed by hand. This is because of limited room to place the power nailer and swing the hammer. The next-to-the-last row can be blind nailed if care is taken. However, the flooring must be brought up tightly by prying between the flooring and the wall. Use a bar to pry the pieces tight at each nail location (*Figure 75–11*).

The last course is installed in a similar manner. However, it must be face nailed. It may need to be ripped to the proper width. If it appears that the installation will end with an undesirable, difficult-to-apply, narrow strip, lay wider strips in the last row (*Procedure 75–A*).

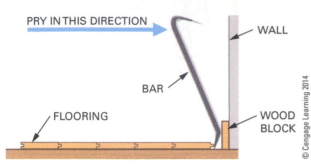

Figure 75–11 The last two courses of strip flooring may be brought tight with a pry bar.

Framing around Obstructions

A much more professional and finished look is given to a strip flooring installation if hearths and other floor obstructions are framed. Use flooring, with mitered joints at the corners, as framework around the obstructions (*Figure 75–12*).

Changing Direction of Flooring

Sometimes it is necessary to change direction of flooring when it extends from a room into another room, hallway, or closet. To do this, face nail the extended piece to a chalk line. Change directions by joining groove edge to groove edge and inserting a spline, ordinarily supplied with the flooring (*Figure 75–13*). For best appearance, avoid bunching short or long strips. Open extra bundles, if necessary, to get the right selection of lengths.

Laying Laminated Strip Flooring

Before laying laminated strip flooring, a ⅛-inch foam underlayment, supplied or approved by the

---

StepbyStep Procedures

Procedure 75–A Installing the Last Strip of Flooring

STEP 1 TO OVERCOME THE DIFFICULTY, FASTEN THE NARROW ENDING STRIP TO THE NEXT TO LAST STRIP BEFORE INSTALLING.

STEP 2 ANOTHER WAY TO OVERCOME THE DIFFICULTY IS TO USE WIDER FLOORING FOR THE ENDING ROW.

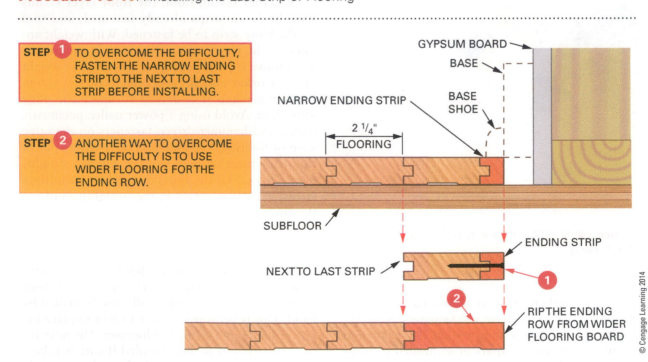

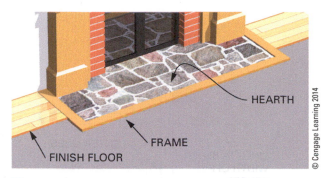

Figure 75-12 Frame around floor obstructions, such as hearths, with strips that are mitered at the corners.

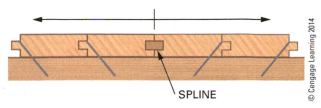

Figure 75-13 The direction of strip flooring can be changed by the use of a spline.

Figure 75-14 A tapping block is used to drive laminated strip flooring boards tight against those already installed.

Figure 75-15 Technique for scribing the last strip on flooring using a scrap piece of flooring.

manufacturer, is applied to the subfloor or slab. The flooring is not fastened or cemented to the floor. However, the boards must be glued to each other along the edges and ends. Apply glue on edges in 8-inch-long beads with 12-inch spaces between them. Apply a full bead across the ends.

The first row is laid in a straight line with end joints glued. The groove edge is placed toward the wall, leaving about a ½-inch expansion space. Subsequent courses are installed. Edges and ends are glued. Each piece is brought tight against the other with a hammer and tapping block. The tapping block should be used only against the tongue. It should never be used against the grooved edge (*Figure 75–14*). Tapping the grooved edge will damage it. Stagger end joints at least 2 feet from those in adjacent courses.

Usually the last course must be cut to width. To lay out the last course to fit, lay a complete row of boards, unglued, tongue toward wall, directly on top of the already installed next to last course. Cut a short piece of flooring for use as a scribing block. Hold the tongue edge against the wall. Move the block along the wall while holding a pencil against the other edge to lay out the flooring. The width of the tongue of the scribing block provides the necessary expansion space (*Figure 75–15*). The last row, when cut, can be glued and wedged tightly in place with a pry bar.

Installing Plank Flooring

Plank flooring is installed like strip flooring. Alternate the courses by widths. Start with the narrowest pieces. Then use increasingly wider courses, and repeat the pattern. Stagger the joints in adjacent courses. Use lengths so they present the best appearance.

Manufacturers' instructions for fastening the flooring vary and should be followed. Generally, the flooring is blind nailed through the tongue of the plank and at intervals along the plank in a manner similar to strip flooring.

INSTALLING PARQUET FLOORING

Procedures for the application of parquet flooring vary with the style and the manufacturer. Detailed installation directions are usually provided with the flooring. Generally, both parquet blocks and strips are laid in *mastic*. Use the recommended type. Apply with a notched trowel. The depth and spacing of the notches are important to leave the correct amount of mastic on the floor. Parquet may be installed either square with the walls or diagonally.

Square Pattern

When laying unit blocks in a square pattern, two layout lines are snapped, at right angles to each other, and parallel to the walls. Blocks are laid with their edges to the lines. Lines are usually laid out so that rows of blocks are either centered on the floor or half the width of a block off center. This depends on which layout produces border blocks of equal and maximum widths against opposite walls.

To determine the location of the layout lines, measure the distance to the center of the room's width. Divide this distance by the width of a block. If the remainder is half or more, snap the layout line in the center. If the remainder is less than half, snap the layout line off center by half the width of the block. Find the location of the other layout line in the same way. It is possible that one of the layout lines will be centered, while the other must be snapped off center.

Other factors may determine the location of layout lines, such as ending with full blocks under a door, or where they meet another type of floor. Regardless of the location, two lines, at right angles to each other, must be snapped.

Place one unit at the intersection of the lines. Position the grooved edges exactly on the lines. Lay the next units ahead and to one side of the first one and along the lines. Install blocks in a pyramid. Work from the center outward toward the walls in all directions. Make adjustments as installation progresses to prevent misalignment (*Figure 75–16*).

Diagonal Pattern

To lay unit blocks in a diagonal pattern, measure an equal distance from one corner of the room along both walls. Snap a starting line between the two marks. At the center of the starting line, snap another line at right angles to it. The location of

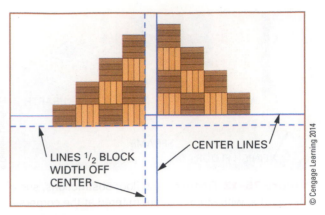

Figure 75–16 Parquet blocks are laid to the center line or off the center lines. This depends on which produces the best size border blocks.

both lines may need to be changed in order to end with border blocks of equal and largest possible size against opposite walls. The diagonal pattern is then laid in a manner similar to the square pattern (*Procedure 75–B*).

Special Patterns

Many parquet patterns can be laid out with square and diagonal layout lines. The *herringbone* pattern requires three layout lines. One will be the 90-degree line used for a square pattern. The other line crosses the intersection at a 45-degree angle. Align the first block or strip with its edge on the diagonal line. The corner of the piece should be lined up at the intersection. Continue the pattern in rows of three units wide, aligning the units with the layout lines (*Figure 75–17*).

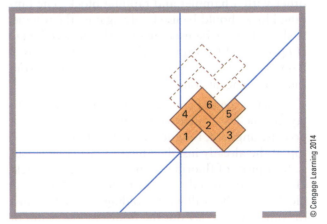

Figure 75–17 The herringbone pattern requires 90- and 45-degree layout lines.

StepbyStep Procedures

Procedure 75–B Laying Out a Diagonal Pattern

STEP 1 LAY OUT EQUAL DISTANCES FROM A SQUARE CORNER.

STEP 2 SNAP A LAYOUT LINE BETWEEN THE TWO POINTS.

STEP 3 SNAP A LAYOUT LINE AT RIGHT ANGLES TO THE FIRST LAYOUT LINE.

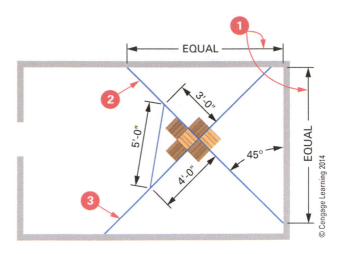

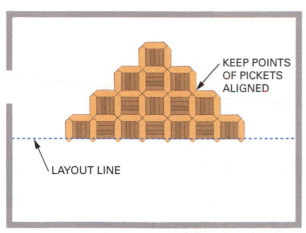

KEEP POINTS OF PICKETS ALIGNED

LAYOUT LINE

SQUARE PATTERN

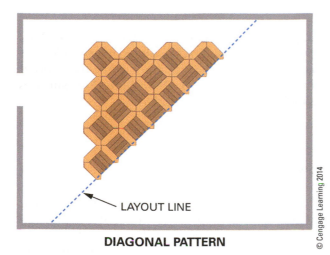

LAYOUT LINE

DIAGONAL PATTERN

Figure 75–18 Layout of the Monticello parquet pattern.

The *Monticello* pattern can be laid square or diagonally in the same way as unit blocks. For best results, lay the parquet in a pyramid pattern. Alternate the grain of the center blocks. Keep *picket* points in precise alignment (*Figure 75–18*).

The *Marie Antoinette* pattern may also be laid square or diagonally. It is started by laying a *band* with its grooved edge and end aligned with the layout lines with the tongued edge to the right. Continue laying the pattern by placing center blocks and bands in a sequence so that bands appear woven (*Figure 75–19*). Tongues of all members face toward the right or ahead. In this pattern, the grain of center blocks runs in the same direction.

Many other parquet patterns are manufactured. With the use of parquet blocks and strips, the design possibilities are almost endless. Competence in laying the popular patterns described in this chapter will enable the student to apply the principles for professional installations of many different parquet floors.

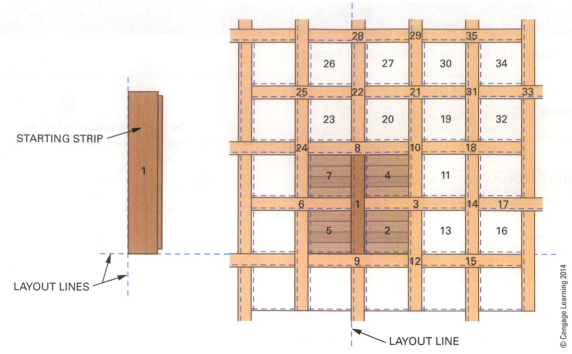

STARTING STRIP

1

LAYOUT LINES

LAYOUT LINE

© Cengage Learning 2014

Figure 75–19 Layout of the Marie Antoinette parquet pattern.

CHAPTER 76

Underlayment and Resilient Tile

Resilient flooring is widely used in residential and commercial buildings. It is a thin, flexible material that comes in sheet or tile form. It is applied on a smooth concrete slab or an underlayment.

UNDERLAYMENT

Plywood, strandboard, hardboard, or particleboard may be used for underlayment. It is installed on top of the subfloor. This provides a base for the application of resilient sheet or tile flooring. The number of underlayment panels required is found by dividing the area to be covered by the area of the panel. *****IRC R503.3.2

Underlayment thickness may range from ¼ to ¾ inch, depending on the material and job

requirements. In many cases, where a finish wood floor meets a tile floor, the underlayment thickness is determined by the difference in the thickness of the two types of finish floor. Both floor surfaces should come flush with each other.

Installing Underlayment

Sweep the subfloor as clean as possible. Cover with asphalt felt lapped 4 inches at the edges. Stagger joints between the subfloor and underlayment. If installation is started in the same corner as the subfloor, the thickness of the walls will be enough to offset the joints. Leave about ¹⁄₃₂ inch between underlayment panels to allow for expansion. ***** *****IRC R503.3.3

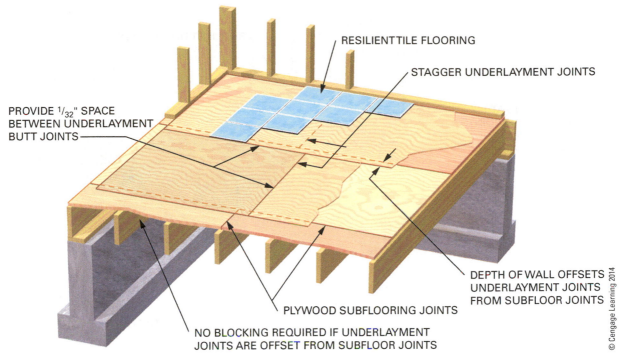

PROVIDE ¹⁄₃₂" SPACE BETWEEN UNDERLAYMENT BUTT JOINTS

RESILIENT TILE FLOORING

STAGGER UNDERLAYMENT JOINTS

DEPTH OF WALL OFFSETS UNDERLAYMENT JOINTS FROM SUBFLOOR JOINTS

PLYWOOD SUBFLOORING JOINTS

NO BLOCKING REQUIRED IF UNDERLAYMENT JOINTS ARE OFFSET FROM SUBFLOOR JOINTS

© Cengage Learning 2014

Figure 76–1 The joints between underlayment and subfloor are offset.

If the underlayment is to go over a board subfloor, install the underlayment with its face grain across the boards. In all other cases, a stiffer and stronger floor is obtained if the face grain is across the floor joists (*Figure 76–1*).

Fasten the first row in place with staples or nails. Underlayment requires more nails than subfloor to provide a squeak-free and stiff floor. Nail spacing and size are shown in *Figure 76–2* for various thicknesses of plywood underlayment.

Similar nail spacing and size are appropriate for other types of underlayment.

Install the remaining rows of underlayment with end joints staggered from the previous courses. The last row of panels is ripped to width to fit the remaining space.

If APA Sturd-I-Floor is used, it can double as a subfloor and underlayment. If square-edged panels are used, blocking must be provided under the joints to support the edges. If tongue-and-groove panels are used, blocking is not required (*Figure 76–3*).

RESILIENT TILE FLOORING

Most resilient floor tiles are made of vinyl. Many different colors, textures, and patterns are available. Tiles come in 12 × 12 squares. They are applied to the floor in a manner similar to that used for applying parquet blocks. The most common tile thicknesses are ¹⁄₁₆ and ⅛ inch. Thicker tiles are used in commercial applications subjected to considerable use.

Long strips of the same material are called feature strips. They are available from the manufacturer. The strips vary in width from ¼ inch to 2 inches. They are used between tiles to create unique floor patterns.

Plywood Thickness (inch)	Fastener Size (approx.) and Type	Fastener Spacing (inches)	
		Panel Edges	Panel Interior
¼	1¼" ring-shank nails	3	6 each way
	18 gauge staples	3	6 each way
⅜ or ½	1¼" ring-shank nails	6	8 each way
	16 gauge staples	3	6 each way
⅝ or ¾	1½" ring-shank nails	6	8 each way
	16 gauge staples	3	6 each way
¼	1¼" ring-shank nails	3	6 each way
	18 gauge staples	3	6 each way

© Cengage Learning 2014

Figure 76–2 Nailing specifications for plywood underlayment.

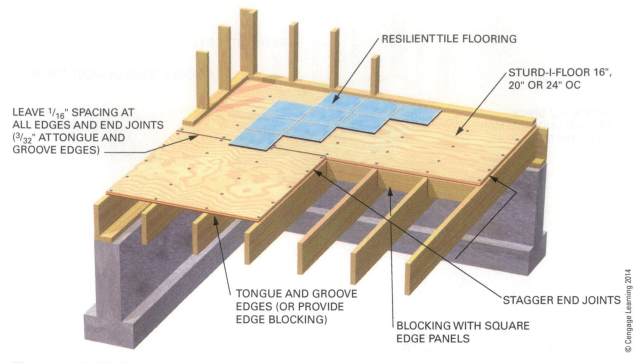

RESILIENT TILE FLOORING

STURD-I-FLOOR 16", 20" OR 24" OC

LEAVE ¹/₁₆" SPACING AT ALL EDGES AND END JOINTS (³/₃₂" AT TONGUE AND GROOVE EDGES)

TONGUE AND GROOVE EDGES (OR PROVIDE EDGE BLOCKING)

BLOCKING WITH SQUARE EDGE PANELS

STAGGER END JOINTS

© Cengage Learning 2014

Figure 76–3 APA Sturd-I-Floor requires no underlayment for the installation of resilient floors.

INSTALLING RESILIENT TILE

CAUTION

If the application involves the removal of existing resilient floor covering, be aware that it, and the adhesive used, may contain asbestos. The presence of asbestos in the material is not easily determined. If there is any doubt, always assume that the existing flooring and adhesive do contain asbestos. Practices for removal of existing flooring or any other building material containing asbestos should comply with standards set by the U.S. Occupational Safety and Health Administration (OSHA) or corresponding authorities in other jurisdictions.

If the application is over an existing resilient floor covering, do not sand the existing surface unless absolutely sure it does not contain asbestos. Inhalation of asbestos dust can cause serious harm.

© Cengage Learning 2014

Figure 76–4 Uneven floor surfaces are patched with a patching compound before tile is installed.

Before installing resilient tile, make sure underlayment fasteners are not projecting above the surface. Fill any open areas, such as splits, with a floor leveling compound (*Figure 76–4*). On underlayment, use a portable disc sander to bring the joints flush. Skim the entire surface with the sander to make sure the surface is smooth. Sweep the floor clean.

Check to see if the spaces between the door stops and casing are sufficient to allow tile to slide under. They may be trimmed using a handsaw on a scrap piece of tile as a gauge (*Figure 76–5*).

Layout

Snap layout lines across the floor at right angles to each other. The lines may be centered if border tiles are a half width or more. If border tiles are less than a half width, layout lines may be snapped half

Figure 76–5 The bottom end of molding may be trimmed to allow the floor tile to slide under.

the width of a tile off center. The important thing is to have two perpendicular lines that serve to guide the joints in the tile.

Applying Adhesive

Adhesive is applied to the entire floor area (*Figure 76–6*). When the adhesive becomes transparent and appears to be dry, it is ready for tile to be installed. It will remain tacky and ready for application for up to seventy-two hours.

A notched trowel is used to spread the adhesive. It must be properly sized according to the manufacturer's recommendation. If the notches are too deep, more adhesive than necessary will be applied. This will result in the adhesive squeezing up through the joints onto the face of the tile and will probably require removal of the application.

Figure 76–6 Floor adhesive is often troweled onto the entire floor area before tile application begins.

Laying Tiles

Start by applying a tile to the intersection of the layout lines with the two adjacent edges on the lines. Lay tiles with edges tight. Work toward the walls (*Figure 76–7*). Watch the grain pattern. It may be desired to alternate the run of the patterns or to lay the patterns in one direction. Some tiles are stamped with an arrow on the back. They are placed so the arrows on all tiles point in the same direction.

Lay tiles in place instead of sliding them into position. Sliding the tile pushes the adhesive through the joint. With most types of adhesives, it may be difficult or impossible to slide the tile.

Applying Border Tiles

Border tiles are often cut using a tile cutter (*Figure 76–8*). It cuts fast, and the cut edges are clean and straight. Each piece is measured, marked, and cut by rotating the handle.

Border tiles may also be cut by scoring with a sharp utility knife and bending. To lay out and score a border tile to fit snugly, first place the tile to be cut directly on top of the last tile installed. Make sure all edges are in line. Place a full tile with its edge against the wall and on top of the one to be fitted. Score the border tile along the outside edge of the top tile (*Figure 76–9*). Bend and break the tile along the scored line.

If the base has not been installed yet, the border tiles are fit roughly into place. Then when the base is installed later, it covers the cut edge. If the base has already been installed, the tile must be fit

Figure 76–7 Tile is applied from the layout lines towards the walls.

Figure 76–8 A tile cutter is often used to install border tile.

Figure 76–10 A propane torch is used to soften the tile for cutting. Use care not to overheat and burn the tile.

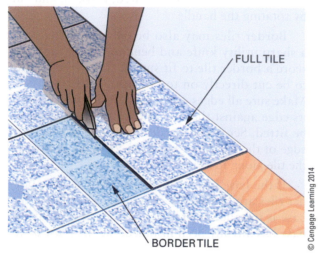

FULL TILE

BORDER TILE

Figure 76–9 Fitting a border tile by placing it under a guide piece of tile while scoring.

Figure 76–11 Fitting a tile around a corner involves two cuts.

closely. Scored cuts may need to be smoothed with a file or sandpaper to improve their look.

For tiles that require curved cuts, a propane torch can be used on the back side of the tile to warm it. This makes the knife cut easily and the resulting cut is smooth (*Figure 76–10*). Use care not to overheat and burn the tile. This is also done when fitting a tile around a corner where two or more cuts are required (*Figure 76–11*).

Applying a Vinyl Cove Base

Many times a vinyl cove base is used to trim a tile floor. A special vinyl base cement is applied to its back. The base is pressed into place (*Figure 76–12*).

Figure 76–12 Vinyl base may be bent to fit around a corner.

ESTIMATING FLOORING

When estimating the amount of materials for finish floors, wood strip flooring, and resilient tiles, remember that if the number of pieces does not come out even, round it up to the next whole number of pieces. The floor area for the rooms of a building may be calculated from a set of prints. Find the room's width and length, then multiply them together.

Wood Strip Flooring

To estimate the amount of hardwood flooring material needed, first determine the area to be covered. Divide this by the square feet per carton of flooring to get the number of cartons. Add 2 percent waste for end matching and normal waste. For example, the area of a room is 384 square feet and a bundle of flooring covers 15 square feet per carton, $384 \div 15 \times 1.02 = 26.1 + 27$ cartons.

Resilient Flooring

To estimate the amount of resilient tile flooring needed, find the area of the room in square feet. For 12×12 tiles, the result is the number of pieces needed. Divide the number of pieces by square feet per box. Rounding up this number to the next whole box usually takes care of any waste factor. For example, if a room is 520 SF and a box covers 36 SF, then the boxes needed is $520 \div 36 = 14.4 \Rightarrow 15$ boxes.

See *Figure 76–13* for a listing of the shorthand version of these estimating techniques, along with another example.

Estimate the materials for flooring materials to cover a 475 SF area in wood strip flooring and also in resilient tiles. Cartons of wood flooring cover 15 SF and resilient tile boxes cover 36 SF.			
Item	**Formula**	**Waste factor**	**Example**
Wood Strip Flooring	area to be covered ÷ SF per carton × waste = cartons	2%	$475 \div 15 \times 102\% = 32.3 \Rightarrow 33$ cartons
Resilient Flooring	area to be covered ÷ SF per carton = boxes	10%	$475 \div 36 \times 110\% = 13.1 \Rightarrow 14$ boxes

Figure 76–13 Example of estimating floor covering materials with formulas.

© Cengage Learning 2014

DECONSTRUCT THIS

Carefully study *Figure 76–14* and think about what is wrong and/or what is right. Consider all possibilities. What construction practice or method is different in your area of the country?

Figure 76–14 This photo shows flooring made with 3-, 4-, and 5-inch-wide white oak boards.

KEY TERMS

herringbone	plank	slat block	unfinished strip
laminated block	plank flooring	sleepers	unit block
laminated strip flooring	prefinished strip	spline	
parquet block	racking the floor	strip	
parquet strip	resilient	strip flooring	

REVIEW QUESTIONS

Select the most appropriate answer.

1. If hardwood flooring is stored in a heated building, the temperature should not exceed

   a. 72°F.
   b. 78°F.
   c. 85°F.
   d. 90°F.

2. Most hardwood finish flooring is made from

   a. Douglas fir.
   b. hemlock.
   c. southern pine.
   d. oak.

3. Bundles of strip flooring may contain pieces over and under the nominal length of the bundle by

   a. 4 inches.
   b. 6 inches.
   c. 8 inches.
   d. 9 inches.

4. No pieces are allowed in bundles of hardwood strip flooring shorter than

   a. 4 inches.
   b. 6 inches.
   c. 8 inches.
   d. 9 inches.

5. The edges of prefinished strip flooring are chamfered to

   a. prevent splitting.
   b. apply the finish.
   c. simulate cracks between adjoining pieces.
   d. obscure any unevenness in the floor surface.

6. The best grade of unfinished oak strip flooring is

   a. prime.
   b. clear.
   c. select.
   d. quarter-sawed.

7. The end seams of strip flooring in adjacent rows should be no closer than

   a. 6 inches.
   b. 8 inches.
   c. 9 inches.
   d. 12 inches.

8. When it is necessary to cut the last strip in a course of flooring, the waste is used to start the next course and should be at least

   a. 8 inches long.
   b. 10 inches long.
   c. 12 inches long.
   d. 16 inches long.

9. The adhesive used for resilient tiles is

   a. applied with a caulking gun.
   b. allowed to dry before tiles are applied.
   c. troweled to the back side of each tile.
   d. all of the above.

10. To change direction when installing strip flooring,

    a. face nail both strips.
    b. turn the extended strip around.
    c. blind nail both strips.
    d. use a spline.

UNIT 27
Interior Doors and Door Frames

CHAPTER 77 Description of Interior Doors
CHAPTER 78 Installation of Interior Doors and Door Frames

Interior doors used in residential and light-commercial buildings are less dense than exterior doors. They are not ordinarily subjected to as much use and are not exposed to the weather. In commercial buildings, such as hospitals and schools, heavier and larger interior doors are specified that meet special conditions.

Doors must be installed level and plumb to operate properly. Otherwise they may not remain in the position last placed in, but instead move on their own.

OBJECTIVES

After completing this unit, the student should be able to:

- describe the sizes and kinds of interior doors.
- make and set interior door frames.
- hang an interior swinging door.
- install locksets on interior swinging doors.
- set a prehung door and frame.
- install bypass, bifold, pocket, and folding doors.

CHAPTER 77

Description of Interior Doors

Interior doors are classified by style as *flush, panel, French, louver,* and *café* doors. Interior flush doors have a smooth surface, are usually less expensive, and are widely used when a plain appearance is desired. Some of the other styles have special uses. Doors are also classified by the way they operate, such as *swinging, sliding,* or *folding.*

INTERIOR DOOR SIZES AND STYLES

For residential and light-commercial use, most interior doors are manufactured in $1\frac{3}{8}$-inch thickness. Some, like café and bifold doors, are also made in $1\frac{1}{4}$- and $1\frac{1}{8}$-inch thicknesses. Most doors are manufactured in 6'-8" heights. Some types may be obtained in heights of 6'-0''' and 6'-6". Door widths range from 1'-0" to 3'-0" in increments of 2 inches. However, not all sizes are available in every style.

Flush Doors

Flush doors are made with *solid* or *hollow* cores. Solid-core doors are generally used as entrance or fire-rated doors. (They have been previously described in Chapter 57, Exterior Doors.) Hollow-core doors are commonly used in the interior except when fire resistance or sound transmission is critical.

A hollow-core door consists of a light perimeter frame. This frame encloses a mesh of thin wood or composition material supporting the faces of the door. Solid wood blocks are appropriately placed in the core for the installation of locksets. The frame and mesh are covered with a thin plywood called a *skin. Lauan* plywood is used extensively for flush door skins. Flush doors are also available with veneer faces of *birch, gum, oak,* and *mahogany,* among others (*Figure 77–1*). When flush doors are to be painted, an overlay plywood or tempered hardboard may be used for the skin.

Panel Doors

Interior panel doors consist of a frame with usually one to eight wood panels in various designs (*Figure 77–2*). They are similar in style to some exterior panel doors. (The construction of panel doors has been previously described in Chapter 57, Exterior Door Installation.)

MESH OR CELLULAR CORE
7 PLY CONSTRUCTION ILLUSTRATED

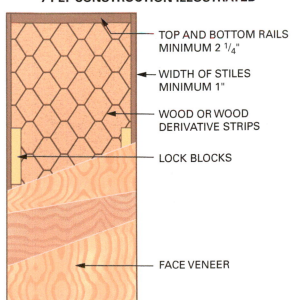

- TOP AND BOTTOM RAILS MINIMUM 2 $\frac{1}{4}$"
- WIDTH OF STILES MINIMUM 1"
- WOOD OR WOOD DERIVATIVE STRIPS
- LOCK BLOCKS
- FACE VENEER

LADDER CORE
7 PLY CONSTRUCTION ILLUSTRATED

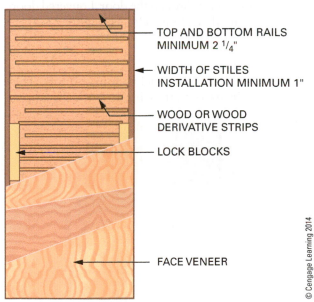

- TOP AND BOTTOM RAILS MINIMUM 2 $\frac{1}{4}$"
- WIDTH OF STILES INSTALLATION MINIMUM 1"
- WOOD OR WOOD DERIVATIVE STRIPS
- LOCK BLOCKS
- FACE VENEER

Figure 77–1 The construction of hollow-core flush doors.

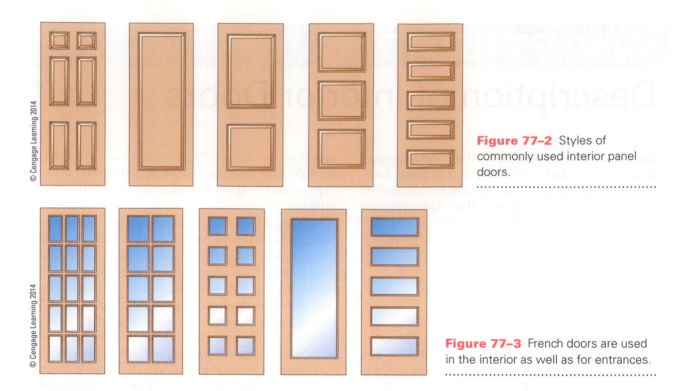

Figure 77–2 Styles of commonly used interior panel doors.

Figure 77–3 French doors are used in the interior as well as for entrances.

French Doors

French doors may contain from one to fifteen lights of glass. They are made in a $1\frac{3}{4}$-inch thickness for exterior doors and $1\frac{3}{8}$-inch thickness for interior doors (*Figure 77–3*).

Louver Doors

Louver doors are made with spaced horizontal slats called louvers used in place of panels. The louvers are installed at an angle to obstruct vision but permit the flow of air through the door. Louvered doors are widely used on clothes closets (*Figure 77–4*).

Café Doors

Café doors are short panel or louver doors. They are hung in pairs that swing in both directions.

They are used to partially screen an area, yet allow easy and safe passage through the opening. The tops and bottoms of the doors are usually shaped in a pleasing design (*Figure 77–5*).

METHODS OF INTERIOR DOOR OPERATION

Doors are also identified by their method of operation. Doors either swing on hinges or slide on tracks. The choice of door operation

Figure 77–4 Louver doors obstruct vision, but permit the circulation of air.

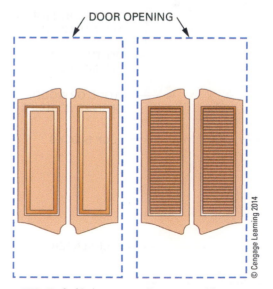

Figure 77–5 Café doors usually are used between kitchens and dining areas.

depends on such factors as convenience, cost, safety, and space.

Swinging Doors

Swinging doors are hinged on one edge. They swing out of the opening. When closed, they cover the total opening. Swinging doors that swing in one direction are called *single-acting doors* (*Figure 77–6*). With special hinges, they can swing in both directions. They are then called *double-acting doors*. Swinging doors are the most commonly used type of door. They have the disadvantage of requiring space for the swing.

Bypass Doors

Bypass doors are commonly used on wide clothes closet openings. A double track is mounted on the header jamb of the door frame. Rollers that ride in the track are attached to the doors so that they slide by each other. A floor guide keeps the doors in alignment at the bottom (*Figure 77–7*). Usually two doors are used in a single opening. Three or more doors may be used, depending on the situation.

The disadvantage of bypass doors is that although they do not project out into the room, access to the complete width of the opening is not possible. They are easy to install, but are not practical in openings less than 6 feet wide.

Pocket Doors

The pocket door is opened by sliding it sideways into the interior of the partition. When opened, only the lock edge of the door is visible (*Figure 77–8*). Pocket doors may be installed as a single unit,

Figure 77–7 Bypass doors are used on wide closet openings.

Figure 77–6 A single-acting swinging door is the most widely used type of interior door.

Figure 77–8 The pocket door slides into the interior of the partition.

sliding in one direction, or as a double unit sliding in opposite directions. When opened, the total width of the opening is obtained, and the door does not project out into the room. Pocket doors are used when these advantages are desired.

The installation of pocket doors requires more time and material than other methods of door operation. A special pocket door frame unit and track must be installed during the rough framing stage (*Figure 77–9*). The rough opening in the partition must be large enough for the door opening and the pocket.

Bifold Doors

Bifold doors are made in flush, panel, louver, or combination panel and louver styles. They are made in narrower widths than other doors. This allows them to operate in a folding fashion on closet and similar type openings (*Figure 77–10*).

Bifold doors consist of pairs of doors hinged at their edges. The doors on the jamb side swing on pivots installed at the top and bottom. Other doors fold up against the jamb door as it is swung open. The end door has a guide pin installed at the top. The pin rides in a track to guide the set when opening or closing (*Figure 77–11*). On very wide openings, the guide pin is replaced by a combination guide and support to keep the doors from sagging.

Bifold doors may be installed in double sets, opening and closing from each side of the opening. They have the advantage of providing access to almost the total width of the opening, yet they do not project out much into the room.

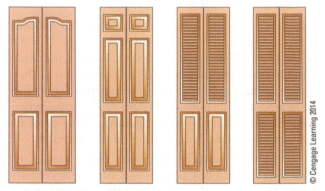

Figure 77–10 Bifold doors are manufactured in many styles.

Figure 77–9 A pocket door frame comes preassembled from the factory. It is installed when the interior partitions are framed.

Figure 77–11 Bifold doors provide access to almost the total width of the opening.

CHAPTER 78

Installation of Interior Doors and Door Frames

Many interior doors come prehung in their frames for easier and faster installation on the job. However, it is often necessary to build and set the door frame, hang the door, and install the locksets.

INTERIOR DOOR FRAMES

Special rabbeted jamb stock or nominal 1-inch square-edge lumber is used to make interior door frames. If square-edge lumber is used, a separate *stop*, if needed, is applied to the inside faces of the door frame.

Checking Rough Openings

The first step in making an interior door frame is to measure the door opening to make sure it is the correct width and height. The rough opening width for single-acting swinging doors should be the width of the door plus twice the thickness of the side jamb, plus $\frac{1}{2}$ inch on each side for shimming between the door frame and the opening. For example, if the thickness of the side jamb beyond the door is $\frac{3}{4}$ inch, the rough opening width is $2\frac{1}{2}$ inches more than the door width. The rough opening height should be the height of the door, plus the thickness of the header jamb, plus $\frac{1}{2}$ inch for clearance at the top, plus the thickness of the finished floor, plus a desired clearance under the door. An allowance of $\frac{1}{2}$ to 1 inch is usually made for clearance between the finished floor and the door.

For example, if the header jamb and finished floor thickness are both $\frac{3}{4}$ inch, the rough opening height should be $2\frac{1}{2}$ inches over the door height, if $\frac{1}{2}$ inch is allowed for clearance under the door (*Figure 78–1*).

The rough opening size for other than single-acting swinging doors, such as bypass and bifold doors, should be checked against the manufacturer's directions. The sizes of the doors and allowances for hardware may differ with the manufacturer.

Making an Interior Door Frame

Interior door frames are constructed like exterior door frames except they have no sill. Interior door frames usually are installed after the interior wall

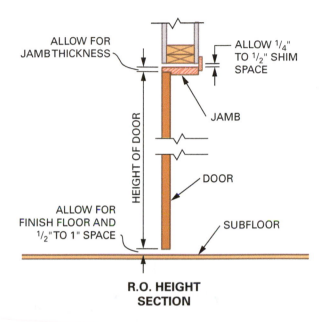

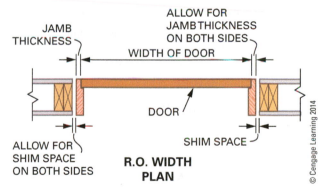

Figure 78–1 The size of rough openings for doors needs to take into account space for jambs and shimming.

covering has been applied. Measure the total thickness of the wall, including the wall covering, to find the jamb width.

Cutting Jambs to Width

If necessary, rip the door jamb stock so its width is the same as the wall thickness. If rabbeted jamb stock is used, cut the edge opposite the rabbet. Plane and smooth both edges to a slight back bevel.

The back bevel permits the door casings, when later applied, to fit tightly against the edges of the door frame in case there are irregularities in the wall (*Figure 78–2*). *Ease* all sharp exposed corners.

Cutting Jambs to Length

On interior door frames, the *head jamb* is usually cut to fit between, and is dadoed into, the *side jambs*. The side jambs run the total height of the rough opening. Cut both side jambs to a length equal to the height of the opening.

Head Jambs for Hinged Doors. If rabbeted jambs are used, the length of the head jamb is the width of the door plus ³⁄₁₆ inch. The extra ³⁄₁₆ inch is for joints of ³⁄₃₂ inch on each side, between the edges of the door and the side jambs. If square-edge lumber is used, its length is the same as a rabbeted head jamb. However, ¹⁄₂ inch is added for dadoing ¹⁄₄ inch deep into each side jamb (*Figure 78–3*). Cut the head jamb to length with both ends square.

Side Jambs for Hinged Doors. Measure up from the bottom ends. Square lines across the side jambs to mark the location of the bottom side of the head jamb. This dimension is the sum of:

- the thickness of the finish floor, if the door frame rests on the subfloor,
- an allowance of ¹⁄₂ inch minimum between the door and the finish floor,
- the height of the door, and
- ³⁄₃₂ inch for a joint between the door and the head jamb;
- on rabbeted jambs, subtract ¹⁄₂ inch for the depth of the rabbet.

Hold a scrap piece of jamb stock to the squared lines. Mark its other side to lay out the width of the dado. Mark the depth of the dadoes on both edges of the side jambs. Cut the dadoes to receive the head jamb. On rabbeted jambs, dado depth is to the face of the rabbet. A dado depth of ¹⁄₄ inch is sufficient on plain jambs (*Figure 78–4*).

Jamb Lengths for Other Types of Doors. The length of head and side jambs for other types of doors, such as bypass and bifold, must be

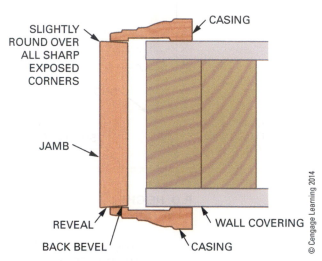

SLIGHTLY ROUND OVER ALL SHARP EXPOSED CORNERS

CASING

JAMB

REVEAL

BACK BEVEL

WALL COVERING

CASING

© Cengage Learning 2014

Figure 78–2 Back bevel jamb edges slightly to permit casings to fit snugly against them.

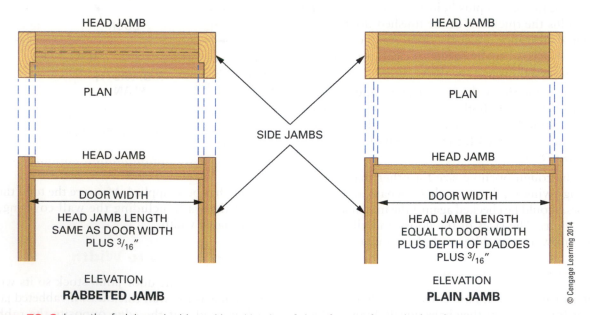

HEAD JAMB

PLAN

HEAD JAMB

DOOR WIDTH

HEAD JAMB LENGTH SAME AS DOOR WIDTH PLUS ³⁄₁₆"

ELEVATION
RABBETED JAMB

SIDE JAMBS

HEAD JAMB

PLAN

HEAD JAMB

DOOR WIDTH

HEAD JAMB LENGTH EQUAL TO DOOR WIDTH PLUS DEPTH OF DADOES PLUS ³⁄₁₆"

ELEVATION
PLAIN JAMB

© Cengage Learning 2014

Figure 78–3 Length of plain and rabbeted head jambs of door frames for swinging doors.

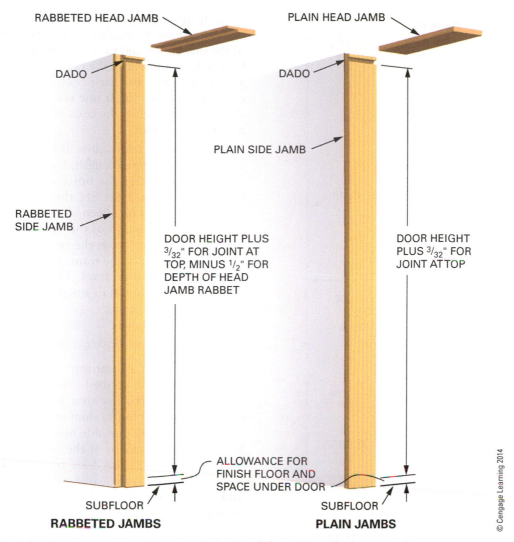

RABBETED HEAD JAMB

PLAIN HEAD JAMB

DADO

DADO

PLAIN SIDE JAMB

RABBETED
SIDE JAMB

DOOR HEIGHT PLUS
$^3/_{32}$" FOR JOINT AT
TOP, MINUS $^1/_2$" FOR
DEPTH OF HEAD
JAMB RABBET

DOOR HEIGHT
PLUS $^3/_{32}$" FOR
JOINT AT TOP

ALLOWANCE FOR
FINISH FLOOR AND
SPACE UNDER DOOR

SUBFLOOR

SUBFLOOR

RABBETED JAMBS

PLAIN JAMBS

© Cengage Learning 2014

Figure 78–4 Laying out plain and rabbeted side jambs of door frames for swinging doors.

determined from instructions provided by the manufacturer of the hardware and the door. Door hardware and door sizes differ with the manufacturer. This affects the length of the door jambs.

Plain, square-edge lumber jambs are used to make door frames for doors other than single-acting swinging doors. The rest of the procedure, such as checking rough openings, cutting jambs to width, assembling, and setting, is the same for all door frames.

ASSEMBLING THE DOOR FRAME

Fasten the side jambs to the head jamb, keeping the edges flush. If there is play in the dado, first wedge the head jamb with a chisel so the face side comes up tight against the dado shoulders before fastening.

Cut a narrow strip of wood. Tack it to the side jambs a few inches up from the bottom so that the frame width is the same at the bottom as it is at the top. This strip is commonly called a spreader (*Figure 78–5*).

SETTING THE DOOR FRAME

Door frames must be set so that the jambs are straight, level, and plumb. They are usually set before the finish floor is laid. If a rabbeted frame is used, determine the swing of the door so that the rabbet is facing toward the correct side.

Cut any *horns* from the top ends of side jambs. Place the frame in the opening. The horns are cut off in case the side jambs need to be shimmed to level the head jamb.

Install shims directly opposite the ends of the header jamb between the opening and the side

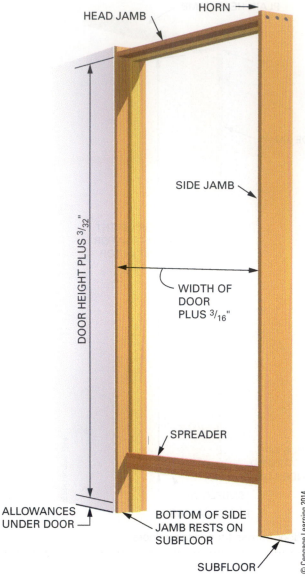

HEAD JAMB

HORN →

DOOR HEIGHT PLUS $3/32$"

SIDE JAMB

WIDTH OF DOOR PLUS $3/16$"

SPREADER

ALLOWANCES UNDER DOOR

BOTTOM OF SIDE JAMB RESTS ON SUBFLOOR

SUBFLOOR

© Cengage Learning 2014

Figure 78–5 An assembled interior rabbeted door frame.

jambs. Shim an equal amount on both sides so that the frame is close to being centered at the top of the opening. Drive shims up snugly not tightly.

Leveling the Header Jamb

Keep the edges of the frame flush with the wall. With the bottom ends of both side jambs resting on the subfloor, check whether the head jamb is level. Level the head jamb, if necessary, by placing shims between the bottom end of the appropriate side jamb and the subfloor. When the header jamb is level, tack the frame in place on both sides, close to the top. Drive fasteners through side jambs and shims into the studs.

If the door frame rests on a finish floor, then a tight joint must be made between the bottom ends of the side jambs and the finish floor. If the floor is level, side jambs will fit the floor if their lengths are exactly the same and their bottom ends have been cut square. Once the frame is set, the head jamb should be level when the ends of the side jamb are resting on the floor.

If the floor is not quite level, or if the side jambs are of unequal length, level the head jamb by shimming under the bottom end of the side jamb on the low side. Set the dividers for the amount the jamb has been shimmed. Scribe that amount on the bottom of the opposite side jamb. Remove the frame from the opening. Cut to the scribed line. Replace the frame in the opening. The head jamb should be level. The bottom ends of both side jambs should fit snugly against the finish floor (*Figure 78–6*).

Plumbing the Side Jambs

Several ways of plumbing door frames have been previously described. An accurate and fast method is with the use of a 6-foot level. When one side jamb is plumb, shim and tack its bottom end in place. Only one side needs to be plumbed. Locate the bottom end of the other side jamb by measuring across. The door frame width at the bottom should be the same as on the top. Shim and tack the bottom end of the other side jamb in place.

Straightening Jambs

Use a 6-foot straightedge against the side jambs. Straighten them by shimming at intermediate points. Besides other points, shims should be placed opposite hinge and lockset locations. Fasten the jambs by nailing through the shims. Header jambs on wide door frames are straightened, shimmed, and fastened in a similar way.

Sighting the Door Frame for a Wind

Before any nails are set, sight the door frame to see if it has a *wind*. The frame must be sighted by eye to make sure that side jambs line up vertically with each other and that the frame is not twisted. This is important when installing rabbeted jambs. The method of checking for a wind in door frames has been previously described in *Figure 56–8*.

If the frame has a wind, move the top or bottom ends of the side jambs slightly until they line up

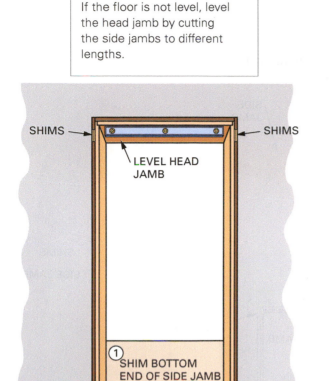

Figure 78–6 Technique for cutting side jambs to make the head jamb level.

SHIMS — LEVEL HEAD JAMB — SHIMS

① SHIM BOTTOM END OF SIDE JAMB UNTIL HEAD JAMB IS LEVEL

② ADJUST SCRIBER TO THE AMOUNT JAMB IS SHIMMED OUT OF LEVEL FINISH FLOOR

③ SCRIBE AMOUNT ON OPPOSITE SIDE JAMB AND CUT

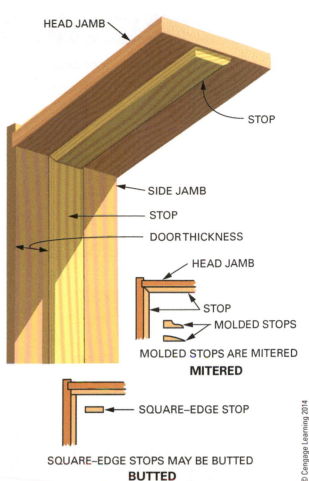

HEAD JAMB
STOP
SIDE JAMB
STOP
DOOR THICKNESS
HEAD JAMB
STOP
MOLDED STOPS
MOLDED STOPS ARE MITERED
MITERED
SQUARE–EDGE STOP
SQUARE–EDGE STOPS MAY BE BUTTED
BUTTED

Figure 78–7 A back miter joint is used for molded stops, and a butt joint is usually used on square-edge stops.

with each other. Fasten top and bottom ends of the side jambs securely. Set the nails (*Procedure 78–A*).

Applying Door Stops

At this time, *door stops* may be applied to plain jambs. The stops are not permanently fastened, in case they have to be adjusted when locksets are installed. A back miter joint is usually made between molded side and header stops. A butt joint is made between square-edge stops (*Figure 78–7*).

HANGING INTERIOR DOORS

The method of fitting and hanging single-acting, hinged, interior doors is similar to that for exterior doors.

Double-Acting Doors

Double-acting doors are installed either with special pivoting hardware installed on the floor and the head jamb or with spring-loaded double-acting hinges. Both types return the door to a closed position after being opened. When opened wide, the doors can be held in the open position. A different type of light-duty, double-acting hardware is used on café doors. To install double-acting door hardware, follow the manufacturer's directions.

Installing Bypass Doors

Bypass doors are installed so they overlap each other by about 1 inch when closed. Cut the track to length. Install it on the header jamb according to the manufacturer's directions.

Installing Rollers. Install pairs of *roller hangers* on each door. The roller hangers may be offset a different amount for the door on the outside

Step**by**Step Procedures

Procedure 78–A Installing a Door Jamb

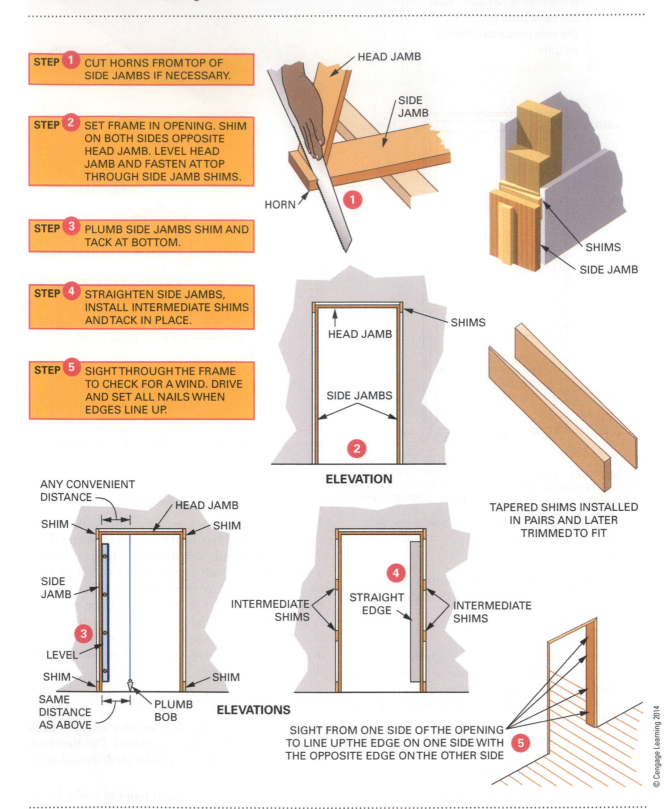

STEP 1 CUT HORNS FROM TOP OF SIDE JAMBS IF NECESSARY.

STEP 2 SET FRAME IN OPENING. SHIM ON BOTH SIDES OPPOSITE HEAD JAMB. LEVEL HEAD JAMB AND FASTEN AT TOP THROUGH SIDE JAMB SHIMS.

STEP 3 PLUMB SIDE JAMBS SHIM AND TACK AT BOTTOM.

STEP 4 STRAIGHTEN SIDE JAMBS, INSTALL INTERMEDIATE SHIMS AND TACK IN PLACE.

STEP 5 SIGHT THROUGH THE FRAME TO CHECK FOR A WIND. DRIVE AND SET ALL NAILS WHEN EDGES LINE UP.

HEAD JAMB

SIDE JAMB

HORN

SHIMS

SIDE JAMB

HEAD JAMB

SHIMS

SIDE JAMBS

ELEVATION

TAPERED SHIMS INSTALLED IN PAIRS AND LATER TRIMMED TO FIT

ANY CONVENIENT DISTANCE

SHIM

HEAD JAMB

SHIM

SIDE JAMB

LEVEL

SHIM

SHIM

SAME DISTANCE AS ABOVE

PLUMB BOB

ELEVATIONS

INTERMEDIATE SHIMS

STRAIGHT EDGE

INTERMEDIATE SHIMS

SIGHT FROM ONE SIDE OF THE OPENING TO LINE UP THE EDGE ON ONE SIDE WITH THE OPPOSITE EDGE ON THE OTHER SIDE

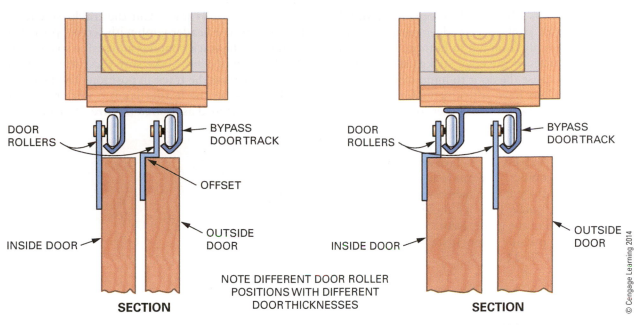

Figure 78–8 Bypass door rollers are offset different distances for use on doors of various thicknesses.

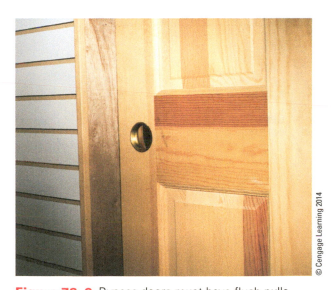

Figure 78–9 Bypass doors must have flush pulls.

than the door on the inside. They are also offset differently for doors of various thicknesses. Make sure that rollers with the same and correct offset are used on each door (*Figure 78–8*). The location of the rollers from the edge of the door is usually specified in the manufacturer's instruction sheet.

Installing Door Pulls.   Mark the location and bore holes for door pulls. Flush pulls must be used so that bypassing is not obstructed (*Figure 78–9*). The proper size hole is bored partway into the door. The pull is tapped into place with a hammer and wood block. The press fit holds the pull in

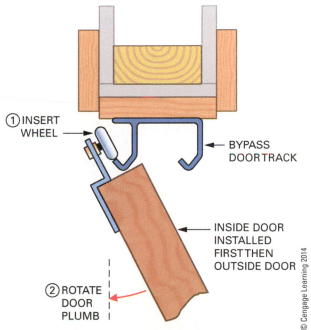

Figure 78–10 Bypass doors are hung on the overhead track by holding the bottom of the door outward.

place. Rectangular flush pulls, also used on bypass doors, are held in place with small recessed screws.

Hanging Doors.   Hang the doors by holding the bottom outward. Insert the rollers in the overhead track. Then gently let the door come to a vertical position. Install the inside door first, then the outside door (*Figure 78–10*).

Fitting Doors.   Test the door operation and the fit against side jambs. Door edges must fit against side jambs evenly from top to bottom. If the top or bottom portion of the edge strikes the side jamb first, it may cause the door to jump from the track. The door rollers have adjustments for raising and lowering. Adjust one or the other to make the door edges fit against side jambs.

Installing Floor Guides.   A floor guide is included with bypass door hardware to keep the doors in alignment. The guide is centered on the lap to steady the doors at the bottom. Mark the location of the guide. Remove the doors. Install the inside section of the guide. Replace the inside door. Replace the outside door. Install the rest of the guide (*Figure 78–11*).

Installing Bifold Doors

Before installing bifold doors, make sure the opening size is as specified by the hardware or door manufacturer. Usually bifold doors come hinged together in pairs. The hardware consists of the track, pivot sockets, pivot pins and guides, door aligners, door pulls, and necessary fasteners (*Figure 78–12*).

Installing the Track.   Cut the track to length. Fasten it to the header jamb with screws provided in the kit. The track contains adjustable sockets for the door *pivot pins*. Make sure these are inserted before fastening the track in position. The position of the track on the header jamb is not critical. It may be positioned as desired (*Figure 78–13*).

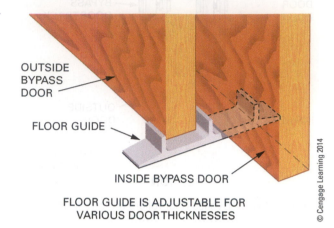

FLOOR GUIDE IS ADJUSTABLE FOR VARIOUS DOOR THICKNESSES

© Cengage Learning 2014

Figure 78–11 A floor guide is installed to keep bypass doors aligned.

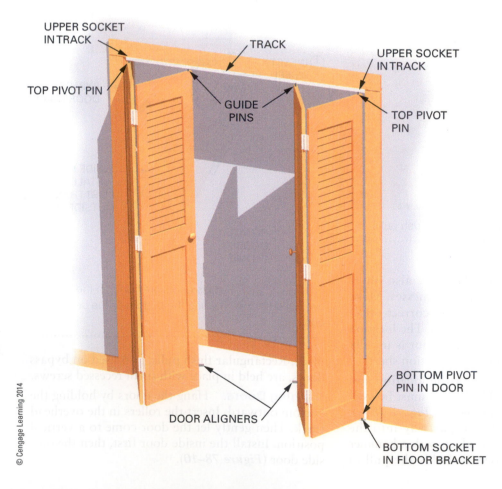

© Cengage Learning 2014

Figure 78–12
Installation of the bifold door requires several kinds of special hardware.

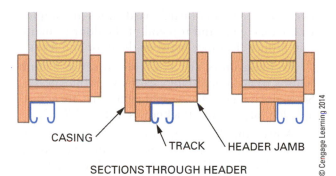

CASING   TRACK   HEADER JAMB

SECTIONS THROUGH HEADER

Figure 78–13 The bifold door track may be located in any position on the header jamb in several ways. Trim conceals the track from view.

Installing Bottom Pivot Sockets. Locate the bottom pivot sockets. Fasten one on each side, at the bottom of the opening. The pivot socket bracket is L-shaped. It rests on the floor against the side jamb. It is centered on a plumb line from the center of the pivot sockets in the track on the header jamb above.

Installing Pivot and Guide Pins. In most cases, bifold doors come with prebored holes for *pivot* and *guide pins*. If not, it is necessary to bore them. Follow the manufacturer's directions as to size and location. Install pivot pins at the top and bottom ends of the door in the prebored holes closest to the jamb. Sometimes the top pivot pin is spring loaded. It can then be depressed for easier installation of the door. The bottom pivot pin is threaded and can be adjusted for height. The guide pin rides in the track. It is installed in the hole provided at the top end of the door farthest away from the jamb.

Hanging the Doors. After all the necessary hardware has been applied, the doors are ready for installation. Loosen the set screw in the top pivot socket. Slide it along the track toward the center of the opening about 1 foot away from the side jamb. Place the door in position by inserting the bottom pivot pin in the bottom pivot socket. Tilt the doors to an upright position. At the same time insert the top pivot pin in the top socket, and the guide pin in the track, while sliding the socket toward the jamb.

Adjusting the Doors. Adjust top and bottom pivot sockets in or out so the desired joint is obtained between the door and the jamb. Lock top and bottom pivot sockets in position. Adjust the bottom pivot pin to raise or lower the doors, if necessary.

If more than one set of bifold doors is to be installed in an opening, install the other set on the opposite side in the same manner. Install knobs in the manner and location recommended by the manufacturer.

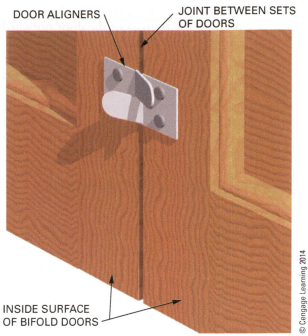

DOOR ALIGNERS   JOINT BETWEEN SETS OF DOORS

INSIDE SURFACE OF BIFOLD DOORS

Figure 78–14 Door aligners are used near the bottom where sets of bifold doors meet.

Where sets of bifold doors meet at the middle of an opening, door aligners are installed, near the bottom, on the inside of each of the meeting doors. The door aligners keep the faces of the center doors lined up when closed (*Figure 78–14*).

Installing Pocket Doors

The pocket door frame, complete with track, is installed when interior partitions are framed. The pocket consists of two ladderlike frames between which the door slides. A steel channel is fastened to the floor. The channel keeps the pocket opening spread the proper distance apart. The frame, which is usually preassembled at the factory, is made of nominal 1-inch stock. The pocket is covered by the interior wall finish. Care must be taken when covering the pocket frame not to use fasteners that are so long that they penetrate the frame. If fasteners penetrate through the pocket door frame, they will probably scratch the side of the door as it is operated or stop its complete entrance into the pocket.

Installing Door Hardware. Attach rollers to the top of the door in the location specified by the manufacturer. Install pulls on the door. On pocket doors an edge pull is necessary, in addition to recessed pulls on the sides of the door. A special pocket door pull contains edge and side pulls. It is mortised in the edge of the door. In most cases, all the necessary hardware is supplied when the pocket door frame is purchased.

Hanging the Door. Engage the rollers in the track by holding the bottom of the door outward in a way similar to that used with bypass doors. Test the operation of the door to make sure it slides easily and butts against the side jamb evenly. Make adjustments to the rollers, if necessary. Stops are later applied to the jambs on both sides of the door. The stops serve as guides for the door. When the door is closed, the stops prevent it from being pushed out of the opening (*Figure 78–15*).

INSTALLING A PREHUNG DOOR

A prehung single-acting, hinged door unit consists of a door frame with the door hinged and casings installed. Some have the casings already applied to one side and another set cut and ready to install. Holes are provided, if locksets have not already been installed. Small cardboard shims are stapled to the lock edge and top end of the door to maintain proper clearance between the door and frame.

Prehung units are available in several jamb widths to accommodate various wall thicknesses. Some prehung units have split jambs that are adjustable for varying wall thicknesses (*Figure 78–16*).

A prehung door unit can be set fairly quickly. If the prehung unit comes without the casing attached, cut and fit casing to one side of the jamb. Set the unit in the opening. Center the unit in the opening, so the door will swing in the desired direction. Be sure the door is closed and spacer shims are in place between the jamb and door. Plumb the door unit. Tack it to the wall through the casing.

Open the door and move to the other side. Install shims between the side jambs and the rough opening at intermediate points, keeping the side

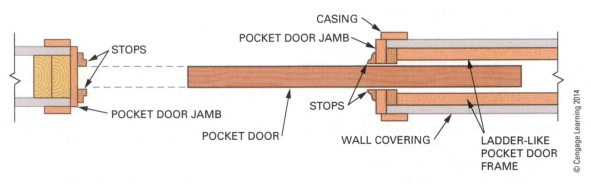

Figure 78–15 Plan view of a pocket door.

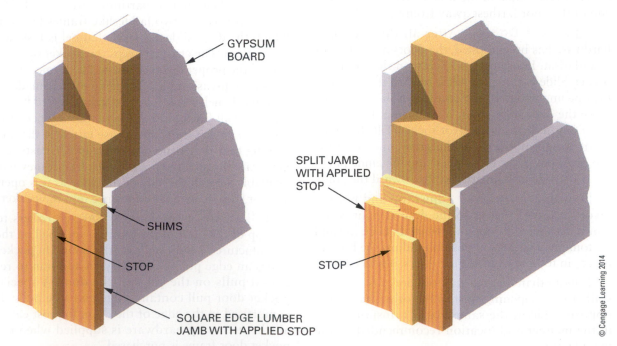

Figure 78–16 Prehung door units come with solid or split jambs.

StepbyStep Procedures

Procedure 78–B Installing a Prehung Door

STEP 1 REMOVE CASINGS FROM ONE SIDE OF UNIT.

STEP 2 PLACE UNIT IN OPENING AND PLUMB SIDE CASING.

STEP 3 FASTEN THROUGH CASINGS INTO WALL.

STEP 4 MOVE TO OTHER SIDE OF DOOR.

STEP 5 INSTALL SHIMS BETWEEN JAMB AND WALL.

STEP 6 FASTEN THROUGH JAMB AND SHIMS. CHECK THAT JAMBS ARE PLUMB AND STRAIGHT.

STEP 7 REPLACE CASINGS THAT WERE PREVIOUSLY REMOVED.

STEP 8 FASTEN THROUGH CASINGS INTO WALL.

DOOR

CASING

HINGE

LEVEL

A A

CASINGS REMOVED FROM THIS SIDE

JAMB

DOOR

CASING

SECTION A-A

DOOR

SHIM

SHIM

JAMB

B B

CASING

DOOR

FASTEN THROUGH JAMB AND SHIMS

SHIMS

SECTION B-B

DOOR

CASING

STOP

JAMB

FASTEN CASING

SECTION C-C

C C

© Cengage Learning 2014

jambs straight. Nail through the side jambs and shims. Remove spacers. Check the operation of the door. Make any necessary adjustments. Replace the previously removed casings. Drive and set all nails (*Procedure 78–B*).

Prehung door units with split jambs are set in a similar manner. However, there is no need to remove the casings. One section is installed as described earlier. The remaining section is inserted into the one already in place.

INSTALLING LOCKSETS

Locksets are installed on interior doors in the same manner as for exterior doors and as described in Chapter 58, Installing Exterior Door Locksets. Although their installation is basically the same, some locks are used exclusively on interior doors.

The privacy lock is often used on bathroom and bedroom doors. It is locked by pushing or turning a button on the room side. On most privacy locks, a turn of the knob on the room side unlocks the door. On the opposite side, the door can be unlocked by a pin or key inserted into a hole in the knob. The unlocking device should be kept close by, in a prominent location, in case the door needs to be opened quickly in an emergency.

The passage lockset has knobs on both sides that are turned to unlatch the door. This lockset is used when it is not desirable to lock the door.

DECONSTRUCT THIS

Carefully study *Figure 78–17* and think about what is wrong and/or what is right. Consider all possibilities. What construction practice or method is different in your area of the country?

© Cengage Learning 2014

Figure 78-17 This photo shows the top edge of an interior door.

KEY TERMS

back bevel	double-acting doors	louver doors	swinging doors
back miter	floor guide	passage lockset	
café doors	french doors	privacy lock	

REVIEW QUESTIONS

Select the most appropriate answer.

1. Most interior doors are manufactured in a thickness of

   a. 1 inch.
   b. $1\frac{3}{8}$ inches.
   c. $1\frac{1}{2}$ inches.
   d. $1\frac{3}{4}$ inches.

2. The height of most interior doors is

   a. 6'-0".
   b. 6'-6".
   c. 6'-8".
   d. 7'-0".

3. Interior door widths usually range from

   a. 1'-6" to 2'-6".
   b. 2'-2 to 2'-8".
   c. 2'-6" to 2'-8".
   d. 1'-0" to 3'-0".

4. Used extensively for flush door skins is

   a. fir plywood.
   b. lauan plywood.
   c. metal.
   d. plastic laminate.

5. The usual distance between the finish floor and the bottom of swinging doors for clearance is

   a. $\frac{1}{4}$ to $\frac{1}{2}$ inch.
   b. $\frac{1}{2}$ to $\frac{3}{4}$ inch.
   c. $\frac{3}{4}$ to 1 inch.
   d. $\frac{1}{2}$ to 1 inch.

6. A disadvantage of bypass doors is that they

   a. project out into the room.
   b. cost more and require more time to install.
   c. are difficult to operate.
   d. do not provide total access to the opening.

7. If the plain jamb stock is $\frac{3}{4}$-inch thick, the rough opening width for a swinging door should be the door width plus

   a. $\frac{3}{4}$ inch.
   b. $1\frac{1}{2}$ inches.
   c. 2 inches.
   d. $2\frac{1}{2}$ inches.

8. If the plain jamb stock and the finished floor are both $\frac{3}{4}$-inch thick and the space under the door is $\frac{1}{2}$ inch, the rough opening height for a 6'-8" swinging door should be

   a. 7'-0".
   b. 6'-11$\frac{1}{2}$".
   c. 6'-10$\frac{1}{2}$".
   d. 6'-9$\frac{1}{2}$".

9. When a plain door frame is made for a swinging door, the header jamb length is the door width plus

   a. twice the dado depth.
   b. twice the dado depth and twice $\frac{3}{32}$" space.
   c. twice the dado depth, twice $\frac{3}{32}$" space, and twice $\frac{1}{2}$" shim space.
   d. none of the above.

10. An accurate and fast method of plumbing side jambs of a door frame is by the use of a

   a. builder's level.
   b. carpenter's 26-inch hand level.
   c. 6-foot level.
   d. straightedge.

UNIT 28
Interior Trim

CHAPTER 79 Description and Application of Molding
CHAPTER 80 Application of Door Casings, Base,
and Window Trim

©Shutterstock, Inc./Eimantas Buzas

*I*nterior trim, also called *interior finish,* involves the application of molding around windows and doors; at the intersection of walls, floor, and ceilings; and to other inside surfaces. Moldings are strips of material, shaped in numerous patterns, for use in a specific location. Wood is used to make most moldings, but some are made of plastic or metal.

Interior trim is among the final materials installed and requires the installer to take care not to mar the finish. Blemishes and dings in the finish may be visible for the life of the building.

OBJECTIVES

After completing this unit, the student should be able to:

- identify standard moldings and describe their use.
- apply ceiling and wall molding.
- apply interior door casings, baseboard, base cap, and base shoe.
- install window trim, including stools, aprons, jamb extensions, casings, and stop beads.
- install closet shelves and closet pole.
- install mantels.

CHAPTER 79

Description and Application of Molding

Moldings are available in many *standard* types. Each type is manufactured in several sizes and patterns. Standard patterns are usually made only from softwood. When other kinds of wood, or special patterns, are desired, mills make *custom* moldings to order. All moldings must be applied with tight-fitting joints to present a suitable appearance.

Molding usually comes in lengths of 8, 10, 12, 14, and 16 feet. Some moldings are available in odd lengths. Door casings, in particular, are available in lengths of 7 feet to reduce waste.

Finger-jointed lengths are made of short pieces joined together. These are used only when a paint finish is to be applied. The joints show through a stained or natural finish.

STANDARD MOLDING PATTERNS

Standard moldings are designated as bed, crown, cove, full round, half round, quarter round, base, base shoe, base cap, casing, chair rail, back band, apron, stool, stop, and others (*Figure 79–1*).

MOLDING SHAPE AND USE

Some moldings are classified by the way they are shaped. Others are designated by location. For example, beds, crowns, and coves are terms related to shape. Although they may be placed in other locations, they are usually used at the intersections

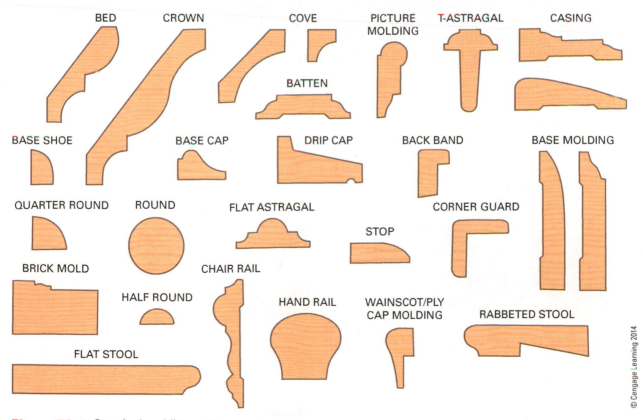

Figure 79–1 Standard molding patterns.

BED   CROWN   COVE   PICTURE MOLDING   T-ASTRAGAL   CASING
BATTEN
BASE SHOE   BASE CAP   DRIP CAP   BACK BAND   BASE MOLDING
QUARTER ROUND   ROUND   FLAT ASTRAGAL   CORNER GUARD
STOP
BRICK MOLD   CHAIR RAIL
HALF ROUND   HAND RAIL   WAINSCOT/PLY CAP MOLDING   RABBETED STOOL
FLAT STOOL

© Cengage Learning 2014

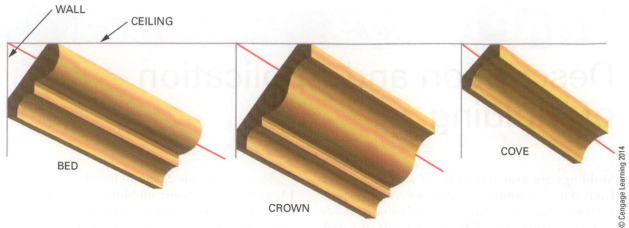

Figure 79-2 Bed, crown, and cove moldings are often used at the intersections of walls and ceilings.

of walls and ceilings (*Figure 79-2*). Also classified by their shape are full rounds, half rounds, and quarter rounds. They are used in many locations. Full rounds are used for such things as closet poles. Half rounds may be used to conceal joints between panels or to trim shelf edges. Quarter rounds may be used to trim outside corners of wall paneling and for many other purposes (*Figure 79-3*).

Designated by location, base, base shoe, and base cap are moldings applied at the bottom of walls where they meet the floor. When square-edge base is used, a base cap is usually applied to its top edge. Base shoe is normally used to conceal the joint between the bottom of the base and the finish floor (*Figure 79-4*).

Casings are used to trim around windows, doors, and other openings. They cover the space between the frame and the wall. Back bands are applied to the outside edges of casings for a more decorative appearance.

Aprons, stools, and stops are parts of window trim. Stops are also applied to door frames. On the same window, aprons should have the same molded shape as casings. Aprons, however, are not *backed out*. They have straight, smooth backs and sharp, square, top edges that butt against the bottom of the stool (*Figure 79-5*).

Corner guards are also called *outside corners*. They are used to finish exterior corners of interior wall finish. Caps and chair rail trim the top edge of wainscoting. (These moldings have been previously

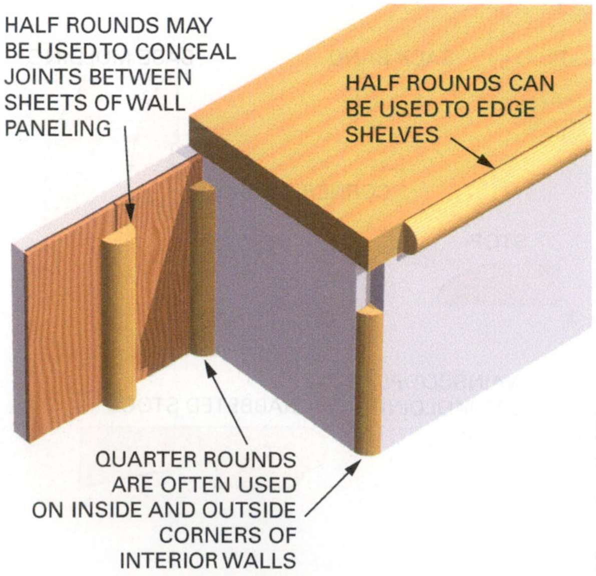

Figure 79-3 Half round and quarter round moldings are used for many purposes.

HALF ROUNDS MAY BE USED TO CONCEAL JOINTS BETWEEN SHEETS OF WALL PANELING

HALF ROUNDS CAN BE USED TO EDGE SHELVES

QUARTER ROUNDS ARE OFTEN USED ON INSIDE AND OUTSIDE CORNERS OF INTERIOR WALLS

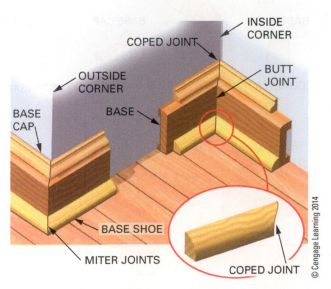

Figure 79-4 Base, base shoe, and base cap are used to trim the bottom of the wall.

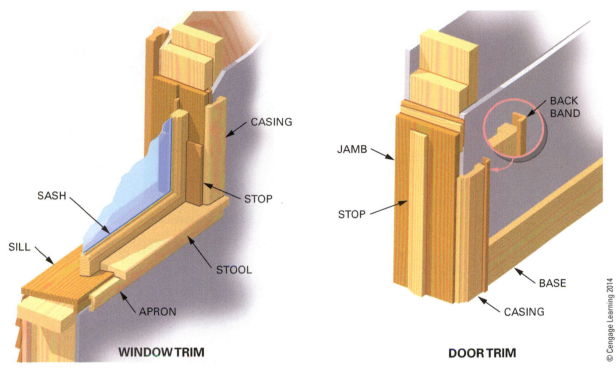

WINDOW TRIM

DOOR TRIM

© Cengage Learning 2014

Figure 79–5 Casing, back bands, and stops are used for window and door trim. Stools and aprons are part of window trim.

described in Chapter 70, Application of Wall Paneling.) Others, such as *astragals, battens, panel,* and *picture* moldings, are used for various purposes.

MAKING JOINTS ON MOLDING

End joints between lengths of ceiling molding may be made square or at a 45-degree angle. Many carpenters prefer to make square joints between moldings because less joint line is shown. Also, the square end acts as a stop when bowing and snapping the last length of molding into place at a corner.

Usually, the last piece of molding along a wall is cut slightly longer. It is bowed outward in the center, then pressed into place when the ends are in position. This makes the joints come up tight. After the molding has been fastened, joints between lengths should be sanded flush, except on prefinished moldings. Failure to sand butted ends flush with each other results in a shadow being cast at the joint line. This gives the appearance of an open joint.

Joints on exterior corners are mitered. Joints on interior corners are usually coped, especially on large moldings. A coped joint is made by fitting the piece on one wall with a square end into the corner. The end of the molding on the other wall is cut to

© Cengage Learning 2014

Figure 79–6 A coped joint is made by fitting the end of one piece of molding against the shaped face of the other piece.

fit against the shaped face of the molding on the first wall (*Figure 79–6*).

Methods of Mitering Using Miter Boxes

Moldings of all types may be mitered by using either hand or power miter boxes. A miter box is a tool that cuts a piece of material at an angle, most commonly, 45 degrees. The hand miter box is metal with a backsaw attached, and is easily adjustable to cut different angles (*Figure 79–7*).

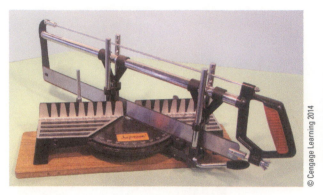

Figure 79-7 Handsaw miter box.

Figure 79-8 A power miter box makes easy work of cutting molding.

SAW HEAD BEVEL ANGLE SET TO ZERO.

THICK EDGE AGAINST FENCE.

BASEBOARD WITH FACE SIDE OUT.

MITER ANGLE SET TO 45°.

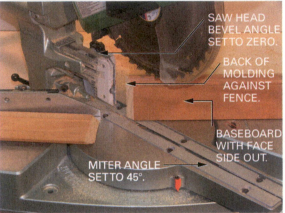

SAW HEAD BEVEL ANGLE SET TO ZERO.

BACK OF MOLDING AGAINST FENCE.

BASEBOARD WITH FACE SIDE OUT.

MITER ANGLE SET TO 45°.

Figure 79-9 The position of molding in a miter box depends on how the saw is set up and where the molding is to be installed.

The most popular way to cut miters and other end cuts on trim is with a power miter box (*Figure 79–8*). With this tool, a carpenter is able to cut virtually any angle with ease, whether it is a simple or a compound miter (one with two angles). Fine adjustments to a piece of trim, $\frac{1}{64}$ inch, can be made with great speed and accuracy. The power miter box is discussed in more depth in Chapter 18.

Positioning Molding in the Miter Box

It is important to visualize the cut ahead of time. Sometimes it is best to hold the back of molding against the fence, sometimes against the base. Placing molding in the correct position in the miter box is essential for accurate mitering. Cut all moldings with their best face or sides up toward the operator so as the saw cuts, it splinters only out the back side, not the face side. Flat miters are cut by holding the molding with its back against the base and its thicker edge against the fence of the miter box. Some moldings, such as baseboard, base cap and base shoe, are

held right side up with their back sides against the fence of the miter box (*Figure 79–9*).

Cutting Crown and Bed Moldings

Cutting bed, crown, and cove molding may be done using two methods. Either method will produce accurate clean fitted moldings.

Upside-Down Method. The first method positions the molding in the meter saw upside down from how it will be installed. The portion of the molding's back side that touches the wall (W) will rest against the fence. The portion of the back side that touches the ceiling (C) will rest on the base of the saw (*Figure 79–10*). Care should be taken to ensure that the molding back sides are resting flat and square against the surfaces of the miter saw. This position is convenient because the bottom edge of this type of molding is the edge that is usually marked for cutting to length. With the bottom edge up, the mark on which to start the cut can be easily seen. This method is inconvenient because

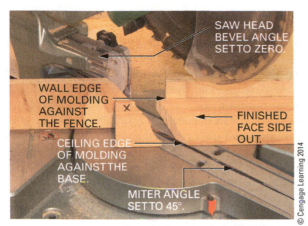

Figure 79–10 Mitering ceiling moldings may be cut with moldings positioned upside. In this method the fence serves as the wall and the base as the ceiling.

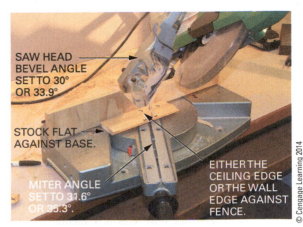

Figure 79–11 Ceiling moldings may be mitered while held flat to miter saw base and cutting at a compound angle.

the piece must be held at the precise angle before cutting.

Flat-Cut Method. The second method of cutting large crown moldings places the broad back side if the molding against the base. The miter saw is set to a compound angle while cutting the molding flat (*Figure 79–11*). Setting the saw to the required angles is easy to do but a little more difficult to understand. High-quality miter saws have detents in the scales to make adjustments easy to find, like those typically at 0 degrees and 45 degrees.

There are three main concepts to cutting molding on the flat in a compound miter saw. First,

determine the angle at which the molding was designed to lay against the wall. Second, determine which side you want the miter saw bevel angle to be tilted, left or right; and third, all cuts are made with the saw at the same bevel angle while changing only the base miter angles.

Crown moldings are typically manufactured to fasten to the wall at 38 degrees, bed moldings at 45 degrees. You can determine this by eye since either it has a symmetrical 45-degree type of look, or it looks like (*Figure 79–12*). Angle settings for the compound miter depend on which configuration the molding was manufactured.

Tilting the saw head left or right to make a bevel cut is entirely user preference. Right-handed

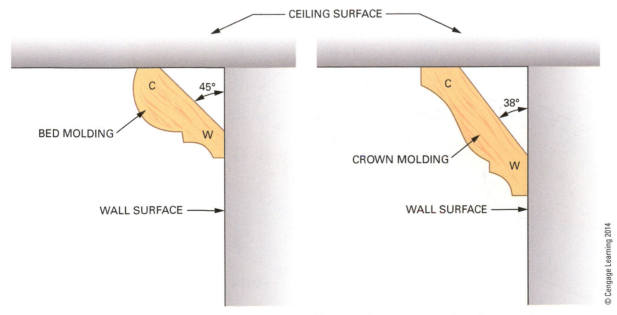

Figure 79–12 The profile of bed moldings and crown molding may be either 38 and 45 degrees.

Molding Design Angle	Base Miter Angle (left or right)	Saw Head Bevel Angle (left or right)
38°	31.6°	33.9°
45°	35.3°	30°

Figure 79–13 The angles to set a compound miter saw to cut crown molding on the flat.

users may want the saw tilted to the right so the blade is fully visible, whereas left-handed users may want the saw tilted to the left. Either setup will produce accurate and clean cuts. The angles for setting the compound angle are shown in (*Figure 79–13*). Begin set up by tilting the saw head in the desired direction and at the correct bevel angle. Then choose the miter angle direction.

Setting the base miter angle direction determines which side of the miter is being cut. There are essentially four types of cuts. Two cuts come from either side of the blade of the left-swing miter, two more from the right-swing miter. Labeling each cut #1, #2, #3, or #4 makes referring to them easier (*Figure 79–14*).

The last step in understanding the compound miter flat-cut method is to note which edge of the molding touches the fence. Either the wall (W) or ceiling (C) side of the molding's cross section touches the fence in each setup (*Figure 79–15*).

Procedure 79–A shows the steps to cut the molding shown in Figure 79–14.

Mitering with a Table Saw

Miters may also be made by using the table saw or the radial arm saw. The use of mitering jigs is helpful when making flat miters on window and door casings. The jigs allow both right- and left-hand miters to be cut quickly and easily without any changes in the setup (*Figure 79–16*). (The construction and use of mitering jigs for saws are more fully described in Unit 7.)

Coped Joints

Coped joints are often done to improve the appearance of a molding corner. Inside miter joints may shrink or open up slightly. They may sometimes not fit perfectly due to the wall corner not being a perfect 90-degree corner. The joint becomes clearly visible and unsightly. Coped joints are able to better hide slight variations because to see into the joint, one's eye must be near the wall. Either piece of an inside miter may be coped, while the other piece is cut square to the corner (*Figure 79–17*).

To cope the end of molding, first make an inside miter cut as shown in Procedure 79–A. Cuts #1 and #2 will produce a molding piece that is ready to be coped. The edge of the cut along the face side forms the profile of the coped cut. You may rub the side of a pencil point lightly along the profile to make the outline more clearly visible.

Hold the molding so it is over the end of a sawhorse. Hold the side of the molding that will touch the wall flat on the top of the sawhorse. This way the cut is done with the coping saw blade plumb.

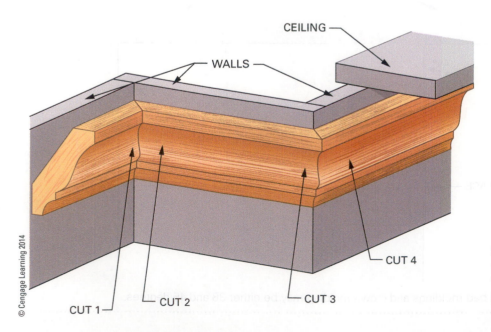

Figure 79–14 There are four types of cuts for miters. Two for inside and two for outside miters.

CEILING

WALLS

CUT 1   CUT 2   CUT 3   CUT 4

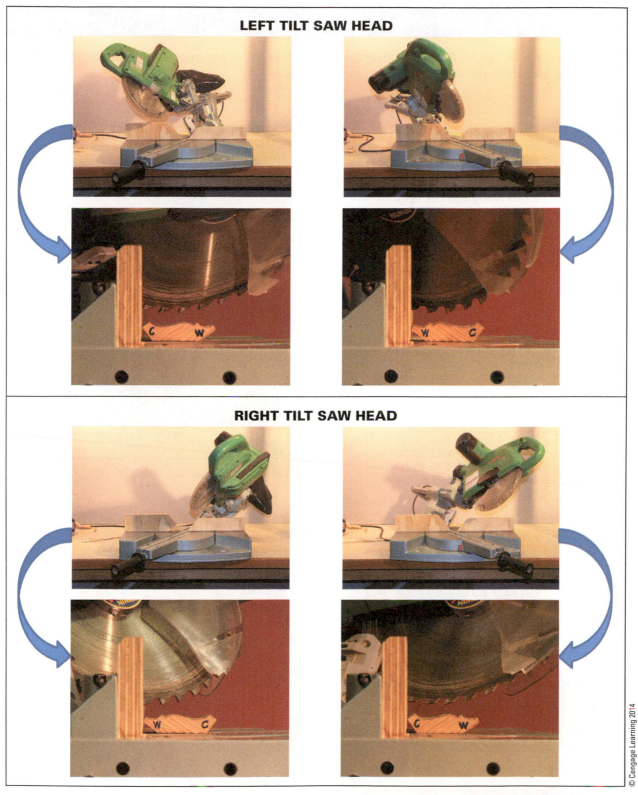

Figure 79–15 To cut large ceiling moldings on the flat, first choose a saw head tilt direction, then make all cuts by changing miter angle direction and which side of molding touches saw fence, ceiling (C) or wall (W).

StepbyStep Procedures

Procedure 79–A Cutting Crown Molding on the Flat. Procedure assumes a right saw head bevel angle and a 38 degree molding profile.

STEP 1 LIGHTLY MARK MOLDING WITH A W AND C. PLACE MOLDING ON SAW BASE WITH W AGAINST THE FENCE. FIRST CUT WILL HAVE #1 ON THE RIGHT AND #4 ON THE LEFT.

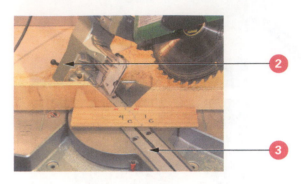

STEP 2 FROM FIGURE 79–13 SET SAW HEAD BEVEL ANGLE TO 33.9°.

STEP 3 SET BASE MITER ANGLE TO 31.6° TO THE RIGHT. CUT PIECE AS NEEDED.

STEP 4 ROTATE MITER ANGLE TO OPPOSITE SIDE AND SET TO THE SAME ANGLE AS BEFORE. DO NOT READJUST THE SAW HEAD BEVEL ANGLE.

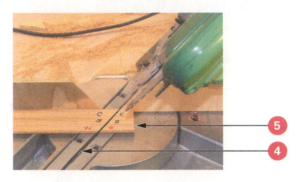

STEP 5 MEASURE AND LIGHTLY MARK NEW PIECE TO CUT #2 AND #3. PLACE NEW MOLDING PIECE ON BASE SO CEILING EDGE OF MOLDING TOUCHES THE FENCE.

STEP 6 MEASURE THE LENGTH AND CUT AT THE SAME ANGLE SETTINGS. THIS SIMPLY MEANS SLIDE THE STOCK TO THE RIGHT FOR A NEW CUT.

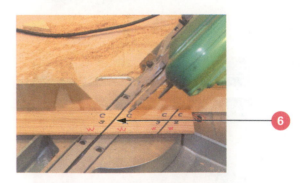

STEP 7 ASSEMBLE THE PIECES.

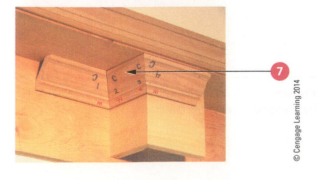

Figure 79–16 With a mitering jig, left- and right-hand miters can be made quickly and easily without changing the setup (*guard has been removed for clarity*).

Figure 79–17 Coped joints form inside angles that look like miters.

Use a fine-tooth blade coping saw, carefully cutting along the outlined profile keeping the saw plumb. A slight undercut of the blade helps ensure that the cut will fit nicely (*Figure 79–18*). It may be

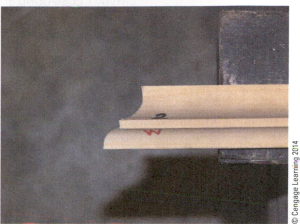

Figure 79–18 To cut a coped joint, first cut an inside miter, place the wall side of molding down on horse and cut along the miter profile.

necessary to touch up the cut with a sharp utility knife or sandpaper.

APPLYING MOLDING

To apply chair rail, caps, or some other type of molding located on the wall, chalk lines should be snapped. This ensures that molding is applied in a straight line. No lines need to be snapped for base moldings or for small-size moldings applied at the intersection of walls and ceiling.

For large-size ceiling moldings, such as beds, crowns, and coves, a chalk line may be snapped. This ensures straight application of the molding and easier joining of the pieces. Without a straight line to guide application, the molding may be forced at different angles along its length. This results in a noticeably crooked bottom edge and difficulty making tight-fitting miters and copes.

Hold a short scrap piece of the molding at the proper angle at the wall and ceiling intersection. Lightly mark the wall along the bottom edge of

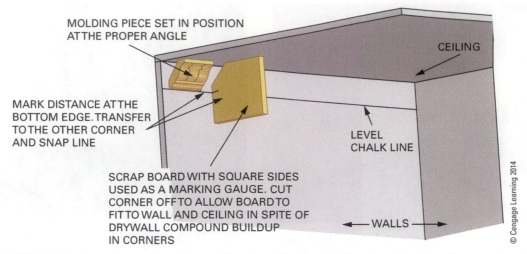

MOLDING PIECE SET IN POSITION AT THE PROPER ANGLE

CEILING

MARK DISTANCE AT THE BOTTOM EDGE. TRANSFER TO THE OTHER CORNER AND SNAP LINE

LEVEL CHALK LINE

SCRAP BOARD WITH SQUARE SIDES USED AS A MARKING GAUGE. CUT CORNER OFF TO ALLOW BOARD TO FIT TO WALL AND CEILING IN SPITE OF DRYWALL COMPOUND BUILDUP IN CORNERS

WALLS

© Cengage Learning 2014

Figure 79–19 Hold a scrap piece of molding against the ceiling and wall to determine the distance from the ceiling to its bottom edge.

the molding. Measure down from the ceiling to the mark. Measure and mark this same distance at the end of each wall to which the molding is to be applied. If the drywall is already finished with joint compound, place a square block with the corner cut off against the wall and ceiling to transfer mark (*Figure 79–19*). This will ensure level molding, Snap lines between the marks. Apply the molding so its bottom edge is to the chalk line.

Apply the molding to the first wall with square ends in both corners. If more than one piece is required to go from corner to corner, install the first piece with both ends square. If mitered joints between lengths are desired, cut a square end into the corner and an inside miter on the other end. On subsequent lengths, make matching end joints until the other corner is reached.

On some moldings, such as quarter rounds and small cove moldings, the straight, back surfaces are not always of equal width. One of the back surfaces of these moldings should be marked with a pencil to ensure positioning them in the miter box the same way each time. Mitering the molding with the same side down each time helps make fitting more accurate (*Figure 79–20*).

If a small-size molding is used, fasten it with finish nails in the center. Use nails of sufficient length to penetrate into solid wood at least 1 inch. If large-size molding is used, fastening is required along both edges (*Figure 79–21*).

Press the molding in against the wall or intersection with one hand while driving the nail almost home. Then set the nail below the surface. Nail at about 16-inch intervals and in other locations as

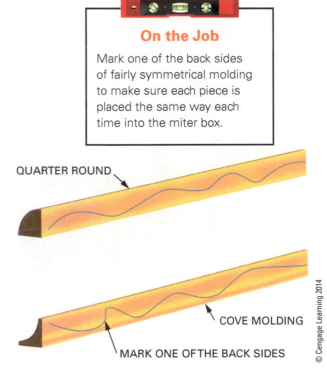

On the Job

Mark one of the back sides of fairly symmetrical molding to make sure each piece is placed the same way each time into the miter box.

QUARTER ROUND

COVE MOLDING

MARK ONE OF THE BACK SIDES

© Cengage Learning 2014

Figure 79–20 Technique for reducing the confusion of working with molding that has a fairly symmetrical cross section.

necessary to bring the molding tight against the surface. End nails should be placed 2 to 3 inches from the end to keep from splitting the molding.

The last piece is easiest to install if only one end fits into a corner. Cut and prepare the end that fits the corner with the miter or coped joint. Place this end into the corner and let the other end overlap

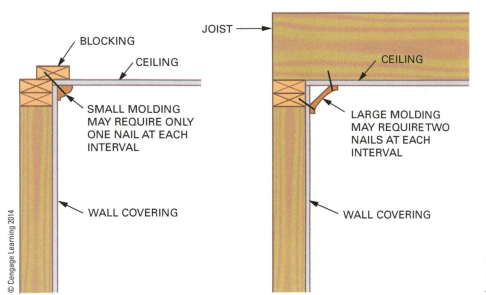

JOIST

BLOCKING

CEILING

CEILING

SMALL MOLDING
MAY REQUIRE ONLY
ONE NAIL AT EACH
INTERVAL

LARGE MOLDING
MAY REQUIRE TWO
NAILS AT EACH
INTERVAL

WALL COVERING

WALL COVERING

© Cengage Learning 2014

Figure 79–21 Fastening
methods for molding.

the first piece installed. Mark the cut at the overlap. This method is more accurate than measuring and then transferring the measurement to the piece. Mark all pieces of interior trim for length in this manner whenever possible. Cope the starting end on each succeeding piece against the face of the last piece installed. Work around the room in one direction, either clockwise or counterclockwise. The end of the last piece installed must be coped to fit against the face of the first piece. It is best to plan where to begin and end the molding. It should be in the least visible place in the room. Good choices include corners behind doors that are usually open, and walls or corners farthest from view.

CHAPTER 80

Application of Door Casings, Base, and Window Trim

In addition to wall and ceiling molding, the application of door casings, base, base cap, base shoe, and window trim is a major part of interior finish work. Care must be taken to avoid marring the work and to make neat, tight-fitting joints.

DOOR CASINGS

Door casings are moldings applied around the door opening. They trim and cover the space between the door frame and the wall. Casings must be applied before any base moldings because the base butts against the edge of door casing (*Figure 80–1*). Door casings extend to the floor,

Design of Door Casings

Moldings or *S4S* stock may be used for door casings. S4S is the abbreviation for surfaced four sides. It is used to describe smooth, square-edge lumber. When molded casings are used, the joint at the head must be mitered unless butted against plinth blocks.

Plinth blocks are small decorative blocks. They are thicker and wider than the door casing. They are used as part of the door trim at the base and at the head (*Figure 80–2*).

When using S4S lumber, the joint may be mitered or butted. If a butt joint is used, the head casing overlaps the side casing. The appearance of S4S

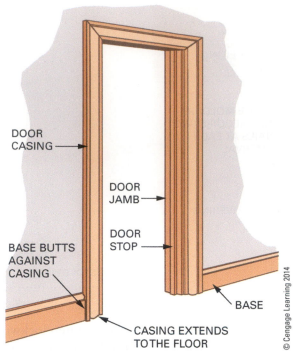

Figure 80–1 Door casings are applied before the base is installed.

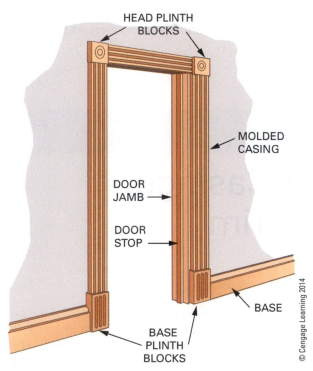

Figure 80–2 Molded casings are mitered at the head unless plinth blocks are used.

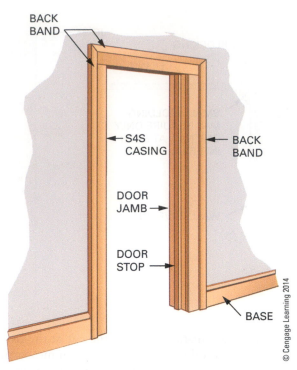

Figure 80–3 Back bands may be applied to improve the appearance of door casings.

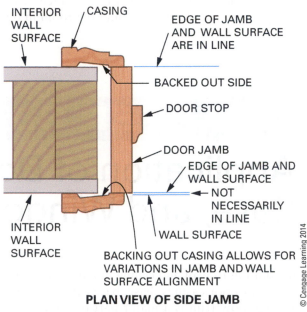

PLAN VIEW OF SIDE JAMB

Figure 80–4 Backing out door casings allows for a tight fit on wall and jamb.

casings and some molded casings may be enhanced with the application of back bands (*Figure 80–3*).

Molded casings usually have their back sides backed out. In cases where the jamb edges and the wall surfaces may not be exactly flush with each other, the backed-out surfaces allow the casing to come up tight on both wall and jamb (*Figure 80–4*). If S4S casings are used, they must be backed out on the job.

Applying Door Casings

Door casings are set back from the inside face of the door frame a distance of about $\frac{5}{16}$ inch. This allows room for the door hinges and the striker

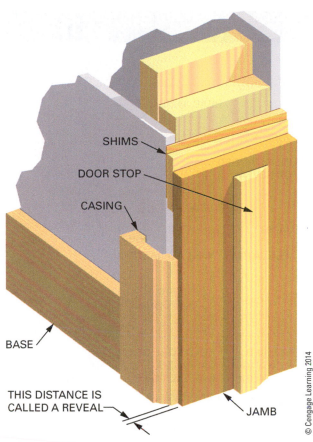

SHIMS

DOOR STOP

CASING

BASE

THIS DISTANCE IS
CALLED A REVEAL

JAMB

© Cengage Learning 2014

Figure 80–5 The setback of the door casing on the jamb is called a reveal.

plate of the door lock. This setback is called a reveal (*Figure 80–5*). The reveal also improves the appearance of the door trim.

Set the blade of the combination square so that it extends $5/16$ inch beyond the body of the square. Gauge lines at intervals along the side and head jamb edges by riding the square against the inside face of the jamb. Let the lines intersect where side and head jambs meet. Mark lightly with a sharp pencil or mark with a utility knife. The knife leaves no pencil lines to erase later.

The following procedure applies to molded door casings mitered at the head. If several door openings are to be cased, cut the necessary number of casings to rough lengths with a miter cut on one end of each piece. Rough lengths are a few inches longer than actually needed. For each interior door opening, four side casings and two head casings are required. Cut side casings in pairs with right- and left-hand miters for use on both sides of the opening.

Applying the Head Casing

Miter one end of the head casing. Hold it against the head jamb of the door frame so that the miter is on the intersection of the gauged lines. Mark the length of the head casing at the intersection of the gauged lines on the opposite side of the door frame. Miter the casing to length at the mark.

Fasten the head casing in position. Its inside edge should be to the gauged lines on the head jamb. The mitered ends should be in line with the gauged lines on the side jambs. Use finish nails along the inside edge of the casing into the header jamb. If the casing edge is thin, use 3d or 4d finish nails spaced about 12 inches apart. Keep the edge of the casing to the gauged lines on the jamb. Straighten the casing as necessary as nailing progresses. Drive nails at the proper angle to keep them from coming through the face or back side of the jamb. Pneumatic finish nailers speed up the job of fastening interior trim.

Fasten the top edge of the casing into the framing. The outside edge is thicker, so longer nails are used, usually 6d or 8d finish nails. They are spaced farther apart, about 16 inches on center (OC) (*Procedure 80–A*). Do not set nails at this time. It may be necessary to move the ends slightly to fit the mitered joint between head and side casings.

Applying the Side Casings

Mark one of the previously mitered side casings by turning it upside down with the point of the miter touching the floor. If the finish floor has not been laid, hold the point of the miter on a scrap block of wood that is equal in thickness to the finish floor. Mark the side casing in line with the top edge of the head casing (*Procedure 80–B*). Make a square cut on the casing at the mark.

Place the side casing in position. Try the fit at the mitered joint. If the joint needs fitting, trim the mitered end of the side casing by planing thin shavings with a sharp block plane. The joint may also be fitted by shimming the casing away from the side of the chop saw and making a thin corrective cut. Shim either near or far from the saw blade as needed to hold the casing at the desired angle (*Figure 80–6*). When fitted, apply a little glue to the joint. Nail the side casing in the same manner as the head casing.

Avoid sanding the joint to bring the casing faces flush. It is difficult to keep from sanding across the grain on one or the other of the pieces. Cross-grain scratches will be very noticeable, especially if the trim is to have a stained finish. Bring the faces flush, if necessary, by shimming between the back of the casing and the wall. Usually, only very thin shims are needed. Any small space between the casing and the wall can be filled later with joint filling compound. Also, the back side of the thicker piece may be planed or chiseled thinner. Most carpenters prefer to do this rather than try to sand the joint.

StepbyStep Procedures

Procedure 80–A Cutting a Head Casing to Fit

STEP 1 MARK LIGHT GAUGE LINES ON THE EDGE OF JAMB TO INDICATE DESIRED REVEAL.

STEP 2 CUT A HEAD CASING PIECE SLIGHTLY LONG WITH ONLY ONE MITER. POSITION IT INTO PLACE.

STEP 3 ALIGN MITER TO THE REVEAL GAUGE LINES.

STEP 4 MARK THE SECOND MITER AT THE REVEAL LINE.

STEP 5 TACK HEAD CASING IN PLACE AND DO NOT SET THE NAILS YET.

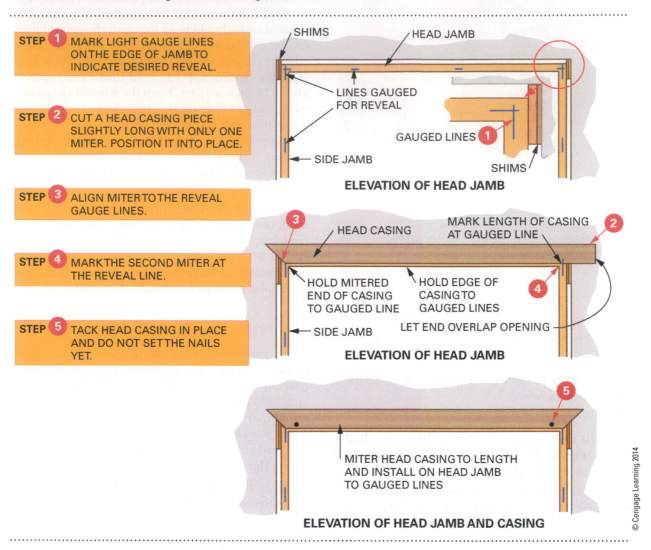

SHIMS   HEAD JAMB

LINES GAUGED FOR REVEAL

GAUGED LINES ①

SIDE JAMB

SHIMS

ELEVATION OF HEAD JAMB

③ HEAD CASING   MARK LENGTH OF CASING AT GAUGED LINE ②

HOLD MITERED END OF CASING TO GAUGED LINE   HOLD EDGE OF CASING TO GAUGED LINES

④

SIDE JAMB   LET END OVERLAP OPENING

ELEVATION OF HEAD JAMB

⑤

MITER HEAD CASING TO LENGTH AND INSTALL ON HEAD JAMB TO GAUGED LINES

ELEVATION OF HEAD JAMB AND CASING

© Cengage Learning 2014

On the Job

Slight adjustments may be made in miter box cuts by shimming molding against fence.

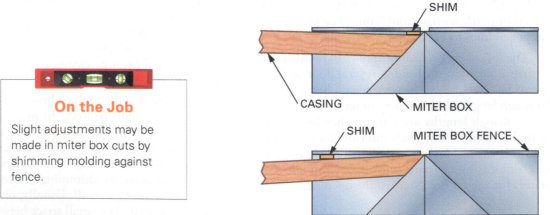

SHIM

CASING   MITER BOX

SHIM   MITER BOX FENCE

PLAN VIEW OF MITER BOX

© Cengage Learning 2014

Figure 80–6 Technique for making small adjustments to the angle of a miter.

StepbyStep Procedures

Procedure 80–B Cutting Side Casings to Fit

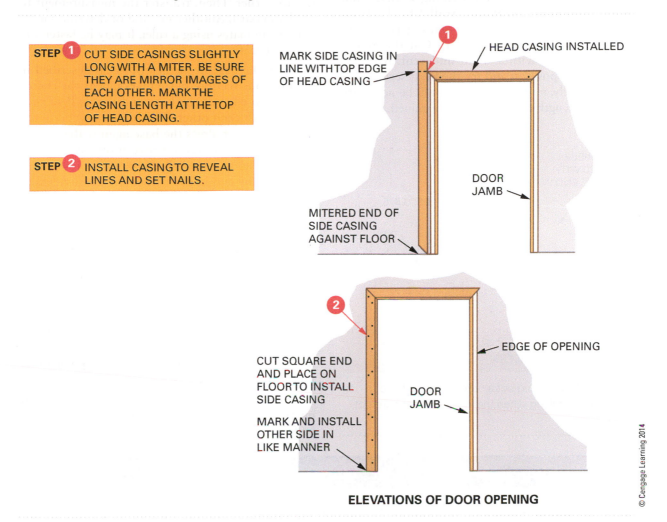

STEP **1** CUT SIDE CASINGS SLIGHTLY LONG WITH A MITER. BE SURE THEY ARE MIRROR IMAGES OF EACH OTHER. MARK THE CASING LENGTH AT THE TOP OF HEAD CASING.

STEP **2** INSTALL CASING TO REVEAL LINES AND SET NAILS.

MARK SIDE CASING IN LINE WITH TOP EDGE OF HEAD CASING

HEAD CASING INSTALLED

DOOR JAMB

MITERED END OF SIDE CASING AGAINST FLOOR

CUT SQUARE END AND PLACE ON FLOOR TO INSTALL SIDE CASING

MARK AND INSTALL OTHER SIDE IN LIKE MANNER

EDGE OF OPENING

DOOR JAMB

ELEVATIONS OF DOOR OPENING

Drive a 4d finish nail into the edge of the casing and through the mitered joint. Then set all fasteners. Keep nails 2 or 3 inches from the end to avoid splitting the casing. If there is danger of splitting, blunt the pointed end or drill a hole slightly smaller in diameter than the nail.

APPLYING BASE MOLDINGS

Molded or S4S stock may be used for base. If S4S base is used, it should be backed out. A base cap should be applied to its top edge. The base cap conforms easily to the wall surface, resulting in a tight fit against the wall. The base trim should be thinner than the door casings against which it butts. This makes a more attractive appearance.

The base is applied in a manner similar to wall and ceiling molding. However, copes are laid out for joints in interior corners. Instead of back-mitering to outline the cope, it is sometimes desired to lay out the cope by scribing (*Figure 80–7*). When placed against the wall, the face of the base may not always be square with the floor. Therefore, if the base is tilted slightly, back-mitering to obtain the outline of the cope will result in a poor fit against it.

Apply the base to the first wall with square ends in each corner. Drive and set two finishing nails, of sufficient length, at each stud location. Nailing blocks previously installed during framing provide solid wood for fastening the ends of the base in interior corners.

Cut the base to go on the next wall about an inch longer than required. Lay the base against the wall by bending it so the end to be scribed lies flat against the wall and against the first base. Set the dividers to scribe about $\frac{1}{2}$ inch. Lay out the cope by riding the dividers along the face of the base on the first wall.

Hold the dividers while scribing so that a line between the two points is parallel to the floor. Twisting the dividers while making the scribe results in an inaccurate layout. Cut the end to the scribed line with a slight undercut. Bend the base back in position and try the fit. If scribed and cut accurately, no adjustments should be necessary. Its overall length must now be determined.

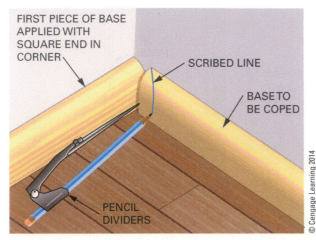

FIRST PIECE OF BASE APPLIED WITH SQUARE END IN CORNER

SCRIBED LINE

BASE TO BE COPED

PENCIL DIVIDERS

© Cengage Learning 2014

Figure 80–7 Laying out a coped joint on base molding by scribing when the walls and floor are not square.

Cutting the Base to Length

The length of a baseboard that fits between two walls may be determined by measuring from corner to corner. Then, transfer the measurement to the baseboard. Another method of determining its length eliminates using a ruler. It may be faster and more accurate.

With the base in the last position described in the previous section, place marks, near the center, on the top edge of the base and the wall so they line up with each other. Place the other end in the opposite corner. Press the base against the wall at the mark. The difference between the mark on the wall and the mark on the base is the amount to scribe off the end in the corner. Set the dividers to this distance. Scribe the end. Cut to the scribed line (*Procedure 80–C*). If a tighter fit is desired, set the dividers slightly less than the distance between the marks. This method of fitting long lengths between walls may be applied to other kinds of trim. However, this works especially well with the base.

Place one end in the corner, and bow out the center. Place the other end in the opposite corner, and press the center against the wall. Fasten in place. Continue in this manner around the room in a previously planned order. Make regular miter joints on outside corners.

If both ends of a single piece are to have regular miters for outside corners, it is imperative that it be fastened in the same position as it was marked.

StepbyStep Procedures

Procedure 80–C Cutting Molding to Fit between Walls

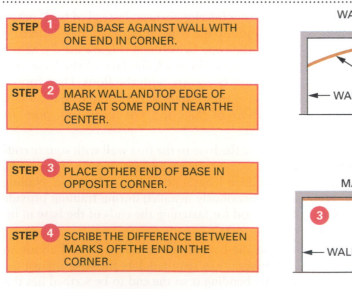

STEP 1 BEND BASE AGAINST WALL WITH ONE END IN CORNER.

STEP 2 MARK WALL AND TOP EDGE OF BASE AT SOME POINT NEAR THE CENTER.

STEP 3 PLACE OTHER END OF BASE IN OPPOSITE CORNER.

STEP 4 SCRIBE THE DIFFERENCE BETWEEN MARKS OFF THE END IN THE CORNER.

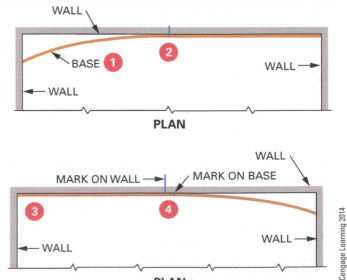

WALL

BASE ①

②

WALL →

← WALL

PLAN

WALL

MARK ON WALL → ← MARK ON BASE

③

④

← WALL

WALL →

PLAN

© Cengage Learning 2014

StepbyStep Procedures

Procedure 80–D Cutting Outside Miters

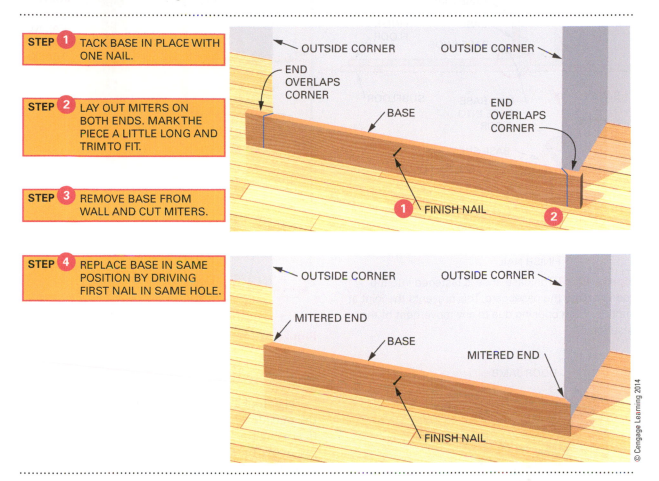

STEP 1 TACK BASE IN PLACE WITH ONE NAIL.

STEP 2 LAY OUT MITERS ON BOTH ENDS. MARK THE PIECE A LITTLE LONG AND TRIM TO FIT.

STEP 3 REMOVE BASE FROM WALL AND CUT MITERS.

STEP 4 REPLACE BASE IN SAME POSITION BY DRIVING FIRST NAIL IN SAME HOLE.

OUTSIDE CORNER OUTSIDE CORNER
END OVERLAPS CORNER
END OVERLAPS CORNER
BASE
FINISH NAIL
1 2

OUTSIDE CORNER OUTSIDE CORNER
MITERED END
BASE
MITERED END
FINISH NAIL

© Cengage Learning 2014

Tack the rough length in position with one finish nail in the center. Mark both ends. Remove, and cut the miters. Installing the piece by first fastening into the original nail hole ensures that the piece is fastened in the same position as marked (*Procedure 80–D*).

Applying the Base Cap and Base Shoe

The base cap is applied in the same manner as most wall or ceiling molding. Cope interior corners and miter exterior corners. The base shoe is also applied in a similar manner as other molding. However, it is ordinarily nailed into the floor and not into the baseboard. This prevents the joint under the shoe from opening should the shrinkage take place in the baseboard (*Figure 80–8*).

Because the base shoe is a small-size molding and has solid backing, both interior and exterior corners are mitered. When the base shoe must be stopped at a door opening or other location, with nothing to butt against, its exposed end is generally back-mitered and sanded smooth (*Figure 80–9*). No base shoe is required if carpeting is to be used as a finish floor.

INSTALLING WINDOW TRIM

Interior window trim, in order of installation, consists of the *stool*, also called *stool cap, apron, jamb extensions, casings,* and *stops* or *stop bead* (*Figure 80–10*). Although the kind and amount of trim may differ, depending on the style of window, the application is basically the same. The procedure described in this chapter applies to most double-hung windows.

Installing the Stool

The bottom side of the stool is rabbeted at an angle to fit on the sill of the window frame so its top side

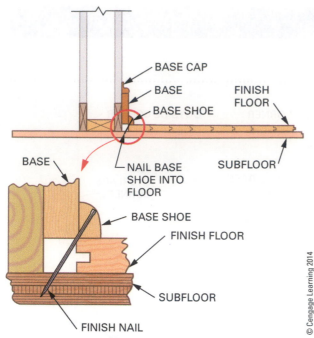

Figure 80–8 The base shoe is fastened into the floor, not into the baseboard. This prevents the joint at the floor from opening due to any movement of the baseboard.

Figure 80–10 Component parts of window trim.

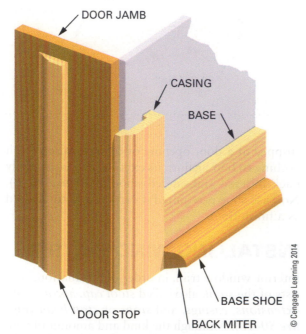

Figure 80–9 The exposed ends of base shoe molding are usually back-mitered and sanded smooth.

will be level. Its final position has the outside edge against the sash. Both ends are notched around the side jambs of the window frame. Each end projects beyond the casings by an amount equal to the casing thickness.

The stool length is equal to the distance between the outside edges of the vertical casing plus twice the casing thickness. On both sides of the window, just above the sill, hold a scrap piece of casing stock on the wall. Its inside edge should be flush with the inside face of the side jamb of the window frame. Draw a light line on the wall along the outside edge of the casing stock. Lay out a distance outward from each line equal to the thickness of the window casing. Cut a piece of stool stock to length equal to the distance between the outermost marks.

Raise the lower sash slightly. Place a short, thin strip of wood under it, on each side, that projects inward to support the stool while it is being laid out (*Figure 80–11*). Place the stool on the strips. Raise or lower the sash slightly so the top of the stool is level. Position the stool with its outside edge against the wall. Its ends should be in line with the marks previously made on the wall.

Square lines, across the face of the stool, even with the inside face of each side jamb of the window frame. Set the pencil dividers or scribers so that, on both sides, an amount equal to twice the casing thickness will be left on the stool. Scribe the stool by riding the dividers along the wall on both sides and along the bottom rail of the window sash (*Figure 80–12*).

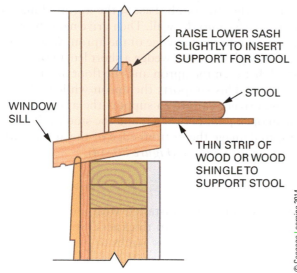

Figure 80–11 Technique to hold a stool for easy layout.

RAISE LOWER SASH SLIGHTLY TO INSERT SUPPORT FOR STOOL

STOOL

WINDOW SILL

THIN STRIP OF WOOD OR WOOD SHINGLE TO SUPPORT STOOL

© Cengage Learning 2014

On the Job

Wood shingles or trim wood strips placed under the window sash support the stool in a level position during layout.

Cut to the lines, using a handsaw. Smooth the sawed edge that will be nearest to the sash. Shape and smooth both ends of the stool the same as the inside edge. Next, close the sash to make sure the stool fits the window. Mark and cut as needed. Apply a small amount of caulking compound to the bottom of the stool along its outside edge. Fasten the stool in position by driving finish nails along its outside edge into the sill. Set the nails.

Applying the Apron

The apron covers the joint between the sill and the wall. It is applied with its ends in line with the outside edges of the window casing. Cut a length of apron stock equal to the distance between the outer edges of the window casings.

Each end of the apron is then *returned upon itself.* This means that the ends are shaped the same as its face. To return an end upon itself, hold a scrap piece on the apron. Draw its profile flush with the end.

Cut to the line with a coping saw. Sand the cut end smooth (*Figure 80–13*). Return the other end upon itself in the same manner.

Place the apron in position with its upper edge against the bottom of the stool. Be careful not to force the stool upward. Keep the top side of the stool level by holding a square between it and the

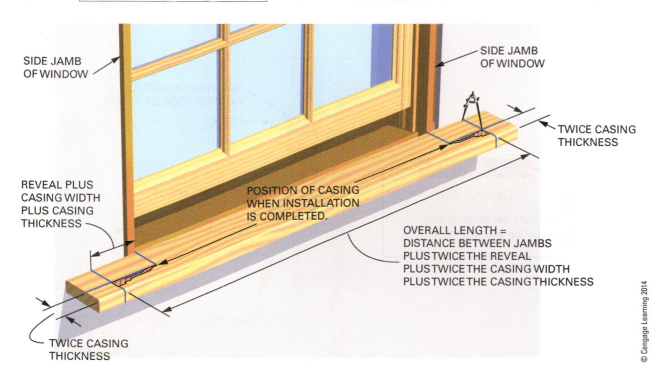

SIDE JAMB OF WINDOW

SIDE JAMB OF WINDOW

TWICE CASING THICKNESS

REVEAL PLUS CASING WIDTH PLUS CASING THICKNESS

POSITION OF CASING WHEN INSTALLATION IS COMPLETED.

OVERALL LENGTH = DISTANCE BETWEEN JAMBS PLUS TWICE THE REVEAL PLUS TWICE THE CASING WIDTH PLUS TWICE THE CASING THICKNESS

TWICE CASING THICKNESS

© Cengage Learning 2014

Figure 80–12 Method of laying out a stool.

Figure 80–13 Returning the end of an apron upon itself.

edge of the side jamb. Fasten the apron along its bottom edge into the wall. Then drive nails through the stool into the top edge of the apron. When nailing through the stool, wedge a short length of 1 × 4 stock between the apron and the floor at each nail location. This supports the apron while nails are being driven. Failure to support the apron results in an open joint between it and the stool. Take care not to damage the bottom edge of the apron with the supporting piece (*Figure 80–14*).

Installing Jamb Extensions

Windows are often installed with jambs that are narrower than the wall thickness. Strips must be fastened to these narrow jambs to bring the inside edges flush with inside wall surface. These strips are called jamb extensions.

Some manufacturers provide jamb extensions with the window unit. However, they are not always applied when the window is installed, but rather when the window is trimmed. Therefore when windows are set, these pieces should be carefully stored and then retrieved when it is time to apply the trim. They are usually precut to length and need only to be cut to width.

Measure the distance from the inside edge of the jamb to the finished wall. Rip the jamb extensions to this width with a slight back-bevel on the inside edge. Cut the pieces to length, if necessary, and apply them to the header and side jambs. Drive finish nails through the edges into the edge of the jambs (*Figure 80–15*).

On the Job

Support the apron when fastening the stool to it.

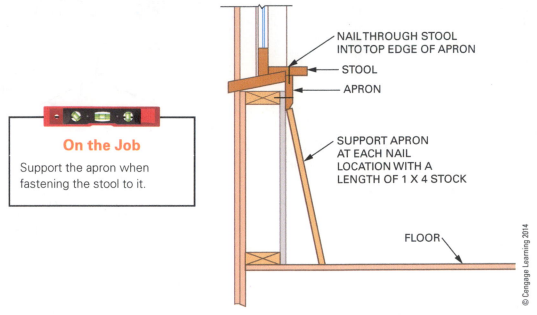

NAIL THROUGH STOOL INTO TOP EDGE OF APRON

STOOL

APRON

SUPPORT APRON AT EACH NAIL LOCATION WITH A LENGTH OF 1 X 4 STOCK

FLOOR

Figure 80–14 Technique for holding an apron in place for nailing.

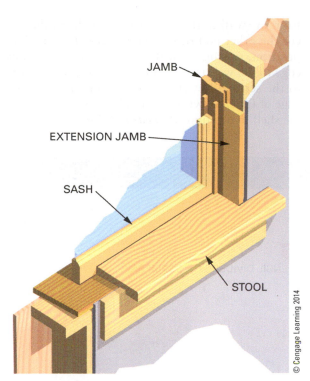

Figure 80–15 Jamb extensions are used to widen the window jamb.

Applying the Casings

Window casings usually are installed with a reveal similar to that of door casings. They also may be installed flush with the inside face. In either case, the bottom ends of the side casings rest on the stool. The window casing pattern is usually the same as the door casings. Window casings are applied in the same manner as door casings.

Cut the number of window casings needed to a rough length with a miter on one end. Cut side casings with left- and right-hand miters. Install the header casing first and then the side casings. Find the length of side casings by turning them upside down with the point of the miter on the stool in the same manner as door casings. Fasten casings with the proper reveal. Make neat, tight-fitting joints at the stool and at the head.

INSTALLING CLOSET TRIM

A simple clothes closet is normally furnished with a shelf and a rod for hanging clothes. Usually a piece of 1 × 5 stock is installed around the walls of the closet to support the shelf and the rod. This piece is called a cleat. The shelf is installed on top of it. The closet pole is installed in the center of it. Shelves are not fastened to the cleat. Rods are installed for easy removal in case the closet walls need refinishing.

Shelves are usually 1 × 12 boards. Rods may be $3/4$-inch steel pipe, $15/16$-inch full round wood poles, or chrome plated rods manufactured for this purpose. On long spans, the rod may be supported in its center by special metal closet pole supports. On each end, the closet pole is supported by plastic or metal closet pole *sockets*. In place of sockets, holes and notches are made in the cleat to support the ends of the closet pole.

For ordinary clothes closets, the height from the floor to the top edge of the cleat is 66 inches (*Figure 80–16*). Measure up from the floor this distance. Draw a level line on the back wall and two end walls of the closet. Ease the bottom outside corner and install the cleat so its top edges are to the line. The cleat is installed in the same manner as baseboard. Butt the interior corners. Fasten with two finish nails at each stud.

Install the closet pole sockets on the end cleats. The center of the socket should be at least 12 inches from the back wall and centered on the width of the cleat. Fasten the sockets through the predrilled holes with the screws provided.

Installing Closet Shelves

For a professional job, fit the ends and back edge of the shelf to the wall. Cut the shelf about $1/2$ inch longer than the distance between end walls.

Place the shelf in position by laying one end on the cleat and tilting the other end up and resting against the wall. Scribe about $1/4$ inch off the end

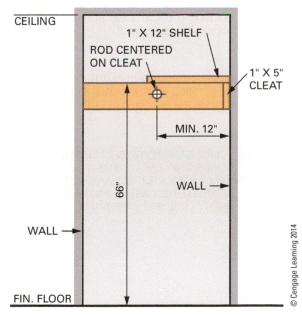

Figure 80–16 Specifications for an ordinary clothes closet.

resting on the cleat. Remove the shelf. Cut to the scribed line. Measure the distance between corners along the back wall. Transfer this measurement to the shelf, measuring from the scribed cut along the back edge of the shelf.

Place the shelf in position, tilted in the opposite direction. Set the dividers to scribe the distance from the wall to the mark on the shelf. Scribe and cut the other end of the shelf. Place the shelf into position, resting it on the cleats. Scribe the back edge to the wall to take off as little as possible. Cut to the scribed line. Ease the corners on the front edge of the shelf with a hand plane. Sand and place the shelf in position (*Procedure 80–E*).

Step**by**Step Procedures

Procedure 80-E Fitting a Closet Shelf

STEP 1 CUT SHELF ABOUT ½" LONGER THAN WIDTH OF CLOSET.

STEP 2 TILT SHELF IN POSITION WITH ONE END ON CLEAT. SCRIBE ABOUT ¼" ON THIS END. REMOVE SHELF AND CUT TO SCRIBED LINE.

STEP 3 FROM SCRIBED END, LAY OUT LENGTH OF SHELF ON BACK EDGE.

STEP 4 REPLACE SHELF IN TILTED POSITION WITH OPPOSITE END ON CLEAT.

STEP 5 SET DIVIDERS FOR DISTANCE FROM WALL TO MARK INDICATING SHELF LENGTH AT BACK EDGE. SCRIBE ALONG END OF SHELF.

STEP 6 REMOVE SHELF AND CUT END TO SCRIBED LINE. REPLACE SHELF WITH BOTH ENDS ON CLEAT.

STEP 7 SCRIBE BACK EDGE TO BACK WALL. SCRIBE ONLY ENOUGH TO FIT SHELF. REMOVE SHELF AND CUT TO SCRIBED LINE. EASE CORNERS ON FRONT EDGE. REPLACE SHELF.

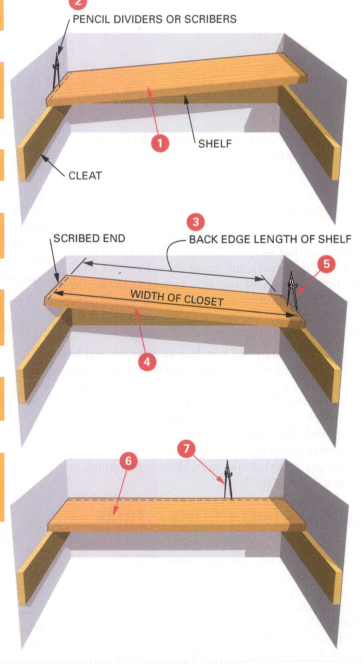

PENCIL DIVIDERS OR SCRIBERS

SHELF

CLEAT

SCRIBED END — BACK EDGE LENGTH OF SHELF

WIDTH OF CLOSET

Installing the Closet Pole

Measure the distance between pole sockets. Cut the pole to length. Install the pole on the sockets. One socket is closed. The opposite socket has an open top. Place one end of the pole in the closed socket. Then rest the other end on the opposite socket.

Linen Closets

Linen closets usually consist of a series of shelves spaced 12 to 16 inches apart. Cleats used to support shelves are $\frac{3}{4} \times 1$ stock, chamfered on the bottom outside corner. A *chamfer* is a bevel on the edge of a board that extends only partway through the thickness of the stock.

Lay out level lines for the top edges of each set of cleats. Install the cleats and shelves in the same manner as described for clothes closets.

MANTELS

Mantels are used to decorate fireplaces and to cover the joint between the fireplace and the wall. Most mantels come preassembled from the factory. They are available in a number of sizes and styles (*Figure 80–17*).

Study the manufacturer's directions carefully. Place the mantel against the wall. Center it on the fireplace. Scribe it to the floor or wall as necessary. Carefully fasten the mantel in place and set all nails.

CONCLUSION

All pieces of interior trim should be sanded smooth after they have been cut and fitted, and before they are fastened. The sanding of interior finish provides a smooth base for the application of stains,

Figure 80–17 Mantels may come preassembled in a number of styles and sizes.

© Cengage Learning 2014

paints, and clear coatings. Always sand with the grain, never across the grain.

All sharp, exposed corners of trim should be rounded over slightly. Use a block plane to make a slight chamfer. Then round over with sandpaper.

If the trim is to be stained, make sure every trace of glue is removed. Excess glue, allowed to dry, seals the surface. It does not allow the stain to penetrate, resulting in a blotchy finish.

Be careful not to make hammer marks in the finish. Occasionally rubbing the face of the hammer with sandpaper to clean it helps prevent it from glancing off the head of a nail.

Make sure any pencil lines left along the edge of a cut are removed before fastening the pieces. Pencil marks in interior corners are difficult to remove after the pieces are fastened in position. Pencil marks show through a stained or clear finish and make the joint appear open. When marking interior trim make light, fine pencil marks.

Note: Layout lines in the illustrations are purposely made dark and heavy only for the sake of clarity.

Make sure all joints are tight fitting. Measure, mark, and cut carefully. Do not leave a poor fit. Do it over, if necessary!

DECONSTRUCT THIS

Carefully study *Figure 80–18* and think about what is wrong and/or what is right. Consider all possibilities. What construction practice or method is different in your area of the country?

FIGURE 80–18 The photo in this figure shows a mitered joint on casing.

KEY TERMS

aprons	caps	full rounds	plinth blocks
back bands	chair rail	half rounds	quarter rounds
base	cleat	jamb extensions	reveal
base cap	coped joint	mantel	scribing
base shoe	corner guards	miter	surfaced four sides
bed	crown	molding	

REVIEW QUESTIONS

Select the most appropriate answer.

1. Bed, crown, and cove moldings are used frequently as

   a. window trim.   c. part of the base.
   b. ceiling molding.   d. door casings.

2. Back bands are sometimes applied to

   a. wainscoting.   c. casings.
   b. exterior corners.   d. interior corners.

3. A stool is part of the

   a. soffit.   c. base.
   b. door trim.   d. window trim.

4. The best joint between moldings of interior corners is usually

   a. coped.
   b. mitered.
   c. butted.
   d. bisected.

5. The setback of casings from the face of the jamb is often referred to as a

   a. gain.
   b. backset.
   c. reveal.
   d. quirk.

6. The length of door side casings is best found by

   a. measuring the distance from floor to the header casing.
   b. marking the length on a scrap strip and transferring it to the side casing.
   c. turning the side casing upside down with the point of the miter against the floor.
   d. holding the side casing with the right end up and marking the miter.

7. If the joint between a head and side casing is not a tight fit, it is best to

   a. plane or refit the mitered surfaces.
   b. fill the gap with glue.
   c. sand the casing face.
   d. nail the casing tighter.

8. The cope on baseboard for a room with walls and floor that are not square is laid out more accurately by

   a. back mitering.
   b. returning it.
   c. a combination square.
   d. scribing.

9. The base shoe is fastened

   a. to the baseboard only.
   b. to both the base and the floor.
   c. to the floor only.
   d. directly to the wall.

10. When the end of a molding has no material to butt against, its end is usually

    a. forward-mitered.
    b. mitered.
    c. returned upon itself.
    d. coped.

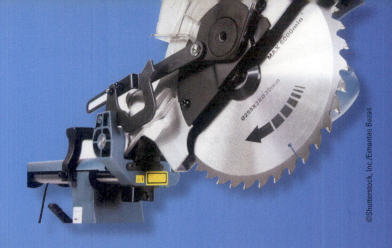

UNIT 29
Stair Finish

The staircase is usually the most outstanding feature of a building's interior. It is a showplace for architectural appeal and carpentry skills. All stair finish work must be done in a first-class manner. Joints between stair finish members must be accurate and tight-fitting. Balustrades are installed with perfectly fitting joints.

OBJECTIVES

After completing this unit, the student should be able to:

- name various stair finish parts and describe their location and function.

- lay out, dado, and assemble a housed-stringer staircase.

- apply finish to open and closed staircases.

- lay out treads for winding steps.

- install a post-to-post balustrade, without fittings, from floor to balcony on the open end of a staircase.

- install an over-the-post balustrade, with fittings, on an open staircase that runs from a starting step to an intermediate landing and, then, to a balcony.

SAFETY REMINDER

Staircases are used by many different-size people. They must be built to safety standards. They are also assembled from many small pieces. Each piece should be examined for defects, looking for strength as well as appearance characteristics. Each must be installed securely enough to provide support and prevent accidents.

CHAPTER 81
Description of Stair Finish

Many kinds of stair finish parts are manufactured in a wide variety of wood species, such as oak, beech, cherry, poplar, pine, and hemlock. It is important to identify each of the parts, know their location, and understand their function when learning to apply stair finish.

TYPES OF STAIRCASES

The stair finish may be separated in two parts: the stair body and the balustrade. Important components of the stair body finish are *treads, risers,* and *finish stringers.* The stair body may be constructed as an open or closed staircase. In an open staircase, the ends of the treads are exposed to view. In a closed staircase, they butt against the wall. Staircases may be open or closed on one or both sides.

Major parts of the balustrade include *handrails, newel posts,* and *balusters.* Balustrades are constructed in either a post-to-post or over-the-post method. In the post-to-post method, the handrail is fitted between the newel posts (*Figure 81–1*). In the over-the-post method, the handrail runs continuously from top to bottom. It requires special curved sections, called *fittings,* where the handrail changes height or direction (*Figure 81–2*).

STAIR BODY PARTS

Many kinds of stair parts are required to finish the stair body.

Risers

Risers are vertical members that enclose the space between treads. They are manufactured in a thickness of $\frac{3}{4}$ inch and in widths of $7\frac{1}{2}$ and 8 inches. ✱IRC ✱ R311.7.5.1

Figure 81–1 A closed staircase with a post-to-post balustrade on a kneewall.

Figure 81–2 An open-one-side staircase with an over-the-post balustrade.

Treads

Treads are horizontal finish members on which the feet are placed when ascending or descending stairs. High-quality treads are made from oak. Others are made from poplar or hard pine. They normally come in $\frac{3}{4}$- and $1\frac{1}{32}$-inch thicknesses, in $10\frac{1}{2}$- and $11\frac{1}{2}$-inch widths, and in lengths from 36 up to 72 inches.✱

✱**IRC**
R311.7.5.2

Nosings.   The outside edge of the tread beyond the riser has a half-round shape. It is called the nosing.✱

✱**IRC**
R311.7.5.3

A *return nosing* is a separate piece mitered to the open end of a tread to conceal its end grain. Return nosings are available in the same thickness as treads and in $1\frac{1}{4}$-inch widths. Treads are available with the return nosing already applied to one end (*Figure 81–3*).

Landing Treads.   Landing treads are used at the edge of landings and balconies. They match the tread thickness at the nosing. However, they are rabbeted to match the finish floor thickness on the landing. They come in $3\frac{1}{2}$- and $5\frac{1}{2}$-inch widths. The wider landing tread is used when newel posts are more than $3\frac{1}{2}$ inches wide (*Figure 81–4*).

Tread Molding.   The tread molding is a small cove molding used to finish the joint under stair and landing treads. The molding should be the same kind of wood as the treads. Its usual size is $\frac{5}{8} \times \frac{13}{16}$.

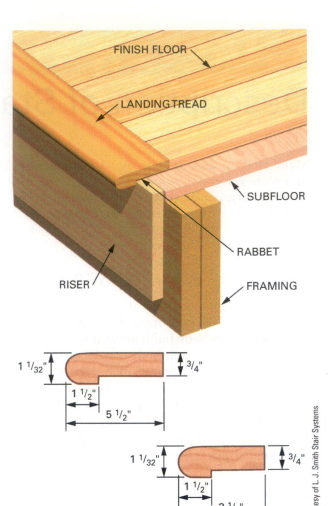

Figure 81–4 Landing treads are rabbeted to match the thickness of the finish floor.

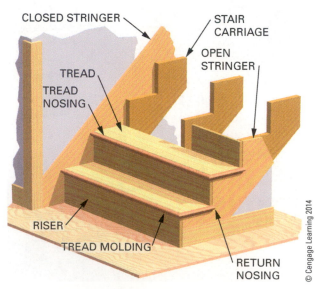

Figure 81–3 Treads and risers may be fastened to supporting stair carriages.

Finish Stringers

Finish stringers are sometimes called *skirt boards*. They are members of the stair body used to trim the intersection of risers and treads with the wall. They are called *closed finish stringers* when they are located above treads that butt the wall. They are termed *open finish stringers* when they are placed on the open side of a stairway below the treads (*Figure 81–5*). Finish stringer lineal stock is available in a $\frac{3}{4}$-inch thickness and in widths of $9\frac{1}{4}$ and $11\frac{1}{4}$ inches.

Starting Steps

A starting step or bull-nose step is the first tread-and-riser unit sometimes used at the bottom of a stairway. The starting step is used when the staircase is open on one or both sides and the handrail curves outward at the bottom. They are available

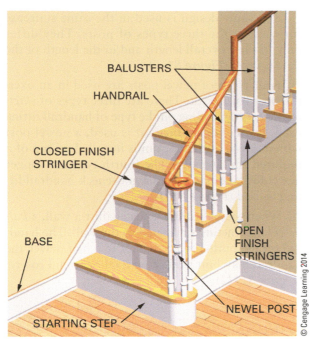

Figure 81–5 Open and closed finish stringers are finished trim pieces used to cover the intersections of the treads and risers with the wall and stair body.

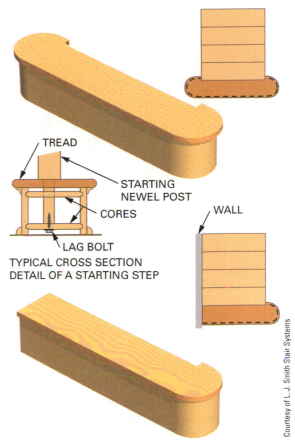

TYPICAL CROSS SECTION
DETAIL OF A STARTING STEP

Figure 81–6 Starting steps are available in a number of styles and sizes.

in a number of styles with *bull-nosed* ends, preassembled, and ready for installation (*Figure 81–6*).

BALUSTRADE MEMBERS

Finish members of the balustrade are available in many designs that are combined to complement each other. Various types of fittings are sometimes joined to straight lengths of handrail when turns in direction are required.

Newel Posts

Newel posts are anchored securely to the staircase to support the handrail. In post-to-post balustrades, the newel posts have flat, square surfaces near the top, against which the handrails are fitted, and also at the bottom for fitting and securing the post to the staircase. In between the flat surfaces, the posts may be *turned* in a variety of designs (*Figure 81–7*).

In over-the-post systems, a round pin at the top of each newel posts fits into the underside of handrail fittings. The posts are tapered toward the top end in a number of turned designs (*Figure 81–8*).

Three types of newel posts are used in a post-to-post balustrade. *Starting newels* are used at the bottom of a staircase. They are fitted against the

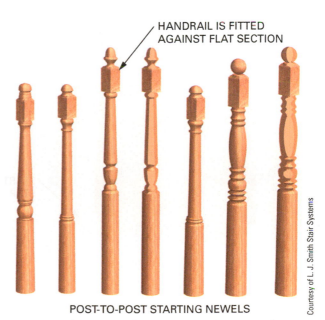

POST-TO-POST STARTING NEWELS

Figure 81–7 Newels in post-to-post balustrades must have flat surfaces where the handrails attach.

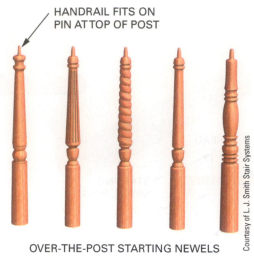

HANDRAIL FITS ON
PIN AT TOP OF POST

OVER-THE-POST STARTING NEWELS

Courtesy of L. J. Smith Stair Systems

Figure 81–8 Newels in over-the-post balustrades are made with a pin at the top.

first or second riser. If fitted against the second riser, the flat, square surface at the bottom must be longer. At the top of the staircase, *second-floor newels* are used. *Intermediate landing newels* are also available. Because part of the bottom end of these newels is exposed, turned *buttons* are available to finish the end.

The same design is used in the same staircase for each of the three types of posts. They differ only in their overall length and in the length of the flat surfaces (*Figure 81–9*).

Four types of newel posts are used in an over-the-post balustrade. There are three types of *starting newels* depending on the type of handrail fitting used. If a *volute* or *turnout* is used, a newel post with a dowel at the bottom is installed on top of a required starting step. The fourth type is a longer newel for landings, where a gooseneck handrail fitting is used (*Figure 81–10*).

When the balustrade ends against a wall, a *half newel* is sometimes fastened to the wall. The handrail is then butted to it. In place of a half newel, the handrail may butt against an oval or round rosette (*Figure 81–11*).

Handrails

The handrail is the sloping finish member grasped by the hand of the person ascending or descending the stairs. It is installed horizontally when it runs along the edge of a balcony. Handrail heights are 34 to 38 inches vertically above the nosing edge of the tread.✱ There should be a continuous 1½-inch finger clearance between the rail and the wall.✱ Several styles of handrails come in

✱IRC
R311.7.8.1

✱IRC
R311.7.8.2

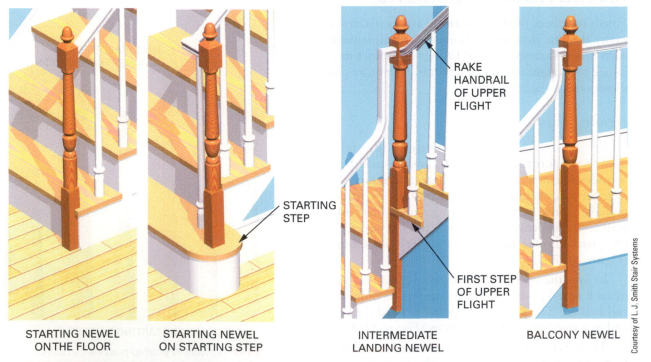

STARTING NEWEL
ON THE FLOOR

STARTING NEWEL
ON STARTING STEP

STARTING STEP

RAKE HANDRAIL OF UPPER FLIGHT

FIRST STEP OF UPPER FLIGHT

INTERMEDIATE LANDING NEWEL

BALCONY NEWEL

Courtesy of L. J. Smith Stair Systems

Figure 81–9 Three types of newel posts are used in a post-to-post balustrade.

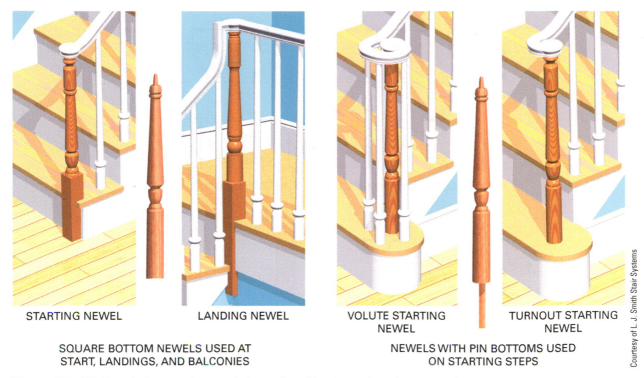

STARTING NEWEL LANDING NEWEL

SQUARE BOTTOM NEWELS USED AT
START, LANDINGS, AND BALCONIES

VOLUTE STARTING
NEWEL

TURNOUT STARTING
NEWEL

NEWELS WITH PIN BOTTOMS USED
ON STARTING STEPS

Courtesy of L. J. Smith Stair Systems

Figure 81–10 Newels for over-the-post balustrades either have pinned or square bottoms.

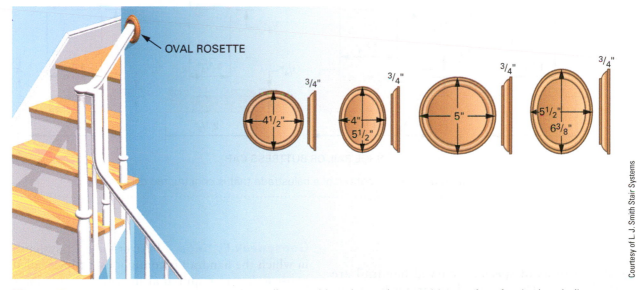

OVAL ROSETTE

Courtesy of L. J. Smith Stair Systems

Figure 81–11 Rosettes are fastened to the wall to provide a decorative attaching surface for the handrail.

lineal lengths that are cut to fit on the job. Some handrails are *plowed* with a wide groove on the bottom side to hold square-top balusters in place (*Figure 81–12*). ✳

✳IRC R311.7.8.3

On closed staircases, a balustrade may be installed on top of a **kneewall** or buttress. In relation to stairs, a kneewall is a short wall that projects

a short distance above and on the same rake as the stair body. A **shoe rail** or buttress cap, which is plowed on the top side, is usually applied to the top of the kneewall on which the bottom end of balusters are fastened (*Figure 81–13*). Narrow strips, called **fillets,** are used between balusters to fill the plowed groove on handrails and shoe rails.

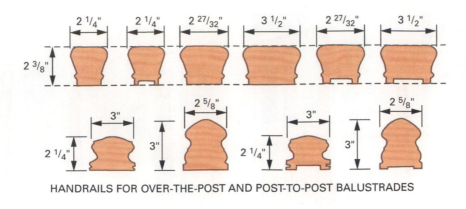

HANDRAILS FOR OVER-THE-POST AND POST-TO-POST BALUSTRADES

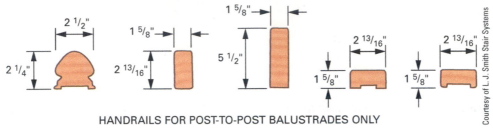

HANDRAILS FOR POST-TO-POST BALUSTRADES ONLY

Courtesy of L. J. Smith Stair Systems

Figure 81–12 Straight lengths of handrail are manufactured in many styles.

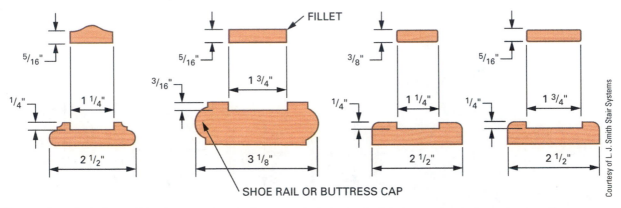

FILLET

SHOE RAIL OR BUTTRESS CAP

Courtesy of L. J. Smith Stair Systems

Figure 81–13 A shoe rail is often used at the bottom of a balustrade that is constructed on a kneewall.

Handrail Fittings

Short sections of specially curved handrail are called fittings. They are used at various locations, joined to straight sections, to change the direction of the handrail. They are classified as *starting, gooseneck,* and *miscellaneous* fittings.

Starting Fittings. To start an over-the-post handrail, starting fittings called volutes, turnouts, or *starting easings* may be used. In a post-to-post system, a straight length of handrail may be used at the bottom. To start with a soft, graceful curve, an *upeasing* is used (*Figure 81–14*).

Gooseneck Fittings. In over-the-post systems, in which the handrail is continuous, fittings called goosenecks are required at intermediate landings and at the top. This is because of changes in the handrail height or direction. In post-to-post systems, goosenecks are not required (*Figure 81–15*).

Goosenecks are available for handrails that continue level or sloping or that turn 90 or 180 degrees right or left. They are made with or without caps for both types of handrail systems.

Miscellaneous Fittings. Among the miscellaneous handrail fittings are *easings* of various

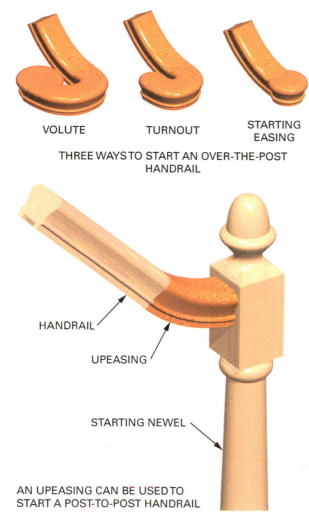

VOLUTE   TURNOUT   STARTING EASING

THREE WAYS TO START AN OVER-THE-POST HANDRAIL

HANDRAIL

UPEASING

STARTING NEWEL

AN UPEASING CAN BE USED TO START A POST-TO-POST HANDRAIL

Courtesy of L. J. Smith Stair Systems

Figure 81–14 A handrail fitting is used to start an over-the-post handrail. An upeasing is sometimes used to start a post-to-post handrail.

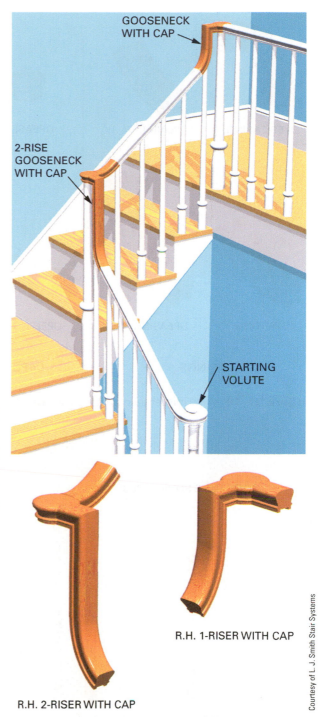

GOOSENECK WITH CAP

2-RISE GOOSENECK WITH CAP

STARTING VOLUTE

R.H. 1-RISER WITH CAP

R.H. 2-RISER WITH CAP

Courtesy of L. J. Smith Stair Systems

Figure 81–15 Gooseneck fittings are used at landings when handrails change direction or height.

kinds, *coped* and *returned ends, quarterturns,* and *caps* (**Figure 81–16**). They are used where necessary to meet the specifications for the staircase. All handrail fittings are ordered to match the straight lengths of handrail being used. When the handrail being used is plowed, a matching plow is specified for the handrail fittings. A special fillet, shaped to fit the plow, comes with the handrail fitting.

Balusters

Balusters are vertical, usually decorative pieces between newel posts. They are spaced close together and support the handrail. On a kneewall, they run

from the handrail to the shoe rail. On an open staircase, they run from the handrail to the treads (see **Figures 81–1** and **81–2**).

Balusters are manufactured in many styles. They should be selected to complement the newel

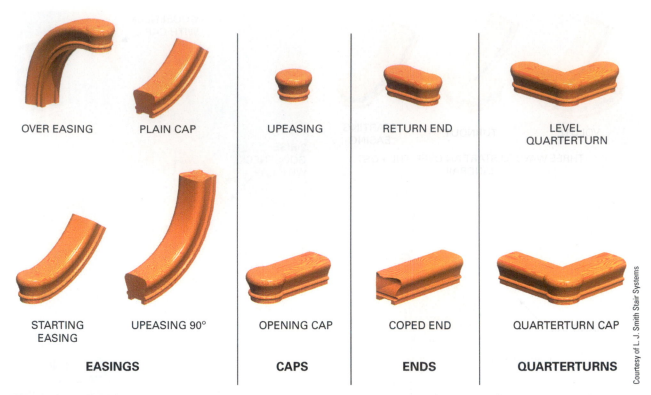

OVER EASING   PLAIN CAP   UPEASING   RETURN END   LEVEL QUARTERTURN

STARTING EASING   UPEASING 90°   OPENING CAP   COPED END   QUARTERTURN CAP

EASINGS   **CAPS**   **ENDS**   **QUARTERTURNS**

Courtesy of L. J. Smith Stair Systems

Figure 81–16 Many miscellaneous fittings are available to meet the needs of any type of handrail installation.

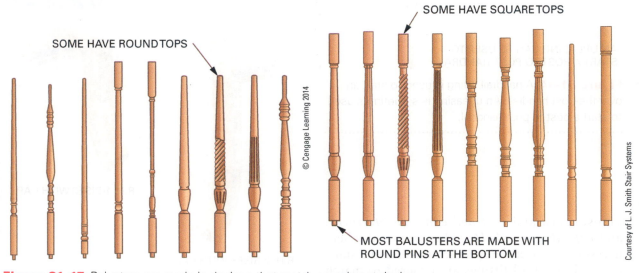

SOME HAVE SQUARE TOPS

SOME HAVE ROUND TOPS

© Cengage Learning 2014

MOST BALUSTERS ARE MADE WITH ROUND PINS AT THE BOTTOM

Courtesy of L. J. Smith Stair Systems

Figure 81–17 Balusters are made in designs that match newel post design.

posts being used (*Figure 81–17*). For example, in a post-to-post staircase, balusters with square tops are usually used. In an over-the-post staircase, balusters that are tapered at the top complement the newel posts.

Most balusters are made in lengths of 31, 34, 36, 39, and 42 inches for use in any part of the stairway. Several lengths of the same style baluster are needed for each tread of the staircase because of the rake of the handrail.

CHAPTER 82

Finishing Open and Closed Staircases

Several methods are used to finish a set of stairs. Treads and risers may be inserted into housed stringers, fastened onto a stair carriage, or a combination of both. Both methods require the riser height (unit rise) and tread width (unit run) to be determined. Example calculations of the unit rise and unit run of the stair may be found in Chapter 47. Review Unit 16 for the description, location, and function of the finish members.

National building codes maintain requirements for variations in dimensions. The maximum variation from largest to smallest riser must be $\frac{3}{8}$ inch. Check local codes. With care, all riser dimensions of a set of stairs can be built with only $\frac{1}{16}$-inch variations. ✱

✱ **IRC R311.7.5.1**

MAKING A HOUSED STRINGER

Housed stringers can be laid out using a stair router template or a pitch board and job-made router template. In either case the layout begins on the face side of the stringer stock. Draw a setback line parallel to and about 2 inches down from the top edge. The intersection of the tread and riser faces will land on this line (*Figure 82–1*).

The 2-inch distance may vary, depending on the width of the stringer stock and the desired height of the top edge of the stringer above the stair treads.

Using a Stair Router Template

Stair routing templates are manufactured to guide a router in making dadoes in stringers for treads and risers. The router must be equipped with a straight bit and a template guide of the correct size. Stair templates are adjustable for different rises and runs. They are easily clamped to the stock for routing the dadoes and then moved.

The template is shaped so the dadoes will be the exact tread width at the nosing and wider toward the back side of the stair. The template has nonparallel sides so the finished dadoes will be tapered. The treads and risers are then wedged tightly against the face side shoulders of the dadoes (*Figure 82–2*).

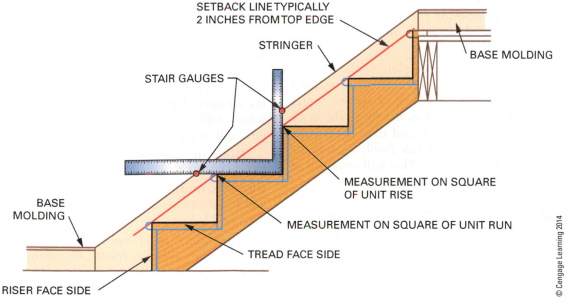

SETBACK LINE TYPICALLY 2 INCHES FROM TOP EDGE

STRINGER

BASE MOLDING

STAIR GAUGES

MEASUREMENT ON SQUARE OF UNIT RISE

BASE MOLDING

MEASUREMENT ON SQUARE OF UNIT RUN

TREAD FACE SIDE

RISER FACE SIDE

© Cengage Learning 2014

Figure 82–1 Layout considerations for a housed stringer.

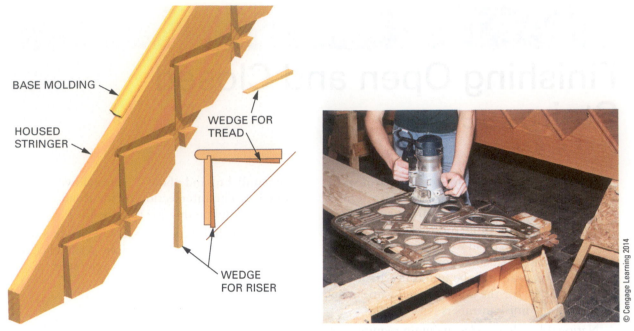

BASE MOLDING

HOUSED STRINGER

WEDGE FOR TREAD

WEDGE FOR RISER

© Cengage Learning 2014

Figure 82–2 A housed stringer is dadoed to accept tread, risers, and wedges.

Full layout of all treads and risers is not necessary while using the template. An alignment gauge is designed to position the template along the setback line according to the unit length of the stair (*Figure 82–3*). Using the Pythagorean theorem, calculate the unit length. Lightly mark squared lines on the setback line, spaced at the unit length distance.

Using a framing square, mark the unit Rise and unit Run on the board. Mark them such that a Rise–Run intersection lands on the setback line. Then, using these lines, mark the thickness of a tread and riser pair. Loosen template shoulder clamp bolts and position the square edges of the template to fit parallel to the tread-riser layout. Retighten shoulder clamps.

Move and clamp the template to the stringer with the alignment gauge on a unit length line. Rout the stringer, about $\frac{1}{4}$ inch deep. Place the router on the template where it does not touch the stock material. Start the router and ease it into the stringer. Press the router guide firmly against the template on all four sides. This will ensure that the dado is completely removed.

> ## CAUTION
> Take great care when removing the router from the template. The butt will damage the template if it is touched, and the operator is at risk from flying metal fragment.

Let the router come to a complete stop before removing it from the template. This will reduce the danger of personal injury and damage to the template and bit. Loosen the template clamp and move the template to the next unit length line. Make sure the template is resting entirely on the face of the stock. Rout the dado and repeat for the remaining treads and risers.

Using a Pitch Board

A pitch board may be used to lay out a housed stringer. This process requires that each tread and riser pair be laid out. A pitch board is a piece of stock, usually $\frac{3}{4}$ inch thick. It is cut to the Rise and Run of the stairs. A strip of wood is fastened to the rake edge of the pitch board. This is used to hold the pitch board against the edge while laying out the stringer (*Figure 82–4*). Care should be taken when making the pitch board because many layouts will be made from it.

Using the pitch board, lay out the risers and treads for each step of the staircase. These lines show the location of the face side of each riser and tread and are the outside edges of the housing (*Figure 82–5*).

After the stringer has been laid out, make a template to guide the router by cutting out the shape of the dadoes from a piece of thin plywood or hardboard. The cut is made slightly larger than the dadoes to allow for the router guide to follow the template. Take care to make cuts smooth and

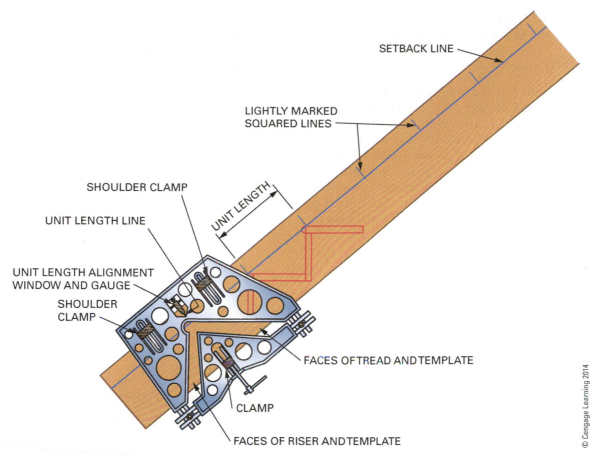

SETBACK LINE

LIGHTLY MARKED
SQUARED LINES

SHOULDER CLAMP

UNIT LENGTH LINE

UNIT LENGTH

UNIT LENGTH ALIGNMENT
WINDOW AND GAUGE

SHOULDER
CLAMP

FACES OF TREAD AND TEMPLATE

CLAMP

FACES OF RISER AND TEMPLATE

© Cengage Learning 2014

Figure 82–3 A stair router template is clamped to the stock at each unit length marking.

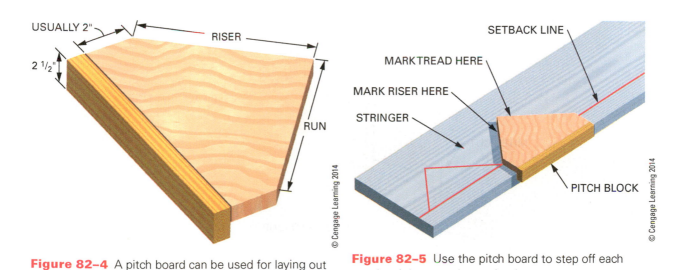

USUALLY 2"

RISER

2 1/2"

RUN

© Cengage Learning 2014

Figure 82–4 A pitch board can be used for laying out a housed stringer.

SETBACK LINE

MARK TREAD HERE

MARK RISER HERE

STRINGER

PITCH BLOCK

© Cengage Learning 2014

Figure 82–5 Use the pitch board to step off each tread and riser on a housed stringer.

clean as the router guide will transfer onto the stringer every deviation in the template.

Completing the Stringer

Cut and fit the top and bottom ends of the stringer to the floor and the top end to the landing. Equalize the bottom riser to account for the finished floor thickness. Make end cuts that will properly join with the baseboard. This joint should be made in a professional manner to provide a continuous line of finish from one floor to the next. Since the stringer is usually S4S stock and the base often is not, the stringer should be cut to allow the base to end against it.

LAYING OUT AN OPEN STRINGER

The layout of an open (or *mitered*) stringer is similar to that of a housed stringer. However, riser and tread layout lines intersect at the top edge of the stringer, instead of against a line in from the edge. The riser layout line is the outside face of the riser. This layout line is mitered to fit the mitered end of the riser. The tread layout is to the face side of the tread. The risers and treads are marked lightly with a sharp pencil (*Figure 82–6*).

To lay out the *miter cut* for the risers, measure in at right angles from the riser layout line a distance equal to the thickness of the riser stock. Draw another plumb line at this point. Square both lines across the top edge of the stringer stock. Draw a diagonal line on the top edge to mark the miter angle (*Figure 82–7*).

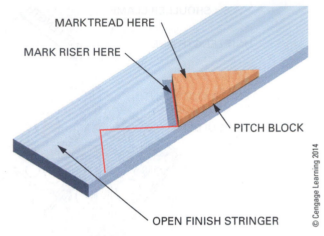

Figure 82–6 Laying out the open finish stringer of a housed staircase.

© Cengage Learning 2014

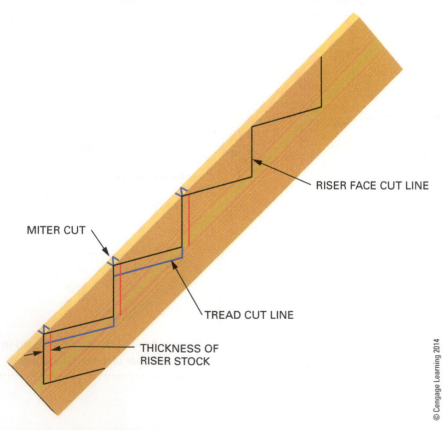

Figure 82–7 Laying out the miter angle on an open finish stringer.

© Cengage Learning 2014

To mark the tread cut on the stringer, measure down from the tread layout line a distance equal to the thickness of the tread stock. Draw a level line at this point for the tread cut. The tread cut is square through the thickness of the stringer. Fit the bottom end to the floor. Fit the top end against the landing. Make the mitered plumb cuts for the risers and the square level cuts for the treads.

Installing Risers and Treads

Cut the required number of risers to a rough length. Determine the face side of each piece. Sometimes a rabbet and a groove are made to increase the strength of the tread and riser joint forming the inside corner. This is done by cutting a $3/8$-inch groove near the bottom edge on the face side of all but the starting riser. The groove is located so its top is a distance from the bottom edge equal to the tread thickness. The groove is made for the rabbeted inner edge of the tread to fit into it (*Figure 82–8*). Rip the risers to width by cutting the edge opposite the groove. Rip the treads to width. Rabbet their back edges to fit in the riser grooves. Cut the risers and tread to exact lengths.

On a closed staircase, the risers are installed with wedges, glue, and screws between housed stringers. On the open side of a staircase, where the riser and open stringer meet, a miter joint is made so no end grain is exposed (*Figure 82–9*). The treads are then installed with wedges, glue, and

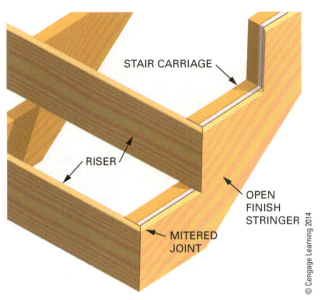

Figure 82–9 A mitered joint is made between the risers and open finish stringer so no end grain is exposed.

screws on the closed side and with screws through screw blocks on the open side. Screw blocks reinforce interior corners on the underside at appropriate locations and intervals.

Applying Return Nosings and Tread Molding

If the staircase is open, a return nosing is mitered to the end of the tread. The back end of the return nosing projects past the riser the same amount as the tread overhangs the riser. The end is returned on itself.

The tread molding is then applied under the overhang of the tread. If the staircase is closed on both sides, the molding is cut to fit between finish stringers. On the open end of a staircase, the molding is mitered around and under the return nosing. It is stopped and returned on itself at a point so the end assembly appears the same as at the edge (*Figure 82–10*).

After the housed-stringer staircase is assembled, it is installed in position. The balustrade is then constructed in a manner similar to that described later in this chapter for framed staircases.

FINISHING THE BODY OF A CLOSED STAIRCASE SUPPORTED BY STAIR FRAMING

The following section describes the installation of finish to a closed staircase in which the supporting stair carriages have already been installed between two walls that extend from floor to ceiling.

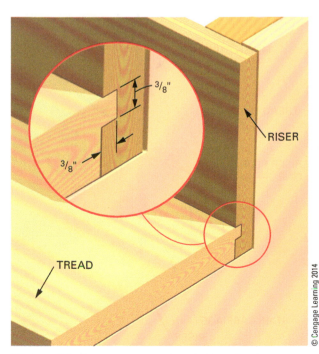

Figure 82–8 A groove can be made in the riser to receive the rabbeted inner edge of the tread.

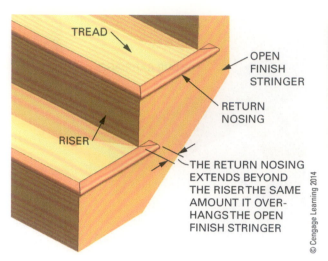

Figure 82–10 A return nosing is mitered to the open end of a tread.

Applying Risers

The first trim members applied to the stair carriages are the risers. Rip the riser stock to the proper width. Cut grooves as described previously for housed-stringer construction. Cut the risers to a length, with square ends, about ¼ inch less than the distance between walls.

Fasten the risers in position with three 2½-inch finish nails into each stair carriage. Start at the top and work down. Remove the temporary treads installed previously as work progresses downward (*Figure 82–11*).

CAUTION

Put up positive barriers at the top and bottom of the stairs so that the stairs cannot be used while the finish is applied. A serious accident can happen if a person, who does not realize that the temporary treads have been removed, uses the stairs.

Laying Out and Installing the Closed Finish Stringer

After the risers have been installed, the *closed finish stringer* is cut around the previously installed risers. Usually 1 × 10 lumber is used. When installed, its top edge will be about 3 inches above the tread nosing. A 1 × 12 may be used if a wider finish stringer is desired.

Tack a length of stringer stock to the wall. Its bottom edge should rest on the top edges of the previously installed risers. Its bottom end should rest on the floor. The top end should extend about 6 inches beyond the landing.

Lay out plumb lines, from the face of each riser, across the face of the finish stringer. If the riser itself is out of plumb, then plumb upward from that part of the riser that projects farthest outward. Then, lay out level lines on the stringer, from each tread cut of the stair carriage and also from the floor of the landing above (*Figure 82–12*).

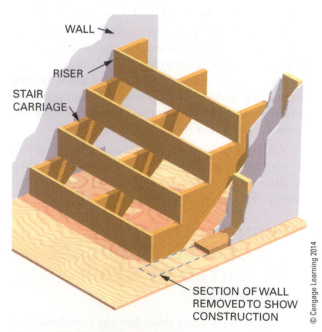

Figure 82–11 Risers are typically the first finish members applied to the stair carriage in a closed staircase.

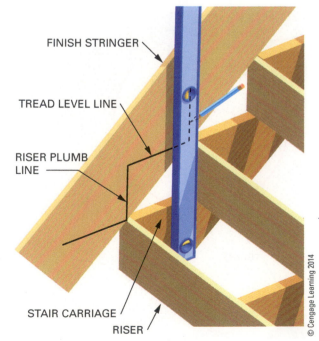

Figure 82–12 The closed finish stringer is laid out using a level to extend plumb and level lines from the stair carriage.

Remove the stringer from the wall. Cut to the layout lines. Follow the plumb lines carefully. Plumb cuts will butt against the face of the risers, so a careful cut needs to be made. Not as much care needs to be taken with level cuts because treads will later butt against and cover them. A circular saw may be used to make most of the cut, and then a handsaw is used to finish the cut.

After the cutouts are made in the finish stringer, tack it back in position. Fit it to the floor. Then, lay out top and bottom ends to join the base that will later be installed on the walls. Remove the stringer. Make the end cuts. Sand the board, and place it back in position. Fasten the stringer securely to the wall with finishing nails. Do not nail too low to avoid splitting the lower end of the stringer. Install the finish stringer on the other wall in the same manner.

Drive shims, at each step, between the back side of the risers and the stair carriage. The shims force the risers tightly against the plumb cut of the finish stringer. Shim at intermediate stair carriages to straighten the risers, from end to end, between walls.

Installing Treads

Treads are cut on both ends to fit snugly between the finish stringers. The nosed edge of the tread projects beyond the face of the riser by $1\frac{1}{8}$ inches (*Figure 82–13*).✳

✳IRC R311.7.5.3

Along the top edge of the riser, measure carefully the distance between finish stringers. Transfer the measurement and square lines across the tread.

Cut in from the nosed edge. Square through the thickness for a short distance. Then undercut slightly. Smooth the cut ends with a block plane. Rub one end with wax. Place the other end in position. Press on the waxed end until the tread lays flat on the stair carriages.

Place a short block on the nosed edge. Tap it until the inner rabbeted edge is firmly seated in the groove of the riser.

If it is possible to work from the underside, the tread may be fastened by the use of screw blocks at each stair carriage and at intermediate locations.

If it is not possible to work from the underside, the treads must be face nailed. Fasten each tread in place with three 8d finish nails into each stair carriage. It may be necessary to drill holes in hardwood treads to prevent splitting the tread or bending the nail. A little wax applied to the nail makes driving easier and helps keep the nail from bending.

Start from the bottom and work up, installing the treads in a similar manner. At the top of the stairs, install a landing tread. If $1\frac{1}{32}$-inch-thick treads are used on the staircase, use a landing tread that is rabbeted to match the thickness of the finish floor (*Figure 82–14*).

Tread Molding

The tread molding is installed under the overhang of the tread and against the riser. Cut the molding to the same length as the treads, using a miter box. Predrill holes. Fasten the molding in place with 4d finish nails spaced about 12 inches apart. Nails are driven at about a 45-degree angle through the center of the molding.

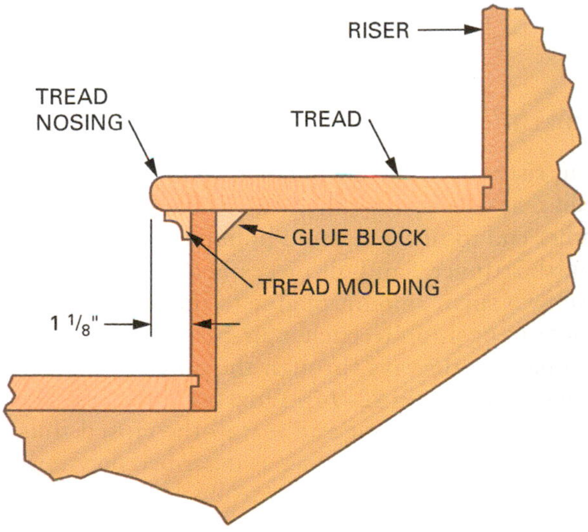

RISER

TREAD NOSING

TREAD

GLUE BLOCK

TREAD MOLDING

$1\frac{1}{8}$"

SECTION

© Cengage Learning 2014

Figure 82–13 Tread and riser details.

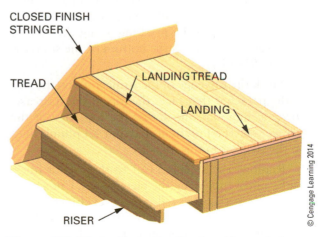

CLOSED FINISH STRINGER

TREAD

LANDING TREAD

LANDING

RISER

© Cengage Learning 2014

Figure 82–14 A rabbeted landing tread is used at the top of the stairway.

FINISHING THE BODY OF AN OPEN STAIRCASE SUPPORTED BY STAIR FRAMING

The following section describes the installation of finish to the stair body of a staircase, supported by stair carriages, which is closed on one side and open on the other side.

Installing the Finish Stringers

The *open finish stringer* must be installed before the *risers* and the *closed finish stringer*. To lay out the open finish stringer, cut a length of finish stringer stock. Fit it to the floor and against the landing. Its top edge should be flush with the top edge of the stair carriage. Tack it in this position to keep it from moving while it is being laid out.

First, lay out level lines on the face of the stringer in line with the tread cut on the stair carriage. Next, plumb lines must be laid out on the face of the finish stringer for making miter joints with risers.

Using a Preacher to Lay Out Plumb Lines.

Use a **preacher** to lay out the plumb lines on the open finish stringer. A preacher is made from a piece of nominal 4-inch stock about 12 inches long. Its thickness must be the same as the riser stock. The preacher is notched in the center. It should be wide enough to fit over the finish stringer. It should be long enough to allow the preacher to rest on the tread cut of the stair carriage when held against the rise cut.

Place the preacher over the stringer and against the rise cut of the stair carriage. Plumb the preacher with a hand level. Lay out the plumb cut on the stringer by marking along the side of the preacher that faces the bottom of the staircase (*Figure 82–15*).

Mark the top edge of the stringer along the side of the preacher that faces the top of the staircase. Draw a diagonal line across the top edge of the stringer for the miter cut. Lay out all plumb lines on the stringer in this manner.

Remove the stringer. Cut to the layout lines. Make miter cuts along the plumb lines. Cut square through the thickness along the level lines. Sand the piece. Fasten it in position. To ensure the piece will be in the same position as it was when laid out, fasten it first in the same holes where the piece was originally tacked.

On the Job

Use a preacher to lay out plumb cuts on an open finish stringer.

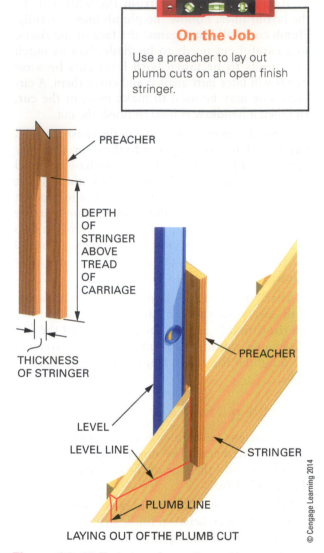

PREACHER

DEPTH OF STRINGER ABOVE TREAD OF CARRIAGE

THICKNESS OF STRINGER

PREACHER

LEVEL

LEVEL LINE

STRINGER

PLUMB LINE

LAYING OUT OF THE PLUMB CUT

© Cengage Learning 2014

Figure 82–15 Technique for easily marking a stringer on both faces.

Installing Risers

Cut risers to length by making a square cut on the end that goes against the wall. Make miters on the other end to fit the mitered plumb cuts of the open finish stringer. Sand all pieces before installation. Apply a small amount of glue to the miters. Fasten them in position to each stair carriage. Drive finish nails both ways through the miter to hold the joint tight (*Figure 82–16*). Wipe off any excess glue. Set all nails. Lay out and install the closed finish stringer in the same manner as described previously.

Installing the Treads

Rip the treads to width. Rabbet the back edges. Make allowance for the rabbet when ripping treads

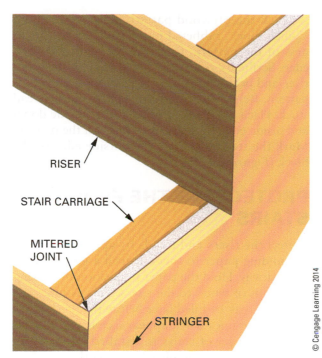

Figure 82–16 Open finish stringers are mitered to receive mitered risers.

On the Job

Make sure nails used to fasten return nosings do not line up with balusters.

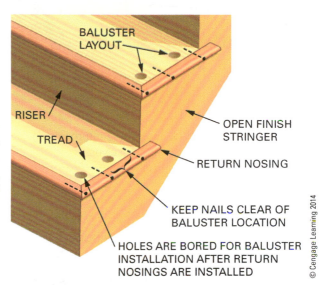

Figure 82–17 Alignment of trim nails should be positioned to avoid future baluster holes.

to width. Cut one end to fit against the closed finish stringer. Make a cut on the other end to receive the return nosing. This is a combination square and miter cut. The square cut is made flush with the outside face of the open finish stringer. The miter starts a distance equal to the width of the return nosing beyond the square cut as shown in Figure 82–9.

Applying the Return Nosings

The return nosings are applied to the open ends of the treads. Miter one end of the return nosing to fit against the miter on the tread. Cut the back end square. Return the end on itself. The end of the return nosing extends beyond the face of the riser, the same amount as its width.

Predrill pilot holes in the return nosing for nails. Locate the holes so they are not in line with any balusters that will later be installed on the treads. Holes must be bored in the treads to receive the balusters. Any nails in line with the holes will damage the boring tool (*Figure 82–17*).

Apply glue to the joint. Fasten the return nosing to the end of the tread with three 8d finishing nails. Set all nails. Sand the joint flush. Apply all other return nosings in the same manner. Treads may be purchased with the return nosing applied in the factory. If used, the closed end of the tread is cut so the nosed end overhangs the finish stringer by the proper amount.

Applying the Tread Molding

The tread molding is applied in the same manner as for closed staircases. However, it is mitered on the open end and returned back onto the open stringer. The back end of the return molding is cut and returned on itself at a point so the end assembly shows the same as at the edge (*Figure 82–18*). Predrill pilot holes in the molding. Fasten it in place. Molding on starting and landing treads should only be tacked in case it needs to be removed for fitting after newel posts have been installed.

INSTALLING TREADS ON WINDERS

Treads on winding steps are especially difficult to fit because of the angles on both ends. A method used by many carpenters involves the use of a pattern and a scribing block. Cut a thin piece of plywood so it fits in the tread space within $\frac{1}{2}$ inch on the ends and back edge. The outside edge should be straight and in line with the nosed edge of the tread when installed.

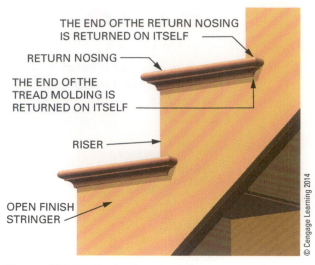

THE END OF THE RETURN NOSING IS RETURNED ON ITSELF

RETURN NOSING

THE END OF THE TREAD MOLDING IS RETURNED ON ITSELF

RISER

OPEN FINISH STRINGER

© Cengage Learning 2014

Figure 82–18 The back ends of the return nosing and molding are returned on themselves.

Tack the plywood pattern in position. Use a ¾-inch block, rabbeted on one side by the thickness of the pattern, to scribe the ends and back edge. Scribe by riding the block against stringers, riser, and post while marking its inside edge on the pattern. Remove the pattern. Tack it on the tread stock. Place the block with its rabbeted side down and inside edge to the scribed lines on the pattern. Mark the tread stock on the outside edge of the block (*Figure 82–19*).

PROTECTING THE FINISHED STAIRS

Protect the risers and treads by applying a width of building paper to them. Unroll a length down the stairway. Hold the paper in position by tacking thin strips of wood to the risers. This assumes the tack hole may be filled later before the finish is applied. Sometimes when appearance is paramount, the finished set of stairs is installed last after all tradespeople have completed their work.

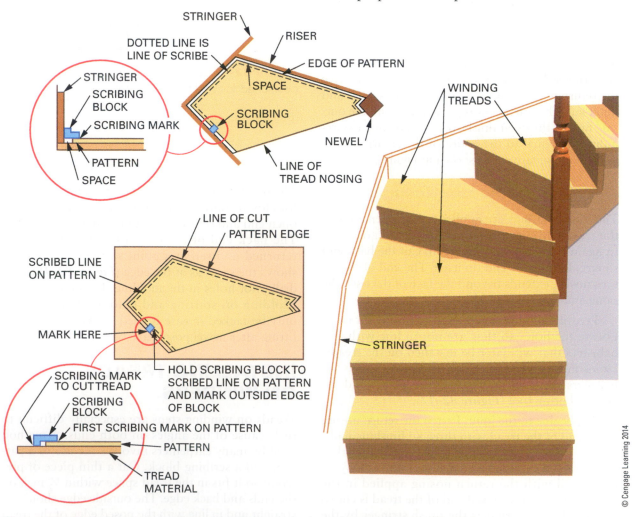

STRINGER

DOTTED LINE IS LINE OF SCRIBE

RISER

EDGE OF PATTERN

SPACE

STRINGER

SCRIBING BLOCK

SCRIBING MARK

PATTERN

SPACE

SCRIBING BLOCK

NEWEL

LINE OF TREAD NOSING

LINE OF CUT

PATTERN EDGE

SCRIBED LINE ON PATTERN

MARK HERE

HOLD SCRIBING BLOCK TO SCRIBED LINE ON PATTERN AND MARK OUTSIDE EDGE OF BLOCK

SCRIBING MARK TO CUT TREAD

SCRIBING BLOCK

FIRST SCRIBING MARK ON PATTERN

PATTERN

TREAD MATERIAL

WINDING TREADS

STRINGER

© Cengage Learning 2014

Figure 82–19 Technique for scribing winding treads to the stringers, risers, and posts.

CHAPTER 83

Balustrade Installation

Balustrades are the most visible and complex component of a staircase (see *Figure I–9*). Installing them is one of the most intricate kinds of interior finish work. This chapter describes installation of post-to-post and over-the-post balustrades. Mastering the techniques described in this chapter will enable the student to install balustrades for practically any situation.

LAYING OUT BALUSTRADE CENTERLINES

For the installation of any balustrade, its centerline is first laid out. On an open staircase, the centerline should be located a distance inward, from the face of the finish stringer, that is equal to half the baluster width. It is laid out on top of the treads. If the balustrade is constructed on a kneewall, it is centered and laid out on the top of the wall (*Figure 83–1*).

Laying Out Baluster Centers

The next step is to lay out the baluster centers. Code requirements for maximum baluster spacing may vary. Check the local building code for allowable spacing. Most codes require that balusters be spaced so that no object 4 inches in diameter or greater can pass through. ✳

On open treads, the center of the front baluster is located a distance equal to half its thickness back from the face of the riser. If two balusters can be used on each tread, the spacing is half the run. If codes require three balusters per tread, the spacing is one-third the run (*Figure 83–2*).

✳ **IRC R312.1.3**

INSTALLING A POST-TO-POST BALUSTRADE

The following procedure applies to a post-to-post balustrade running, without interruption, from floor to floor.

Laying Out the Handrail

Clamp the handrail to the tread nosings. Use a short bar clamp from the bottom of the finish stringer to the top of the handrail. Clamp opposite a nosing to avoid bowing the handrail. Use only enough pressure to keep the handrail from moving. Protect the

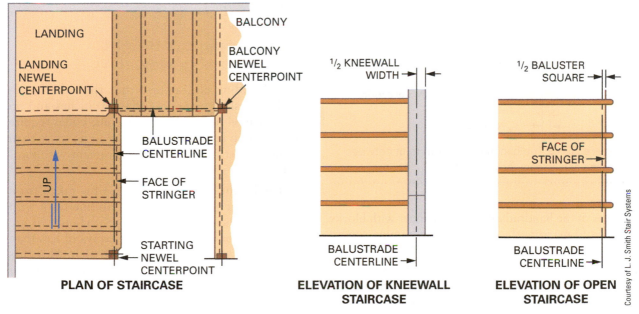

Figure 83–1 The centerline of the balustrade is laid out on a kneewall or open treads.

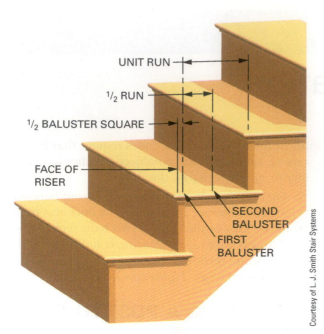

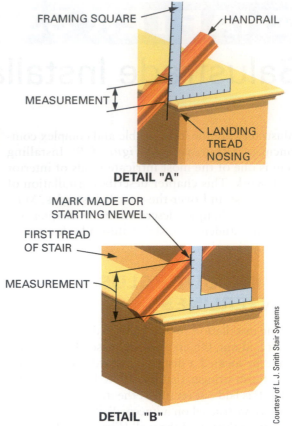

Figure 83–2 Layout of baluster centers on open treads.

DETAIL "A"

DETAIL "B"

Figure 83–4 Determine the two measurements shown and record for future use.

measure the vertical thickness of the rake handrail. Also, at the bottom, measure the height from the first tread to the top of the handrail where it butts the newel post. Record and save the measurements for later use (*Figure 83–4*).

Determining the Height of the Starting Newel

Most building codes state that the rake handrail height shall be no less than 30 inches and no more than 34 inches. However, some codes require heights of no less than 34 inches and no more than 38 inches. Verify the height requirement with local building codes. ✳

✳**IRC R311.7.8.1**

The height of the stair handrail is taken from the top of the tread along a plumb line flush with the face of the riser (*Figure 83–5*). Handrails are required on only one side in stairways of less than 44 inches in width. Stairways wider than 44 inches require a handrail on both sides. A center handrail must be provided in stairways more than 88 inches wide. Handrail heights are 34 to 38 inches

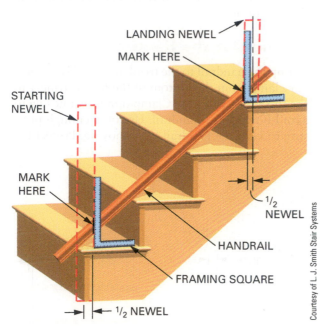

Figure 83–3 The handrail is laid out to fit between starting and balcony newel posts.

edges of the handrail and finish stringer with blocks to avoid marring the pieces. Use a framing square to mark the handrail where it will fit between starting and balcony newel posts (*Figure 83–3*).

While the handrail is clamped in this position, use a framing square at the landing nosing to

Courtesy of L. J. Smith Stair Systems

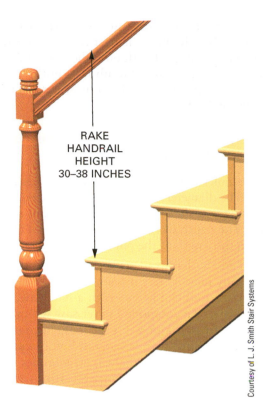

Courtesy of L. J. Smith Stair Systems

Figure 83–5 The rake handrail height is the vertical distance from the tread nosing to the top of the handrail.

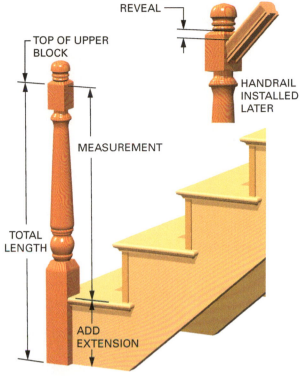

Courtesy of L. J. Smith Stair Systems

TO DETERMINE MEASUREMENT, ADD THE DIFFERENCE BETWEEN DISTANCES SHOWN IN DETAILS "A" AND "B" IN FIGURE 83–4 TO THE HANDRAIL HEIGHT. ADD 1" MORE FOR BLOCK REVEAL.

Figure 83–6 Determining the height of the starting newel.

vertically above the nosing edge of the tread. There should be a continuous $1\frac{1}{2}$-inch finger clearance between the rail and the wall. ✳

✳IRC R311.7.8.2

If a turned starting newel post is used, add the difference between the two previously recorded measurements (see previous paragraph) to the required rake handrail height. Add 1 inch for a block reveal. The block reveal is the distance from the top of the handrail to the top of the square section of the post. This sum is the distance from the top of the first tread to the top of the upper block. To this measurement add the height of the turned top and the distance the newel extends below the top of the first tread to the floor (*Figure 83–6*). Cut the starting newel to its total length.

Installing the Starting Newel

The starting newel is notched over the outside corner of the first step. One-half of its bottom thickness is left on from the front face of the post to the face of the riser. In the other direction, it is notched

so its centerline will be aligned with the balustrade centerline (*Figure 83–7*). The post is then fastened to the first step with lag screws. The lag screws are counterbored and later concealed with wood plugs.

Newel posts must be set plumb. They must be strong enough to resist lateral force applied by persons using the staircase. The post may be slightly out of plumb after it is fastened. If so, loosen the lag screws slightly. Install thin shims, between the post and riser or finish stringer, to plumb the post. On one or both sides, install the shims, near the bottom or top of the notch as necessary, to plumb the post. When plumb, retighten the lag screws.

Installing the Balcony Newel

Generally, codes require that balcony rails for homes be no less than 36 inches. For commercial or public structures, the rails are required to be no less than 42 inches. Check local codes for requirements. ✳

✳IRC R312.1.2

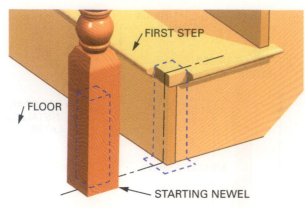

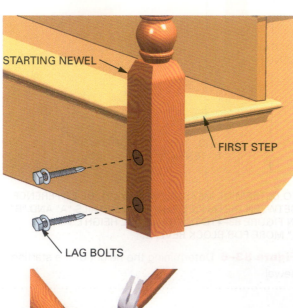

FIRST STEP

FLOOR

STARTING NEWEL

STARTING NEWEL

FIRST STEP

LAG BOLTS

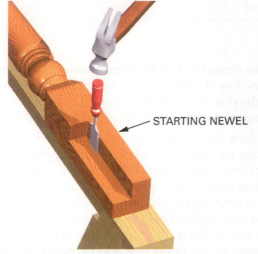

STARTING NEWEL

Courtesy of L. J. Smith Stair Systems

Figure 83–7 The starting newel is notched to fit over the first step.

The height of the balcony newel is determined by finding the sum of the required balcony handrail height, a block reveal of 1 inch, the height of the turned top, and the distance the newel extends below the balcony floor.

Trim the balcony newel to the calculated height. Notch and fit it over the top riser with its centerlines aligned with both the rake and balcony handrail centerlines. Plumb it in both directions. Fasten it in place with counterbored lag bolts (*Figure 83–8*).

Preparing Treads and Handrail for Baluster Installation

Bore holes in the treads at the center of each baluster. The diameter of the hole should be equal to the diameter of the pin at the bottom end of the baluster. The depth of the hole should be slightly more than the length of the pin (*Figure 83–9*).

Cut the handrail to fit between starting and balcony newels. Lay it back on the tread nosings. The handrail can be cut with a handsaw, a radial arm saw, or a compound miter saw with the blade tilted. If using a power saw, make a practice cut to be sure the setup is correct, before cutting the handrail. Transfer the baluster centerlines from the treads to the handrail (*Figure 83–10*).

Turn the handrail upside down and end for end. Set it back on the tread nosings with the starting newel end facing up the stairs. Bore holes at baluster centers at least $3/4$ inch deep, if balusters with round tops are to be used (*Figure 83–11*).

Installing Handrail and Balusters

Prepare the posts for fastening the handrail by counterboring and drilling shank holes for lag bolts through the posts. Place the handrail at the correct height between newel posts. Drill pilot holes. Temporarily fasten the handrail to the posts (*Figure 83–12*).

Cut the balusters to length. Allow $3/4$ inch for insertion in the hole in the bottom of the handrail. The handrail may have to be removed for baluster installation and then fastened permanently. The bottom pin is inserted in the holes in the treads. The top of the baluster is inserted in the holes in the handrail bottom.

If *square-top balusters* are used, they are trimmed to length at the rake angle. They are inserted into a *plowed handrail*. The balusters are then fastened to the handrail with finish nails and glue (*Figure 83–13*). Care must be taken to keep the handrail in a straight line from top to bottom when fastening square-top balusters. Care must also be taken to keep each baluster in a plumb line. Install fillets in the plow of the handrail, between the balusters.

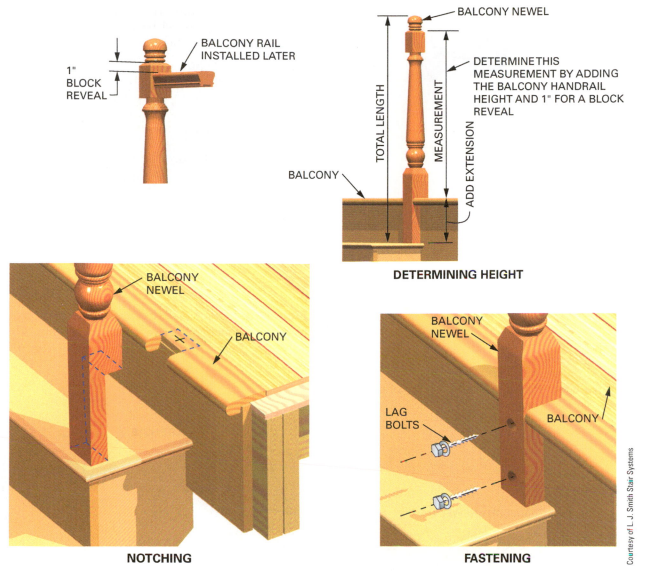

Figure 83–8 The height of the balcony newel post is calculated, notched at the bottom, and fastened in place.

Courtesy of L. J. Smith Stair Systems

Installing the Balcony Balustrade

Cut a half newel to the same height as the balcony newel. Temporarily place it against the wall. Mark the length of the balcony handrail (*Figure 83–14*). Cut the handrail to length. Fasten the half newel to one end of it. Temporarily fasten the half newel to the wall and the other end of the handrail to the landing newel, if they must be removed to install the balcony balusters.

If the balcony handrail ends at the wall against a rosette, first fasten the rosette to the end of the handrail. Hold the rosette against the wall. Mark the length of the handrail at the landing newel.

Cut the handrail to length. Temporarily fasten it in place (*Figure 83–15*).

Spacing and Installing Balcony Balusters

The balcony balusters are spaced by adding the thickness of one baluster to the distance between the balcony newel and the half newel. The overall distance is then divided into spaces that equal, as close as possible, the spacing of the rake balusters (*Figure 83–16*). The balcony balusters are then installed in a manner similar to the rake balusters.

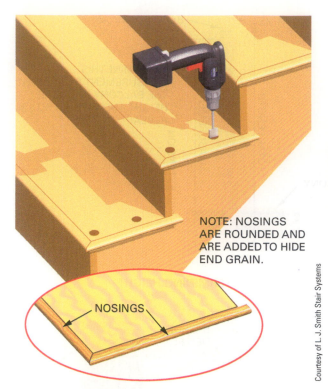

NOTE: NOSINGS ARE ROUNDED AND ARE ADDED TO HIDE END GRAIN.

NOSINGS

Courtesy of L. J. Smith Stair Systems

Figure 83–9 Holes are bored in the top of the treads at each baluster center point.

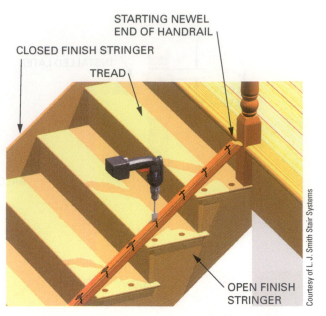

STARTING NEWEL END OF HANDRAIL

CLOSED FINISH STRINGER

TREAD

OPEN FINISH STRINGER

Courtesy of L. J. Smith Stair Systems

Figure 83–11 Rotate and invert rake handrail to bore holes in the bottom to receive round-top balusters.

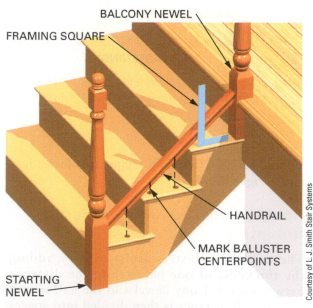

BALCONY NEWEL

FRAMING SQUARE

HANDRAIL

MARK BALUSTER CENTERPOINTS

STARTING NEWEL

Courtesy of L. J. Smith Stair Systems

Figure 83–10 The handrail is fitted between newel posts and then the baluster centers are transferred to it.

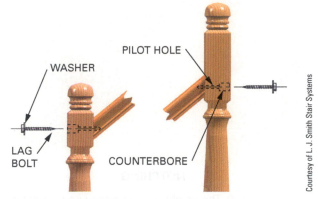

PILOT HOLE

WASHER

LAG BOLT

COUNTERBORE

Courtesy of L. J. Smith Stair Systems

Figure 83–12 The handrail is fastened to newel posts with lag bolts. The use of nails when constructing balustrades is discouraged.

INSTALLING AN OVER-THE-POST BALUSTRADE

The following procedures apply to an over-the-post balustrade running from floor to floor with an intermediate landing (see *Figure 81–2*). An over-the-post balustrade is more complicated. Handrail fittings are required to be joined to straight sections to construct a continuous handrail from start to end. The procedure consists of constructing the

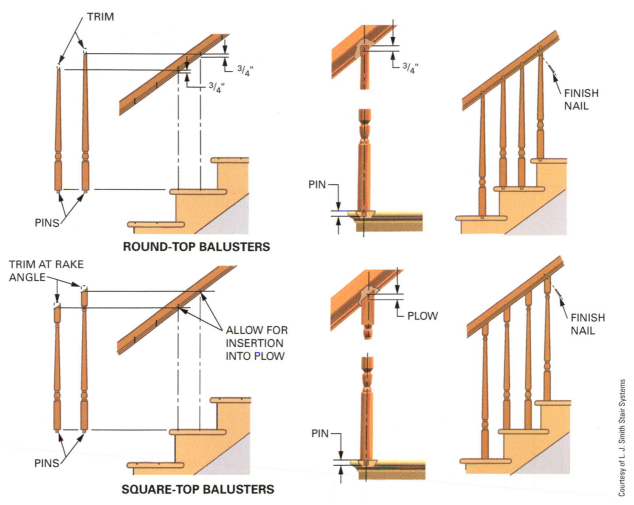

ROUND-TOP BALUSTERS

SQUARE-TOP BALUSTERS

TRIM

3/4"

3/4"

PINS

TRIM AT RAKE ANGLE

ALLOW FOR INSERTION INTO PLOW

PINS

3/4"

PIN

FINISH NAIL

PLOW

PIN

FINISH NAIL

Courtesy of L. J. Smith Stair Systems

Figure 83–13 Balusters are cut to length and installed between handrail and treads.

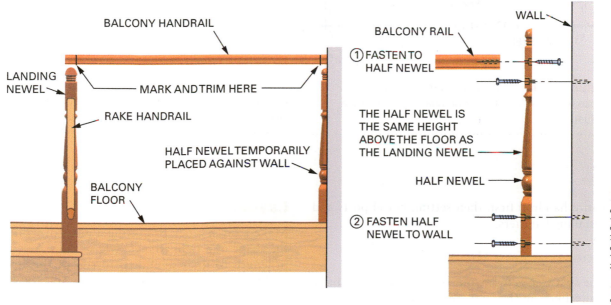

BALCONY HANDRAIL

LANDING NEWEL

MARK AND TRIM HERE

RAKE HANDRAIL

HALF NEWEL TEMPORARILY PLACED AGAINST WALL

BALCONY FLOOR

WALL

BALCONY RAIL

① FASTEN TO HALF NEWEL

THE HALF NEWEL IS THE SAME HEIGHT ABOVE THE FLOOR AS THE LANDING NEWEL

HALF NEWEL

② FASTEN HALF NEWEL TO WALL

Courtesy of L. J. Smith Stair Systems

Figure 83–14 The balcony rail is fitted between the landing newel and a half newel placed against the wall.

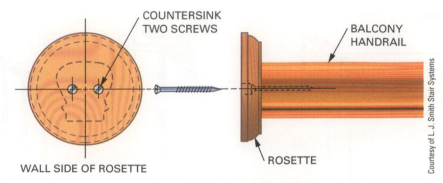

Figure 83–15 A rosette is sometimes used to end the balcony handrail instead of a half newel.

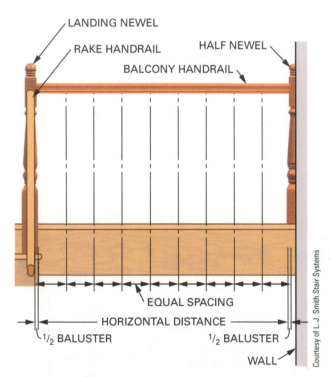

Figure 83–16 Balcony balusters are installed as close as possible to the same spacing as the rake balusters.

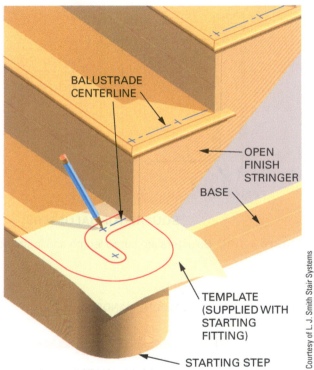

Figure 83–17 Baluster and newel post centers are laid out on a starting step with a template.

entire handrail first, then setting newel posts and installing balusters.

Constructing the Handrail

The first step is to lay out balustrade and baluster centerlines on the stair treads as described previously. If a starting step is used, lay out the baluster and starting newel centers using a template provided with the starting fitting (*Figure 83–17*).

Laying Out the Starting Fitting

Make a pitch block by cutting a piece of wood in the shape of a right triangle whose sides are equal in length to the rise and tread run of the stairs. Hold the cap of the starting fitting and the Run side of the pitch block on a flat surface. The rake edge of the pitch block should be against the bottom side of the fitting. Mark the fitting at the tangent point, where its curved surface touches the pitch block. A straight line, tangent to a curved line, touches the curved line at one point only.

Then turn the pitch block so its Rise side is on the flat surface. Mark the fitting along the rake edge of the pitch block. Place the fitting in a power miter box, supported by the pitch block. Cut it to the layout line (*Procedure 83–A*).

> ## CAUTION
>
> When cutting handrail fittings supported by a pitch block in a power miter box, clamp the pitch block and fitting securely to make sure they will not move when cut. If either one moves during cutting, a serious injury could result. Even if no injury occurs, the fitting would most likely be ruined.

Joining the Starting Fitting to the Handrail

The starting fitting is joined to a straight section of handrail by means of a special handrail bolt. The bolt has threads on one end designed for fastening into wood. The other end is threaded for a nut. Holes must be drilled in the end of both the fitting and the handrail in a manner that ensures their alignment when joined.

To mark the hole locations, make a template by cutting about a $\frac{1}{8}$-inch piece of handrail. Drill a $\frac{1}{16}$-inch hole centered on the template width and $\frac{15}{16}$ inch from the bottom side. Mark one side *rail* and the other side *fitting*. This will ensure that the template is facing in the right direction when making the layout. If the hole in the template is drilled

---

StepbyStep Procedures

Procedure 83–A Cutting the Starting Fitting

STEP 1 MAKE A PITCH BLOCK USING UNIT RISE AND UNIT RUN OF STAIRS.

STEP 2 MARK TANGENT POINT WHERE PITCH BLOCK TOUCHES RAIL.

STEP 3 ROTATE PITCH BLOCK AND MARK ANGLE ALONG RAKE EDGE OF PITCH BLOCK.

STEP 4 TRIM CONNECTING EASEMENT USING PITCH BLOCK AS A GUIDE. BE SURE TO SECURE RAIL BEFORE CUTTING.

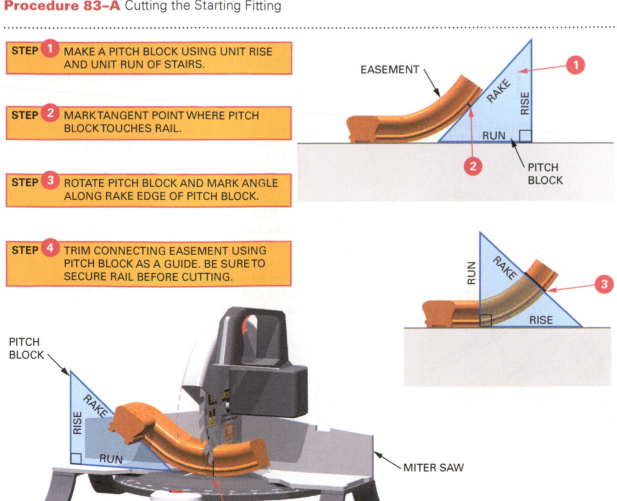

Courtesy of L. J. Smith Stair Systems

slightly off center and the template turned when making the layout, the handrail and fitting will not be in alignment when joined. Use the template to mark the location of the rail bolt on the end of the fitting and the handrail.

Drill all holes to the depth and diameter shown in *Figure 83–18*. Double nut the rail bolt and turn it into the fitting. Remove the nuts, place the handrail on the bolt. Install a washer and nut, and tighten to join the sections. Clamp the assembly to the tread nosings. The newel center points on the fitting, and the starting tread should be in a plumb line (*Figure 83–19*). Note the measurement in Detail C of Figure 83–19.

Installing the Landing Fitting

The second flight of stairs turns at the landing, so a right-hand, two-riser gooseneck fitting is used.

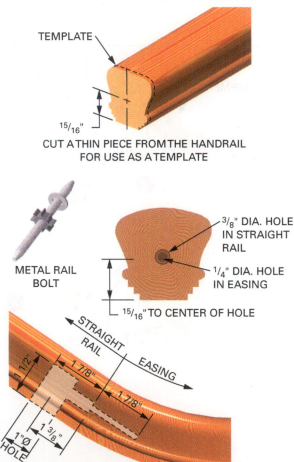

TEMPLATE

$15/16"$

CUT A THIN PIECE FROM THE HANDRAIL FOR USE AS A TEMPLATE

METAL RAIL BOLT

$3/8"$ DIA. HOLE IN STRAIGHT RAIL

$1/4"$ DIA. HOLE IN EASING

$15/16"$ TO CENTER OF HOLE

STRAIGHT RAIL

EASING

1 1/2"

1 7/8"

1 7/8"

1 3/8"

1" Ø HOLE

Figure 83-18 A template is made to mark the ends of handrails and fittings for joining with rail bolts. The ends are drilled to specific depths and diameters.

© Cengage Learning 2014

It is laid out and joined to the bottom end of the handrail of the second flight in a similar manner as the starting fitting (*Procedure 83–B*). The assembled fitting and rail are then clamped to the nosings of the treads in the second flight. Position the rail so that the *newel cap* of the fitting is in a plumb line with the baluster centerlines of both flights (*Figure 83–20*).

Joining First- and Second-Flight Handrails

In preparation for laying out the easement used to join the first- and second-flight handrails, tack a piece of plywood about 5 inches wide to the bottom side of the gooseneck fitting and the handrail of the first flight. These pieces are used to rest the connecting easement against when laying out the joint.

Clamp the handrails back on the treads. Make sure the newel cap centers are plumb with starting and landing newel post centers. Rest the connecting easement on the plywood blocks. Level its upper end. Mark the gooseneck in line with the upper end. Mark the lower end of the easement at a point tangent with the block under the handrail (*Procedure 83–C*). Note the measurement in Detail D of *Procedure 83–C*.

Make a square cut at the mark laid out on the lower end of the gooseneck. Join the gooseneck and easement with a rail bolt. Lay out the lower end of the easement with the pitch block. Cut it using the pitch block for support (*Procedure 83–D*).

Place the assembled landing fitting and handrail back on the stair nosings. Mark the handrail of the lower flight where it meets the end of the fitting. Cut the handrail square at the mark. Join it to the landing fitting with a rail bolt. Clamp the entire handrail assembly to the tread nosings with newel cap centers plumb with newel post centerlines (*Figure 83–21*).

Installing the Balcony Gooseneck Fitting

A one-riser balcony gooseneck fitting is used when balcony rails are 36 inches high. Two-riser fittings are used for rails 42 inches high. In this case, a one-riser fitting is used at the balcony. Laying out and fitting a two-riser fitting at the landing has been previously described.

Hold the fitting so the center of its cap is directly above the balcony newel post centerline. Hold it against the handrail of the upper flight.

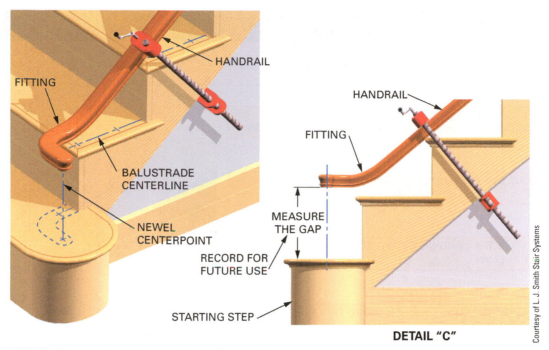

Courtesy of L. J. Smith Stair Systems

DETAIL "C"

Figure 83–19 The starting fitting and handrail assembly are clamped to the tread nosings of the first flight of stairs.

StepbyStep Procedures

Procedure 83–B Cutting the Landing Fitting to the Second-Flight Handrail

STEP 1 USING PITCH BOARD, MARK TANGENT POINT.

STEP 2 ROTATE PITCH BOARD TO MARK ANGLE ALONG RAKE EDGE.

STEP 3 TRIM CONNECTING EASEMENT. MAKE SURE HANDRAIL IS SECURE BEFORE CUTTING.

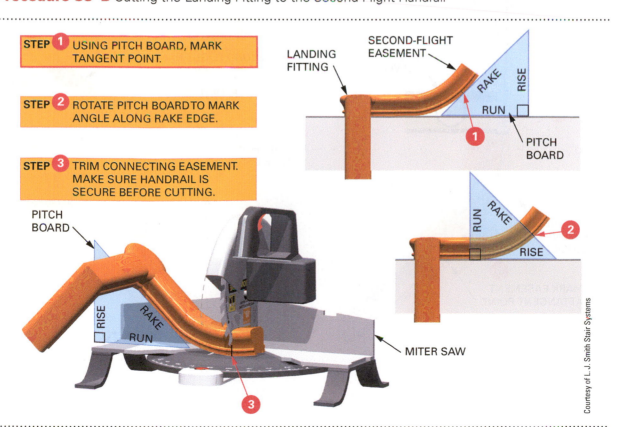

Courtesy of L. J. Smith Stair Systems

StepbyStep Procedures

Procedure 83–C Marking the Landing Fitting

STEP 1 PLACE EASEMENT BLOCKS UNDER RAILING AND AGAINST STRINGER AND ON TOP OF TREAD NOSINGS.

STEP 2 ADJUST EASEMENT SO TOP END IS LEVEL.

STEP 3 MARK THE LANDING FITTING AT THE TOP OF THE EASEMENT.

STEP 4 MARK THE LOWER TANGENT POINT OF EASEMENT AND THE BLOCK.

STEP 5 MEASURE AND RECORD THE HEIGHT OF THE LANDING FITTING.

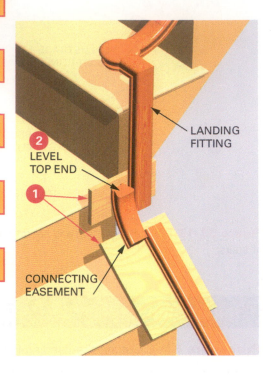

2 LEVEL TOP END

1 CONNECTING EASEMENT

LANDING FITTING

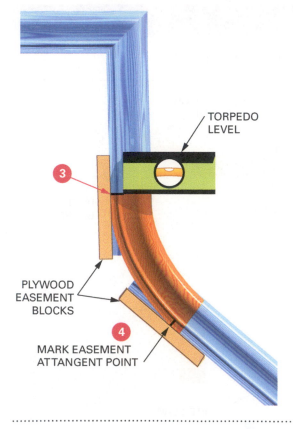

TORPEDO LEVEL

3

PLYWOOD EASEMENT BLOCKS

4 MARK EASEMENT AT TANGENT POINT

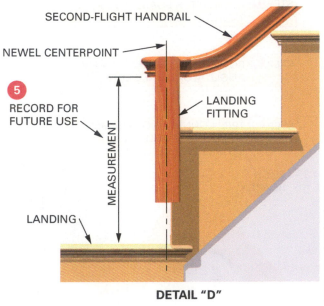

SECOND-FLIGHT HANDRAIL

NEWEL CENTERPOINT

5 RECORD FOR FUTURE USE

LANDING FITTING

MEASUREMENT

LANDING

DETAIL "D"

Courtesy of L. J. Smith Stair Systems

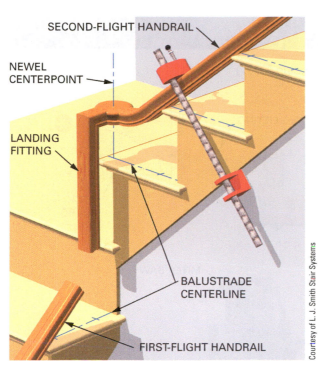

SECOND-FLIGHT HANDRAIL

NEWEL CENTERPOINT

LANDING FITTING

BALUSTRADE CENTERLINE

FIRST-FLIGHT HANDRAIL

Courtesy of L. J. Smith Stair Systems

Figure 83–20 The second-flight handrail is clamped to the tread nosings after the landing fitting is joined to it.

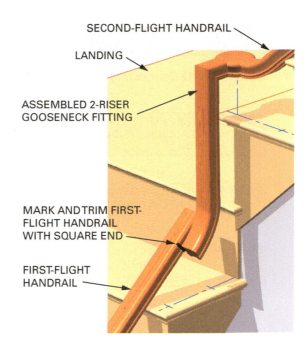

SECOND-FLIGHT HANDRAIL

LANDING

ASSEMBLED 2-RISER GOOSENECK FITTING

MARK AND TRIM FIRST-FLIGHT HANDRAIL WITH SQUARE END

FIRST-FLIGHT HANDRAIL

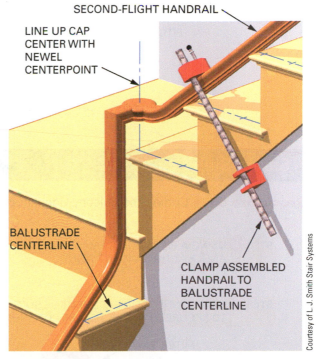

SECOND-FLIGHT HANDRAIL

LINE UP CAP CENTER WITH NEWEL CENTERPOINT

BALUSTRADE CENTERLINE

CLAMP ASSEMBLED HANDRAIL TO BALUSTRADE CENTERLINE

Courtesy of L. J. Smith Stair Systems

Figure 83–21 Lower and upper handrail assemblies are joined. They are then clamped to the tread nosings.

Mark it and the handrail at the point of tangent (*Figure 83–22*). Lay out and make the cut on the gooseneck with the use of a pitch block (*Procedure 83–E*). Cut the handrail square at the mark. Join the gooseneck fitting and the handrail with a rail bolt.

Clamp the entire rail assembly back on the nosing of the treads in line with the balustrade centerlines and the three newel post centers. Use a framing square to transfer the baluster centers from the treads to the side of the handrails. Remove the handrail assembly out of the way of newel post installation.

INSTALLING NEWEL POSTS

In this staircase, the starting newel is installed on a starting step. The height of the rake handrail is calculated, from the height of the starting newel, to make sure the handrail will conform to the height required by the building code.

From the height of the starting newel to be used, subtract the previously recorded distance between the starting fitting and the starting tread as shown in Detail C of *Figure 83–19*. Then, add the vertical thickness of the rake handrail shown in *Figure 83–4*, Detail A. The result is the rake handrail height (*Procedure 83–F*). If the height does not

conform to the building code, the starting newel post height must be changed.

Installing the Starting Newel Post

Before installing the starting step, measure the diameter of the dowel at the bottom of the starting newel post. At the centerpoint of the newel post,

StepbyStep Procedures

Procedure 83–D Connecting the Easement and Landing Fitting

STEP 1 SQUARE CUT THE LANDING FITTING AND ATTACH THE EASEMENT FITTING.

STEP 2 MARK THE ANGLE ALONG THE RAKE AT THE TANGENT POINT.

STEP 3 TRIM CONNECTING EASEMENT. MAKE SURE THE RAIL IS SECURE BEFORE CUTTING.

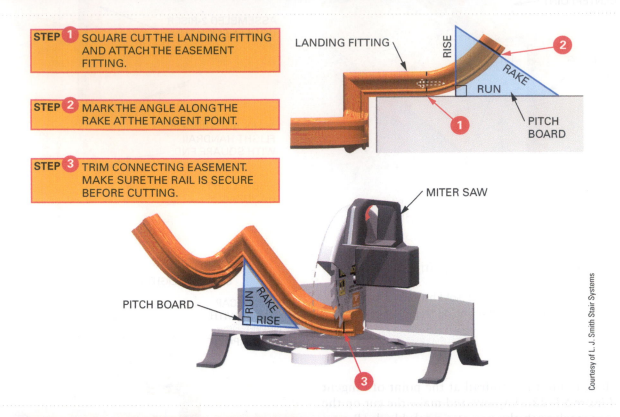

Courtesy of L. J. Smith Stair Systems

StepbyStep Procedures

Procedure 83–E Cutting the Balcony Gooseneck Fitting

STEP 1 USING A PITCH BOARD, MARK THE ANGLE AT THE TANGENT POINT.

STEP 2 TRIM CONNECTING EASEMENT. BE SURE TO SECURE HANDRAIL BEFORE CUTTING.

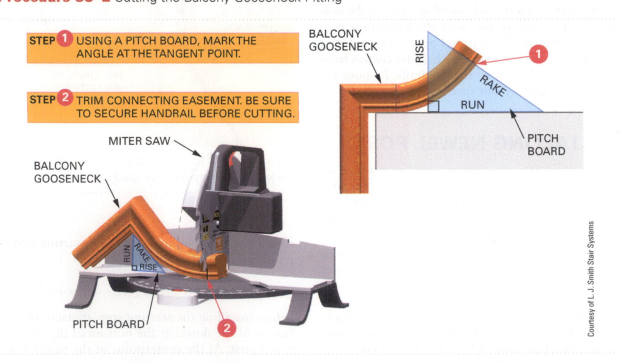

Courtesy of L. J. Smith Stair Systems

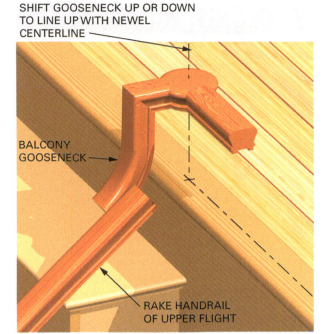

SHIFT GOOSENECK UP OR DOWN TO LINE UP WITH NEWEL CENTERLINE

BALCONY GOOSENECK

RAKE HANDRAIL OF UPPER FLIGHT

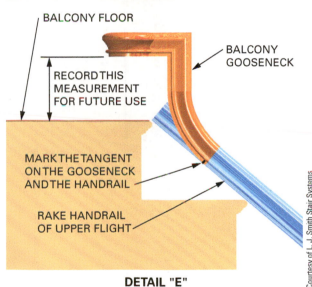

BALCONY FLOOR

RECORD THIS MEASUREMENT FOR FUTURE USE

MARK THE TANGENT ON THE GOOSENECK AND THE HANDRAIL

BALCONY GOOSENECK

RAKE HANDRAIL OF UPPER FLIGHT

DETAIL "E"

Courtesy of L. J. Smith Stair Systems

Figure 83–22 Method used to mark the balcony gooseneck fitting and handrail of the upper flight.

bore a hole for the dowel through the tread and floor. Install the post. Wedge it under the floor. Wedges are driven in a through mortise cut in the doweled end of the post.

An alternate method is used when there is no access under the floor. Bore holes only through the tread and upper riser block of the starting step. Cut the dowel to fit against the lower riser block so the newel post rests snugly on the tread. Fasten the end of the dowel with a lag screw through the lower

riser block (*Figure 83–23*). Fasten the assembled starting newel and starting step in position.

Installing the Landing Newel Post

The height of the landing newel above the landing is found by subtracting the previously recorded distance between the starting fitting and the starting tread (*Figure 83–19*, Detail C) from the height of the starting newel. Then add the distance between the landing fitting and the landing as previously recorded and shown in *Procedure 83–C*, Detail D. To this length, add the distance that the landing newel extends below the landing.

Notch the landing newel to fit over the landing and the first step of the upper flight. Fasten the post in position with lag bolts in counterbored holes (*Procedure 83–G*).

Installing the Balcony Newel Post

The height of the balcony handrail must be calculated before the height of the balcony newel can be determined. The height of the balcony handrail is found by subtracting the previously recorded distance between the starting fitting and the starting tread (*Figure 83–19*, Detail C) from the height of the starting newel. Then add the previously recorded distance between the balcony gooseneck fitting and the landing (*Figure 83–22*, Detail E). Then, add the thickness of the balcony handrail. The balcony handrail height must conform to the building code. If not, substitute a two-riser gooseneck fitting instead of a one-riser fitting.

The height of the balcony newel above the balcony floor is found by subtracting the handrail thickness from the handrail height. To this length, add the distance the post extends below the floor. Notch and install the post over the balcony riser in line with balustrade centerlines (*Figure 83–24*).

Installing Balusters

Bore holes for balusters in the tread and bottom edge of the handrail. No holes are bored in the handrail if square-top balusters are used. Install the handrail on the posts. Cut the balusters to length. Install them in the manner described previously for post-to-post balustrades.

Installing the Balcony Balustrade

Cut a half newel to the same height as the balcony newel extends above the floor. Install it against the wall on the balustrade centerline. Cut an opening cap so it fits on top of the half newel. Join the

Step**by**Step Procedures

Procedure 83–F Determining the Height of the Starting Newel Post

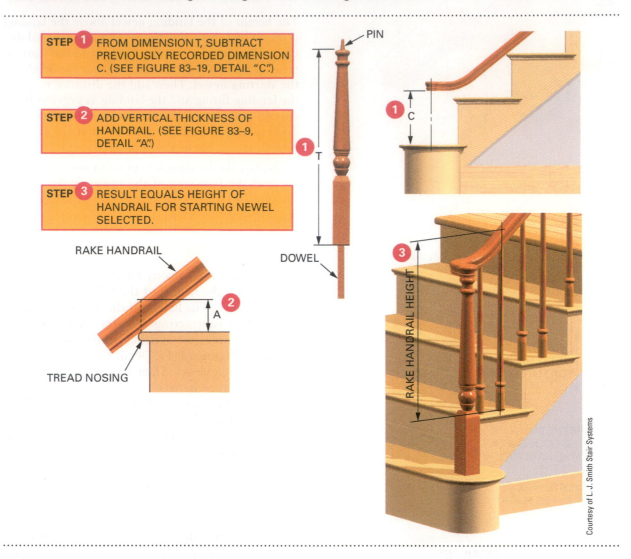

STEP 1 FROM DIMENSION T, SUBTRACT PREVIOUSLY RECORDED DIMENSION C. (SEE FIGURE 83–19, DETAIL "C".)

STEP 2 ADD VERTICAL THICKNESS OF HANDRAIL. (SEE FIGURE 83–9, DETAIL "A".)

STEP 3 RESULT EQUALS HEIGHT OF HANDRAIL FOR STARTING NEWEL SELECTED.

PIN

C

T

DOWEL

RAKE HANDRAIL

A

TREAD NOSING

RAKE HANDRAIL HEIGHT

Courtesy of L. J. Smith Stair Systems

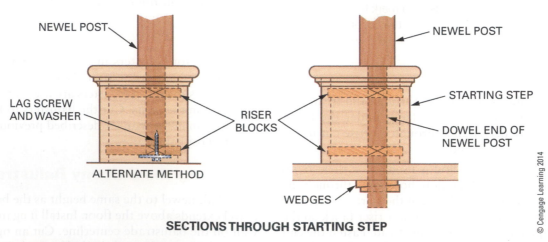

NEWEL POST

NEWEL POST

LAG SCREW AND WASHER

RISER BLOCKS

STARTING STEP

DOWEL END OF NEWEL POST

ALTERNATE METHOD

WEDGES

SECTIONS THROUGH STARTING STEP

© Cengage Learning 2014

Figure 83–23 Methods of installing the starting newel on a starting step.

StepbyStep Procedures

Procedure 83–G Determining the Height of the Landing Newel

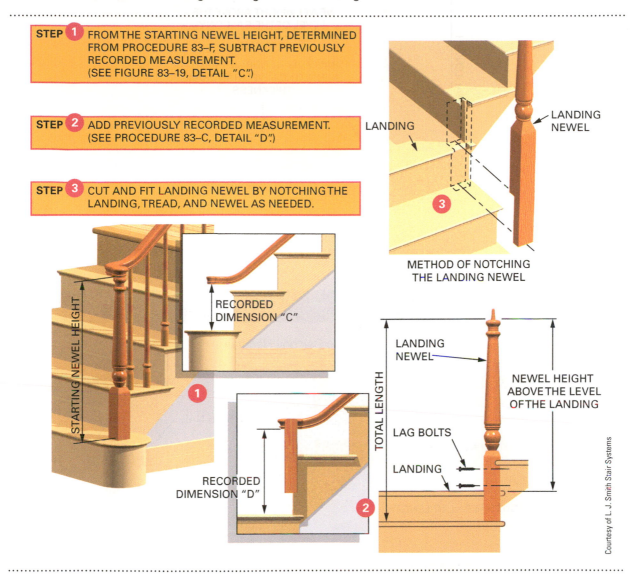

STEP 1 FROM THE STARTING NEWEL HEIGHT, DETERMINED FROM PROCEDURE 83–F, SUBTRACT PREVIOUSLY RECORDED MEASUREMENT. (SEE FIGURE 83–19, DETAIL "C".)

STEP 2 ADD PREVIOUSLY RECORDED MEASUREMENT. (SEE PROCEDURE 83–C, DETAIL "D".)

STEP 3 CUT AND FIT LANDING NEWEL BY NOTCHING THE LANDING, TREAD, AND NEWEL AS NEEDED.

LANDING

LANDING NEWEL

METHOD OF NOTCHING THE LANDING NEWEL

STARTING NEWEL HEIGHT

RECORDED DIMENSION "C"

RECORDED DIMENSION "D"

TOTAL LENGTH

LANDING NEWEL

LAG BOLTS

LANDING

NEWEL HEIGHT ABOVE THE LEVEL OF THE LANDING

Courtesy of L. J. Smith Stair Systems

cap to the end of the balcony handrail with a rail bolt. Place the cap on the half newel. Mark the length of the handrail at the balcony newel. Cut the handrail and join it, on one end, to the balcony newel post and, on the other end, to the half newel (*Procedure 83–H*).

A rosette may be used against the wall instead of the half newel and opening cap. The procedure for installing a rosette has been previously described (see *Figure 83–15*).

Installing Balcony Balusters

Balcony balusters are laid out and installed in the same manner as described previously for post-to-post balustrades. Balconies with a span of 10 feet or more should have intermediate balcony newels installed every 5 or 6 feet.

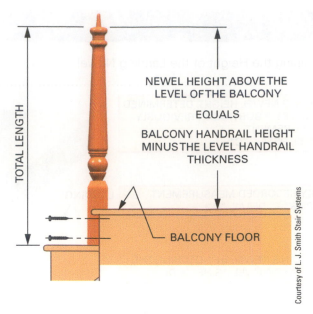

Figure 83–24 The height of the balcony newel is determined. The post is then fastened in place.

Courtesy of L. J. Smith Stair Systems

Step**by**Step Procedures

Procedure 83–H Installing a Half Newel and the Balcony Handrail

STEP 1 INSTALL HALF NEWEL TO WALL.

STEP 2 TRIM CAP TO CENTERLINE.

STEP 3 MARK AND TRIM RAIL TO LENGTH.

STEP 4 JOIN RAIL TO GOOSENECK AND HALF NEWEL.

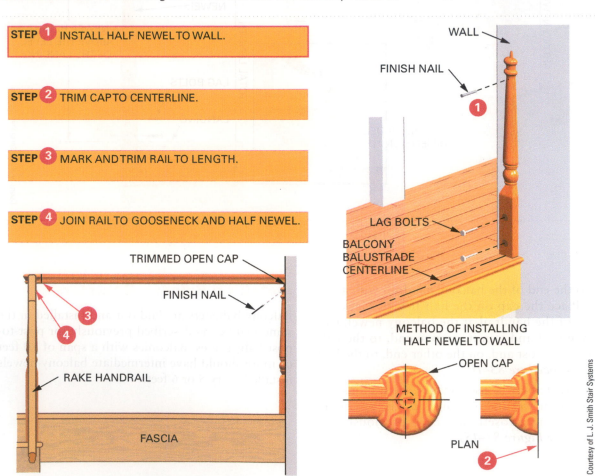

Courtesy of L. J. Smith Stair Systems

DECONSTRUCT THIS

Carefully study **Figure 83–25** and think about what is wrong and/or what is right. Consider all possibilities. What construction practice or method is different in your area of the country?

Figure 83–25 The photo in this figure shows a set of carpeted stairs with a handrail.

KEY TERMS

balustrade	goosenecks	over-the-post	shoe rail
closed	handrail bolt	pitch block	stair body
dowel	kneewall	post-to-post	starting step
fillets	landing treads	preacher	tread molding
finish stringers	newel posts	return nosings	turnouts
fittings	open	rosette	volutes

REVIEW QUESTIONS

Select the most appropriate answer.

1. The rounded outside edge of a tread that extends beyond the riser is called a

   a. housing.
   b. turnout.
   c. coving.
   d. nosing.

2. Finish boards between the stairway and the wall are called

   a. returns.
   b. balusters.
   c. stringers.
   d. casings.

3. Treads are sometimes rabbeted on their back edge to fit into

   a. risers.
   b. housed stringers.
   c. return nosings.
   d. newel posts.

4. An open stringer is

   a. housed to receive risers.
   b. mitered to receive risers.
   c. housed to receive treads.
   d. mitered to receive treads.

5. A volute is part of a

   a. tread.
   b. baluster.
   c. newel post.
   d. handrail.

6. The entire rail assembly on the open side of a stairway is called a

   a. baluster.
   b. balustrade.
   c. guardrail.
   d. finish stringer assembly.

7. In a framed staircase, the treads and risers are supported by

   a. stair carriages.
   b. housed stringers.
   c. each other.
   d. blocking.

8. One of the first things to do when trimming a staircase is

   a. check the rough framing for Rise and Run.
   b. block the staircase so no one can use it.
   c. straighten the stair carriages.
   d. install all the risers.

9. Return nosings project beyond the face of the stringer

   a. the same as tread nosing distance.
   b. $1/4$ inch less than tread nosing distance.
   c. $1/4$ inch more than tread nosing distance.
   d. at least $1\frac{1}{2}$ inches.

10. Newel posts are notched around the stairs so that their centerline lines up with the

    a. centerline of the stair carriage.
    b. centerline of the balustrade.
    c. outside face of the open stringer.
    d. outside face of the stair carriage.

UNIT 30
Cabinets and Countertops

Chapter 84 Cabinet Design and Installation
Chapter 85 Countertop and Cabinet Construction

©Shutterstock, Inc./Eimantas Buzas

Cabinets and countertops usually are purchased in preassembled units and may be installed by a carpenter. Manufactured cabinets are often used because of the great variety and shorter installation time than job-built cabinets. Cabinets can be custom-made to meet the specifications of almost any job, but they are usually made in a cabinet shop. Countertops, cabinet doors, and drawers may be customized in a wide variety of styles and sizes.

OBJECTIVES

After completing this unit, the student should be able to:

- state the sizes and describe the construction of typical base and wall kitchen cabinet units.
- plan, order, and install manufactured kitchen cabinets.
- construct and install a cabinet
- construct, laminate, and install a countertop.
- identify cabinet doors and drawers according to the type of construction and method of installation.
- identify overlay, lipped, and flush cabinet doors and proper drawer construction.
- apply cabinet hinges, pulls, and door catches.

SAFETY REMINDER

Kitchen cabinets are installed in large units that are cumbersome and often heavy to move. Watch that other wall and floor finishes are not marred when positioning the cabinets. Also, remember to lift with your legs and not your back.

CHAPTER 84

Cabinet Design and Installation

Manufactured kitchen and bath cabinets come in a wide variety of styles, materials, and finishes (*Figure 84-1*). The carpenter must be familiar with the various kinds, sizes, uses, and construction of the cabinets to know how to plan, order, and install them.

CABINET DESIGN

For commercial buildings, many kinds of specialty cabinets are manufactured. They are designed for specific uses in offices, hospitals, laboratories,

Figure 84-1 Kitchen cabinets are available in a wide variety of styles and sizes.

schools, libraries, and other buildings. Most cabinets used in residential construction are manufactured for the kitchen or bathroom. All cabinets, whether for commercial or residential use, consist of a case that is fitted with shelves, doors, and/or drawers. Cabinets are manufactured and installed in essentially the same way. Designs vary considerably with the manufacturer, but sizes are close to the same.

Kinds and Sizes

One method of cabinet construction utilizes a face frame. This frame provides openings for doors and drawers. Another method, called *European* or frameless, eliminates the face frame (*Figure 84-2*). Face-framed cabinets usually give a traditional look. Frameless cabinets are used when a contemporary appearance is desired.

The two basic kinds of kitchen cabinets are the wall unit and the base unit. The surface of the countertop is usually about 36 inches from the floor. Wall units are installed about 18 inches above the countertop. This distance is enough to accommodate such articles as coffeemakers, toasters, blenders, and mixers. Yet it keeps the top shelf within reach, not over 6 feet from the floor. The usual overall height of a kitchen cabinet installation is 7'-0" (*Figure 84-3*).

Wall Cabinets. Standard wall cabinets are 12 inches deep. They normally come in heights of 42,

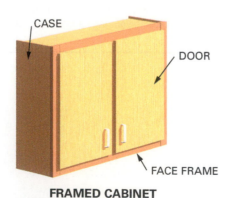

FRAMED CABINET

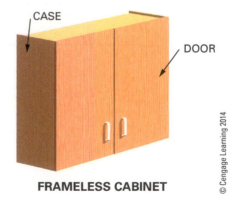

FRAMELESS CABINET

Figure 84-2 Two basic methods of cabinet construction.

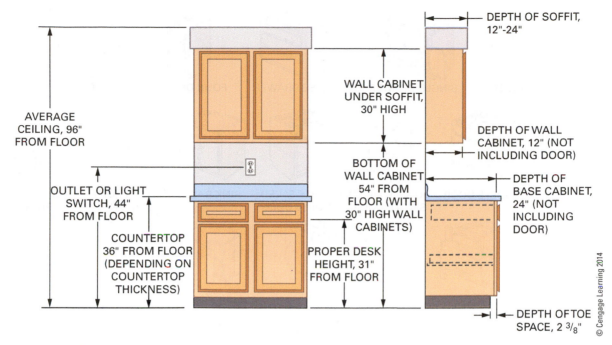

Figure 84–3 Common kitchen cabinet heights and dimensions.

30, 24, 18, 15, and 12 inches. The standard height is 30 inches. Shorter cabinets are used above sinks, refrigerators, and ranges. The 42-inch cabinets are for use in kitchens without soffits where more storage space is desired. A standard height wall unit usually contains two adjustable shelves.

Usual wall cabinet widths range from 9 to 48 inches in 3-inch increments. They come with single or double doors depending on their width.

Single-door cabinets can be hung so doors can swing in either direction.

Wall corner cabinets make access into corners easier. **Double-faced cabinets** have doors on both sides for use above island and peninsular bases. Some wall cabinets are made 24 inches deep for installation above refrigerators. A microwave oven case, with a 30-inch-wide shelf, is available (*Figure 84–4*).

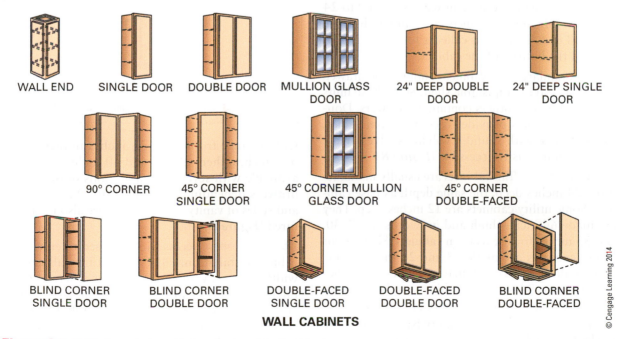

WALL CABINETS

Figure 84–4 Kinds and sizes of manufactured wall cabinets.

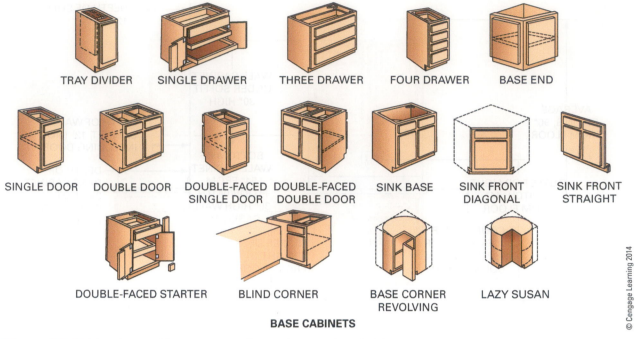

TRAY DIVIDER   SINGLE DRAWER   THREE DRAWER   FOUR DRAWER   BASE END

SINGLE DOOR   DOUBLE DOOR   DOUBLE-FACED SINGLE DOOR   DOUBLE-FACED DOUBLE DOOR   SINK BASE   SINK FRONT DIAGONAL   SINK FRONT STRAIGHT

DOUBLE-FACED STARTER   BLIND CORNER   BASE CORNER REVOLVING   LAZY SUSAN

BASE CABINETS

Figure 84–5 Most base cabinets are manufactured to match wall units.

Base Cabinets.

Most base cabinets are manufactured 34½ inches high and 24 inches deep. By adding the usual countertop thickness of 1½ inches, the work surface is at the standard height of 36 inches from the floor. Base cabinets come in widths to match wall cabinets. Single-door cabinets are manufactured in widths from 9 to 24 inches. Double-door cabinets come in widths from 27 to 48 inches. A recess called a toe space is provided at the bottom of the cabinet.

The standard base cabinet contains one drawer, one door, and an adjustable shelf. Some base units have no drawers; others contain all drawers. Double-faced cabinets provide access from both sides. Corner units, with round revolving shelves, make corner storage easily accessible (*Figure 84–5*).

Tall Cabinets.

Tall cabinets are usually manufactured 24 inches deep, the same depth as base cabinets. Some utility cabinets are 12 inches deep. They are made 66 inches high and in widths of 27, 30, and 33 inches for use as oven cabinets. Single-door utility cabinets are made 18 and 24 inches wide. Double-door pantry cabinets are made 36 inches wide (*Figure 84–6*). Wall cabinets with a 24-inch depth are usually installed above tall cabinets.

Vanity Cabinets.

Most vanity base cabinets are made 31½ inches high and 21 inches deep. Some are made in depths of 16 and 18 inches. Usual

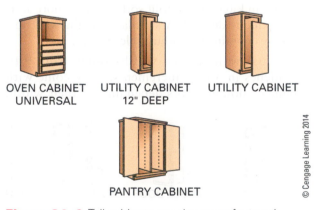

OVEN CABINET UNIVERSAL   UTILITY CABINET 12" DEEP   UTILITY CABINET

PANTRY CABINET

Figure 84–6 Tall cabinets are also manufactured.

widths range from 24 to 36 inches in increments of 3 inches, then 42, 48, and 60 inches. They are available with several combinations of doors and drawers depending on their width. Various sizes and styles of vanity wall cabinets are also manufactured (*Figure 84–7*).

Accessories.

Accessories are used to enhance a cabinet installation. Filler pieces fill small gaps in width between wall and base units when no combination of sizes can fill the existing space. They are cut to necessary widths on the job. Other accessories include cabinet end panels, face panels for dishwashers and refrigerators, open shelves for cabinet ends, and spice racks.

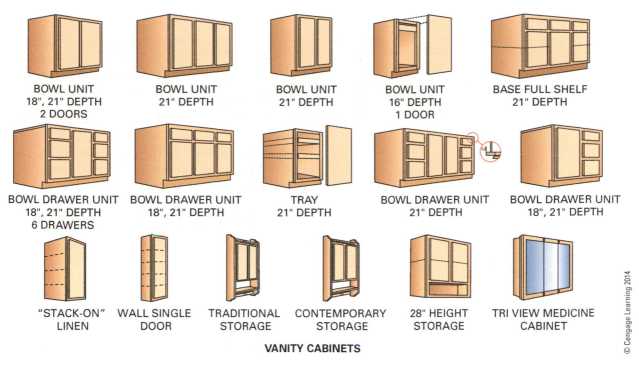

BOWL UNIT
18", 21" DEPTH
2 DOORS

BOWL UNIT
21" DEPTH

BOWL UNIT
21" DEPTH

BOWL UNIT
16" DEPTH
1 DOOR

BASE FULL SHELF
21" DEPTH

BOWL DRAWER UNIT
18", 21" DEPTH
6 DRAWERS

BOWL DRAWER UNIT
18", 21" DEPTH

TRAY
21" DEPTH

BOWL DRAWER UNIT
21" DEPTH

BOWL DRAWER UNIT
18", 21" DEPTH

"STACK-ON"
LINEN

WALL SINGLE
DOOR

TRADITIONAL
STORAGE

CONTEMPORARY
STORAGE

28" HEIGHT
STORAGE

TRI VIEW MEDICINE
CABINET

VANITY CABINETS

© Cengage Learning 2014

Figure 84–7 Vanity cabinets are made similar to kitchen cabinets, but differ in size.

CABINET MATERIALS

Cabinets are typically made from wood and wood products. A wide variety of wood species are available to satisfy the creative spirits of woodworkers and their customers. Oak, maple, cherry, birch, and hickory make up some of the more popular species. In general any wood species could be used; it is all a matter of taste. Some wood species are more workable than others, which has an effect on what species are most often chosen. *Figure 1–4* and *Figure 1–5* list a variety of hardwoods and softwoods with their characteristics.

Wood panels are also an important part of cabinet construction. They are used for most of the cabinet; sides, back, bottom, and sometimes the doors. Types of panels include hardwood and softwood plywood, particleboard, and fiberboards. They may be surfaced veneered with any of the different wood species. This lessens the expense of the cabinet because only the top surface is made from the more costly wood species. Chapter 4 gives more information on these products.

CABINET LAYOUT

The blueprints for a building contain plans, elevations, and details that show the cabinet layout. Architects may draw the layout. But they may not specify the size or the manufacturer's identification for each individual unit of the installation.

In residential construction, particularly in remodeling, no plans are usually available to show the cabinet arrangement. In addition to installation, it becomes the responsibility of the carpentry contractor to plan, lay out, and order the cabinets, in accordance with the customer's specifications.

The first step is to measure carefully and accurately the length of the walls on which the cabinets are to be installed. A plan is then drawn to scale. It must show the location of all appliances, sinks, windows, and other necessary items (*Figure 84–8*).

Next, draw elevations of the base cabinets, referring to the manufacturer's catalog for sizes. Always use the largest size cabinets available instead of two or three smaller ones. This reduces the cost and makes installation easier.

Match up the wall cabinets with the base cabinets, where feasible. If filler strips are necessary, place them between a wall and a cabinet or between cabinets in the corner. Identify each unit on the elevations with the manufacturer's identification (*Figure 84–9*). Make a list of the units in the layout. Order from the distributor.

Computer Layouts

Computer programs are available to help in laying out manufactured kitchen cabinets. When the required information is fed into the computer, a number of different layouts can be quickly made. When an acceptable layout is made, it can be printed with

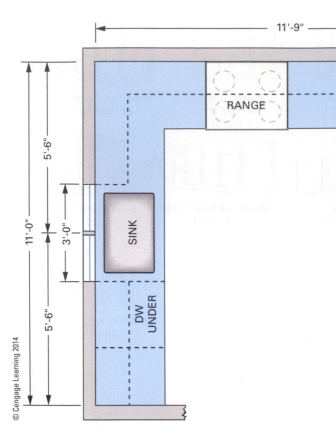

Figure 84–8 Typical plan of a kitchen cabinet layout showing location of walls, windows, and appliances.

each of the cabinets in the layout identified and priced. Most large kitchen cabinet distributors will supply computerized layouts on request.

CABINET INSTALLATION

Cabinets must be installed level and plumb even though floors are not always level and walls not always plumb. Level lines are first drawn on the wall for base and wall cabinets. To level base cabinets that set on an unlevel floor, either shim the cabinets from the high point of the floor or scribe and fit the cabinets to the floor from the lowest point on the floor. Shimming the base cabinets leaves a space that must be later covered by a molding. Scribing and fitting the cabinets to the floor eliminates the need for a molding. The method used depends on various conditions of the job. If shimming base cabinets, lay out the level lines on the wall from the highest point on the floor where cabinets are to be installed. If fitting cabinets to the floor, measure up from the low point.

Laying Out the Wall

Measure $34\frac{1}{2}$ inches up the wall. Draw a level line to indicate the tops of the base cabinets. Use the most accurate method of leveling available (described in Chapter 25, Leveling and Layout Tools). Another level line must be made on the wall 54 inches from the floor. The bottoms of the wall units are installed to this line. It is often faster to measure $19\frac{1}{2}$ inches up from the first level line and snap lines parallel to it than to level another line.

The next step is to mark the stud locations in a framed wall. (Cabinet mounting screws will be driven into the studs.) An electronic stud finder works well to locate framing. The other method is to lightly tap on and across a short distance of the wall with a hammer. Drive a finish nail in at the point where a solid sound is heard. Drive the nail where holes are later covered by a cabinet. If a stud is found, mark the location with a pencil. If no stud is found, try a little over to one side or the other.

Measure at 16-inch intervals in both directions from the first stud to locate other studs. Drive a finish nail to test for solid wood. Mark each stud location. If studs are not found at 16-inch centers, try 24-inch centers. At each stud location, draw

SINK WALL ELEVATION

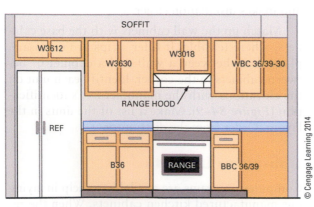

Figure 84–9 Elevations of the installation are sometimes drawn and the cabinets identified.

plumb lines on the wall. Mark the outlines of all cabinets on the wall to visualize and check the cabinet locations against the layout (*Figure 84–10*).

Installing Wall Units

Many installers prefer to install the soffit and wall cabinets first so the work does not have to be done leaning over the base units. The soffit may be framed using any of several methods (*Figure 84–11*). One method uses drywall to cover 2 × 2 framing. Another uses paneling or wood strips to cover

a 2 × 2 frame. In either case, the ceiling and wall drywall should be installed completely to the corner beforehand. This makes the house more airtight.

A cabinet lift may be used to hold the cabinets in position for fastening to the wall. If a lift is not available, the doors and shelves may be removed to make the cabinet lighter and easier to clamp together. If possible, screw a 1 × 3 strip of lumber so its top edge is on the level line for the bottom of the wall cabinets. This is used to support the wall units while they are being fastened. If it is not possible to screw to the wall, build a

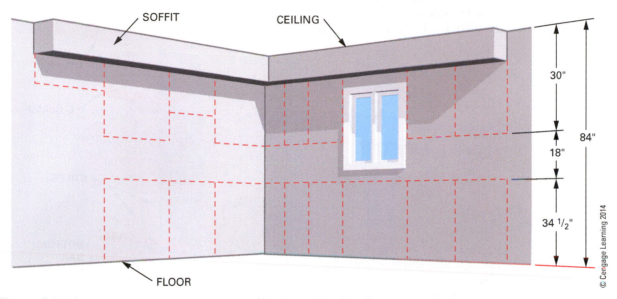

Figure 84–10 The wall is laid out with outlines for the cabinet locations.

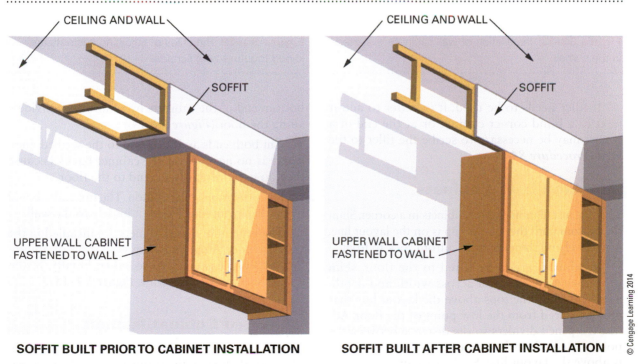

SOFFIT BUILT PRIOR TO CABINET INSTALLATION

SOFFIT BUILT AFTER CABINET INSTALLATION

Figure 84–11 Two methods of finishing a soffit.

stand on which to support the unit near the line of installation.

Start the installation of wall cabinets in a corner. On the wall, measure from the line representing the outside of the cabinet to the stud centers. Transfer the measurements to the cabinets. Drill shank holes for mounting screws through mounting rails usually installed at the top and bottom of the cabinet. Place the cabinet on the supporting strip or stand it so its bottom is on the level layout line. Wood and steel framing typically do not need predrilled screw holes. Concrete screws used in masonry walls do need to have predrilled holes. This is normally done with the cabinet held in place. Make sure holes are sufficiently deep to prevent

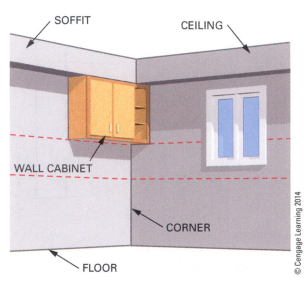

Figure 84–12 Installation of wall cabinets is started in the corner.

the screw from bottoming out in the hole. Fasten the cabinet in place with mounting screws of sufficient length to hold the cabinet securely. Do not fully tighten the screws (*Figure 84–12*).

The next cabinet is installed in the same manner. Align the adjoining stiles so their faces are flush with each other. Clamp them together with C-clamps. Screw the stiles tightly together (*Figure 84–13*). Continue this procedure around the room. Tighten all mounting screws.

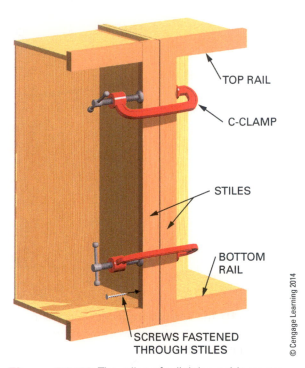

Figure 84–13 The stiles of adjoining cabinets are joined together with screws.

If filler needs to be used, it is better to add it next to a blind corner cabinet or at the end of a run. It may be necessary to scribe the filler to the wall (*Procedure 84–A*).

Installing Base Cabinets

Start the installation of base cabinets in a corner. Shim the bottom until the cabinet top is on the layout line. Then level and shim the cabinet from back to front.

If cabinets are to be fitted to the floor, shim until their tops are level across width and depth. This will bring the tops above the layout line that was measured from the low point of the floor. Adjust the pencil dividers so the distance between the points is equal to the amount the top of the unit is above the layout line. Scribe this amount on the

bottom end of the cabinets by running the dividers along the floor (*Figure 84–14*).

Cut both ends and toeboard to the scribed lines. There is no need to cut the cabinet backs because they do not, ordinarily, extend to the floor.

Place the cabinet in position. The top ends should be on the layout line. Fasten it loosely to the wall.

The remaining base cabinets are installed in the same manner. Align and clamp the stiles of adjoining cabinets. Fasten them together. Finally, fasten all units securely to the wall (*Figure 84–15*).

Layout of Cabinet Islands

Islands are cabinet units not attached to a wall. They stand alone in the room and may be accessed

StepbyStep Procedures

Procedure 84–A A Scribing a Filler Piece Using a Block

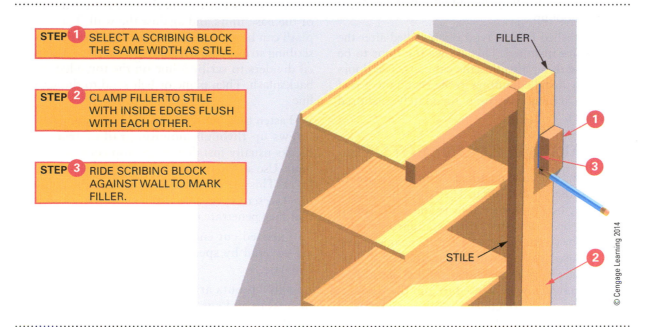

STEP 1 SELECT A SCRIBING BLOCK THE SAME WIDTH AS STILE.

STEP 2 CLAMP FILLER TO STILE WITH INSIDE EDGES FLUSH WITH EACH OTHER.

STEP 3 RIDE SCRIBING BLOCK AGAINST WALL TO MARK FILLER.

FILLER

STILE

© Cengage Learning 2014

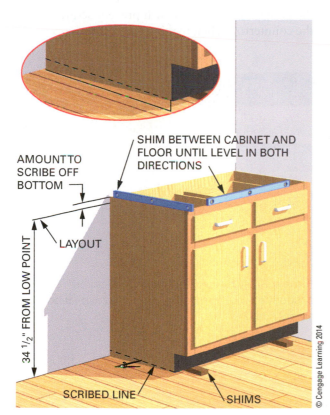

AMOUNT TO SCRIBE OFF BOTTOM

SHIM BETWEEN CABINET AND FLOOR UNTIL LEVEL IN BOTH DIRECTIONS

LAYOUT

34 ½" FROM LOW POINT

SCRIBED LINE

SHIMS

© Cengage Learning 2014

Figure 84–14 Method of scribing base cabinets to the floor.

© Cengage Learning 2014

Figure 84–15 Base cabinets are secured with screws to wall studs.

from all sides. Some are backed up and attached to short walls; others are free standing with doors and drawers on both long sides. This allows for a wider countertop as a possibility (*Figure 84–16*).

The walking space around the island should be at least 36 inches. This allows a minimum of space to access all parts of the island and other cabinets. Increasing this space to 42 or 48 inches provides more space for multiple occupants of the kitchen.

Islands with a sink must be plumbed for water and drains. Typically the plumbing rough-in is done before the installation of cabinets. The carpenter must then position the island cabinets over the pipes in such a way the plumber may later perform the hookup.

The finished floor is typically installed after the cabinets are installed. This allows the floor to be replaced without affecting the cabinets. Tile floors are usually installed butting to the island toe board after the cabinets have been adequately anchored to the floor.

INSTALLING COUNTERTOPS

Custom and pre-manufactured *countertops* are installed in the same fashion. They both need to be cut and scribed to fit the cabinets and wall. Manufactured tops come in various standard lengths. They can be cut to fit any installation against walls. They are also available with one end precut at a 45-degree angle for joining with a similar one at corners. Special hardware is used to join the sections. The countertops are covered with a thin, tough *high-pressure plastic laminate*. This is generally known as mica. It is available in many colors and patterns. The countertops are called postformed countertops. This term comes from the method of forming the mica to the rounded edges and corners

of the countertop (*Figure 84–17*). Postforming is bending the mica with heat to a radius of $\frac{3}{4}$ inch or less. This can be done only with special equipment.

After the base units are fastened in position, the countertop is cut to length. It is fastened on top of the base units and against the wall. The backsplash can be scribed, limited by the thickness of its scribing strip, to an irregular wall surface. Use pencil dividers to scribe a line on the top edge of the backsplash. Then plane or belt sand to the scribed line.

Fasten the countertop to the base cabinets with screws up through mounting rail or triangular blocks usually installed in the top corners of base units. Use a stop on the drill bit when drilling pilot holes. This prevents drilling through the countertop. Use screws of sufficient length, but not so long that they penetrate the countertop.

Exposed cut ends of postformed countertops are covered by specially shaped pieces of plastic laminate.

Sink cutouts are made by carefully outlining the cutout and cutting with a saber saw. The cutout pattern usually comes with the sink. Use a fine-tooth blade to prevent chipping out the face of the mica beyond the sink. Some duct tape applied to the base of the saber saw will prevent scratching of the countertop when making the cutout.

Figure 84–16 Islands are free standing cabinets.

Figure 84–17 A section of a manufactured postformed countertop. The edges and interior corner are rounded.

CHAPTER 85

Countertop and Cabinet Construction

Cabinets are usually purchased from cabinet shops and installed by carpenters on the job. Occasionally they are made by the carpenter who installs them. Nevertheless, all carpenters should understand how cabinets are constructed.

CABINET CONSTRUCTION

Cabinets are constructed by precutting, machining, and shaping pieces from dimensions on a set of prints. The process begins with the sides, bottom, and the back. They are then joined and assembled as a unit (*Figure 85–1*). The height of kitchen base cabinet is normally 34 ½ inches. This allows for the countertop to be 1½ inch thick, making the overall height 36 inches. Wall units may be 30 or 36 inches high depending on customer's desire. The depth of these cabinets is found on the set of prints.

The sides and bottom pieces are usually made from ¾-inch plywood. The quality of material for the more visible side pieces should be the highest of the material being used because of their visual prominence. Pieces are joined together in a dado joint (*Figure 85–2*). This joint offers superior strength to the cabinet. The bottom is fastened to the sides with pocket screws from the bottom side or with corner glue blocks. The concern is to hide the fasteners from normal view and still have a strong joint.

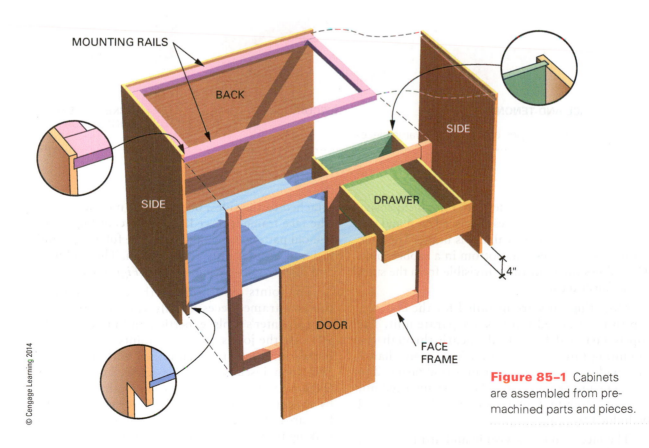

MOUNTING RAILS

BACK

SIDE

SIDE

DRAWER

4"

DOOR

FACE FRAME

Figure 85–1 Cabinets are assembled from premachined parts and pieces.

© Cengage Learning 2014

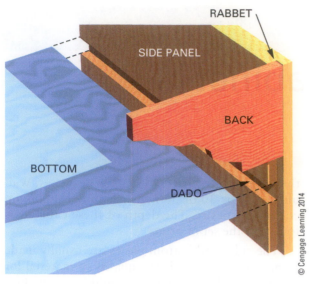

Figure 85-2 Cabinet sides are rabbeted and dadoed to accept bottom and back panels.

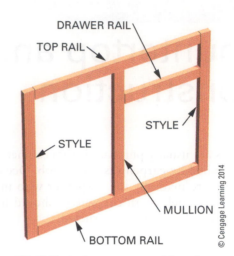

Figure 85-3 Typical arrangment of face frame pieces.

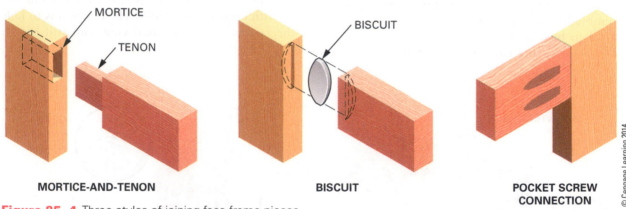

MORTICE-AND-TENON   **BISCUIT**   **POCKET SCREW CONNECTION**

Figure 85-4 Three styles of joining face frame pieces.

The back may be made from thin, inexpensive plywood because little stress is placed on it and it is difficult to see after the cabinet is in use. The back is fitted to the sides and bottom in a rabbet joint. This allows the joint to be invisible from the sides of the finished cabinet.

Mounting rails are installed for the countertop to be attached later as a separate unit. The top is fastened from underneath through the mounting rails. These are nothing more than 2 inch wide strips of wood or plywood fastened to cabinet's upper perimeter. These strips need not be made from high quality material as they will be hidden from view.

The face frame is a visible and attractive part of a cabinet. It is often assembled as a unit, then attached to the sides, bottom, and mounting rail. The face frame is pieced together according to an age-old pattern, where the styles are full length and top and bottom rails butt to them. The mullions and drawer rails are joined later (*Figure 85-3*).

The joints used to make the connections in the face frame pieces usually vary according to the carpenter's tools, expertise, and preference. In any case the joints are glued. Joints appear to be butted but they are often joined with a mortise-and-tenon, biscuits, or pocket-screwed connections (*Figure 85-4*). This gives more strength to the joint by making a pining effect and increasing the surface area of the glued wood. It is important when making these joints to keep the face piece surfaces aligned and flat. They are sanded after assembly to make the surfaces smooth. Frames are then attached to the mounting rails using glue and pocket screws from the back side.

Mounting rails are secured in a dado joint with an angled screw into the cabinet side panels. Rails are butted and attached to the back panel and face frame top rail with pocket screws. This makes a secure mounting rail connection that will resist pull-out when the countertop is later attached.

COUNTERTOP CONSTRUCTION

A laminated countertop is made from two layers of ¾ or ⅝-inch panels, either plywood, particle board, or MDF (medium density fiberboard). The top layer is full width and length of the countertop. The bottom layer is only few inches wide and along the perimeter. This gives the illusion the countertop is 1 ½" thick. Splices in the top pieces are reinforced with a bottom layer pieces glued and screwed. The overall width of the countertop should be about 25 inches to allow for an overhang of the front of surface of the cabinet.

Fitting the Countertop

After the top panel has been cut to size and the bottom layer attached it should be checked to see if it fits to the wall. If the wall has an irregular surface it may need to be scribed to the top to the wall. Place the top panel assembly on the base cabinets, against the wall. Its outside edge should overhang the face

frame the same amount along the entire length. Open the pencil dividers or scribers to the amount of overhang. Scribe the back edge of the countertop to the wall. Cut the countertop to the scribed line. Place it back on top of the base cabinets. The ends should be flush with the ends of the base cabinets. The front edge should be flush with the face of the face frame (*Figure 85–5*). Install a 1 × 2 on the front edge and at the ends if an end overhang is desired. Keep the top edge flush with the top side of the countertop.

Applying the Backsplash

If a backsplash is used, rip a 4-inch-wide length of ¾-inch stock the same length as the countertop. Use lumber for the backsplash, if lengths over 8 feet are required, to eliminate joints. Fasten the backsplash on top of and flush with the back edge of the countertop by driving screws up through the countertop and into the bottom edge of the backsplash (*Figure 85–6*). In corners, fasten the ends of the backsplash together with screws. Backsplash should be installed after the laminate is attached.

LAMINATING A COUNTERTOP

Countertops may be covered with plastic laminate. Before laminating a countertop, make sure all surfaces are flush. Check for protruding nailheads. Fill in all holes and open joints. Lightly hand or power sand the entire surface, making sure joints are sanded flush.

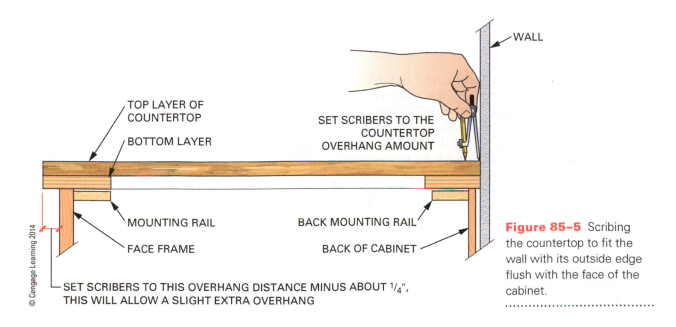

TOP LAYER OF COUNTERTOP

BOTTOM LAYER

MOUNTING RAIL

FACE FRAME

SET SCRIBERS TO THIS OVERHANG DISTANCE MINUS ABOUT ¼", THIS WILL ALLOW A SLIGHT EXTRA OVERHANG

SET SCRIBERS TO THE COUNTERTOP OVERHANG AMOUNT

WALL

BACK MOUNTING RAIL

BACK OF CABINET

Figure 85–5 Scribing the countertop to fit the wall with its outside edge flush with the face of the cabinet.

© Cengage Learning 2014

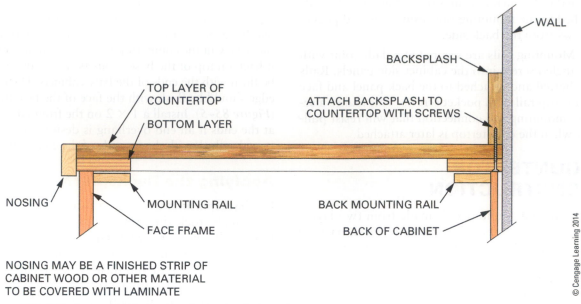

NOSING MAY BE A FINISHED STRIP OF
CABINET WOOD OR OTHER MATERIAL
TO BE COVERED WITH LAMINATE

Figure 85–6 Technique for fastening the backsplash to the countertop.

Laminate Trimming Tools and Methods

Pieces of laminate are first cut to a **rough size**, about ¼ to ½ inch wider and longer than the surface to be covered. A strip is then cemented to the edge of the countertop. Its edges are flush trimmed even with the top and bottom surfaces. Laminate is then cemented to the top surface, overhanging the edge strip. The overhang is then bevel trimmed even with the laminated edge. A laminate trimmer or a small router fitted with laminate trimming bits is used for rough cutting and flush and bevel trimming of the laminate (*Figure 85–7*).

Cutting Laminate to Rough Sizes

Sheets of laminate are large, thin, and flexible. This makes them difficult to cut on a table saw. One method of cutting laminates to rough sizes is by clamping a straightedge to the sheet. Cut it by guiding a laminate trimmer with a flush-trimming bit along the straightedge (*Figure 85–8*). It is easier to run the trimmer across the sheet than to run the sheet across the table saw. Also, the router bit leaves a smooth, clean-cut edge. Use a solid carbide trimming bit, which is smaller in diameter than one with ball bearings. It makes a narrower cut. It is easier to control and creates less waste. With this method, cut all the pieces of laminate needed to a rough width and length. Cut the narrow edge strips from the sheet first.

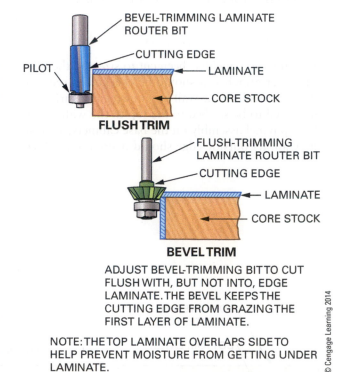

FLUSH TRIM

BEVEL TRIM

ADJUST BEVEL-TRIMMING BIT TO CUT FLUSH WITH, BUT NOT INTO, EDGE LAMINATE. THE BEVEL KEEPS THE CUTTING EDGE FROM GRAZING THE FIRST LAYER OF LAMINATE.

NOTE: THE TOP LAMINATE OVERLAPS SIDE TO HELP PREVENT MOISTURE FROM GETTING UNDER LAMINATE.

Figure 85–7 The laminate trimmer is used with flush and bevel bits to trim overhanging edges of laminate.

Another method for cutting laminate to rough size is to use a carbide-tipped hand cutter (*Figure 85–9*). It scores the laminate sufficiently in three passes to be able to break the piece away. Make

On the Job

Clamp the laminate to a straightedge. Cut rough sizes with a laminate trimmer.

Figure 85-8 Technique for using a laminate trimmer to cut laminate to size.

the cut on the face side and then bend the piece up to create the cleanest break. When used with a straightedge, the cutter is fast and effective.

Using Contact Cement

Contact cement is used for bonding plastic laminates and other thin, flexible material to surfaces. A coat of cement is applied to the back side of the laminate and to the countertop surface. The cement must be dry before the laminate is bonded to the core. The bond is made on contact without the need to use clamps. A contact cement bond may fail for several reasons:

- Not enough cement is applied. If the material is porous, like the edge of particleboard or plywood, a second coat is required after the first coat dries. When enough cement has been applied, a glossy film appears over the entire surface when dry.

- Too little time is allowed for the cement to dry. Both surfaces must be dry before contact is made. To test for dryness, lightly press your finger on the surface. Although it may feel sticky, the cement is dry if no cement remains on the finger.

- The cement is allowed to dry too long. If contact cement dries too long (more than about two hours, depending on the humidity), it will not bond properly. To correct this condition, merely apply another coat of cement and let it dry.

- The surface is not rolled out or tapped after the bond is made. Pressure must be applied to the entire surface using a 3-inch J-roller or by tapping with a hammer on a small block of wood.

CAUTION

Some contact cements are flammable. Apply only in a well-ventilated area around no open flame. Avoid inhaling the fumes.

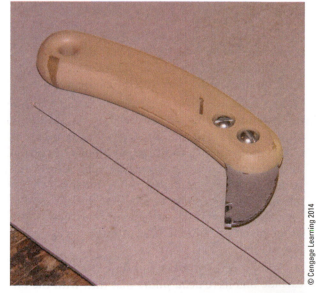

Figure 85-9 A hand cutter is often used to cut laminate to a rough size.

Figure 85–10 Applying laminate to the edge of the countertop.

Figure 85–11 Flush trimming the countertop edge laminate.

Laminating the Countertop Edges

Remove the backsplash from the countertop. Apply coats of cement to the countertop edges and the back of the edge laminate with a narrow brush or small paint roller. After the cement is dry, apply the laminate to the front edge of the countertop (*Figure 85–10*). Position it so the bottom edge, top edge, and ends overhang. A permanent bond is made when the two surfaces make contact. A mistake in positioning means removing the bonded piece—a time-consuming, frustrating, and difficult job. Roll out or tap the surface.

Apply the laminate to the ends in the same manner as to the front edge piece. Make sure that the square ends butt up firmly against the back side of the overhanging ends of the front edge piece to make a tight joint.

Trimming Laminated Edges

The overhanging ends of the edge laminate must be trimmed before the top and bottom edges. If the laminate has been applied to the ends, a bevel trimming bit must be used to trim the overhanging ends.

Using a Bevel-Trimming Bit.   When using a bevel-trimming bit, the router base is gradually adjusted to expose the bit so that the laminate is trimmed flush with the first piece but not cutting into it. The bevel of the cutting edge allows the laminate to be trimmed without cutting into the adjacent piece (see Figure 85–7). A flush-trimming bit cannot be used when the pilot rides against another piece of laminate because the cutting edge may damage it.

Ball bearing trimming bits have *live pilots*. Solid carbide bits have *dead pilots* that turn with the bit. When using a trimming bit with a dead pilot, the laminate must be lubricated where the pilot will ride. Rub a short piece of white candle or some solid vegetable shortening on the laminate to prevent marring the laminate with the bit.

Using the bevel-trimming bit, trim the overhanging ends of the edge laminate. Then, using the flush-trimming bit, trim off the bottom and top edges of both front and end edge pieces (*Figure 85–11*). To save the time required to change and adjust trimming bits, some installers use two laminate trimmers, one with a flush bit and the other with a bevel bit.

Use a belt sander or a file to smooth the top edge flush with the surface. Sand or file *flat* on the countertop core so a sharp square edge is made. This ensures a tight joint with the countertop laminate. Sand or file *toward* the core to prevent chipping the laminate. Smooth the bottom edge. Ease the sharp outside corner with a sanding block.

Laminating the Countertop Surface

Apply contact bond cement to the countertop and the back side of the laminate. Let dry. To position large pieces of countertop laminate, first place thin strips of wood or metal venetian blind slats about a foot apart on the surface. Lay the laminate to be bonded on the strips or slats. Then position the laminate correctly (*Figure 85–12*).

Make contact on one end. Gradually remove the slats one by one until all are removed. The laminate should then be positioned correctly with no costly errors. Roll out the laminate (*Figure 85–13*). Trim the overhanging back edge with a flush-trimming bit. Trim the ends and front edge with a bevel-trimming bit (*Figure 85–14*). Use a flat file to smooth the trimmed edge. Slightly ease the sharp corner.

Figure 85–12 Position the laminate on the countertop using strips before allowing cemented surfaces to contact each other.

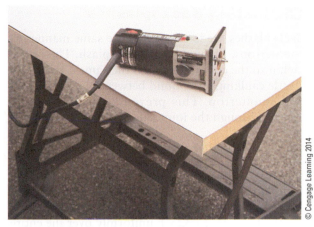

Figure 85–14 The outside edge of the countertop laminate is bevel trimmed flush to the edge laminate.

Figure 85–13 Rolling out the laminate with a J-roller is required to ensure a proper bond.

LAMINATING A COUNTERTOP WITH TWO OR MORE PIECES

When the countertop is laminated with two or more lengths, tight joints must be made between them. Tight joints can be made by clamping the two pieces of laminate in a straight line on some strips of $\frac{3}{4}$-inch stock. Butt the ends together or leave a space less than $\frac{1}{4}$ inch between them.

Using one of the strips as a guide, run the laminate trimmer, with a flush-trimming bit installed, through the joint. Keep the pilot of the bit against the straightedge. Cut the ends of both pieces at the same time to ensure making a tight joint (*Figure 85–15*). Bond the sheets as previously described. Apply seam-filling compound, specially made for laminates, to make a practically invisible joint. Wipe off excess compound with the recommended solvent.

A GUIDE STRIP FOR ROUTER MUST BE INSTALLED ON OPERATOR'S RIGHT WHEN PULLING ROUTER THROUGH CUT.

BOTH PIECES ARE HELD SECURELY AND CUT AT THE SAME TIME.

SUPPORTING STRIPS

On the Job

Tight joints are required between the ends of two lengths of laminate.

Figure 85–15 Making a tight laminate butt seam.

Laminating Backsplashes

Backsplashes are laminated in the same manner as countertops. Laminate the backsplash. Then reattach it to the countertop with the same screws. Use a little caulking compound between the backsplash and countertop. This prevents any water from seeping through the joint (*Figure 85–16*).

Laminating Rounded Corners

If the edge of a countertop has a rounded corner, the laminate can be bent. Strips of laminate can be cold bent to a minimum radius of about 6 inches. Heating the laminate to 325°F uniformly over the entire bend will facilitate bending to a minimum radius of about 2½ inches. Heat the laminate carefully with a *heat gun*. Bend it until the desired radius is obtained (*Figure 85–17*). Experimentation may be necessary until success in bending is achieved.

CAUTION

Keep fingers away from the heated area of the laminate. Remember that the laminate retains heat for some time.

Figure 85–16 Apply the laminate to the backsplash and then fasten it to the laminated countertop.

Figure 85–17 A heat gun makes laminate bend easily.

KINDS OF DOORS

Cabinet doors are classified by their construction and also by the method of installation. Sliding doors are occasionally installed, but most cabinets are fitted with hinged doors that swing.

Hinged cabinet doors are classified as overlay, lipped, and flush, based on the method of installation (*Figure 85–18*). The overlay method of hanging cabinet doors is the most widely used.

Overlay Doors

The **overlay door** laps over the opening, usually ³⁄₈ inch on all sides. However, it may overlay any amount. In many cases, it may cover the entire face frame. The overlay door is easy to install. It does not require fitting in the opening, and the face frame of the cabinet acts as a stop for the door. *European-style* cabinets omit the face frame. Doors completely overlay the front edges of the cabinet (*Figure 85–19*).

Lipped Doors

The **lipped door** has rabbeted edges that overlap the opening by about ³⁄₈ inch on all sides. Usually the ends and edges are rounded over to give a more pleasing appearance. Lipped doors and drawers are easy to install. No fitting is required, and the rabbeted edges stop against the face frame of the cabinet. However, a little more time is required to shape the rabbeted edges.

Flush Type

Flush-type doors fit into and flush with the face of the opening. They are a little more difficult to hang because they must be fitted in the opening. A fine joint,

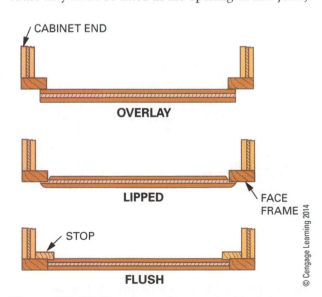

Figure 85–18 Plan views of types of cabinet doors.

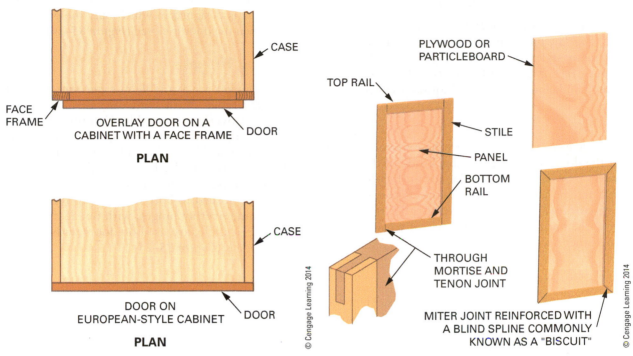

FACE FRAME

CASE

OVERLAY DOOR ON A CABINET WITH A FACE FRAME

DOOR

PLAN

CASE

DOOR ON EUROPEAN-STYLE CABINET

DOOR

PLAN

© Cengage Learning 2014

PLYWOOD OR PARTICLEBOARD

TOP RAIL

STILE

PANEL

BOTTOM RAIL

THROUGH MORTISE AND TENON JOINT

MITER JOINT REINFORCED WITH A BLIND SPLINE COMMONLY KNOWN AS A "BISCUIT"

© Cengage Learning 2014

Figure 85–19 Overlay doors lap the face frame by varying amounts.

Figure 85–21 Panel doors of simple design can be made on the job.

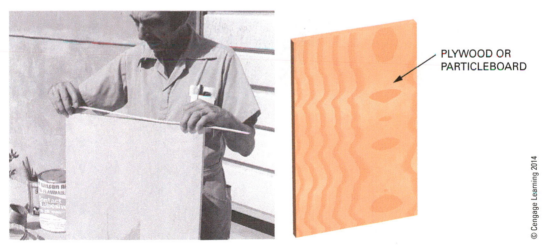

PLYWOOD OR PARTICLEBOARD

© Cengage Learning 2014

Figure 85–20 Solid doors may be made of plywood or particleboard and then laminated.

about the thickness of a dime, must be made between the opening and the door. Stops must be provided in the cabinet against which to close the door.

Door Construction

Doors are also classified, by their construction, as solid or paneled. Solid doors are made of plywood, particleboard, or solid lumber. Particleboard doors are ordinarily covered with plastic laminate. Matched boards with V-grooves and other designs, such as those used for wall paneling, are often used to make solid doors (*Figure 85–20*). Designs may

be grooved into the face of the door with a router. Small moldings may be applied for a more attractive appearance.

Paneled doors have an exterior framework of solid wood with panels of solid wood, plywood, hardboard, plastic, glass, or other panel material. Many complicated designs are manufactured by millworkers with specialized equipment. With the equipment available, carpenters can make paneled doors of simple design only (*Figure 85–21*). Both solid doors and paneled doors may be hinged in overlay, lipped, or flush fashion.

TYPES OF HINGES

Several types of cabinet hinges are *surface, offset, overlay, pivot,* and *butt.* For each type, there are many styles and finishes (*Figure 85–22*). Some types are *self-closing* hinges that hold the door closed and eliminate the need for door catches.

Surface Hinges

Surface hinges are applied to the exterior surface of the door and frame. The back side of the hinge leaves may lie in a straight line for flush doors. One leaf may be offset for lipped doors (*Figure 85–23*). The surface type is used when it is desired to expose the hardware, as in the case of wrought iron and other decorative hinges.

Offset Hinges

Offset hinges are used on lipped doors. They are called *offset surface* hinges when both leaves are fastened to outside surfaces. The *semiconcealed offset* hinge has one leaf bent to a $\frac{3}{8}$-inch offset that is screwed to the back of the door. The other leaf screws to the exterior surface of the face frame. A *concealed offset* type is designed in which only the pin is exposed when the door is closed (*Figure 85–24*).

Overlay Hinges

Overlay hinges are available in *semiconcealed* and *concealed* types. With semiconcealed types, the amount of overlay is variable. Certain concealed overlay hinges are made for a specific amount of overlay, such as $\frac{1}{4}$, $\frac{5}{16}$, $\frac{3}{8}$, and $\frac{1}{2}$ inch. European-style hinges are completely concealed. They are not usually installed by the carpenter because of the equipment needed to bore the holes to receive the hinge. Some overlay hinges, with one leaf bent at a 30-degree angle, are used on doors with reverse beveled edges (*Figure 85–25*).

Pivot Hinges

Pivot hinges are usually used on overlay doors. They are fastened to the top and bottom of the door and to the inside of the case. They are frequently used when there is no face frame and the door completely covers the face of the case (*Figure 85–26*).

Courtesy of Amerock, A Newell Rubbermaid Company

Figure 85–22 Cabinet door hinges come in many styles and finishes.

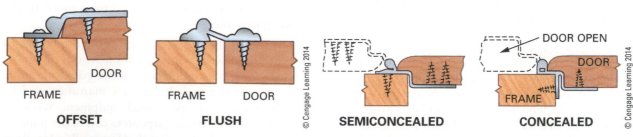

© Cengage Learning 2014

| OFFSET | FLUSH | | SEMICONCEALED | CONCEALED |

Figure 85–23 Surface hinges.

Figure 85–24 Offset hinges.

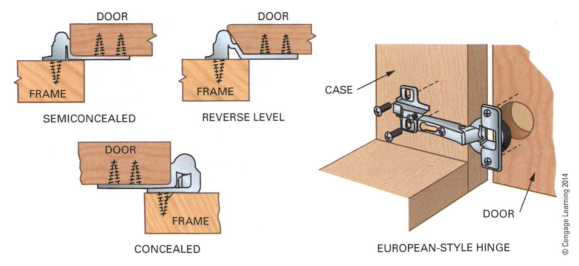

Figure 85–25 Overlay hinges.

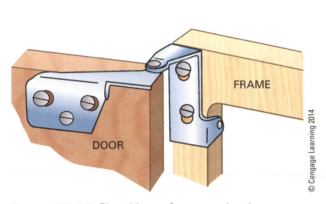

Figure 85–26 Pivot hinges for an overlay door.

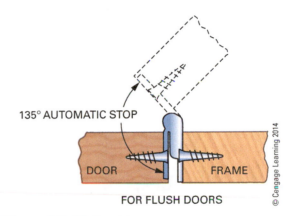

Figure 85–27 Butt hinges.

Butt Hinges

Butt hinges are used on flush doors. Butt hinges for cabinet doors are a smaller version of those used on entrance doors. The leaves of the hinge are set into **gains** in the edges of the frame and the door, in the same manner as for entrance doors. Butt hinges are used on flush doors when it is desired to conceal most of the hardware. They are not often used on cabinets because they take more time to install than other types (*Figure 85–27*).

HANGING CABINET DOORS

Surface Hinges

To hang cabinet doors with surface hinges, first apply the hinges to the door. Then shim the door in the opening so an even joint is obtained all around. Screw the hinges to the face frame.

Semiconcealed Hinges

For semiconcealed hinges, screw the hinges to the back of the door. Then center the door in the opening. Fasten the hinges to the face frame. When more than one door are to be installed side by side, clamp a straightedge to the face frame along the bottom of the openings for the full length of the cabinet. Rest the doors on the straightedge to keep them in line (*Figure 85–28*).

Concealed Hinges

When installing concealed hinges, first screw the hinges on the door. Center the door in the opening. Press or tap on the hinge opposite the face frame. Small projections on the hinge make indentations to mark its location on the face frame.

Open the door. Place the projections of the hinges into the indentations. Screw the hinges to the face frame.

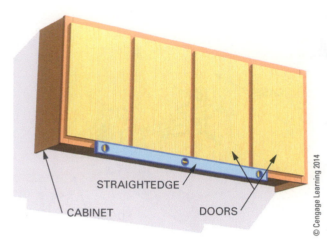

Figure 85–28 When installing doors, use a straightedge to keep them in line.

Figure 85–29 The VIX bit is a self-centering drill stop used for drilling holes for cabinet hinges.

Butt Hinges

Hanging flush cabinet doors with butt hinges is done in the same manner as hanging entrance doors. Drill pilot holes for all screws so they are centered on the holes in the hinge leaf. Drilling the holes off center throws the hinge to one side when the screws are driven. This usually causes the door to be out of alignment when hung. Many carpenters use a self-centering tool, called a *VIX bit*, when drilling pilot holes for screw fastening of cabinet door hinges of all types (*Figure 85–29*). The tool centers a twist drill on the hinge leaf screw hole. It also stops at a set depth to prevent drilling through the door or face frame.

INSTALLING PULLS AND KNOBS

Cabinet pulls or knobs are used on cabinet doors and drawers. They come in many styles and designs. They are made of metal, plastic, wood, porcelain, or other material (*Figure 85–30*).

Pulls and knobs are installed by drilling holes through the door. Then fasten them with machine screws from the inside. When two screws are used to fasten a pull, the holes are drilled slightly oversize in case they are a little off center. This allows the pulls to be fastened easily without crossthreading the screws. Usually $3/16$-inch-diameter holes are drilled for $1/8$-inch machine screws. To drill holes quickly and accurately, make a template from scrap wood that fits over the door. The template can be made so that holes can be drilled for doors that swing in either direction (*Figure 85–31*).

DOOR CATCHES

Doors without self-closing hinges need catches to hold them closed. Many kinds of catches are available (*Figure 85–32*). Catches should be placed where they are not in the way, such as on the bottom of shelves, instead of the top.

Figure 85–30 A few of the many styles of pulls and knobs used on cabinet doors and drawers.

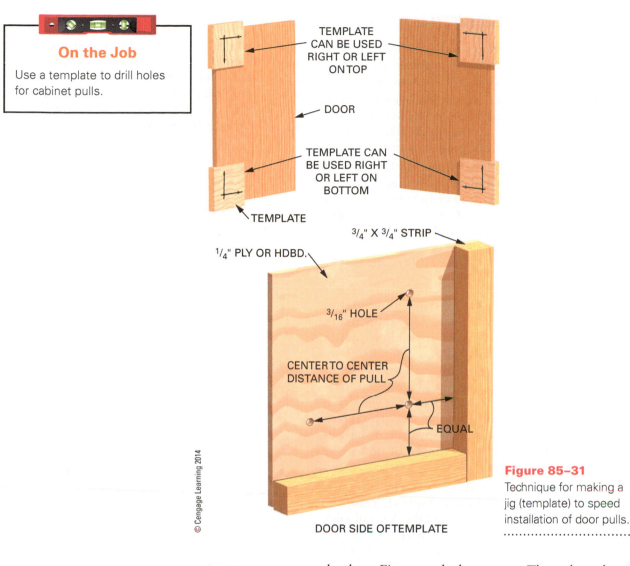

On the Job

Use a template to drill holes for cabinet pulls.

TEMPLATE CAN BE USED RIGHT OR LEFT ON TOP

DOOR

TEMPLATE CAN BE USED RIGHT OR LEFT ON BOTTOM

TEMPLATE

³/₄" X ³/₄" STRIP

¹/₄" PLY OR HDBD.

³/₁₆" HOLE

CENTER TO CENTER DISTANCE OF PULL

EQUAL

DOOR SIDE OF TEMPLATE

Figure 85–31
Technique for making a jig (template) to speed installation of door pulls.

Figure 85–32 Several types of catches are available for use on cabinet doors.

Courtesy of Amerock, A Newell Rubbermaid Company

the door. First attach the magnet. Then place the plate on the magnet. Close the door and tap it opposite the plate. Projections on the plate mark its location on the door. Attach the plate to the door where marked. Try the door. Adjust the magnet, if necessary.

Friction catches are installed in a similar manner to that used for magnetic catches. Fasten the adjustable section to the case and the other section to the door.

Elbow catches are used to hold one door of a double set. They are released by reaching to the back side of the door. These catches are usually used when one of the doors is locked against the other.

Bullet catches are spring loaded. They fit into the edge of the door. When the door is closed, the catch fits into a recessed plate mounted on the frame.

Magnetic catches are widely used. They are available with single or double magnets of varying holding power. An adjustable magnet is attached to the inside of the case. A metal plate is attached to

DRAWER CONSTRUCTION

Drawers are classified as overlay, lipped, and flush in the same way as doors. In a cabinet installation, the drawer type should match the door type.

Drawer fronts are generally made from the same material as the cabinet doors. Drawer sides and backs are generally ½ inch thick. They may be made of solid lumber, plywood, or particleboard.

Medium-density fiberboard with a printed wood grain is also manufactured for use as drawer sides and backs. The drawer bottom is usually made of ¼-inch plywood or hardboard. Small drawers may have ⅛-inch hardboard bottoms.

Drawer Joints

Typical joints between the front and sides of drawers are the *dovetail, lock,* and *rabbet* joints. The dovetail joint is used in higher-quality drawer construction. It takes a longer time to make, but is the strongest. Dovetail drawer joints may be made using a router and a dovetail template (*Figure 85–33*). The lock joint is simpler. It can be easily made using a table saw. The rabbet joint is the easiest to make. However, it must be strengthened with fasteners in addition to glue (*Figure 85–34*).

Joints normally used between the sides and back are the *dovetail, dado and rabbet, dado,* and *butt* joints. With the exception of the dovetail joint, the drawer back is usually set in at least ½ inch from the back ends of the sides to provide added strength. This helps prevent the drawer back from being pulled off if the contents get stuck while opening the drawer (*Figure 85–35*).

Figure 85–33 Dovetail joints can be made with a router and a dovetail template.

Drawer Bottom Joints

The drawer bottom is fitted into a groove on all four sides of the drawer (*Figure 85–36*). In some cases, the drawer back is made narrower, the four sides assembled, the bottom slipped in the groove, and its back edge fastened to the bottom edge of the drawer back (*Figure 85–37*).

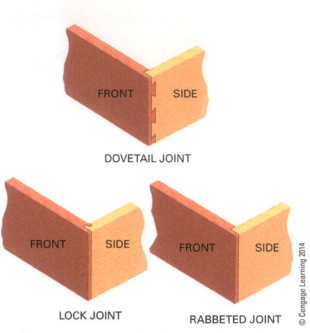

Figure 85–34 Various joints between drawer front and side.

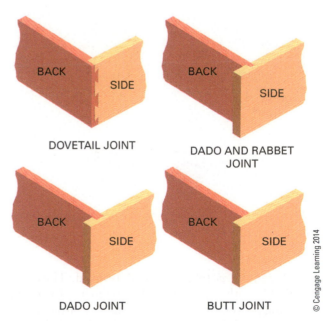

Figure 85–35 Various joints between drawer back and side.

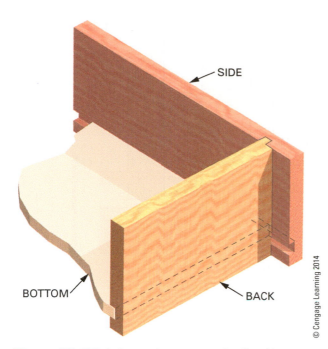

Figure 85–36 A drawer bottom may be fitted into a groove at the drawer back.

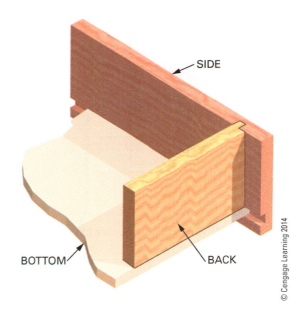

Figure 85–37 A drawer bottom may be fastened to bottom edge of drawer back.

DRAWER GUIDES

There are many ways of guiding drawers. Drawer guides are designed to make the drawer easy to pull out and return. They are often pre-manufactured metal guides. Sometimes the carpenter will build them of hardwood.

Metal Drawer Guides

Metal guides come in many styles and configurations. They are made for light-, medium-, and heavy-duty situations. Most kitchen cabinets will need to be medium duty. Some have a single track mounted on the bottom center of the opening. These have a nylon wheel that rides in a metal track mounted on each side of the drawer guide (*Figure 85–38*). Others are self-contained guides with steel rollers that ride in tracks. These may be normal extension or full extension where the back of the drawer rides out to the edge of the face frame (*Figure 85–39*).

Instructions for installation differ with each type and manufacturer. Read the instructions before making the drawer so that proper allowances for the drawer guide can be made. Most guides need only 1/2 inch on either side of the drawer box and face frame.

Wood Guides

The type of drawer guide selected affects the size of the drawer. The drawer must be supported level and guided sideways. It must also be kept from tilting down when opened. Probably the simplest wood guide is the center strip. It is installed in the bottom center of the opening from front to back (*Figure 85–40*). The strip projects above the bottom of the opening about $\frac{1}{4}$ inch. The bottom edge of the drawer back is notched to ride in the guide.

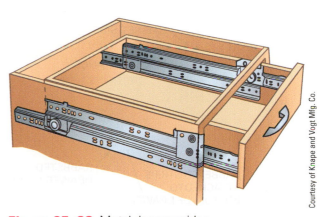

Figure 85–38 Metal drawer guides.

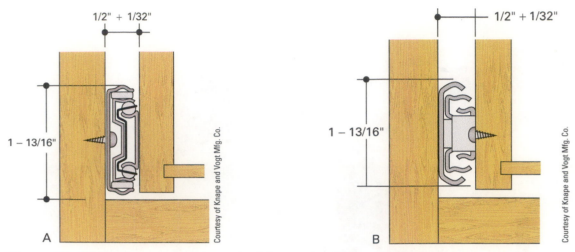

1/2" + 1/32"

1 − 13/16"

A

Courtesy of Knape and Vogt Mfg. Co.

1/2" + 1/32"

1 − 13/16"

B

Courtesy of Knape and Vogt Mfg. Co.

Figure 85–39 Metal drawer guides may be full extension (A) or standard (B).

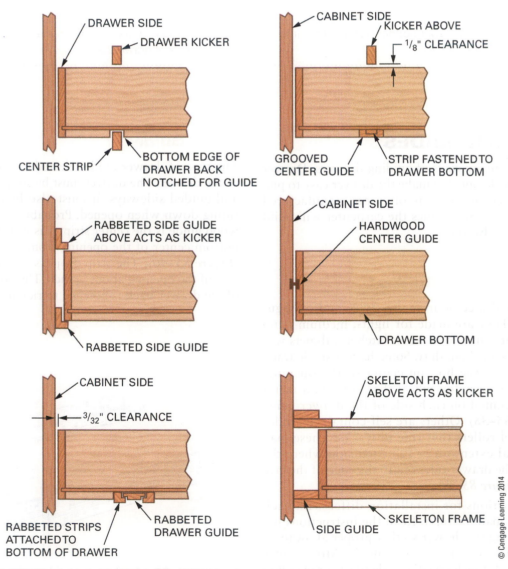

DRAWER SIDE

DRAWER KICKER

CENTER STRIP

BOTTOM EDGE OF DRAWER BACK NOTCHED FOR GUIDE

CABINET SIDE

KICKER ABOVE

1/8" CLEARANCE

GROOVED CENTER GUIDE

STRIP FASTENED TO DRAWER BOTTOM

RABBETED SIDE GUIDE ABOVE ACTS AS KICKER

RABBETED SIDE GUIDE

CABINET SIDE

HARDWOOD CENTER GUIDE

DRAWER BOTTOM

CABINET SIDE

3/32" CLEARANCE

RABBETED STRIPS ATTACHED TO BOTTOM OF DRAWER

RABBETED DRAWER GUIDE

SKELETON FRAME ABOVE ACTS AS KICKER

SIDE GUIDE

SKELETON FRAME

© Cengage Learning 2014

Figure 85–40 Wood drawer guides are installed in several ways.

A kicker is installed. It is centered above the drawer to keep it from tilting downward when opened.

Another type of wood guide is the grooved center strip. The strip is placed in the center of the opening from front to back. A matching strip is fastened to the drawer bottom. In addition to guiding the drawer, this system keeps it from tilting when opened, eliminating the need for drawer kickers.

Another type of wood guide is a rabbeted strip. Strips are used on each side of the drawer opening. The drawer sides fit into and slide along the rabbeted pieces. Sometimes these guides are made up of two pieces instead of rabbeting one piece. A kicker above the drawer is necessary with this type of guide.

DECONSTRUCT THIS

Carefully study **Figure 85–40** and think about what is wrong and/or what is right. Consider all possibilities. What construction practice or method is different in your area of the country?

© Cengage Learning 2014

Figure 85-40 The photo in this figure shows a set of kitchen cabinets being installed.

KEY TERMS

accessories	filler	mica	postforming
backsplash	flush type door	offset hinges	rough size
base unit	frameless	overlay door	seam-filling compound
cabinet lift	gains	overlay hinges	solid
catches	joints	paneled	surface hinges
double-faced cabinets	J-roller	pilot	toe space
face frame	lipped door	pivot	wall unit

REVIEW QUESTIONS

Select the most appropriate answer.

1. The vertical distance between the base unit and a wall unit is usually
   a. 12 inches.
   b. 15 inches.
   c. 18 inches.
   d. 24 inches.

2. The distance from the floor to the surface of the countertop is usually
   a. 30 inches.
   b. 32 inches.
   c. 36 inches.
   d. 42 inches.

3. To accommodate sinks and provide adequate working space, the width of the countertop is
   a. 25 inches.
   b. 28 inches.
   c. 30 inches.
   d. 32 inches.

4. Standard wall cabinet height is
   a. 24 inches.
   b. 30 inches.
   c. 32 inches.
   d. 36 inches.

5. The height of most manufactured base kitchen cabinets is
   a. $30\frac{3}{4}$ inches.
   b. $32\frac{1}{2}$ inches.
   c. $34\frac{1}{2}$ inches.
   d. $35\frac{1}{4}$ inches.

6. A drawer front or door with its edges and ends rabbeted to fit over the opening is called
   a. an overlay type.
   b. a flush type.
   c. a lipped type.
   d. a rabbeted type.

7. The offset hinge is used on
   a. paneled doors.
   b. flush doors.
   c. lipped doors.
   d. overlay doors.

8. Butt hinges are used on
   a. flush doors.
   b. lipped doors.
   c. overlay doors.
   d. solid doors.

9. The joint used on high-quality drawers is the
   a. dado joint.
   b. dado and rabbet joint.
   c. dovetail joint.
   d. rabbeted joint.

10. A wood strip installed in cabinets with wood drawer guides to prevent a drawer from tilting downward when opened is called a
    a. top guide.
    b. kicker.
    c. sleeper.
    d. tilt strip.

Accessories in cabinetry, items used to enhance cabinet installation such as end panels, face panels for dishwashers and refrigerators, open shelves for cabinet ends, and spice racks

Acoustical tile a fiberboard ceiling tile whose surface consists of small holes and/or fissures that act as sound traps

Adjustable wrench wrench with jaws that move to various widths by turning a screw

Admixture material used in concrete or mortar to produce special qualities

Aggregates materials, such as sand, rock, and gravel, used to make concrete

Air-dried lumber lumber that has been seasoned by drying in the air

Air-entrained cement cement that has undergone the process of having tiny air bubbles added to it to increase its resistance to freeze and thaw cycles

Anchor a device used to fasten structural members in place

Annular ring the rings seen when viewing a cross section of a tree trunk; each ring constitutes one year of tree growth

Apron a piece of the window trim used under the stool

Apron flashing shingle courses that are applied under the flashing on the lower side of a chimney

Architect's scale ruler used to draw and read measurements in various proportions or scales

Arch windows windows that have a curved top or head that make them well suited to be joined in combination with a number of other types of windows or doors

Areaway below-grade, walled area around basement windows

Argon gas a gas used to fill the space between layers of insulated glass to increase the R-value

Asphalt felt a building paper saturated with asphalt for waterproofing

Asphalt shingles a type of shingle surfaced with selected mineral granules and coated with asphalt to provide weather-proofing qualities

Astragal a semicircular molding often used to cover a joint between doors

Auger bits wood-boring bit with a piloting screw tip

Aviation snips metal shears with lever-action handles to increase cutting power

Awning window a type of window in which the sash is hinged at the top and swings outward

Back band molding applied around the sides and tops of windows and doors

Back bevel a bevel on the edge or end of stock toward the back side

Backing strips or blocks installed in walls or ceilings for the purpose of fastening or supporting trim or fixtures

Backing board a type of gypsum board designed to be used as a base layer in multilayer systems

Backing the hip beveling the top edge of a hip rafter to line it up with adjacent roof surfaces

Back miter an angle cut starting from the end and coming back on the face of the stock

Backset the distance an object is set back from an edge, side, or end of stock, such as the distance a hinge is set back from the edge of a door

Backsplash a raised portion on the back edge of a countertop to protect the wall

Balloon frame a type of frame in which the studs are continuous from foundation to roof

Baluster vertical members of a stair rail, usually decorative and spaced closely together

Balustrade the entire stair rail assembly, including handrail, balusters, and posts

Base see *base shoe*

Baseboard finish board used to cover the joint at the intersection of wall and floor

Base cap a molding applied to the top edge of the baseboard

Base shoe a molding applied between the baseboard and floor

Base unit one of two basic kinds of kitchen cabinets; the base unit is on the bottom

Batten a thin, narrow strip usually used to cover joints between vertical boards

Batter board a temporary framework erected to hold the stretched lines of a building layout

Bay window a window, usually three-sided, that projects out from the wall line

Beam pocket an indentation in a foundation wall where a girder rests

Bearer horizontal members of a wood scaffold that support scaffold plank

Bearing the surface area of a structural member where weight or load is transferred

Bearing partition an interior wall that supports the floor above

Bed a standard molding often used at the intersections of walls and ceilings

Belt sander a two-handed electric sanding tool with a looped strip of sandpaper

Benchmark a reference point for determining elevations during the construction of a building

Bench plane iron cutting portion of a bench plane

Bench planes hand tool used to shave thin layers of wood

Bevel the sloping edge or side of a piece for which the angle formed by the slope is not a right angle

Bevel ripping term used to describe cutting wood with the grain and at an angle

Bifold door doors that are hinged to each other in pairs as well as one being hinged to the jamb

Bird's mouth a notch cut in the underside of a rafter to fit on top of the wall plate

Bit brace two-handed drilling tool used to spin auger bits

Blind nail a method of fastening that conceals the nails

Blind stop part of a window finish applied just inside the exterior casing

Blocking short pieces of lumber installed in floor, ceiling, or wall construction to provide weather tightness, fire-stopping, or support for parts of the structure

Blockouts the forms used for providing the larger openings in concrete walls

Block planes a small hand tool with a low-blade angle used to shave thin layers of wood

Blueprinting older process of creating copies of construction drawings where the result is blue with white lines and letters

Board lumber usually 8 inches or more in width and less than 2 inches thick

Board and batten a kind of wood siding suitable for only vertical applications

Board on board a kind of wood siding suitable for only vertical applications

Board-on-board fence similar to a picket fence, however the boards are alternated from side to side so that the fence looks the same from both sides

Board foot a volume of wood that measures 1 foot square and 1 inch thick or any equivalent lumber volume

Board paneling paneling used on interior walls when the warmth and beauty of solid wood is desired

Boring term used to mean drilling of large holes

Boring jig a tool frequently used to guide bits when boring holes for locksets

Bow a type of warp in which the side of lumber is curved from end to end

Bow window window units that project out from the building, often creating more floor space

Box cornice A cornice with overhang that helps protect the sidewalls from the weather and provides shade for windows; probably the most common cornice

Box header in platform construction, framing members that cap the ends of the floor joists; also called a *band joist, rim joist,* or *joist header*

Box nail a thin nail with a head, usually coated with a material to increase its holding power

Braces diagonal members of a scaffold that prevent the poles from moving or buckling

Brad a thin, short, finishing nail

Brake a tool used to bend sheet metal

Bridging diagonal braces or solid wood blocks between floor joists used to distribute the load imposed on the floor

Bright term used to describe uncoated nails

Btu the abbreviation for British thermal unit

Buck a rough frame used to form openings in concrete walls

Bucks in masonry construction, wood pieces that are fastened to the sides of a window opening against which the window may be installed

Builder's level a telescope to which a spirit level is mounted

Building codes see *national building codes*

Building permits proof of permission to build a structure granted by the local authorities

Bulls eye bubble a leveling device requiring a liquid spirit bubble to be centered within a circle

Butt the joint formed when one square cut piece is placed against another; also, a type of hinge

Butt hinge a hinge composed of two plates attached to abutting surfaces of a door and door jamb, joined by a pin

Butt hinge template a metal template positioned and tacked into place with pins that is used as a guide to rout hinge gains on doors and frames

Butt markers tool used to outline a mortise for butt hinges

Butterfly roof an inverted gable roof

Bypass doors doors, usually two for a single opening, that are mounted on rollers and tracks so that they slide by each other

Cabinet lift a lift used to hold cabinets in position for fastening to the wall

Café doors short panel or louver doors hung in pairs that swing in both directions; used to partially screen an area yet allow easy and safe passage through the opening

Cambium layer a layer just inside the bark of a tree where new cells are formed

Capillary action the movement of a liquid into fine cracks and spaces due to molecular attraction

Caps molding used to finish the top edge of a molding application

Carbide blade see *Tungsten carbide-tipped blade*

Carbide-tipped refers to cutting tools that have small, extremely hard pieces of carbide steel welded to the tips

Carpenter's level 2-foot-long tool with spirit vials used to determine level and plumb

Carriage bolt a machine-threaded bolt having a rounded head and no screw slot

Casement window a type of window in which the sash is hinged at the edge and usually swings outward

Casing molding used to trim around doors, windows, and other openings

Casing nails similar to finish nails; preferred by many carpenters to fasten exterior finish

Catches hardware used to keep cabinet doors shut when they do not have self-closing hinges

Caulk puttylike mastic used to seal cracks and crevices

Cement board a panel product with a core of portland cement reinforced with a glass fiber mesh embedded in both sides

Cements adhesives that bond asphalt roofing products and flashings

Ceramic tile hard, brittle, heat-resistant, and corrosion-resistant tiles made by firing a nonmetallic mineral, such as clay; ceramic tiles come in a variety of geometric shapes, sizes, patterns, colors, and designs

Chair rail molding applied horizontally along the wall to prevent chair backs from marring the wall

Chalk line string in a box with colored powder used to establish straight lines

Chamfer a beveled edge, such as that machined between the face side and edges of strip flooring prior to finishing to obscure any unevenness after installation

Chase wall two closely-spaced, parallel walls constructed for the running of plumbing, heating and cooling ducts, and similar items

Check lengthwise split in the end or surface of lumber, usually resulting from more rapid drying of the end than the rest of the piece

Cheek cut a compound miter cut on the end of certain roof rafters

Clamps tools used to temporarily hold material together

Claw hammer carpenter's tool used to drive and pull nails

Cleat a piece of $1'' \times 5''$ stock that is installed around the walls of a closet to support the shelf and the rod; also, blocks fastened to the stair carriage to support the treads

Closed a type of staircase where the treads butt against the wall and are not exposed to view

Closed stairway a type of stair installed between walls where the ends of treads and risers are hidden from view

Closed valley a roof valley in which the roof covering meets in the center of the valley, completely covering the valley

Close-grained wood in which the pores are small and closely spaced

Coatings a product used to resurface old roofing or metal that has become weathered

Coil stock aluminum stock sold in 50-foot-long rolls of various widths ranging from 12 to 24 inches

Cold-rolled channels (CRCs) channels formed from 54-mil steel that are often used in suspended ceilings and as bridging for lateral support in walls

Collar tie a horizontal member placed close to the ridge at right angles to the plate

Combination blade circular saw blade designed to cut across and with wood grain

Combination pliers pliers with adjustable jaws

Combination square a squaring tool with a sliding blade

Common nails fasteners used for rough work such as framing

Common rafter extends from the wall plate to the ridge board, where its run is perpendicular to the plate

Common screw multi-purpose slot-headed fastener

Compass saw short hand saw with a tapered blade used to cut curves or irregular shapes

Competent person an individual who supervises and directs scaffold erection, and who has the authority to take corrective measures to ensure that scaffolding is safe to use

Composites veneer faces bonded to reconstituted wood cores

Compression glazing a term used to describe various means of sealing monolithic and insulating glass in the supporting framing system with synthetic rubber and other elastomeric gasket materials; this method virtually eliminates the possibility of any water seeping through the joints

Compressive strength the quality of being resistant to crushing; concrete has high compressive strength

Concrete a building material made from portland cement, aggregates, and water

Condensation when water, in a vapor form, changes to a liquid due to cooling of the air; the resulting drops of water that accumulate on the cool surface

Conduction the transfer of heat by contact

Conductor a vertical member used to carry water from the gutter downward to the ground; also called *downspout* or *leader*

Conical screw hollow wall fastener with deep threads used as a holder for screws

Coniferous cone-bearing tree; also known as *evergreen* tree

Connectors term used to describe a large metal fastener used to join framing members

Construction heart the most suitable and most economical grade of California redwood for deck posts, beams, and joists

Contact adhesives adhesives used for laminating gypsum boards to each other; the board cannot be moved after contact has been made

Contact cement an adhesive used to bond plastic laminates or other thin material; so called because the bond is made on contact, eliminating the need for clamps

Contour interval scale used between lines of a contour map

Contour line lines on a drawing representing a certain elevation of the land

Control joints metal strips with flanges on both sides of a $\frac{1}{4}$-inch, V-shaped slot placed in large drywall areas to relieve stresses induced by expansion or contraction

Convection fluid (air) movement caused by differences in warm and cool fluid (air) densities; denser cool air is pulled downward by gravity displacing the less dense warm air

Coped joint a type of joint between moldings in which the end of one piece is cut to fit the molded surface of the other

Coping saw hand saw with short thin blade held by a bow-shaped frame used to cut irregular shapes in trim

Cordless nailing gun tool that drives fasteners by use of a fuel cell instead of air pressure

Corner bead metal trim used on exterior corners of walls to trim and reinforce them

Corner boards boards used to trim corners on the exterior walls of a building

Corner brace diagonal member of the wall frame used at the corners to stiffen and strengthen the wall

Corner guards molding applied to wall corners

Corner post built-up stud used in the corner of a wall frame

Cornice the entire finished assembly where the walls of a structure meet the roof; sometimes called the eaves

Cornice return used to change the level box cornice to the angle of the roof

Counterboring boring a larger hole partway through the stock so that the head of a fastener can be recessed

Countersink to make a flared depression around the top of a hole to receive the head of a flathead screw; also, the tool used to make the depression

Counter Flashing flashing installed on the sides of a chimney where the top edge is bedded into mortar joint and the bottom edge covers step flashing

Course a continuous row of building material, such as brick, siding, roofing, or flooring

Cove a concave-shaped molding

Coverage the number of overlapping layers of roofing and the degree of weather protection offered by roofing material

Crawl space foundation type creating a space under the first floor that is not tall enough to allow a full basement

Cricket a small, false roof built behind a chimney or other roof obstacle for the purpose of shedding water; also called *saddle*

Cripple jack rafter a common rafter cut shorter that does not contact either a top plate or a ridge

Crook a type of warp in which the edge of lumber is not straight

Crosscut a cut made across the grain of lumber

Crosscut circular saw blade saw blade designed to cut best when cutting across wood grain

Crosscut saws cutting tools designed to cut across wood grain

Cross tees metal pieces that come in 2- and 4-foot lengths with connecting tabs on each end; the tabs are inserted and locked into main runners and other cross tees in a suspended ceiling grid

Crown a kind of standard molding; see *bed*

Crown usually referred to as the high point of the crooked edge of joists, rafters, and other framing members

Cup a type of warp in which the side of a board is curved from edge to edge

Cylindrical lockset hardware for shutting or locking a door that is often called a *key-in-the-knob lockset*

Dado a cut, partway through and across the grain of lumber

Deadbolt door-locking bolt operated by a key from the outside and by a handle or key from the inside

Dead load Dead Load is the weight of the structural members as well as all material that is fastened into place.

Deadman a T-shaped wood device used to support ceiling drywall panels and other objects

Deciduous trees that shed leaves each year

Deck the wood roof surface to which roofing materials are applied

Deck board lumber that covers the joist and girder platform to provide the surface and walking area of the deck

Deflection is the amount a structural member will sag when under load.

Detail a drawing showing a close-up or zoomed-in view of part of another drawing

Dew point temperature at which moisture in the air condenses into drops

Diagonal at an angle, usually from corner to corner in a straight line

Dimension lumber wood used for framing having a nominal thickness of 2 inches

Door casings the molding used to trim the sides and top of a door

Door jack a manufactured or job-made jack that holds a door steady during installation

Door schedules informational chart found on a set of prints providing pertinent information on doors of the building

Door stop molding fastened to door jambs for the door to stop against

Dormer a structure that projects out from a sloping roof to form another roofed area to provide a surface for the installation of windows

Double-acting doors doors that swing in both directions or the hinges used on these doors

Double coursing when two layers of shingles are applied to sidewalls in one course

Double coverage a concealed-nail type of roll roofing

Doubled-faced cabinets cabinets that have doors on both sides for use above island and peninsular bases

Double-hung window a window in which two sashes slide vertically by each other

Double pole a type of wood scaffold used when the scaffolding must be kept clear of the wall for the application of materials or for other reasons

Double-strength glass (DS) glass about $\frac{1}{8}$ inch thick and used for larger lights of glass

Dovetail a type of interlocking joint resembling the shape of a dove's tail

Dowel hardwood rods of various diameters

Downspout see *conductor*

Draft see *firestop*

Draft-stop blocking see *fire-stop blocking*

Drilling term uses to describe cutting holes with turning bits

Drip that part of a cornice or a course of horizontal siding that projects below another part; also, a channel cut in the underside of a windowsill that causes water to drop off instead of running back and down the wall

Drip cap a molding placed on the top of exterior door and window casings for the purpose of shedding water away from the units

Drip edge metal edging strips placed on roof edges to provide a support for the overhang of the roofing material

Drive anchor a solid wall fastener that is set with a hammer

Drop-in anchor a solid wall fastener that requires a setting tool used to support bolt threads

Dropping the hip increasing the depth of the hip rafter seat cut so that the centerline of its top edge will lie in the plane of adjacent roof surfaces

Dry rot dry, powdery residue of wood left after fungus destruction of wood due to excessive moisture

Drying potential a building constructed so that building materials can dry easily after exposure to severe weather

Drywall a type of construction usually referred to as the installation of gypsum board

Drywall cutout tool a portable electric tool used to make cutouts in wall panels for electrical boxes, ducts, and similar objects

Drywall foot lifter a tool used to lift gypsum panels in the bottom course up against the top panels

Dry well gravel- or stone-filled excavation for catching water so it can be absorbed into the earth

Duplex nail a double-headed nail used for temporary fastening such as in the construction of wood scaffolds

Eased edge an edge of lumber whose sharp corners have been rounded

Easement the curved piece that joins the first- and second-flight handrails in a balustrade

Eaves that part of a roof that extends beyond the sidewall

Edge the narrow surface of lumber running with the grain

Edge grain boards in which the annular rings are at or near perpendicular to the face; sometimes called vertical grain

Elastomeric membrane a material having the elastic properties of natural rubber

Electrolysis the decomposition of the softer of two unlike metals in contact with each other in the presence of water

Elevation a drawing in which the height of the structure or object is shown; also, the height of a specific point in relation to another point

End the extremities of a piece of lumber

End lap the horizontal distance that the ends of roofing in the same course overlap each other

Engineered lumber products (ELPs) manufactured lumber substitutes, such as wood I-beams, glue-laminated beams, laminated veneer lumber, parallel strand lumber, and laminated strand lumber

Engineered panels human-made products in the form of large reconstituted wood sheets

Equilibrium a state of balance; heat energy is thought to move from warmer materials into cooler ones in an attempt to reach this state

Equilibrium moisture content the point at which the moisture content of wood is equal to the moisture content of the surrounding air

Escutcheon protective plate covering the knob or key hole in doors

Expanded polystyrene a type of rigid foam plastic insulation that will absorb moisture if it comes in contact with water; also called *white bead board*

Expansion anchors hollow wall fastener that spreads out behind the drywall to secure screws

Expansive bit wood-drilling tool that adjusts to various diameters

Exposure the amount that courses of siding or roofing are exposed to the weather

Extension jambs extensions applied to the inside edge of window jambs to accommodate various wall thicknesses

Extension ladder a ladder whose length can be extended by pulling on the rope and raising it to the desired height

Extruded polystyrene a type of rigid foam plastic insulation with a closed-cell structure that will not absorb water

Face the best-appearing side of a piece of wood or the side that is exposed when installed, such as finish flooring

Face frame a framework of narrow pieces on the face of a cabinet containing door and drawer openings

Face nail a method of driving a fastener nearly perpendicular to the material surface

Faceplate marker a tool used to lay out the mortise for the latch faceplate

Fall protection an OSHA requirement when using scaffolding where workers are at heights above 10 feet; consists of either guardrails or a personal fall protection system

Fancy butt shingles that are available in a variety of designs; used in combination with square butts or other types of siding, they provide interesting patterns

Fascia a vertical member of the cornice finish installed on the tail end of rafters

Feather boards guide tool used to secure material against a fence while it is being fed into stationary power tools

Feather edge the edge of material brought down in a long taper to a very thin edge, such as a wood shingle tip

Fiberboard building material made from fine wood chips pressed into sheets

Fiber-saturation point the moisture content of wood when the cell cavities are empty but the cell walls are still saturated

Filler in cabinetry, pieces that are used to fill small gaps in width between wall and base units when no combination of sizes can fill the existing space

Fillets narrow strips used between balusters to fill the plowed groove on handrails and shoe rails

Finish carpentry that part of the carpentry trade involved with the application of exterior and interior finish

Finger-jointed joints made in a mill used to join short lengths together to make long lengths

Finish nail a thin nail with a small head designed for setting below the surface of finish material

Finish schedules informational chart found on a set of prints providing pertinent information on interior design of a building

Finish stringer the finish board running with the slope of the stairs and covering the joint between the stairs and the wall; also called a *skirt board*

Finishing coat the third and final coat of drywall compound applied to cover wall joints, metal trims, and fasteners

Fire-stop blocking material installed to slow the movement of fire and smoke within smaller cavities of a building frame during a fire; also called *draft-stop blocking*

Firestop material used to fill air passages in a frame to prevent the spread of fire; and hot gases; also called *draft stop*

Firsts and Seconds the best grade of hardwood lumber

Fissured irregularly shaped grooves made in material, such as ceiling tile, for acoustical purposes

Fittings short sections of specially curved handrail

Fixed windows windows that consist of a frame in which the sash is fitted in a fixed position

Flange size the steel term that is similar to the thickness of a piece of wood

Flashing material used at intersections such as roof valleys, dormers, and above windows and doors to prevent the entrance of water

Flat bar general-purpose prying tool with flatten claws

Flat grain grain in which the annular rings of lumber lie close to parallel to the sides; opposite of edge grain

Flatwork a term used to describe concrete slabs, driveways, and sidewalks.

Flexible insulation a type of insulation manufactured in blanket or batt form

Floor guide hardware installed at the bottom of bypass doors to keep the doors in alignment

Floor joists horizontal members of a frame that rest on and transfer the load to sills and girders

Floor plans pages of a set of construction drawings showing the walls as viewed from above

Flush term used to describe when the surface joint between two materials is perfectly aligned

Flush door a door with a smooth, flat surface of wood veneer or metal

Flush type door a type of cabinet door that fits into and is flush with the face of the opening

Flute closely spaced concave grooves in lumber for decorative purposes on a column, post, or pilaster

Foamed-in-place a urethane foam insulation produced by mixing two chemicals that are injected into place to expand on contact with surfaces

Footing a foundation for a column, wall, or chimney made wider than the object it supports, to distribute the weight over a greater area

Foundation that part of a wall on which the major portion of the structure is erected

Foundation wall the supporting portion of a structure below the first floor construction, or below grade, including footings

Frameless a type of cabinet construction that does not use a face frame; also called *European method*

Framing hammer heavier (20–32 ounce) hammers used in framing

Framing square L-shaped steel tool, 24 inches long, used to lay out rafters and stairs

French door a door, usually one of a pair, of light construction with glass panes extending for most of its length

Frieze a part of the exterior finish applied at the intersection of an overhanging cornice and the wall

Frost line the depth to which frost penetrates into the ground in a particular area; footings must be placed below this depth

Full rounds a round piece of molding often used for such things as closet poles

Full sections pages of a set of construction drawings showing the cross section of the building

Furring channels hat-shaped pieces made of 18- and 33-mil steel that are applied to walls and ceilings for the screw attachment of gypsum panels

Furring strips strips that are usually attached directly to exposed joists to provide a surface for fastening ceiling tiles

Gable end the triangular-shaped section on the end of a building formed by the rafters in a common or gable roof and the top plate line

Gable end studs studs that form the wall closing in the triangular area under a gable roof

Gable roof a type of roof that pitches in two directions

Gables the triangular areas formed by the rake rafters and the wall plate at the ends of the building

Gable studs a stud whose bottom end is cut square and fit against the top of the wall plate, and whose top end fits snugly against the bottom edge and inside face of the end rafter to frame the gable

Gain a cutout made in a piece to receive another piece, such as a cutout for a butt hinge

Galvanized protected from rusting by a coating of zinc

Galvanized steel steel that is coated with zinc to help protect against corrosion

Gambrel roof a type of roof that has two slopes of different pitches on each side of center

Girder a heavy timber or beam used to support vertical loads

Glass bead small molding used to hold lights of glass in place

Glazier a person who installs glass in a frame

Glazing the act of installing glass in a frame

Glazing compound a soft, plastic-type material, similar to putty, used for sealing lights of glass in a frame

Glazing points small, triangular, or diamond-shaped pieces of metal used to secure and hold lights of glass in a frame

Glue-laminated lumber large beams or columns made by gluing smaller-dimension lumber together side to side; commonly called *glulam*

Gooseneck a curved section of handrail used when approaching a landing; also, an outlet in a roof gutter

Grade the level of the ground; also identifies the quality of lumber

Grade rod the height of the instrument minus the control point elevation

Grain in wood, the design on the surfaces caused by the contrast, spacing, and direction of the annular rings

Green a) a term applied to concrete that has not fully cured, b) concept of building with renewable, energy efficient and cost effective measures

Green lumber lumber that has not been dried to a suitable moisture content

Green space areas of a building site devoted to natural vegetation

Grilles on a window, false muntins applied as an overlay to simulate small lights of glass

Grit material of sandpaper that actually provides the sanding action

Groove a cut, partway through, and running with the grain of lumber

Ground strips of wood placed at the base of walls and around openings and used as a guide for the application of an even thickness of plaster; also, a system used for electrical safety

Ground Fault Circuit Interrupter (GFCI) device used in electrical circuits for protection against electrical shock; it detects a short circuit instantly and shuts off the power automatically

Grouting the process of filling in small space with a thick paste of cement

Guardrails rails installed on all open sides and ends of scaffolds that are more than 10 feet in height

Gusset plate see *scab*

Gutter a wood or metal trough used at the roof edge to carry off rainwater and water from melting snow

Gypsum a chalky type rock that is the basic ingredient of plaster

Gypsum board a sheet product made by encasing gypsum in a heavy paper wrapping

Gypsum lath a type of gypsum board used as a base to receive conventional plaster

Hacksaws saws designed to cut metal

Half-round a molding with its end section in the shape of a semicircle

Hammer-drills electric drilling tools that also provide an impact action to speed drilling holes in masonry

Handrail a railing on a stairway intended to be grasped by the hand to serve as a support and guard

Handrail bolt a bolt with threads on one end designed for fastening into wood and threads for a nut on the other end; used to join the starting fitting to a straight section of handrail on the balustrade

Hand scraper   tool designed to remove thin layers of material from various surfaces

Hardboard   a building product made by compressing wood fibers into sheet form

Hardwoods   the wood of broadleafed dicotyledonous trees (as distinguished from the wood of conifers)

Head casing   the top member of a window unit's exterior casing

Headers   pieces placed at right angles to joists, studs, and rafters to form and support openings in a wood frame

Head flashing   the flashing on the upper side of a chimney

Head jamb   the top horizontal member of a window frame

Head lap   the distance from the bottom edge of an overlapping shingle to the top of a shingle two courses under, measured up the slope

Hearth   an area near a fireplace, usually paved and extending out into a room, around which a wood floor installation must be framed

Heartwood   the wood in the inner part of a tree, usually darker and containing inactive cells

Heel   the back end of objects, such as a handsaw or hand plane

Height of the instrument   measurement found by placing the leveling rod on the benchmark, then adding the rod reading to the elevation of the benchmark

Herringbone   a pattern used in parquet floors

High-speed steel blade   saw blade with no carbide cutting tip

Hinge gain   the recessed area for the hinge on a door

Hinge mortise   see *hinge gain*

Hip jack rafter   a shortened common rafter that spans from the wall plate to a hip rafter

Hip rafter   extends diagonally from the corner of the plate to the ridge at the intersection of two surfaces of a hip roof

Hip roof   a roof that slopes upward toward the ridge from four directions

Hip-valley cripple jack rafter   a short rafter running parallel to common rafters, cut between hip and valley rafters

Hole saws   drills that cut with saw teeth along its perimeter

Hopper window   a type of window in which the sash is hinged at the bottom and swings inward

Horizontal circle scale   a scale that is divided into quadrants of 90 degrees each and remains stationary as the telescope is turned

Horizontal vernier   a scale that rotates with the telescope and is used to read minutes of a degree

Hose bibbs   external water faucets of a building

Housed stringer   a finished stringer that is dadoed to receive treads and risers of a stairway

Housewrap   type of building paper with which the entire sidewalls of a building are covered

Hydration   chemical reaction between cement and water causing concrete or mortar to harden

Ice and water shield flashing   flashing applied whenever there is a possibility of ice dams forming along the eaves and causing a backup of water

Ice dam   ice that forms on an overhang, causing water buildup behind it to back up under roofing material

Impact noise   noise caused by the vibration of an impact, such as dropped objects or footsteps on a floor

Impact noise rating (INR)   a rating that shows the resistance of various types of floor-ceiling construction to impact noises

Independent slab   a concrete slab that is separate from the foundation

Inserts   lead and plastic anchors commonly used for fastening lightweight fixtures to masonry walls

Insulated glass   double- or triple-glazed windows; improved insulated glass design uses argon gas between two panes of glass

Insulation   material used to restrict the passage of heat or sound

Intersecting roof   the roof of irregularly shaped buildings; valleys are formed at the intersection of the roofs

Isometric   a drawing in which three surfaces of an object are seen in one view, with the base of each surface drawn at a 30-degree angle

Jack rafter   part of a common rafter, shortened for framing a hip rafter, a valley rafter, or both

Jack studs   shortened studs that support the headers

Jamb   the sides and top of window and door frames

Jamb extension   narrow strips of wood fastened to the edge of window jambs to increase their width

J-channel   a siding system accessory

Jig   any type of fixture designed to hold pieces or guide tools while work is being performed

Jigsaw   a small hand-held electric saw that cuts with a stroking action

Joint compound   a substance similar to plaster used to cover joints or the heads of screws or nails in plasterboard

Joint   as a verb, denotes straightening the edge of lumber; as a noun, means the place where parts meet and unite

Jointer plane   largest of the bench planes used to finish wood by removing thin strips of wood

Joint reinforcing tape   material used to cover, strengthen, and provide crack resistance to drywall joints

Joist   horizontal framing members used in a spaced pattern that provide support for the floor or ceiling system

Joist hanger   metal stirrups used to support the ends of joists

Joist header   see *box header*

J-roller   a 3-inch-wide rubber roller used to apply pressure over the surface of contact cement-bonded plastic laminates

Juvenile wood   the portion of wood that contains the first seven to fifteen growth rings of a log; they are located in the pith

Kerf   the width of a cut made with a saw

Keyhole saw   thinner version of a compass saw

Keyway   a groove made in concrete footings for tying in the concrete foundation wall

Kicker   in drawer construction, a piece centered above the drawer to keep it from tilting downward when opened

Kiln-dried   lumber dried by placing it in huge ovens called kilns

Kneewall   in relation to stairs, a short wall that projects a short distance above and on the same rake as the stair body

Knot   a defect in lumber caused by cutting through a branch or limb embedded in the log

Ladder jacks   metal brackets installed on ladders to hold scaffold planks

Lag screws   large threaded screw with a hex head

Lag shield   anchor used to secure lags in solid walls

Laminate   a thin layer of plastic often used as a finished surface for countertops

Laminated block   a type of parquet block generally made of three-ply laminated oak in a $3/_8$-inch thickness

Laminated glass   a type of safety glass that has two or more layers of glass with inner layers of transparent plastic bonded together

Laminated strand lumber (LSL)   lumber manufactured by bonding thin strands of wood, up to 12 inches long, with adhesive and pressure

Laminated strip flooring   a five-ply prefinished wood assembly

Laminated veneer lumber (LVL)   lumber manufactured by laminating many veneers of plywood with the grain of all running in the same direction

Laminate trimmer   small router used to cut and fit plastic laminate in cabinet making

Landing   an intermediate-level platform between flights

Landing treads   treads used at the edge of landings and balconies

Laser a concentrated, narrow beam of light; laser-equipped devices are used in building construction

Laser level a level that emits a red beam and can rotate through a full 360 degrees, creating a level plane of light

Lateral bracing temporary or permanent bracing that runs perpendicular to the braced members

Lattice thin strips of wood, spaced apart and applied in two layers at angles to each layer resulting in a kind of grillwork

Lattice work frequently used to close the space between the deck and the ground; also sometimes used to fill in the rail spaces above the deck

Ledger a horizontal member of a wood scaffold that ties the scaffold posts together and supports the bearers; a temporary or permanent supporting member for joists or other members running at right angles

Level line any line on the rafter that is horizontal when the rafter is in position

Level horizontal or perpendicular to the force of gravity

Lever handle used in place of knobs on locksets likely to be used by handicapped persons or in other situations where a lever is more suitable than a knob

Light a pane of glass or an opening for a pane of glass

Lignin the natural glue in wood that holds together the wood cells and fibers

Line length the length of a rafter along a measuring line without consideration to the width or thickness of the rafter

Line level device suspended from a string to determine level

Linear feet a one-dimensional measurement of length in foot increments

Linear measure a measurement of length

Lintel horizontal load-bearing member over an opening; also called a *header*

Lipped door a type of cabinet door that has rabbeted edges that overlap the opening by about $3/8$ inch on all sides

Live load Live load is weight on a structure that is moveable or temporary.

Load-bearing partitions (LBP) partitions that support the inner ends of ceiling or floor joists

Lock rail a rail situated at lockset height, usually 38 inches from the finish floor to its center

Lockset a set of hardware for shutting or locking a door

Lookout horizontal framing pieces in a cornice, installed to provide fastening for the soffit

Loose-fill insulation a type of insulation usually composed of materials in bulk form and supplied in bags or bales

Louver an opening for ventilation consisting of horizontal slats installed at an angle to exclude rain, light, and vision, but to allow the passage of air

Louver door a door made with spaced horizontal slats called louvers used in place of panels; the louvers are installed at an angle to obstruct vision but permit the flow of air through the door

Louvered fence a panel fence with vertical boards set at an angle

Low emissivity the term used to designate a type of glazing that reflects heat back into the room in winter and blocks heat from entering in the summer; window glass that has low emissivity is called LoE glass

Low-emissivity coating a coating on the inner glass surface of a window that allows ultraviolet rays from the sun to pass but reflects the inside infrared (radiant) rays back into the building

Lumber wood that is cut from the log to form boards, planks, and timbers

Lumber grades numbers and letters used to rank wood according to quality

Machine bolt hex-headed threaded fastener typically used with a nut and washer

Magazine a container in power nailers and staplers in which the fasteners to be ejected are placed

Main runners metal pieces shaped in the form of an upside-down T that are the primary support of a suspended ceiling's weight; main runners extend from wall to wall

Major span width of the larger portion of a building with six or eight corners

Mansard roof a type of roof that has two different pitches on all sides of the building, with the lower slopes steeper than the upper

Mantel the ornamental finish around a fireplace, including the shelf above the opening

Masonry any construction of stone, brick, tile, concrete plaster, and similar materials

Masonry drill bits drill bits with a carbide tip designed primarily to drill masonry products

Masonry nails hardened steel fasteners used in cementitious materials

Mastic a thick adhesive

Matched boards boards that have been finished with tongue-and-grooved edges

Medullary ray bands of cells radiating from the cambium layer to the pith of a tree to transport nourishment toward the center

Metal corner tape material used to protect corners on arches, windows with no trim, and other locations; applied by embedding in joint compound

Metes and bounds boundaries established by distances and compass directions

Mica a thin, tough high-pressure plastic laminate

Millwork any wood products that have been manufactured, such as moldings, doors, windows, and stairs for use in building construction; sometimes called *joinery*

Minor span width of the smaller portion of a building with six or eight corners

Miter boxes a fixed or adjustable tool for guiding handsaws in cutting miter joints or in making crosscuts

Mitered joint the joining of two pieces by cutting the end of each piece by bisecting the angle at which they are joined

Miter gauge a guide used on the table saw for making miters and square ends

Miters angular cuts at any angle other than a right angle made by a miter box

Mobile scaffold scaffold whose components include casters and horizontal diagonal bracing; also called a *rolling tower*

Modular measurement process of designing structures to best fit standard material sizes

Moisture content the amount of moisture in wood expressed as a percentage of the dry weight

Moisture meter a device used to determine the moisture content of wood

Molding decorative strips of wood used for finishing purposes

Monolithic slab a combined slab and foundation; also referred to as a thickened edge slab

Mortared using a mixture of portland cement, lime, sand, and water to bond masonry units together

Mortise a rectangular cavity cut in a piece of wood to receive a tongue or tenon projecting from another piece

Mud sill typically a 2×10 board approximately 18 to 24 inches long upon which scaffold baseplates rest

Mullion a vertical division between windows or panels in a door

Multispur bit a power-driven bit, guided by a boring jig, that is used to make the hole in a door for the lockset

Muntin slender strips of wood between lights of glass in windows or doors

Nail claw prying tool used to remove nails that have been driven all the way

Nailers general term use to describe power tools used to drive nails

Nail set tool used to drive nail heads below the surface

National building codes rules and regulations guiding the construction industry as set by national agencies

Newel posts a post to which the end of a stair railing or balustrade is fastened; also, any post to which a railing or balustrade is fastened

No. 1 common a lower grade of hardwood lumber

Nominal size the stated size of the thickness and width of lumber even though it differs from its actual size; the approximate size of rough lumber before it is surfaced

Nonconforming term used to describe buildings that do not fit the local zoning laws

Non-load-bearing partitions (NLBPs) partitions built to divide a space into rooms of varying size carrying only the load of the partition material itself and no structural load from the rest of the building

Nosing the rounded edge of a stair tread projecting over the riser

Nylon plug used for a number of hollow-wall and some solid-wall applications

Offset hinges a type of cabinet hinge used on lipped doors

On center (OC) the distance from the center of one structural member to the center of the next one

Open a type of staircase where the ends of the treads are exposed to view

Open cornice a cornice with no soffit

Open-grained a texture quality of wood where wood cells or pores are open to the surface

Open stairway type of stairway with the ends of the treads and risers visible

Open valley a roof valley in which the roof covering is kept back from the centerline of the valley

Optical levels commonly used instruments for leveling, plumbing, and angle layout; includes the builder's level and transit-level

Oriented strand board (OSB) common name for nonveneered panels

Orthographic multiview drawings

Overlay door a type of cabinet door that laps over the opening, usually $^3/_8$ inch on all sides

Overlay hinges a type of cabinet hinge that is available in either semiconcealed or concealed

Over-the-post a balustrade system that utilizes fittings to go over newels for an unbroken, continuous handrail

Panel a large sheet of building material

Panel door a door that consists of a frame with usually one to eight wood panels in various designs

Paneled a type of cabinet door construction that has an exterior framework of solid wood with panels of solid wood, plywood, hardboard, plastic, glass, or other panel material

Panel siding siding that comes in 4-foot widths and lengths of 8 and 10 feet, and sometimes 9 feet; it is usually applied vertically, but may also be applied horizontally

Parallel strand lumber (PSL) lumber manufactured by bonding strands of structural lumber, up to 8 feet long, with adhesive, heat, and pressure

Parquet a floor made with strips or blocks to form intricate designs

Parquet block a type of flooring consisting of square or rectangular blocks; sometimes installed in combination with strips, to form mosaic designs

Parquet strip a type of flooring with short strips that are laid to form various mosaic designs

Partial section elevations orthographic drawing showing only one side of the outside of the building at a distance of about 100 feet

Particleboard a building product made by compressing wood chips and sawdust with adhesives to form sheets

Parting strip a small strip of wood separating the upper and lower sash of a double-hung window

Partition an interior wall separating one portion of a building from another

Passage lockset a lockset with knobs on both sides that are turned to unlatch the door; used when it is not desirable to lock the door

Patio door a door unit normally consisting of two or three sliding or swinging glass doors completely assembled in the frame

Penny (d) a term used in designating nail sizes

Perforated material that has closely spaced holes in a regular or irregular pattern

Perlite loose-fill insulation sometimes mixed into concrete to make it lighter in weight and a better insulator

Phillips head a type of screw head with a cross-slot

Picket fence a fence commonly used on boundary lines or as barriers for pets and small children

Pier a column of masonry, usually rectangular in horizontal cross-section, used to support other structural members

Pilaster column built within and usually projecting from a wall to reinforce the wall

Pile concrete, metal, or wood pillar forced into the earth or cast in place as a foundation support

Pilot a guide on the end of edge-forming router bits used to control the amount of cut

Pilot hole a small hole drilled to receive the threaded portion of a wood screw

Pitch the amount of slope to a roof expressed as a ratio of the total Rise to the span

Pitch block a piece of wood cut in the shape of a right triangle; used as a pattern for laying out stair stringers, rafters, and handrail fittings

Pitch board a block of wood that is cut to form a right triangle using the rise and run dimensions of the stair

Pitch pocket an opening in lumber between annular rings containing pitch in either liquid or solid form

Pith the small, soft core at the center of a tree

Pivot revolving around a point

Plain-sawed a method of sawing lumber that produces flat-grain

Plan view blueprint that shows the building from above looking down

Plancier the finish member on the underside of a box cornice; also called *soffit*

Plank lumber that is 6 or more inches in width and from $1^1/_2$ to 6 inches in thickness

Plank flooring a type of solid wood flooring that may be laid using alternating widths

Plaster a mixture of portland cement, sand, and water used for covering walls and ceilings of a building

Plastic laminate a very tough, thin material in sheet form used to cover countertops; available in a wide variety of colors and designs

Plastic toggle hollow wall anchor use to secure screws

Plate top or bottom horizontal member of a wall frame

Platform frame method of wood frame construction in which walls are erected on a previously constructed floor deck or platform

Plies sheets (thin pieces) of veneer used to make most plywood panels

Plinth block a small, decorative block, thicker and wider than a door casing, used as part of the door trim at the base and at the head

Plot plan a drawing showing a bird's-eye view of the lot, the position of the building, and other pertinent information; also called *site plan*

Plow a wide groove cut with the grain of lumber

Plumb vertical; at right angles to level

Plumb bob a pointed weight attached to a line for testing plumb

Plumb line any line on the rafter that is vertical when the rafter is in position

Plywood a building material in which thin sheets of wood are glued together with the grain of adjacent layers at right angles to each other

Pneumatic powered by compressed air

Pocket door a door that slides sideways into the interior of a partition; when opened, only the lock edge of the door is visible

Pocket tapes measuring device made of a steel self-retracting strip

Pocket a recess in a wall to receive a piece, such as a recess in a concrete foundation wall to receive the end of a girder

Point of beginning mark on plot plan indicating the start point for laying out the lot; usually a large object that is unlikely to move during construction such as a large rock or tree is used

Poles the vertical members of a scaffold

Polyurethane a type of rigid foam insulation made with a facing of foil or building paper

Portland cement an improved type of cement that is a fine gray powder

Post-and-rail fence a basic and inexpensive style of fencing normally used for long boundaries

Postforming method used to bend plastic laminate to small radii

Post-to-post a balustrade system where the handrail is cut and attached between newels

Powder-actuated drivers tool that uses gunpowder to drive fasteners into metal and masonry

Power miter saw tool used to cut trim material at various angles

Preacher a small piece of wood of the same thickness as stair risers; it is notched in the center to fit over the finish stringer and rests on the tread cut of the stair carriage; it is used to lay out open finished stringers in a staircase

Prefinished strip solid wood flooring that is sanded, finished, and waxed at the factory; it cannot be sanded after installation

Prehung door a door that comes already fitted and hinged in the door frame with the lockset and outside casings installed

Pressure-treated treatment given to lumber that applies preservative under pressure to penetrate the total piece

Prism a device used to reflect a light beam back along the exact path form which it arrived

Privacy lock a lock often used on bathroom and bedroom doors that functions by pushing or turning a button on the room side of the doorknob

Pullsaws hand saws that cut while being drawn toward the user

Pump jack a type of scaffold that consists of 4 × 4 poles, a pump jack mechanism, and metal braces for each pole

Purlin knee wall supporting timbers at which rafters may intersect each other in a gambrel roof

Putty glazing compound used to cover and seal around the edges of glass panes

Pythagorean theorem a theorem that can be used to determine the diagonal of any right triangle: $a^2 + b^2 = c^2$

Quarter round a type of molding, an end section of which is in the form of a quarter-circle

Quarter-sawed a method of sawing lumber parallel to the medullary rays to produce edge-grain lumber; see *edge grain*

Rabbet a cutout along the edge or end of lumber

Rabbeted jamb a type of side door jamb in which the rabbet acts as the stop; on a double-rabbeted jamb, one rabbet is used as a stop for the main door and the other as a stop for storm and screen doors

Racking term used to describe the lateral movement of a wall where the 90-degree angle between the studs and plates is affected

Racking the floor the practice of laying out seven or eight loose rows of flooring, end to end, after the second course of flooring has been fastened; done to save time and material

Radial arm saw older style of stationary saw used for crosscuting and ripping

Radiation a general word referring to electromagnetic radiation, which includes microwaves, radio waves, infrared, visible light, ultraviolet light, X rays, and cosmic rays

Radius edge decking specially shaped lumber that is usually used to provide the surface and walking area of the deck

Rafter a sloping structural member of a roof frame that supports the roof sheathing and covering

Rafter tables information found printed on the body of a framing square; used to calculate the lengths of various components of a roof system

Rail the horizontal member of a frame

Rake the sloping portion of the gable ends of a building

Rake cornice the exterior finish applied to the sloped portion of the end rafters

Rake rafter the first and last rafter of a gable roof, usually having a finish or trim applied to it; also called *barge* or *fly rafter*

Rebar steel reinforcing rod used in concrete

Reciprocating a back-and-forth action, as in certain power tools

Reciprocating saw general term used to describe saws that cut in a back-and-forth or stroking motion

Reinforced concrete concrete that has been reinforced with steel bars

Relative humidity percentage of moisture suspended in air compared to the maximum amount it could hold at the same temperature

Resilient a type of thin, flexible flooring material widely used in residential and commercial buildings that comes in sheet or tile form

Return a turn and continuation for a short distance of a molding, cornice, or other kind of finish

Return nosing a separate piece mitered to the open end of a stair tread for the purpose of returning the tread nosing

Reveal the amount of setback of the casing from the face side of window and door jambs or similar pieces

Ribbon a narrow board let into studs of a balloon frame to support floor joists

Ribbon coursing an alternative to straight-line courses in which both layers are applied in straight lines; the outer course is raised about 1 inch above the inner course

Ridge the highest point of a roof that has sloping sides

Ridgeboard a horizontal member of a roof frame that is placed on edge at the ridge and into which the upper ends of rafters are fastened

Rim joist floor member that is nailed perpendicular to the ends of joists

Rip fence guide tool used to support and maintain a straight cut in material being cut on by stationary power tools

Ripping sawing lumber in the direction of the grain

Ripsaw blade saw blade designed to cut best when cutting with the grain of the wood

Ripsaws saws designed to cut in the direction that is parallel to wood grain

Rise in stairs, the vertical distance of the step; in roofs, the vertical distance from plate to ridge; may also be the vertical distance through which anything rises

Riser the finish member in a stairway covering the space between treads

Roll roofing a type of roofing made of the same material as asphalt shingles that comes in rolls that are 36 inches wide

Roofing brackets a metal bracket used when the pitch of the roof is too steep for carpenters to work on without slipping

Roofing nails large-headed galvanized fastener used to attach roofing materials

Roof window a window for the roof that contains an operating sash to provide light and ventilation

Rosette a round or oval decorative wood piece used to end a handrail against a wall

Rough carpentry that part of the trade involved with construction of the building frame or other work that will be dismantled or covered by the finish

Rough opening opening in a wall in which windows or doors are placed

Rough sills members that form the bottom of a window opening, at right angles to the studs

Rough size approximate size; for example, cutting countertop laminate about $\frac{1}{4}$ to $\frac{1}{2}$ inch wider and longer than the surface to be covered

Router hand-held electric tool used to shape material with high spinning cutters

Run the horizontal distance over which rafters, stairs, and other like members travel

R-value a number given to a material to indicate its resistance to the passage of heat

Saddle see *cricket*

Safety glass glass constructed, treated, or combined with other materials to minimize the possibility of injuries resulting from contact with it

Sapwood the outer part of a tree just beneath the bark containing active cells

Sash that part of a window into which the glass is set

Sawhorses frameworks for holding material that is being laid out or cut to size

Sawyer a person whose job is to cut logs into lumber

Scab a length of lumber applied over a joint to stiffen and strengthen it

Scaffold an elevated, temporary working platform

Scale the proportional reduction of each line in a drawing of a building to a size that clearly shows the information and can be handled conveniently

Screed strips of wood, metal, or pipe secured in position and used as guides to level the top surface of concrete

Screwdriver bits interchangeable tips or various driving-style heads

Screwdrivers tools used to turn fasteners into and out of materials

Screwguns electric drills with a special nose piece to drive wallboard screws to an exact depth

Scribe laying out woodwork to fit against an irregular surface

Scribing fitting woodwork to an irregular surface; with moldings, cutting the end of one piece to fit the molded face of the other at an interior angle to replace a miter joint

Scuttle attic access or drain through a parapet wall

Seam-filling compound a compound made especially for laminates used to make a practically invisible joint

Seasoned lumber lumber that has been dried to a suitable moisture content

Seat cut a cut on the rafter that is a combination of a level cut and a plumb cut; also called the *bird's mouth*

Section view building plan that shows a cross section of the building as if it were sliced to reveal its skeleton

Self-tapping screw screws designed to drill a pilot hole as it is turned

Selvage the ungranulated or unexposed part of roll roofing covered by the course above

Setbacks distance buildings must be kept from property lines

Shake a defect in lumber caused by a separation of the annular rings; also, a type of wood shingle

Shank hole a hole drilled for the thicker portion of a wood screw

Shark tooth saw handsaw with aggressive teeth that cuts wood fast in both directions

Shear walls wall bracing, framed of 2 × 6s anchored to the slab, with OSB heavily nailed to them; used in regions of severe seismic activity

Sheathing boards or sheet materials that are fastened to roofs and exterior walls and on which the roof covering and siding are applied

Shed roof a type of roof that slopes in one direction only

Sheet paneling paneling, such as prefinished plywood and hardboard, that is usually applied to walls with long edges vertical

Shim a thin, wedge-shaped piece of material used behind pieces for the purpose of straightening them, or for bringing their surfaces flush at a joint

Shingle butt the bottom exposed edge of a shingle

Shingling hatchet a lightweight hatchet with both a sharp blade and a heel that is used to apply wood shakes and shingles

Shoe rail a plowed, lineal molding designed to receive the bottom square of a baluster

Shortened valley rafter a valley rafter that runs from the plate to the supporting valley rafter

Side the wide surfaces of a board, plank, or sheet

Side casings the side members of a window unit's exterior casing

Side flashing flashing tucked in between shingles along the sides of a chimney

Side jambs the vertical sides of a window frame

Siding exterior sidewall finish covering

Sidelight a framework containing small lights of glass placed on one or both sides of the entrance door

Sill horizontal timbers resting on the foundation supporting the framework of a building; also, the lowest horizontal member in a window or door frame

Single coursing when shingles are applied to walls with a single layer in each course, in a way similar to roof application

Single-hung window a window in which the lower sash slides vertically, but the upper sash is fixed

Single pole a type of wood scaffold used when it can be attached to the wall and does not interfere with the work

Single-strength (SS) glass glass about $3/_{32}$ inch thick used for small lights of glass

Skilsaw general term use to describe portable electric circular saws

Skylight a fixed-sash window for the roof that provides light only, no ventilation

Slab-on-grade foundation a solid concrete building base used instead of a foundation because it saves on material and labor

Slat block a type of parquet block made by joining many short, narrow strips together in small squares of various patterns

Sleeper strips of wood laid over a concrete floor to which finish flooring is fastened

Sleeve anchor solid wall fastener used to secure bolts

Sliding a type of exterior door in which the doors are opened by sliding the panels along a track horizontally

Sliding gauge a gauge used for fast and accurate checking of shingle exposure that permits several shingle courses to be laid at a time without snapping a chalk line

Sliding windows windows that have sashes that slide horizontally in separate tracks located on the header jamb and sill

Slope term used to indicate the steepness of a roof; stated as unit Rise on unit Run, for example, 6 on 12

Slump test a test given to concrete to determine its consistency

Snaplock punch a special tool used to produce raised lugs on a piece of siding so that it will snap snugly into the trim when fitting around windows

Snap tie a metal device to hold concrete wall forms the desired distance apart

Snub cornice a cornice with no rafter projection beyond the wall; chosen primarily to cut down the cost of material and labor

Soffit the underside trim member of a cornice or any such overhanging assembly

Softboard a low-density fiberboard

Softwood wood from coniferous (cone-bearing) trees

Sole plate the bottom horizontal member of a wall frame

Solid a type of cabinet door construction with doors made of plywood, particleboard, or solid lumber

Solid sheathing usually APA-rated panels over which shingles or shakes may be applied

Solid wall anchors metal fasteners that tighten inside a hole as a bolt or screw is turned

Sound-deadening insulating board a sound resistance system using a gypsum board outer covering and resilient steel channels placed at right angles to the studs to isolate the gypsum board from the stud

Sound transmission class (STC) a rating given to a building section, such as a wall, that measures the resistance to the passage of sound

Spaced sheathing usually 1 × 4 or 1 × 6 boards over which shingles or shakes may be applied

Specifications written or printed directions of construction details for a building, sometimes referred to as specs

Specifications guide used by spec writers for complex commercial projects; developed by the Construction Specifications Institute (CSI)

Specifications writer a person who writes supplemental information for construction projects to include any information that cannot be communicated in drawings or schedules

Speed Square tool used for quick layout of angles, particularly for rafters

Spline a thin, flat strip of wood inserted into the grooved edges of adjoining pieces

Split fast anchor fastener that tightens in a masonry hole by the opposing pressure of shaft legs

Spray foams a type of foam insulation in an aerosol can that is sprayed into gaps to expand and create an airtight seal

Spreader a narrow strip of wood tacked to the side jambs of a door frame a few inches up from the bottom so that the frame width is the same at the bottom as it is at the top

Spud vibrator an immersion type vibrator with a metal tube on its end that is commonly used to vibrate and consolidate concrete

Square the amount of roof covering that will cover 100 square feet of roof area

Staggered coursing coursing in which the butt lines of the shingles are alternately offset below, but not above, the horizontal line

Staging planks planks that rest on scaffold bearers with edges close together to form a tight platform

Stair body the part of the stair consisting mainly of the treads, risers, and finish stringers

Stair carriage the main strength of a set of stairs supporting the treads and the risers. Also called a stair horse

Staircase term used to refer to an entire set of stairs, which include treads, risers, stringers, and balustrade

Stair horse See stair carriage

Stairwell an opening in the floor for climbing or descending stairs, or the space of a structure where the stairs are located

Staple U-shaped fasteners used to secure a variety of materials

Staplers tools that drive U-shaped fasteners by either squeezing a spring-loaded lever or swinging like a hammer

Stapling guns tools that drive U-shaped fasteners by use of power provided by electricity, air pressure, or a fuel cell

Starter course usually used in reference to the first row of shingles applied to a roof or wall

Starter strip a furring strip of the same thickness and width of the siding headlap that is first fastened along the bottom edge of the sheathing

Starting step the first step in a flight of stairs

Steel buck a metal door frame using all steel studs to frame around door openings

Steel tapes term used to describe long measuring tools that are rewound by hand

Step ladder a folding portable ladder hinged at the top

Stickering machine that makes moldings or a thin strip placed between layers of lumber to create an air space for drying

Sticking the molded inside edge of the frame of a panel door

Stile the outside vertical members of a frame, such as in a paneled door

Stockade fence a higher fence with picket edges applied tightly together to provide privacy

Stool cap a horizontal finish piece covering the stool or sill of a window frame on the interior; also called *stool*

Stool the bottom member of a door or window frame; also called a *sill*

Stop a strip of wood applied vertically to the door jamb upon which the door rests when closed

Stop bead a vertical member of the interior finish of a window against which the sash butts or sides

Storm sash an additional sash placed on the outside of a window to create a dead air space to prevent the loss of heat from the interior in cold weather

Storm window a secondary window attached over the usual window to protect against the wind and cold

Story the distance between the upper surface of any floor and the upper surface of the floor above

Story pole a narrow strip of wood used to lay out the heights of members of a wall frame or courses of siding

Stove bolt bolt with a machine thread and a screw-slot head

Straightedge a length of wood or metal having at least one straight edge to be used for testing straight surfaces

Straight tin snips shearlike tool used for all-purpose cutting of sheet material

Striated finish material with random and finely spaced grooves running with the grain

Striker plate a plate installed on the door jamb against which the latch on the door engages when the door is closed

Stringer the finish material applied to cover the stair carriage

Strip see *strip flooring*

Strip flooring a type of widely-used solid wood flooring most often tongue-and-grooved on edges and ends to fit precisely together

Strongback a member placed on edge and fastened to others to help support them

Structural insulated panels (SIPs) panels that consist of two outer skin panels and an inner core of an insulating material

Stub joists joists that are shorter and run perpendicular to the normal joists

Stud vertical framing member in a wall running between plates

Subfascia a horizontal framing member fastened to the rafter tails; sometimes called the *false fascia* or *rough fascia*

Supporting valley rafter a rafter that runs from the plate to the ridge of the main roof

Surfaced four sides smooth, square-edge lumber used for door casings; abbreviated as *S4S*

Surface hinges a type of cabinet hinge that is applied to the exterior surface of the door and frame

Swing direction an installed door will open

Swinging doors doors that are hinged on one end and swing out of an opening; when closed, they cover the total opening

Table saw stationary circular power saw with a large bed or table

Tack to fasten temporarily in place; also, a short nail

Tail cut a cut on the extreme lower end of a rafter

Tail joist short joist running from an opening to a bearing

Taper becoming thinner from one end to the other

Taper ripping jig a wide board with the length and amount of taper cut out of one edge that is used to cut tapered pieces

Tempered treated in a special way to be harder and stronger

Tempered glass a type of safety glass treated with heat or chemicals that cause the entire piece of glass to immediately disintegrate into a multitude of small granular pieces when broken

Template a pattern or a guide for cutting or drilling

Tensile strength the greatest longitudinal stress a substance can bear without tearing apart

Termites insects that live in colonies and feed on wood

Termite shield metal flashing plate over the foundation to protect wood members from termites

Thermal envelope the part of a building that creates the boundary between conditioned and unconditioned air

Thin set a mortar-type adhesive often used to set ceramic tile; used when tile is likely to get soaked with water

Threshold a piece with chamfered edges placed on the floor under a door; also called a *sill*

Tile square or rectangular blocks placed side by side to cover an area

Tile grout a thin mortar, mixed from a powder and water, to form a paste that is worked into tile joints with a trowel

Timber large pieces of lumber over 5 inches in thickness and width

Toe the forward end of tools, such as a hand saw and hand plane

Toeboard a strip of material located at the back of the toe space under a base cabinet; also the bottom horizontal member of a scaffold guardrail

Toenailing technique of driving nails diagonally to fasten the end of framing

Toe space a recess provided at the bottom of a cabinet

Toggle bolts hollow wall anchor with spring-loaded wings to secure a machine-treaded bolt

Top lap the height of the shingle or other roofing minus the exposure

Topography a detailed description of the land surface

Top plates the top members of a wall frame

Total Rise the vertical distance that the roof rises from plate to ridge

Total Run the total horizontal distance over which the rafter slopes

Total span the horizontal distance covered by the roof

Total station a surveying instrument that uses a light beam to measure distances and angles accurately

Trammel points tool with sharp points that is clamped to a strip of wood to lay out arcs

Transit-level similar to a builder's level, but with a telescope that can be moved up and down 45 degrees in each direction

Transom small sash above a door

Tread horizontal finish members in a staircase on which the feet of a person ascending or descending the stairs are placed

Tread molding a small cove molding used to finish the joint under stair and landing treads

Trestle similar to a sawhorse used to support scaffold plank

Trimmer joist full-length joists that run along the inside of an opening

Trimmers members of a frame placed at the sides of an opening running parallel to the main frame members

Truss an engineered assembly of wood or wood and metal members used to support roofs or floors

Tungsten carbide-tipped blade a wood-cutting blade tipped in tungsten carbide, a material used to make cutting edges stay sharper longer

Turnout a type of handrail fitting

Twist lumber defect in wood

Twist drills term used for drilling bits typically used for making holes in metal

Undercourse the bottom layer of less expensive wood shingles applied when double coursing sidewalls

Undercut a cut made through the thickness of finished material at slightly less than 90 degrees, such that butt joints in the material will fit tightly on the face sides

Underlayment material placed on the subfloor to provide a smooth, even surface for the application of resilient finish floors. It is also the base material installed from rolls placed on the roof deck before the roof finish.

Undersill trim a siding system accessory; also known as *finish trim*

Unfinished strip solid wood flooring that is milled with square, sharp corners at the intersections of the face and edges; once laid, any unevenness in the faces of adjoining pieces is removed by sanding the surface

Unit block a parquet block made up of smaller strips of wood

Unit Length the length of a stair stringer or rafter per unit of Run

Unit Rise the amount a stair or rafter rises per unit of Run

Unit Run a horizontal distance of a stair tread or horizontal segment of the total Run of a rafter

Unit Span twice the unit run for a common rafter

Urethane glue a high-performance glue that is strong, flexible, and durable, and is used for exterior applications

Utility knife all-purpose cutting tool, typically with a retractable blade

Valley the intersection of two roof slopes at interior corners

Valley cripple jack rafter a rafter running between two valley rafters

Valley jack rafter a rafter running between a valley rafter and the ridge

Valley rafter the rafter placed at the intersection of two roof slopes in interior corners

Vapor normally invisible, a cool gaseous state of water

Vapor barrier plastic sheet used to prevent moisture from penetrating the building surface

Vapor retarder a material used to prevent the passage of vapor; also called *vapor barrier*

Variance a notion granted by the Zoning Board of Appeals in a community to change the zoning code due to hardships imposed by zoning regulations

Veneer a very thin sheet or layer of wood

Vermiculite a mineral closely related to mica with the ability to expand on heating to form a lightweight material with insulating qualities

Vertical arc a scale, attached to a telescope, that measures vertical angles to 45 degrees above and below the horizontal

Vertical grain siding in which the annual growth rings, viewed in cross section, must form an angle of 45 degrees or more with the surface of the piece

Vinyl-clad windows whose exposed wood parts are encased in vinyl

Volute a spiral fitting at the beginning of a handrail

Wainscoting a wall finish applied partway up the wall from the floor

Waler horizontal or vertical members of a concrete form used to brace and stiffen the form and to which ties are fastened

Wall angle metal L-shaped pieces that are fastened to the wall to support the ends of main runners and cross tees in a suspended ceiling grid

Wallboard saw hand saw designed to cut gypsum board

Wall sheathing exterior wall covering that may consist of boards, rated panels, fiberboard, gypsum board, or rigid foam board

Wall unit one of two basic kinds of kitchen cabinets; the wall unit is installed about 18 inches above the countertop

Wane bark, or lack of wood, on the edge of lumber

Warp any deviation from straightness in a piece of lumber

Waste factor an amount added to a material order beyond the exact calculation to ensure sufficient material supply for job completion

Water table finish work applied just above the foundation that projects beyond it and sheds water away from it

Weather stripping narrow strips of thin metal or other material applied to windows and doors to prevent the infiltration of air and moisture

Web wood or metal members connecting top and bottom chords in trusses; also, the center section of a wood or steel I-beam

Web stiffener a wood block that is used to reinforce the web of an I-joist, often at locations where the I-joist is supported in a hanger and the sides of the hanger do not extend up to the top flange

Wedge anchor fastener used to secure bolts in solid walls

Whet the sharpening of a tool on a sharpening stone by rubbing the tool on the stone

Wind a defect in lumber caused by a twist in the stock from one end to the other

Winder a tread in a stairway, wider on one end than the other, that changes the direction of travel

Window flashing a piece of metal as long as the head casing that is bent to fit over the head casing and against the exterior wall to prevent the entrance of water at this point; also called a *drip cap*

Window frame the stationary part of a window unit; the window sash fits into the window frame

Window schedules drawings that give information about the location, size, and kind of windows to be installed in the building

Wing dividers compasslike tool use to lay out circles and perform incremental step-offs in various layouts

Wire nails general term for most nails that are made from coils of wire

Wood chisel metal tool designed to be driven by a hammer, and that is used to make mortises and other rectangular holes in wood

Wood shingles a common roof covering most often produced from western red cedar

Woven valley valleys flashed by applying asphalt shingles on both sides of the valley and alternately weaving each course over and across the valley

Wrecking bar large pry bar used in demolition; often called a *crow* bar

Yoking applying vertical members (2 × 4s) that hold outside form corners together

Zones areas communities are divided into to separate the types of buildings that can be built in that area

Zoning regulations keeps buildings of similar size and purpose in areas for which they have been planned

Accessories (Accesorios) para los armarios, artículos utilizados para mejorar la instalación de estos muebles, tales como paneles laterales, paneles frontales para lavavajillas y refrigeradores, estanterías abiertas para los extremos de los armarios y especieros

Acoustical tile (Losetas acústicas) losetas de fibra para techos cuya superficie presenta grietas u orificios pequeños que atrapan el sonido

Adjustable wrench (Llave inglesa ajustable) llave inglesa con mordazas que varían su ancho de apertura al girar un tornillo roscado

Admixture (Aditivo) material que se agrega al concreto o mortero para lograr cualidades especiales

Aggregates (Áridos) materiales como arena, rocas o grava que se utilizan para hacer concreto

Air-dried lumber (Madera secada al aire) madera que ha sido curada mediante su exposición al aire

Air-entrained cement (Cemento celular) cemento al que se han agregado pequeñas burbujas de aire para incrementar su resistencia a los ciclos de congelación y descongelación

Anchor (Anclaje) dispositivo que se usa para sujetar dos partes de una estructura en su lugar

Annular ring (Anillo de crecimiento) los anillos que se ven en una sección transversal de un tronco de árbol; cada anillo constituye un año de crecimiento del árbol

Apron (Zócalo) pieza del contramarco de la ventana que se coloca bajo la repisa

Apron flashing (Vierteaguas para chimeneas) hileras de tejas que se colocan debajo de los vierteaguas en el lado más bajo de una chimenea

Architect's scale (Escalímetro) regla usada para dibujar y tomar medidas en varias proporciones o escalas

Arch windows (Ventanas de arco) ventanas cuya sección superior es curva, lo que las hace aptas para combinarlas con diferentes tipos de ventanas o puertas

Areaway (Acceso al sótano) área amurallada bajo el nivel del suelo que rodea las ventanas del sótano

Argon gas (Gas argón) gas usado para rellenar el espacio entre las capas de un cristal aislante para aumentar el valor de R

Asphalt felt (Fieltro asfaltado) papel aislante saturado con asfalto que se utiliza como impermeabilizante

Asphalt shingles (Tejas de asfalto) tipo de tejas cuya superficie contiene una capa de gránulos minerales seleccionados y que están cubiertas con asfalto para impermeabilizarlas

Astragal (Astrágalo) moldura semicircular que se utiliza para cubrir la unión entre puertas

Auger bits (Brocas salomónicas) broca para perforar madera con punta en forma de hélice

Aviation snips (Tijeras de aviación) tijeras metálicas cuyo mango tiene mecanismo de palanca para aumentar la fuerza de corte

Awning window (Ventana abatible hacia afuera) tipo de ventana cuyo marco está abisagrado en la parte superior y se abre hace afuera

Back band (Banda trasera) moldura que se aplica alrededor de los lados y los extremos de puertas y ventanas

Back bevel (Bisel dorsal) biselado hecho sobre el borde o extremo del material hacia el lado trasero

Backing (Refuerzo) bandas o bloques colocados en las paredes o techos para sujetar o sostener instalaciones o accesorios

Backing board (Panel de refuerzo) tipo de panel de yeso diseñada para usarse como capa base en sistemas multicapa

Backing the hip (Recortar el borde de una limatesa) biselar el borde superior de un cabio de limatesa para alinearlo con las superficies adyacentes del techo

Back miter (Inglete) ángulo cortado desde el extremo hacia la superficie del material

Backset (Distancia de entrada) distancia entre un objeto y un borde, lado o extremo de un material, tal como la distancia entre una bisagra y el borde de una puerta

Backsplash (Protector de pared contra salpicaduras) parte elevada del borde trasero de una encimera de cocina para proteger la pared

Balloon frame (Armadura sin rigidez) tipo de armadura en el que los puntales se extienden ininterrumpidamente desde los cimientos hasta el techo

Baluster (Balaústre) piezas verticales de un pasamano de escalera que se usan con frecuencia de manera decorativa y se ubican uno al lado del otro con una pequeña separación entre sí

Balustrade (Barandaje) toda la estructura del pasamanos de una escalera, lo que incluye baranda, balaústres y postes

Base (Remate) ver *base shoe* (**moldura de remate**)

Baseboard (Zócalo) placa de terminación usada para cubrir la unión en la intersección de la pared y el suelo

Base cap (Moldura de zócalo) moldura que se coloca en el borde superior del zócalo

Base shoe (Moldura de remate) moldura que se coloca entre el zócalo y el piso

Base unit (Armario base) uno de dos tipos básicos de armarios de cocina; el armario base se ubica abajo

Batten (Listón) banda delgada y angosta que se usa para cubrir las uniones entre paneles verticales

Batter board (Tabla de excavación) tabla provisoria que se coloca erguida para sostener las líneas extendidas de la disposición de la construcción

Bay window (Ventana mirador estilo bay window) ventana que generalmente tiene 3 hojas y sobresale de la línea de la pared

Beam pocket (Caja para viga) abertura en la pared de cimentación donde reposa la viga maestra

Bearer (Soporte) parte horizontal de un andamio de madera que sostiene el tablón de éste

Bearing (Apoyo) área de la superficie de una pieza estructural donde se transfiere el peso o la carga

Bearing partition (Pared divisoria con carga) pared interior que sostiene el peso del techo

Bed (Lecho) moldura estándar que se usa generalmente en la intersección de paredes y techos

Belt sander (Lijadora de correa) herramienta lijadora eléctrica con una cinta giratoria de papel de lija

Benchmark (Punto de referencia) referencia para determinar elevaciones durante la construcción de un edificio

Bench plane iron (Hierro para cepillo de banco)   parte cortante de un cepillo de banco

Bench planes (Cepillos de banco)   herramienta manual para cortar capas finas de madera

Bevel (Bisel)   el borde o lado de una pieza que forma ángulo oblicuo y no recto

Bevel ripping (Corte al hilo en bisel)   término utilizado para describir el tallado de la madera en sentido de la veta y en ángulo

Bifold door (Puertas plegables)   puertas que están unidas en pares por bisagras y otra que está abisagrada a la jamba

Bird's mouth (Muesca tipo "pico de pájaro")   corte en forma de V hecho en una viga para ensamblarla en la parte superior de la pared

Bit brace (Berbiquí)   herramienta perforadora con dos agarraderas usada para hacer orificios

Blind nail (Clavado invisible)   método de sujeción con clavos en el que los clavos quedan ocultos

Blind stop (Tope invisible)   parte de la terminación de una ventana que se coloca dentro del marco exterior de ésta

Blocking (Refuerzo)   piezas cortas de madera instaladas en el piso, el techo o las paredes que proveen aislamiento climático, actúan como contrafuegos o sostienen las partes de la estructura

Blockouts (Blockouts)   formas usadas para lograr las aberturas más grandes en paredes de concreto

Block planes (Cepillos de contrafibra)   herramienta manual pequeña con una cuchilla de ángulo bajo que se usa para cortar capas finas de madera

Blueprinting (Copia azul / Blueprinting)   antiguo proceso para copiar planos de construcción cuyo resultado era de color azul con líneas y letras blancas

Board (Tabla)   pieza de madera que en general tiene 8 pulgadas o más de ancho y menos de 2 pulgadas de grosor

Board and batten (Entablado y listonado)   tipo de revestimiento de madera que solamente se puede colocar de manera vertical

Board on board (Tablas superpuestas)   tipo de revestimiento de madera que solamente se puede colocar de manera vertical

Board-on-board fence (Cerca de tablas superpuestas)   similar a una cerca de estacas, pero las tablas se colocan alternadas, una de un lado y otra del otro, para que la cerca se vea igual de ambos lados

Board foot (Pie de tabla)   medida de volumen de madera que equivale a 1 pie de ancho por 1 pie de largo y 1 pulgada de grosor o cualquier volumen de madera equivalente

Board paneling (Paneles)   tablas utilizadas en las paredes interiores cuando se desea mantener la calidez y la belleza de la madera sólida

Boring (Taladrado)   término usado para referirse a la perforación de orificios profundos

Boring jig (Matriz de taladrado)   herramienta que se usa generalmente para guiar las brocas cuando se perforan orificios para juegos de cerraduras

Bow (Comba)   tipo de alabeo en el que un trozo de madera se curva de un extremo al otro

Bow window (Ventana de mirador tipo Bow window)   ventana que sobresal del edificio y, por lo general, crea más espacio en el suelo

Box cornice (Cornisa en caja)   cornisa con voladizo que ayuda a proteger las paredes laterales del clima y proporciona sombra a las ventanas; probablemente sea el tipo de cornisa más común

Box header (Cabezal de cajón)   en la construcción de plataformas, partes del marco que tapan los extremos de las vigas del piso; también llamadas *vigas perimetrales*, *viguetas de apoyo perimetrales*, o *cabezales de viga*

Box nail (Clavo de encajonar)   clavo delgado con cabeza que por lo general está recubierto de un material para aumentar su fuerza de sujeción

Braces (Riostras)   parte diagonal de un andamio que previene que los parales se muevan o doblen

Brad (Puntilla o clavo corto)   clavo de terminación delgado y corto

Brake (Plegadora)   herramienta plegadora de chapa metálica

Bridging (Riostras diagonales)   riostas o bloques de madera sólida diagonales puestos entre las vigas del suelo que se usan para distribuir el peso impuesto sobre el suelo

Bright (Brillante)   término usado para describir clavos sin recubrimiento

Btu (Btu, por sus siglas en inglés)   abreviación de unidad térmica británica

Buck (Marco)   estructura compacta que se usa para formar aberturas en paredes de concreto

Bucks (Marcos de ventanas)   en la construcción de mampostería, piezas de madera que se sujetan a los lados del hueco de las aberturas para ventanas y sobre las que éstas se instalan

Builder's level (Nivel para constructor)   telescopio al que se le coloca un nivel de burbuja

Building codes (Normas de construcción)   ver *national building codes (normas nacionales de construcción)*

Building permits (Autorización para construcción)   prueba de autorización otorgada por las autoridades locales para construir una estructura

Bulls eye bubble (Burbuja de nivel)   herramienta de nivelación que para su funcionamiento utiliza una burbuja de líquido que debe centrarse dentro de un círculo

Butt (Tope)   junta que se forma cuando una pieza cortada en forma de cuadrado se coloca junto con otra; también es un tipo de bisagra

Butt hinge (Bisagra de tope)   bisagra formada por dos placas pegadas a las superficies lindantes de una puerta y de una jamba y que están unidas por un pasador

Butt hinge template (Plantilla de bisagra de tope)   plantilla metálica posicionada y asegurada con espigas que se usa para guiar las muescas de las bisagras en las puertas y marcos

Butt markers (Delineadores de tope)   herramienta usada para delinear las muescas para bisagras de tope

Butterfly roof (Techo tipo mariposa)   techo a dos aguas invertido

Bypass doors (Puertas corredizas)   puertas, usualmente dos en una sola abertura, que tienen rueditas y que se montan sobre carriles para que puedan deslizarse una al lado de la otra

Cabinet lift (Elevadores de gabinetes)   elevador usado para sostener los armarios en posición para poder así atornillarlos a la pared

Café doors (Puertas de bar)   puertas persiana o de paneles cortos que se sujetan de a pares y se abren en ambas direcciones; se usan para dividir un área de manera parcial pero sin evitar un paso fácil y seguro a través de la abertura

Cambium layer (Capa de cambium)   capa que se encuentra dentro de la corteza de un árbol donde se forman las células nuevas

Capillary action (Acción capilar)   movimiento del líquido dentro de grietas o espacios pequeños debido a la atracción molecular

Caps (Molduras de terminación)   molduras usadas para darles un acabado a los bordes superiores de otras molduras

Carbide blade (Hoja de sierra de carburo)   ver *Tungsten carbide-tipped blade (Cuchilla con punta de carburo de tungsteno)*

Carbide-tipped (Herramienta con puntas de carburo)   hace referencia a herramientas cortantes que tienen piezas pequeñas y extremadamente duras de acero soldadas a las puntas

Carpenter's level (Nivel de carpintero)   herramienta de 2 pies de largo con nivel de burbuja que se usa para determinar nivel y plomada

Carriage bolt (Bulón de cabeza de hongo y cuello cuadrado)   bulón fileteado mecánicamente que tiene cabeza redonda y no tiene la ranura típica de los tornillos

Casement window (Ventana batiente) tipo de ventana cuyo marco está abisagrado en los bordes y usualmente se abre hacia afuera

Casing (Marco) moldura que se coloca alrededor de las puertas, ventanas y otras aberturas

Casing nails (Puntilla de contramarcos) similar a los alfilerillos; muchos carpinteros la prefieren para sujetar el acabado exterior

Catches (Pestillos) herraje utilizado para mantener las puertas de un armario cerradas cuando éstas no tienen bisagras de cierre automático

Caulk (Sellador) sellador tipo masilla que se utiliza para tapar rajaduras y hendiduras

Cement board (Placa de cemento) panel con núcleo de cemento pórtland reforzado con una malla de fibra de vidrio en ambos lados

Cements (Cementos) adhesivos que unen los productos de asfalto para el techo y los vierteaguas

Ceramic tile (Teja cerámica) tejas duras, frágiles, resistentes al calor y a la corrosión que se fabrican al quemar un mineral no metálico como la arcilla; vienen en gran variedad de formas geométricas, tamaños, diseños, colores y modelos

Chair rail (Guardasilla) moldura que se coloca de manera horizontal a lo largo de la pared para evitar que los respaldos de las sillas choquen contra la pared

Chalk line (Tendel) soga con polvo de color que se usa para marcar líneas rectas

Chamfer (Chaflán) borde biselado, como esos hechos entre la cara superior y los bordes de un piso sólido de tablones de canto antes de terminarlo para ocultar cualquier desnivel luego de la instalación

Chase wall (Muro ducto) estructura formada por dos paredes paralelas que se ubican muy cerca una de la otra y que se fabrican para contener las tuberías, los ductos de calefacción y refrigeración y artículos similares

Check (Grieta) Separación en el extremo o a lo largo de la superficie de una madera que resulta luego de que una parte de ésta se haya secado más rápido que el resto de la pieza

Cheek cut (Corte biselado tipo cheek cut) corte a inglete compuesto hecho en el extremo de ciertas vigas de techo

Clamps (Mordazas) herramientas utilizadas para sostener materiales juntos temporalmente

Claw hammer (Martillo de uña) herramienta de carpintero para clavar o remover clavos

Cleat (Ristrel) pieza de medida de 1″ × 5″ de madera que se instala alrededor de las paredes de un guardarropas para sostener estantes y la percha; también

son bloques que se ajustan al soporte de una escalera para sostener los escalones

Closed (Escalones cerrados) tipo de escalera en la que los topes de los escalones están contra la pared y no a la vista

Closed stairway (Escalera encerrada) tipo de escalera que se instala entre dos paredes y cuyos topes de escalones y contrahuellas están ocultos

Closed valley (Limahoya cerrada) limahoya del techo en el que su recubrimiento se une en el centro de la limahoya y la cubre totalmente

Close-grained (De grano fino o cerrado) madera cuyos poros son pequeños y están muy juntos

Coatings (Revestimiento) producto que se usa para recubrir la superficie de techos o metales viejos intemperizados

Coil stock (Material en espiral) material de aluminio que se vende en rollos de 50 pulgadas de largo y con diferentes anchos que van desde 12 hasta 24 pulgadas

Cold-rolled channels (CRCs) (Perfiles laminados en frío (CRC)) perfiles formados por 54 milipulgadas de acero que se usan en cielos rasos colgantes y como crucetas para soporte lateral de paredes

Collar tie (Falso tirante) parte horizontal que se coloca cerca del camellón en ángulo recto con la solera

Combination blade (Hoja de sierra combinada) hoja de sierra circular diseñada para cortar la madera de manera transversal o en el sentido de las vetas

Combination pliers (Pinzas ajustables) pinzas con mordazas ajustables

Combination square (Escuadra de combinación) escuadra con hoja deslizable

Common nails (Clavos comunes) sujetadores usados para trabajos como las armaduras

Common rafter (Cabio común) se extiende desde el panel de la pared hasta la viga de caballete, en el que su paso es perpendicular al panel

Common screw (Perno común) sujetador de cabeza ranurada que se usa con diferentes funciones

Compass saw (Serrucho de canal) serrucho manual pequeño con hoja ahusada que se usa para hacer cortes curvos o con formas irregulares

Competent person (Persona competente) individuo que supervisa y dirige la construcción del andamiaje y que tiene la autoridad para tomar medidas correctivas para garantizar su seguridad

Composites (Compuestos) placas de madera sujetadas a un núcleo de madera reconstituido

Compression glazing (Compresión de vidriería) término utilizado para describir varias formas de sellar cristal monolítico o aislante en el sistema de estructuras de apoyo con caucho sintético y otras empaquetaduras elastoméricas; este método elimina toda posibilidad de que se filtre agua a través de las juntas

Compressive strength (Resistencia a la compresión) cualidad de ser resistente al aplastamiento; el concreto tiene alta resistencia a la compresión

Concrete (Concreto) material de construcción que se hace con cemento pórtland, áridos y agua

Condensation (Condensación) cuando el agua en estado gaseoso cambia a líquido porque se enfría; las gotas de aguas resultantes se acumulan en la superficie fría

Conduction (Conducción) transferencia de calor por contacto

Conductor (Conductor) elemento vertical utilizado para transportar agua desde la canaleta hasta el suelo; también *se denomina tubo conductor o guía*

Conical screw (Tarugo) sujetador de pared hueco con roscas profundas que sirve para sostener tornillos

Coniferous (Coníferas) árbol cuyos frutos tienen forma de cono; también conocido como *árbol de hoja perenne*

Connectors (Conectores) término usado para describir sujetadores de metal largos usados para unir las partes de una armadura

Construction heart (Corazón de la construcción) el tipo más apto y económico de las secuoyas de California que usan para hacer postes para plataformas, vigas y juntas

Contact adhesives (Adhesivos de contacto) adhesivos usados para pegar placas de yeso entre sí; no se pueden mover las placas una vez pegadas

Contact cement (Cemento de contacto) adhesivo utilizado para unir laminados plásticos y otros materiales delgados; se llama así ya que la unión se hace al contacto y no se necesitan abrazaderas

Contour interval (Intervalos de contorno) escala usada entre líneas de un mapa topográfico

Contour line (Curva de nivel) líneas de un mapa que representan cierta elevación del suelo

Control joints (Juntas de control) tiras metálicas con aletas de $1/4$ pulgadas y con una ranura en forma de V que se coloca en paredes de yeso grandes para reducir la presión por la expansión y la contracción

Convection (Convección) movimiento fluido (del aire) causado por las diferencias entre las densidades de fluido cálido y frío (del aire); la gravedad empuja el aire frío más denso hacia abajo y desplaza el aire cálido menos denso

Coped joint (Corte en falsa escuadra) tipo de junta ubicada entre molduras en la que el extremo de una se corta para encajar la superficie moldeada de la otra

Coping saw (Serrucho de calar) sierra manual de hoja delgada y corta sujetada por un marco arqueado que se usa para cortar formas irregulares en ebanistería

Cordless nailing gun (Pistola de clavos inalámbrica) herramienta para colocar clavos que funciona con una pila de combustible en vez de presión de aire

Corner bead (Guardavivo) vestidura metálica que se usa en las esquinas exteriores de las paredes para decorarlas y protegerlas

Corner boards (Tablas para esquinas) Tablas usadas para decorar las esquinas de las paredes exteriores de un edificio

Corner brace (Riostra de esquina) pieza diagonal del marco de la pared que se usa en las esquinas para endurecer y reforzar la pared

Corner guards (Esquinero) moldura que se coloca en las esquinas de las paredes

Corner post (Poste esquinero) montante armado que se usa en la esquina de un panel de pared

Cornice (Cornisa) ensamble entero terminado donde las paredes de una estructura se unen con el techo; también denominada alero o sofito.

Cornice return (Vuelta de cornisa) usada para cambiar la caja de cornisa nivelada según el ángulo del techo

Counterboring (Ensanchamiento) hacer un orificio más grande en una pequeña porción a través del material para embutir la cabeza de un sujetador

Countersink (Abocardo) hundir un poco la superficie de un orificio para recibir la cabeza de un perno de cabeza chata; también se lo llama así a la herramienta para realizar dicha acción

Counter Flashing (Contraplancha de escurrimiento) Planchas de escurrimiento ubicadas en los lados de una chimenea donde el borde superior está apoyado en una junta de mortero y el borde inferior cubre el vierteaguas escalonado

Course (Hilera) fila continua de materiales de construcción como ladrillos, revestimiento, tejas o pisos

Cove (Ensenada) moldura de forma cóncava

Coverage (Cobertura) número de capas superpuestas de tejas y el grado de protección del clima que estas ofrecen

Crawl space (Hueco sanitario) tipo de cimientos que dejan un espacio debajo del primer piso y que no tiene la altura suficiente para usarse como sótano

Cricket (Banquillo) techo falso pequeño construido detrás de una chimenea u otro obstáculo para el techo y que tiene como propósito repeler el agua; también se lo conoce como *chaflán*

Cripple jack rafter (Cabrio corto de limatesa) cabrio corto que no hace contacto ni con la solera superior ni con el camellón

Crook (Combadura) tipo de alabeo en el que el borde de la madera está doblado

Crosscut (Aserrado transversal) corte en sentido transversal a las vigas de la madera

Crosscut circular saw blade (Hoja circular de sierra de trozar) hoja de sierra apropiada para realizar cortes transversales a las vetas de la madera

Crosscut saws (Sierra de trozar) herramienta diseñada para cortar en sentido transversal a las vetas de la madera

Cross tees (Perfiles en T) piezas metálica que viene en largos de 2 y 4 pies y que tienen lengüetas de conexión en cada extremo; estas lengüetas se insertan y se traban en las correderas principales y en otras cruces en T suspendidas en la rejilla del techo

Crown (Moldura de corona) tipo de moldura estándar; ver *bed (lecho)*

Crown (Vértice) usualmente indica el punto superior del borde doblado de vigas, limatesas y otros elementos de armaduras

Cup (Desviación de canto) tipo de comba en el que el lado de una placa está curvada de borde a borde

Cylindrical lockset (Cerradura cilíndrica) herraje para cerrar o trabar una puerta al que se lo conoce como *cerradura de llave en la perilla (del inglés key-in-the-knob)*

Dado (Dado) corte parcial o perpendicular a las vetas de la madera

Deadbolt (Cerrojo muerto) cerradura que se abre y se cierra con una llave desde el exterior y con una manija o llave desde adentro

Dead load (Peso muerto) peso de todos los componentes estructurales y de todos los materiales que se sujetan en su lugar

Deadman (Macizo de anclaje) dispositivo de madera en forma de T que se usa para sostener los paneles de yeso del techo y otros objetos

Deciduous (Árboles de hojas caducas) árboles cuyas hojas se caen todos los años

Deck (Plataforma) superficie de madera o metal a la cual se aplican los materiales para el techado

Deck board (Placas para la plataforma) madera que cubre perfiles y vigas para proporcionar un área de superficie y de paso en la plataforma

Deflection (Desviación) grado de comba que presentará un componente estructural cuando tenga peso encima.

Detail (Detalle) vista en primer plano de un plano o una sección de plano

Dew point (Punto de rocío) temperatura en la que la humedad en el aire se condensa en gotas

Diagonal (Diagonal) en un ángulo, generalmente de esquina a esquina en una línea recta

Dimension lumber (Madera dimensionada) madera utilizada para enmarcar que tiene un espesor nominal de 2 pulgadas

Door casings (Marcos de la puerta) la moldura utilizada para cortar los lados y la parte superior de una puerta

Door jack (Gato para la puerta) gato fabricado o improvisado que mantiene una puerta firme durante la instalación

Door schedules (Lista de puertas) gráfico informativo que se encuentra en un conjunto de impresiones que brindan información pertinente sobre las puertas del edificio

Door stop (Tope de puerta) moldura fijada a las jambas de la puerta para que la puerta se detenga contra ésta

Dormer (Buhardilla) estructura que se proyecta hacia afuera de un techo con vertiente para formar otra zona techada que brinda una superficie para la instalación de ventanas.

Double-acting doors (Puertas de vaivén) puertas que oscilan en ambas direcciones o las bisagras utilizadas en esas puertas

Double coursing (Hilera doble) cuando se aplica dos capas de tejas a las paredes laterales en una hilera

Double coverage (Cobertura doble) tipo de techo en rollo con clavos ocultos

Doubled-faced cabinets (Armarios con frente doble) armarios que tienen puertas en ambos lados para usar sobre bases en forma de isla o península

Double-hung window (Ventana de guillotina) ventana en la que dos hojas se deslizan verticalmente hacia arriba y hacia abajo

Double pole (Poste doble) tipo de andamio de madera utilizado cuando se debe mantener el andamiaje alejado de la pared para aplicar materiales o por otras razones.

Double-strength glass (DS) (Vidrio de doble resistencia) vidrio de aproximadamente $\frac{1}{8}$ de pulgada de espesor que se utiliza para hojas de vidrio más grandes

Dovetail (Cola de milano) tipo de junta entrelazada semejante a la forma de la cola de una paloma

Dowel (Espiga) varas de madera noble de varios diámetros

Downspout (Canaleta) ver *conductor (conductor)*

Draft (Corriente) ver *firestop (contrafuegos)*

Draft-stop blocking (Bloqueador de corriente) ver *fire-stop blocking (bloqueador contra incendio)*

Drilling (Perforación) término utilizado para describir la realización de pozos con brocas torneadas

Drip (Goterón) parte de una cornisa o de una capa de panel de revestimiento horizontal que se proyecta debajo de otra parte; también, corte en canal en la parte inferior de un alféizar que hace que el agua caiga en vez de retroceder y deslizarse por la pared

Drip cap (Cubeta de goteo) moldura colocada en la parte superior de los marcos de puertas y ventanas exteriores cuya finalidad es alejar el agua de las unidades

Drip edge (Reborde de escurrimiento) tiras marginales de metal colocadas en los bordes de los techos para brindar soporte al excedente del material del techo

Drive anchor (Anclaje de conducción) sujetador mural sólido que se coloca con un martillo

Drop-in anchor (Anclaje embutido) sujetador mural sólido que requiere una herramienta de inserción y ajuste para sostener las tuercas del perno

Dropping the hip (Profundizar la limatesa) aumentar la profundidad del corte de asiento del cabio de limatesa para que la línea central del borde superior se apoye en el plano de las superficies adyacentes del techo

Dry rot (Pudrición en seco) residuo seco y arenoso de madera que queda después de la destrucción de la madera por parte de hongos debido a una humedad excesiva

Drying potential (Potencial de secado) edificio construido para que los materiales de construcción puedan secarse fácilmente después de la exposición a climas extremos

Drywall (Panel de yeso) tipo de construcción que se refiere, generalmente, a la instalación de placas de yeso

Drywall cutout tool (Herramienta para cortar paneles de yeso) herramienta eléctrica portátil utilizada para hacer cortes en paneles de pared para cajas eléctricas, conductos y objetos similares

Drywall foot lifter (Elevador de panel de yeso) herramienta utilizada para elevar paneles de yeso en la capa inferior contra los paneles superiores

Dry well (Pozo seco) excavación rellena de grava o rocas para atrapar agua a fin de que pueda ser absorbida en la tierra

Duplex nail (Clavo dúplex) clavo de dos cabezas utilizado para sujeciones temporales, como en la construcción de andamios de madera

Eased edge (Borde liso) borde de madera cuyas esquinas filosas han sido redondeadas

Easement (Curva de transición) pieza curvada que une la primera y segunda barandilla en una balaustrada

Eaves (Alero) parte de un techo que se extiende por fuera de la pared lateral

Edge (Borde) superficie estrecha de madera que se une a la veta

Edge grain (Veta del borde) tablas en las que los anillos de crecimiento están en posición perpendicular o casi perpendicular al frente; también denominada veta vertical

Elastomeric membrane (Membrana elastomérica) material que tiene las propiedades elásticas del caucho natural

Electrolysis (Electrólisis) descomposición del más suave de dos metales diferentes en contacto entre sí ante la presencia de agua

Elevation (Elevación) dibujo en el que se muestra la altura de la estructura o del objeto; también, la altura de un punto específico en relación a otro punto

End (Extremo) las extremidades de un pedazo de madera

End lap (Junta de solape) distancia horizontal en la que los extremos del techo en la misma hilera se superponen entre sí

Engineered lumber products (ELPs) (Productos de madera industrializados) sustitutos elaborados de madera como vigas en L de madera, vigas laminadas encoladas, madera revestida laminada, madera de tiras paralelas y madera de tiras laminadas

Engineered panels (Paneles industrializados) productos hechos por el hombre en la forma de hojas de madera reconstituidas grandes

Equilibrium (Equilibrio) estado de equilibrio; la energía calórica pasa de materiales más calientes a materiales más fríos con el propósito de alcanzar este estado

Equilibrium moisture content (Contenido de humedad de equilibrio) el punto en el que el contenido de humedad de la madera es igual al contenido de humedad del aire circundante

Escutcheon (Bocallave) placa de protección que cubre el pomo o el ojo de la cerradura en las puertas

Expanded polystyrene (Poliestireno expandido) un tipo de aislante de plástico espumado rígido que absorberá la humedad si entra en contacto con agua; también llamado *placa decorativa blanca*

Expansion anchors (Anclajes de expansión) sujetador mural hueco que se expande detrás del panel de yeso para asegurar los tornillos

Expansive bit (Broca de expansión) herramienta para perforar madera que se ajusta a varios diámetros

Exposure (Exposición) la cantidad de superficie de panel de revestimiento o techo que está expuesta a la intemperie

Extension jambs (Jambas extensibles) extensiones aplicadas al borde interno de las jambas de la ventana para acomodar varios espesores de pared

Extension ladder (Escalera extensible) escalera cuya longitud puede extenderse tirando de la cuerda y levantándola a la altura deseada

Extruded polystyrene (Poliestireno extruido) tipo de aislante de plástico espumado rígido con un tipo de estructura de célula cerrada que no absorberá el agua

Face (Frente) el lado con mejor apariencia de un pedazo de madera o el lado que queda expuesto cuando se lo instala, como el piso acabado

Face frame (Marco del frente) estructura de pedazos estrechos en el frente de un armario que contiene aberturas para puerta y cajón

Face nail (Claveteado delantero) método para colocar un sujetador casi perpendicularmente a la superficie del material

Faceplate marker (Marcador de placa frontal) herramienta utilizada para disponer la muesca para la placa frontal del pestillo

Fall protection (Protección contra caídas) requisito de OSHA cuando se utilizan andamios donde los trabajadores están a alturas superiores a los 10 pies; consta de barandillas o un sistema de protección contra caídas

Fancy butt (Tejas decorativas) tejas disponibles en diferentes diseños; utilizadas en combinación con tejas cuadradas u otros tipos de revestimiento; ofrecen patrones atractivos

Fascia (Imposta) membrana vertical del acabado de la cornisa instalada en la parte trasera de las vigas

Feather boards (Tablas de canto biselado) herramienta de guía utilizada para asegurar material contra una valla mientras se lo trabaja en herramientas eléctricas fijas

Feather edge (Canto biselado) el canto de un material reducido a una astilla larga a un canto muy delgado, como la punta de una teja de madera

Fiberboard (Tablero de fibra) material de construcción hecho de virutas finas de madera prensadas en láminas

Fiber-saturation point (Punto de saturación de la fibra) el contenido de humedad de la madera cuando las cavidades de la célula están vacías pero las paredes celulares aún están saturadas

Filler (Relleno) en la fabricación de muebles, pedazos que se utilizan para llenar espacios pequeños en espesor entre unidades en pared y unidades en piso cuando ninguna combinación de tamaños puede llenar el espacio existente

Fillets (Filetes) tiras angostas utilizadas entre balaústres para llenar la ranura labrada en barandillas

Finish carpentry (Carpintería acabada) la parte de la industria de la carpintería involucrada en la aplicación de acabados exteriores e interiores

Finger-jointed (Juntas dentadas) juntas hechas en una fábrica para unir longitudes cortas para hacer longitudes largas

Finish nail (Clavo de acabado) clavo delgado con una cabeza pequeña diseñado para ser colocado debajo de la superficie de materiales acabados

Finish schedules (Lista de acabados) gráfico informativo que se encuentra en un conjunto de impresiones que brindan información pertinente sobre el diseño interno de un edificio

Finish stringer (Plano acabado) tabla acabada que se une a la pendiente de las escaleras y que cubre la junta entre las escaleras y la pared; también llamado *borde lateral de faldón*

Finishing coat (Última mano) tercera y última mano de compuesto de yeso aplicado para cubrir las juntas de la pared, arreglos de metal y sujetadores

Fire-stop blocking (Bloqueador contra incendio) material instalado para reducir el movimiento del fuego y humo dentro de cavidades más pequeñas de la estructura de un edificio durante un incendio; también llamado *bloqueador de corriente*

Firestop (Contrafuegos) material utilizado para llenar los conductos de aire en una estructura para evitar la propagación del fuego y gases calientes; también llamado *contracorriente*

Firsts and Seconds (Primeros y segundos) la mejor calidad de madera noble

Fissured (Agrietado) ranuras moldeadas irregulares hechas de material, como losa de techo, para fines acústicos

Fittings (Accesorios) secciones cortas de barandillas especialmente curvadas

Fixed windows (Ventanas fijas) ventanas que constan de un marco en el que la hoja está colocada en una posición fija

Flange size (Tamaño del reborde) término de acero que es similar al espesor de un pedazo de madera

Flashing (Tapajuntas) material utilizado en intersecciones como valles del techo, claraboyas, y arriba de ventanas y puertas para evitar que el agua ingrese

Flat bar (Barra plana) llave de palanca multipropósito con pinzas planas

Flat grain (Veta plana) veta en la que los anillos de crecimiento de la madera están casi de forma paralela a los laterales; opuesto a veta del borde

Flatwork (Planchado) término utilizado para describir losas, caminos y aceras de concreto

Flexible insulation (Aislamiento flexible) tipo de aislamiento fabricado en forma de lámina o panel

Floor guide (Guía de piso) herraje instalado en la base de puertas corredizas para mantener las puertas alineadas

Floor joists (Vigas de piso) miembros horizontales de un marco que se apoyan sobre y transfieren la carga a umbrales y vigas

Floor plans (Planta) páginas de un conjunto de dibujos de construcción que muestran las paredes como vistas desde arriba

Flush (A ras) término utilizado para describir cuando la junta superficial entre dos materiales está perfectamente alineada

Flush door (Puerta a ras) puerta con una superficie suave y plana de laminado de madera o de metal

Flush type door (Puerta tipo a ras) tipo de puerta de armario que se encaja en y está alineada con el frente de la abertura

Flute (Estría) ranuras cóncavas cercanas entre sí de madera para fines decorativos en una columna, un poste o una pilastra

Foamed-in-place (Espumado in situ) aislamiento de espuma de uretano producido al mezclar dos químicos que se inyectan en el lugar para expandirse en contacto con superficies

Footing (Base de apoyo) cimiento para una columna, pared o chimenea que es más amplio que el objeto que sostiene, para distribuir el peso sobre una área mayor

Foundation (Cimiento) parte de una pared sobre la cual se erige la porción mayor de la estructura

Foundation wall (Pared de cimentación) porción de apoyo de una estructura debajo de la construcción del primer piso, o debajo del nivel, incluidas las bases de apoyo

Frameless (Sin marco) tipo de construcción de armario que no usa marco delantero; también llamado *método europeo*

Framing hammer (Martillo de carpintero) martillos más pesados (20 a 32 onzas) utilizados en la realización de marcos

Framing square (Escuadra de enmarcar) herramienta de acero en forma de L, de 24 pulgadas de largo, utilizada para disponer vigas y escaleras

French door (Puerta francesa) puerta, generalmente una de un par, de construcción liviana con hojas de vidrio que se extienden en casi toda su longitud

Frieze (Friso) parte del acabado exterior en la intersección de una cornisa sobresaliente y la pared

Frost line (Profundidad de penetración del congelamiento) profundidad a la que el hielo penetra en el suelo en un área particular; las bases de apoyo deben colocarse debajo de este nivel de profundidad

Full rounds (Redondos completos) pedazo redondo de moldura que por lo general se utiliza como postes para armarios

Full sections (Secciones completas) páginas de un conjunto de dibujos de construcción que muestran la sección transversal del edificio

Furring channels (Canales enrasados) pedazos en forma de sombrero hechos de acero de 18 y 33 mm que se aplican a paredes y techos para fijar los tornillos de paneles de yeso

Furring strips (Tiras para enrasar) tiras que por lo general se fijan directamente a vigas expuestas para brindar una superficie de ajuste para losas de techo

Gable end (Hastial) sección triangular al final de un edificio formada por las vigas en un techo común o a dos aguas y la línea del techo

Gable end studs (Pernos del hastial) pernos que forman la pared que se cierra en el área triangular debajo del techo a dos aguas

Gable roof (Techo a dos aguas) tipo de techo con vertiente en dos direcciones

Gables (Faldones) áreas triangulares formadas por las vigas inclinadas y la placa de pared en los extremos del edificio

Gable studs (Perno del faldón) perno cuyo extremo inferior es cuadrado y se fija contra la parte superior de la placa de pared, y cuyo extremo superior se fija contra el borde inferior y lado interno de la viga para enmarcar el faldón

Gain (Ganancia) recorte hecho en una pieza para recibir otra pieza, como un recorte para una bisagra plana

Galvanized (Galvanizado) protegido de la oxidación mediante una capa de zinc

Galvanized steel (Acero galvanizado) acero que está revestido con zinc para ayudar a protegerlo de la corrosión

Gambrel roof (Techo abuhardillado) tipo de techo que tiene dos inclinaciones diferentes en cada lado del centro

Girder (Viga maestra) madera o viga pesada que se usa para sostener cargas verticales

Glass bead (Perla de vidrio) moldura pequeña utilizada para sostener una hoja de vidrio en su lugar

Glazier (Vidriero) persona que instala vidrios en un marco

Glazing (Encristalar) instalar vidrios en un marco

Glazing compound (Pasta para instalar cristales) material suave de plástico similar a la masilla que sirve para sellar los cristales a un marco

Glazing points (Puntas de vidriar) piezas de metal pequeñas, triangulares o con forma de diamante que se usan para asegurar y sostener el cristal en el marco

Glue-laminated lumber (Maderas laminadas encoladas) columnas o vigas largas que se hacen al pegar maderas de menor dimensión una al lado de la otra; popularmente se las conoce como *laminado de madera encolada*

Gooseneck (Cuello de cisne) parte curva de un pasamanos ubicada al llegar a un descanso; también es una salida de una canaleta de techo

Grade (Nivel) nivel del piso; también identifica la calidad de la madera

Grade rod (Lectura para rasante) la altura del instrumento menos la elevación del punto de control

Grain (Vetas) en la madera, el diseño de las superficies causado por el contraste, el espaciado y la dirección de los anillos anulares

Green (Verde) a) (del inglés green concrete), concreto fresco que aún no está completamente curado; b) concepto que se deriva de construir con energías eficientes y renovables y tomando medidas económicas efectivas

Green lumber (Madera verde) madera que aún no se ha secado a su punto justo

Green space (Espacios verdes) áreas de una construcción destinadas a la vegetación

Grilles (Rejas) en una ventana, parteluz falso colocado delante de ésta para simular pequeños cristales

Grit (Arenilla) material del papel de lija que realiza la acción de lijar

Groove (Rebajo) corte parcial perpendicular y paralelo a las vetas de la madera

Ground (Ristrel) listones de madera que se colocan en la base de paredes y alrededor de las aberturas y se usan como guías para la aplicación de una capa pareja de yeso; también (del término inglés ground) puesta a tierra

Ground Fault Circuit Interrupter (GFCI) (Disyuntor diferencia (GFCI, por sus siglas en inglés)) dispositivo usado en circuitos eléctricos para la protección contra descargas eléctricas; detecta el cortocircuito instantáneamente y corta la corriente de manera automática

Grouting (Enlechado) proceso de relleno de pequeños espacios con una pasta espesa de cemento

Guardrails (Barandas para andamios) barandas instaladas en todos los lados abiertos y extremos de andamios que están a más de 10 pies de altura

Gusset plate (Escuadra de ensamble) ver *scab (costanera)*

Gutter (Canaleta) canal de madera, metal o plástico utilizado en la orilla del techo para escurrir el agua de lluvia y la nieve derretida

Gypsum (Aljez) tipo de piedra que es el ingrediente principal del yeso

Gypsum board (Panel de yeso) placa que se hace revistiendo un tipo de papel pesado con yeso

Gypsum lath (Papel enyesado) tipo de placa de yeso que se usa para como base para colocar yeso convencional

Hacksaws (Sierra cortametales) sierra diseñada para cortar metales

Half-round (Moldura de media caña) moldura cuyo extremo tiene forma de semicírculo

Hammer-drills (Perforadora de martillo) herramienta perforadora eléctrica que, mediante el impacto, hace orificios en mampostería a gran velocidad

Handrail (Pasamanos) baranda de escaleras de donde las personas se agarran para subir o bajar y que tiene la función de proporcionar apoyo y seguridad

Handrail bolt (Perno para pasamanos) perno que en un extremo tiene un espiral para atornillarse en madera y en el otro tiene un espiral para colocar tuercas; se usa para unir los ajustes con una sección derecha del pasamanos en la balaustrada

Hand scraper (Raspador manual) herramienta diseñada para quitar capas finas de material de diferentes superficies

Hardboard (Tablero de aglomerado) producto para la construcción que se hace comprimiendo fibras de madera para formar planchas

Hardwoods (Madera dura) madera de árboles dicotiledóneos de hoja ancha (para distinguirla de la madera de las coníferas)

Head casing (Chambrana de dintel) parte superior del revestimiento exterior de una ventana

Headers (Travesaños) partes colocadas en ángulo recto a los cabrios, montantes y limatesas para formar y sostener las aberturas en los marcos de madera

Head flashing (Tapajuntas superior) tapajuntas ubicado en la parte superior de una chimenea

Head jamb (Jamba superior) el miembro horizontal superior del marco de una ventana

Head lap (Solape superior) distancia entre el borde inferior de una teja superpuesta a la parte superior de una teja dos hileras abajo, a la altura de la inclinación

Hearth (Hogar) área cerca de una chimenea, generalmente enlosada y que se extiende hacia afuera en una habitación, alrededor de la cual se debe colocar una instalación de piso de madera

Heartwood (Duramen) la madera de la parte interna de un árbol, generalmente más oscura y que contiene células inactivas

Heel (Talón) parte trasera de los objetos, como una sierra de mano o un cepillo de mano

Height of the instrument (Altura del instrumento) medida que se obtiene colocando la barra de nivelación en el punto de referencia y sumando después la lectura de la barra a la elevación del punto de referencia

Herringbone (Espiguilla) patrón utilizado en pisos de parqué

High-speed steel blade (Hoja de acero de alta velocidad) hoja de sierra sin punta cortante de carburo

Hinge gain (Ganancia de bisagra) área cóncava para la bisagra en una puerta

Hinge mortise (Muesca de bisagra) ver *hinge gain (ganancia de bisagra)*

Hip jack rafter (Cabio corto de limatesa) un cabio común acortado que cruza desde la placa de pared a un cabio de limatesa

Hip rafter (Cabio de limatesa) se extiende diagonalmente desde la esquina de la placa al camellón en la intersección de dos superficies de un techo a cuatro aguas

Hip roof (Techo a cuatro aguas) techo que asciende hacia el camellón desde cuatro direcciones

Hip-valley cripple jack rafter (Cabio de limatesa a limahoya) cabio corto que se extiende de forma paralela a cabios comunes, cortado entre cabios de limatesa y limahoya

Hole saws (Sierras perforadoras) taladros que cortan con dientes para sierra a lo largo de su perímetro

Hopper window (Ventana de hoja basculante) tipo de ventana en la cual la hoja está abisagrada en la base y oscila hacia adentro

Horizontal circle scale (Escala circular horizontal) escala que se divide en cuadrantes de 90 grados cada uno y que permanece fija a medida que se gira el telescopio

Horizontal vernier (Pie de rey horizontal) escala que rota con el telescopio y se utiliza para leer los minutos de un grado

Hose bibbs (Grifos de manguera) grifos de agua externos de un edificio

Housed stringer (Zanca cubierta) zanca acabada que está frisada para recibir escalones o contrahuellas de una escalera

Housewrap (Membrana microporosa) tipo de papel de construcción con el que están cubiertas todas las paredes laterales de un edificio

Hydration (Hidratación) reacción química entre el cemento y el agua que hace el concreto o el mortero se endurezcan

Ice and water shield flashing (Tapajuntas protectora contra hielo y agua) tapajunta aplicada siempre que existe la posibilidad de que se formen acumulaciones de hielo a lo largo de aleros y que provoquen el retroceso de agua

Ice dam (Acumulación de hielo) hielo que se forma en un alero y que hace que la acumulación de agua detrás de éste quede retenida debajo del material del techo

Impact noise (Ruido por impacto) ruido causado por la vibración de un impacto, como objetos caídos o pasos en el suelo

Impact noise rating (INR) (Clasificación de ruido por impacto) clasificación que muestra la resistencia de varios tipos de construcciones suelo-techo para impactar los ruidos

Independent slab (Losa independiente) losa de concreto que está separada de los cimientos

Inserts (Encastres) anclajes plásticos y de plomo que se usan comúnmente para sujetar accesorios livianos en paredes de mampostería

Insulated glass (Vidrio aislante) Ventanas con doble o triple vidriado; los vidrios aislantes mejorados contienen gas argón entre sus paneles

Insulation (Aislación) material usado para restringir el paso del calor o el sonido

Intersecting roof (Cubierta a múltiples aguas) techo de edificios con diseños irregulares; limahoyas que se forman en la intersección de los techos

Isometric (Isométrico) plano en el que se aprecian tres superficies de un objeto y la base de cada superficie está dibujada a un ángulo de 30 grados respecto de las otras

Jack rafter (Cabrio corto o secundario) parte de un cabrio común que se corta para ensamblarse en una limatesa, limahoya o en ambas

Jack studs (Travesaño corto) travesaño acortado que sostiene los travesaños

Jamb (Jamba) lados y parte superior del marco de una puerta o ventana

Jamb extension (Extensión de jamba) listones angostos de madera sujetados al borde de las jambas de una ventana para aumentar su ancho

J-channel (Canal J) accesorio del revestimiento

Jig (Guía) cualquier tipo de accesorio diseñado para sostener piezas o para guiar herramientas mientras se realiza el trabajo

Jigsaw (Sierra caladora) pequeña sierra eléctrica manual que corta con un movimiento de vaivén

Joint compound (Compuesto para juntas) sustancia similar al yeso usada para cubrir juntas o las cabezas de pernos o clavos en las placas de yeso

Joint (Estirar / Junta) como verbo (estirar), significa enderezar el

extremo de una madera; como sustantivo (junta), denota el lugar donde las partes se juntan y se unen

Jointer plane (Cepillo juntera) el más largo de los cepillos de banco que se usa para darle un acabado a la madera quitándole tiras finas

Joint reinforcing tape (Cinta reforzadora de juntas) material usado para cubrir las juntas de yeso, reforzarlas y brindarles resistencia contra las grietas

Joist (Vigueta) parte horizontal de la estructura distribuida de manera espaciada que da soporte al sistema de piso o techo

Joist hanger (Colgaderos de viguetas) estribos de metal que se utilizan para sostener los extremos de las viguetas

Joist header (Cabezal de vigueta) ver*cabezal de cajón*

J-roller (Rodillo en J) rodillo de caucho de 3 pulgadas de ancho usado para aplicar presión sobre la superficie de contacto de laminados plásticos adheridos con cemento

Juvenile wood (Madera joven) porción de la madera que contiene los primeros siete a 15 anillos de crecimiento de un tronco; se encuentra en la médula

Kerf (Corte o sección) ancho de un corte hecho con una sierra

Keyhole saw (Sierra de punta) versión más delgada de un serrucho de calar

Keyway (Adaraja) ranura hecha en la cimentación de concreto para sujetarla a la pared de cimientos de concreto

Kicker (Taco) en la construcción de cajones, pieza centrada arriba del cajón que impide que este se incline y se caiga cuando se abre

Kiln-dried (Secado al horno) madera secada en hornos grandes llamados "hornos de secar"

Kneewall (Pared de buhardilla) pared corta que, en relación con las escaleras, se proyecta hacia arriba con la misma inclinación que la estructura de la escalera

Knot (Nudo) defecto en la madera causado por seccionarla a través de una rama unida al tronco

Ladder jacks (Ménsulas para escaleras) ménsulas metálicas instaladas en las escaleras para sostener los escalones

Lag screws (Tirafondos) tornillos largos roscados con cabeza hexagonal

Lag shield (Tirafondos de expansión) anclaje usado para asegurar los tirafondos en las paredes sólidas

Laminate (Laminado) capa fina de plástico usada comúnmente como superficie de terminación de encimeras de cocina

Laminated block (Plancha laminada) tipo de plancha para parqué hecha generalmente de roble laminado de tres capas con un espesor de $^3/_8$ pulgadas

Laminated glass (Vidrio laminado) tipo de vidrio de seguridad de dos o más capas de cristal y capas interiores de plástico transparente pegadas entre sí

Laminated strand lumber (LSL) (Tablas de virutas laminadas (LSL, por sus siglas en inglés)) madera que se fabrica al pegar virutas finas de madera de hasta 12 pulgadas de largo con adhesivo y haciendo presión

Laminated strip flooring (Pisos de tablones de cantos laminados) tablones de madera preacabados de 5 capas

Laminated veneer lumber (LVL) (Madera de chapa laminada (LVL, por sus siglas en inglés)) madera fabricada mediante el laminado de muchas chapas de madera contrachapada cuyas vetas corren todas en la misma dirección

Laminate trimmer (Recortadora de laminado) pequeño contorneador usado para cortar y encastrar laminados plásticos en el armado de los armarios

Landing (Descanso) plataforma de nivel intermedio ubicada entre los tramos de la escalera

Landing treads (Escalón de descanso) escalones ubicados en el borde de los descansos o balcones

Laser (Láser) el láser produce un rayo de luz concentrada; en la construcción de edificios se usan dispositivos equipados con láser

Laser level (Nivel láser) nivel que emite un rayo de luz rojo que rota 360 grados y crea, así, un plano nivelado

Lateral bracing (Abrazaderas laterales) abrazaderas temporales o permanentes que se ubican de manera perpendicular a los elementos sujetados

Lattice (Celosía) listones delgados de madera dispuestos de manera espaciada uno al lado del otro y en dos capas en ángulo una respecto de la otra similares a un enrejado

Lattice work (Panel de celosía) usado con frecuencia para cerrar espacios entre la plataforma y el suelo; también se usa para rellenar espacios en las cercas sobre la plataforma

Ledger (Larguero) parte horizontal de un andamio de madera que sostiene los postes juntos y los soportes; pieza de soporte temporal para vigas y otras partes que se colocan en ángulos rectos

Level line (Línea de nivel) cualquier línea en el cabio que queda horizontal cuando este está en posición

Level (Nivel) horizontal o perpendicular a la fuerza de gravedad

Lever handle (Mango de palanca) se usa en lugar de las perillas en las cerraduras que pueden ser usadas por personas con capacidades reducidas o en otras situaciones en las que sean más apropiadas que una perilla

Light (Hoja de vidrio) panel de cristal o la abertura para éste

Lignin (Lignina) pegamento natural de la madera que une las células de la madera y las fibras

Line length (Largo de línea) medida a lo largo de un cabio sin considerar su ancho ni espesor

Line level (Nivel de cuerda) dispositivo suspendido desde una cuerda para determinar un nivel

Linear feet (Pies lineales) medida de longitud en pies de una dimensión

Linear measure (Medida de longitud) medida de largo

Lintel (Dintel) pieza horizontal que sostiene la carga sobre una abertura; también llamado *cabezal*

Lipped door (Puerta con reborde) tipo de puerta de armario con bordes rebajados que se superponen a las aberturas en una medida de $3/8$ pulgadas en todos los lados

Live load (Carga viva) carga movible o temporaria sobre una estructura.

Load-bearing partitions (LBP) (Tabiques de carga (LBP, por sus siglas en inglés)) tabiques que sostienen los extremos internos de las vigas de los techos o pisos

Lock rail (Peinazo de la cerradura) peinazo situado a la altura de la cerradura, usualmente a 38 pulgadas del suelo hacia su centro

Lockset (Cerradura) herraje usado para cerrar y trabar las puertas

Lookout (Soporte de sofito) armadura horizontal en una cornisa que se instala para sujetar el sofito

Loose-fill insulation (Aislamiento de relleno suelto) tipo de aislamiento que se compone de materiales por bultos y se vende en bolsas o fardos

Louver (Rejilla de ventilación) abertura para ventilación formada por listones horizontales instalados en ángulo para evitar el paso de la lluvia, la luz y la visión pero que dejan paso al aire

Louver door (Puerta de persiana) puerta hecha con listones horizontales llamados rejilla que reemplazan los paneles; los listones están instalados en ángulo, lo que obstruye la visión pero permite el paso del aire a través de la puerta

Louvered fence (Cerca tipo persiana) Panel de cercado con tablas verticales dispuestas en ángulo

Low emissivity (Vidrio de baja emisividad) término utilizado para designar un tipo de vidrio que atrapa el calor y calienta los cuartos en invierno, y no permite su paso en verano; se los llama vidrios de baja emisividad

Low-emissivity coating (Revestimiento de baja emisividad) tipo de revestimiento de la superficie interior del vidrio de una ventana que permite el paso de los rayos ultravioleta del sol pero que refleja los rayos infrarrojos (radiantes) hacia adentro del edificio

Lumber (Madera) material que se extrae del tronco de un árbol con el que se hacen planchas, tablones y madera para la construcción

Lumber grades (Grados de la madera) números y letras utilizados para designar la madera según su calidad

Machine bolt (Perno ordinario) sujetador roscado de cabeza hexagonal que se usa con tuerca y arandela

Magazine (Depósito de clavos) contenedor de clavadoras eléctricas y engrapadoras donde se colocan los clavos que van a ser clavados

Main runners (Canales principales) piezas de metal con forma de T invertida que forman el soporte principal del peso del cielo raso colgante; se extienden de pared a pared

Major span (Tramo mayor) ancho de la porción más grande de un edificio con seis u ocho esquinas

Mansard roof (Techo de mansarda) tipo de techo que tiene dos inclinaciones distintas en todos los lados del edificio y las pendientes más bajas son más empinadas que las más altas

Mantel (Repisa de chimenea) la terminación de adorno alrededor de una chimenea que incluye el estante sobre la abertura

Masonry (Mampostería) cualquier construcción de piedra, ladrillo, tejas, yeso de concreto o materiales similares

Masonry drill bits (Brocas de perforación para mampostería) broca de perforación con punta de carbón diseñada especialmente para perforar productos de mampostería

Masonry nails (Clavos para mampostería) sujetadores de acero endurecido que se usan en materiales cementosos

Mastic (Masilla) adhesivo espeso

Matched boards (Planchas machihembradas) planchas que han sido terminadas con bordes a ranura y lengüeta

Medullary ray (Rayos medulares) bandas de células que parten del centro de la capa de cambium hasta la médula de un árbol para transportar nutrientes hacia el centro

Metal corner tape (Cinta metálica para esquinas) material usado para proteger esquinas de arcos, ventanas sin contramarcos y otros; se aplica incrustándola en el compuesto para juntas

Metes and bounds (Trazado) límites establecidos por las distancias y la dirección de la brújula

Mica (Mica) laminado plástico delgado duro de alta presión

Millwork (Carpintería) cualquier producto de madera manufacturado, como molduras, puertas, ventanas y escaleras que se usan en la construcción; a veces se la llama *ebanistería*

Minor span (Tramo menor) ancho de la porción más pequeña de un edificio de seis u ocho esquinas

Miter boxes (Caja de ingletes) herramienta fija o ajustable que se usa para guiar las sierras manuales a la hoja de cortar juntas a ingletes o cuando se hacen cortes transversales

Mitered joint (Juntas a ingletes) unión de dos piezas al cortarles los extremos biseccionando el ángulo por el que están unidas

Miter gauge (Calibre de inglete) guía usada en la sierra de mesa para hacer ingletes y extremos con forma cuadrada

Miters (Ingletes) cortes angulares en cualquier ángulo que no sea recto hechos con una caja de ingletes

Mobile scaffold (Andamio móvil) andamio cuyos componentes incluyen ruedas y abrazaderas diagonales horizontales; también llamado *torre rodante*

Modular measurement (Medida modular) proceso de diseño de estructuras para que éstas encajen de la manera más adecuada con el tamaño estándar de los materiales

Moisture content (Contenido de humedad) cantidad de humedad de una madera expresado como porcentaje del peso seco

Moisture meter (Higrómetro) dispositivo usado para determinar la cantidad de agua de la madera

Molding (Moldura) listones de madera decorativos usado para hacer acabados

Monolithic slab (Losa monolítica) losa combinada con cimientos; también llamada losa engrosada

Mortared (Relleno de mortero) uso de una mezcla de cemento pórtland, cal, arena y agua para unir piezas de mampostería

Mortise (Entalladura) cavidad rectangular cortada en una pieza de madera para recibir la lengüeta o espiga de otra pieza

Mud sill (Zapata de asiento) una placa de 2 3 10 de aproximadamente 18 a 24 pulgadas de largo sobre la que se apoyan los andamios

Mullion (Entreventana) división vertical entre ventanas o paneles de una puerta

Multispur bit (Broca de puntas múltiples) broca mecánica guiada por un patrón para hacer orificios que se usa para hacer el agujero para insertar la cerradura

Muntin (Parteluz) listones delgados de madera que ponen entre las hojas de vidrio de una ventana o puerta

Nail claw (Sacaclavos) herramienta de palanca usada para remover los clavos que ya están totalmente clavados

Nailers (Clavadoras) término generalizado que se usa para nombrar a las herramientas eléctricas que clavan sujetadores

Nail set (Embutidor) herramienta usada para empujar las cabezas de los clavos debajo de la superficie

National building codes (Normas nacionales de construcción) normas y disposiciones que dan pautas sobre la industria de la construcción según lo estipulan los organismos nacionales

Newel posts (Poste de barandilla) poste al que se une el final de la baranda de una escalera o balaustrada; también se llama así a cualquier poste al que se une cualquier barandilla o balaustrada

No. 1 common (N.° 1 común) grado inferior de madera dura

Nominal size (Tamaño nominal) el grado estipulado de grosor y ancho de la madera aunque difiera con su tamaño real; el tamaño aproximado de la madera sin labrar antes de que sea trabajada

Nonconforming (Antirreglamentario) término utilizado para describir a aquellos edificios que no se ajustan a las leyes de zonificación locales

Non-load-bearing partitions (NLBPs) (Tabiques sin carga (NLBP, por sus siglas en inglés)) tabiques construidos para dividir los espacios dentro de cuartos de diferentes tamaños que solamente soportan su propio peso y no el peso estructural del resto del edificio

Nosing (Vuelo del escalón) borde redondeado de un peldaño de la escalera que sobresale del contrapeldaño

Nylon plug (Tarugo) se usa para aplicaciones en paredes huecas y en algunas paredes sólidas

Offset hinges (Bisagras acodadas) tipo de bisagras de armario que se usan en las puertas con reborde

On center (OC) (de centro a centro (OC, por sus siglas en inglés)) distancia desde el centro de un miembro estructural al centro del siguiente

Open (De ojo) tipo de escalera en la que los extremos de los escalones están expuestos a la vista

Open cornice (Cornisa abierta) cornisa sin sofito

Open-grained (Madera porosa) calidad de textura de una madera en la que sus células están abiertas a la superficie

Open stairway (Escalera de ojo) tipo de escalera en la que los extremos de los escalones y contrapeldaños están a la vista

Open valley (Limahoya abierta) tipo de limahoya en la que la cubierta del techo se mantiene detrás de la línea central de la limahoya

Optical levels (Nivel óptico) instrumentos usados para nivelar, aplomar y determinar la disposición de los ángulos; incluye nivel para constructor y nivel de tránsito

Oriented strand board (OSB) (Tablero de partículas orientadas (OSB, por sus siglas en inglés)) denominación común de paneles no laminadas

Orthographic (Ortográfico) planos de vistas múltiples

Overlay door (Puerta superpuesta) tipo de puerta de armario que se superpone a la abertura por un tramo de $3/8$ pulgadas en todos los ángulos

Overlay hinges (Bisagra superpuesta) tipo de bisagra de armario disponible en versión oculta o semioculta

Over-the-post (Pasamanos tipo Over-the-post) sistema de balaustrada en el que se utilizan ajustes sobre los postes de la escalera para lograr un pasamanos continuo y sin cortes

Panel (Panel) Placa larga de material de construcción

Panel door (Puerta ensamblada) puerta hecha con un marco y de uno a ocho paneles con distintos diseños

Paneled (Puerta de paneles) tipo de puerta de armario construida con un marco exterior de madera sólida y paneles de madera sólida, madera contrachapada, madera prensada, plástico, vidrio y otro material

Panel siding (Revestimiento de paneles) revestimiento de 4 pies de ancho y de 8 y 10 pies de largo, algunos son de 9 pies; generalmente se colocan de manera vertical pero también pueden usarse de manera horizontal

Parallel strand lumber (PSL) (Perfiles de astillas paralelas (PSL, por sus siglas en inglés)) madera fabricada uniendo astillas de madera estructural de hasta 8 pies de largo con adhesivo, calor y presión

Parquet (Parqué) suelo hecho con tiras o bloques para formar diseños intrincados

Parquet block (Bloque de parqué) tipo de material para suelos que consiste en bloques cuadrados o rectangulares; a veces se combinan con listones para lograr diseños de mosaicos

Parquet strip (Listón de parqué) tipo de material para piso de listones cortos que se colocan de varias formas para lograr diseños en mosaicos

Partial section elevations (Alzada en corte parcial) plano ortográfico que muestra solamente un lado exterior del edificio a una distancia de 100 pies aproximadamente

Particleboard (Tablero prensado) producto para la construcción que se hace comprimiendo astillas de madera y aserrín con adhesivos para formar placas

Parting strip (Listón separador) pequeño listón de madera que separa los bastidores superiores e inferiores de una ventana de guillotina

Partition (Pared de separación) pared interior que separa una parte del edificio de otra

Passage lockset (Cerradura sin llave) cerradura con perillas de los dos lados que se giran para alzar el pestillo de la puerta; se usan cuando no se desea trabar la puerta

Patio door (Puerta corrediza) puerta que consiste en dos o tres hojas de puerta de vidrio deslizables ensambladas en un marco

Penny (d) (Clavos Penny (d)) término usado para designar los tamaños de clavos

Perforated (Perforado) material con orificios de corta separación entre sí distribuidos en patrones regulares o irregulares

Perlite (Perlita) aislamiento de relleno suelto que a veces se mezcla con concreto para hacerlo más liviano y mejorar su función aislante

Phillips head (Tornillo con cabeza Phillips) tipo de tornillo que tiene una ranura en cruz en la cabeza

Picket fence (Cerca puntiaguda) tipo de cerca que se usa comúnmente como barrera para mascotas y niños pequeños

Pier (Pilar) columna de mampostería que, por lo general, es rectangular en secciones cruzadas de manera horizontal, que se usan para sostener partes estructurales

Pilaster (Pilastra) columna interna que usualmente se proyecta desde la pared para reforzarla

Pile (Pilote) pilar de concreto, metal o madera hundido en la tierra o moldeado en el lugar como soporte de los cimientos

Pilot (Guía) guía en el extremo de broca rebajadora cortadora de bordes que se usa para controlar el corte

Pilot hole (Orificio piloto) pequeño orificio hecho para recibir la porción roscada del tornillo de madera

Pitch (Inclinación) el grado de pendiente de un techo expresado en proporción con la elevación del tramo

Pitch block (Bloque de ajuste) pieza de madera cortada en forma de triángulo rectángulo; se usa como patrón para instalar largueros de escaleras, cabios y accesorios para pasamanos

Pitch board (Tablón de inclinación) bloque de madera que se corta con forma de triángulo rectángulo usando las dimensiones de huella y contrahuella de la escalera

Pitch pocket (Bolsa de resina) abertura en la madera entre los anillos de crecimiento que contiene resina líquida o sólida

Pith (Médula) núcleo suave y pequeño del centro de un árbol

Pivot (Pivote) punto de rotación

Plain-sawed (Aserrado simple) método de aserrado de madera que produce vetas planas

Plan view (Vista en planta) copia heliográfica que muestra el edificio desde arriba con vista hacia abajo

Plancier (Apoyo de corona) parte del acabado de la parte inferior de una caja de cornisa; también llamado *sofito*

Plank (Madero) madera que tiene 6 o más pulgadas de ancho y desde $1\frac{1}{2}$ d 6 pulgadas de espesor

Plank flooring (Entablonado) tipo de suelo de madera sólida que puede colocarse alternando los anchos

Plaster (Yeso) mezcla de cemento pórtland, arena y agua usando para cubrir las paredes y techos de un edificio

Plastic laminate (Laminado plástico) material duro y delgado en capas usado para cubrir encimeras de cocina; se consigue en muchos colores y diseños

Plastic toggle (Fiador de plástico) anclaje hueco de pared para asegurar tornillos

Plate (Solera) parte horizontal inferior o superior de la estructura de la pared

Platform frame (Estructura de plataforma) método de construcción de estructuras de madera en el cual las paredes se levantan sobre una plataforma construida previamente

Plies (Capas) planchas (piezas finas) de chapa de madera utilizado para hacer la mayoría de los paneles de madera laminada

Plinth block (Bloque de plinto) bloque pequeño y decorativo, más grueso y ancho que el marco de la puerta, que se usa como parte del contramarco de la puerta en la base y el cabezal

Plot plan (Plano de terreno) plano que muestra la vista aérea de lote, la posición del edificio y otros datos relevantes; también llamado *plano logístico*

Plow (Ranura) ranura profunda hecha en el sentido de las vetas de la madera

Plumb (A plomo) vertical; en ángulos rectos al nivel

Plumb bob (Plomada) peso con punta amarrado a una línea para evaluar el aplomo

Plumb line (Tranquil) cualquier línea de un cabio que queda vertical cuando éste está en su lugar

Plywood (Madera laminada) material de construcción que se forma pegando capas finas de madera en las que las vetas de las capas adyacentes quedan en ángulos rectos entre sí

Pneumatic (Neumático) accionado por aire comprimido

Pocket door (Puerta embutida) puerta que se desliza de manera lateral dentro de una pared; cuando está abierta, solamente se ve el borde de la cerradura

Pocket tapes (Cinta de bolsillo) dispositivo retráctil para medir de acero

Pocket (Cavidad de pared) cavidad en una pared para recibir una pieza, como la cavidad en los cimientos de concreto para recibir el extremo de la viga maestra

Point of beginning (Punto de partida) marca de un plano que indica el comienzo de la disposición del terreno; suele ser un objeto que no es probable que se mueva durante la construcción, como una roca grande o un árbol

Poles (Paral) partes verticales de un andamio

Polyurethane (Poliuretano) tipo de aislación de espuma rígida que se hace con un revestimiento de hoja o con papel aislante

Portland cement (Cemento pórtland) tipo de cemento mejorado en forma de polvo gris

Post-and-rail fence (Cerco de postes y barandales) estilo básico y económico de cerca que se usa para rodear distancias largas

Postforming (Postformación) método usado para doblar laminados plásticos en radios pequeños

Post-to-post (Poste a poste) sistema de balaustrada donde el pasamanos se corta y se encastra entre los postes

Powder-actuated drivers (Clavadoras accionadas a pólvora) herramienta que utiliza pólvora para clavar los sujetadores en metal o en mampostería

Power miter saw (Sierra ingleteadora eléctrica) herramienta usada para cortar ebanistería en varios ángulos

Preacher (Guía preacher) pieza pequeña de madera que tiene el mismo espesor que el contrapeldaño de la escalera; tiene un corte rectangular en el centro para encajar sobre el larguero y se apoya sobre el corte de escalón de la zanca de la escalera; se usa para delimitar los largueros abiertos de una escalera

Prefinished strip (Listón pre acabado) pisos de madera sólida que han sido lijados, terminados y encerados en fábrica; no pueden lijarse luego de la instalación

Prehung door (Puertas premontadas) puerta que ya viene encastrada y abisagrada al marco y que contiene cerradura y contramarco exterior

Pressure-treated (Tratada a presión) tratamiento que recibe la madera en el que se le aplican conservantes a presión para que penetren toda la pieza

Prism (Prisma) dispositivo que se usa para reflejar un rayo de luz y devolverlo por el mismo camino por donde viene

Privacy lock (Cerradura de dormitorio) cerradura que se usa con frecuencia en baños y dormitorios que funciona al apretar o girar un botón que tiene la perilla del lado de adentro del cuarto

Pullsaws (Sierra de tracción) sierra manual que corta al hacer presión sobre la madera hacia atrás y hacia adelante

Pump jack (Guimbalete) tipo de andamio que está formado por parales 4 3 4, un mecanismo de balancín y abrazaderas metálicas para cada paral

Purlin knee wall (Vigueta de pared de buhardilla) maderas de soporte en las que pueden intersectarse los cabios entre sí en un techo abuhardillado

Putty (Masilla) compuesto para vidrios que se usa para cubrir y sellar los bordes de paneles de vidrio

Pythagorean theorem (Teorema de Pitágoras) teorema que se puede usar para calcular la diagonal de cualquier ángulo recto: $a^2 + b^2 = c^2$

Quarter round (Cuarto bocel) tipo de moldura, extremo de una parte que tiene forma de cuadrante de círculo

Quarter-sawed (Aserrado por cuartos) método de aserrar madera paralela a los rayos medulares; ver *edge grain (veta del borde)*

Rabbet (Rebajo) corte a lo largo del borde o extremo de una madera

Rabbeted jamb (Jamba reducida) tipo de jamba de puerta lateral en la que la superficie reducida actúa como tope; en una jamba reducida doble, una reducción actúa como tope para la puerta principal y la otra como tope para canceles y puertas de tela metálica

Racking (Deformación) término utilizado para describir el movimiento lateral de una pared donde el ángulo de 90 grados entre el montante y las placas está afectado

Racking the floor (Escalonar el piso) práctica en la que se disponen siete u ocho filas sueltas de lamas de suelo uniendo extremo con extremo después de que la segunda fila de lamas de suelo ha sido abulonada; se hace para ahorrar tiempo y material

Radial arm saw (Sierra de brazo radial) tipo antiguo de sierra fija que se usa para troceo y corte al hilo

Radiation (Radiación) término general que se refiere a la radiación electromagnética, que incluye microondas, ondas de radio, infrarrojas, luz visible, luz ultravioleta, rayos X y rayos cósmicos

Radius edge decking (Tablas redondeadas para terrazas) madera con forma especial que se usa como superficie y área de paso de una terraza, piso o plataforma

Rafter (Cabio)   parte estructural oblicua de un techo que sostiene la cubierta y el entarimado de un techo

Rafter tables (Tabla de cabios) información impresa en el cuerpo de una escuadra de ajustar; se usa para calcular el largo de varios componentes de un sistema de techado

Rail (Peinazo)   miembro horizontal de un marco

Rake (Inclinación)   parte oblicua de los hastiales de un edificio

Rake cornice (Cornisa inclinada) terminación exterior aplicada a la parte oblicua del cabio

Rake rafter (Cabio inclinado)   el primer y último cabio de un techo a dos aguas que usualmente presenta una terminación o contramarco; también llamado *cabio* volante

Rebar (Barra de armadura)   barra de refuerzo de acero usada en concreto

Reciprocating (Recíproco)   acción alternativa, como sucede con algunas herramientas eléctricas

Reciprocating saw (Sierra alternativa)   término general usado para describir las sierras que cortan con un movimiento de vaivén

Reinforced concrete (Concreto reforzado)   concreto reforzado con barras de acero

Relative humidity (Humedad relativa)   porcentaje de humedad suspendida en el aire comparado con la cantidad máxima presente en la misma temperatura

Resilient (Elástico)   tipo de suelo flexible y delgado que se usa con mucha frecuencia en viviendas y edificios comerciales y que se consigue en planchas o baldosas

Return (Vuelta)   vuelta y continuación corta de una moldura, cornisa u otro tipo de acabado

Return nosing (Vuelta de vuelo de escalón)   pieza separada ingletada al extremo abierto de un escalón para retraer el vuelo del escalón

Reveal (Derrame de la puerta o ventana)   distancia del marco a la cara lateral de las jambas de la puerta o ventana o piezas similares

Ribbon (Cinta)   listón delgado puesto entre montantes de una armadura sin rigidez para sostener las vigas del suelo

Ribbon coursing (Hilera de cintas superpuestas)   alternativa a las hileras derechas en las que ambas capas se aplican en líneas rectas; la hilera de encima se coloca 1 pulgada por encima de la hilera de abajo

Ridge (Camellón)   punto más alto de un techo que tiene laterales inclinados

Ridgeboard (Tabla de cumbrera o camellón)   parte horizontal del marco de un techo que se colocan de canto en el camellón y al que se ajustan los extremos superiores de los cabios

Rim joist (Vigueta de apoyo periférica)   parte del techo que se clava en sentido perpendicular al extremo de las vigas

Rip fence (Guía de corte)   guía usada para sostener y mantener un corte derecho en el material trabajado con herramientas eléctricas fijas

Ripping (Aserrar a lo largo)   aserrar madera en la dirección de las vetas

Ripsaw blade (Hoja de sierra de hender)   hoja diseñada para cortar mejor cuando lo hace según las vetas de la madera

Ripsaws (Sierras de hender)   sierra diseñada para cortar en sentido paralelo a las vetas de la madera

Rise (Elevación)   en escaleras, la altura vertical del escalón; en techos, la altura vertical desde la placa de la pared al camellón; también puede ser cualquier altura en la que algo se eleva

Riser (Contrapeldaño)   terminación de una escalera que cubre el espacio entre escalones

Roll roofing (Techado prearmado) tipo de techado que se hace del mismo material que las tejas de asfalto y que se consigue en rollos de 36 pulgadas de ancho

Roofing brackets (Cartelas para andamios de techar)   Abrazadera metálica que se usa cuando la inclinación del techo es muy pronunciada para evitar que los carpinteros se resbalen

Roofing nails (Clavos de techado) sujetadores galvanizados de cabeza grande que se usan para clavar el techo

Roof window (Claraboya)   ventana del techo que contiene una hoja operativa para la entrada de luz y ventilación

Rosette (Roseta)   pieza de madera decorativa de forma redonda u oval que se usa para terminar un pasamanos contra la pared

Rough carpentry (Obra preliminar de carpintería)   parte del trabajo relacionado con la construcción o con otro trabajo que será desmantelado o cubierto por los acabados

Rough opening (Abertura preliminar)   abertura en la pared en la que se colocan las puertas y las ventanas

Rough sills (Solera preliminar) partes que forman el piso de una abertura de ventana, en ángulos rectos con los parantes

Rough size (Tamaño bruto)   tamaño aproximado; por ejemplo, cortar el laminado de la encimera de cocina de $\frac{1}{4}$ a $\frac{1}{2}$ pulgadas más ancho y largo que la superficie a cubrir

Router (Contorneador)   herramienta eléctrica manual que tienen cuchillas que rotan a gran velocidad usadas para darles forma a los materiales

Run (Huella)   distancia horizontal sobre la que corren las vigas, escaleras y otras partes similares

R-value (Valor R)   número que se le asigna a un material para indicar su resistencia al paso del calor

Saddle (Asiento)   ver *cricket* (*banquillo*)

Safety glass (Vidrio inastillable) vidrio construido, tratado o combinado con otros materiales para disminuir la posibilidad de lesiones por el contacto con él

Sapwood (Albura)   parte exterior de un árbol que se encuentra justo debajo de la corteza y que contiene células activas

Sash (Hoja de la ventana)   parte de la ventana en la que se coloca el vidrio

Sawhorses (Caballete de aserrado) caballetes para sostener el material que se está marcando o cortando

Sawyer (Aserrador)   persona cuyo trabajo es cortar troncos para obtener la madera

Scab (Costanera)   largo de la madera colocada sobre una junta para endurecerla y reforzarla

Scaffold (Andamio)   plataforma temporaria elevada para trabajar

Scale (Escala)   reducción proporcional de cada línea en el plano de un edificio a un tamaño que muestra con claridad la información y que puede usarse de manera conveniente

Screed (Listón guía)   listones de madera, metal o caños asegurados en su lugar y usados como guías para nivelar la cara externa superior del concreto

Screwdriver bits (Broca desatornilladora)   brocas o cabezales varios intercambiables

Screwdrivers (Destornilladores) herramientas usadas para colocar los sujetadores dentro de los materiales mediante un movimiento giratorio concéntrico

Screwguns (Destornillador eléctrico)   taladro eléctrico con una punta especial para atornillar sujetadores de pared a una distancia exacta

Scribe (Trazado)   Contornear madera para encajarla contra una superficie irregular

Scribing (Ajuste)   Encaje de la madera en una superficie irregular; en molduras, corte de uno de los extremos de una pieza para calzar la cara moldeada con la otra en un ángulo interior y para, así, reemplazar un inglete

Scuttle (Escotillón)   acceso al ático o drenaje a través de un parapeto

Seam-filling compound (Compuesto para rellenar)   compuesto hecho especialmente para laminados que deja juntas casi invisibles

Seasoned lumber (Madera curada) madera que se ha secado y tiene la humedad apropiada

Seat cut (Corte de asiento) corte en del cabio que es una combinación de un corte de nivel y un corte vertical; también llamado *"pico de pájaro"*

Section view (Vista de sección) plano de construcción que muestra una sección transversal de un edificio como si estuviese cortado para mostrar su esqueleto

Self-tapping screw (Tornillo autorroscante) tornillos diseñados para agujerear un orificio guía mientras se atornillan

Selvage (Orillo) parte no granulada o no expuesta de un techado prearmado tapada por la cubierta de este

Setbacks (Retroceso) distancia que deben mantener los edificios de los límites de la propiedad

Shake (Venteada) defecto en la madera causado por una separación de los anillos de crecimiento; también, del inglés shake, teja de madera

Shank hole (Orificio para espigas) orificio hecho para la parte más ancha de un tornillo de madera

Shark tooth saw (Sierra tipo "dientes de tiburón") sierra manual con dientes agresivos que corta madera en ambas direcciones

Shear walls (Pared de corte) abrazaderas de pared, con estructura de 2 3 6s ancladas a la losa, con tablas de partículas orientadas clavados muy firmes; se usan en regiones de grandes sismos

Sheathing (Entablado) planchas o placas de material que se sujetan al techo y a las paredes exteriores y sobre las cuales se coloca la cubierta del techo y el revestimiento

Shed roof (Techo de un agua) tipo de techo que se inclina en una sola dirección

Sheet paneling (Paneles en planchas) paneles de madera contrachapada o madera dura que se aplican en las paredes con bordes verticales largos

Shim (Calza) material delgado con forma de cuña que se usa detrás de las piezas para enderezarlas o para nivelar y unir sus superficies

Shingle butt (Tope de teja) borde inferior expuesto de una teja

Shingling hatchet (Hachuela para tejamaniles) hacha liviana con una cuchilla afilada y taco que se usa para colocar tejas de madera o tejas comunes

Shoe rail (Barandilla inferior) moldura lineal labrada diseñada para sostener los extremos inferiores cuadrados de la balaustrada

Shortened valley rafter (Limahoyas acortadas) limahoya que va desde la placa de la pared a la limahoya de soporte

Side (Cara) superficie ancha de una tabla, plancha o placa

Side casings (Marcos laterales) partes laterales del marco exterior de una ventana

Side flashing (Tapajuntas laterales) tapajuntas puestos entre las tejas alrededor de los lados de una chimenea

Side jambs (Jambas laterales) lados verticales de un marco de ventana

Siding (Revestimiento) cobertura de terminación exterior de una pared)

Sidelight (Cristaleras laterales) marco que contiene hojas de vidrio pequeñas de uno o de los dos lados de una puerta de entrada

Sill (Durmiente) pieza de madera horizontal que descansa sobre los cimientos y sostiene la estructura de un edificio; también, alféizar, la parte horizontal más baja del marco de una ventana o puerta

Single coursing (Hilera simple) cuando las tablas se colocan en las paredes con una capa simple en cada hilera, de manera similar a como se colocan las tejas en el techo

Single-hung window (Ventana de guillotina simple) ventana cuya hoja inferior se desliza verticalmente pero la hoja superior queda fija

Single pole (Andamio de poste sencillo) tipo de andamio de madera que se usa se puede adherir a la pared y que no interfiere con el trabajo

Single-strength (SS) glass (Vidrio delgado (SS, por sus siglas en inglés) vidrio de $^3/_{32}$ pulgadas de ancho aproximadamente usado en hojas de vidrio pequeñas

Skilsaw (Sierra circular eléctrica) término general utilizado para llamar a las sierras circulares eléctricas portables

Skylight (Claraboya) ventana fija del techo que solamente permite el paso de la luz y no del aire

Slab-on-grade foundation (Cimientos de losa a nivel de tierra) base de construcción sólida de concreto usada en vez de los cimientos ahorra material y trabajo

Slat block (Parqué de bloque de listones) tipo de bloque de parqué que se hace uniendo muchos listones angostos y cortos de madera en cuadrados pequeños con varios patrones

Sleeper (Traviesa) listones de madera distribuidos sobre un piso de concreto en donde se asegurará el suelo de terminación

Sleeve anchor (Ancla de manguito) sujetador de pared sólida que se usa para asegurar los pernos

Sliding (Puerta corrediza) tipo de puerta exterior en la que sus paneles se deslizan de manera horizontal a través de un carril

Sliding gauge (Calibre "pie de rey") calibre usado para controlar de manera rápida y precisa la exposición de una teja, lo que permite hacer muchas hileras de tejas de una sola vez sin necesidad de usar un tendel

Sliding windows (Ventanas corredizas) ventanas de hojas deslizables que se abren arrastrándolas de manera horizontal en carriles separados ubicados en la jamba superior y el alféizar

Slope (Inclinación) término utilizado para indicar la pendiente de un techo; se calcula como unidad de elevación sobre unidad de recorrido, por ejemplo, 6 sobre 12

Slump test (Prueba de asentamiento) prueba que se hace sobre el concreto para determinar su consistencia

Snaplock punch (Punzón de cerradura de resorte) herramienta especial usada para hacer lengüetas elevadas en una pieza de revestimiento para que ésta calce estrechamente en el contramarco alrededor de las ventanas

Snap tie (Varilla de tensión) dispositivo metálico para sostener paredes de concreto a la distancia deseada

Snub cornice (Cornisa oculta) cornisa cuyo cabio no se proyecta más allá de la pared; se elige principalmente para reducir el costo de los materiales y el trabajo

Soffit (Sofito) parte inferior del contramarco de una cornisa o cualquier ensamble sobresaliente

Softboard (Cartón comprimido) cartón fibra de baja densidad

Softwood (Madera blanda) madera de coníferas (árboles cuya punta tiene forma de cono)

Sole plate (Solera inferior) parte horizontal inferior de la estructura de la pared

Solid (Sólidas) tipo de construcción de puertas de armario que se hacen con madera dura, aglomerado o madera sólida

Solid sheathing (Entablado sólido) paneles clasificados por ATA sobre los que se colocan tejas

Solid wall anchors (Anclajes de paredes sólidas) sujetadores de metal que se ajustan dentro de una pared mientras se atornilla un perno o tornillo

Sound-deadening insulating board (Paneles aislantes de sonido) sistema de aislación de sonido formado por una cubierta exterior de placas de yeso y canales de acero flexible en ángulos rectos a los parantes para aislar la placa de yeso de los mismos

Sound transmission class (STC) (Clasificación de la transmisión de sonido (STC, por sus siglas en inglés)) clasificación dada a una sección de un edificio, como una pared, que mide la resistencia al paso del sonido

Spaced sheathing (Entablado espaciado) placas que usualmente miden 1 3 4 o 1 3 6 sobre las cuales se aplican las tejas

Specifications (Especificaciones) pautas escritas o impresas sobre los detalles de la construcción de un edificio

Specifications guide (Guía de especificaciones) usado por los redactores de especificaciones para proyectos comerciales complejos; desarrolladas por el Instituto de las Especificaciones para la Construcción (CSI, por sus siglas en inglés)

Specifications writer (Redactor de especificaciones) persona que escribe información complementaria para proyectos de construcción para incluir cualquier información que no puede incluirse en planos o programas

Speed Square (Escuadra triangular de acero) herramienta utilizada para trazar ángulos de manera rápida, especialmente para cabios

Spline (Lengüeta postiza) listón fino y delgado de madera insertado en los extremos ranurados de las piezas adyacentes

Split fast anchor (Anclaje tipo split fast) sujetador que se ajusta en el orificio de la mampostería por la presión opuesta de las solapas de su eje

Spray foams (Espumas de aislación en aerosol) tipo de espuma de aislación en aerosol que se pulveriza en los espacios de los materiales para expandir y crear un sello hermético

Spreader (Separador) listón angosto de madera clavado a las jambas laterales de un marco de puerta unas pulgadas por encima del borde inferior para que el ancho del marco sea el mismo en el borde inferior y el en superior

Spud vibrator (Vibrador de concreto con punta) vibrador de inmersión con un tubo metálico en el extremo que se usa comúnmente para hacer vibrar y consolidar el concreto

Square (Cuadro de cubierta de techo) cantidad de cubierta de techo que equivale a 100 pies cuadrados de techado

Staggered coursing (Hileras alternadas) hileras en las que los topes de las tejas están retraídos de manera alternada por debajo, pero no por arriba, de la línea horizontal

Staging planks (Tablas de andamiaje) tablas que descansan sobre los travesaños de los andamios cuyos extremos están muy juntos para formar una plataforma firme

Stair body (Cuerpo de la escalera) parte de la escalera formada por los escalones, los contrapeldaños y los largueros acabados

Stair carriage (Soporte de la escalera) principal estructura de soporte de un conjunto de escaleras sobre la cual descansan los peldaños y los contrapeldaños. También llamado *stair horse (caballete de escalera)*

Staircase (Armazón completa de escalera) término usado para nombrar la estructura completa de una escalera, que incluye escalones, contrapeldaños, soportes y balaustrada

Stair horse (Caballete de escalera) ver *stair carriage (soporte de escalera)*

Stairwell (Agujero de escalera) abertura en el piso para escaleras ascendentes o descendentes o el espacio de una estructura donde se colocan las escaleras

Staple (Grampa) sujetadores con forma de U que se usan para asegurar algunos materiales

Staplers (Engrapadoras) herramientas que clavan grampas al presionar una palanca accionada por resorte o con movimiento de martillo

Stapling guns (Pistolas engrapadoras) herramientas que clavan grampas accionadas con electricidad, a presión de aire o con baterías

Starter course (Primera fila) hilera que se aplica como referencia para las otras capas de tejas de un techo o pared

Starter strip (Primer listón) listón de enrasado del mismo espesor y ancho que la superposición del revestimiento que se sujeta primero a lo largo del extremo inferior del entablado

Starting step (Primer escalón) el primer escalón de una escalera

Steel buck (Marco de acero) marco de puerta de metal con montantes de acero para formar el marco en la abertura de la puerta

Steel tapes (Cinta de acero) término usado para nombrar herramientas de medición largas que se enrollan a mano

Step ladder (Escalera de tijera) escalera plegable portable que está abisagrada en la parte superior

Stickering (Máquina de empalillar) máquina que hace molduras o un listón delgado que se ubica entre capas de madera para crear un espacio que facilite el secado

Sticking (Contorno) borde moldeado interior de un marco de una puerta ensamblada

Stile (Montante) miembros verticales exteriores de un marco, como los de una puerta ensamblada

Stockade fence (Cerca de estacas) cercado alto con extremos puntiagudos que se colocan muy juntos para brindar privacidad

Stool cap (Cubierta de repisa) pieza horizontal de acabado que cubre la repisa o alféizar del marco de una ventana en el interior; también llamada *repisa*

Stool (Repisa) miembro inferior del marco de una puerta o ventana; también llamado *alféizar*

Stop (Tope) listón de madera colocado de manera vertical a la jamba sobre la que descansa la puerta cuando está cerrada

Stop bead (Batiente) parte vertical del acabado interior de una ventana sobre la cual la hoja hace tope

Storm sash (Contraventana) hoja adicional ubicada en el exterior de una ventana para crear un espacio de aire sin circulación con el propósito de prevenir la pérdida de calefacción del interior durante periodos de clima frío

Storm window (Guardaventana) ventana secundaria sujeta a la ventana regular para protección contra el viento y el frío

Story (Piso) distancia entre la superficie superior de cualquier suelo y la superficie superior del suelo de abajo

Story pole (Regla calibrada) listón angosto de madera usado para trazar las alturas de las partes de un marco de pared o de hileras de revestimiento

Stove bolt (Perno de ranura) perno con espiral y cabeza con muesca de tornillo

Straightedge (Escantillón) corte de madera o metal que tiene al menos un lado recto que se usa para controlar superficies rectas

Straight tin snips (Tijeras de hojalatero) herramienta tipo tijeras para cortar planchas de material con propósitos múltiples

Striated (Estriado) material de acabado con ranuras aleatorias y poco espaciadas en el sentido de las vetas

Striker plate (Placa del cerradero) placa instalada en la jamba de la puerta en la que encaja el pestillo cuando se cierra la puerta

Stringer (Larguero) material de terminación que se aplica para cubrir el cuerpo de la escalera

Strip (Tablones de canto) ver *strip flooring (piso sólido de tablones de canto)*

Strip flooring (Piso sólido de tablones de canto) tipo de piso de madera sólida muy utilizado que tiene ranuras y lengüetas en los bordes para que los extremos de los tablones encajen perfectamente unos con otros

Strongback (Carrera) elemento que se ubica en las aristas y se ajusta a otros para ayudar a sostenerlos

Structural insulated panels (SIPs) (Paneles estructurales aislantes (SIP, por sus siglas en inglés)) paneles que consisten en dos tablas de revestimiento externas y un núcleo interno de material aislante

Stub joists (Viguetas) vigas que son más cortas y que se ubican en forma perpendicular a las vigas comunes

Stud (Puntal) elemento vertical del armazón de una pared que va de solera a solera

Subfascia (Imposta inferior) elemento horizontal del armazón que se ajusta a la parte trasera de las vigas; a veces se la llama *falsa imposta* o *imposta preliminar*

Supporting valley rafter (Viga de lima hoya) viga que se extiende de la solera hasta la cumbrera de la cubierta

Surfaced four sides (Cepillado por los cuatro lados) madera suave y escuadrada que se utiliza para los marcos de la puerta; abreviatura *S4S (por sus siglas en inglés)*

Surface hinges (Bisagras de superficie) tipo de bisagra de un armario que se coloca en la superficie exterior de la puerta y del marco

Swing (Oscilación) dirección en la que se abre una puerta instalada

Swinging doors (Puertas de vaivén) puertas que están fijas con bisagras en un extremo y que oscilan desde una abertura; cuando están cerradas, cubren la totalidad de la abertura

Table saw (Sierra de mesa) sierra circular fija con un lecho o mesa

Tack (Clavetear) sujetar algo con clavos temporariamente en su lugar; también, clavo pequeño

Tail cut (Corte de cola) corte en el extremo inferior de un cabio

Tail joist (Cabecero) vigueta corta que se extiende desde una abertura hasta un apoyo

Taper (Ahusado) que se vuelve más delgado desde un extremo al otro

Taper ripping jig (Patrón para aserrado ahusado) tabla ancha que tiene la longitud y la magnitud de estrechamiento necesarias para cortar piezas ahusadas

Tempered (Templado) material al que se le da un tratamiento especial para hacerlo más duro y resistente

Tempered glass (Vidrio templado) tipo de vidrio de seguridad que se trata con calor o con productos químicos que producen que el vidrio se desintegre inmediatamente en múltiple piezas granulares cuando se rompe

Template (Plantilla) patrón o guía para cortar o taladrar

Tensile strength (Resistencia a la tracción) la mayor tensión longitudinal que una sustancia puede soportar sin romperse

Termites (Termitas) insectos que viven en colonias y se alimentan de madera

Termite shield (Barrera antitermitas) placa tapajuntas metálica que se coloca sobre los cimientos para proteger los elementos de madera contra las termitas

Thermal envelope (Revestimiento térmico) parte de un edificio que crea el límite entre el aire acondicionado y no acondicionado

Thin set (Adhesivo de capa delgada) pegamento tipo mortero que se utiliza a menudo para colocar tejas de cerámica; se utiliza cuando es probable que la teja se moje con agua

Threshold (Umbral) pieza con bordes biselados que se ubica en el piso debajo de la puerta; también llamado *alféizar*

Tile (Teja) bloques cuadrados o rectangulares que se ubican lado a lado para cubrir una superficie

Tile grout (Lechada) mortero delgado que se obtiene de la mezcla de polvo y agua, y que forma una pasta que se coloca entre las juntas de las tejas con una paleta

Timber (Madera de construcción) piezas grandes de madera de alrededor de 5 pulgadas de ancho y de espesor

Toe (Punta) extremo delantero de las herramientas, como el de una sierra de mano o el de una lijadora de mano

Toeboard (Rodapié) a tabla de material que se ubica detrás del espacio libre inferior que se encuentra debajo de la base de un armario; también elemento horizontal de la parte inferior de la barandilla de un andamio

Toenailing (Sujetar con clavos oblicuos) técnica para clavar clavos diagonalmente para sujetar el extremo del armazón

Toe space (Espacio libre inferior) hueco que se encuentra en la parte inferior de un armario

Toggle bolts (Tornillos de fiador) anclaje para pared hueca con alas de resorte que sujetan un bulón fileteado mecánicamente

Top lap (Solape de remate) altura de la teja o de otro componente del techado menos la exposición

Topography (Topografía) descripción detallada de la superficie del terreno

Top plates (Soleras superiores) elementos superiores del marco de una pared

Total Rise (Elevación total) distancia vertical de elevación del techo que va desde la solera hasta la cumbrera

Total Run (Huella total) distancia horizontal total sobre la que cae un cabio

Total span (Envergadura total) distancia horizontal cubierta por el techo

Total station (Estación total) instrumento de agrimensura que utiliza un haz de luz para medir distancias y ángulos con precisión

Trammel points (Puntas para compás de vara) herramienta con puntas afiladas que se fija a una tabla de madera para diseñar arcos

Transit-level (Nivel de tránsito) similar a un nivel para constructor, pero con un telescopio que puede moverse 45 grados hacia arriba y hacia abajo en cada dirección

Transom (Travesaño) hoja pequeña sobre la puerta

Tread (Peldaño) elementos horizontales de terminación de una escalera sobre los cuales coloca los pies una persona que asciende la escalera o desciende de ella

Tread molding (Moldura del peldaño) moldura cóncava pequeña que se utiliza para terminar la junta debajo de los peldaños de la escalera

Trestle (Caballete) similar a un caballete de aserrado, se lo utiliza para sostener los tablones de los andamios

Trimmer joist (Viga recortada) vigas de longitud completa que corren a lo largo del interior de una abertura

Trimmers (Vigas de embrochalado) elementos de un marco que se ubican en los laterales de una abertura y que corren paralelos a los elementos del marco principal

Truss (Armadura) conjunto de elementos de madera o de madera y metal que se utiliza para sostener los techos o los pisos

Tungsten carbide-tipped blade (Cuchilla con punta de carburo de tungsteno) cuchilla para cortar madera que tiene una punta de carburo de tungsteno, material utilizado para que los bordes de corte permanezcan filosos por más tiempo

Turnout (Apartadero) a tipo de accesorio de un pasamanos

Twist (Torcedura) defecto de la madera

Twist drills (Brocas helicoidales) término utilizado para barrenas que se utilizan típicamente para hacer orificios en metal

Undercourse (Hilada inferior) capa inferior compuesta por tejas de madera más económicas que se aplica cuando se realiza una hilada doble en las paredes laterales

Undercut (Rebajo) corte algo menor de 90 grados que se realiza a través del espesor de un material terminado de modo que los topes de las juntas del material encajen perfectamente

Underlayment (Contrapiso) material que se coloca por debajo del piso y que proporciona una superficie suave y pareja para la aplicación de pisos elásticos. También es el material de base que se coloca desde el prearmado del techo hacia la plataforma del techo antes de terminarlo.

Undersill trim (Moldura del contramarco) accesorio de revestimiento exterior; también conocido como *acabado del contramarco*

Unfinished strip (Tabla sin acabar) piso de madera sólida que está tallado y que tiene vértices cuadrados y puntiagudos en las intersecciones de la superficie con los bordes; una vez colocada, se elimina cualquier imperfección en la superficie de las piezas adjuntas mediante un proceso de arenado

Unit block (Bloque unitario) bloque de parqué hecho de pequeñas tablas de madera

Unit Length (Proyectura) la longitud del larguero o del cabio de una escalera por unidad de huella

Unit Rise (Altura de peldaño) valor de elevación de una escalera o cabio por unidad de huella

Unit Run (Unidad de huella) distancia horizontal del peldaño de una escalera o segmento horizontal de la huella total o de un cabio

Unit Span (Unidad de envergadura) dos veces la unidad de huella para un cabio normal

Urethane glue (Cola de uretano) cola de alto rendimiento que es resistente, flexible y duradera y que se utiliza para aplicaciones en exteriores

Utility knife (Navaja para uso general) herramienta cortante multiuso, típicamente con una cuchilla retráctil

Valley (Lima hoya) intersección de dos pendientes del techo en los ángulos internos

Valley cripple jack rafter (Cabio corto o secundario) cabio que se extiende entre dos cumbreras

Valley jack rafter (Cabio de lima hoya) cabio que se extiende desde la cumbrera hasta la techumbre

Valley rafter (Cumbrera) cabio que se coloca en la intersección de dos pendientes en los ángulos internos

Vapor (Vapor) estado gaseoso del agua, comúnmente invisible

Vapor barrier (Barrera de vapor) lámina de plástico utilizada para evitar que la humedad penetre en la superficie de la construcción

Vapor retarder (Retardador de vapor) material utilizado para evitar el pasaje de vapor de agua; también se lo denomina *barrera de vapor*

Variance (Varianza) noción otorgada por la Junta de Apelaciones de Zonificación (del inglés Zoning Board of Appeals) a una comunidad para cambiar el código de zonificación debido a dificultades impuestas por las regulaciones de zonificación

Veneer (Chapa de madera) lámina o capa de madera muy delgada

Vermiculite (Vermiculita) mineral parecido a la mica que tiene la capacidad de expandirse con el calor para formar un material liviano con propiedades aislantes

Vertical arc (Arco vertical) báscula unida a un telescopio que se utiliza para medir ángulos verticales a 45 grados por encima y por debajo de la horizontal

Vertical grain (Aserrado por cuartos) revestimiento en el que los anillos de crecimiento anual, vistos desde un corte transversal, forman un ángulo de 45 grados o más en la superficie de la pieza

Vinyl-clad (Ventana revestida en vinilo) ventanas cuyas partes de madera expuestas se cubren con vinilo

Volute (Voluta) a accesorio espiralado que se coloca al comienzo de un pasamanos

Wainscoting (Friso) acabado de una pared que se coloca a mitad de la pared desde el piso

Waler (Larguero de entibación) elementos horizontales o verticales de una estructura de concreto que se utilizan para apuntalarla y fortalecerla y en los cuales se ajustan los amarres

Wall angle (Ángulo de pared) piezas de metal con forma de L que se ajustan a la pared para sostener los extremos de los canales principales y de los perfiles en T de una rejilla para cielo raso colgante

Wallboard saw (Sierra para paneles) sierra de mano diseñada para cortar paneles de yeso

Wall sheathing (Revestimientos para paredes) cobertura exterior de una pared que puede consistir en tableros, paneles, cartón fibra, tableros de yeso o tableros de espuma rígida

Wall unit (Módulo mural) uno de los dos tipos básicos de armarios para cocinas; se lo instala a 18 pulgadas por encima del mostrador

Wane (Gema) corteza o pérdida de material en el extremo de una madera

Warp (Alabeo) desviación de la rectitud de una pieza de madera

Waste factor (Factor de desperdicio) cantidad de material que se suma al cálculo exacto necesario para asegurar la provisión suficiente de material para completar un trabajo

Water table (Retallo de derrames) obra de terminación que se coloca por encima de los cimientos y que se proyecta más allá de éstos para escurrir el agua

Weather stripping (Burlete) tiras estrechas de metal delgado u otro material que se coloca en las ventanas y las puertas para evitar la infiltración de aire y humedad

Web (Tejido) elementos de madera o de metal que conectan las cuerdas superiores e inferiores de las armaduras; también, parte central de una viga en L de madera o de acero

Web stiffener (Montantes de refuerzo) bloque de madera que se utiliza para reforzar el tejido de una vigueta en L, con frecuencia en ubicaciones en donde

una barra de suspensión sostiene la vigueta y los laterales de la barra no se extienden hasta la parte superior del reborde

Wedge anchor (Anclaje en cuña) sujetador que se utiliza para apretar los pernos en paredes sólidas

Whet (Afilar) saca filo a una herramienta en una piedra de afilado al frotar la herramienta contra dicha piedra

Wind (Desviación) defecto de la madera que se produce por una torsión de un extremo al otro en el tallo leñoso

Winder (Peldaño radial) peldaño de una escalera, más ancho en un extremo que en el otro, que cambia la dirección del recorrido

Window flashing (Lámina de escurrimiento para ventanas) pieza metálica a lo largo de la chambrana de dintel que se dobla sobre la chambrana de dintel y contra la pared exterior para evitar el ingreso de agua en este punto; también se la denomina *cubierta de drenado*

Window frame (Marco de ventana) parte fija de la ventana; la hoja de la ventana encaja en el marco de la ventana

Window schedules (Plano de ventana) esquemas que proporcionan información sobre la ubicación, el tamaño y el tipo de ventanas que se instalarán en una construcción

Wing dividers (Compás con arco) herramienta parecida a un compás que se utiliza para diseñar círculos e intervalos crecientes de diversos motivos

Wire nails (Clavo francés) término general que se utiliza para denominar a la mayoría de los clavos hechos con rollos de alambre

Wood chisel (Cincel para madera) herramienta metálica diseñada para ser impulsada con un martillo, lo que se utiliza para hacer muescas y otros orificios rectangulares en la madera

Wood shingles (Tejuelas de madera) cubierta del techo muy utilizada que se produce habitualmente a partir de cedro rojo occidental

Woven valley (Lima hoya tejida) limahoyas recubiertas mediante la aplicación de tejas de asfalto en ambos lados y el tejido alternado de cada hilada por encima y a través de la limahoya

Wrecking bar (Barra de demolición) barreta grande que se utiliza en demolición; a veces también se la llama *barra de uñas*

Yoking (Yuntar) colocar los elementos verticales (2 3 4s) que se sostienen juntos en los rincones

Zones (Zonas) áreas en las que se dividen las comunidades para separar el tipo de construcción que se puede levantar en ella

Zoning regulations (Regulaciones de zonificación) mantienen a las construcciones del mismo tamaño en las áreas para las cuales fueron planeadas

Index